Metal-Organic Frameworks (MOFs) as Catalysts

Shikha Gulati
Editor

Metal-Organic Frameworks (MOFs) as Catalysts

Springer

Editor
Shikha Gulati
Department of Chemistry
Sri Venkateswara College
University of Delhi
New Delhi, India

ISBN 978-981-16-7958-2 ISBN 978-981-16-7959-9 (eBook)
https://doi.org/10.1007/978-981-16-7959-9

This Springer imprint is published by the registered company Springer Nature Singapore Pte Ltd.
The registered company address is: 152 Beach Road, #21-01/04 Gateway East, Singapore 189721, Singapore

Dedicated to My beloved Parents

Foreword

Make everything as simple as possible, but not simpler
Albert Einstein

The above quote writely and adequately justifies the emerging and the exciting field of Metal-Organic Frameworks (MOFs). MOFs are one of the most important classes of materials which involve strong bonding of molecular components into crystalline porous extended architectures. As a result, MOFs offer extremely high surface areas and very low crystalline densities. However, the real beauty of MOFs lies in their *simplicity* in preparing outstanding architectures by the judiciously selection of the straightforward molecular components which simply self-assemble following the geometrical concepts. MOFs offer noteworthy applications in a variety of fields including gas and vapor storage, gas and vapor separation, carbon dioxide storage and capture, ion conduction, drug delivery, catalysis, luminescence, batteries, water treatment, and photo- and electrocatalysis.

Out of varied applications, catalysis is not only fundamentally important but is also paramount for the industrial and medicinal applications. The exponential growth in the field of MOFs in the last few decades has tremendously enhanced our understanding both about their synthesis and structures. While a number of review articles as well as books are available about the synthetic concepts and structures of MOFs; there is a noticeable lack of a general book for the undergraduate and the graduate studies that not only covers the underlying principles of MOFs but also focuses on their catalytic applications. In this context, it is overwhelming to learn about the new contribution "Metal-Organic Frameworks (MOFs) as Catalysts" in the Molecular Catalysis Book Series published by the Springer Nature. This book is not only timely but also enriches the readers about various concepts and examples in the ever-gowing field of MOF-catalysis. I applaud the contribution of Dr. Shikha Gulati, as the Editor, and all contributing authors to fill this void with this book on *Metal-Organic Frameworks (MOFs) as Catalysts*". This book covers 28 chapters for showcasing the development of several MOFs and MOF-based materials as the heterogeneous materials for a wide range of applications.

This book has been organized in such a way that it may be useful as a comprehensive text with sixteen parts devoted to the synthesis, characterization, properties, functionalization, and applications of MOFs. While the initial chapters focus on the synthesis and the functionalization strategies to aid in scaling-up while also addressing difficulties that arise during the said process; the subsequent chapters are devoted to assorted applications. Every chapter provides a good number of chemical drawings and structures making the whole book a useful resource for the stakeholders working in the field. Throughout the book, images of molecular structures are shown while their electronic files have been made available online to enhance comprehension.

MOF-based catalytic systems are likely to have a significant impact on the future growth as evidenced by the examples outlined in this book. While there are many obstacles to overcome in terms of selectivity, compatibility, cost, and robustness in the field of heterogeneous catalysis; this book is likely to provide an ideal platform for the development of innovative catalytic materials based on MOFs and MOF-based materials.

Prof. Rajeev Gupta
Department of Chemistry
University of Delhi
New Delhi, India

Preface

The field of metal-organic frameworks (MOFs) is growing rapidly and steadily. The method of combining easily accessible metal precursors with organic linkers to create an extensive range of protracted frameworks evidently captivated the thoughts of chemists and materials scientists all over the world back then, and it continues to do so now. There is a significant relationship between the chemistry of MOFs and Inorganic chemistry from a fundamental approach. MOF's chemistry permits chemists to link formerly prevailing coordination complexes to form a conceptual connection into the field of materials chemistry in diverse ways. This relationship has now evolved to allow for a wide range of applications, the most common of which is catalysis. The purpose of this book is to provide a comprehensive understanding of a field where new research is emerging that is of interest to a larger scientific audience to non-specialist readers, whether in academia or industry. As a result, it gives me great pleasure to be able to put together a multi-author book with impressive contributions from topmost research laboratories in the field of Metal-organic frameworks.

My aim is to deliver insight into how the area of MOFs came to be, as well as an explanation of the basics that describe where it has come from and where it can go in the future. These possibilities have manifested themselves in critical developments and significant advancements described in the chapters of the sixteen parts. Part One provides an overview of MOFs, highlighting their potential state-of-the-art performance in catalytic applications. The Kinetic Stability of Robust Metal-Organic Frameworks (MOFs) in Catalytic Reactions is demonstrated in Part Two. Part Three then focuses on the designing and synthesis of MOFs, as well as their functionalization procedures, which allow MOFs to be reasonably modified with organic groups for a variety of applications. Part Four shows how to describe these frameworks using various analytical approaches, while Part Five provides an overview of MOF applications in catalyzing various sorts of reactions. Making optimal use of greenhouse gases into valuable compounds is a major problem for humanity. In the system of gas storage, separation, and conversion into fine chemicals, MOFs have developed as possibly effective platforms for addressing this dilemma. As a result, Part Six focuses on how MOFs interact with gas molecules (such as CO_2) and how these frameworks might be used into capture and utilization strategies. Part Seven explores the use of covalent-organic frameworks as catalysts in a variety of processes. Part Eight then presents a quick overview of the heterogeneous nature of the most important MOFs in catalysis. Part Nine deals with the rising subject of environmental monitoring using MOFs, and some groups' work indicates how these frameworks might be constructed to act as sensor platforms. The significance of MOFs in the capture and degradation of chemical warfare weapons is explained in detail in Part Ten. This part also discusses how computational skills in chemistry can be used to better comprehend and eventually support the design of MOFs, with precise applications in the detection and catalytic degradation of chemical warfare chemicals. Asymmetric catalysis also relies heavily on chiral MOFs. Asymmetric organic transformations with MOFs are discussed in Part Eleven. MOFs are described as catalysts for methane storage in Part Twelve. Parts Thirteen and Fourteen detail some of the ways MOFs can be used to catalyze photochemical and bio-catalytic reactions, respectively. Nanotechnology is a rapidly growing field that includes nanostructures, nanomaterials, and nanoparticles, among other subdisciplines. These materials have attracted a lot of interest in science because of their size, structure, and potential utility. Part Fifteen looks at how these MOFs encase nanoparticles, which play a key role in catalytic applications. Finally, Part Sixteen concludes with Concluding Remarks and Future Perspectives on MOFs, allowing researchers interested in this topic to plan their future work and progress.

The contributing authors have chosen carefully selected references to help the reader over the rich literature, making the area available to a broad and diverse audience. Contributions also provide a forecast for future advancements in the topic. As an editor, it will bring me great satisfaction if this book aids those working on the study and development of catalytic systems based on MOFs.

New Delhi, India

Dr. Shikha Gulati

Acknowledgements

I thank God, the Almighty, for his blessings throughout my life and conceding me the potential to complete the book successfully. Next, I would like to express my sincere gratitude to Prof. C. Sheela Reddy, Principal, Sri Venkateswara College, University of Delhi for her generous and fruitful advice, kind encouragement, selfless support, readiness to help all the time, and facilitating appropriate milieu in the College. I also acknowledge her to provide well equipped ICT Laboratory and library which helped me a lot during the work. I express my gratitude to my Beloved parents, my daughter and all my mentors for their encouragement and inspiration.

I would like to thank all the Renowned contributing authors without whose efforts this book could not have been written. Besides this, numerous individuals have significantly and innocently facilitated me in the efficacious accomplishment of this book so I would also say a great thanks to them.

We would like to thank the staff at Springer, for their help and support.

Finally, I'd like to express my gratitude to Springer Nature Singapore for allowing us to publish this book.

Dr. Shikha Gulati

About This Book

The emerging and exciting area of MOF inspired me to write the book *Metal-Organic Frameworks (MOFs) as Catalysts*. This book brings together a global and interdisciplinary group of leading specialists in the fields of metal-organic framework synthesis, functionalization, characterisation, and applications (MOFs). The authors present a unique perspective on MOF research that provides a solid foundation in basic methodologies and instructs the reader on how to think about MOFs. This book can be used as a starting point for theoretical or applied research by researchers who are new to MOFs. There are twenty-eight chapters in the book, which are organised into sixteen divisions. This book covers everything from MOFs' synthetic and functionalization strategies to their use as sensors and their association with relevant encapsulated metal nanoparticles in CO_2 capture and fixation, chemical warfare agent capture and degradation, environmental monitoring, and water purification, etc. Because of the wide range of possibilities that MOFs offer in terms of structure, network topology, integration of active species and post-synthetic modifications, among other things, the potential of MOFs in these areas is presently developing at a rapid pace. The purpose is to establish future research objectives that focus on important peculiarities in this class of materials as a whole. Scientists and researchers who are interested in MOFs will find the book interesting.

Contents

MOFs as Catalysts: Introduction and Prospects

Introduction to Metal-Organic Frameworks (MOFs) 3
Rajender S. Varma, Arikta Baul, Rachit Wadhwa, and Shikha Gulati

Stability of MOFs and Kinetics of MOFs Catalyzed Reactions

Kinetic Stability of Robust Metal–Organic Frameworks (MOFs) in Catalytic Reactions 45
Sanjay Kumar, Sweta Kumari, Shikha, Shefali Shukla, and Shikha Gulati

Strategies for the Synthesis and Functionalization of MOFs

Strategies to Synthesize Diverse Metal–Organic Frameworks (MOFs) 69
Chetna Gupta, Parul Pant, and Himanshu Rajput

Functionalization Strategies of Metal–Organic Frameworks (MOFs): Diverse Ways to Versatile MOFs 99
Sanjay Kumar, Sneha Vijayan, Kartika Goyal, Mansi Kathuria, and Shikha Gulati

Characterization Techniques of MOFs

Spectroscopic and Microscopic Techniques: Tools for Characterizing Nanoscale Metal–Organic Frameworks (NMOFs) 127
Parul Pant, Aadya Jaipuria, and Chetna Gupta

Metal–Organic Frameworks (MOFs) Characterization Using Different Analytical Methods 165
Dipeshkumar D. Kachhadiya, Amol Vijay Sonawane, and Z. V. P. Murthy

Reactions Catalyzed by MOFs and Prospects for Applications

Versatile Metal–Organic Frameworks: Perspectives on Contribution in Reaction Catalysis and Applications 183
Sanjay Kumar, Kartika Goyal, Mansi, Sweta Kumari, and Shikha Gulati

MOFs as Catalysts for CO_2 Capture and Fixation

Metal–Organic Frameworks as Promising Catalysts for CO_2 Capture and Fixation 207
Anand Prakash and Rakesh Kumar Sharma

Theoretical Study on Catalytic Capture and Fixation of Carbon Dioxide by Metal–Organic Frameworks (MOFs) 237
Upasana Issar and Richa Arora

Covalent Organic Frameworks as Catalysts

Covalent Organic Frameworks (COFs) as Catalysts: An Overview 267
Arti Jain and Priti Malhotra

Recent Advances in the Synthesis of Covalent Organic Frameworks for Heterogeneous Catalysis 285
Subodh and Dhanraj T. Masram

Designing, Synthesis, and Applications of Covalent Organic Frameworks (COFs) for Diverse Organic Reactions 319
Shefali Shukla, Abhay Gaur, and Shikha Gulati

MOFs as Heterogeneous Catalysts

Metal–Organic Frameworks (MOFs) as Heterogeneous Catalysts: An Overview 355
Sushma Yadav and Priti Malhotra

Recent Trends of Metal–Organic Frameworks in Heterogeneous Catalysis 369
Manoj Trivedi, Sanjay Kumar, Aryan Arora, and Kartika Goyal

MOFs as Sensors

Metal–Organic Frameworks (MOFs) as Sensors for Environmental Monitoring 393
Sushma Yadav and Priti Malhotra

Metal–Organic Frameworks for Pesticide Sensing: Trend in the Recent Years 411
Navin Kumar Mogha and Dhanraj T. Masram

MOFs as Catalysts for the Capture and Degradation of Chemical Warfare Agents

Computational Approach Toward Identification and Catalytic Degradation of Chemical Warfare Agents Using MOFs 431
Richa Arora and Upasana Issar

Metal–Organic Frameworks (MOFs) as Versatile Detoxifiers for Chemical Warfare Agents (CWAs) 453
Laishram Saya and Sunita Hooda

Chiral MOFs for Asymmetric Catalysis

Chiral Metal-Organic Frameworks for Asymmetrical Catalysis 493
Bhavna Dwivedi, Pawan K. Mishra, and Vivek Mishra

MOFs as Catalysts for the Storage of Methane

Metal–Organic Frameworks: Promising Materials for Methane Storage 517
Pooja Rani, Ahmad Husain, and Girijesh Kumar

Photocatalysis by MOFs

Photocatalysis by Metal–Organic Frameworks (MOFs): An Overview 553
Ankur Srivastava, Manju Yadav, Bani Mahanti, Arun Kumar, and Mrituanjay D. Pandey

Metal Organic Frameworks as Photocatalyst for Water Purification 561
Naveen Goyal and Pragya Arora

Metal-Organic Framework as a Photocatalyst: Recent Growth in Environmental Applications 595
Anita Yadav and Rakesh Kumar Sharma

Metal–Organic Frameworks (MOFs) Catalyzed Photodegradation of Organic Dyes: Syntheses, Designing Strategies, and Mechanism 645
Ayushi Singh and Abhinav Kumar

Role of MOFs in Bio-catalysis

Multifaceted Metal–Organic Frameworks: An Emerging Platform for Biocatalytic Reactions 683
Shikha Gulati, Kartika Goyal, Nandini Sharma, Sanjay Kumar, and Kanchan Batra

MOFs Encapsulated Metal Nanoparticles as Catalysts

Metal–Organic Frameworks (MOFs) Encapsulated Nanoparticles: Potential Catalysts for Diverse Organic Reactions 705
Gunjan Purohit and Diwan S. Rawat

Recent Progress in the Synthesis and Electrocatalytic Application of Metal–Organic Frameworks Encapsulated Nanoparticle Composites 731
Naveen Goyal and Rajeev Kumar Rai

Concluding Remarks and Future Perspectives of MOFs as Catalysts

Concluding Remarks About Metal–Organic Frameworks (MOFs): From Properties to Potential Applications 767
Sanjay Kumar, Aryan Arora, Kartika Goyal, Shikha Gulati, and Manoj Trivedi

About the Editor

Dr. Shikha Gulati (M.Sc., Ph.D.) is working as an Assistant Professor of Chemistry in Sri Venkateswara College, University of Delhi. She has expertise in Inorganic Chemistry, Nano-materials, Metal-organic frameworks, Green Chemistry, Catalysis, and Analytical Chemistry. Dr. Shikha has authored several research papers in International Journals of repute and written numerous books as well as chapters in diverse books which attest to her research aptitude, and good writing skills. Her books are referred to in diverse Universities across India for different undergraduate courses. Dr. Gulati has also been awarded the Young Researcher Award 2020 for her work in the field of Nanotechnology. Her knowledge in the field of Inorganic Chemistry, Metal-organic frameworks and Nanomaterials are paid off immeasurably to this book.

Abbreviations

°C	Degree Celsius
1D	One dimensional
2-CP	2-chlorophenol
2D	Two dimensional
3D	Three dimensional
4-bpmh	N,N′-bis-pyridin-4-ylmethylene-hydrazine
4-MAP	4-methylaminopyridine
4-NP	4-nitrophenol
5-HT	5-hydroxy-tryptamine
Å	Angstrom
AB	Amido-black
ABDC	2-amino-1,4-benzenedicarboxylic acid
ACh	Acetyl choline
AChE	Acetylcholinesterase
ADA	Adamantanediacetic acid
AFM	Atomic Force Microscopy
AIM	ALD in MOFs
ALD	Atomic Layer Deposition
ANNs	Artificial neural networks
AO	Acid orange
AOP	Advanced Oxidation Process
AR3R	Acid Red 3R
ARO	Asymmetric Ring Opening
ASU	Air Separation Unit
ATA	2-aminoterephthalic acid
atm	Atmospheric pressure
AuNPs	gold nanoparticles
AY	Acid Yellow
BASF	Baden Aniline and Soda Factory
BB41	Basic Blue 41
BBTA	1H,5H-benzo(1,2-d:4,5-d′)bistriazole

BDBA	Benzene-1,4-diboronic acid
BDC	1,4 Benzenedicarboxylic Acid
bdp	1,4-benzenedipyrazolate
BE	Back Scattered electrons
BET	Brunauer-Emmett-Teller
BFDC	2,2′-bifuran-5,5′-dicarboxylate
BiOBr	Bismuth oxybromide
BPA	Bisphenol A
BPDA	N,N′-bis(4-pyridinyl)-1,4-benzenedicarboxamide
BPDC	Biphenyl-4,4′-dicarboxylate
BPEI	Branched poly (ethylenimine)
BPHZ	1,2-bis(4-[pirydyl-methylene)hydrazine
BPTC	3,3′,5,5′-biphenyltetracarboxylate
bpy	2,2′-bipyridine
bpydc	2,2′-bipyridine-5,5′-dicarboxylate
BPZ	Bipyrazole
BSE	Back Scattered Electrons
BSL2	*Bacillus subtilis* lipase
BTB	1,3,5-Tris (4-carboxyphenyl)benzene
BTC	Bis(trichloromethyl)carbonate
BTDC	2,2′-bithiophene-5,5′-dicarboxylate)
BTDD	Bis(1,2,3-triazolato-[4,5-b],[4′,5′-i])dibenzo-[1,4]-dioxin
BTP	Benzenetripyrazolate
BTT	1,2,5-benzenetristetrazolate
BTTA	2,5-di(1H-1,2,4-triazol-1-yl)terephthalate
CAL	Coordinative Alignment
CALF	Calgary Framework
CAU-1	$[Al_4(OH)_2(OCH_3)_4(H_2N\text{-}BDC)_3]\cdot xH_2O$ MOF
CB	Conduction Band
CBMC	Configurational-Bias Monte Carlo
CCS	CO_2 Capture and Storage
CCs	Cyclic carbonates
CEES	2-chloroethyl ethyl sulfide
CEESO	2-chloroethyl ethyl sulfoxide
CFSE	Crystal-field stabilization energy
CIF	Cobalt imidazolate framework
CIP	Ciprofloxacin
CL	Chemiluminescence
CMOF	Catalytic metal organic framework
CN	2-chloroacetophenone
CNT	Carbon Nanotubes
COC	Cyclic olefin copolymer
COFs	Covalent organic frameworks
CoRE	Computation-ready
CORR	Carbon dioxide reduction reaction

Co-TAPP	Cobalt(II) 5,10,15,20-tetrakis(ρ-tetraphenyl amino) porphyrin
	Cp*: Pentamethylcyclopentadienyl
CPL	Coordination polymer with pillared Layer structure
CP-MAS	Cross-Polarization Magic Angle Spinning
CPTPY	Bis(4′-(4-carboxyphenyl)-terpyridine
CPU	CO_2 Processing Unit
CQDs	Carbon quantum dots
CR	Congo Red
CS	2-(2-chlorobenzylidene) malononitrile
CTAB	Cetyl trimethyl ammonium bromide
CTF	Covalent triazine framework
CTPY	4-carboxy-4,2′,6′,4″-terpyridine
CuNWs	Copper nanowires
CUS	Coordination Unsaturated Sites
CV	Crystal violet
CWAs	Chemical Warfare Agents
Cyt c	Cytochrome Complex
DA	2,5-dihydroxyterephthalaldehyde
DABCO	1,4-diazabicyclo[2.2.2]octane
DCM	Dichloromethane
DCP	Diethyl chloro phosphonate
DCS	Differential Scanning Calorimetry
DEF	Diethyl formamide
DENP	Diethyl 4-nitrophenyl phosphate
DESH	Diisopropylamino ethyl mercaptan
DETH	2,5-diethoxy-terephthalohydrazide
DFPase	Diisopropylfluorophosphatase
DFT	Density Functional Theory
DHBDC	2,5-dihydroxybenzenedicarboxylate
DHTA	2,5-dihydroxyl-terephthalaldehyde
DIFP	Diisopropylfluorophosphate
DLS	Dynamic Light Scattering
DMA	Dimethylamine
DMA	N,N-di-methylacetamide
DMF	N, N-dimethyl formamide
DMNP	Dimethyl (4-nitrophenyl) phosphate
DMSO	Dimethyl sulfoxide
DMTA	2,5-dimethoxyterephthalaldehyde
DOBDC	2,5-dioxido-1,4-benzene-dicarboxylate
DOX	Doxorubicin
DPBIB	(S)-4- 7-diphenyl-2-(pyrrolidin-2-yl)-1H-benzo[d]imidazole
DPDS	4,4′-dipyridyldisulfide
dpNDI	N,N’ -di(4-pyridyl)-1,4,5,8-naphthalenediimide
DPP	1,3-di(4-pyridyl) propane
DQTP	2,6-diaminoanthraquinone-2,4,6-triformylphloroglucinol

DR23	Direct Red 23
DRS	Diffuse reflectance spectroscopy
DTDAO	dibenzo[b,d]thiophene-3,7-dicarboxylate 5,5-dioxide
DUT	Dresden University of Technology
E. coli	Escherichia coli
EA-2192	S-[2(diisopropyla mino) ethyl]methyl phosphonic acid
EBSD	Electron Backscatter Diffraction
ECSA	Electrochemical active surface area
EDS	Energy Dispersive Spectroscopy
EDX	Energy Dispersive X-Ray Analysis
EELS	Electron Energy Loss Spectrometry
EIS	Electrochemical impedance spectroscopy
EMD	Equilibrium molecular dynamics
EMPA	Ethyl methyl phosphonic acid
EtOAc	Ethyl acetate
FAO	Food and Agriculture Organization of the United Nations
FCS	Fluorescence Correlation Spectroscopy
FDA-DAM	Hexafluoropropane dianhydride-diaminomesitylene
FDCA	9-fluorenone-2,7-dicarboxylate
FDM	Fudan materials
FE-SEM	Field Emission Scanning Electron Microscopy
FET	Field effect transistor
FF	Fill factor
FMOFs	Functional metal organic frameworks
FTIR	Fourier Transform Infrared Spectroscopy
FTO	Fluorine-Doped Tin Oxide
g-C_3N_4	Graphitic Carbon Nitride
GC	Gas chromatography
GCMC	Grand Canonical Monte Carlo
GF	Glufosinate
GGA	Generalized Gradient Approximation
GMC	Gibbs Ensemble Monte Carlo
GO	Graphene Oxide
GOD	Glucose Oxidase
GP	Glyphosate
GR	Graphene
H_2BBTA	1H,5H-benzo(1,2-d:4,5-d')bistriazole
H_2BDIM	1,5-dihydrobenzo[1,2-d:4,5-d']diimidazole
H_2CBPTZ	3-(4-carboxylbenzene)-5-(2-pyrazinyl)-1H-1,2,4-triazole
H_2HFBBA	4,4′-(hexafluoroisopropylidene) bis(benzoic acid)
H_2Me2BPZ	3,3′-dimethyl-1H,1′H-4,4′-bipyrazole
H_2SBPDC	2,2′-sulfone-4,4′-biphenyldicarboxylate
H_3BTP	1,3,5-tris(1H-pyrazol-4-yl)benzene
H_4TBAPy	4,4' ,4", 4"'- (pyrene-1,3,6,8-tetrayl) tetrabenzoic acid
H_4TPP	5,10,15,20-tetra (1H-pyrazol-4-yp porphyrin)

HATZ	3-amino-1,2,4- triazole
HER	Hydrogen evolution reactions
HF	Hartree-Fock
HKUST	Hong Kong University of Science and Technology
HMF	Hydroxymethylfurfural
HN1	Bis-(2-chloro-ethyl)-ethyl-amine
HN2	2-chloro-ethyl)-methyl-amine
HN3	Tris-(2-chloro-ethyl)-amine
HOMO	Highest Occupied Molecular Orbital
HPLC	High-performance liquid chromatography
HRP	Horseradish Peroxidase
HR-TEM	High-Resolution Transmission Electron Microscopy
HSAB	Hard-soft acid-base
ICP-MS	Inductively Coupled Plasma-Mass-Spectrometry
ICP-OES	Inductively Coupled Plasma Optical Emission Spectroscopy
ID50	Lethal Dose 50
IDA	Imidazole-4,5- dicarboxylic acid
ILAG	Ion-and-liquid assisted grinding
IMPA	Isopropyl methyl phosphonic acid
INA	Isonicotinate
INT	Intermediate
IPCC	Intergovernmental Panel on Climate Change
^{i}PrOH	Isopropanol
IRMOF	IsoReticular Metal-Organic Frameworks
K	Kelvin
KAUST	King Abdullah University of Science and Technology
LAG	Liquid-assisted grinding
LbL	Layer-by-layer
LDH	Layered double hydroxides
LED	Light-emitting diode
LIFM	Lehn Institute of Functional Materials
LMCT	Ligand to Metal Charge Transition
LMOF	Luminescent Metal-Organic Frameworks
Ln-ZMOFs	Lanthanide Zeolite-like Metal-Organic Frameworks
LOD	Limit of detection
LPA	Lysophosphatidic Acid
LSD	Lysergic acid diethylamide
LSV	Linear sweep voltammetry
LUMO	Lowest unoccupied molecular orbital
MAF	Metal Azolate Framework
MB	Methylene Blue
M-BDC	m-benzenedicarboxylate
MCM	Mobil Composition of Matter
MCR	Multi-component reactions
MD	Molecular Dynamics

MeOH	Methanol
MEPY	Methyl pyridine
MFM	Manchester Framework Material
MIL	Materials Institute Lavoisier
ML	Machine Learning
MLCT	Metal to ligand charge transfer
MMSV	Morse–Morse–spline–van der Waals
MNPs	Metal Nanoparticles
MNPs@MOFs	Encapsulated/anchored metal nanoparticles to MOFs
MO	Methyl Orange
MOF	Metal-Organic Frameworks
MOP	Metal-Organic Polyhedra
MOT	Molecular orbital theory
MP-11	Microperoxidase-11
MPA	Methylphosphonic acid
MPa	Mega pascal
MPFA	Methyl phosphonofluoridic acid
M-PMOF	Polyoxometalate-Metalloporphyrin Organic Frameworks
MRI	Magnetic Resonance Imaging
MTV	Multi variate
MTV-MOFs	Multivariate metal-organic frameworks
MW	Microwave
MWCVD	Microwave-assisted chemical-vapour deposition
NADH	Nicotinamide adenine dinucleotide hydrogenase
NDC	2,6-Naphthalene dicarboxylate
NDPA	5,5' -(naphthalene-2,7-diyl)diisophthalate
NFT	Nitrofurantoin
NG	Neat Grinding
NH_2-BDC	2-aminobenzenedicarboxylic acid
NHPI	N-Hydroxyphthalimide
Ni/Cu-BTC	Nickel/Copper-benzene tri-carboxylic acid
Ni-BDC	Nickel-benzene dicarboxylic acid
NiO	Nickel Oxide
NIR	Near-Infrared
NJU-Bai	Nanjing University Bai's group
NMO	4-Methylmorpholine-N-oxide
NMOF	Nanoscale Metal organic frameworks
NMP	N-Methyl-2-Pyrrolidine
NMR	Nuclear Magnetic Resonance
NOTT	University of Nottingham
NP	Nanoparticles
NRR	Nitrogen reduction reaction
NS	Nanosheet
NTA	Nitrilotriacetate
NTR	Nitroreductase

NZF	Nitrofurazone
OD Cu/C	Oxide-derived Cu/carbon
OER	Oxygen evolution reactions
OMS	Open metal site
OP	Organophosphorus
OPAA	Organophosphorus acid anhydrolase
OPH	Organophosphorus hydrolase
OPs	Organophosphorous pesticides
ORR	Oxygen reduction reactions
PA-6	Polyamide-6 nanofiber
PAC	Poly-aluminium chloride
PAET	Poly(3-acetoxyethylthiophene
PAH	Polycyclic aromatic hydrocarbons
PAHC	Poly-aluminium hydroxy chloride
PAN	Polyacrylonitrile
PBS buffer	Phosphate-buffered saline buffer
PCN	Porous Coordination Network
PCO	Photocatalytic oxidation
PCP	Porous Coordination Polymers
Pd/C	Palladium on charcoal
PDA	Palladium diacetate
PDFT	Periodic Density Functional Theory
PE	Primary electrons
PEI	Polyethyleneimine
PFIB	Perfluoro isobutylene
PGM	Pt group metal
PHEN	1,10 phenanthroline
PI	Polyimide
PIM	Polymers of intrinsic microporosity
PL	Photoluminescence
PMOF	Polyoxometalate metalloporphyin Metal Organic Framework
PMPA	Pinacolyl methyl phosphonic acid
PNP	p-nitrophenol
PNPDPP	p-nitrophenyl diphenyl phosphate
PNRR	Photocatalytic nitrogen reduction reaction
POM	Polyoxometallate
POP	Porous Organic Polymer
ppm	Part per million
PSA	Pressure oscillatory absorption
PSD	Post-synthetic deprotection
PSE	Post-synthetic exchange
PSM	Post-synthetic modifications
PTE	Phosphotriesterase
PXRD	Powder X-ray diffraction
PYDC	Pyridinedicarboxylic acid

QM/MM	Quantum Mechanics/Molecular Mechanics
QSAR	Quantitative structure-activity relationship
R6G	Rhodamine 6G
RB5	Reactive Black 5
rGO	Reduced graphene oxide
RH	Relative Humidity
RhB	Rhodamine B
ROSs	Reactive oxygen species
RT	Room temperature
SAED	Selected area electron Diffraction
SALE	Solvent Assisted Ligand Exchange
SALI	Solvent-Assisted Ligand Incorporation
SBU	Secondary Building Unit
SC-XRD	Single-crystal X-ray diffraction
SE	Scattered Electrons
SEM	Scanning Electron Microscopy
SERS	Surface-enhanced Raman scattering
SFG	Solvent free grinding
SLI	Sequential Linker Installation
S-OMC	Sulfur doped organic mesoporous carbon
SPM	Scanning Probe Microscopy
SPR	Surface Plasmon Resonance
SSNMR	Solid State Nuclear Magnetic Resonance
STEM	Scanning Transmission Electron Microscopy
STP	Standard Temperature and Pressure
TA	Terephthalaldehyde
TAA	1,3,5,7-Tetraaminoadamantane
TAM	Tetrakis-(4-aminobenzyl)methane
TAPB	1,3,5-Tri(4-aminophenyl)benzene
TAPT	1,3,5-tris-(4-aminophenyl) triazine
TATB	4,4′,4″-s-Triazine-2,4,6-triyl-tribenzoic acid
TBA	Triboronic acid
TBAB	Tetra-butylammonium bromide
TBAPY	Pyrene tetra-p-benzoate
TBHP	*Tert*-butyl hydroperoxide
TBPS	Tetra(4-dihydroxyborylphenyl)silane
TC	Tetracycline
TCA	Tricarboxy Triphenyl Amine
TCPP	Tetra kis(4-carboxyphenyl porphyrin)
TDM	Tetrakis[(3,5-dicarboxyphenyl)oxamethyl]methane
TED	Triethylenediamine
TEM	Transmission Electron Microscopy
TEMPO	2,2,6,6-Tetramethylpiperidinyloxy
TEOS	Tetraethyl orthosilicate
TEPA	Tetraethylene pentaamine

TFA	Trifluoroacetic acid
TFB	1,3,5-triformylbenzene
TFPB	1,3,5-tris(4-formylphenyl)benzene
TFPT	1,3,5-tris-(4-formyl-phenyl)triazine
TGA	Thermogravimetric Analysis
THC	Tetrahydrocannabinol
THF	Tetrahydrofuran
TIPA	Tris-(4-imidazolylphenyl)amine
TMAOH	Tetramethylammonium hydroxide
TMB	Tetramethylbenzidine
TMS	Trimesic acid
TMSCN	Trimethylsilyl cyanide
TNP	2,4,6 Nitrophenol
TOF	Turn Over Frequency
TON	Turn Over Number
Tp	2,4,6-trihydroxybenzene-1,3,5-tricarbaldehyde
TPA	1,3,5-triformylphloroglucinol
TPD	Temperature-programmed Desorption
TPDC	p-terphenyl-4,4′-dicarboxylate
TPO	Temperature-programmed Oxidation
TPSS	Tao-Perdew-Staroverov-Scuseria
TPVT	2,4,6-tris(2-(pyridine-4-yl)vinyl)-1,3,5-triazine
TS	Transition state
TSA	Temperature oscillatory absorption
TTTPC	Tris(2,4,6-trimethylbenzene-1,3,5-triyl)-tris(methylene)-tris(pyridine-4-carboxylic acid)
Tz	4,4′,4″-(1,3,5-Triazine-2,4,6-triyl)trianiline
TZC	Tetrazole-5-carboxylate
TZPA	5-(4-(tetrazol-5-yl)phenyl)isophthalate
UFF	Universal Force Field
UV	Ultraviolet
UV–Vis DRS	UV-visible Diffuse Reflectance Spectroscopy
VB	Valence band
VOCs	Volatile Organic Compounds
VSA	Vacuum oscillatory absorption
XPS	X-ray Photoelectron Spectroscopy
XRD	X-ray Diffraction
ZIFs	Zeolitic Imidazolate Frameworks
Zr-MOFs	Zirconium based Metal-Organic Frameworks

MOFs as Catalysts: Introduction and Prospects

Introduction to Metal-Organic Frameworks (MOFs)

Rajender S. Varma, Arikta Baul, Rachit Wadhwa, and Shikha Gulati

Contents

1 Introduction ... 4
2 Metal–organic Frameworks (MOFs): A Promising Class of Crystalline Heterogeneous Catalysts with Promising Properties ... 6
3 Structural Analysis of MOFs ... 8
4 Strategies for Synthesis of Metal–Organic Frameworks ... 11
4.1 Solvothermal Synthesis ... 11
4.2 Microwave(MW)-assisted Synthesis ... 11
4.3 Electrochemical Synthesis ... 11
4.4 Solvent-Free Synthesis ... 12
5 Post-Synthetic Modification (PSM) Approaches for Enhanced Functionality of Synthesized MOFs ... 13
6 Comparative Study of MOFs Catalysis with Traditional Heterogeneous Catalysts like Zeolites and Mesoporous Silica ... 14
7 Strategies for Incorporation of Specific Catalytic Activity into a MOF ... 18
8 Different Catalytic Applications of MOFs: Appropriate Properties and Opportunities ... 20
8.1 Heterogeneous Catalysis ... 20
8.2 Photocatalysis ... 20
8.3 Electrocatalysis ... 21
9 Some Examples of the Common MOFs-Catalyzed Reactions ... 22
9.1 CO/CO_2 Hydrogenation ... 24
9.2 Alkyne Hydrogenation ... 24
9.3 Dehydrogenation ... 24
9.4 Epoxidation ... 24
9.5 Oligomerization ... 25
9.6 Hydrolysis, Dehydration, Esterification, and Condensation ... 26
9.7 Isomerization and Alkylation Reactions ... 27
9.8 Photocatalysis ... 29
10 Some Other Important Applications of MOFs ... 29

R. S. Varma
Regional Centre of Advanced Technologies and Materials, Czech Advanced Technology and Research Institute, Palacký University in Olomouc, Šlechtitelů 27, 783 71 Olomouc, Czech Republic
e-mail: Varma.Rajender@epa.gov

A. Baul · R. Wadhwa · S. Gulati (✉)
Department of Chemistry, Sri Venkateswara College, University of Delhi, Delhi 110021, India

S. Gulati (ed.), *Metal-Organic Frameworks (MOFs) as Catalysts*,
https://doi.org/10.1007/978-981-16-7959-9_1

10.1 Luminescent Sensing ... 29
10.2 CO_2 Capture and Photocatalytic Reduction ... 29
10.3 Biomedical Applications ... 30
11 Limitations in Catalytic Properties of MOFs ... 31
11.1 Transport Limitations in Catalysis by MOFs ... 31
11.2 Minority and Defect Structures in MOFs ... 31
11.3 Stability Limitations in MOFs ... 32
12 Conclusion and Future Outlook ... 32
Abbreviations ... 33
References ... 34

Abstract In this era of incessant development and progress in the production of diverse class of chemical compounds, heterogeneous catalysis has become an integral part of the expeditious and efficient synthesis in view of the non-reusability of the homogenous catalysts. The quality and efficiency of the heterogeneous catalysts is essentially determined by their stability and availability of their well-defined active sites. In this regard, synthetically created metal–organic frameworks (MOFs) have garnered immense interest in the past decade, bringing favorable characteristics for their application as heterogeneous catalysts over other materials, such as zeolites. They comprise special organometallic coordination polymers having a cationic center and linkage with organic ligands giving rise to a stable crystalline porous cage-like structure. Because of their flexibility in redefining surface area, crystallinity, tunability, rich topology, porosity, and dispersity of active sites, they find numerous applications, catalysis being the more promising one. In this chapter, the basics of MOFs are introduced, their synthesis methods, typical characteristics along with limelight of their promising appliances as catalysts. Further, the salient advantages of MOFs over homogenous counterparts and conventional heterogeneous catalysts has been deliberated with limitations, and their future prospects in the field of catalysis.

Keywords Metal–Organic Frameworks (MOFs) · Zeolites · Porosity · Crystallinity · Catalysis · Synthesis

1 Introduction

A catalyst is an entity capable of increasing the rate of a chemical reaction towards equilibrium, without getting itself consumed in the process [19]. With almost 90% of the chemical processes linked to a catalytic reaction, it can be said that catalysis has an enormous impact on the global economy [28]. Catalysis contributes immensely to accomplishing the desired reaction(s) with higher selectivity and rate under relatively mild conditions. Nowadays, a major chunk of all industrial chemicals produced in industries around the globe deploys catalysts within the manufacturing process, with the global catalyst market close to 35 billion USD.

Conventionally, catalysts were differentiated into two types- homogeneous (where catalysis operates in the same phase as the reactants and heterogeneous (where the phase of catalysis and reactants are different). The proposal of a rational theory for the effect of nitrogen oxides as homogeneous catalysts in the lead chamber process for the synthesis of sulphuric acid dates back to the eighteenth century; it has been in use for thousands of years, like fermentation. However, homogeneous catalysis has limitations which include difficulty in its recovery from the reaction medium, environmentally compromised, and corrosive nature. In order to reutilize homogeneous catalysts, the distillation of reaction products or elimination of precipitating counter-ion is typically required, which often leads to the deactivation of the catalyst.

Heterogeneous catalysis has offered significant advantages over homogeneous ones, the major one being the simple and inexpensive separation options and reuse of heterogeneous catalysts over the homogeneous counterparts due to distinction in the phases of catalysis and reaction medium. Their preferability in industrial chemical production can be attributed to their robustness and lower operational costs. MOFs, mesoporous silica, and zeolites are the three types of porous materials that have garnered enormous attention as heterogeneous catalysts for chemical industries over the years.

In the past two decades, metal–organic frameworks (MOFs), have successfully bloomed as a relatively new class of crystalline porous materials, drawing significant attention globally with potential applications in various fields, ranging from gas sorption and separation [46, 71, 121, 122], to luminescence [23], biomedical [54], and electrochemical applications [10, 146], among others [70], [98].

MOFs serve as the gateway to interesting possibilities in the field of catalysis as their well-defined single active sites contribute to overcoming the typical drawbacks of both, the homogeneous and crystalline heterogeneous catalysts [83]. In the domain of heterogeneous catalysis, MOFs are a relatively new entry with the first report of their catalytic application in 1994 when Fujita et al. [40] produced a 2-dimensional network of $Cd(4,4'\text{-}bpy)_2\ (NO_3)_2$ through the combination of $Cd(NO_3)_2$ and (4,4'-bpy), deploying them as heterogeneous catalysts in the cyanosilylation of aldehydes. Early patents in 2003 [6, 100, 119], [126], with BASF developing MOFs catalysts at that initial stage. The initial reports kindled research devoted to the exploration of heterogeneous catalysis in a variety of reactions, later transcending towards the finding of specific catalytic applications. Over the years, the field of heterogeneous catalysis has been explored for other catalytic applications, like photocatalysis, and electrocatalysis, etc.; Fig. 1.1 shows the timeline of progress for MOFs in catalytic applications.

Although the proposal of these materials with high expectations in catalytic applications has seen quite a few years [40], MOFs only have experienced a substantial activity in last decade. Easy recovery and recyclability are some of the appealing properties which make them favorable candidates for replacement of conventional Bronsted [59] or Lewis acids as well as other homogeneous acids utilized in classical synthesis [1]. Their tunable adsorption properties, pore size, and topology, etc., all point to their potential applicability in heterogeneous catalysis [37, 83, 108]. The main driver of the growing recognition of MOFs can be attributed to the advantages that it offers over traditional porous materials like zeolites and mesoporous silica. Its structural diversity, tunability, well-defined structures, mutable pore environments, etc. make them ideal candidates for catalytic applications.

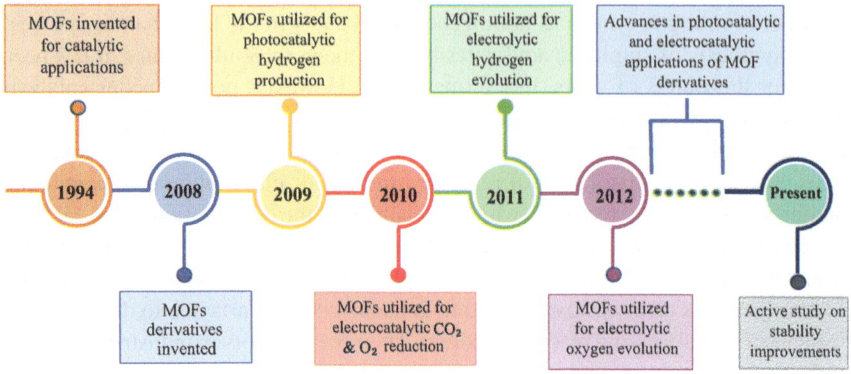

Fig. 1 Timeline of progress for MOFs in catalytic applications. (i) MOFs invented for catalytic applications [40]; (ii) MOFs derivatives invented [78]; (iii) MOFs utilized for photocatalytic hydrogen production [107]; (iv) MOFs utilized for electrocatalytic CO_2 & O_2 reduction [68, 140]; (v) MOFs utilized for electrolytic hydrogen evolution [95]; (vi) MOFs utilized for electrolytic oxygen evolution [17]; (vii) Advances in photocatalytic and electrocatalytic applications of MOF derivatives [24, 82, 88, 89, 135], [144, 147], [149]; (viii) Active study on stability improvements of MOF catalysts

Furthermore, although the acid/base sites of organic linkers, and metal sites of MOFs are limited, the structural tailorability of MOFs greatly compensates for its limitations leading to the origination of more active sites. The possibility to graft various functional groups on their metal nodes and linkers through post-synthetic modification approaches, and to capture into their pores and defects, the species like metal NPs, enzymes, etc. that are catalytically active, greatly enhances their catalytic potential.

2 Metal–organic Frameworks (MOFs): A Promising Class of Crystalline Heterogeneous Catalysts with Promising Properties

Metal–organic frameworks (MOFs) are inorganic–organic hybrid materials that represent an advanced class of solid, crystalline materials; inorganic building units like metal ions/clusters or metal–oxygen clusters are interlinked with the help of organic linkers [65], [110].

Since the arrival of MOFs as a promising novel organic–inorganic material, their potential applications have seen strikingly revolutionary, from being used in gas storage, separation, molecular sensing, optoelectronic devices, food packaging materials, to being exploited as heterogeneous catalysts, owing to their chemical mutability, pore structures, high metal content, and large internal surface areas, and flexibility in designing the active sites.

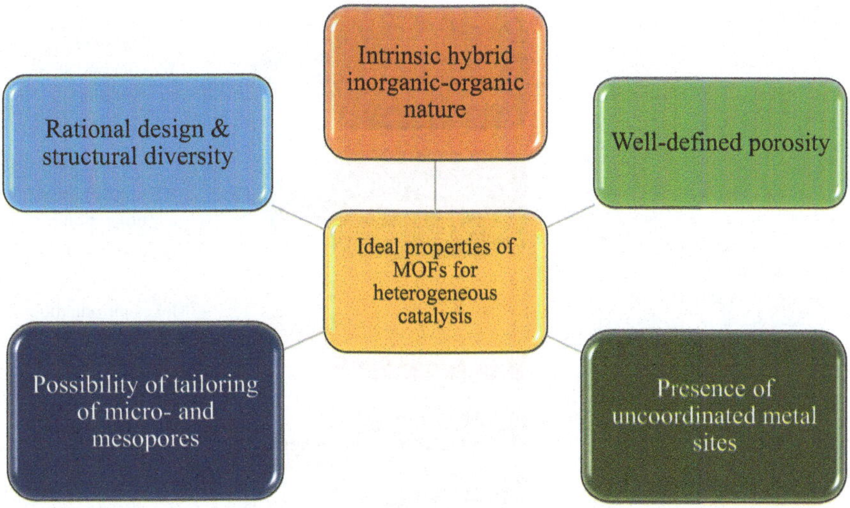

Fig. 2 Ideal properties of MOFs that make them exceptional candidates for heterogeneous catalysis

In the Metal–organic frameworks (MOFs), inorganic clusters or metal ions are linked by organic ligands, thereby creating solid hybrid structures that are crystalline and have varied dimensionalities [41]. The chemical building components form the basis for designing these materials by presenting specific structures and properties [141]. The structural and compositional variability favors the applicability of MOFs in various fields. Moreover, the post-modification ability of the organic linkers results in the formation of compatible and flexible MOFs [97, 134]. Functional organic groups can further decorate the nanoscale features of MOFs for specific interactions with biomolecules, thus making them an attractive candidate for the stabilization of enzymes in catalytic applications [47, 76, 87].

Figure 2 shows several properties of MOFs that make them exceptional candidates for heterogeneous catalysis.

MOFs comprise diverse frameworks types because of their finely tunable pore size, dimensionality, and chemical environment, which is totally a win–win over other crystalline porous solids that lack diversity in their framework types. Its pore network can be used to discriminate reagents or direct reactivity by manipulating the size and shape of the pore spaces available within a MOF. Excellent mass transport can be facilitated by tailoring the pore network, which can be further enhanced by structuralizing the MOF onto a hierarchically porous support. Furthermore, the high surface areas offer a high density of active sites and potentially a more effective catalyst. Synthesis or post-modification methods can easily control the chemical environment and host–guest interactions.

MOF's environment can be rationally designed along with the active sites with great precision [48]. The catalytic function can be implemented at the organic [51] or at the inorganic [127] component directly, or by post-synthetic modifications

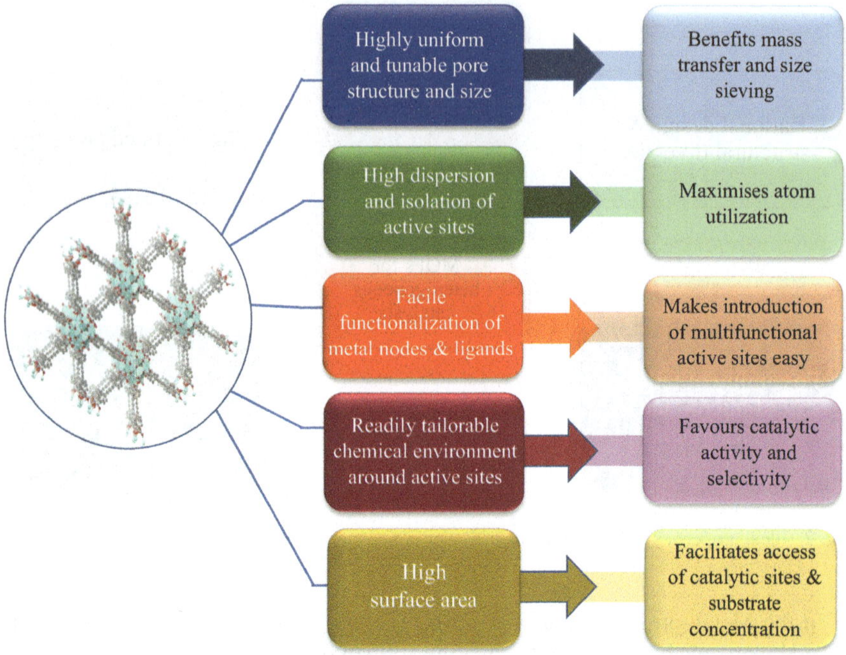

Fig. 3 Favorable structural and chemical properties of MOFs and their corresponding advantages

post [114, 128, 134, 136]. Catalytically active species (e.g., metal nanoparticles, metal coordination complexes, etc.) can also be captured in their pore system [60, 93], thereby providing space for the chemical reactions to occur. Figure 3 lists the properties of MOFs and their corresponding advantages.

3 Structural Analysis of MOFs

MOFs are composed of two types of units, organic and inorganic (Fig. 4). The inorganic part, which comprises metal ions or clusters forms the core. The resulting porous topology of MOFs with respect to the desired application is decided by secondary building units (SBUs) which enable the formation of extended networks feasible. On the other hand, the organic units play the role of linking the molecule together and are often organic ligands consisting of carboxylates, phosphate, sulfonate, etc. This combination of metal nodes interlinked with organic linkers results in the formation of a crystalline material that is highly porous, i.e., MOFs (Fig. 5). Based on the number of extension points, different secondary building

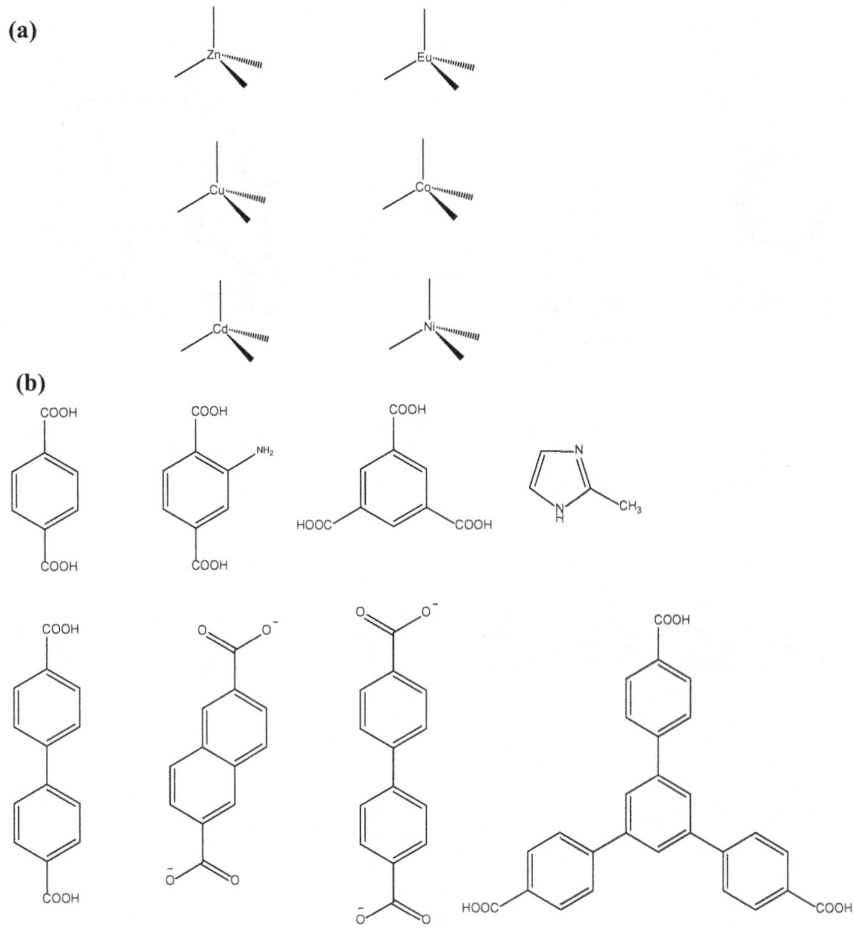

Fig. 4 Some examples of common (A) Metal ions and (B) Carboxylate and other organic linkers, and their coordination geometry

unit (SBU) geometries can be observed in MOFs namely, triangle, square paddle-wheel, trigonal prism, and octahedron with 3, 4, 6, and 6 points of extension respectively (Fig. 6). The porous and open structure of MOFs makes renders them suitable candidates for catalysis [64].

The possible diversification in the structure of MOFs with the usage of certain inorganic and organic units for diverse applications becomes their vital asset. MOFs also comes with the flexibility of further tuning of special cavities by post-functionalization methods in order to increase its functionality spectrum, eventually assisting in specific target-oriented catalysis applications [56].

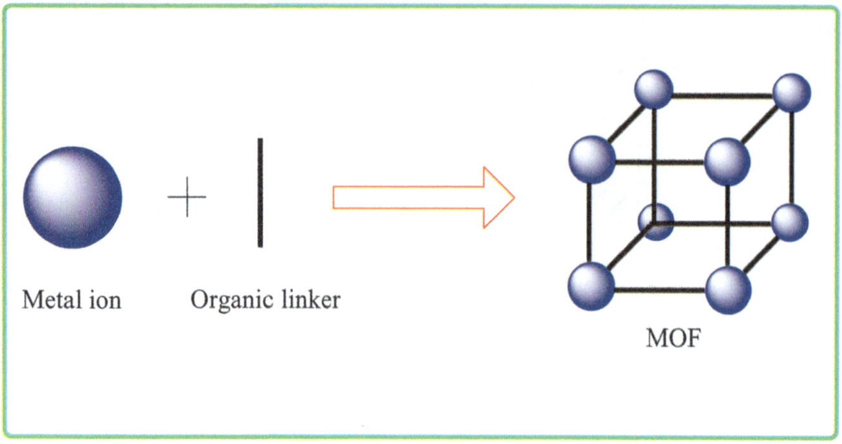

Fig. 5 Simplified representation of Metal–Organic Frameworks (MOFs)

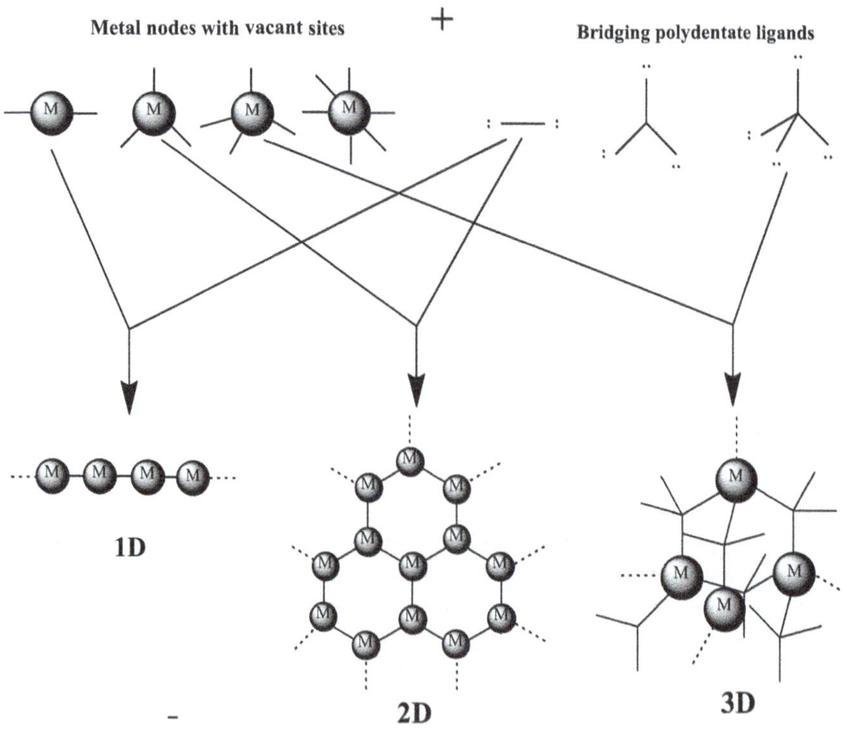

Fig. 6 Combinatorial possibilities for MOFs originating from different metal nodes and bridging ligands

4 Strategies for Synthesis of Metal–Organic Frameworks

4.1 *Solvothermal Synthesis*

The conventional approach for the synthesis of MOFs via heating is known as solvothermal synthesis. The process is initiated by the self-assembling of MOFs typically by the inorganic unit (metal-ion) and the organic linkers. HKUST-1 [112] and MOF-5 remain as the two benchmarks in MOFs chemistry reported back in 1999. In HKUST-1, the porous cubic networks are coordinated via 1,3,5-benzenetricarboxylate (BTC). As for MOF-5, they're composed of Zn_4O clusters linked with linear BDC (benzene dicarboxylate) linkers. This approach has been inherited by many researchers in order to synthesize versatile MOFs for different applications. To name a few, MIL-[85, 130] MIL-101 [85][39][52][49], MIL-53 [39], MIL-74 [25, 91, 109], etc., among others.

4.2 *Microwave(MW)-assisted Synthesis*

During this process, MW irradiation serves the purpose of providing energy for the synthesis of MOFs. The mutual interaction between mobile electric charges (e.g., ions in solution) and electromagnetic waves derives the principle which encapsulates this approach. The MW-assisted synthetic approach proves to be vastly advantageous over other conventional methods as it has higher efficiency, phase selectivity, morphology control, and flexible particles size reduction [53, 62, 79, 117, 120]. Many MOFs have been prepared using MW-assisted method and one of the most popular ones is Cr-MIL-101 which was synthesized in 2011 through MW heating at 210 °C [66]. Sabouni et al. [113], first reported a facile method that generated CPM-5 with a high surface area of 2187 m^2g^{-1} in about 10 min, thereby exhibiting high adsorption of CO_2 [113].

4.3 *Electrochemical Synthesis*

In 2005, HKUST-1 MOFs have been synthesized using the electrochemical synthesis approach as reported by BASF. The report stated an aim of exclusion of anions for the production of MOFs at a large scale. Since then, the electrochemical synthetic strategy has created a revolution in MOF chemistry, as many Zn, Cu, and Al-based MOFs are being synthesized using this approach [90]. Electrochemical synthesis usually takes place under high temperature and pressure conditions where electrochemical deposition takes place. Even though the electrochemical synthesis route has been highly recognized for the production of MOF layers, anodic electrodeposition comes with a downside which is, the possibility of synthesis of only those MOFs

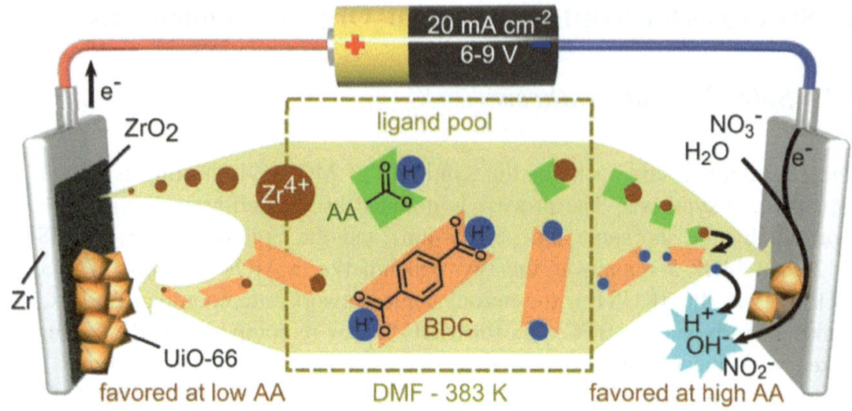

Fig. 7 Schematic of the anodic and cathodic electrochemical deposition mechanisms. Reprinted with permission from [118]. Copyright 2015 American Chemical Society

that have the same metal as the substrate. Figure 7 is a schematic representation of electrochemical deposition mechanisms depicted by Stassen et al. [118].

4.4 Solvent-Free Synthesis

There had been extensive research in the field of MOFs over the past decade regarding the exploration of solvent-free routes for the synthesis of MOFs which could eventually replace the solvent-based reaction between the organic ligands and metal ions. Greener synthesis methods have always been preferred over conventional methods which can cause harm to the environment by the release and usage of toxic chemicals [7, 45]. Under this solvent-free category, there are a number of mechanochemical synthesis methods which allow clean synthesis of MOFs with minimal generation of by-products. They particularly are less time-consuming reactions which a highly efficient yield rate. There are several branches of mechanochemical synthesis of MOFs: (i) by NG (neat grinding or no solvent), (ii) by LAG (liquid-assisted grinding) in which a few drops of solvent are utilized to increase the reactivity of reagents, (iii) ILAG (ion-and-liquid assisted grinding in which minimal amounts of solvents and salts), and (iv) via extrusion [33, 124]. The first solvent-free mechanochemical synthesis of microporous 3-dimensional MOF [$Cu(INA)_2$], has been reported by grinding together copper acetate and isonicotinic acid for 10 min [105].

5 Post-Synthetic Modification (PSM) Approaches for Enhanced Functionality of Synthesized MOFs

The addition of various chemical functionalities, post-synthesis, enables the properties of MOFs to be tuned in different ways to channel their utilization for different applications (Fig. 8). Attempts to bring functionalities simultaneous to the synthesis of MOFs is not feasible in many cases as it risks chemical and thermal stability, solubility, functional group compatibility, unwanted interference in the interactions of metal ions and organic linkers during the assembly of MOFs, etc. Post-synthetic approaches address these problems and bring functionalities to the pre-synthesized MOFs. Investigation on several of these post-synthetic approaches has been accomplished, namely, post-synthetic modification (PSM) [18, 125, 134], post-synthetic deprotection (PSD) [38], and post-synthetic exchange (PSE) [26, 34]. Adding functionalities after the synthesis helps to maintain the structure and stability of MOFs which is difficult in the direct synthesis of MOFs. A few of these approaches are presented here briefly.

One of the most common routes of post-synthetic modification is the use of covalent modifiers; usually, amino groups are used for functional modification. Amino-functionalized linkers show high chemical compatibility with a number of MOFs such

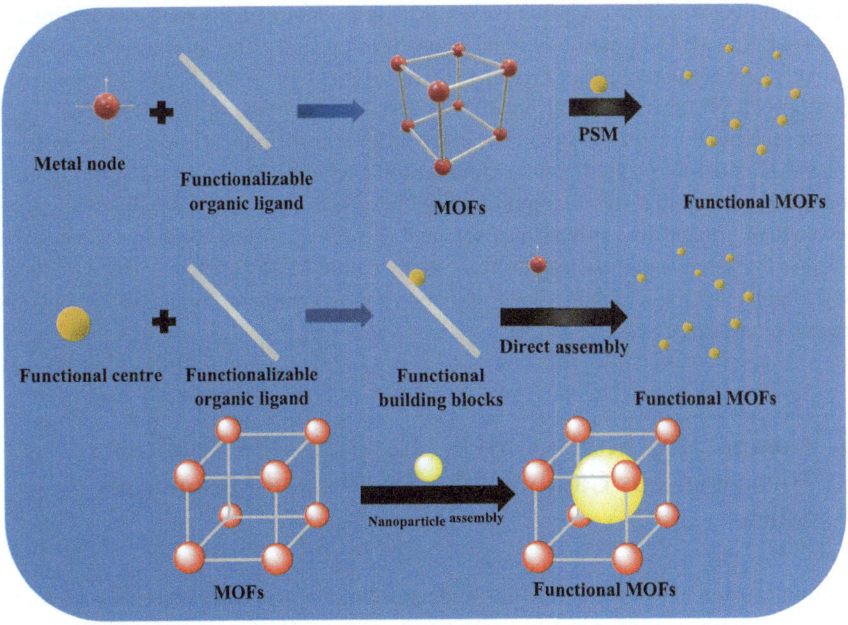

Fig. 8 Assembly of MOFs having active sites for catalysis by post-synthetic modification or direct assembly of a (i) linker or (ii) a metal node or (iii) nanoparticle encapsulation

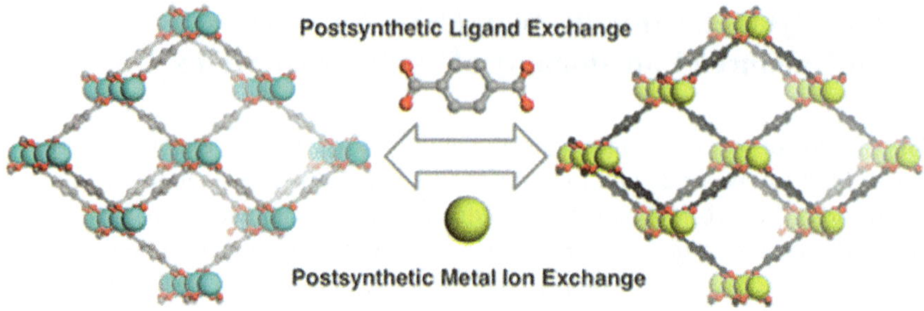

Fig. 9 Ligand and cation exchange for post-synthetic modifications in robust MOFs. Reprinted with permission from [67]. Copyright 2012 American Chemical Society

as MIL-53 [8, 20], MIL-101 [90], UiO-66 [43], and CAU-1 [3]. By deploying aldehydes, anhydrides, alkyl bromides, isocyanates, and additional such metal–organic complexes PSMs of amino groups have been reported [3, 43, 129].

Besides covalent modifications, another post-synthesis approach has been opted, via the metalation of MOFs with the assistance of soft metal ions. Metalation of Zr-based MOFs in PSM showed good results of the product as an organic catalyst for a broad range of conversions and reactions.

Apart from amino groups, different additional ligands can also be used to modify the metal sites which are coordinatively vacant or unsaturated. Solvent-assisted ligand incorporation (SALI) methods have been developed for the alteration of coordinatively unsaturated cluster $[Zr_6(\mu_3\text{-}O)_4(\mu_3\text{-}OH)_4(OH)_4(H_2O)_4(COO)_8]$ cluster by perfluoroalkyl carboxylate entities for enhanced CO_2 uptake in Nu-1000 [26].

Another method that has demonstrated a difference in application-based modifications is the post-synthetic ligand and metal-ion exchange [67] (Fig. 9). Although the metallic bonding in a stable complex of MOF has shown a rather subdued behavior towards the ligand and metal-ion exchange, however, further studies revealed the overestimation of these assumptions as many stable MOFs such as ZIFs, UiO, and MIL family MOFs, etc. have been realized by post-synthetic ligand and metal-ion exchange processes.

6 Comparative Study of MOFs Catalysis with Traditional Heterogeneous Catalysts like Zeolites and Mesoporous Silica

Rational design of MOF frameworks with coordinatively unsaturated metal sites and controlled pore size for selective catalytic properties offers a powerful platform for the development of heterogeneous systems in comparison to other porous materials like zeolites [77], [103]

A promising solution to green and sustainable chemistry is the translation of homogeneous catalysis into heterogeneous catalysis. As a result of the several difficulties associated with the utilization of conventional homogeneous catalysts for organic reactions, such as complications in separation, recovery, and disposal of spent catalysts, the use of heterogeneous catalysts in the liquid phase is highly preferable.

In this regard, MOFs as heterogeneous catalysts are expected to see major growth in the near future, as a replacement for not-so-environment-friendly and homogenous catalysts. The tunable nature, permanent microporosity, and well-defined reaction environmental conditions of MOFs make them preferable candidates for new innovations in catalyst design and application. In addition to their excellent robustness and reusability, these materials showcase size selectivity relative to their homogeneous counterparts. One major highlight of using MOFs is the fact that MOFs as heterogeneous catalysts are easily separable from the reaction mixture by centrifugation and can be recovered and reused multiple times without significantly losing their catalytic activity. Unlike homogeneous catalysis, homochiral MOFs can also be employed for enantiomeric selectivity in product formation. Although MOFs prove to be inferior to zeolites when it comes to thermal and hydrolytic stability, they exhibit other favorable features like extensive tunability, extremely regular catalytic sites, etc. where zeolites do not seem to be a preferable option.

MOFs, mesoporous silica, and zeolites are the three porous materials utilized significantly in heterogeneous catalysis. Zeolites are an important group of catalysts with their limited pore size and appreciable stability, have been increasingly utilized for many years under harsh conditions, however, they prove to be quite inadequate when the reactants' molecular dimensions are larger than their pores. Over time, with the discovery of MCM-41 [9, 69], mesoporous materials created a new buzz in this domain with their favorable properties such as large surface area and pore volume and pore dimensions that could be tuned, thus enabling catalytic reactions with bulky products. However, the amorphous walls of mesoporous materials provided lower stability and catalytic activity when compared to crystalline pore walls of zeolites. In contrast, MOFs have been garnering a lot of attention as they offer a large variety of frameworks with varied structures and compositions [92], [96]. These attractive features of MOFs form the basis for the large number of reviews published on MOF catalysis [31, 44, 80, 83, 106, 115]. However, their low thermal and chemical stability still poses a problem, and researchers are actively working for improvements in this context.

A detailed comparison of the structure and catalytic behavior is presented in Table 1) [75]

(a) Comparison of the structures of Zeolites, Mesoporous Silica, and MOFs

Structural comparison between MOFs, mesoporous silica, and zeolites is provided in Fig. 10.

(b) Comparison of the catalytic behavior of MOFs, Zeolites, and Mesoporous Silica

(i) Gas-phase reactions

Table 1 Depicts different structural features of MOFs, mesoporous silica, and zeolites and their advantages

Property	MOFs	Mesoporous silica	Zeolites
Pore walls	Crystalline porous structures prove to be advantageous for determining the composition and positioning of atoms in space	It has amorphous pore walls, suitable for facilitating the incorporation of heteroelements, thereby leading to sites with different local environments. However, major disadvantages are its low thermal and hydrothermal stability	Its crystalline porous structure offers advantages similar to those offered by MOFs
Pore size	MOFs have a pore size between mesoporous silica and zeolites which consequently bridge the gap between the two	With a pore size of 1 nm, mesoporous silica is suitable for catalytic reactions involving bulky products	Zeolites, with pore size larger than 1 nm prove to be an advantageous option for shape selectivity
Stability	The thermal and chemical stability of MOFs is comparatively lesser than inorganic porous solids, so much so that some of them even decompose in air, moisture, etc	They showcase good thermal stability, which can be effectively improved by increasing the wall thickness. Although their hydrothermal stability is much lower than that of zeolites	Zeolites generally display excellent thermal and hydrothermal stability. High silica zeolites can show stability up to 800 degrees celsius
Surface area	MOFs possess a high surface area, thereby permitting higher concentrations of active sites. The regular active sites are an added advantage. They behave as single-site catalysts	Although mesoporous silica has a high surface area, the distribution of active sites is inhomogeneous because of their amorphous walls	The distribution of active sites is although homogeneous, but random, which proves to be less advantageous than single-site catalysts

Zeolites, with their unique structural, chemical, and thermal stability, prove to be the ideal heterogeneous catalysts for gas-phase reactions carried out at temperatures > 300 degrees celsius. In petrochemical industries, zeolite-based catalysts are widely utilized for many well-established processes like aromatic alkylation, fluid catalytic cracking, hydrocracking, etc.

The success of zeolites in this domain although cannot just be attributed to their stability, but to their microporous structures and flexible chemical composition. The main advantage that zeolites show is that they can easily be reactivated to remove heavier residual products; unlike heterogeneous catalysts that tend to become deactivated as a result of adsorption or formation of residual products.

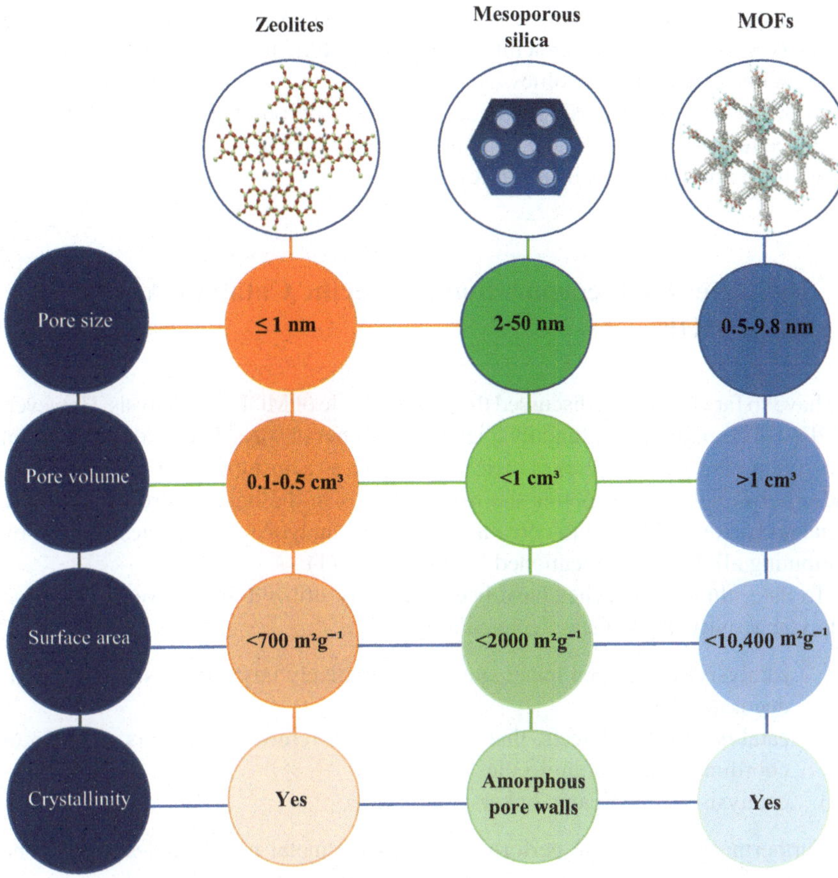

Fig. 10 A comparison of the structural features of MOFs, mesoporous silica, and zeolites

Therefore, it is still difficult for MOFs and mesoporous silica to compete with zeolites that still dominate this field.

(ii) **Liquid-phase reactions**

Unlimited pore size and lower thermal and chemical stability provide mesoporous silica and MOFs an edge over zeolites for liquid-phase reactions; liquid-phase reactions are typically carried out at temperatures < 200 °C. Mesoporous silica and MOFs complement zeolites in the production of fine chemicals that proceed in a liquid phase under mild conditions.

Zeolites have a propensity to open the epoxide ring easily, while mesoporous silica shows good potential for Friedel–Crafts acylation requiring lower acidic levels, and MOFs seem suitable for cycloaddition of CO_2 into epoxides because of possessing both, the acidic and basic sites.

In Knoevenagal condensation, interactions between substrates, intermediates, or products with active sites are facilitated by the catalytic behavior of MOFs, relative to negligible influence by zeolites.

In the case of alkene oxidation reactions, mesoporous silica with larger pores shows better performance in the oxidation of bulky alkenes while zeolites are more suited for oxidation of small substrates.

7 Strategies for Incorporation of Specific Catalytic Activity into a MOF

We have so far elaborately discussed the potential role of MOFs in catalysis. However, MOFs can be catalytically functionalized in a number of ways. For example, they can encapsulate the active species in their pores; serve as a host for metal NPs, enzymes, and other moieties; incorporate the catalytic sites in the organic struts linking the metal nodes; or by PSM of the organic linkers by catalytic metal species; or even by combining all the above-mentioned features (Fig. 11).

To date, three approaches have been primarily utilized to achieve the intrinsic catalytic activity of MOFs as depicted in Fig. 12.

(i) catalysis on the open metal sites (coordinatively unsaturated sites) and non-decorated MOFs.
(ii) catalysis on the defects, that arise from the removal of linkers, clusters of coordinated solvent molecules.
(iii) catalysis on the organic linkers.

Furthermore, MOFs can perform as a host for metal nanoparticles (NPs). The catalytic centers in MOFs are either metal ions that act as Lewis acids or organic linkers. The centers can be modified or tuned for them to act as supports for NPs.

Due to the high porosity and crystallinity of these composites, NPs are highly dispersed within the pores and are easily accessible as active sites.

In such composites, MOFs play numerous roles.

(i) They stabilize the metal NPs within pores.
(ii) They help in the selectivity of the reaction.
(iii) By controlling charge transfer and electronic density between the metal nanoparticle and MOF, they modify the electronic properties of metal NPs.
(iv) They can catalyze one-pot tandem reactions.

The catalytic performance of metal nanoparticle@MOF composites is capable of covering a wide range of reactions.

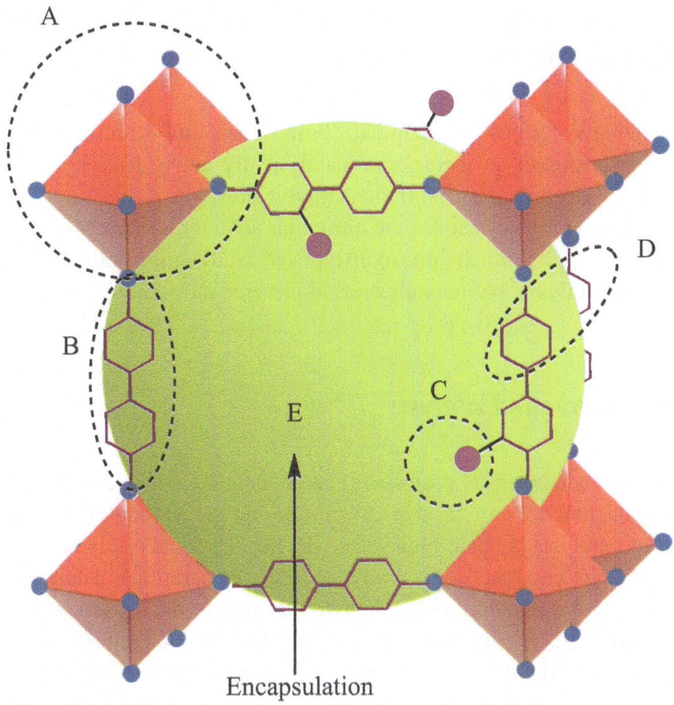

Fig. 11 Coarse representation of catalytically active sites in MOFs; (A) Sites on metal clusters, (B) Organic linkers (ligands such as BDC), (C) PSM grafted sites, (D) Catalytically active site, (E)Encapsulated MOF composite sites

Fig. 12 Three main strategies for achieving intrinsic catalytic activity in a MOF

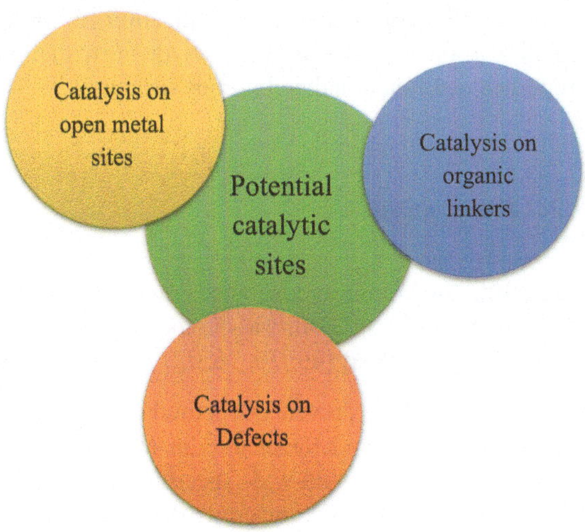

8 Different Catalytic Applications of MOFs: Appropriate Properties and Opportunities

With the extensive ranges of chemical composition and structures, MOFs offer substantial opportunities to incorporate catalytically active sites by tailoring pore structures and environments around the catalytic sites.

The following section discusses the particular advantages that MOFs provide for assorted catalytic applications; three main types of catalysis are deliberated namely—Heterogeneous catalysis, photocatalysis, and electrocatalysis.

8.1 Heterogeneous Catalysis

MOFs have proven to be a very promising class of heterogeneous catalysts for organic transformations, that act as a "bridge" by combining the advantages offered by both homogeneous catalysts (such as high selectivity, activity, and easily accessible sites), and heterogeneous catalysts (reusability and easy separation). Their uniform and tunable pore size and structure make them highly preferable for size-selective catalysis. Their high porosity and high surface area benefit the contact and interaction between reactants and catalytic sites, thereby improving the catalytic efficiency.

Their pore space provides the chiral environment, appropriate hydrophobic and hydrophilic pore nature, electron-rich and deficit circumstances, that greatly affect the reactivity and interaction of reactant molecules. The possibility to graft desired active sites through post-modification approaches provides an extra edge. All in all, MOF-derived porous catalysts exhibit high desirability for heterogeneous catalysis.

Figure 13 depicts the broad utilization of MOFs for the synthesis of cyclic organic carbonates (COCs) from CO_2, owing to their favorable physical and chemical properties and functionalized porous structure.

8.2 Photocatalysis

Apart from the common advantages mentioned above for heterogeneous catalysis, MOFs show superb suitability as photocatalysts, owing to the numerous opportunities that they have to offer: [29, 30, 36, 72, 139, 148], [152].

- High crystallinity prevents structural defects.
- Organic linkers and metal ions in MOFs can be rationally designed as light-harvesting centers with wide spectrum absorption.
- Metal clusters provide uniformly distributed catalytically active sites.
- There is a possibility of introducing photosensitizers like polyoxometalates (POMs), metal nanoparticles, semiconductors, etc., owing to the high surface

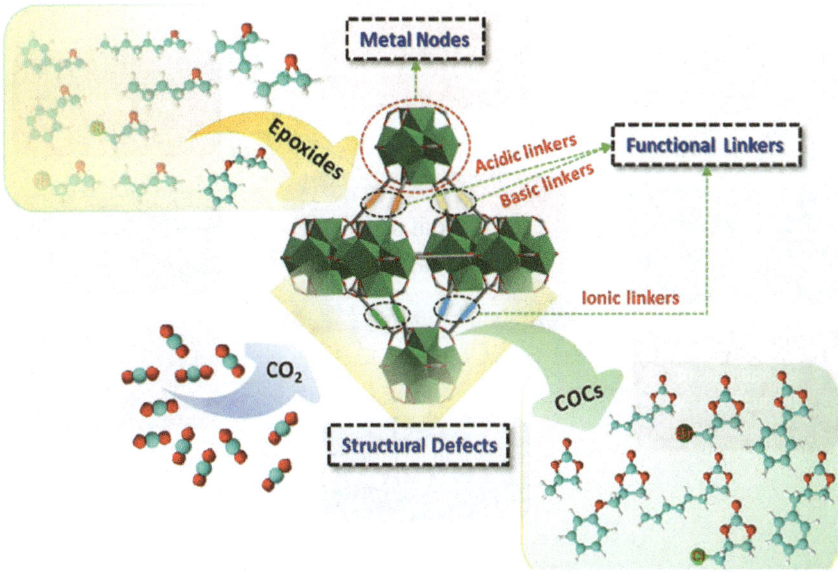

Fig. 13 Schematic representation of MOFs as heterogeneous catalysts for cycloaddition into organic carbonates, of CO_2 with epoxides. Reprinted from Coordination Chemistry Reviews, 387, Wen-Gang Cui, Guo-Ying Zhang, Tong-Liang Hu, Xian-He Bu, Metal–organic framework-based heterogeneous catalysts for the conversion of C1 chemistry: CO, CO2 and CH4, 42, Copyright (2019), with permission from Elsevier [22]

area and porous structures of MOFs, which further helps in achieving improved photocatalytic efficiency.

- CO_2 photoreduction reactions can be accelerated as their excellent CO_2 adsorption capacity may lead to a high concentration of CO_2 around their active sites [63, 72].
- MOF-derived porous materials possess exceptionally higher performance than their bulkier counterparts.

8.3 Electrocatalysis

In recent years, MOFs have emerged as the desirable materials for electrocatalytic applications [81, 138], [151]. MOFs and MOF derivatives are capable of driving significant electrochemical reactions like hydrogen evolution reactions (HER), oxygen reduction reactions (ORR), oxygen evolution reactions (OER), carbon dioxide reduction reactions (CO_2RR), among others. The large surface area, tailorable structures, and tunable pore sizes are highly desirable for electrocatalysis.

MOF-derived materials like doped porous carbons, metal oxides, nitrides, phosphides, etc., and their composites, easily obtained by chemical or thermal conversion have been studied intensively for electrocatalytic applications.

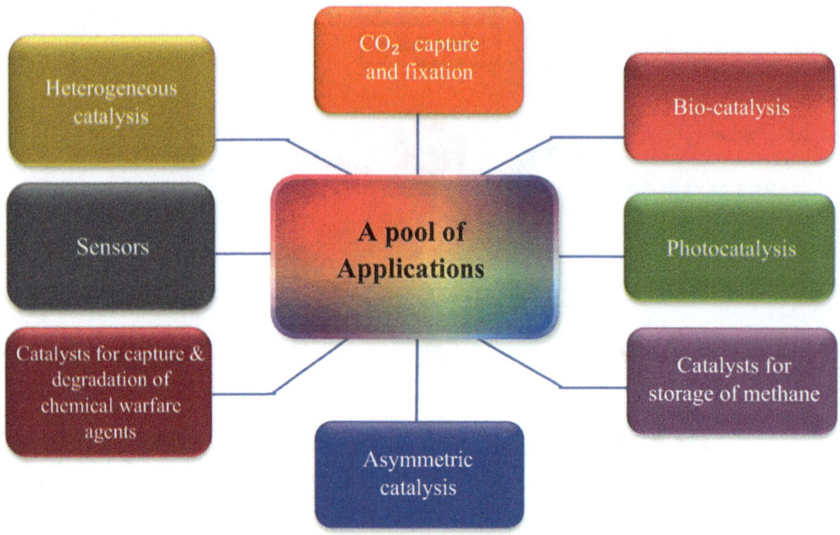

Fig. 14 Assorted applications of MOFs in catalysis discussed in the following chapters

This is only a glimpse of MOFs prowess in the field of catalysis. The relevance of MOFs has been realized in a number of applications, such as bio-catalysis, catalysis for CO_2 capture and fixation, sensors, asymmetric catalysis, catalysis for capture and degradation of chemical warfare agents (Fig. 14). A detailed discussion has been undertaken in the chapters ahead.

9 Some Examples of the Common MOFs-Catalyzed Reactions

Owing to their exceptional properties, MOFs tend to catalyze a variety of reactions. A good amount of these reactions can be catalyzed on a limited scale due to reactants being of higher value and under preferably mild conditions. Some of the examples include hydrogenation of ethylene, hexene, octene, and reactions of alcohols/CO_2 with epoxide ring opening. These catalysis reactions have added value to the literature on MOFs providing outcomes and deriving correlations between the structure and performance of the catalysts. Here, some of the enticing reactions catalyzed by MOFs are discussed (Table 2).

Table 2 Illustrative examples of some MOFs for assorted catalyzed reactions

MOFs	Node component (clustered core)	Organic ligand/linker	Pore aperture (Å)	Catalyzed reaction	Reference
UiO-66	$Zr_6O_4(OH)_4$	benzene-1,4-dicarboxylic acid	6	Dehydrohalogenation	[12, 80, 86, 127, 131, 143]
				Ring opening reaction	
				Cyclization	
				Hydrogenation, condensation	
UiO-67	$Zr_6O_4(OH)_4$	4,4'-biphenyldicarboxylic acid	8	Hydrogenation	[50]
MOF-808	$Zr_6O_3(OH)_3$	1,3,5-benzene tricarboxylate	14	Isomerization	[58]
Zr-MOF	$Zr_6O_4(OH)_4$	dibenzoate-substituted 2,2′-bipyridine, dibenzoate-substituted 2-phenylpyridine	4.8, 8.8	Oxidation	[132]
Pd-MOF	Pd	2-hydroxypyrimidinolate	3.4	Hydrogenation	[84]
IRMOFs (MOF-5)	Zn_4O	benzene-1,4-dicarboxylic acid	18	Alkylation	[104]
				Hydrogenation	
MIL-101	Cr_3O	benzene-1,4-dicarboxylic acid	23	Dehydrogenation	[61]
HKUST-1	Cu	1,3,5-benzene tricarboxylate	10, 14	Esterification	[123]
NU-1000	$Zr_6O_4(OH)_4$	tetratopic 1,3,6,8-tetrakis(p-benzoate)pyrene	12, 30	Hydrolysis	[11, 26, 74, 99, 102, 142]
				Hydrogenation	
				Oligomerization	
ZIF-8	Zn	2-methylimidazolate	8	Hydrogenation	[131]

9.1 CO/CO_2 Hydrogenation

MOFs have been used as catalysts in hydrogenation reactions for the production of various hydrocarbons, alcohols, and many other oxygenates in presence of CO, making it of great industrial importance;

However, in the case of CO_2 hydrogenation reactions, there are limited industrial applications but they have been actively investigated by catalysis researchers due to the availability, economic and environmental prospects; CO_2 being a greenhouse gas accumulated in the atmosphere. Table 1 illustrates some examples, such as UiO-67 MOFs [50] applied as catalysts for hydrogenation reactions using CO_2; additional cases can be found [5, 57, 111].

9.2 Alkyne Hydrogenation

One of the most important industrial reactions is the purification of alkynes such as ethylene and propylene by semi-hydrogenation reaction (Fig. 15). Conversion of a family of acyclic alkynes to alkenes has been catalyzed by NU-1000 MOFs having Rh-Ga bimetallic sites has been reported with very high stereoselectivity (greater than 99%) [27]. Although in many cases, limiting issues pertain to the stability and regenerability of catalysts.

9.3 Dehydrogenation

Another example of an extensively used industrial process is the dehydrogenation of alkanes. Some examples of industrial applications such as, dehydrogenation of propane reveal that the conditions under which they are undertaken aren't suitable for MOFs due to deployment of extreme conditions; temperatures can be > 500 °C. However, research has shown that dehydrogenation of propane at 230 °C using NU-1000 MOFs-supported Co clusters is feasible as the catalyst has sustained stability for 20 h (Fig. 16) [73].

9.4 Epoxidation

Epoxidation is another process that is amongst the important ones in the industrial context. MOFs are often used as supporters of the main catalyst involved in the epoxidation reaction as exemplified by Hupp et al. [99] using NU-1000-supported molybdenum oxide as the catalyst. Unlike conventional catalysts, the reaction didn't require extreme conditions and occurred only at 60 °C in presence of oxygen gas.

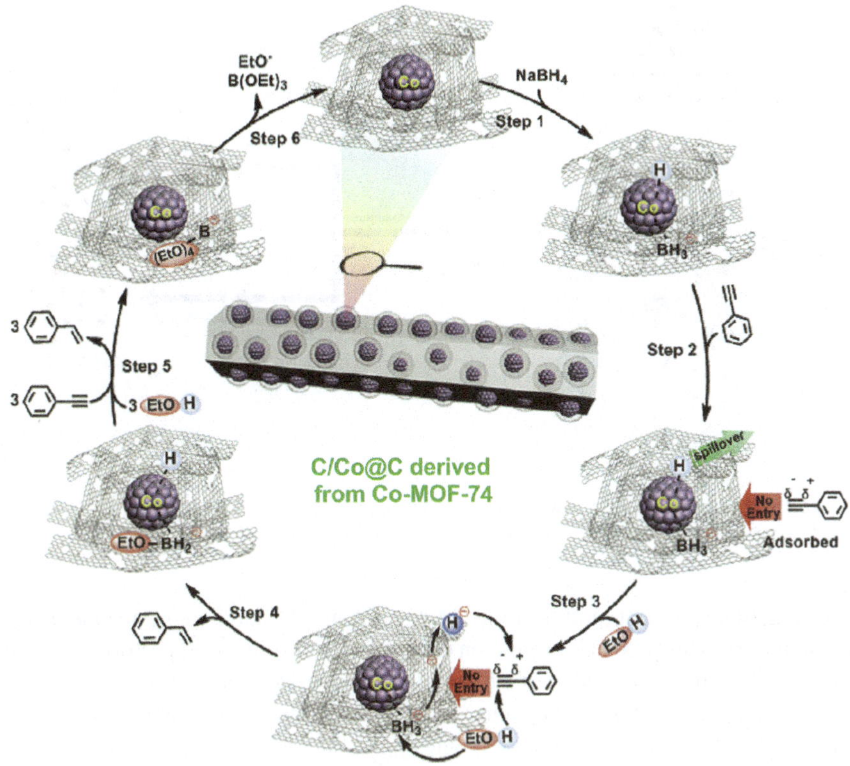

Fig. 15 Reaction mechanism for semi-hydrogenation of phenylacetylene to styrene using Co-MOF-74-derived catalyst by Caia et al. Reprinted from Carbon, 160, Jingyu Cai, Yi Chen, Haotian Song, Linxi Hou, Zhaohui Li, MOF derived C/Co@C with a "one-way-valve"-like graphitic carbon layer for selective semi-hydrogenation of aromatic alkynes, 7, 2020, with permission from Elsevier [13]

Another research group has reported the epoxidation of cyclopentene catalyzed by Mn-exchanged MOF-5 at room temperature. Zhou et al. gave a generally graphical representation of epoxidation of olefins using Cu-BTC MOF [150].

9.5 *Oligomerization*

In the polymer industries, processes such as oligomerization of olefins to produce monomers, polymers, and motor components are common. These reactions are usually catalyzed by metal complexes or acids. Presently, a number of MOFs have demonstrated their reliability for the oligomerization reaction. As an example, the dimerization of ethene to form 1-butene has been accomplished using MIL-101(Fe)-anchored nickel as a highly selective catalyst [14].

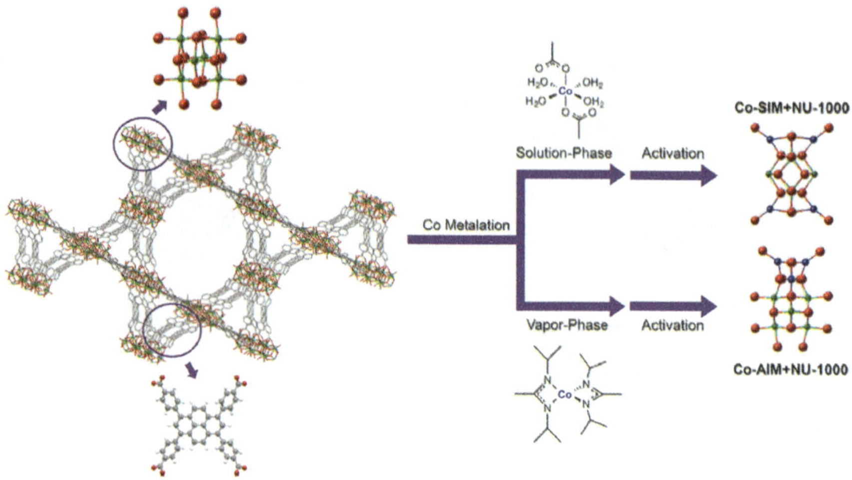

Fig. 16 Structural representation of NU-1000 MOF support and preparative routes for the MOF-based Co catalyst for oxidative dehydrogenation of propane. Reprinted from [73]

Coming back to post modifications, Metzger et al. [94] have suggested the incorporation of Ni sites in MFU-4 by cation exchange for the enhanced performance and selectivity for the dimerization of ethylene to 1-butene. Polymerization reactions for isobutylene, 1-hexene, and many other olefins have shown exceptional results under such mild conditions suggesting a bright future for MOFs being deployed in the mainstream of polymer industries.

9.6 Hydrolysis, Dehydration, Esterification, and Condensation

These are some of the most common reactions in the industrial context wherein MOFs can play crucial role in numerous applications such as the manufacture of various olefins and ethers. Water is an integral part of all these reactions and is catalyzed by active sites which are acidic in nature. For example, UiO-66 feasibly catalyzes the reaction of dehydration of ethanol to diethyl ether at 200 °C. However, the dehydration of alcohols to give alkene (olefins) requires temperature conditions > 300 °C where MOFs fail to assist due to stability reasons. In contrast to dehydration, MOFs show high activity in esterification reactions which occur at temperatures 60–80°[15, 58]. Figure 17 presents the esterification of benzyl alcohol and methanol carried out by Fan et al. [35]. The hydrolysis and condensation reactions occurred comfortably at room temperature and under mild conditions. Thus, MOFs are expected to have fascinating prospects once adequate studies are performed on their stability under realistically harsh conditions.

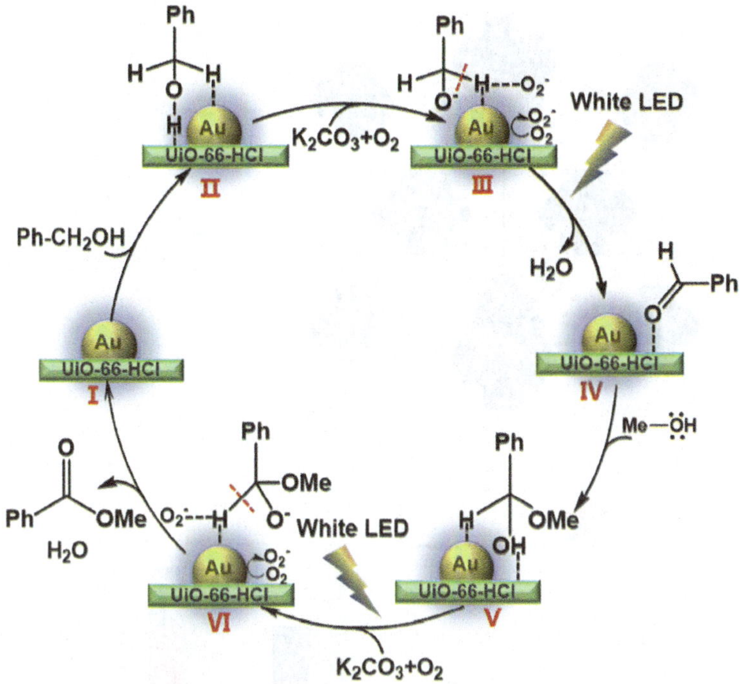

Fig. 17 Schematic representation of Au/UiO-66-HCl catalyzed (MOF-based catalyst) oxidative esterification of benzyl alcohol and methanol to the corresponding ester by fan et al. Reprinted from Catalysis Communications, 140, Chaoyang Fan, Ruiyi Wang, Peng Kong, Xiaoyu Wang, Jie Wang, Xiaochao Zhang, Zhanfeng Zheng, Modification of Au nanoparticles electronic state by MOFs defect engineering to realize highly active photocatalytic oxidative esterification of benzyl alcohol with methanol, 1, 2020, with permission from Elsevier [35]

9.7 Isomerization and Alkylation Reactions

These acid-catalyzed reactions are of immense importance in the industry. Skeletal isomerization reactions of olefins and kinds of paraffin aren't feasible for MOFs to catalyze because of the recurring drawback of not being stable at elevated temperatures. On the other hand, MOFs offer some prospects for reactions of cyclic hydrocarbons as they take place at relatively lower temperatures. Take an example of methylcyclopentane whose conversions take place at 150–200 °C [58]. It has been reported that NU-1000-supported catalyst could perform the isomerization reaction for o-xylene to m-xylene at very high temperatures, compromising with the stability of MOFs [2]. Figure 18 shows the graphical representation and comparison of results for isomerization of glucose and fructose.

Similar extreme temperature conditions are required for the alkylation reactions of paraffin and olefins which may utilize MOFs as potential catalysts. However, the

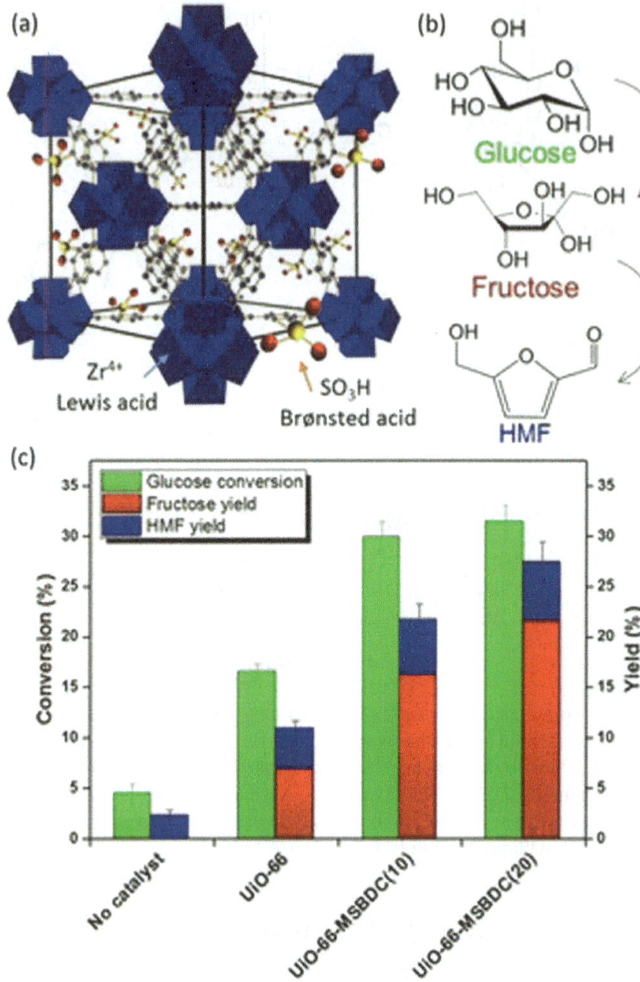

Fig. 18 **a** Structure of UiO-66 framework unit. **b** Isomerization of Glucose and Fructose to HMF (Hydroxymethylfurfural). **c** Graphical comparison of the efficiency of reaction using no catalyst, UiO-66 MOF, and its derivatives. Reprinted with permission from Oozeerally, R., Burnett, D.L., Chamberlain, T.W., Walton, R.I., Degirmenci, V., 2018. Exceptionally Efficient and Recyclable Heterogeneous Metal–Organic Framework Catalyst for Glucose Isomerization in Water. ChemCatChem 10, 706–709. https://doi.org/10.1002/cctc.201701825. Copyright (2018) American Chemical Society [101]

alkylation reaction of alcohols and benzene takes place at relatively lower temperatures of 100–200 °C which makes a number of MIL-series MOFs potential catalysts for limited periods of time.

9.8 *Photocatalysis*

In order to harness solar energy, photocatalysis provides a direct and sustainable path to convert it into clean chemical energy. MOFs owing to their versatile properties on different post-modification treatments tend to be photo-responsive by showing adsorption of light, an aspect broadly looks into in the following chapters.

10 Some Other Important Applications of MOFs

Luminescent MOF (LMOF) is a class of MOFs that have high potential in different sensing applications as described below.

10.1 *Luminescent Sensing*

Many substances such as organic solvents, water, aromatics, ions, etc. have been used to exploit the high-level performance of LMOFs with the transition metals in metal-ion sensing. Due to the interference of metallic ions (Cu^{2+}, Ag^{+}, Zn^{2+}, etc.), a reduction in the intensity of fluorescence of lanthanide-based MOF has been observed. The fluorescence spectrum of MOFs changed from showing multiple peaks to a single peak with the introduction of Ag^{+} ion in the coordination site. These alterations in the luminescence of the MOFs indicate the presence of metal ions.

Furthermore, not only in metal-ion but MOFs also have prospects in various other types of sensing such as biosensing, electrochemical sensing, aromatic sensing, among others which are discussed in the following chapters.

10.2 *CO_2 Capture and Photocatalytic Reduction*

The versatility of MOFs brings out exceptional adsorptive properties as well (Fig. 19). There have been many routes that have established an industry acceptance but are not yet as efficient in economic terms; MOFs have become breakthrough candidates for the adsorption of CO_2 [21].

The porosity of MOFs becomes an asset in giving them a large CO_2 uptake capacity and as a result, the level of CO_2 on the surface also increases to attain intimation between the gas and the catalytic sites.

Semiconductors have been the most explored candidates for the photocatalytic reduction of CO_2 where MOFs have made inroads only recently. In view of the low photogenerated charge and low charge separation efficiency, MOFs have not been exceptional as photocatalysts. However, with pre-and post-synthetic modification,

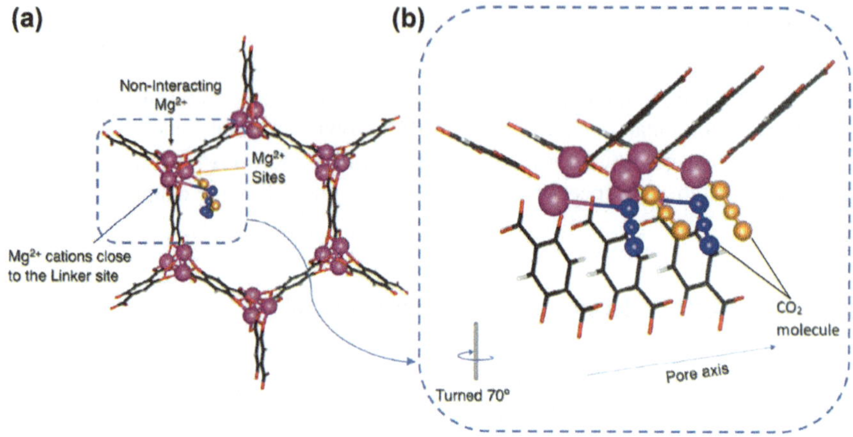

Fig. 19 3-D representation of adsorption of CO_2 molecules on Mg^{2+} and Linker sites on Mg-MOF-74 pore. Reprinted with permission from Alonso, G., Bahamon, D., Keshavarz, F., Giménez, X., Gamallo, P., Sayós, R., 2018. Density functional theory-based adsorption isotherms for pure and flue gas mixtures on Mg-MOF-74. Application in CO2 capture swing adsorption processes. J. Phys. Chem. C 122, 3945–3957. https://doi.org/10.1021/acs.jpcc.8b00938. Copyright (2018) American Chemical Society [4]

MOFs with high potential in efficient and low-cost photoreduction have attracted the interest; target-improved efficiency and chemical stability of MOFs being the two major attributes.

10.3 Biomedical Applications

High surface area, tunability of pores, and many other physical and chemical properties allow MOFs to be exploited in diverse biological and medical applications including biosensing, biomimetic catalysis, drug release, etc. Chemical stability is an essential and standout property that allows its usage in these biomedical applications. The robustness of MOFs plays an important role in order to show resistance against collapse due to hydrolysis under physiological environments.

Among a few examples, two MIL-series MOFs, MOF-68 and MOF-100 with a Fe core, showed colorimetric biosensor behavior for the detection of hydrogen peroxide [145]. MIL-53 with Fe and Cr cores (MOF-53(Fe) and MOF-53(Cr)) have been applied for the drug release of *ibuprofen* [55]. Owing to the flexibility, the slow and complete drug release was accomplished in a duration of 3 weeks. The use of biomolecular ligands such as adenine in order to enhance biocompatibility has also been a part of biomedical applications of MOFs.

11 Limitations in Catalytic Properties of MOFs

Numerous discoveries and novel demonstrations of the catalytic properties of MOFs made so far indicate the potential that researchers and scientists see in MOFs as catalysts, however, their application as catalysts is still lacking.

In order to put MOFs to feasible catalytic use, knowing their limitations is of utmost importance. Some of the major limitations of using MOFs as catalysts are discussed here.

11.1 Transport Limitations in Catalysis by MOFs

Narrow MOFs pores may inhibit the access of reactants inside the interior part of MOFs by offering transport resistance. When the transportation restrictions in pores are sufficient enough to influence the rate of reactions, measured rates become less than that of intrinsic ones, the kinetics of catalytic reactions are disguised, and hindrance in measurement rates of reactions in fundamental terms occurs. Corma, Garcia, and co-workers [84] demonstrated an example of reactant shape selectivity by choosing 1-octene and cyclododecene as reactants for hydrogenation in the MOF incorporating nodes of Pd ions (2 pymo = 2-hydroxypyrimidinolate). They reported that due to the pores of cyclododecene being too large to enter its pores, MOFs did not show activity for cyclododecene because of its failure in accessing the catalytically active Pd; it did exhibit sufficient activity for the hydrogenation of 1-octene. Another group [131] found that the platinum NPs in ZIF-8 did not exhibit activity for hydrogenation of cyclooctene because its critical diameter (5.5 Å) was large enough to fit into the MOF with an aperture diameter of 3.4 Å.

Synthesizing MOFs under harsh conditions has been considered as the solution to this limitation which may result in the generation of hierarchical pores and it is expected that the catalytic activity of the MOFs will increase by doing so.

11.2 Minority and Defect Structures in MOFs

Due to the non-uniformity in the MOF structures, identifying and counting the active sites has proven to be the central challenge in MOF catalysis. The major catalytic sites, as in MOFs having metal nodes, can be straightforwardly determined. But catalytic sites in some of them have defects in the crystalline structures. Vacancies like node sites, not linker bonded are one example. Some could have the linkers missing [137], while some may have node binding sites not occupied by liners, but other ligands formed during synthesis [42, 133].

In some cases, defects may be generated by modulators competing with linkers for bonding to nodes [116]. Nodes may be missing in some defects [16]; while missing linkers could be a case too. The Kieslich and Fischer group [32] showed that missing node defects are the consequence of the distorted distribution of linker defects within the MOF framework.

In MOF-catalyzed processes, investigations of defects have been mostly detailed in the context of catalysis on MOF nodes by the Lewis acid sites. However, there are still opportunities for future research on the investigation of other types of defect sites, for example, the -OH groups on defect sites.

11.3 Stability Limitations in MOFs

One major limitation of MOFs is their relatively low thermal and chemical stability. The thermal stability of a MOF is usually established by thermogravimetric analysis (TGA) in which the temperature is raised enormously (typically up to 850-1000 K) up to the point where the framework is completely destroyed. This temperature is high enough to carry out MOF-catalyzed liquid-phase reactions (typically carried out at temperatures < 473 K). However, structures that exhibit stability over 573 K might undergo extensive damage at lower temperatures when exposed for longer periods of time.

Another major issue that can restrict their stability is the mechanical resistance of MOFs. Upon submission to mechanical compression, some MOFs might lose their crystallinity, robustness, and specific surface area and just collapse after a limiting pressure. However, the stability of MOFs as catalysts is not only influenced by their intrinsic resistance, but also by their compatibility with reactants such as O_2 and other compounds that can break the metal-linked bonds.

12 Conclusion and Future Outlook

Over the past two decades, MOF-related catalytic applications are advancing at a tremendous rate. These MOFs, endowed with a variety of catalytic sites, provide a vast range of chemical compositions and structures for catalysis. Since their advent, researchers and scientists have believed in their competence in the field of catalysis, owing to the extensive opportunities that they have to offer, such as high metal content, large surface areas, chemical mutability, structural diversity, and well-defined porosity, among others. Furthermore, the post-modification ability of its organic linkers and the possibility to incorporate catalytically active species such as metal NPs, enzymes, etc. into its pores, enrich their catalytic activity considerably, making them potential competitors to traditional crystalline porous materials for multitudinous catalytic applications.

However, despite the striking advances made in recent years, in order to put them into practical application as catalysts, a few challenges need to be circumvented: (i) cheap and large-scale high-yield production of MOFs is yet to be achieved; (ii) the thermal, chemical and mechanical stability of MOFs is still unsatisfactory, which further makes high-temperature regeneration process challenging; (iii) the mass transfer resistance to large substrates due to the microporous structures of MOFs is problematic; (iv) Poor conductivity of MOF catalysts affect the charge transfer in photocatalysis and electrocatalysis. Thus, it can be justifiably said that research on MOFs has entered a new phase where challenges and opportunities coexist, and there is enormous scope for future research and improvement in this field. Undoubtedly, by virtue of the wide-ranging strengths and opportunities of MOF structures, they have the potential to be exploited for practical applications in industry in the near future.

Abbreviations

MOF	Metal-Organic Frameworks
BASF	Baden Aniline and Soda Factory
NPs	Nanoparticles
SBU	Secondary Building Unit
HKUST	Hong Kong University of Science and Technology-series MOFs
BTC	1,3,5-Benzenetricarboxylate
MIL-series	Matériaux de l'Institut Lavoisier-series
NG	Neat Grinding
LAG	Liquid-assisted grinding
ILAG	Ion-and-liquid assisted grinding
INA	1,5-Bis(n-benzyloxycarbonyl-l-leucinyl)carbohydrazide
PSM	Post-synthetic modifications
PSE	Post-synthetic exchange
PSD	Post-synthetic deprotection
UiO	Universitetet i Oslo-series MOFs
CAU-1	$[Al_4(OH)_2(OCH_3)_4(H_2N\text{-}BDC)_3]\cdot xH_2O$ MOF
SALI	Solvent-Assisted Ligand Incorporation
ZIFs	Zeolitic imidazolate frameworks
MCM	Mobil Composition of Matter
COC	Cyclic organic carbonates
POMs	Polyoxometalates
HER	Hydrogen evolution reactions
ORR	Oxygen reduction reactions
OER	Oxygen evolution reactions
CO_2RR	Carbon dioxide reduction reactions
NU-1000	Zirconium-based MOF
HMF	Hydroxymethylfurfural
LMOF	Luminescent Metal–Organic Frameworks

References

1. A. Dhakshinamoorthy, HG (2014) Metal organic frameworks as solid catalysts for the synthesis of nitrogen-containing heterocycles Table of Content Ph Ph O This Critical Review summarizes the recent developments in the use of metal-organic frameworks as catalysts for the synthesis of nit
2. Ahn S, Nauert SL, Buru CT, Rimoldi M, Choi H, Schweitzer NM, Hupp JT, Farha OK, Notestein JM (2018) Pushing the limits on metal-organic frameworks as a catalyst support: NU-1000 supported tungsten catalysts for o-xylene isomerization and disproportionation. J Am Chem Soc 140:8535–8543. https://doi.org/10.1021/jacs.8b04059
3. Ahnfeldt T, Guillou N, Gunzelmann D, Margiolaki I, Loiseau T, Férey G, Senker J, Stock N (2009) [Al4(OH)2(OCH3)4(H 2N- Bdc)3]·xH2O: A 12-connected porous metal-organic framework with an unprecedented aluminum-containing brick. Angew Chemie—Int Ed 48:5163–5166. https://doi.org/10.1002/anie.200901409
4. Alonso G, Bahamon D, Keshavarz F, Giménez X, Gamallo P, Sayós R (2018) Density functional theory-based adsorption isotherms for pure and flue gas mixtures on Mg-MOF-74. Application in CO2 capture swing adsorption processes. J Phys Chem C 122:3945–3957. https://doi.org/10.1021/acs.jpcc.8b00938
5. An B, Zhang J, Cheng K, Ji P, Wang C, Lin W (2017) Confinement of ultrasmall Cu/ZnOx nanoparticles in metal-organic frameworks for selective methanol synthesis from catalytic hydrogenation of CO2. J Am Chem Soc 139:3834–3840. https://doi.org/10.1021/jacs.7b00058
6. Arbor A (2003) (12) United States Patent 1
7. Ayoub G, Karadeniz B, Howarth AJ, Farha OK, Lilović I, Germann LS, Dinnebier RE, Užarević K, Friščić T (2019) Rational synthesis of mixed-metal microporous metal-organic frameworks with controlled composition using mechanochemistry. Chem Mater 31:5494–5501. https://doi.org/10.1021/acs.chemmater.9b01068
8. Bauer S, Serre C, Devic T, Horcajada P, Férey G, Stock N (2008) High-Throughput assisted rationalization of the formation iron (III) aminoterephthalate solvothermal system. Inorg Chem 9:7568–7576
9. Beck JS, Vartuli JC, Roth WJ, Leonowicz ME, Kresge CT, Schmitt KD, Chu CTW, Olson DH, Sheppard EW, McCullen SB, Higgins JB, Schlenker JL (1992) A new family of mesoporous molecular sieves prepared with liquid crystal templates. J Am Chem Soc 114:10834–10843. https://doi.org/10.1021/ja00053a020
10. Beitollahi H, Van Le Q, Farha OK, Shokouhimehr M, Tajik S, Nejad FG, Kirlikovali KO, Jang HW, Varma RS (2020) Recent electrochemical applications of metal-organic framework-based materials. Cryst Growth Des 20:7034–7064. https://doi.org/10.1021/acs.cgd.0c00601
11. Bernales V, Yang D, Yu J, Gümüşlu G, Cramer CJ, Gates BC, Gagliardi L (2017) Molecular rhodium complexes supported on the metal-oxide-like nodes of metal organic frameworks and on zeolite HY: Catalysts for ethylene hydrogenation and dimerization. ACS Appl Mater Interfaces 9:33511–33520. https://doi.org/10.1021/acsami.7b03858
12. Cai G, Jiang HL (2017) A modulator-induced defect-formation strategy to hierarchically porous metal-organic frameworks with high stability. Angew. Chemie - Int. Ed. 56:563–567. https://doi.org/10.1002/anie.201610914
13. Cai J, Chen Y, Song H, Hou L, Li Z (2020) MOF derived C/Co@C with a "one-way-valve"-like graphitic carbon layer for selective semi-hydrogenation of aromatic alkynes. Carbon N. Y. 160:64–70. https://doi.org/10.1016/j.carbon.2020.01.006
14. Canivet J, Aguado S, Schuurman Y, Farrusseng D (2013) MOF-supported selective ethylene dimerization single-site catalysts through one-pot postsynthetic modification. J Am Chem Soc 135:4195–4198. https://doi.org/10.1021/ja312120x
15. Caratelli C, Hajek J, Cirujano FG, Waroquier M, Llabrés i Xamena FX, Van Speybroeck V (2017) Nature of active sites on UiO-66 and beneficial influence of water in the catalysis of Fischer esterification. J Catal 352:401–414. https://doi.org/10.1016/j.jcat.2017.06.014

16. Cliffe MJ, Wan W, Zou X, Chater PA, Kleppe AK, Tucker MG, Wilhelm H, Funnell NP, Coudert FX, Goodwin AL (2014) Correlated defect nanoregions in a metal-organic framework. Nat Commun 5:1–8. https://doi.org/10.1038/ncomms5176
17. Clough AJ, Yoo JW, Mecklenburg MH, Marinescu SC (2015) Two-dimensional metal-organic surfaces for efficient hydrogen evolution from water. J Am Chem Soc 137:118–121. https://doi.org/10.1021/ja5116937
18. Cohen SM (2012) Postsynthetic methods for the functionalization of metal-organic frameworks. Chem Rev 112:970–1000. https://doi.org/10.1021/cr200179u
19. Comyns AE (2001) Chemical process technology. JA Moulijn, M Makkee, A van Diepen, John Wiley and Sons Ltd, Chichester, 2001. xii?+?453 pages. 27.50 ISBN 0–471–63062–4 (paperback). Appl Organomet Chem 15:956–956. https://doi.org/10.1002/aoc.202
20. Couck S, Denayer JFM, Baron GV, Rémy T, Gascon J, Kapteijn F (2009) An amine-functionalized MIL-53 metal-organic framework with large separation power for CO2 and CH4. J Am Chem Soc 131:6326–6327. https://doi.org/10.1021/ja900555r
21. Crake A, Christoforidis KC, Kafizas A, Zafeiratos S, Petit C (2017) CO2 capture and photocatalytic reduction using bifunctional TiO2/MOF nanocomposites under UV–vis irradiation. Appl Catal B Environ 210:131–140. https://doi.org/10.1016/j.apcatb.2017.03.039
22. Cui W, Zhang G, Hu T, Bu X (2019) Metal-organic framework-based heterogeneous catalysts for the conversion of C1 chemistry: CO, CO 2 and CH 4 q. Coord Chem Rev 387:79–120. https://doi.org/10.1016/j.ccr.2019.02.001
23. Cui Y, Yue Y, Qian G, Chen B (2012) Luminescent functional metal à organic frameworks 1126–1162
24. Dekrafft KE, Wang C, Lin W (2012) Metal-organic framework templated synthesis of Fe 2O 3/TiO 2 nanocomposite for hydrogen production. Adv Mater 24:2014–2018. https://doi.org/10.1002/adma.201200330
25. Deng H, Grunder S, Cordova KE, Valente C, Furukawa H, Hmadeh M, Gándara F, Whalley AC, Liu Z, Asahina S, Kazumori H, O'Keeffe M, Terasaki O, Stoddart JF, Yaghi OM (2012) Large-pore apertures in a series of metal-organic frameworks. Science (80-.). 336:1018–1023. https://doi.org/10.1126/science.1220131
26. Deria P, Mondloch JE, Tylianakis E, Ghosh P, Bury W, Snurr RQ, Hupp JT, Farha OK (2013) Perfluoroalkane functionalization of NU-1000 via solvent-assisted ligand incorporation: synthesis and CO2 adsorption studies. J Am Chem Soc 135:16801–16804. https://doi.org/10.1021/ja408959g
27. Desai SP, Ye J, Zheng J, Ferrandon MS, Webber TE, Platero-Prats AE, Duan J, Garcia-Holley P, Camaioni DM, Chapman KW, Delferro M, Farha OK, Fulton JL, Gagliardi L, Lercher JA, Penn RL, Stein A, Lu CC (2018) Well-defined rhodium-gallium catalytic sites in a metal-organic framework: promoter-controlled selectivity in alkyne semihydrogenation to e-alkenes. J Am Chem Soc 140:15309–15318. https://doi.org/10.1021/jacs.8b08550
28. Deutschmann O, Knözinger H, Kochloefl K, Turek T (2011) Heterogeneous catalysis and solid catalysts, 1. Fundamentals. Ullmann's Encycl Ind Chem. https://doi.org/10.1002/14356007.a05_313.pub3
29. Dhakshinamoorthy A, Asiri AM, García H (2016) Metal-organic framework (MOF) compounds: photocatalysts for redox reactions and solar fuel production. Angew. Chemie - Int. Ed. 55:5414–5445. https://doi.org/10.1002/anie.201505581
30. Dhakshinamoorthy A, Li Z, Garcia H (2018) Catalysis and photocatalysis by metal organic frameworks. Chem Soc Rev 47:8134–8172. https://doi.org/10.1039/c8cs00256h
31. Dhakshinamoorthy A, Opanasenko M, Čejka J, Garcia H (2013) Metal organic frameworks as heterogeneous catalysts for the production of fine chemicals. Catal Sci Technol 3:2509–2540. https://doi.org/10.1039/c3cy00350g
32. Dissegna S, Epp K, Heinz WR, Kieslich G, Fischer RA (2018) Defective metal-organic frameworks. Adv Mater 30:1–23. https://doi.org/10.1002/adma.201704501
33. Do JL, Friščić T (2017) Mechanochemistry: a force of synthesis. ACS Cent Sci 3:13–19. https://doi.org/10.1021/acscentsci.6b00277

34. Evans JD, Sumby CJ, Doonan CJ (2014) Post-synthetic metalation of metal-organic frameworks. Chem Soc Rev 43:5933–5951. https://doi.org/10.1039/c4cs00076e
35. Fan C, Wang R, Kong P, Wang X, Wang J, Zhang X, Zheng Z (2020) Modification of Au nanoparticles electronic state by MOFs defect engineering to realize highly active photocatalytic oxidative esterification of benzyl alcohol with methanol. Catal Commun 140:106002. https://doi.org/10.1016/j.catcom.2020.106002
36. Fang Y, Ma Y, Zheng M, Yang P, Asiri AM, Wang X (2018) Metal–organic frameworks for solar energy conversion by photoredox catalysis. Coord Chem Rev 373:83–115. https://doi.org/10.1016/j.ccr.2017.09.013
37. Farrusseng D, Aguado S, Pinel C (2009) Metal-organic frameworks: opportunities for catalysis. Angew. Chemie - Int. Ed. 48:7502–7513. https://doi.org/10.1002/anie.200806063
38. Fei H, Shin J, Meng YS, Adelhardt M, Sutter J, Meyer K, Cohen SM (2014) Reusable oxidation catalysis using metal-monocatecholato species in a robust metal-organic framework. J Am Chem Soc 136:4965–4973. https://doi.org/10.1021/ja411627z
39. Férey C, Mellot-Draznieks C, Serre C, Millange F, Dutour J, Surblé S, Margiolaki I (2005) Chemistry: a chromium terephthalate-based solid with unusually large pore volumes and surface area. Science (80-.). 309:2040–2042. https://doi.org/10.1126/science.1116275
40. Fujita M, Kwon J (1994) Preparation, clathration ability, and catalysis of a two-dimensional square network material composed of cadmium(11) and 4,Y-Bipyridine 1151–1152. https://doi.org/10.1021/ja00082a055
41. Furukawa H, Cordova KE, O'Keeffe M, Yaghi OM (2013) The chemistry and applications of metal-organic frameworks. Science (80-.). 341. https://doi.org/10.1126/science.1230444
42. Furukawa H, Gándara F, Zhang YB, Jiang J, Queen WL, Hudson MR, Yaghi OM (2014) Water adsorption in porous metal-organic frameworks and related materials. J Am Chem Soc 136:4369–4381. https://doi.org/10.1021/ja500330a
43. Garibay SJ, Cohen SM (2010) Isoreticular synthesis and modification of frameworks with the UiO-66 topology. Chem Commun 46:7700–7702. https://doi.org/10.1039/c0cc02990d
44. Gascon J, Corma A, Kapteijn F, Llabrés I, Xamena FX (2014) Metal organic framework catalysis: Quo vadis? ACS Catal 4:361–378. https://doi.org/10.1021/cs400959k
45. Germann LS, Katsenis AD, Huskić I, Julien PA, Užarević K, Etter M, Farha OK, Friščić T, Dinnebier RE (2020) Real-time in situ monitoring of particle and structure evolution in the mechanochemical synthesis of UiO-66 metal-organic frameworks. Cryst Growth Des 20:49–54. https://doi.org/10.1021/acs.cgd.9b01477
46. Getman RB, Bae Y, Wilmer CE, Snurr RQ, Carlo M (2012) Review and analysis of molecular simulations of methane , hydrogen , and acetylene storage in metal à organic frameworks 703–723.
47. Gkaniatsou E, Sicard C, Ricoux R, Mahy JP, Steunou N, Serre C (2017) Metal-organic frameworks: a novel host platform for enzymatic catalysis and detection. Mater. Horizons 4:55–63. https://doi.org/10.1039/c6mh00312e
48. Goesten MG, Kapteijn F, Gascon J (2013) Fascinating chemistry or frustrating unpredictability: observations in crystal engineering of metal-organic frameworks. CrystEngComm 15:9249–9257. https://doi.org/10.1039/c3ce41241e
49. Gu ZY, Yan XP (2010) Metal-organic framework MIL-101 for high-resolution gaschromatographic separation of xylene isomers and ethlbenzene. Angew. Chemie - Int. Ed. 49:1477–1480. https://doi.org/10.1002/anie.200906560
50. Gutterød ES, Øien-ØDegaard S, Bossers K, Nieuwelink AE, Manzoli M, Braglia L, Lazzarini A, Borfecchia E, Ahmadigoltapeh S, Bouchevreau B, Lønstad-Bleken BT, Henry R, Lamberti C, Bordiga S, Weckhuysen BM, Lillerud KP, Olsbye U (2017) CO2 hydrogenation over Pt-containing UiO-67 Zr-MOFs—the base case. Ind Eng Chem Res 56:13206–13218. https://doi.org/10.1021/acs.iecr.7b01457
51. Hasegawa S, Horike S, Matsuda R, Furukawa S, Mochizuki K, Kinoshita Y, Kitagawa S (2007) Three-dimensional porous coordination polymer functionalized with amide groups based on tridentate ligand: Selective sorption and catalysis. J Am Chem Soc 129:2607–2614. https://doi.org/10.1021/ja067374y

52. Henschel A, Gedrich K, Kraehnert R, Kaskel S (2008) Catalytic properties of MIL-101. Chem Commun 4192–4194. https://doi.org/10.1039/b718371b
53. Horcajada P, Chalati T, Serre C, Gillet B, Sebrie C, Baati T, Eubank JF, Heurtaux D, Clayette P, Kreuz C, Chang JS, Hwang YK, Marsaud V, Bories PN, Cynober L, Gil S, Férey G, Couvreur P, Gref R (2010) Porous metal-organic-framework nanoscale carriers as a potential platform for drug deliveryand imaging. Nat Mater 9:172–178. https://doi.org/10.1038/nmat2608
54. Horcajada P, Gref R, Baati T, Allan PK, Maurin G, Couvreur P (2012) Metal À organic frameworks in biomedicine 1232–1268
55. Horcajada P, Serre C, Maurin G, Ramsahye NA, Balas F, Vallet-Regí M, Sebban M, Taulelle F, Férey G (2008) Flexible porous metal-organic frameworks for a controlled drug delivery. J Am Chem Soc 130:6774–6780. https://doi.org/10.1021/ja710973k
56. Hu ML, Safarifard V, Doustkhah E, Rostamnia S, Morsali A, Nouruzi N, Beheshti S, Akhbari K (2018) Taking organic reactions over metal-organic frameworks as heterogeneous catalysis. Microporous Mesoporous Mater 256:111–127. https://doi.org/10.1016/j.micromeso.2017.07.057
57. Isaeva VI, Eliseev OL, Kazantsev RV, Chernyshev VV, Davydov PE, Saifutdinov BR, Lapidus AL, Kustov LM (2016) Fischer-Tropsch synthesis over MOF-supported cobalt catalysts (Co@MIL-53(Al)). Dalt. Trans. 45:12006–12014. https://doi.org/10.1039/c6dt01394e
58. Jiang J, Gándara F, Zhang YB, Na K, Yaghi OM, Klemperer WG (2014) Superacidity in sulfated metal-organic framework-808. J Am Chem Soc 136:12844–12847. https://doi.org/10.1021/ja507119n
59. Jiang J, Yaghi OM (2015) Brønsted acidity in metal-organic frameworks. Chem Rev 115:6966–6997. https://doi.org/10.1021/acs.chemrev.5b00221
60. Juan-Alcañiz J, Gascon J, Kapteijn F (2012) Metal-organic frameworks as scaffolds for the encapsulation of active species: state of the art and future perspectives. J Mater Chem 22:10102–10119. https://doi.org/10.1039/c2jm15563j
61. Juan-Alcañiz J, Ramos-Fernandez EV, Lafont U, Gascon J, Kapteijn F (2010) Building MOF bottles around phosphotungstic acid ships: one-pot synthesis of bi-functional polyoxometalate-MIL-101 catalysts. J Catal 269:229–241. https://doi.org/10.1016/j.jcat.2009.11.011
62. Jung DW, Yang DA, Kim J, Kim J, Ahn WS (2010) Facile synthesis of MOF-177 by a sonochemical method using 1-methyl-2-pyrrolidinone as a solvent. Dalt. Trans. 39:2883–2887. https://doi.org/10.1039/b925088c
63. Kajiwara T, Fujii M, Tsujimoto M, Kobayashi K, Higuchi M, Tanaka K, Kitagawa S (2016) Photochemical reduction of low concentrations of CO 2 in a porous coordination polymer with a ruthenium(II)–CO complex. Angew. Chemie 128:2747–2750. https://doi.org/10.1002/ange.201508941
64. Kalmutzki MJ, Hanikel N, Yaghi OM (2018) Secondary building units as the turning point in the development of the reticular chemistry of MOFs. Sci Adv 4. https://doi.org/10.1126/sciadv.aat9180
65. Kee MO, Yaghi OM (2012) Deconstructing the crystal structures of metal à organic frameworks and related materials into their underlying nets 675–702
66. Khan NA, Kang IJ, Seok HY, Jhung SH (2011) Facile synthesis of nano-sized metal-organic frameworks, chromium-benzenedicarboxylate, MIL-101. Chem Eng J 166:1152–1157. https://doi.org/10.1016/j.cej.2010.11.098
67. Kim M, Cahill JF, Fei H, Prather KA, Cohen SM (2012) Postsynthetic ligand and cation exchange in robust metal-organic frameworks. J Am Chem Soc 134:18082–18088. https://doi.org/10.1021/ja3079219
68. Kortlever R, Shen J, Schouten KJP, Calle-Vallejo F, Koper MTM (2015) Catalysts and reaction pathways for the electrochemical reduction of carbon dioxide. J Phys Chem Lett 6:4073–4082. https://doi.org/10.1021/acs.jpclett.5b01559
69. Kresge et al (1992) Ordered mesoporous molecular sieves synthesized by a liquid-crystal template mechanism. Nature 359:167–169
70. Kurmoo M (2009) Magnetic metal–organic frameworks. https://doi.org/10.1039/b804757j

71. Li J, Sculley J, Zhou H (2012) Metal À organic frameworks for separations 869–932
72. Li R, Zhang W, Zhou K (2018) Metal–organic-framework-based catalysts for photoreduction of CO2. Adv Mater 30:1–31. https://doi.org/10.1002/adma.201705512
73. Li Z, Peters AW, Bernales V, Ortuño MA, Schweitzer NM, Destefano MR, Gallington LC, Platero-Prats AE, Chapman KW, Cramer CJ, Gagliardi L, Hupp JT, Farha OK (2017) Metal-organic framework supported cobalt catalysts for the oxidative dehydrogenation of propane at low temperature. ACS Cent Sci 3:31–38. https://doi.org/10.1021/acscentsci.6b00290
74. Li Z, Schweitzer NM, League AB, Bernales V, Peters AW, Getsoian AB, Wang TC, Miller JT, Vjunov A, Fulton JL, Lercher JA, Cramer CJ, Gagliardi L, Hupp JT, Farha OK (2016) Sintering-resistant single-site nickel catalyst supported by metal-organic framework. J Am Chem Soc 138:1977–1982. https://doi.org/10.1021/jacs.5b12515
75. Liang J, Liang Z, Zou R, Zhao Y (2017) Heterogeneous catalysis in zeolites, mesoporous silica, and metal-organic frameworks. Adv Mater 29:1–21. https://doi.org/10.1002/adma.201701139
76. Lillerud KP, Olsbye U, Tilset M (2010) Designing heterogeneous catalysts by incorporating enzyme-like functionalities into MOFs. Top Catal 53:859–868. https://doi.org/10.1007/s11244-010-9518-4
77. Liqing Ma CA, WL (2009) Enantioselective catalysis with homochiral metal—organic frameworks. https://doi.org/10.1039/b807083k
78. Liu B, Shioyama H, Akita T, Xu Q (2008) Metal-organic framework as a template for porous carbon synthesis. J Am Chem Soc 130:5390–5391. https://doi.org/10.1021/ja7106146
79. Liu HK, Tsao TH, Zhang YT, Lin CH (2009) Microwave synthesis and single-crystal-to-single-crystal transformation of magnesium coordination polymers exhibiting selective gas adsorption and luminescence properties. Cryst Eng Comm 11:1462–1468. https://doi.org/10.1039/b819559e
80. Liu J, Chen L, Cui H, Zhang J, Zhang L, Su CY (2014) Applications of metal-organic frameworks in heterogeneous supramolecular catalysis. Chem Soc Rev 43:6011–6061. https://doi.org/10.1039/c4cs00094c
81. Liu J, Zhu DD, Guo CX, Vasileff A, Qiao SZ (2017) Design strategies toward advanced mof-derived electrocatalysts for energy-conversion reactions. Adv Energy Mater 7:1–26. https://doi.org/10.1002/aenm.201700518
82. Liu Y, Su Y, Quan X, Fan X, Chen S, Yu H, Zhao H, Zhang Y, Zhao J (2018) Facile ammonia synthesis from electrocatalytic N2 reduction under ambient conditions on N-doped porous carbon. ACS Catal 8:1186–1191. https://doi.org/10.1021/acscatal.7b02165
83. Llabre FX (2010) Engineering metal organic frameworks for heterogeneous catalysis 4606–4655
84. Llabrés i Xamena FX, Abad A, Corma A, Garcia H (2007) MOFs as catalysts: activity, reusability and shape-selectivity of a Pd-containing MOF. J Catal 250:294–298. https://doi.org/10.1016/j.jcat.2007.06.004
85. Llewellyn PL, Bourrelly S, Serre C, Vimont A, Daturi M, Hamon L, Weireld GD, Chang J, Hong D, Hwang YK, Jhung SH (2008) Llewellyn 2008:7245–7250
86. Luz I, Rösler C, Epp K, Llabrés I Xamena FX, Fischer RA (2015) Pd@UiO-66-Type MOFs prepared by chemical vapor infiltration as shape-selective hydrogenation catalysts. Eur J Inorg Chem 2015:3904–3912. https://doi.org/10.1002/ejic.201500299
87. Lykourinou V, Chen Y, Wang XS, Meng L, Hoang T, Ming LJ, Musselman RL, Ma S (2011) Immobilization of MP-11 into a mesoporous metal-organic framework, MP-11@mesoMOF: a new platform for enzymatic catalysis. J Am Chem Soc 133:10382–10385. https://doi.org/10.1021/ja2038003
88. Ma S, Goenaga GA, Call AV, Liu DJ (2011) Cobalt imidazolate framework as precursor for oxygen reduction reaction electrocatalysts. Chem. - A Eur. J. 17:2063–2067. https://doi.org/10.1002/chem.201003080
89. Ma TY, Dai S, Jaroniec M, Qiao SZ (2014) Metal-organic framework derived hybrid Co3O4-carbon porous nanowire arrays as reversible oxygen evolution electrodes. J Am Chem Soc 136:13925–13931. https://doi.org/10.1021/ja5082553

90. Martinez Joaristi A, Juan-Alcañiz J, Serra-Crespo P, Kapteijn F, Gascon J (2012) Electrochemical synthesis of some archetypical Zn 2+, Cu 2+, and Al 3+ metal organic frameworks. Cryst Growth Des 12:3489–3498. https://doi.org/10.1021/cg300552w
91. McDonald TM, Lee WR, Mason JA, Wiers BM, Hong CS, Long JR (2012) Capture of carbon dioxide from air and flue gas in the alkylamine-appended metal-organic framework mmen-Mg 2(dobpdc). J Am Chem Soc 134:7056–7065. https://doi.org/10.1021/ja300034j
92. Meek ST, Greathouse JA, Allendorf MD (2011) Metal-organic frameworks: a rapidly growing class of versatile nanoporous materials. Adv Mater 23:249–267. https://doi.org/10.1002/adma.201002854
93. Meilikhov M, Yusenko K, Esken D, Turner S, Van Tendeloo G, Fischer RA (2010) Metals@MOFs—loading MOFs with metal nanoparticles for hybrid functions. Eur J Inorg Chem 3701–3714. https://doi.org/10.1002/ejic.201000473
94. Metzger ED, Brozek CK, Comito RJ, Dinca M (2016) Selective dimerization of ethylene to 1-butene with a porous catalyst. ACS Cent Sci 2:148–153. https://doi.org/10.1021/acscentsci.6b00012
95. Miner EM, Fukushima T, Sheberla D, Sun L, Surendranath Y, Dincə M (2016) Electrochemical oxygen reduction catalysed by Ni3 (hexaiminotriphenylene)2. Nat Commun 7:1–7. https://doi.org/10.1038/ncomms10942
96. Mohamed Eddaoudi David B. Moler, H.L.I.B.C.T.M.R.M.O, Yaghi OM (2001) Modular chemistry: secondary building units as a basis forbuilding units as a basis for robust metal-organic carboxylate frameworks. Acc Chem Res 34:319–330
97. MSc. Khan AH (2020) Solid-state NMR study of nitric oxide adsorption in carboxylate based MOFs
98. Natarajan S, Kim K, Yoon M, Suh K, Natarajan S, Kim K (2013) Proton conduction in metal—organic frameworks and related modularly built porous solids 2688–2700. https://doi.org/10.1002/anie.201206410
99. Noh H, Cui Y, Peters AW, Pahls DR, Ortuno MA, Vermeulen NA, Cramer CJ, Gagliardi L, Hupp JT, Farha OK (2016) An exceptionally stable metal-organic framework supported molybdenum(VI) oxide catalyst for cyclohexene epoxidation. J Am Chem Soc 138:14720–14726. https://doi.org/10.1021/jacs.6b08898
100. Omar M, Arbor A, Arbor A (2003) (12) United States Patent 122, 12487–12496
101. Oozeerally R, Burnett DL, Chamberlain TW, Walton RI, Degirmenci V (2018) Exceptionally efficient and recyclable heterogeneous metal-organic framework catalyst for glucose isomerization in water. Chem Cat Chem 10:706–709. https://doi.org/10.1002/cctc.201701825
102. Otake KI, Cui Y, Buru CT, Li Z, Hupp JT, Farha OK (2018) Single-atom-based vanadium oxide, catalysts supported on metal-organic frameworks: selective alcohol oxidation and structure-activity relationship. J Am Chem Soc 140:8652–8656. https://doi.org/10.1021/jacs.8b05107
103. Garcia-Garcia P, Muller M, AC (2015) MOF catalysis in perspective to their homogeneous counterparts and conventional solid catalysts. J Mater Chem C 3:10715–10722. https://doi.org/10.1039/b000000x
104. Pan D, Xi C, Li Z, Wang L, Chen Z, Lu B, Wu M (2013) Electrophoretic fabrication of highly robust, efficient, and benign heterojunction photoelectrocatalysts based on graphene-quantum-dot sensitized TiO2 nanotube arrays. J Mater Chem A 1:3551–3555. https://doi.org/10.1039/c3ta00059a
105. Pichon A, Lazuen-Garay A, James SL (2006) Solvent-free synthesis of a microporous metal-organic framework. Cryst Eng Comm 8:211–214. https://doi.org/10.1039/b513750k
106. Ranocchiari M, Bokhoven JAV (2011) Catalysis by metal-organic frameworks: Fundamentals and opportunities. Phys Chem Chem Phys 13:6388–6396. https://doi.org/10.1039/c0cp02394a
107. Rimoldi M, Nakamura A, Vermeulen NA, Henkelis JJ, Blackburn AK, Hupp JT, Stoddart JF, Farha OK (2016) A metal-organic framework immobilised iridium pincer complex. Chem Sci 7:4980–4984. https://doi.org/10.1039/c6sc01376g

108. Rodenas T, Van Dalen M, García-Pérez E, Serra-Crespo P, Zornoza B, Kapteijn F, Gascon J (2014) Visualizing MOF mixed matrix membranes at the nanoscale: towards structure-performance relationships in CO2/CH4 separation over NH2-MIL-53(Al)@PI. Adv Funct Mater 24:249–256. https://doi.org/10.1002/adfm.201203462
109. Rowsell JLC, Yaghi OM (2006) Effects of functionalization, catenation, and variation of the metal oxide and organic linking units on the low-pressure hydrogen adsorption properties of metal-organic frameworks. J Am Chem Soc 128:1304–1315. https://doi.org/10.1021/ja056639q
110. Rowsell JLC, Yaghi OM (2004) Metal—organic frameworks: a new class of porous materials 73:3–14. https://doi.org/10.1016/j.micromeso.2004.03.034
111. Rungtaweevoranit B, Baek J, Araujo JR, Archanjo BS, Choi KM, Yaghi OM, Somorjai GA (2016) Copper nanocrystals encapsulated in Zr-based Metal-organic frameworks for highly selective CO2 hydrogenation to methanol. Nano Lett 16:7645–7649. https://doi.org/10.1021/acs.nanolett.6b03637
112. Chui SSY, Lo SMF, Charmant JPH, Orpen AG, Williams ID (1999) A chemically functionalizable nanoporous material [Cu3(TMA)2(H2O)3]n. Science (80-.). 283:1148
113. Sabouni R, Kazemian H, Rohani S (2012) Microwave synthesis of the CPM-5 metal organic framework. Chem Eng Technol 35:1085–1092. https://doi.org/10.1002/ceat.201100626
114. Seo JS, Whang D, Lee H, Jun SI, Oh J, Jeon YJ, Kim K (2000) A homochiral metal-organic porous material for enantioselective separation and catalysis. Nature 404:982–986. https://doi.org/10.1038/35010088
115. Sharma RK, Yadav P, Yadav M, Gupta R, Rana P, Srivastava A, Zbořil R, Varma RS, Antonietti M, Gawande MB (2020) Recent development of covalent organic frameworks (COFs): synthesis and catalytic (organic-electro-photo) applications. Mater. Horizons 7:411–454. https://doi.org/10.1039/c9mh00856j
116. Shearer GC, Chavan S, Bordiga S, Svelle S, Olsbye U, Lillerud KP (2016) Defect engineering: tuning the porosity and composition of the metal-organic framework UiO-66 via modulated synthesis. Chem Mater 28:3749–3761. https://doi.org/10.1021/acs.chemmater.6b00602
117. Shulman A, Zanghellini E, Palmqvist A (2010) Reversible sorption of water in the crystalline microporous semiconductor K-SBC-1. Microporous Mesoporous Mater 132:128–131. https://doi.org/10.1016/j.micromeso.2010.02.009
118. Stassen I, Styles M, Van Assche T, Campagnol N, Fransaer J, Denayer J, Tan JC, Falcaro P, De Vos D, Ameloot R (2015) Electrochemical film deposition of the zirconium metal-organic framework uio-66 and application in a miniaturized sorbent trap. Chem Mater 27:1801–1807. https://doi.org/10.1021/cm504806p
119. States U (2004) (12) Patent Application Publication (10) Pub . No .: US 2004/0092606A1 1, 2002–2005
120. Stock N, Biswas S (2012) Synthesis of metal-organic frameworks (MOFs): routes to various MOF topologies, morphologies, and composites. Chem Rev 112:933–969. https://doi.org/10.1021/cr200304e
121. Suh MP, Park HJ, Prasad TK, Lim D (2017) Hydrogen storage in metal à organic frameworks 782–835
122. Sumida K, Rogow DL, Mason JA, Mcdonald TM, Bloch ED, Herm ZR, Bae T, Long R (2012) Carbon dioxide capture in metal à organic frameworks 724–781. https://doi.org/10.1021/cr2003272
123. Sun CY, Liu SX, Liang DD, Shao KZ, Ren YH, Su ZM (2009) Highly stable crystalline catalysts based on a microporous metal-organic framework and polyoxometalates. J Am Chem Soc 131:1883–1888. https://doi.org/10.1021/ja807357r
124. Tan D, García F (2019) Main group mechanochemistry: from curiosity to established protocols. Chem Soc Rev 48:2274–2292. https://doi.org/10.1039/c7cs00813a
125. Tanabe KK, Cohen SM (2011) Postsynthetic modification of metal–organic frameworks—a progress report. Chem Soc Rev 40:498–519. https://doi.org/10.1039/c0cs00031k
126. Ulrich Muller, Hermann Putter, Michael Hesse, M.S (2007) Method For electrochemical production of a crystalline porous material organic skeleton material

127. Valvekens P, Vermoortele F, De Vos D (2013) Metal-organic frameworks as catalysts: the role of metal active sites. Catal Sci Technol 3:1435–1445. https://doi.org/10.1039/c3cy20813c
128. Venkataraman D, Moore JS, Gardner GB, Lee S (1995) Zeolite-like behavior of a coordination network. J Am Chem Soc 117:11600–11601. https://doi.org/10.1021/ja00151a034
129. Volkringer C, Cohen SM (2010) Generating reactive MILs: isocyanate- and isothiocyanate-bearing mils through postsynthetic modification. Angew Chemie 122:4748–4752. https://doi.org/10.1002/ange.201001527
130. Volkringer C, Popov D, Loiseau T, Férey G, Burghammer M, Riekel C, Haouas M, Taulelle F (2009) Synthesis, single-crystal X-ray microdiffraction, and NMR characterizations of the giant pore metal-organic framework aluminium trimesate MIL-100. Chem Mater 21:5695–5697. https://doi.org/10.1021/cm901983a
131. Wang B, Liu W, Zhang W, Liu J (2017) Nanoparticles@nanoscale metal-organic framework composites as highly efficient heterogeneous catalysts for size-and shape-selective reactions. Nano Res 10:3826–3835. https://doi.org/10.1007/s12274-017-1595-2
132. Wang C, Wang JL, Lin W (2012) Elucidating molecular iridium water oxidation catalysts using metal-organic frameworks: a comprehensive structural, catalytic, spectroscopic, and kinetic study. J Am Chem Soc 134:19895–19908. https://doi.org/10.1021/ja310074j
133. Wang TC, Vermeulen NA, Kim IS, Martinson ABF, Fraser Stoddart J, Hupp JT, Farha OK (2016) Scalable synthesis and post-modification of a mesoporous metal-organic framework called NU-1000. Nat Protoc 11:149–162. https://doi.org/10.1038/nprot.2016.001
134. Wang Z, Cohen SM (2009) Postsynthetic modification of metal–organic frameworks. Chem Soc Rev 38:1315–1329. https://doi.org/10.1039/b802258p
135. Wu HB, Xia BY, Yu L, Yu XY, Lou XW (2015) Porous molybdenum carbide nano-octahedrons synthesized via confined carburization in metal-organic frameworks for efficient hydrogen production. Nat Commun 6:1–8. https://doi.org/10.1038/ncomms7512
136. Wu CD, Hu A, Zhang L, Lin W (2005) A homochiral porous metal-organic framework for highly enantioselective heterogeneous asymmetric catalysis. J Am Chem Soc 127:8940–8941. https://doi.org/10.1021/ja052431t
137. Wu H, Chua YS, Krungleviciute V, Tyagi M, Chen P, Yildirim T, Zhou W (2013) Unusual and highly tunable missing-linker defects in zirconium metal-organic framework UiO-66 and their important effects on gas adsorption. J Am Chem Soc 135:10525–10532. https://doi.org/10.1021/ja404514r
138. Xia W, Mahmood A, Zou R, Xu Q (2015) Metal-organic frameworks and their derived nanostructures for electrochemical energy storage and conversion. Energy Environ Sci 8:1837–1866. https://doi.org/10.1039/c5ee00762c
139. Xiao JD, Jiang HL (2018) Metal-organic frameworks for photocatalysis and photothermal catalysis. Acc Chem Res. https://doi.org/10.1021/acs.accounts.8b00521
140. Xu C, Liu H, Li D, Su JH, Jiang HL (2018) Direct evidence of charge separation in a metal-organic framework: Efficient and selective photocatalytic oxidative coupling of amines via charge and energy transfer. Chem Sci 9:3152–3158. https://doi.org/10.1039/c7sc05296k
141. Yaghi OM, O'Keeffe M, Ockwig NW, Chae HK, Eddaoudi M, Kim J (2003) Reticular synthesis and the design of new materials. Nature 423:705–714. https://doi.org/10.1038/nature01650
142. Yang D, Odoh SO, Wang TC, Farha OK, Hupp JT, Cramer CJ, Gagliardi L, Gates BC (2015) Metal-organic framework nodes as nearly ideal supports for molecular catalysts: NU-1000- and UiO-66-supported iridium complexes. J Am Chem Soc 137:7391–7396. https://doi.org/10.1021/jacs.5b02956
143. Yang D, Ortuño MA, Bernales V, Cramer CJ, Gagliardi L, Gates BC (2018) Structure and dynamics of Zr6O8 metal-organic framework node surfaces probed with ethanol dehydration as a catalytic test reaction. J Am Chem Soc 140:3751–3759. https://doi.org/10.1021/jacs.7b13330
144. Zhang H, Wang T, Wang J, Liu H, Dao TD, Li M, Liu G, Meng X, Chang K, Shi L, Nagao T, Ye J (2016) Surface-plasmon-enhanced photodriven co2reduction catalyzed by metal-organic-framework-derived iron nanoparticles encapsulated by ultrathin carbon layers. Adv Mater 28:3703–3710. https://doi.org/10.1002/adma.201505187

145. Zhang JW, Zhang HT, Du ZY, Wang X, Yu SH, Jiang HL (2014) Water-stable metal–organic frameworks with intrinsic peroxidase-like catalytic activity as a colorimetric biosensing platform. Chem Commun 50:1092–1094. https://doi.org/10.1039/c3cc48398c
146. Zhang K, Kirlikovali KO, Le QV, Jin Z, Varma RS, Jang HW, Farha OK, Shokouhimehr M (2020) Extended metal-organic frameworks on diverse supports as electrode nanomaterials for electrochemical energy storage. ACS Appl. Nano Mater. 3:3964–3990. https://doi.org/10.1021/acsanm.0c00702
147. Zhang M, Huang YL, Wang JW, Lu TB (2016) A facile method for the synthesis of a porous cobalt oxide-carbon hybrid as a highly efficient water oxidation catalyst. J Mater Chem A 4:1819–1827. https://doi.org/10.1039/c5ta07813j
148. Zhang T, Lin W (2014) Metal-organic frameworks for artificial photosynthesis and photocatalysis. Chem Soc Rev 43:5982–5993. https://doi.org/10.1039/c4cs00103f
149. Zhao C, Dai X, Yao T, Chen W, Wang X, Wang J, Yang J, Wei S, Wu Y, Li Y (2017) Ionic exchange of metal-organic frameworks to access single nickel sites for efficient electroreduction of CO2. J Am Chem Soc 139:8078–8081. https://doi.org/10.1021/jacs.7b02736
150. Zhou Z, Li X, Wang Y, Luan Y, Li X, Du X (2020) Growth of Cu-BTC MOFs on dendrimer-like porous silica nanospheres for the catalytic aerobic epoxidation of olefins. New J Chem 44:14350–14357. https://doi.org/10.1039/d0nj02672g
151. Zhu B, Xia D, Zou R (2018) Metal-organic frameworks and their derivatives as bifunctional electrocatalysts. Coord Chem Rev 376:430–448. https://doi.org/10.1016/j.ccr.2018.07.020
152. Zhu J, Li PZ, Guo W, Zhao Y, Zou R (2018) Titanium-based metal–organic frameworks for photocatalytic applications. Coord Chem Rev 359:80–101. https://doi.org/10.1016/j.ccr.2017.12.013

Stability of MOFs and Kinetics of MOFs Catalyzed Reactions

Kinetic Stability of Robust Metal–Organic Frameworks (MOFs) in Catalytic Reactions

Sanjay Kumar, Sweta Kumari, Shikha, Shefali Shukla, and Shikha Gulati

Contents

1 Introduction 46
2 Kinetic Factors Affecting MOFs Topology 47
 2.1 The Rigidity of the Linker 47
 2.2 Coordination Bond 48
 2.3 Surface Hydrophobicity 49
 2.4 Framework Interpenetration 49
3 Strategies for Stabilizing the Kinetics of Porous MOFs 50
 3.1 Metal Ion Substitution Kinetics 50
 3.2 Strengthening the Bonds for Stabilizing the Porous Phase in Contrast to the Transition State 52
 3.3 Steric and Hydrophobicity 54
 3.4 Linkers and Nodes Connectivity 54
4 MOFs Stability in Gases and Vapors Phase 55
 4.1 Effect of Different Gases 55
 4.2 Effect of Water and Vapors 56
5 Applications of Kinetically Stable MOFs 58
 5.1 Adsorption and Separation 58
 5.2 Fluorescence Sensing 60
 5.3 Heterogeneous Catalysis 60
 5.4 Biological and Pharmaceutical Application 61
6 Conclusion and Future Perspectives 61
Abbreviations 62
References 63

Abstract In the past few decades, researchers have ventured to search for kinetically stable framework materials as conventional ones were expensive and lack chemical robustness, which hinders their practical application. In this context, metal–organic frameworks (MOFs) have gained a lot of attention as metastable design catalysts for a broad range of chemical alterations including catalysis, gas storage, separation, and conversion, etc. MOFs are one the fastest emerging class of organic–inorganic hybrid crystalline porous material assembled with metal ions/clusters as nodes with organic linkers. Although MOFs inherit a unique set of properties including high porosity, ultra-high surface area, diverse composition, and versatile functionality,

S. Kumar · S. Kumari · Shikha · S. Shukla · S. Gulati (✉)
Department of Chemistry, Sri Venkateswara College, University of Delhi, Delhi 110021, India

S. Gulati (ed.), *Metal-Organic Frameworks (MOFs) as Catalysts*,
https://doi.org/10.1007/978-981-16-7959-9_2

however, for these materials to be industrially viable, the kinetic stability is necessary. In order to attain improved stability, kinetic factors such as coordination number, surface hydrophobicity, as well as framework interpenetration are considered to give rise to numerous robust MOFs, broadening their applicability rate and benefiting in industrialization. The present chapter focuses on the kinetic factors affecting the topology of the framework, different strategies to stabilize the porous phase, their stability in the gas and vapors phase has been highlighted, and their role in various application prospectus. Apart from this, significant emphasis has been given to techniques to improve the kinetic stability of existing frameworks using the contrasting methodology.

Keywords Metal–organic frameworks (MOFs) · Kinetics · Stability · Catalyst

1 Introduction

The porosity of any material is of eminent importance for its application in catalysis, gas storage, sensing, drug delivery, etc. Over the past several years, researchers have tended to develop a material with high porosity. To this, metal–organic frameworks (MOFs) have generally been considered to be among the foremost groups of porous crystalline structures. MOFs distinguish themselves from all other typical porous materials such as zeolites and carbon materials in terms of their chemical tunability and adaptable tailorability. MOFs are organic–inorganic hybrid materials, constructed via metal ions/clusters that serve as nodes and organic ligands as linkers, bonded by coordinate bonds. MOFs regulated porosity, high surface area, and crystallinity are some of the unique features that help to control the topology of the material through molecular design. The property of porosity creates opportunity at micro-, meso-, and nanoscale to design facile MOFs with desirable characteristics by incorporating various guest species [44].

A variety of frameworks with almost the same topology but with enhanced pore size, wider dimension, indeed be prepared by adopting a variety of techniques. Unfortunately, industrial applications of these materials are not beneficial as they lack reliable performance. Many efforts have been taken to address this challenge. One of the primary issues in this context is the stability of MOFs, particularly kinetic, chemical, thermal, and mechanical [10]. In general, the stability of MOFs is determined by several factors including thermodynamic, kinetic, chemical, and other operating environmental factors. Even when the metal cluster and framework topology remain the same, the chemical stability of certain MOFs diminishes with linker extension and pore widening. This is generally attributable to kinetic variables that are intrinsically linked to the structural stiffness, coordination bond, hydrophobicity of framework surface, strength of interpenetration, etc. [2]. The first porous moisture-labile MOF(MOF-5) was developed by Li et al. in the year 1999. It was observed that this MOF may preserve its crystalline structure even after 24 h of continuous heating at 300 °C [9]. Numerous potential approaches to modify MOF stability, particularly kinetic stability, have been established which be the crucial need for their potential utilization. In this context, various approaches to understanding the stability of MOFs on exposure to different gases or water have been studied by researchers around the globe. Herein, the MOFs are subjected to coordinating, acidic, and oxidizing agent in gases and vapors phase, and in addition to this, its effect on MOFs is used to propose strategies to synthesize more firm MOFs [44]. The kinetics of perforated MOFs must be improved by enhancing the energy barrier between the porous and denser layers of the transition state, and this can be done by adopting various techniques. Owing to their unique set of properties and highly ordered crystalline structure, it is vital to highlight that the position and degree of modification in MOFs are controllable, so that they may be used for a variety of purposes, as discussed in detail in the various sections of this chapter.

The current chapter focuses on the distinctive factors affecting the kinetic stability of MOFs. Furthermore, from an industrial application viewpoint, a section is devoted to strategies to enhance the kinetic stability of MOFs and their endurance, on exposure to different gases and water phases. In addition to this, various applications of stabilized MOFs are elaborated.

2 Kinetic Factors Affecting MOFs Topology

MOFs are a special variant of porous materials, constructed via coordination bonding between metal ions/clusters and organic ligands as linkers. Over the past few years, the stability of these materials are of great concern about their utilization in catalysis, adsorption, separation, drug discovery, etc. The kinetic stability of MOFs, in particular, describes the degree of decomposition of the MOF at any temperature, relative humidity (RH), and exposure interval [33]. To investigate the compatibility of MOFs over conventional materials, several kinetic variables that impact the topology of the framework are shown in Fig. 1.

2.1 The Rigidity of the Linker

The rigidity of the ligand linkers is crucial to determine the endurance of the framework. It is usually observed that the dense and rigid frameworks built from robust and concentrated constituents, typically demonstrate excellent stability. In this context, Song and co-workers used two MOFs to illustrate transmetalation. The synthesized MOFs are Zn-HKUST-1 and Zn-PMOF-2. These MOFs possess identical metal centers of Zn with varying ligand linker sizes. It was revealed that the large organic linkers of flexible Zn-PMOF-2 support proper transmetalation, whereas small organic linkers of rigid Zn-HKUST-1 framework favor limited transmetalation [47]. It should be noted here that metal centers linked by a flexible organic linker seem to be more reactive than metal centers linked by a firm organic linker. As metal centers on the

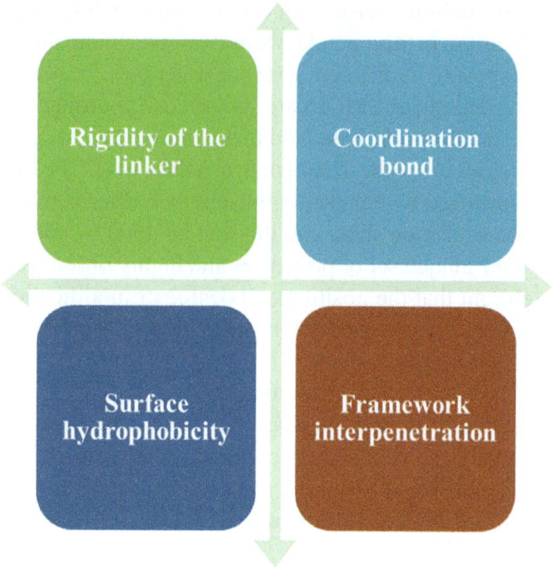

Fig. 1 Different kinetic factors that affect the topology of MOFs

periphery of a particle are in a more flexible environment compared to others present in the core, which is in a more rigid environment, therefore, the latter is more stable than the former.

2.2 Coordination Bond

In order to understand the ligand-rigidification mechanism, it is very crucial to study coordination bonds formed between the metal cluster and the ligand. To increase the kinetic stability of MOFs, it is necessary to robust M-L coordination bond in the material. Herein, it should be kept in mind that although strong coordination bonds are essential for stability it is not sufficient. For example, a series of Zr-MOFs were having quite a strong coordination bond, but still some of them are found to be unstable [8]. When water substitution causes dissociation on one side of a ligand attached to a metal cluster, the ligand poses rotational flexibility to exit the initial coordination sites. The change in coordination offers various possibilities of different types of flexibility for MOFs [35]. These coordination changes in MOFs can change/improve MOFs properties. There are different ways by which coordination changes can occur in MOFs:

- Coordination geometry changes
- Formation of open metal sites non-topological changes
- Topological changes.

These changes bring change in their structure and proved to have better adsorption for gases, magnetic behavior, etc. Mendes et al. discussed various examples of MOFs in which transformation brought improvement in MOFs [31]. The ligand is not able to migrate much in the framework as structural deformation of stiff ligands is limited by other coordination sites. As a result, the lability of the metal–ligand coordination bond at some positions may cause fast structural repair. The bond between a metal cluster and a ligand can dissociate for flexible ligands without having any effect on other coordination sites, which leads to the reformation of a slower and more complex ligand–metal bond. As a result, the defect ratio in the system will rise, resulting in the demise of the framework [26].

2.3 Surface Hydrophobicity

Surface hydrophobicity inhibits water from adsorbing into the pores of metal clusters, and this improves the stability of MOFs in the presence of water even at the higher temperature. The approach of increasing the bond durability of metal–ligand interaction is used to enhance the hydrophobicity of MOFs. This stabilization approach can be examined by analyzing the adsorption isotherm of the MOF to demonstrate that water is prevented from trapping into the framework by hydrophobic groups. It has been extensively studied that stability of MOF under humid circumstances might be enhanced by adding hydrophobically fluorinated and/or alkyl functional groups to the ligand site [2].

2.4 Framework Interpenetration

Metal–organic frameworks (MOFs) interpenetration is a fascinating phenomenon that has critical structural, porous, and functional effects on the framework. It is also called a framework catenation. With a considerable rise in the number of documented MOFs, framework interpenetration is now becoming a prevalent study. It is only possible when the pore space of the framework is sufficiently wide enough to incorporate host material. In this context, the interpenetrated motifs greatly improve the framework stability by filling the space, and this also forms a repulsive force that helps to prevent framework collapse [17]. Some of the contrasting features of framework interpenetration that strike to enhance the stability of MOFs are:

- 2D or 3D networks are used to form motifs with identical structure.
- The finite number of interpenetrated motifs is used.
- Same dimensionalities are there in the single network and final structure.
- In order to form the final structure, each network is interwoven with all the others.

In a study, Jasuja and co-workers used the strategy of pillaring to catenate two MOFs (MOF-508 and DMOF) [16]. Their study demonstrated that on exposure to

90% RH, the non-interpenetrated DMOF showed a reduction of surface area from 1980 to 7 $m^2 g^{-1}$, whereas twofold interpenetrated MOF-508 blocked water adsorption and maintained their stability after the same treatment. With an outstanding unique set of features that makes MOFs robust, it should be noted that interpenetration has a detrimental impact on the porosity of open frames caused by limiting the size of open pores [17]. In order to reduce this impact and to build a highly porous MOF with extraordinary surface area, the framework interpenetration needs to be controlled. To this, various strategies to suppress interpenetration are as follows:

- By regulating reaction temperature and concentration
- By implementing a template to avoid multiple net interpenetrations
- By selectively modifying induced ligand design.

3 Strategies for Stabilizing the Kinetics of Porous MOFs

Porous MOFs are the best examples of unique hollow-structure materials that can provide a whole raft of excellent features. Porous MOFs are found to be metastable as compared to their denser phase. The denser phase of MOFs is an imaginary component of the same constituents, but its porosity is negligible. In the case of all silica zeolites, it can be easily conceptualized and achieved by simple heating, but in the case of MOFs, it is a difficult task due to the directionality of ligands [22]. However, it is worth noting that some MOFs have been achieved thermally in their denser and amorphous phase. As mentioned above, the porous MOFs are less stable when compared to their denser phase, so their stability can be enhanced kinetically only [44]. For kinetically stabilizing the porous MOFs, the energy barrier for transition state between the porous and denser phase needs to be increased which can be done in two ways:

1. By enhancing the energy of transition state
2. By reducing the porous phase energy.

In this section, we will discuss different approaches to stabilize different MOFs, as shown in Fig. 2. There are four methods that can increase the stability of porous MOFs kinetically either by enhancing the energy of the transition state or by reducing the porous phase energy, and they are mentioned below:

3.1 Metal Ion Substitution Kinetics

By this method, the energy of the transition state can be increased. When the transition from porous phase to denser phase occurs, the substitution of ligands takes place around the metal ion. The complexes with higher ligands exchange rates are considered to be liable and with lower ligands exchange rates are considered to be inert, represented in Fig. 3 [36]. For example, the rate of substitution of ligands in

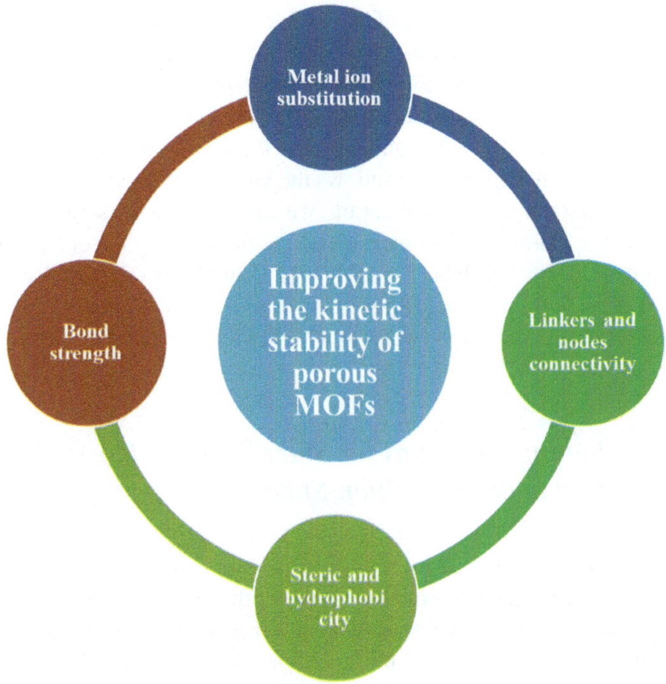

Fig. 2 Different strategies for improving the kinetic stability of MOFs

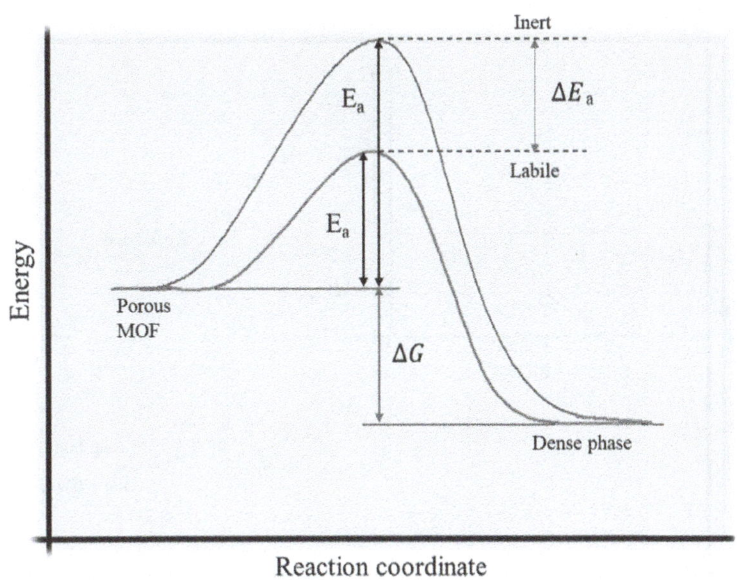

Fig. 3 Metal ion substitution kinetics

octahedral complexes for Cu^{2+} and Cr^{2+} is 5.9 × 109 s^{-1}, which are liable, and for inert Ir^{3+}, its value is 1.1 × 10–10 s^{-1}. Cr^3 + which has a ligand exchange rate slower than 1 s^{-1} most widely used in MOFs synthesis because it forms a carboxylate framework that is stable on water and steam exposure [19]. The stability trend within the family of triazolate MOFs and an impact of the energy of transition state which is an outcome of the metal ion's size follows the kinetic metal-aquo substitution rate. Thus, by increasing the inertness of cation, we can increase the stability.

Ni^{2+} is another example of an inert metal cation showing excellent stability. Li et al. replaced native Zn^{2+} with Ni^{2+} in MOF-5 [23] and Jaio et al. partially substituted Ni^{2+} in Mg-MOF-74 framework [18]. Herein, in both cases, the new MOF was found to have more stability toward the water at a higher temperature.

3.2 Strengthening the Bonds for Stabilizing the Porous Phase in Contrast to the Transition State

In this approach, the ligands with higher donating ability are being used to enhance the porous MOFs stability. The weak metal–ligand bonding of porous MOFs is the reason for their poor stability. If the heterolytic metal–ligand bond strength is increased, it results in enhancing the stability of porous MOFs toward polar substances or it can lower the energy of porous MOFs, demonstrated in Fig. 4. An increase in the donating ability of ligands especially in the case of late transition metals can be beneficial,

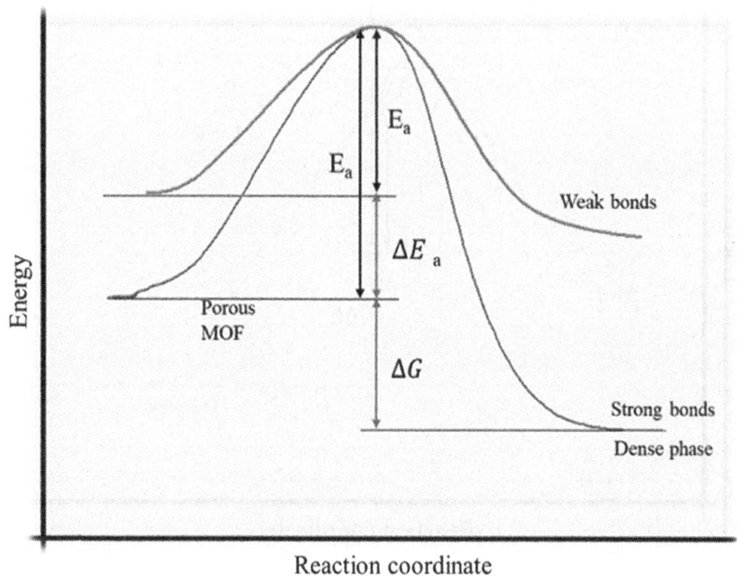

Fig. 4 Effect of strength of M-L bond on kinetics

and this will also enhance the chemical stability. Choi et al. used pyrazolate-bridged MOFs and combined it with tetrahedral Zn^{2+} [5, 6]. It was found to be highly stable toward boiling water, organic solvents, and even in an acidic medium.

Another strategy to enhance the metal–ligand bond strength is by increasing the valency of metal ions. With the same ligand and coordinating atmosphere, metal ions having higher valency with high charge densities might form a stable framework. Metals with higher valency like Ti^{4+}, Zr^{4+}, and Cr^{3+} when combined with carboxylates are found to have stronger linkage [53]. There is no doubt in the fact that by increasing the bond strength of metal and ligands, the porous MOFs' stability can be enhanced, but in this approach, the stability of the denser phase also increases at the same rate. So, the net energy for the transition from porous to denser phase is almost the same.

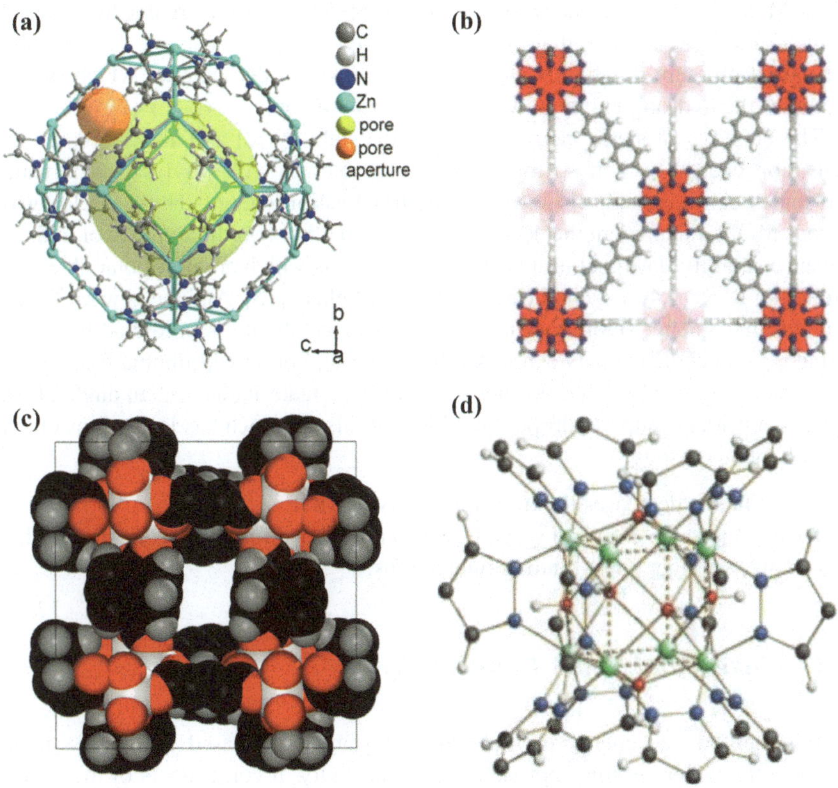

Fig. 5 Some MOFs blocks with high kinetic stability **a** Zn(2-methylimidazolate)2 (ZIF-8). Reused with permission from [42]. **b** $Zr_6O_4(OH)_4(BDC)_6$(UiO-66). Reused with permission from [4]. **c** Al(OH)(isophthalate) (CAU-10). Reused with permission from [43]. **d** $Ni_8(OH)_4(H_2O)_2(BDP)_6$. Reused with permission from [30]

3.3 Steric and Hydrophobicity

This is a simple method that involves steric shielding of the metal–ligand bond. For a hydrolysis reaction to proceed, there are two required conditions. Firstly, water molecules should come close enough, so that interaction can take place, and secondly, there should be an available minimum amount of energy, so that energy of activation can be overcome. In the first case, framework hydrophobicity can be an important factor to increase the MOFs stability which is possible by substituting hydrophobic alkyl and fluorinated functional groups with ligands. Nguyen et al. used the post-synthetic method and modified isoreticular metal–organic frameworks (IRMOFs) by substituting longer alkyl groups [38]. Most of the modified IRMOFS were found to have better stability, even after exposure to ambient atmosphere for few weeks. In the year 2013, a series of Ni(II)-pyrazolate based MOFs was prepared by Paidel et al. These MOFs were introduced with a fluorinated methyl group on organic linkers [39]. In the results, it was found that fluorinated methyl groups are capable of increasing the partial pressure at which water vapor condenses in the material which further increases the hydrophobicity of MOFs to a greater extent.

The presence of steric factors enhances the stability [2] and also increases the activation energy barrier, so that they cannot interact with MOFs even if water molecules come enough close, due to high activation energy. Thus, shielding of porous MOFs can prevent the access of water, and other coordinating vapor and can enhance stability. Jasuja et al. placed methyl groups which are non-polar shielding groups on BDC (1,4-benzenedicarboxylic acid) linkers [15]. And the new MOF was found to have better stability. The extent of hydrophobicity is predicted by calculating the contact angle between water and material. For a hydrophobic framework, the contact angle is between 90° and 150°, and the material having an angle above 150° is considered superhydrophobic [29]. But this approach has limitations due to the following reasons:

- Sometimes it decreases the overall porosity of MOFs.
- It can impede the access of sorbates which are needed for framework sites as they can give stronger guest binding interactions.

3.4 Linkers and Nodes Connectivity

In this method, the energy of the transition state is increased as it does in the chelate effect. Here in this approach, the connectivity of constituents of the framework is increased. This kinetic stabilization can be increased by increasing the number of a special kind of metal ligands bond. And during the phase transition or collapsing of pores, those bonds should rearrange or dissociates. Mondloch using this approach targeted the MOF made from hexa-ZrIV nodes and tetratopic linkers [34]. He used carboxylates linkers and did many changes and formed the new MOF having the formula $Zr_6O_4(OH)_4(HCOO)_4(TBAPy)_2$(NU-1000; $TBAPy^{4-}$ = pyrene tetra-p-benzoate). The metal–oxo–hydroxo clusters which are frameworks composed of secondary building units (SBUs) with greater connectivity have proved to have increased stability. For example, MOFs are made from Zr^{4+} oxo–hydroxo nodes, which are linked by 6, 8, 10, or 12 carboxylate groups, with node connectivity the stability usually increases [4]. A pyrazolate-based porphyrinic MOF with ftw-a topology, namely PCN-601 was synthesized by Wang et al. [50]. It was connected by Ni8 cluster and 5,10,15,20-tetra(1H-pyrazol-4-yl)-porphyrin (H_4TPP)PCN-601 and it is found to be highly stable even in a saturated solution of aqueous NaOH. Cadiau et al. [3] reported (KAUST)-7 which is also known as NbOFFIVE-1Ni is found to have extreme stability in aqueous solution for 6 months and even stable in an atmosphere of 95% humidity and 10% H_2S [9]. It was connected by $(NbOF_5)^{2-}$, and assembled by a square-shaped Ni-pyrazine (4,4′) square grid.

4 MOFs Stability in Gases and Vapors Phase

The bond strength between the metal cluster and the linker is an essential factor, understanding the kinetic stability of MOFs. Change in ligand connectivity and binding typically results in reductions of effective surface area and porosity. This damage is generally observed during the reaction processor in the regeneration of the framework. It also reduces the effectiveness of sorbent [25]. One of the sources of such disturbances is the exposure of the framework to different gases and vapors. In this section, we will discuss different approaches for assessing the sustainability of various MOFs. The tolerance of numerous the framework is studied either on exposure to profuse gases or water, and its out-turn is elaborated below.

4.1 Effect of Different Gases

- **Coordinating Gases: H_2O, NH_3, and H_2S**

The MOF stability trends toward polar gases are heavily influenced by the acidity and nucleophilicity of the gas. Accordingly, the operative reaction mechanisms include the following:

- Substitution of ligand
- Hydrolysis of metal–ligand bonds
- Ligand reconfiguration mediated by coordination site

The aforementioned variables witness the relationship between the kinetics of metal–ligand bonding strength and the robustness of MOFs in response to coordinating gas exposure [44]. In a study by Peterson et al., it was concluded that when ammonia combines with water, it produces NH_4OH, which is extremely abrasive and results in metal–ligand interaction hydrolysis. Herein, the MOF $Cu_3(BTC)_2$

evidenced persistent breakdown of structure and porosity, particularly with ammonia under humid circumstances [41].

- **Acidic and Oxidizing Gases: SO_x, and NO_x**

The development of materials that can resistant to acidified as well as oxidizing gases including sulfur dioxide (SO_2), and nitrogen oxides (NO_x) presents distinct problems. The substitution kinetics can be significantly be altered by oxidizing the metal center using an oxidizing gas. Although, several composites are robust towards toxic pollutants in single component investigations, also by the addition of moist air introduces a slew of additional difficulties due to possible side reactions that generate powerful acids [13]. The apparent instability of MOFs in humid environments is due to their unexplored interactions with SO_x and NO_x. Herein, SO_2 reacts with water vapor to produce sulfurous acid (H_2SO_3), which can then react with oxygen to form SO_3 and sulfuric acid (H_2SO_4). Therefore, in order to mitigate this issue, Elder and co-workers investigated the function of metal center on exposure of MOFs to humid SO_2 atmosphere. They consummated zinc benzenedicarboxylate (ZnBDC) to be the most stable to date on 80 ppm-days exposure in a moist SO_2 environment [11]. Also, adsorption of SO_2 in humid air is difficult, because the protonated linker produces metal sulfites or sulfates.

However, it should be noted that NO_x capture poses extra complications. Comparable to SO_2, under humid circumstances, both NO and NO_2 produce strong acids such as nitrous acid (HNO_2) and nitric acid (HNO_3), this can also protonate linkers and deteriorate the framework. Even nitrogen oxides can go through many redox reactions including NO disproportionation, NO oxidation reactions, or dimerization/disproportionation of NO_2 [1].

4.2 Effect of Water and Vapors

Porous MOFs have been extensively investigated for catalysis, and they are used in two ways as a catalyst directly or as a scaffold. Some catalytic reactions might not involve water, while in many reactions water is required as it is considered as a novel solvent for catalysis due to environmental, economical, and safety concerns. Similarly, MOFs are widely investigated for luminance and drug delivery too where water can also be present [2]. So, to increase the use of MOFs in these areas, it has become an important need of MOFs to be water stable. There is a high diversity of MOFs out of which some can be highly hydrolytic stable while others may not be. Different MOFs show different behavior on exposure to water or humidity, and it also depends on what type of water exposure it is. There are six levels proposed in Fig. 6 to discuss the harshness of moisture exposure.

To discuss the hydrolytic stability of water, it is necessary to know what type of water exposure is there and some of the types of water exposures are as follows:

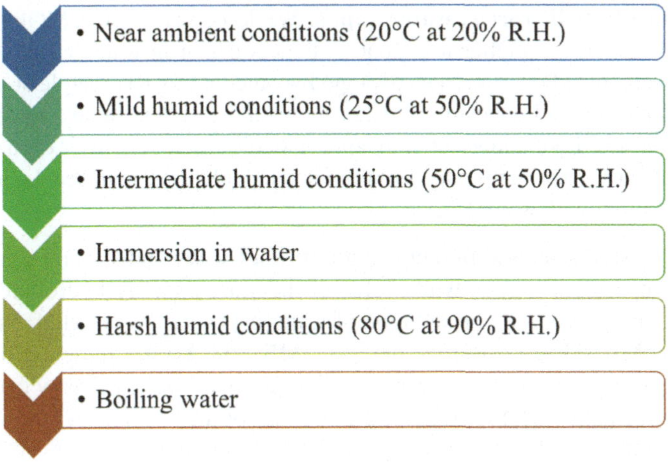

Fig. 6 Several levels of harshness on moisture exposure

- **Exposure to ambient air**: This condition is considered a level 1 condition. Only testing MOFs for ambient conditions is not a good method for stating whether MOF is stable or not because it is much dependent on local conditions existing during testing and it includes ambient relative humidity also. The ambient relative humidity is used to predict the amount of water vapor in the air based on external temperature and pressure conditions [45]. MOF is not fully activated, then it may have some remaining guests in pores which can prohibit the excess of water and impart extra stability to MOFs. And if MOFs are kept in a closed system, then equilibrium will exist between water absorbed in MOFs and water vapor in a surrounding which might result in different relative humidity.
- **Immersion in liquid water**: It is another method of testing the hydrolytic stability of MOFs. Generally, it is considered that if MOF is synthesized in the water, then it will be stable toward the water which is not always true [7]. There can be different reasons why MOF is not soluble in water like it might be kinetically insoluble in water or there can be some other constituents available like organic solvents or salts which result in decreasing the solubility [40].
- **Uncertain relative humidity exposure**: Another method for testing hydrolytic stability of MOFs is the use of humidity with unknown concentration by using vial in vial method at a fixed temperature [37]. Since internal pressure is unknown, the amount of water content cannot be measured in this method which can show uncertainties. The main advantage of this method it provides a fast-screening way for predicting whether MOF has the capability of stability toward water and vapors or not.
- **Certain relative humidity exposure**: In this method, the MOF is exposed to various degrees of water at different temperatures for different time periods [37]. The important advantage of this method is that the stability of internal pores can be tested.

- **Exposure to bulk versus water vapor**: While describing the MOFs stability, it should be clear whether it is done in bulk water or in water vapor. Adhesion and entropy decide the surface coverage by water at low RH. While at saturation point, adhesion and cohesion of water molecules can determine hydrophobicity of the surface. Like a material may appear stable when the only external surface is exposed to water and it may not when it is in bulk because internal pores might be interacting with water [2].

Different MOFs show different stability toward water and vapors, some MOFs are highly hydrolytic stable, while some of them are partially hydrolytic stability and their stability depends on the type of water exposure. Weber et al. studied the stability of MIL-53(Al) toward water [51]. MIL -53(Al) has quite good stability toward the water, but in boiling water, a slight decrease in its porosity can be seen due to the formation of a thin layer of AlO(OH) on it. And if it is further boiled, then a thick layer is formed, and it completely loses its porosity. According to a study HKUST-1, $[Cu_3(Btc)_2]$ when kept in water for 24 h at 50 °C, then there was a decrease in its surface area and porosity was retained by 48%. According to another study by Majano et al. when it was exposed to humidity, there was a complete loss of porosity and crystallinity [28]. This example shows that a change in the type of water exposure changes the stability. In another study, CALF-28 was studied, when it was exposed to 80% to 90% RH, then a decrease in its surface area was found, and its bimodal pore distribution changed into unimodal but further increase in humidity did not show any effect. Similar to this, there are many other MOFs which are there who show decreases in surface area to a specific extent but on further increase the exposure to humidity or water they do not show effect [14].

5 Applications of Kinetically Stable MOFs

MOF catalysts are being extensively researched for their use in precision chemical synthesis. MOFs with a variety of functions may be systematically developed and built for a particular purpose. Figure 7 depicts some of the most notable uses of stabilized MOFs, which are briefly addressed below:

5.1 Adsorption and Separation

In the recent few years, MOFs have made an important place in novel materials for adsorption and gas separations because of their large surface area, plus tunable porous framework. These frameworks are ideal candidates for the storage of gases like H_2, CO_2, CH_4, and noble gases because of their large surface areas and highly ordered porous structures. Large surface areas of MOFs help in the efficient adsorption of gas, but the adsorption of pure MOFs is not satisfactory in practical applications [32].

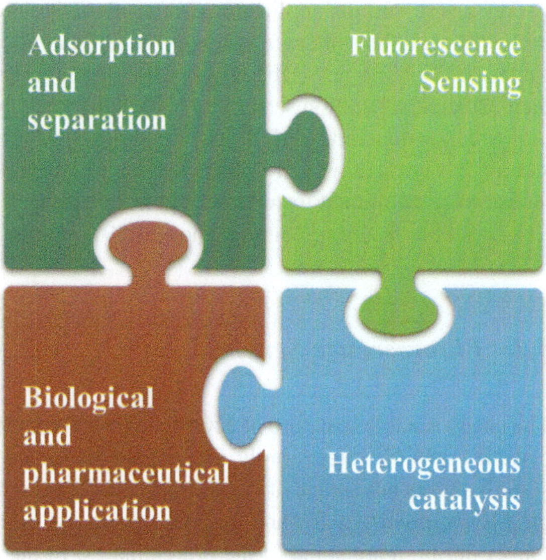

Fig. 7 Some of the applications of kinetically stable MOFs

So, some modifications are mandatory, for instance, confinement of certain superior metal nanoparticles in MOFs by also merging the stronger adsorption capacity of metal nanoparticles as well as extremely wide surface areas of MOFs. The resulted material shows excellent adsorption capacity. Palladium is a well-known metal for hydrogen storage. According to a study, by Li et al. when these Pd nanocrystals were combined with MOFs then the resulted material showed a great increase in the storage capacity of hydrogen [21]. Even when compared with the PC curves of MOFs, then completely reversible absorption/desorption response was observed for Pd@HKUST-1 while simple Pd shows hysteresis behavior.

In practical applications, water, humidity, and other harsh conditions are present which are unavoidable. So, the MOFs should have enough hydrophobicity to increase practical applications. The MOFs show a high assortment in their pore size, structures, inorganic clusters, and their chemical activity, so their water adsorption properties lie between high diverse profiles of adsorption–desorption isotherms [46]. Highly porous MOFs can show great potential in industrial applications, for example, hydrolytic stable MOFs show highly water adsorption profiles for heat pump and chillers.

Along with gas adsorption, MOFs also show high potential in gas separation too which can be ascribed to the following two characteristics of MOFs:

1. By regulating inorganic metal ions and organic linkers, their pores can be controlled.
2. By various methods, their pore surface can be improved with a lot of functional groups.

The gas separation applications are attributed to the shape and size of gas mixture molecules and their interactions. For industrial applications on large scale, also over substrates, consistent MOFs membranes of superior quality are coated, which can increase their separation selectivity [10]. Song et al. synthesized a core porous structure which is denoted by MOF@ZIF using a UiO-66-NH_2 core having a higher pore size of 0.6 nm and a shell of ZIF-8 [48]. The newly formed MOF@ZIF nanomaterials were combined to form hybrid polysulfone (PSF) membranes for capturing CO_2. And the resulting material showed 50% higher selectivity for CO_2 and N_2.

5.2 *Fluorescence Sensing*

The easily tunable structure of MOFs serves it to be an exceptional platform for fluorescence sensing. Interestingly, fluorescent MOFs are way more sensitive to external stimuli, taking advantage of this attribute MOF's can be excellent sensors. Moreover, the MOFs sensitivity toward analyte, selectivity, time of response, and stability for longer duration are some of the prerequisites. In the application perspective, Wang et al. used a topological design strategy to synthesis Zr(IV)-based MOFs which showed outstanding fluorescent properties [49]. The produced MOF helps to identify and eliminate organic contaminants from water waste which include antibiotics and organic explosives. Also, it showed an exceptional capacity for detecting nitrofurazone (NZF), nitrofurantoin (NFT), 2,4,6-trinitrophenol (TNP), and 4-nitrophenol (4-NP) in aqueous solutions with minimal detection limit, the result of the combined electron, and energetic exchange effect. In contrast to all chemical sensors, the triazolate MAF-2 Cu(I) was developed by Liu et al. as an oxygen-gassing sensor [24]. It was based on high humidity stability, rapid absorption and sorption kinetics of oxygen, and other optical characteristics. It is worth pointing out that for the first time soft membrane oxygen sensor was produced by the counter-diffusion crystal-growth method with astonishing oxygen sensing performance (detection limit = 0.047 mbar) and remarkable kinetic stability.

5.3 *Heterogeneous Catalysis*

One of the prime demonstrated implementations of MOFs is heterogeneous catalysis. Kinetically stabilized MOFs have made tremendous progress toward their prospective application in catalysis over the last two decades. However, the presence of metal cluster active sites or organic MOF connectors makes them stable and reliable for heterogeneous catalysis [10]. In addition to this, MOF or MOF-supported catalysts have attracted a lot of interest in applications, such as organocatalysis, photocatalysis, and electrocatalysis due to their outstanding catalyst characteristics. In a study, for the first time, Lv et al. used Mn^{3+} porphyrin ligand as a base and constructed PCN-602(Mn) framework for halogenation of C-H bond in basic medium. The as-constructed MOF shows outstanding performance with a high yield of 92% chlorocyclohexane [27]. To beneficially utilize the outstanding porous feature of the framework, Kornienko and co-workers introduced cobalt porphyrin MOF to form Al-TCPP-Co composite. The selectivity rate of this MOF was found to be more than 76% for electrochemical reduction of CO_2 to CO with the stability of more than 7 h and 1400 turnover numbers [20].

5.4 Biological and Pharmaceutical Application

Metal–organic frameworks (MOFs) are widely used for drug delivery owing to their unique properties like tunability and ultra-high surface areas. They have demonstrated tremendous potential and distinct benefits in the domain of pharmacy. Broadly, the applications of kinetically stabilized MOFs include the following:

- Biosensing
- Drug delivery
- Biomimetic catalysis
- Antimicrobial enzyme
- Biological imaging.

Interestingly, Cu-TCA, a biosensing sensor was built using $Cu_2(O_2CR)_4$ paddle-wheel units and tricarboxy-triphenyl amine (TCA) as linkersa by Wu et al. [52]. Herein, when NO was added, the coordination association with NO and Cu(II) complexes converts Cu^{2+} to Cu^{+} ions, resulting in the restoration of luminescence from Cu-TCA in an aqueous medium. In another study by Horcajada et al., two flexible frameworks, MIL-53(Cr) and MIL-53(Fe), were chosen as bearer frameworks for transporting ibuprofen. It was a prior effort to utilize MOFs for drug transport. Both prepared MOFs, MIL-53(Cr) and MIL-53(Fe), were able to achieve a progressive plus continuous release of ibuprofen beyond a three-week period [12].

6 Conclusion and Future Perspectives

Metal–organic frameworks (MOFs) offer abundant applications in a variety of fields due to their exceptional structural, inherited porosity, and chemical characteristics. This is justified as MOFs have become a high priority in the scientific community whose interest and development in the domain are escalating. However, there are several issues with these MOFs that must be addressed by alterations. The primary issue that MOFs confront is maintaining their porosity and stability, which can be simply explained by their dynamics.

In this chapter, we have thoroughly put highlights on various methods to stabilize the MOF units and the involved challenges. Despite their numerous potential uses, MOFs have pertinent issues, as a result, different techniques for further improvement have to be proposed. More detailed studies are required in various domains of MOFs to solve the current challenges. These characteristics point to exploring new opportunities for achieving a more stable framework for reactions that require extreme RH value at relatively high temperatures on prolonged exposure.

Important Links

1. https://www.azom.com/article.aspx?ArticleID=17851
2. https://www.nanowerk.com/mof-metal-organic-framework.php
3. https://www.nature.com/articles/s41578-019-0140-1#:~:text=MOFs%20formed%20with%20kinetically%20inert,30%2C31%2C32
4. https://novomof.com/metal-organic-frameworks/?hsCtaTracking=ec8597ef-7650-4f00-9975-26a88ba24130%7C8d8099f6-9030-4abd-8c17-bc155c09fe0a
5. https://chemistry-europe.onlinelibrary.wiley.com/doi/abs/10.1002/celc.201700931.

Abbreviations

MOF	Metal-Organic Framework
RH	Relative Humidity
PMOF	Polyoxometalate metalloporphyin Metal–Organic Framework
HKUST	Hong Kong University of Science and Technology
M-L	Metal-Ligand
DMOF	DABCO Metal–Organic Framework
IRMOFs	Isoreticular Metal–Organic Framework
BDC	1,4 Benzenedicarboxylic Acid
MIL	Materials Institute Lavoisier
NU	Northwestern University
TBAPY	Pyrene tetra-p-benzoate
SBU	Secondary Building Units
PCN	Porous Coordination Network
KAUST	King Abdullah University of Science and Technology
H_4TPP	5,10,15,20-Tetra (1H-pyrazol-4-yp porphyrin
CAU	Christian-Albrechts-University
ZIF	Zeolitic Imidazolate Framework
BTC	1,3,5Benzenedicarboxylate
PSF	Polysulfone
CALF	Calgary Framework
NZF	Nitrofurazone
NFT	Nitrofurantoin
TNP	2,4,6 Nitrophenol
NP	Nitrophenol
MAF	Metal Azolate Framework

TCPP Tetra kis(4-carboxyphenyl porphyrin)
TCA Tricarboxy Triphenyl Amine

References

1. Brozek CK, Miller JT, Stoian SA, Dinca M (2015) NO Disproportionation at a mononuclear site-isolated Fe2+ center in Fe2+-MOF-5. J Am Chem Soc 137(23):7495–7501
2. Burtch NC, Jasuja H, Walton KS (2014) Water stability and adsorption in metal-organic frameworks. Chem Rev 114(20):10575–10612
3. Cadiau A et al (2016) A Metal-Organic framework-based splitter for separating propylene from propane downloaded from. http://science.sciencemag.org/
4. Cavka JH et al (2008) A new zirconium inorganic building brick forming metal organic frameworks with exceptional stability. J Am Chem Soc 130(42):13850–13851
5. Choi HJ, Dincă M, Dailly A, Long JR (2010) Hydrogen storage in water-stable metal-organic frameworks incorporating 1,3- and 1,4-benzenedipyrazolate. Energy Environ Sci 3(1):117–123
6. Colombo V et al (2011) High thermal and chemical stability in pyrazolate-bridged metal-organic frameworks with exposed metal sites. Chem Sci 2(7):1311–1319
7. Dang D et al (2012) Assembly of polyoxometalate-based metal-organic frameworks with silver(I)-schiff base coordination polymeric chains as building blocks. Crystal Growth Design 12(8):3856–3867
8. Decoste JB et al (2013) Stability and degradation mechanisms of metal-organic frameworks containing the Zr6O4(OH)4 secondary building unit. J Mater Chem A 1(18):5642–5650
9. "Design and synthesis of an exceptionally stable and highly" (1999) 402(November):276–79
10. Ding M, Cai X, Jiang HL (2019) Improving MOF stability: approaches and applications. Chem Sci 10(44):10209–10230
11. Elder AC, Bhattacharyya S, Nair S, Orlando TM (2018) Reactive adsorption of humid SO2 on metal-organic framework nanosheets. J Phys Chem C 122(19):10413–10422
12. Horcajada P et al (2008) Flexible porous metal-organic frameworks for a controlled drug delivery. J Am Chem Soc 130(21):6774–6780
13. Hungerford J et al (2018) DMOF-1 as a representative MOF for SO2 adsorption in both humid and dry conditions. J Phys Chem C 122(41):23493–23500
14. Isaeva VI, Tarasov AL, Chernyshev VV, Kustov LM (2015) Control of morphology and size of microporous framework MIL-53(Al) crystals by synthesis procedure. Mendeleev Commun 25(6):466–467
15. Jasuja H, Huang Y, Walton KS Adjusting the stability of metal-organic frameworks under humid conditions by ligand functionalization. http://pubs.acs.org
16. Jasuja H, Walton KS (2013) Effect of catenation and basicity of pillared ligands on the water stability of MOFs. Dalton Trans 42(43):15421–15426
17. Jiang HL, Makal TA, Zhou HC (2013) Interpenetration control in metal-organic frameworks for functional applications. Coordination Chem Rev 257(15–16):2232–2249
18. Jiao Y et al (2015) Tuning the kinetic water stability and adsorption interactions of Mg-MOF-74 by partial substitution with Co or Ni. Industrial Eng Chem Res 54(49):12408–12414
19. Kang IJ, Khan NA, Haque E, Jhung SH (2011) Chemical and thermal stability of isotypic metal-organic frameworks: effect of metal ions. Chem—A Euro J 17(23):6437–6442
20. Kornienko N et al (2015) Metal-organic frameworks for electrocatalytic reduction of carbon dioxide. J Am Chem Soc 137(44):14129–14135
21. Li G et al (2014) Hydrogen storage in pd nanocrystals covered with a metal-organic framework. Nature Mater 13(8):802–806
22. Li H et al. Design and construction of chemically stable metal-organic frameworks. www.worldscientific.com

23. Li H et al (2012) Enhanced hydrostability in Ni-doped MOF-5. Inorganic Chem 51(17):9200–9207
24. Liu SY et al (2014) Porous Cu(I) triazolate framework and derived hybrid membrane with exceptionally high sensing efficiency for gaseous oxygen. Adv Function Mater 24(37):5866–5872
25. Low JJ et al (2009) Virtual high throughput screening confirmed experimentally: porous coordination polymer hydration. J Am Chem Soc 131(43):15834–15842
26. Lv XL et al (2019) Ligand rigidification for enhancing the stability of metal-organic frameworks. J Am Chem Soc 141(26):10283–10293
27. Lv XL et al (2016) J Am Chem Soc, Just Accepted Manuscript • Publication Date A Base-Resistant Metalloporphyrin MOF for C-H Bond Halogenation A Base-Resistant Metalloporphyrin MOF for C-H Bond Halo-Genation. http://pubs.acs.org
28. Majano G et al (2014) Solvent-mediated reconstruction of the metal-organic framework HKUST-1 (Cu3(BTC)2). Adv Function Mater 24(25):3855–3865
29. Makal TA, Wang X, Zhou HC (2013) Tuning the moisture and thermal stability of metal-organic frameworks through incorporation of pendant hydrophobic groups. Crystal Growth Design 13(11):4760–4768
30. Masciocchi N et al (2010) Cubic octanuclear Ni(II) clusters in highly porous polypyrazolyl-based materials. J Am Chem Soc 132(23):7902–7904
31. Mendes RF, Almeida Paz FA (2015) Transforming metal-organic frameworks into functional materials. Inorganic Chem Frontiers 2(6):495–509
32. Meng J et al (2020) Advances in metal-organic framework coatings: versatile synthesis and broad applications. Chem Soc Rev 49(10):3142–3186
33. Ming Y et al (2015) Kinetic stability of MOF-5 in humid environments: impact of powder densification, humidity level, and exposure time. Langmuir 31(17):4988–4995
34. Mondloch JE et al (2013) Vapor-phase metalation by atomic layer deposition in a metal-organic framework. J Am Chem Soc 135(28):10294–10297
35. Morris RE, Brammer L (2017) Coordination change, lability and hemilability in metal-organic frameworks. Chem Soc Rev 46(17):5444–5462
36. Muthaiah S, Bhatia A, Kannan M (2020) "Stability of metal complexes." In Stability and applications of coordination compounds, IntechOpen
37. Nagarkar SS, Das R, Poddar P, Ghosh SK (2012) Bistable dynamic coordination polymer showing reversible structural and functional transformations. Inorganic Chem 51(15):8317–8321
38. Nguyen JG, Cohen SM (2010) Moisture-resistant and superhydrophobic metal-organic frameworks obtained via postsynthetic modification. J Am Chem Soc 132(13):4560–4561
39. Padial NM et al (2013) Highly hydrophobic isoreticular porous metal-organic frameworks for the capture of harmful volatile organic compounds. Angewandte Chemie—International Edition 52(32):8290–8294
40. Park KS et al (2006) Exceptional chemical and thermal stability of zeolitic imidazolate frameworks. www.pnas.orgcgidoi10.1073pnas.0602439103
41. Peterson GW et al (2009) Ammonia vapor removal by Cu 3(BTC) 2 and its characterization by MAS NMR. J Phys Chem C 113(31):13906–13917
42. Railey P, Song Y, Liu T, Li Y (2017) Metal organic frameworks with immobilized nanoparticles: synthesis and applications in photocatalytic hydrogen generation and energy storage. Mater Res Bull 96:385–394
43. Reinsch H et al (2013) "Structures, sorption characteristics, and nonlinear optical properties of a new series of highly stable aluminum MOFs." Chem Mater 17–26
44. Rieth AJ, Wright AM, Dincă M (2019) Kinetic stability of metal-organic frameworks for corrosive and coordinating gas capture. Nature Rev Mater 4(11):708–725
45. Serhan M et al (2019a). "Total iron measurement in human serum with a smartphone." In AIChE annual meeting, conference proceedings, American Institute of Chemical Engineers
46. Serhan M et al (2019b) "Total iron measurement in human serum with a smartphone." In AIChE Annual Meeting, Conference Proceedings, American Institute of Chemical Engineers

47. Song X, Jeong S, Kim D, Lah MS (2012) Transmetalations in two metal-organic frameworks with different framework flexibilities: kinetics and core-shell heterostructure. Cryst Eng Comm 14(18):5753–5756
48. Song Z et al (2017) Dual-channel, molecular-sieving core/shell ZIF@MOF architectures as engineered fillers in hybrid membranes for highly selective CO2 separation. Nano Lett 17(11):6752–6758
49. Wang B et al (2016) Highly stable Zr(IV)-based metal-organic frameworks for the detection and removal of antibiotics and organic explosives in water. J Am Chem Soc 138(19):6204–6216
50. Wang K et al (2016) Pyrazolate-based porphyrinic metal-organic framework with extraordinary base-resistance. J Am Chem Soc 138(3):914–919
51. Weber G et al (2016) Mechanism of water adsorption in the large pore form of the gallium-based MIL-53 metal-organic framework. Microporous Mesoporous Mater 222:145–152
52. Wu P et al (2012) Luminescent metal-organic frameworks for selectively sensing nitric oxide in an aqueous solution and in living cells. Adv Function Mater 22(8):1698–1703
53. Yuan S, Qin JS, Lollar CT, Zhou HC (2018) Stable metal-organic frameworks with group 4 metals: current status and trends. ACS Central Sci 4(4):440–450

Strategies for the Synthesis and Functionalization of MOFs

Strategies to Synthesize Diverse Metal–Organic Frameworks (MOFs)

Chetna Gupta, Parul Pant, and Himanshu Rajput

Contents

1 Introduction ... 70
2 History of MOFs ... 71
3 Categories and Names of Metal–Organic Frameworks ... 72
4 Reticular Synthesis of MOFs ... 73
5 Properties and Parameters Considered During Synthesis of MOFs ... 75
6 Different Methods of Synthesis of MOFs ... 80
7 Microwave-Assisted MOF Synthesis ... 83
8 Electrochemical Synthesis ... 86
9 Mechanochemical Synthesis ... 88
10 Ultrasonic Synthesis ... 92
11 Conclusions and Future Perspectives ... 94
Abbreviations ... 94
References ... 95

Abstract Metal–organic frameworks (MOFs) are remarkable platforms with numerous applications. These coordination polymer networks of inorganic metals blended with organic ligands in a 3-D crystalline fashion, result in porous, thermally, and chemically stable, robust polymers. Functionalized MOFs have been widely researched and used for energy storage, catalysis, and separations. This chapter focuses on the different strategies adopted for the synthesis and functionalization of MOFs. Various synthetic routes of MOFs are discussed in detail that includes the solvothermal method, ultrasonic method, microwave method, electrochemical method, and mechanical method of synthesis. Parameters governing

C. Gupta · P. Pant (✉)
Department of Chemistry, Hansraj College, University of Delhi, Delhi 110007, India
e-mail: parulpant@hrc.du.ac.in

C. Gupta
e-mail: chetnagupta@hrc.du.ac.in

H. Rajput
Department of Chemistry, University of Delhi, Delhi 110007, India

S. Gulati (ed.), *Metal-Organic Frameworks (MOFs) as Catalysts*,
https://doi.org/10.1007/978-981-16-7959-9_3

different shapes and sizes of MOFs, like reaction temperature, solvent, pH are also briefly covered in this chapter. Lastly, the advantages of the functionalization of MOFs, their applications in several fields like catalysis, drug delivery, sensors, etc., are also highlighted.

Keywords Metal–organic frameworks (MOFs) · Synthetic routes · Reaction temperature · Solvent · pH

1 Introduction

Over the last decade, the synthesis, characterization, and analysis of metal–organic frameworks (MOFs) have increased manifolds. These structures are formed by combining metal-containing units with organic ligands and generating open crystalline frameworks with persistent porosity employing strong bonds (reticular synthesis). Their modular design provides for a wide range of synthetic tuning options, including precise chemical and structural control. Porosity, stability, particle shape, and conductivity are all adjusted for specific purposes through innovative synthetic synthesis. More than 20,000 distinct MOFs have been reported and investigated in the last decade due to the ease with which the materials' shape, size, and functionality may be changed. MOFs have surface areas that vary from 1000 to 10,000 m^2 per gram, much surpassing those of conventional synthetic polymers like zeolites and carbons [12].

MOFs are a type of porous inorganic–organic hybrid component made up of metal oxo clusters coupled by organic ligands. MOFs have both organic and metal cores in their structure. Organic links are regarded as organic SBUs (Secondary Building Units) that operate as "struts" in MOF structures, whereas metal centers are called inorganic SBUs that act as "joints." (Fig. 1) The structure of the framework, inorganic metal centers, and organic ligands are the three primary components of MOF. Ligands for organic compounds are usually molecules with one or more N-

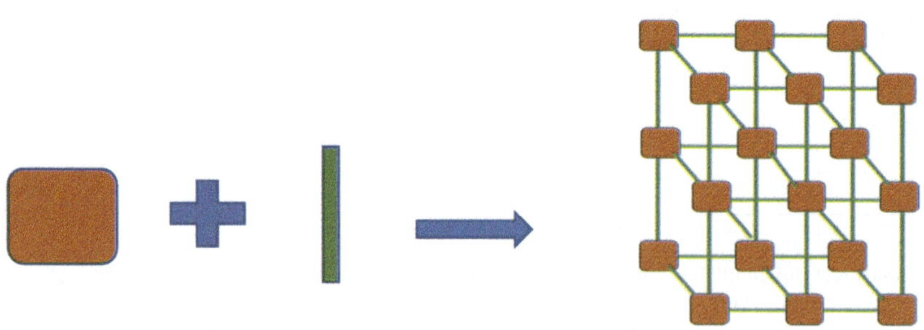

Inorganic SBU(Metal) **Organic SBU(Ligand)** **MOF**

Fig. 1 A general structure of metal–organic frameworks (MOFs) illustrating the bonding between organic ligands and metals or O-donor atoms. These molecules act as connecting linkers that connect the metal ions. Ligands like carboxylates, pyridyl, polyamines (particularly those derived from benzene, imidazole, oxalic acid), and cyano groups are most commonly used for the purpose.

MOFs have a wide range of applications, such as gas storage, separation, and catalysis. Fuel cells, supercapacitors, and catalytic conversions, in particular, have been the subject of substantial research, industrial-scale manufacture, and application due to their uses in energy technology. Metal–organic frameworks feature porous topologies that are comparable to zeolites, but they can be manufactured in an infinite number of configurations with various surface chemistries and pore topologies.

MOFs' easy tunability also allows them to be used as precursors and templates in the production of synthetic polymers with specific chemical compositions and topologies. Materials produced from MOFs, such as metallic compounds, permeable carbons, and their nanomaterials, are being studied extensively. Chemical and structural control may be achieved by adjusting MOF initial composition and modifying conversion processes.

2 History of MOFs

MOFs like structures were initially reported by Tomic et al., in the 1960s [35]. Later, in the 1990s, a porous crystal with persistent porosity was developed. Omar M. Yaghi was the scientist who coined the term "Metal–Organic Framework" to describe these porous crystals. MOF research work has exploded since then, notably when Yaghi and his research team rediscovered MOF-based porous materials [27].

Professor Yaghi's invention of MOF-5 (Fig. 2) ushered in a new era for metal–organic frameworks. This MOF-5 shattered the previous world record for porosity at

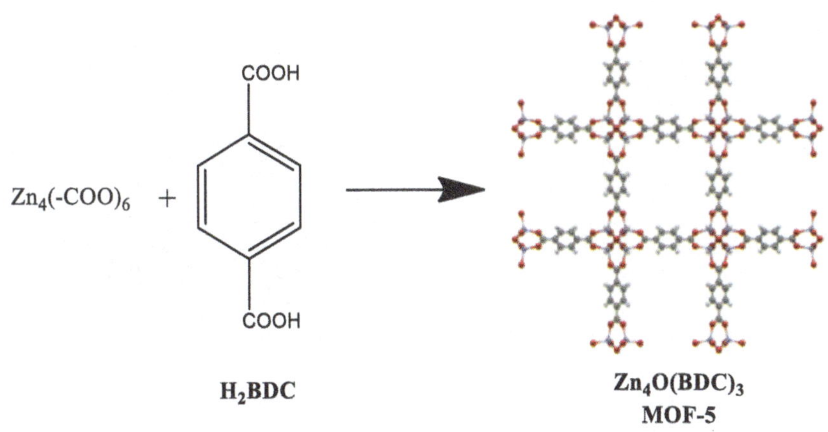

Fig. 2 Synthesis of MOF-5

the time, with a surface area of $6500m^2g^{-1}$. Many experts see MOF-5's discovery as the most notable example of a metal–organic framework. The synthesis of MOF-5 marked a significant step forward in MOF Chemistry [17].

3 Categories and Names of Metal–Organic Frameworks

MOFs are an abbreviation for metal–organic frameworks, which is normally used as a generic term for a class of substances (Fig. 3). When MOF is followed by ordinal numbers, it refers to a particular metal–organic framework structure (Table 1, lines 1–2). The study of MOF properties and structures aid in the development of framework structures that have desirable characteristics, such as the same symmetry as the family of isoreticular metal–organic frameworks abbreviated as IRMOFs (Table 1, lines 3–4). The other large class of MOFs is the ones having zeolite-like geometry and arrangement. Similarly, the zeolite imidazolate framework (ZIF) (Table 1, line 8) can be used to generate a spectrum of neutral frameworks. ZIFs are a class of microporous materials where metal ions such as Fe, Co, Cu, Zn, and others are surrounded by nitrogen-based tetrahedra and connected by imidazole rings, with different functional

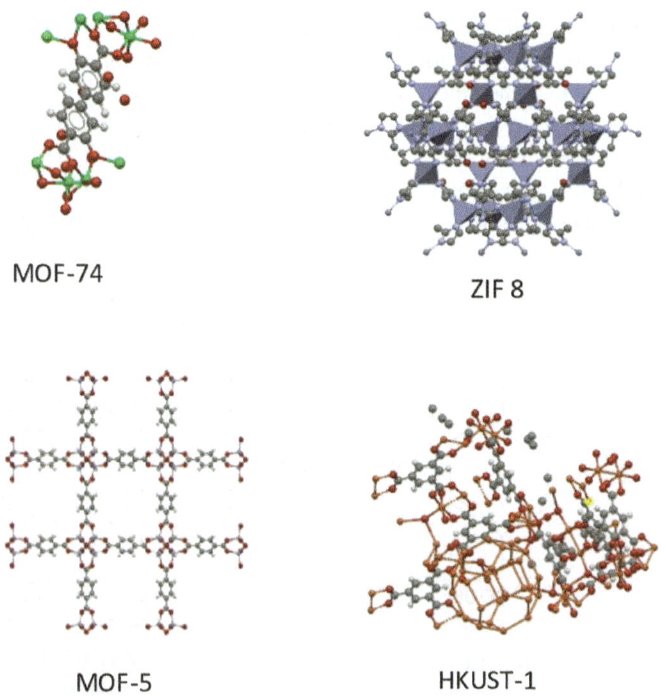

Fig. 3 Some representative MOF structures

Table 1 Examples of MOFs (names and their composition)

S.No	Abbreviated name of MOF	Formula	Abbreviation meaning
1	MOF-101	$Cu_2(BDC\text{-}Br)_2(H_2O)_2$	Metal–organic frameworks
2	MOF-253	Al(OH)(BPYDC)	Metal–organic frameworks
3	IRMOF-16	$Zn_4O(TPDC)_3$. 17DEF.$2H_2O$	Isoreticular metal–organic frameworks
4	IRMOF-1	$Zn_4O(BDC)_3$. 7DEF.$3H_2O$	Isoreticular metal–organic frameworks
5	UiO-67	Zr_6O_6(BPDC)6	Universitetet i Oslo
6	MIL-53	Al(OH)(BDC)	Materials of Institut Lavoisier
7	HKUST-1	$Cu_3(BTC)_2$	Hong Kong University of Science and Technology
8	ZIF-8	$Zn(MIM)_2$	Zeolite Imidazolate Framework
9	MOP-1	$Cu_{24}(m\text{-}BDC)_{24}(DMF)_{14}(H_2O)_{10}$	Metal–organic Polyhedra
10	CPL-2	$Cu_2(PZDC)_2$(4,4'-BPY)	Coordination polymer with pillared layer structure

Note BDC is benzenedicarboxylate; BPYDC is 2,2'-bipyridine-5,5'-dicarboxylate; TPDC is p-terphenyl-4,4'-dicarboxylate, BPDC is biphenyl-4,4'-dicarboxylate; MIM is 2-methylimidazolate; m-BDC is m-benzenedicarboxylate; PZDC is pyrazine-2,3-dicarboxylate, 4,4'-BPY is 4,4'-bipyridine

groups. CPL, UiO, HKUST, and MOP are among the other classifications used by research groups to classify synthesized MOFs.

4 Reticular Synthesis of MOFs

The most accepted mechanism which is followed by the different types of MOF synthetic routes is reticular synthesis. Reticular synthesis is the process of combining specially constructed rigid molecular building blocks into prescribed ordered networks held together by strong bonds. In reticular synthesis, the secondary building units (SBUs) are connected by strong bonds to produce crystalline solid-state materials with a defined topology [3]. The creation of novel structures known as metal–organic frameworks has resulted from the chemistry of linking building components together with strong chemical bonds to form extended frameworks [36]. The utilization of predefined reaction conditions in reticular synthesis in the in-situ generation of rigid and very well-structured molecules that preserve their structure and strength throughout the synthesis.

As a result of reticular synthesis, a wide spectrum of MOFs is made from polyatomic inorganic metal clusters, generally, the first-row transition metal (inorganic SBUs), connected by polytopic chelating linkers (the organic SBU). Both inorganic and organic SBUs can have a range of geometric structures, ranging from linear to polygonal.

Secondary building units (SBUs) are molecular structures and cluster entities in which polytopic linkers such as 1,4-benzenedicarboxylate, 1,3,5,7-adamantanetetracarboxylate, etc., can be used to convert fragments into extended porous networks using ligand coordination sites and metal coordination environments. The geometric and chemical properties of the SBUs and linkers are used to predict the framework topology, which leads to the design and synthesis of MOFs [7].

So, we can say that SBUs give MOF:

- building directionality
- as well as framework durability.

As a result, the SBU idea is useful for rationalizing MOF structure topologies. A vast number of geometric structures can be created using structurally varied inorganic and organic SBUs.

If the reaction circumstances are appropriately recognized to generate an SBU in-situ with a specific shape, the predesigned network can be formed. As a result, reticular chemistry entails designing, synthesizing, and predictably analyzing the structure. This form of crystal chemistry allows MOF to design structure and characteristics to a far larger extent than zeolite, which uses the silicate tetrahedron as the basic building block rather than the SBU.

Different functionalized organic linkers are directly introduced as a reactant into the syntheses of MOFs, and the metal-based units assemble in-situ during crystallization. As a result, using organic linkers with the same geometric arrangements of bonding groups (commonly carboxylates and amines) but different organic molecules, it is possible to generate a series of compounds with the same topology but different composition and dimension, both in principle and in practice.

The creation of desired types of pores is feasible if the inorganic moiety (Inorganic SBUs) and organic linkers (organic SBUs) are properly selected. Furthermore, the ability to chemically alter the organic linker via organic synthesis enables the predictable production of functionalized MOFs.

5 Properties and Parameters Considered During Synthesis of MOFs

Over the past few years, MOFs have gained much interest due to the properties depicted in Fig. 4.

- The enormous porosity and surface area of MOFs are arguably their most well-known feature. Hierarchically porous MOFs—frameworks with a mix of micropores and mesopores—create even more strategic channels and pore space.
- By choosing the right ligands and metal sites, the framework's pore size and structure may be fine-tuned to get atomic-level structural uniformity.
- Isoreticular MOFs have a lot of control over pore size and the chemical environment.
- MOFs have great thermal stability and mechanical stability due to their whole composition of strong bonds.
- A wide range of construction components and connecting types allows for a wide range and variety of MOFs.
- The network flexibility can be easily achieved by tuning how they are connected.

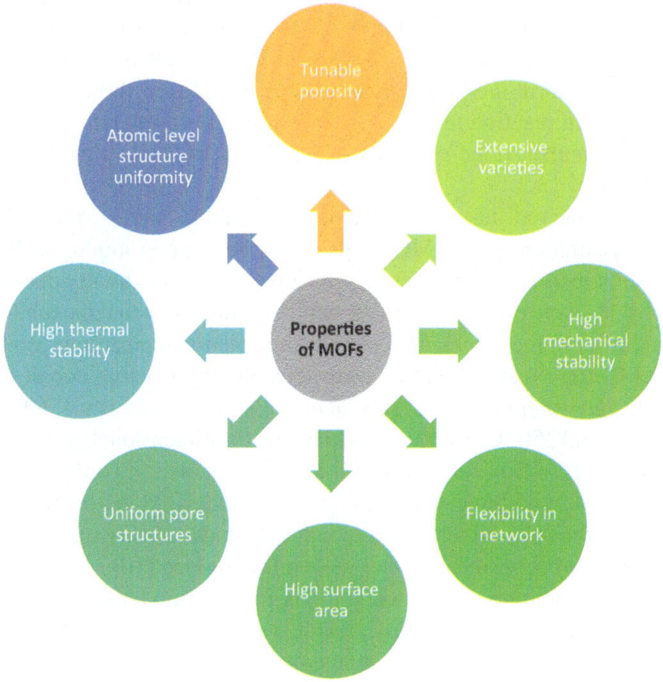

Fig. 4 Properties of MOFs

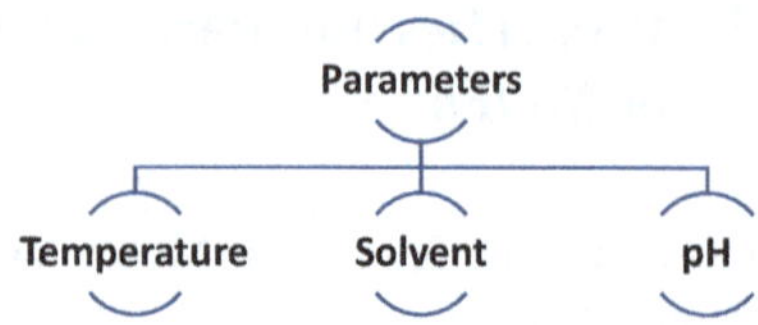

Fig. 5 Some important parameters to be considered for the synthesis of MOFs

Adjusting the synthesis conditions to reach suitable quantities for real-life application is the major challenge for MOF synthesis. While preparing the MOFs with controlled morphology and structure, the most considerable parameters to be taken into account are temperature, size, solvent, and pH (Figure 5).

These parameters are discussed below:

(i) **Temperature**

In the synthesis of metal–organic frameworks, one of the most important factors is the reaction temperature. Despite the fact that the numerous experimental parameters do not all agree, reaction temperature has been proven to have a significant impact on MOF formation and structure, particularly in terms of controlling topology and dimensionality. In reaction thermodynamics and reaction kinetics, the reaction temperature has a direct effect on the reaction energy barrier. Significant temperature produces high reaction pressure in the sealed system, which affects the assembly and final architecture of MOFs in hydro/solvothermal processes, which are widely employed in the synthesis of MOFs. It has been shown that raising the hydro/solvothermal reaction temperature increases the coordination number of the central metal ion and the dimensionality of the MOFs while lowering the coordinated solvent molecules.

It has been observed that as the reaction temperature is raised, the dimensionality and concentration of the MOFs also increase. For example, with the same starting reaction mixture but a different reaction temperature, totally distinct Co(II)-succinates MOFs were produced, and the results demonstrate that the dimensionality and concentration of the MOFs increased as the reaction temperature was elevated. Five distinct Co(II)-succinate MOFs were separated by reacting cobalt hydroxide, succinic acid, and water in a 1:1:28 ratio at different temperatures viz. 60 °C, 100 °C, 150 °C, 190 °C, and 250 °C. At 60 °C and 100 °C, the dimensionality of the resultant Co(II)-MOFs evolves from one-dimensional (1D) chains to two-dimensional (2D) networks at 150 °C, and three-dimensional (3D) frameworks at 190 °C and 250 °C. Furthermore, as the reaction temperature was raised, the density of the MOFs grew, while the number of coordinated water molecules dropped [10].

At higher temperatures, the effect of classical thermodynamic considerations can be seen in the shift from highly hydrated complexes at low temperatures to metal hydroxide complexes at higher temperatures. A mixture of $Mn(OAC)_2.4H_2O$, 4,4‘-oxy bis(benzoic acid), imidazole, and NaOH were dissolved in water and heated at different temperatures. By raising the temperature, three novel manganese oxy-bis(benzoate) compounds, $[Mn(H_2O)_3\ (C_{12}H_8O)\ (COO)_2]H_2O$ (a), $[Mn(H_2O)$

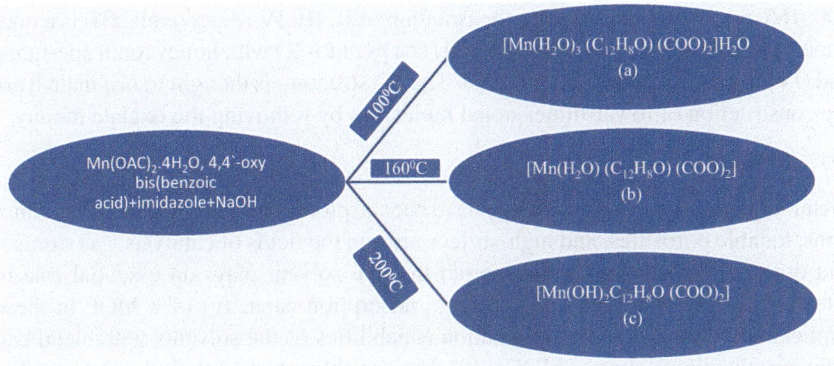

Fig. 6 Formation of 3 different structures with different temperature conditions for same starting materials

$(C_{12}H_8O)$ $(COO)_2]$ (b), were generated as pure phases (Fig. 6). The synthesis of a higher dimensional structure $[Mn(OH)_2C_{12}H_8O$ $(COO)_2]$(c) was achieved by the removal of certain aqua ligands at a higher temperature. This is thought to be a thermodynamically entropy-driven dehydration pathway that reduced the Mn–Mn distances inside the framework. The structures' dimensionality was determined to be a pseudo 2D for (a) at 100 °C and two 2D layer structures for (b) at 160 °C and (c) at 220 °C [22].

Another research that displayed when the reaction temperature rises, the total dimensionality of the MOFs rises, was studied by Dan et al. Effect of temperature on the dimensionality of zinc(II) oxalate complex MOF was done by using a zero-dimensional (0D) dimeric zinc (II) oxalate complex. Dan and Rao [6] observed a gradual rise in dimensionality with MOF $(C_4N_2H_{12})_3[Zn_2(C_2O_4)_5]0.8H_2O$ (I) (where $C_4N_2H_{12}$ = piperazine di-cation). MOFs $(C_4N_2H_{12})_2[Zn_2(C_2O_4)_4]0.3H_2O$ (II), $(C_4N_2H_{12})_3[Zn_2(C_2O_4)_7]$ $0.4H_2O$ (III), as well as$(C_4N_2H_{12})$ $[Zn_2(C_2O_4)_3]$ (IV) were successfully synthesized (Fig. 7). Heating (I) in the presence of piperazine (PIP) at

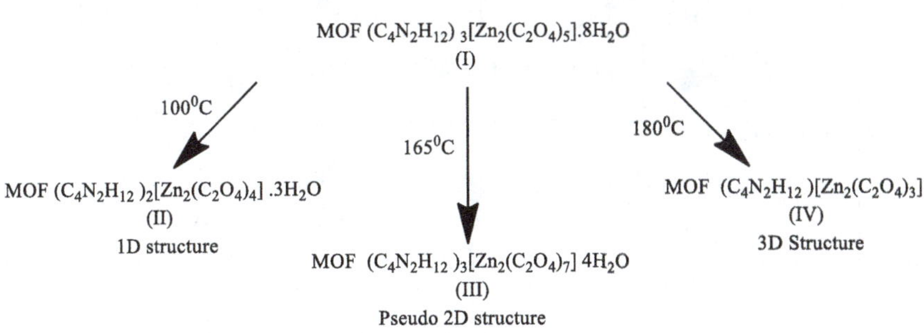

Fig. 7 Formation of different dimensionalities of zinc(II) oxalate complex MOFs with a variation of temperature

100, 165, and 180 °C, resulted in the isolation of II, III, IV respectively. (II) is a one-dimensional endless chain structure, (III) is a pseudo-2D with honeycomb apertures, and (IV) is a three-dimensional design. The 3D structure is thought to originate from the construction of lower-dimensional molecules by removing the oxalate moiety.

(ii) **Solvent**

Metal–organic frameworks (MOFs) have been explored for their customizable functions, tunable porosities, and high-surface area, in the fields of catalysis, gas storage, and drug delivery. It has been reported that the solvent plays an essential role in affecting the stability and function (e.g., adsorption capacity) of a MOF in these applications. The different coordination capabilities of the solvents with metal can influence the dimensions of MOFs [40]. Various solvents (and their derivatives) that have been used over the years for MOF synthesis are depicted in Fig. 8.

To study the effect of solvents employed in the reaction combination, Pachfule et al. [25] observed that by reacting $Cu(NO_3)_2.3H_2O$ with 4,4' -(hexafluoroisopropylidene) bis(benzoic acid) ($C_{17}H_{10}F_6O_4$, H2hfbba) and terminal monodentate ligand 3-methyl pyridine (3-picoline/3-mepy) solvothermally, and taking N, N-dimethyl formamide (DMF) and N,N-diethyl formamide (DEF) as solvents different MOF structures were obtained for MOF $Cu_2(hfbba)_2(3\text{-}mepy)_2](DMF)_2(3\text{-}mepy)]$ and, $[Cu_2(hfbba)_2(3\text{-}mepy)_2]$.

Banerjee's group [2] also studied the effects of a mixed solvent system on a group of MOFs. They used a mixture of DMF, MeOH, EtOH, and H2O to create four Mg-based coordination networks: three 3D $[Mg_3(3,5\text{-}PDC)_3\ (DMF)_3]$. DMF,

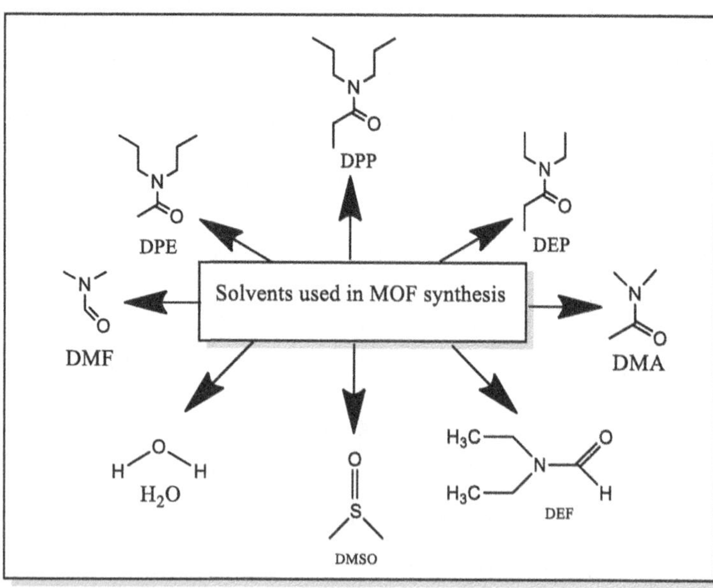

Fig. 8 Various solvents used in MOF synthesis

$[Mg(3,5\text{-}PDC)(H_2O)].(H_2O)$, $[Mg_4(3,5\text{-}PDC)_4(DMF)_2(H_2O)_2]2DMF$. 4.5H2O, 2D $[Mg(3,5\text{-}PDC)(H_2O)_2]$ (PDC-pyridine dicarboxylate). The dimensions of these four MOFs were found to be affected by the relative coordination capabilities of the above-mentioned solvents with the metal. It was observed that H2O has a strong propensity for coordinating with metal centers. When MeOH/H_2O and EtOH/DMF combinations were employed, DMF has a lower affinity for Mg than H_2O, but MeOH and EtOH had no affinity for the metal core.

In order to study the solvent effect on drug delivery, Gomar et al. [13] researched on ZIFs (Zeolite Imidazolate Frameworks, ZIF-6, ZIF-3) and ethanol interaction at different pressures. They discovered that when ethanol is present at 300 K and 105 Pa, drug molecules are better adsorbed at the ZIF-ethanol interface. As a result, at atmospheric pressure, the solvent plays a crucial role in drug adsorption, but it was found that at lower pressures, it has no effect. This study implied that solvent plays a crucial role in the adsorption of small molecules in porous materials. So, the impact of solvent on small molecules should be considered when designing new materials investigation, both experimentally and computationally.

(iii) **pH**

The acidity/basicity of the reaction medium has a significant impact on the crystallization and development of inorganic–organic hybrid materials. On the basis of the acid–base concept, the amount of deprotonation of an organic ligand and, in certain cases, the creation of an OH-ligand in an aqueous solution with regard to the pH of the reaction medium would favor the connection of ligand (e.g., polycarboxylate) to the metal ion.

It has been observed that:

(a) During MOF synthesis, higher pH values encourage the production of interpenetrating networks, whereas lower pH values encourage the construction of simple un-interpenetrated 3D frameworks.
(b) Distinct composition and dimensionality are influenced by the pH value of the solution.
(c) pH values can alter the color of MOF.

To study the role of pH, Yuan et al. [38] and his colleagues used $CdCl_2.2.5H_2O$, 4-carboxy-4,2‘,6‘,4“-terpyridine(CTPY), and oxalic acid to make two MOFs by varying the pH levels. At pH 7.5 and 5.5, frameworks of the formula $[Cd_2(CTPY)_4]_n2nH_2O]$ (a) and $[Cd_2(CTPY)_2(ox)]_n$ (b) were produced. It was observed that CTPY adopts multiple coordination modes based on pH. Oxalic acid's role as a templating agent was also affected by pH. It was found that oxalic acid can function as a medium for the synthesis of (a) at a higher pH of 7.5, resulting in a 3D interpenetrating framework, but it is coordinated with the Cd(II) atom in (b) at a lower pH of 5.5 (Fig. 9) So, it was concluded that higher pH values favor the production of interpenetrating networks, whereas lower pH values favor the formation of a simple un-interpenetrated three-dimensional framework.

Similarly, Zhang and his research group [39] systematically examined the ability of the flexible ligand 1,3-adamantanediacetic acid (H_2ADA)'s to react with cobalt

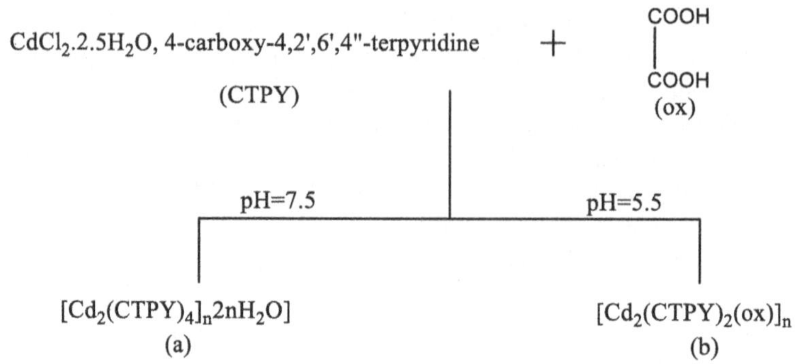

Fig. 9 Formation of different MOFs $[Cd_2(CTPY)_4]_n 2nH_2O]$ **a** and $[Cd_2(CTPY)_2(ox)]_n$ at different pH conditions

salt for the formation of different MOFs, reveals that the varied coordination preferences are mostly influenced by the pH value of the solution. Three neutral MOFs, $[Co(HADA)_2(bpp)]_n$, $[Co\ (ADA)(bpp)(CH_3OH)](H_2O)_n$ and $[Co_2(ADA)_2(bpp)]_n$, were obtained at pH = 5, pH = 6, and pH = 7, each with its own composition and dimensionality.

The pH value was also studied by Luo et al. [21] to check its effect on the color of the Co-MOFs formed by taking 1,3,5 benzene tricarboxylate (BTC) as ligand. The pH value was adjusted to produce three distinct colored Co-BTC-3,3‘,5,5‘-tetra(1H-imidazole-1-yl)-1,1’ -biphenyl (L) compounds. At pH = 5, pink-colored crystals of $[Co(L)\ (HBTC)_2(\mu_2\text{-}H_2O)(H_2O)_2]0.3H_2O$ were produced, as were purple-colored crystals of $[Co_3(L)_2(BTC)_2]0.4H_2O$ at pH = 7, and brown-colored crystals of $[Co_2(L)(BTC)(\mu_2\text{-}OH)(H_2O)_2]0.2H_2O$ at the pH = 9. It was observed that pH = 5, BTC was slightly deprotonated, while at pH 7 and pH 9, it was completely deprotonated. The pH value of the reaction system was found to control the BTC ligand's coordination mode as well as the color of the compounds.

6 Different Methods of Synthesis of MOFs

MOFs are now regarded as a significant category of porous compounds due to their unique functional and structural characteristics. Bridging organic linkers and metal ion clusters and metal ions are used to generate these frameworks (Fig. 10).

It becomes very important to discuss the synthesis of the MOFs employing different methods in order to understand the advantages and limitations of each method of synthesis.

It is critical to employ a variety of synthesis techniques because different procedures can lead to new "frameworks" that would otherwise be impossible to obtain. Alternative methods of synthesis can result in compounds of various particle sizes and sizes. Morphologies, as well as distributions, can also have an impact depending

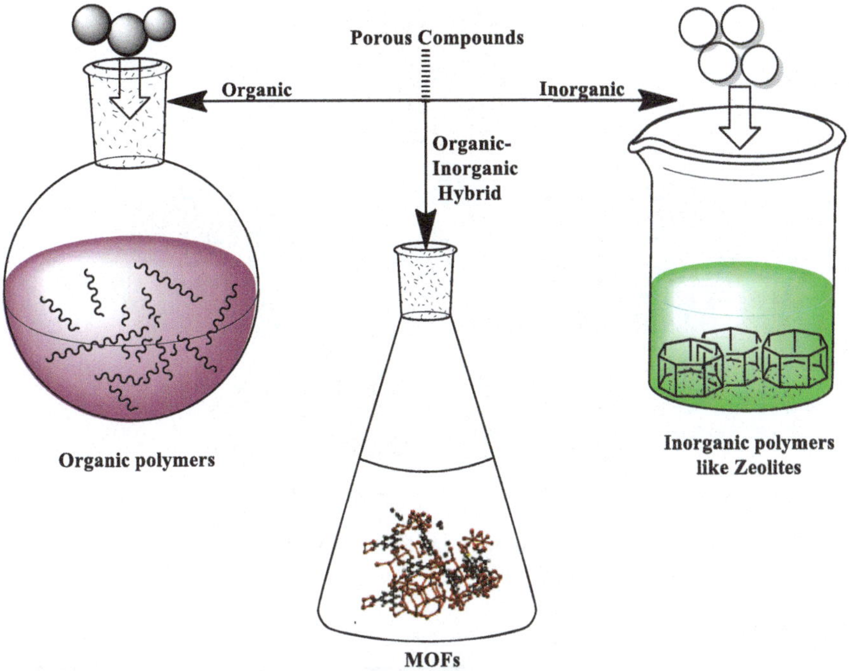

Fig. 10 MOFs are organic–inorganic hybrids

on the material's qualities. Different particle sizes, for example, can influence the migration of guest molecules in MOFs which has a direct impact on catalytic reactions or catalytic reactions in general molecular adsorption and separation.

Figure 11 summarizes the scope of this topic. The MOF synthesis processes like hydrothermal (solvothermal), electrochemical mechanochemical, microwave aided, heating, and ultrasonic methods that have been used over the years are discussed under this topic.

(a) **Solvothermal Synthesis**

Metal–organic frameworks (MOFs) are formed when a metal-containing precursor reacts with an organic linker in an organic solvent at a high temperature (more than that of the solvent), and this process is known as solvothermal synthesis (Fig. 12). Here, the MOFs are synthesized through electrical heating on small scale either in NMR tubes or small vials. The procedure generally uses thermal energy (353–453 K), and it might take anywhere from 48 to 96 h to complete. DMF (dimethylformamide), DEF (diethyl formamide), ethanol, methanol, acetone, and other organic solvents were often employed in solvothermal processes. In the end washing and drying are done for obtaining pure MOFs.

Different starting materials exhibit different solubility so either a single solvent or combination of solvents have been used to carry out solvothermal reactions. Depending on the reaction, solvothermal reactions can be performed at a variety

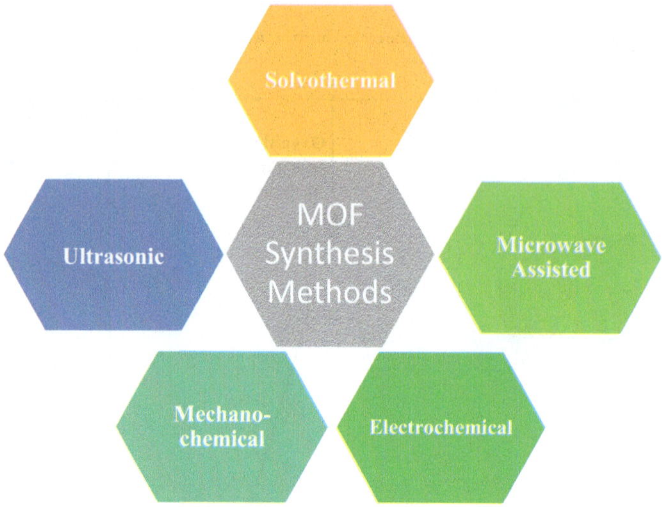

Fig. 11 Different types of synthesis methods employed for MOF synthesis

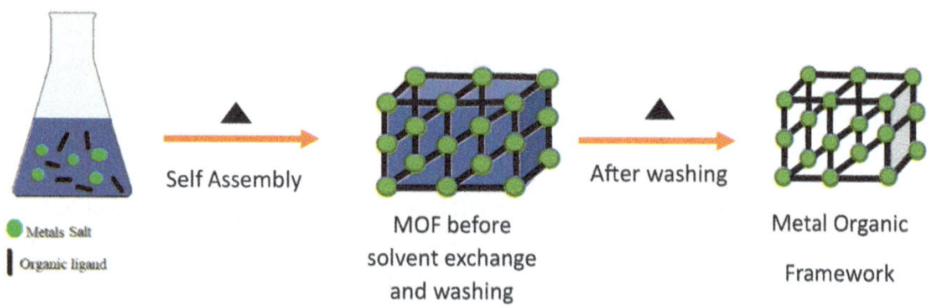

Fig. 12 Steps involved in solvothermal synthesis

of temperatures. As the self-assembly interaction between metal ions and organic ligands occurs easily in the liquid phase, solvothermal synthesis is the most often utilized method.

For example, different MOFs were obtained depending on the solvents used in the reaction combination by employing solvothermal synthesis in the following way. Inorganic salt $Cu(NO_3)_2.3H_2O$ with 4,4' -(hexafluoroisopropylidene) bis(benzoic acid) ($C_{17}H_{10}F_6O_4$, H_2hfbba) was reacted with terminal monodentate ligand 3-methyl pyridine (3-picoline/3-mepy) solvothermally, taking N,N-dimethyl formamide (DMF) and N,N-diethyl formamide (DEF) as solvents respectively to produce two different MOF structures Cu_2(hfbba)$_2$(3-mepy)$_2$](DMF)$_2$(3-mepy)] called F-MOF-4 and, [Cu_2(hfbba)$_2$(3-mepy)$_2$] called Cu-F-MOF-4B [26].

Similarly, Férey and co-workers [9] described the synthesis of MIL-101 (Matérial Institut Lavoisier-101) by solvothermal method (taking water as a solvent and the process also known as Hydrothermal synthesis). They reacted terephthalic acid with

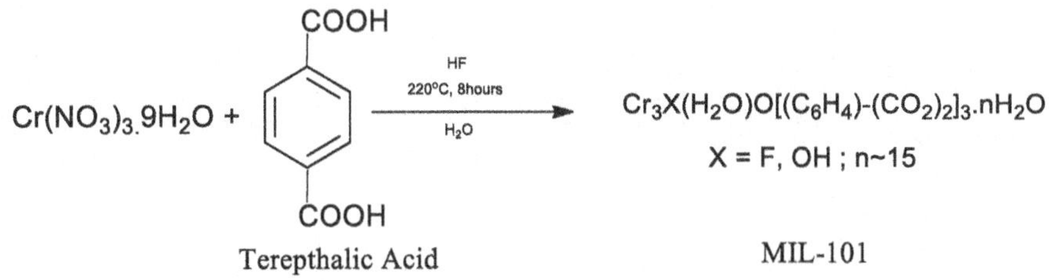

Fig. 13 Steps involved in the preparation of MIL-101

$Cr(NO_3)_3.9H_2O$, fluorohydric acid, and H_2O in a hydrothermal vessel and heated the resultant mixture for about 8 h at 220 °C. The reaction mixture produced crystallized chromium terephthalate (MIL-101) on completion (Fig. 13).

In a similar way, Clausen et al. [5] the Zn–terephthalic acid (benzene-1,4-dicarboxylic acid, H_2BDC)–dimethyl formamide (DMF) system yielded two novel crystal structures via solvothermal synthesis. The two structures reported by them were obtained by reacting a solution of benzene-1,4-dicarboxylic acid in DMF with a solution of $Zn(NO_3)_2.6H_2O$ in DMF. They heated the reactant mixture in an autoclave for 4 days at a temperature of about 381 K. When the tank was brought to room temperature, two different sorts of macroscopically indistinguishable colorless single crystals (1) $Zn(C_8H_4O_4)(C_3H_7NO)$ and (2) $Zn(HCO_2)_3(C_2H_8N)$ were obtained.

7 Microwave-Assisted MOF Synthesis

The technique relies on electromagnetic waves interacting with any substance having mobile electric charges, such as polar molecules in a liquid or conducting ions in a solid. Unlike traditional solvothermal techniques, which transmit thermal energy from the heat source to the solution via the reaction vessel, Microwave irradiations interact directly with the reactants, resulting in more efficient and quicker heating. Furthermore, in microwave synthesis, crystallization happens in hot spots formed by direct heating of the solvent, rather than the reactor vessel's wall, as with traditional heating techniques. As a result, it is considerably quicker and produces smaller particles.

Microwave technique comprises of treating the reactants in a suitable solvent inside a sealed Teflon vessel placed in the microwave unit, where it is heated for the required time at the set temperature. The microwave technique leads to rapid heating of the liquid phase because of the interaction of the applied oscillating electric field is with the permanent dipole moment of the molecules in the synthesis medium. The final product is obtained by washing, followed by heating or vacuum drying. Figure 14 depicts various steps involved in microwave-assisted MOF synthesis.

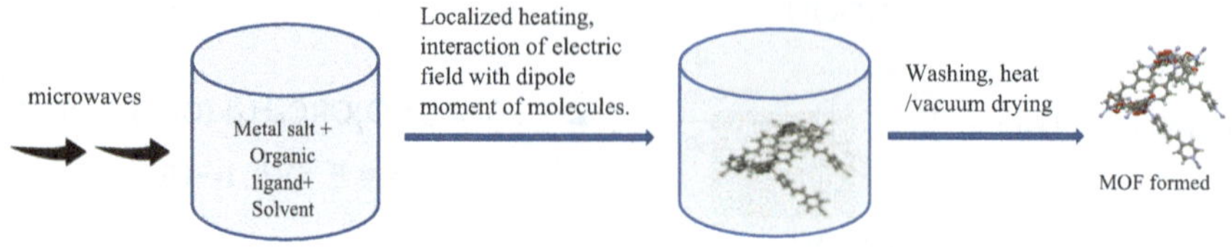

Fig. 14 Schematic representation of microwave-assisted MOF synthesis

Some of the important aspects of microwave synthesis are listed below:

- Microwaves may be used to increase the temperature of the solution for an hour or longer in order to generate metal nanosized crystals.
- Microwave synthesis is an essential technique for doing high-speed synthesis.
- The form and size of the resultant particles may be precisely controlled using Microwave synthesis. Because uniform nanocrystals are easily produced, MW irradiation is an appealing technique for synthesizing MOFs with biological applications, such as iron-carboxylate MOFs.
- Microwave irradiation (MI) is an innovative approach for increasing MOF scalability because MI syntheses depend on targeted heating rather than heating the entire solution, microwave heating uses less energy than standard solvothermal techniques.

As a result, MW synthesis can significantly reduce the synthesis cycle for a variety of operational purposes while also regulating the form and size of the crystals. Another advantage of microwave synthesis is that microwave irradiation results in the formation of crystals in minutes of reaction time, whereas the conventional method required hours for producing the same. The ideal microwave settings produce regular cubic crystals in certain cases. Aside from rapid crystallization, phase selectivity, narrow particle size distribution, and easy morphological control are potential benefits of this method for the synthesis of porous materials [33].

Jhung et al. [14] described the water-based synthesis of the chromium trimesate MIL-100 MOF by microwave synthesis. They reacted Cr salts with H_3BTC (benzene tricarboxylic acid) in presence of fluorohydric acid and water. The reaction mixture was sealed in an autoclave and kept in a microwave oven and heated to about 220 °C for a specific period of time. For reaction durations after 4 h, the data revealed the presence of reacted metallic chromium species. After 4 h, the crystal yield was 44 percent, which is equivalent to the 45 percent obtained in a 4-day traditional synthesis.

The same group also researched the Microwave synthesis chromium terepthalate of Cr-MIL-101 [14]. They reacted $Cr(NO_2)_3 \cdot 9H_2O$, H_2BDC (benzene dicarboxylate) fluorohydric acid, and H_2O. The reacting mixture was sealed in a Teflon autoclave and kept inside a microwave. The microwave oven was preheated to about 210 °C. Then, after keeping the reacting mixture in the microwave oven, it was heated at the same temperature (210 °C) for about 1–60 min. MIL-101 was produced in this study utilizing microwave irradiation at a significantly lower temperature than previously reported. The substance was further purified using a two-step procedure that comprised double filtering and ethanol extraction. The purified MIL-101 remained stable in air for months, even after being handled with different organic solvents at high temperatures or boiling water during purification.

Similarly, Ni et al. [23] reported the MW synthesis of different isoreticular metal–organic frameworks (IRMOFs): IRMOF-1, IRMOF-2, and IRMOF-3 in less than 2 min obtaining microcrystals with very uniform size and similar cubic shape. For IRMOF-1, they reacted $Zn(NO_3)_2.6H_2O$ and 1,4-benzene dicarboxylate acid (H_2BDC) and dissolved them in N,N′-diethylformamide (DEF). The dissolved solution was sealed in a pyrex vial and heated in a microwave synthesizer at 150 W for 25 s. Microsized cubic crystals of $Zn_4O(BDC)_3.7DEF.3H_2O$ IRMOF-1 were obtained after the completion of the process. A similar case was seen in IRMOF-2 and IRMOF-3 but with a difference of micro-units in size. They also demonstrated that by varying the concentration of the starting material, and the crystal size can be changed from micrometer to submicrometer.

MW irradiation has been used as a successful technique for synthesizing MOFs with biological applications, such as iron-carboxylate MOFs. Taylor and coworkers [34] for example, published the MW synthesis of iron-MIL-101 (Fe_3-(μ_3-O)$Cl(H_2O)_2(BDC)_3$) MOF nanoparticles and their amino-functionalized form in 2009. The reactants $FeCl_3$ and terephthalic acid (BDC) were dissolved in DMF and heated with microwave at 150 °C for about 10 min. The product formed Fe-MIL-101 was isolated and washed with DEF and C_2H_5OH. Post-synthetic modifications were done on Fe-MIL-101 and the product formed by incorporating 2-amino terephthalic acid (NH_2-BDC) that produced MOF-88B (Fe-based MOF) used in biological applications.

MW irradiation has been used as a successful technique for synthesizing MOFs with biological applications, such as iron-carboxylate MOFs. For example, Vilela et al. [37] synthesized quality single crystals of the microporous cationic $[Ce_2(pydc)_2(Hpydc)(H_2O)_2]Cl$ (where pydc corresponds to the 2,5-pyridinedicarboxylic acid) by using MW heating at 200 °C for 20 min. They synthesized a series of lanthanide (Eu, Gd and Tb) bisphosphonates in a similar way, using conventional hydrothermal synthesis (180 °C, 3 days), MW-assisted heating (40 °C, 5 s). They proved the efficacy of MW synthesis for the above synthesis over conventional hydrothermal synthesis with their research work.

Bag and co-workers [1] also synthesized $[La(TTTPC)(O)_2(NO_2)_2(Cl)] \cdot (H_2O)_{10}$ by reacting $La(NO_3)_3{\cdot}6H_2O$, H_3TTTPC, in a mixture (1:1) of H_2O/EtOH. The reactant species were taken in a microwave tube and heated by microwaves at 85 °C for about 5 min. The same MOF was produced solvothermally at the same temperature in 2 days. Bag et al., demonstrated the gram-scale manufacture of nine other isostructural microporous lanthanide MOFs in 5 min using a microwave. The same synthesis using a traditional solvothermal process took seven days to generate the same materials with a similar yield.

8 Electrochemical Synthesis

The study of the interconversion of chemical and electrical energy is referred to as electrochemistry. It is a branch of chemistry that deals with chemical changes produced by an electrical current.

MOF powders are produced on a large scale using an electrochemical method (Fig. 15). When compared to solvothermal synthesis, this technique has several advantages, including the avoidance of anions such as nitrates from metal salts, lower reaction temperatures, and highly rapid synthesis. As a result, as compared to solvothermal approaches, electrochemical synthesis methods provide greater fine tuning parameters owing to simple voltage adjustments or the imposition of specific signals (e.g., pulses).

The electrochemical synthesis of MOFs has evoked a lot of interest as it offers a lot of benefits. One of them is the ability to carry out MOF synthesis in a continuous manner. It also enables them to be synthesized under gentler temperatures than traditional solvothermal or microwave syntheses, which reduces reaction time. Electrochemical techniques usually create MOF material in minutes or hours, whereas solvothermal synthesis might take hours or days. Furthermore, the electrochemistry technique allows for direct control of MOF synthesis throughout the process by adjusting the passing current or applied voltage. In addition, the electrochemistry technique allows for the creation of homogenous thin films or coatings.

The two major approaches for electrosynthesis of MOFs are:

(i) anodic deposition, which was the first way patented by BASF, and
(ii) cathodic deposition.

(i) Anodic Deposition

In this process, an applied electric potential causes metal ions from the electrode to be released, which subsequently react with an organic linker present in the solution to produce a MOF film. The use of a metallic electrode (rather than metal salts) as the source of metal cations in this environment prevents the production of any corrosive anions (mostly nitrate and acetate anions) or by-products. Anodic dissolution is normally done in a two-electrode setup without a reference electrode, and protic

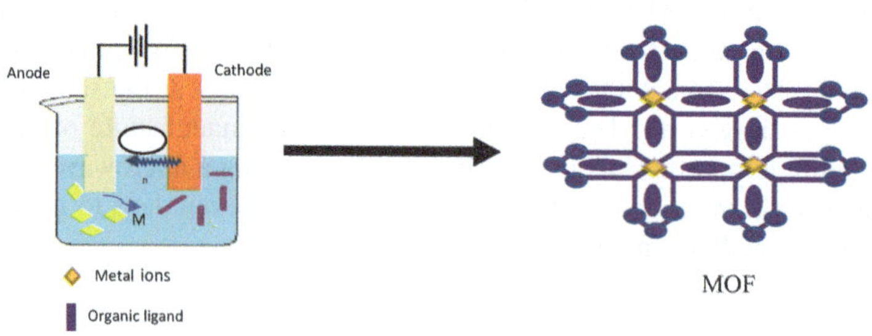

Fig. 15 Representation of MOF synthesis by electrochemical method

solvents are usually required to assure hydrogen evolution and minimize metal ion reduction at the counter electrode. Furthermore, the use of a selectively reduced sacrificial compound (e.g., acrylonitrile, acrylic, or maleic esters) or a counter electrode with a sufficient over potential for hydrogen evolution is also advisable for getting the best results.

The company BASF first patented the use of electrochemical synthesis to produce MOFs in 2005. BASF was the first to prepare and patent HKUST-1 (HKUST = "Hong Kong University of Science and Technology"), by the process of anodic deposition. The synthesis consisted of immersing a copper plate in a solution containing the organic linker, 1,3,5-benzenetricarboxylic acid (BTC), and an electrolyte, taking advantage of the potential of electrochemistry to synthesize materials and their large experience in the domain. In this case, the source of Cu(II) ions was a copper plate that served as the electrode. Cu(II) ions were released from the copper electrode into the solution and interacted with the dissolved linker when a certain current or voltage was applied. After applying a voltage of 12–19 V and a current of 1.3 A for 150 min, a powder of electrochemically generated HKUST-1 which is also called MOF-199, with octahedral crystals (size: 0.5–5 mm) was created. This synthesized HKUST-1 had a surface area of 1820 $m^2\ g^{-1}$, which is greater than the solvothermally synthesized HKUST-1 (1550 $m^2\ g^{-1}$).

Many attempts have been made to study and optimize the conditions of electrochemical synthesis of HKUST-1 (by BASF) utilizing anodic dissolution since it was first patented. Campagnol et al. [4] suggested a four-phase process for the anodic dissolution synthesis of HKUST-1:

(a) Initial nucleation;
(b) growth of HKUST-1 islands;
(c) intergrowth; and
(d) crystal detachment.

(a) Initial nucleation: Initially in the solution, Cu (II) ions are released where the organic ligand is already present. After reaching the critical concentration of the reagent, the formation of nuclei takes place in and on the surface.
(b) Growth of Islands: On the surface, these nuclei develop to micrometer-sized crystals adjacent to and on top of each other.
(c) Intergrowth: New crystals continue to form and develop, producing an intergrown layer.
(d) Crystal Detachment: Finally, due to the internal pressure in the MOF layer and the depression of the crystals, portions of the MOF surface break interaction with the substrate and thus are discharged into the solution.

Schafer et al. [31] characterized the anodic dissolving process from a chemical perspective at the same time. According to these researchers, the chemical species implicated in the electrochemical synthesis of HKUST-1, include $Cu(I)_2O$, which is produced on oxidation of copper plate in the presence of H_2O or O_2. Thus, on further oxidation of $Cu(I)_2O$ to Cu(II)O, these may subsequently interact with the organic ligands to form HKUST-1 crystals.

Similarly, anodic dissolution electrochemical synthesis of nanostructures of rod-like MOF-5 crystals were reported by Li et al. [19] to make dense and robust MOF-5 films, zinc electrodes were utilized in an aqueous phase that contained H_2BDC (BDC = 1,4-benzenedicarboxylate) and the salt ammonium fluoride as electrolyte and the voltage of about 2 Volts was applied at 65 °C.

(ii) **Cathodic Deposition**

In this technique, a cathodic surface is contacted with a solution comprising the organic linker, metal ions, and a so-called probase. In this method, MOF film deposition is caused by raising the pH at the cathodic surface, where the probase is electrochemically reduced. The nitrite ions formed by the reduction of nitrates are an example of a probase since they can deprotonate the organic linker and create the MOF.

As previously mentioned, the second main technique for the electrochemical synthesis of MOFs is cathodic deposition. In 2011, cathodic deposition of MOF was investigated by Li and Dinca [18] to overcome two major anodic dissolution limitations:

1. The deposition surface (anode surface) is continually degraded during the synthesis process because it is used to produce metal cations.
2. Since the anode is also used as a metal resource, the metal option for the anode is limited.

In the cathodic deposition, the metal salt is used as the metal precursor, which is incorporated in the electrolyte solution along with the organic linker and the probase. Dinca et al. proved the method's efficiency by synthesizing HKUST-1 and MOF-5 in less than 15 min at room temperature. Fluorine doped tin oxide (FTO) was used as working electrodes. Ag/Ag(cryptand) (Cryptands are a class of synthetic bicyclic and polycyclic multidentate ligands that may bind to a wide range of cations.) was used as a reference electrode and the electrolyte fluid was a mixture of DMF and water in the ratio 100:1 that included organic linkers and metal salts. The kind of metal salt used has been revealed to have a key function in these syntheses. This was because of probase character of counter anions, which can either hinder or favor the creation of the desired MOF. Because the production of $Zn_5(OH)_8(H_2O)_2(NO_3)_2$ is the first step in the synthesis of MOF-5 (by reacting zinc nitrate and benzene dicarboxylic acid (BDC)), employing chlorine anions can prevent this step by producing $Zn_5(OH)_8(Cl)_2(H_2O)_2$. By serving as the probase and contributing to the synthesis of the intermediate species, the use of nitrate anions can help in its production.

9 Mechanochemical Synthesis

Mechanochemical synthesis is a well-known technique in metallurgy and mineral extraction, but it has expanded into a variety of chemical disciplines in recent years, including catalysis, inorganic chemistry, and pharmaceutical synthesis. Chemical reactions and a range of physical phenomena can be induced by mechanical force (mechanophysics). Mechanical breakdown in mechanochemical synthesis is preceded by breaking intramolecular connections. In multicomponent (ternary and higher) procedures, this process has a wide history to generate pharmaceutically active co-crystals, and also in conventional inorganic chemistry (solid-state), polymer science, organic synthesis, and many other disciplines, mechanochemistry has played a great role. Mechanically induced MOF synthesis is appealing for a variety of reasons. Without the need for organic solvents, reactions may also be carried out at an ambient temperature, which again is beneficial for many reasons when organic solvents are unavailable. Short reaction durations (10–16 min) make it possible to make products with few constituents and precise yields.

In the mechanochemical synthesis of MOFs, the organic linkers and metal nodes are mixed physically (grinding, ball-milling, etc.) rather than solvents or prolonged heating, which reduces the expense and chemical waste. Mechanochemical syntheses may also be utilized to produce mixed metal MOF in large amounts by altering the constituent reagents. The cost of production can be further reduced by using cheap metal sources such as metal salts, oxides, hydroxides, or carbonates.

The main concept behind this technique of synthesis is to improve chemical reactions by grinding or milling particles with or without the use of minimal solvent. In this approach, conventional solvothermal synthesis of MOF reactors is completely replaced by either a mortar and pestle or mechanized ball mills in a mechanical process. Generally, mechanical milling consumes higher energy and ensures production uniformity. This technique allows for faster and more efficient production of MOFs with measurable yields, including the use of MOF substrates with low solubility, for example, oxides, hydroxides, and carbonates, in contrast to the solvent-free conditions.

Some shortcomings of the mechanochemical method include.

(i) Because mechanosynthesis is essentially a batch processing method with a restricted rate of output, expanding it up is challenging
(ii) Despite almost 'solvent-free production, purification may still be necessary, which may necessitate the need for a solvent MOFs are produced using three distinct mechanochemical methods.

Nevertheless, this synthesis technique is the most ecologically responsible way of manufacturing MOFs, which might result in considerable cost savings.

Mechanochemical Synthesis can be applied by using the following two types:

1. Solvent-free grinding (SFG). This is the simplest method that eliminates the need for solvents
2. Liquid-assisted grinding (LAG). This method is more flexible and faster because it employs catalytic quantities of liquid phases, which improves reagent mobility
3. Ion and liquid assisted grinding.

The synthesis of nearly all MOF classes has been proven using these methods, and the following part will discuss some examples of each type.

(1) **Solvent-free grinding**

For various reasons, solvent-free synthesis of metal–organic frameworks (MOFs) is of importance. It might, for example, provide information about the functions of solvent molecules in the synthesizing of microporous structures, as well as access to environmentally friendly large-scale manufacturing procedures and more efficient lab-scale sample preparation methods. Solvent-free production of porous MOFs is a major step forward.

Pichon and colleagues [28] utilized the SFG technique by milling a dry combination of copper acetate and isonicotinic acid (Hina) powder for about 10 min to synthesize copper(II) isonicotinate ($Cu(INA)_2$) MOF with acetic acid and water molecules clogged in pores (Fig. 16). The formation of the product was observed when the color changed from green to blue with a characteristic order of acetic acid (by-product). Surprisingly, they also discovered that in their solvent-free synthesis, grinding was only required to start the reaction; it was not essential to grind further to complete it. Even if the process was ground only for one minute, the reaction nevertheless continued in quantitative yield to provide the product, although the entire reaction took a longer time to complete, taking 6 h total.

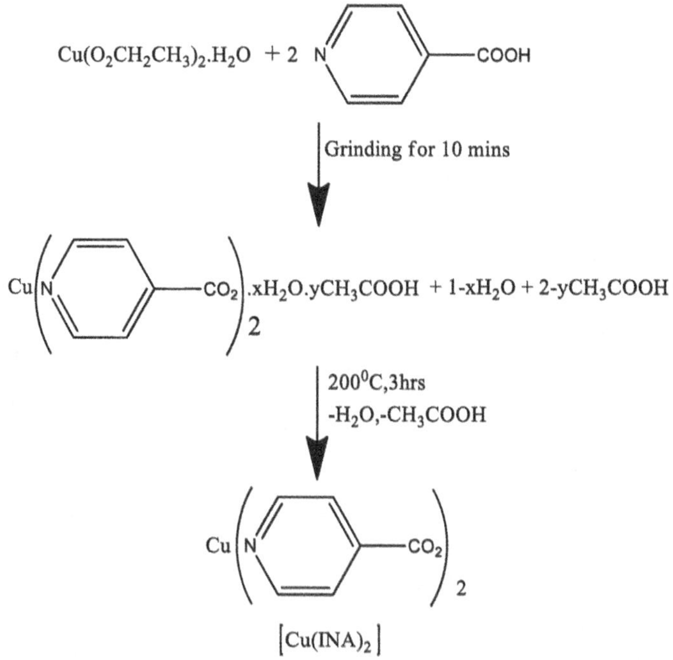

Fig. 16 Solventless preparation of $Cu(INA)_2$ MOF

The same group used the same approach to screen sixty distinct combinations of twelve different divalent metal salts consisting of copper, nickel, and zinc with five different carboxylate organic linkers for 15 min. HKUST-1 and $Cu(INA)_2$, two microporous metal–organic frameworks, were produced as a consequence. The capacity to synthesize MOFs with mere water as a by-product, eliminating the need for a purification process altogether, is a key advantage of this technique. This is achieved by combining protons generated by the organic ligand with metal substrates such as hydroxides or oxides to form H_2O.

MIL-101(Cr) was recently synthesized without the need for solvent or hydrofluoric acid, according to Leng and colleagues [16]. The chromium salt was ground with terephthalic acid at ambient temperature for 30 min before being placed in an oven at 220 °C for 4 h, giving a product with a BET surface area of 3517 $m^2\ g^{-1}$ and a smaller particle size than the batch procedure.

Niraj and colleagues [24] reported the production of the yttrium-based MIL-78 MOF under totally liquid-free circumstances by using a metal hydride as a reactant material and generating hydrogen as a by-product. They observed that on reacting trimesic acid (TMS, $C_6H_3(COOH)_3$) with yttrium hydride (YH_3), only MIL-78 and Hydrogen as a product were produced as a result of solvent-free grinding (Fig. 17).

(2) **Liquid assisted grinding**

In liquid-assisted grinding, a little (catalytic) volume of liquid is added to the grinding mixture in the LAG (or kneading) method. Unlike conventional grinding, LAG employs the addition of a small amount of liquid phase to aid or enable the mechanochemical reaction. LAG products often have a high level of crystallinity, implying a way to prevent the amorphous impurities that are common in neat grinding.

For example, Klimakow et al. [15] synthesized the well-known HKUST-1 and its benzenetribenzoate(H_3BTB) based analog MOF-14 using the LAG technique in 2010. They ground copper acetate monohydrate and benzenetribenzoate(H_3BTB)

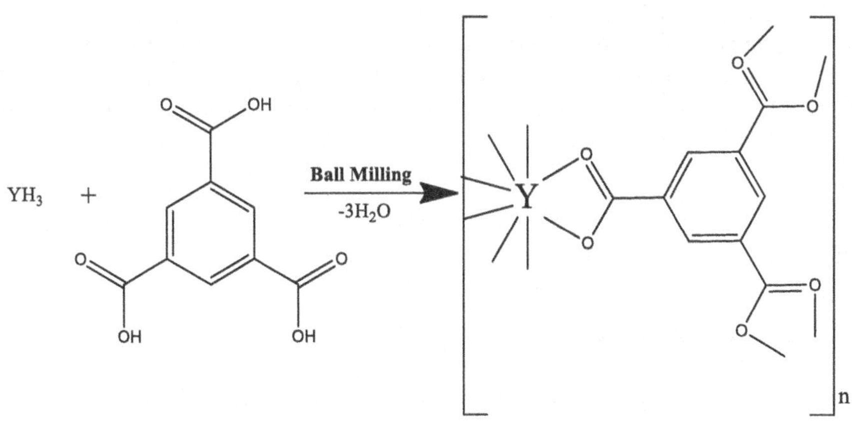

Fig. 17 MIL-78 synthesis by solvent-free grinding

powders in a ball mill with two balls. The mixture powder, at a frequency of about 25 Hz was ground for ten minutes. The mixture was milled at 40 Hz after adding ethanol for another 15 min. The change in the wet powder's color changed from deep green to blue during the reaction process, followed by an odor of acetic acid odor. The color of the product turned to a more greenish-blue after drying in the air.

Similarly, Pichon and co-workers [29] showed that introducing tiny amounts of liquid by-products created using the SFG technique before the mechanical process might speed up the synthesis. The production of $Cu(INA)_2$ MOF (INA = isonicotinic acid) was substantially increased by adding tiny quantities of acetic acid to the reactant mixtures, but there was no enhancement for HKUST-1 due to the reduced solubility of the trimesic acid.

(3) **Ion and liquid assisted grinding (ILAG)**

Over the past years, the ion and liquid-assisted grinding (ILAG) technique has been utilized to aid the mechanochemical production of MOFs by combining ions and liquids. ILAG induces, enhances, and directs mechanochemical production by adding catalytic quantities of simple ionic salts to the metal oxides.

Friscic et al. [11] showed that utilizing ZnO as a metal substrate and H_2BDC (BDC = 1,4-benzenedicarboxylate) and DABCO (diazabicyclo[2.2.2]octane) as a mixture of ligands, the catalytic quantity of simple salts can speed up MOF formation. Through the effects of the template, the salts cause the development of product structure. Under mechanochemical circumstances, salts are trapped in the pores of MOFs, whereas ion templates may play an important factor in the formation of neutral MOFs. They then investigated the three imidazoles (imidazole, methyl imidiazole ethylimidazole) in the presence of various solvents (DMF, DEF, and EtOH) and salts (NH_4NO_3, $NH_4CH_3SO_3$, and $(NH_4)_2SO_4$) with ZnO as the metal substrate. They discovered through the liquid is absent, the ammonium salt enhanced ZIF production. The reaction between imidazole and ZnO is facilitated by the substoichiometric amount of ammonium salt, and quantifiable yields were achieved in most instances.

Pilloni et al. [30] also synthesized Fe-MOF by ILAG. They placed H_3BTC(1,3,5-benezenetricarboxylic acid) and $Fe(NO_3)_3.9H_2O$ (iron(III) nitrate nanohydrate) in Teflon coated stainless-steel grinding jar followed by aqueous TMAOH (tetramethylammonium hydroxide) solution. A vibrating ball-mill was used to grind the mixture, and then after 1 h of grinding, a thick slurry was obtained. After that the washed and air-dried light-orange colored sample of Fe-MOF (MIL-100(Fe) was obtained.

10 Ultrasonic Synthesis

Ultrasound is now considered one of the most effective methods for the production of metal–organic frameworks as it leads to the high efficiency of energy. The process involves the phenomenon of acoustic cavitation, which is caused by strong ultrasound radiation (in the range of 20 kHz–10 MHz) in the field of sonochemistry which causes molecules to undergo chemical reactions. Cavitation consists of formation, development, and sudden uncontrollable breaking down of bubbles in a liquid, resulting in the formation of local hot spots of about 5000 °C temperature and pressure of around 500 atm within microseconds. As a consequence, these severe circumstances can aid in the development of nanoscale structures, mostly by increasing the number of crystallization nuclei.

A variety of chemical reactions may now be performed at room temperature with the help of ultrasonic synthesis at room temperature, even those that were earlier difficult to perform with traditional methods. High-intensity ultrasound may also be employed to develop nano/micro-scaled coordination complex compounds with novel morphologies and properties in a quick, easy, cost-effective, consistent, and environmentally friendly manner.

Feldblyum et al. [8], synthesized Zn-HKUST-1 ($[Zn_3(BTC)_2(H_2O)_3]$ ultrasonically (BTC = benzenetricarboxylic acid) in DMF by taking 1,3,5-benzenetricarboxylic acid (H_3BTC) and zinc(II) acetate dihydrate as reactants. They observed that the ultrasonic synthesis produced MOF nanorods in 30 min at ambient temperature and pressure, whereas single crystals of the framework were synthesized using a conventional heating method technique. When compared to traditional thermal synthesis of the similar chemical Zn-HKUST-1, which is performed at 85 °C for 12 h, ultrasonic synthesis is observed to be a very efficient method (Fig. 18).

Liu et al. [20] also carried out the ultrasonic synthesis of $Ce(1,3,5\text{-}BTC)(H_2O)_6$ by adding a $Ce(NO_3)_3$ solution to a 1,3,5-H_3BTC (benzenetricarboxylic acid) solution and treating it at room temperature. It was observed that depending upon the reacting species, the size of MOFs formed also varied. All the products obtained exhibited a bundle-like structure made up of perfectly aligned nanorods. When the

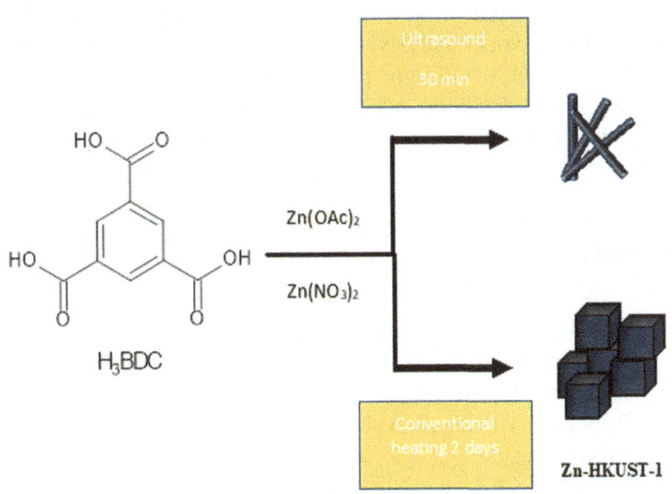

Fig. 18 Preparation of Zn-HKUST-1

concentrations of reactant species 1,3,5-H_3BTC and $Ce(NO_3)_3$ were increased, the MOFs of significantly smaller sizes were obtained.

MOF-5 crystals were synthesized by Son et al. [32] using the ultrasonic synthesis method and 1-methyl-2- pyrolidine (NMP) was utilized as the solvent. As a typical prerequisite of synthesis, MOF-5 was synthesized by reacting zinc nitrate hexahydrate $Zn(NO_3)_2.6H_2O$ with H_2BDC in the presence of NMP(solvent). They dissolved the reactant species in NMP, and the solution was further processed for about 10–75 mins to ultrasonic treatment to get MOF-5 crystals at different power levels. They also investigated the effects of various sonochemical synthesis parameters on MOF-5, and then their parameters were compared to those of a sample synthesized using traditional solvothermal synthesis. It was found that in the case of ultrasonic synthesis the maximum time required for the formation of MOF-5 was 75 minutes but the same MOF-5 synthesis took 12–24 hours by the solvothermal conventional method of synthesis.

11 Conclusions and Future Perspectives

MOFs are more attractive than traditional materials like zeolites, COFs, and polymers because they combine different metals and organic linkers to produce a diverse range of materials with varied crystal structures and chemical compositions. MOFs are paving the way for systematic studies of fundamental polymerization reactions and the development of next-generation materials, from supporting organometallic moieties as heterogeneous analogs of molecular catalysts to employing them for optics, ion exchange, gas storage, gas separation, sensing, polymerization, and drug delivery, molecular electronics, and so on. Due to the tuneability and versatility of metal-organic framework catalysts, as well as the large range of polymerization reactions already proven, the design space and opportunity for future research are broad. So, using logical designing methodologies and diverse synthetic methods of MOF-based materials will be very rewarding in a multitude of ways.

Abbreviations

ADA	Adamantanediacetic acid
BDC	Benzenedicarboxylate
BET	Brunauer-Emmett-Teller
BPDC	Biphenyl-4,4′-dicarboxylate
BPY	Bipyridine
BPYDC	Bipyridine-5,5′-dicarboxylate
BTB	Benzene tribenzoate
BTC	Benzene tricarboxylate
COF	Covalent Organic Framework

CPL	Coordination polymer with pillared layer structure
CTPY	4-Carboxy-4,2′,6′,4′′-terpyridine
DABCO	Diazabicyclo[2.2.2]octane
DEF	Diethyl formamide
DMF	Dimethyl Formamide
FTO	Fluorine-Doped Tin Oxide
H_2HFBBA	4,4′ -(Hexafluoroisopropylidene) bis(benzoic acid)
HKUST	Hong Kong University of Science and Technology
ILAG	Ion Liquid-Assisted Grinding
INA	Isonicotinate
IRMOF	IsoReticular Metal–Organic Frameworks
LAG	Liquid-Assisted Grinding
M-BDC	M-benzenedicarboxylate
MEPY	Methyl pyridine
MIL	Matérial Institut Lavoisier
MIM	Methylimidazolate
MOP	Metal-Organic Polyhedra
NMP	N-Methyl-2- Pyrrolidine
PYDC	Pyridinedicarboxylic acid
PZDC	Pyrazine-2,3-dicarboxylate
SBU	Secondary Building Blocks
SFG	Solvent free grinding
TMAOH	Tetramethylammonium hydroxide
TMS	Trimesic acid
TPDC	P-terphenyl-4,4′-dicarboxylate
TTTPC	Tris(2,4,6-trimethylbenzene-1,3,5-triyl)-tris(methylene)-tris(pyridine-4-carboxylic acid)
UiO	Universitetet i Oslo
ZIF	Zeolite Imidazolate Framework

References

1. Bag PP, Wang XS, Cao R (2015) Microwave-assisted large scale synthesis of lanthanide metal-organic frameworks (Ln-MOFs), having a preferred conformation and photoluminescence properties. Dalt Trans 44:11954–11962. https://doi.org/10.1039/c5dt01598g
2. Banerjee D, Finkelstein J, Smirnov A et al (2011) Synthesis and structural characterization of magnesium based coordination networks in different solvents. Cryst Growth Des 11:2572–2579. https://doi.org/10.1021/cg200327y
3. Banerjee T (2010) Thesis: impact of nickel doping on hydrogen storage in porous metal-organic frameworks. Adsorpt J Int Adsorpt Soc
4. Campagnol N, Van Assche TRC, Li M et al (2016) On the electrochemical deposition of metal-organic frameworks. J Mater Chem A 4:3914–3925. https://doi.org/10.1039/c5ta10782b

5. Clausen HF, Poulsen RD, Bond AD et al (2005) Solvothermal synthesis of new metal organic framework structures in the zinc-terephthalic acid-dimethyl formamide system. J Solid State Chem 178:3342–3351. https://doi.org/10.1016/j.jssc.2005.08.013
6. Dan M, Rao CNR (2006) A building-up process in open-framework metal carboxylates that involves a progressive increase in dimensionality
7. Eddaoudi M, Moler DB, Li H et al (2001) Modular chemistry: secondary building units as a basis for the design of highly porous and robust metal-organic carboxylate frameworks. Acc Chem Res 34:319–330. https://doi.org/10.1021/ar000034b
8. Feldblyum JI, Liu M, Gidley DW, Matzger AJ (2011) Reconciling the discrepancies between crystallographic porosity and guest access as exemplified by Zn-HKUST-1. J Am Chem Soc 133:18257–18263. https://doi.org/10.1021/ja2055935
9. Férey G (2005) Erratum: a chromium terephthalate-based solid with unusually large pore volumes and surface area (Science (September 23) (2040)). Science (80-.). 310:1119
10. Forster PM, Burbank AR, Livage C et al (2004) The role of temperature in the synthesis of hybrid inorganic°Corganic materials: the example of cobalt succinates. Chem Commun 4:368–369. https://doi.org/10.1039/b311156c
11. Friscić T (2010) New opportunities for materials synthesis using mechanochemistry. J Mater Chem 20:7599–7605. https://doi.org/10.1039/c0jm00872a
12. Furukawa H, Cordova KE, O'Keeffe M, Yaghi OM (2013) The chemistry and applications of metal-organic frameworks. Science (80-.). 341
13. Gomar M, Yeganegi S (2019) Adsorption of 5-Fluorouracil and Thioguanine drugs into ZIF-1, ZIF-3 and ZIF-6 by simulation methods. Mater Sci Eng C 97:461–466. https://doi.org/10.1016/j.msec.2018.12.068
14. Jhung SH, Lee JH, Yoon JW, et al (2007) Microwave synthesis of chromium terephthalate MIL-101 and its benzene sorption ability. Adv Mater 121–124
15. Klimakow M, Klobes P, Thünemann AF et al (2010) Mechanochemical synthesis of metal-organic frameworks: a fast and facile approach toward quantitative yields and high specific surface areas. Chem Mater 22:5216–5221. https://doi.org/10.1021/cm1012119
16. Leng K, Sun Y, Li X et al (2016) Rapid synthesis of metal-organic frameworks MIL-101(Cr) without the addition of solvent and hydrofluoric acid. Cryst Growth Des 16:1168–1171. https://doi.org/10.1021/acs.cgd.5b01696
17. Li H, Eddaoudi M, O'Keeffe M, Yaghi OM (1999) Design and synthesis of an exceptionally stable and highly porous metal- organic framework. Nature 402:276–279. https://doi.org/10.1038/46248
18. Li M, Dincă M (2011) Reductive electrosynthesis of crystalline metal-organic frameworks. J Am Chem Soc 133:12926–12929. https://doi.org/10.1021/ja2041546
19. Li WJ, Lü J, Gao SY et al (2014) Electrochemical preparation of metal-organic framework films for fast detection of nitro explosives. J Mater Chem A 2:19473–19478. https://doi.org/10.1039/c4ta04203d
20. Liu K, You H, Jia G et al (2010) Hierarchically nanostructured coordination polymer: facile and rapid fabrication and tunable morphologies. Cryst Growth Des 10:790–797. https://doi.org/10.1021/cg901170j
21. Luo L, Lv GC, Wang P et al (2013) PH-Dependent cobalt(II) frameworks with mixed 3,3′,5,5′-tetra(1H-imidazol-1-yl)-1,1′-biphenyl and 1,3,5-benzenetricarboxylate ligands: Synthesis, structure and sorption property. CrystEngComm 15:9537–9543. https://doi.org/10.1039/c3ce41056k
22. Mahata et al (2007) The role of temperature on the structure and dimensionality of MOFs: an illustrative study of the formation of manganese oxy-bis(benzoate) structures
23. Ni Z, Masel RI (2006) Rapid production of metal-organic frameworks via microwave-assisted solvothermal synthesis. J Am Chem Soc 128:12394–12395. https://doi.org/10.1021/ja0635231
24. Niraj SK, Meennakshi H, Viktor BP (2012) Mechanochemistry: fundamentals and applications in synthesis. Chem Commun 48:11289–11291
25. Pachfule P, Das R, Poddar P, Banerjee R (2011) Solvothermal synthesis, structure, and properties of metal organic framework isomers derived from a partially fluorinated link. Cryst Growth Des 11:1215–1222. https://doi.org/10.1021/cg101414x

26. Pachfule P, Das R, Poddar P, Banerjee R (2011) Solvothermal synthesis, structure, and properties of metal organic framework isomers derived from a partially fluorinated link. Cryst Growth Des 11:1215–1222
27. Peplow M (2015) Materials science: the hole story. Nature 520:148–150
28. Pichon A, Lazuen-Garay A, James SL (2006) Solvent-free synthesis of a microporous metal-organic framework. CrystEngComm 8:211–214. https://doi.org/10.1039/b513750k
29. Pichon et al (2006) Solvent-free synthesis of a microporous metal-organic farmework
30. Pilloni et al (2015) Liquid assisted mechanochemical synthesis of an iron carboxylate metal-organic framework and its evaluation in diesel fuel desulfurization. Adv Mater 14:629–638. https://doi.org/10.1002/1521-4095(20020503)14:9%3c629::AID-ADMA629%3e3.0.CO;2-B
31. Schäfer P, Van Der Veen MA, Domke KF (2016) Unraveling a two-step oxidation mechanism in electrochemical Cu-MOF synthesis. Chem Commun 52:4722–4725. https://doi.org/10.1039/c6cc00534a
32. Son WJ, Kim J, Kim J, Ahn WS (2008) Sonochemical synthesis of MOF-5. Chem Commun 6336–6338. https://doi.org/10.1039/b814740j
33. Sung HJ, Lee JH, Chang JS (2005) Microwave synthesis of a nanoporous hybrid material, chromium trimesate. Bull Korean Chem Soc 26:880–881. https://doi.org/10.5012/bkcs.2005.26.6.880
34. Taylor-Pashow KML, Della Rocca J, Xie Z et al (2009) Postsynthetic modifications of iron-carboxylate nanoscale metal-organic frameworks for imaging and drug delivery. J Am Chem Soc 131:14261–14263. https://doi.org/10.1021/ja906198y
35. Tomic EA (1965) Thermal stability of coordination polymers. J Appl Polym Sci 9:3745–3752. https://doi.org/10.1002/app.1965.070091121
36. Tranchemontagne DJ, Tranchemontagne JL, O'keeffe M, Yaghi OM (2009) Secondary building units, nets and bonding in the chemistry of metal–organic frameworks. Chem Soc Rev 38:1257–1283. https://doi.org/10.1039/b817735j
37. Vilela SMF, Ananias D, Fernandes JA et al (2014) Multifunctional micro-and nanosized metal-organic frameworks assembled from bisphosphonates and lanthanides. J Mater Chem C 2:3311–3327. https://doi.org/10.1039/c3tc32114b
38. Yuan F, Xie J, Hu HM et al (2013) Effect of pH/metal ion on the structure of metal-organic frameworks based on novel bifunctionalized ligand 4′-carboxy-4,2′:6′,4″- terpyridine. CrystEngComm 15:1460–1467. https://doi.org/10.1039/c2ce26171e
39. Zhang B, Zhang J, Liu C et al (2015) Solvent determines the formation and properties of metal-organic frameworks. RSC Adv 5:37691–37696. https://doi.org/10.1039/c5ra02440d
40. Zhang WH, Wang YY, Lermontova EK et al (2010) Interaction of 1, 3-adamantanediacetic acid (H2ADA) and ditopic pyridyl subunits with cobalt nitrate under hydrothermal conditions: PH influence, crystal structures, and their properties. Cryst Growth Des 10:76–84. https://doi.org/10.1021/cg900285b

Functionalization Strategies of Metal–Organic Frameworks (MOFs): Diverse Ways to Versatile MOFs

Sanjay Kumar, Sneha Vijayan, Kartika Goyal, Mansi Kathuria, and Shikha Gulati

Contents

1 Introduction 100
2 Metal–Organic Frameworks: General Characteristics, Structure and Synthesis 101
 2.1 Synthetic Methods of MOFs 102
3 Functionalization of MOF Backbone (Internal-Surface Functionalization) 103
 3.1 Inorganic Nodes 103
 3.2 Organic Linkers 104
4 Strategies for Functionalization 106
 4.1 In Situ Functionalization 106
 4.2 Pre-Synthetic Functionalization 107
 4.3 Post-Synthetic Modifications 108
5 Immobilization of Guest Species in MOF Matrices and Pores 112
6 Coating MOFs with Functional Materials 114
7 Applications of Functionalized MOFs 116
 7.1 Chemical Sensing 117
 7.2 Heterogeneous Catalysis 117
 7.3 Gas Storage and Separation 117
 7.4 Biomedical Applications 118
8 Conclusion and Future Outlook 119
Abbreviations 119
References 120

Abstract Recently, metal–organic frameworks have been regarded as one of the major classes of porous crystalline materials. Each year witnesses new and novel MOF structures. The as-developed novel structures possess a distinct set of hallmarks, i.e., permanent porosity, diverse composition, and unique surface properties. In structural aspects, MOFs are often constructed from metal ions/clusters coordinated by organic linkers. Any modifications in the adjustment of linker geometry, length, ratio, or functional group can tune the size, shape, and internal surface properties of a MOF for a targeted application. Available literature significantly proved the improved properties like high thermal stability of the MOFs upon functionalization. In this chapter, we have focused on the recent advances in MOF synthesis through various strategies of functionalization to attain unique properties. Finally, the possible

---

S. Kumar · S. Vijayan · K. Goyal · M. Kathuria · S. Gulati (✉)
Department of Chemistry, Sri Venkateswara College, University of Delhi, Delhi 110021, India

S. Gulati (ed.), *Metal-Organic Frameworks (MOFs) as Catalysts*,
https://doi.org/10.1007/978-981-16-7959-9_4

future development of functional MOFs toward different practical applications is presented.

Keywords Metal–organic frameworks · Functionalization · Synthesis · Applications

1 Introduction

In the last few decades, particularly since the 1990s, a novel family of porous, organic–inorganic supramolecular hybrids, known as the metal–organic frameworks (MOFs), have been in the limelight of extensive and fast-growing research. So far, over 20,000 distinct metal–organic frameworks, differing in morphology, size, functionality, and structure, have been reported [20]. These materials rely on Werner's coordination chemistry, wherein metal ions (or clusters) are coordinately connected by bridging ligands, to form infinite porous frameworks [16, 37]. The concept of isoreticularity, deals with the idea of building different solids having identical framework topology, but differing in morphology and other features like the size of the pores. Isoreticular synthesis of MOFs is a very effective approach to tune the porosity, increase the surface area, and introduce functional groups within the framework [19, 21]. To put it simply, given a starting framework geometry, by elongating the organic ligands, a number of frameworks with identical, unaltered topology but with bigger dimensions, wider pore aperture, increased window sizes, and different functional groups, can be prepared. With the gradual addition of more atoms in the organic linkers, precise control of the pore size of the frameworks can be achieved at the Angstrom level [57]. Moreover, due to the obtainability of a wide array of building units of metal clusters and bridging ligands, there is an ever-expanding scope of preparing infinitely many new MOFs with various structures, porosity, morphologies, and other characteristics.

With their outstanding, unique set of properties, MOFs form an intriguing class of porous materials, distinguishing them from other materials such as mesoporous silica, porous carbon, and microporous fully inorganic zeolites. Precisely tunable pore size, high surface area which provides a more available surface for interaction with guest species, chemical tailorability, and synthetic flexibility are some of the noteworthy properties of MOFs that set them apart. Most importantly, the structure and functionality of MOFs can be tuned during their synthesis, following the isoreticular principle [58]. MOFs can be functionalized by the addition of substituent functional groups on the backbone of the bridging ligands during synthesis. However, this strategy is limited as many of the desired functionalities cannot withstand the reaction conditions of MOF synthesis [7]. Additionally, even a small change in the reaction parameters, chemical nature of the linker, sterics, electronic configuration of the metal ion and the ligand, results in drastic changes in the MOF thus formed and we may not get the target MOF [58]. To overcome these challenges and limitations of pre-synthetic functionalization, an alternative strategy, known as post-synthetic modification, can

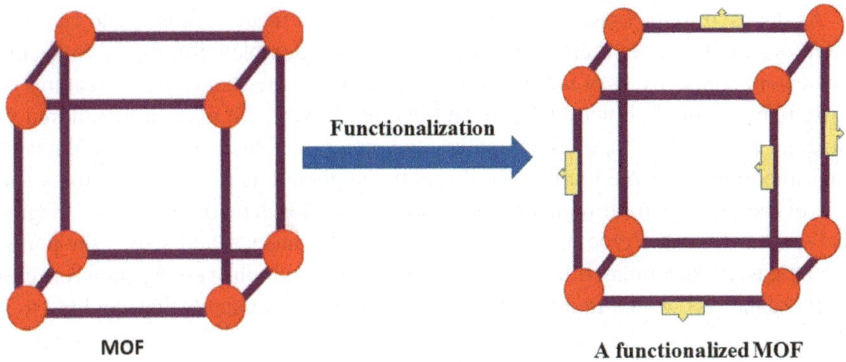

Fig. 1 Schematic diagram of functionalization of MOF

be employed. It is an approach that leads to diverse functionalities without affecting the structural stability, porosity, and crystallinity of the as-synthesized framework [7, 35]. Owing to their unique set of properties and highly ordered crystalline structure, it is of paramount significance to mention that the position, as well as the degree of functionalization of MOFs, are controllable. Figure 1 represents the functionalization process. For the representative purpose, we have shown a ligand functionalized MOF, but the functionalization strategies are not limited to it and are explained in detail in the different sections of this chapter.

In this chapter, we will present the various methods to impart functionality and elaborate on the strategies of functionalization. Furthermore, from the application perspective, to gain insights into the enhanced features of the functional MOFs and to understand the host–guest interactions and synergistic effects, we will devote a section to the applications of functionalized MOFs, wherein we cite examples from recent research works to illuminate the concept.

2 Metal–Organic Frameworks: General Characteristics, Structure and Synthesis

Metal–organic frameworks (MOFs), or Porous Coordination Polymers (PCPs), are an impressive class of crystalline, hybrid materials of organic and inorganic units, which have gained immense attention since the 1990s. They are composed of secondary building units, SBUs, or nodes (metal clusters or ions), and organic bridging ligands, connected by coordination bonds. They possess an outstanding set of properties such as ultrahigh porosity (about 90% free volume), large specific surface area (as high as 10,000 m^2/g), flexibility, and chemical tunability [46]. Owing to these remarkable features, MOFs have been meticulously explored for a wide variety of applications, such as heterogeneous catalysis, gas separation and storage, chemical sensing, optical luminescence, water remediation, biomedicine, and many more.

As opposed to the classical porous materials like activated carbons and zeolites, MOFs stand out due to their chemical tunability and flexible tailorability. Since the framework connectivity of the building blocks governs the characteristic features of a MOF, by prudent choice of SBUs and organic linkers, MOFs can be tailored to have the desired pore size, functionality, and structure. This feature of MOFs is of vital importance for the realization of specific applications mentioned before. The size of the pores can be regulated by choosing organic ligands of various lengths and spaces. Also, the type of metal sites in the MOF, surface, and size of the pores, can be wisely manipulated by judicious choice of metal ions (giving preference to its coordination), which in turn determines the number of ligands that can bind to it [12].

MOFs are characterized by their permanent porosity, which is highly favorable for incorporating various guest species. From the application perspective, MOFs with greater pore size, capable of accommodating diverse guest species such as metal complexes, metal nanoparticles, single-metal atoms, polymers, organic dyes, and small enzymes, are immensely useful [50]. The serious concerns regarding MOFs for their practical applications are their low chemical and thermal stabilities. Many efforts have been taken to address this challenge. In this regard, increasing the strength of the coordination bonds connecting the organic ligands and the metal ions is a verified strategy. Combining metal ions having high oxidation states with hard bases like carboxylate ligands is another successful strategy to enhance their acid/base stability. The thermal stability can be improved by using high-valence metal ions like Al^{3+}, Ln^{3+}, Ti^{4+}, etc. [28].

2.1 Synthetic Methods of MOFs

An interesting feature of MOFs is that even a slight variation in the reaction conditions including the precursors' ratio, temperature, solvent, can result in striking changes in the structural properties like pore size, topology, etc. Also, the same starting material can give rise to different products when different synthetic strategies are employed. The commonly used preparation strategies are described in brief:

- **Solvothermal/Hydrothermal Method:** It is the most commonly employed synthetic strategy for the preparation of MOFs. In this method, the crystals are grown from solutions of organic-linker source and mixed-metal source by self-assembly in closed vessels, at high temperature (higher than the boiling point of the solvent) and high pressure [9, 46]. Special care has to be taken in adjusting the reaction environment and parameters, such as the pH and temperature, the solubility of the precursors in the selected solvent, and the metal salt-organic linker ratio, lest the structure of the MOF get altered. One of the major limitations of this method is that the sensitive functional groups cannot withstand the harsh reaction conditions, making it a tough task to incorporate them into the MOF.

Nonetheless, the sensitive functional groups can be incorporated into the MOFs at milder conditions via post-synthetic modifications.

- **Microwave-Assisted Synthesis:** This method utilizes microwave irradiation as the source of energy for the reaction, and facilitates the quick synthesis of smaller crystals, plausibly beneficial in industrial applications. In the course of the process of irradiation, the energy is generated within an oscillating electromagnetic field. A special metal tube carries the microwaves to the vial containing the reaction mixture. However, regulating the power and time of irradiation is a daunting task [46].
- **Seeded-Growth Method:** In this method, the growth of a new crystal having well-ordered morphology, is attained by nucleation of a small crystal that is used as the seed. This synthetic route was first reported by Falcaro et al. [15]. They also demonstrated that this strategy can simultaneously facilitate confinement of nanoparticles (NPs) in the pores of a MOF, by the introduction of the NPS to the nucleating seeds [15]. Therefore, this approach also helps prevent the issues of disintegration of the framework and outer surface functionalization of the MOF, attached with the conventional synthetic methods of host–guest MOF composites.

Electrochemical, mechanochemical, and sonochemical syntheses are alternative strategies for MOF preparation and are yet to be explored more. A serious issue during MOF preparation is that the solvent molecules most often retain in the MOF pores, and it is necessary to remove them for activation of the MOF. Activation by heating at high temperatures is avoided, as there are chances for decomposition of the sample. Instead, activation at lower temperatures is preferred.

3 Functionalization of MOF Backbone (Internal-Surface Functionalization)

3.1 Inorganic Nodes

The inorganic/metal nodes in various MOFs are in coordination with labile ligands, which can be eliminated during the activation stage, resulting in the formation of coordination unsaturated sites (CUSs) in the metal nodes [52]. Nevertheless, the organic linkers block the coordination sphere over the metal nodes in most cases, leading to poor MOF activity [55]. For partially exposing the inorganic nodes of MOFs with defects, two synthetic strategies have been applied:

(i) "de novo" synthesis
(ii) Post-synthetic treatment.

In "de novo" synthesis, synthetic conditions are modified in order to purposely induce defects in the parent framework while retaining its structural integrity. In such techniques, metal ion vacancies can occur, partially exposing them [17]. Post-synthetic modification is another effective technique to expose the metal nodes, which involves the heterogeneous treatment after MOF synthesis. Moreover, the inorganic nodes can be electronically modified by employing a mixed linker approach in which the ligand of the parent framework is substituted by a ligand with distinct functional groups. Sometimes, it is possible to enhance the properties of MOFs by replacing their metal species. For example, in MOF-5, Botas and co-workers substituted zinc with cobalt to obtain CoZn-MOF-5. The obtained composite showed higher uptake capacity for H_2, CO_2, and CH_4 [4]. Some strategies have been developed to partially break the association of metal nodes and linkers, which also help in the creation of defects in the inorganic nodes. These include:

- Mechanical treatment
- Treatment with acid or base
- Solvent-assisted ligand exchange (SALE)
- Harsh activation procedure.

The inorganic nodes can also be functionalized by the immobilization of active species, which can lead to the prevention of intermolecular deactivation [44, 60].

3.2 *Organic Linkers*

Organic molecules containing functional groups are used as ligands for the preparation of MOFs. Linkers consisting of carboxylate groups are the most preferred ones, owing to their high chemical and thermal stability, and flexibility [40]. Organic ligands can be more conveniently functionalized as compared to metal centers. A comprehensive study has demonstrated that the inclusion of functional moieties into the ligands is a suitable method for the functionalization of MOFs. For example, chromogenic groups can be linked to the backbone of the ligand in order to enhance the photo-responsive properties of MOFs [43]. Some of the commonly functionalized organic ligands are shown in Table 1. Functionalization of ligands also can also lead to the improvement in the water stability of MOFs. In a study, Nguyen et al. [39] showed that IRMOF-3, earlier which displayed hydrophilic properties, became hydrophobic after its integration with the medium-long alkyl group [39]. However, sometimes the immobilization of functional groups on the organic ligands can result in problematic solubility. Therefore, alternative strategies like post-synthetic methods are generally employed for the introduction of functional groups into MOFs [55]. The various methods involved are stated below:

- Post-synthetic modification of the ligand
- Post-synthetic exchange of ligand
- Post-synthetic deprotection.

The immobilization of functional groups of organic ligands with the metal ions leads to the development of some properties, which facilitate various applications of MOFs, such as catalysis. Both precious, as well as earth-abundant metal ions, have been successfully immobilized on the organic linkers of MOFs by:

Table 1 Some of the commonly used organic ligands in functionalization

IUPAC name	Abbreviation	Chemical structure
1,4-benzenedicarboxylate	bdc	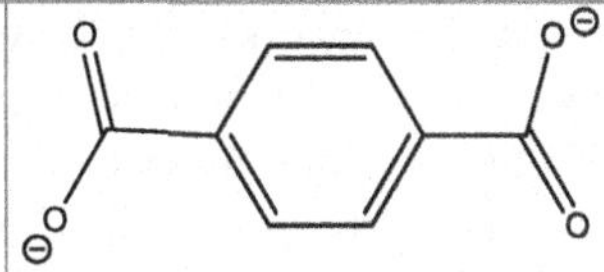
4,4'-bipyridyl	bpy	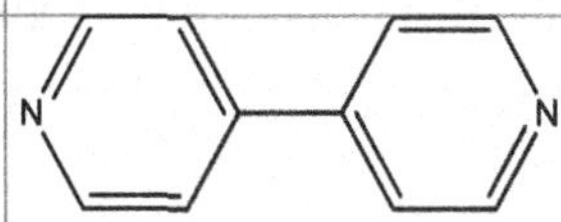
Pyrrole-2-carboxylate	PyC	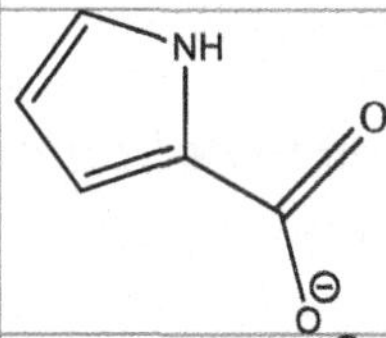
1,3,5-benzenetricarboxylate	btc	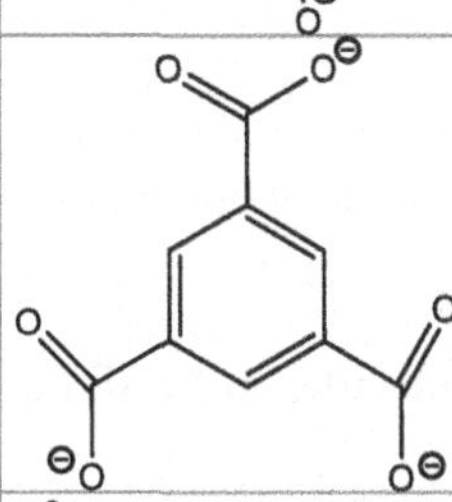
p,p'-terphenyldicarboxylate	tpdc	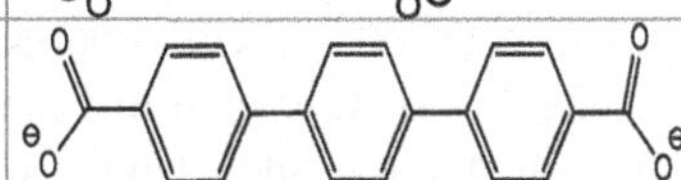
4,4'-biphenyldicarboxylate	bpdc	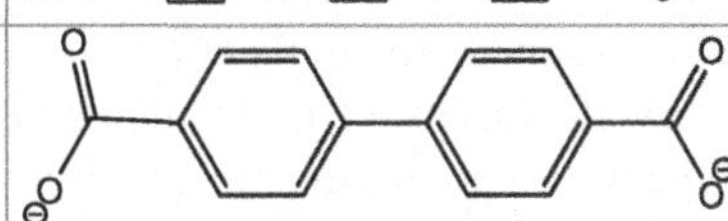
2,5-dioxobenzene-1,4-dicarboxylate	dobdc	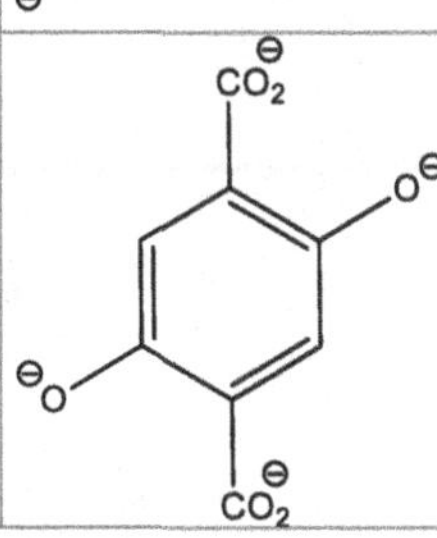

(i) Employing pre-metallated organic linkers
(ii) Post-synthetic metalation of the functionalized ligands.

In addition, the immobilization of metal complexes on organic linkers of MOFs can prevent mutual deactivation by isolating the active sites [55].

4 Strategies for Functionalization

Metal–organic frameworks can be synthesized in countless ways out of which few were reported in the previous section. Though these synthesis methods allow the targeted synthesis of porous structures with tunable pore sizes yet they do not represent the full synthetic potential of MOFs. For this, researchers across the globe suggested the functionalization of MOFs prior to their synthesis, or through in situ functionalization, or post-synthesis functionalization [58]. In this section, we will discuss different approaches for functionalization as represented in Fig. 2.

4.1 In Situ Functionalization

The introduction of guest species during the MOF synthesis is referred to as in situ functionalization. These guest species can be introduced either inside the pores or within MOF matrices. The various guest molecules are different organic molecules, metal clusters, metal nanoparticles, etc. Briefly, these can be achieved in two ways.

Entrapping Molecules in MOF Pores

For trapping the species within pores of a MOF framework, the pore aperture must be considerably smaller than the diameter of the guest species and the framework must have the binding sites. Under fulfillment of either of the conditions, molecules will be trapped efficiently [6].

A wide variety of molecules of interest can be immobilized by this approach, including but not limited to, quantum dots, polyoxometallates, dyes, drug molecules, and enzymes. Recently, a polyoxometalate (POM), $H_3PW_{12}O_{40}$, was encapsulated

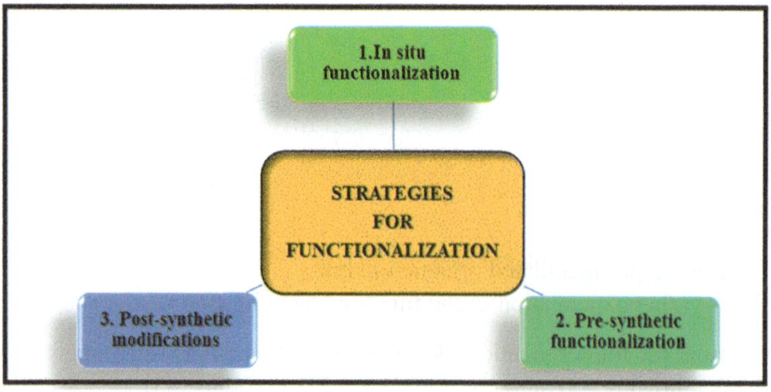

Fig. 2 Different strategies for functionalization

within the pores of UiO-67 during the framework formation, to form a remarkably capable heterogeneous catalyst for oxidative desulfurization reaction [41]. The suitable cage size and confined windows of the MOF help prevent the leaching of the POM as well as facilitates the interaction with the commonly found sulfur compounds in fuels while retaining the structure of the MOF. On a similar note, by the immobilization of different guest species, a plethora of applications can be facilitated and improved.

Encapsulation of Nanoparticles in MOF Matrix

This technique is known to prevent the Ostwald ripening of nanoparticles, resulting in immense potential for a plethora of applications. In various studies, the increased activity of MOF-catalyst composite is witnessed due to synergic effects [58].

In a recent study, copper nanoparticles were encapsulated into the single crystals of UiO-66 MOF that upgraded the activity as well as selectivity of the catalyst in CO_2 hydrogenation. This is due to the strong interaction between the nanoparticles and the SBUs of the MOF [45].

4.2 *Pre-Synthetic Functionalization*

In another way, functionality can be imparted by the conversion of the organic linkers to their derivatives before the MOF synthesis. It is important to wisely choose the substituents attached to the linker to ensure that they do not affect the formation of the desired MOF. Linkers substituted with amino or bipyridyl groups are generally preferred because they offer suitable sites for the corresponding functionalization. Post-synthetic deprotection is often done to take the advantage of these functionalities [32].

To briefly explain the functionalization of the bridging ligand prior to MOF synthesis and its effect on the properties, we consider and compare two isoreticular MOFs- $\{[Cd(bdc)(4\text{-}bpmh)]\}_n.2n(H_2O)$ and its amino-functionalized analog, $\{[Cd(2\text{-}NH_2bdc)(4\text{-}bpmh)]\}_n.2n(H_2O)$ [48]. The amino-functionalized MOF was prepared by replacing the bdc linker (benzene dicarboxylic acid) with 2-NH_2bdc. The adsorption experiments revealed that the CO_2 uptake ability of NH_2-functionalized MOF was almost twice that of the parent MOF, which can be attributed to the formation of a complex between the electron donor NH_2 groups and the electron acceptor CO_2 molecules. Moreover, the selectivity for CO_2 over methane was higher in the analog (7.0 wt% at 298 K) than the non-functionalized homolog (4.3 wt%). Furthermore, the amino-functionalized MOF could more efficiently adsorb and release I_2, and offers a promising application in the management of nuclear waste.

4.3 *Post-Synthetic Modifications*

The so far discussed methods are not that effective as they offer various constraints as MOFs undergo subtle changes in reaction parameters. These limitations for functionalization made it necessary to functionalize the MOFs post-synthesis. These approaches often deal with reactions involving the exchange of metal ions or solvent molecules, and the incorporation of guest species in the pores [58]. These can be further divided into three parts briefly based on interactions involved.

Functionalization Involving Weak Interaction

This approach takes benefit of coordinative interactions of MOF framework with guest molecules imparting functionality. Citing a few examples, here we briefly explain the various methods employed.

Besides the in situ functionalization by which various large functional molecules are trapped in the matrices (or pores), small molecules may be trapped within the pores by means of adsorption, however, due to the weak interactions, the molecules cannot be effectively trapped by spatial confinement [21]. Often, it is difficult to immobilize already synthesized metal NPs or bigger molecules directly, due to the small pore aperture. Alternatively, precursors for the formation of NPs can be diffused into the pores by methods like solution impregnation or chemical vapor deposition, wherein the NPs are formed within the framework by reduction. In a study, MIL-101 was loaded with a precursor complex of palladium $[(\eta^5\text{-}C_5H_5)Pd(\eta^3\text{-}C_3H_5]$ via the method of vapor deposition, to synthesize Pd@MIL-101 for application as catalysts [25].

Lewis acidic open metal sites are formed when terminal ligands get dissociated from the SBUs during the removal of solvent molecules coordinated to metal at high temperatures or in vacuum conditions [42]. Coordinative functionalization of such open metal sites is another way of imparting new properties to MOFs. The polar Lewis acidic nature of the metal sites increases the sorption capacity of MOFs, particularly for nonpolar gases like H_2, CO_2, etc. In this regard, Kang and co-workers grafted the well-exposed unsaturated metal sites of a framework resembling MOF-74 with diamines of varying lengths and arrangements, to tune the CO_2 uptake abilities [29]. The principle is that while one end of the inserted diamine group stays free, the other end binds to the metal ion and the Lewis acidic metal sites couple with the oxygen in CO_2. The experimental results revealed significant information regarding the relationship between the structure and functionalization potential. The framework of Mg_2 (dondc) which had narrow channel sizes, exhibited relatively low grafting of diamine groups to the open metal sites. Whereas frameworks with larger channel sizes could accommodate more diamines, resulting in enhanced CO_2 uptake.

Another strategy renders the linkers functionalized via the pre-synthetic approach to be "coordinatively functionalized," wherein metal ions are coordinated to such linkers. Long et al. presented the first MOF (MOF-253) with open 2,2'-bipyridine ligand sites [3]. The Al(OH)(bpydc) (known as MOF-253) was prepared by the reaction of $AlCl_3.6H_2O$ with H_2bpydc (2,2'-bipyridine-5,5'-dicarboxylic acid), and they demonstrated that it reacts with $Cu(BF_4)_2$ and $PdCl_2$, filling 97% and 83% of the 2,2'-bipyridyl sites, respectively. The resulting material of the complexation of Cu^{2+} showed an enhanced adsorption selectivity for CO_2 over N_2 under typical flue gas conditions.

Functionalization Involving Strong Interaction

The various commonly employed strategies for functionalization involving strong interactions are enlisted in Fig. 3.

Atomic layer deposition (ALD) refers to a typical synthetic strategy generally adopted for hydroxyl-containing surfaces of exclusively inorganic substances. ALD when applied in MOFs is popularly known by the acronym "AIM," this is possible because the metal nodes (or SBUs) within some zirconium MOFS, could be chemically considered as small fragments of zirconia. Kim and co-workers reported an example of the coordinative functionalization of the SBU by AIM. The existence of -OH and -OH_2 functional groups at the surface of the Zr_6 node, makes NU-1000 MOF an ideal candidate for AIM. They treated NU-1000 with the ALD precursors, $Al(CH_3)_3$ and $In(CH_3)_3$, and successfully demonstrated that Al^{3+} and In^{3+}, respectively, bond to the zirconium secondary building units by the terminal oxygen atoms and make up heterometallic clusters [31]. Therefore, it can be inferred that the terminal ligands on Zr SBUs act as Lewis bases and coordinate to additional metal ions.

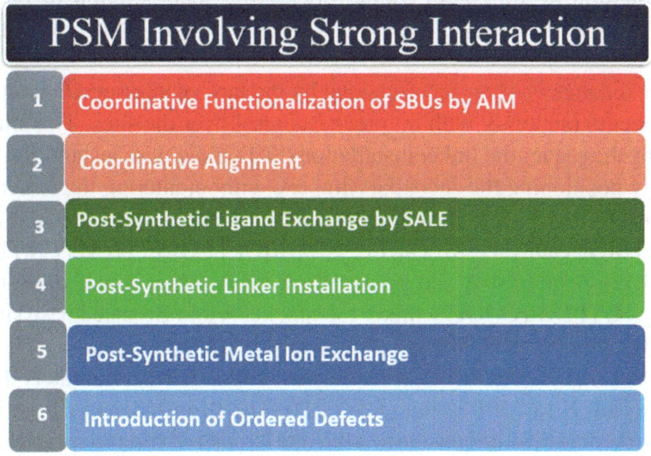

Fig. 3 Different strategies for post-synthetic modifications involving strong interactions

Coordinative alignment (CAL) is a strategy by which various organic molecules are attached to MOF backbones that are chiral. This is of particular interest in single-crystal X-ray diffraction (XRD), an analytical technique for the precise determination of chemical structures, which is often hindered by the orientational disorders of the molecules. In this context, crystallization of the molecules in the pores of the chiral MOF helps reduce their motional degrees of freedom. Moreover, the chiral metal–organic framework backbone acts as a reference in the sample solution for the accurate identification of the absolute configuration of the aligned molecules. An exemplary example is that Lee and co-workers coordinatively aligned 16 different molecules of 4 common functional groups (phenols, vicinal diols, primary alcohols, and carboxylic acids) to the chiral MOF-520 [33]. With this strategy, they could accurately distinguish the gibberellins, A_1 and A_3, which differed only in the presence of a single bond and a double bond. Of significance is the fact that they could enantioselectively crystallize and obtain the structure of the racemic jasmonic acid.

Another strategy for obtaining preferred functionalized frameworks, which cannot be achieved by direct synthesis, is the post-synthetic exchange of the ligand. It involves the exchange (partially or completely) of a terminal ligand of the framework by a different ligand of a distinct functional group or length, while retaining the MOF topology. In this regard, the solvent-assisted ligand exchange (SALE), involving the soaking of a metal–organic framework in a concentrated solution of the chosen ligand molecule for a long duration, at elevated temperatures, is a commonly employed method. Very recently, Gharib et al. [23] adopted a SALE pathway for obtaining new MOFs with improved catalytic efficiency in reactions involving epoxide ring-opening. They exchanged a malonamide pillar ligand containing acidic hydrogen, N1,N3-di(pyridine-4-yl) malonamide linker (S), with the acylamide pillars in three amide-functionalized MOFs (TMU-49, TMU-50, and TMU-51). The resultant MOFs were isostructural to the parent frameworks. After incorporation of the S linker, TMU-50S showed a 98% efficiency in catalysis, much greater than the parent TMU-50.

Ligand exchange is often accompanied by the lack of site-specificity, that is, the exchange occurs randomly with no precise control over the sites. This limitation is overcome by the sequential linker installation (SLI), a stepwise installation technique for precisely positioning the secondary linkers, without altering the structure of the parent MOF. Recently, Liu et al., employed SLI to prepare a Bronsted acid–base bifunctional MOF catalyst [26]. The PCN-700 MOF has two "pockets" with different lengths, pocket A of 16.4 Å and pocket B of 7.0 Å. They installed Bronsted acid site-bearing $H_2TPDC\text{-}(COOH)_2$ linker into pocket A and Bronsted base site-bearing $H_2BDC\text{-}NH_2$ into pocket B. The installation followed single-crystal-to-single-crystal transformation and as a result, the acid and bases sites could be accurately determined by single-crystal X-ray diffraction (SC-XRD). The Bronsted acid carboxyl groups, catalyze the de-acetalization reaction, while the Bronsted base group, $\text{-}NH_2$, catalyze the Knoevenagel condensation reaction, thus rendering the resultant MOF with an outstanding acid–base catalytic ability for the one-pot tandem reaction.

Post-synthetic metal ion exchange is another strategy to prepare desired MOFs that cannot be prepared directly via conventional methods. In this strategy, a metal center within a MOF is exchanged for another, and can also be used for the preparation of MOFs with mixed-metal SBUs. Denysenko et al. [13] successfully carried out post-synthetic metal ion exchange by replacing four of the five zinc centers in $[Zn_5Cl_4(BTDD)_3]$ with cobalt centers, by treating the MOF with $CoCl_2$ in DMF. The isostructural replacement of Zn^{2+} ions gave rise to the formation of the redox-active $[ZnCo_4Cl_4(BTDD)_3]$ MOF possessing catalytically active metal sites, which cannot be prepared directly. The resultant MOF having open (or unsaturated) metal sites exhibited reversible gas-phase oxidation properties, as revealed from the cyclic temperature-programmed oxidation (TPO) results.

By the controlled removal of linkers and/or metal ions, ordered vacancies can be created at specific locations, and the topology can be altered to a desired one. The introduction of such ordered defects bestows the MOF with superior properties like increased pore size, which can be exploited for several potential applications. Furthermore, the deliberately created vacancies can be filled with new functional groups to prepare more complex and useful MOFs, which cannot be synthesized by conventional methods. As a proof-of-concept, Tu, Pang, and co-workers introduced ordered vacancies in a MOF constituting Zinc(II) and pyrazolecarboxylic acid, $Zn_4O(Pyc)_3$ [51]. The deliberate elimination of half of the PyC (4-pyrazolecarboxylate) ligands and a quarter of the Zn^{2+} ions, introduced ordered vacancies and changed the topology of the framework from **pcu** to **srs**. The increased pore size of the resultant framework allows bigger dye molecules to occupy the pores. Further, they filled the metal vacancies with different metal ions such as Li^+, Cd^{2+}, etc., and functionalized linkers such as NH_2-PyC and CH_3-PyC were introduced into the vacancies formed by the removal of the linker.

Functionalization Involving Covalent Interaction

The covalent post-synthetic modification was the first successfully reported post-synthetic functionalization technique [30]. So far, covalent transformations of the organic linkers are the most extensively studied PSM type in MOF chemistry, such as amide coupling, imine condensation, bridging hydroxyl groups within the SBUs, etc. This approach often relies on the existence of functional groups in the ligands, which act as the site for modification. Here, we throw light into the principles of this approach with the help of examples.

Wang and Cohen [54] came up with the terminology "post-synthetic modification," and pioneered in demonstrating a post-synthetic covalent modification by an amine to amide transformation. They carried out the reaction of IRMOF-3, the isoreticular NH_2-functionalized analog of MOF-5, with acetic anhydride in chloroform or dichloromethane, to form an amide-functionalized IRMOF-3(AM1). The Cohen group further demonstrated that the degree of amidation decreases with increasing chain length [49]. This implies that the size of the reactant with respect to the pore aperture is a crucial factor. By this strategy of amide coupling, chiral centers can be introduced to an achiral MOF, by the reaction with chiral anhydrides [22]. In more recent work, Stock and co-workers treated MIL-101(Cr)-NH_2 with p-phenylazobenzoylchloride to produce Cr-MIL-101_amide and further probed into the photoisomerism of the azo groups [38].

"Click chemistry" mainly comprises the Huisgen cycloaddition reactions of alkynes with azido group (RN_3) catalyzed by Cu(I), yielding a triazole compound. MOFs with organic linkers having a terminal azide can chemically react with alkynes in cycloaddition reactions and, those with linkers bearing an alkyne moiety can react with azides. A noteworthy advantage of this reaction is its low steric demand, which enables a broad range of MOFs to undergo this reaction. Jiang et al. [27] reported covalent PSM via click chemistry of the remarkably stable isoreticular MOF series- PCN-56, PCN-57, PCN-58, and PCN-59. The azide group in PCN-58 ($Zr_3O_2(OH)_2(TPDC\text{-}(CH_2N_3)_2)$) and PCN-59 ($Zr_3O_2(OH)_2(TPDC\text{-}(CH_2N_3)_4)$) undergoes click reaction with acetylenes post-synthesis to produce triazole species. By varying the ratio of the azide group-containing ligands during the MOF synthesis, the loading of the azide groups can be precisely regulated. This in turn permits the anchoring of various functional groups to the pore walls, which influences their gas sorption selectivity.

When multiple PSMs are carried out in a sequence, the process is referred to as "tandem functionalization." Besides the strategies mentioned above, some of the other strategies for PSM involving covalent interactions include tandem functionalization of amino-functionalized metal–organic frameworks, cycloaddition reactions, bridging hydroxyl groups within SBUs of MOFs, etc.

5 Immobilization of Guest Species in MOF Matrices and Pores

Taking advantage of the characteristic high permanent porosity of MOFs and well-ordered pore structure, various guest species can be encapsulated within the pores of MOF, including metal nanoparticles, metal oxides, quantum dots, polyoxometalates, polymers, enzymes, dyes, and biomolecules. Besides enhancing the activity of the guest molecules, this strategy also endows the resulting composites with superior properties that are not found in the individual components, on account of synergistic effect. In contrast, the deposition of the guest species on the outer surface of MOFs would lead to their self-deactivation and leaching, and hence, is avoided [55]. The presence of binding sites within the pores in the MOF framework for interaction and fixing of the guest molecules, the size of the pore aperture being smaller than the diameter of the introduced species, are some of the necessary requirements for trapping of the guest molecules [58]. Also, the host framework should be inert toward the embedded precursors or free ligands of the precursors. By virtue of the caging effect or the spatial confinement of the cavities of the MOF, the growth of the guest molecules is limited to the size of the corresponding pore diameter.

There are several characteristic properties of MOFs that make them ideal platforms for the immobilization/encapsulation of different guest molecules [10]. Tunable pore environments and pore size are noteworthy features of MOFs, because of which preparation of isoreticular MOFs (MOFs with the same topology) with various functionalities or different pore sizes, respectively, is possible. This is of significance as it helps in determining the influence of the pore environment on the performance of the immobilized guest molecules or helps in understanding the diffusivity of guest species in MOFs for gaining insights into their potential applications. The pore size can be readily tuned by changing the length of the bridging ligands; increasing the length of the linkers would result in larger pore apertures. The uniform pore structure, that is, the long-range order distribution of uniform cavities throughout the three-dimensional structure, facilitates an ordered arrangement of the guest molecules in the lattice. This is particularly useful for studying the host–guest interactions by various crystallographic methods. Typical cage-like MOFs, characterized by large pores and narrow windows, permit the encapsulation of guest molecules with no constraints on their degree of freedom and prevent their leaching and aggregation. Moreover, it is highly effective in catalytic applications, as the pore windows facilitate easy diffusion of substrates in and out of the cavities to get to the catalytically active sites which are embedded within the MOF.

The conventional strategies for the immobilization of guest species within MOFs are described in brief:

- **Ship-in-bottle strategy:** It can be accomplished by pre-assembly incorporation or post-assembly incorporation [10]. In the former, the guest species precursors are bound to the organic linkers before the assembly of the MOF. Whereas, in the latter, the precursors are dispersed within the pores of the MOF matrix. This strategy has some limitations such as difficulty in regulating the size, shape, internal location, and composition of the guest species within the framework.
- **Bottle-around-ship strategy:** In this method, the guest species, say metal nanoparticles, are pre-synthesized and, MOF is assembled or grown in situ around the metal NPs [11]. The use of surfactants or capping agents is essential, for enhancing the stability of the guest species and assisting the overgrowth of MOF on them. Controlling the overgrowth of the MOF is often difficult and is a limitation of this strategy.
- **Sandwich assembly strategy:** This approach facilitates embedding the guest species between the layers of MOF. Initially, a MOF core is synthesized, followed by the deposition of guest species. Further, the growth of a MOF shell with controllable thickness is achieved on the core. MOFs with similar crystal topologies are preferred to MOFs with mismatched topologies, in view of attaining conformal outgrowth of fracture-free shells on the core.
- **In situ encapsulation or One-pot synthesis:** It facilitates a quick, cost-effective, scalable preparation technique of MOF composites, by simultaneous synthesis of MOF and the nanoentity [10]. It is accomplished by direct mixing of the precursors of the MOF and the nanoentity as well as the other reagents, to form a solution. In this approach, MOF shells grow precisely on the surface of the guest species ruling out self-nucleation. To ensure the assembly of both the components into a

single nanostructure, the reaction parameters (such as the temperature, precursors, solvents, surfactants, etc.) must be kept in check.

In bottle-around-ship strategy and in situ encapsulation, it is essential to maintain mild reaction conditions, for retaining the stability and activity of the guest molecules.

The MOF composites formed by the immobilization of guest molecules have excellent potential in numerous applications. For example, Zhao and co-workers reported novel "sandwich" type composites, MIL-101@Pt@MIL-101, by sandwiching platinum NPs between a core and an outer shell composed of MIL-101 MOF, consisting of metal nodes of either Fe^{3+} or Cr^{3+} or both [61]. The composites were employed as selectivity regulators in the hydrogenation reaction of α, β-unsaturated aldehydes to unsaturated alcohols, because of their ability to selectively tune catalysis. In practice, hydrogenation of the carbon–carbon group is favored thermodynamically, as opposed to that of the carbon–oxygen group. However, preferential interactions of the MOF metal site with the carbon–oxygen group make the hydrogenation of it (and not carbon–carbon group) by the Pt NPs a thermodynamically favored reaction. Thus, with the new catalytic pathway, the "sandwich" structures were successfully tested as suitable catalysts for the selective reduction of α, β-unsaturated aldehydes.

In another study, Let et al. [34] successfully prepared a dye@MOF luminescent composite for chemical recognition application for water remediation. They prepared Bio-MOF-1@RhB composite by encapsulating luminescent, cationic RhB dye (Rhodamine B) in the porous channels of the water stable, anionic Bio-MOF-1, via the ion-exchange process. The composite exhibited selective and sensitive recognition/detection of Fe^{3+} ions in water even in the presence of other metal ions, via a luminescence quenching mechanism. The quenching coefficient value (K_{SV}) of the composite was as high as $5.5 \times 10^4\ M^{-1}$, making it an excellent sensory probe. Similarly, a plethora of applications can be realized using such MOF composites and there is an endless scope of research in this field.

6 Coating MOFs with Functional Materials

Functionalization of MOFs can also be accomplished by adopting any of the three approaches: (1) coating the MOF with a functional shell (2) encapsulation of MOF in functional matrices (3) deposition of MOF on functional supports [55]. Figure 4 shows a general representation of MOF coated with a functional material. However, these strategies are relatively underexplored, and more extensive research is anticipated. In these strategies, the type of interface between the MOF and the functional support (such as oxides of metals, polymers, carbon, other MOFs, etc.) is pivotal in controlling the assembly process, which involves growth kinetics, nucleation, and alignment. Furthermore, the interaction at the interphase between the MOF and the functional materials influences the properties of the composites formed and also triggers the synergistic effects for enhanced properties [5].

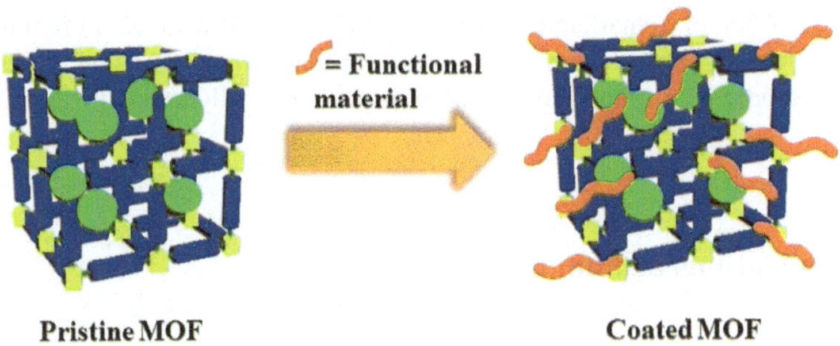

Fig. 4 An illustration of coating a MOF with functional material

The various strategies for the preparation of functional material-coated MOF composites are essentially based on the three general methods of preparation of composites: top-down synthesis, bottom-up synthesis, and one-pot synthesis. The growth of MOFs on substrates is the approach in top-down synthesis. However, it is less preferable due to the intrinsic limitation of difficulty in precise controlling of the size and morphology of the nanostructure. Bottom-up synthesis involves the coating of MOF crystals with protective functional shells, wherein the shell can be another MOF too. One-pot synthesis accomplishes operationally simple, low-cost preparation of the MOF/functional material composites.

Besides improving the physicochemical properties, morphology, and stability of MOFs, these strategies also introduce several new, desirable functionalities such as mechanical strength, magnetic properties, conduction of electricity, hydrophilicity or hydrophobicity, etc., Silica, monoliths, carbon, alumina, and polymers are some of the commonly used substrates for integration with or deposition of MOFs. This method of imparting functionality to MOFs is an effective strategy to significantly improve properties and bring about better properties, specific to different applications. For example, combining biopolymer interfaces with MOFs which have exhibited potential in diagnostics and drug delivery, helps improve their biocompatibility and solubility in an aqueous medium.

As a proof-of-concept of this functionalization strategy, Liang and co-workers prepared an enzyme-coated MOF shell, for attaining adaptive cell survival in the simulated oligotrophic environment [36]. They covered living eukaryotic cells with the bioactive shell, and the cells could thrive in the nutrient-deficient medium even with other unfavorable conditions such as ultraviolet irradiation and the presence of toxic agents. This was possible because the enzyme/MOF coating generated the nutrients required by the cells and the MOF allowed the transport of nutrients, as well as protected the cells and enzymes from the toxic agents by providing a selective barrier. Furthermore, on the disposal of the biocomposite MOF covering, the cells regained their full growth capacity.

In another work, Wang et al. coated UiO-66 MOF crystals with a luminescent biocompatible polymer, BDP-imine. The composite was prepared by epitaxial growth of the polymer on the surface of UiO-MOFs seeds. The resulting composite displayed a set of remarkable properties such as well-retained pore integrity, enhanced luminescence, low cytotoxicity, and available functional sites, making it a potential candidate for bio-imaging and improved intracellular uptake applications [53].

7 Applications of Functionalized MOFs

MOFs find a wide range of applications due to their unique properties like high permanent porosity, extraordinary functionality, tunable pore structures, and ultrahigh surface areas. Moreover, combining MOFs with other suitable materials can further improve their properties [1]. MOFs with diverse functionalities can be designed and constructed for certain applications [56]. Some of the important applications of functionalized MOFs are listed in Fig. 5 and are discussed below briefly.

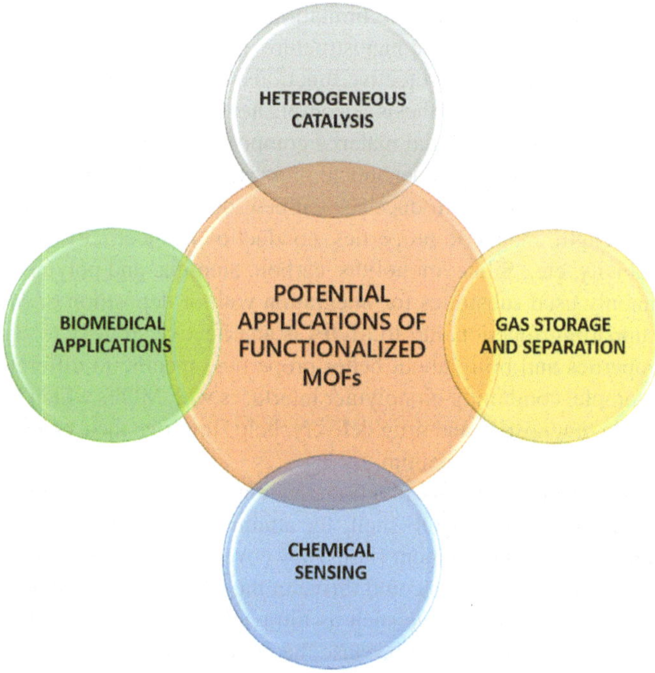

Fig. 5 Some of the potential applications of functionalized MOFs

7.1 Chemical Sensing

MOFs serve as excellent platforms for constructing chemical sensors owing to their highly ordered and tunable structures. The incorporation of a guest species often leads to the change in luminescence for the MOF hosts [28]. The inclusion of various chemical functionalities into the coordination nano space of MOFs has shown great sensing applications. Water sensing holds a great significance in chemical industries. Douvali et al. reported a flexible MOF $[Mg(H_2dhtp)(H_2O)_2]\cdot DMAc$ with a three-dimensional open framework structure. It showed remarkable breathing behavior upon the incorporation of guest DMAc molecules and was found to exhibit the excellent potential to detect the trace water in several organic solvents [14]. The research concerning the potential application of MOFs in electronic devices, like sensors is still in its infancy, probably because the fabrication of MOFs with well-designed structures represents a big challenge.

7.2 Heterogeneous Catalysis

Heterogeneous catalysis is among the earliest reported applications of MOFs. Functionalized MOFs have presented expeditious progress toward their prospective application in catalysis in the past two decades. MOF-based catalysts were firstly reported by Fujita et al. [18] in 1994. They developed a 3D Cd(II)-MOF with 4,4'-bipyridine which was found to have catalytic properties for the cyanosilylation of aldehydes. Since then, in a quite short period of time, these materials have been found useful as heterogeneous solid catalysts for various organic processes like ring-opening reactions, transesterification, coupling reactions, Mukaiyama–aldol reaction, oxidation, Knoevenagel condensation, etc. Besides these, owing to their excellent catalytic properties, MOF-supported catalysts have gained immense interest in energy-related applications like photocatalysis and electrocatalysis [28]. In order to explore new visible-light-promoted photocatalysts with increased H_2 evolution, Han et al. [24] investigated the photocatalytic properties of hybrid materials obtained from MIL-125. They reported the first example of methylthio-functionalized MOF decorated with Pt co-catalyst for effective production of H_2. The H_2 production rate of this hybrid material was found to be as high as 3814 $\mu molg^{-1}\ h^{-1}$.

7.3 Gas Storage and Separation

MOFs are excellent candidates for the storage of gases like H_2, CO_2, CH_4 because of their large surface areas and high permanent porosity. Large surface areas of MOFs help in the efficient adsorption of gas [28]. The functionalization of MOF materials is a primary and effective strategy for altering the interaction between the MOF framework and the guest species in order to obtain high gas storage capacity [56]. Because of hydrogen's gaseous form and low density, its storage is a critical aspect in determining its practical use as transportation, stationary, and portable fuel. In the past decade, functionalized MOFs have been enormously explored for hydrogen storage.

Anthropogenic CO_2 emissions are causing disastrous environmental issues, leading to an adverse impact on human lives. Carbon dioxide, in comparison to hydrogen, forms stronger intermolecular interactions with the functionalities of MOFs, allowing it to be stored more effectively. While gas storage emphasizes storing a pure gas, gas separation requires MOFs to have high selective adsorption of a particular gas [28]. The low partial pressure of CO_2 is responsible for making its separation from the gaseous mixture challenging. This is because it significantly lowers its adsorption by MOF [28]. Open metal sites and amine functionalities can be introduced in the MOF adsorbents in order to enhance the adsorption capacity as well as the selectivity of CO_2 over other gases. Recently, Bahamon et al. [2] reviewed the effect of the introduction of amine functionalities in MOFs for improved capture and separation of CO_2.

7.4 Biomedical Applications

MOFs have presented great potential and unique advantages in the field of biomedicine. Thanks to the rapid growth of synthetic methods and surface functionalized strategies, smart MOF-based composites with superior bio-related properties have been designed to fulfill the rising demands of MOF materials in biomedical applications [59]. In this domain, functionalized MOFs are finding various applications such as:

- Drug delivery
- Bioimaging
- Antimicrobial
- Biosensing
- Biocatalysis.

Metal–organic frameworks (MOFs) are widely used for drug delivery because of their unique properties like tunability and large surface areas. Their high surface area enables the loading of small molecules as well as biomacromolecules. Recently, Cai and co-workers employed mono substituents for the functionalization of MOF-5, and studied its effect on drug delivery. A substantial effect on the drug delivery behavior of MOF upon the incorporation of single substituents has been reported. Moreover, the obtained composites were found to possess good biocompatibility and negligible cytotoxicity [8].

Till now, various MOFs have been employed for the preparation of MRI contrast agents [59]. For example, in 2017, Sene et al. decorated the MIL-100(Fe) with γ-Fe_2O_3 for MRI imaging and cancer treatment. The obtained composite, after being loaded with DOX, showed high antitumor activity. Also, it showed low cytotoxicity and great biocompatibility, and hence, could be used as an efficient MRI contrast agent [47].

8 Conclusion and Future Outlook

Progressing interest and development in the domain of MOF studies is continued which is justified enough by vast applications of exceptional size dimensions, gas storage, and separation, owing to the inherent pores and their volume. Despite the large potential applications, there are pertinent challenges that MOFs face in their conventional aspects, and hence to further improve various strategies have been suggested. Functionalization is commonly utilized to exploit the pore structure of MOF increasing its tunability by many folds. In this chapter, we have thoroughly put highlights on various methods to functionalize the MOF units and the involved challenges.

More detailed studies are required in various domains of MOFs to solve the current challenges to utilize more effectively for current applications. For instance, the porosity of most MOFs makes them potential agents for gas storage application but the size exclusion approach requires fine-tuning of pore size for separation of gases with similar kinetic diameters that is often a challenging task. Indeed, size exclusion is difficult for separating molecules of similar size and polarizability. Hence, this is one such area that needs future exploration by researchers across the globe.

Important Websites

1. https://www.nanowerk.com/mof-metal-organic-framework.php
2. http://wuttkescience.com/research/functionalization-of-materials-backbone/
3. https://www.sigmaaldrich.com/IN/en/technical-documents/technical-article/materials-science-and-engineering/photovoltaics-and-solar-cells/metal-organic-frameworks.

Abbreviations

MOF	Metal-Organic Framework
PCP	Porous Coordination Polymers
SBU	Secondary Building Unit
CUS	Coordination Unsaturated Sites
bdc	1,4-Benzenedicarboxylate
bpy	4,4'-Bipyridyl
PyC	Pyrrole-2-carboxylate

btc	1,3,5-Benzenetricarboxylate
tpdc	P,p'-terphenyldicarboxylate
bpdc	4,4'-Biphenyldicarboxylate
dobdc	4,4'-Biphenyldicarboxylate
4-bpmh	N,N'-bis-pyridin-4-ylmethylene-hydrazine
bpydc	2,2'-Bipyridine-5,5'-dicarboxylate
BTDD	Bis(1,2,3-triazolato-[4,5-b],[4',5'-i])dibenzo-[1,4]-dioxin
POM	Polyoxometallate
NPs	Nanoparticles
UiO	Universitetet i Oslo (University of Oslo)
MIL	Materials Institute Lavoisier
ALD	Atomic Layer Deposition
AIM	ALD in MOFs
CAL	Coordinative Alignment
XRD	X-ray diffraction
SALE	Solvent-Assisted Ligand Exchange
SLI	Sequential Linker Installation
SC-XRD	Single-crystal X-ray diffraction
TPO	Temperature-programmed Oxidation
IRMOF	Isoreticular MOF
PCN	Porous Coordination Network
PSM	Post-synthetic Modifications
RhB	Rhodamine B
DMAc	N,N-dimethylacetamide
MRI	Magnetic Resonance Imaging
DOX	Doxorubicin

References

1. Ahmed I, Jhung SH (2014) Composites of metal-organic frameworks: preparation and application in adsorption. Mater Today 17(3):136–146. https://doi.org/10.1016/j.mattod.2014.03.002
2. Bahamon D et al (2021) Effect of amine functionalization of MOF adsorbents for enhanced CO_2 capture and separation: a molecular simulation study. Front Chem 8(January):1–11
3. Bloch ED et al (2010) Metal insertion in a microporous metal-organic framework lined with 2,2′-Bipyridine. J Am Chem Soc 132(41):14382–14384
4. Botas JA, Calleja G, Sánchez-Sánchez M, Gisela Orcajo M (2010) Cobalt doping of the MOF-5 framework and its effect on gas-adsorption properties. Langmuir 26(8):5300–5303. https://doi.org/10.1021/la100423a
5. Bradshaw D, Garai A, Huo J (2012) Metal–organic framework growth at functional interfaces: thin films and composites for diverse applications. Chem Soc Rev 41(6):2344–2381. http://xlink.rsc.org/?DOI=C1CS15276A
6. Bromberg L et al (2012) Bromberg2012 (III)
7. Burrows AD (2013) Metal-organic frameworks as heterogeneous catalysts. Post-synthetic modification of MOFs

8. Cai M et al (2020) Functionalization of MOF-5 with mono-substituents: effects on drug delivery behavior. RSC Adv 10(60):36862–36872
9. Chen J, Li Y (2016) The road to MOF-related functional materials and beyond: desire, design, decoration, and development. Chem Rec 16(3):1456–1476. https://doi.org/10.1002/tcr.201500304
10. Chen L, Luque R, Li Y (2017) Controllable design of tunable nanostructures inside metal-organic frameworks. Chem Soc Rev 46(15):4614–4630. https://doi.org/10.1039/C6CS00537C
11. Chen L et al (2020) The function of metal-organic frameworks in the application of MOF-based composites. Nanoscale Advances 2(7):2628–2647
12. Consideration General (2020) Metal organic frameworks. MDPI. http://www.mdpi.com/books/pdfview/book/2140
13. Denysenko D et al (2012) Reversible gas-phase redox processes catalyzed by co-exchanged MFU-4l(Arge). Chem Commun 48(9):1236–1238
14. Douvali A et al (2015) Turn-on luminescence sensing and real-time detection of traces of water in organic solvents by a flexible metal-organic framework. Angew Chem Int Ed 54(5):1651–1656
15. Falcaro P et al (2011) A new method to position and functionalize metal-organic framework crystals. Nat Commun 2(1):237. http://www.ncbi.nlm.nih.gov/pubmed/21407203
16. Fang QR, Makal TA, Young MD, Zhou HC (2010) Recent advances in the study of mesoporous metal-organic frameworks. Comments Inorg Chem 31(5):165–195
17. Fang Z, Bueken B, De Vos DE, Fischer RA (2015) Defect-engineered metal-organic frameworks. Angew Chem Int Ed 54(25):7234–7254. https://doi.org/10.1002/anie.201411540
18. Fujita M, Washizu S, Ogura K, Kwon YJ (1994) Preparation, clathration ability, and catalysis of a two-dimensional square network material composed of Cadmium(II) and 4, 4′-Bipyridine. J Am Chem Soc 116(3):1151–1152
19. Furukawa H et al (2011) Isoreticular expansion of metal-organic frameworks with triangular and square building units and the lowest calculated density for porous crystals. Inorg Chem 50(18):9147–9152
20. Furukawa H, Cordova KE, O'Keeffe M, Yaghi OM (2013) The chemistry and applications of metal-organic frameworks. Science 341(6149)
21. Garibay SJ, Cohen SM (2010) Isoreticular synthesis and modification of frameworks with the UiO-66 topology. Chem Commun 46(41):7700–7702
22. Garibay SJ, Wang Z, Tanabe KK, Cohen SM (2009) Postsynthetic modification: a versatile approach toward multifunctional metal-organic frameworks. Inorg Chem 48(15):7341–7349
23. Gharib M, Esrafili L, Morsali A, Retailleau P (2019) Solvent-assisted ligand exchange (SALE) for the enhancement of epoxide ring-opening reaction catalysis based on three amide-functionalized metal-organic frameworks. Dalton Trans 48(24):8803–8814
24. Han SY et al (2018) A methylthio-functionalized-MOF photocatalyst with high performance for visible-light-driven H_2 evolution. Angew Chem Int Ed 57(31):9864–9869
25. Hermannsdörfer J, Kempe R (2011) Selective palladium-loaded MIL-101 catalysts. Chem Eur J 17(29):8071–8077
26. Hu XJ et al (2020) Designing a bifunctional brønsted acid-base heterogeneous catalyst through precise installation of ligands on metal-organic frameworks. CCS Chem 2(1):616–622
27. Jiang H-L et al (2012) Pore surface engineering with controlled loadings of functional groups via click chemistry in highly stable metal–organic frameworks. J Am Chem Soc 134(36):14690–14693. https://doi.org/10.1021/ja3063919
28. Jiao L et al (2019) Metal–organic frameworks: structures and functional applications. Mater Today 27(August):43–68
29. Kang M, Kang DW, Hong CS (2019) Post-synthetic diamine-functionalization of MOF-74 type frameworks for effective carbon dioxide separation. Dalton Trans 48(7):2263–2270
30. Kiang YH et al (1999) Variable pore size, variable chemical functionality, and an example of reactivity within porous phenylacetylene silver salts. J Am Chem Soc 121(36):8204–8215
31. Kim IS et al (2015) Targeted single-site MOF node modification: trivalent metal loading via atomic layer deposition. Chem Mater 27(13):4772–4778

32. Lee H, Varma CM et al (1988) Proc Natl Acad Sci USA 37(January):714
33. Lee S, Kapustin EA, Yaghi OM (2016) Coordinative alignment of molecules in chiral metal-organic frameworks. Science 353(6301):808–811
34. Let S, Samanta P, Dutta S, Ghosh SK (2020) A dye@MOF composite as luminescent sensory material for selective and sensitive recognition of Fe(III) ions in water. Inorg Chim Acta 500(III):119205. https://doi.org/10.1016/j.ica.2019.119205
35. Lian X et al (2017) Enzyme-MOF (metal-organic framework) composites. Chem Soc Rev 46(11):3386–3401
36. Liang K et al (2017) An enzyme-coated metal-organic framework shell for synthetically adaptive cell survival. Angew Chem 129(29):8630–8635
37. Meek ST, Greathouse JA, Allendorf MD (2011) Metal-organic frameworks: a rapidly growing class of versatile nanoporous materials. Adv Mater 23(2):249–267
38. Modrow A, Zargarani D, Herges R, Stock N (2012) Introducing a photo-switchable Azo-functionality inside Cr-MIL-101-NH_2 by covalent post-synthetic modification. Dalton Trans 41(28):8690–8696
39. Nguyen JG, Cohen SM (2010) Moisture-resistant and superhydrophobic metal—organic frameworks obtained via postsynthetic modification 4560–4561
40. Ozer D (2020) Fabrication and functionalization strategies of MOFs and their derived materials 'MOF architecture'. In: Applications of metal–organic frameworks and their derived materials. Wiley, pp 63–100. https://onlinelibrary.wiley.com/. https://doi.org/10.1002/9781119651079.ch3
41. Peng YL et al (2018) A size-matched POM@MOF composite catalyst for highly efficient and recyclable ultra-deep oxidative fuel desulfurization. Inorg Chem Front 5(7):1563–1569
42. Pentyala V et al (2016) Carbon dioxide gas detection by open metal site metal organic frameworks and surface functionalized metal organic frameworks. Sens Actuators, B Chem 225:363–368. https://doi.org/10.1016/j.snb.2015.11.071
43. Qian B, Chang Z, Bu X-H (2020) Functionalized dynamic metal–organic frameworks as smart switches for sensing and adsorption applications. Top Curr Chem 378(1):5. https://doi.org/10.1007/s41061-019-0271-2
44. Rogge SMJ et al (2017) Metal-organic and covalent organic frameworks as single-site catalysts. Chem Soc Rev 46(11):3134–3184
45. Rungtaweevoranit B et al (2016) Copper nanocrystals encapsulated in Zr-based metal-organic frameworks for highly selective CO_2 hydrogenation to methanol. Nano Lett 16(12):7645–7649. https://doi.org/10.1021/acs.nanolett.6b03637
46. Sahoo B (2017) Synthesis and characterizations of novel metal-organic frameworks (MOFs), pp 109–126. https://doi.org/10.1007/978-3-319-48350-4_5
47. Sene S et al (2017) Maghemite-NanoMIL-100(Fe) bimodal nanovector as a platform for image-guided therapy. Chem 3(2):303–322
48. Sriparshamoni N, Sanda S, Jena HS, Konar S (2015) Tuning CO_2 uptake and reversible iodine adsorption in two isoreticular MOFs through ligand functionalization. Chem Asian J 10(3):653–660
49. Tanabe KK, Wang Z, Cohen SM (2008) Systematic functionalization of a metal-organic framework via a postsynthetic modification approach. J Am Chem Soc 130(26):8508–8517
50. Tranchemontagne DJ, Tranchemontagne JL, O'keeffe M, Yaghi OM (2009) Secondary building units, nets and bonding in the chemistry of metal–organic frameworks. Chem Soc Rev 38(5):1257–1283
51. Tu B et al (2014) Ordered vacancies and their chemistry in metal-organic frameworks. J Am Chem Soc 136(41):14465–14471
52. Valvekens P, Vermoortele F, De Vos D (2013) Metal–organic frameworks as catalysts: the role of metal active sites. Catal Sci Technol 3(6):1435. http://xlink.rsc.org/?DOI=c3cy20813c
53. Wang L et al (2017) Nanoscale fluorescent metal-organic framework@microporous organic polymer composites for enhanced intracellular uptake and bioimaging. Chem Eur J 23(6):1379–1385

54. Wang Z, Cohen SM (2007) Postsynthetic covalent modification of a neutral metal−organic framework. J Am Chem Soc 129(41):12368–12369. https://doi.org/10.1021/ja0743660
55. Xu C et al (2019) Functional metal-organic frameworks for catalytic applications. Coord Chem Rev 388:268–292. https://doi.org/10.1016/j.ccr.2019.03.005
56. Xue DX, Wang Q, Bai J (2019) Amide-functionalized metal-organic frameworks: syntheses, structures and improved gas storage and separation properties. Coord Chem Rev 378:2–16. https://doi.org/10.1016/j.ccr.2017.10.026
57. Yaghi OM et al (2003) Reticular synthesis and the design of new materials. Nature 423(6941):705–714
58. Yaghi OM, Kalmutzki MJ, Diercks CS (2019) Functionalization of MOFs. In: Introduction to reticular chemistry, pp 145–176
59. Yang J, Yang YW (2020) Metal–organic frameworks for biomedical applications. Small 16(10):1–24
60. Yuan S, Qin J-S, Lollar CT, Zhou H-C (2018) Stable metal-organic frameworks with group 4 metals: current status and trends. ACS Cent Sci 4(4):440–450. http://www.ncbi.nlm.nih.gov/pubmed/29721526
61. Zhao M et al (2016) Metal–organic frameworks as selectivity regulators for hydrogenation reactions. Nature 539(7627):76–80. https://doi.org/10.1038/nature19763

Characterization Techniques of MOFs

Spectroscopic and Microscopic Techniques: Tools for Characterizing Nanoscale Metal–Organic Frameworks (NMOFs)

Parul Pant, Aadya Jaipuria, and Chetna Gupta

Contents

1 Introduction 128
2 Microscopic Characterizations 132
  2.1 Visible Light Microscopy 132
  2.2 Electron Microscopy 133
  2.3 Scanning Probe Microscopy 141
  2.4 Scanning Tunneling Microscopy (STM) 143
  2.5 Atomic Force Microscopy (AFM) 144
3 Spectroscopic Characterizations 146
  3.1 NMR Spectroscopy 149
  3.2 UV–Visible Differential Reflectance Spectroscopy 152
  3.3 Infrared and Raman Spectroscopy 153
  3.4 Fluorescence Spectroscopy 156
  3.5 Fluorescence Correlation Spectroscopy 157
  3.6 Dynamic Light Scattering 157
  3.7 Inductively Coupled Plasma Optical Emission Spectrometry (ICP-OES) 159
  3.8 Photoluminescence Spectroscopy 159
  3.9 Powder X-Ray Diffraction (PXRD) 160
Abbreviations 162
References 162

Abstract Nanoscale metal–organic frameworks are well known for their ease of synthesis, surface modification, stability, porosity, and thus are found to have great potential in numerous industrial and biomedical applications such as in gas storage, molecular separations, sensors, catalysis, and drug delivery. This chapter focuses on various characterization techniques that are required to study the properties of synthesized and functionalized nanoscale metal–organic frameworks (NMOFs). To understand the morphological properties of NMOFs, microscopic characterization such as scanning electron microscopy (SEM), transmission electron microscopy (TEM), scanning probe microscopy (SPM), scanning tunneling microscopy (STM),

P. Pant · C. Gupta (✉)
Department of Chemistry, Hansraj College, University of Delhi, Delhi, India
e-mail: chetnagupta@hrc.du.ac.in

A. Jaipuria
Massachusetts College of Pharmacy and Health Sciences, Boston, USA

S. Gulati (ed.), *Metal-Organic Frameworks (MOFs) as Catalysts*,
https://doi.org/10.1007/978-981-16-7959-9_5

and atomic force microscopy (AFM) are indispensable tools. However, to gain insights into the different functional groups present inside the framework and in determining its elemental and chemical composition, spectroscopic characterizations such as solid-state nuclear magnetic resonance (SS-NMR), fluorescence spectroscopy, powder X-ray diffraction, infrared and Raman spectroscopy are employed. In this chapter, various microscopic and spectroscopic techniques that provide in-depth knowledge of the size, stability, porosity, and functionality of NMOFs will be elaborated on with some examples cited from the literature.

Keywords Nanoscale metal–organic frameworks · Surface · Modification · Microscopic characterization · Spectroscopic techniques

1 Introduction

Metal–organic frameworks (MOF) are an extremely important emerging class of organic–inorganic hybrid, crystalline compounds with a porous and modular structure. Metal–organic frameworks comprise of two components: metal ion nodes or inorganic clusters, which are referred to as secondary building units, and organic linker molecules or spacers. The variety of combinations of nodes and spacers as well as the flexibility with which their size, functionality, and geometry can be varied aids in the preparation of a huge number of metal–organic frameworks (MOFs). These compounds are of tremendous interest to scientists owing to their unique features like large surface area, easily tunable structure, high pore size and volume, porosity, and surface functionality. MOFs are widely used in the field of drug delivery, catalysis, optics, sensing, diagnosis, and storage [1].

In addition, the size of MOFs can be controlled at the nano level to generate nanoscale metal–organic frameworks (NMOFs). At the nanoscale level, the properties of NMOF depend on both its inner and outer surface due to its larger surface area to volume proportion. Hence, NMOF properties and behavior are considerably size and shape-dependent [2]. Depending upon the synthesis procedure, NMOFs can exist in diverse morphologies resulting in unique physicochemical properties. A slight variation in morphology can lead to changes in their properties. Hence, the characterization of NMOFs plays a crucial role and is requisite for their future applications.

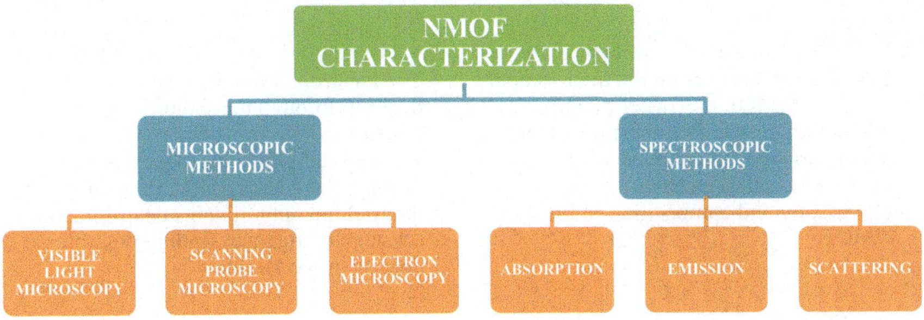

Fig. 1 Different methods used for the characterization of NMOFs

This chapter reviews the different microscopic and spectroscopic techniques for the characterization of NMOFs (Fig. 1). These techniques are employed to study morphology, functional groups, crystal structure, and chemical composition of NMOFs.

Microscopic Characterization

It is used for the purpose of magnification of NMOFs and visualizing what cannot be seen with the naked eye. These techniques are invaluable for imaging and observing the structural features of NMOFs. There are mainly three types of microscopes available. Visible light microscopes, and electron microscopes, scanning probe microscopes (Fig. 2) [3].

Visible Light Microscopy uses light in the visible range to visualize NMOFs. Due to a larger wavelength of photons, the images formed do not have a high resolution and cannot be magnified further. Thus, to overcome the disadvantages of optical microscopes, electron microscopes are used.

Electron Microscopes replace photons with electrons. Owing to the wave nature of electrons and a smaller de-Broglie wavelength, images of higher atomic resolution and greater magnification can be produced [4]. Thus, the characterization tools that employ electron microscopes are (Fig. 3): scanning electron microscopy (SEM) and transmission electron microscopy (TEM) [5].

Scanning Probe Microscopy does not involve light or electron beams but instead scans the NMOF using a sharp probe kept at a constant distance of a few nanometres.

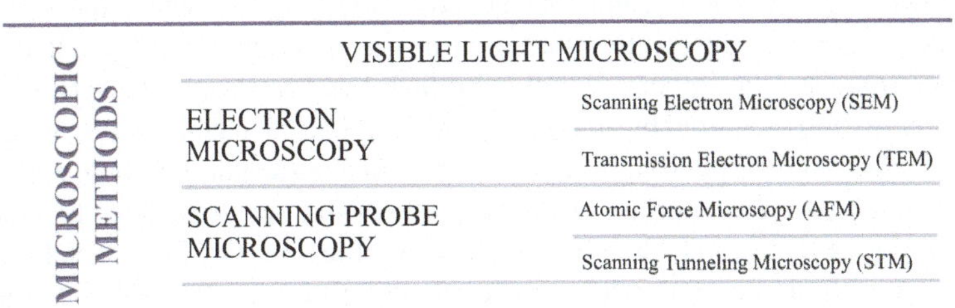

Fig. 2 Various microscopic characterization methods used for NMOFs

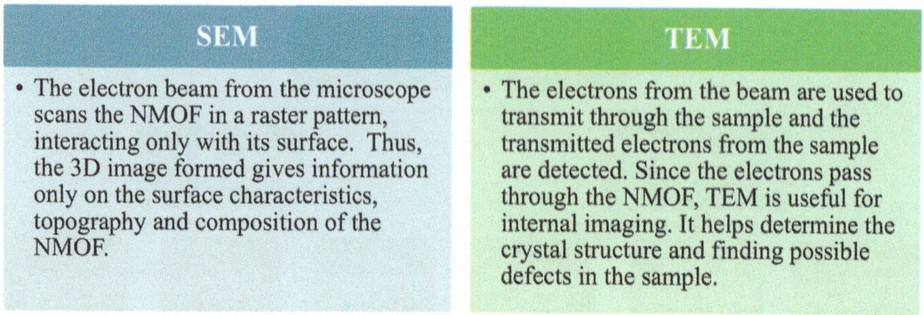

Fig. 3 Basic principles and uses of scanning electron microscopy and transmission electron microscopy in NMOF characterization

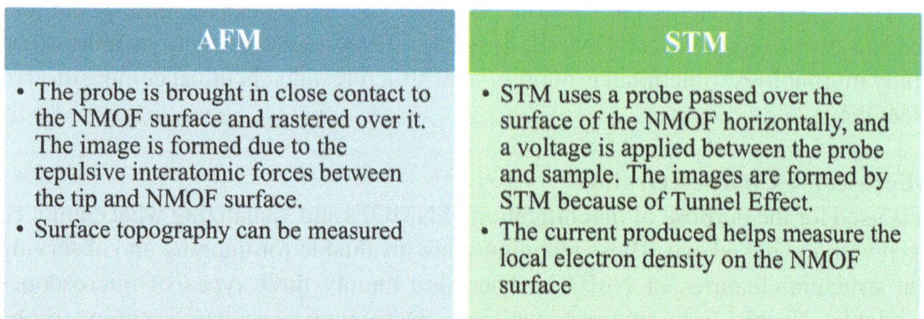

Fig. 4 Basic principles and uses of atomic force microscopy and scanning tunneling microscopy in NMOF characterization

Due to this direct interaction, 3D topographical images of higher magnification can be produced. SPM is the most efficient microscopic method for generating high-resolution images of NMOFs [3]. It can be broadly classified into atomic force microscopy (AFM) and scanning tunneling microscopy (STM) (Fig. 4) [6].

Spectroscopic Characterizations

The interaction of electromagnetic radiation with matter gives rise to various phenomena, and hence, different spectroscopic techniques are required to understand the spectra generated. Electrons in an atom or molecule, principally reside in the lowest energy state (ground state) and when a photon strikes the surface, it transfers its energy to the electron that in turn rises to a higher level (excited state). Each electronic state is associated with vibrational and rotational states (Fig. 5).

The energy thus acquired by an electron can be released in many ways, and hence, different spectra corresponding to different phenomena would be obtained such as absorption, emission, scattering, or fluorescence. Thus, spectroscopic characterization techniques (Fig. 6) are helpful in analyzing the structure of a metal–organic framework with respect to the different functional groups present inside the framework and in determining its elemental and chemical composition. The details are discussed subsequently in the chapter.

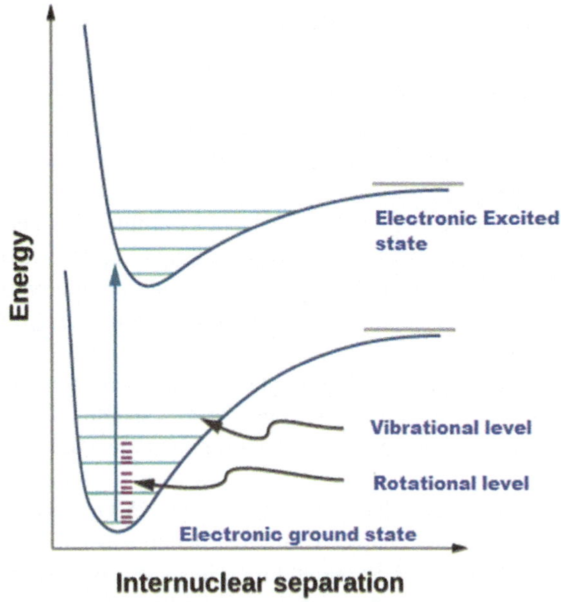

Fig. 5 Energy level diagram of a molecule depicting various levels in it

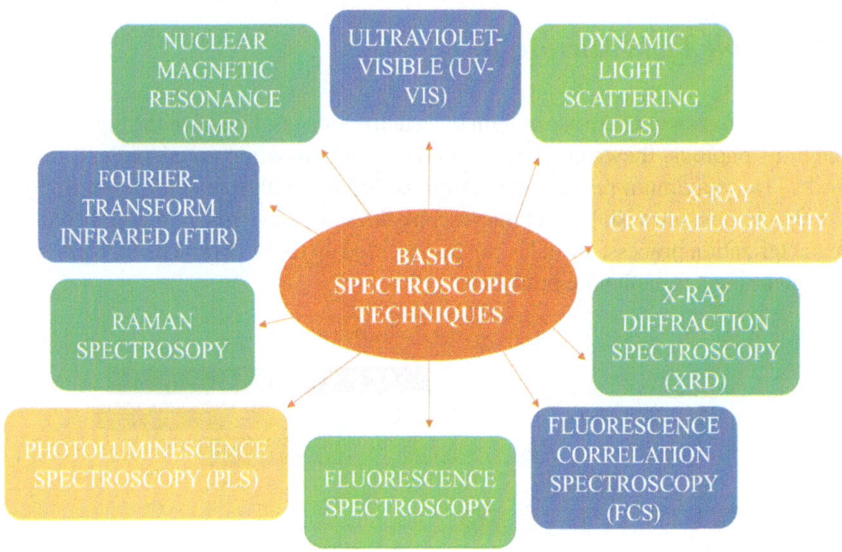

Fig. 6 Basic spectroscopic characterization techniques used for NMOFs

2 Microscopic Characterizations

Microscopic characterization techniques are used for the physical characterization and imaging of NMOFs. Microscopy is advantageous in providing images of objects with good resolution. Figure 7 compares the resolutions achieved by the different kinds of microscopies [7].

Human eyes can resolve images up to 1/10th of a millimeter. For better resolution, optical microscopes are required that can resolve images up to a micrometer due to a higher wavelength of visible light. However, to study the morphological properties of MOFs, even higher resolution is required for which electron microscopes are employed. Electron microscopes, scanning electron microscopes, and transmission electron microscopes can resolve images up to the nanometre level, with high-resolution transmission electron microscopes being capable of atomic resolution. Another powerful tool that can provide resolution up to the atomic level is scanning probe microscopy (SPM). All these techniques are widely implemented; for example, SEM is employed when the surface topography of an NMOF needs to be analyzed, and TEM is advantageous for characterization of features inside the compound, i.e., below the surface. SPM is the best method for when surface sensitivity is needed.

The three microscopic methods are discussed in the further section.

2.1 *Visible Light Microscopy*

Optical microscopes make use of light to visualize images. Owing to the higher wavelength of photons, they cannot fully interact with nanoscale metal–organic frameworks. The maximum resolution possible is 200 nm with a magnification of 1000 × only. As a result, visible light microscopy does not serve much importance in the characterization process. Particles can be observed up to the micron level, but beyond that, clear images cannot be produced.

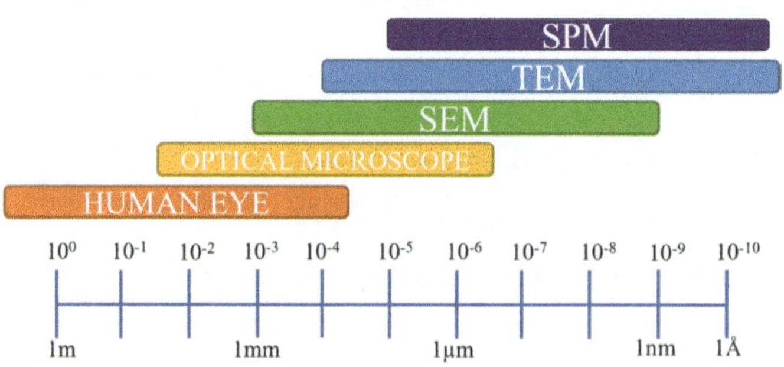

Fig. 7 Image resolution achieved in various kinds of microscopes

Fig. 8 Optical microscope image at a resolution of 500. Reprinted with permission [8]. Copyright (2013) Elsevier

The resolution of a microscope can be described as its ability to produce clear images. Higher the resolution, better will be the microscope's ability to separate two features in proximity as different parts of the image.

Salmi et al. [8] took optical images of MOF-5 (Fig. 8) using an Olympus PX51 light microscope. They observed the crystallization of MOF-5 films when kept under 300 °C under 60% controlled humidity. This observation contrasted with their original hypothesis that MOF-5 was stable under moist conditions.

2.2 *Electron Microscopy*

Electron microscopy is one of the most widely used techniques for the visual characterization of nanoscale metal–organic frameworks. Unlike light microscopy, electron microscopes can provide images of higher resolution and greater magnification. This advantage is possible by replacing photons in a light microscope with an electron beam. As compared to photons, electrons have a smaller de-Broglie wavelength; therefore, they can strongly interact with the nanoscale compounds.

An electron gun is used to produce electrons in an electron microscope. These electrons follow de Broglie's relation, $\lambda = h/p$, i.e., the wavelength of an electron is directly proportional to its momentum. High energy electrons having smaller wavelengths can be generated using higher accelerating voltages. An electron gun applying voltages of 1–300 kV can produce electrons with wavelengths in the range of 1–40 pm. This is useful in generating images with atomic resolution down to < 0.1 nm and magnified up to 1 million times depending on the microscope [5].

Interaction of Electrons

The electrons falling on the metal–organic framework from the microscope are known as primary electrons. When primary electrons interact with the specimen, it causes scattering of electrons from the sample along with the other interactions (Fig. 9).

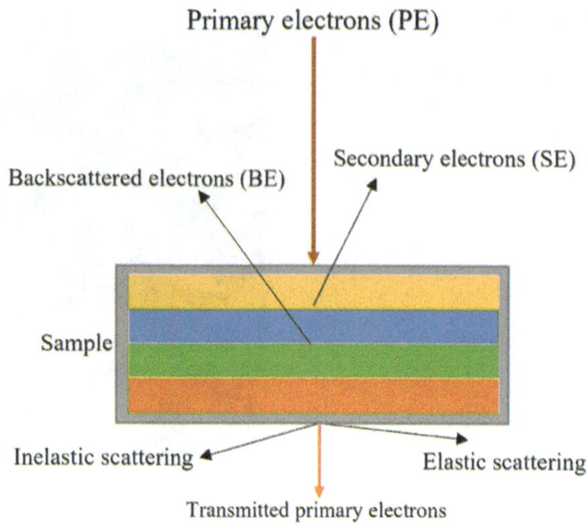

Fig. 9 Some of the valuable signals generated when a focused beam of electrons impinges on the nanoscale sample

Secondary electrons (SEs) are produced because of the inelastic scattering of electrons. This causes a transfer of kinetic energy from primary electrons to the loosely bound valence electrons on the sample surface. The SEs generated have low energy, 2–5 eV, and get deflected at low angles from the sample.

Elastic scattering occurs due to repulsion between the primary electrons and electrons in the sample. As a result, primary electrons get deflected from the specimen and are now called backscattered electrons (BEs). BEs are scattered at higher angles and are comparatively higher in energy than SEs. They are originated from deep within the sample and reach the surface due to their high energy [9].

The two most generic electron microscopes used to study the morphological details of a specimen are scanning electron microscopes (SEM) and transmission electron microscopes (TEM). While SEM is used to characterize the surface of an NMOF, TEM is primarily used for the analysis of internal structure.

Scanning Electron Microscopy

SEM is one of the most useful microscopic techniques to gain insight into the NMOF surface. The scanning electron microscope uses an electron beam to scan the sample in a raster pattern, interacting only with atoms on the sample surface. The standard electron energy of the beam is in the range of 1–30 keV; therefore, to generate a beam of higher energy, a field emission source is employed. The scattered electron and signals emitted from the sample at each location are collected and detected using an Everhart–Thornley (E–T) detector. The corresponding signal intensity regulates the pixel of the image. The entire setup needs to be kept under high vacuum pressure [10].

Scanning electron microscopy is especially valuable for studying the surface topography, composition, and properties like the electrical conductivity of an NMOF. SEM can also be useful for analyzing specific point locations on the surface of the sample due to the raster manner of the scan. SEM is usually coupled with energy dispersive X-ray (EDS) technique to help qualitatively determine the chemical composition of the sample [11].

Brief Description of the Scanning Electron Microscope

The scanning electron microscope was developed by Dr. Charles Oatley in 1948. The microscope consists of the following parts:

- electron gun and converging lenses
- coils that enable raster scanning
- a deflection system, and
- an electron detector.

A standard scanning electron microscope generates the electron beam by a thermionic process (using a heated tungsten filament). The electron beam thus generated has lower energy and is incapable of generating clear images of nanoscale substances. A field emission scanning electron microscope (FESEM) employs electro magnetization (using needle-shaped tungsten) instead to produce the electron beam. This beam has a higher electron density and is more suitable to produce higher resolution images for the particle. In addition, field emission sources are brighter and more stable [12]. A series of electromagnetic lenses are applied to shape and converge the beam onto the sample, adjust astigmatism, move the beam across the specimen to scan the NMOF, and generate images.

The magnification of the image is the ratio of the linear size of the viewing screen to the corresponding length of the specimen area being scanned. Due to the variable nature of the scannable area, magnifications of 10X to 300,000X can be achieved in FESEM [5].

An **Everhart–Thornley (E–T) detector,** a scintillator–photomultiplier combination, is used to collect SEs and BSEs. Depending on the angle of incidence of the beam with the sample, the interaction area increases and so does the intensity of electrons emitted. Thus, flat surfaces have a lower number of electrons emitted as compared to angled surfaces, ridges, slopes, and edges. The brightness of the signal is subject to the number of electrons striking the detector. Hence, the irregularities on the sample surface tend to be brighter than the flat areas. This results in 3D images with resolution < 0.5 nm [13].

By detecting the X-ray emitted by the specimen, its elemental composition can be determined. This is termed energy dispersive X-ray spectroscopy (EDS).

Sample Preparation

Due to the electronic nature of the microscopic analysis, the sample under observation must be electrically conducting and able to withstand high-pressure conditions. The

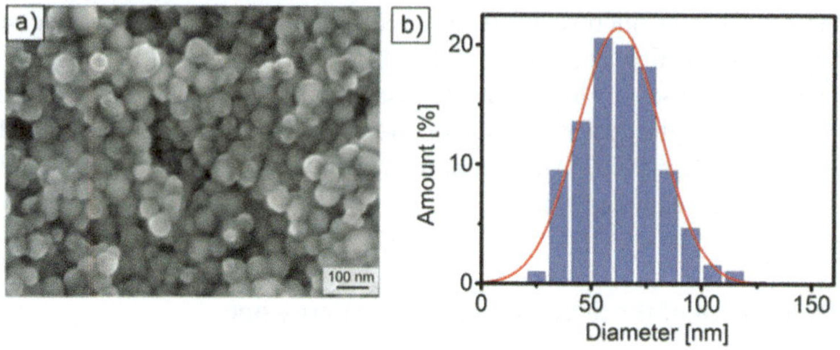

Fig. 10 **a** Microscopic image generated from scanning electron microscope (SEM); **b** Histogram of diameter measured from a 1000 NMOF particles

primary electrons are supposed to be absorbed by the specimen creating a dissipating current, the absence of which makes imaging difficult and produces faulty images.

Most NMOFs are non-conducting in nature and hence get damaged when impinged by an electron beam for a long duration. To avoid damage, NMOFs require a coating with a thin layer of conductive material like gold or platinum, deposited under high vacuum conditions [11].

SEM Analysis of Zr–Fumarate NMOFs

Hirschle et al. [14] prepared zirconium fumarate nanoscale MOFs and coated them with a thin layer of carbon to make the surface conducting. When SEM analysis was performed on the NMOF, a spherical morphology was revealed. Figure 10 shows the microscopic image generated from the microscope. To measure the particle size, the diameter of 1000 NMOFs (Fig. 11) was measured and plotted in a histogram. This histogram was fitted using a Gaussian function and the average diameter of the Zr–fum NMOF was found to be 62.0 ± 18.9 nm. Scanning electron microscopic analysis was thus used for the determination of the average particle size of Zr–fum NMOF.

Transmission Electron Microscopy

Transmission electron microscopy (TEM) is a versatile technique used for NMOF characterization. It is most widely used for the determination of NMOF crystallographic arrangement, chemical composition, and atomic dimensions. TEM imaging, diffraction patterns, and energy spectrum are all produced using transmission electron microscopes. High-resolution TEM imaging can capture the finest details like surface facets and possible defects in the crystalline structure. High-resolution TEM (HRTEM) can also measure at the nanometric scale and is the best technique for NMOF characterization [4]. TEM is beneficial to obtain high-resolution images of NMOFs. This is possible due to the presence of elastically scattered electrons originating from deep inside the sample that is highly localized. With energy loss,

Fig. 11 A 1000 NMOF particles for size determination. Images [14] are being reused under permission of a Creative Commons License

localization increases, and TEM can filter out electrons providing lattice resolution of up to 0.3–0.5 nm [15].

TEM presents high magnification of images, ranging from 50 to 106. Both image and diffraction patterns can be detected using TEM, which is crucial for determining NMOF crystallinity.

Despite the effectiveness of TEM for MOF characterization, there is a paucity in the number of TEM studies. This is due to the instability of NMOFs toward the electron beam used in the microscope.

Working Principle of TEM

In HRTEM, a beam of electrons is transmitted through a thin specimen. This results in interactions between the primary beam and the electrons of the sample causing elastic and inelastic scattering. The scattered electrons are then transmitted through imaging lenses (objective and magnifying). The enlarged image is finally focused onto a screen where the images are observed. The main signal used for bright images in TEM is elastically scattered electrons due to their origin from deep inside the specimen. However, inelastically scattered electrons can be detected using electron energy loss spectroscopy (EELS) for information on the NMOF composition [16].

Brief Description of Transmission Electron Microscope

A transmission electron microscope is an electron microscope capable of producing magnified images of thin specimens, usually with a magnification of the range 10^3–10^6. The electron–optical system of the microscope consists of the following [17]:

- An **electron gun** responsible for electron beam generation.

- **Illumination System** consists of the electron gun and a couple of condenser lenses used to converge the electron beam onto the sample. It is responsible for the electron beam diameter and thus the intensity of the final image.
- **Specimen Stage** is the location of the specimen slide. It is necessary to keep the sample stable and hence produce images with better spatial resolution.
- **Imaging System** comprises lenses responsible for magnification and spatial resolution. Depending on the imaging mode, TEM applications vary.
- **Screen/Computer** where the images are seen or recorded such as a fluorescent screen or photographic film layer.

Selected Area Electron Diffraction

TEM can combine diffraction patterns with various imaging techniques and is hence a superior characterization technique. The diffraction pattern of NMOFs can provide information on their crystal structure. Selected area electron diffraction (SAED) and high-resolution TEM (HRTEM) combination are employed when identifying the crystal structure and defects in an NMOF.

Transmission electron microscopes are also capable of analyzing spectroscopic information for electron transmissions. (Fig. 12).

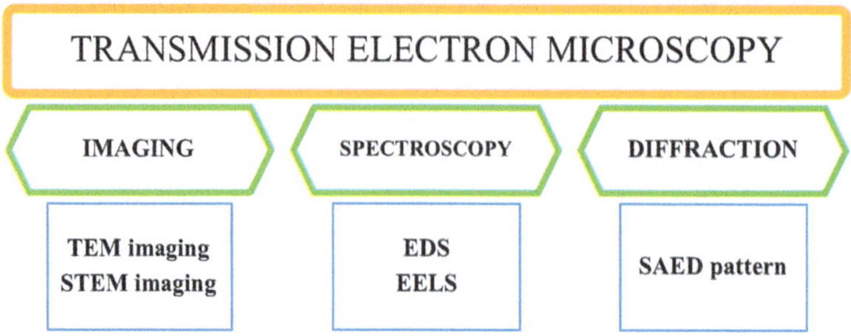

Fig. 12 Various techniques that can be used to get data from a transmission electron microscope

1. Energy-Dispersive X-Ray Spectrometry (EDS)
   An X-ray spectrum is useful for the determination of element concentration of a specific portion in the NMOF sample.
2. Electron Energy-Loss Spectrometry (EELS)
   EELS is useful for the determination of the chemical composition of the NMOF. EELS analysis is also valuable in the determination of the distribution of lighter atoms like B, C, N, and O in the sample [18].

These spectroscopies are highly localized and useful in the combination of SEM and TEM, i.e., scanning transmission electron microscopy (STEM). STEM is useful for the high spatial resolution of the chemical maps [19].

The first systematic TEM analysis was performed by Lebedev et al. [20] in 2005. It was conducted on the MIL-101(Cr) sample comprised of trimeric chromium(III) octahedral clusters interconnected by 1,4-benzenedicarboxylates. The electron diffraction (ED) patterns were recorded in multiple crystallographic orientations along with HRTEM images in [111] and [011] orientations. The HRTEM images were interpreted by comparing them with simulations. Both these data helped reveal that MIL-101 was crystalline and octahedral in shape. They also seemed to have [111]-like surface facets.

TEM Analysis of Zr–fumarate NMOFs

Hirschle et al. [2] prepared a carbon grid on which a droplet of the Zr–fum NMOF suspension had been evaporated. They analyzed this sample under a transmission electron microscope. The TEM micrograph of the sample showed the interconnection between the particles in the sample.

As shown in Fig. 13, Zr–fum NMOF particles are interconnected via necks. When the diameter of 1000 particles (Fig. 14b) was measured, the following histogram was

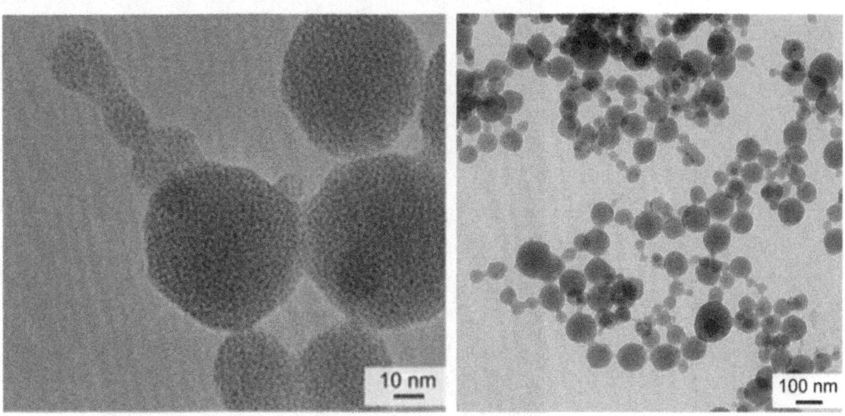

Fig. 13 TEM imaging at a resolution of **a** 10 nm, **b** 100 nm. Image [14] is being reused under permission of a Creative Commons License

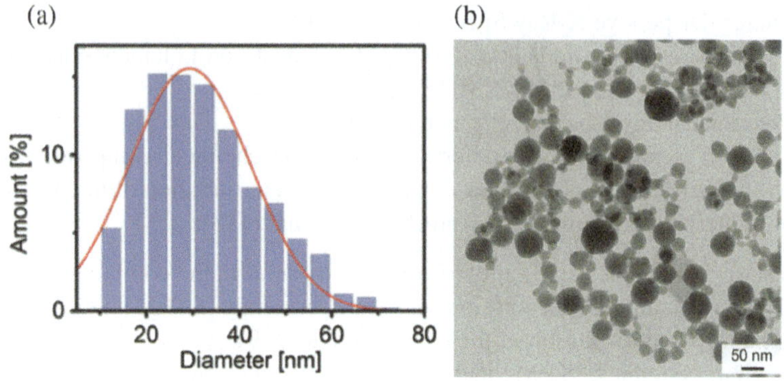

Fig. 14 **a** Histogram of diameters measure from 1000 particles, **b** Zr–fum particles measured for TEM size determination. Figure [14] is being reused under permission of a Creative Commons License

drawn (Fig. 14a). This distribution when adjusted with a curve following the normal law, the average NMOF diameter was calculated as 29 ± 12.9 nm.

The ED pattern of Zr–fum particles is shown in Fig. 15. This is helpful in determining the structure of the sample. Zr–fum is crystalline in nature as proved by the ED pattern. The radial distance in the image gives information on the lattice distance of the particle. When compared with literature values for a crystalline compound, the sample's values agree with the crystal structure. The Debye–Scherrer rings in Fig. 15 are proof enough for crystallinity.

Fig. 15 ED pattern of the Zr–fum sample. Figure [14] is being reused under permission of a Creative Commons License

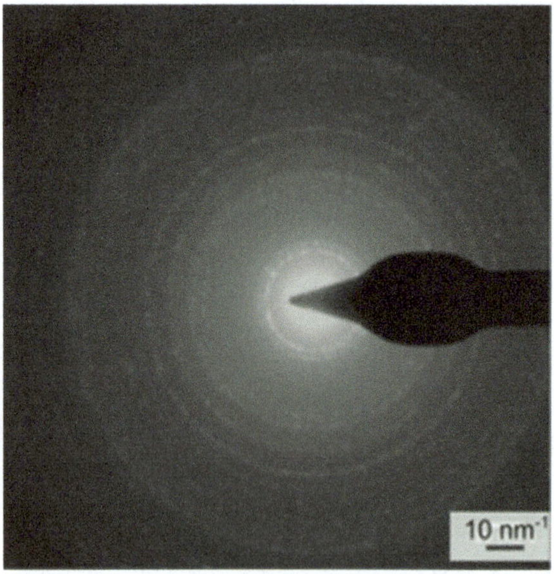

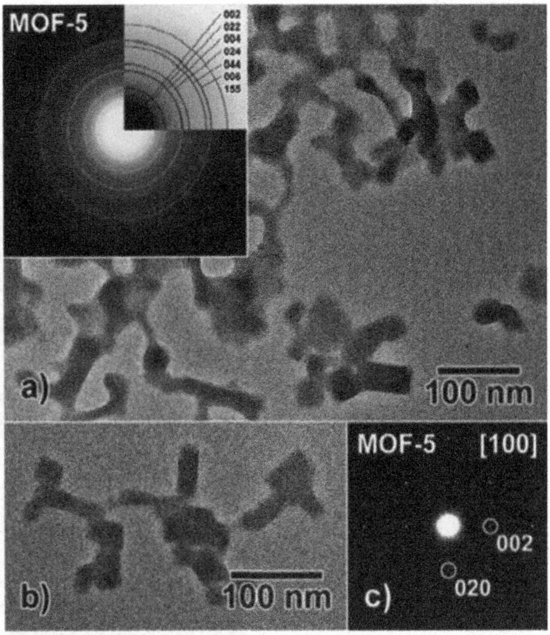

Fig. 16 **a** and **b** Low magnification BFTEM of MOF-5 crystals. The inset in (**a**) shows ED ring pattern of MOF-5 crystals. **c** SAED pattern of a single MOF-5 crystal

Wiktor et al. [21] performed TEM analysis on MOF-5 ($[Zn_4O(bdc)_3]$) at temperatures as low as -200 °C and applying very low doses of electron exposure. Since MOF-5 was found to be highly unstable, HRTEM images had to be recorded using the first electron exposure itself. TEM images showed MOF-5 to contain cuboidal crystals (Fig. 16). MOF constituents were located by conducting HRTEM simulations and comparing them with the HRTEM images generated (Fig. 17) [21].

2.3 Scanning Probe Microscopy

Scanning probe microscopy (SPM) is a microscopic tool commonly employed for the characterization of substances down to the atomic level. The principle of SPM is to use a sharp tip to scan the surface of a sample and since the properties at nano or atomic scale. The most important types of scanning probe microscopes are the scanning tunneling microscope (STM) and the atomic force microscope (AFM).

Scanning tunneling microscopy (STM) was the first kind of scanning probe microscopy developed. It was invented by Binnig and Rohrer in 1981. STM uses the tunneling property of current to produce images of greater resolution. Here, the scanning tip is adjusted in a way to keep the tunneling current as well as the tip–sample distance constant. However, scanning tunneling microscopy can only be conducted for conductive material due to the principle of tunneling current. Atomic force microscopy (AFM) developed in 1985 by Binnig, Quate, and Gerbe resolves

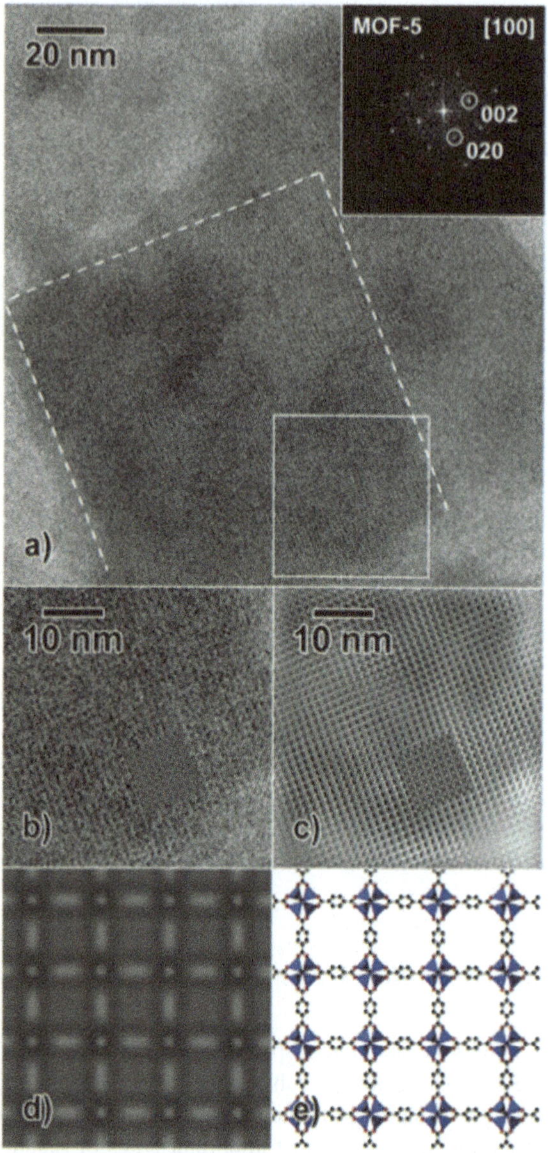

Fig. 17 **a** HRTEM images of MOF-5. **b** Magnified image of area marked in (**a**). **c** Further magnified and filtered images of the same area. **d** Magnified and noise-free version of the same image. **e** Crystal structure of MOF-5. Reprinted with permission [21] Copyright (2012) Elsevier

this disadvantage of STM. AFM, alternatively known as scanning force microscopy (SFM), can also be used on non-conducting/insulating samples. In AFM, the sharp tip is attached over a cantilever with is oscillated over the sample. The forces between the tip and the surface of the sample are sensed by the deflection of the cantilever which can be detected.

SPM is highly important for molecular imaging to elucidate the arrangements in arrays, orientations, and intramolecular structures of molecules. SPM can be used for substances in a high vacuum as well as in a solution [22].

2.4 Scanning Tunneling Microscopy (STM)

Scanning tunneling microscopy is one of the most advanced microscopic techniques for the atomic characterization of NMOFs. In STM, a very sharp metal wire tip is used to raster scan the sample surface. Once the tip is brought closer to the sample surface and a voltage is applied, the electric circuit is completed and one can image the sample surface down to the atomic scale, i.e., individual atoms. Scanning tunneling microscopy works on the principle of tunneling effect and piezoelectric effect.

Tunneling effect is a quantum mechanical effect allowing us to produce images of the specimen. This effect is based on the wave nature of electrons. The **piezoelectric effect** ensures high resolution of the images and precision.

A feedback loop is maintained to monitor the tunneling current being produced. The current coming from between the tip and the surface helps analyze the electron density and topography of the sample beneath the tip (Fig. 18). However, scanning tunneling microscopy has its disadvantages. Being surface sensitive, it requires a clean surface to operate and hence has a low success rate as compared to TEM. In addition, scanning tunneling microscopy can only detect conductive samples, thereby making it difficult to characterize insulating NMOFs. But scanning tunneling microscopy is preferred to explore the structures of conductive materials like carbon nanotubes and graphene.

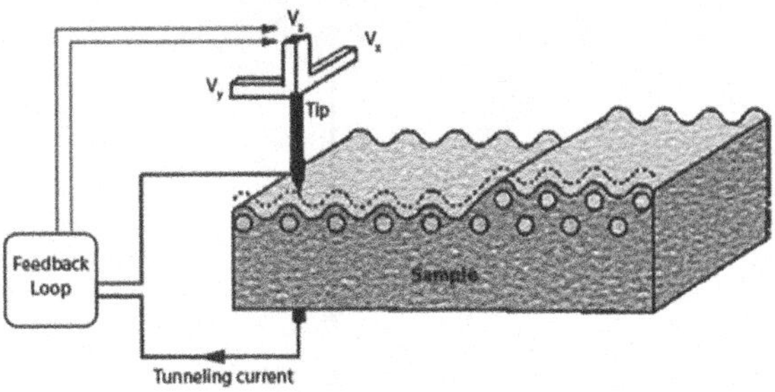

Fig. 18 Schematic diagram of an STM [23]

2.5 *Atomic Force Microscopy (AFM)*

Atomic force microscopy measures the forces between the sharp tip and sample to obtain the topography of the sample. The AFM probes the sample with a sharp tip located at the free end of a cantilever. The tip used has a diameter of < 100 Å and the cantilever can be as long in length as 500 μm. Unlike scanning tunneling microscopy, in AFM, the tip and the sample are in direct contact, therefore, instead of tunneling current, interactive forces between the tip and the sample are measured.

The forces between the tip and the sample cause the deflection of the cantilever. These deflections are then measured by the detector as and when they happen. The detection is carried out using a laser beam and a split photodiode. These signals are further used by a computer to generate a map of the surface topography (Fig. 19) [7].

Van der Waals' forces of attraction play the most important role in the cantilever deflection. The forces are at the nanoscale level, measuring < 1 nanonewton. When the distance is > 0.5 nm, the forces are attractive in nature (Van der Waals) and at shorter ranges, become repulsive. Owing to this, atomic force microscopy can be employed for the study of insulators and semiconductors as well as electrically conductive materials.

When the tip is further away from the surface, the forces between the two are negligible. In contrast, at very small distances, a repulsive force is observed between the tip and the surface due to the repulsion of electrons (Fig. 20).

The ideal distance as shown in the figure is between points 1 and 2. At this distance, the sharp tip, and the surface experience attractive forces, mainly Van der Waals. Attractive forces are negative. This distance must be maintained in order to conduct atomic force microscopy [7].

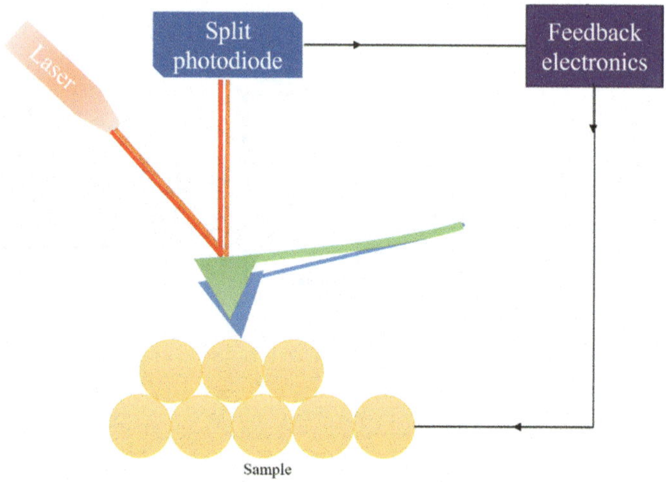

Fig. 19 Schematics of atomic force microscopy

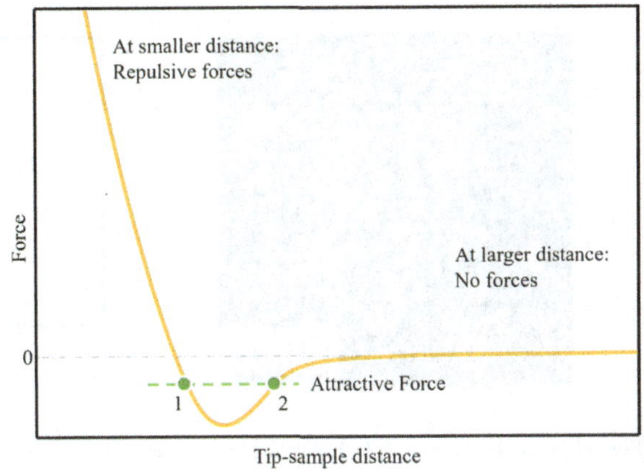

Fig. 20 Forces between the tip and the sample surface at varying distances

Atomic force microscopy has its own advantages over other microscopic methods. One of its main advantages is its ability to characterize any material, be it conductive, insulating or in solution. Magnetic materials can also be characterized using AFM unlike in SEM and TEM. [6]

Imaging Modes

Depending on the scanning pattern of the cantilever on the sample, there are three imaging modes in AFM.

(a) **Contact Mode (C-AFM)**

In contact mode AFM, the tip makes physical contact with the sample surface. As the tip gently scans the surface of the sample, the Van der Waals forces cause the cantilever to bend or deflected in accordance with the topographical changes. (Howland and [24] Here, the motion of the scanner is used to generate the images.

The contact mode is the easiest method for NMOF topography imaging; however, it has the highest wear and tear rate due to friction from direct contact. This method is also more prone to tip failure.

(b) **Non-Contact Mode (NC-AFM)**

In non-contact AFM, the cantilever is vibrated to a frequency of 100–400 kHz and the tip hovers 1–10 nm over the sample, i.e., with an amplitude ranging from 10–100 Å.

The image is generated by detecting the change in the amplitude of the vibrating cantilever. The more sensitive the detector in the AFM, the better is the resolution (down to angstrom scale) of the image.

In non-contact AFM modes, the tip or sample does not undergo degradation effects. NC-AFM is also preferred over contact AFM for analyzing soft samples.

(c) **Tapping/Intermittent Contact mode (IC-AFM)**

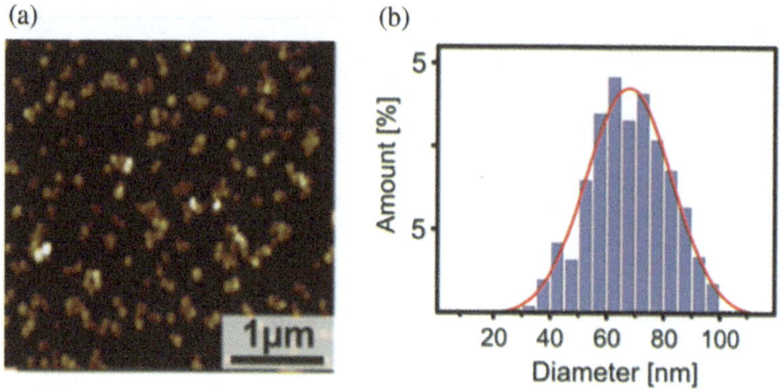

Fig. 21 **a** AFM Micrograph. **b** Particle size distribution of Zr–fum MOFs from AFM images

In tapping mode, the vibrating cantilever oscillates over the sample surface and taps/hits the sample at a constant interval of time. In comparison to NC-AFM, in the tapping mode, the cantilever oscillates with a higher amplitude. This leads to a greater deflection and hence easier signal detection. The tapping mode is like the non-contact mode, except the cantilever just taps the sample and makes no contact further. The life of the tip in the IC-AFM mode is less due to the tapping nature of the tip. It is more prone to wear and tear and hence has a smaller life.

AFM Imaging of Zr–Fum NMOFs

Hirchle et al. [14] performed a closed loop IC-AFM in air on Zr–fum NMOFs. The sample was prepared by making an ethanolic dispersion of the NMOF and drying it on a SiO_2 slide. The cantilever was excited close to its resonance frequency with the help of a piezoelectric device. With changing distance, the forces (Van der Waals attractive forces and repulsive Coulomb forces) between the tip and sample varied, leading to a change in the amplitude of the cantilever vibrations. The feedback loop maintained a constant oscillation amplitude. The topographical image was produced by detecting the variations in the amplitude of the vibrations.

The AFM micrograph and height distribution obtained from scanning probe imaging process is shown in Fig. 21. The average diameter of the Zr–fum NMOFs was calculated to be 68 nm with a standard deviation of 15 nm (Fig. 22).

3 Spectroscopic Characterizations

Spectroscopic characterizations play a significant role in knowing the constituents present in each material. When electromagnetic radiations (light) interact with matter, different phenomenon could take place. The interaction results in either the absorption of photons from light, by the material or the promotion of an electron to a higher

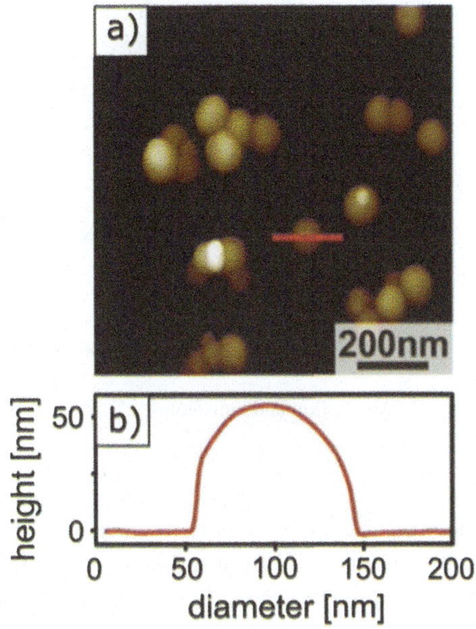

Fig. 22 **a** Zoomed in AFM micrograph. **b** Cross section of a single particle. Images [14] are being reused under permission of a Creative Commons License

level after acquiring energy from the incident photon, and consequently releasing excess energy in the form of heat [25] or by emitting a photon spontaneously or non-spontaneously. The spontaneous energy emission leads to a scattering phenomenon, where the emitted photon is identical to the absorbed photon except for its direction of propagation [26] or Raman scattering if energy is released spontaneously but in two stages. Here, two photons at a longer wavelength are released per photon absorbed. The energy can also be released after some relaxation times that may vary from nanoseconds to hours and is non-spontaneous then the phenomenon is called fluorescence. The absorption or emission requires the difference in absolute energies between the final and the initial state as $\Delta E = E_{\text{final}} - E_{\text{initial}}$ (Fig. 23).

The different phenomena give rise to different spectra that generally require:

1. a source of light
2. a disperser to separate light based on wavelength
3. a detector to detect the light after dispersion.

The instrument consisting of these three parts is called a spectrometer, and the spectrum obtained from it is unique to the element, that is, each spectrum is the fingerprint of each element.

The light consists of photons that carry discrete amounts of energy called quanta. Absorption of a photon by an atom or molecule could take place in any region of the EMR spectrum and depending on which part of the spectrum is absorption taking place, different structural information can be obtained from the spectra; for example, absorption of a photon in UV or visible have sufficient energy to excite the electrons and hence reveals more about the electronic structure of a molecule.

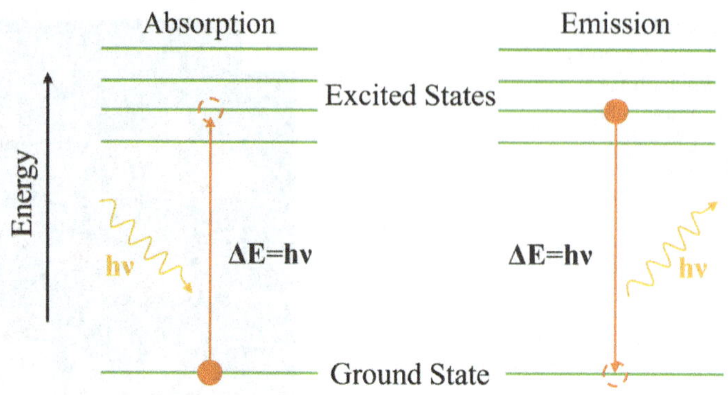

Fig. 23 Absorption and emission

However, lower energy IR radiations do not possess sufficient energy to excite the electrons but are sufficient to cause the vibrations in a molecule; therefore, the IR spectrum obtained from this would give insights into the types of chemical bonds present in the molecule. Other spectroscopies based on absorption phenomena such as nuclear magnetic resonance (NMR) spectroscopy, X-ray diffraction and those based on scattering phenomena, diffuse reflectance spectroscopy, X-ray crystallography, and Raman scattering are equally significant in analytical applications. The emission spectroscopy is also an important tool by which the elemental and chemical composition of MOFs can be determined. In this technique, the frequency of photons emitted by atoms or molecules is examined when an electron jumps from a higher energy level to a lower energy level. The energy of the emitted photon is equal to the energy difference between the two levels. Depending upon the type of source that is employed to excite the electrons, there are different techniques that are called as such as fluorescence spectroscopy (or fluorimetry) where fluorescence of the sample is analyzed, and the ultraviolet light is used to excite the electrons in the sample. Most of the time, radiations are emitted in the visible region of the electromagnetic spectrum. Another technique is inductively coupled plasma optical emission spectrometry (ICP-OES) which is used to determine the elemental composition of MOFs. This employs inductively coupled plasma to produce excited atoms and ions that give off radiations at a wavelength characteristic of a specific element. Energy-dispersive X-ray spectroscopy (EDS or EDX) uses the source that emits the X-rays from the sample, and the energy of the emitted X-rays is characteristic of the difference in energy between the shells and hence of a particular atom. X-ray fluorescence (XRF) is closely related to the EDS technique but is more sensitive in terms of reading and reporting concentration levels in the parts per million (ppm) range. In this high energy, X-rays are being used for excitation that give off secondary or fluorescent X-rays from the sample. The spectral data obtained by employing these spectral techniques and the corresponding spectra that we obtain is revealing different information about the molecule. To have complete information about the structure and composition of the material, all spectroscopic characterizations play a vital role.

We would be discussing each in detail with reference to metal–organic frameworks which are loaded with numerous functional groups and metal ions.

3.1 NMR Spectroscopy

The most powerful spectroscopic technique used to characterize NMOFs is solid-state NMR (SSNMR). It is extremely useful for the study of NMOF dynamics and how the adsorbed molecules react once inside the framework. Thus, solid-state NMR is an asset in the structure-to-function relationship deduction of NMOFs. This chapter will investigate the application of SSNMR for studying the absorbate interactions inside the NMOF pores [27].

Brief Description of Solid-State NMR

NMR spectroscopy is dependent on the magnetic property of atomic nuclei. Those atomic nuclei with both nonzero electric quadrupole and electric dipole can detect slight changes in the local electric and magnetic field. The local magnetic field around the nucleus depends largely on the electron dipole surrounding it. The external magnetic field is generated using a superconducting magnet. This induces an electric current in the surrounding electronic cloud of the nucleus, which results in a local magnetic field shielding the external one, i.e., chemical shift and shielding. This is the reason why NMR spectra of the nucleus largely depend on its neighboring atoms in the first and second coordination, their nature, bond strength, and bond angle.

Dipole interactions (dipolar coupling) from neighboring nuclei having non-zero magnetic moments are the second-important interaction that contributes to the local magnetic field of the main atomic nucleus. Due to the non-zero nature of the magnetic moment, those atomic nuclei also act like tiny magnets generating a magnetic field of their own. The strength of this interaction decreases with an increase in distance thereby providing information on interatomic distances. The local electric field around the nucleus depends upon the electric charges present around it and their arrangement. This electric field gradient is detected only for non-spherical nuclei. These are termed quadrupolar nuclei and have spin quantum numbers $> \frac{1}{2}$. This sensitivity can help give information on the symmetry of their environment. Measuring the strength of the quadrupolar interaction can also provide information on the dynamics of the species of which the atomic nucleus is a part. In this case, the dynamics of the metal–organic framework.

The factors on which an NMR spectrum depends are shown in Fig. 24.

The chemical shift helps distinguish environments around the atomic nucleus, i.e., between CH_4, CH_3OH due to the different environments around ^{13}C nuclei. Dipolar coupling enables signal enhancement as well as distance measurement, while quadrupole coupling is useful for the detection of symmetry in the environment.

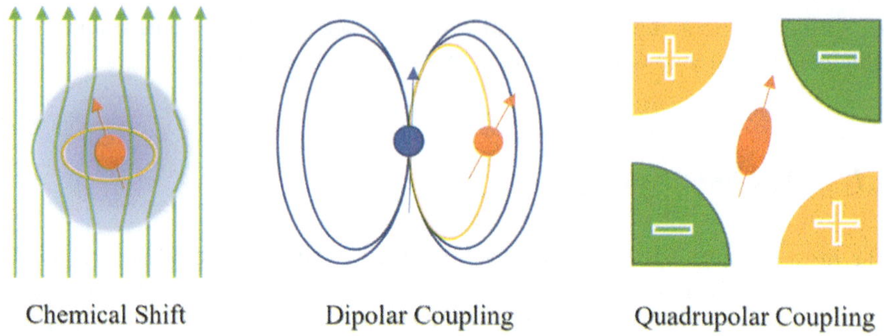

Fig. 24 Factors that influence NMR spectroscopy

Fig. 25 Showing NMR active nuclei in NMOFs

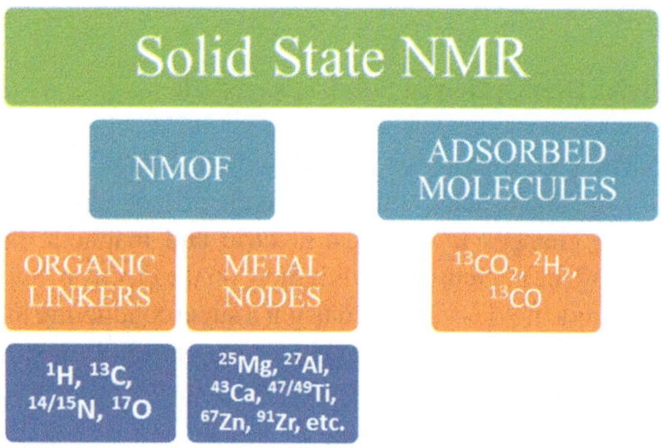

Nanoscale metal–organic frameworks can include several NMR active nuclei, some of which are shown in Fig. 25.

Measuring ^{13}C and ^{1}H NMR spectra is useful for determining the organic linker and attached functional groups present in the NMOF. Solid-state NMR can give an insight into the dynamics of the linkers as well. Aluminum is by far the most studied metal center in NMOFs. Carrying out ^{27}Al MAS NMR can help us get important information on the framework. Quite often, NMOFs may contain paramagnetic metal centers making NMR spectroscopy a little difficult. Despite having magnetic moments, the unpaired electrons have very strong interactions with the nuclei, and hence, NMR spectra cannot be measured. However, ^{13}C and ^{1}H NMR spectra can be measured from the linkers without any difficulty.

Solid-state NMR is extremely useful for studying molecules adsorbed within the NMOF pores. The most common adsorbed species studied is water since it can give important insight into the NMOFs stability.

Benzaqui et al. [28] used ^{27}Al MAS NMR and 2D double-quantum–single-quantum (DQ–SQ) MAS NMR to study the effect on Al environment upon hydration and dehydration of MIL-96. The changes in the Al3 site of the metal–organic framework were denoted by the splitting of the Al3 signal (Al3 and Al3') in the spectrum

(Fig. 26). The ^{27}Al chemical shifts and the quadrupole coupling constants were in agreement with the density functional theory (DFT) calculations that provided a

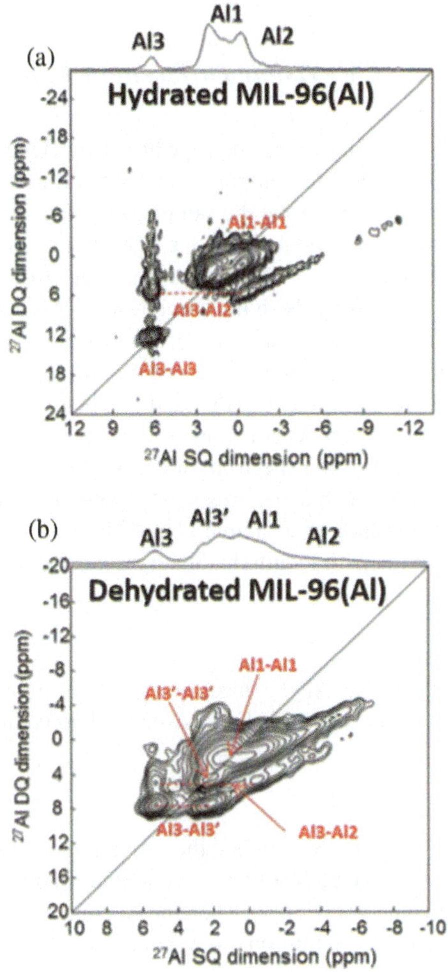

Fig. 26 ^{27}Al DQ–SQ spectrum of MIL-96 when **a** Hydrated; **b** Dehydrated. Reprinted with permission [28]. Copyright (2017) American Chemical Society

structural model of the interactions between water molecules and MIL-96. Therefore, SSNMR can be used to detect highly subtle structural changes in the NMOF and can provide valuable information when used in combination with DFT [29].

3.2 UV–Visible Differential Reflectance Spectroscopy

Differential reflectance spectroscopy (DRS) is a surface analytical technique used for solid-state materials. It uses the ultraviolet (UV) range of the electromagnetic spectrum to probe the sample. UV–visible spectroscopy is most frequently used to study the optical properties of NMOFs. Since NMOFs contain organic linkers with unsaturated bonds, they have a π-electron cloud with high conjugation and hence is UV active. NMOFs can absorb light in the UV region and provide a spectrum. This spectrum aids in the calculation of the maximum wavelength at which the absorbance is most. The maximum wavelength is useful for further carrying out photoluminescence (PL) and to obtain the PL emission spectrum. This technique uses the UV–visible part of the electromagnetic spectrum, 200–800 nm, to record the absorption spectrum of the NMOF sample.

UV–visible spectroscopy follows Beer Lambert's law. This law states that the absorbance of a sample is directly proportional to the concentration of the sample and the path length that the light must travel.

$$A\alpha(b)(c)$$

where A = Absorbance; b = length of light path and c = concentration.

The proportionality constant is known as molar absorptivity (ε).

$$A = \varepsilon bc$$

Li et al. [30] studied the activation process of Ni-doped MOF-5 by observing the color change. MOF-5 is white and gradually turns green upon nickel doping. The green color deepens when the Ni content increases in the sample. The sample turned purple upon introducing vacuum, i.e., degassing. DRUV-vis spectra were performed to confirm the color change thus observed. (Fig. 27) The Ni-doped MOF spectrum showed 2 peaks at 420 and 720 nm which resembled the literature Ni(II) values in an octahedrally coordinated system. The degassed MOF-5 exhibited 3 bands in the UV spectrum and was purple in color. Upon introducing air back in the environment, the original color and spectrum were observed, proving the possibility of reversible change in the Ni(II) coordination environment.

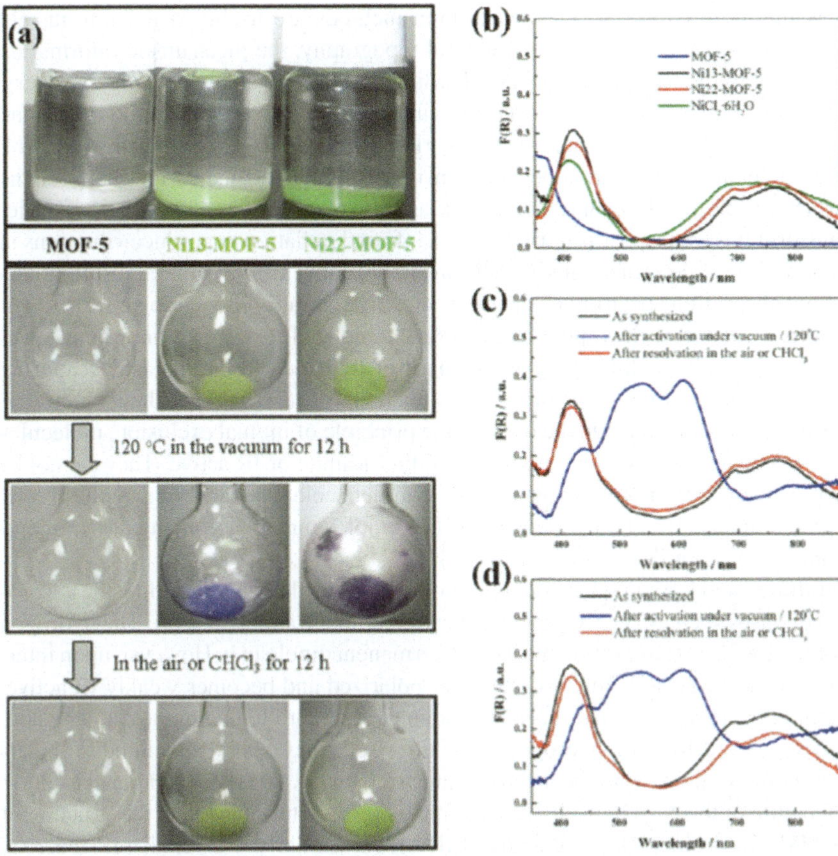

Fig. 27 **a** Color changes observed in the specimen upon Ni-doping and introducing different environments, **b** DRUV-vis spectra of MOF-5, Ni13-MOF-5, and Ni22-MOF-5. **c** DRUV-vis spectra of Ni13-MOF-5. **d** DRUV-vis spectra of Ni22-MOF-5. Reprinted with permission [30]. Copyright (2012) American Chemical Society

3.3 *Infrared and Raman Spectroscopy*

Infrared and Raman spectroscopy involves the interaction of radiation with molecular vibrations that yield information about the molecular structure. These two techniques are part of vibrational spectroscopy that provide complementary information about the nature and strength of chemical bonds present in the molecule. They differ in the way they transfer photon energy to the molecule by changing its vibrational state [31]. IR spectroscopy measures absolute frequencies at which a molecule absorbs radiation, and it responds to the change in the dipole moment while Raman spectroscopy responds to the polarizability changes in the molecule and measures the relative frequencies at which a sample scatters radiation.

While the microscopic characterization methods are useful for gaining insight into the local geometry and the structural topography, the mechanistic information pertaining to the working of an NMOF while engaging in catalysis, drug discovery, etc., is difficult to obtain. However, the vibrational spectroscopic methods, IR, and Raman are widely used to analyze the workings of an NMOF while interacting with other molecules as they are especially sensitive to detect the interactions between the surface of the metal–organic framework and guest molecules [32]. FT-IR (Fourier transform infrared) spectroscopy employs infrared radiation to produce vibrations in the sample. While some radiation gets absorbed, some are transmitted resulting in a unique absorption spectrum of the molecule. The vibration is said to be IR active if there is a change in the dipole moment of the molecule. However, in Raman spectroscopy, there is a shift in the wavelength of light due to the inelastic scattering phenomenon. A vibration is termed as Raman active if it causes a change in polarization in the molecule. On the basis of the principle of mutual exclusion, molecules having a center of symmetry are usually either Raman or IR active. They cannot be both. However, in molecules with other symmetry elements, they can either be IR active, Raman active, both, or none. Lastly, complex molecules having no symmetry elements are both IR and Raman active [33].

Infrared and Raman spectroscopic studies are crucial to understanding the adsorption of gases such as H_2, CO_2 into the metal–organic framework [32]. The hydrogen molecule is IR inactive due to the lack of permanent dipole in it. However, upon interaction with the NMOF, the molecule gets polarized and becomes weakly IR active. During the process, a redshift is observed in the H–H stretching modes. Nijem et al. [34] suggested that the hydrogen stretching frequency shifts' magnitudes and its nature of interaction with the metal–organic framework largely depend on the chemical nature of the NMOF itself. The organic linkers, metal center, and the structure of the NMOF dominate the magnitude of IR shifts. Similarly, CO_2 is also IR-inactive, being a linear molecule and having a center of symmetry, and when it interacts with the MOF framework, the adsorption band changes and becomes IR active. Depending upon the nature of metal ions present in the framework, various red shifts and blue shifts are observed.

A lot of studies have been carried out in exploring MOFs as adsorption materials for H_2 and CO_2; however, some of the MOF's are found to lose their stability in the presence of a small amount of water. Tan et al. [35] investigated the behavior of metal–organic framework upon exposure to water vapor with the help of infrared and Raman spectroscopic studies and concluded that the nature of central metal ion in MOF plays important role in stability and decomposition pathways in humid environments. Different vibrational spectra of the frameworks $M(bdc)(ted)_{0.5}$ [M = Cu, Zn, Ni, Co; bdc = 1,4-benzenedicarboxylate; ted = triethylenediamine] were recorded in controlled humidity to observe the structural changes in the MOF upon gas uptake.

The hydrated spectra of different MOF's were found to have distinct bands as depicted in Fig. 28. The stretching frequency of the ligand, bdc in MOF framework in all the samples yield information about the different responses of MOF to hydration, that is, it redshifts to 1010 cm^{-1} when M = Ni, Co and blue shifts to 1027 cm^{-1}

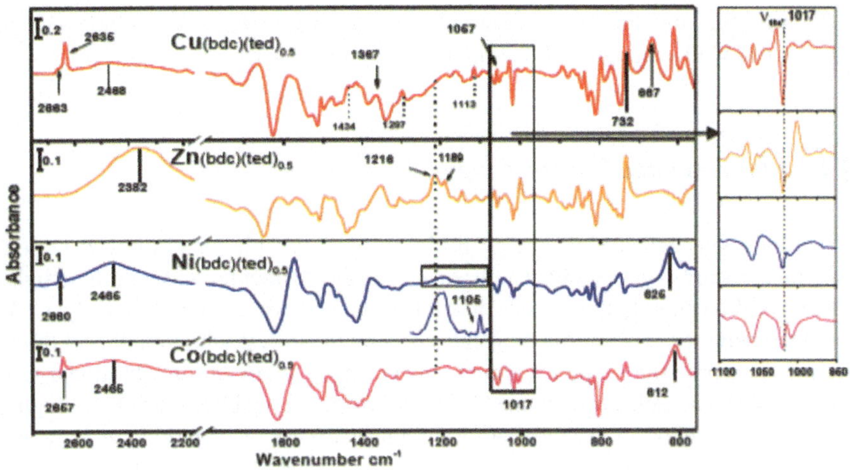

Fig. 28 IR adsorption spectra of hydrated M(bdc)(ted)$_{0.5}$, referenced to activated MOF in a vacuum after the introduction of 9.5 Torr D_2O vapor and evacuation of gas phase D_2O vapor. Reprinted with permission [35]. Copyright (2012) American Chemical Society

when M = Cu. However, Zn(bdc)(ted)$_{0.5}$, observed to undergo a larger red shift to 1001 cm^{-1}. These observations were applied for CO_2 adsorption and dehydration processes in MIL-53 framework as reported in the literature.

Nijem et al. [34] investigated the interaction of CO_2 with the MOF framework and observed C–C inter-ring of the bpdc ligand in $Zn_2(bpdc)_2(bpee)$ gets weakened. The Raman spectra in Fig. 29 supported this observation by depicting red shift in

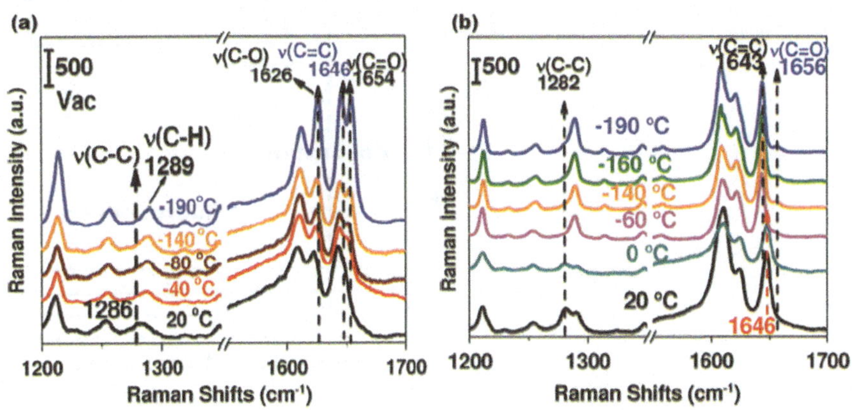

Fig. 29 Raman spectra of activated $Zn_2(bpdc)_2(bpee)$ (**a**) in a vacuum and **b** after the introduction of 1 atm of CO_2 at room temperature. Reprinted with permission [34]. Copyright (2011) American Chemical Society

C–C inter-ring stretching in the ligand by approximately 3 cm^{-1} and blue shift in coordinated C-O symmetric stretching by approximately 10 cm^{-1}.

3.4 Fluorescence Spectroscopy

When the molecules absorb energy in the infrared, visible, or ultraviolet region, outermost electrons get excited which when returning to their ground state, emit light, which is called photoluminescence, and this is the basis of fluorometric analysis. At room temperature, most of the molecules occupy the lowest vibrational level of the ground electronic state. The excitation may occur up to any of the vibrational sub-levels within each electronic state. Once the molecule reaches one of the higher vibrational levels of an excited state it dissipates excess of vibrational energy and falls to the lowest vibrational level of the excited state [36]. The further dropping down to any of the vibrational levels of the ground state emits energy in the form of fluorescence (Fig. 30).

As far as Metal–organic frameworks are concerned they are generally reported to be luminescent, owing to the presence of various aromatic or conjugated groups in the organic linkers [37]. Hao and Yan [38] developed a luminescent lanthanide-organic framework to capture toxic and carcinogenic cadmium ions that are being widely used in industrial and agricultural fields. They encapsulated Eu^{3+} cations into the pores of luminescent Ln-MOF and recorded its emission spectrum that confirmed the successful encapsulation of Eu^{3+} inside the framework. In this way,

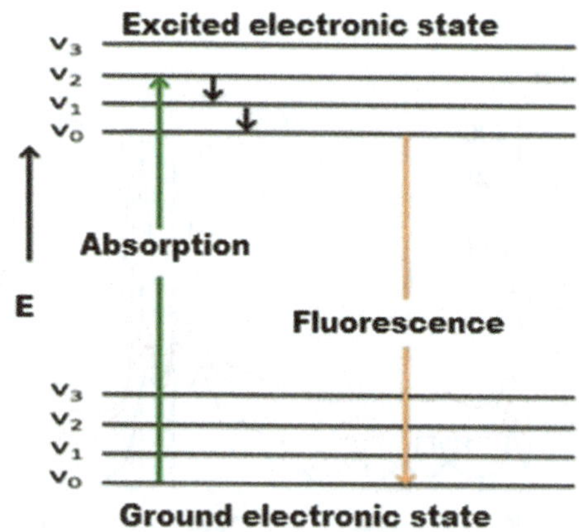

Fig. 30 Phenomenon of Fluorescence as explained by Jablonski diagram

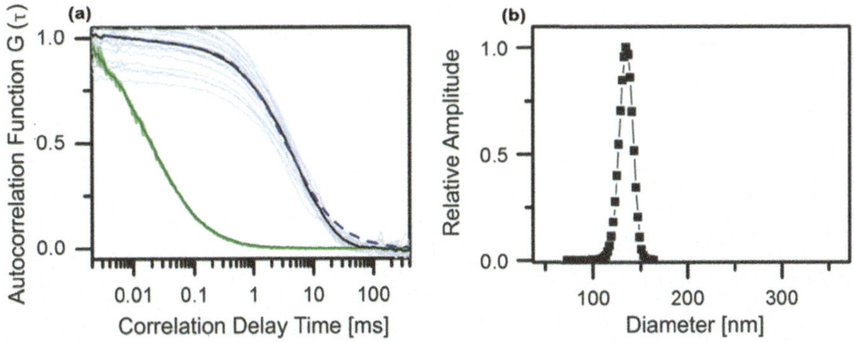

Fig. 31 FCS autocorrelation curves of **a** the free dye (green) and Zr-fum NMOFs in water (black), **b** GDM curve peaking at 135 nm representing average NMOF size. Figures are being reused under permission of a Creative Commons License

3.5 *Fluorescence Correlation Spectroscopy*

Fluorescence correlation spectroscopy is a statistical tool that gives the concentration and size of the particles by measuring the fluorescence intensity of the fluctuating particles. Due to the Brownian motion of the particles, there is continuous diffusion of particles or molecules in and out of the tiny observed volume that results in the fluctuations of fluorescent particles and hence variable fluorescence intensity [39]. In this technique, the sample is being focused on by laser and fluctuating fluorescence intensity is recorded using a photodiode which is then used to calculate the time autocorrelation function. Further, the use of Stokes–Einstein relation gives the hydrodynamic diameter of the particles. Hirschle et al. used Fluorescence Correlation spectroscopy (FCS) to determine the hydrodynamic diameter of Zr-fum MOF NP's. The three samples of Zr-fum-MOF NP's were labeled with dye Alex Flour 488 and examined with FCS (Fig. 31).

The autocorrelation function of Zr-fum MOF NP is shifted toward higher correlation times with respect to the free dye that shows the slower diffusion and hence larger hydrodynamic diameter of MOF NP's as compared to free dye. Using the Gaussian Distribution Model (GDM), the hydrodynamic diameter was calculated as 135 nm.

3.6 *Dynamic Light Scattering*

Dynamic light scattering (DLS) is another tool that is used to measure the hydrodynamic diameter of particles like FCS. Here, the intensity fluctuations are measured, which are caused by interference of laser light that is scattered by diffusing particles. The fluctuations are observed due to the particles undergoing Brownian motion and so there is constant change in the distance between the scattered particles with time.

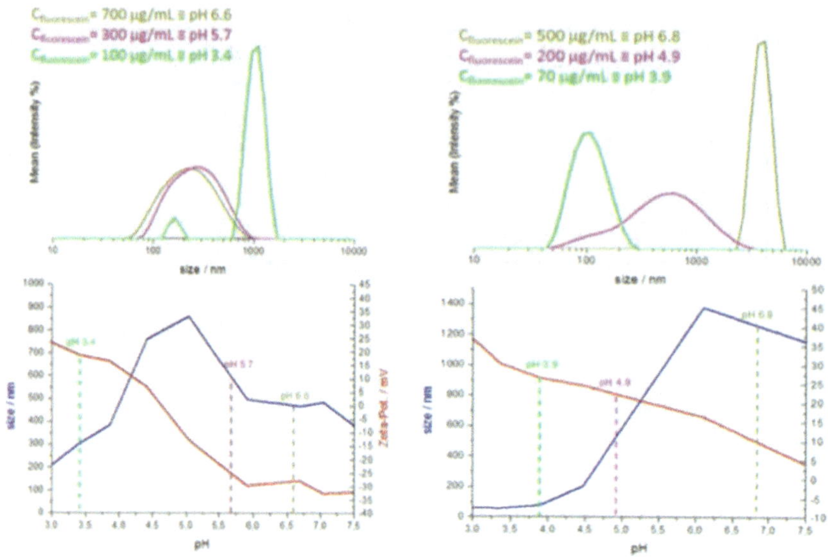

(a) MIL-101(Cr) (b) MIL-100(Fe)

Fig. 32 DLS (blue) and zeta-potential (red) measurements of **a** MIL-101(Cr) and **b** MIL-100(Fe). The figure is being reused under an open access Creative Common CC BY license [40]

It is then correlated to the diffusion coefficient of particles that gives the size of the particle after plotting a second-order autocorrelation function.

Preiß et al. [40] studied the uptake and release of fluorescein into the pores of MOF framework and carried out DLS studies on two representatives of MOF NP's, MIL-101 (Cr) and MIL-100 (Fe), and obtained data on their diffusive behaviors. The data was being analyzed to calculate the hydrodynamic diameter of these two MOF NP's. Further, to study the pH dependency of effective particle size in a well-defined system the DLS was performed with concurrent zeta potential experiments (Fig. 32).

The calculated size of MIL-100(Fe) and MIL-101 (Cr) to be about 200 nm and 50 nm respectively. It was also observed that the size of MIL-100 (Fe) varies with the concentration of fluorescein that indicated another observation that NPs tend to aggregate and depend on pH and about 50 nm for MIL-101(Cr). Thus, the finding that particle size is pH-dependent was taken into consideration in their theoretical model.

Hirschle et al. [14] calculated the hydrodynamic diameter of Zr-fum MOF NP's in water and ethanol (Fig. 33) and found it comparable to that obtained by FCS measurements. The hydrodynamic diameter of Zr-fum NPs in water and ethanol was calculated by DLS to be 42 nm and 130 nm respectively.

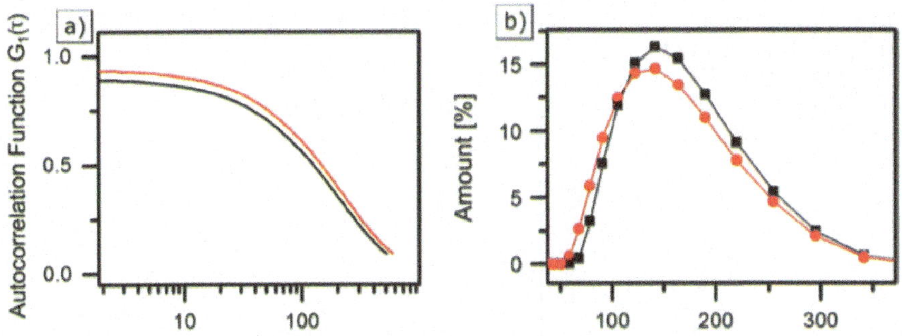

Fig. 33 DLS correlation data (**a**) and size distribution (**b**) of Zr-fum MOF NPs in ethanol (red) and water (black). Figure [14] is being reused under permission of a Creative Commons License

3.7 Inductively Coupled Plasma Optical Emission Spectrometry (ICP-OES)

This technique uses emission spectra to analyze the elemental composition of a sample.

However, for solid MOF's its use is restricted because it is necessary to ensure the complete dissolution of a sample which is always not so easy even after treating MOF with nitric acid, sulphuric acid, or even hydrofluoric digestion. That's why the other technologies that analyze the elemental composition are widely used in comparison to ICP-OES.

3.8 Photoluminescence Spectroscopy

Photoluminescence spectroscopy has found its wide applicability in the characterization of complex molecules in biochemistry as well as in characterizing optoelectronic properties of semiconductors. The material absorbs radiation when excited by a light from a xenon lamp and dissipates the excess of energy through the process, called Photoluminescence. This gives the energy difference between the orbitals (HOMO–LUMO gap in molecules) or bands (bandgap in semiconductors) [41]. Photoluminescence spectroscopy is an informative tool regarding defects present in semiconductors. Sofi et al. [42] studied the photoluminescence response of two heterostructures, HKUST-1 and Ag/Ag_3PO_4/HKUST-1 in order to photocatalytic activity of the MOF's. The PL spectra of HKUST-1 and Ag/Ag_3PO_4/HKUST-1 are depicted in Fig. 34 that shows that the less intense band in the latter indicating the decrease in the electron–hole recombination due to the significant separation of charge carriers.

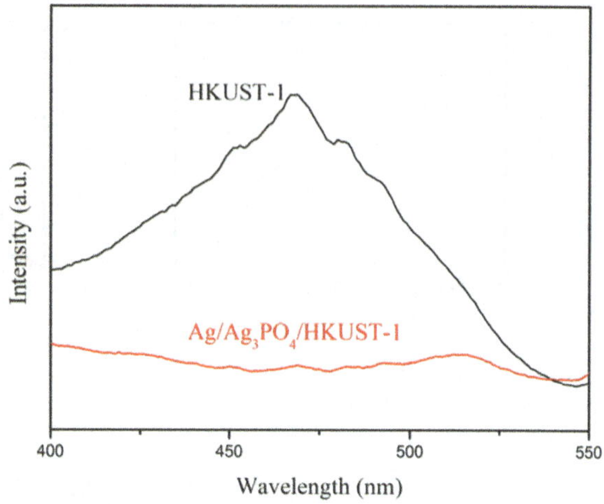

Fig. 34 PL spectra of (**a**) Pristine HKUST-1 and (**b**) Ag/Ag3PO4/HKUST-1 catalyst. Reprinted with permission [42] Copyright (2020) Elsevier

3.9 Powder X-Ray Diffraction (PXRD)

The powder X-Ray Diffraction method is used in the study of NMOF to analyze its crystallinity and structural parameters. PXRD aids in the determination of the NMOF's crystalline structure, the crystallinity percentage, and further identification of lattice parameters, unit cell size, and crystallite size.

PXRD employs an incident X-ray beam that interacts with the powdered sample to obtain a diffraction pattern. X-rays are produced in a cathode tube by heating electrons, which are then accelerated toward the target (Cu is the most commonly used target material) by applying a voltage. This results in the target being bombarded with electrons and the production of X-Rays. These X-rays are collimated, passed through a monochromator, and then directed toward the powdered sample. (Fig. 35) [12].

If the specimen under observation is crystalline, sharp, well-defined, significant, and narrow peaks are observed in the diffraction pattern. Amorphous compounds on the other hand give noise signals and bumpy peaks. Upon identifying the peaks in the diffraction pattern, the crystallite site is calculated using Scherrer's equation.

Venna et al. used a Bruker D8-Discover diffractometer with Cu Kα radiation to record PXRD patterns of ZIF-8 to observe its formation kinetics. The diffraction pattern shows the gradual increase in crystallinity of the sample. The crystallinity slowly increases in the initial stages and then shows a rapid increase after 30–40 min. After 50 min, the crystallinity doesn't increase and remains practically constant (>90%). 100% crystallinity is reached after 24 h as shown in Fig. 36 (Venna, Jasinski, and Carreon 2010).

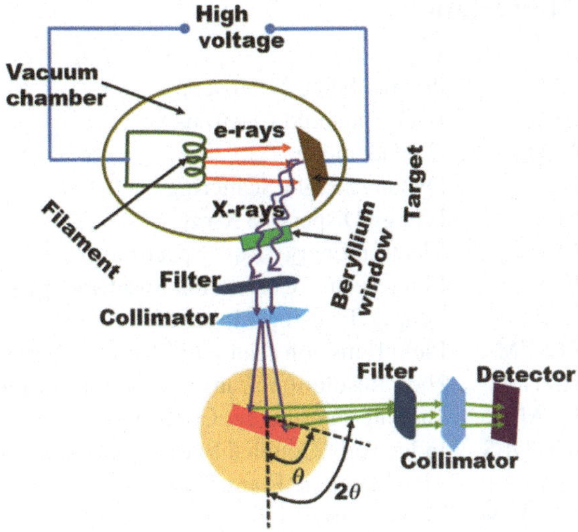

Fig. 35 Schematic diagram of PXRD technique. Reprinted with permission [12] Copyright (2017) Elsevier

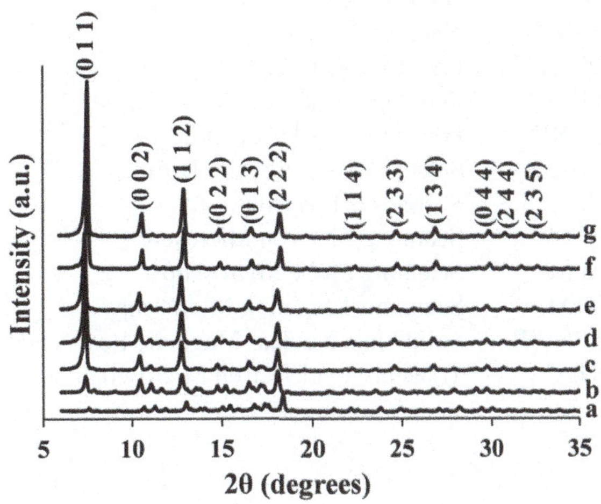

Fig. 36 PXRD patterns of formation of ZIF-8 as a function of time. **a** 20 min; **b** 30 min; **c** 40 min; **d** 50 min; **e** 1 h; **f** 12 h; **g** 24 h. Reprinted with permission from Venna et al. (2010). Copyright (2010) Journal of American Chemical Society

In addition to PXRD, various other techniques also employ X-Rays for the characterization of NMOFs. X-Ray absorption fine structure (XAFS) is used for the analysis of the number of atoms in a particular local environment and their type. X-Ray absorption near-edge structure (XANES) provides data on the bond angles, valence state, and energy bandwidth of the molecule [41].

Abbreviations

AFM	Atomic Force Microscopy
BE	Back Scattered Electrons
C-AFM	Contact mode
DLS	Dynamic Light Scattering
EDS	Energy Dispersive X-ray
EELS	Electron Energy Loss Spectrometry
FCS	Fluorescence Correlation Spectroscopy
FTIR	Fourier Transform Infrared
FESEM	Field Emission scanning Electron Microscopy
HRTEM	High Resolution Transmission Electron Microscopy
IC-AFM	Tapping/Intermittent Contact Mode
ICP-OES	Inductively Coupled Plasma Optical Emission Spectroscopy
MOF	Metal Organic Frameworks
NC-AFM	Non-Contact Mode
NMOF	Nanoscale Metal Organic Frameworks
NMR	Nuclear Magnetic Resonance
NP	Nanoparticles
PE	Primary Electrons
PL	Photoluminescence
PXRD	Powder X-ray Diffraction
SAED	Selected Area Electron Diffraction
SE	Secondary Electrons
SEM	Scanning Electron Microscopy
SPM	Scanning Probe Microscopy
STM	Scanning Tunneling Microscopy
SSNMR	Solid State Nuclear Magnetic Resonance
TEM	Transmission Electron Microscopy

References

1. Shi X, Shan Y, Meng D, Pang H (2021) Synthesis and application of metal-organic framework films. Coord Chem Rev 444(October):214060. https://doi.org/10.1016/J.CCR.2021.214060
2. Hirschle P, Preiß T, Auras F, Pick A, Völkner J, Valdepérez D, Witte G, Parak WJ, Rädler JO, Wuttke S (2014) From chip-in-a-lab to lab-on-a-chip: towards a single handheld electronic system for multiple application-specific lab-on-a-chip (ASLOC), vol 18, pp 4359. https://doi.org/10.1039/c6ce00198j
3. 2.3: Instruments of Microscopy-Biology LibreTexts. n.d. https://bio.libretexts.org/Bookshelves/Microbiology/Microbiology_(OpenStax)/02%3A_How_We_See_the_Invisible_World/2.03%3A_Instruments_of_Microscopy
4. Mayeen A, Shaji LK, Nair AK, Kalarikkal N (2018) Morphological characterization of nanomaterials. In: Characterization of nanomaterials. Elsevier, pp 335–364. https://doi.org/10.1016/B978-0-08-101973-3.00012-2

5. Inkson BJ (2016) Scanning electron microscopy (SEM) and transmission electron microscopy (TEM) for materials characterization. In: Materials characterization using nondestructive evaluation (NDE) methods. Elsevier Inc., pp 17–43. https://doi.org/10.1016/B978-0-08-100040-3.00002-X
6. Barhoum A, Luisa García-Betancourt M (2018) Physicochemical characterization of nanomaterials: size, morphology, optical, magnetic, and electrical properties. In: Emerging applications of nanoparticles and architectural nanostructures: current prospects and future trends. Elsevier, pp 279–304. https://doi.org/10.1016/B978-0-323-51254-1.00010-5
7. Voigtländer B (2016) Scanning probe microscopy: atomic force microscopy and scanning tunneling microscopy. MRS Bull 41(2):165–166. https://doi.org/10.1557/mrs.2016.18
8. Salmi LD, Heikkilä MJ, Puukilainen E, Sajavaara T, Grosso D, Ritala M (2013) Studies on atomic layer deposition of MOF-5 thin films. Microporous Mesoporous Mater 182(December):147–154. https://doi.org/10.1016/j.micromeso.2013.08.024
9. Fujii K, Lazuen Garay A, Hill J, Sbircea E, Pan Z, Xu M, Apperley DC, James SL, Harris KDM (2010) Direct structure elucidation by powder X-ray diffraction of a metal-organic framework material prepared by solvent-free grinding. Chem Commun 46(40):7572–7574. https://doi.org/10.1039/c0cc02635b
10. Goldstein JI, Newbury DE, Echlin P, Joy DC, Lyman CE, Lifshin E, Sawyer L, Michael JR (2003) Electron beam–specimen interactions. In: Scanning electron microscopy and X-ray microanalysis. Springer US, pp 61–98. https://doi.org/10.1007/978-1-4615-0215-9_3
11. Carreon MA, Venna SR (2020) Metal-organic frameworks history and structural features. In: Metal-organic framework membranes for molecular gas separations, pp 1–29. https://doi.org/10.1142/9781786346735_0001
12. Samui A, Sahu SK (2020) Characterizations of MOFs for biomedical application. In: Metal-organic frameworks for biomedical applications, pp 277–295. https://doi.org/10.1016/B978-0-12-816984-1.00015-9
13. Goldstein JI, Newbury DE, Echlin P, Joy DC, Lyman CE, Lifshin E, Sawyer L, Michael JR (2003) The SEM and its modes of operation. In: Scanning electron microscopy and X-ray microanalysis. Springer US, pp 21–60. https://doi.org/10.1007/978-1-4615-0215-9_2
14. Hirschle P, Preiß T, Auras F, Pick A, Völkner J, Rädler JO, Wuttke S (2016) Exploration of MOF nanoparticle sizes using various physical characterization methods—is what you measure what you get?, pp 4359–4368. https://doi.org/10.1039/c6ce00198j
15. Reimer L (1997) Elements of a transmission electron microscope, pp 79–142. https://doi.org/10.1007/978-3-662-14824-2_4
16. Smith DJ (2006) High resolution transmission electron microscopy. In Handbook of microscopy for nanotechnology. Springer, Boston, MA, pp 427–453. https://doi.org/10.1007/1-4020-8006-9_14
17. Williams DB, Barry Carter C (1996) The transmission electron microscope. In: Transmission electron microscopy. Springer International Publishing, Cham, pp 3–17. https://doi.org/10.1007/978-1-4757-2519-3_1
18. Fultz B, Howe J (2013). The TEM and its optics, pp 59–115. https://doi.org/10.1007/978-3-642-29761-8_2
19. Wiktor C, Meledina M, Turner S, Lebedev OI, Fischer RA (2017) Transmission electron microscopy on metal-organic frameworks-a review. J Mat Chem A. https://doi.org/10.1039/c7ta00194k
20. Lebedev OI, Millange F, Serre C, Van Tendeloo G, Férey G (2005) First direct imaging of giant pores of the metal-organic framework MIL-101. Chem Mater 17(26):6525–6527. https://doi.org/10.1021/cm051870o
21. Wiktor C, Turner S, Zacher D, Fischer RA, Van Tendeloo G (2012) Imaging of intact MOF-5 nanocrystals by advanced TEM at liquid nitrogen temperature. Microporous Mesoporous Mater 162(November):131–135. https://doi.org/10.1016/j.micromeso.2012.06.014
22. Kunitake M, Uemura S (2020) Construction and scanning probe microscopy imaging of two-dimensional nanomaterials. Chem Lett. https://doi.org/10.1246/cl.200080

23. Scanning Tunneling Microscopy—Nanoscience Instruments. n.d. Nanoscience instruments. https://www.nanoscience.com/techniques/scanning-tunneling-microscopy/
24. Howland R, Benatar L (1996) A practical guide to scanning probe microscopy, p 87. http://web.mit.edu/cortiz/www/AFMGallery/PracticalGuide.pdf
25. Hofmann A, Griffith Research Online (2010) Spectroscopic techniques: I spectrophotometric techniques author book title principles and techniques of biochemistry and molecular biology link to published version. http://hdl.handle.net/10072/34561. http://www.cambridge.org/catalogue/catalogue.asp?isbn=9780521731676&ss=fro.
26. Chapter 3 Absorption, emission, reflection, and scattering 3.1. Absorption and emission
27. Mali G (2016) Looking into metal-organic frameworks with solid-state NMR spectroscopy. In: Metal-organic frameworks. IntechOpen. https://doi.org/10.5772/64134
28. Benzaqui M, Pillai RS, Sabetghadam A, Benoit V, Normand P, Marrot J, Menguy N et al. (2017) Revisiting the aluminum trimesate-based MOF (MIL-96): from structure determination to the processing of mixed matrix membranes for CO_2 capture. Chem Mater 29(24):10326–10338. https://doi.org/10.1021/ACS.CHEMMATER.7B03203
29. Brunner E, Rauche M (2020) Solid-state NMR spectroscopy: an advancing tool to analyse the structure and properties of metal–organic frameworks. Chem Sci 11(17):4297–4304. https://doi.org/10.1039/d0sc00735h
30. Li H, Shi W, Zhao K, Li H, Bing Y, Cheng P (2012) Enhanced hydrostability in Ni-doped MOF-5. Inorg Chem 51(17):9200–9207. https://doi.org/10.1021/IC3002898
31. Larkin P (2011) Introduction. In: Infrared and Raman spectroscopy. Elsevier, pp 1–5. https://doi.org/10.1016/b978-0-12-386984-5.10001-1
32. Tan K, Chabal YJ (2016) Interaction of small molecules within metal organic frameworks studied by in situ vibrational spectroscopy. In Metal-organic frameworks, October. https://doi.org/10.5772/64906
33. Raman Spectroscopy, IR Spectroscopy. Comparison of Raman and IR spectroscopy. http://www.chemvista.org/ramanIR4.html. Accessed 31 Aug 2021
34. Nijem N, Thissen P, Yao Y, Longo RC, Roodenko K, Wu H, Zhao Y et al (2011) Understanding the preferential adsorption of CO_2 over N_2 in a flexible metal-organic framework. J Am Chem Soc 133(32):12849–12857. https://doi.org/10.1021/ja2051149
35. Tan K, Nijem N, Canepa P, Gong Q, Li J, Thonhauser T, Chabal YJ (2012) Stability and hydrolyzation of metal organic frameworks with paddle-wheel SBUs upon hydration. Chem Mater 24(16):3153–3167. https://doi.org/10.1021/cm301427w
36. Gooijer C (2000) Introduction to fluorescence spectroscopy. Anal Chim Acta 419(1):116–117. https://doi.org/10.1016/s0003-2670(00)01086-2
37. Allendorf MD, Bauer CA, Bhakta RK, Houk RJT (2009) Luminescent metal-organic frameworks. Chem Soc Rev 38(5):1330–1352. https://doi.org/10.1039/b802352m
38. Hao JN, Yan B (2015) A water-stable lanthanide-functionalized MOF as a highly selective and sensitive fluorescent probe for Cd^{2+}. Chem Commun 51(36):7737–7740. https://doi.org/10.1039/c5cc01430a
39. Yu L, Lei Y, Ma Y, Liu M, Zheng J, Dan D, Gao P (2021) A comprehensive review of fluorescence correlation spectroscopy. Front Phys. https://doi.org/10.3389/fphy.2021.644450
40. Preiß T, Zimpel A, Wuttke S, Rädler JO (2017) Kinetic analysis of the uptake and release of fluorescein by metal-organic framework nanoparticles. Materials 10(2). https://doi.org/10.3390/ma10020216
41. Bedia J, Muelas-Ramos V, Peñas-Garzón M, Gómez-Avilés A, Rodríguez JJ, Belver C (2019) A review on the synthesis and characterization of metal organic frameworks for photocatalytic water purification. Catalysts. Multidisciplinary Digital Publishing Institute. https://doi.org/10.3390/catal9010052
42. Sofi FA, Majid K, Mehraj O (2018) The visible light driven copper based metal-organic-framework heterojunction: HKUST-1@Ag-Ag_3PO_4 for plasmon enhanced visible light photocatalysis. J Alloy Compd 737(March):798–808. https://doi.org/10.1016/j.jallcom.2017.12.141

Metal–Organic Frameworks (MOFs) Characterization Using Different Analytical Methods

Dipeshkumar D. Kachhadiya, Amol Vijay Sonawane, and Z. V. P. Murthy

Contents

1 Introduction 166
2 Characterization Techniques 167
2.1 FE-SEM and EDX 167
2.2 TGA 168
2.3 XRD 169
2.4 DLS 170
2.5 FTIR 171
3 Characterizations of Some Prepared MOFs 172
3.1 ZIF-8 172
3.2 ZIF-67 174
3.3 CuBTC and GO@CuBTC 175
4 Conclusions 177
Abbreviations 177
References 178

Abstract Metal–organic frameworks (MOFs) are a novel form of porous material with a wide range of present and prospective applications in a variety of fields, including catalysis, gas storage, ion exchange, sensing, molecular recognition, separation, drug delivery, and so on. Different characterization techniques play a vital role in the structural characterization to understand the interactions of MOFs with other materials. The field emission scanning electron microscopy with energy dispersive X-rays (FESEM-EDX), Thermogravimetry analysis (TGA), Dynamic light scattering (DLS), X-ray diffraction analysis (XRD), and Fourier transform infrared spectroscopy (FTIR) characterization techniques are used to analyze the surface morphology, thermal stability, particle size distribution, crystal structure and crystallinity, and chemical composition, respectively, of the MOFs. This chapter gives brief information about different characterization tools used to characterize different MOFs.

D. D. Kachhadiya · A. V. Sonawane · Z. V. P. Murthy (✉)
Department of Chemical Engineering, Sardar Vallabhbhai National Institute of Technology, Surat, Gujarat 395007, India
e-mail: zvpm@ched.svnit.ac.in

S. Gulati (ed.), *Metal-Organic Frameworks (MOFs) as Catalysts*,
https://doi.org/10.1007/978-981-16-7959-9_6

Keywords MOFs · Characterizations · Thermogravimetry · Diffraction · X-rays · Scattering · Spectroscopy

1 Introduction

Nanostructures have gotten a lot of press as a rapidly evolving class of materials with a wide variety of applications. Various techniques have been used to determine the crystallinity, crystal structure, size and shape, elemental composition, thermal stability, surface area, and pore volume, and various other physical dynamics of nanoparticles (NPs) [1]. In many cases, some physical parameters may be approximated using more than one method. Some physical properties may be estimated using more than one approach in many circumstances. The most critical factors, size, and shape are examined first in the characterization of NPs. We may also assess the surface chemistry and quantify the size distribution, surface area, degree of aggregation, and particle charge [2]. The physicochemical properties of nanomaterials are important to assess the quality control of manufacturing processes and the effect of nanomaterials on the environment and human beings [3]. The summary of characterization techniques that are used for nanoparticles characterizations is listed in Table 1. Table 2 shows that parameters must be analyzed for nanoparticles and which characterization technique is best for them.

Table 1 List of characterization techniques and their key findings

Technique	Key findings
XRD	Crystal structure, crystal grain structure, composition
FE-SEM	Morphology, shape, size, the structure of NPs
EDX	Elemental composition, dispersion of NPs within cell/support
TGA	Thermal stability
DLS	Particle size, detection of agglomeration
BET	Surface area, pore volume
FTIR	Ligand bindings, surface composition, functional groups
TEM	NP size, structure, shape, growth kinetics, aggregation state
AFM	3D shape and size, lateral dimensions, dispersion of NPs in cell/support
UV–vis	Size, concentration, and also hint on the shape of the particles
ICP-MS	Elemental information, NP concentration, shape and size distribution

Table 2 List of parameters required to be analyzed and the corresponding characterization method

Parameter	A suitable technique for characterization
Size	SEM, AFM, TEM, XRD, DLS, UV–Vis
Shape	TEM, HRTEM, AFM,
Elemental and chemical composition	SEM–EDX, ICP-MS, XRD, XPS, NMR
Crystal structure	XRD, STEM, HRTEM
Size distribution	DLS, DCS, ICP-MS
Ligand bindings, surface composition	FTIR, NMR, XPS
Surface charge	BET, NMR
Concentration	ICP-MS, UV–Vis, DCS
3D visualization	AFM, SEM, TEM
Dispersion and detection of NPs	SEM, AFM, TEM, EBSD

2 Characterization Techniques

Various characterization techniques such as FTIR, FE-SEM, XRD, TGA, and DLS characterization techniques are used to analyze the chemical composition, surface morphology, crystal structure and crystallinity, thermal stability, and particle size distribution of the MOFs, respectively.

2.1 FE-SEM and EDX

The SEM is an instrument that scans the specimen's surface with a high-energy electron beam. Traditional light microscopes bend light waves via a series of glass lenses to magnify an image. On the other hand, the SEM magnifies images by using electrons rather than light waves [4]. Basic components used in SEM are electron gun, condenser lenses, electron column, anode, scanning coils, specimen holder, electron detectors, and vacuum system. The schematic diagram of the SEM instrument is shown in Fig. 1. Signals in the form of distinctive X-rays, secondary electrons (SE), and backscattered electrons (BSE) are created when electrons contact the top surface of a sample and interrelate with the atoms of the sample, including information about the surface topography, composition, and so on. In its principal detection mode, the SEM can create high-resolution pictures of a specimen, showing features as small as 1–5 nm by SE imaging. Another most frequent imaging approach for SEM is distinctive X-rays. Using an energy dispersive X-ray method (EDX), these distinct X-rays are utilized to determine the sample's elemental composition. BSE can also be utilized to create an image of samples. In analytical SEM, BSE images and spectra

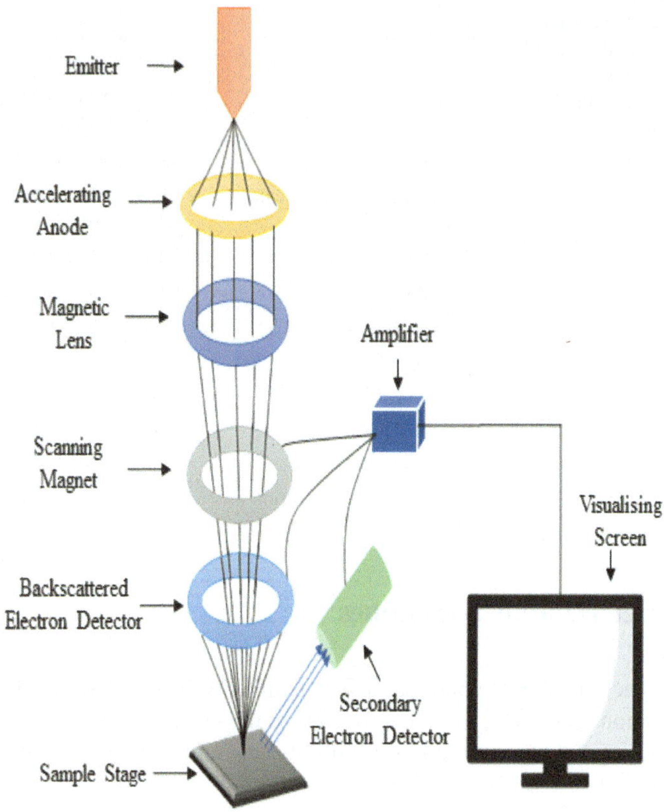

Fig. 1 Schematic representation of SEM [4]

derived from distinctive X-rays are frequently utilized to determine the elemental analysis of the material.

2.2 *TGA*

TGA measures the amount of weight loss and rate of change with temperature and time in the gaseous atmosphere in the materials. Generally, TGA is used to determine the composition of the material and its thermal stability at 1000 °C temperature. Using TGA, one can know whether the material undergoes loss or gain of weight and is due to decomposition, oxidation, or dehydration. In the analysis, with the help of microbalance, the sample is placed in a platinum crucible. This whole system is kept in a controlled temperature system like a small oven. In analysis, the temperature is increased gradually with inert gas flow, and the weight loss versus temperature graph is plotted. Figure 2 represents the schematic diagram of the TGA instrument.

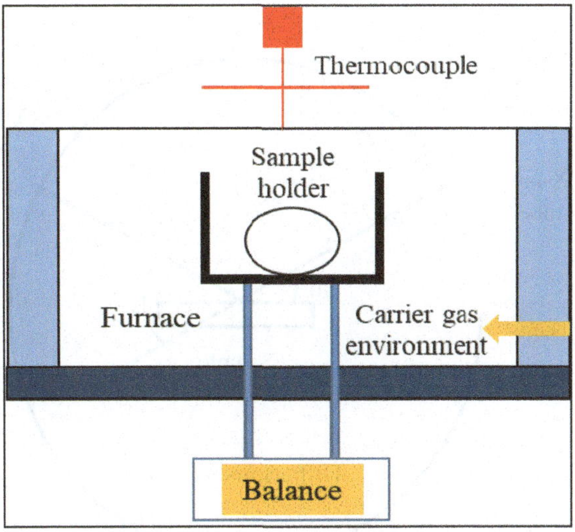

Fig. 2 Schematic diagram of TGA [5]

TGA gives essential information on material selection, product prediction, and improved quality for the product. It is generally used to measure the composition of a multi-component system, estimated lifetime of the product, catalyst information regarding moisture content, mass change during calcination/reaction condition environment, and effect of reactive corrosive atmosphere on the material. TGA does not give any information about which process is responsible for mass change.

2.3 XRD

XRD is a useful tool for identifying a material's crystal structure. It can identify crystalline materials with crystal domains larger than 3–5 nm. The X-ray tube, goniometer, incident-beam optics, receiving-slits, sampler, and detector are the major components of the XRD instrument. It is used to determine the bulk crystal structure and phase identification. Figure 3 represents a schematic diagram of the XRD. The specimens are examined in the diffractometer using a single wavelength X-ray beam. Constantly changing the incidence angle of the X-ray beam captures a spectrum of intensity vs diffraction angle between incident and diffraction beam. Bragg's law, which is given by Eq. 1, is the connection by which diffraction occurs [6].

$$n\lambda = 2d\sin\theta \tag{1}$$

where d is the lattice spacing, θ is an angle between the wave vector of the incident plane wave, k_o, and the lattice planes, λ its wavelength, and n is an integer.

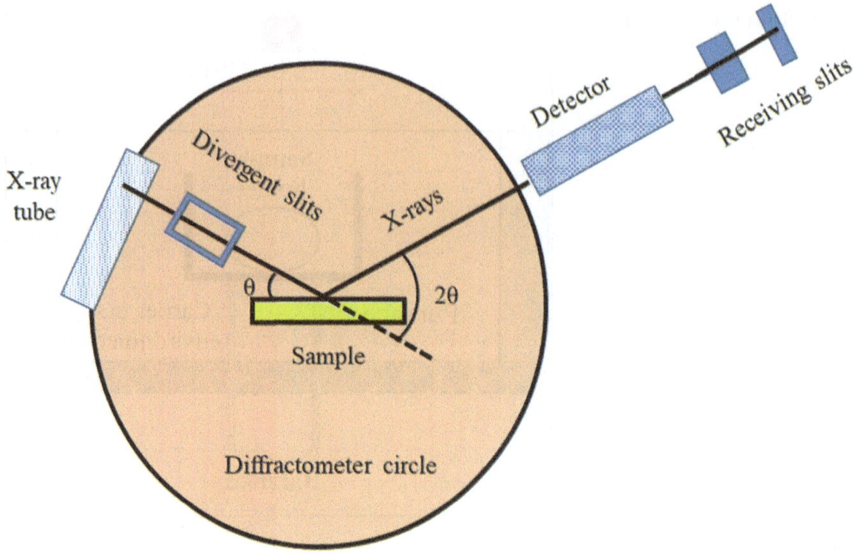

Fig. 3 Schematic diagram of X-ray diffractometer. Adapted from https://pubs.usgs.gov/of/2001/of01-041/htmldocs/xrpd.htm

2.4 DLS

DLS, also known as photon correlation spectroscopy (PCS), is an intra-operative, well-known system for determining the size of particles in the range of submicron and nanosize. Figure 4 represents a schematic diagram of DLS. Brownian motion occurs when particles, emulsions, and molecules are suspended. The bombardment of solvent molecules causes them to move, which they do because of their thermal energy. Assume that a laser is used to light the particles or molecules. The smaller particles are propelled further and move faster by the solvent molecules, the intensity of the scattered light varies at a rate that is proportional to particle size. The Brownian motion velocity and the particle size (radius r_k) from these intensity fluctuations are calculated using the Stokes–Einstein relationship [7].

$$r_k = \frac{kT}{6\pi \eta D} \tag{2}$$

where k is the Boltzmann's constant, η is the solvent viscosity, T is the temperature in K, and D is the diffusion coefficient.

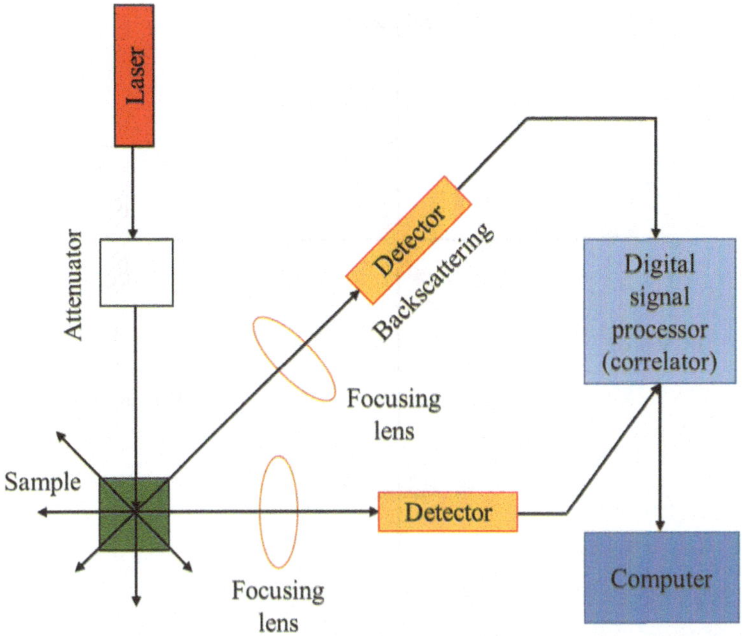

Fig. 4 Schematic diagram of DLS instrument [7]. Reproduced with permission from Elsevier Inc.

2.5 *FTIR*

The vibrational stretch frequency of metal–oxygen bonds is revealed by FTIR [8]. To create light across a wide range of infrared wavelengths, a continuous light source is utilized. A half-silvered mirror splits the light from this continuous source into two directions. Light is focused onto the material of interest in FTIR Spectroscopy, and the intensity is measured using an infrared detector. The resultant graph shows the Fourier transform of the intensity of light as a function of wavenumber when the intensity of light is measured and plotted as a function of the location of the moveable mirror. A Michelson Interferometer is utilized as a radiation source. Figure 5 depicts the fundamental structure of the device. The phenomenon of numerous internal reflections is utilized to increase the sensitivity of semiconductors. The edges of the sample are polished with this procedure, and the light is delivered at an angle. A total of 30–50 bounces occur when the light bounces around and inside the sample. This improves sensitivity by roughly a factor of 30–50, allowing researchers to detect the absorption of molecules in less than one monolayer on a surface.

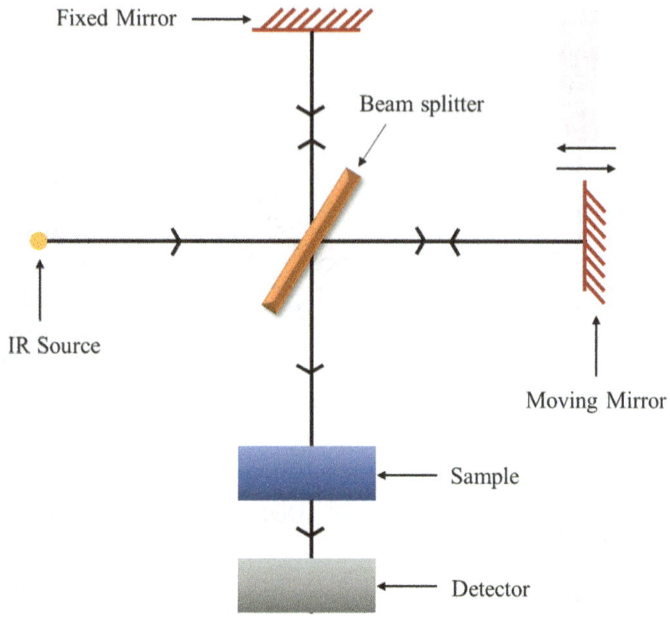

Fig. 5 Schematic diagram of FTIR instrument. Adapted from https://lab-training.com/2014/09/system-features-ft-ir

3 Characterizations of Some Prepared MOFs

3.1 ZIF-8

The characterizations of the in-house prepared ZIF-8 NPs are carried out using SEM, XRD, DLS, FTIR, and TGA (Figs. 6, 7 and 8). Figure 6a and b shows the

Fig. 6 SEM image of ZIF-8 particles, **a** 1 μm, and **b** 500 nm

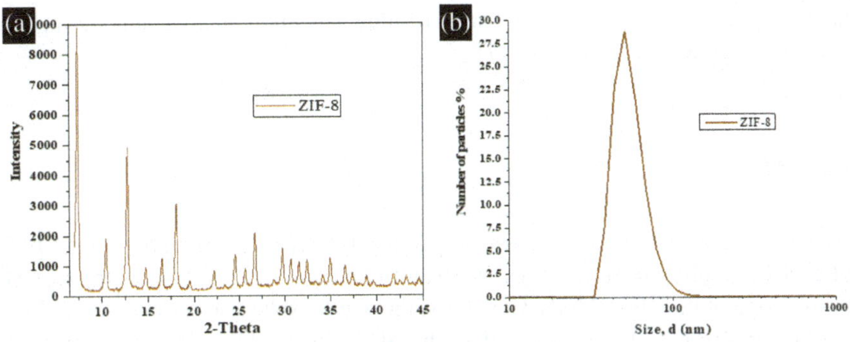

Fig. 7 **a** XRD spectra and **b** DLS curve of ZIF-8 particles

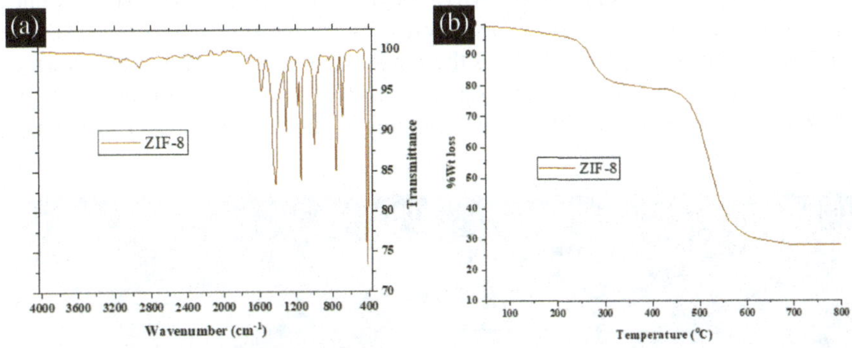

Fig. 8 **a** FTIR and **b** TGA analysis of ZIF-8 particles

surface morphology of the ZIF-8 NPs. Regular homogenous nano-sized and octahedral crystal morphology of most of ZIF-8 NPs are observed in the SEM images [9]. 2θ values of 7.38° (011), 10.42° (022), 12.76° (112), 14.74° (022), 18.08° (222), 22.16° (114), 24.54° (233), 26.72° (134), 29.7° (004), 30.64° (334), and 31.54° (244) were obtained using XRD for the ZIF-8 (Fig. 7a). These 2θ values validated the formation of ZIF-8 NPs, as described by various researchers [10, 11]. The synthesized ZIF-8 NPs size falls in the range of 40–80 nm (Fig. 7b). Various researchers have reported the same size for ZIF-8 particles [12]. Figure 8a shows the FTIR analysis of prepared ZIF-8 NPs. Significant bands of 1595 (C = N stretching of Hmim), 1458 (entire ring stretching of Hmim), 1421, 1307, 1145, 994 (in plane bending of Hmim), 752 and 592 (out of plane bending of Hmim), and 421 (Zn–N stretching) cm^{-1} were confirmed for ZIF-8 NPs from the FTIR spectrum. The various group of researchers could confirm these [13–15]. The weight loss percentage of ZIF-8 particles was obtained by TGA in the range of RT to 800 °C with airflow (Fig. 8b). The 72% weight loss of the ZIF-8 NPs is obtained from the evaporation of methanol, bounded moisture, unreacted 2-Hmim, and organic linker molecules from the cavities and surfaces of the ZIF-8 particles. The final white solid residue was ZnO remain stable during the heating at high temperatures [16, 17].

3.2 ZIF-67

The characterizations of the in-house prepared ZIF-8 NPs are carried out using SEM, XRD, DLS, FTIR, and TGA (Figs. 9, 10 and 11). Figure 9a and b show the surface morphology of the ZIF-67 NPs. Generally, the regular spherical crystal morphology of most of ZIF-67 NPs is observed in the SEM images [18]. 2θ values of 7.40° (011), 10.20° (002), 12.80° (112), 14.80° (022), 18.10° (222), 22.18° (114), 24.56° (233), 25.68° (224), 26.72° (134), 29.70° (044), 30.62° (334), 31.52° (244), and 32.42° (235) were obtained using XRD for the ZIF-67 (Fig. 10a). These values validated the formation of ZIF-67, as reported by various researchers [19–21]. The synthesized ZIF-67 NPs size falls in the range of 40–100 nm (Fig. 10b). Various researchers have reported the same size for ZIF-67 [20]. Figure 11a represents the FTIR analysis of prepared ZIF-67 NPs. Significant bands of 3367 (NH- residual of

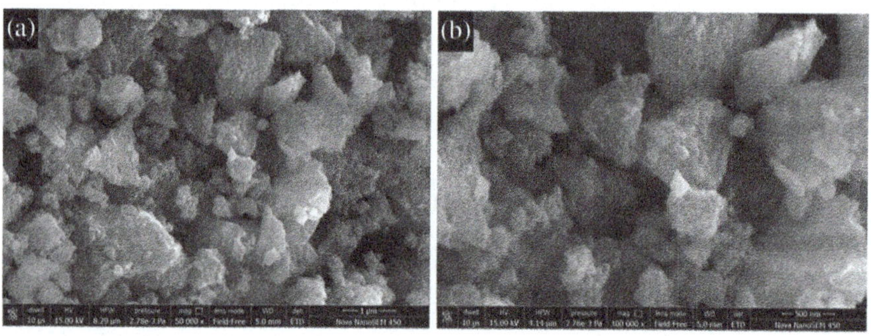

Fig. 9 SEM image of ZIF-67 particles, **a** 1 μm, and **b** 500 nm

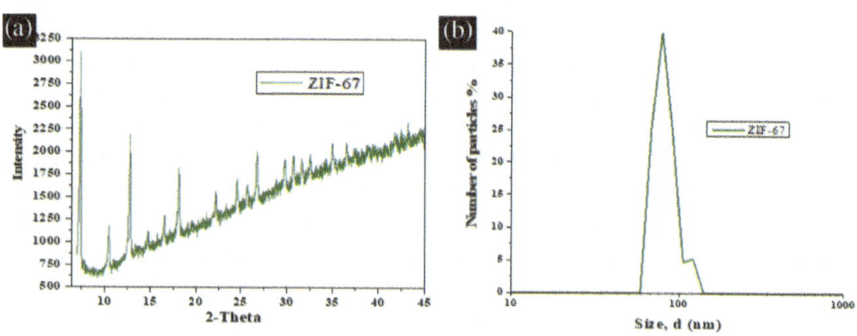

Fig. 10 **a** XRD spectra and **b** DLS curve of ZIF-67 particles

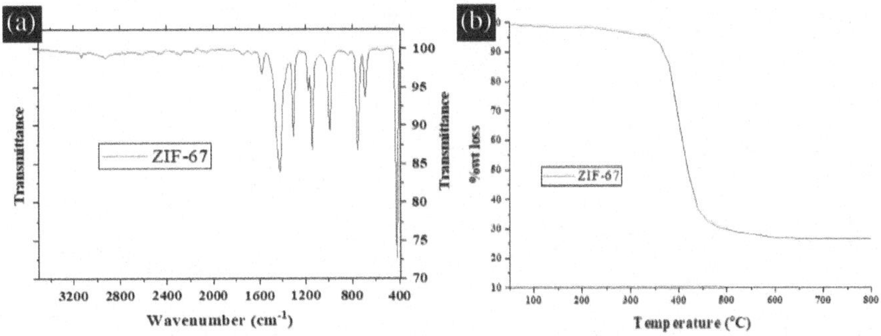

Fig. 11 **a** FTIR and **b** TGA analysis of ZIF-67 particles

Hmim), 2927 and 3133 (aliphatic and aromatic C-H stretching vibration of Hmim), 1584 (C = N stretching of Hmim), 1458 (entire ring stretching of Hmim), peaks ranging from 600 to 1500 cm^{-1} (bending and stretching of the imidazole), and 425 (Co–N stretching) cm^{-1} were confirmed for ZIF-67 NPs from the FTIR spectrum. These could be confirmed by the various group of researchers [20, 21]. The weight loss percentage of ZIF-67 particles was obtained by TGA in the range of RT to 800 °C with airflow (Fig. 11b). When the MOF precursor is heated after 385 °C, the ZIF-67 will disintegrate and carbonize.

3.3 *CuBTC and GO@CuBTC*

The characterizations such as SEM, XRD, DLS, FTIR, and TGA of the CuBTC and GO@CuBTC nanoparticles are carried out by many researchers and shown in Figs. 12, 13 and 14. SEM images of CuBTC and GO@CuBTC were displayed in Fig. 12a and b. CuBTC has an octahedral structure with a smooth surface, which is consistent with the literature [22–24]. In comparison to the clean CuBTC

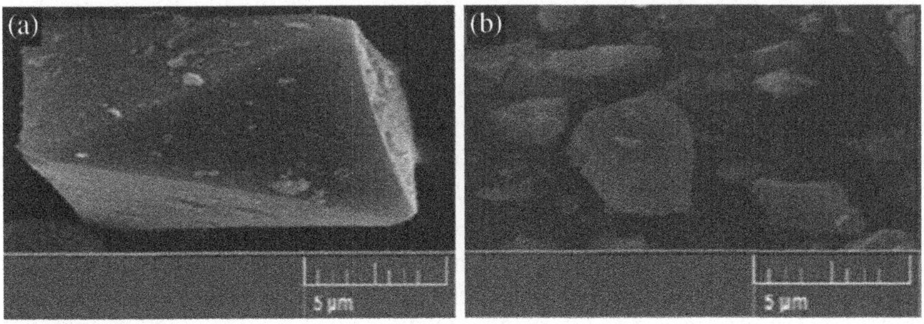

Fig. 12 SEM image of **a** CuBTC and **b** GO@CuBTC [31]. Reproduced with permission from Elsevier Inc.

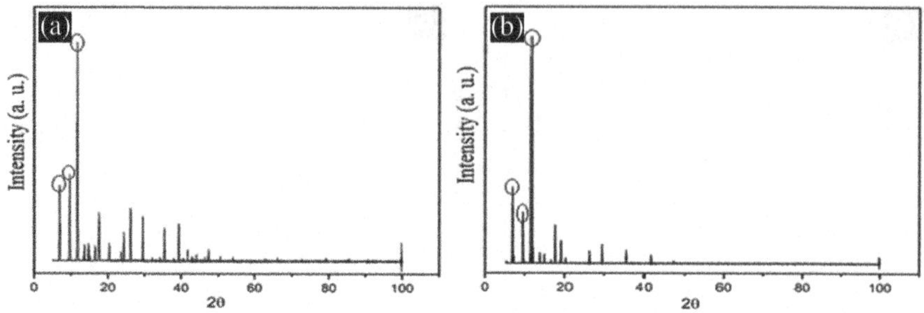

Fig. 13 XRD spectra of **a** CuBTC and **b** GO@CuBTC [31]. Reproduced with permission from Elsevier Inc.

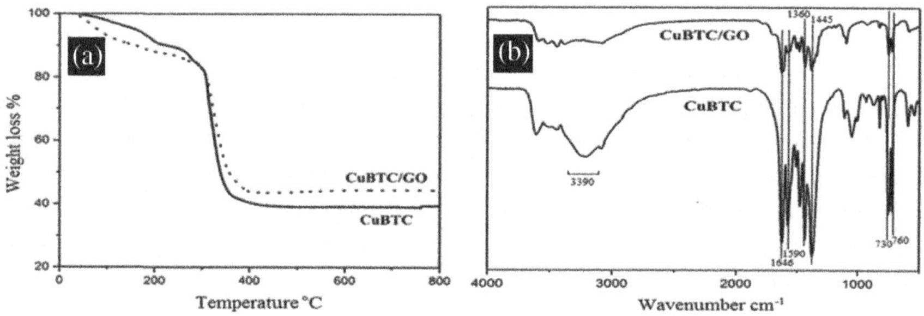

Fig. 14 **a** FTIR and **b** TGA analysis of CuBTC and GO@CuBTC [31]. Reproduced with permission from Elsevier Inc.

crystal size, the composite CuBTC crystal size was reduced. 2θ values of 6.5°, 9.4°, 11.7°, and 13.8° were obtained using XRD for the CuBTC NPs. A similar XRD pattern with lower intensities was observed for GO-coated CuBTC (Fig. 13a). The characteristic peak of GO was observed at a 2θ value of 11.4° (Fig. 13b). These values validated the formation of CuBTC and GO@CuBTC, as reported by various researchers [25, 26]. Figure 14a shows the FTIR analysis of synthesized CuBTC and GO@CuBTC nanoparticles. Significant bands of 3290 (O–H stretching of carboxylic acid), 1646 and 1590 (asymmetric stretching of carboxylate), 1360 and 1445 (symmetric stretching of carboxylate), 1110 (C–O–Cu stretching of CuBTC), and peaks at 760 and 730 cm^{-1} (Cu substitution in benzene group) were detected for CuBTC and GO@CuBTC particles from the ATR-FTIR spectrum [27, 28]. Figure 14b shows the thermal stability of CuBTC and GO@CuBTC NPs in a nitrogen atmosphere. A similar trend was observed for both nanoparticles. Here weight loss was observed in three stages. In the first stage, bounded moisture was evaporated. Hydrolyzed ethanol was separated from the BTC linker in the second stage. And at last, considerable weight loss was due to the decomposition of linkers and transformation of CuBTC to Cu_2O and CuO [29, 30]. The major weight loss for CuBTC and GO@CuBTC was observed at 330 °C and 350 °C, respectively.

4 Conclusions

Nanotechnology, as a well-known field, has a wide range of applications, including the pharma sector, cosmetic industries, energy sector, petroleum industries, defense and security, electronics sector, and many more. Nanotechnology involves the fundamentals of synthesis, large-scale production, and modification of nanoparticles. Characterization of nanoparticles is equally necessary to understand the properties and their application. This chapter gives brief information about some major basic characterization techniques such as FE-SEM, XRD, TGA, DLS, and FTIR. We have discussed in detail the characterization of ZIF-8, ZIF-67, CuBTC, and GO@CuBTC by using FE-SEM, XRD, TGA, DLS, and FTIR. Here we have discussed the size, shape, crystal nature, present functional groups, and thermal stability of the different MOFs. Nanotechnology has a massive prospective for future generations in various fields of applications.

Acknowledgements The authors (DDK and AVS) acknowledge the Ministry of Education, Government of India, for a doctoral fellowship.

The authors acknowledge the Materials Research Centre (MRC), Malaviya National Institute of Technology Jaipur (MNIT) for providing membrane characterization facilities.

Abbreviations

AFM	Atomic Force Microscopy
BET	Brunauer-Emmett-Teller
BSE	Back Scattered Electrons
DCS	Differential Scanning Calorimetry
DLS	Dynamic Light Scattering
EBSD	Electron Backscatter Diffraction
EDX	Energy Dispersive X-Ray Analysis
FE-SEM	Field Emission-Scanning Electron Microscopy
FTIR	Fourier Transform Infrared Spectroscopy
GO	Graphene Oxide
HR-TEM	High-Resolution Transmission Electron Microscopy
ICP-MS	Inductively Coupled Plasma-Mass-Spectrometry
MOFs	Metal-Organic Frameworks
NMR	Nuclear Magnetic Resonance
NPs	Nanoparticles
SE	Scattered Electrons
STEM	Scanning Transmission Electron Microscopy
TEM	Transmission Electron Microscopy
TGA	Thermogravimetric Analysis
UV-vis	Ultraviolet-Visible Spectrophotometry
XPS	X-ray Photoelectron Spectroscopy

XRD X-ray Diffraction Analysis
ZIFs Zeolitic Imidazolate Frameworks

References

1. Mourdikoudis S, Pallares RM, Thanh NTK (2018) Characterization techniques for nanoparticles: comparison and complementarity upon studying nanoparticle properties. Nanoscale 10:12871–12934. https://doi.org/10.1039/C8NR02278J
2. Upadhyay S, Parekh K, Pandey B (2016) Influence of crystallite size on the magnetic properties of Fe_3O_4 nanoparticles. J Alloys Compd 678:478–485. https://doi.org/10.1016/j.jallcom.2016.03.279
3. Bushell M, Beauchemin S, Kunc F, Gardner D, Ovens J, Toll F, Kennedy D, Nguyen K, Vladisavljevic D, Rasmussen PE, Johnston LJ (2020) Characterization of commercial metal oxide nanomaterials: crystalline phase, particle size and specific surface area. Nanomaterials 10:1812. https://doi.org/10.3390/nano10091812
4. Sharma S, Jaiswal S, Duffy B, Jaiswal A (2019) Nanostructured materials for food applications: spectroscopy, microscopy and physical properties. Bioengineering 6:26. https://doi.org/10.3390/bioengineering6010026
5. Arumugasamy SK, Selvarajoo A, Tariq MA (2020) Artificial neural networks modelling: gasification behaviour of palm fibre biochar. Mater Sci Energy Technol 3:868–878. https://doi.org/10.1016/j.mset.2020.10.010
6. Anderoglu O (2004) Residual stress measurement using X-ray diffraction. https://www.researchgate.net/publication/26898174%0AResidual
7. Bhattacharjee S (2016) DLS and zeta potential—What they are and what they are not? J Control Release 235:337–351. https://doi.org/10.1016/j.jconrel.2016.06.017
8. Smith BC (2011) Fundamentals of Fourier transform infrared spectroscopy, 2nd edn. CRC Press, Boca Raton. https://doi.org/10.1201/b10777
9. Tsalaporta E, MacElroy JMD (2020) A comparative study of the physical and chemical properties of pelletized HKUST-1, ZIF-8, ZIF-67 and UiO-66 powders. Heliyon 6:e04883. https://doi.org/10.1016/j.heliyon.2020.e04883
10. Vatanpour V, Khorshidi S (2020) Surface modification of polyvinylidene fluoride membranes with ZIF-8 nanoparticles layer using interfacial method for BSA separation and dye removal. Mater Chem Phys 241:122400. https://doi.org/10.1016/j.matchemphys.2019.122400
11. Rao Y, Ni F, Sun Y, Zhu B, Zhou Z, Yao Z (2020) Efficient recovery of the volatile aroma components from blackberry juice using a ZIF-8/PDMS hybrid membrane. Sep Purif Technol 230:115844. https://doi.org/10.1016/j.seppur.2019.115844
12. Denny MS, Cohen SM (2015) In situ modification of metal-organic frameworks in mixed-matrix membranes. Angew Chemie Int Ed 54:9029–9032. https://doi.org/10.1002/anie.201504077
13. Hu Y, Kazemian H, Rohani S, Huang Y, Song Y (2011) In situ high pressure study of ZIF-8 by FTIR spectroscopy. Chem Commun 47:12694. https://doi.org/10.1039/c1cc15525c
14. Low Z-X, Yao J, Liu Q, He M, Wang Z, Suresh AK, Bellare J, Wang H (2014) Crystal transformation in zeolitic-imidazolate framework. Cryst Growth Des 14:6589–6598. https://doi.org/10.1021/cg501502r
15. Jian M, Liu B, Zhang G, Liu R, Zhang X (2015) Adsorptive removal of arsenic from aqueous solution by zeolitic imidazolate framework-8 (ZIF-8) nanoparticles. Colloids Surfaces A Physicochem Eng Asp 465:67–76. https://doi.org/10.1016/j.colsurfa.2014.10.023
16. Yang S, Xia N, Li M, Liu P, Wang Y, Qu L (2019) Facile synthesis of a zeolitic imidazolate framework-8 with reduced graphene oxide hybrid material as an efficient electrocatalyst for nonenzymatic H_2O_2 sensing. RSC Adv 9:15217–15223. https://doi.org/10.1039/C9RA02096A

17. Han Y, Qi P, Li S, Feng X, Zhou J, Li H, Su S, Li X, Wang B (2014) A novel anode material derived from organic-coated ZIF-8 nanocomposites with high performance in lithium ion batteries. Chem Commun 50:8057–8060. https://doi.org/10.1039/C4CC02691H
18. Li X, Qiao Y, Wang C, Shen T, Zhang X, Wang H, Li Y, Gao W (2019) MOF-derived Co/C nanocomposites encapsulated by $Ni(OH)_2$ ultrathin nanosheets shell for high performance supercapacitors. J Alloys Compd 770:803–812. https://doi.org/10.1016/j.jallcom.2018.08.164
19. Truong T, Hoang TM, Nguyen CK, Huynh QTN, Phan NTS (2015) Expanding applications of zeolite imidazolate frameworks in catalysis: synthesis of quinazolines using ZIF-67 as an efficient heterogeneous catalyst. RSC Adv 5:24769–24776. https://doi.org/10.1039/C4RA16168H
20. Khan A, Ali M, Ilyas A, Naik P, Vankelecom IFJ, Gilani MA, Bilad MR, Sajjad Z, Khan AL (2018) ZIF-67 filled PDMS mixed matrix membranes for recovery of ethanol via pervaporation. Sep Purif Technol 206:50–58. https://doi.org/10.1016/j.seppur.2018.05.055
21. Nadar SS, Rathod VK (2019) One pot synthesis of α-amylase metal organic framework (MOF)-sponge via dip-coating technique. Int J Biol Macromol 138:1035–1043. https://doi.org/10.1016/j.ijbiomac.2019.07.099
22. Yang Y, Dong H, Wang Y, He C, Wang Y, Zhang X (2018) Synthesis of octahedral like Cu-BTC derivatives derived from MOF calcined under different atmosphere for application in CO oxidation. J Solid State Chem 258:582–587. https://doi.org/10.1016/j.jssc.2017.11.033
23. Kaur R, Kaur A, Umar A, Anderson WA, Kansal SK (2019) Metal organic framework (MOF) porous octahedral nanocrystals of Cu-BTC: synthesis, properties and enhanced adsorption properties. Mater Res Bull 109:124–133. https://doi.org/10.1016/j.materresbull.2018.07.025
24. Shang S, Tao Z, Yang C, Hanif A, Li L, Tsang DCW, Gu Q, Shang J (2020) Facile synthesis of CuBTC and its graphene oxide composites as efficient adsorbents for CO_2 capture. Chem Eng J 393:124666. https://doi.org/10.1016/j.cej.2020.124666
25. Zhao Q, Zhu L, Lin G, Chen G, Liu B, Zhang L, Duan T, Lei J (2019) Controllable synthesis of porous Cu-BTC@polymer composite beads for iodine capture. ACS Appl Mater Interfaces 11:42635–42645. https://doi.org/10.1021/acsami.9b15421
26. Szczęśniak B, Phuriragpitikhon J, Choma J, Jaroniec M (2020) Mechanochemical synthesis of three-component graphene oxide/ordered mesoporous carbon/metal-organic framework composites. J Colloid Interface Sci 577:163–172. https://doi.org/10.1016/j.jcis.2020.05.080
27. Misdan N, Ramlee N, Hairom NHH, Ikhsan SNW, Yusof N, Lau WJ, Ismail AF, Nordin NAHM (2019) CuBTC metal organic framework incorporation for enhancing separation and antifouling properties of nanofiltration membrane. Chem Eng Res Des 148:227–239. https://doi.org/10.1016/j.cherd.2019.06.004
28. Mazani M, Aghapour Aktij S, Rahimpour A, Tavajohi Hassan Kiadeh N (2019) Cu-BTC metal–organic framework modified membranes for landfill leachate treatment. Water 12:91. https://doi.org/10.3390/w12010091
29. Domán A, Madarász J, László K (2017) In situ evolved gas analysis assisted thermogravimetric (TG-FTIR and TG/DTA–MS) studies on non-activated copper benzene-1,3,5-tricarboxylate. Thermochim Acta 647:62–69. https://doi.org/10.1016/j.tca.2016.11.013
30. Li Y, Miao J, Sun X, Xiao J, Li Y, Wang H, Xia Q, Li Z (2016) Mechanochemical synthesis of Cu-BTC@GO with enhanced water stability and toluene adsorption capacity. Chem Eng J 298:191–197. https://doi.org/10.1016/j.cej.2016.03.141
31. Azizi A, Feijani EA, Ghorbani Z, Tavasoli A (2021) Fabrication and characterization of highly efficient three component CuBTC/graphene oxide/PSF membrane for gas separation application. Int J Hydrogen Energy 46:2244–2254. https://doi.org/10.1016/j.ijhydene.2020.10.025

Websites

32. https://pubs.usgs.gov/of/2001/of01-041/htmldocs/xrpd.htm
33. https://lab-training.com/2014/09/system-features-ft-ir

Reactions Catalyzed by MOFs and Prospects for Applications

Versatile Metal–Organic Frameworks: Perspectives on Contribution in Reaction Catalysis and Applications

Sanjay Kumar, Kartika Goyal, Mansi, Sweta Kumari, and Shikha Gulati

Contents

1 Introduction 184
2 Potential of Metal–Organic Frameworks (MOFs) Over Other Conventional Catalysts 185
3 Fundamental Properties of MOFs Contributing Towards Catalysis 187
4 MOFs as Robust Host for Various Nanoparticles in Heterogeneous Catalysis 190
5 Reactions Catalyzed by MOFs and Prospects for Applications 192
5.1 Polymerization Reactions 193
5.2 Oxidation Reactions 193
5.3 Coupling and Cross-Coupling Reactions 194
5.4 CO_2 Conversion Reactions 195
6 Significance of MOFs as Lewis Acid Catalysts 197
6.1 In Condensation Reactions 197
6.2 In Cyanosilylation Reaction 198
6.3 In Friedel–Crafts Reaction 199
7 Conclusion and Future Perspectives 200
References 201

Abstract In the past decade, scientists have struggled to look for potential catalysts as the available ones were highly expensive with poor durability. Hence it is desirable to develop durable catalysts with low cost and more abundantly available resources. To this, Metal–Organic Frameworks (MOFs) stand out as a promising platform. MOFs as an emerging class of stable hybrid materials with multimodal structures, unique surface properties, high porosity, diverse composition, and crystallinity have become potential candidates for reaction catalysis in industrial applications. They are known to accelerate the reactions of high interest that even

S. Kumar · K. Goyal · Mansi · S. Kumari · S. Gulati (✉)
Department of Chemistry, Sri Venkateswara College, University of Delhi, New Delhi, Delhi 110021, India

S. Gulati (ed.), *Metal-Organic Frameworks (MOFs) as Catalysts*,
https://doi.org/10.1007/978-981-16-7959-9_7

surpass the shortcomings associated with homogeneous catalysts. Compared with homogeneous catalysts, MOF catalyst stands out in terms of recyclability and reusability for multiple cycles.. In view of the essential need of structural and chemical uniformity at the meso, nano, and atomic-scale level MOFs have attracted attention as model catalysts or catalyst-supports for a wide array of chemical transformations. The present chapter focuses on the limitations associated with available catalysts, fundamental properties of MOF, their role as a robust host for nanoparticles, reactions catalyzed, and application prospects. Besides, special attention has been given to the reactions that are of high industrial interest demonstrating the significant role of MOFs in lewis acid and heterogeneous catalysis.

Keywords Reaction catalysis · MOFs · Porous material · Robust host material · Surface properties

1 Introduction

Metal–Organic Frameworks (MOFs) are keenly structured with the coordination bonds between inorganic metal nodes and organic ligands (Fig. 1). Pronounced by high porosity, tunable functionality, and concentrated metal sites, they are convincing enough to act as the adsorbent and catalytic materials for the generation to come. Specifically, the existence of unsaturated metal centers and electron-deficient groups make them competent of acting as Lewis acid sites, which has further made MOFs highly potential candidates in catalysis applications [43].

These materials rely on Werner's coordination chemistry, wherein metal ions (or clusters) are coordinately connected by bridging ligands, to form infinite porous

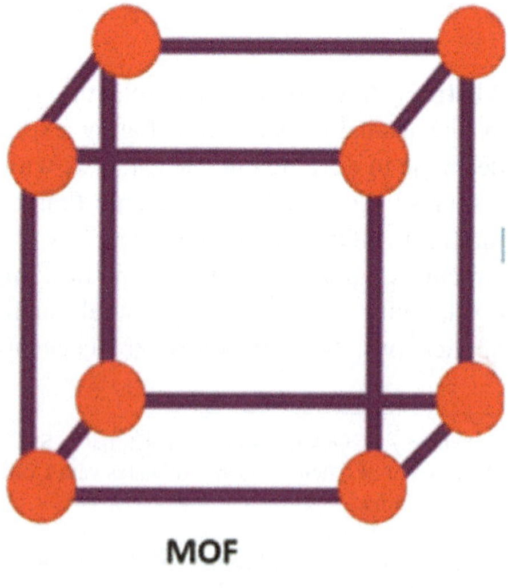

Fig. 1 Basic structural unit of MOF

frameworks. Prior to the mid-1990s, researchers mainly focused on two types of porous materials, either purely inorganic materials or carbon materials. MOF is an unprecedented class of porous material that combines the feature of both two types. Similar to zeolites, MOFs are generally prepared via hydrothermal or solvothermal methods, which permits for steady nucleation of crystal seed from hot solutions. However, MOFs use organic linkers to bridge metal clusters and form two or three-dimensional lattices, which offers enhanced structural diversity than classical porous materials. Zeolites have found broad application for catalysis in the chemical industry owing to their uniform porosity, but the types of reactions are limited to acid-catalyzed reactions [26].

With their outstanding, unique set of properties, MOFs form an intriguing class of porous materials, distinguishing them from other materials such as mesoporous silica, porous carbon, and microporous fully inorganic zeolites. Precisely tunable pore size, high surface area which provides a more available surface for interaction with guest species, chemical tailorability, and synthetic flexibility are some of the noteworthy properties of MOFs that set them apart. Most importantly, the structure and functionality of MOFs can be tuned during their synthesis, following the isoreticular principle [61]. MOFs can be functionalized by the addition of substituent functional groups on the backbone of the bridging ligands during synthesis. However, this strategy is limited as many of the desired functionalities cannot withstand the reaction conditions of MOF synthesis [4]. Additionally, even a small change in the reaction parameters, chemical nature of the linker, sterics, electronic configuration of the metal ion and the ligand, results in drastic changes in the MOF thus formed and we may not get the target MOF [61]. To overcome these challenges and limitations of pre-synthetic functionalization, an alternative strategy, known as post-synthetic modification, can be employed. It is an approach that leads to diverse functionalities without affecting the structural stability, porosity, and crystallinity of the as-synthesized framework [4, 36]. Owing to their unique set of properties and highly ordered crystalline structure, it is of paramount significance to mention that the position, as well as the degree of functionalization of MOFs, are controllable.

In this chapter, we will present how imparting functionality can further accelerate the reaction and elaborate the array of reactions catalyzed by MOFs. Furthermore, from the application perspective, we have highlighted their use as lewis acid catalyst, wherein we cite examples from recent research works to illuminate the concept.

2 Potential of Metal–Organic Frameworks (MOFs) Over Other Conventional Catalysts

MOFs are a new class of organic–inorganic hybrid crystalline porous materials composed of metal ions/clusters as nodes and organic ligands as linkers, first synthesized in 1999 [24]. These materials possess a number of significant

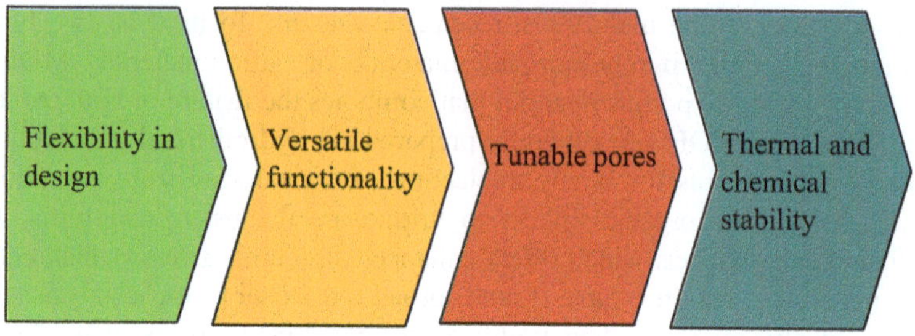

Fig. 2 The fascinating properties of MOFs

advantages over conventional catalysts (e.g., zeolites, silicates, organometallic complexes, transition metals, activated carbon, another homogeneous catalyst, etc.). The fascinating characteristics of MOFs that make them ideal for their application in various reactions are shown in Fig. 2.

This is worth noting that the presences of a large surface area and porous structure functionalize the guest species to be introduced into the pores and permit the substrates to approach the internal active sites. This facilitates mass transport and enables the encapsulation of precursors at the atomic level into the pores of MOFs to design functional materials with enhanced morphology [26].

The MOFs acts as active centers in catalysis that makes them capable of a range of organic transformation that could not be successfully catalyzed by the traditional catalyst. Besides other conventional catalysts, MOFs bridge the metal clusters with the utilization of organic linkers and this provide a broad range of MOFs configurations including nanocubes, nanoframes, nanowires with a precise combination of atomic metal ligands, it reflects the flexibility in the design and increases the structural assortment of MOFs than traditional materials. Also, theoretical studies indicate that thousands of MOFs are possible, to date more than 88,000 have been reported and the possibility to create new structures is even greater [35].

The thermal and chemical stability of catalysts plays an essential role in catalysis. Thermal stability is the ability of MOF compounds to resist any change in their physical structure and chemical properties upon heating to relatively high temperatures. In particular, the key attributes for determining the thermal stability of MOFs are:

- the arrangement of functional units
- ligand/nodes stability
- a coordinated solvent molecule's presence.

Generally, MOFs can only be heated up to 150–300 °C but according to a study by Ma et al. after interpenetration, MOFs are stable up to 400 °C maintaining their framework integrity, and after double interpenetration Yb MOF show thermal stability up to 500 °C with enhanced gas-adsorption feature [39]. In a recent study, the most extensively analyzed MOFs in catalysis are MOF-74, ZIF-8, CPO-27, UiO-66, CuBDC (HKUST-1), MIL-53, and MIL-101 which can withstand prolonged heating [56].

The chemical stability of MOFs, refers to their ability to resist the consequences of exposure to numerous chemicals in their environment, for example, moisture, solvents, acids, bases, and binary compound solutions. Additionally, the chemical stability of MOFs have been studied broadly in following three medium:

- Acidic medium
- Basic medium
- Hydrolytic medium.

The chemical stability of MOFs is significantly influenced by their intrinsic structure, which includes metal ion charge density, ions/clusters connectivity, basicity, and orientation, as well as hydrophobicity of ligand. Despite the fact that many MOFs are susceptible to structural degradation even in the ambient atmosphere due to the lability of coordination bonds between metal ions and ligands, in recent years an increasing number of MOFs with excellent chemical stability have been discovered by primarily using two methods. The first is to create stable unknown MOFs by de novo synthesis, while the second is to increase the stability of existing MOFs [15].

As a result, MOFs are ideal for their application in gas separation, chemical sensing, and numbers of catalysis reactions. Thus, all these factors contribute to the advancement in broader applications of MOFs over zeolites and other conventional catalysts.

3 Fundamental Properties of MOFs Contributing Towards Catalysis

Metal–Organic Frameworks (MOFs) are suitably heterogeneous catalysts that belong to a class of porous crystalline materials featuring a series of unique properties demonstrated in Fig. 3.

Currently, there are several types of catalytic processes which include biocatalysis, chemical catalysis, photocatalysis, electrocatalysis, and much more heterogeneous catalysis, where it is generally observed that different structural uniqueness of MOFs is beneficial for its utilization in these processes.

One of the most prominent characteristics of MOFs is their high porosity which is the pore space at micro-and mesoscale in frameworks, even sufficiently large pores allow access to harness the pore chemistry. In MOFs the pore spaces range from a few Angstroms to tens of nanometers, also a variety of pore sizes can be observed within the same material. MOFs can achieve size and shape selectivity due to their adjustable and well-defined pores as the voids facilitate the mass transfer from the exterior to the interior of materials which significantly also increases the surface area [35].

Furukawa et al. synthesized four MOFs, named MOF-180, MOF-200, MOF-205, MOF-210 respectively using $Zn_4O(CO_2)_6$ units with one or two types of

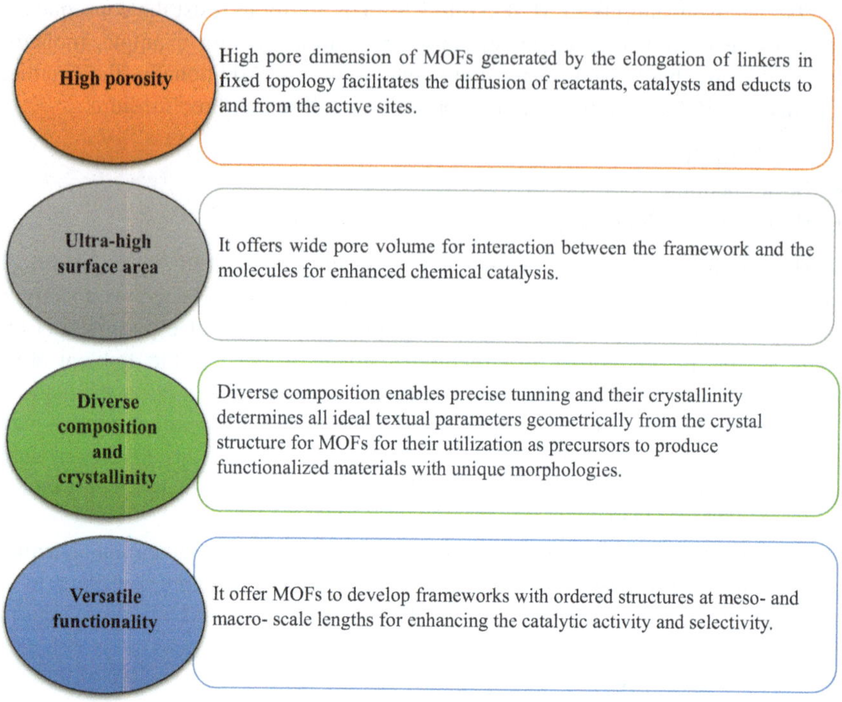

Fig. 3 Unique properties of MOFs

organic linkers. Their study highlighted the defining porosity property of MOFs, which is an important considering factor in gas storage application as MOF-200 possess the lowest known crystal density (0.22 g/cm^3) with 90% of the total volume is empty and this makes them porous material [21]. Porosity information of various MOFs is given in Table 1.

Metal–Organic Frameworks (MOFs) exhibits ultra-high surface area due to their highly organized pore structure and these offer more space for molecular adsorption as well as in chemical reactions. Also, various catalytic active sites/centers of MOFs such as open metal sites, bifunctional acid–base sites, and introduction of multiple pores inside the MOFs by post-synthetic modification (PSM), increases the surface area which is conducive for the improved properties of materials [14]. According to a study by Li and co-workers, Th-MOFs were prepared by is reticular synthesis method without changing the comprehensive topology of MOFs. The synthesized product Th-SINAP-13 exhibited the largest surface area of 3396.5 m^2/g with a void space of 74% among thorium materials [35].

The uniform and crystalline structure of MOFs makes the reactant accessible to active sites via open channels, which increases the selectivity of the ligands and modifies the components, this provides the possibility to synthesis a limitless number of MOFs materials. As a result, fundamental materials can be tuned

Table 1 Porosity information of several MOFs

MOF	Linker	Void volume (%)	References
MOF-200	4,4′,4″-(benzene-1,3,5 triyltribenzene-4,1-diyl) Tribenzene	90	Furukawa et al. [21]
MOF-205	2,6-napthalendicarboxylate and 4,4′,4″-benzene-1,3,5 triyltribenzene	85	Furukawa et al. [21]
MOF-210	Biphenyl-4,4′-dicarboxylate and 4,4′,4″ (benzene-1,3,5 triyl-tris ethylyne-2, 1-diyl) tribenzoate	89	Furukawa et al. [21]
UMCM-2	Thieno[3, 2-b]thiophene-5-dicarboxylate and 4,4′,4″-benzene-1,3,5 Triyltribenzene	83	Koh et al. [33]
Cr-MIL-101	1,4-benzene dicarboxylate	83	Férey et al. [18]

according to their specific application to provide unique structural features and desired chemical compositions. They offer better opportunities to explore structure–activity correlations and undergo computational investigations. Recently, the multi-metal catalyst has been developed to provide a substantial degree of freedom in composition design, further promoting the synergistic effect between components to improve its chemical performance [42].

Versatile functionalities of MOFs include the following:

- Catalysis
- Gas and liquid adsorption
- Gas storage and treatment
- Conductivity
- Food
- Sensoring and detection
- Textile reform.

The term catalysis describes a process wherein the reaction rate and results are altered by the presence of catalysts which usually don't get consumed during the reaction and are often removed after reaction to avoid the existence of impurity in the end product. Whereas the process of adsorption refers to the binding of the molecules on the surface, basically it is used to separate mixtures (liquid or gas) into their constituent components. Also, MOFs improve the effectiveness of membranes and filters by increasing the accessible surface area that binds molecules [35].

It is widely used in the storage of compressed gases, separation, removal of impurities and odors. It also, open new horizons for addressing conductivity challenges by developing electrodes and electrolyte composites. In the food industry, MOFs can be utilized in a variety of applications such as quality assurance, management of the shelf life, agrochemical distribution, etc. MOFs present the possibility for sensing and detecting small molecules, medical diagnostics, explosives, etc. because of the presence of conciliated porous structure. The fabrics reformed by MOFs decrease the odor and protect the wearer from hazardous chemicals [35].

4 MOFs as Robust Host for Various Nanoparticles in Heterogeneous Catalysis

The various unique properties possessed by MOFs as discussed earlier enable them to show the potential in heterogeneous catalysis. However, in order to further enhance their catalytic activity, MOFs can be made to undergo controlled integration with functional materials like metal nanoparticles (MNPs), polyoxometallates (POMs), quantum dots (QDs), enzymes, etc. Metal Nanoparticles are rapidly gaining researchers' interest as one of the most promising guest species. Because of the high surface energies of MNPs, they are thermodynamically unstable and tend to aggregate during the catalytic reaction, which generally leads to loss of catalytic activity [1, 63].

In this regard, the use of various surface capping agents has been recognized as a solution but such kind of surface contamination has a negative impact on the catalytic activity of MNPs. To this end, the immobilization of NPs inside porous materials like zeolites, porous silica, has been demonstrated as an effective approach. Amongst the porous materials, MOFs have been found to be the most promising candidates, because of the various reasons which are reported below [1, 7, 63]:

- They possess unique properties like diverse chemical composition, ultra-high surface area, and permanent porosity.
- They are versatile with tunable pore structures for meeting the desired requirements of MNPs.
- They are highly adaptable to catalytic design.
- Aggregation and leaching issues are eventually minimized due to confinement and electronic effects offered by MOFs.
- The inner pore surfaces of MOFs can be easily modified.

Broadly, there are three strategies for the preparation of MNP@MOF composites, as listed in Fig. 4 and discussed below in brief:

1. **Ship-in-bottle method**: In this approach, for the synthesis of MNPs, the preformed MOF is placed into a metal precursor followed by further treatment in order to reduce the metal precursor into its metallic state [24]. It includes procedures such as:

   (i) Solution impregnation
   (ii) Chemical vapor deposition

Fig. 4 Preparation of MNP@MOF composites

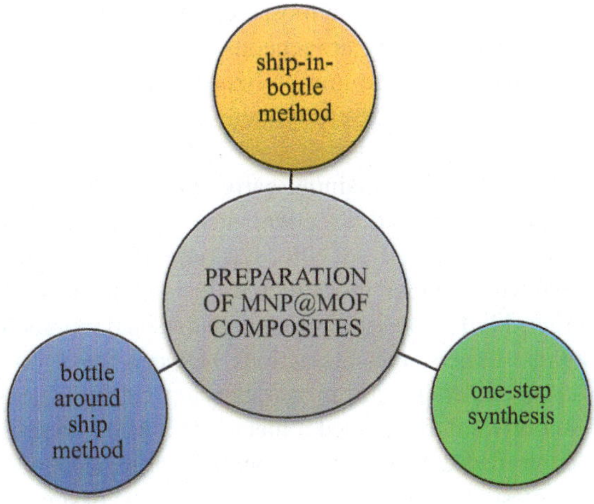

  (iii) Double solvent approach (DSA)
  (iv) Thermal decomposition.

2. **"Bottle around ship" method**: In this approach, the preformed MNPs are added during the synthesis of MOF. Due to this fact, better location and size control of MNPs can be obtained, as compared to the ship-in-bottle process [24]. However, such procedures are not suitable for all types of MOFs, UiO-66 and ZIF-8 have been mostly reported so far.
3. **One-step synthesis**: As the name suggests, this method involves one-step procedure where both MNPs and MOF precursors are mixed together to form the corresponding MNP@MOF composite.

Chen et al. developed an advanced Pd-MOF catalyst using a novel and efficient strategy. They encapsulated palladium precursors within the pores of UiO-67 prior to the MOF assembly [7]. This strategy allowed the Pd nanoparticles to evenly distribute within the cavities of MOFs. Moreover, the obtained composite showed significantly enhanced catalytic activity and stability compared to those prepared by the traditional impregnation method.

The combination of MOFs and MNPs for enhanced catalytic activity has attracted immense interest in recent years. This is because, in MNP@MOF composites, the multiple advantages of both the components are integrated due to the synergistic effect between MOFs and MNPs while the shortcomings of MNPs and MOFs are mitigated [58, 63].

The combination of MOFs and MNPs for enhanced catalytic activity has attracted immense interest in recent years. This is because, in MNP@MOF composites, the multiple advantages of both the components are integrated due to the synergistic effect between MOFs and MNPs while the shortcomings of MNPs and MOFs are mitigated [58, 63].

MOFs are one of the most excellent supports of noble metal NPs, especially for the application in heterogeneous catalysis. It is worth noting that the catalytically active metal ions are immobilized by the functional groups present on the organic linkers of MOFs, therefore, there is no loss of expensive metals [38]. Zhang et al., for the first time, reported a versatile strategy for the incorporation of noble metal NPs into MOFs containing carboxylic acid-based ligands. The resulting NP/MOF composites showed excellent shape selectivity in various reactions like olefin hydrogenation and aqueous reaction in the reduction of 4-nitrophenol. Also, it exhibited a higher molecular diffusion rate in CO oxidation [67]. As discussed earlier, MOFs consist of organic linkers and inorganic nodes. Due to the presence of organic linkers, the interactions between metal NPs and the inorganic nodes are weaker which results in unsatisfying catalytic activity. To overcome this issue, Tsumori et al. developed a metal/quasi-MOF composite, Au/MIL-101(Cr), through a controlled deligandation process in order to expose the inorganic Cr–O nodes to the guest Au NPs. This composite showed significantly enhanced catalytic performance in the low-temperature oxidation of CO [53].

Noble metal NPs have been extensively used in the field of heterogeneous catalysis. However, the practical use of these precious metals is limited due to their high costs and global reserve scarcities. Therefore, the development of non-noble MNP@MOF composites is important which possess similar catalytic activity as noble MNP@MOF composite [56]. To this, transition metals come up as a sustainable alternative owing to their low cost and abundance. In a recent study, Habib and co-workers have discussed the applications of non-noble MNP@MOF composites in heterogeneous catalysis. These composites are found to be useful in CO and CO_2 conversion reactions [24].

The field of application of MOF-based composites in heterogeneous catalysis is still in its developing stage. Some large-scale synthetic strategies need to be developed in order to synthesis MNP@MOF composites at affordable costs for their practical applications. It is expected that these composites will have a bright future if persistent efforts are made towards such challenges [63].

5 Reactions Catalyzed by MOFs and Prospects for Applications

In recent years, MOFs have gained tremendous attention as potential catalysts. Catalysis is one of the most successful implementations of these materials. Earlier studies were focused mainly on demonstrating the catalytic sites possessed by MOFs which are required to catalyze a certain process. Now, with the advancement in the research, the catalytic applications of MOFs range from conventional catalytic implementations to photocatalysis, and electrocatalysis too [2]. Different MOFs have been used to catalyze a wide variety of chemical transformations. Some of the important reactions catalyzed by MOFs are discussed below.

5.1 Polymerization Reactions

Catalysts enable the rapid, and efficient formation of desired polymers from the corresponding monomers. Though various molecular catalysts have been found effective for these reactions, their separation from the reaction products is quite difficult. To overcome this issue, the use of heterogeneous catalysts has been suggested. MOFs have attracted attention in this field owing to their fundamental desirability of chemical and structural uniformity, high porosity, and well-defined crystallography [23].

Polyalkenes are among the most ubiquitous polymers. They are used in electronic connections, building, and construction, toys, etc. Li and co-workers prepared a new class of PCP catalysts for ethylene polymerization, using Zr^{4+} ions and a tritopic phenoxy-imine linker. These PCP catalysts were found to exhibit moderate to high activity. The study nicely demonstrated that a prospective catalyst can be used as a building block for developing a porous network, which is functional for ethylene polymerization [34].

Dienes also play an important part in everyday life, since they are widely utilized in the automotive and roofing industries due to their resistance to ozone, ultraviolet light, and heat. Russell et al. demonstrated that MOF compounds containing two sets of different lanthanide elements (Nd^{3+}, Tb^{3+}/Eu^{3+}) can be used for luminescent polymer production. Neodymium was used because of its well-known catalytic properties for dienes polymerization. With the aim to provide luminescent properties, a second transition element, terbium or europium, was added to the MOF structure. Various MOFs satisfying the aforementioned criteria were prepared and used for the catalysts for the polymerization of isoprene [49].

5.2 Oxidation Reactions

Recently, various research groups have attempted to use MOFs for oxidation reactions. Torbina et al. investigated chromium-based MOFs, MIL-100, and MIL-101, for the oxidation of propylene glycol, using *tert*-butyl hydroperoxide (TBHP) as an oxidant. Hydroxyacetone was obtained as the main oxidation product while acetaldehyde and acetic acid were the minor oxidation products. MIL-101 catalyst could preserve its structure during a minimum of three catalytic cycles. Moreover, it could be recycled without much deterioration of activity and selectivity [52].

Kholdeevaa et al. prepared Fe, and Cr containing MOFs, MOF-100, and MOF-101, and compared their catalytic activities in allylic oxidation of alkenes with molecular oxygen and oxidation of anthracene with *tert*-butyl hydroperoxide (TBHP). In the oxidation of anthracene, 100% selectivity was observed for both Cr-MOFs and Fe-MIL-101, with 92–100% anthracene conversion. In the case of oxidation of alkenes, Cr-based MOFs gave unsaturated alcohols while Fe-containing

MOFs produced unsaturated alcohols. The stability of MOFs was found to increase in the following order: Fe-MIL-101 < Fe-MIL-100 < Cr-MIL-100, Cr-MIL-101 [32].

It is worth noting that the polyoxometallate (POM)-based MOFs are efficient catalysts for the oxidation of various sulfides into sulfoxides, with high conversion (90–100%) and selectivity (95–100%). Heterogeneous catalysts like PW-MOF are more active catalysts as compared to H_2O_2, Also, these could be recovered easily and reused five times, without much loss of catalytic activity [25].

5.3 Coupling and Cross-Coupling Reactions

Coupling reactions are widely opted by organic chemists for carrying out various organic transformations and chemical synthesis. Nguyen and co-workers developed a metal–organic framework $Fe_3O_4(BPDC)_3$ and used it to catalyze the direct C–N coupling of azoles with ethers by oxidative C–H activation, to produce azole derivatives. In the study, it was demonstrated that the contribution of the leached active iron species to the production of the preferred azole product was negligible. The catalyst could be reused several times without a significant change in the catalytic efficiency [41]. Some other MOFs which have been used to catalyze coupling reactions are summarized in Table 2.

Cross-coupling reactions, which produce new carbon–carbon, and carbon-heteroatom bonds, are performing well in the pharmaceutical industry. Homogeneous catalysts are popular in this field because of their relative stability in air and water. However, their industrial applications are limited because these are:

- prone to aggregation
- difficult to recycle.

Table 2 Various MOF catalysts used for coupling reactions

MOF catalyst used	Coupling reaction	References
Cu-BDC MOF	Suzuki coupling	Rostamnia et al. [47]
UiO-68Se	Aerobic cross-dehydrogenative coupling	Zhang et al. [68]
Pd(II)-porphyrinic MOF	Heck coupling	Chen and Jiang [8]
Pd NPs supported on UiO-66-NH_2	Suzuki cross-coupling	Kardanpour et al. [31]
Pd@MIL-101	Biginelli and Hantzsch coupling	Rostamnia and Morsali [48]
NPC-Pd MOF	Suzuki–Miyaura coupling	Zhang et al. [66]
Transition metal-free MOF	Oxidative coupling of amines	Qiu et al. [44]
$Cu(4\text{-ba})_2\{(\text{solvent})\}_n$	Chan–Lam coupling, Suzuki–Miyoura coupling, Heck coupling	Wang et al. [55]
Cu^{2+}-M′MOF ZJU-22	Cross-dehydrogenative coupling	Yang et al. [64]

MOFs have been gradually applied in such organic transformations, taking advantage of being a heterogeneous catalytic material for easy catalyst recovery [30]. Li et al. prepared a Pd(II)@UiO-67 composite, using a mixed ligand approach. They used a direct incorporation strategy to immobilize the organic palladium complex Pd (H_2bpydc)Cl_2 on a porous MOF, UiO-67. The obtained composite showed high efficiency for the catalytic conversion of aryl chlorides. High yields were obtained for the Heck and Suzuki–Miyaura coupling reactions of aryl chlorides. Moreover, the catalytic efficiency of the catalyst remained almost unaffected after at least five cycles [6].

5.4 CO_2 Conversion Reactions

The rising quantity of CO_2 in the atmosphere is mostly due to the combustion of fossil fuels which leads to the emission of greenhouse gases and this is well known to have a substantial influence on climate change causing global warming. Consequently, the reduction of CO_2 to commodity products is a recent critical study in green chemistry. But, the conversion of CO_2 into valuable chemicals is a major challenge because of the thermodynamic stability of CO_2. Considering the major challenge in reduction due to the complexity of the multielectron step involved, MOFs have attracted intense research interest as it could become a replacement for currently used commercial techniques of CO_2 sorption due to their selectivity, structural features for efficient formation of desirable products [56].

MOFs offer multiple advantages such as active surface area, robust pore distribution, crystallinity, etc. These materials are considered to mitigate the harmful impact of CO_2 through fixation and conversion of CO_2 using different chemical reduction reaction approaches. Also, the careful selection of linkers as well as by adjustment of pores shape/size and surface area considerably increases the gas-adsorption efficiency [11].

Various catalytic processes such as photocatalytic or electrocatalytic reduction as well as hydrogenation are used for the conversion of CO_2 into value-added products. General products of catalytic conversion of CO_2 are carbon monoxide (CO), formic acid (HCOOH), methanol (CH_3OH), ethanol (C_2H_5OH), methane (CH_4), and others. Therefore, CO_2 reduction occurs through various processes and reduction conditions to develop targeted products using appropriate MOFs [67].

Photoreduction

Interestingly, photocatalytic CO_2 reduction is well known as artificial photosynthesis which is used to convert CO_2 to carbon-based fuels through the utilization of solar energy. This is a straightforward and ecologically beneficial method of converting solar energy into chemical energy. The aforementioned process was first discovered by Inoue et al. in 1979, combining a suspended semiconductor photocatalyst with Xe lamp irradiation [28].

It's important to note that photocatalytic reduction of CO_2 can not only reduce the atmospheric CO_2 concentration but also alleviate the energy constraints. However, the traditional photocatalysts endure some inevitable defects as these materials are inert in chemical reactivity and are difficult to recollect and recycle for reuse. Hence MOFs are ideal for utilization in photoreduction reactions as they inherit excellent adsorption capability toward CO_2 transformation and unique structural characteristics [10]. As MOFs are formed from metal or metal cluster nodes linked to multi-dented organic components, the metal nodes can initiate photocatalysis. Based on the studies of controlling the modifications on the metal ions or organic linkers, it is easier to tune the light adsorption capacity of MOFs.

In this regard, for the very first time, Li et al. reported that a Ti-containing MOF material (NH_2-MIL-125(Ti)) was prepared on replacing the linker in MIL-125(Ti). It was revealed that $HCOO^-$ was the only produced product under visible light irradiation [19].

Electroreduction

The major by-product of fuel burning is CO_2 which is a relatively low energy molecule, so its reduction necessitates large reduction potentials. Electrocatalytic CO_2 reduction occurs in the cathode, which uses MOFs as an electrode, and involves the absorption of CO_2 on the cathode's surface, electron transfer from the cathode to the absorbed CO_2, and product desorption from the cathode's surface. Simultaneously, the oxygen evolution process takes place in the anode compartments, releasing H^+ that migrates to the cathode to facilitate CO_2 reduction. Copper (Cu) materials are recommended for this process as they advance strong electrochemical activity towards multi-carbon products [37].

Han and co-workers reported the first paper on the reduction of CO_2 electrochemically by utilizing MOFs as cathode and ionic liquid (ILs) as an electrolyte to produce multielectron reduction products, i.e., methane. In another study by Wang et al. zeolitic imidazolate framework (ZIF-8) nanomaterial was synthesized and used as a framework for electrocatalytic CO_2 reduction and yielded 65% CO [57].

It's worth noting that in a comparative study by Yadav and co-workers, differently shaped tin(Sn) catalysts were synthesized using the new solar electro-deposition method. The selective formation of HCOOH with faradaic efficiency of 94.5% at 1.6 V versus Ag/AgCl using the prepared Sn electrocatalyst [60]. It is proposed that Sn-based catalysts are generally used in CO_2 electroreduction to produce HCOOH is because of their intrinsic activity. To enhance this activity, a strategy of node doping in MOFs for CO_2 electroreduction was introduced by Geng et al. Herein, an active tin (Sn) node was doped into zeolitic imidazolate framework-8 (ZIF-8) using an ion-exchange method, resulting in the highest faradaic efficiency of 74% with a current density of 27 mA cm^{-2} at −1.1 V versus RHE for production of HCOOH [22].

Hydrogenation

Hydrogenation is a reduction process in which molecular hydrogen is added to an element, compound, or molecule, usually in the presence of a catalyst. The most commonly used catalysts are metal nickel, platinum, and palladium [4]. Using a renewable source of H_2 to capture CO_2 and convert it to methanol is a potential technique to minimize net CO_2 emissions while providing valuable fuels. In a study by Ye et al. a catalyst consisting of microporous metal–organic framework (UiO-67) further functionalized with catalytically active lewis pair, the functional group was synthesized by density functional theory. This catalyst was capable of producing methanol from CO_2 and H_2 [65].

6 Significance of MOFs as Lewis Acid Catalysts

As mentioned earlier in this chapter, MOFs are a class of hybrid crystalline materials, where metal ions/clusters are erected as nodes and organic ligands as linkers. Generally, these nodes are eliminated from crystal lattice by heating at higher temperatures to form an' Open metal site' (OMS). This OMS mainly constitutes transition metals that can serve as Lewis acid [27]. In the year 1994, for the very first time, Fujita et al. reported a Cd-bipyridine MOF that acts as a lewis acid catalyst [20]. Thereafter, researchers have widely studied various reactions where MOFs act as lewis acid. Some of the examples of these reactions are condensation reactions, cyanosilylation reactions, and Friedel–crafts reactions.

6.1 In Condensation Reactions

Dhakshinamoorthy et al. reported the Claisen–Schmidt Condensation catalyzed by Metal–Organic Frameworks wherein in particular they take Metal–organic framework [Fe(BTC) (BTC = 1,3,5-benzenetricarboxylic acid)] to prepare selectively different chalcone derivatives bearing various functionalities [13].

In another similar report, N-Hydroxyphthalimide (NHPI) was assimilated on a Fe III-based metal–organic framework [NHPI/Fe(BTC), BTC: 1,3,5-benzenetricarboxylate)]. This obtained structure is further utilized as a heterogeneous catalyst for the aerobic oxidation of benzylamine and its derivatives to synthesize the corresponding benzyl imines under typical conditions with molecular oxygen as the only oxidant at 1008 °C. The catalytic activity of NHPI/Fe (BTC) compared with other heterogeneous catalysts (supported gold nanoparticles) proves to be superior in terms of percentage conversion and selectivity [12].

In a recent pH-sensitive study of different structural materials, 3D MOF materials stand out for aldol condensation reactions of various aromatic aldehydes with acetone under heterogeneous conditions. pH-controlled synthesis of MOFs involving magnesium and pyrazole-3,5-dicarboxylic acid (H3L) was done, which permits a progressive increase of dimensionality from zero-dimensional to three-dimensional frameworks. The porous 3D compound has been found to effectively catalyze aldol condensation reactions [50].

6.2 In Cyanosilylation Reaction

Cyanosilylation is a nucleophilic addition reaction in which nitrile and silyl groups are added across double or triple bonds of carbonyl compounds to produce cyanohydrins using Lewis acids as catalysts. It's noteworthy, during the reaction procedure, the reagent trimethylsilylcyanide (TMSCN) is preferred because it is easy to handle and poses low dissociation energy for Si@C bond. Also, it has remarkable conversion efficiency without causing adverse effects [59].

- The mechanism of cyanosilylation reaction involves three steps
  1. Aldehyde oxygen coordinates onto Lewis acid sites.
  2. Nitrile groups attack carbonyl groups.
  3. Silyl groups isomerize to form cyanohydrin.

The cyanosilylation reaction indicates certain important characteristics for effective catalytic activity in MOFs. They are reported below:

- Pore size
- Active sites number
- Dimension of MOFs
- Size of particle
- Condition of solvent.

The large pore size of the MOF allows the easy accessibility of the substrates to the exposed metal sites in the cages or channels, which is very important for the reaction to proceed smoothly. In MIL-101(Cr), the presence of wider pores makes it possible to perform product separation and incrementation relative to MIL-47, MIL-53(Al), and UiO-66 [51].

The mass transfer capability and open metal sites play a prominent role in cyanosilylation reaction. In a view to enhancing these properties, Hu et al. used the strategy of mixed ligand to synthesize HP-CuBTC framework, developed by integrating the BTC ligand with defective linkers. In comparison to pristine Cu-BTC, the latter exhibits significantly improved lewis acid catalytic activity in cyanosilylation reaction for the conversion of benzaldehyde to cyanohydrins [51].

Recently, for practical application of chiral MOFs as asymmetric cyanation catalysts has been studied. To this, Zhu and co-workers reported VO-MOF complex which demonstrated improved stereoselectivity and cyanation of aldehyde [69].

6.3 In Friedel–Crafts Reaction

The alkylation process has been widely employed by chemical and petrochemical industries to produce different commodities and fine products. Conventionally, this process was carried out with help of zeolites such as ZSM-5, Y-Zeolite at high temperatures. But this high temperature usually deactivates the catalyst leading to poor yields of product. To this instance, various MOFs such as MOF-5, IRMOF-3, or MOF-69 °C were employed for tert-butylation of toluene and biphenylene at 170 °C. Cu-MOF-74 is also used to perform the alkylation of anisole at low temperatures [45].

The catalytic behavior of Metal–Organic Frameworks of different structures (Fe (BTC), MIL-100 (Fe), MIL-100(Cr), and $Cu_3(BTC)_2$) was investigated in annulation reaction between 2-methyl-3- buten-2-ol and phenols differing in size (phenol, 2-naphthol). The highest conversions of phenols (45 and 75% after 1300 min of TOS for phenol and 2-naphthol, respectively) and selectivity (45 and 65% at 16% of phenol and 2-naphthol conversion, respectively) to target benzopyran were achieved over MIL-100 (Fe) possessing intermediate Lewis acidity, perfect crystalline structure, and the highest SBET surface area [3]. Table 3 summarizes the various MOF catalysts used for Friedel–crafts reactions.

Table 3 Various MOF catalysts used for Friedel–crafts reactions

S. No.	Type of MOF catalyst	Friedel–crafts reaction	References
1	Cu-MOF (MOF-891)	Synthesis of bis(indolyl)methanes	Nguyen et al. [40]
2	HPW@Zr-BTC	Acylation of anisole with benzoyl chloride	Ullah et al. [54]
3	ZnMOF	Alkylation of indoles with nitroalkenes	Rao and Mandal [46]
4	Al-based MIL-53	Alkylation of benzene with ethanol	Rahmani and Rahmani [45]
5	MIL-53(Al)	Acylation of indole with benzoyl chloride	Yan et al. [62]
6	Cu-MOF-74	Acylation of anisole	Calleja et al. [5]
7	Zr-MOF	Benzoylation of arenes	Doan et al. [16]
8	MIL-53(Al) @SiO_2@Fe_3O_4	Acylation of 2-methylindole with benzoyl chloride	Jiang et al. [29]
9	Zn MOF	Alkylation of toluene with benzyl bromide	Farzaneh and Mortazavi [17]
10	Sulfonated Zr-MOF	Acylation of p-xylene with benzoyl chloride	Chung et al. [9]
11	Cu4I4-MOF	Alkylation of indoles with acetals	Zhu et al. [70]

7 Conclusion and Future Perspectives

Progressing interest and development in the domain of MOF studies is continued which is justified enough by various applications of MOF material. With remarkable porosity and distinguishable functional properties, they have been identified as materials of high interest for the application of heterogeneous catalysis. Besides the large array of successful applications, there are pertinent challenges that MOFs face in their conventional aspects, and hence to further improving various strategies have been suggested. Functionalization is commonly utilized for exploiting the catalytic activity by many folds. In this chapter, we have thoroughly summarized various reactions that are accelerated by the MOF units.

Besides the edge of heterogeneous catalysts, various large-scale industrial processes still rely on homogeneous catalysis (e.g., Wacker oxidation, hydroformylation, ethylene oligomerization), owing to the lack of compositional and electronic control of heterogeneous catalysts. Collaborative efforts from industry experts and researchers from different fields can help in overcoming the obstacles and ensure advancement in the field of Metal–Organic Frameworks.

List of Abbreviations

MOF	Metal-Organic Framework
HKUST	Hong Kong University of Science and Technology
IRMOFs	Isoreticular Metal-Organic Framework
BDC	1,4 Benzenedicarboxylic Acid
MIL	Materials Institute Lavoisier
H_4TPP	5,10,15,20-tetra (1H-pyrazol-4-yp porphyrin)
ZIF	Zeolitic Imidazolate Framework
BTC	1,3,5benzenedicarboxylate
TCA	Tricarboxy Triphenyl Amine
TBHP	*tert*-butyl hydroperoxide
POM	polyoxometallate
OMS	Open metal site
NHPI	N-Hydroxyphthalimide
TMSCN	trimethylsilylcyanide
PCP	Porous Coordination Polymers
CUS	Coordination Unsaturated Sites
Bpy	4,4′-bipyridyl
PyC	Pyrrole-2-carboxylate
Bpdc	4,4′-biphenyldicarboxylate
Bpydc	2,2′-bipyridine-5,5′-dicarboxylate
NPs	Nanoparticles
MNPs	Metal Nanoparticles
UiO	Universitetet i Oslo (University of Oslo)

CAL Coordinative Alignment
PCN Porous Coordination Network
PSM Postsynthetic Modifications

Important Links

1. https://www.google.com/amp/s/blog.novomof.com/blog/gasseparation-zeolites-vs-mofs%3fhs_amp=true
2. https://www.nanowerk.com/mof-metal-organic-framework.php
3. https://novomof.com/metal-organic-frameworks/?hsCtaTracking=ec8597ef-7650-4f00-9975-26a88ba24130%7C8d8099f6-9030-4abd-8c17-bc155c09fe0a
4. https://www.nature.com/articles/s42004-019-0184-6
5. https://www.sigmaaldrich.com/IN/en/applications/chemistry-and-synthesis/synthetic-methods/photocatalysis
6. https://www.nanowerk.com/news2/green/newsid=55082.php
7. https://www.thoughtco.com/definition-of-hydrogenation-604530
8. https://www.azom.com/article.aspx?ArticleID=16233.

References

1. Aijaz A, Xu Q (2014) Catalysis with metal nanoparticles immobilized within the pores of metal-organic frameworks. J Phys Chem Lett 5(8):1400–1411
2. Alhumaimess MS (2020) Metal–organic frameworks and their catalytic applications. J Saudi Chem Soc 24(6):461–473. https://doi.org/10.1016/j.jscs.2020.04.002
3. Anugrah S (2015) 2015. No Title طرق تدريس اللغة العربية. *Экономика Региона*: 32
4. Burrows AD (2013) Metal organic frameworks as heterogeneous catalysts. Post-synthetic modification of MOFs
5. Calleja G et al (2014) Copper-based MOF-74 material as effective acid catalyst in Friedel-crafts acylation of anisole. Catal Today 227:130–137. https://doi.org/10.1016/j.cattod.2013.11.062
6. Chen L, Rangan S et al (2014) A molecular Pd(Ii) complex incorporated into a MOF as a highly active single-site heterogeneous catalyst for C–Cl bond activation. Green Chem 16 (8):3978–3985
7. Chen L, Chen H, Luque R, Li Y (2014) Metal−organic framework encapsulated Pd nanoparticles: towards advanced heterogeneous catalysts. Chem Sci 5(10):3708–3714. https://doi.org/10.1039/C4SC01847H
8. Chen, YZ, Jiang HL (2016) Porphyrinic metal-organic framework catalyzed heck-reaction: fluorescence 'turn-on' sensing of Cu(II) ion. Chem Mater 28(18):6698–6704
9. Chung YM, Kim HY, Ahn WS (2014) Friedel-crafts acylation of p-xylene over sulfonated zirconium terephthalates. Catal Lett 144(5):817–824
10. Dao XY et al (2019) Solvent-free photoreduction of CO_2 to CO catalyzed by Fe-MOFs with superior selectivity. Inorg Chem 58(13):8517–8524
11. De D et al (2016) A versatile CuII metal-organic framework exhibiting high gas storage capacity with selectivity for CO_2: conversion of CO_2 to cyclic carbonate and other catalytic abilities. Chem Eur J 22(10):3387–3396

12. Dhakshinamoorthy A, Alvaro M, Garcia H (2010) Aerobic oxidation of benzyl amines to benzyl imines catalyzed by metal-organic framework solids. ChemCatChem 2(11):1438–1443
13. Dhakshinamoorthy A (2010) Claisen-Schmidt condensation catalyzed by metal-organic frameworks. Adv Synth Catal 352(4):711–717
14. Dhakshinamoorthy A, Li Z, Garcia H (2018) Catalysis and photocatalysis by metal organic frameworks. Chem Soc Rev 47(22):8134–8172
15. Ding M, Cai X, Jiang HL (2019) Improving MOF stability: approaches and applications. Chem Sci 10(44):10209–10230
16. Doan TLH et al (2016) An efficient combination of Zr-MOF and microwave irradiation in catalytic Lewis acid Friedel-crafts Benzoylation. Dalton Trans 45(18):7875–7880
17. Farzaneh F, Mortazavi SS (2017) Zn metal organic framework as a heterogeneous catalyst for the alkylation of toluene with benzyl bromide. React Kinet Mech Catal 120(1):333–344
18. Férey G et al (2005) A chromium terephthalate-based solid with unusually large pore volumes and surface area to cite this version: HAL Id: Hal-03182295 a chromium terephthalate—based solid with unusually large pore volumes and surface area. Science 309:2040–2042
19. Fu Y et al (2012) An amine-functionalized titanium metal-organic framework photocatalyst with visible-light-induced activity for CO_2 reduction. Angew Chem—Int Ed 51(14):3364–3367
20. Fujita M, Washizu S, Ogura K, Kwon YJ (1994) Preparation, clathration ability, and catalysis of a two-dimensional square network material composed of cadmium(II) and 4, 4′-bipyridine. J Am Chem Soc 116(3):1151–1152
21. Furukawa H et al (2010) Ultrahigh porosity in metal-organic frameworks. Science 329 (5990):424–428
22. Geng W et al (2020) Induced CO_2 electroreduction to formic acid on metal−organic frameworks via node doping. Chemsuschem 13(16):4035–4040
23. Goetjen TA et al (2020) Metal-organic framework (MOF) materials as polymerization catalysts: a review and recent advances. Chem Commun 56(72):10409–10418
24. Habib, NR, Asedegbega-Nieto E, Mengesha A, Diaz I (2021) Non-noble MNP@MOF: synthesis and applications to heterogenous catalysis. Dalton Trans
25. Haddadi H, Hafshejani SM, Farsani MR (2015) Selective and reusable oxidation of sulfides to sulfoxides with hydrogen peroxide catalyzed by organic-inorganic polyoxometalate-based frameworks. Catal Lett 145(11):1984–1990
26. Hailong Q, Xiuping Y (2020) Recent advances in the application of metal-organic frameworks for sample pretreatment. Chin J Chromatogr (Se Pu) 38(1):22–27
27. Hu Z, Zhao D (2017) Metal-organic frameworks with Lewis acidity: synthesis, characterization, and catalytic applications. CrystEngComm 19(29):4066–4081
28. Inoue T, Fujishima A, Konishi S, Honda K (1979) Photoelectrocatalytic reduction of carbon dioxide in aqueous suspensions of semiconductor powders [3]. Nature 277(5698):637–638
29. Jiang S, Yan J, Habimana F, Ji S (2016) Preparation of magnetically recyclable MIL-53(Al) @SiO_2@Fe_3O_4 catalysts and their catalytic performance for Friedel-crafts acylation reaction. Catal Today 264:83–90. https://doi.org/10.1016/j.cattod.2015.10.003
30. Kang YS et al (2019) Metal–organic frameworks with catalytic centers: from synthesis to catalytic application. Coord Chem Rev 378:262–280
31. Kardanpour R et al (2014) Highly dispersed palladium nanoparticles supported on amino functionalized metal-organic frameworks as an efficient and reusable catalyst for Suzuki cross-coupling reaction. J Organomet Chem 761:127–133. https://doi.org/10.1016/j.jorganchem.2014.03.012
32. Kholdeeva OA et al (2014) Hydrocarbon oxidation over Fe- and Cr-containing metal-organic frameworks MIL-100 and MIL-101-a comparative study. Catal Today 238:54–61. https://doi.org/10.1016/j.cattod.2014.01.010
33. Koh K, Wong-Foy AG, Matzger AJ (2009) A porous coordination copolymer with over 5000 m^2/g BET surface area. J Am Chem Soc 131(12):4184–4185

34. Li, H et al (2015) Highly active self-immobilized FI-Zr catalysts in a PCP framework for ethylene polymerization. Chem Commun 51(93):16703–16706
35. Li ZJ et al (2020) Modulated synthesis and isoreticular expansion of Th-MOFs with record high pore volume and surface area for iodine adsorption. Chem Commun 56(49):6715–6718
36. Lian X et al (2017) Enzyme-MOF (metal-organic framework) composites. Chem Soc Rev 46 (11):3386–3401
37. Lu Q, Jiao F (2016) Electrochemical CO_2 reduction: electrocatalyst, reaction mechanism, and process engineering. Nano Energy 29:439–456
38. Luo S et al (2019) Metal organic frameworks as robust host of palladium nanoparticles in heterogeneous catalysis: synthesis, application, and prospect. ACS Appl Mater Interfaces 11 (36):32579–32598
39. Ma S, Wang XS, Yuan D, Zhou HC (2008) A coordinatively linked Yb metal-organic framework demonstrates high thermal stability and uncommon gas-adsorption selectivity. Angew Chem—Int Ed 47(22):4130–4133
40. Nguyen HTD, Nguyen TT, Nguyen PTK, Tran PH (2020) A highly active copper-based metal-organic framework catalyst for a Friedel-crafts alkylation in the synthesis of bis(indolyl) methanes under ultrasound irradiation. Arab J Chem 13(1):1377–1385. https://doi.org/10.1016/j.arabjc.2017.11.009
41. Nguyen KD et al (2016) Direct C–N coupling of azoles with ethers via oxidative C–H activation under metal-organic framework catalysis. J Ind Eng Chem 44:136–145. https://doi.org/10.1016/j.jiec.2016.08.025
42. Paquin F et al (2015) Multi-phase semicrystalline microstructures drive exciton dissociation in neat plastic semiconductors. J Mater Chem C 3(207890):10715–10722
43. Pascanu V, Miera GG, Ken Inge A, Martín-Matute B (2019) Metal-organic frameworks as catalysts for organic synthesis: a critical perspective. J Am Chem Soc 141(18):7223–7234
44. Qiu X, Len C, Luque R, Li Y (2014) Solventless oxidative coupling of amines to imines by using transition-metal-free metal-organic frameworks. Chemsuschem 7(6):1684–1688
45. Rahmani E, Rahmani M (2018) Al-Based MIL-53 metal organic framework (MOF) as the new catalyst for Friedel-crafts alkylation of benzene. Ind Eng Chem Res 57(1):169–178
46. Rao PC, Mandal S (2017) Friedel–crafts alkylation of indoles with nitroalkenes through hydrogen-bond-donating metal-organic framework. ChemCatChem 9(7):1172–1176
47. Rostamnia S, Alamgholiloo H, Liu X (2016) Pd-grafted open metal site copper-benzene-1,4-dicarboxylate metal organic frameworks (Cu-BDC MOF's) as promising interfacial catalysts for sustainable Suzuki coupling. J Colloid Interface Sci 469:310–317. https://doi.org/10.1016/j.jcis.2016.02.021
48. Rostamnia S, Morsali A (2014) Basic isoreticular nanoporous metal-organic framework for Biginelli and Hantzsch coupling: IRMOF-3 as a green and recoverable heterogeneous catalyst in solvent-free conditions. RSC Adv 4(21):10514–10518
49. Russell S, Loiseau T, Volkringer C, Visseaux M (2015) Luminescent lanthanide metal organic frameworks for Cis-selective isoprene polymerization catalysis. Inorganics 3(4):467–481
50. Sen R, Saha D, Koner S (2012) Controlled construction of metal-organic frameworks: hydrothermal synthesis, X-Ray structure, and heterogeneous catalytic study. Chem Eur J 18 (19):5979–5986
51. Serhan M et al (2019) Total iron measurement in human serum with a smartphone. In: AIChE annual meeting, conference proceedings 2019-November
52. Torbina VV et al (2016) Propylene glycol oxidation with tert-butyl hydroperoxide over Cr-containing metal-organic frameworks MIL-101 and MIL-100. Catal Today 278:97–103. https://doi.org/10.1016/j.cattod.2016.04.008
53. Tsumori N et al (2018) Quasi-MOF: exposing inorganic nodes to guest metal nanoparticles for drastically enhanced catalytic activity. Chem 4(4):845–856. https://doi.org/10.1016/j.chempr.2018.03.009
54. Ullah L et al (2018) 12-Tungstophosphoric acid niched in Zr-based metal-organic framework: a stable and efficient catalyst for Friedel-crafts acylation. Sci China Chem 61(4):402–411

55. Wang B, Yang P, Ge ZW, Li CP (2015) A porous metal-organic framework as active catalyst for multiple C–N/C–C bond formation reactions. Inorg Chem Commun 61:13–15. https://doi.org/10.1016/j.inoche.2015.08.010
56. Wang Q, Astruc D (2020) State of the art and prospects in metal-organic framework (MOF)-based and MOF-derived nanocatalysis. Chem Rev 120(2):1438–1511
57. Wang Y, Hou P, Wang Z, Kang P (2017) Zinc imidazolate metal-organic frameworks (ZIF-8) for electrochemical reduction of CO_2 to CO. ChemPhysChem 18(22):3142–3147
58. Xiang W, Zhang Y, Lin H, Liu CJ (2017) Nanoparticle/metal-organic framework composites for catalytic applications: current status and perspective. Molecules 22(12)
59. Yadav A, Kanoo P (2019) Metal-organic frameworks as platform for Lewis-acid-catalyzed organic transformations. Chem Asian J 14(20):3531–3551
60. Yadav VSK, Noh Y, Han H, Kim WB (2018) Synthesis of Sn catalysts by solar electro-deposition method for electrochemical CO_2 reduction reaction to HCOOH. Catal Today 303(June):276–281
61. Yaghi OM, Kalmutzki MJ, Diercks CS (2019) Functionalization of MOFs. Introduction to reticular chemistry, pp 145–176
62. Yan J et al (2015) Metal-organic framework MIL-53(Al): synthesis, catalytic performance for the Friedel-crafts acylation, and reaction mechanism. Sci China Chem 58(10):1544–1552
63. Yang Q, Xu Q, Jiang HL (2017) Metal-organic frameworks meet metal nanoparticles: synergistic effect for enhanced catalysis. Chem Soc Rev 46(15):4774–4808. https://doi.org/10.1039/C6CS00724D
64. Yang XL et al (2014) A stable microporous mixed-metal metal-organic framework with highly active Cu^{2+} sites for efficient cross-dehydrogenative coupling reactions. Chem Eur J 20 (5):1447–1452
65. Ye J, Karl Johnson J (2016) Catalytic hydrogenation of CO_2 to methanol in a Lewis pair functionalized MOF. Catal Sci Technol 6(24):8392–8405
66. Zhang, L et al (2015) Palladium nanoparticle supported on metal-organic framework derived N-decorated nanoporous carbon as an efficient catalyst for the Suzuki coupling reaction. Catal Commun 61:21–25. https://doi.org/10.1016/j.catcom.2014.12.004
67. Zhang W et al (2014) A family of metal-organic frameworks exhibiting size-selective catalysis with encapsulated noble-metal nanoparticles. Adv Mater 26(24):4056–4060
68. Zhang WQ et al (2016) Robust metal-organic framework containing benzoselenadiazole for highly efficient aerobic cross-dehydrogenative coupling reactions under visible light. Inorg Chem 55(3):1005–1007
69. Zhu C et al (2016) Chiral metal–organic framework as a platform for cooperative catalysis in asymmetric cyanosilylation of aldehydes. ACS Catal 6(11):7590–7596
70. Zhu NX et al (2016) Micro-Cu4I4-MOF: reversible iodine adsorption and catalytic properties for tandem reaction of Friedel-crafts alkylation of indoles with acetals. Chem Commun 52 (86):12702–12705

MOFs as Catalysts for CO_2 Capture and Fixation

Metal–Organic Frameworks as Promising Catalysts for CO_2 Capture and Fixation

Anand Prakash and Rakesh Kumar Sharma

Contents

1 Introduction 208
2 Metal–Organic Frameworks (MOFs) for Capturing CO_2 210
2.1 Oxy-Fuel Combustion-Based CO_2 Capture 212
2.2 CO_2 Capture from the Air and Natural Gas 214
2.3 CO_2 Separation by MOF-Based Membranes 215
2.4 Pre-combustion CO_2 Capture 217
2.5 Post-combustion CO_2 Capture 218
2.6 Photocatalytic Conversion of CO_2 220
2.7 Electrochemical and Electrocatalytic Conversion of CO_2 221
3 Factors Affecting the Efficiency of MOFs for CO_2 Capture 222
3.1 Heteroatoms 223
3.2 Hydrophobicity 224
3.3 The Presence of Unsaturated Metal Sites 225
3.4 Interactions with Functional Groups of SBUs 225
4 Regeneration of MOFs 226
5 Summary 226
Abbreviations 228
References 229

Abstract Alarming CO_2 emission is a cause of global warming, and its capture is nowadays a topical issue in mitigating global warming, sustainable development, and environmental protection. Considerable work has been done on the manufacturing of novel and promising adsorbents for CO_2 capture. Microporous and nanoporous adsorbents such as zeolites and metal–organic frameworks (MOFs) are reported in the literature for capturing CO_2. Very interestingly, MOFs usually offer exceptional water stability, high surface area, thermal stability, porous nature, and ease of surface modification. MOFs act as a potential candidate for storage, catalysis, separation, and other widespread potential applications. MOFs offer high selectivity for CO_2 capture not only due to high surface area but also due to high heat of adsorption. However, the number of MOFs developed to date is extremely high and acts as an ideal candidate for high-performance CO_2 capture. This chapter represents the summarized study of

A. Prakash · R. K. Sharma (✉)
Department of Chemistry, University of Delhi, New Delhi, Delhi 110007, India
e-mail: sharmark@chemistry.du.ac.in

S. Gulati (ed.), *Metal-Organic Frameworks (MOFs) as Catalysts*,
https://doi.org/10.1007/978-981-16-7959-9_8

MOFs for CO_2 capture. Additionally, various strategies including functionalization of MOFs, N-doping, and metal-doping in MOFs are discussed for enhancing capture performance.

Keywords Metal–organic frameworks · CO_2 capture · Adsorption · MOF membranes · Photocatalytic reduction

1 Introduction

An increasing amount of greenhouse gas, i.e., carbon dioxide (CO_2), primarily arises from industries and the burning of fossil fuels including petroleum, coal, and natural gas which affects the global climate system. Around 85% of the world's energy demand is fulfilled by fossil fuels and the burning of fossil fuels generates CO_2 on large scale in the atmosphere. It is the worrisome issue, and one of the major environmental concerns leads to extreme weather, ocean acidification, and species extinction [107, 122]. The alarming increase in the concentration of CO_2 in the atmosphere causes global warming [30, 33, 56]. Basically, these greenhouse gases are blaming for capturing the solar radiation in the atmosphere, and this process is called the greenhouse effect which gives rise to global warming. However, rapid increase in the population and rise in the number of industries are the main cause of CO_2 emission in the atmosphere.

Intergovernmental Panel on Climate Change (IPCC) predicted that by 2100, the atmospheric CO_2 concentration will reach up to 950 ppm (Climate Change 2007, no date) [147]. As a result, the capture and fixation of CO_2 is a pressing task and became a necessity nowadays. For this, there is a need to pay immediate attention to develop or fabricate some advanced material that can capture CO_2 efficiently. In the past few years, numerous efforts have been made for capturing and fixation of CO_2 using ionic liquids, porous materials including zeolites, organic polymers, and metal–organic frameworks (MOFs) (Climate Change 2007, no date) [8, 30, 41, 54, 103].

The chemistry of MOFs plays a significant role in addressing environmental concerns. MOFs are the porous and new class of crystalline materials fabricated from the secondary building units (SBUs) and an organic linker [79]. The feature of tuning the physical and chemical properties of MOFs makes them potential materials for wide applications including heterogeneous catalysis, drug delivery, gas storage, gas separation, and sensing [151]. Particularly, for CO_2 capture and fixation, MOFs have emerged as a potential candidate for adsorption and catalysis and offers unique features like:

1. Chemical tunability and structural designability allow the MOFs to capture.
2. MOFs can form hybrid structures with other materials and can act as templates to show exceptional chemical and physical properties.
3. MOFs have enough strengths for catalysis having combinatorial features of heterogeneous and homogeneous catalysts including stability, reusability, high catalytic separation, and facile separation.

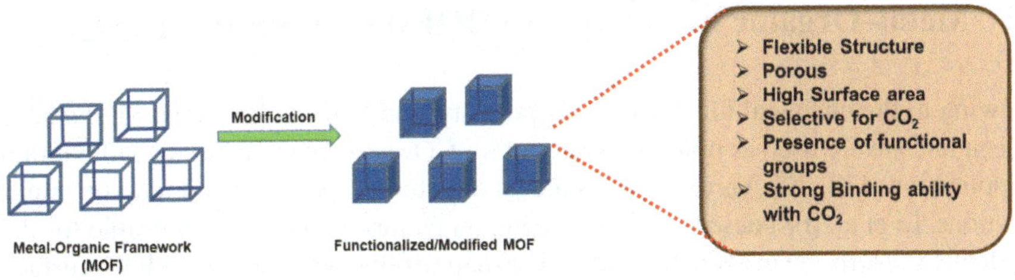

Fig. 1 Characteristics of functionalized MOFs for CO_2 capture

4. The tailored and well-defined structure of MOFs enabled a great understanding of property and structural relationships in MOF-based catalysts.
5. By understanding the chemistry of MOFs, one can modulate the MOF structure to achieve desired properties required for the desired application.
6. The high surface area and ultrahigh porosity of MOFs make them hold huge quantities of gases for an extendable period of time in a confined space.

These are some unique features of MOFs that make them a suitable candidate for CO_2 capture and fixation as shown in Fig. 1 [30].

The utilization of MOFs in capturing and fixation of carbon dioxide has gone through three different stages. The first stage focused on the tuning of selectivity of MOFs and CO_2 adsorption capacity. In order to enhance the interactions between MOFs and CO_2, various strategies were developed and reported in the literature [68, 74, 139, 142]. The second stage is more focused on the production and use of MOFs-based material for the fixation of CO_2 into organic products. Specifically, CO_2 cycloaddition with epoxides has become one of the trendings and well-explored MOF-catalyzed CO_2 conversion reactions [46, 77]. It is important to note that many MOFs have limited reactive sites. MOFs composites produced from a combination of MOF with other materials having high stabilities and multiple active sites are considered as potential materials for CO_2 capture and fixation [39, 60, 136]. The most recent and third stage is focused on exploring and expanding the MOF-based materials involved in CO_2 transformation reactions in addition to optimization of its catalytic performance. Owing to their unique functionalities, MOF-based materials can serve as a potential catalyst for CO_2 conversion reactions, particularly electro-catalytic and photocatalytic CO_2 reduction [39, 82, 125, 126, 131, 132, 137]. With all these aspects and stages in mind, more advancement can be done in the near future in the optimization of the synergistic effect of CO_2 adsorption and conversion in MOF-based materials.

This chapter basically summarizes the MOFs and MOF-based materials for efficient capture of CO_2 and its fixation to mitigate environmental concerns. MOF-based catalysis including heterogeneous conversion of CO_2 into organic products, photocatalytic and electrocatalytic reduction, and hydrogenation is well-discussed. An overview of strategies of CO_2 capture with special emphasis on the interaction between CO_2 and MOFs is described. Additionally, the future perspectives and existing challenges in the field of CO_2 capture and fixation are highlighted.

2 Metal–Organic Frameworks (MOFs) for Capturing CO_2

Owing to the chemical tunability, high porosity, and high surface area, MOFs have been intensively used as potential material for CO_2 adsorption, capture, and fixation applications [32, 70]. Various MOFs are studied for CO_2 capture and fixation applications. Li et al. [68] described that three main factors are solely responsible for the selective adsorption in MOFs includes size/shape exclusion, adsorbate-MOFs surface interactions, and simultaneous corporation of both factors [68]. For example, based on the molecular sieving effect, many MOFs are utilized for selective adsorption of different gases [31, 81, 97]. The kinetic diameter of various gases is summarized in Table 1. The kinetic diameter of gas is defined as the parameter applicable to atoms and molecules of gas which tells the probability of collision of gas molecules with other molecules. Only the gas molecules having suitable pore kinetic diameters can go through the pores of the MOFs.

However, the framework's interaction with CO_2 molecules plays a crucial role in CO_2 adsorption and capture applications. Upon increasing the strength of interaction between framework and CO_2 molecules, especially at low pressure, the MOFs CO_2 uptake capacity increases. One can increase the strength of this interaction by tailoring the physicochemical properties of the framework. The physicochemical properties of the framework can be tuned based on its synthesis and design or by post-synthetic modifications. Recently, post-synthetic modifications have been considered as the method with a high degree of precision and extensively applied for achieving the desired modifications. Our goal is to design the framework in such a way that it can show a high affinity for CO_2 molecules.

Among porous materials, MOFs have surpassed the conventional/traditional materials like zeolites, hybrid ultra-porous materials, and ionic liquids. MOFs excelled in the storage and capturing of methane and carbon dioxide, uptake of hydrogen upon physisorption [58, 99, 129]. Table 2 shows MOFs strength and working capacity among other materials used for adsorption of different gases. It has been shown that MOFs have high working capacity among other materials.

Wang et al. [127] reported that the pre-combustion CO_2 capture approach can be considered best among other approaches for gasification plants, whereas post-combustion CO_2 capture is well-suited for retrofitting power plants. Additionally, oxy-fuel combustion is frequently recommended for newly-built power plants [127].

Table 1 Kinetic diameters of different gases [110]

Gas molecule	Kinetic diameter (in Å)
Hydrogen	2.89
Oxygen	3.46
Methane	3.8
Carbon dioxide	3.3
Nitrogen	3.64
Water vapor	2.65

Table 2 Strengths and weaknesses of MOFs among other materials [89]

	Ionic liquids	Amine grafted MOFs	Ultra-porous materials (Hums)	Zeolites	Amine grafted inorganics	MOFs
Kinetics	F	M	F	M	M	M
Material Cost	L	H	L	L	M	M/H
Process Cost	M	M	L	L	M	M
Selectivity	H	H	VH	L	H	L
Stability	H	M	M	H	H	L
Recycling cost	M/H	M	L	H	M	H
Working capacity	L	M	M	M	M	H

F—fast; H—high; VH—very high; M—medium; L—low

Zou and Zhu [152] showed that ideal MOFs have unique features including function control, pore size, polar functional groups, open metal sites, and grafting of alkylamines onto the framework. Owing to these unique functionalities, MOFs are ideal potential materials with the high capturing ability of carbon dioxide gas due to the high uptake and adsorption of CO_2 molecules over other gases like nitrogen and methane [152]. Kang et al. [57] mentioned that the most demanding and challenging process is the gas separation process for industries. In the field of gas separation, MOFs showed outstanding performance and capabilities [57]. In another study, Millward et al. [88] showed that a gas cylinder containing MOF-177 ($Zn_4O(BTB)_2$; BTB = 4,4′,4″-benzene-1,3,5-triyl-tribenzoate) has more CO_2 capturing and storage ability (about nine times) when compared with the same gas cylinder without MOF [88]. Table 3 summarizes the description of MOFs and MOFs-based materials utilized for capturing CO_2 gas.

From Table 3, we observed that upon increment in temperature, there is a decrement in CO_2 adsorption capacity. Chen et al. [22] stated that the temperature and saturated adsorption capacity relationship can be best explained by the exponential function. As temperature increases, the adsorption capacity decreases [21]. In short, the adsorption temperature must be low in order to achieve high adsorption capacity.

Different ways for capturing CO_2 using frameworks to mitigate CO_2 from the atmosphere include oxy-fuel combustion, pre-combustion, and post-combustion capture. Methods for capturing CO_2 (listed in Fig. 2) can be selected on the basis of its merits, demerits, and feed input requirements like concentration of CO_2, partial pressure in the flue gas, and temperature.

Table 3 Summary of MOFs and MOF-based materials for CO_2 capture [33]

MOFs	CO_2 uptake (g^{-1})	Temperature (°C)	Pressure	References
1D-MOF	4.0 mmol	78	1 bar	Bataille et al. [7]
ZIF-20	70 mL	0	760 torr	Hayashi et al. [45]
MIL-53 M = Al, Cr	10 mmol	30.85	30 bar	Sandrine [106]
PCN-5	210 mg	78.15	760 torr	Shengqian [80]
Mg-MOF-74	8.0 mmol	23	1 bar	Caskey et al. [18]
2D-MOF	2.9 mmol	0	1 bar	Yan [135]
$Ni_3(BTC)_2$	3.0 mmol	40	1 bar	Wade and Dincă [121]
core–shell MOF	41 mmol	0	1 bar	Li et al. [69]
PCN-124	9.1 mmol	0	1 bar	Park et al. [95]
Zn doped Ni-ZIF-8	4.3 mmol	0	1 bar	Yu and Balbuena [140]
Zn(II)-based MOFs	9.2 mmol	25	1 bar	Masoomi et al. [86]
MOF with PEI	4.2 mmol	78	0.15 bar	Lin et al. [72]
$[Ni(bpe)_2 (N(CN)_2)] (N(CN)_2)$	35 mL	78.15	1p/p	Maji et al. [83]
HCM-$Cu_3(BTC)_2$–3	2.8 mmol	25	1 bar	Qian et al. [100]
MOF-5/graphite oxide	1.1 mmol	25	4 bar	Zhao et al. [148]
Cu(bdc) $(4,40\text{-bpy})_{0.5}$	70 mL	24.85	0.1–0.2 MPa	Kitaura et al. [61]
MIL-53 with BNH_x	4.5 mmol	0	0.1 bar	Bataille et al. [7]
UMCM-1-NH_2-MA	19.8 mmol	25	18 bar	Xiang et al. [130]

2.1 Oxy-Fuel Combustion-Based CO_2 Capture

The supreme and one of the prominent technologies for capturing CO_2 is oxy-fuel combustion which involves the combustion of fuel with pure O_2 rather than air. The temperature can be controlled by diluting the oxygen instead of diluting the nitrogen of the flue gas. For capturing the CO_2 from the coal-fired power plant, the main focus is on the production of flue gas consisting of a high amount of CO_2 and water vapor. After that, CO_2 can be separated from the produced flue gas by the process of low-temperature desulfurization and by dehydration processes [17, 111].

The oxy-fuel combustion process mainly comprised of three major sections which are as follows and also shown in Fig. 3:

1. The first section involves the production of oxygen, i.e., air separation unit (ASU).
2. The second section involves the fuel combustion and production of heat.

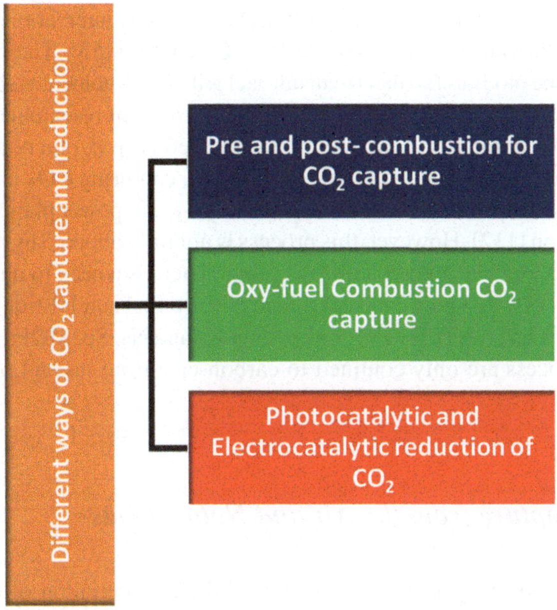

Fig. 2 Different ways of CO_2 capture and reduction

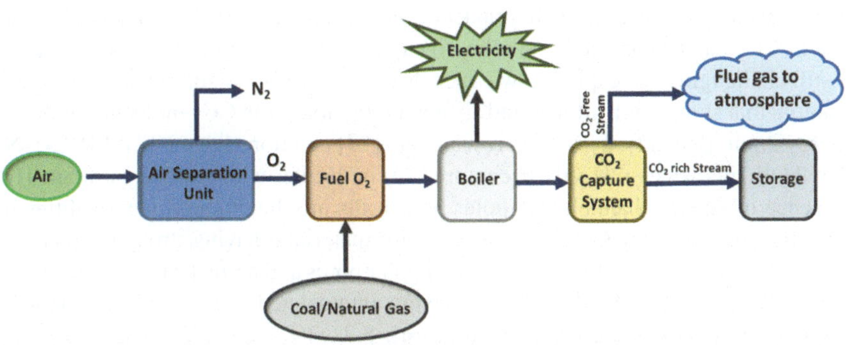

Fig. 3 Schematic view of the sections involved in oxy-fuel combustion process for carbon capture and storage system

3. The third section is mainly focused on cleaning flue gas known as flue gas processing, and CO_2 is purified for capturing and storage. The CO_2 purification unit is named as CO_2 processing unit (CPU).

Hu et al. [53] reported that among several adsorbents including zeolites, MOFs, and activated carbon, MOFs are the potential materials for direct CO_2 capture and storage from the air. All these adsorbents are applied to pre-combustion, oxy-fuel combustion, post-combustion, and CO_2 capture processes. Except for MOFs, other adsorbents such as zeolites, activated carbon showed weak interactions with CO_2 and hence are the materials of lower interest for CO_2 capture [53]. The structure of the framework can be modified so that it can interact with CO_2 molecules strongly and can be used in capturing CO_2 directly from the air. A group of researchers[65] mentioned that the oxy-fuel combustion process is advantageous over others as this process involves the flue gas consisting entirely of CO_2. So, capturing CO_2 in this process is an easy task. This capturing process is extensively used in power plants because of its ease of separation [112]. However, this process is not fully adopted by power plants as mentioned by laboratory and theoretical studies. There is a need to understand some design parameters which play a crucial role in the oxy-fuel combustion process. Thus, MOFs have high selectivity for CO_2 rather than H_2 and N_2. So, MOFs in the oxy-fuel combustion process are only confined to carbon-capturing instead of nitrogen and hydrogen [42].

2.2 *CO_2 Capture from the Air and Natural Gas*

Humans are mainly responsible for increasing the concentration of CO_2 in the air due to respiration and the burning of fossil fuels. Increasing concentration of CO_2 in the air leads to deep unresponsiveness, pathological state of inactivity, sleepiness, and headaches. Gradually, CO_2 build-up in sealed space including submarines and spacecraft may create a negative impact on the occupants [65]. The high concentration of CO_2 can be captured easily but to capture CO_2 present in low concentration is still challenging. To capture low concentration, there is a requirement to design a material that shows high affinity and strong interactions with CO_2 molecules without energy input. Recently, a group of researchers [27] mentioned that NbOFFIVE-1-Ni [$NiNbOF_5(pyrazine)_2 \cdot 2H_2O$] can capture CO_2 molecules with low energy input. The presence of square channels and polar pore walls are the unique functionalities in this MOF, for capturing CO_2 at low levels. This material is having strong interactions with CO_2 molecules due to the presence of polar pores and perfect sites. NbOFFIVE-1-Ni consists of 2D Ni(II) pyrazine layers intercalated with $(NbOF_5)_2^-$anions. NbOFFIVE-1-Ni has the capacity to adsorb 8.2 wt% $CO_2$with 1% CO_2in a dry nitrogen atmosphere. Upon changing the medium of study, i.e., under 75% relative humidity, the CO_2 adsorption capacity of the material decreased to 5.6 wt% [11]. This material shows outstanding performance among other MOFs in addition to the low energy input required for its regeneration. With these perspectives in mind, we can also design new MOFs with higher CO_2 uptake capacity than the already existing MOFs. The higher CO_2 uptake capacity can also be achieved by modifying the MOFs by introducing some polar groups on their surface so that they can show strong binding with the CO_2 molecules present in the air.

Carbon dioxide needs to be separated from methane because, in the humid environment, CO_2 is corrosive in nature and results in lowering energy efficiency [19, 38]. CO_2 and CH_4 have different (more or less) kinetic pore diameters (from Table 1). So, by taking this advantage, one can separate the CO_2 from CH_4 by adjusting the pore size of the material [134]. Additionally, the material should be selective in nature so that it can show more selectivity toward CO_2 instead of CH_4. For example,

UTSA-49 ($Zn(mtz)_2$ where mtz is 5-methyl-1*H*-tetrazolate) contains the tetrazolate framework and shows selectivity of 33.7 for CO_2 in an equimolar mixture of CO_2 and CH_4 at 298 K [134]. The main reasons for showing higher selectivity toward CO_2 instead of CH_4 are-

(1) Small pore openings which can capture only CO_2 molecules and
(2) Presence of uncoordinated heteroaromatic nitrogen atoms.

Based on these reasons, many MOFs including polymorphic MOFs [$Cu(quinoline\text{-}5\text{-}carboxylate)_2$], mixed-metal ZIFs, etc. are proved to be more selective and efficacious toward CO_2 capture. ZIFs can capture CO_2 selectively from a ternary mixture of gases including CH_4, CO_2, and H_2O. In ternary mixture, the concentration of CO_2 is low, i.e., 1.6 wt%, and interestingly, the humid environment had not shown any negative effect on CO_2 capture [91]. Also, in another example, the presence of a humid atmosphere had not affected the performance of polymorphic MOFs. It is mentioned that polymorphic MOFs, composed of copper ions, showed size-selective CO_2 adsorption. This material showed pore size tuning capabilities for efficient capture of CO_2 over CH_4 even in a humid environment for low concentration CO_2 [23]. Much more research is focused nowadays on constructing modified materials having more unsaturated uncoordinated sites for enhancing the selectivity of CO_2 over CH_4.

Over the past decades, the humid environment is known to have an adverse effect on the selectivity of MOFs. But now the issue has been addressed with time. Another main challenging task is the removal of natural gas impurities like H_2S. To date, only limited research is focused on the stability of MOFs in the presence of H_2S [9, 43, 44, 119]. But this challenge needs to be addressed and explored well before the implementation of MOFs for capturing CO_2 from the air and natural gas.

2.3 CO_2 Separation by MOF-Based Membranes

In contrast to pressure and temperature-based adsorption, membrane-based technology for the gas separation process is of high importance and of particular interest because of low infrastructure costs, easy mode of operation, and low energy consumption [143]. Few years ago, only polymers were used because of their low production cost, ease of processability, and mechanical flexibility [101]. But, the demerits of using polymeric membrane are their limited chemical and thermal stabilities, and short lifetime [104]. Additionally, plasticization of polymers was observed when these membranes were subjected to CO_2 separation under natural or flue gas conditions [63, 102]. MOFs are viable and promising materials and have been extensively used in gas separation technologies because of their stability, and they have the capability to overcome the drawbacks of polymeric membranes [120]. Therefore, MOFs-based membranes are trending nowadays for gas separation and research is focused mainly on increasing the adsorption and selectivity of CO_2. MOFs-based membranes can be categorized into two types, i.e., pure-MOF-based membranes and mixed-MOF-based membranes.

Pure-MOF-Based Membrane

MOFs are promising candidates for membrane-based gas separation technologies. MOFs are advantageous for gas separation technologies because of their structural features and using different synthetic strategies, one can fabricate the continuous membrane in a well-oriented manner which shows beneficial properties for the separation process. They usually take the advantage of their small pore aperture feature and have high CO_2 affinity. The first MOF-based membrane that consists of MOF-5 was fabricated in 2005 and proved best for its use in gas separation processes [49, 73]. In another study, ZIF-8 [Zn(2-mIm)$_2$]-based membrane was produced on alumina substrate and showed high CO_2 permeance in CO_2/CH_4 mixture rather than CO_2 single-component permeation [76]. ZIF-69 (Zn(cbIm)(nIm) where nIm is 2-nitroimidazolate)-based membrane showed high selectivity and permeation of CO_2 in equimolar binary mixture of gases, i.e., CO_2/CH_4 and CO_2/N_2 compared to single gas component system. In 2012, bio-MOF-1 [$Zn_8(ad)_4(BPDC)_6O{\cdot}2Me_2NH_2$]-based continuous membrane was produced on porous stainless-steel substrate [3, 13]. This MOF-based membrane showed the highest CO_2 permeance and selectivity in CO_2/CH_4 gas mixture. With this MOF-based membrane in mind, another bio-MOF-13 [$Co_2(ad)_2(butyrate)_2$] membrane was designed to have small pore apertures, high CO_2 uptake capacity, and high surface area [133]. In short, research is more focused on MOF-based membrane for targeting the CO_2 separation from a mixture of gases.

Mixed-MOF-Based Membrane

Considerable success has been achieved with the pure MOF-based membrane including high selectivity of CO_2 in the mixture of gases. But, according to the industrial standards, satisfactory achievements have not been achieved that is combining high selectivity and high permeability of CO_2 in a mixture of gases. The gas molecules can easily pass through the membranes because of pinholes and cracks which create defects in the membrane. Additionally, grain boundary defects are also responsible for creating the defects in the pure MOF-based membrane. However, the cracks in the membrane can be filled by incorporating MOF nanoparticles within a pure MOF-based membrane. The resultant mixed-MOF-based membrane has better gas separation properties and mechanical stability than the pure ones. In 2004, the first mixed-MOF (Cu BPDC-TED/PAET where TED is triethylenediamine, PAET = poly(3-acetoxyethylthiophene))-based membrane was produced and subject to single gas permeation measurements [138]. The membrane consists of Mg-MOF-74 and polyimide that showed enhancement in CO_2 removal efficiency [6]. In another example, ZIF-90/6FDA-DAM (6FDA-DAM-2,2-bis(3,4-carboxyphenyl) hexafluoropropane dianhydride-diaminomesitylene)-based membrane was produced, and binary gas permeation studies were performed at 2 bar, 298 K, and 1:1 CO_2/CH_4. It also showed exceptional selectivity of CO_2 and enhanced CO_2 permeability [115]. Additionally, to meet industrial demands, the gas separation performance of mixed-MOF-based membranes could be further improved by various efforts including post-modifications, tailoring the morphology, and selecting proper MOF fillers and its perfect loading on the membrane to fill the cracks [2, 144, 149].

2.4 *Pre-combustion CO_2 Capture*

As its name suggests, this technology refers to the capture of carbon from biomass, coal gasification, and natural gas, before completion of the combustion process [55, 78, 150]. Although the separation of individual gas such as CO_2 and H_2 from the binary mixture can be attained by several efforts, pre-combustion CO_2 capture technology is more advantageous than others because of low energy requirements and lower cost. It is mentioned that MOFs are potential candidates than other porous materials including zeolites for pre-combustion CO_2 capture benefited from a porous and flexible structure, enhancing CO_2 uptake even at moderately high pressures [145]. MOFs are promising adsorbents when added to pre-combustion CO_2 capture, showed high CO_2 uptake which reduces the concentration of CO_2 in the atmosphere and can be of particular interest for industries. According to in-silico discovery, top MOFs with outstanding performance were chosen and produced. These MOFs were then activated and showed high CO_2 working capacity and CO_2/H_2 selectivity [26]. In another example, the authors synthesized ultra-microporous Ni-(4-pyridylcarboxylate) MOF with high working capacity (3.95 mmol/g), and selectivity and utilized it for pre-combustion CO_2 capture. In this work, MOFs showed almost double CO_2 self-diffusion property than zeolites [90].

In another work, the authors reported that pre-combustion CO_2 combustion is low-cost technology. The turbine used in this study presented has high thermal efficiency compared to the latest technology (Gough et al., no date). MOFs are widely explored for CO_2 capture and conversion applications. However, many factors are responsible mainly for capturing and conversion applications. Efforts can be made to enhance the interactions between MOFs and CO_2 molecules. In this way, the adsorption capacity and selectivity of CO_2 gas molecules can be tuned greatly [30]. Now, many MOFs are present which shows strong interactions and can capture CO_2 for CO_2 separation. For example, Ni-4PyC MOF can act as an ideal and potential candidate and shows pre-combustion CO_2 capture at 35 bar pressure with a negligible amount of adsorbed hydrogen [30]. A different study reported that Mg_2 (dobdc) and Cu-BTTri have high concentrations of metal sites showed higher efficacy of pre-combustion CO_2 capture when compared with the traditional adsorbents [48]. Asgari et al. [4] mentioned that activated carbons and zeolites have not shown considerable improvements in the separation of CO_2 from a mixture of gases because of the restrictions or limited tunability in pore size, shape, and surface functionalities [33].

MOFs showed overwhelming CO_2 capture and adsorption capability in the pre-combustion capture process. For example, NU-11 showed the highest CO_2 adsorption with the uptake of 856 cm^3/g of NU-11 at 30 bar and 25 °C. Long et al. proved that MOF-based membranes (either mixed-MOF-based or pure-MOF-based) have an outstanding capacity for pre-combustion CO_2 capture. The driving force for this high performance is pre-combustion gas mixture pressure, which can separate CO_2 and H_2 [33, 112]. However, MOFs owing to their high surface area, open metal sites, structural flexibility, and tunability can show enhancement in binding efficacy of CO_2 with MOFs with the strong interactions.

2.5 *Post-combustion CO_2 Capture*

The post-combustion CO_2 capture technology is an expensive technology used in the CO_2 capture field and is subjected to various conditions, including solvents, sorbents, and membranes [33]. Certain limitations and demerits of this technology regulate its usage [14, 34, 118]. In this process, carbon capture, i.e., CO_2 separation from flue gas, takes place after combustion [14]. Here, pre-treatment of flue gas is required in order to remove all impurities and corrosive substances. The temperature of flue gas obtained after combustion is very high and in the range of 120–180 °C. So, before pre-treatment, there is a necessity for an energy-intensive cooling system. Also, the volume of flue gas obtained is very high, and partial pressure of CO_2 is low, and hence, large size apparatus is required. This additional requirement further increases the cost of this technology. Fossil fuels combustion leads to the 60% (approx.) CO_2 emission worldwide. Post-combustion CO_2 plays a crucial role in capturing CO_2 from the flue gas obtained from coal-fired power plants [52]. For example, a study reported that flue gas obtained after combustion consists of 15% CO_2, 75% N_2, moisture, and impurities at 1 bar and 30 °C. To capture and separate this 15% CO_2 from flue gas, and economic post-combustion CO_2 strategy is of utmost importance which can capture CO_2 with low cost and with no effect of moisture [37, 52].

Researchers[112] designed three different MOFs, i.e., ZIF-8, HKUST-I, and MIL-53(Al), and modified them with wet impregnation of tetraethylenepentamine (TEPA) molecules. These MOFs were subjected to CO_2 capture under the post-combustion process. It was found that amino-grafted ZIF-8 showed higher adsorption and CO_2 uptake capacity (104 mg CO_2/gads) under post-combustion conditions even in the presence of moisture. This enhancement is due to the combination of physical and chemical adsorptions in contrast to TEPA-grafted microporous materials. This improvement is achieved because of the presence of amino groups which interact strongly with CO_2 molecules [84]. Pai et al. [94] demonstrated several diamine-fabricated MOFs and utilized them for post-combustion carbon capture from emitted flue gas. These MOFs showed S-shaped isotherm upon adsorption of CO_2 molecules. Mmen-M_2 (dobpdc) ($M_2 = Mg_2, Mn_2$) MOFs were modified by the researchers to reduce parasitic energy and to enhance productivity. Note that S-shape CO_2 isotherm and low N_2 affinity are solely responsible for lower energy dissipation [94].

MOFs are classified as emergent solid adsorbents for capturing CO_2 molecules and can show higher uptake and more selectivity of CO_2 over N_2 gas. It is mentioned that MOFs are suitable material for remarkable adsorption of CO_2 for post-combustion CO_2 capture. For applying MOFs for post-combustion CO_2 capture from flue gas, many efforts need to be done. Researchers reported that selectivity of CO_2 over N_2 also limits their performance for CO_2 capture [47, 105]. Singh et al.[116] reported three different water-stable microporous MOFs, two covalent organic frameworks (COFs), and single-wall carbon nanotubes for post-combustion CO_2 capture under flue gas conditions [87]. It is found that ZIF-8 out of several other MOFs and other solid adsorbents showed better efficiency for post-combustion CO_2 capture and the order of these is as follows

$$\text{ZIF}-8 > \text{COF}-108 > \text{UiO}-66 > \text{COF}-300 > \text{InOF}-1 > \text{SWCNT}$$

Jiang et al.[117] characterized and utilized lithium cation-exchanged MOFs (Li^+-MOF) for post-combustion CO_2 capture theoretically. Molecular simulations have been applied to investigate the $CO_2/N_2$and CO_2/H_2 separation from the binary gas mixture. Density functional theory (DFT) studies have been applied for the optimization of atomic charges of cations and their locations within the framework. Compared with non-ionic and nanoporous solid adsorbents including MOFs, the selectivity of Li^+-MOF is higher, i.e., 550 for CO_2/H_2 binary mixture and 60 for CO_2/N_2 gas mixture. It is found that the charge and location of cation within the framework had a great impact on the selectivity of CO_2 over N_2 and H_2. By modulating the charges of the frameworks, the order of selectivity decreased. Additionally, the hydration of this MOF affects its adsorption properties. Hydration of this MOF resulted in lower adsorption capacity because of a decrement in free volume [5]. Photoresponsive frameworks are also studied nowadays for post-combustion CO_2 capture. Mg-IRMOF-74-III, a photoresponsive MOF, consists of azopyridine units and is utilized for CO_2 capture application. Simulations showed that the trans-to-cis isomerization feature of photoresponsive MOFs affects the CO_2 uptake to a great extent [96]. These results are the outcomes of computational studies because before starting any actual and new experimental work, it is better to take a blueprint of any new material through computational studies. However, research is more focused on the MOFs with fine-tunable structures, distinct functionalities, which show exceptional properties and can open up windows for the investigation of high-performance materials and frameworks for post-combustion CO_2 capture application [124]. It is reported that post-combustion CO_2 capture technology is the most commonly used technology among others for power generation [33]. This technology captures 90% CO_2 from flue gas and is of low cost. Interestingly, a new class of MOFs is designed which shows no adsorption of N_2 gas. In this work, MIL-101 was modified by decorating the framework with alkylamine molecules which binds to the unsaturated Cr (III) centers and is used for post-combustion CO_2 capture. This modified Cr-based MIL-101 showed higher CO_2 adsorption and no adsorption for N_2 gas. This is because of strong interactions between the amine groups present in the framework and CO_2 molecules [51]. In another example, diethylenetriamine modified MIL-101 showed high CO_2/N_2 selectivity, exceptional stability, mild regeneration energy, and extraordinary CO_2 adsorption capacity, makes it a promising material for post-combustion CO_2 capture. This material can benefit industries where a higher amount of CO_2 is generated.

2.6 Photocatalytic Conversion of CO_2

The photocatalytic conversion of CO_2 consists of multielectron steps, generates methanol, formaldehyde, methane, carbon monoxide, higher-order hydrocarbons, and formic acid. However, due to the formation of mentioned products, it is very difficult to control the selectivity of CO_2 in this process. Additionally, hydrogen production or evolution along with CO_2 reduction and low solubility of CO_2 in the aqueous phase is still a challenging task. So, to eliminate and address these issues and for better understanding and elucidation of the mechanism of this process, the design of suitable material that can address these issues is of utmost importance.

Photoreduction of CO_2 to CO was investigated with modified UiO-67 ($Zr_6O_4(OH)_4$(BPYDC)) by substituting the BPDC linker with BPYDC (2,2'-bipyridine-5,5'-dicarboxylate) (4.2 wt%) which consists of $Re(CO)_3Cl$ unit (acts as a catalyst) conjugated to N-atoms [123]. This photoresponsive MOF has the CO_2/H_2 selectivity of factor 10 and reduced the carbon dioxide to carbon monoxide in 6 h with a turnover number of 5. This study was done in the presence of ultraviolet light but others can think of extending this study in visible light. Although with 13% Re-based catalyst loading on UiO-67, the framework showed good results, and activity got enhanced seven times by decoration of Re-based MOF on silver nanocubes [25]. This enhancement of the activity of photoactive MOF is due to the surface plasmons. In a similar way, other Ti, Fe, Cd, Gd, Y, Al-based photoactive MOFs are studied with some additional photoactive metal sites. (106–113) UiO-66 and UiO-67 are zirconium-based MOFs utilized for the reduction of CO_2 to formate. UiO-67 MOF modified with Ru- and Mn-based photosensitizer showed a similar turnover number as that of Re-based catalyst loaded UiO-67 MOF. But this Ru- and Mn-based photosensitizer modified UiO-67 was less selective for reduction of CO_2 to $HCOO^-$ [35]. The turnover number of the material can be reduced upon CO_2 dilution with Ar. It was observed that along with a decrement in turnover number, selectivity can rise drastically. Hence, these types of materials are promising materials for high selectivity of CO_2 from binary or ternary mixture of gases. For extending the application of these photoactive MOFs to the visible region, linkers present in UiO-66 and UiO-67 MOFs were substituted with amino groups bonded to the phenyl ring of the linker. This MOF acts as a template for catalytically active metal and metal oxide clusters [114]. The performance of these MOFs can be enhanced by the post-synthetic introduction of titanium and the replacement of Zr along with amino group incorporation [66, 113]. The post-synthetic exchange of Zr by Ti helps in the modulation of band structure, and it can be a promising material for CO_2 reduction to formatephotoactively.NNU-28 [$Zr_6O_4(OH)_4(L)_6$ where L = 4,4'-(anthracene-9,10-diylbis(ethyne-2,1-diyl))dibenzoate] is the most active Zr-based MOF reported for photoreduction of CO_2 to formate without any additional photosensitizer [20]. This MOF works in the visible region with a turnover number of 18 in 10 h. The enhancement in reduction property is due to the combined effects of catalytic sites of linker and cluster which together are responsible for the production of the photoactive framework.

Apart from the formation of formate, which is the product formed by photoreduction of CO_2, other products can also be obtained including methane, methanol, formaldehyde, and hydrogen. It is reported that Al-based MOF consists of porphyrin ring which was utilized for the photoreduction of CO_2 to methanol instead of formate [75]. Here, the porphyrin ring contains the catalytic active sites for the photoreduction process. The complete reduction of CO_2 to methane can be achieved by using MOF composite with metal nanoparticles. For example, the complete reduction of CO_2 to methane photocatalytically was achieved by Ti-based MIL-125-NH_2 MOF decorated with gold nanoparticles [59]. On the other hand, core–shell MOFs can also play important role in achieving the complete reduction of CO_2. It was observed that by coating the HKUST-1 MOF with the TiO_2 nanoparticles, activity was enhanced. Here HKUST-1 acts as core and TiO_2 as a shell. This material composite shows high selectivity for CO_2 because of HKUST-1, and decreased the formation of hydrogen resulted from charge separation between MOF and nanoparticles. The studies with this core–shell material were performed in the gas phase rather than the aqueous phase by taking the high CO_2 uptake feature of MOF into consideration [67].

In conclusion, from a future point of view, MOFs show potential for photocatalysis as observed from different studies. Various strategies can be implemented to improve the photoreduction efficiency of MOFs. This includes the composite of MOFs with nanoparticles, modification of linkers, and forming core–shell MOFs. It is observed that the incorporation of photosensitizer along with co-catalyst has the highest photoreduction capability in terms of turnover number. Core–shell MOFs can also be the material of interest for photoreduction due to unique physicochemical properties arising from core and shell material.

2.7 *Electrochemical and Electrocatalytic Conversion of CO_2*

For mitigation of greenhouse gas generated from fossil fuel burning, the effective clean and remarkable approach is an electrochemical and electrocatalytic conversion of CO_2 into value-added substances and hydrocarbons. This process is based on half-reactions which involve 2–14 electron exchange phenomenon. The challenging task in this process is high overpotential which leads to the production of CO_2 radical, having a high energy barrier [64]. The material should be designed in such a way it should be acid and water stable and can be operated at low overpotential to enhance the selectivity and limit the production of hydrogen. Cu(II) rubeanate framework was the first MOF utilized for CO_2 reduction electrocatalytically having unsaturated metal active sites [128]. In this study, CO_2 was reduced to formic acid with a high turnover number compared to Cu electrode. Cu (II) sites present in the MOF showed weak adsorption when compared with metallic copper attributed to the high selectivity of CO_2. Apart from high selectivity, a massive amount of hydrogen is generated as a side product with 30% faradaic efficiency. Here, faradaic efficiency is defined as the efficiency of transferring the electrons in the system to facilitate the electrochemical reaction. HKUST-1 is also utilized for electrocatalytic reduction of CO_2 to oxalic acid with 90% selectivity in DMF solvent instead of an aqueous medium [108]. Here, the faradaic efficiency was 51% which is low and Cu(II) is reduced to Cu(I) and electron transfer takes place from framework to bonded CO_2.

In another study, cobalt–porphyrin complex is conjugated with one-dimensional aluminum oxide rods to form $Al_2(OH)_2$TCPP-Co. This structure proves the merits of combining both homogeneous and heterogeneous catalysis, and finally, thin films of this MOF were produced which allows the transportation of both charge and mass transport. This transportation can be controlled by tuning the thickness of MOF film [62]. The benefit of using the thin film for electrocatalytic reduction of CO_2 is that the CO_2 molecules remain in contact with the electrode during the process and due to which the performance increases. It was observed that CO_2 reduces selectively to CO in 7 h with 76% faradaic efficiency and turnover number of 1400 in aqueous media. During the reduction process, the Co(II) present in cobalt–porphyrin complex was reduced to Co(I). Further, the strategy of producing the thin film was extended, and thin films of MOF-525 ($Zr_6O_4(OH)_4$(TCPP-Fe)$_3$) were prepared. This Zr-based MOF film reduced CO_2 to CO in DMF solvent in 4 h with a turnover number of 272. In addition to CO, hydrogen was also evolved in 1:1 ratio with CO [50]. Upon addition of weak Bronsted acid-like 2,2,2-trifluoroethanol, increase in current densities along with the increase in the turnover number of 1520 due to the presence of Fe(0) active center within the framework during the process.

3 Factors Affecting the Efficiency of MOFs for CO_2 Capture

Although MOFs are a promising material for CO_2 capture, reduction, and fixation, their working capacity can be enhanced by doing some modifications to the frameworks. This includes the introduction of some heteroatoms, introducing the hydrophobicity to the framework, incorporation of some coordinatively unsaturated metal active sites, and the presence of some additional functional groups which are part of secondary building units and are listed in Fig. 4. In this way, one can enhance the efficiency of MOFs for capturing CO_2 to mitigate the environmental issues caused by CO_2.

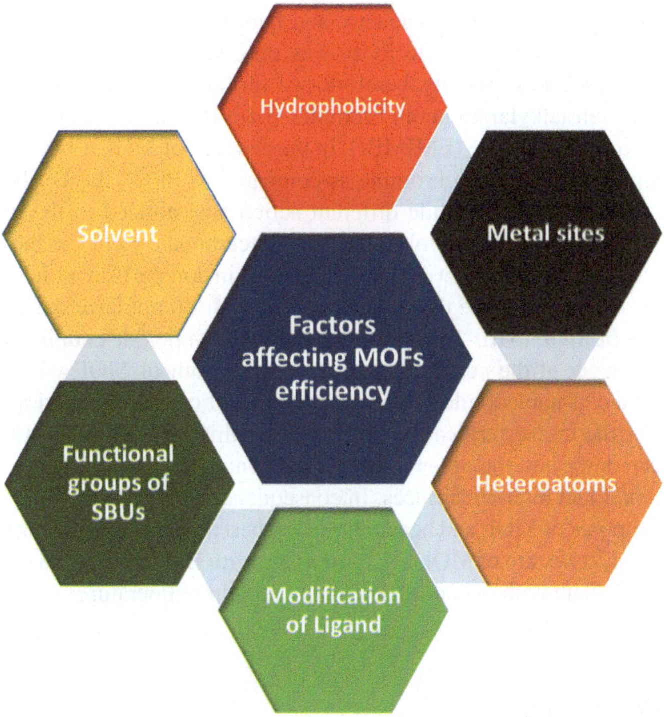

Fig. 4 Factors affecting the efficiency of MOFs for CO_2 capture

3.1 Heteroatoms

The type of organic halide which has been utilized and combined with Lewis acid shows a huge effect on the performance of the catalytic process upon cycloaddition of CO_2 to epoxides [98, 116, 117]. The activity of the catalyst and its selectivity is dependent on the type of halide acting as a nucleophile and cation present in the organic halide salt [98]. The organic halide is different for every catalyst. MIL-100(Cr) is one of the most favorable candidates combined with anion and cation to establish high catalytic activity. The two classes of organic halides which can be combined with Lewis acid complexes are tetrabutylammonium halides (Bu_4NX) (halides = Cl, Br, I) and bis(triphenylphosphine)iminium halides (PPNX) (halides = Cl, Br, I). The best activity was shown by halides is in the order Br > I > Cl. Bromide shows higher nucleophilicity as well as better leaving capacity in comparison with chloride [98]. The nucleophilic character eases the epoxide ring-opening followed by the ring closure to form cyclic carbonate with good leaving ability. PPNX shows better activity than Bu_4NX due to weak interactions present between PPN^+ and the halide ion. The charge is delocalized over cation and halide which enhances the attack of halide ion on epoxide ring and initiates the reaction [28].

Chlorides possess poor leaving capacity and hence lead to expansion rather than ring closure. Here, complete selectivity was observed toward styrene carbonate. Bromine shows the best activity,therefore, we further explored other two organic bromides, i.e., tetraalkylammonium bromides (Et_4NBr) and ionic liquid 1-ethyl-3-methylimidazolium bromide (EMIMBr). In the absence of MOF, Et_4N^+ and Me_4N^+ show strong interactions with bromine as compared to PPN^+ and Bu_4N^+ suggests the poor activity of halide. While different action was noticed in the presence of MOF, the MOF shows a crucial role by increasing carbonate yield and the effect was more applicable to EMIMBr, Et_4NBr, and Me_4NBr in comparison with Bu_4NBr and PPNBr. This may be because the PPNBr and Bu_4NBr do not have the access to go into the pores of MOF (MIL-100(Cr)). Hence, the CO_2-epoxide coupling reaction is presumed to occur at the surface and at the pore mouth of MOF when PPNX or Bu_4NX are used as nucleophile sources. Another reaction was reported where ZIF-8 is utilized for the transesterification of oil [24]. In this case, Me_4NBr, Et_4NBr, and EMIMBr can enter into the pores of MOF, and hence, reaction is assumed to occur at both external and internal surfaces. In consequence, EMIMBr is small enough to enter into the pores of MOF and hence shows the best activity out of the other halides used. The catalyst system of MOF (MIL-100(Cr)) with EMIMBr gave a yield of 94% of styrene carbonate with enhanced selectivity at mild temperature.

3.2 Hydrophobicity

The selectivity and uptake of CO_2 can be enhanced by introducing polar groups which can show strong binding affinity with CO_2 molecules. Additionally, hydrophobicity is also responsible for enhancing the CO_2 capture process by the removal of water molecules from the pores. For example, studies reported that the zeolitic imidazolate frameworks (ZIFs) class, i.e., ZIF-300, ZIF-301, ZIF-302, are hydrophobic in nature and can work equally for CO_2 capture in dry as well as in the presence of moisture (80%). The additional benefit of using hydrophobic ZIFs is their recyclability even after consecutive trials without altering their performance. Other than ZIFs, for introducing hydrophobicity in the MOFs, post-synthetic strategy can be of great interest. In one study, it is reported that MOF-5 (zinc-based MOF) was post-modified with polynapthylene. Polynapthylene can be conjugated with MOF-5 by using monomer 1,2-diethynylbenzene which can be loaded into the pores of the framework. After that, Bergman cyclization and radical polymerization happened to get polyethylene after heating. This modified framework showed almost double CO_2 uptake capacity and enhanced the selectivity of CO_2/N_2 by 25 times. In the presence of moisture, this framework showed more than 90% CO_2 uptake capacity [29, 146]. Particularly, the frameworks that primarily depend on only hydrophobicity for CO_2 capture selectively experienced a lack of strong binding sites for CO_2 and showed poor uptake over water as compared with the frameworks having other structural functionalities. From the practical point of view, combining the hydrophobicity with other structural functionalities within one framework can be of great use in comparison with only using the hydrophobic framework for CO_2 capture. However, one can design the promising framework by introducing hydrophobicity in addition to heteroatoms, polar functional groups, that can work efficiently for CO_2 selective uptake.

3.3 The Presence of Unsaturated Metal Sites

MIL-100(Cr) acts as a favorable catalyst for the cycloaddition of CO_2 to styrene oxide [12]. MIL-100 can also be prepared using other metals which shows good Lewis acidity. Therefore, to compare the activities, two other MIL-100(M) frameworks were fabricated, i.e., MIL-100(Fe) and MIL-100(V) having surface areas 2230 and 2133 m^2g^{-1}, respectively. Taking Bu_4NBr and Bu_4NI as the nucleophilic source, the activity was best shown in the order MIL-100(Cr) > MIL-100(V) > MIL-100(Fe) with the best carbonate yield based on their Lewis acid strength. This trend can be related to the CO_2 adsorption capacity of these materials. Carbon dioxide is a linear polar molecule and has partial positive and negative charges on carbon and oxygen atoms, respectively. CO_2 interacts with Lewis bases and can also adsorb on Lewis acid sites, provided by the metal present in MOF. MIL-100(Cr) showed more CO_2 adsorption ability as compared to MIL-100(V) [16]. The results showed that Bu_4NBr gives the best carbonate yield with MIL-100(Cr), while Bu_4NI showed the best performance with MIL-100(V) and MIL-100(Fe). The trend or order can be inverted for other catalytic systems and is dependent upon the halide anion which maintains the balance with its nucleophilicity, its coordination tendency to the metal center, and also its leaving ability during cyclic carbonate formation [98].

3.4 Interactions with Functional Groups of SBUs

Apart from the charge on unsaturated metal sites, non-metallic functionalities of secondary building blocks are also responsible for making interactions with CO_2 gas. It is reported that MOFs, for example, SIFSIX-1-Cu and SIFSIX-2-Cu-I, showed CO_2 adsorption efficiency of 19.1 and 19.2 wt%, respectively, at ambient pressure and temperature. These MOFs do not contain unsaturated metal sites. But irrespective of metal sites, these MOFs showed high CO_2 uptake because of functional groups of secondary building units [15, 92]. It was observed that monodentate hydroxide units are also advantageous for strong adsorption of CO_2 even in humid conditions and act as capping ligands [71]. Particularly, MAF-X27ox [Co(II)Co(III)(OH)Cl_2(bbta], MAF-X27 [$Co(II)_2Cl_2$(bbta)], MAF-X25ox [Mn(II)Mn(III)(OH)Cl_2(bbta)], and MAF-X25 ($Mn(II)_2Cl_2$(bbta) where bbta is 1*H*,5*H*-benzo(1,2-*d*:4,5-*d*')bistriazolate) contains monodentate hydroxide moieties, due to which there is enhancement in 50% CO_2 uptake compared to parent MOF. MAF-X27ox showed equal CO_2 uptake capacity in both dry and humid conditions. Additionally, it exhibited selective and strong bonding with CO_2 molecules [71].

Additional and more efforts are required to investigate the recyclability of MOFs with SBU-based units so that they can fulfill the long-term CO_2 capturing requirements.

4 Regeneration of MOFs

The CO_2 adsorption and capture technologies require a suitable material that can capture CO_2 efficiently. The key parameter for selecting the adsorbent for CO_2 capture is the ease of regeneration of the adsorbent. Regeneration of MOFs is one of the predominant characteristics when we apply MOFs for CO_2 adsorption. It is recommended that after multiple CO_2 adsorptions and desorption, there should not be any change in uptake capacity, structure, and properties of the MOFs [85]. Based on physicochemical and structural properties and by varying the temperature and pressure, the adsorption and desorption cycles can be performed. Recently, the adsorption regeneration technologies have more focused on vacuum oscillatory absorption (VSA), pressure oscillatory absorption (PSA), and temperature oscillatory absorption (TSA) processes [93, 141].

In VSA and PSA, adsorption of CO occurred at a particular pressure and reduces the given pressure. In VSA, CO is adsorbed at ambient pressure, and in PSA, it is adsorbed at increased pressure (pressure range of 8–28 bar). Keeping this view in mind, we can say that PSA is better for the separation of CO in the pre-combustion capture process and VSA is for post-combustion capture process. MOFs like MOF-177, Mg-MOF-74, CuBTTri, Co(BDP) (BDP = 1.4 benzene dipyrazole) and BeBTB [$Be_{12}(OH)_{12}$ $(BTB)_4$] were used for PSA process for separating CO_2 and H_2 components up to the pressure of 40 bar [48]. Mg-MOF-74 and CuBTTri showed an excellent working capacity of carbon capture as compared to activated carbon and zeolites. These MOFs contain unsaturated metal sites which strongly interact with the CO_2 molecules. With MIL-53 (Al) MOF, 93% of methane was recovered with a flow rate of 13% CO_2 and 87% CH_4 with a purity of 99.4% and desorption pressure of 0.1 bar at 303 k [36, 109]. Another technique used for the regeneration of MOF is TSA. In this technique, the adsorbent is heated to detach/remove the adsorbed molecules [10, 85].

5 Summary

The potential of MOFs for adsorption, CO_2 capture, membrane separation, photocatalytic conversion, and electrocatalytic reduction of CO_2 to various other substances including formate, methanol, methane is well-discussed in this chapter (given in Table 4).

Owing to porous nature, tunable structure, high surface area, ease of functionalization, MOFs are the materials of particular interest for the scientific community. Due to large surface area, porosity, and ordered structure, MOFs are promising

Table 4 Summary of achievements by MOFs in the carbon capture system

Process of capture and reduction of CO_2	Accomplishment using MOFs	Future perspectives and challenges
MOFs membrane for separation	• Both pure and mixed MOF membranes are permselective (an important membrane property to evaluate the quality of membrane separation) than polymeric membranes • Better gas separation properties	• Improvement in processability and permeability • Restrict defects in the membrane
Post-combustion capture RE	• High uptake capability and selectivity of CO_2 over N_2 • Equal performance in the presence and absence of moisture • Tunability in structure to enhance the capture of CO_2	• New MOFs with high capacity and high selectivity in humid conditions • Reduction in cost
Photocatalytic reduction	• Optimization of band gaps by modifying the framework • Highly selective for $HCOO^-$ and CO • Enhancement in catalytic active sites	• Reduction of CO_2 to hydrocarbons • Higher performance
Electrocatalytic reduction	• Low over potential with a high Turnover number • Higher mass and charge transport	• Reduction of CO_2 to hydrocarbons • Proper understanding of electrocatalytic mechanisms • Extension to MOFs other than porphyrin containing MOFs
Regeneration of MOFs	• Low processability cost compared to conventional materials • Recovery achieved with proof of concepts	• Exploration of more MOFs for the regeneration process • Retainment in working capacity after regeneration

catalysts or adsorbents for CO_2 adsorption and are in massive demand for separation, capture, and reduction processes. The three basic technologies for capturing CO_2 including oxy-fuel-based combustion, pre-combustion, and post-combustion are briefly explained. Out of these three techniques, Post-combustion for CO_2 capture is the most extensively used and adopted technique in industries and the carbon capture sector. Although some points need to be addressed and keep in mind before implementation of MOFs for post-combustion carbon capture technology, these include low cost, thermal stability, selectivity, permeability, uptake capacity, regeneration, and reusability of MOFs after multiple operations.

For bulk processes, scalability of the process should be considered and the availability of raw materials required for the synthesis of MOFs should be checked. The thermal and chemical stabilities of MOFs are a very important parameter for their usage in capturing and fixation of CO_2. The thermal stability of MOFs will enable its use under high temperature and pressure conditions. If MOF is not thermally stable, then one cannot subject it to capture CO_2 at high temperature and pressure conditions. But the research is rolling on improving the properties of MOFs, and in that way, it will be the material of the future for capturing, separating, adsorption, and reducing CO_2. In membrane fabrication technology, the MOFs fill the cracks that arise due to the defects. Otherwise, gas molecules can easily pass through the cracks present in them. Another factor that we should think of is to generate economic MOFs for filling these cracks. Various factors to enhance the CO_2 uptake capacity including the incorporation of heteroatoms, unsaturated metal sites, hydrophobicity is well-described briefly in this chapter. The MOFs can work equally to capture CO_2 even in the presence and absence of humid conditions. In conclusion, MOFs broaden up the scope, and by introducing the functionalities which can strengthen the binding of CO_2 with MOFs, it will benefit the scientific community and helps in mitigation of the adverse and toxic effects created by CO_2.

Abbreviations

MOFs	Metal–organic frameworks
IPCC	Intergovernmental Panel on Climate Change
ppm	Part per million
SBUs	Secondary building units
Å	Angstrom
ASU	Air separation unit
CPU	CO_2 processing unit
ZIF	Zeolitic imidazolate framework
TED	Triethylenediamine
PAET	Poly(3-acetoxyethylthiophene
FDA-DAM	Hexafluoropropane dianhydride-diaminomesitylene
TEPA	Tetraethylenepentamine
COF	Covalent organic frameworks
MIL	Matériaux de l'Institut Lavoisier
CO_2	Carbon dioxide
N_2	Nitrogen
CH_4	Methane
H_2	Hydrogen
CO	Carbon monoxide
CH_3OH	Methanol
HCOOH	Formic acid
C_2H_4	Ethylene
HCHO	Formaldehyde
$HCOO^-$	Formate
VSA	Vacuum oscillatory absorption
PSA	Pressure oscillatory absorption
TSA	Temperature oscillatory absorption

References

1. *Climate change 2007. Synthesis report. Contribution of Working Groups I, II and III to the fourth assessment report (Book)* | ETDEWEB (no date). Available at: https://www.osti.gov/etdeweb/biblio/944235 Accessed 1 Aug 2021
2. Al-Maythalony BA et al (2017) Tuning the interplay between selectivity and permeability of ZIF-7 mixed matrix membranes. ACS Appl Mater Interfaces 9(39):33401–33407. https://doi.org/10.1021/ACSAMI.6B15803
3. An J et al (2009) Synthesis, structure, assembly, and modulation of the CO_2 adsorption properties of a zinc-adeninate macrocycle. J Am Chem Soc 131(24):8401–8403. https://doi.org/10.1021/JA901869M
4. Asgari M, Queen W (2018) Carbon Capture in etal–Organic rameworks; Wiley: Hoboken, NJ, USA, 1–78
5. Babarao R, Jiang JW (2010) Cation characterization and CO_2 capture in Li^+-exchanged metal−organic frameworks: from first-principles modeling to molecular simulation†. Ind Eng Chem Res 50(1):62–68. https://doi.org/10.1021/IE100214A
6. Bae T-H, Long JR (2013) CO_2/N_2 separations with mixed-matrix membranes containing Mg_2(dobdc) nanocrystals. Energy Environ Sci 6(12):3565–3569. https://doi.org/10.1039/C3EE42394H
7. Bataille T et al (2012) Solvent dependent synthesis of micro- and nano- crystalline phosphinate based 1D tubular MOF: structure and CO_2 adsorption selectivity. CrystEngComm 14(21):7170–7173. https://doi.org/10.1039/C2CE26138C
8. Bates, B.C., Z.W. Kundzewicz S Wu JP Palutikof E (2008) Climate change and water: climate change 2007: synthesis report. Available at: http://www.ipcc.ch/pdf/assessment-report/ar4/syr/ar4_syr_full_report.pdf Accessed: 1 August 2021
9. Belmabkhout Y et al (2017) Metal–organic frameworks to satisfy gas upgrading demands: fine-tuning the soc-MOF platform for the operative removal of H_2S. J Mater Chem A 5(7):3293–3303. https://doi.org/10.1039/C6TA09406F
10. Berger AH, Bhown AS (2011) Comparing physisorption and chemisorption solid sorbents for use separating CO_2 from flue gas using temperature swing adsorption. Energy Procedia 4:562–567. https://doi.org/10.1016/J.EGYPRO.2011.01.089
11. Bhatt PM et al (2016) A fine-tuned fluorinated MOF addresses the needs for trace CO_2 removal and air capture using physisorption. J Am Chem Soc 138(29):9301–9307. https://doi.org/10.1021/JACS.6B05345
12. Bhattacharjee S, Chen C, Ahn W-S (2014) Chromium terephthalate metal–organic framework MIL-101: synthesis, functionalization, and applications for adsorption and catalysis. RSC Adv 4(94):52500–52525. https://doi.org/10.1039/C4RA11259H
13. Bohrman JA, Carreon MA (2012) Synthesis and CO_2/CH_4 separation performance of Bio-MOF-1 membranes. Chem Commun 48(42):5130–5132. https://doi.org/10.1039/C2CC31821K
14. Breeze P (2015) Carbon capture and storage. Coal-Fired Gener, pp 73–86. https://doi.org/10.1016/B978-0-12-804006-5.00013-7
15. Burd SD et al (2012) Highly selective carbon dioxide uptake by [Cu(bpy-n)2(SiF6)] (bpy-1 = 4,4′-bipyridine; bpy-2 = 1,2-bis(4-pyridyl)ethene). J Am Chem Soc 134(8):3663–3666. https://doi.org/10.1021/JA211340T
16. Cabello CP et al (2014) Enhanced CO_2 adsorption capacity of amine-functionalized MIL-100(Cr) metal–organic frameworks. CrystEngComm 17(2):430–437. https://doi.org/10.1039/C4CE01265H
17. Carpenter SM Long HA (2017) 13 Integration of carbon capture in IGCC systems. Integr Gasification Combined Cycle (IGCC) Technol, pp 445–463. https://doi.org/10.1016/B978-0-08-100167-7.00036-6
18. Caskey SR, Wong-Foy AG, Matzger AJ (2008) Dramatic tuning of carbon dioxide uptake via metal substitution in a coordination polymer with cylindrical pores. J Am Chem Soc 130(33):10870–10871. https://doi.org/10.1021/JA8036096

19. Cavenati S, Grande CA, Rodrigues* AE (2006) Removal of carbon dioxide from natural gas by vacuum pressure swing adsorption. Energy Fuels 20(6):2648–2659. https://doi.org/10.1021/EF060119E
20. Chen D et al (2016) Highly efficient visible-light-driven CO_2 reduction to formate by a new anthracene-based zirconium MOF via dual catalytic routes. J Mater Chem A 4(7):2657–2662. https://doi.org/10.1039/C6TA00429F
21. Chen S, Jin L, Chen X (2011) The effect and prediction of temperature on adsorption capability of coal/CH_4. Procedia Eng 26:126–131. https://doi.org/10.1016/J.PROENG.2011.11.2149
22. Chen W, Zhang Z, Hou L, Yang C, Shen H, Yang K, Wang Z (2020) Metal-organic framework MOF-801/PIM-1mixed-matrix membranes for enhanced CO2/N2 separation performance. Sep Purif Technol 250:117198
23. Chen KJ et al (2016) Tuning pore size in square-lattice coordination networks for size-selective sieving of CO_2. Angew Chem (International ed. in English), 55(35):10268–10272. https://doi.org/10.1002/ANIE.201603934
24. Chizallet C et al (2010) Catalysis of transesterification by a nonfunctionalized metal–organic framework: acido-basicity at the external surface of ZIF-8 probed by FTIR and ab Initio calculations. J Am Chem Soc 132(35):12365–12377. https://doi.org/10.1021/JA103365S
25. Choi KM et al (2016) Plasmon-enhanced photocatalytic CO_2 conversion within metal-organic frameworks under visible light. J Am Chem Soc 139(1):356–362. https://doi.org/10.1021/JACS.6B11027
26. Chung YG et al (2016) In silico discovery of metal-organic frameworks for precombustion CO_2 capture using a genetic algorithm Sci Adv 2(10). https://doi.org/10.1126/SCIADV.1600909
27. Dallbauman LA., Finn JE (1999) Adsorption and its applications in industry and environmental protection. In: Dabrowski A (ed) Applications in environmental protections, Part B, vol. 120. Elsevier, pp 455–471
28. Decortes A, Castilla AM, Kleij AW (2010) Salen-complex-mediated formation of cyclic carbonates by cycloaddition of CO_2 to epoxides. Angew Chem Int Ed 49(51):9822–9837. https://doi.org/10.1002/ANIE.201002087
29. Ding N et al (2016) Partitioning MOF-5 into confined and hydrophobic compartments for carbon capture under humid conditions. J Am Chem Soc 138(32):10100–10103. https://doi.org/10.1021/JACS.6B06051
30. Ding M et al (2019) Carbon capture and conversion using metal–organic frameworks and MOF-based materials. Chem Soc Rev 48(10):2783–2828. https://doi.org/10.1039/C8CS00829A
31. Dybtsev DN † et al (2003) Microporous manganese formate: a simple metal–organic porous material with high framework stability and highly selective gas sorption properties. J Am Chem Soc 126(1):32–33. https://doi.org/10.1021/JA038678C
32. Eddaoudi M, Li H, Yaghi OM (2000) Highly porous and stable metal-organic frameworks: structure design and sorption properties. https://doi.org/10.1021/ja9933386
33. Elhenawy SEM et al (2020) Metal-organic frameworks as a platform for CO_2 capture and chemical processes: adsorption, membrane separation, catalytic-conversion, and electrochemical reduction of CO_2. Catalysts 2020 10(11):1293. https://doi.org/10.3390/CATAL10111293
34. Favre E (2011) Membrane processes and postcombustion carbon dioxide capture: challenges and prospects. Chem Eng J 171(3):782–793. https://doi.org/10.1016/J.CEJ.2011.01.010
35. Fei H et al (2015) Photocatalytic CO_2 reduction to formate using a Mn(I) molecular catalyst in a robust metal-organic framework. Inorg Chem 54(14):6821–6828. https://doi.org/10.1021/ACS.INORGCHEM.5B00752
36. Ferreira AFP et al (2015) Methane purification by adsorptive processes on MIL-53(Al). Chem Eng Sci 124:79–95. https://doi.org/10.1016/J.CES.2014.06.014
37. Figueroa JD et al (2008) Advances in CO_2 capture technology—The U.S. Department of Energy's Carbon Sequestration Program. Int J Greenhouse Gas Control 2(1):9–20. https://doi.org/10.1016/S1750-5836(07)00094-1

38. Férey G et al (2011) Why hybrid porous solids capture greenhouse gases? Chem Soc Rev 40(2):550–562. https://doi.org/10.1039/C0CS00040J
39. Gadipelli S et al (2014) A thermally derived and optimized structure from ZIF-8 with giant enhancement in CO_2 uptake. Energy Environ Sci 7(7):2232–2238. https://doi.org/10.1039/C4EE01009D
40. Gough C et al (no date) Biomass energy with carbon capture and storage (BECCS): unlocking negative emissions
41. Guo Z et al (2016) Hydroxyl-exchanged nanoporous ionic copolymer toward low-temperature cycloaddition of atmospheric carbon dioxide into carbonates. ACS Appl Mater Interfaces 8(20):12812–12821. https://doi.org/10.1021/ACSAMI.6B02461
42. Günther C, Weng M, Kather A (2013) Restrictions and limitations for the design of a steam generator for a coal-fired oxyfuel power plant with circulating fluidised bed combustion. Energy Procedia 37:1312–1321. https://doi.org/10.1016/J.EGYPRO.2013.06.006
43. Hamon L et al (2009) Comparative study of hydrogen sulfide adsorption in the MIL-53(Al, Cr, Fe), MIL-47(V), MIL-100(Cr), and MIL-101(Cr) metal−organic frameworks at room temperature. J Am Chem Soc 131(25):8775–8777. https://doi.org/10.1021/JA901587T
44. Hamon L et al (2011) Molecular insight into the adsorption of H_2S in the flexible MIL-53(Cr) and rigid MIL-47(V) MOFs: infrared spectroscopy combined to molecular simulations. J Phys Chem C 115(5):2047–2056. https://doi.org/10.1021/JP1092724
45. Hayashi H et al (2007) Zeolite A imidazolate frameworks. Nat Mater 6(7):501–506. https://doi.org/10.1038/nmat1927
46. He H et al (2016) Metal-organic frameworks for co_2 chemical transformations. Small 12(46):6309–6324. https://doi.org/10.1002/SMLL.201602711
47. Hedin N et al (2013) Adsorbents for the post-combustion capture of CO_2 using rapid temperature swing or vacuum swing adsorption. Appl Energy 104:418–433. https://doi.org/10.1016/J.APENERGY.2012.11.034
48. Herm ZR et al (2011) Metal−organic frameworks as adsorbents for hydrogen purification and precombustion carbon dioxide capture. J Am Chem Soc 133(15):5664–5667. https://doi.org/10.1021/JA111411Q
49. Hermes S et al (2005) Selective nucleation and growth of metal-organic open framework thin films on patterned COOH/CF_3-terminated self-assembled monolayers on Au(111). J Am Chem Soc 127(40):13744–13745. https://doi.org/10.1021/JA053523L
50. Hod I et al (2015) Fe-Porphyrin-based metal-organic framework films as high-surface concentration, heterogeneous catalysts for electrochemical reduction of CO_2. ACS Catal 5(11):6302–6309. https://doi.org/10.1021/ACSCATAL.5B01767
51. Hu Y et al (2014) Alkylamine-tethered stable metal-organic framework for CO_2 capture from flue gas. Chemsuschem 7(3):734–737. https://doi.org/10.1002/CSSC.201301163
52. Hu Z et al (2015) Ionized Zr-MOFs for highly efficient post-combustion CO_2 capture. Chem Eng Sci 124:61–69. https://doi.org/10.1016/J.CES.2014.09.032
53. Hu Z et al (2019) CO_2 Capture in metal-organic framework adsorbents: an engineering perspective. Adv Sustain Syst 3(1):1800080. https://doi.org/10.1002/ADSU.201800080
54. Hudson MR et al (2012) Unconventional, highly selective CO_2 adsorption in zeolite SSZ-13. J Am Chem Soc 134(4):1970–1973. https://doi.org/10.1021/JA210580B
55. Jansen D et al (2015) Pre-combustion CO_2 capture. Int J Greenhouse Gas Control 40:167–187. https://doi.org/10.1016/J.IJGGC.2015.05.028
56. Ji G et al (2016) Hierarchically mesoporous o-hydroxyazobenzene polymers: synthesis and their applications in CO_2 capture and conversion. Angew Chem Int Ed 55(33):9685–9689. https://doi.org/10.1002/ANIE.201602667
57. Kang Z, Fan L, Sun D (2017) Recent advances and challenges of metal–organic framework membranes for gas separation. J Mater Chem A 5(21):10073–10091. https://doi.org/10.1039/C7TA01142C
58. Kayal S, Sun B, Chakraborty A (2015) Study of metal-organic framework MIL-101(Cr) for natural gas (methane) storage and compare with other MOFs (metal-organic frameworks). Energy 91:772–781. https://doi.org/10.1016/J.ENERGY.2015.08.096

59. Khaletskaya K et al (2015) Fabrication of gold/titania photocatalyst for CO_2 reduction based on pyrolytic conversion of the metal-organic framework NH2-MIL-125(Ti) loaded with gold nanoparticles. Chem Mater 27(21):7248–7257. https://doi.org/10.1021/ACS.CHEMMATER.5B03017
60. Kitao T et al (2017) Hybridization of MOFs and polymers. Chem Soc Rev 46(11):3108–3133. https://doi.org/10.1039/C7CS00041C
61. Kitaura R et al (2003) Porous coordination-polymer crystals with gated channels specific for supercritical gases. Angew Chem Int Ed 42(4):428–431. https://doi.org/10.1002/ANIE.200390130
62. Kornienko N et al (2015) Metal–organic frameworks for electrocatalytic reduction of carbon dioxide. J Am Chem Soc 137(44):14129–14135. https://doi.org/10.1021/JACS.5B08212
63. Koros WJ, Mahajan R (2000) Pushing the limits on possibilities for large scale gas separation: which strategies? J Membr Sci 175(2):181–196. https://doi.org/10.1016/S0376-7388(00)00418-X
64. Kumaravel V, Bartlett J, Pillai SC (2020) Photoelectrochemical conversion of carbon dioxide (CO_2) into fuels and value-added products. ACS Energy Lett, pp 486–519. https://doi.org/10.1021/ACSENERGYLETT.9B02585
65. Law J et al (2010) In-flight carbon dioxide exposures and related symptoms: association, susceptibility, and operational implications. Available at: http://citeseerx.ist.psu.edu/viewdoc/summary?doi=10.1.1.694.8269 Accessed: 7 September 2021
66. Lee Y et al (2015) Photocatalytic CO_2 reduction by a mixed metal (Zr/Ti), mixed ligand metal–organic framework under visible light irradiation. Chem Commun 51(26):5735–5738. https://doi.org/10.1039/C5CC00686D
67. Li R et al (2014) Metal–organic frameworks: integration of an inorganic semiconductor with a metal-organic framework: a platform for enhanced gaseous photocatalytic reactions (Adv Mater 28/2014). Adv Mater 26(28):4907–4907. https://doi.org/10.1002/ADMA.201470193
68. Li J-R, Kuppler RJ, Zhou H-C (2009) Selective gas adsorption and separation in metal–organic frameworks. Chem Soc Rev 38(5):1477–1504. https://doi.org/10.1039/B802426J
69. Li T, Sullivan JE, Rosi NL (2013) Design and preparation of a core-shell metal–organic framework for selective CO_2 capture. J Am Chem Soc 135(27):9984–9987. https://doi.org/10.1021/JA403008J
70. Li H et al (1999) Design and synthesis of an exceptionally stable and highly porous metal-organic framework. Nature 402(6759):276–279. https://doi.org/10.1038/46248
71. Liao P-Q et al (2015) Monodentate hydroxide as a super strong yet reversible active site for CO_2 capture from high-humidity flue gas. Energy Environ Sci 8(3):1011–1016. https://doi.org/10.1039/C4EE02717E
72. Lin Y et al (2013) Polyethyleneimine incorporated metal-organic frameworks adsorbent for highly selective CO_2 capture. Sci Rep 3(1):1–7. https://doi.org/10.1038/srep01859
73. Liu Y et al (2009) Synthesis of continuous MOF-5 membranes on porous α-alumina substrates. Microporous Mesoporous Mater 1–3(118):296–301. https://doi.org/10.1016/J.MICROMESO.2008.08.054
74. Liu J et al (2012) Progress in adsorption-based CO_2 capture by metal–organic frameworks. Chem Soc Rev 41(6):2308–2322. https://doi.org/10.1039/C1CS15221A
75. Liu Y et al (2013) Chemical adsorption enhanced CO_2 capture and photoreduction over a copper porphyrin based metal organic framework. ACS Appl Mater Interfaces 5(15):7654–7658. https://doi.org/10.1021/AM4019675
76. Liu Y et al (2015) Remarkably enhanced gas separation by partial self-conversion of a laminated membrane to metal-organic frameworks. Angew Chem Int Ed 54(10):3028–3032. https://doi.org/10.1002/ANIE.201411550
77. Liu T-T et al (2019) Salen-Co(III) insertion in multivariate cationic metal–organic frameworks for the enhanced cycloaddition reaction of carbon dioxide. Chem Commun 55(28):4063–4066. https://doi.org/10.1039/C8CC10268F
78. Looyd PJD (2005) Precombustion technologies to aid carbon capture. Greenhouse Gas Control Technol, pp 1957–1961. https://doi.org/10.1016/B978-008044704-9/50249-4

79. Lu W et al (2014) Tuning the structure and function of metal–organic frameworks via linker design. Chem Soc Rev 43(16):5561–5593. https://doi.org/10.1039/C4CS00003J
80. Ma S et al (2007) Metal−organic framework based on a trinickel secondary building unit exhibiting gas-sorption hysteresis. Inorg Chem 46(9):3432–3434. https://doi.org/10.1021/IC070338V
81. Ma S et al (2009) Microporous lanthanide metal-organic frameworks containing coordinatively linked interpenetration: syntheses, gas adsorption studies, thermal stability analysis, and photoluminescence investigation. Inorg Chem 48(5):2072–2077. https://doi.org/10.1021/IC801948Z
82. Mahmood A et al (2016) Metal-organic framework-based nanomaterials for electrocatalysis. Adv Energy Mater 6(17):1600423. https://doi.org/10.1002/AENM.201600423
83. Maji TK, Matsuda R, Kitagawa S (2007) A flexible interpenetrating coordination framework with a bimodal porous functionality. Nat Mater 6(2):142–148. https://doi.org/10.1038/nmat1827.
84. Martínez F, Sanz R, Orcajo G, Briones D, Yángüez V (2016) Amino-impregnated MOF materials for CO_2 capture at post-combustion conditions. Chem Eng Sci 142:55–61. https://doi.org/10.1016/J.CES.2015.11.033
85. Mason JA et al (2011) Evaluating metal–organic frameworks for post-combustion carbon dioxide capture via temperature swing adsorption. Energy Environ Sci 4(8):3030–3040. https://doi.org/10.1039/C1EE01720A
86. Masoomi MY et al (2014) Selective CO_2 capture in metal-organic frameworks with azine-functionalized pores generated by mechanosynthesis. Cryst Growth Des 14(5):2092–2096. https://doi.org/10.1021/CG500033B
87. Maurya M, Singh JK (2019) Effect of ionic liquid impregnation in highly water-stable metal-organic frameworks, covalent organic frameworks, and carbon-based adsorbents for post-combustion flue gas treatment. Energy Fuels 33(4):3421–3428. https://doi.org/10.1021/ACS.ENERGYFUELS.9B00179
88. Millward AR, Yaghi* OM (2005) Metal−organic frameworks with exceptionally high capacity for storage of carbon dioxide at room temperature. J Am Chem Soc 127(51):17998–17999. https://doi.org/10.1021/JA0570032
89. Mukherjee S, Kumar A, Zaworotko MJ (2019) Metal-organic framework based carbon capture and purification technologies for clean environment. Metal-Organic Frameworks (MOFs) Environ Appl, pp 5–61. https://doi.org/10.1016/B978-0-12-814633-0.00003-X
90. Nandi S et al (2015) A single-ligand ultra-microporous MOF for precombustion CO_2 capture and hydrogen purification. Sci Adv 1(11). https://doi.org/10.1126/SCIADV.1500421
91. Nguyen NTT et al (2016) Mixed-metal zeolitic imidazolate frameworks and their selective capture of wet carbon dioxide over methane. Inorg Chem 55(12):6201–6207. https://doi.org/10.1021/ACS.INORGCHEM.6B00814
92. Nugent P et al (2013) Porous materials with optimal adsorption thermodynamics and kinetics for CO_2 separation. Nature 495(7439):80–84. https://doi.org/10.1038/NATURE11893
93. Olajire AA (2010) CO_2 capture and separation technologies for end-of-pipe applications—a review. Energy 35(6):2610–2628. https://doi.org/10.1016/J.ENERGY.2010.02.030
94. Pai KN, Baboolal, JD, Sharp DA, Rajendran A (2019) Evaluation of diamine-appended metal-organic frameworks for post-combustion CO_2 capture by vacuum swing adsorption. Sep Purif Technol 211:540–550. https://doi.org/10.1016/J.SEPPUR.2018.10.015
95. Park J et al (2012) A versatile metal–organic framework for carbon dioxide capture and cooperative catalysis. Chem Commun 48(80):9995–9997. https://doi.org/10.1039/C2CC34622B
96. Park J, Suh BL, Kim J (2020) Computational design of a photoresponsive metal-organic framework for post combustion carbon capture. J Phys Chem C 124(24):13162–13167. https://doi.org/10.1021/ACS.JPCC.0C01878
97. Parkes MV et al (2013) Screening metal–organic frameworks for selective noble gas adsorption in air: effect of pore size and framework topology. Phys Chem Chem Phys 15(23):9093–9106. https://doi.org/10.1039/C3CP50774B

98. Pescarmona PP, Taherimehr M (2012) Challenges in the catalytic synthesis of cyclic and polymeric carbonates from epoxides and CO_2. Catal Sci Technol 2(11):2169–2187. https://doi.org/10.1039/C2CY20365K
99. Qasem NAA, Ben-Mansour R, Habib MA (2018) An efficient CO_2 adsorptive storage using MOF-5 and MOF-177. Appl Energy 210:317–326. https://doi.org/10.1016/J.APENERGY.2017.11.011
100. Qian D et al (2012) Synthesis of hierarchical porous carbon monoliths with incorporated metal-organic frameworks for enhancing volumetric based CO_2 capture capability. ACS Appl Mater Interfaces 4(11):6125–6132. https://doi.org/10.1021/AM301772K
101. Robeson LM (1991) Correlation of separation factor versus permeability for polymeric membranes. J Membr Sci 62(2):165–185. https://doi.org/10.1016/0376-7388(91)80060-J
102. Robeson LM (2008) The upper bound revisited. J Membr Sci 320(1–2):390–400. https://doi.org/10.1016/J.MEMSCI.2008.04.030
103. Rochelle GT (2009) Amine scrubbing for CO_2 capture. Science 325(5948):1652–1654. https://doi.org/10.1126/SCIENCE.1176731
104. Rui Z et al (2016) Metal-organic framework membrane process for high purity CO_2 production. AIChE J 62(11):3836–3841. https://doi.org/10.1002/AIC.15367
105. Samanta A et al (2011) Post-combustion CO_2 capture using solid sorbents: a review. Ind Eng Chem Res 51(4):1438–1463. https://doi.org/10.1021/IE200686Q
106. Sandrine Bourrelly, † et al (2005) Different adsorption behaviors of methane and carbon dioxide in the isotypic nanoporous metal terephthalates MIL-53 and MIL-47. J Am Chem Soc 127(39):13519–13521. https://doi.org/10.1021/JA054668V
107. Sanz-Pérez ES et al (2016) Direct capture of CO_2 from ambient air. Chem Rev 116(19):11840–11876. https://doi.org/10.1021/ACS.CHEMREV.6B00173
108. Senthil Kumar R, Senthil Kumar S, Kulandainathan MA (2012) Highly selective electrochemical reduction of carbon dioxide using Cu based metal organic framework as an electrocatalyst. Electrochem Commun 25:70–73. https://doi.org/10.1016/j.elecom.2012.09.018
109. Serra-Crespo P et al (2015) Preliminary design of a vacuum pressure swing adsorption process for natural gas upgrading based on amino-functionalized MIL-53. Chem Eng Technol 38(7):1183–1194. https://doi.org/10.1002/CEAT.201400741
110. Songolzadeh M et al (2014) Carbon dioxide separation from flue gases: a technological review emphasizing reduction in greenhouse gas emissions. The Sci World J 2014. https://doi.org/10.1155/2014/828131
111. Stanger R et al (2015) Oxyfuel combustion for CO_2 capture in power plants. Int J Greenhouse Gas Control 40:55–125. https://doi.org/10.1016/J.IJGGC.2015.06.010
112. Sumida K, Rogow DL, Mason JA, McDonald TM, Bloch ED, Herm ZR, Bae T-H, Long JR (2012) Carbon dioxide capture in metal-organic frameworks. Chem Rev 112(2):724–781. https://doi.org/10.1021/CR2003272
113. Sun D et al (2015) Introduction of a mediator for enhancing photocatalytic performance via post-synthetic metal exchange in metal–organic frameworks (MOFs). Chem Commun 51(11):2056–2059. https://doi.org/10.1039/C4CC09407G
114. Sun D et al (2013) Studies on photocatalytic CO_2 reduction over NH2-Uio-66(Zr) and its derivatives: towards a better understanding of photocatalysis on metal–organic frameworks. Chem—A Eur J 19(42):14279–14285. https://doi.org/10.1002/CHEM.201301728
115. Tae-Hyun B et al (2010) 'A high-performance gas-separation membrane containing submicrometer-sized metal-organic framework crystals. Angew Chem (International ed. in English), 49(51):9863–9866. https://doi.org/10.1002/ANIE.201006141
116. Taherimehr M et al (2013) High activity and switchable selectivity in the synthesis of cyclic and polymeric cyclohexene carbonates with iron amino triphenolate catalysts. Green Chem 15(11):3083–3090. https://doi.org/10.1039/C3GC41303A
117. Taherimehr M et al (2015) New iron pyridylamino-bis(phenolate) catalyst for converting CO_2 into cyclic carbonates and cross-linked polycarbonates. Chemsuschem 8(6):1034–1042. https://doi.org/10.1002/CSSC.201403323

118. Tillman DA (2018) The development of postcombustion control technology. Coal-Fired Electricity Emissions Control, pp 237–276. https://doi.org/10.1016/B978-0-12-809245-3.00009-4
119. Vellingiri K, Deep A, Kim K-H (2016) Metal–organic frameworks as a potential platform for selective treatment of gaseous sulfur compounds. ACS Appl Mater Interfaces 8(44):29835–29857. https://doi.org/10.1021/ACSAMI.6B10482
120. Venna SR, Carreon MA (2015) Metal organic framework membranes for carbon dioxide separation. Chem Eng Sci 124:3–19. https://doi.org/10.1016/J.CES.2014.10.007
121. Wade CR, Dincă M (2012) Investigation of the synthesis, activation, and isosteric heats of CO_2 adsorption of the isostructural series of metal–organic frameworks M3(BTC)2 (M = Cr, Fe, Ni, Cu, Mo, Ru). Dalton Trans 41(26):7931–7938. https://doi.org/10.1039/C2DT30372H
122. Wang Q et al (2010) CO_2 capture by solid adsorbents and their applications: current status and new trends. Energy Environ Sci 4(1):42–55. https://doi.org/10.1039/C0EE00064G
123. Wang C et al (2011) Doping metal-organic frameworks for water oxidation, carbon dioxide reduction, and organic photocatalysis. J Am Chem Soc 133(34):13445–13454. https://doi.org/10.1021/JA203564W
124. Wang Q et al (2015) Finely tuning MOFs towards high-performance post-combustion CO_2 capture materials. Chem Commun 52(3):443–452. https://doi.org/10.1039/C5CC07751F
125. Wang C, An B, Lin W (2018) Metal–organic frameworks in solid-gas phase catalysis. ACS Catal 9(1):130–146. https://doi.org/10.1021/ACSCATAL.8B04055
126. Wang S, Wang X (2015) Multifunctional metal-organic frameworks for photocatalysis. Small 11(26):3097–3112. https://doi.org/10.1002/SMLL.201500084
127. Wang Y, Zhao L, Otto A, Robinius M, Stolten D (2017) A review of post-combustion CO_2 capture technologies from coal-fired power plants. Energy Procedia 114:650–665. https://doi.org/10.1016/J.EGYPRO.2017.03.1209
128. Wang Y et al (2020) Reticular chemistry in electrochemical carbon dioxide reduction. Sci China Mater 63(7):1113–1141. https://doi.org/10.1007/S40843-020-1304-3
129. Wei W et al (2013) A heterometallic microporous MOF exhibiting high hydrogen uptake. Microporous Mesoporous Mater 165:20–26. https://doi.org/10.1016/J.MICROMESO.2012.07.036
130. Xiang Z et al (2011) CNT@$Cu_3(BTC)_2$ and metal-organic frameworks for separation of CO_2/CH_4 mixture. J Phys Chem C 115(40):19864–19871. https://doi.org/10.1021/JP206959K
131. Xiao J-D, Jiang H-L (2017) Thermally stable metal-organic framework-templated synthesis of hierarchically porous metal sulfides: enhanced photocatalytic hydrogen production. Small 13(28):1700632. https://doi.org/10.1002/SMLL.201700632
132. Xiao J-D, Jiang H-L (2018) Metal–organic frameworks for photocatalysis and photothermal catalysis. Acc Chem Res. https://doi.org/10.1021/ACS.ACCOUNTS.8B00521
133. Xie Z et al (2013) Alumina-supported cobalt–adeninate MOF membranes for CO_2/CH_4 separation. J Mater Chem A 2(5):1239–1241. https://doi.org/10.1039/C3TA14058J
134. Xiong S et al (2014) A new tetrazolate zeolite-like framework for highly selective CO_2/CH_4 and CO_2/N_2 separation. Chem Commun 50(81):12101–12104. https://doi.org/10.1039/C4CC05729E
135. Yan Q et al (2013) Designed synthesis of functionalized two-dimensional metal-organic frameworks with preferential CO_2 capture. ChemPlusChem 78(1):86–91. https://doi.org/10.1002/CPLU.201200270
136. Yao J, Wang H (2014) Zeolitic imidazolate framework composite membranes and thin films: synthesis and applications. Chem Soc Rev 43(13):4470–4493. https://doi.org/10.1039/C3CS60480B
137. Yi F-Y et al (2017) Metal–organic frameworks and their composites: synthesis and electrochemical applications. Small Methods 1(11):1700187. https://doi.org/10.1002/SMTD.201700187
138. Yin H et al (2014) A highly permeable and selective amino-functionalized MOF CAU-1 membrane for CO_2–N_2 separation. Chem Commun 50(28):3699–3701. https://doi.org/10.1039/C4CC00068D

139. Yu J et al (2017) CO_2 capture and separations using MOFs: computational and experimental studies. Chem Rev 117(14):9674–9754. https://doi.org/10.1021/ACS.CHEMREV.6B00626
140. Yu J, Balbuena PB (2013) Water effects on postcombustion CO_2 capture in Mg-MOF-74. J Phys Chem C 117(7):3383–3388. https://doi.org/10.1021/JP311118X
141. Yu C-H, Huang C-H, Tan C-S (2012) A review of CO_2 capture by absorption and adsorption. Aerosol Air Quality Res 12(5):745–769. https://doi.org/10.4209/AAQR.2012.05.0132
142. Zhai Q-G et al (2017) Pore space partition in metal-organic frameworks. Acc Chem Res 50(2):407–417. https://doi.org/10.1021/ACS.ACCOUNTS.6B00526
143. Zhang Y et al (2013) Current status and development of membranes for CO_2/CH_4 separation: a review. Int J Greenhouse Gas Control 12:84–107. https://doi.org/10.1016/J.IJGGC.2012.10.009
144. Zhang C et al (2014) Highly scalable ZIF-based mixed-matrix hollow fiber membranes for advanced hydrocarbon separations. AIChE J 60(7):2625–2635. https://doi.org/10.1002/AIC.14496
145. Zhang Z et al (2014) Perspective of microporous metal–organic frameworks for CO_2 capture and separation. Energy Environ Sci 7(9):2868–2899. https://doi.org/10.1039/C4EE00143E
146. Zhang Z et al (2016) Polymer–metal–organic frameworks (polyMOFs) as water tolerant materials for selective carbon dioxide separations. J Am Chem Soc 138(3):920–925. https://doi.org/10.1021/JACS.5B11034
147. Zhang L, Zhao Z-J, Gong J (2017) Nanostructured materials for heterogeneous electrocatalytic CO_2 reduction and their related reaction mechanisms. Angew Chem Int Ed 56(38):11326–11353. https://doi.org/10.1002/ANIE.201612214
148. Zhao Y, Ding H, Zhong Q (2013) Synthesis and characterization of MOF-aminated graphite oxide composites for CO_2 capture. Appl Surf Sci 284:138–144. https://doi.org/10.1016/J.APSUSC.2013.07.068
149. Zhenggong W et al (2016) Interfacial design of mixed matrix membranes for improved gas separation performance. Adv Mater (Deerfield Beach, Fla.) 28(17):3399–3405. https://doi.org/10.1002/ADMA.201504982
150. Zhong DL et al (2016) Precombustion CO_2 capture using a hybrid process of adsorption and gas hydrate formation. Energy 102:621–629. https://doi.org/10.1016/J.ENERGY.2016.02.135
151. Zhou H-C "Joe", Kitagawa S (2014) Metal–organic frameworks (MOFs). Chem Soc Rev 43(16):5415–5418. https://doi.org/10.1039/C4CS90059F
152. Zou X, Zhu G (2019) CO_2 capture with MOF membranes. Microporous Mater Sep Membr, pp 323–359. https://doi.org/10.1002/9783527343997.CH10

Theoretical Study on Catalytic Capture and Fixation of Carbon Dioxide by Metal–Organic Frameworks (MOFs)

Upasana Issar and Richa Arora

Contents

1 Introduction 238
2 Metal–Organic Frameworks 239
3 Theoretical Studies on the Capture of CO_2 by MOFs 241
4 Theoretical Studies on Fixation of CO_2 by MOFs 249
5 Conclusion 254
Abbreviations 254
References 256

Abstract Over the years, the production of carbon dioxide (CO_2) has been on the rise owing to industrialization, and globalization. The presence of excess CO_2 in the atmosphere has been causing a lot of problems lately; the major one being global warming due to its severe impact on the ozone layer. In the last few decades, the main problem bogging scientists and environmentalists is how to capture and store excessive CO_2 and stop it from entering the carbon cycle. Further, since CO_2 is made of essential elements like carbon and oxygen, it would be highly economical and environment-friendly, if somehow this could be converted to some other form that could be used as a fuel. The tricky part is to capture CO_2 gas and for that, various methods have been employed, adsorption being quite efficient and inexpensive among them. There are many adsorbents in use for capturing CO_2, but metal–organic frameworks (MOFs) have piqued the interest of scientists owing to their valuable properties like high surface area, resilience to water and chemicals, economic viability, and environment-friendliness. MOFs have also been used as a catalyst to reduce CO_2 into other energy-rich compounds. To understand the mechanism of action of MOF for CO_2 adsorption, molecular modeling and simulation techniques have been in use. One major advantage of MOFs is that it has organic ligands connecting metal ions, wherein functionalities of the organic groups can be changed to increase its catalytic and adsorption power. Such a feature of changing metal ions and organic functionalities to develop better MOFs can easily be studied

U. Issar
Department of Chemistry, Kalindi College, University of Delhi, New Delhi, Delhi, India

R. Arora (✉)
Department of Chemistry, Shivaji College, University of Delhi, New Delhi, Delhi, India

S. Gulati (ed.), *Metal-Organic Frameworks (MOFs) as Catalysts*,
https://doi.org/10.1007/978-981-16-7959-9_9

by using computational methods along with experiments. This chapter addresses various computational techniques used to study MOFs which could not only capture CO_2 but reduce them to form valuable energy-rich substances.

Keywords MOF · CO_2 capture · CO_2 fixation · DFT · Dispersion · Machine learning

1 Introduction

With each passing day, the human race is increasing in huge numbers. This puts pressure on already depleting natural resources. To overcome this huge gap between demand and supply, mankind has been working tremendously toward the advancement of technology. Such a growing population with ever-increasing energy demand along with the use of advanced technology has impacted our environment in a huge way. One such menacing problem is the release of greenhouse gases like carbon dioxide into the environment. These greenhouse gases have resulted in dramatic shifts in climate leading to global warming, a major environmental problem faced by humans today.

The main reason behind the focus on CO_2 is that it remains in the atmosphere for quite long periods of time, thereby causing serious damage to the environment and global warming [6]. With the increasing use of fossil fuels, even though a major shift has come to use other renewable sources of energy, the CO_2 levels in the atmosphere are at an all-time high of approximately 414 ppm (parts per million) as was recorded in August 2020 [34]. This has also led to abnormally warm temperatures around the world [123]. One major evidence of the damaging effect of high CO_2 levels in the atmosphere is the increase in the water level of the ocean which has led to a loss of shoreline and coastal lands [34, 72]. Such a dramatic change in climate and environment poses a great threat to the existence of life on earth and hence, demands quick and effective measures to curtail the increasing levels of CO_2.

Now, there are mainly three measures that can be taken up to solve the CO_2 menace. The first measure is the alteration of human and industrial activities so as to reduce the CO_2 emission in the atmosphere. This involves less dependence and usage of CO_2 producing fossil fuels and more use of sustainable and clean renewable resources for energy production. It also involves improving fuel efficiency and the use of recycled products. But this option is far from being feasible as complete reliance on alternate sources and methods is not possible because of huge energy demand and also because the use of renewable sources for energy production and alternate methods are way more costly than the simple use of fossil fuels. The second measure to contain CO_2 levels is the CO_2 capture and storage (CCS) or CO_2 sequestration method. This involves the capturing of CO_2 as soon as it is released in the atmosphere, then safely transporting the compressed CO_2 and finally storing them for future uses [41]. The main issue with this method is again the huge cost involved. This method would sustain if somehow the transportation and storage cost is reduced and chances of leakage of CO_2 and accidents during transportation and storage are reduced [75]. The last method is the complete depletion of CO_2 by converting it into some useful products [123].

Over the years, researchers have been constantly working on finding suitable methods and materials for CO_2 capture, separation, storage, and conversion. The next important task for researchers is to find a suitable porous material that could be used as an adsorbent for CO_2. The best adsorbent would be the one that is specific to the desired target molecule. The choice of adsorbent also depends on the nature of the adsorbent like its size, porous nature, structure, adsorption process, and compatibility of the adsorbent with the adsorbate. Since CO_2 has a high quadrupole moment, the adsorbent chosen must have high electric field gradients [37].

Out of many useful materials, metal–organic frameworks (MOFs) have emerged as a new and effective class of solid porous material, which could be used for the capture, separation, and conversion of CO_2. This application of MOFs could be attributed to many factors like their diverse structures, ease of preparation, highly porous nature, flexible, and adjustable size, metals, organic linkers, which work in favor of MOFs and gives them an advantage over other porous materials [122]. But again, cost remains one of the big factors in designing and devising any material to control the CO_2 problem. Since the development of any method and/or material requires a lot of effort in terms of manpower as well as money, cost-cutting in this regard is also of paramount importance. So, scientists over a period of time have made a swift shift toward the use of computers, software, and modeling techniques so as to reduce the cost (to some extent) involved in research purposes. Although money is a huge factor for inventing new materials and then making use of them for CO_2 capture and conversion to some extent, money could be saved while researching for new materials. With this aim in mind, this chapter exhaustively focuses on the use of various theoretical, computational, and simulation methods in use for designing new MOF materials for capturing and converting CO_2.

2 Metal–Organic Frameworks

MOFs were developed in the 1990s, by Omar Yaghi, who is touted as a forefather of MOF development [26]. MOFs have received a great deal of consideration from scientists due to their properties like huge surface area, stability, porosity, and diverse functionalities [119]. Moreover, large databases of MOFs are available which were created theoretically using advanced computational techniques and tools. Such hypothetical MOFs can be manufactured experimentally [98].

MOFs are multi-dimensional and porous nanomaterials. They constitute metal ions coordinated to organic molecules called linkers (Fig. 1). The metal ion and the organic moiety used determines not only the shape and structure but also application of MOF. Nearly all metal ions can be used to prepare MOFs. Some of the commonly used organic linkers are carboxylates and azoles moiety. Major research has been going on for A-MOF-74 (A = Mg and Ni), MOF-5 (e.g., Zn_4O tetrahedra linked

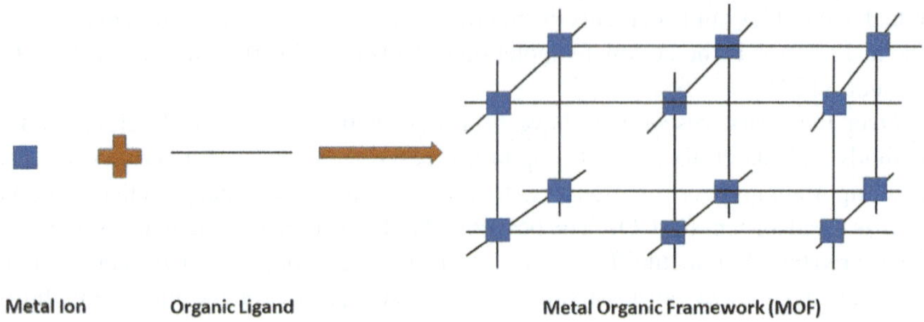

Fig. 1 Metal–organic framework (MOF)

by 1,4-benzenedicarboxylate (BDC) molecules), and X-BTT (X = Ca, Fe, Mn, Cu, Co, and BTT = 1,2,5-benzenetristetrazolate) families. The A-MOF-74 has 54 atoms in its primitive unit cell and contains carboxyl and hydroxyl groups. The MOF-5 is cubic and has 424 and 106 atoms in its primitive unit cell. The application of MOF as an adsorbent for CO_2 depends upon the type of metal ion present, the organic group present, and the separation of organic groups from each other [13].

Computational studies have been quite useful in understanding the importance and interactions of metal ions with organic linkers and how changing or modifying either of these would impact the properties and hence, the functionality of the resulting MOF. The first ever theoretical study on MOF was done in 2006. The properties of MOF-5 were analyzed, and it was proposed that the energy levels are majorly determined by the organic linkers and the metal–oxygen moiety which includes Zn_4O clusters, structurally similar but electronically different from zinc oxide cubic cell [17]. This was later confirmed experimentally [35] as well. This prompted researchers to combine theoretical and experimental studies to understand the electronic and structural properties of MOFs. Another such study involving combined experimental and theoretical techniques involves Zr-BDC MOF (UiO-66) [102] and its isoreticular parent MOF, UiO-67 [14]. The theoretical studies confirmed the experimental findings that the coordination number of Zr atoms ($Zr_6O_4(OH)_4$) is modified along with complete phase transformation of UiO-66 from its hydroxylated to dehydroxylated form. One more example which confirms the importance of theoretical investigation is an analysis of bipyrazole-based (BPZ) MOFs like M-BPZ MOF having different divalent metal metals (M = Ni, Cu, Zn) [3, 8] and $Ni_3(BTP)_2$ (BTP = benzenetripyrazolate) [90]. The structure and different forms (having different porosity) of MIL-53(Al) (MIL = Matérial Institut Lavoisier) were explained using ab initio calculations. This particular MOF has two temperature-driven stable forms: the one having narrow pores is stable at low temperature and the other having large pores is stable at high temperature. The stability of both forms at different temperatures was explained on the basis of dispersion and entropic factors. The low-temperature structure is stable due to dispersion interactions between phenyl rings and the high-temperature structure is stable owing to entropy which also leads to an increase in the size of pores [70, 104].

A vast variety of MOFs with diverse structures and properties are available. The leading properties of MOFs are their strong and flexible 3D structure (leading to their high surface area) and porosity. Also, they are capable of maintaining their structural integrity while removing water or solvent molecules leading to the generation of pores (open metal or unsaturated spots are created), which can be used for adsorption. The flexibility of MOF is in terms of using different metal ions, functional groups of the organic linkers, and their pore size, making them useful in diverse fields like catalysis, as sensors, for gas storage, etc. [33].

The MOFs can be rigid as well as flexible. The rigid MOFs have strong and permanent pores while the structure of flexible ones can be easily changed [54]. For CO_2 adsorption, mainly rigid MOFs are in use, but flexible ones are also gaining importance owing to their extraordinary properties [37].

3 Theoretical Studies on the Capture of CO_2 by MOFs

CO_2 capture has over time, emerged as an alternative method to control the CO_2 concentration in the atmosphere. CO_2 can be captured and put to good use like in the food industry as dry ice or to prepare carbonated drinks by converting it into some useful carbon-containing products and also to provide artificial CO_2 atmosphere for efficient growth of plants. CO_2 can be captured directly from the atmosphere or from some CO_2 sources like fuel gas, natural gas, etc., but retrieving CO_2 from the former is quite tricky [142].

Using computational simulation techniques like density functional theory (DFT), periodic density functional theory (PDFT), molecular dynamics (MD) simulation, and grand canonical Monte Carlo (GCMC) simulation, the separation and adsorption of CO_2 can be studied efficiently against MOF. Theoretical calculations can also work with hypothetical MOFs, which could tell us which types of modifications are better suited to prepare a selective and efficient adsorbent for CO_2.

DFT study could explain the interaction of CO_2 with $[Zn_2(OX)(ATZ)_2].H_2O_{0.5}$ (OX = oxalate, HATZ = 3-amino-1,2,4-triazole) [101]. According to X-ray crystallographic studies performed, there are two different sites for CO_2 adsorption and according to DFT calculations, the binding energies of both sites are approximately the same (39.6 and 38.1 kJ mol^{-1}). In the first adsorption site, the carbon atom of CO_2 interacts with the oxalate group, and oxygen atoms of CO_2 forms hydrogen bonds with the $–NH_2$ group. For the second adsorption site, the interactions were found between the oxygen atom of CO_2 and carbon atom of oxalate and nitrogen atom of $–NH_2$ group. Another such interaction between CO_2 and MOF involves $[Cu(TZC)(DPP)_{0.5}]_n.1.5H_2O$, (PCN (porous coordination network)—200, TZC = tetrazole-5-carboxylate, DPP = 1,3-di(4-pyridyl) propane) [115], wherein GCMC and DFT calculations revealed the presence of one CO_2 molecule in each of the cavity, and they show multiple interactions with MOF [142].

Another study that again emphasizes the importance of nitrogen-containing groups in MOF involves $[Cu_3L_2(H_2O)_5]$.xGuest (NJU-Bai3: Nanjing University Bai group), with multidentate ligand, 5-(4-carboxy-benzoylamino)-isophthalic acid [25], having an amide group. Owing to the presence of amide groups, the surface area of MOF rises to 2690 m^2 g^{-1} which increases its selectivity and activity for CO_2. Two MOFs having acylamide-functionalized linkers, $[Cu_3(BTB^{6-}]_n$ (BTB = 1, 3, 5—Tris (4-carboxyphenyl)benzene) and $[Cu_3(TATB^{6-})]_n$ (TATB = 4,4′,4″-s-Triazine-2,4,6-triyl-tribenzoic acid), exhibited greater CO_2 capture capacity amounting to 157% [140]. This has been further confirmed by GCMC simulation and first-principles calculations [122]. Adenine can also be used as functionality to prepare an effective MOF for capturing CO_2. The key advantage of adenine is that it is a rigid molecule with different nitrogen sites and all the atoms of adenine are in one plane, promoting inter-ligand interactions. Computational studies have also reiterated the importance of the presence of adenine in MOF [37].

DFT studies revealed the interaction of CO_2 (Lewis acid) with $-NH_2$ group (lewis base) of $[Zn_2(NH_2\text{-}BDC)_2(dpNDI)]_n$ (NH_2-BDC = 2-aminoterephthalic acid, dpNDI = N,N′-di(4-pyridyl)-1,4,5,8-naphthalenediimide) [93], amide group of $\{[Zn_4(BDC)_4(BPDA)_4]0.5DMF.3H_2O\}_n$ (BPDA = N,N′-bis(4-pyridinyl)-1,4-benzenedicarboxamide) [48], polar imide group of $\{[Zn_2(BDC)_2(bpNDI)].4DMF\}_n$ [92] to be the prime reason for stronger interactions of CO_2 with nitrogen-containing MOFs. GCMC simulation also proposed similar interactions between oxalamide groups containing MOFs like, $[Cu_2(H_2O)_2L]0.4H_2O.2DMA$ (H_4L = oxalylbis(azanediyl)diisophthalic acid; DMA = N,N-di-methylacetamide) (NOTT-125) and CO_2, mainly emphasizing dipole–quadrupole interactions and hydrogen bonding between the N atom of amide and the O atoms of CO_2 [5]. DFT calculations on Mg-BTT with tetrazole linker showed strong electrostatic interactions (owing to larger negative partial charges on tetrazole) and Van der Waals interaction between MOF and CO_2 [80]. Molecular simulations and experimental studies on three MOFs, NTU-111, NTU-112, and NTU-113 formed by integrating Cu(II) ions with nitrogen-rich tetra-carboxylate ligands, [57], showed Cu(II) and nitrogen atoms to be the sites of adsorption for CO_2 [7]. A modified MIL like NH_2-MIL-101 based on Cr(III) and 2-aminobenzene-1,4-dicarboxylate linker, offers more binding sites if further modification with tetraethylenepentamine (TEPA) is done using the wet impregnation method. MD simulations showed that the incorporation of TEPA in MOF increased its CO_2 adsorption capacity because of the presence of abundant amine groups [43]. N,N′-dimethylethylenediamine-functionalized MOF having DOPDC group (DOPDC = 4,4′-dioxidobiphenyl-3,3′-dicarboxylate) [69] for CO_2 capture was studied using spectroscopic, diffraction, and computational techniques. CO_2 can insert itself between metal (M = Mg, Mn, Fe, etc.) and amine (bonded together) leading to the conversion of the amine group into ammonium carbamate [53]. Van der Waals interaction between organic ligands of MOF and CO_2 has also been found to positively impact the selectivity and activity for CO_2 capture. For example, silver-triazolate MOFs, $\{Ag_3[Ag_5(\mu\text{-}3,5\text{-}^tBu_2tz)_6](BF_4)_2\}_n$, and $\{Ag_3[Ag_5(\mu\text{-}3,5\text{-}Ph_2tz)_6](NO_3)_2\}_n$ have been prepared and studied experimentally and theoretically. DFT studies showed the Van der Waals interactions between CO_2 and the aromatic and aliphatic hydrocarbons are the major reason for its activity [126].

Apart from nitrogen-containing functional groups, other functional groups have been used to prepare MOF for CO_2 capture. A sulfone-functionalized MOF, analog of DUT-5, was prepared, namely [Al(OH)(SBPDC)] (H_2SBPDC = 2,2′-sulfone-4,4′-biphenyldicarboxylate) [86]. The resulting MOF was studied experimentally (gas sorption method) and theoretically (using the DFT method) and showed enhanced CO_2 capture behavior because of the presence of the sulfone group [46]. Another functional group, integration of which in MOF enhances its applicability for CO_2 uptake is carboxylate. Strong electrostatic interactions and Van der Waals interactions are responsible for robust interaction of CO_2 with HKUST-1 (having Cu(II) ions and BTC organo-ligand, BTC = 1,3,5-benzenetricarboxylate) and MIL-101 (Cr(III)-terephthalic acid), shown using GCMC simulation and experimental results [99]. A 3D, porous MOF, $\{[Co_2(TZPA)(OH)(H_2O)_2].DMF\}_n$, where TZPA = 5-(4-(tetrazol-5-yl)phenyl)isophthalate, was built. This MOF has OH groups, O atoms of the carboxylate group, and Co(II) metal sites, due to which it is able to interact with CO_2 much more than CH_4 and CO owing to its large quadrupole moment and polarizability, proved by GCMC simulations [46, 106]. GCMC simulations showed that UiO-66(Zr)-$(COOH)_2$ is quite effective for CO_2 capture from a mixture of gases like CH_4 and N_2 [7, 116]. Few other MOFs suitable to separate CO_2 from other gases include UiO-67, Zr-BTDC (BTDC = 2,2′-bithiophene-5,5′-dicarboxylate), Zr-BFDC (BFDC = 2,2′-bifuran-5,5′-dicarboxylate). They have the same $Zr_6(\mu_3\text{-}O)_4(\mu_3\text{-}OH)_4$ core. GCMC simulations showed that Zr-BFDC adsorb maximum CO_2 at low pressures and Zr-BTDC take maximum CO_2 at high pressures [42]. Due to the strong interactions between the carbonyl and sulfone groups in BUT-10 ($[Zr_6O_4(OH)_4(FDCA)_6]$; FDCA = 9-fluorenone-2,7-dicarboxylate) and BUT-11 ($[Zr_6O_4(OH)_4(DTDAO)_6)]$; DTDAO = dibenzo[b,d]thiophene-3,7-dicarboxylate 5,5-dioxide), their CO_2 uptake capacity is enhanced from their parent MOF (UiO-67), revealed by GCMC simulation [79, 105]. Apart from metal ions, MOFs with different functional groups can be prepared and used for CO_2 capture. One such study was done by Luo and co-workers, wherein MOFs were prepared, by using Zn(II) ion, carboxamide, and carboxylate ligands [66]. Different functional groups, namely, amino, hydroxyl, nitro, and carboxyl groups were integrated. Again, MOFs with amide and acylamide groups were found to have the best affinity for CO_2, according to DFT studies [37]. One more study involving different metal ions was done for M/DOBDC (DOBDC = 2,5-dioxido-1,4-benzenedicarboxylate) series of MOFs (M = Ca and first transition metal elements), MOF-74 [78]. DFT calculations showed the highest binding energy for Ti- and V-/DOBDC MOFs against CO_2, owing to their strong electric fields and also effective overlapping of their empty d-orbitals with CO_2 lone-pair orbitals [7].

Although carboxylate is important functionality to modify MOF for CO_2 capture, it is quite unstable in water [56]. This problem could be solved if MOF is integrated with azolyl group along with the carboxyl group. The presence of azolyl group, because of strong M–N bonds, leads to stability of the resulting MOF, not only in water but also in various organic solvents and in acidic as well basic pH conditions. A polar MOF ($[Mn_2(HCBPTZ)_2(Cl)(H_2O)]$Cl.DMF.0.5

CH_3CN) having specific and strong interaction with CO_2 consists of pyrazinyl triazolyl carboxyl ligand, H_2CBPTZ (3-(4-carboxylbenzene)-5-(2-pyrazinyl)-1H-1,2,4-triazole) and Mn(II) ions. The adsorption studies and GCMC simulations revealed strong interaction of CO_2 with MOF, and it was found that CO_2 interacts with MOF through uncoordinated N atoms, Cl^- ions, and aryl ring [107]. A stable kag-MOF ($[Zn(HCN_4)_2.(H_2O)_{2/9}]0.1.4H_2O$) has been prepared using Zn(II) ions and tetrazole-5-carboxylate organic framework. In this MOF, Zn(II) has an octahedral structure, coordinated to six nitrogen atoms of tetrazole ligands. GCMC simulations showed that CO_2 interacts with the hydrogen atoms attached to carbon and nitrogen atoms of the tetrazole group [46, 71]. A rigid carboxylate containing MOF like PCN-26 ($Cu_4(H_2O)_4(TDM).xS$, where S = non-coordinated solvent molecules and TDM = tetrakis[(3,5-dicarboxyphenyl)oxamethyl]methane have also shown (experimental and theoretical studies) to exhibit enhanced affinity for CO_2 than other gases like N_2 and CH_4. It contains octahedral and cuboctahedral cages [37, 143]. Functionalization of MIL-53 (Al) MOF was done with polar groups like hydroxyl, amino, and carboxyl, and DFT and GCMC simulations were performed. The electrostatic interactions and hydrogen bonding interactions were attributed to be the main reason for the interaction of CO_2 with such polar functional groups [7, 100]. A Cu-based MOF, FJI-H14 having 2,5-di(1H-1,2,4-triazol-1-yl)terephthalate (BTTA) [62] demonstrated open Cu(II) sites and oxygen atoms of carboxyl groups as probable binding sites for CO_2, according to GCMC simulation studies. The said MOF is water and pH-resistant because of abundant free nitrogen atoms [53]. Another set of water-resistant carboxylate-based mixed ligand MOF is $Zn(BDC\text{-}X)\text{-}(DABCO)_{0.5}$ series (DABCO = 1,4-diazabicyclo[2.2.2]octane) manufactured by Jasuja's group using ligands having a polar group like fluorine and nonpolar group like methyl [44]. The molecular simulation was performed which showed improved stability due to the presence of tetramethyl group. Another water-stable MOF was prepared from Cu(II) ions and phosphonate monoesters (1,3,5-tri(4-phosphonato)benzene-tris(monoethylester)) [36]. The experimental and computational results showed its improved stability against water and also its improved activity against CO_2 owing to the presence of the ester group [37].

One more interesting approach is to integrate multiple functional groups in a single MOF. A large number of computationally generated multivariate metal–organic frameworks (MTV-MOFs) having multiple linkers and functional groups were screened computationally for their selectivity and activity for CO_2 capture using GCMC simulations [58]. The GCMC simulation demonstrated enhanced CO_2 capture ability of MOFs having multiple functional groups and small pores compared to the single functional group and large pores in MOF. The type of functional groups, number of functional groups, and pore size play a significant role in determining its activity for CO_2 uptake. In another combined experimental and computational study [63] of zirconium-based MIL-140 frameworks, revealed the importance of pore size. It was found that smaller pores and nitrogen-containing functional groups would enhance the overlap between the adsorbent and adsorbate and hence, better CO_2 uptake capacity [46]. A similar study was done on zirconium-based MOFs. Three different topologies (csq, ftw, and scu) were used, which were further functionalized by three functional groups, –F, $-NH_2$, and $-OCH_3$. GCMC simulations revealed that ftw-MOFs exhibited the strongest affinity toward CO_2. $-NH_2$ functionalized MOFs in csq and scu again reiterated the importance of pore size [60]. MOFs having carboxyl and amino groups, $\{[Cd(NH_2\text{-}BDC)(BPHZ)_{0.5}]$ $DMF.H_2O\}_n$ (NH_2-BDC = 2-aminobenzenedicarboxylic acid, BPHZ = 1,2-bis(4-[pirydyl-methylene)hydrazine) were prepared. One isomer of this MOF has a rigid framework and the other one has a flexible framework. DFT calculations showed interactions between $-NH_2$ and aromatic ring of the ligand with CO_2 [7, 38].

Usually, one CO_2 molecule binds with one metal ion, but there have been studies conducted wherein it was revealed that CO_2 can bind more than one metal ion of MOF, like in the case of $[Cu_2(NDPA)]$ (PCN-88; NDPA = 5,5′-(naphthalene-2,7-diyl)diisophthalate) [55], as shown by DFT studies. Another such example involves $[Cu_2(\mu\text{-}OH)_2(BDIM)]0.5.4H_2O$ (MAF-35; H_2BDIM = 1,5-dihydrobenzo[1,2-d:4,5-d′]diimidazole) [141]. It has $Cu_2(\mu\text{-}OH)_2(BDIM)_4$ clusters exposing both sides of square-planar coordinated Cu(II) ions for CO_2. Again, GCMC simulation showed that each copper ion can bind two CO_2 molecules, and even two adjacent Cu(II) ions, can bind one CO_2 molecule [142]. A unique MOF-74 isomer, UTSA-74, $[Zn_2(DOBDC)(H_2O)]0.0.5H_2O$ [67] has one Zn(II) ion tetrahedrally coordinated and another one, octahedrally coordinated. X-ray crystal structure and molecular modeling studies showed that each Zn-site binds one CO_2 molecule, forming a bridge between two Zn-sites [64].

The determination of which metal ions and which functional groups need to be integrated into MOF to improve its adsorption behavior toward CO_2 using experiments is a very costly affair. A computational technique called high-throughput computational screening comes to the rescue for this. Large-scale screening of a huge number of MOFs can be done so as to ascertain the factors absolutely necessary for selectivity and activity against CO_2 and then new MOFs could be proposed without doing any experimental work. Along these lines, a computational screening of around 41,825 MOFs was done [83, 84]. Eight best MOFs consisting of Mg-MOF-74, Ni-MOF-74, Co-MOF-74, Zn-MOF-74, Al-MIL-53, Cr-MIL-53, UiO-66, and UiO-67, for CO_2 capture were determined and then modified using amine functional groups, $-NH_2$, $-NHCH_2CH_2NH_2$, –NHCOH, and $-NHCOCH_3$. Out of 48 functionalized MOFs for Al-MIL-53, Cr-MIL-53, UiO-66, and UiO-67, the best performing MOF for CO_2 capture was $(NHCOH)_n$-Al-MIL-53. For the rest of the MOFs, 80 functionalized forms were made using the same amine functional groups, and the best amine functional group was found to be $NHCH_2CH_2NH_2$. Then, parallel experiments were conducted to verify the computational results [83]. Another similar study involves more than 1,30,000 hypothetical MOFs [113], wherein a strong correlation between structural properties like pore size, surface area, and chemical properties of MOFs was found. In another screening study using GCMC simulation, 4764 computation-ready, experimental MOFs (CoRE-MOFs), wherein the importance of lanthanide ions in MOF for CO_2 capture came forward rather than usual alkali and alkaline earth metal ions [84]. Computational screening using MC and MD simulation along with machine learning (ML) yields good results in selecting the best MOF for CO_2 from a large dataset [21]. A total of 6,013 CoRE-MOFs were analyzed to find the correlation between adsorption and diffusion selectivity for CO_2 and the MOF descriptors at very low CO_2 concentrations. Random forest was found to be the best ML algorithm which showed that the major factor for capturing CO_2 at a low concentration of air is diffusion [21]. Several other studies using different descriptors and different machine learning algorithms were done [11, 16, 31]. In machine learning, the first step is to analyze the structural diversity among MOFs selected. Then, geometric properties like pore size, shape, surface area, volume, etc., of different MOFs have to be analyzed using various descriptors, which affect the adsorption behavior of MOFs. ML methods can be used to design new and better MOFs, but the results of ML depend on the experimental conditions of MOFs preparation as well as the function and diversity of MOFs used. One such study was done by Burns and co-workers wherein they used molecular simulation, process modeling, and ML to evaluate more than 1,600 MOFs for CO_2 [12]. The use of ML not only speeds up the process but also increases efficiency [85]. One interesting study was done by [30] using 1,37,953 hypothetical MOFs. A universal ML algorithm was proposed based on basic chemical properties like the type of atoms (determined using Python software) rather than building blocks. By doing so, the number of MOFs required to perform such screening was reduced as building blocks could be infinite but atom types are finite and accuracy is also increased [124]. Artificial neural networks (ANNs) (a machine learning method) along with multi-objective genetic algorithm method for optimization can be used to study MOF properties for adsorption. One such study includes UiO-66, Zr-MOF. The numerical data were used to evaluate the heat of adsorption of gases like CO_2 and CH_4 using the Dubinin Astakhov model for $-NO_2$ and $-NH_2$ functionalized MOFs [135]. Computational and experimental screening of hypothetical and real ethylenediamine-functionalized CPO-27-Mg (Mg-DOBDC), was done for their CO_2 capture property [10, 46]. High-throughput GCMC screening of approximately 3000 existing MOFs, taken from the CoRE MOF 2014-DDEC database, was carried out to study their capturing power for CO_2, even in the presence of water. Various filters and selection criteria like pore size, adsorption activity, adsorption sites, their distribution, and surface area were considered. Thirteen promising MOF structures, especially those containing Zn(II) ions and nitrogen atoms in ligand with small channels, were revealed by the screening for their selectivity for CO_2, even in the presence of other gases. The best MOF came out to be $MnH_6C_{12}(NO_2)_2$ [88].

Equilibrium molecular dynamics (EMD) simulations were performed to study the interaction of CO_2 with MOF-5 ($C_{6.40}H_{3.20}O_{3.47}Zn_{1.07}$). Charge–charge interactions between CO_2 and MOF revealed selective and strong adsorption of CO_2 [94]. DFT method was used to study CO_2 adsorption on CPO-27-M (M = Mg, Mn, Fe, etc.) [133]. Geometry optimizations of the MOFs and CO_2-adsorbed MOFs were done using generalized gradient approximation (GGA) exchange–correlation functional of the DFT method. Single-point energy calculations at the MP2 level were done to calculate the binding energies, which were further corrected using a quantum mechanics/molecular mechanics (QM/MM) approach. In QM/MM, the QM energies were calculated using the MP2-level calculations and the MM energies using Dreiding force field or Universal force field (UFF) [133]. ab initio calculations were performed for the adsorption of CO_2 in CPO-27-Mg [15]. A Buckingham-type potential (the Carra–Konowalow potential) was used to calibrate a Morse–Morse–spline–Van der Waals (MMSV) interaction potential and then the optimization of the MMSV potential was done using some known ab initio data obtained using a double-hybrid density functional with empirical dispersion correction, B2PLYP-D2 with multi-objective genetic algorithm. This optimized force field was used to perform GCMC simulations, and the results matched with the experimental ones [18]. Hartree–Fock (HF) method with 6-31G* basis set and Gibbs Ensemble Monte Carlo (GMC) and the MD simulations were performed to study the adsorption of CO_2 on MIL-127(Fe). This MOF consists of Fe(III) and carboxylate ligands. It was found that apart from metal ions, adsorption sites were found at the center of MOF, and adsorption capacity is high at low concentrations of CO_2 and it decreases at high concentrations of CO_2 because of an increase in CO_2–CO_2 repulsions [82]. A novel MOF with heterofullerene ($C_{48}B_{12}$) as a linker (with and without lithium doping) was investigated for its adsorption properties using DFT and GCMC simulations. Li-modified MOF was found to have good selectivity for CO_2. Few more examples of fullerene-based MOF include C_{60}-impregnated derivatives of MOF-177, MOF-180, and MOF-200. GCMC simulations showed that such C_{60} modified MOFs exhibit better CO_2 separation and capture tendency [120, 121]. Another combined GCMC simulation and DFT study on adsorption characteristics of CO_2 on $[Zn(BDC)(DPDS)]_n$ (BDC = benzenedicarboxylate (anionic linker), DPDS = 4,4′-dipyridyldisulfide (neutral linker)) was carried out and showed excellent results in correspondence with experimental results. Simulation studies revealed that the major adsorption site for CO_2 comprises two DPDS ligands [65]. A combined GCMC simulation and DFT screening method involving charges obtained from ChelpG method and charge equilibration method has been proved to be an effective technique for studying MOF for their CO_2 adsorption properties [114]. A GCMC simulation and DFT study using unrestricted B3LYP functional and ChelpG method include the study of catenated and non-catenated isoreticular metal–organic frameworks (IRMOFs), consisting of organic dicarboxylate linkers. The catenated forms demonstrated better CO_2 adsorption power [128].

A unique phenomenon of co-adsorption of SO_2 and CO_2 was found in Ni/DOBDC (CPO-27 or MOF-74). The molecular simulations (DFT and GCMC) found that from a gas mixture of CO_2, N_2, and SO_2 at standard temperature and pressure (STP), CO_2 capture by MOF increases from 0.76 to 1.04 mmol g^{-1} with an increase in SO_2 concentration to 4% but decreases on further increase in the concentration of SO_2 up to 10%, attributed to strong SO_2–CO_2 interactions [23]. Zeolitic imidazolate frameworks (ZIFs) are nanomaterials with properties of both MOF (having metal ions and imidazolate linkers) and zeolites (cage-like structure) integrated together [51]. Although ZIFs have more affinity for SO_2, in some of ZIFs like ZIF-68, ZIF-69, ZIF-10, and ZIF-71, GCMC simulations revealed interactions between CO_2 and SO_2 and hence, does not majorly impact their interaction with CO_2 [24]. Another research group studied the CO_2 capture ability of $-NH_2$, –OH, and –Br functionalized UiO-66 in the presence of other gases like H_2O, SO_2, and NO_2 using molecular simulations [9]. Again, a major impact on CO_2 uptake was found due to H_2O and SO_2 but in the case of UiO-66-Br, CO_2 uptake was optimum because of hydrophobic interactions and also because water interacts with SO_2 rather than MOF [9]. Mg-MOF-74 was found to be a better adsorbent for CO_2 from a mixture of gases having CH_4, N_2, H_2, revealed from Configurational-Bias Monte Carlo (CBMC) and MD simulations [47]. GCMC simulations on IRMOF series were performed to gauge the effect of porosity and interpenetration on CO_2 capture [40]. It contains Zn_4O metal clusters interconnected via organic linkers like naphthalene in IRMOF-8 and biphenyl in IRMOF-9. At low pressures, the long-interpenetrated linkers result in more CO_2 uptake [7].

As CO_2 interaction with different MOFs involves a complex interplay of electrostatic, orbital overlap, and dispersion interactions, involving empty d-orbitals of metal ions of MOF and lone pairs of CO_2, theoretical studies must include dispersion effects in order to understand long-range CO_2–MOF interactions. For example, in a Ca(II) ion-containing MOF–CO_2 system, dispersion interactions were evaluated by representing binding energy as a function of the cut-off distance [81]. Different groups have performed complex computational calculations [81, 87] using dispersion corrected DFT functionals like Van der Waals density functionals [vdW-X (X = DF, DF_2], and the results obtained have a good correlation with experiments. A hybrid MP2:B3LYP-D functional used for Mg-, Zn-, and Ni-DOBDC (CPO-27 CO_2 adsorption systems gave results at par with vdW-X, again with results matching that of experiments [103]. In MP2:B3LYP-D hybrid, MP2 calculations are done on the adsorption site (a cluster model) and long-range dispersion corrections are added later using B3LYP-D. The use of only B3LYP-D functional did not yield good results [13]. Another such combined experimental and ab initio study (using B3LYP-D*) amino-functionalized UiO-66 having ligands, BDC, and 2-amino-1,4-benzenedicarboxylic acid (ABDC) [29]. With the advancement of computational infrastructure and supercomputers, efficiently parallelized codes like CRYSTAL [28], theoretical calculations on MOFs as large as MIL-100 (M) having 2788 atoms in the unit cell [20] are possible. DFT-D study of MIL-100(M), where M(III) = Al, Sc, Cr, Fe, predicted the maximum activity of Cr(III) containing MOF for CO_2 exhibiting the shortest M-CO_2 distance, which could be attributed to its electrostatic potential and partially occupied d-orbitals of Cr(III). It was a complex study as there are 136 possible sites for CO_2 adsorption. There was a slight deviation between experimental and theoretical results of MIL-100(Cr) because the real structure is defective with uneven distribution of the adsorption sites [20]. Another dual B3LYP-D3 level DFT and experimental study of CO_2 adsorption capacity of the M(Me_2BPZ) (M = Co, Zn and H_2Me_2BPZ = 3,3′-dimethyl-1H,1′H-4,4′-bipyrazole) MOFs and its non-methylated M(BPZ) parents were also done. The adsorption energies of methylated families were found to be more than their non-methylated counterparts as supported by theoretical calculations [70, 73]. A set of six different MOFs, with varying pore geometry, organic linking groups, and metals, $Zn_2(BDC)_2$(DABCO), ZIF-69, Zn_2(DHBDC) (DHBDC = 2,5-dihydroxybenzenedicarboxylate), MOF-177, $Cu_3(BTC)_2$, and Cu_2(BPTC) (BPTC = 3,3′,5,5′-biphenyltetracarboxylate) were studied using GCMC simulations and DFT method (using PBE functional) was used to calculate charges. Dispersion energies of CO_2–MOFs were calculated and it was observed that at low pressures, the CO_2–MOF interaction is strong and at high pressures, CO_2–CO_2 repulsions are predominant [112]. M-MOF-74 with different metal ions like Mg, Ti, V, Cr, etc. [49] were prepared and investigated for their interaction with components of flue gas. Since d-orbitals are involved, dispersion interactions are included in DFT calculations by using a non-local Van der Waals density functional, vdW-DF2. According to DFT calculations, the highest binding energy for CO_2 was found to be for Cu-MOF-74 [37].

4 Theoretical Studies on Fixation of CO_2 by MOFs

CO_2 is a very stable molecule as it consists of two strong C=O bonds. Therefore, its reduction is a complex multi-step process that requires either an electrocatalyst or a photocatalyst. The final product of CO_2 reduction varies from compounds like methanol (CH_3OH), carbon monoxide (CO), formic acid (HCOOH) to hydrocarbons like methane (CH_4), ethylene (C_2H_4), etc. [98] and chemicals like carbonates [2], carbamates [97] as well as hydrocarbon fuel [137] (Fig. 2) [117].

Cyclic carbonates (CCs) are stable, non-toxic, and highly efficient solvents and extractants. They are also used as precursors for the manufacturing of other important chemicals like polycarbonates [32]. CO_2 can be artificially fixed as CCs by inserting it in epoxides (Fig. 3). This conversion is very beneficial as it is 100% economical [19, 74].

Out of various catalysts for the artificial fixation of CO_2, MOFs have emerged as very efficient ones. DMA-MF {[$(CH_3)_2NH_2$][$M(COOH)_3$] (DMA = dimethylamine)} is one such MOF, wherein, metal ions were varied as Mn, Co, Ni, and Zn. Out of these, DMA-MnF exhibited maximum catalytic ability. NH_3-TPD (temperature-programmed desorption) profile was used to propose the mechanism of action of DMA-MF. CO_2 is activated by DMA, which acts as a Lewis base. The epoxide first binds metal ions (Lewis acid). The oxygen atom of the activated CO_2 then attacks the β-carbon of the activated epoxide resulting in the formation of an intermediate which further gives cyclic carbonate product. DFT calculations using the M06 method were done using Los Alamos effective core potential (LANL2DZ)/6–31 G(d, p) basis set to confirm the mechanism [117]. Wu et al. [118] further investigated Mn-based MOFs for CO_2 coupling with epoxides using DFT studies. The MOFs studied were, [(CH_3NH_3)][$Mn(COOH)_3$], [$(CH_3CH_2NH_3)$][$Mn(COOH)_3$], and [$C_3H_5N_2$][$Mn(COOH)_3$]. polycarbonates The MOF having $C_3H_5N_2$ group was found to exhibit maximum catalytic activity [118].

Another MOF studied for CO_2 fixation with epoxides involves MIL-101(Cr)/tetrabutylammonium bromide (TBAB). A DFT study was carried out to study the mechanism of the catalytic activity of MOF. CO_2 fixation was studied using MIL-101, TBAB, and MIL-101(Cr)/(TBAB) as catalysts, and a non-catalyzed CO_2 fixation was also analyzed for comparison. The activation barrier of the rate-determining step involving the addition of CO_2 to epoxide (propylene oxide) was found to be lowest for

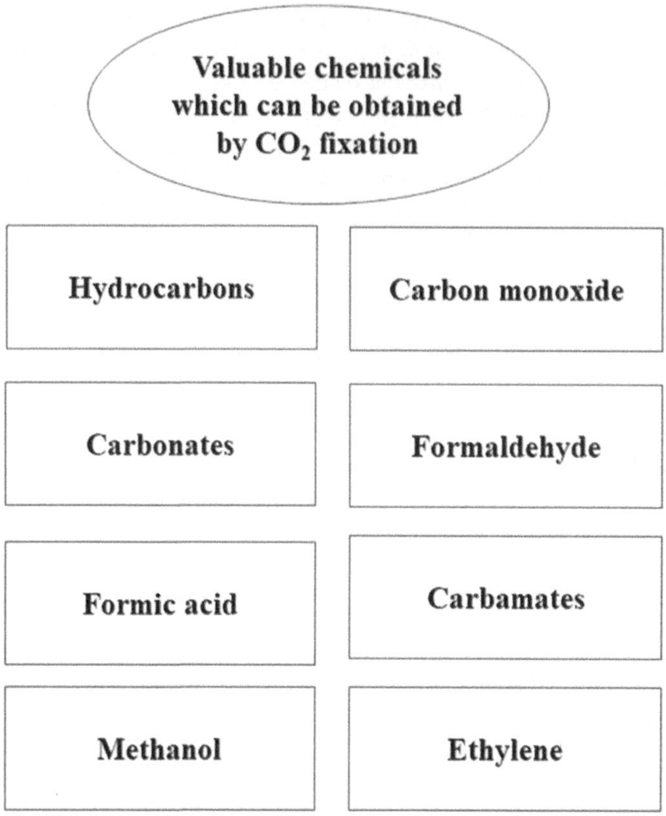

Fig. 2 Various chemicals that can be obtained by reduction or fixation of CO_2

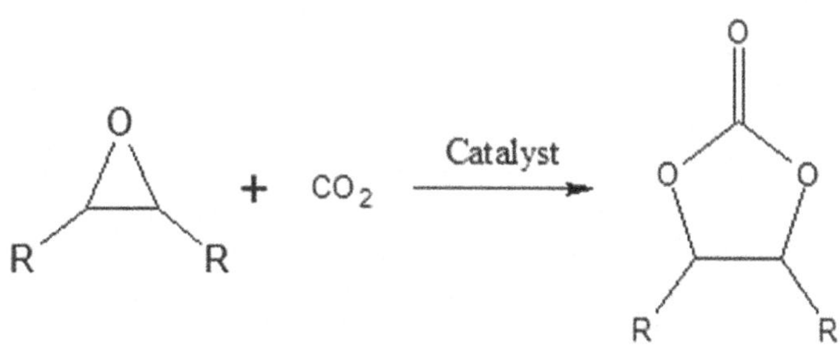

Fig. 3 Insertion of CO_2 in an epoxide molecule

MIL-101/TBAB (18.11 kcal mol^{-1}) followed by TBAB-catalyzed (26.86 kcal mol^{-1}) reaction and MIL-101-catalyzed (46.89 kcal mol^{-1}) reaction. The non-catalyzed reaction was found to have a very high activation barrier of 57.67 kcal mol^{-1} [42]. Another TBAB co-catalyzed conversion of CO_2 to cyclic carbonates (propylene carbonate) involves the use of Zn(II) ion-containing MOFs. Such a MOF is called 1-Zn and contains dimeric paddlewheel ($Zn_2(CO_2)_4$) and tetrameric ($Zn_4(O)(CO_2)_6$) units [144]. This MOF could be further processed to introduce Cu(II) and Co(II) ions to the dimeric paddlewheel group using the ion-exchange method. The catalytic efficiency was found to be maximum for 1-Zn, followed by Co(II) and Cu(II) containing MOF. To explain this order, molecular dynamics simulations were performed. The calculations showed that the higher activity of 1-Zn is due to the lowest energy gap between the highest occupied molecular orbital (HOMO) of epoxide attached to the metal site (propylene oxide) and the lowest unoccupied molecular orbital (LUMO) of CO_2. Also, the binding energy of CO_2 to metal ions was calculated and found to be maximum for 1-Zn [144]. Sulfonate-based MOF, like TMOF-1 has also been shown to catalyze the conversion of CO_2 to cyclic carbonate via addition to epoxide. TMOF-1 has sulfonate, 4,4′-bipyridine (BPY) ligand, and Cu(II) ions. TMOF-1 exhibited a surface area of 256 ± 5 m^2 g^{-1} and CO_2 uptake of 32.4 cm^3 g^{-1} at STP. It was found to perform better than its corresponding carboxylate-based MOF. Greater efficiency of sulphonate-based MOF over carboxylate-based MOF was further proved using DFT calculations [61, 136]. Amino group functionalized MOFs were prepared and were shown to exhibit better catalytic activity than the one without amino group for cycloaddition of CO_2 to epoxide using experimental and DFT studies (DFT-D2 method) [50]. MIL-68(In)-NH_2 and MIL-68(In) were prepared. MIL-68(In)-NH_2 was found to have a 1-D rod-shaped structure having hexagonal large and triangular tight channels resulting in a chain structure bridged by hydroxyl groups [76]. ML technique was used to study MOFs so as to understand the factors responsible for their efficiency to act as a catalyst for fixing carbon dioxide into cyclic carbonate. 1311 hypothetical MOFs were created from the experimental results available in the literature. Six best metal ions and four best ligands were revealed and combined to give 24 hypothetical MOFs having CO_2 fixation ability [59].

CO_2 can be transformed to CO using MOF-derived catalysts. One such catalyst is the (ZIF)-derived carbons obtained from the carbonization of MOF. One such study, using ZIF as catalyst was done, wherein four Co/Zn-ZIF-derived nanocarbons were prepared, containing N-doped carbons with Co-N_2, Co-N_3, Co-N_4, and Co-NPs (nanoparticles) active sites [108]. The ZIF having Co-N_2 was found to have the maximum catalytic activity. DFT calculations were used to calculate free energy for CO_2 to CO conversion using all four ZIFs. The endothermic free energy value of adsorption of CO_2 on ZIF was found to be the least for Co-N_2 [108]. Another MOF-derived catalyst for the same CO_2 conversion is the one containing nickel, prepared by MOF carbonization. DFT calculations were done to study the catalytic activity of two such MOFs, one is manufactured by Ni^{2+} and 2-aminoterephthalic acid in carbon black (contains nickel nanoparticles as active sites) [45] and another MOF has single nickel atoms as active sites [134]. CO_2 reduction to CO can be realized by polyoxometalate (POM)-metalloporphyrin organic frameworks (M-PMOF) [111]. Both POM and Co-porphyrin play a crucial role in CO_2 reduction. According to DFT calculations, Co-PMOF shows relatively lower values of adsorption free energies for the intermediates like COOH and CO. It was proposed that Co-porphyrin provides a favorable active site and POM acts as an electron-rich site [111]. Single-atom catalysts involving Co and Fe integrated on ZIF-8 have also been explored as catalysts for CO_2 fixation [132]. Experimental and theoretical calculations demonstrated that M–N–C active sites on the surface of ZIF are more exposed and available than those embedded in the ZIF. It was also proposed that the metal centers with dangling bonds adjacent to N are the actual active sites that adsorb the intermediates like CO and OH [77]. To enhance the activity and stability of molecular catalysts like $ReL(CO)_3Cl$ (L = 2,2-bipyridine-5,5′-dicarboxylic acid) [131], it can be incorporated with a MOF. The resulting Re-MOF-based electrocatalyst displayed superior catalytic activity for the conversion of CO_2 to CO. The mechanism was proposed to involve the reduction of Re^+ to Re^0. DFT studies, using Grimme's D3 dispersion (PBE-D3) and the TZP Basis Set, revealed that the orientation of the resulting MOFs favors charge transfer between the electrode and the catalyst [89, 131]. A series of isostructural honeycomb-like MOFs ($[Co_2(\mu\text{-}Cl)_2(BBTA)]$ (H_2BBTA = 1H,5H-benzo(1,2-d:4,5-d′)bistriazole), $[Co_2(\mu\text{-}OH)_2(BBTA)]$ and $[Co_2(DOBDC)]$ were prepared for photocatalytic conversion of CO_2 to CO. PDFT calculations revealed strong CO_2 binding power of $\mu\text{-}OH^-$ ligands (being a good hydrogen bond donor) present near cobalt center in MAF-X27-OH, stabilizing the Co–CO_2 adduct, which promotes the reduction process [110]. Such strong CO_2 binding power was also seen in Ni-MOF monolayers leading to highly selective photoreduction of CO_2 to CO [39, 52].

CO produced by CO_2 can be further reduced to prepare valuable compounds like methanol. For this, through-hole carbon nanofibers having copper single atoms were used, prepared from polyacrylonitrile (PAN) fibers-integrated Cu/ZIF-8 NPs containing Cu–N_4 moieties as active sites. DFT method was used to reveal the mechanism of action. The free energy value of the desorption reaction of CO from the catalyst was found to be very low (endergonic) and hence, increases its probability to be further converted to methanol [1, 127]. An oxide-derived Cu/carbon (OD Cu/C) catalyst was prepared by carbonization of Cu-based MOF (HKUST-1). The resulting catalyst effectively reduced CO_2 to methanol and ethanol owing to the synergistic effect of copper oxides/copper and porous carbon. The infrared (IR) spectroscopy and DFT studies showed that the CO molecule adsorbs on OD Cu/C catalyst, and then carbon (having synergistic interaction with Cu) promotes C–C coupling resulting in the formation of C_2 product [89, 138]. Another set of DFT simulations were performed on Lewis pair (LP) functionalized MOF (UiO-67) which could reduce CO_2 to methanol [22, 130].

CO_2 can be converted to other useful products like HCOO by hydrogenation. DFT calculations were performed using M06-L functional to study the mechanism of catalytic conversion of CO_2 to HCOOH by Cu-alkoxide-functionalized MOF-5 (Cu-MOF-5). Out of the two probable mechanisms, concerted and stepwise, the stepwise mechanism was found to be more energetically favorable having lower activation barriers of 24.2 and 18.3 kcal mol^{-1}, respectively, for two steps than the concerted mechanism (activation barrier of 67.2 kcal mol^{-1}) [68]. Frustrated Lewis pairs functionalized MOFs (UiO-66-P-BF_2) have also been used for the reduction of CO_2 to HCOOH. DFT calculations using CP2K code with Grimme's D3 dispersion corrections were done which showed the stability of such unique MOFs before and after catalyzing CO_2. DFT studies proposed that both H_2 and CO_2 have to be chemisorbed on the MOF catalyst for efficient interaction between the two [19, 129].

CO_2 can be catalytically hydrogenated to form CH_4 called CO_2 methanation or the Sabatier reaction. Thermodynamically, CO_2 methanation is favorable but has kinetic limitations as it is an eight-electron reduction process. Again, MOFs and modified MOFs have emerged as useful catalysts for the Sabatier reaction [95]. One such MOF reported in the literature is Ni NPs-encapsulated MIL-101. DFT studies showed that CO_2 is first chemisorbed on the modified MOF catalyst to form CO. Then, adsorbed CO species are converted to CHO and then CH_4, which is released from the catalyst [19, 139].

A cobalt-containing ZIF9 was prepared and integrated on the surface of Co_3O_4 nanowires for photoelectrochemical CO_2 fixation. DFT studies using Tao-Perdew-Staroverov-Scuseria (TPSS) functional, with LanL2DZ basis set for Co atom and all-electron 6–31G (d, p) basis set for rest of the atoms, showed that CO_2 is adsorbed on the Co sites of ZIF9 and CO_2 is reduced to formate [91]. CO_2 can be converted to formate by photocatalytic reduction, thereby using solar energy. One such catalyst is Ti-doped NH_2-Uio-66(Zr) MOF [96]. DFT studies were performed to decipher the impact of Ti on its activity. The density of states and their band structure revealed that pure NH_2-Uio-66(Zr) has a valence band comprising of lone pairs of nitrogen atoms and the conduction band is due to the π^* states of the ligand, 2-aminoterephthalic acid (ATA). The Ti introduces new energy bands in the conduction band of Ti-doped MOF. This leads to a favorable electron transfer from the excited ligand to the Ti leading to the formation of the excited state, $(Ti^{3+}/Zr^{4+})_6O_4(OH)_4$. Ti^{3+} in the excited moiety acts as an electron donor to Zr^{4+} forming Zr^{3+} which leads to enhancement of its catalytic power [4, 96].

DFT studies were done on Ru-based metal ligand as a photosensitizer for MOFs used for CO_2 reduction. Three metal-based (Ni, Co, and Cu) MOFs have been prepared [109]. Out of these, Ni–based MOF exhibited maximum activity for the photoreduction of CO_2 to CO owing to its strong CO_2 binding, the low free energy value of CO reduction. They used Ru-based metalloligand, $[Ru(BPY)_3]Cl_2.6H_2O$, as a photosensitizer, which transfers an electron to MOF because of their matching LUMOs [109]. Another study involving $Ru(PHEN)_3$-derived tricarboxylate acid metalloligand (PHEN = 1,10 phenanthroline) as a photosensitizer for MOF having dinuclear $Eu(III)_2$ clusters. The resulting MOF could reduce CO_2 to HCOOH with high selectivity. An efficient electron transfer occurs from Ru metalloligands to $Eu(III)_2$ initiating the photoreduction reaction [125]. A photoactive Zr-MOF with $[Ru(CPTPY)_2]$ complex (CPTPY = bis(4′-(4-carboxyphenyl)-terpyridine)) was prepared to exhibit high CO_2 to HCOOH photoreduction owing to efficient charge transfer from Ru-based photosensitizer to MOF showed by DFT studies using B3LYP/6-311G(d,p) exchange–correlation functional with LanL2DZ basis set [27, 52].

5 Conclusion

The mounting levels of CO_2 in the atmosphere because of urbanization and industrialization have been a source of worry for people all over the world. The capturing of CO_2 or its conversion to some useful carbon-containing compounds is the only viable option available to deal with this problem. MOFs have emerged as a great adsorbent for CO_2 whether one talks about its capture or fixation. This is probably because of the highly porous nature and highly efficient binding sites of MOF (metal ions and organic linkers) and the fact that MOFs can be fine-tuned by changing metal ions and organic and inorganic groups and also by integrating them with other nanomaterials. But to manufacture and study different MOFs for their adsorption properties is a costly affair. In this chapter, we have highlighted the importance of integration of computational techniques like DFT, ab initio, and simulations with experimental methods to effectively and completely study MOFs–CO_2 interaction including the mechanism of action of MOF. Computational techniques like high-throughput screening and ML also help in screening huge datasets of real as well as hypothetical MOFs with the help of which we can factor out the best MOFs and also the properties responsible for best adsorption behavior against CO_2 under different environmental conditions. This could help in zeroing on a few MOFs which could be further synthesized and used for CO_2 adsorption which reduces the cost effectively. But the use of computational methods does not rule out the importance of experiments, which are absolutely necessary for giving concrete results.

For CO_2 adsorption, there are multiple binding sites on MOFs like metal ions and atoms of organic groups. Usually, MOFs are effective for CO_2 at low pressures (lower CO_2 concentration) as at high pressures, adsorbate–adsorbate mutual repulsions are predominant. Also, the effective interaction of CO_2 with MOFs is due to hydrogen bonding, Van der Waals interaction, electrostatic interactions, and dispersion interactions between them. Dispersion effects can be readily studied using dispersion corrected functionals of DFT like B3LYP-D, vdW-X, etc., and electrostatic interactions are studied by calculating charges from DFT calculations. The charge is something that cannot be evaluated using an experimental technique.

Abbreviations

MOFs	Metal–Organic Frameworks
Ppm	Parts per million
CCS	CO_2 Capture and Storage
BDC	1,4-Benzenedicarboxylate
BTT	1,2,5-Benzenetristetrazolate
BPZ	Bipyrazole
BTP	Benzenetripyrazolate
MIL	Matérial Institut Lavoisier

DFT	Density Functional Theory
PDFT	Periodic Density Functional Theory
MD	Molecular Dynamics
GCMC	Grand Canonical Monte Carlo
OX	Oxalate
HATZ	3-Amino-1,2,4- triazole
PCN	Porous Coordination Network
TZC	Tetrazole-5-carboxylate
DPP	1,3-Di(4-pyridyl) propane
NJU-Bai3	Nanjing University Bai group
BTB	1, 3, 5—Tris (4-carboxyphenyl)benzene
TATB	4,4′,4″-S-Triazine-2,4,6-triyl-tribenzoic acid
NH_2-BDC	2-Aminoterephthalic acid
dpNDI	N,N′-di(4-pyridyl)-1,4,5,8-naphthalenediimide
BPDA	N,N′-bis(4-pyridinyl)-1,4-benzenedicarboxamide
DMA	N,N-di-methylacetamide
TEPA	Tetraethylenepentamine
DOPDC	4,4′-Dioxidobiphenyl-3,3′-dicarboxylate
H_2SBPDC	2,2′-Sulfone-4,4′-biphenyldicarboxylate
BTC	1,3,5-Benzenetricarboxylate
TZPA	5-(4-(Tetrazol-5-yl)phenyl)isophthalate
BTDC	2,2′-Bithiophene-5,5′-dicarboxylate)
BFDC	2,2′-Bifuran-5,5′-dicarboxylate
FDCA	9-Fluorenone-2,7-dicarboxylate
DTDAO	Dibenzo[b,d]thiophene-3,7-dicarboxylate 5,5-dioxide
DOBDC	2,5-Dioxido-1,4-benzenedicarboxylate
H_2CBPTZ	3-(4-Carboxylbenzene)-5-(2-pyrazinyl)-1H-1,2,4-triazole
TDM	Tetrakis[(3,5-dicarboxyphenyl)oxamethyl]methane
BTTA	2,5-Di(1H-1,2,4-triazol-1-yl)terephthalate
DABCO	1,4-Diazabicyclo[2.2.2]octane
MTV-MOFs	Multivariate metal–organic frameworks
NH_2-BDC	2-Aminobenzenedicarboxylic acid
BPHZ	1,2-Bis(4-[pirydyl-methylene)hydrazine
NDPA	5,5′-(Naphthalene-2,7-diyl)diisophthalate
H_2BDIM	1,5-Dihydrobenzo[1,2-d:4,5-d′]diimidazole
CoRE-MOFs	Computation-ready, Experimental MOFs
ML	Machine Learning
ANNs	Artificial neural networks
EMD	Equilibrium molecular dynamics
GGA	Generalized Gradient Approximation
QM/MM	Quantum Mechanics/Molecular Mechanics
UFF	Universal Force Field
MMSV	Morse–Morse–spline–van der Waals
HF	Hartree-Fock
GMC	Gibbs Ensemble Monte Carlo

DPDS	4,4′-Dipyridyldisulfide
IRMOFs	Isoreticular metal–organic frameworks
STP	Standard Temperature And Pressure
ZIFs	Zeolitic Imidazolate Frameworks
CBMC	Configurational-Bias Monte Carlo
ABDC	2-Amino-1,4-benzenedicarboxylic acid
H_2Me_2BPZ	3,3′-Dimethyl-1H,1′H-4,4′-bipyrazole
DHBDC	2,5-Dihydroxybenzenedicarboxylate
BPTC	3,3′,5,5′-Biphenyltetracarboxylate
CCs	Cyclic carbonates
DMA	Dimethylamine
NH_3-TPD	Temperature-programmed Desorption
TBAB	Tetra-butylammonium bromide
HOMO	Highest Occupied Molecular Orbital
LUMO	Lowest Unoccupied Molecular Orbital
BPY	4,4′-Bipyridine
NPs	Nanoparticles
POM	Polyoxometalate
M-PMOF	Polyoxometalate-Metalloporphyrin Organic Frameworks
H_2BBTA	1H,5H-benzo(1,2-d:4,5-d′)bistriazole
PAN	Polyacrylonitrile
OD Cu/C	Oxide-derived Cu/carbon
TPSS	Tao-Perdew-Staroverov-Scuseria
ATA	2-Aminoterephthalic acid
PHEN	1,10 Phenanthroline
CPTPY	Bis(4′-(4-carboxyphenyl)-terpyridine

References

1. Abdelkader-Fernández VK, Fernandes DM, Freire C (2020) Carbon-based electrocatalysts for CO_2 electroreduction produced via MOF, biomass, and other precursors carbonization: a review. J CO_2 Util 42:101350
2. Abimanyu H, Kim CS, Ahn BS, Yoo KS (2007) Synthesis of dimethyl carbonate by transesterification with various MgO-CeO_2 mixed oxide catalysts. Catal Lett 118(1–2):30–35
3. Albanese E, Civalleri B, Ferrabone M, Bonino F, Galli S, Maspero A, Pettinari C (2012) Theoretical and experimental characterization of pyrazolato-based Ni(II) metal-organic frameworks. J Mater Chem 22(42):22592–22602
4. Alkhatib II, Garlisi C, Pagliaro M, Al-Ali K, Palmisano G (2020) Metal-organic frameworks for photocatalytic CO_2 reduction under visible radiation: a review of strategies and applications. Catal Today 340:209–224
5. Alsmail NH, Suyetin M, Yan Y, Cabot R, Krap CP, Lü J, Easun TL, Bichoutskaia E, Lewis W, Blake AJ, Schröder M (2014) Analysis of high and selective uptake of CO_2 in an oxamide-containing $\{Cu_2(OOCR)_4\}$-based metal-organic framework. Chem Eur J 20(24):7317–7324

6. Anderson TR, Hawkins E, Jones PD (2016) CO_2, the greenhouse effect and global warming: from the pioneering work of Arrhenius and Callendar to today's Earth system models. Endeavour 40(3):178–187
7. Andirova D, Cogswell CF, Lei Y, Choi S (2016) Effect of the structural constituents of metal organic frameworks on carbon dioxide capture. Micropor Mesopor Mat 219:276–305
8. Baima J, Macchieraldo R, Pettinari C, Casassa S (2015) Ab initio investigation of the affinity of novel bipyrazolate-based MOFs towards H_2 and CO_2. CrystEngComm 17(2):448–455
9. Balbuena PB, Yu J (2015) How impurities affect CO_2 capture in metal-organic frameworks modified with different functional groups. ACS Sustain Chem Eng 3(1):117–124
10. Bernini MC, García Blanco AA, Villarroel-Rocha J, Fairen-Jimenez D, Sapag K, Ramirez-Pastor AJ, Narda GE (2015) Tuning the target composition of amine-grafted CPO-27-Mg for capture of CO_2 under post-combustion and air filtering conditions: a combined experimental and computational study. Dalton Trans 44(43):18970–18982
11. Borboudakis G, Stergiannakos T, Frysali M, Klontzas E, Tsamardinos I, Froudakis GE (2017) Chemically intuited, large-scale screening of MOFs by machine learning techniques. Npj Computat Mater 3(1):40
12. Burns TD, Pai KN, Subraveti SG, Collins SP, Krykunov M, Rajendran A, Woo TK (2020) Prediction of MOF performance in vacuum swing adsorption systems for postcombustion CO_2 capture based on integrated molecular simulations, process optimizations, and machine learning models. Environ Sci Technol 54(7):4536–4544
13. Cazorla C (2015) The role of density functional theory methods in the prediction of nanostructured gas-adsorbent materials. Coord Chem Rev 300:142–163
14. Chavan S, Vitillo JG, Gianolio D, Zavorotynska O, Civalleri B, Jakobsen S, Nilsen MH, Valenzano L, Lamberti C, Lillerud KP, Bordiga S (2012) H_2 storage in isostructural UiO-67 and UiO-66 MOFs. Phys Chem Chem Phys 14:1614–1626
15. Chen L, Morrison CA, Düren T (2012) Improving predictions of gas adsorption in metal-organic frameworks with coordinatively unsaturated metal sites: model potentials, ab initio parameterization, and gcmc simulations. J Phys Chem C 116(35):18899–18909
16. Chung YG, Gómez-Gualdrón DA, Li P, Leperi KT, Deria P, Zhang H, Vermeulen NA, Stoddart JF, You F, Hupp JT, Farha OK, Snurr RQ (2016) In silico discovery of metal-organic frameworks for precombustion CO_2 capture using a genetic algorithm. Sci Adv 2:e1600909
17. Civalleri B, Napoli F, Noël Y, Roetti C, Dovesi R (2006) Ab-initio prediction of materials properties with CRYSTAL: MOF-5 as a case study. CrystEngComm 8(5):364–371
18. Coudert FX, Fuchs AH (2016) Computational characterization and prediction of metal-organic framework properties. Coord Chem Rev 307:211–236
19. Cui WG, Zhang GY, Hu TL, Bu XH (2019) Metal-organic framework-based heterogeneous catalysts for the conversion of C1 chemistry: CO, CO_2 and CH_4. Coord Chem Rev 387:79–120
20. D'Amore M, Civalleri B, Bush IJ, Albanese E, Ferrabone M (2019) Elucidating the interaction of CO_2 in the giant metal-organic framework MIL-100 through large-scale periodic ab initio modeling. J Phys Chem C 123(47):28677–28687
21. Deng X, Yang W, Li S, Liang H, Shi Z, Qiao Z (2020) Large-scale screening and machine learning to predict the computation-ready, experimental metal-organic frameworks for CO_2 capture from air. Appl Sci 10(2):569
22. Din IU, Usman M, Khan S, Helal A, Alotaibi MA, Alharthi AI, Centi G (2021) Prospects for a green methanol thermo-catalytic process from CO_2 by using MOFs based materials: a mini-review. J CO_2 Util 43:101361
23. Ding L, Yazaydin AÖ (2012) Subscriber access provided by UNIV OF CALGARY how well do metal-organic frameworks tolerate flue gas impurities? J Phys Chem C
24. Ding L, Yazaydin AO (2013) The effect of SO_2 on CO_2 capture in zeolitic imidazolate frameworks. Phys Chem Chem Phys 15(28):11856–11861
25. Duan J, Yang Z, Bai J, Zheng B, Li Y, Li S (2012) Highly selective CO_2 capture of an agw-type metal-organic framework with inserted amides: experimental and theoretical studies. Chem Comm 48(25):3058–3060

26. Eddaoudi M, Moler DB, Li H, Chen B, Reineke TM, O'Keeffe M, Yaghi OM (2001) Modular chemistry: secondary building units as a basis for the design of highly porous and robust metal-organic carboxylate frameworks. Acc Chem Res 34(4):319–330
27. Elcheikh Mahmoud M, Audi H, Assoud A, Ghaddar TH, Hmadeh M (2019) Metal-organic framework photocatalyst incorporating Bis(4′-(4-carboxyphenyl)-terpyridine)ruthenium(II) for visible-light-driven carbon dioxide reduction. J Am Chem Soc 141(17):7115–7121
28. Erba A, Baima J, Bush I, Orlando R, Dovesi R (2017) Large-scale condensed matter DFT simulations: performance and capabilities of the CRYSTAL code. J Chem Theory Comput 13(10):5019–5027
29. Ethiraj J, Albanese E, Civalleri B, Vitillo JG, Bonino F, Chavan S, Shearer GC, Lillerud KP, Bordiga S (2014) Carbon dioxide adsorption in amine-functionalized mixed-ligand metal-organic frameworks of UiO-66 topology. Chemsuschem 7(12):3382–3388
30. Fanourgakis GS, Gkagkas K, Tylianakis E, Froudakis GE (2020) A universal machine learning algorithm for large-scale screening of materials. J Am Chem Soc 142(8):3814–3822
31. Fernandez M, Barnard AS (2016) Geometrical properties can predict CO_2 and N_2 adsorption performance of metal-organic frameworks (MOFs) at low pressure. ACS Comb Sci 18(5):243–252
32. Fukuoka S, Kawamura M, Komiya K, Tojo M, Hachiya H, Hasegawa K, Aminaka M, Okamoto H, Fukawa I, Konno S (2003) A novel non-phosgene polycarbonate production process using by-product CO_2 as starting material. Green Chem 5(5):497–507
33. Furukawa H, Cordova KE, O'Keeffe M, Yaghi OM (2013) The chemistry and applications of metal-organic frameworks. Science 341(6149):1230444
34. Garba MD, Usman M, Khan S, Shehzad F, Galadima A, Ehsan MF, Ghanem AS, Humayun M (2021) CO_2 towards fuels: a review of catalytic conversion of carbon dioxide to hydrocarbons. J Environ Chem Eng 9(2):104756
35. Gascon J, Hernández-Alonso MD, Almeida AR, van Klink GPM, Kapteijn F, Mul G (2008) Isoreticular MOFs as efficient photocatalysts with tunable band gap: an operando FTIR study of the photoinduced oxidation of propylene. Chemsuschem 1:981–983
36. Gelfand BS, Huynh RPS, Collins SP, Woo TK, Shimizu GKH (2017) Computational and experimental assessment of CO_2 uptake in phosphonate monoester metal-organic frameworks. Chem Mater 29(24):10469–10477
37. Ghanbari T, Abnisa F, Wan Daud WMA (2020) A review on production of metal organic frameworks (MOF) for CO_2 adsorption. Sci Total Environ 707:135090
38. Haldar R, Reddy SK, Suresh VM, Mohapatra S, Balasubramanian S, Maji TK (2014) Flexible and rigid amine-functionalized microporous frameworks based on different secondary building units: supramolecular isomerism, selective CO_2 capture, and catalysis. Chem Eur J 20(15):4347–4356
39. Han B, Ou X, Deng Z, Song Y, Tian C, Deng H, Xu YJ, Lin Z (2018) Nickel metal-organic framework monolayers for photoreduction of diluted CO_2: metal-node-dependent activity and selectivity. Angew Chemie Int Ed 57(51):16811–16815
40. Han SS, Jung DH, Heo J (2013) Interpenetration of metal organic frameworks for carbon dioxide capture and hydrogen purification: good or bad? J Phys Chem C 117(1):71–77
41. Haszeldine RS (2009) Carbon capture and storage: how green can black be? Science 325:1647–1652
42. Hu J, Liu Y, Liu J, Gu C, Wu D (2018) Effects of incorporated oxygen and sulfur heteroatoms into ligands for CO_2/N_2 and CO_2/CH_4 separation in metal-organic frameworks: a molecular simulation study. Fuel 226:591–597
43. Huang X, Lu J, Wang W, Wei X, Ding J (2016) Experimental and computational investigation of CO_2 capture on amine grafted metal-organic framework NH2-MIL-101. Appl Surf Sci 371:307–313
44. Jasuja H, Burtch NC, Huang YG, Cai Y, Walton KS (2013) Kinetic water stability of an isostructural family of zinc-based pillared metal-organic frameworks. Langmuir 29(2):633–642

45. Jia M, Choi C, Wu TS, Ma C, Kang P, Tao H, Fan Q, Hong S, Liu S, Soo YL, Jung Y, Qiu J, Sun Z (2018) Carbon-supported Ni nanoparticles for efficient CO_2 electroreduction. Chem Sci 9(47):8775–8780
46. Kazemi S, Safarifard V (2018) Carbon dioxide capture in MOFs: the effect of ligand functionalization. Polyhedron 154:236–251
47. Krishna R, van Baten JM (2011) Investigating the potential of MgMOF-74 membranes for CO_2 capture. J Membr Sci 377(1–2):249–260
48. Lee CH, Huang HY, Liu YH, Luo TT, Lee GH, Peng SM, Jiang JC, Chao I, Lu KL (2013) Cooperative effect of unsheltered amide groups on CO_2 adsorption inside open-ended channels of a zinc(II)-organic framework. Inorg Chem 52(7):3962–3968
49. Lee K, Howe JD, Lin LC, Smit B, Neaton JB (2015) Small-molecule adsorption in open-site metal-organic frameworks: a systematic density functional theory study for rational design. Chem Mater 27(3):668–678
50. Lescouet T, Chizallet C, Farrusseng D (2012) The origin of the activity of amine-functionalized metal-organic frameworks in the catalytic synthesis of cyclic carbonates from epoxide and CO_2. ChemCatChem 4(11):1725–1728
51. Lewis DW, Ruiz-Salvador AR, Gómez A, Rodriguez-Albelo LM, Coudert F-X, Slater B, Cheetham AK, Mellot-Draznieks C (2009) Zeolitic imidazole frameworks: structural and energetics trends compared with their zeolite analogues. CrystEngComm 11:2272–2276
52. Li D, Kassymova M, Cai X, Zang SQ, Jiang HL (2020) Photocatalytic CO_2 reduction over metal-organic framework-based materials. Coord Chem Rev 412:213262
53. Li H, Li L, Lin R-B, Zhou W, Zhang Z, Xiang S, Chen B (2019) Porous metal-organic frameworks for gas storage and separation: status and challenges. EnergyChem 1(1):100006
54. Li JR, Ma Y, McCarthy MC, Sculley J, Yu J, Jeong HK, Balbuena PB, Zhou HC (2011) Carbon dioxide capture-related gas adsorption and separation in metal-organic frameworks. Coord Chem Rev 255(15–16):1791–1823
55. Li JR, Yu J, Lu W, Sun LB, Sculley J, Balbuena PB, Zhou HC (2013) Porous materials with pre-designed single-molecule traps for CO_2 selective adsorption. Nat Commun 4:1538
56. Li P, Chen J, Zhang J, Wang X (2015) Water stability and competition effects toward CO_2 adsorption on metal organic frameworks. Sep Purif Rev 44(1):19–27
57. Li PZ, Wang XJ, Zhang K, Nalaparaju A, Zou R, Zou R, Jiang J, Zhao Y (2014) 'Click'-extended nitrogen-rich metal–organic frameworks and their high performance in CO_2-selective capture. Chem Comm 50(36):4683–4685
58. Li S, Chung YG, Simon CM, Snurr RQ (2017) High-throughput computational screening of multivariate metal-organic frameworks (MTV-MOFs) for CO_2 capture. J Phys Chem Lett 8(24):6135–6141
59. Li S, Zhang Y, Hu Y, Wang B, Sun S, Yang X, He H (2021) Predicting metal-organic frameworks as catalysts to fix carbon dioxide to cyclic carbonate by machine learning. J Materiomics 7(5):1029–1038
60. Li W, Li S (2018) CO_2 adsorption performance of functionalized metal-organic frameworks of varying topologies by molecular simulations. Chem eng Sci 189:65–74
61. Liang J, Huang YB, Cao R (2019) Metal–organic frameworks and porous organic polymers for sustainable fixation of carbon dioxide into cyclic carbonates. Coord Chem Rev 378:32–65
62. Liang L, Liu C, Jiang F, Chen Q, Zhang L, Xue H, Jiang HL, Qian J, Yuan D, Hong M (2017) Carbon dioxide capture and conversion by an acid-base resistant metal-organic framework. Nat Commun 8(1):1233
63. Liang W, Babarao R, Church TL, D'Alessandro DM (2015) Tuning the cavities of zirconium-based MIL-140 frameworks to modulate CO_2 adsorption. Chem Comm 51(56):11286–11289
64. Lin RB, Xiang S, Li B, Cui Y, Qian G, Zhou W, Chen B (2019) Our journey of developing multifunctional metal-organic frameworks. Coord Chem Rev 384:21–36
65. Lu SI, Liao JM, Huang XZ, Lin CH, Ke SY, Wang CC (2017) Probing adsorption sites of carbon dioxide in metal organic framework of $[Zn(bdc)(dpds)]_n$: a molecular simulation study. Chem Phys 497:1–9

66. Luo F, Wang MS, Luo MB, Sun GM, Song YM, Li PX, Guo GC (2012) Functionalizing the pore wall of chiral porous metal–organic frameworks by distinct –H, –OH, $-NH_2$, $-NO_2$, –COOH shutters showing selective adsorption of CO, tunable photoluminescence, and direct white-light emission. Chem Comm 48(48):5989–5991
67. Luo F, Yan C, Dang L, Krishna R, Zhou W, Wu H, Dong X, Han Y, Hu TL, O'Keeffe M, Wang L, Luo M, Lin RB, Chen B (2016) UTSA-74: A MOF-74 isomer with two accessible binding sites per metal center for highly selective gas separation. J Am Chem Soc 138(17):5678–5684
68. Maihom T, Wannakao S, Boekfa B, Limtrakul J (2013) Production of formic acid via hydrogenation of CO_2 over a copper-alkoxide-functionalized MOF: a mechanistic study. J Phys Chem C 117(34):17650–17658
69. McDonald TM, Mason JA, Kong X, Bloch ED, Gygi D, Dani A, Crocellà V, Giordanino F, Odoh SO, Drisdell WS, Vlaisavljevich B, Dzubak AL, Poloni R, Schnell SK, Planas N, Lee K, Pascal T, Wan LF, Prendergast D, Neaton JB, Smit B, Kortright JB, Gagliardi L, Bordiga S, Reimer JA, Long JR (2015) Cooperative insertion of CO_2 in diamine-appended metal-organic frameworks. Nature 519(7543):303–308
70. Mercuri G, Giambastiani G, di Nicola C, Pettinari C, Galli S, Vismara R, Vivani R, Costantino F, Taddei M, Atzori C, Bonino F, Bordiga S, Civalleri B, Rossin A (2021) Metal–organic frameworks in Italy: from synthesis and advanced characterization to theoretical modeling and applications. Coord Chem Rev 437:213861
71. Mohideen MIH, Pillai RS, Adil K, Bhatt PM, Belmabkhout Y, Shkurenko A, Maurin G, Eddaoudi M (2017) A fine-tuned MOF for gas and vapor separation: a multipurpose adsorbent for acid gas removal, dehydration, and BTX sieving. Chem 3(5):822–833
72. Moomaw WR, Chmura GL, Davies GT, Finlayson CM, Middleton BA, Natali SM, Perry JE, Roulet N, Sutton-Grier AE (2018) Wetlands in a changing climate: science, policy and management. Wetlands 38(2):183–205
73. Mosca N, Vismara R, Fernandes JA, Casassa S, Domasevitch KV, Bailon-Garcia E, Maldonado-Hodar FJ, Pettinari C, Galli S (2017) CH_3-tagged Bis(pyrazolato)-based coordination polymers and metal-organic frameworks: an experimental and theoretical insight. Cryst Growth Des 17(7):3854–3867
74. North M, Pasquale R, Young C (2010) Synthesis of cyclic carbonates from epoxides and CO_2. Green Chem 12(9):1514–1539
75. Oraee-Mirzamani B, Cockerill T, Makuch Z (2013) Risk assessment and management associated with CCS. Energy Procedia 37:4757–4764
76. Pal TK, De D, Bharadwaj PK (2020) Metal–organic frameworks for the chemical fixation of CO_2 into cyclic carbonates. Coord Chem Rev 408:213173
77. Pan F, Zhang H, Liu K, Cullen DA, More KL, Wang M, Feng Z, Wang G, Wu G, Li Y (2018) Unveiling active sites of CO_2 reduction on nitrogen coordinated and atomically dispersed iron and cobalt catalysts. ACS Catal 8(4):3116–3122
78. Park J, Kim H, Han SS, Jung Y (2012) Tuning metal-organic frameworks with open-metal sites and its origin for enhancing CO_2 affinity by metal substitution. J Phys Chem Lett 3(7):826–829
79. Petrovic B, Gorbounov M, Masoudi Soltani S (2021) Influence of surface modification on selective CO_2 adsorption: a technical review on mechanisms and methods. Micropor Mesopor Mater 312:110751
80. Poloni R, Smit B, Neaton JB (2012a) Ligand-assisted enhancement of CO_2 capture in metal-organic frameworks. J Am Chem Soc 134(15):6714–6719
81. Poloni R, Smit B, Neaton JB (2012b) CO_2 capture by metal-organic frameworks with van der Waals density functionals. J Phys Chem A 116(20):4957–4964
82. Pongsajanukul P, Parasuk V, Fritzsche S, Assabumrungrat S, Wongsakulphasatch S, Bovornratanaraks T, Chokbunpiam T (2017) Theoretical study of carbon dioxide adsorption and diffusion in MIL-127(Fe) metal organic framework. Chem Phys 491:118–125
83. Qiao Z, Wang N, Jiang J, Zhou J (2016a) Design of amine-functionalized metal-organic frameworks for CO_2 separation: the more amine, the better? Chem Comm 52(5):974–977
84. Qiao Z, Zhang K, Jiang J (2016b) In silico screening of 4764 computation-ready, experimental metal-organic frameworks for CO_2 separation. J Mater Chem A 4(6):2105–2114

85. Rahimi M, Moosavi SM, Smit B, Hatton TA (2021) Toward smart carbon capture with machine learning. Cell Rep Phys Sci 2(4):100396
86. Raja DS, Chang IH, Jiang YC, Chen HT, Lin CH (2015) Enhanced gas sorption properties of a new sulfone functionalized aluminum metal-organic framework: synthesis, characterization, and DFT studies. Micropor Mesopor Mater 216:20–26
87. Rana MK, Koh HS, Hwang J, Siegel DJ (2012) Comparing van der Waals density functionals for CO_2 adsorption in metal organic frameworks. J Phys Chem C 116(32):16957–16968
88. Rogacka J, Seremak A, Luna-Triguero A, Formalik F, Matito-Martos I, Firlej L, Calero S, Kuchta B (2021) High-throughput screening of metal–organic frameworks for CO_2 and CH_4 separation in the presence of water. Chem Eng J 403:126392
89. Shao P, Yi L, Chen S, Zhou T, Zhang J (2020) Metal-organic frameworks for electrochemical reduction of carbon dioxide: the role of metal centers. J Energy Chem 40:156–170
90. Shearer GC, Colombo V, Chavan S, Albanese E, Civalleri B, Maspero A, Bordiga S (2013) Stability vs. reactivity: Understanding the adsorption properties of $Ni_3(BTP)_2$ by experimental and computational methods. Dalton Trans 42(18):6450–6458
91. Shen Q, Huang X, Liu J, Guo C, Zhao G (2017) Biomimetic photoelectrocatalytic conversion of greenhouse gas carbon dioxide: Two-electron reduction for efficient formate production. Appl Catal B Environ 201:70–76
92. Sikdar N, Bonakala S, Haldar R, Balasubramanian S, Maji TK (2016) Dynamic entangled porous framework for hydrocarbon (C2–C3) storage, CO_2 capture, and separation. Chem Eur J 22(17):6059–6070
93. Singh Dhankhar S, Sharma N, Kumar S, Dhilip Kumar TJ, Nagaraja CM (2017) Rational design of a bifunctional, two-fold interpenetrated ZnII-metal–organic framework for selective adsorption of CO_2 and efficient aqueous phase sensing of 2,4,6-trinitrophenol. Chem Eur J 23(64):16204–16212
94. Skoulidas AI, Sholl DS (2005) Self-diffusion and transport diffusion of light gases in metal-organic framework materials assessed using molecular dynamics simulations. J Phys Chem B 109(33):15760–15768
95. Su X, Xu J, Liang B, Duan H, Hou B, Huang Y (2016) Catalytic carbon dioxide hydrogenation to methane: a review of recent studies. J Energy Chem 25(4):553–565
96. Sun D, Liu W, Qiu M, Zhang Y, Li Z (2015) Introduction of a mediator for enhancing photocatalytic performance via post-synthetic metal exchange in metal-organic frameworks (MOFs). Chem Comm 51(11):2056–2059
97. Tamura M, Honda M, Noro K, Nakagawa Y, Tomishige K (2013) Heterogeneous CeO_2-catalyzed selective synthesis of cyclic carbamates from CO_2 and aminoalcohols in acetonitrile solvent. J Catal 305:191–203
98. Tekalgne MA, Do HH, Hasani A, van Le Q, Jang HW, Ahn SH, Kim SY (2020) Two-dimensional materials and metal-organic frameworks for the CO_2 reduction reaction. Mater Today Adv 5:100038
99. Teo HWB, Chakraborty A, Kayal S (2017) Evaluation of CH_4 and CO_2 adsorption on HKUST-1 and MIL-101(Cr) MOFs employing Monte Carlo simulation and comparison with experimental data. Appl Therm Eng 110:891–900
100. Torrisi A, Bell RG, Mellot-Draznieks C (2013) Predicting the impact of functionalized ligands on CO_2 adsorption in MOFs: a combined DFT and grand canonical Monte Carlo study. Micropor Mesopor Mater 168:225–238
101. Vaidhyanathan R, Iremonger SS, Shimizu GKH, Boyd PG, Alavi S, Woo TK (2010) Direct observation and quantification of CO_2 binding within an amine-functionalized nanoporous solid. Science 330:650
102. Valenzano L, Civalleri B, Chavan S, Bordiga S, Nilsen MH, Jakobsen S, Lillerud KP, Lamberti C (2011a) Disclosing the complex structure of UiO-66 metal organic framework: a synergic combination of experiment and theory. Chem Mater 23(7):1700–1718
103. Valenzano L, Civalleri B, Sillar K, Sauer J (2011b) Heats of adsorption of CO and CO_2 in metal-organic frameworks: quantum mechanical study of CPO-27-M (M = Mg, Ni, Zn). J Phys Chem C 115(44):21777–21784

104. Walker AM, Civalleri B, Slater B, Mellot-Draznieks C, Corà F, Zicovich-Wilson CM, Román-Pérez G, Soler JM, Gale JD (2010) Flexibility in a metal-organic framework material controlled by weak dispersion forces: the bistability of MIL-53(Al). Angew Chem Int Ed 49(41):7501–7503
105. Wang B, Huang H, Lv XL, Xie Y, Li M, Li JR (2014) Tuning CO_2 selective adsorption over N_2 and CH_4 in UiO-67 analogues through ligand functionalization. Inorg Chem 53(17):9254–9259
106. Wang HH, Hou L, Li YZ, Jiang CY, Wang YY, Zhu Z (2017) Porous MOF with highly efficient selectivity and chemical conversion for CO_2. ACS Appl Mater Interfaces 9(21):17969–17976
107. Wang HH, Shi WJ, Hou L, Li GP, Zhu Z, Wang YY (2015) A cationic MOF with high uptake and selectivity for CO_2 due to multiple CO_2-philic sites. Chem Eur J 21(46):16525–16531
108. Wang X, Chen Z, Zhao X, Yao T, Chen W, You R, Zhao C, Wu G, Wang J, Huang W, Yang J, Hong X, Wei S, Wu Y, Li Y (2018a) Regulation of coordination number over single co sites: triggering the efficient electroreduction of CO_2. Angew Chem Int Ed 57(7):1944–1948
109. Wang XK, Liu J, Zhang L, Dong LZ, Li SL, Kan YH, Li DS, Lan YQ (2019) Monometallic catalytic models hosted in stable metal-organic frameworks for tunable CO_2 photoreduction. ACS Catal 9(3):1726–1732
110. Wang Y, Huang NY, Shen JQ, Liao PQ, Chen XM, Zhang JP (2018b) Hydroxide Ligands cooperate with catalytic centers in metal-organic frameworks for efficient photocatalytic CO_2 reduction. J Am Chem Soc 140(1):38–41
111. Wang YR, Huang Q, He CT, Chen Y, Liu J, Shen FC, Lan YQ (2018c) Oriented electron transmission in polyoxometalate-metalloporphyrin organic framework for highly selective electroreduction of CO_2. Nat Commun 9(1)
112. Wells BA, Liang Z, Marshall M, Chaffee AL (2009) Modeling gas adsorption in metal organic frameworks. Energy Procedia 1(1):1273–1280
113. Wilmer CE, Farha OK, Bae YS, Hupp JT, Snurr RQ (2012) Structure-property relationships of porous materials for carbon dioxide separation and capture. Energy Environ Sci 5(12):9849–9856
114. Wilmer CE, Snurr RQ (2011) Towards rapid computational screening of metal-organic frameworks for carbon dioxide capture: calculation of framework charges via charge equilibration. Chem Eng J 171(3):775–781
115. Wriedt M, Sculley JP, Yakovenko AA, Ma Y, Halder GJ, Balbuena PB, Zhou HC (2012) Low-energy selective capture of carbon dioxide by a pre-designed elastic single-molecule trap. Angew Chem Int Ed 51(39):9804–9808
116. Wu D, Maurin G, Yang Q, Serre C, Jobic H, Zhong C (2014) Computational exploration of a Zr-carboxylate based metal-organic framework as a membrane material for CO_2 capture. J Mater Chem A 2(6):1657–1661
117. Wu Y, Song X, Xu S, Yu T, Zhang J, Qi Q, Gao L, Zhang J, Xiao G (2019a) $[(CH_3)_2NH_2][M(COOH)_3]$ (M=Mn, Co, Ni, Zn) MOFs as highly efficient catalysts for chemical fixation of CO_2 and DFT studies. Mol Catal 475:110485
118. Wu Y, Song X, Zhang J, Xu S, Gao L, Zhang J, Xiao G (2019b) Mn-based MOFs as efficient catalysts for catalytic conversion of carbon dioxide into cyclic carbonates and DFT studies. Chem Eng Sci 201:288–297
119. Xia W, Mahmood A, Zou R, Xu Q (2015) Metal–organic frameworks and their derived nanostructures for electrochemical energy storage and conversion. Energy Environ Sci 8(7):1837–1866
120. Xu G, Meng Z, Guo X, Zhu H, Deng K, Xiao C, Liu Y (2019a) Molecular simulations on CO_2 adsorption and adsorptive separation in fullerene impregnated MOF-177, MOF-180 and MOF-200. Comput Mater Sci 168:58–64
121. Xu G, Meng Z, Liu Y, Guo X, Xiao C, Deng K (2019b) Theoretical study of heterofullerene-linked metal–organic framework with lithium doping for CO_2 capture and separation from CO_2/CH_4 and CO_2/H_2 mixtures. Micropor Mesopor Mater 284:385–392
122. Xue DX, Wang Q, Bai J (2019) Amide-functionalized metal–organic frameworks: syntheses, structures and improved gas storage and separation properties. Coord Chem Rev 378:2–16

123. Yaashikaa PR, Senthil Kumar P, Varjani SJ, Saravanan A (2019) A review on photochemical, biochemical and electrochemical transformation of CO_2 into value-added products. J CO_2 Util 33:131–147
124. Yan Y, Zhang L, Li S, Liang H, Qiao Z (2021) Adsorption behavior of metal-organic frameworks: from single simulation, high-throughput computational screening to machine learning. Comput Mater Sci 193:110383
125. Yan ZH, Du MH, Liu J, Jin S, Wang C, Zhuang GL, Kong XJ, Long LS, Zheng LS (2018) Photo-generated dinuclear {Eu(II)}2 active sites for selective CO_2 reduction in a photosensitizing metal-organic framework. Nat Commun 9(1):3353
126. Yang G, Santana JA, Rivera-Ramos ME, García-Ricard O, Saavedra-Arias JJ, Ishikawa Y, Hernández-Maldonado AJ, Raptis RG (2014) A combined experimental and theoretical study of gas sorption on nanoporous silver triazolato metal-organic frameworks. Micropor Mesopor Mater 183:62–68
127. Yang H, Wu Y, Li G, Lin Q, Hu Q, Zhang Q, Liu J, He C (2019) Scalable production of efficient single-atom copper decorated carbon membranes for CO_2 electroreduction to methanol. J Am Chem Soc 141(32):12717–12723
128. Yang Q, Xu Q, Liu B, Chongli Z, Berend S (2009) Molecular simulation of CO_2/H_2 mixture separation in metal-organic frameworks: effect of catenation and electrostatic interactions. Chin J Chem Eng 17(5):781–790
129. Ye J, Johnson JK (2015) Design of Lewis pair-functionalized metal organic frameworks for CO_2 hydrogenation. ACS Catal 5(5):2921–2928
130. Ye J, Johnson JK (2016) Catalytic hydrogenation of CO_2 to methanol in a Lewis pair functionalized MOF. Catal Sci Technol 6(24):8392–8405
131. Ye L, Liu J, Gao Y, Gong C, Addicoat M, Heine T, Wöll C, Sun L (2016) Highly oriented MOF thin film-based electrocatalytic device for the reduction of CO_2 to CO exhibiting high faradaic efficiency. J Mater Chem A 4(40):15320–15326
132. Ye Y, Cai F, Li H, Wu H, Wang G, Li Y, Miao S, Xie S, Si R, Wang J, Bao X (2017) Surface functionalization of ZIF-8 with ammonium ferric citrate toward high exposure of Fe-N active sites for efficient oxygen and carbon dioxide electroreduction. Nano Energy 38:281–289
133. Yu D, Yazaydin AO, Lane JR, Dietzel PDC, Snurr RQ (2013) A combined experimental and quantum chemical study of CO_2 adsorption in the metal-organic framework CPO-27 with different metals. Chem Sci 4(9):3544–3556
134. Yuan CZ, Liang K, Xia XM, Yang ZK, Jiang YF, Zhao T, Lin C, Cheang TY, Zhong SL, Xu AW (2019) Powerful CO_2 electroreduction performance with N-carbon doped with single Ni atoms. Catal Sci Technol 9(14):3669–3674
135. Yulia F, Chairina I, Zulys A, Nasruddin (2021) Multi-objective genetic algorithm optimization with an artificial neural network for CO_2/CH_4 adsorption prediction in metal–organic framework. Thermal Sci Eng Process 25:100967
136. Zhang G, Wei G, Liu Z, Oliver SRJ, Fei H (2016) A robust sulfonate-based metal-organic framework with permanent porosity for efficient CO_2 capture and conversion. Chem Mater 28(17):6276–6281
137. Zhang J, Fu J, Chen S, Lv J, Dai K (2018) 1D carbon nanofibers@TiO_2 core-shell nanocomposites with enhanced photocatalytic activity toward CO_2 reduction. J Alloys Compd 746:168–176
138. Zhao K, Liu Y, Quan X, Chen S, Yu H (2017) CO_2 electroreduction at low overpotential on oxide-derived Cu/carbons fabricated from metal organic framework. ACS Appl Mater Interfaces 9(6):5302–5311
139. Zhen W, Gao F, Tian B, Ding P, Deng Y, Li Z, Gao H, Lu G (2017) Enhancing activity for carbon dioxide methanation by encapsulating (1 1 1) facet Ni particle in metal–organic frameworks at low temperature. J Catal 348:200–211
140. Zheng B, Yang Z, Bai J, Li Y, Li S (2012) High and selective CO_2 capture by two mesoporous acylamide-functionalized rht-type metal-organic frameworks. Chem Comm 48(56):7025–7027

141. Zhou DD, He CT, Liao PQ, Xue W, Zhang WX, Zhou HL, Zhang JP, Chen XM (2013) A flexible porous Cu(II) bis-imidazolate framework with ultrahigh concentration of active sites for efficient and recyclable CO_2 capture. Chem Comm 49(100):11728–11730
142. Zhou DD, Zhang XW, Mo ZW, Xu YZ, Tian XY, Li Y, Chen XM, Zhang JP (2019) Adsorptive separation of carbon dioxide: from conventional porous materials to metal–organic frameworks. EnergyChem 1(3):100016
143. Zhuang W, Yuan D, Liu D, Zhong C, Li JR, Zhou HC (2012) Robust metal-organic framework with an octatopic ligand for gas adsorption and separation: combined characterization by experiments and molecular simulation. Chem Mater 24(1):18–25
144. Zou R, Li PZ, Zeng YF, Liu J, Zhao R, Duan H, Luo Z, Wang JG, Zou R, Zhao Y (2016) Bimetallic metal-organic frameworks: probing the Lewis acid site for CO_2 conversion. Small 12(17):2334–2343

Covalent Organic Frameworks as Catalysts

Covalent Organic Frameworks (COFs) as Catalysts: An Overview

Arti Jain and Priti Malhotra

Contents

1 Introduction 269
2 Chemistry and Design of COFs 269
3 Catalytic Applications of COFs 271
4 Engaging COF for Catalytic Splitting of Water Molecules Irradiated by Visible Light 272
5 COF Catalyzed Suzuki Coupling Reaction [22, 23] 272
6 Degradation of 4-Nitrophenol by the Ultra-Stable Covalent Organic Framework–Au Nanoparticles System [24] 272
7 Covalent Organic Frameworks Based on Azine as Metal-Free Photocatalysts for CO_2 Reduction with H_2O [26] 273
8 Imine-Based Covalent Organic Frameworks for Selective Olefin Oxidation [27] 274
9 A Novel Hybrid COF for Photocatalytic Removal of Organic Dye and Cr(VI) from Wastewater [28] 274
10 A Covalent Organic Framework–Cadmium Sulfide System for Visible Light-Initiated Hydrogen Gas Production [29] 275
11 Electrocatalytic Reduction of Carbon Dioxide on Active Sites of COF [30] 275
12 Ultra-Stable Covalent Organic Frameworks with Benzoxazole for Photocatalysis [31] 276
13 COF Efficiency in the Production of Solar Fuel from CO_2 [32] 276
14 Covalent Organic Frameworks for Serotonin Detection Using Luminol Chemiluminescence [33] 276
15 Degradation of Organic Pollutants by Covalent Organic Framework-Silver Nanoparticles@Sand Hybrid [34] 277
16 CdS/COF Heterostructure for Enhanced Photocatalytic Decomposition of Bisphenol-A [36] 278
17 MnO_2-loaded Ultrathin COF Nanosheets as Cathode Catalysts for Li-CO_2 Batteries [37] 278
18 Magnetic COFs for Efficient Fenton-Like Degradation of Pharmaceutical Waste Sulfamethazine [38] 279
19 Triazine Functionalized Porous Covalent Organic Framework for Photo-Organocatalytic E–Z Isomerization of Olefins [39] 279
20 Conclusion 280
Abbreviations 280
References 281

A. Jain · P. Malhotra (✉)
Department of Chemistry, Daulat Ram College, University of Delhi, Delhi 110007, India

S. Gulati (ed.), *Metal-Organic Frameworks (MOFs) as Catalysts*,
https://doi.org/10.1007/978-981-16-7959-9_10

Abstract Covalent organic frameworks (COFs) are currently evolving as a novel family of materials categorized as electively designed and tailored crystalline porous polymers. Both its structure and composition are fabricated along with their tunable functionality and porosity giving rise to a new class of efficient and promising catalysts. Potential insights into COFs skeleton and their availability for effortless bonding with metals and various functional groups endow them with multifold efficacious characteristics. Over the recent past, the substantial focus has been laid on the design, synthesis, and exploration of applications of COFs in diverse fields. Owing to their vast surface areas, ease of surface modifications, and structural adaptability, they serve as extraordinary platforms for various catalytic processes. COFs are excellent mimics of nature's porous materials whose applications have always been intriguing the scientific community and has inspired their focus towards exploring their catalytic efficacy in wide areas ranging from organic synthesis, redox reactions, coupling reactions to enzyme supports, polymer supports, and templates for the formation of various metal composites. The current chapter is aimed at evaluating the significance and dynamics of COFs as they are proving their enormous versatility in a wide range of catalytic applications, gas storage and separation, chemical sensing, proton conductivity, optoelectronics, medicinal chemistry, and chromatographic techniques.
Graphical Abstract

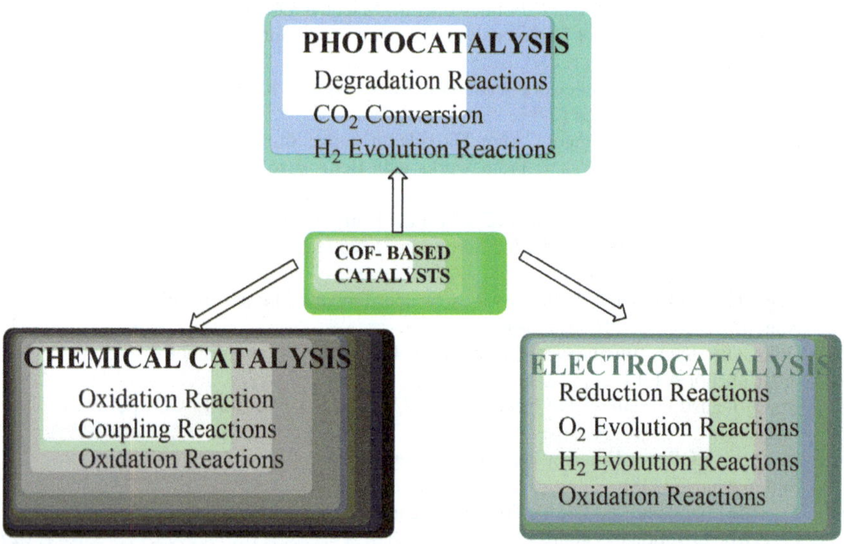

Keywords Covalent organic framework · Catalyst · Redox · Coupling · Enzyme · Polymer

1 Introduction

For a decade, chemists have been fascinated by the reticular chemistry of building blocks and connecting molecules through covalent bonds. Featuring among one of the potential organic crystalline porous materials, they demonstrate extraordinary behavior owing to the presence of covalently bonded carbon atoms along with elements such as N, O, B, and H. Extending both over two-dimensional and three-dimensional frameworks, they exhibit highly adaptable porosity and remarkable thermal stability making them widely significant within long-range ordered crystalline materials. Yaghi et al. first synthesized them in 2005 as novel crystalline porous organic polymers using building blocks of lighter elements that are bonded through strong covalent bonds formed by condensation reactions that were reversible [1]. In contrast to MOFs where metals are bonded through coordinate bonds [2–4], COFs have exceptional chemical stability which can be seen in their high-grade integrity in organic solvents and strong basic and acidic conditions. This escalated order of stability proves the mighty presence of purely covalent bonds along with hydrogen bonds and $\pi-$ stacking interactions which renders strength to the porous skeleton and resistance towards hydrolysis and disintegration in other solvents. Catalytic performance generally gets enhanced with larger pore sizes of the catalyst as the diffusion of the reactants and desorption of the products both get facilitated which further justifies their remarkable role as a significant catalyst with high order selectivity [5, 6].

Covalent organic frameworks (COFs) are also well known for providing catalytic efficiency and durability to nanomaterials and other well-known surface-active materials. Their specialty arises from their lightweight porosity suitable for gas storage applications, semiconducting and photoconductive characteristics along catalytic support. Although the earlier known porous materials, viz. zeolites, dendrimers, charcoal, polymers, and mesoporous silica, etc., have been used in the past for catalyzing many chemical reactions their low order stability hampered their reusability and efficacy. On the contrary, we'll order crystallinity and porosity of COF and COF docked nanoparticles proved a better versatility. Several nanoparticles@COFs has been employed as a catalyst for C–H activation, Suzuki coupling, Knoevenagel condensation, nitro reduction, glycerol oxidation, etc. [7–17].

2 Chemistry and Design of COFs

COFs have emerged as robust new catalytic materials which can be easily tailored for suitable purposes because a flexible range of tunable functionality can be provided along with pliable pore sizes. The crystalline nature of the pores along with their nano size and their block-by-block building from organic polymer further adorns them with highly precise and efficient heterogeneous catalysts. Depending on their dimensions and nature of building blocks, the construction of COFs can lead to

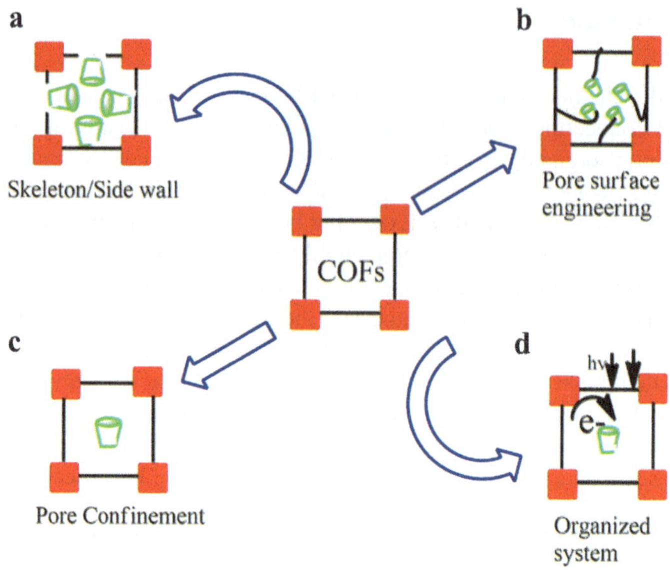

Fig. 10.1 Scheming COFs for application as heterogeneous catalysis on the basis of **a** skeleton and sidewall, **b** pore surface engineering, **c** pore confinement, and **d** organized systems

either two- (2D) or three-dimensional (3D) porous networks. In the 2D category, the covalently bound framework makes a two-dimensional sheet-like surface having repeated linear columns. It involves stacking of 2D polymer networks resulting in layered skeletons having the one-dimensional open channel in which every layer is separated by 3–5 Å and their separation is supported by π–π interactions. Such an arrangement allows easy transport of electrons, excitons, and electronic holes and provides an excellent arrangement for promoting photocatalysis and electrocatalysis. (Fig. 10.1) Although a few catalytic sites get hidden due to stacking nevertheless, 2D COFs offer plenty of catalytic centers at the skeletons where pores are developed by π columns.

On the other hand, the stacking assembly would give coverage to the catalytic sites and leaves only those on the surface layers reachable to reactions; this replicates that 2D COFs is not appropriate to rightly fit in the catalytic sites to the π columns, particularly reconnoitering the focal point of the π backbones as catalytic centers There are also examples where monomers are employed to construct 2D COFs, and then fix the catalytic sites by joining pores thereby providing a technology of surface pores for heterogeneous catalysis. Similarly, the design of 3D COFs also prevents the direct involvement of the skeleton for catalytic intervention which is further hampered by small pore size and also restricts the merging of pores for creating catalytic platforms [18, 19]. The ordered imine-based 2D COF were prepared by Smith et al. and have reportedly given way to many porphyrin-containing 2D COFs with special properties due to the hydroxy groups. Their performance apart from catalysis is also well established in optoelectronics.

The porosity of covalent organic frameworks provide exceptional catalytic sites due to facile mass transit and tunable catalytic centers offering heterogeneous catalysis through the following different ways:

(i) Metal Nanoparticles or clusters can be conveniently supported on the porous frameworks through coordination which gives extra stability to the resulting MOFs.
(ii) Due to the induction of heteroatoms at the organic sites and simultaneous Immobilization of metal coordinated within the molecule, the benefit of a typical heterogeneous catalytic platform is created.
(iii) Another way by which suitable solid catalysts are provided is by those possessing intrinsic catalytic properties in the absence of metals and pure organic frameworks to suffice for heterogenous catalysis either due to the presence of heteroatoms or common functional groups contributing to organocatalysis.

3 Catalytic Applications of COFs

Inspired by the inorganic zeolites-driven catalytic reactions, the high order crystallinity and porosity of COFs were also explored by various scientists in plentiful industrial and photochemical events. Chemical reactions highly depend on heterogeneous catalysis and whether exclusive COFs skeletons or their metal pendant couples, all of their analogs provide devoted sites for numerous applications. Heterogeneous catalysts have large benefits as they suffice for reusable material and are used to design cyclable surfaces multidimensional applications. Such a platform can be explored for tailoring suitable and effective transformations leading to hybrid systems which can offer wide applications and ensure a promising future for covalent organic frameworks (COFs). Their crystalline nature and porosity can be combined with organic networks which can eventually integrate organic units into defined and ordered skeletons resulting in an insoluble and robustly stable platform for employing them successfully into various heterogeneous catalytic events. The uniqueness of COFs and COF-supported entities can be witnessed in a huge range of strategically designed platforms based on four different types of pathways where COFs are employed as walls, skeletons, pores, and ordered systems boosting stability, reusability, and efficiency to the catalytic performances. Therefore, designer approaches are developed to provide immense benefits to the catalytic features of COFs through interactions of electrons, holes, and molecular interfaces all reinforcing their heterogeneous catalytic efficacy.

Some of such breakthrough applications and COF-based activities have been reported in the recent past and have been reviewed in the following text.

4 Engaging COF for Catalytic Splitting of Water Molecules Irradiated by Visible Light

COF s catalytic endowments have proved their versatility in many areas which also include their prominent role in being able to drive both the two half-reactions of water splitting under visible light irradiation. This is attributed to the high crystallinity of COFs and their definite structures which support photocatalytic mechanism-based interventions. Semiconducting organic molecules which support π-conjugated graphene-like systems were designed which led to the formation of a family of COF and offered a successful solution to the catalytic splitting of water molecules. A 2 D pyridinylic COF has reportedly enabled the splitting of water under visible light [20, 21].

5 COF Catalyzed Suzuki Coupling Reaction [22, 23]

Spherical shaped covalent organic frameworks (COFs) consisting of 2,4,6-trihydroxybenzene-1,3,5-tricarbaldehyde (Tp) and p-phenylenediamine (Pa) encapsulating palladium (Pd) nanoparticles generated a yolk-shell structure of (Pd/C)@TpPa COFs (Fig. 10.2). The unique composites (Pd/C)@TpPa COFs possess remarkable catalytic efficacy reportedly exhibited in catalyzing the Suzuki reaction. The catalytic performance observed a high conversion and repeated recovery of the catalyst up to five consecutive cycles with simultaneous retention of the original activity.

6 Degradation of 4-Nitrophenol by the Ultra-Stable Covalent Organic Framework–Au Nanoparticles System [24]

When COF-supported Gold [Au(0)] nanoparticles were synthesized via infiltration method [25], it not only immobilized the nanoparticles it also gave rise to a stable Au(0)@TpPa-1 catalyst showing high efficiency (Fig. 10.3). They proved to possess high order recyclability and superior reactivity for reduction of nitrophenol as compared to $HAuCl_4$,3H2O. It has been reported that Au(0)@TpPa-1 demonstrates a good amount of stability in aqueous and common organic solvents owing to its ordered crystalline architecture with the significant association of oxygen and nitrogen atoms in the framework contributing immensely to its high catalytic efficacy and optimal stability of nanoparticles in acidic, alkaline and aqueous media. Their recyclability is also commendable as the catalytic efficacy for the reduction of nitrophenol lasts for six cycles proving the immense stability and catalytic efficacy of Au(0)@TpPa-1.

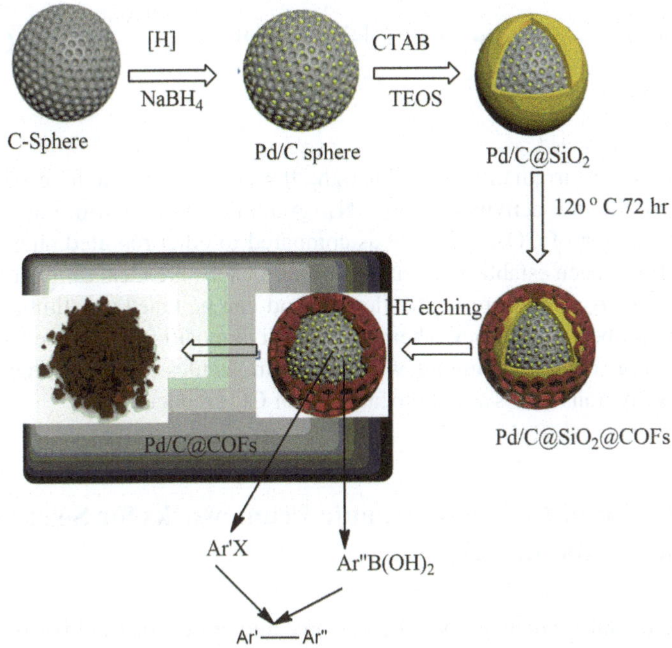

Fig. 10.2 Diagrammatic representation showing the synthesis of Pd/C)@TpPa COFs and their catalytic application

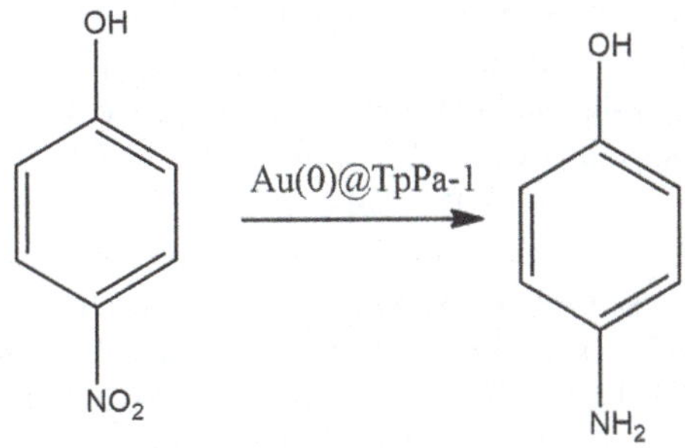

Fig. 10.3 Scheme showing reduction of nitrophenol

7 Covalent Organic Frameworks Based on Azine as Metal-Free Photocatalysts for CO_2 Reduction with H_2O [26]

Covalent organic frameworks (COFs), as they exhibit an outstanding thermal and chemical stability due to their covalent bonds, also lead to an interesting new class of materials in the form of azine-based COFs. Azine-based COFs feature among such photoactive materials which have added benefits of capturing CO_2 while promoting heterogeneous photocatalysis so that CO_2 is easily converted to various products like

$$\text{TAPT} + \text{DHTA} \longrightarrow \text{TAPT-DHTA-COF} \xrightarrow{\text{Cu-acetate}} \text{Cu-COF}$$

Fig. 10.4 Synthesis of Cu-COF

methanol under the irradiation of visible light. It was observed that these azine-based COFs showed higher activity than g-C_3N_4 (graphitic carbon nitride) in the photocatalytic reduction of CO_2 with H_2O as compared to other reported photocatalytic reactions. It has been established that azine-based COFs are ideal for photocatalytic reduction of CO_2 as they are highly flexible and can be tuned constitutionally and structurally for better efficiency. Their metal-free composition is an added advantage that allows more options for linking with organic molecules thereby producing a new class of energy transducers via photocatalysis of CO_2 reduction.

8 Imine-Based Covalent Organic Frameworks for Selective Olefin Oxidation [27]

Imine-linked COFs which are two dimensional and are bonded to hydroxyl groups showed high thermal stability, porosity. They displayed optimum crystal structure to anchor copper acetate which could readily coordinate with imine and the hydroxyl end of COF. This type of arrangement where copper supported COF provides larger surface area and greater pore size along with imine linkage which escalated catalytic efficacy proving to be excellent catalyst converting styrene to benzaldehyde. Cu-COFHX demonstrated outstanding catalytic performance with a high grade of recyclability and reusability for a selective reaction involving the oxidation of styrene to benzaldehyde.

Herein, 2-D COFs comprising hydroxyl and imine units (TAPT-DHTA-COF) were synthesized by the use of 2,5-dihydroxyl-terephthalaldehyde (DHTA) present at the vertices and 1,3,5-tris-(4-aminophenyl) triazine (TAPT,) at edges. (Fig. 10.4) Under reflux and solvothermal situations, 2-D imine-linked COFs, TAPT-DHTA-COFHX, and TAPT-DHTA-COFDMF were accomplished respectively and employed as the platform for docking copper acetate. Next, copper-containing COFs represented as Cu-COFDMF and Cu-COFHX were arranged and verified in the selective oxidation of styrene to benzaldehyde.

9 A Novel Hybrid COF for Photocatalytic Removal of Organic Dye and Cr(VI) from Wastewater [28]

A novel BiOBr/TzDa covalent organic framework (COF) composite was designed as a 2D/2D Z-scheme heterostructure photocatalyst and was explored for reduction of

Cr (VI) and organic dye Rhodamine B (RhB). It turned out to be the most significant research where solar energy was trapped for the degradation of water contaminants to purify the wastewater. The photocatalytic activity of BiOBr/TzDa COF was highly important as it was able to completely reduce Cr (Vl) at pH 2.1 with a 90 min irradiation and prove to be challenging for purifying wastewater with solar energy. Alongside a novel 2D/2D Z-scheme heterocatalyst of BiOBr/TzDa covalent organic framework (COF) (BTDC) was tailored and its photocatalytic behavior was observed under the influence of visible light for the degradation of Rhodamine (B) and Cr (Vl) both first individually and then in a combined mixture. It was observed that better performance was exhibited in the case of the mixture than individually because of the synergistic effect resulting in 97% removal at 2.1 pH and within a span of 20–40 min. The enhanced efficacy of BTDC could have been possible due to the raised transportation of charges, wide separation between the charges, and large surface area which has an increased impact. Various studies have indicated a photocatalytic mechanism for the efficient role of BTDC in removing the pollutants such as heavy metals and organic contaminants from water.

10 A Covalent Organic Framework–Cadmium Sulfide System for Visible Light-Initiated Hydrogen Gas Production [29]

As compared to the solo CdS nanoparticles when the hybrid model in which CDS was embedded in COF matrix was studied, it proved multifold high efficacy in the photocatalytic production of hydrogen. For a varied content of COF, investigations revealed that even with 1% (by wt) inclusion of COF deposited on a highly stable, two-dimensional (2D) covalent organic framework (COF) matrix a ten times hiked photocatalytic efficacy was witnessed. Such a steep rise was attributed to the COF providing a supportive backbone through a p- conjugation and a two-dimensional hetero-interface including the extraordinarily high surface area.

11 Electrocatalytic Reduction of Carbon Dioxide on Active Sites of COF [30]

The reticular structure of COF possessing an active surface of porphyrin was found to be responsible for electrocatalytic initiation of the reduction process of CO_2 to form CO (Fig. 10.5). The electronic character of porphyrin active sites for electrocatalytic reduction of CO_2 to CO. The reduction of CO_2 to CO was enabled by charge transfer between the functional groups and the backbone of COF which consists of the porphyrin ring acting as the site for the reduction and the reduction could be

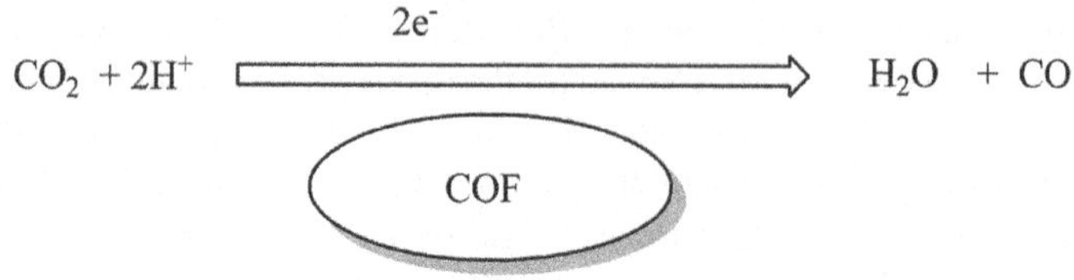

Fig. 10.5 Schematic representation reduction of carbon dioxide

carried out at very low potential (~550 mV). Furthermore, a high level of efficiency and selectivity was generated due to the COF matrix.

12 Ultra-Stable Covalent Organic Frameworks with Benzoxazole for Photocatalysis [31]

The significance of COF could be further emphasized when their metal-free series of super stable benzoxazole-linked COFs were used for photocatalytic applications. The π-conjugated system of benzoxazole molecule on COF was highly stable and had improved ability for the absorption of visible light. The organocatalyst had a unique potential of recyclability during photocatalysis as the catalyst could be reused many times.

13 COF Efficiency in the Production of Solar Fuel from CO_2 [32]

COF based on triazine also designated as covalent triazine framework (CTF) were successfully employed as efficient photoactive materials for generating solar fuels trapping CO_2 and visible light (Fig. 10.6). It is a high environment-friendly pathway as it reduces carbon footprints simultaneously opening new avenues for solar devices thereby conserving energy. Perylene based on triazine containing COF led to artificial photosynthesis and provided an extraordinary scaffold for an efficient photocatalyst engaged in harvesting light energy to produce solar fuels from CO_2.

14 Covalent Organic Frameworks for Serotonin Detection Using Luminol Chemiluminescence [33]

The chemiluminescence (CL) of the luminol system in the covalent organic frameworks (COFs) is used to determine the serotonin concentration in such a way that as

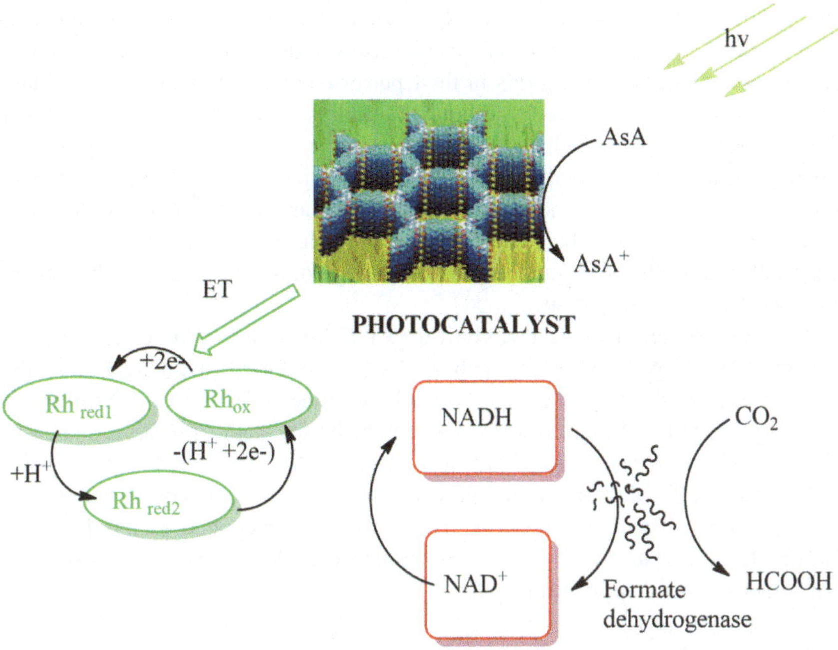

Fig. 10.6 Schematic diagram of CTF film photocatalyst-enzyme coupled system involved in exclusive production of formic acid from CO_2. Rh_{ox} = [Cp * Rh(bipy)H2O]2 +, Rh_{red1} = Cp * Rh(bipy), Rh_{red2} = [Cp * Rh(bipy)H] +; Cp* = pentamethylcyclopentadienyl, bpy = 2,2′-bipyridine

the serotonin concentration increases the CL of the luminol-COF system decreases. Coupling it with flow injection techniques the sensing of serotonin was observed by CL measurement of a highly sensitive COF-based luminol system. Such a CL sensing technique using COF-based materials gave excellent results and further emphasized the role of metal-free efficient photocatalytic systems. Serotonin (5-hydroxy-tryptamine, 5-HT) has a significant presence in the human body which also includes the central nervous system and its optimum content is a marker for well being of many body parts and physiological systems.

15 Degradation of Organic Pollutants by Covalent Organic Framework-Silver Nanoparticles@Sand Hybrid [34]

Nanoparticles of noble metals like silver anchored on the sand and supported by a covalent organic framework have exhibited high-grade reduction efficacy towards the reduction of organic contaminants. Methylene blue, Congo red, and 4-nitrophenol are some of the pollutants which have reportedly been degraded and reduced through heterogenous catalysis offered by Ag Nps@sand-COF and have also shown exceptional reusable characteristics as they can be recovered and reused for many subsequent runs without any alterations in their performance. Following both the static technique and the flow injected technique the catalysts proved their versatility with more than 90% efficacy which was accounted for by not only the Ag-COF hybrid but also the contribution of sand due to its high hydrophilicity and resistance to leaching during treatment as they have a larger size. Therefore sand supported nanocomposites provided added benefits by being cost-effective, enabling the fast flow of the contaminated solution for the continuous removal and degradation of organic pollutants by COFs-Ag NPs@Sand) system.

Another report has disclosed that Fe_3O_4@COF-Au systems which were prepared via immobilizing Au Nps in the matrix of COF-supported magnetic Fe_3O_4 also exhibited excellent catalytic efficacy for a high-grade reduction of 4-Nitrophenol and methylene blue. The catalyst was found to be highly stable, reusable, and easily separable as it demonstrated the fine surface of Au Nps combined with magnetic properties of Fe_3O_4 and crystalline porosity of COF thus representing an extraordinary potential of synergic benefits. The strategy of integrated materials having nanostructured surfaces and variable functionality could be proven successful for many other remedial applications [35].

16 CdS/COF Heterostructure for Enhanced Photocatalytic Decomposition of Bisphenol-A [36]

Nanocomposites of the type where CdS loaded on the surface of COF forming a heterogeneous structure CDS/COF were tested for the photocatalytic degradation of Bisphenol-A (BPA) and a successful removal rate of 85.68% was observed as compared to 40.61% of the rate observed in case of isolated CdS. This enhanced rate is complemented by reduced irradiation time and the overall performance of the photochemical degradation is due to augmented by the addition of surface properties and benefits of COF. The coupling of CDS with COF facilitates the degradation of one of the most toxic pollutants BPA (0.3 g) at ph = 7 and 0.5%wt of the catalyst.

17 MnO_2-loaded Ultrathin COF Nanosheets as Cathode Catalysts for Li-CO_2 Batteries [37]

With the advent of rechargeable Li-CO_2 batteries, a novel green strategy has evolved for not only solving energy problems but has also found its green solution by fixing the atmospheric CO_2 into useful energy and thereby reducing carbon footprints. The strategy involves exfoliation of COF-supported quinone into super-thin MnO_2/ 2,6-diaminoanthraquinone-2,4,6-triformylphloroglucinol (DQTP)- COF-nanosheet

(NS) type conjugated materials which serve as a potential catalyst at the cathode of Li-CO_2 batteries. It emphasizes the significant role of porous and crystalline substances for catalyzing the fixation of CO_2 and providing extraordinary electrode materials for a powerful green strategy and environment-friendly technique ensuring the sustainability of the environment.

18 Magnetic COFs for Efficient Fenton-Like Degradation of Pharmaceutical Waste Sulfamethazine [38]

Pharmaceutical wastes in today's times are overwhelming the water bodies and are hugely contaminating them. Sulfamethazine is one of such commonly found pharmaceutical waste which was observed to be degraded by magnetic covalent organic framework (COFs) (Fe_2O_3@COFs) nanocomposite. These composites were able to play the role of Fenton-like catalysts (in H_2O_2 medium) employed in the catalytic degradation of an aqueous solution of sulphamethazine and was attributed to the restriction of the agglomeration and growth of Fe_3O_4 Nps while embedded in the COF channels and exposing the crystallinity and porosity of COFs for exceptional results. 100% removal of sulphamethazine was witnessed as the magnetic nature of Fe_2O3 based COFs facilitated complete adsorption of the toxic organic contaminant from aqueous solution at acidic pH and the reusability of the catalyst was shown in subsequent five runs of the degradation reaction.

19 Triazine Functionalized Porous Covalent Organic Framework for Photo-Organocatalytic E–Z Isomerization of Olefins [39]

Many synthetic drugs and polymers have alkenes or olefins as their major constituent and they make essential building blocks for numerous significant polymers. Stereo regular olefins have wider prospects as they are significant anti-cancer agents incorporated into many anti-cancer drugs, industrial dyes, and many chromatic materials including lasers. As the synthesis of Z form (cis) is always tedious a convenient pathway will be to synthesize the less stable E for (trans) first and then convert E to Z using isomerization reaction via photocatalytic initiation. Such an efficient photocatalyst was developed using COF-supported catalyst in which all the benefits derived from their porosity, crystallinity, ultra stability and reusability could be harnessed successfully. The photoactive nature of triazine and β-ketoenamine were useful as building units for such a covalent organic framework which could be utilized for catalyzing E to Z transformation reaction of the olefins upon light irradiation.

Other applications of COF and its hybrid molecules are humongous and include solid–solid reaction synthesis of the covalent organic framework as a stable and highly active photocatalyst for degradation of sulfathiazole in industrial wastewater [40]. It has also been reported that metal nanoparticles when dispersed in COF containing in thioether are capable of many impressive catalytic applications. [41] Pyrazine-supported COF has exhibited Lewis acid behavior in their catalytic events [42], and Thiazolo[5,4-d] thiazole-linked Covalent Organic Framework is capable of Sustained Solar H2 Evolution from water [43].

20 Conclusion

Thus COF and COF-supported hybrid systems have been under researchers' sharp eye in the current scenario and have gained immense attention from scientists. Covalent organic frameworks provide a molecular template where the organic molecules integrate with other components and simultaneously create ordered and well-organized 2 to 3-dimensional lattices possessing pores of varied sizes. This unique structure facilitates the tailoring and designing of various polymeric COF composites suitable for high-grade catalytic activity. The synergic effect in hybrid COFs further enhances their capabilities and opens a new field for exploring a huge range of combinations for humongous applications. There have been numerous explorations of catalytic activities of COF and COF-based diverse composites exposing a large number of active sites and functionality for multiple heterogeneous catalysis and other benefits. The above insight can provide a novel perspective to scientists working on heterogeneous catalysis and open new avenues for future research on crystalline porous materials and their advantages.

Abbreviations

COFs	Covalent organic frameworks
Tp	2,4,6-Trihydroxybenzene-1,3,5-tricarbaldehyde
Pa	p-phenylenediamine
g-C_3N_4	Graphitic carbon nitride
DHTA	2,5-Dihydroxyl-terephthalaldehyde
TAPT	1,3,5-Tris-(4-aminophenyl) triazine
Tz	4,4′,4″-(1,3,5-Triazine-2,4,6-triyl)trianiline
Da	2,5-Dihydroxyterephthalaldehyde
BTDC	BiOBr/TzDa COF
CTF	Covalent triazine framework
Rh_{ox}	[Cp * Rh(bipy)H_2O]2+
Rh_{red1}	Cp * Rh(bipy)
Rh_{red2}	[Cp * Rh(bipy)H] +
Cp*	Pentamethylcyclopentadienyl
bpy	2,2′-Bipyridine

CL	Chemiluminescence
5-HT	5-Hydroxy-tryptamine
BPA	Bisphenol-A
DQTP	2,6-Diaminoanthraquinone-2,4,6-triformylphloroglucinol
NS	Nanosheet

References

1. Côté AP, Benin AI, Ockwig NW, O'Keeffe M, Matzger AJ, Yaghi OM (2005) Porous, crystalline, covalent organic frameworks. Science 310:1166–1170
2. Yang D, Gates BC (2019) Catalysis by metal organic frameworks: perspective and suggestions for future research. ACS Catal 9:1779–1798
3. Dhakshinamoorthy A, Li Z, Garcia H (2018) Catalysis and photocatalysis by metal organic frameworks. Chem Soc Rev 47:8134–8172
4. Dhakshinamoorthy A, Opanasenko M, Cejka J, Garcia H (2013) metal organic frameworks as heterogeneous catalysts for the production of fine chemicals. Catal Sci Technol 3:2509–2540
5. Feng X, Ding X, Jiang D (2012) Covalent organic frameworks. Chem Soc Rev 41:6010–6022
6. Geng K, He T, Liu R, Dalapati S, Tan KT, Li Z, Tao S, Gong Y, Jiang Q, Jiang D (2020) Covalent organic frameworks: design, synthesis, and functions. Chem Rev. https://doi.org/10.1021/acs.chemrev.9b00550
7. Hou YX, Zhang XM, Sun JS (2015) Good Suzuki-coupling reaction performance of Pd immobilized at the metal-free porphyrin-based covalent organic framework. Micropor Mesopor Mater 214:108–114
8. Yao Y, Hu Y, Hu H (2019) Metal-free catalysts of graphitic carbon nitride-covalent organic frameworks for efficient pollutant destruction in water. J Colloid Interface Sci 554, 376–387
9. Chen X, Xia L, Pan R (2020) Covalent organic framework mesocrystals through dynamic modulator manipulated mesoscale self-assembly of imine macrocycle precursors. J Colloid Interface Sci 568:76–80
10. Kang ZX, Peng YW, Qian YH (2016) Mixed matrix membranes (MMMs) comprising exfoliated 2D covalent organic frameworks (COFs) for efficient CO_2 separation. Chem Mater 28:1277–1285
11. Fan HW, Mundstock A, Feldhoff A (2018) Covalent organic framework–covalent organic framework bilayer membranes for highly selective gas separation. J Am Chem Soc 140:10094–10098
12. Li Y, Hu T, Chen R (2020) Novel thiol-functionalized covalent organic framework as adsorbent for simultaneous removal of BTEX and mercury (II) from water. Chem Eng J 398:125566
13. Yu Y, Li G, Liu J (2020) A recyclable fluorescent covalent organic framework for exclusive detection and removal of mercury (II). Chem Eng J 401:126139
14. Mal A, Mishra RK, Praveen VK (2018) Supramolecular reassembly of self-exfoliated ionic covalent organic nanosheets for label-free detection of double-stranded DNA. Angew Chem Int Ed 57:8443–8447
15. Li ZP, Huang N, Lee KH (2018) Light-emitting covalent organic frameworks: fluorescence improving via pinpoint surgery and selective switch-on sensing of anions. J Am Chem Soc 140:12374–12377
16. Fang Q, Wang J, Gu S (2015) 3D porous crystalline polyimide covalent organic frameworks for drug delivery. J Am Chem Soc 137:8352–8355
17. Biswal BP, Chandra S. Kandambeth S (2013) Mechanochemical synthesis of chemically stable isoreticular covalent organic frameworks. J Am Chem Soc 135 5328–5331
18. Fang Q, Gu S, Zheng J, Zhuang Z, Qiu S, Yan Y (2014) 3D microporous base-functionalized covalent organic frameworks for size-selective catalysis. Angew Chem Int Ed 53:2878–2882

19. Li H, Pan Q, Ma Y, Guan X, Xue M, Fang Q, Yan Y, Valtchev V, Qiu S (2016) Three-dimensional covalent organic frameworks with dual linkages for bifunctional cascade catalysis. J Am Chem Soc 138:14783–14788
20. McCrory CCL, Jung SH, Peters JC, Jaramillo TF (2013) Benchmarking heterogeneous electrocatalysts for the oxygen evolution reaction. J Am Chem Soc 135:16977–16987
21. Zhang B, Zheng XL, Voznyy O, Comin R, Bajdich M, Garcia-Melchor M, Han LL, Xu JX, Liu M, Zheng LR, de Arquer FPG, Dinh CT, Fan FJ, Yuan MJ, Yassitepe E, Chen N, Regier T, Liu PF, Li YH, De Luna P, Janmohamed A, Xin HLL, Yang HG, Vojvodic A, Sargent EH (2016) Homogeneously dispersed multimetal oxygen-evolving catalysts. Science 352:333–337
22. Li Y, Pei B, Chen J, Bing S, Hou L, Sun Q, Xu G, Yao Z, Zhang L (2021) Hollow nanosphere construction of covalent organic frameworks for catalysis: (Pd/C)@TpPa COFs in Suzuki coupling reaction. J Colloid Interface Sci 591:273–280
23. Ding SY, Gao J, Wang Q (2011) Construction of covalent organic framework for catalysis: Pd/COF-LZU1 in Suzuki-Miyaura coupling reaction. J Am Chem Soc 133:19816–19822
24. Pachfule P, Kandambeth S, Díaz Díaz D, Banerjee R (2014) Highly stable covalent organic framework–Au nanoparticles hybrids for enhanced activity for nitrophenol reduction. Chem Commun 50:3169
25. Kandambeth S, Mallick A, Lukose B, Mane MV, Heine T, Banerjee R (2012) Construction of crystalline 2D covalent organic frameworks with remarkable chemical (acid/base) stability via a combined reversible and irreversible route. J Am Chem Soc 134:19524
26. Fua Y, Zhua X, Huangb L, Zhanga X, Zhanga F, Zhua W (2018) Azine-based covalent organic frameworks as metal-free visible light photocatalysts for CO_2 reduction with H_2O. Catal B Environ 239:46–51
27. Mu M, Wang Y, Qin Y, Yan X, Li Y, Chen L (2017) Two-dimensional imine-linked covalent organic frameworks as a platform for selective oxidation of olefins. ACS Appl Mater Interfaces 9:22856–22863
28. Zhang Y., Chen Z., Shi Z., Lua T., Chen D., Wang Q., Zhan Z (2021) A direct Z-scheme BiOBr/TzDa COF heterojunction photocatalyst with enhanced performance on visible-light driven removal of organic dye and Cr(VI). Sep Purif Technol 275:119216
29. Thote J, Aiyappa HB, Deshpande A, DiazDíaz D, Kurungot S, Banerjee R (2014) A covalent organic framework–cadmium sulfide hybrid as a prototype photocatalyst for visible-light-driven hydrogen production. Chem Eur J 20:15961–15965
30. Diercks CS, Lin S, Kornienko KEA, Nichols EM, Zhu C, Zhao Y, Chang CJ, Yaghi OM (2018) Reticular electronic tuning of porphyrin active sites in covalent organic frameworks for electrocatalytic carbon dioxide reduction. J Am Chem Soc 140:1116–1122
31. Wei PF, Qi MZ, Wang ZP, Ding SY, Yu W, Liu Q, Wang LK, Wang HZ, Wan-Kai An WK, Wang W (2018) Benzoxazole-linked ultrastable covalent organic frameworks for photocatalysis. J Am Chem Soc 140:4623–4631
32. Yadav RK, Kumar A, Park NJ, Kong KJ, Baeg JO (2016) A highly efficient covalent organic framework film photocatalyst for selective solar fuel production from CO_2. J Mater Chem A 4:9413–9418
33. Kong M, Jin P, Wei W, Wang W, Qin H, Chen H, He J (2021, January) Covalent organic frameworks (COF-300-AR) with unique catalytic performance in luminol chemiluminescence for sensitive detection of serotonin. Microchemical J 160(Part A):105650. https://doi.org/10.1016/j.microc.2020.105650
34. Pana F, Xiaob F, Wanga N (2021, November 1) Towards application of a covalent organic framework-silver nanoparticles@sand heterostructure as a high-efficiency catalyst for flow-through reduction of organic pollutants. Appl Surf Sci 565:150580
35. Xu Y, Shi X, Hua R, Zhang R, Yao Y, Zhao B, Liu T, Zheng J, Lu G (2020) Remarkably catalytic activity in reduction of 4-nitrophenol and methylene blue by Fe_3O_4@COF supported noble metal nanoparticles. Appl Catal B Environm 260:118142. https://doi.org/10.1016/j.apcatb.2019.118142
36. Sun C, Gurusamy KL, Yang HJ, Liu CH, Dong J, Wu JJ (2021, September15) Facile sonochemical synthesis of CdS/COF heterostructured nanocomposites and their enhanced photocatalytic degradation of Bisphenol-A. Sep Purif Technol 271:118873

37. Jiang C, Zhang Y, Zhang M, Ma NN, Gao GK, Wang JH, Zhang MM, Chen Y, Li S, Lan YQ (2021, April 21) Exfoliation of covalent organic frameworks into MnO2-loaded ultrathin nanosheets as efficient cathode catalysts for Li-CO_2 batteries. Cell Rep Phys Sci 2:100392
38. Zhuang S, Wang J (2021) Magnetic COFs as catalyst for Fenton-like degradation of sulfamethazine. Chemosphere 264:128561
39. Bhadra M, Kandambeth S, Sahoo MK, Addicoat M, Balaraman E, Banerjee R (2019) Triazine functionalized porous covalent organic framework for photo-organocatalytic E-Z isomerization of olefins. J Am Chem Soc 141:6152–6156
40. Niu L., Zhao X., Wu F., Lv H., Tang Z. Liang W., Wang X., Giesy J. (2021) Hotpots and trends of covalent organic frameworks (COFs) in the environmental and energy field: bibliometric analysis. Sci Total Environm. 783:146838
41. Lu S, Hu Y, Wan S, McCaffrey R, Jin Y, Gu H, Zhang W (2017) Synthesis of ultrafine and highly dispersed metal nanoparticles confined in a thioether-containing covalent organic framework and their catalytic applications. J Am Chem Soc 139:17082–17088
42. Ma Y, Liu X, Guan X, Li H, Yusran Y, Xue M, Fang Q, Yan Y, Qiua S, Valtchev V (2019) One-pot cascade syntheses of microporous and mesoporous pyrazine-linked covalent organic frameworks as Lewis-acid catalysts. Dalton Trans 48:7352
43. Biswal BP, Vignolo-Gonzaĺez HA, Banerjee T, Grunenberg L, Gottschling GK, Nuss J, Ochsenfeld C, Lotsch BV (2019) Sustained solar H2 evolution from a thiazolo[5,4-d]thiazole-bridged covalent organic framework and nickel-thiolate cluster in water. J Am Chem Soc 141:11082–11092

Recent Advances in the Synthesis of Covalent Organic Frameworks for Heterogeneous Catalysis

Subodh and Dhanraj T. Masram

Contents

1 Introduction 286
2 Design of Building Blocks for the Synthesis of COFs 288
3 COF as a Support Material for Catalysis 289
4 Methods of Synthesis 290
4.1 Ionothermal Trimerization 290
4.2 Microwave-Assisted (MW) Synthesis 291
4.3 Sonochemical Synthesis 292
4.4 Mechanochemical Synthesis 293
4.5 Schiff's Base Reaction 293
4.6 Synthesis of COFs via One-Pot Multi-component Reactions (MCRs) 295
4.7 Synthesis of Magnetic COFs 296
4.8 Friedel–Crafts Reaction 297
4.9 Coupling Reactions 299
5 Application of Covalent Organic Frameworks in Heterogeneous Catalysis 300
5.1 C–C Coupling Reactions 301
5.2 Nitrophenol Reduction 301
5.3 Asymmetric Catalysis 302
5.4 Cycloaddition of CO_2 to Epoxides 305
5.5 Condensation and Multi-Component Reactions 306
5.6 Dye Degradation 309
5.7 Photocatalytic Applications 310
5.8 Electrocatalysis 311
5.9 Other Catalytic Applications 312
6 Conclusion 313
Abbreviations 314
References 314

Abstract Covalent organic frameworks (COFs) are periodically well-organized polymeric skeletons of organic monomer units connected through strong covalent linkages to form stable crystalline materials. The specific bonding among the organic monomer unit constructs a regular and porous skeleton, and therefore, these COFs are considered as highly ordered and uniform materials. COFs have recently appeared as heterogeneous catalysts for various organic transformations and photocatalytic reactions. These properties are arisen due to the reticular design of these materials, and this

Subodh · D. T. Masram (✉)
Department of Chemistry, University of Delhi, Delhi 110007, India

S. Gulati (ed.), *Metal-Organic Frameworks (MOFs) as Catalysts*,
https://doi.org/10.1007/978-981-16-7959-9_11

reticular design comes from the diversity of organic building blocks, stable geometry, and reversibility of dynamic covalent reactions. These materials are usually formed by organic building blocks consisting of heteroatoms and obtained in situ during the synthesis of 2/3D-ordered porous materials. The COFs are acclaimed as functional materials due to their crystallinity, porous nature, both chemical and thermal stability, low skeleton density, high surface area, diverse and easy synthetic methodologies, flexibility, insolubility, cheaper substrates, and highly simplified for functional modifications. Additionally, COFs can be used as appropriate carriers to lock metal ions or by using their functional in-built sites as a catalyst. This chapter summarizes the noteworthy progress in the synthesis of COFs material and their potential catalytic applications with an emphasis on their property as support material for heterogeneous catalysis.

Keywords Covalent organic framework · Catalysis · Nanoparticles · Photocatalysis · Triazine

1 Introduction

There the design and synthesis of porous covalent organic frameworks (COFs) have resulted in vital advancements in the field of porous materials. COFs are a type of porous material that, unlike supramolecular and coordination polymers, is made up of covalent bonds between organic precursors. These are categorized as a class of highly cross-linked materials having an amorphous to crystalline nature, substantially formed by strong covalent linkage [1]. These materials are usually formed by organic building blocks consisting of heteroatoms and obtained in situ during the synthesis of 2/3D-ordered porous materials (Fig. 11.1a) [2]. In the recent research, COFs have been widely employed and categorized as covalent triazine framework (CTF) [3], covalent organic frameworks (COF) [4], polymers of intrinsic microporosity (PIMs) [5], covalent organic polymers (COPs), Schiff's base networks, porous aromatic framework, porous polymer framework, pyrene-derived benzimidazole-based polymer, and conjugated micro/mesoporous polymers (CMPs) [6].

The idea of synthetic coordination polymers was first introduced and discovered in 1920 by Herman Staudinger, who proposed that the molecular weight measured for natural rubber was higher which could be the result of many small organic molecules that were covalently linked together. Later, this work led him to the 1953 Nobel Prize [7]. Recently, chemists around the globe have recognized the paramount significance of these functional organic frameworks. As per synthetic perspective, there are various synthetic methods such as Friedel–Crafts arylation, Cross-coupling reaction, electrophilic substitution reaction, nucleophilic substitution reaction, oxidative coupling, Schiff's base reaction, ionothermal trimerization, cyclotrimerization, microwave-assisted synthesis, phenazine ring fusion reaction, etc. The irreversible nature of these useful organic reactions results in chemically stable polymers, in

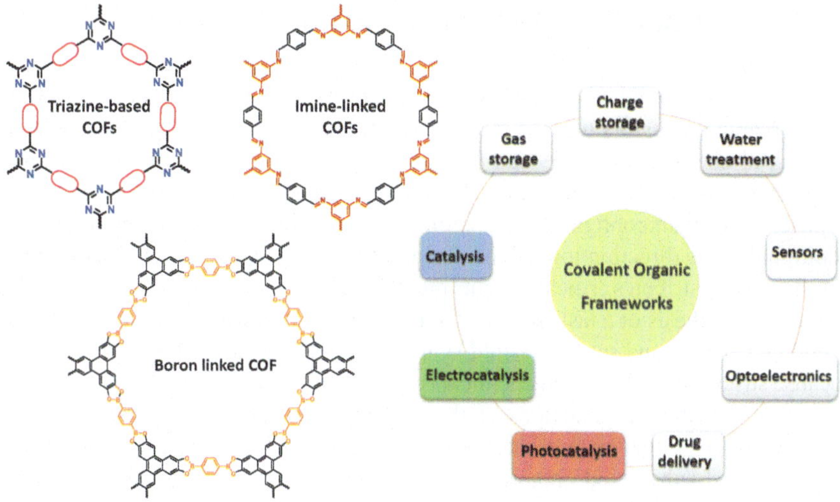

Fig. 11.1 Types of some usual COFs and their various potential applications

most cases, lacks orientation and crystallinity. The low crystallinity characteristic of these polymers drives initiation by chemists to explore the design of polymeric systems with controllable pore sizes or shapes. Later, this challenge was handled by Prof. Omar Yaghi and co-workers, who accomplished the investigation of crystalline polymers with defined porous properties. This new class of material was termed as covalent organic frameworks (COF) [8]. In the last decade, a huge number of research reports for covalent organic frameworks suggested that significant efforts had been made in this new area of porous materials.

The COFs are acclaimed as functional materials due to their crystallinity, porous nature, both chemical and thermal stability, low skeleton density, high surface area, diverse and easy synthetic methodologies, flexibility, insolubility, cheaper substrates, and highly simplified for functional modifications. Relative to other materials, COFs appear to offer outstanding utility in a variety of applications, including sensing [10], catalysis [11], drug delivery [12], gas storage and separation [9], electrocatalysis [13], organic pollutants control, luminescence, and photocatalysis (Fig. 11.1) [14]. COFs structures can be two dimensional or three dimensional depending upon the interactions between the covalent bond formation and non-covalent interactions which ultimately control the structure, morphology, and other properties. Conclusively, research around COFs and their materials has made impressive growth in the last decade.

Herein, we present comprehension of the primary ideologies and principles for the strategy, synthesis, utilization of COFs as catalysts, and some of the challenging issues in the context of chemistry and materials.

2 Design of Building Blocks for the Synthesis of COFs

There are various methods to construct covalent organic frameworks by incorporating organic precursors into the main chain. In this context, the experience gained by material chemists in MOFs can be applied to conceive the synthesis of COFs. However, the assembled architecture of building molecular units of crystalline MOFs is much easier than configuring crystalline COFs via dynamic covalent chemistry. Additionally, the direction toward the design of inorganic materials such as zeolites significantly contributes critical insights into the architectural synthesis of COFs. The design and synthesis of a task-specific COF primarily start with the rationale choice of building blocks and set the desired synthesis conditions accordingly (Fig. 11.2). Because structural features and chemical reactivity are significantly dependent upon the nature and the side reactions of the monomers. There are at least two reactive functional groups such as haloarenes, boronic acids, nitriles, aromatic aldehydes, ethynyl-substituted arenes, and substituted anilines/amines that should be there on the building block either similar or dissimilar to construct the COF architecture. Two different types of monomeric units or building blocks couple with each other to create the polymeric structure while self-condensation of a single type of monomer has also been reported. The monomeric units may consist of arenes, fused rings, phenyl ethylene, and hetero/macrocyclic units. Subsequently, a wide range of building blocks with diverse functional groups bequeaths COFs with distinctive structure and functionality. By varying the building block, the porosity, specific surface area, morphology, and functionality can be easily modified.

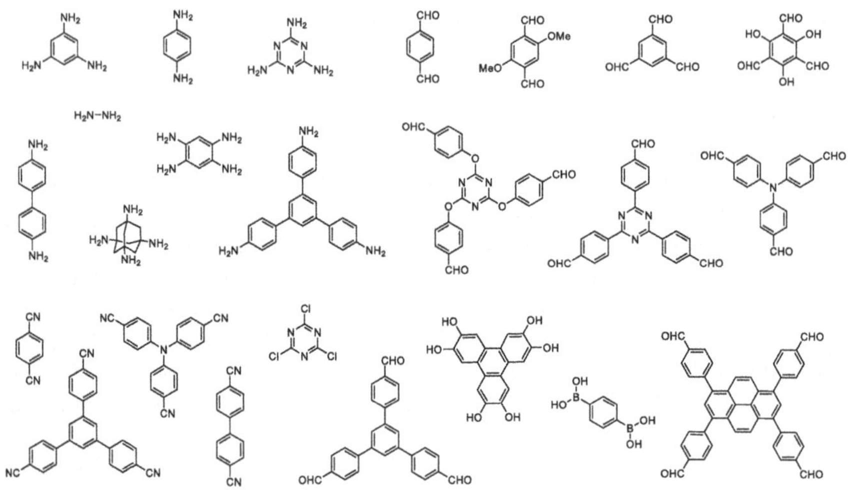

Fig. 11.2 Some of the most widely used building blocks for the synthesis of COFs

3 COF as a Support Material for Catalysis

What is support material and why do we need support material for catalysis? In material chemistry, catalytic support can be described as a material with some primary features like high surface area, optimal binding sites, both thermal and chemical stability, and adequate morphology to which a catalyst or nanoparticles are affixed with fine dispersion [15]. The most important types of heterogeneous catalysts are supported catalysts where catalysis occurs at surface atoms [16]. Nano-catalysts' remarkable progress in diverse chemical transformations has got a lot of attention. Because of their selectivity and efficacy, immobilized catalysts are of critical relevance for catalytic methods. However, there are some downsides related to unsupported nanocatalysis including the agglomeration of particles or active sites, inconsistency in catalytic activity, and their recovery and reusability [17]. The problem of agglomeration of nanoparticles can be avoided by the immobilization of MNPs on a suitable support material due to their synergistic effect [18]. The catalyst's selectivity and performance are influenced by several functionalities on the support material, which can be inert or active in the catalytic process. Therefore, the motive goes around the dispersion of appropriate support material to enhance the stability of the catalytic nanoparticles and to achieve the finest catalytic efficiency, which consequently decreases the expense of costly metal and eventually making it an economical process at the industry level.

Among various support materials reported for heterogeneous catalytic process, covalent organic frameworks, a member of the highly crystalline material family is known as a substantial support material due to its outstanding surface characteristics such as crystalline architect, porous nature, significant surface area, non-toxicity, and chemical as well as thermal stability which increase the loading of active catalytic sites [19]. COF as a support material has shown high chemical stability in various organic and aqueous solvents. Support materials such as COF and its modified variants are well known, and several reports are already available in literature where COF has been utilized as an extensive support material for various catalytically active species [20]. Presence of several heteroatoms such as N, O, S, B helps in the immobilization of nanoparticles, and it might function as a more selective supporting material, bridging the electron transfer. As a result of the immobilization of NPs over COFs, they become more proficient and recoverable [21]. Furthermore, these NPs can aggregate to produce bigger size particles and snag the catalysis process, and this can be avoided by immobilization of NPs over the surface of COFs to improve the stability of the catalyst. Using COF as support has its own advantages such as cheaper, availability, amphiphilic nature, and high dispersibility in aqueous or organic solvents. Owing to their outstanding synergistic activity, COFs are reorganized utilizing nanoparticles settled over their exterior to use as heterogeneous catalysts for many industrial catalytic applications. Thus, COF are best-suited heterogeneous supports catalysts in numerous catalytic organic transformations.

4 Methods of Synthesis

Various methods are reported to synthesize covalent organic frameworks, including the ionothermal cyclotrimerization, condensations of different aldehyde, amino groups, boronic acids, alcohols, and other functional groups, cross-coupling reactions, nucleophilic substitutions, ring fusion reactions, etc. There are also solvent-free techniques that have been described. A brief description has been given below for various methods of synthesis.

4.1 Ionothermal Trimerization

Very first in 2008, Thomas et al. stated ionothermal trimerization to prepare very stable triazine-based frameworks by heating a combination of $ZnCl_2$ and nitrile monomer in quartz ampules, under the reaction conditions of very high temperature around 400–600 °C, where molten $ZnCl_2$ acted as a solvent and catalyst as well (Scheme 11.1a) [3a].

Molten $ZnCl_2$ under a high-temperature environment satisfies all the conditions to construct a porous polytriazine network. Primarily, nitriles have shown remarkable solubility in molten $ZnCl_2$ owing to strong Lewis's acid–base interactions among them; secondly $ZnCl_2$ acts as an effectively good catalyst for most of the aromatic and heterocyclic nitriles. FT-IR spectroscopy and ^{13}C and ^{15}N solid-state CP-MAS NMR spectroscopy were used to determine the structure and synthetic development of the polymeric networks. The elemental presence and their chemical bonding correlation were well confirmed using XPS analysis. By the vigilant variety of monomeric units or building blocks, the porosity, specific surface area, structure, morphology, and functionality are easily alterable. There are various symmetrical and unsymmetrical building blocks counting aromatic and other heterocyclic molecules. Furthermore, the porosity of the resulting polymer is determined by altering reaction parameters such as reaction temperature, solvent, the quantity of $ZnCl_2$, and reaction time. Afterward, in 2012, Cooper et al. modified this method using CF_3SO_3H in place of $ZnCl_2$, which offers very mild reaction conditions of the microwave or room temperature (Scheme 11.1b) [22]. Additionally, this method was reported advantageous in terms of preventing the use of $ZnCl_2$, which contaminates the final product, breakdown of nitriles, and C-H bonds. Though, even this method is associated with its own limitations of CF_3SO_3H is an expensive, corrosive, and eye irritating chemical, low reaction temperature is required for reaction handling, and ice-cold brine solution is necessary for neutralization of acidic medium due to CF_3SO_3H.

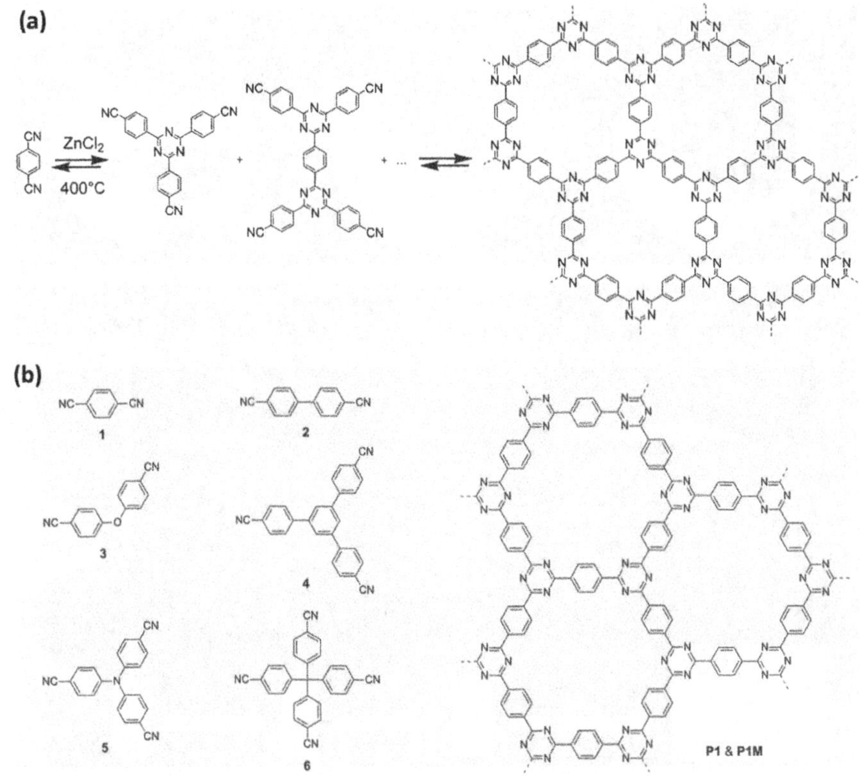

Scheme 11.1 **a** Schematics of preparation of triazine-based precursor and then further polymerization to form triazine-based framework (CTF-1) [3a] and **b** preparation of CTFs using different nitrile precursors [22]

4.2 Microwave-Assisted (MW) Synthesis

The alternative source of energy in place of traditional energy sources is highly recommended in recent times. A typical thermal reaction can be simplified by using microwave radiations, which enables the handling to ease. Cooper et al. published the first synthesis technique for 2/3D COF in 2009, utilizing benzene-1,4-diboronic acid (BDBA) and hydroxy analog (HHTP) after 20 min of microwave irradiation at 100 °C (Fig. 11.3a) [23]. The reaction was reported as 200 times faster than conventional methods, besides this method also results in enhancement of the surface area of COF. Recently, Wenhua et al. reported microwave-assisted preparation of a dioxin-linked TH-COF via nucleophilic substitution (Fig. 11.3b) [24]. The microwave-assisted TH-COF exhibits better stability, higher surface area, and crystallinity and is thus utilized as covering for SPME fiber for detection of traces of PFASs in the water sample. The TH-COF-SPME fiber via UPLC-MS/MS method exhibits low LODs and good reusability.

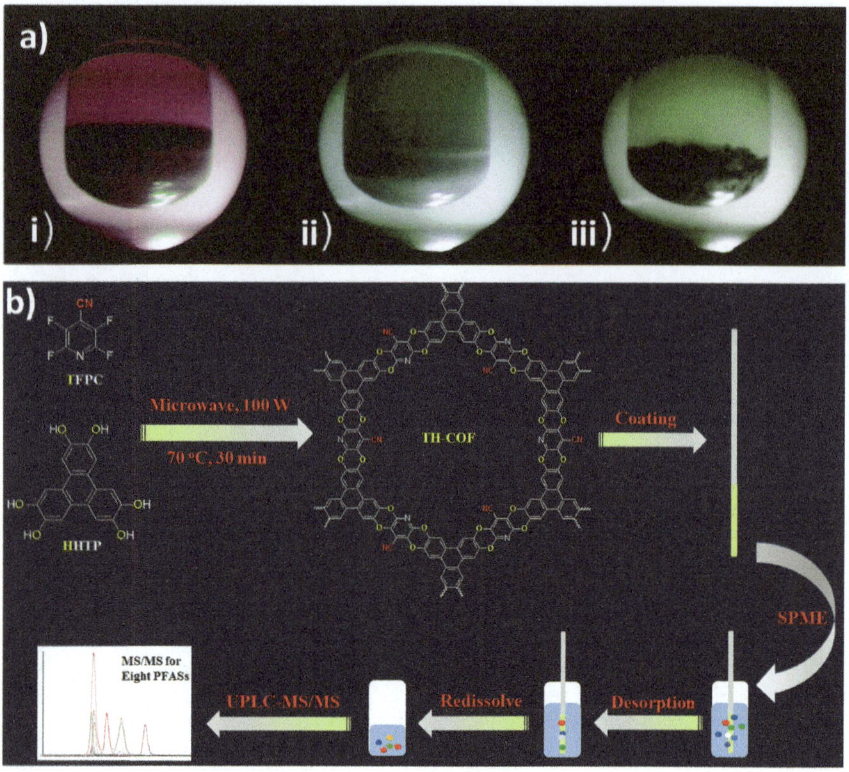

Fig. 11.3 **a** Digital images of microwave-assisted synthesis of COF-5 and its purification: (i) initially formed gray–purple COF-5 containing impurities, (ii) removal of impurities by MW extraction method, and (iii) final pure extracted gray COF-5 [23] and **b** schematic representation of the preparation of TH-COF-SPME fiber and its application [24]

4.3 Sonochemical Synthesis

Likewise, sonochemical synthetic approaches are even of great importance as these methods are comparatively quicker and cost effective. Therefore, in 2012 Ahn et al. described the preparation of two COF materials named COF-1 and COF-5 using benzene-1,4-diboronic acid as a basic unit (Fig. 11.4) [25]. Both the COFs were reported as powder as well as in film form with almost the same surface area, while, as far as their physicochemical properties are concerned, somewhat superior characteristics are shown as compared to traditional methods. The sonication aids in the solubility of reactants in reaction medium or solvent, and therefore, COF-1 was much smaller (~400 times smaller) than that of the conventional process of preparation.

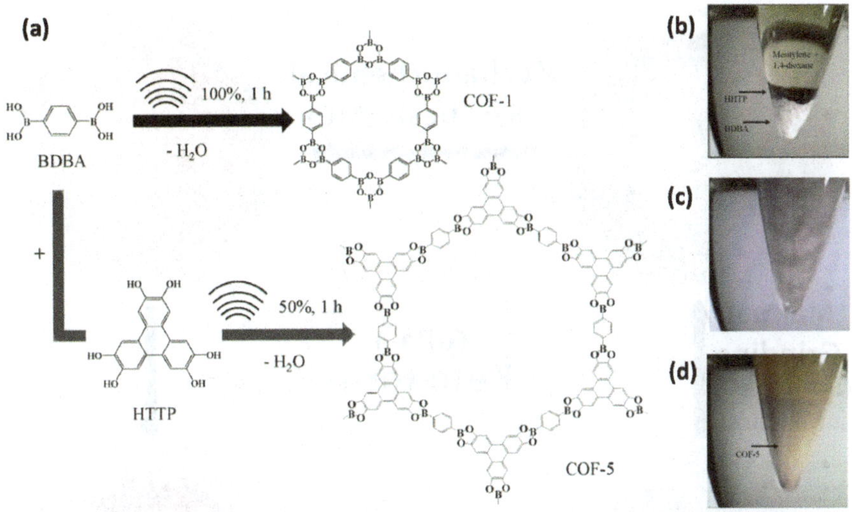

Fig. 11.4 **a** Sonochemical preparation of COF-1 and COF-5. Digital images of preparation of COF-1 showing **b** setup, **c** during sonication, and **d** after completion of the reaction [25]

4.4 Mechanochemical Synthesis

The simplest synthetic route for any chemical transformation is considered as mechanochemical synthesis for simple, faster, economic, greener in terms of solvent-free and lower-temperature methodology just obtained as a result of manual grinding. In 2013, Bishnu et al. reported three imine-linked COF materials TpPa-1, TpPa-2, and TpBD [26]. The naked eye's significant color change confirmed the completion of the reaction. Unlike solvothermal methods, these COF exhibited sheet-like layered morphology which may be ascribed to the depilation of COF layers in the mechanochemical process (Fig. 11.5).

4.5 Schiff's Base Reaction

The most commonly employed synthetic methodology for COFs is Schiff's base condensation reaction. In general, by taking melamine as an amine precursor, this synthetic method has reported numerous COFs using different aldehyde substrates. The biggest advantage of these methods comes from their non-catalytic processing, and the reaction needs moderate reaction conditions with cost-effective precursors. Numerous melamine-based COFs were synthesized using inexpensive precursors, and high nitrogen content (>40 wt.%) was reported, as well as high stability and insoluble in aqueous and organic solvents. In 2017, Manman et al. synthesized two polymeric frameworks containing –OH groups, TAPT-DHTA-COF_{HX}, and

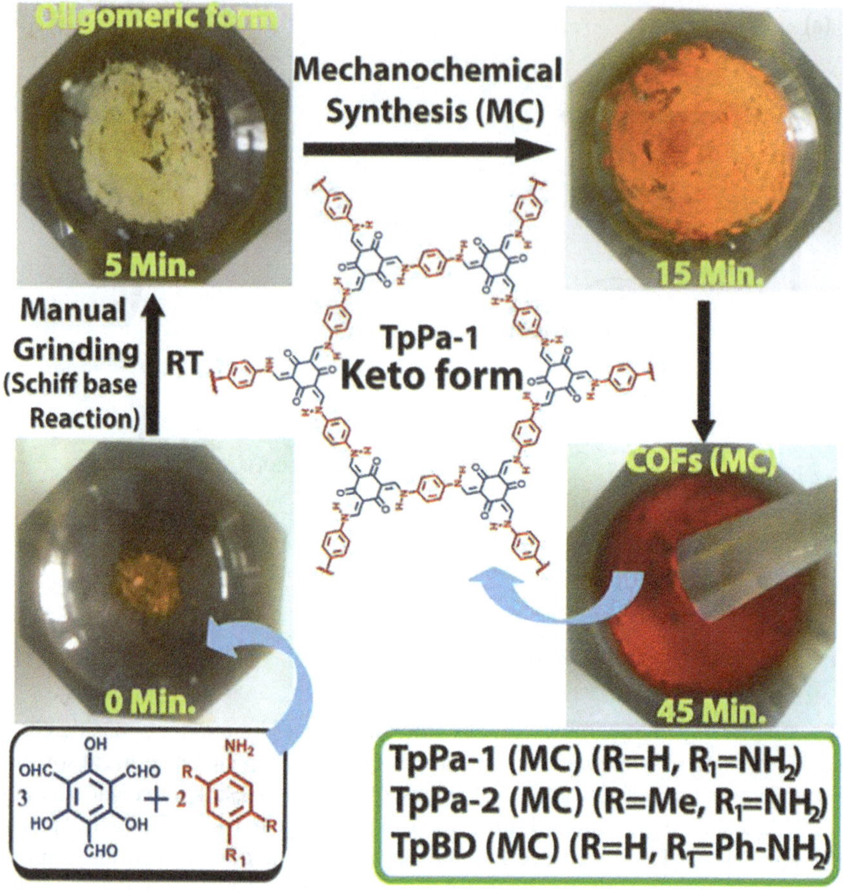

Fig. 11.5 Digital images of mechanochemical preparation of TpPa-1, TpPa-2, and TpBD [26]

TAPT-DHTA-COF_{DMF} (Scheme 11.2a) [27]. The condensation reaction takes place between a triazine precursor 1,3,5-tris-(4-aminophenyl)triazine and 2,5-dihydroxyl-terephthalaldehyde in different solvents. The PXRD data confirm the eclipsed stacking of layers in the framework besides TAPT-DHTA-COF_{HX} possesses an exceptionally high surface area of 2238 m^2g^{-1}. Numerous imine-linked COFs have come into notice, which was prepared using triazine-containing amine precursors other than melamine and amines [28]. In 2017, Mohammad et al. demonstrated the preparation of a N-rich COF material via Schiff's base condensation reaction between two triazine-containing building blocks which were designed to make the architect N-rich (Scheme 11.2b) [29]. Under the solvothermal reaction conditions, the synthesis was performed which gives N-rich COF with high surface area and good crystallinity.

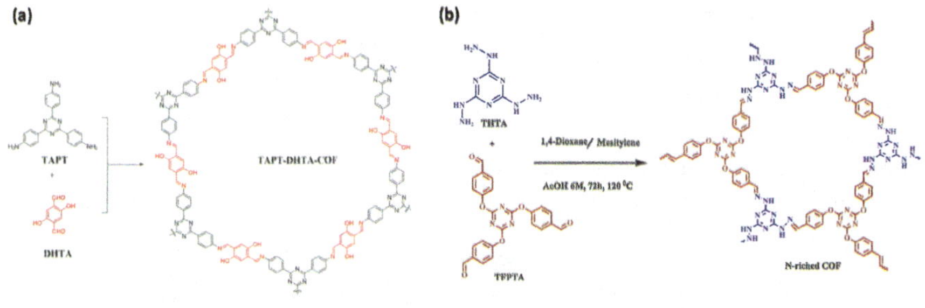

Scheme 11.2 **a** Schematics for preparation of TAPT-DHTA-COF [27] and **b** N-rich COF [29]

4.6 Synthesis of COFs via One-Pot Multi-component Reactions (MCRs)

Commonly, COFs are synthesized by the co-condensation between a linker and a knot, which can be termed as traditional [1 + 1] 2 component reactions. Undeniably, two component synthetic methods have their own pros; however, the three-component synthetic methods are highly recommended if a multifunctional COF material is required. Therefore, a multi-component reaction was used to prepare structurally diverse COF with exceptional properties. Besides, those chemical linkages which cannot be made through conventional synthetic methods are constructed using the robust method of MCR.

In 2019, Peng et. al. demonstrated the synthesis of ultra-stable imidazole-linked COFs. A very famous multi-component reaction named the Debus–Radziszewski reaction was employed for construction (Fig. 11.6i) [30]. The reaction took place in one pot through a greater degree of intricacy and accuracy in covalent architecture. The study claims an interesting cultured reversible/irreversible multi-component reaction for the creation of porous crystalline frameworks.

Amitava et al. presented a combined cyclotrimerization and aldol condensation process in a pot to make vinylene-bridged COF in 2020 [31]. The complete process was simplified in this study by mixing two or more successive reactions in a pot. In this report, the synthesis was initiated by cyclotrimerization of acetonitrile and subsequently followed by the aldol condensation with BPDA aldehyde (Fig. 11.6ii). Further, the applicability was extended for additional multi-topic aryl-aldehyde units, resulting in the formation of some new frameworks.

In 2020 only, Keiwei et al. reported a robust technique to synthesize highly stable covalent organic frameworks (COFs) through a robust three-component reaction between aldehydes, amines, and elemental sulfur (S_8) (Scheme 11.3) [32]. Under metal-free catalytic conditions, the reaction was started by C-H activation of aldehyde followed by oxidative annulation. Effectively, five COFs (thiazole linked) were synthesized which exhibited outstanding physicochemical stability, highly crystalline, and porous nature. Owning these excellent properties, these COFs were employed as photocatalysts in sacrificial hydrogen evolution reactions.

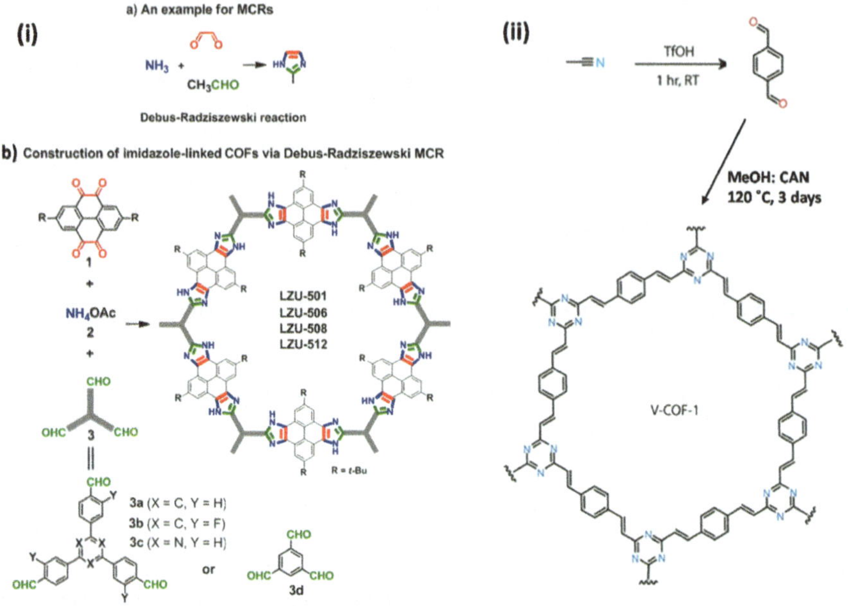

Fig. 11.6 (i) (a) Preparation of 2-methylimidazole via Debus – Radziszewski MCR reaction, (b) one-pot synthesis of imidazole-linked COFs [30], and (ii) schematics for preparation of V-COF-1 [31]

4.7 Synthesis of Magnetic COFs

Magnetism introduces the most interesting feature of a catalyst which is almost 100% recovery from the reaction medium which ultimately makes the catalyst cost effective and less polluting. Thus, catalysts with magnetic properties can be cast into a variety of chemical transformations including industrial applications. Owing to this fact, introducing the magnetic character to COF can bring a lot more fascinating materials of chemical use. In order to introduce the magnetic character to the COF backbone, it has to be attached to the magnetic nanoparticles; a fact introduces the idea of synthesis of magnetic COFs. In the early research, in 2017, Guo et al. published a magnetic COF material utilizing Fe_3O_4 as the magnetic core shell (Fig. 11.7a) [33]. Initially, the Fe_3O_4 nanoparticles were stabilized by citrate ligands using the solvothermal method. Consequently, the carboxylate-rich surface of modified Fe_3O_4 nanoparticles enables the coverage of COFs network through template-controlled TAPB and TPA precipitation polymerization over the exterior of the Fe_3O_4 core shell.

Similarly, in 2017, Sijing et al. published a straightforward technique for making a bouquet-shaped magnetic nanocomposite of a COF-TpPa-1 and surface improved Fe_3O_4 nanoparticles [34]. The synthetic procedure starts from the synthesis of amino-functionalized surface-modified Fe_3O_4 nanoparticles which were then reacted with

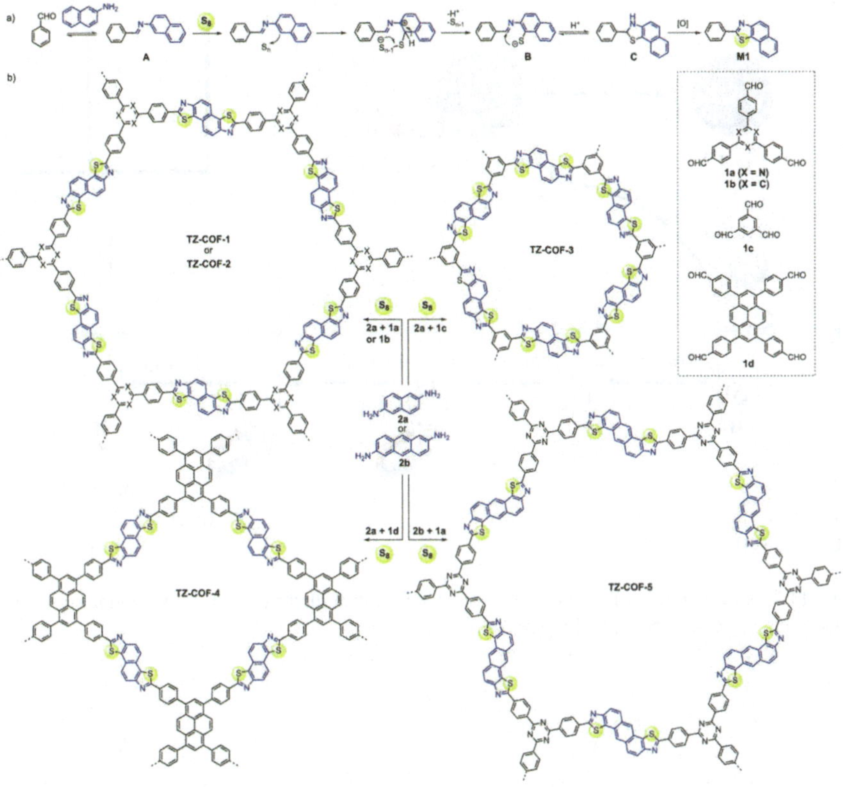

Scheme 11.3 Schematics for **a** probable mechanism for synthesis of thiazole containing precursor and **b** preparation of thiazole-linked COFs (TZ-COFs) via MCR [32]

1,3,5-triformylphloroglucinol (Tp) (Fig. 11.8a); lastly, these imine-linked nanoparticles were then reacted with the mixture of Tp and p-phenylenediamine (Pa-1) to formulate "TpPa-1" a hybrid magnetic nanocomposite. TpPa-1 reportedly possesses high surface area and porosity with super magnetism.

Similarly, in the same year, Rong et. al reported a magnetic adsorbent-based covalent organic framework LZU1 [35]. The polymeric framework LZU1 was immobilized upon the exterior of polyethyleneimine functionalized Fe_3O_4 nanoparticles (magnetic core shell) and termed as COF-LZU1@PEI@Fe_3O_4 (Fig. 11.8b).

4.8 *Friedel–Crafts Reaction*

Chang et al. prepared new triazine-based COPs in 2012 via Friedel–Crafts reaction of cyanuric chloride with benzene, biphenyl, and terphenyl in DCM using $AlCl_3$ as a catalyst (Scheme 11.4a) [36]. The electron deficiency of carbon centers

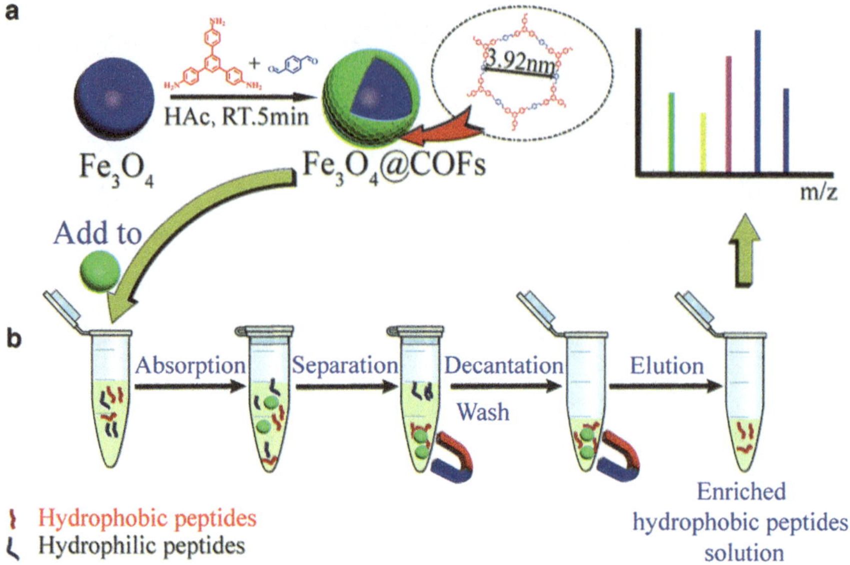

Fig. 11.7 **a** Preparation of magnetic Fe_3O_4@COFs and **b** their application in peptide separation process is depicted schematically [33]

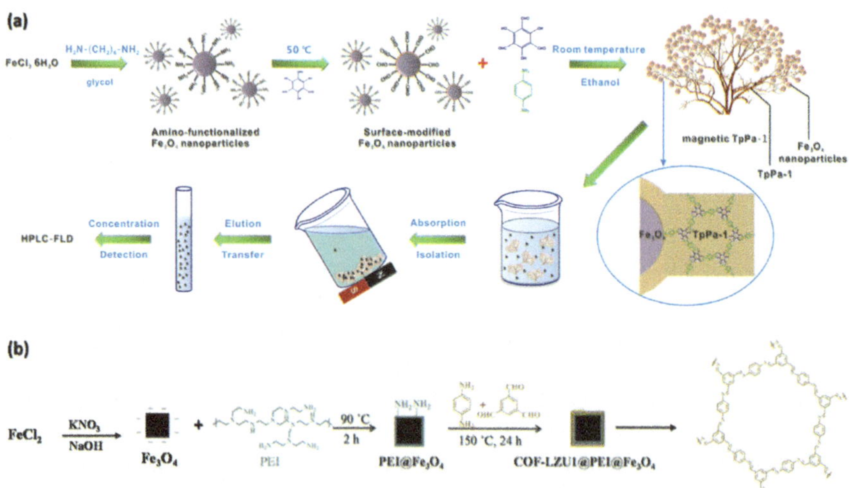

Fig. 11.8 **a** Preparation of magnetic COF-TpPa-1 and its use in the extraction of organic targets are depicted schematically [34] and **b** synthesis of magnetic COF-LZU1@PEI@Fe_3O_4 [35]

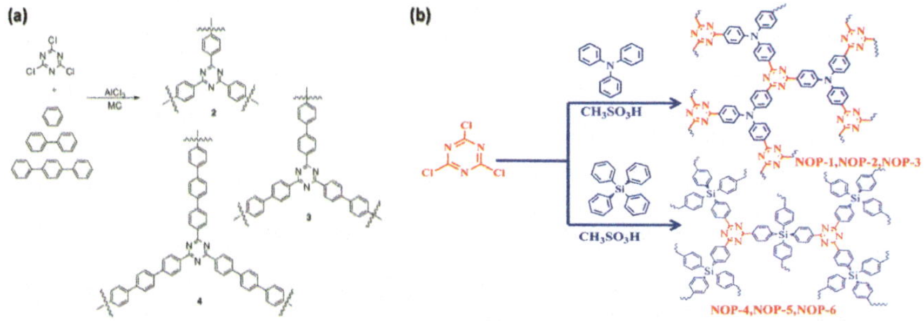

Scheme 11.4 Schematic representation of the preparation of triazine-based COFs via Friedel–Crafts reactions **a** microporous polymers [36] and **b** NOP networks [37]

of triazine ring facilitates the electrophilic substitution by an aromatic compound. A vital synthetic challenge is an easy and straight protocol to synthesize triazine-based COP under mild reaction conditions such as adequate range of temperature, cheaper starting materials, and catalyst. So, in this regard, Pan et al. prepared new nanoporous frameworks in 2014 using a simple approach of modified Friedel–Crafts reaction between 2,4,6-trichloro1,3,5-triazine and substituted aromatic substrates (Scheme 11.4b) [37]. The reaction was formed using CH_3SO_3H as a catalyst and anhydrous toluene as solvent at 140 °C for 24 h. All the polymeric materials synthesized were found in high yields, without utilizing any metal catalyst. While comparing with $AlCl_3$ catalyzed Friedel–Crafts method, it was noted that the precursors were highly miscible into the liquid catalyst (methanesulfonic acid), which subsequently increased the reaction efficiency [38]. Besides, methanesulfonic acid has low toxicity and corrosivity, good thermal steadiness, no harmful volatiles and decomposes into sulfate, carbon dioxide, water, and biomass in an environmentally benign manner; therefore, methanesulfonic acid has been proposed as an excellent catalyst for the Friedel–Crafts method.

4.9 Coupling Reactions

After the coupling reactions were discovered, techniques for forming the C–C bond have become exceedingly simple and advantageous. These transition metal-catalyzed chemical transformations enhanced the exponential growth of pharmaceuticals and industrial chemicals. In this context, in 2012 Cooper et al. developed a method of synthesis of triazine-based COFs using Sonogashira–Hagihara cross-coupling [39]. The reaction was performed via 2,4,6- tris(4-bromophenyl)-1,3,5-triazine with various di- and tri-acetylenes, in the presence of $(PPh_3)_4$, CuI, Et_3N, and anhydrous DMF as solvent (Scheme 11.5). These newly synthesized materials possess remarkably high porosity and are found advantageous in terms of stability. These polymeric networks were found structurally similar to benzene-linked CMPs, besides optical

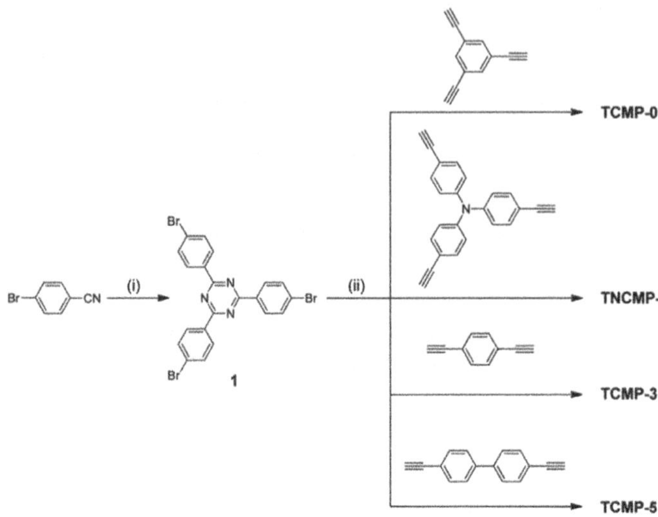

Scheme 11.5 Preparation of triazine-based conjugated microporous polymers via Sonogashira–Hagihara cross-coupling is depicted schematically [39]

band gap of these materials was altered using an appropriate selection of monomeric units. Remarkably, they synthesized a polymer termed TNCMP-2 that contains both an electron donor (triphenylamine) and an electron acceptor (1,3,5-triazine) part and consequently results in fascinating optoelectronic behavior. COFs with the specified sort of building block might be beneficial in photovoltaics or photocatalysis.

5 Application of Covalent Organic Frameworks in Heterogeneous Catalysis

In the field of catalysis and synthetic transformations, heterogeneous catalysis has simplified the process as one of the green, sustainable, and efficient gateway [40]. Covalent organic framework and its hybrid composites are well-known heterogeneous catalysts. Multifunctional COFs possess high potentials as efficient robust catalysts in various types of organic transformations. As a heterogeneous catalyst, these COFs can be used as an appropriate carrier to lock metal ions or by using their functional in-built sites as a catalyst. Large surface area, high porosity, and N-rich sites provide multifunctional heterocatalytic media for catalysis. The core backbone of COFs is made up of entirely organic molecules that are held together through covalent bonding which consequently deliver high stability. Advancements and quick expansion of synthetic methods have ensued into the synthesis of an array of COF-supported metal oxide composites with variable morphologies, chemical, and physical properties. Here, we summarize the representative catalytic application of COF materials.

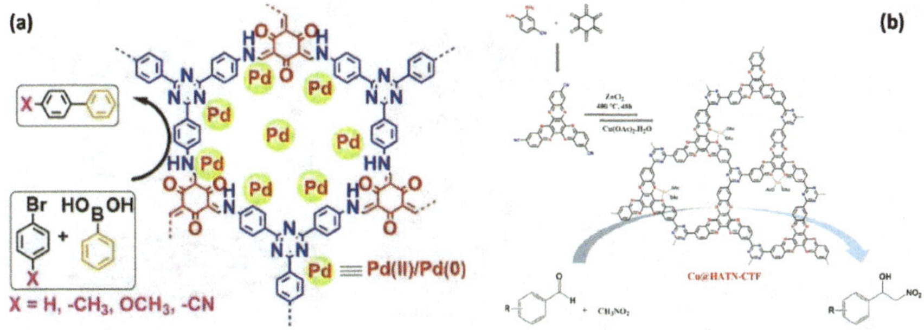

Fig. 11.9 **a** Synthetic procedure of Pd immobilized imine β-ketoenamine-linked COF and its use in coupling reactions [41] and **b** schematic representation of the synthesis of Cu@HATN-CTF and its catalytic application [42]

5.1 C–C Coupling Reactions

Cross-coupling reactions catalyzed by transition metals have long been regarded as critical for the production of C–C bonds in industrial and medicinal applications. Sonogashira, Heck, and Suzuki–Miyaura C– coupling reactions are well known for their potential uses. Metal immobilized COF catalysts are well known for C–C coupling reactions. Numerous reports have been published demonstrating sustainable catalytic systems for the same in terms of recyclability, better heterogeneity, reaction time, solvent, and temperature. In 2017, Dhananjayan et al. reported two imines and β-ketoenamine-linked COFs named TAT-DHBD and TAT-TFP [41]. The high nitrogen content enabled the excellent affinity for Pd NPs; subsequently, the newly synthesized catalyst was utilized for Suzuki–Miyaura cross-coupling between aromatic boronic acid and bromoarenes in DMF as solvent at 100 °C in 24 h with recyclability up to five cycles (Fig. 11.9a).

Recently, in 2019, Norini et al. reported a functionalized porous organic polymeric framework HATN-CTF showing high surface area, consequently subject to metalation using $Cu(OAc)_2$ to synthesize an effective heterogeneous catalyst Cu@HATN-CTF for Henry reaction of aromatic aldehydes and nitromethane in EtOH as solvent at 70 °C in 12 h (Fig. 11.9b) [42]. An excellent catalytic activity and higher TON were reported which may attribute to the electronic configuration effect.

5.2 Nitrophenol Reduction

The day-by-day progressive industrialization has been directed toward the highly polluted aquatic environment over the last few decades leading to tremendous pollution in the aquatic environment. Unprocessed industrial wastewater contaminates water sources resulting in impacts on marine animals and human health [43]. Among

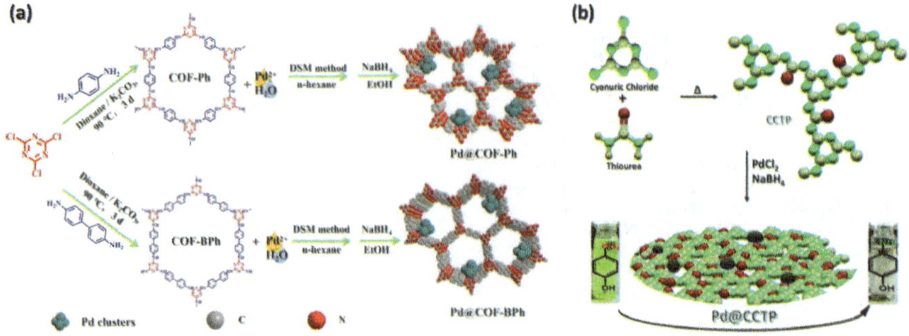

Fig. 11.10 **a** Preparation of COF-Ph and COF-BPh is shown diagrammatically and [46] **b** graphical illustration of solvent-free preparation of COP Pd@CCTP and its catalytic application in the hydrogenation of nitroaromatics [47]

them, nitroaromatics predominantly exist, which are very toxic, stable, and non-biodegradable [44], while their respective amino substituents are valuable in pharmaceuticals [45]. To simplify nitroaromatic reduction, various metal-based heterogeneous catalysts have been designed, among them metal immobilized triazine COFs are our prime focus.

In 2019, Mengying et al. synthesized a triazine-based polymeric material using cyanuric chloride with *p*-phenylenediamine/4,4′- diamino diphenyl, through nucleophilic substitution method and after the Pd immobilization, and the newly synthesized Pd@COF-Ph/ Pd@COF-BPh were utilized into hydrogenation of *p*-nitrophenol (Fig. 11.10a) [46]. More accurately, Pd@COF-BPh exhibited better catalytic efficiency due to larger pores. This catalytic investigation displayed high catalytic recyclability up to ten cycles for both catalysts. Similarly, in 2020, Yadav et al. presented their extended study on the Pd metal immobilized triazine-based COP (Fig. 11.10b). A new sustainable method was designed and practiced for the preparation of a polymer CCTP using thiourea and cyanuric chloride under solvent and additive-free sustainable reaction conditions [47]. After the Pd metal immobilization, the Pd@CCTP was utilized in catalytic reduction of nitroaromatics for experimental and bulk scale reactions. Additionally, thermodynamic and kinetic parameters for the hydrogenation of p-nitrophenol were also computed, indicating that the process is non-spontaneous and endothermic. Surfactants such NH_4OH, FA, and N_2 were also tested for their influence on response performance.

5.3 Asymmetric Catalysis

Asymmetric organocatalysis is a predominantly useful tool for organic chemists; however, being expensive and highly polluting and homogenous in nature are some major downsides associated with it. Asymmetric organocatalysts with recycling properties and effective cost can serve the best for this purpose. Therefore, designing of highly effective heterogeneous asymmetric organocatalyst was highly desirable to overcome the aforementioned drawbacks. There are chiefly two approaches for design and construction; the first one includes the integration or incorporation of chiral auxiliaries upon heterogenous support, while the second one includes direct synthesis of heterogenous asymmetric organocatalysts using functional organic building blocks. The first one has a limitation of uneven and insufficient loading of chiral auxiliaries; while for the second one, it is very difficult to construct directly a chiral heterogeneous network. But, upon literature survey, it was found that chiral heterogeneous organocatalysts possess high stability, porosity, and crystalline properties.

In this context, Zhang et al. explored the multivariate strategy for the synthesis of COFs containing chiral pyrrolidine and imidazolidine catalysts for heterogeneous asymmetric organic transformations (Scheme 11.6) [48]. A series of highly stable and crystalline porous 2D CCOFs was synthesized using a dialdehyde (DMTA) and a mixture of triamine with and without chiral organocatalysts. The systematic arrangement of chiral organocatalysts to the channel walls of porous skeleton acts as a three-component condensation system and served as a highly efficient

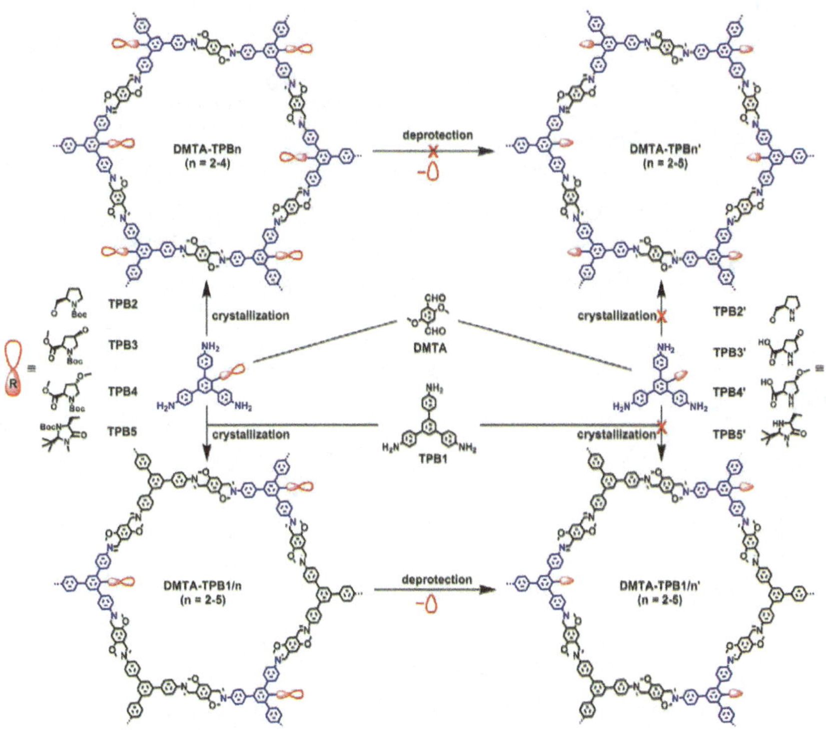

Scheme 11.6 Graphical illustration of the preparation of seven chiral covalent organic frameworks (CCOFs) [48]

hetero-organocatalyst for asymmetric aldol, Diels–Alder, and aminoxylation reactions even under harsh conditions. These metal-free CCOFs catalysts show high stereoselectivity and diastereoselectivity over their organocatalytic building blocks.

In 2017, Hui et al. published the preparation of a homochiral triazine-based COF via S-(+)-2-methylpiperazine and cyanuric chloride through nucleophilic substitution reaction [49]. This chiral framework (CCOF) was immobilized with Pd nanoparticles (NPs) and termed as Pd@CCOF-MPC (Fig. 11.11a). This newly synthe-

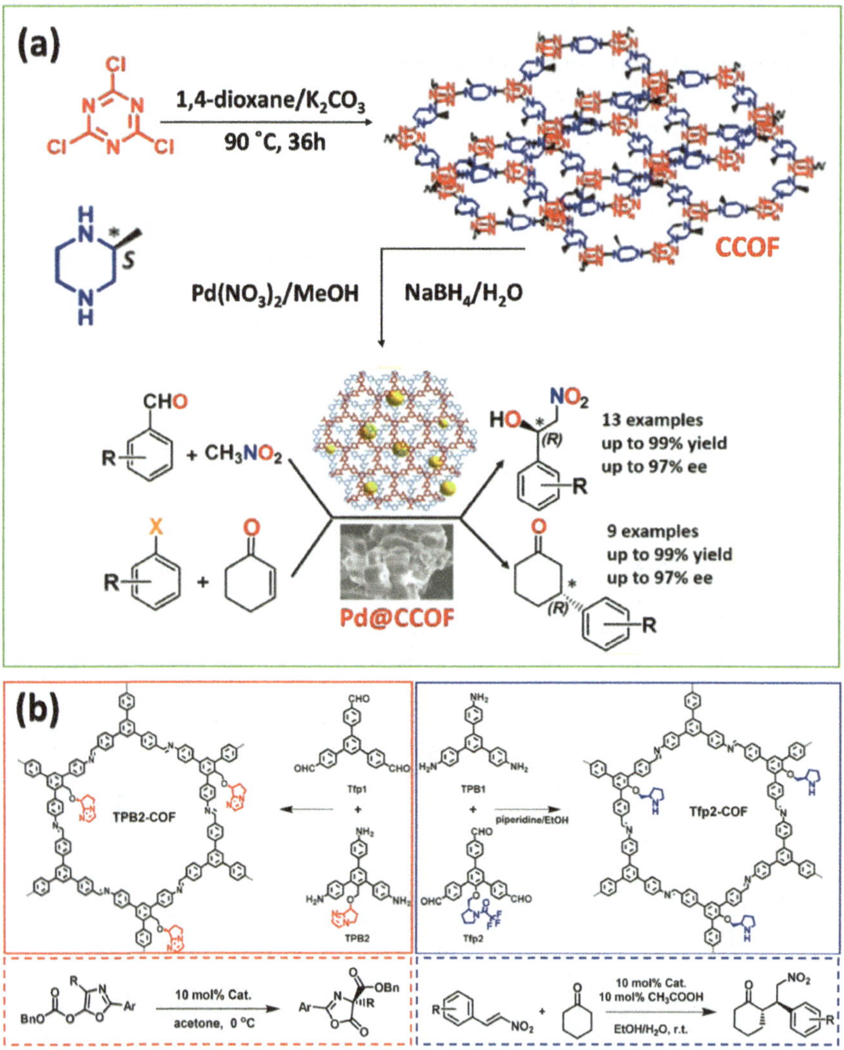

Fig. 11.11 **a** Schematics for the formation of CCOF and Pd@CCOF; asymmetric Henry reactions of different aldehydes with nitromethane [49] and **b** schematic illustration of chiral DHIP and pyrrolidine-based TPB2-COF and Tfp2-COF; their catalytic applications [50]

sized Pd@CCOF-MPC was utilized as an asymmetric heterogeneous catalyst for Henry reaction and reductive Heck reaction. The nitrogen-rich skeleton of CCOF substantiates the loading and stability of Pd NPs such transformations.

Recently in 2019, Jie. et al. utilized chiral organocatalysts as precursors to prepare two chiral COFs named TPB2-COF and Tfp2-COF as shown in Fig. 11.11b [50]. The systematic arrangement of chiral catalytic sites to the porous skeleton served as a highly efficient hetero-organocatalyst for asymmetric Steglich rearrangement and Michael addition reaction. The metal-free chiral catalysts were found highly efficient for both the reactions in terms of activity, enantioselectivity, diastereoselectivity, and reusability. This is the first report demonstrating the asymmetric Steglich rearrangement with a solid heterogeneous catalyst with conversion around 85% and enantiomeric excess (ee) around 80%. The asymmetric Michael addition took place with high stereoselectivity and diastereoselectivity.

5.4 *Cycloaddition of CO_2 to Epoxides*

The rising CO_2 releases from various sources triggering toward environmental illness so better technical assistance is much needed effective utilization of pollutants [51]. The direct utilization of CO_2 gas for the beneficial chemical synthesis of epoxides has drawn substantial attention in terms of atom economical synthesis and abundant source of CO_2.[52] The cycloaddition of CO_2 to epoxides results in industrially significant cyclic carbonate products having tremendous applications in various fields [53]. The high activation energy barrier associated with CO_2 activation needs a suitable catalyst, so a simple and heterogenous catalyst with appropriate functional moieties is extremely worthy to be explored [54]. In 2017, Yongfeng et al. designed two metal-free covalent organic frameworks using triazine-based precursors and 2,5-dihydroxyterephthalaldehyde as the connector and named them COF-JLU6 and COF-JLU7 (Fig. 11.12a) [55]. The in-built catalytic sites due to N, O-rich character and abundant H-bonding provided very high catalytic efficiency with a broad substrate scope under mild reaction conditions. The recyclability examination confirmed the high recyclability of the newly synthesized catalyst up to seven runs with retention of its catalytic performance under the optimized reaction conditions. Besides, in 2018 only, Tao-Tao et al. reported a variety of cationic CTFs synthesized by ionothermal method from 1,3- Bis(4-cyanophenyl)imidazolium chloride (BCI-Cl) (Fig. 11.12b) [56]. Due to this exclusive ionic character, they exhibited enhanced CO_2 uptake capacity and effective catalysis in the cycloaddition of epoxides and CO_2 without the need for any additive.

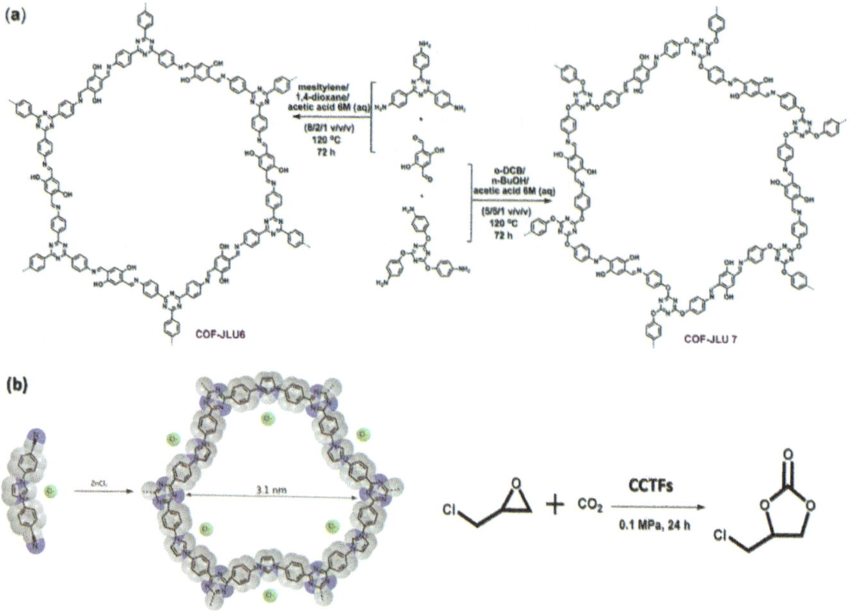

Fig. 11.12 **a** Synthetic procedure of COF-JLU6 and COF-JLU7 and **b** synthesis of the imidazolium-based CTFs and their application in CO_2 fixation [56]

5.5 Condensation and Multi-Component Reactions

MCRs (one-pot multi-component reactions) are an emergent class of reactions having numerous applications in the pharmaceuticals, medical, and chemical industries [57]. The very simplistic one-pot process converts two or more starting materials into the product in such a way that the majority of atoms are found in the final product without isolating the intermediate. Recently in 2020, Zare et al. fabricated a triazine-based COF with magnetic properties by covering it over Fe_3O_4 nanoparticles coated with chitosan (Fig. 11.13) [58]. The magnetically modified COF was finally embedded with Cu^{2+} to synthesize Fe_3O_4/CS/COF/Cu nanocomposite with magnetic properties. The material was utilized as an excellent heterogeneous catalyst in the unsymmetric Hantzsch process to produce polyhydroquinoline derivatives. High product yield, selectivity, recoverability, and reusability are some advantages of this modified COF.

Recently, Yao et al. reported a quinoline-linked COF via Povarov reaction of 2,5-dimethoxy terephthalaldehyde (DMTPA), 1-vinylimidazole (VIM), and 1,3,5-tri(4-aminophenyl)-benzene (TAPB) [59]. The COF-IM was further reacted with 1,3-propanesultone to obtain ionic liquid functionalities (Fig. 11.14). COF-IM-SO_3H serves as an efficient heterogeneous catalyst for multi-component Biginelli reactions because of these sulfonic acid catalytic sites. A broad range of products was synthesized by varying aromatic aldehydes under solvent-free reaction conditions with high efficiency (Fig. 11.14).

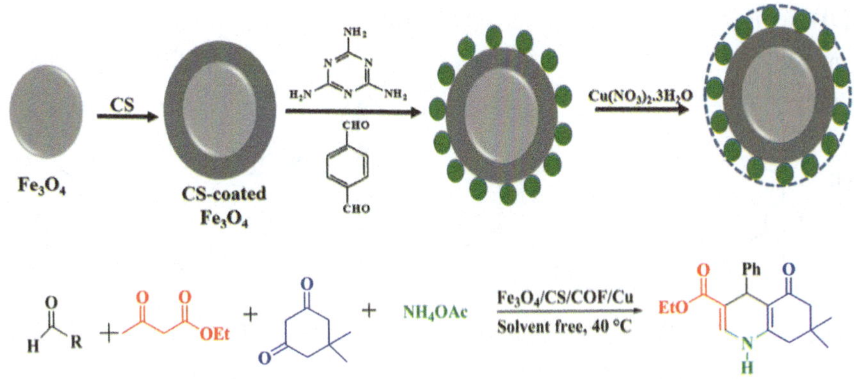

Fig. 11.13 Preparation of Fe_3O_4/CS/COF/Cu and its utilization in Hantzsch reaction [58]

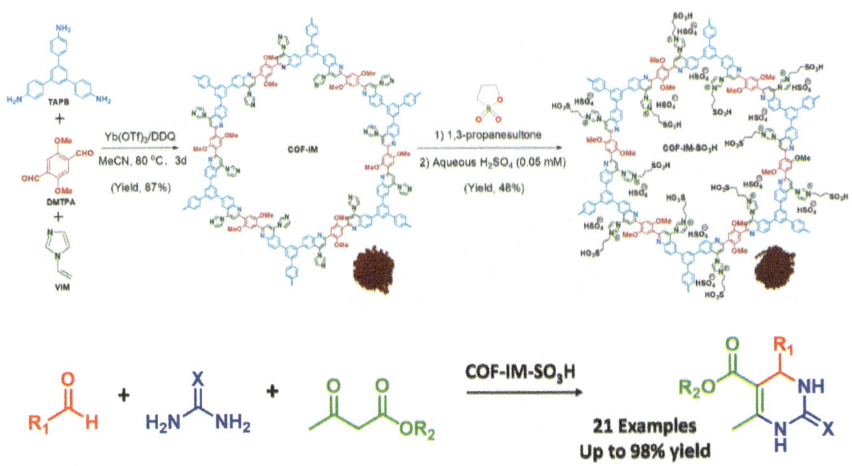

Fig. 11.14 Graphical illustration of the preparation of COF-IM and COF-IM-SO_3H and its application in Biginelli reaction [59]

Using a tetrahedral alkyl amine and two triangular precursors, triformylphloroglucinol (TFP) and 1,3,5-triformylbenzene, Fang et al. synthesized two bases functionalized three-dimensional (3D) COFs in 2014 (Fig. 11.15a) [60]. With 96% and 98% conversion, the 3D BF-COF-1 and BF-COF-2 are used as highly effective size-selective organocatalysts for metal-free the Knoevenagel condensation process.

Likewise, Sun et al. demonstrated the catalytic potential of a bifunctional COF in aerobic oxidation–Knoevenagel condensation reactions [61]. A pyridine containing COF has been synthesized using 2,5-diaminopyridine and triformylphloroglucinol, and the resultant COF-TpPa-Py was partially immobilized with Pd species over pyridine units (Fig. 11.15b). The resultant porous and chemically stable catalyst having the advantage of Pd metal and basic pyridine functionalities exhibits superior catalytic

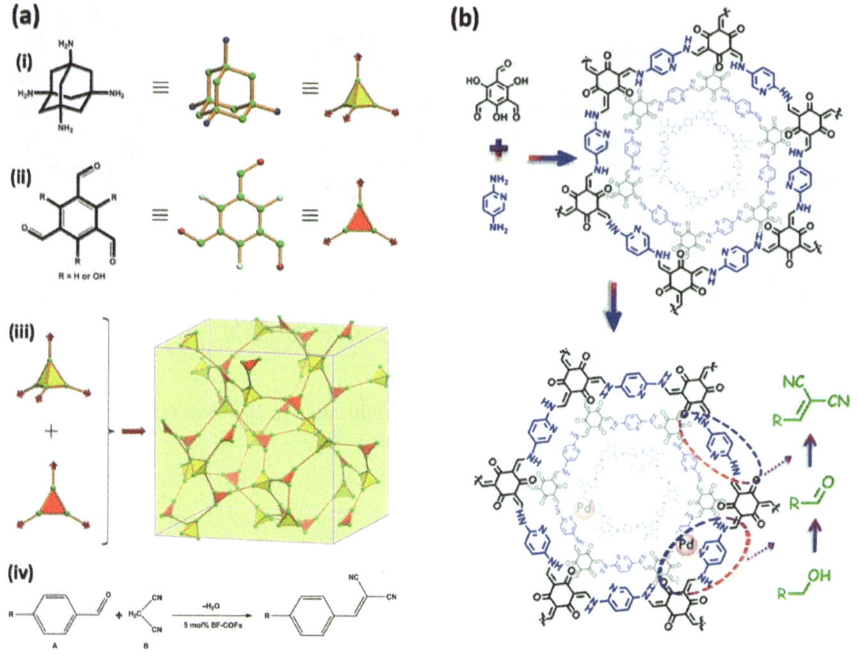

Fig. 11.15 **a** Graphical illustration of preparation of 3D COFs: (i) 3D tetrahedral analog of 1,3,5,7-tetraaminoadamantane, (ii) 3D triangular analog 1,3,5-triformylbenzene, (iii) 3D representation of BF-COF-1 or BF-COF-2, and (iv) their utilization in Knoevenagel condensation reaction and [60] **b** schematic representation of preparation of Pd/COF-TpPa-Py and its application [61]

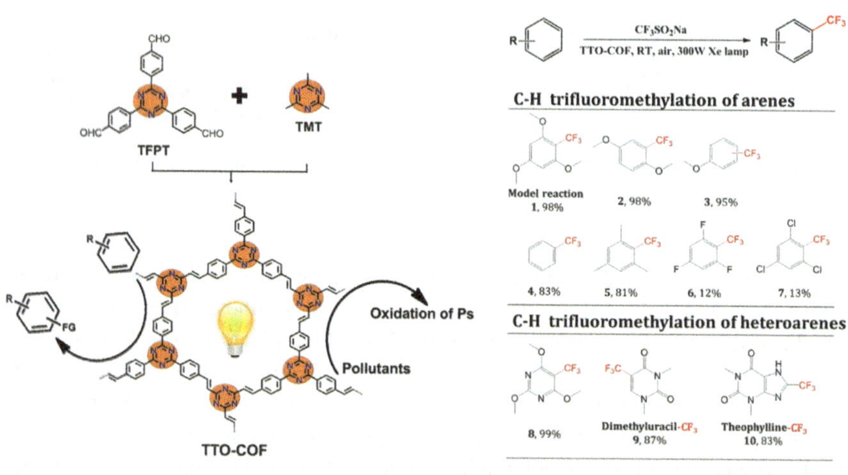

Fig. 11.16 Graphical representation of the preparation of TTO-COF and its application in photocatalytic C-H activation [68]

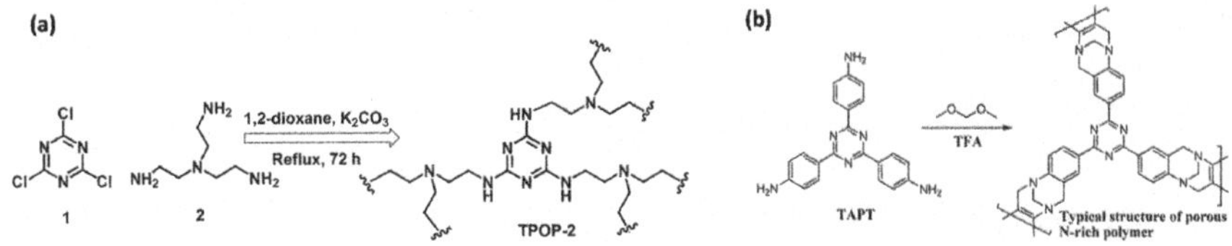

Scheme 11.7 **a** Synthetic procedure of TPOP-2 [62] and **b** synthetic procedure of N-rich polymer [63]

activity using Pd for conversion of alcohols to aldehyde, followed by Knoevenagel condensation with the remaining pyridine units.

In 2015, Bhaumik et al. synthesized a new triazine polymer (TPOP-2) using cyanuric chloride and tris(2-aminoethyl)amine, with high N-content due to amine and triazine functionalities (Scheme 11.7a) [62]. The surface basicity, insoluble nature, and high thermal stability provided excellent catalytic activity as an organocatalyst for the condensation reaction of an aromatic aldehyde with phenols and malononitrile to produce 2-amino-chromenes.

Likewise, in 2018, Yuanzheng et al. used 2,4,6-tris(4-aminophenyl)-s-triazine to prepare a porous N-rich polymer (Scheme 11.7b). With high reusability and recyclability, the polymer was used in the Knoevenagel condensation process of benzaldehyde to benzylidene malononitrile [63].

5.6 *Dye Degradation*

The developing textile and other industries including paper and leather have also increased the usage of synthetic dyes. Dyes are causing serious environmental threats due to their poor biodegradation. Therefore, the systematic removal of these harmful dyes is of utmost priority. Dyes after mixing into water sources obstruct the photosynthesis process of marine animals. Toward this end, various methods have been proposed such as sedimentation, biodegradation, advanced oxidation, carbon adsorption, and activated sludge, but these were reported with their low efficiency and high cost [64].

In 2018, Ning et al. synthesized a triazine-linked covalent organic polymer (COP-NT) using triazine substrate and naphthalene imide derivative through Schiff's condensation method (Scheme 11.8) [65]. This newly synthesized polymeric material exhibited well visible light absorption and semiconductor properties. The COP-NT was used to efficiently photocatalyze the degradation of methyl orange, rhodamine B, and methylene blue with their respective E_a values 9.40 $kJmol^{-1}$, 30.94 kJ^{-1} mol, or 17.54 $kJmol^{-1}$. Similarly, in 2020, Chen et al. used Knoevenagel condensation to prepare a vinylene-linked covalent triazine polymer (sp^2c-CTP-4) (Scheme 11.8b) [66]. The as-prepared sp^2c-CTP-4 exhibited considerable high semiconducting properties. Because of these characteristics, sp2c-CTP-4 was revealed to be an efficient

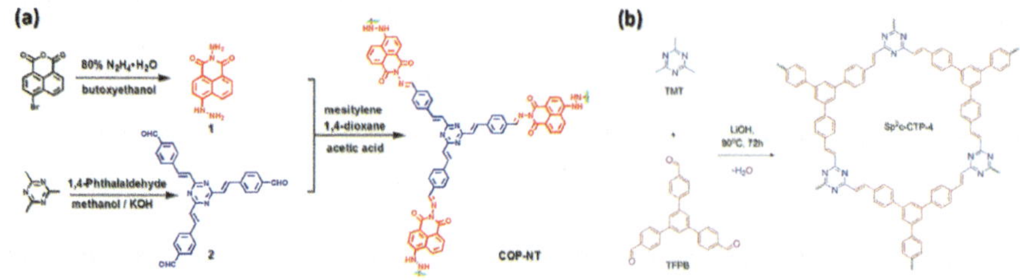

Scheme 11.8 **a** Preparation of COP-NT [65] and **b** schematic representation of the synthesis of sp^2c-CTP-4 [66]

photocatalyst for methylene blue degradation. Furthermore, sp^2c-CTP-4's greater chemical stability led to its high recyclability and photocatalytic retention up to five cycles.

5.7 Photocatalytic Applications

Owning to various key features of COFs, viz. everlasting porosity, highly crystalline nature, high thermal and chemical stability, periodic channels for charge conduction, and columnar p-arrays, COFs have been revealed to be an excellent material for various photocatalytic applications.

Using Schiff's base condensation process with a triazine precursor and 2,5-dimethoxyterephthaldehyde, Yongfeng et al. synthesized a 2D-COF in 2017 (Scheme 11.9) [67]. The high π-electron density, periodic nanochannels, and persis-

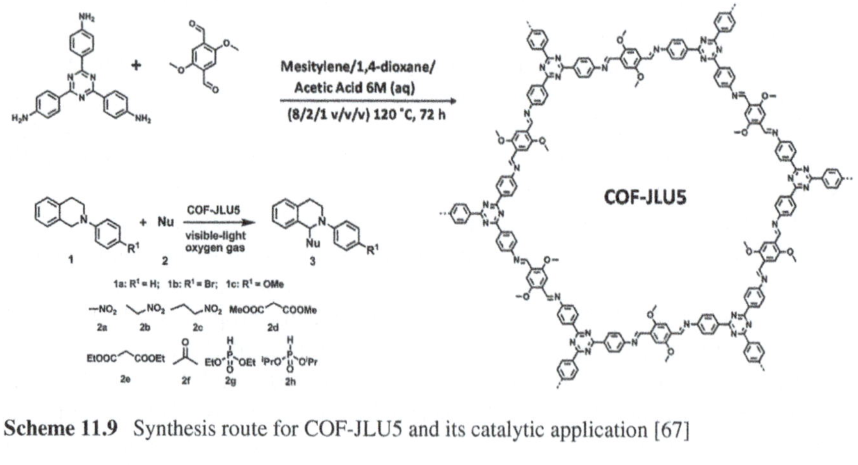

Scheme 11.9 Synthesis route for COF-JLU5 and its catalytic application [67]

tent porosity provided COF-JLU5 excellent photocatalytic activity for visible light-mediated oxidative C–H bond functionalization. The methodology was found advantageous in terms of photocatalytic sustainability, highly efficient metal-free approach, a broad scope of substrates, and recyclability.

In 2020, Yongliang et al. utilize an aldol condensation reaction to synthesize an olefin-linked triazine functionalized COF material using triazine building blocks (Fig. 11.16) [68]. The synthesis was accelerated by trifluoroacetic acid (TFA); thereafter, a completely sp^2 carbon-containing TTO-COF was obtained which was then found as a highly photosensitive semiconductor having an optical band gap of 2.46 eV. As a result, photocatalytic C–H bond functionalization of different arenes, as well as photodegradation of organic dyes, were carried out. The proficiency was reported in terms of higher efficiency and reusability in both applications.

5.8 *Electrocatalysis*

The increasing demand for energy from conventional energy sources has come up with hazardous environmental challenges. Therefore, a promising alternative to these challenges is electrochemical/photoelectrochemical reactions. In this regard, COFs are again extremely capable materials possessing the integrative effect of crystalline nature, high surface area, and unique porous skeleton.

Wenjie et al. demonstrated a cobalt–porphyrin-based COF synthesized under solvothermal conditions via Schiff's base condensation reaction subsequently followed by pyrolysis in N_2 atmosphere at 900 °C [69]. This newly synthesized material Co-COF-900 was then utilized for oxygen reduction reaction (ORR) in basic conditions (Scheme 11.10a). This catalyst was advantageous in terms of its comparatively better performance with commercial Pt/C and deprived of methanol crossover effect. Co-COF-900 verifies itself as an industry viable electrocatalyst by eliminating the use of precious noble metal Pt with high efficiency.

In 2018, Cang et al. prepared a new series of three cobalt immobilized COF materials using cobalt–phthalocyanine and boronic acid linker as basic building blocks (Scheme 11.10b) [70]. Among these three, the Co-Pc-PBBA was used for large-scale CO_2 utilization through an integrated effect of gas storage and catalytic reduction. The porous architect of Co-Pc-PBBA is highly competitive in CO_2 storage; therefore, the high concentration of CO_2 around the catalytic sites enhances the mass transport of CO_2 and delivers high efficacy in the reduction reaction. The comparative analysis reveals that Co-Pc-PBBA displays an over potential of 0.27 V and remarkably high productivity rate approximately increased by 97.7 times compared with aqueous CO making rate 97.7 times advanced than in aqueous medium.

In 2020, Royuela et. al. synthesized a new naphthalene-based diimide-linked covalent organic framework NDI-COF (Scheme 11.10c) [71]. The newly synthesized NDI-COF was efficaciously exfoliated into COF nanosheets architect (CONs) and utilized for oxygen reduction reaction (ORR) in basic conditions. The major highlights were reported as metal and pyrolysis-free ORR process, and this report

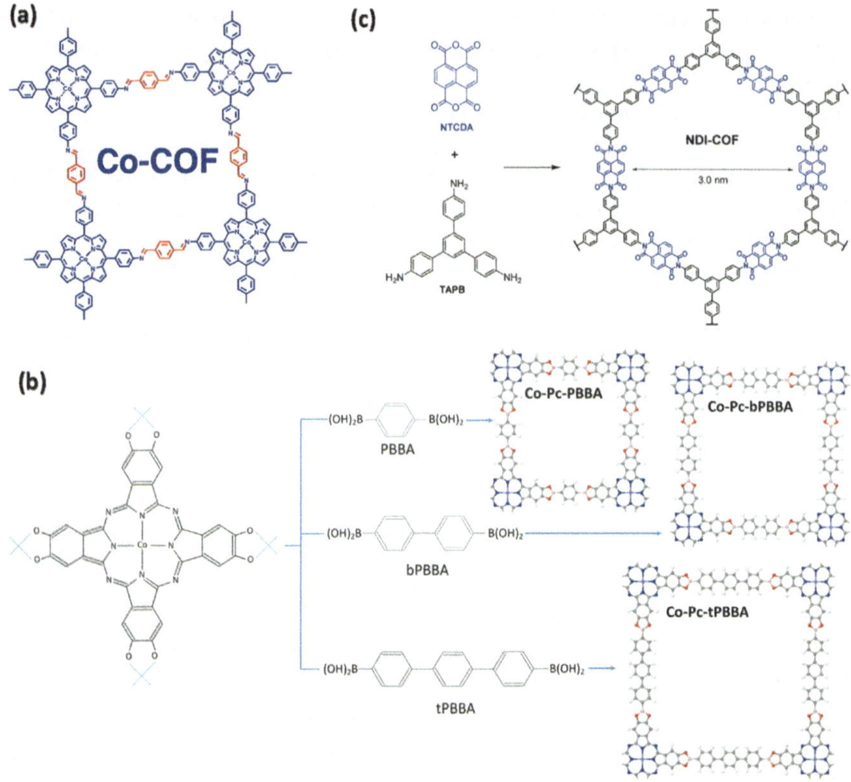

Scheme 11.10 **a** Schematic illustration of Co-COF [69], **b** synthesis of Co immobilized COFs and different boronic acid linkers as basic building blocks, and [70] **c** synthesis of NDI-COF [71]

demonstrates that an electroactive part of the COF material is solely accountable for its electrocatalytic ORR activity.

5.9 Other Catalytic Applications

In 2019, Vardhan et al. illustrate the synthesis of two vanadium decorated covalent organic frameworks by attaching VO to their hydroxyl functionalities [72]. Both the hybrid COFs were explored for their catalytic potential in Mannich-type reactions, Prins reaction, and oxidation of sulfides. The results for catalytic transformations substantiate their significance as highly efficient heterogeneous catalysts. The robust crystalline skeleton, stability, and reusability of these hybrid COFs assure their place in the industrial world (Fig. 11.17).

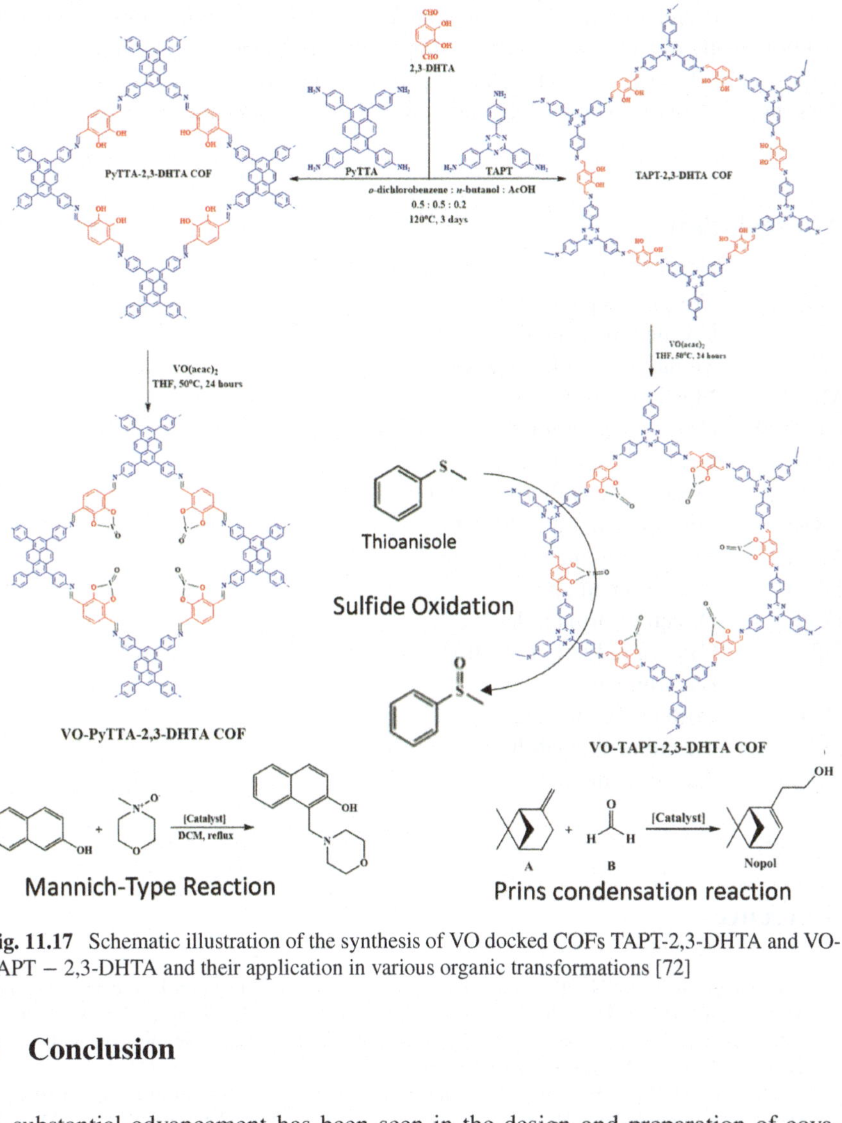

Fig. 11.17 Schematic illustration of the synthesis of VO docked COFs TAPT-2,3-DHTA and VO-TAPT – 2,3-DHTA and their application in various organic transformations [72]

6 Conclusion

A substantial advancement has been seen in the design and preparation of covalent organic frameworks and their applications. The literature survey of these hybrid materials has substantiated their significance in various applications. Catalytic organic transformation is one of the highly vibrant fields in the chemical industry, and new hybrid catalysts are continuously mounting to accomplish this demand through heterogeneous catalysis. Covalent organic frameworks and their functionalized derivatives are acclaimed support materials, and several reports are already available in literature where COFs have been utilized as extensive support materials for various metal/metal oxides. In trend, a diverse range of covalent organic frameworks has already been reported for heterogeneous catalysis applications. A broad scope of various organic transformations has been successfully implemented. In addition, many of COFs were used as support materials, showing significant advantages over traditional methods. This chapter has the primary objective to highlight the synthesis of new hybrid COF materials and their application in catalysis.

Abbreviations

COF	Covalent organic framework
CTF	Covalent triazine framework
MOF	Metal–organic framework
MNPs	Metal nanoparticles
CP-MAS	Cross-polarization magic angle spinning
NMR	Nuclear magnetic resonance
BDBA	Benzene-1,4-diboronic acid
PXRD	Powder X-ray diffraction
MCR	Multi-component reactions
DCM	Dichloromethane
DMF	Dimethylformamide
ORR	Oxygen reduction reaction
eV	Electronvolt
TFA	Trifluoroacetic acid
2D/3D	Two dimensional/three dimensional
TON	Turn over number

References

1. (a) Zhang Y, Riduan SN (2012) Functional porous organic polymers for heterogeneous catalysis. Chem Soc Rev 41(6):2083–2094; (b) Puthiaraj P, Lee Y-R, Zhang S, Ahn W-S (2016) Triazine-based covalent organic polymers: design, synthesis and applications in heterogeneous catalysis. J Mater Chem A 4(42):16288–16311
2. Fischbach DM, Rhoades G, Espy C, Goldberg F, Smith BJ (2019) Controlling the crystalline structure of imine-linked 3D covalent organic frameworks. Chem Commun 55(25):3594–3597
3. (a) Kuhn P, Antonietti M, Thomas A (2008) Porous, covalent triazine-based frameworks prepared by ionothermal synthesis. Angew Chem Int Ed 47(18):3450–3453. https://doi.org/10.1002/anie.200705710; (b) Roeser J, Kailasam K, Thomas A (2012) Covalent triazine frameworks as heterogeneous catalysts for the synthesis of cyclic and linear carbonates from carbon dioxide and epoxides. ChemSusChem 5(9):1793–1799
4. Zeng Y, Zou R, Zhao Y (2016) covalent organic frameworks for CO_2 capture. Adv Mater 28(15):2855–2873
5. McKeown NB, Budd PM (2006) Polymers of intrinsic microporosity (PIMs): organic materials for membrane separations, heterogeneous catalysis and hydrogen storage. Chem Soc Rev 35(8):675–683

6. Cao H-L, Huang H-B, Chen Z, Karadeniz B, Lü J, Cao R (2017) Ultrafine silver nanoparticles supported on a conjugated microporous polymer as high-performance nanocatalysts for nitrophenol reduction. ACS Appl Mater Interfaces 9(6):5231–5236
7. Mülhaupt R (2004) Hermann staudinger and the origin of macromolecular chemistry. Angew Chem Int Ed 43(9):1054–1063
8. Côté AP, Benin AI, Ockwig NW, Keeffe M, Matzger AJ, Yaghi OM (2005) Porous, crystalline, covalent organic frameworks. Science 310(5751):1166
9. (a) Gomes R, Bhanja P, Bhaumik A (2015) A triazine-based covalent organic polymer for efficient CO_2 adsorption. Chem Commun 51(49):10050–10053; (b) Xiang Z, Zhou X, Zhou C, Zhong S, He X, Qin C, Cao D (2012) Covalent-organic polymers for carbon dioxide capture. J Mater Chem A 22(42):22663–22669
10. (a) Subodh, Prakash K, Masram DT (2020) A reversible chromogenic covalent organic polymer for gas sensing applications. Dalton Trans 49(4):1007–1010; (b) Subodh, Prakash K, Masram DT (2020) Chromogenic covalent organic polymer-based microspheres as solid-state gas sensor. J Mater Chem C 8(27):9201–9204
11. (a) Subodh, Prakash K, Chaudhary K, Masram DT (2020) A new triazine-cored covalent organic polymer for catalytic applications. Appl Catal A 593:117411; (b) Yadav D, Awasthi SK (2020) An unsymmetrical covalent organic polymer for catalytic amide synthesis. Dalton Trans 49(1):179–186
12. Shi Y, Fu Q, Li J, Liu H, Zhang Z, Liu T, Liu Z (2020) Covalent organic polymer as a carborane carrier for imaging-facilitated boron neutron capture therapy. ACS Appl Mater Interfaces 12(50):55564–55573
13. Patra BC, Khilari S, Manna RN, Mondal S, Pradhan D, Pradhan A, Bhaumik A (2017) A Metal-free covalent organic polymer for electrocatalytic hydrogen evolution. ACS Catal 7(9):6120–6127
14. Jejurkar VP, Yashwantrao G, Saha S (2020) Tröger's base functionalized recyclable porous covalent organic polymer (COP) for dye adsorption from water. New J Chem 44(28):12331–12342
15. Julkapli NM, Bagheri S (2015) Graphene supported heterogeneous catalysts: an overview. Int J Hydrogen Energy 40(2):948–979
16. Subodh, Mogha NK, Chaudhary K, Kumar G, Masram DT (2018) Fur-imine-functionalized graphene oxide-immobilized copper oxide nanoparticle catalyst for the synthesis of xanthene derivatives. ACS Omega 3(11):16377–16385
17. (a) Chaudhary K, Subodh, Prakash K, Mogha NK, Masram DT (2020) Fruit waste (Pulp) decorated CuO NFs as promising platform for enhanced catalytic response and its peroxidase mimics evaluation. Arab J Chem 13(4):4869–4881; (b) Kumar G, Mogha NK, Kumar M, Subodh, Masram DT (2020) NiO nanocomposites/rGO as a heterogeneous catalyst for imidazole scaffolds with applications in inhibiting the DNA binding activity. Dalton Trans 49(6):1963–1974
18. Subodh, Chaudhary K, Prakash K, Masram DT (2020) TiO_2 nanoparticles immobilized organo-reduced graphene oxide hybrid nanoreactor for catalytic applications. Appl Surf Sci 509:144902
19. (a) Liu Y, Zhou W, Teo WL, Wang K, Zhang L, Zeng Y, Zhao Y (2020) Covalent-organic-framework-based composite materials. Chem 6(12):3172–3202; (b) Yadav D, Awasthi SK (2020) A Pd NP-confined novel covalent organic polymer for catalytic applications. New J Chem 44(4):1320–1325
20. Hu H, Yan Q, Ge R, Gao Y (2018) Covalent organic frameworks as heterogeneous catalysts. Chin J Catal 39(7):1167–1179
21. Yadav D, Dixit AK, Raghothama S, Awasthi SK (2020) Ni nanoparticle-confined covalent organic polymer directed diaryl-selenides synthesis. Dalton Trans 49(35):12266–12272
22. Ren S, Bojdys MJ, Dawson R, Laybourn A, Khimyak YZ, Adams DJ, Cooper AI (2012) Porous, fluorescent, covalent triazine-based frameworks via room-temperature and microwave-assisted synthesis. Adv Mater 24(17):2357–2361

23. Campbell NL, Clowes R, Ritchie LK, Cooper AI (2009) Rapid microwave synthesis and purification of porous covalent organic frameworks. Chem Mater 21(2):204–206
24. Ji W, Guo Y-S, Xie H-M, Wang X, Jiang X, Guo D-S (2020) Rapid microwave synthesis of dioxin-linked covalent organic framework for efficient micro-extraction of perfluorinated alkyl substances from water. J Hazard Mater 397:122793
25. Yang S-T, Kim J, Cho H-Y, Kim S, Ahn W-S (2012) Facile synthesis of covalent organic frameworks COF-1 and COF-5 by sonochemical method. RSC Adv 2(27):10179–10181
26. Biswal BP, Chandra S, Kandambeth S, Lukose B, Heine T, Banerjee R (2013) Mechanochemical synthesis of chemically stable isoreticular covalent organic frameworks. J Am Chem Soc 135(14):5328–5331
27. Mu M, Wang Y, Qin Y, Yan X, Li Y, Chen L (2017) Two-Dimensional imine-linked covalent organic frameworks as a platform for selective oxidation of olefins. ACS Appl Mater Interfaces 9(27):22856–22863
28. Bai L, Phua SZF, Lim WQ, Jana A, Luo Z, Tham HP, Zhao L, Gao Q, Zhao Y (2016) Nanoscale covalent organic frameworks as smart carriers for drug delivery. Chem Commun 52(22):4128–4131
29. Dinari M, Hatami M (2019) Novel N-riched crystalline covalent organic framework as a highly porous adsorbent for effective cadmium removal. J Environ Chem Eng 7(1):102907
30. Wang P-L, Ding S-Y, Zhang Z-C, Wang Z-P, Wang W (2019) Constructing robust covalent organic frameworks via multicomponent reactions. J Am Chem Soc 141(45):18004–18008
31. Acharjya A, Longworth-Dunbar L, Roeser J, Pachfule P, Thomas A (2020) Synthesis of vinylene-linked covalent organic frameworks from acetonitrile: combining cyclotrimerization and aldol condensation in one pot. J Am Chem Soc 142(33):14033–14038
32. Wang K, Jia Z, Bai Y, Wang X, Hodgkiss SE, Chen L, Chong SY, Wang X, Yang H, Xu Y, Feng F, Ward JW, Cooper AI (2020) Synthesis of stable thiazole-linked covalent organic frameworks via a multicomponent reaction. J Am Chem Soc 142(25):11131–11138
33. Lin G, Gao C, Zheng Q, Lei Z, Geng H, Lin Z, Yang H, Cai Z (2017) Room-temperature synthesis of core–shell structured magnetic covalent organic frameworks for efficient enrichment of peptides and simultaneous exclusion of proteins. Chem Commun 53(26):3649–3652
34. He S, Zeng T, Wang S, Niu H, Cai Y (2017) Facile synthesis of magnetic covalent organic framework with three-dimensional bouquet-like structure for enhanced extraction of organic targets. ACS Appl Mater Interfaces 9(3):2959–2965
35. Wang R, Chen Z (2017) A covalent organic framework-based magnetic sorbent for solid phase extraction of polycyclic aromatic hydrocarbons, and its hyphenation to HPLC for quantitation. Microchim Acta 184(10):3867–3874
36. Lim H, Cha MC, Chang JY (2012) Preparation of microporous polymers based on 1,3,5-Triazine units showing high CO_2 adsorption capacity. Macromol Chem Phys 213(13):1385–1390
37. Xiong S, Fu X, Xiang L, Yu G, Guan J, Wang Z, Du Y, Xiong X, Pan C (2014) Liquid acid-catalysed fabrication of nanoporous 1,3,5-triazine frameworks with efficient and selective CO_2 uptake. Polym Chem 5(10):3424–3431
38. (a) Tian Y, Meng X, Duan J-y, Shi L (2012) A novel application of methanesulfonic acid as catalyst for the alkylation of olefins with aromatics. Ind Eng Chem Res 51(42):13627–13631; (b) Gernon MD, Wu M, Buszta T, Janney P (1999) Environmental benefits of methanesulfonic acid . comparative properties and advantages. Green Chem 1(3):127–140
39. Ren S, Dawson R, Laybourn A, Jiang J-X, Khimyak Y, Adams DJ, Cooper AI (2012) Functional conjugated microporous polymers: from 1,3,5-benzene to 1,3,5-triazine. Polym Chem 3(4):928–934
40. Mizuno N, Misono M (1998) Heterogeneous catalysis. Chem Rev 98(1):199–218
41. Kaleeswaran D, Antony R, Sharma A, Malani A, Murugavel R (2017) Catalysis and CO_2 capture by palladium-incorporated covalent organic frameworks. ChemPlusChem 82(10):1253–1265
42. Tahir N, Wang G, Onyshchenko I, De Geyter N, Leus K, Morent R, Van Der Voort P (2019) High-nitrogen containing covalent triazine frameworks as basic catalytic support for the Cu-catalyzed Henry reaction. J Catal 375:242–248

43. Hartikainen H (2005) Biogeochemistry of selenium and its impact on food chain quality and human health. J Trace Elem Med Biol 18(4):309–318
44. Balasubramanian P, Balamurugan TST, Chen S-M, Chen T-W (2019) Simplistic synthesis of ultrafine $CoMnO_3$ nanosheets: an excellent electrocatalyst for highly sensitive detection of toxic 4-nitrophenol in environmental water samples. J Hazard Mater 361:123–133
45. Ansari A, Badhe RA, Garje SS (2019) Preparation of CdS–TiO_2-based palladium heterogeneous nanocatalyst by solvothermal route and its catalytic activity for reduction of nitroaromatic compounds. ACS Omega 4(12):14937–14946
46. Fan M, Wang WD, Zhu Y, Sun X, Zhang F, Dong Z (2019) Palladium clusters confined in triazinyl-functionalized COFs with enhanced catalytic activity. Appl Catal B 257:117942
47. Yadav D, Awasthi SK (2020) A Pd confined hierarchically conjugated covalent organic polymer for hydrogenation of nitroaromatics: catalysis, kinetics, thermodynamics and mechanism. Green Chem 22(13):4295–4303
48. Zhang J, Han X, Wu X, Liu Y, Cui Y (2017) Multivariate chiral covalent organic frameworks with controlled crystallinity and stability for asymmetric catalysis. J Am Chem Soc 139(24):8277–8285
49. Ma H-C, Kan J-L, Chen G-J, Chen C-X, Dong Y-B (2017) Pd NPs-loaded homochiral covalent organic framework for heterogeneous asymmetric catalysis. Chem Mater 29(15):6518–6524
50. Zhang J, Han X, Wu X, Liu Y, Cui Y (2019) Chiral DHIP- and pyrrolidine-based covalent organic frameworks for asymmetric catalysis. ACS Sustain Chem Eng 7(5):5065–5071
51. Siriwardane RV, Shen M-S, Fisher EP (2003) Adsorption of CO_2, N_2, and O_2 on natural zeolites. Energy Fuels 17(3):571–576
52. Lenden P, Ylioja PM, González-Rodríguez C, Entwistle DA, Willis MC (2011) Replacing dichloroethane as a solvent for rhodium-catalysed intermolecular alkyne hydroacylation reactions: the utility of propylene carbonate. Green Chem 13(8):1980–1982
53. (a) Decortes A, Castilla AM, Kleij AW (2014) Salen-complex-mediated formation of cyclic carbonates by cycloaddition of CO_2 to epoxides. Angew Chem Int Ed 49(51):9822–9837; (b) Kim SH, Ahn D, Go MJ, Park MH, Kim M, Lee J, Kim Y (2014) Dinuclear aluminum complexes as catalysts for cycloaddition of CO_2 to epoxides. Organometallics 33(11):2770–2775
54. (a) Huang J-W, Shi M (2003) Chemical fixation of carbon dioxide by NaI/PPh_3/PhOH. J Org Chem 68(17):6705–6709; (b) Yamaguchi K, Ebitani K, Yoshida T, Yoshida H, Kaneda K (1999) Mg–Al mixed oxides as highly active acid–base catalysts for cycloaddition of carbon dioxide to epoxides. J Am Chem Soc 121(18):4526–4527; (c) Yasuda H, He L-N, Takahashi T, Sakakura T (2006) Non-halogen catalysts for propylene carbonate synthesis from CO_2 under supercritical conditions. Appl Catal A 298:177–180; (d) Kim HS, Kim JJ, Lee SD, Lah MS, Moon D, Jang HG (2003) New mechanistic insight into the coupling reactions of CO_2 and epoxides in the presence of zinc complexes. Chem Eur J 9(3):678–686
55. Verma S, Kumar G, Ansari A, Kureshy RI, Khan N-UH (2017) A nitrogen rich polymer as an organo-catalyst for cycloaddition of CO_2 to epoxides and its application for the synthesis of polyurethane. Sustain Energy Fuels 1(7):1620–1629
56. Liu T-T, Xu R, Yi J-D, Liang J, Wang X-S, Shi P-C, Huang Y-B, Cao R (2018) Imidazolium-based cationic covalent triazine frameworks for highly efficient cycloaddition of carbon dioxide. ChemCatChem 10(9):2036–2040
57. Ramón DJ, Yus M (2005) Asymmetric multicomponent reactions (AMCRs): the new frontier. Angew Chem Int Ed 44(11):1602–1634
58. Zare E, Rafiee Z (2020) Magnetic chitosan supported covalent organic framework/copper nanocomposite as an efficient and recoverable catalyst for the unsymmetrical hantzsch reaction. J Taiwan Inst Chem Eng 116:205–214
59. Yao B-J, Wu W-X, Ding L-G, Dong Y-B (2021) Sulfonic acid and ionic liquid functionalized covalent organic framework for efficient catalysis of the Biginelli reaction. J Org Chem 86(3):3024–3032
60. Fang Q, Gu S, Zheng J, Zhuang Z, Qiu S, Yan Y (2014) 3D microporous base-functionalized covalent organic frameworks for size-selective catalysis. Angew Chem 126(11):2922–2926

61. Sun Q, Aguila B, Ma S (2017) A bifunctional covalent organic framework as an efficient platform for cascade catalysis. Mater Chem Front 1(7):1310–1316
62. Kundu SK, Bhaumik A (2015) A triazine-based porous organic polymer: a novel heterogeneous basic organocatalyst for facile one-pot synthesis of 2-amino-4H-chromenes. RSC Adv 5(41):32730–32739
63. Cui Y, Du J, Liu Y, Yu Y, Wang S, Pang H, Liang Z, Yu J (2018) Design and synthesis of a multifunctional porous N-rich polymer containing s-triazine and Tröger's base for CO_2 adsorption, catalysis and sensing. Polym Chem 9(19):2643–2649
64. Yadav D, Awasthi S-K (2021) Ni nanoparticle-immobilized imine-linked microspherical covalent organic polymer for degradation studies of organic dyes. ACS Appl Polym Mater 3(11):5460–5469
65. Xu N, Wang R-L, Li D-P, Meng X, Mu J-L, Zhou Z-Y, Su Z-M (2018) A new triazine-based covalent organic polymer for efficient photodegradation of both acidic and basic dyes under visible light. Dalton Trans 47(12):4191–4197
66. Xu C, Xie Q, Zhang W, Xiong S, Pan C, Tang J, Yu G (2020) A vinylene-bridged conjugated covalent triazine polymer as a visible-light-active photocatalyst for degradation of methylene blue. Macromol Rapid Commun 41(7):2000006
67. Zhi Y, Li Z, Feng X, Xia H, Zhang Y, Shi Z, Mu Y, Liu X (2017) Covalent organic frameworks as metal-free heterogeneous photocatalysts for organic transformations. J Mater Chem A 5(44):22933–22938
68. Yang Y, Niu H, Xu L, Zhang H, Cai Y (2020) Triazine functionalized fully conjugated covalent organic framework for efficient photocatalysis. Appl Catal B 269:118799
69. Ma W, Yu P, Ohsaka T, Mao L (2015) An efficient electrocatalyst for oxygen reduction reaction derived from a Co-porphyrin-based covalent organic framework. Electrochem commun 52:53–57
70. Yao C-L, Li J-C, Gao W, Jiang Q (2018) An integrated design with new metal-functionalized covalent organic frameworks for the effective electroreduction of CO_2. Chem Eur J 24(43):11051–11058
71. Royuela S, Martínez-Periñán E, Arrieta MP, Martínez JI, Ramos MM, Zamora F, Lorenzo E, Segura JL (2020) Oxygen reduction using a metal-free naphthalene diimide-based covalent organic framework electrocatalyst. Chem Commun 56(8):1267–1270
72. (a) Vardhan H, Hou L, Yee E, Nafady A, Al-Abdrabalnabi MA, Al-Enizi AM, Pan Y, Yang Z, Ma S (2019) Vanadium docked covalent-organic frameworks: an effective heterogeneous catalyst for modified Mannich-type reaction. ACS Sustain Chem Eng 7 (5):4878–4888; (b) Vardhan H, Verma G, Ramani S, Nafady A, Al-Enizi AM, Pan Y, Yang Z, Yang H, Ma S (2019) Covalent organic framework decorated with vanadium as a new platform for Prins reaction and sulfide oxidation. ACS Appl Mater Interfaces 11(3):3070–3079

Designing, Synthesis, and Applications of Covalent Organic Frameworks (COFs) for Diverse Organic Reactions

Shefali Shukla, Abhay Gaur, and Shikha Gulati

Contents

1 Introduction 320
2 Diverse Strategies of Synthesis of COFs Using Alternate Energy Sources 321
3 Reticular Chemistry: Next Step in Chemistry 323
  3.1 Types of Linkages in the Coupling Reactions for Synthesizing COFs 325
4 Post-Synthetic Modification 334
5 Application of COFs in Catalysis 337
6 Conclusion and Future Perspectives 343
Abbreviations 344
References 345

Abstract Covalent Organic Frameworks (COFs), a new emerging class of crystalline porous materials fabricated by connecting the organic building blocks with strong covalent bonds to form extended structures. COFs are amongst the few synthetic superstructures which have a highly crystalline nature thus leading to ordered and well-arranged pores which are easy to characterize through common analytical techniques like Brunauer–Emmett–Teller (BET) and Powder X-Ray Diffraction (PXRD). COFs have emerged as an important and fascinating research topic in the last one and a half decades owing to their lower density, large surface area, versatile and robust nature due to strong and intact covalent linking, and easy-to-tune pore size. Thus, crystalline porous structures, with long-range orderliness and larger surface area, have made these functional-tailored materials attractive to the present researchers for the development of promising catalytic materials. COFs can act as nanoreactors which reduce the activation energies of the reactants by bringing them closer and providing the proper environment to react. Being light and insoluble in many solvents, they form an ideal contestant for heterogeneous catalysis. COFs are an excellent host to the metal NPs with minimum aggregation for effective cooperative catalytic activity. In the present chapter the basic concepts used in designing, the recent advancements in the field of synthesis, and the properties shown by these structures are discussed in detail. Furthermore, the applications of these highly porous catalysts in diverse organic transformations are described.

S. Shukla (✉) · A. Gaur · S. Gulati
Department of Chemistry, Sri Venkateswara College, University of Delhi, Delhi 110021, India

S. Gulati (ed.), *Metal-Organic Frameworks (MOFs) as Catalysts*,
https://doi.org/10.1007/978-981-16-7959-9_12

Keywords Covalent organic frameworks · Catalysts · Organic transformations · Crystalline porous structures · Reticular chemistry

1 Introduction

Covalent Organic Frameworks (COFs), the latest and emerging class of microporous materials, are built by integrating organic building blocks via strong covalent bonding to form extended 2D and 3D superstructures [23, 39, 54]. COFs are light materials composed of atoms like C, B, O, N and Si and show unique and unprecedented properties like robust chemical bonding, structural regularity, easy and tunable functionalization, high thermal stability, well-defined pores, larger surface areas, and many more. The specific conformations and morphologies produce a confined molecular space that can accommodate a variety of reactive species like photons, electrons, holes, ions, excitons, spins, and small molecules. This specificity provides novel macromolecular platforms for structural designs and functional development [39, 47]. These microporous materials have positional control over their molecular organic building blocks, which ascertain the synthesis of molecules with high specificity and regularity. Besides synthesizing the channels and micro spaces, COFs are explored extensively for wide applications in fields of catalysis, gas storage, adsorption of metal ions, sensors, drug delivery, optoelectronics, etc. Therefore, these unique properties have made COFs very popular not only among material scientists but also among specialized chemists and physicists. Apart from the application part, the latest research also focuses on unraveling new methods of tailoring novel COF systems and understanding the mechanism of arrangement and action [47, 52].

Creating precise and ordered multidimensional superstructures has been a challenge to science as rightly quoted by the Noble Laureate Prof. R. sn "In two or three dimensions, it's a synthetic wasteland" [44]. Such covalently linked multidimensional and precise superstructures are quite evident. Biopolymers like nucleic acids and proteins are formed as a result of covalent and non-covalent interactions to form controlled superstructures in which the covalent bonds define the primary structure and the non-covalent interactions precisely shape up the higher-ordered secondary structures [17, 38, 48, 60]. To date, the synthesis of molecules up to five monomers has been successfully achieved, but it still looks impossible to mimic the higher-ordered arrangements of biological superstructures through non-covalent interactions. Even polymerization has been of little help for creating such ordered and crystalline materials as the conventional techniques of organic polymerization lead to amorphous materials. Such tailor-made synthetic materials had always been a great challenge to the scientists [36, 95, 111]. Moreover, extended covalent networks are found throughout nature like diamonds but none was synthetically prepared until the phenomenal work from the group of Prof. Yaghi published in 2005 who synthesized a COF using light atoms like B, O, N. These COFs have emerged as a ground-breaking invention in the past 15 years which enforced this goal by incorporating covalent and non-covalent interactions in the polymerization systems [16]. COFs are the first kind of highly crystalline porous materials made up entirely of covalent bonds where the multidentate organic blocks are aligned in well-ordered sites with high precision, thus showing long-range ordered crystallinity. The crystallinity of COFs is well documented with experimental evidence mainly through PXRD data unlike other porous organic polymers (POPs) which are found to be amorphous. In 2018, the first structure of crystalline COF was also reported using single-crystal X-ray diffraction. The findings revealed the high-ordered crystalline arrangement in COFs which make them fascinating for the exploration of unique applications [82]. Since most of the COFs are insoluble in solvents, therefore, a salient feature of COF catalysts is easy separation using simple physical techniques like filtration and centrifugation followed by reactivation [24]. In the past one and a half-decade, this field has expanded a lot to the novel synthetic methods, generation and tuning of skeletal pores, and prospective applications including gas storage, catalysis, semi-conduction, sensing, etc. [37, 72, 73, 87, 129, 135]. This chapter provides a concise overview of the building blocks utilized and various applications, particularly in catalysis, of crystalline COFs to date. The pores and pore walls of these tailor-made COF skeletons are useful to act as domains for designing reaction-specific catalytic sites. These porous crystalline frameworks also offer a suitable platform for designing heterogeneous catalysis where the open nanopores provide a specific and confined space that has the potential to act as nanoreactors. As a result of the broad structural diversity and robust bonding, COFs allow the amalgamation of a wide range of tolerant catalytic sites. Asymmetric catalysis, photocatalysis, electrocatalysis, metal-based catalysis, and other types of catalysis have been developed and have shown the great potential of COF as the porous platform for exploring catalytic systems [14, 129, 130].

2 Diverse Strategies of Synthesis of COFs Using Alternate Energy Sources

There are several strategies to synthesize COFs using Alternate energy sources (Fig. 12.1).

The conventional methods to synthesize COFs were solvothermal where extended reaction time and high energies were required to push the reaction towards thermodynamically stable crystalline materials. Presently, with technical advancements, these reactions can be carried out with alternate sources of energy viz. microwave, ultrasound, mechanical agitation, photochemical conditions, and electron beam irradiation methods. These techniques not only enhance the overall kinetics of the reaction but also help in getting better yields [6, 36, 74, 107, 124].

1. **Microwave (MW)-Assisted Solvothermal Synthesis:** COF synthesis through Microwaves has attracted the attention of researchers as a cleaner alternate that provides better yields, demands lower energy, and lesser time. Therefore, a wide array of these crystalline materials can be synthesized more efficiently as compared to the conventional solvothermal processes [99, 100].

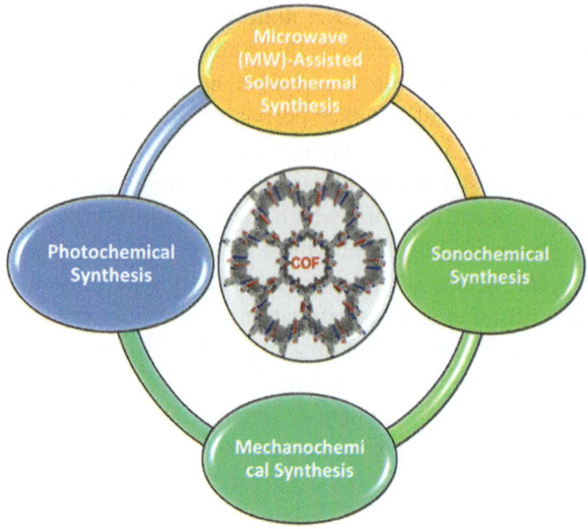

Fig. 12.1 Energy efficient alternate strategy to synthesize COFs

In 2009, the MW-assisted synthesis of boronated ester-linked COFs, COF-5, and COF-102, was reported by using microwave irradiations. As compared to the COFs synthesized by the conventional solvothermal process, the products formed by the MW process took about 200 times less time and had better surface area also allowing the trapped impurities to be removed with ease using a potent microwave extraction process [10]. Wei et al. first synthesized a β-ketoenamine linked COF (TpPa-COF-MW) using the Schiff base reaction between 1,3,5-triformylphloroglucinol (TPG) and p- phenylenediamine (Pa) under the microwave irradiation [120]. The reaction time reported for completion was 1 h. This method resulted in exceptionally high yields, better crystallinity, and higher surface area of the material which was also observed through the Brunauer–Emmett–Teller (BET) analysis. This β-ketoenamine linked COF showed higher adsorption of CO2 as compared to its solvothermal counterpart.

2. **Sonochemical Synthesis**: Synthesis using ultrasounds accelerates crystallization due to the acoustic cavitation which creates a high temperature and high-pressure condition in solutions [3, 43]. This alternate source of energy also features low energy requirements, faster reaction rates, and the overall low cost of the reaction. In 2012, the first ultrasonic process of COF synthesis of boron-based COFs (COF-1 and COF-5) was reported. The reaction reported was completed only in one hour. The resultant crystalline polymer exhibited a better BET area and 100 folds decrease in crystal size when compared with the COF obtained by sonothermal process [56].
3. **Mechanochemical Synthesis**: Mechanochemical synthesis involves manual grinding in solvent-free conditions [63]. That is the reason why this method is considered an eco-friendly and cost-effective alternative to conventional methodologies like solvothermal process [113, 118, 131]. In 2013, the first solventless COFs, TpPa-1, and TpPa-2, were synthesized by the manual grinding

method. Using this method, the reaction got completed in less than an hour and the products showed exceptional stability in acidic, alkaline, and boiling water conditions [6]. Later, by using the Liquid Assisted Grinding approach various COFs having linkages like β-ketoenamine, hydrazine, and imine were also synthesized and reported [4, 93, 128]. These fabricated COFs exhibited outstanding water adsorption property which was even better than commercial zeolites.

4. **Photochemical Synthesis**: In the last decade, the synthesis of organic material using high-energy photons has been studied extensively. The reaction rates observed in these processes were superfast and the energy requirements of the reactions were exceptionally low [107]. The first photochemical synthesis of COF-5, a sea urchin-shaped COF, was reported using UV irradiation in 2017. The reported reaction was performed at room temperature with a 48-fold increase in the kinetics [59]. Later, the sunlight-stimulated synthesis of pyrazine-fused COF (hcc-COF), electron beam irradiation-based synthesis of imine-linked EB-COF-1, and 2D COFs were also reported. It has been proposed and reported in the literature that the irradiation by electron beam probably "froze" the EB-COF-1 in its most stable conformation and thus led to enhanced crystallinity in the COF as compared to the one synthesized by the solvothermal process [58, 134].

3 Reticular Chemistry: Next Step in Chemistry

The word 'Reticulate' simply means something arranged like a net and 'Reticular Chemistry' or 'Reticular Synthesis' is primarily concerned with the controlled formation as well as cleavage of bonds in the molecules. As defined by Prof. Yaghi, 'Reticular Chemistry' can be described as the chemistry of joining modular building blocks and chemical linkers with strong bonds, like covalent linkages, to make extended crystalline structures [23]. In Reticular Chemistry, the groups are bonded together to form supermolecules in the form of extended networks or nets just like how graphite forms 2D layers and diamond contains tetrahedral units extended in 3D [127]. The developments in the area of building complex structures began with Gilbert Lewis' millennial paper "The Atom and the Molecules" which introduced the concept of covalent bonds [16, 23, 31, 64]. Since then, chemists have been exploiting the concepts to design molecules in zero-dimension with varying structures and complexities with incredible precision using retrosynthetic analysis. One milestone in chemistry was achieved when 1D polymers were synthesized. These polymers opened up the area of performance polymers. But these amorphous polymers could not provide extension into 2D and 3D because of lack of long-range order. One solution came out as supramolecular chemistry, where the idea of non-covalent interactions like hydrogen bonds, halogen bonds, van der Waal interactions was utilized as a guiding tool to form self-assembled superstructures [16, 21, 29, 79, 126]. However, they were found to suffer from the serious drawback of high sensitivity and lack of structural integrity. This posed a serious hurdle for their application in materials. However, nature also uses such a self-organization strategy to construct complex molecular architectures like proteins and nucleic acids. However, nature doesn't rely solely on weak non-covalent interactions, rather such complex architectures are built upon strong covalently bonded 1D-chain which is responsible to maintain the structural integrity on which the molecular building blocks like amino acids and saccharides are decorated. This inspired chemists to better look for a strategy that assembles the molecular building blocks in geometrically ordered arrangement and then links them by strong covalent bonds. Once done, then it becomes feasible to introduce various functionalities and complexities [34, 52, 84, 124, 126, 127] (Huang et al. 2016b). The work in the area of Reticular Chemistry started by connecting the metal ions to the charged organic linkers, e.g., carboxylates, which led to the development of metal organic frameworks where various building blocks were linked together in precise geometrical and spatial arrangements [25, 101]. Another development in the area was linking the organic building blocks to form extended 2D and 3D COFs (Fig. 12.2) [41, 79, 83]. This kind of chemistry is a remarkable feat in developing the discipline and moving closer to mimic nature to control the order and synthesis of extended crystalline organic supermolecules like diamond and graphite. In a broader sense, reticular chemistry is the next step forward in developing our understanding to synthesize materials in the form of frameworks. The developments in reticular chemistry, mainly in the past 15 years, have brought the synthesis of 2D and 3D extended frameworks to a logical approach for engineering materials using predesigned building blocks and linkers [126]. Owing to the relevance in academia and potential for applications, this field though new has seen tremendous growth in the last 15 years with more than a thousand publications that

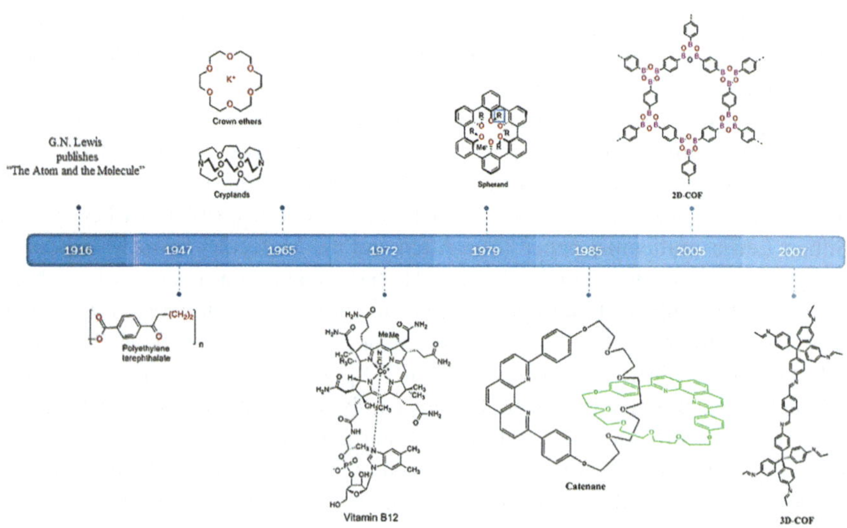

Fig. 12.2 Timeline: From the lewis structures to covalent organic framework

appeared in the last year respectively with the exploitation of COFs in a variety of applications and creative ways to make more and more complex structures.

3.1 Types of Linkages in the Coupling Reactions for Synthesizing COFs

The remarkable crystallinity in COFs is ensured by carrying out the synthesis via coupling of organic building blocks and organic linkers with strong but slightly reversible covalent bonding. The covalent bonds being strong keep the modular chemical blocks and linkers intact, thus providing robustness to the whole extended framework [34, 126, 127]. On the other end, the slight reversibility allows the possibility for checking errors and then rearranging the extended networks by cleaving and reforming the covalent bonds. This ensures the proper stacking of building blocks in an ordered fashion thus stabilizing the long-range ordered crystalline structure. A balance between thermodynamics and kinetics plays a crucial role in imparting crystallinity to the molecular framework [23, 52, 54]. Such ordered assembly led researchers to explore the reversible coupling reactions for synthesizing COFs with desired properties.

In the quest to design new COFs with desired properties, various monomers have been investigated which can be strategized to impart different types of chemical linkages in the framework and thus give variation in structures, pore size, and functionality. Amongst the wide range of COFs, those bearing boroxines and boronic ester, hydrazone, azine, imine, and β-ketoenamine are of particular interest. In the first COF, the reversibility of the five-membered cyclic boronic ester linkages and six-membered boroxines were reported (Fig. 12.3). Since then, various reactions have been used in the successful synthesis of novel COFs [39, 74]. As boroxines and boronic esters were found to be sensitive for hydrolysis, other strategies were explored like a cocondensation of primary aldehydes and amines to form Schiff bases which were more resistant to hydrolysis. Another scheme that was reported to enhance the stability towards hydrolysis included the formation of β-ketoenamine [55]. β-ketoenamine can be obtained by cocondensing primary amines with 1,3,5-triformylphloroglucinol (TPG) which subsequently undergoes enol-keto tautomerization which is an irreversible step. This yields a highly stable network that is resistant to acid/base hydrolysis. In recent years, scientists have explored various schemes to build COFs from organic building blocks and linkages as discussed below:

1. **Boroxines and Boronic ester linkages**: The very first COFs were synthesized by (i) cocondensing boronic acids with functionalized catechol to yield five-membered boronic ester rings and (ii) boronic esters undergo self-condensation to form a six-membered analogy of cyclohexane boroxine. These reactions provided the required reversibility hence ensuring the thermodynamic control over the kinetic in the coupling reaction. Following the strategy for the formation

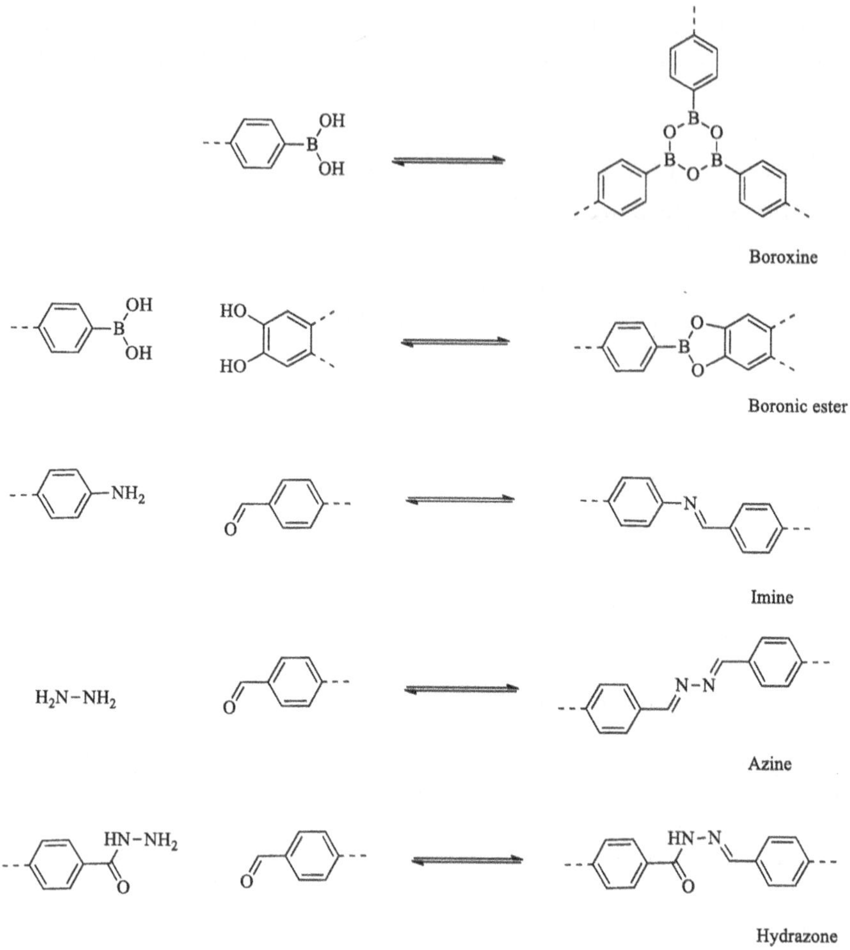

Fig. 12.3 Common coupling reactions for the synthesis of COFs

of the boronic acid linkages, Tilford et al. synthesized COF-18A by cocondensing triboronic acid (TBA) and 1,2,4,5-tetrahydroxybenzene (Fig. 12.4) [114]. 1,4-benzenediboronic acid (BDBA) condensed to form boroxine linkage via thermally assisted dehydration yielding COF-1 (Fig. 12.5) [20]. Though the eclipsed AA stacking was the most common one observed in the 2D materials, COF-1 formed a staggered AB packing in presence of solvents and transformed into eclipsed AA when solvent molecules were removed [16].

Further, a 3D-COF network was synthesized by coupling the tetrahedral boronic acid tetra (4-dihydroxyborylphenyl) silane (TBPS) via self-condensation as well as by condensing with HHTP, and these COFs were found to have larger surface areas [29]. Later, other researchers synthesized and reported a large number of boronic ester-linked COFs containing various

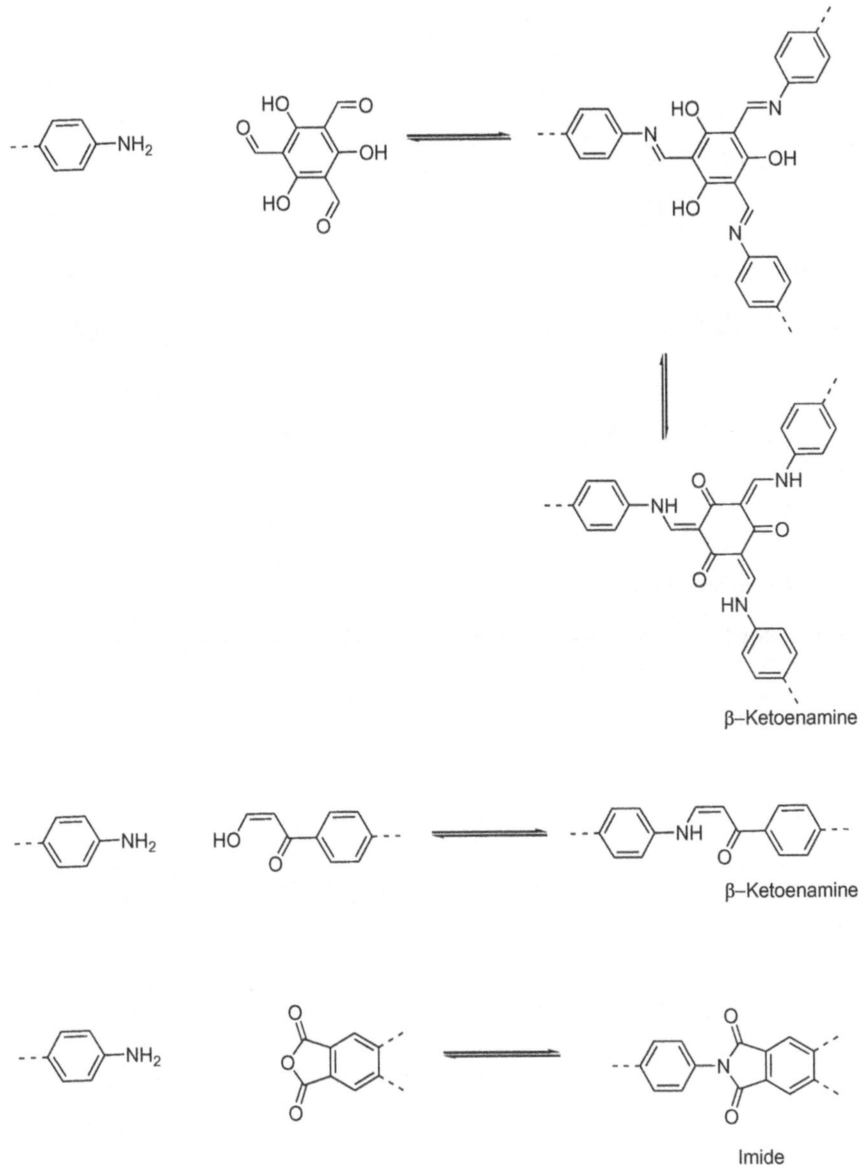

Fig. 12.3 (continued)

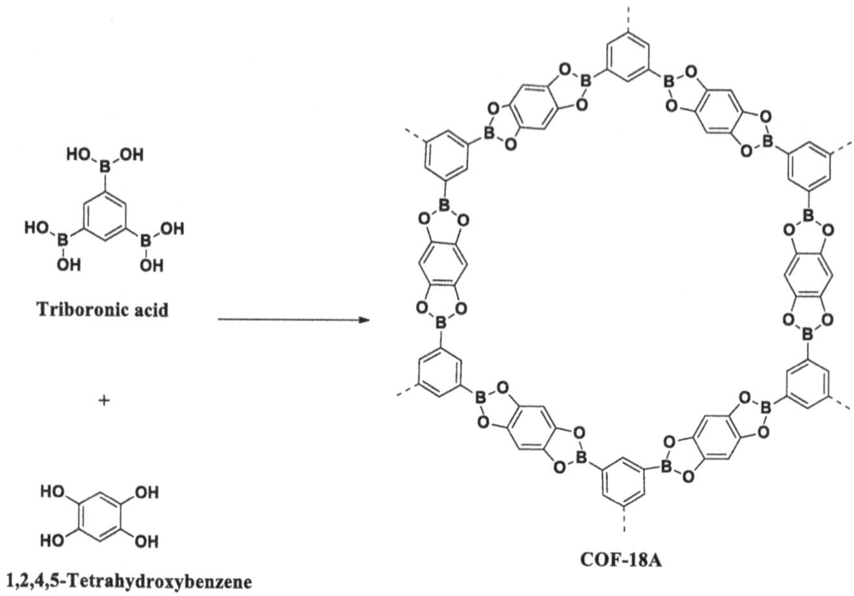

Fig. 12.4 Synthesis of COF-18A, a boroxine linked 2D-COF. Reproduced with permission from [114]. Copyright American Chemical Society 2006

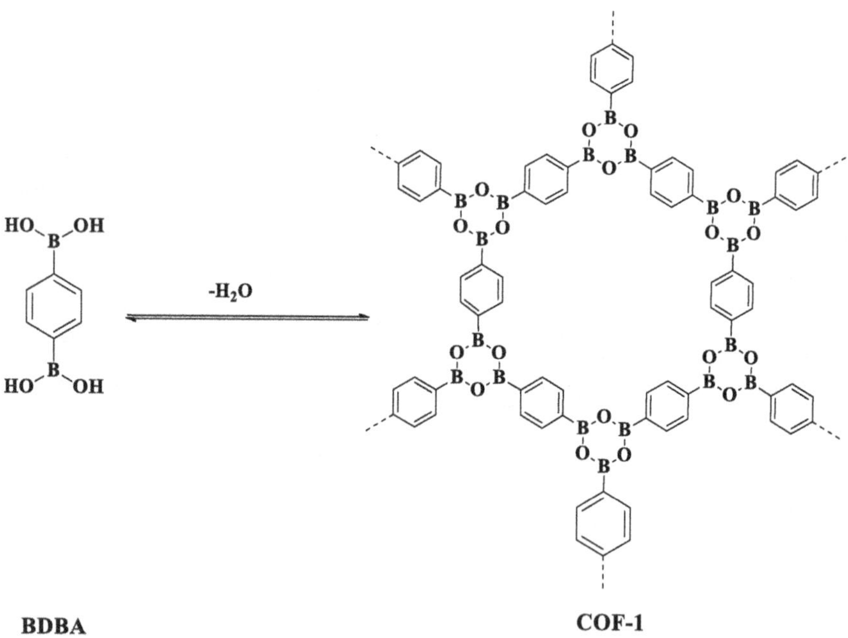

Fig. 12.5 Synthesis of COF-1 by the polymerization of diboronic acid BDBA to form boroxine linkages. Reproduced with permission from [20]. Copyright American Chemical Society 2011

heteroatoms like N, S, Se, etc. in the boronic acid backbone. These COFs were found to have improved properties and wider application potential [68]. For example, incorporating thiadiazole units in the diboronic acid linker yielded a framework with an electron donor (thiadiazole) and electron acceptor (HHTP) sites [33].

2. **Imine linkages**: These are formed by cocondensing an amine and a primary aldehyde. The introduction of imine linkage has been the most common strategy in the synthesis of COFs. These offer much better resistance to hydrolysis without compromising the crystallinity of the material. The first example of the synthesis of such COF was reported by Uribe-Romo et al., i.e., COF-300, which was made by cocondensing tetra-(4-anilyl) methane and tetraphthaldehyde (TA). The COF was crystallized in tetrahedral units to give a 3D COF [115]. Later, extended aldehydes like (3,3'-bipyridine)-6,6'-dicarbaldehyde were utilized as linkers to form 3D frameworks with flexible pores featuring the size-tuning capabilities upon the adsorption and removal of gas [68]. Liu et al. reported a simple room-temperature solution-phase method for fabricating mesoporous TPB-DMTP-COF by Schiff base chemistry of 1,3,5-tri-(4-aminophenyl)benzene (TAPB) with 2,5-dimethoxyterephthalaldehyde (DMTA) (Fig. 12.6) [71].

   Just like boronic ester COFs, various electron acceptor and donor groups like phthalocyanine, thiophene, and pyrene were incorporated by linear linkages to

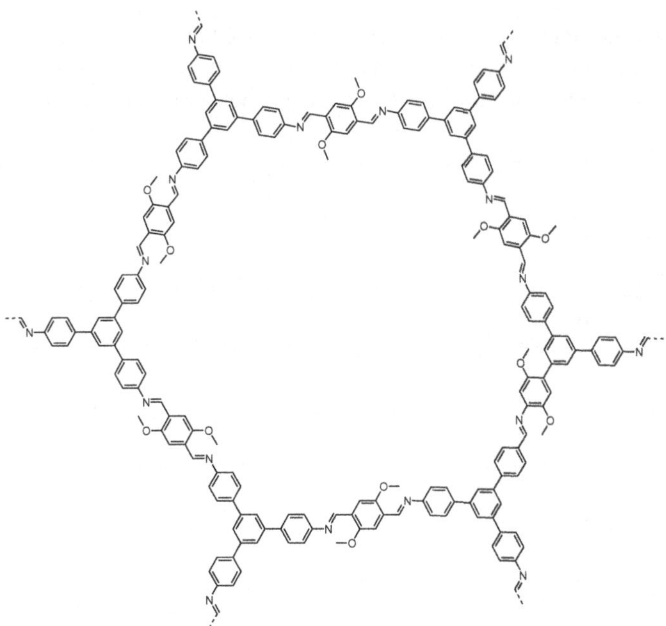

Fig. 12.6 Structure of mesoporous TPB-DMTP-COF depicting imine Linkages. Reproduced with permission from [71] Copyright Elsevier 2019

realize various applications related to charging carriers and optical properties [39, 74].

3. **Hydrazones and Azines**: Hydrazones, azines, and imines are similar linkages that are collectively referred to as Schiff bases [104]. These are reversible under milder acidic conditions and are formed as a result of condensation between aldehydes and the respective primary amines, e.g., COF-42 [55]. Hydrazones are obtained when hydrazides reversibly cocondense with aldehydes (Fig. 12.2).

The first hydrazone COF was synthesized using 2,5-diethoxyterephthalohydrazide and TFB (COF-42) and TFPB (COF-43) [116]. These COFs have been extensively studied for a wide range of applications specific to sensing and catalysis [12, 107, 119, 129]. When hydrazine is used as a linker to join two building blocks, a functionality called azine is established. This was first published by Dalapati et al. for the synthesis of rhombic py-azine (Figure 12.7) COF from hydrazine and 1,3,5,8-tetrakis(4-formylphenyl) pyrene. Azine linkage holds the advantage of providing a variety of pore geometries [19]. It can lead to the most common hexagonal type and may also result in the formation of heterogeneous dual pore structures. The combination of hexaphenylbenzene aldehyde with hydrazine yielded COF with triangular pores of only 1 nm [1]. A variety of such other knots were also reported, which included substituted benzene, triphenyl benzene, triphenyl triazine, and pyrene monomers. These monomers led to the synthesis of COFs with different morphologies [110].

4. **β-Ketoenamines**: Enhancing the stability of COFs is an important task and one of the strategies developed takes the advantage of the reversibility of imines which are synthesized by the cocondensation between aldehydes and 2o-amines

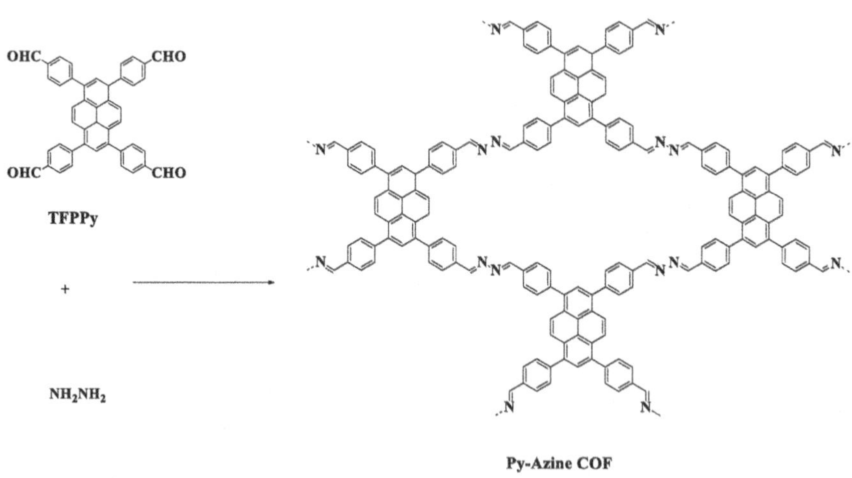

Fig. 12.7 Py-azine COF Synthesized from TFPPy and hydrazine. Reproduced with permission from [19]. Copyright American Chemical Society 2013

to connect organic linkers and the modular organic blocks in well-ordered assemblies. For example, in COF synthesis using TPG as an aldehyde linker where a hydroxyl group is present adjacent to the formyl groups, an imine is obtained. This imine tautomerizes irreversibly via enol-keto method to form an irreversible β-ketoenamine (Fig. 12.8). Though, lesser crystalline behavior was observed which was probably because of the irreversible step introduced

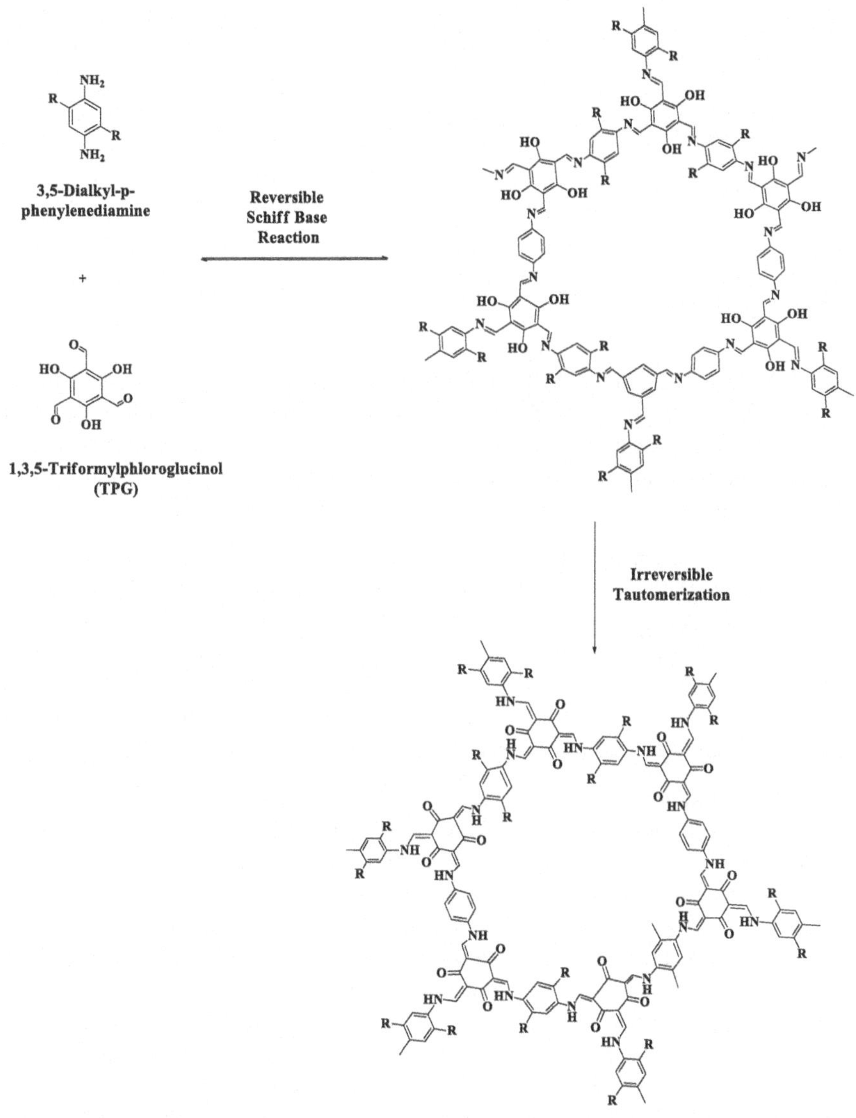

Fig. 12.8 Steps of formation of COF with β-ketoamine linkage. Reproduced with Permission from [55]. Copyright American Chemical Society 2012

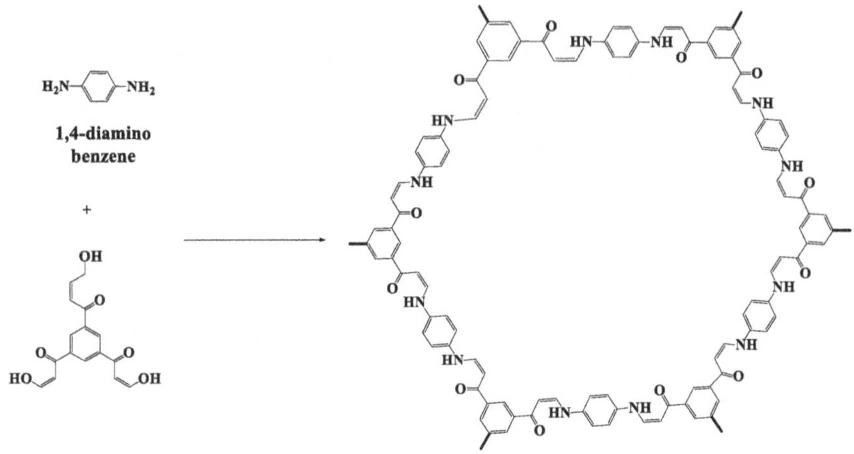

Fig. 12.9 Synthesis of a COF 3PD by forming the β-ketoamine linkage through a one-step michael addition–elimination reaction between amine and β-Ketone. Reproduced with permission from [98]. Copyright American Chemical Society 2017

in the COF formation which would have obstructed the error corrections in the COF lattice [55]. In literature, a structural series is reported with various amines forming β-ketoenamine. Another approach towards the synthesis of these COFs included mechanical synthesized by a simple mortar grinding which results in the formation of nanosheets that are just a few layers thick [6, 11, 56].

Another formation of β-ketoenamine linkage motif utilized a direct one-step formation through the pathway of Michael's addition-elimination reaction by where amines are condensed with β-ketoenols to yield β-ketoenamines (Fig. 12.9) [98]. These β-ketoenamines have added advantage of extra stabilization which arises from intramolecular hydrogen bonding. These β-ketoenamine COFs have found numerous applications in reversible redox processes in acidic electrolytes to energy storage, as proton conducting electrolytes, in reduction of NO_2 to NH_2, etc.

5. **Other linkages**: Besides the mainly used linkages, researchers have developed several other bonding strategies. In 2008, Hunt et al. reported the 3D COF having borosilicate linkages formed by the condensation between boronic acid, tetra (4-dihydroxyboryl-phenyl) methane, and tert-butyl silanetriol which resulted in the formation of borosilicate bonds [50]. In another report, nitrile trimerization reactions were used in the synthesis of crystalline triazine framework (CTF-1) using 1,4-dicyanobenzene, 2,6-naphthalenedicarbonitrile, [1,1′-biphenyl]-4,4′-dicarbonitrile, and 5′-(4-cyanophenyl)-[1,1′:3′,1″-terphenyl]-4,4″-dicarbonitrile [61]. Though most of the triazine polymers were amorphous these were found to be highly stable and useful in various applications like catalysis, polysulfide sorption in Li–S batteries, and these also function as membrane materials in the separation of gases [4, 9, 99]. Exploring further possibilities, the making of COFs with inorganic linkages, Jackson et al. reported the synthesis

of a COF where the borazine linkages were formed by the polymerization between melamine and cyanuric chloride and later by the decomposition of 1,3,5-(p-aminophenyl) benzene- borane under thermal conditions [51]. Another important linkage was formed as a result of squaraine-linked COF in a reaction between squaraic acid and aromatic amines [89, 125]. In further investigation, Pyles et al. introduced a novel strategy to form COF by benzobisoxazole linkages formed by catalytic condensation between 2,5-diamino-1,4-benzenediol and trigonal aldehydes [97], and Nandi et al. utilized benzimidazole linkage in the synthesis of COF [91]. Spiro borate-linked COFs were obtained from macrocycles which were diol functionalized and were reacted with trimethyl-borate to yield the sp^3 hybridized COF with spiroborate linkage. These frameworks were found to have a large surface area and high H_2 absorption potential. In addition, this spiroborate COF showed Li^+ transportation capacity which suggested them as the potential candidate for energy storage applications [28].

Benzoylation, a common lab reaction for amide preparation was also utilized to assemble and thus effectively reticulate building blocks and linkers. Li et al. synthesized two 2D-crystalline COFs 1,2,3-trimesoyl chloride with p-phenyldiamine (Fig. 12.10) and ethylene diamine. These COFs made by amide linkages were further studied for heavy metal recovery [69]. Therefore, various groups of researchers

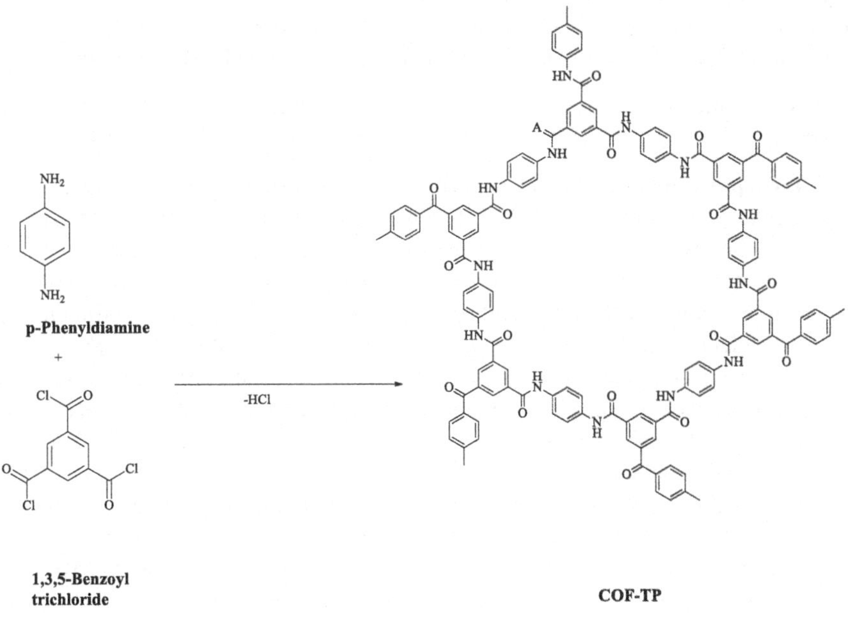

Fig. 12.10 Synthesis of an amide linked COF, COF-TP by benzoylation. Reproduced with permission from [69]. Copyright Elsevier 2019

have used different building blocks to synthesize a wide range of crystalline and amorphous COFs to design the framework with desired properties [13, 39, 129, 130].

4 Post-Synthetic Modification

The COF properties and applications are based on their interaction of the internal surface with the external compounds. This is further dependent on the structure, chemical environment, and the size of the voids or the pores [112]. The respective COF properties can be modulated by reinforcing the pore size and environment. There are three primary approaches adopted to carry out such modifications in COFs viz. pre-synthesis, post-synthesis, or in situ [105]. The possibility of Post-Synthetic Modifications in COFs makes them unique in the category of crystalline polymeric compounds. The extended porous networks and high surface area of COFs allow easy post-synthetic functional group modifications by employing simple synthetic methodologies. As the size of the pore, structure, and chemical environment can be easily modified this further proves the versatility of COFs in a wide range of applications [11, 15]. The general approaches employed to post-modify COFs are:

(i) Incorporation of an external molecule
(ii) Use of metal atom or ion as intercalators or complexing agents
(iii) Formation of new covalent bonds between the ordered COF network.

Some examples of both types of post-synthetic modification in COF are discussed below:

1. **Incorporation of External molecule**: Thermal instability and easy hydrolysis are the major drawbacks of COFs having borate ester linkage. These COFs when post-modified by incorporating the pyridine units, showed enhanced thermal stability and higher resistance for hydrolysis. Through the experimental and computational studies, the authors evidenced the preferential coordination to the defect sites. This strategy greatly enhanced the utility of COFs for gas absorption and storage purposes [27]. People have also tried to incorporate metals as well as various compounds to modify COFs after the polymerization [25, 76, 101, 139]. In this direction, Kalidindi et al. investigated the incorporation of organometallics into the COFs. Ferrocene was found to form ferrocene-COF intercalation compounds where the cyclopentadienyl units of ferrocene were found to interact with cyclopentadienyl units via. π–π interactions [53].
2. **Use of metal atoms or ions as intercalators or complexing agents**: The COFs containing imine linkages can coordinate with specific metals and can act as ligands for metal complexes. This complexation fixes the metal centers at specific positions of the porous scaffold which leads to the defined and separate reaction centers. These complexes, where metal acts as a complexing agent, have been found quite favorable for catalytic applications. It was first exemplified by

Ding et al., who synthesized the first 2D-material COF LZU-1 (using TFB and PDA) and later modified it by incorporating with Pd $(OAc)_2$ (Fig. 12.11) [24].

There are examples in the literature where specific metals are used to bind adjacent 2D layers of COFs and form an interlayer bridge. For example, molybdenum is incorporated post synthetically in a COF having hydrazone linkages, where TPG was used as an aldehyde linker, by employing $MoO_2(acac)_2$. The COF showed remarkable catalytic activity towards oxidation reactions and was utilized for the selective epoxidation of alkenes [132]. In another report, Xiong et al. used the vicinal hydroxyl groups in 2,3-dihyroxylterephthaldehyde to dock the immobilized vanadium as V=O using $VO(acac)_2$ [122]. $Co(OAc)_2$ was also used to incorporate Co(II) ions in the bipyridyl units of the β-ketoenamine linkage COF-TpBpy (made from TPG and 5,5′-diamino- 2,2′-bipyridine). Also, the same COF was further modified post synthetically by using $Cu(OAc)_2$ which embedded the Co(II) ions in it [5]. To mimic the natural heme complexes, Seo

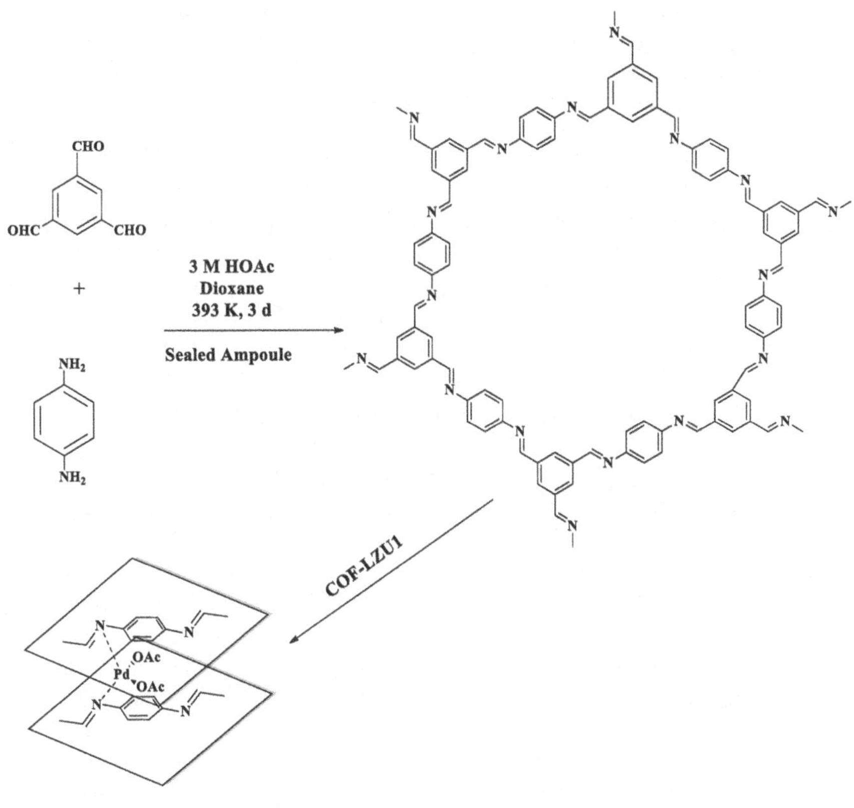

Fig. 12.11 Synthesis of COF-LZU1 Followed by its Post-Functionalization to Incorporate Pd Metal thus Forming Pd/COF-LZU1. Reproduced with Permission from [24]. Copyright from American Chemical Society 2011

et al. reported Dha-Tph-COF made from 2,3-dihydroxy-1,4-dicarbaldehyde and 5,10,15,20-tetrakis(4-aminophenyl)porphyrin. It was modified by immersing the COF in $FeCl_2{\cdot}4H_2O$ solution which immobilized the Fe species in it [106]. Similarly, chloro(1,5-cyclooctadiene)nickel(II), $Ni(COD)_2$ was used to link Ni ions to the dehydrobenzoannulene units of a 3D boronic ester COF with a nominal loss in the porosity [2]. The examples discussed so far involved the incorporation of a single metal ion in the framework. Leng et al. investigated the possibility to achieve a coordination of two different metals incorporated in a COF. For this purpose, they synthesized Schiff base COFs using 2,2′-bipyridyl-5,5′-dialdehyde and 4,4'-biphenyldicarboxaldehyde as the aldehyde source and 4,4′,4″,4‴-(pyrene-1,3,6,8-tetrayl)tetraaniline as an amine having the pyrene core. Two nitrogen sites: the bipyridine and the imine act as coordination-docking sites. Two metal complexes, using Rh and Pd, of different sizes were then used to achieve the complexation of both the metals. The moiety Rh(COD)Cl was found only on the outer surface as it was too large to occupy the interlayer positions between the imines. In the subsequent step, these imine positions were packed by using $Pd(OAc)_2$, yielding a bimetallic Pd and Rh functionalized COF [62].

3. **Formation of new covalent bonds between the ordered COF network**

    (i) The first covalent PSM was first shown by Nagai and co-workers who synthesized an azide appended BDBA, a boronic ester COF. This azide appended COF, N_3-COF-5 reacted with functionalized alkynes using Cu catalyzed click reaction. This approach also led to the development of strategies to append various groups to the COFs (even fullerenes!). In the reversed click version, the reaction first the COF is synthesized with an alkyne-functionalized linker and then it is later reacted with an azide [90]. This click reaction is used to tune the storage capacity of COFs towards CO_2 by introducing functional groups like -COOH, -COOMe, -OH, and $-NH_2$ by post-synthetic modifications, where 50% NH_2 displayed the highest capacity [46]. In another example, the pyrrolidine azide group was incorporated post synthetically to generate a catalytic center in the framework [123, 124]. COFs which can be used for the storage of electrochemical energy can be made by modifying the COF with Ni-porphyrin units yielding TEMPO–NiP-COF where TEMPO-COF was made from 1,3,5-tris(4-aminophenyl) benzene and 2,5-bis(2-propynyloxy) terephthalaldehyde and covalent PSM by employing 4-azido-2,2,6,6-tetramethyl- 1-piperidinyloxy, an azide or an organic radical [49]. The same radical was introduced by using a different strategy (truncation approach) to create covalent docking sites in a COF by Bunck and Dichtel. They mixed a truncated boronic acid node having an allyl group with the tetragonal linker, tetra(4-dihydroxyborylphenyl) methane, which was later reacted with propanethiol in a thiol-ene reaction [8].

(ii) In 2016, a two-step post-synthetic modification was reported where a ketoenamine COF $TpBD(NO_2)_2$ was synthesized from TPG and 3,3′-dinitrobenzidine. The nitro group present in the COF was first reduced to primary amines. As amine functionality is present in the building block of the COF also, thus only post-synthetic modification could lead to the amino-pore-wall functionalization in the COF [75]. In the next step, these amino groups can be further modified with some specific reagents like acetyl chloride.

(iii) The Hydroxyl groups represent another class of functions that can be used to carry out post-synthetic modifications. The hydroxyl groups are modified easily because of their high reactivity towards various functional groups [45]. For example, a hydroxyl group-containing COF (made from 5,10,15,20-tetrakis(p-tetraphenyl amino) porphyrin and a mixture of 2,5-dihydroxyterephthalaldehyde and 1,4-phthalaldehyde) easily reacts with succinic anhydride which undergoes a ring-opening reaction to yield a carboxylic acid group. The formation of carboxylic acid significantly improved the CO_2 storage capacity of the COF under study. Several other groups have suggested modification in the COF with hydroxyl group through reaction with linkers molecules. Another OH-functionalized COF was synthesized by condensing 1,3,5-tris(4-formylphenyl)benzene with 2,5-diamino-1,4-dihydroxybenzenedihydrochloride. In the next step, these -OH groups present in the framework were oxidized with O_2 and Et_3N, to yield the quinone form, and thus a COF was obtained which was otherwise could not be synthesized via direct condensation of the diaminoquinone with the aldehyde linker [122, 136].

In another PSM, the layers of the 2D network were converted into the 3D structure via topochemical [4 + 4] cycloaddition reaction. In this reaction, the anthracene moieties of the adjacent 2D layers get dimerized which resulted in the formation of the 3D network [45]. This PSM strategy involving cycloaddition is further exemplified by a β-ketoenamine COF (using TPG and 2,6-diaminoanthracene) undergoing a [4 + 2] cycloaddition with N-hexylmaleimide. In this case, the cycloaddition being reversible provided a further correction to form a crystalline 3D-COF [57].

5 Application of COFs in Catalysis

Various building blocks and post-synthetic modifications have enriched COFs with the possibility to have diversified catalytic action. To explore the catalytic potential of these COFs, several researchers have linked numerous substituents to these crystalline frameworks which have resulted in expedient catalysts. In the following section, we have discussed the organocatalytic (e.g., coupling reactions, redox reactions, condensation reactions, addition reactions, and isomerization reactions), electrocatalytic and photocatalytic properties of these porous [101] supermolecules. In

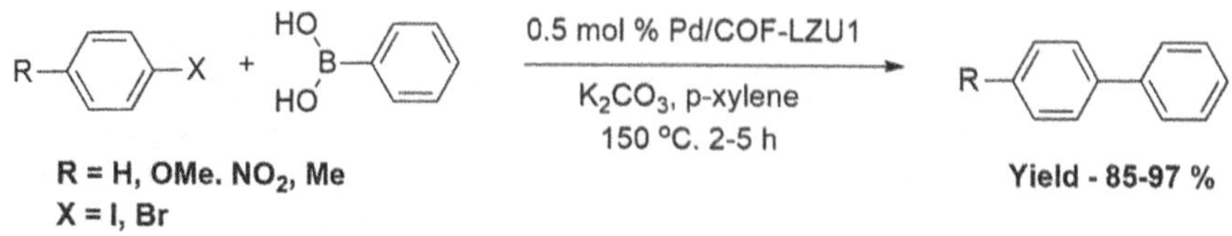

Fig. 12.12 Suzuki–Miyaura coupling catalyzed by Pd/COF-LZU1. Reproduced with Permission from [24]. Copyright American Chemical Society 2011

these applications, the COF surfaces are either directly used, or have been utilized or modified to be employed as coordination locations for various metal species.

1. **COFs as Organocatalysts**:

   (i) **Coupling Reactions:** The first example of COF-mediated Suzuki-Miyaura coupling reaction was reported by Ding et al. in the year 2011. The reported imine-linked COF was designed, synthesized, and was modified post synthetically for the desired catalytic activity. The imine-based COF was synthesized by using Schiff base reaction between 1,3,5-triformylbenzene and 1,4-diaminobenzene and the resultant 2D layered sheets had close nitrogen atoms in eclipsed form (at a distance of ~3.7 Å), which offered easy incorporation of Pd ions using $Pd(OAc)_2$. Furthermore, the presence of regular channels facilitated the efficient access of reactants to the active sites and speedy diffusion of the products. This imine-linked COF material exhibited excellent catalytic activity in the Suzuki–Miyaura coupling with a long catalytic life, i.e., it can be used for multiple cycles. COF derivatives have been reported to function as heterogeneous catalysts for a variety of coupling reactions [24]. Gonçalves et al. synthesized COF-300 utilizing tetrakis-(4-aminobenzyl)methane (TAM) with terephthalaldehyde (TA) followed by post-synthetic treatment with Pd(OAc)2. This catalyst revealed 100% selectivity in the Suzuki- Miyaura coupling (Fig. 12.12), and later provided an alternative for various cross-coupling reactions including Heck, Sonogashira, and Heck-Matsuda [40]. Typically, some heterogeneous COFs have also been developed to build the carbon–nitrogen bonding in Chan–Evans-Lam coupling reactions [108]. More recently, a polyimide-supported Cu catalyst Cu@PI-COF has been developed by Lu et al. which exhibited a notable catalytic activity toward the Chan–Lam coupling reaction of phenyl and aryl boronic acids with aniline and other amines [42].

   Owing to the large surface area and sue generis characteristics, metal nanoparticles have grabbed the attention of researchers as potential candidates for catalytic applications. However, these nanoparticles often tend to aggregate on support surfaces. In this regard, Lu et al. synthesized a β-ketoamine COF fabricated with thioether linkages. These thioethers anchored the metal nanoparticles via the interactions between metal and sulfur. The PdNPs@COF and PtNPs@COF have also shown superior catalytic activity for the Suzuki-Miyaura coupling along with easy reusability [77].

(ii) **Redox Reactions:** Pachfule et al. first attempted to accommodate the Au nanoparticles into a porous matrix of TpPa-1 COF. The Au modified heterogeneous catalyst Au(0)@TpPa-1 exhibited a remarkable catalytic activity. For example- Reduction of 4-nitrophenol to 4- aminophenol could be completed within 13 min. This could be because the Au nanoparticles (about 5 ± 3 nm) were finely distributed at the surface of the COFs [94]. COF complexes have been widely studied for selective oxidation reactions also. For instance, in 2015, Mu et al. reported the synthesis of copper-docked novel 2D TAPT-DHTA-COF using 2, 5-dihydroxyl-terephthalaldehyde (DHTA) and 1, 3, 5- tris-(4-aminophenyl) triazine (TAPT) [88]. The 2D COFs catalyst provided a wonderful catalytic activity for the oxidation of styrene to benzaldehyde. Not only for the simple oxidation reactions, but the COFs are also the rightful candidates for some advanced oxidations, such as oxidation of organic contaminants, Fenton reaction, etc. [78, 102, 103] (Liu et al. 2020c). Preet et al. have reported the novel sulfur-doped ordered mesoporous carbons (S-OMCs) catalyst which could activate persulfate for the degradation of p-nitrophenol (PNP) to smaller organic molecules where the observed removal rate of PNP was up to 82% [96]. Li et al. used a two-step controlled calcination method to prepare the heterogeneous, magnetic porous Fenton-like catalyst, Fe_3O_4/carbon. It was reported that in contrast to the singular Fe_3O_4 nanoparticles from precipitation, the porous Fe_3O_4/carbon showed enhanced catalytic performance towards the oxidation of organic pollutants. The activity of the framework was analyzed by the oxidation of toxic organic dyes like methylene blue which were oxidized to small molecules [67].

(iii) **Condensation Reactions**: It should be addressed that these emerging classes of porous materials validated a preferable catalytic activity over other conventional methods and procedures of catalysis in the reactions like Knoevenagel, Aldol, and Mannich condensation. In this regard, two novel 3D functionalized imine-based COFs have been synthesized by Fang et al., BF-COF-1, and BF-COF-2, by condensing 1,3,5,7-tetraaminoadamantane (TAA), a tetrahedral alkyl amine, and 1, 3, 5-triformylbenzene (TFB) or tri-formylphloroglucinol (TFP), planar triangular building blocks. In BET, the surface areas of the BF-COFs, when evaluated using nitrogen sorption isotherm were calculated to be 730 $m^2\ g^{-1}$ and 680 $m^2\ g^{-1}$ for BF-COF-1 and BF-COF-2 respectively. Both heterogeneous catalysts exhibited remarkable conversion (96% for BF-COF-1 and 98% for BF-COF-2), high sensitivity and selectivity to size and more importantly, good catalytic life span was documented [30]. Similarly, Wang et al. also reported two imine-based

COFs, named LZU-72 and LZU-76, synthesized through a condensation reaction, and further their catalytic potential in the asymmetric aldol reaction was investigated. Both this crystalline COFs presented chiral functional sites, hence the selectivity towards the reactants was observed and the transport of products was also facilitated. Accordingly, when these heterogeneous frameworks were compared with the conventional asymmetric catalysts like (S)-4- 7-diphenyl-2-(pyrrolidin-2-yl)-1H-benzo[d]imidazole (DPBIP), the LZU-76 displayed equivalent enantioselectivity. The vanadium-docked COFs (VO-TAPT-2, 3-DHTA and VO-PyTTA-2, 3-DHTA) are used in the Mannich reactions between β-naphthol and 4-methylmorpholine N-oxide (NMO) using CH_2Cl_2 as the solvent. These highly stable and crystalline COFs exhibited remarkable catalytic efficacy and were utilized for a wide range of substrates, including both electron-donating as well as electron-withdrawing substituents [117]. Furthermore, these stable heterogeneous catalysts were found to be easily recyclable and could be employed multiple times without any loss of structural integrity and activity. Besides, these COFs have shown tremendous achievement in many other reactions such as the Henry reaction. Ma et al. had designed and synthesized a sophisticated homochiral COF by using S-(+)-2-methylpiperazine and cyanuric chloride. The resultant framework was post-modified by using palladium nanoparticles within the porous COF skeleton. This modification efficiently immobilized the metal sites in the void of the COF which is used to catalyze Henry's reaction of nitromethane with different aromatic aldehydes. The reaction produced high yields (80 – 99%) with an excellent enantiomeric excess (R-configuration, 79 – 97%) [81].

(iv) **Addition Reactions**: Just like other organic transformations, COFs have been advertised as ideal solid organocatalysts to carry out addition reactions as well such as Michael addition and cycloaddition. These are considered ideal because of the presence of unique channel structure, porosity, multifunctionality, and recyclability [130] (Liu et al. 2020b; Cheng and Wang 2021b). Xu et al. incorporated both chirality (through chiral centers) and organocatalytic species in the COF skeletons to develop a class of novel chiral COFs, namely [(S)-Py]-TPB-DMTP- COF, as a candidate catalyst for Michael addition reaction between cyclohexanone and β-nitrostyrene [124]. The reaction proceeded effortlessly and exhibited excellent catalytic activity with 100% conversion and 92% enantioselectivity in 12 h of reaction time. In addition, Lyu et al. synthesized the olefin-linked COFBF3 ⊂ COF-701 by immobilizing strong Lewis acid $BF_3{\cdot}OEt_2$ in the pores of the matrix. The framework showed remarkable stability and crystallinity under both strongly acidic as well as alkaline conditions. Apart from those mentioned above, very interesting results of the catalyst were reported in the Diels – Alder cycloaddition reactions [80]. In another case, Dou et al. reported a simple hyper crosslinking method to fabricate a series of Fe(III) porphyrin-based polymers, terming FePp-H-HCP, FePp-Br-HCP, FePp-OMe- HCP, and FePo-OMe-HCP. In the cyclo-addition of benzaldehyde with 2,3-dimethyl-1,3-butadiene, excellent catalytic performances with good yields of 87–97% for all these specially designed COFs were reported. All these catalysts were found to be recyclable and had applicability in the Friedel–Crafts alkylation reaction also [26].

2. **Photocatalysts**: Researchers have incorporated various other components in the COFs which successively led to modifications and introduction of extraordinary properties such as large surface areas, long-range orderliness, tunable bandgaps, and excellent absorbance of visible light. Due to these properties, these hybrid composites have received special attention from various research groups for exploring their photocatalytic applications [25, 66, 76, 101]. Certain promising heterogeneous photoactive 2D frameworks have now been developed to be used in visible-light-driven organic transformations. The observed photocatalytic activity of COFs is because of (i) the special structural design which features excellent visible-light absorption, and fast separation and transfer of electron–hole; (ii) their large crystalline and porous surface area which accelerates the charge transport to the surface (iii) incorporation of photoactive units during synthesis or post-synthesis which prolongs the lifespan of the excited states; and (iv) presence of extended π-conjugation in the structural framework which also adds up to charge carrier mobility [7, 18, 92, 107, 138]. Stegbauer et al. were the first to report the visible-light-driven hydrogen generation through a COF in the presence of Pt as a proton reduction catalyst. The hydrazine-based COF could be synthesized by condensation of 1,3,5-tris-(4-formyl-phenyl)triazine (TFPT) and 2,5-diethoxy-terephthalohydrazide (DETH) building blocks under solvothermal reaction conditions. The COF formed had a layered structure with a honeycomb type lattice framework which featured mesopores of 3.8 nm in diameter [109]. When sodium ascorbate was employed as the sacrificial electron donor, the TFPT–COF/Pt photocatalytic system could produce hydrogen continuously from water without being degraded. A wide range of COFs with a potential application as photosensitizers and photocatalysts have been designed and developed by various groups of researchers. In this respect, Wei et al. also constructed a series of highly stable imine-based COFs (LZU- 190, LZU-191, and LZU-192) which had been used for the visible-light-driven oxidative hydroxylation of aryl boronic acids to the corresponding phenols [121]. Moreover, some of the crystalline COFs have been tailored with an optimum bandgap which is used for the photocatalytic oxidation of the C-H bond. In this context, Zhi et al. reported a solvothermal synthesis of a 2D-COF, COF-JLU5, by condensation of 1,3,5-tris-(4-aminophenyl) triazine and 2,5-dimethoxyterephthaldehyde. The COF was used to catalyze the aerobic cross dehydrogenative coupling of phenyl tetrahydroisoquinolines with various nucleophiles [138]. Zhang et al. have reported a modification in the 2D COF by covalently attaching NH_2-UiO-66 onto the surface of TpPa-1-COF. This tailored

material exhibited high porosity, crystallinity, and large surface area and was used for the visible-light-driven photocatalytic evolution of H_2. Because of the well-matched band gaps and effective separation of charges between NH_2-UiO-66 and TpPa-1-COF, the molecular hybrid presented high H_2 evolution rates [133]. Because COFs have an organic moiety with high covalent interactions, these exhibit various advantages over MOF [65, 133, 140]. Fu et al. had designed chemically and thermally stable metal-free azine-based COFs which demonstrated high photocatalytic potential and electronic performance [35]. Two 'Green' photocatalytic systems were developed employing azine-linked 2D COFs, ACOF-1, and N_3-COF, for the reduction of CO_2 into CH_3OH using water as the reductant [35]. More recently, Bhadra et al. have also synthesized a heterogeneous COF with triazine and keto linkages using melamine/1,3,5-triazine-2,4, 6-triamine (Tt) and 2, 4, 6- triformylphloroglucinol (TPG) aldehyde. The designed COF was employed for the facile isomerization reaction of trans (E) to cis (Z) stilbene with about 90% yield at an exposure time of 18 h [4].

3. **Electrocatalysts**: It is a catalytic process that involves the direct transfer of electrons to carry out oxidation and reduction reactions. An efficient electrocatalyst was synthesized from cobalt-porphyrin-based COF (Co-COF) by Ma and co-workers in 2015. These were synthesized by cocondensation of terephthalaldehyde (TA) and cobalt (II) 5,10,15,20-tetrakis(ρ-tetraphenyl amino) porphyrin (Co-TAPP), followed by post-pyrolysis at 900 °C in N_2 atmosphere. The COFs possessed a 2D graphite-like layered structure with a uniform distribution of cobalt nanoparticles throughout the framework. Excellent electrocatalytic performance was observed towards redox reactions in the alkaline media [86]. Diercks et al. also reported a successful synthesis of a series of electrocatalysts 2D-COFs COF-366-Co, COF-366-$(OMe)_2$-Co, COF-366-F-Co, and COF-366-$(F)_4$-Co used in the reduction of CO_2. These metal impregnated catalysts exhibited better performance even as compared to the mono-dispersed Au nanoparticles, which were reported to have the highest current density [22]. Lin et al. have identified crystal-field stabilization energy (CFSE) as an important tool as an activity descriptor for predicting oxygen reduction reaction/ oxygen evolution reaction (ORR/OER) activities of the transition metal (e.g., Sc, Ti, V, Cr, Mn, Fe, Co, Ni, Cu, and Zn) and alkaline earth metals (AM = Be, Mg, Ca, Sr, Ba) incorporated COFs with porphyrin units. The energy barrier to be overcome (the overpotential) in ORR/OER is usually taken as the measure of catalytic activity. A framework with lower overpotential would exhibit better catalytic performance. Among the series of all transition metals, the Fe-porphyrin-COF presents the lowest overpotential. One of the proposed potential applications of the Fe-porphyrin-COF was in the production of H_2O_2, the 'green oxidizer' which is currently synthesized using a high-energy anthraquinone process at the industrial level [70].
4. **Other catalytic applications**: Apart from the above-discussed catalytic application, COFs have also been used for other catalytic applications, like shape-selective catalysis [30, 32]. Shape selectivity could be achieved by modifying

the building blocks which are responsible for their well-defined conformation and morphology giving rise to specific pore sizes and cavities. This is also responsible for thermal and hydrothermal stability and the possibility of incorporation of a wide range of active sites in the walls of the framework. Recently, Ma et al. have synthesized structural isomorphs featuring highly ordered amide groups in the channels which could assign size, regio, and stereoselectivity to the organic transformation. Various aromatic aldehydes were included for testing the catalytic activity of the framework in the Knoevenagel condensation reaction. 4-phenylbenzaldehyde being more linear ($0.43 \times 1.042\ nm^2$), yielded 90% of the product as compared to 70% yield 1-naphthaldehyde ($0.68 \times 0.62\ nm^2$) and zero condensation with tert-butyl cyanoacetate ($0.58 \times 1.03\ nm^2$) [85]. These owing to highly ordered pore size the framework exhibited the shape selectivity towards the incorporation of guest molecules [137].

6 Conclusion and Future Perspectives

Covalent Organic Frameworks have brought a lot of excitement amongst the fraternity of researchers and have motivated them to explore the promising catalytic potential of these supermolecules. Since the ground-breaking report of COFs from Yaghi and co-workers in 2005, these molecules came into the limelight because of their exceptional properties such as large surface area, tunable porosity, low density, high stability, and easily modifiable functionality. All these properties have evidenced their probable potential in a wide range of applications. In the current chapter, we have covered a comprehensive review of the various energy efficient methods of preparation of these 'designer catalysts' with a special emphasis on its catalytic potential including photocatalytic reactions, electrocatalytic reactions, and organic reactions developed in recent years.

As shown by various groups of researchers, these crystalline porous frameworks have demonstrated their effectiveness in a wide range of reactants and have produced excellent yields in most of the catalytic processes. But these developments are still in their early stages and these potential COFs are expected to illustrate extraordinary performance in catalysis and other areas in the near future. Nevertheless, before the industrial applications of these materials can be achieved there are many challenges that need to be addressed, as summarised below:

- In catalysis, there are many more factors that affect the catalytic potential like structural defects, morphology, and dimensions, etc. In fact, very little research has been reported so far about these aspects. Thus, there is a lot of potential in the rational designing and investigation which leads to diversified COF structures.
- To catalyze the chemical reactions with special requirements, a wide range of active components have been introduced like chiral groups, metals at various sites (via metal nanoparticles/metallic oxide), and even ionic liquid co-catalyst. These COFs besides having catalytic active sites for catalyzing chemical reactions, also express ample electronic, photochemical, and redox properties. Hence there

is a lot of scopes to exploit these crystalline porous frameworks based on the functional groups attached and active sites present in a catalytic system.

- Most of the COFs present to date are synthesized in organic solvents. There is a great opportunity in the field of greener designing in synthesis by making use of 'Green Solvents' and Solventless approaches.
- There is an urgent need to develop new synthetic methodologies employing milder reaction conditions along with the economic considerations for the industrial production of these COF networks.
- Also, when these frameworks are used in the diversified catalytic reaction, achieving an efficient catalyst with excellent stability which could provide good yields of the desired product is an ever-lasting objective.
- While predesigning the functionalities to be incorporated into COFs, the mechanism of the reactions to be carried out plays a major role. This can be achieved by making use of computational modeling of COF materials which would provide an insight into the theoretical information regarding the characterization and potential application of COF under study. In summary, within the past two decades of development, COFs have grown enormously. Their wide range of applications has lengthened well beyond the boundaries of the conventional catalytic reaction of chemical manufacturing. We are certain that the research will not be limited to what all has been reported so far, but there is a great scope in the development of novel COFs with other exciting performances by exploring the inventory of synthetic materials. Thus, more active and selective COFs based catalysts will be designed and their intriguing and sparkling performances in another catalysis would be pursued. Thus, we strongly believe that these COFs materials with diverse functionalities have a bright future toward industrial and commercial applications.

Abbreviations

BDBA	1,4-Benzenediboronic acid
BET	Brunauer–Emmett–Teller isotherm
CFSE	Crystal-field stabilization energy
COF	Covalent Organic Framework
Co-TAPP	Cobalt(II) 5,10,15,20-tetrakis(ρ-tetraphenyl amino) porphyrin
CTF	Covalent Triazine Framework
DETH	2,5-Diethoxy-terephthalohydrazide
DHTA	2, 5-Dihydroxyl- terephthalaldehyde
DMTA	2,5-Dimethoxyterephthalaldehyde
DPBIB	(S)-4- 7-diphenyl-2-(pyrrolidin-2-yl)-1H-benzo[d]imidazole
HHTP	Hexahydroxytriphenylene
MOF	Metal Organic Framework
MW	Microwave

NMO	4-Methylmorpholine-N-oxide
NP	Nano particles
OER	Oxygen Evolution Reaction
ORR	Oxygen Reduction Reaction
Pa	p-phenylenediamine
PDA	Palladium diacetate
PI	Polyimide
PNP	p-nitrophenol
POP	Porous Organic Polymer
PSM	Post-synthetic Modification
PXRD	Powder X-Ray Diffraction
S-OMC	Sulfur doped organic mesoporous carbon
TA	Terephthalaldehyde
TAA	1,3,5,7-Tetraaminoadamantane
TAM	Tetrakis-(4-aminobenzyl)methane
TAPB	1,3,5-Tri(4-aminophenyl)benzene
TAPT	1, 3, 5- Tris-(4-aminophenyl)triazine
TBA	Triboronic acid
TBPS	Tetra(4-dihydroxyborylphenyl)silane
TEMPO	2,2,6,6-Tetramethylpiperidinyloxy
TFB	1,3,5-Triformylbenzene
TFPB	1,3,5-Tris(4-formylphenyl)benzene
TFPT	1,3,5-Tris-(4-formyl-phenyl)triazine
TPA	1,3,5-Triformylphloroglucinol

References

1. Alahakoon SB, Thompson CM, Nguyen AX et al (2016) An azine-linked hexaphenylbenzene based covalent organic framework. Chem Commun 52:2843–2845. https://doi.org/10.1039/c5cc10408d
2. Baldwin LA, Crowe JW, Pyles DA, McGrier PL (2016) Metalation of a mesoporous three-dimensional covalent organic framework. J Am Chem Soc 138:15134–15137. https://doi.org/10.1021/jacs.6b10316
3. Bang JH, Suslick KS (2010) Applications of ultrasound to the synthesis of nanostructured materials. Adv Mater 22:1039–1059. https://doi.org/10.1002/adma.200904093
4. Bhadra M, Kandambeth S, Sahoo MK et al (2019) Triazine functionalized porous covalent organic framework for photo-organocatalytic E- Z isomerization of olefins. J Am Chem Soc 141:6152–6156. https://doi.org/10.1021/jacs.9b01891
5. Bhadra M, Sasmal HS, Basu A et al (2017) Predesigned metal-anchored building block for in situ generation of Pd nanoparticles in porous covalent organic framework: application in heterogeneous tandem catalysis. ACS Appl Mater Interfaces 9:13785–13792. https://doi.org/10.1021/acsami.7b02355
6. Biswal BP, Chandra S, Kandambeth S et al (2013) Mechanochemical synthesis of chemically stable isoreticular covalent organic frameworks. J Am Chem Soc 135:5328–5331. https://doi.org/10.1021/ja4017842

7. Bottecchia C, Noël T (2019) Photocatalytic modification of amino acids, peptides, and proteins. Chem A Eur J 25:26. https://doi.org/10.1002/chem.201803074
8. Bunck DN, Dichtel WR (2013) Postsynthetic functionalization of 3D covalent organic frameworks. Chem Commun 49:2457–2459. https://doi.org/10.1039/c3cc40358k
9. Burns DA, Benavidez A, Buckner JL, Thoi VS (2021) Maleimide-functionalized metal-organic framework for polysulfide tethering in lithium-sulfur batteries. Mater Adv 2:2966–2970. https://doi.org/10.1039/d1ma00084e
10. Campbell NL, Clowes R, Ritchie LK, Cooper AI (2019) Rapid microwave synthesis and purification of porous covalent organic frameworks. Chem Mater 21:204–206. https://doi.org/10.1021/cm802981m
11. Chandra S, Kandambeth S, Biswal BP et al (2013) Chemically stable multilayered covalent organic nanosheets from covalent organic frameworks via mechanical delamination. J Am Chem Soc 135:17853–17861. https://doi.org/10.1021/ja408121p
12. Chen L, Wu Q, Gao J et al (2019) Applications of covalent organic frameworks in analytical chemistry. TrAC Trends Anal Chem 113:182–193. https://doi.org/10.1016/j.trac.2019.01.016
13. Chen X, Geng K, Liu R et al (2020) Covalent organic frameworks: chemical approaches to designer structures and built-in functions. Angew Chem Int Ed 59:5050–5091. https://doi.org/10.1002/anie.201904291
14. Cheng HY, Wang T (2021) Covalent organic frameworks in catalytic organic synthesis. Adv Synth Catal 363:144–193. https://doi.org/10.1002/adsc.202001086
15. Cooper AI (2013) Covalent organic frameworks. CrystEngComm 15:1483–1483. https://doi.org/10.1039/c2ce90122f
16. Côté AP, Benin AI, Ockwig NW et al (2005) Chemistry: porous, crystalline, covalent organic frameworks. Science 310:1166–1170. https://doi.org/10.1126/science.1120411
17. Crick F, Watson J (1953) Molecular structure of nucleic acids: a structure for deoxyribose nucleic acid. Nature 171:737–738. https://doi.org/10.1038/171737a0
18. Dai X, Xu X, Li X (2013) Applications of visible light photoredox catalysis in organic synthesis. Chin J Org Chem 33:2031–2045. https://doi.org/10.6023/cjoc201304026
19. Dalapati S, Jin S, Gao J et al (2013) An azine-linked covalent organic framework. J Am Chem Soc 135:17310–17313. https://doi.org/10.1021/ja4103293
20. Dienstmaier JF, Gigler AM, Goetz AG et al (2011) Synthesis of well-Oodered COF monolayers: surface growth of nanocrystalline precursors versus direct on-surface polycondensation. ACS Nano 5:9737–9745. http://https://doi.org/10.1021/nn2032616
21. Diercks C, Kalmutzki M, Yaghi O (2017) Covalent organic frameworks—organic chemistry beyond the molecule. Molecules 22:1575. https://doi.org/10.3390/molecules22091575
22. Diercks CS, Lin S, Kornienko N et al (2018) Reticular electronic tuning of porphyrin active sites in covalent organic frameworks for electrocatalytic carbon dioxide reduction. J Am Chem Soc 140:1116–1122. https://doi.org/10.1021/jacs.7b11940
23. Diercks CS, Yaghi OM (2017) The atom, the molecule, and the covalent organic framework. Science 355:6328. https://doi.org/10.1126/science.aal1585
24. Ding SY, Gao J, Wang Q et al (2011) Construction of covalent organic framework for catalysis: Pd/COF-LZU1 in Suzuki-Miyaura coupling reaction. J Am Chem Soc 133:19816–19822. https://doi.org/10.1021/ja206846p
25. Dong J, Han X, Liu Y et al (2020) Metal-covalent organic frameworks (MCOFs): a bridge between metal-organic frameworks and covalent organic frameworks. Angew Chem 132:13826–13837. https://doi.org/10.1002/ange.202004796
26. Dou Z, Xu L, Zhi Y et al (2016) Metalloporphyrin-based hypercrosslinked polymers catalyze hetero-diels–alder reactions of unactivated aldehydes with simple dienes: a fascinating strategy for the construction of heterogeneous catalysts. Chem Eur J 22:9919–9922. https://doi.org/10.1002/chem.201601151
27. Du Y, Mao K, Kamakoti P et al (2012) Experimental and computational studies of pyridine-assisted post-synthesis modified air stable covalent-organic frameworks. Chem Commun 48:4606–4608. https://doi.org/10.1039/c2cc30781b

28. Du Y, Yang H, Whiteley JM et al (2015) Ionic covalent organic frameworks with spiroborate linkage. Angew Chem Int Ed 55:1737–1741. https://doi.org/10.1002/anie.201509014
29. El-Kaderi HM, Hunt JR, Mendoza-Cortés JL et al (2007) Designed synthesis of 3D covalent organic frameworks. Science 316:268–272. https://doi.org/10.1126/science.1139915
30. Fang Q, Gu S, Zheng J et al (2014) 3D microporous base-functionalized covalent organic frameworks for size-selective catalysis. Angew Chem 126:2922–2926. https://doi.org/10.1002/ange.201310500
31. Fang Q, Gu S, Zheng J et al (2020) Chemistry of covalent organic frameworks. J Mater Chem A 48:3053–3063. https://doi.org/10.1021/acs.accounts.5b00369
32. Fang Q, Zhuang Z, Gu S et al (2014) Designed synthesis of large-pore crystalline polyimide covalent organic frameworks. Nat Commun 5:1–8. https://doi.org/10.1038/ncomms5503
33. Feng X, Chen L, Honsho Y et al (2012) An ambipolar conducting covalent organic framework with self-sorted and periodic electron donor-acceptor ordering. Adv Mater 24:3026–3031. https://doi.org/10.1002/adma.201201185
34. Freund R, Canossa S, Cohen SM et al (2021) 25 Years of reticular chemistry. Angew Chem Int Ed. https://doi.org/10.1002/anie.202101644
35. Fu Y, Zhu X, Huang L et al (2018) Azine-based covalent organic frameworks as metal-free visible light photocatalysts for CO_2 reduction with H_2O. Appl Catal B 239:46–51. https://doi.org/10.1016/j.apcatb.2018.08.004
36. Fukui T, Kawai S, Fujinuma S et al (2017) Control over differentiation of a metastable supramolecular assembly in one and two dimensions. Nat Chem 9:493–499. https://doi.org/10.1038/nchem.2684
37. Furukawa H, Yaghi OM (2009) Storage of hydrogen, methane, and carbon dioxide in highly porous covalent organic frameworks for clean energy applications. J Am Chem Soc 131:4570–4571. https://doi.org/10.1021/ja9015765
38. Gardel ML (2013) Synthetic polymers with biological rigidity hidden is more. Nature 493(7434):618–619. https://doi.org/10.1038/nature11855
39. Geng K, He T, Liu R et al (2020) Covalent organic frameworks: design, synthesis, and functions. Chem Rev 120:8814–8933. https://doi.org/10.1021/acs.chemrev.9b00550
40. Gonçalves RSB, Deoliveira ABV, Sindra HC et al (2016) Heterogeneous catalysis by covalent organic frameworks (COF): Pd(OAc)2@COF-300 in cross-coupling reactions. ChemCatChem 8:743–750. https://doi.org/10.1002/cctc.201500926
41. Guo J, Xu Y, Jin S et al (2013) Conjugated organic framework with three-dimensionally ordered stable structure and delocalized π clouds. Nat Commun 4:2736. https://doi.org/10.1038/ncomms3736
42. Han Y, Zhang M, Zhang YQ, Zhang ZH (2018) Copper immobilized at a covalent organic framework: an efficient and recyclable heterogeneous catalyst for the Chan-Lam coupling reaction of aryl boronic acids and amines. Green Chem 20:4891–4900. https://doi.org/10.1039/c8gc02611d
43. Hinman JJ, Suslick KS (2017) Nanostructured materials synthesis using ultrasound. Top Curr Chem 375:12. https://doi.org/10.1007/s41061-016-0100-9
44. Hoffmann R (1993) How should chemists think? Sci Am 268:66–73
45. Huang N, Chen X, Krishna R, Jiang D (2015) Two-dimensional covalent organic frameworks for carbon dioxide capture through channel-wall functionalization. Angew Chem 127:3029–3033. https://doi.org/10.1002/ange.201411262
46. Huang N, Krishna R, Jiang D (2015) Tailor-made pore surface engineering in covalent organic frameworks: systematic functionalization for performance screening. J Am Chem Soc 137:7079–7082. https://doi.org/10.1021/jacs.5b04300
47. Huang N, Wang P, Jiang D (2016) Covalent organic frameworks: a materials platform for structural and functional designs. Nat Rev Mater 1:16068. https://doi.org/10.1038/natrevmats.2016.68
48. Huggins ML (1954) The structure of collagen. J Am Chem Soc 76:4045–4046. https://doi.org/10.1021/ja01644a065

49. Hughes BK, Braunecker WA, Bobela DC et al (2016) Covalently bound nitroxyl radicals in an organic framework. J Phys Chem Lett 7:3660–3665. https://doi.org/10.1021/acs.jpclett.6b01711
50. Hunt JR, Doonan CJ, LeVangie JD et al (2008) Reticular synthesis of covalent organic borosilicate frameworks. J Am Chem Soc 130:11872–11873. https://doi.org/10.1021/ja805064f
51. Jackson KT, Reich TE, El-Kaderi HM (2012) Targeted synthesis of a porous borazine-linked covalent organic framework. Chem Commun 48:8823–8825. https://doi.org/10.1039/c2cc33583b
52. Jiang D, Science M, Science N et al (2020) Donglin Jiang answers questions about 15 years of research on covalent organic frameworks. Nat Commun 11:10–11. https://doi.org/10.1038/s41467-020-19302-x
53. Kalidindi SB, Yusenko K, Fischer RA (2011) Metallocenes@COF-102: organometallic host-guest chemistry of porous crystalline organic frameworks. Chem Commun 47:8506–8508. https://doi.org/10.1039/c1cc11450f
54. Kandambeth S, Dey K, Banerjee R (2019) Covalent organic frameworks: chemistry beyond the structure. J Am Chem Soc 141:1807–1822. https://doi.org/10.1021/jacs.8b10334
55. Kandambeth S, Mallick A, Lukose B et al (2012) Construction of crystalline 2D covalent organic frameworks with remarkable chemical (acid/base) stability via a combined reversible and irreversible route. J Am Chem Soc 134:19524–19527. https://doi.org/10.1021/ja308278w
56. Karak S, Kandambeth S, Biswal BP et al (2017) Constructing ultraporous covalent organic frameworks in seconds via an organic terracotta process. J Am Chem Soc 139:1856–1862. https://doi.org/10.1021/jacs.6b08815
57. Khayum MA, Kandambeth S, Mitra S et al (2016) Chemically delaminated free-standing ultrathin covalent organic nanosheets. Angew Chem 128:15833–15837. https://doi.org/10.1002/ange.201607812
58. Kim S, Choi HC (2019) Light-promoted synthesis of highly-conjugated crystalline covalent organic framework. Commun Chem 2:60. https://doi.org/10.1038/s42004-019-0162-z
59. Kim S, Park C, Lee M, et al (2017) Rapid photochemical synthesis of sea-urchin-shaped hierarchical porous COF-5 and its lithography-free patterned growth. Adv Funct Mater 27. https://doi.org/10.1002/adfm.201700925
60. Kouwer PHJ, Koepf M, le Sage VAA et al (2013) Responsive biomimetic networks from polyisocyanopeptide hydrogels. Nature 493:651–655. https://doi.org/10.1038/nature11839
61. Kuhn P, Antonietti M, Thomas A (2008) Porous, covalent triazine-based frameworks prepared by ionothermal synthesis. Angew Chem Int Ed 47:3450–3453. https://doi.org/10.1002/anie.200705710
62. Leng W, Peng Y, Zhang J et al (2016) Sophisticated design of covalent organic frameworks with controllable bimetallic docking for a cascade reaction. Chem Eur J 22:9087–9091. https://doi.org/10.1002/chem.201601334
63. Leonardi M, Villacampa M, Menéndez JC (2018) Multicomponent mechanochemical synthesis. Chem Sci 9:2042–2064. https://doi.org/10.1039/C7SC05370C
64. Lewis GN (1916) The atom and the molecule. J Am Chem Soc 38:762–785. https://doi.org/10.1021/ja02261a002
65. Li Z, Feng X, Zou Y et al (2014) A 2D azine-linked covalent organic framework for gas storage applications. Chem Commun 50:13825–13828. https://doi.org/10.1039/c4cc05665e
66. Li Z, Han S, Li C et al (2020) Screening metal-free photocatalysts from isomorphic covalent organic frameworks for the C-3 functionalization of indoles. J Mater Chem A 8:8706–8715. https://doi.org/10.1039/d0ta02164d
67. Li W, Wu X, Li S et al (2018) Magnetic porous Fe_3O_4/carbon octahedra derived from iron-based metal-organic framework as heterogeneous Fenton-like catalyst. Appl Surf Sci 436:252–262. https://doi.org/10.1016/J.APSUSC.2017.11.151
68. Li X, Yang C, Sun B et al (2020) Expeditious synthesis of covalent organic frameworks: a review. J Mater Chem A 8:16045–16060. https://doi.org/10.1039/d0ta05894g

69. Li G, Ye J, Fang Q, Liu F (2019) Amide-based covalent organic frameworks materials for efficient and recyclable removal of heavy metal lead (II). Chem Eng J 370:822–830. https://doi.org/10.1016/j.cej.2019.03.260
70. Lin CY, Zhang L, Zhao Z, Xia Z (2017) Design principles for covalent organic frameworks as efficient electrocatalysts in clean energy conversion and green oxidizer production. Adv Mater 29:1606635. https://doi.org/10.1002/adma.201606635
71. Liu L, Meng WK, Li L et al (2019) Facile room-temperature synthesis of a spherical mesoporous covalent organic framework for ultrasensitive solid-phase microextraction of phenols prior to gas chromatography-tandem mass spectrometry. Chem Eng J 369:920–927. https://doi.org/10.1016/j.cej.2019.03.148
72. Liu J, Wang N, Ma L (2020) Recent advances in covalent organic frameworks for catalysis. Chem Asian J 15:338–351. https://doi.org/10.1002/asia.201901527
73. Liu S, Lai C, Li B, et al (2020) Role of radical and non-radical pathway in activating persulfate for degradation of p-nitrophenol by sulfur-doped ordered mesoporous carbon. Chem Eng J 384:123304. https://doi.org/10.1016/j.cej.2019.123304
74. Lohse MS, Bein T (2018) Covalent organic frameworks: structures, synthesis, and applications. Adv Func Mater 28:1705553. https://doi.org/10.1002/adfm.201705553
75. Lohse MS, Stassin T, Naudin G et al (2016) Sequential pore wall modification in a covalent organic framework for application in lactic acid adsorption. Chem Mater 28:626–631. https://doi.org/10.1021/acs.chemmater.5b04388
76. Lopez-Magano A, Jiménez-Almarza A, Aleman J, Mas-Ballesté R (2020) Metal–organic frameworks (MOFS) and covalent organic frameworks (COFS) applied to photocatalytic organic transformations. Catalysts 10:720. https://doi.org/10.1021/acs.chemmater.5b04388
77. Lu S, Hu Y, Wan S, McCaffrey R, Jin Y, Gu H, Zhang W (2017) Synthesis of ultrafine and highly dispersed metal nanoparticles confined in a thioether-containing covalent organic framework and their catalytic applications. J Am Chem Soc 139(47):17082–17088. https://doi.org/10.1021/jacs.7b07918
78. Lv H, Zhao X, Niu H et al (2019) Ball milling synthesis of covalent organic framework as a highly active photocatalyst for degradation of organic contaminants. J Hazard Mater 369:494–502. https://doi.org/10.1016/j.jhazmat.2019.02.046
79. Lyle SJ, Waller PJ, Yaghi OM (2019) Covalent organic frameworks: organic chemistry extended into two and three dimensions. Trends Chem 1:172–184. https://doi.org/10.1016/j.trechm.2019.03.001
80. Lyu H, Diercks CS, Zhu C, Yaghi OM (2019) Porous crystalline olefin-linked covalent organic frameworks. J Am Chem Soc 141:6848–6852. https://doi.org/10.1021/jacs.9b02848
81. Ma HC, Kan JL, Chen GJ et al (2017) Pd NPs-loaded homochiral covalent organic framework for heterogeneous asymmetric catalysis. Chem Mater 29:6518–6524. https://doi.org/10.1021/acs.chemmater.7b02131
82. Ma T, Kapustin EA, Yin SX et al (2018) Single-crystal x-ray diffraction structures of covalent organic frameworks. Science 361:48–52. https://doi.org/10.1126/science.aat7679
83. Ma YX, Li ZJ, Wei L et al (2017) A dynamic three-dimensional covalent organic framework. J Am Chem Soc 139:4995–4998. https://doi.org/10.1021/jacs.7b01097
84. Ma DL, Qian C, Qi QY et al (2021) Effects of connecting sequences of building blocks on reticular synthesis of covalent organic frameworks. Nano Res 14:381–386. https://doi.org/10.1007/s12274-020-2723-y
85. Ma L, Wang X, Deng D et al (2015) Five porous zinc(II) coordination polymers functionalized with amide groups: cooperative and size-selective catalysis. J Mater Chem A 3:20210–20217. https://doi.org/10.1039/c5ta06248a
86. Ma W, Yu P, Ohsaka T, Mao L (2015) An efficient electrocatalyst for oxygen reduction reaction derived from a co-porphyrin-based covalent organic framework. Electrochem Commun 52:53–57. https://doi.org/10.1016/j.elecom.2015.01.021
87. Medina DD, Sick T, Bein T (2017) Photoactive and conducting covalent organic frameworks. Adv Energy Mater 7:1700387. https://doi.org/10.1002/aenm.201700387

88. Mu M, Wang Y, Qin Y et al (2017) Two-dimensional imine-linked covalent organic frameworks as a platform for selective oxidation of olefins. ACS Appl Mater Interfaces 9:22856–22863. https://doi.org/10.1021/acsami.7b05870
89. Nagai A, Chen X, Feng X et al (2013) A squaraine-linked mesoporous covalent organic framework. Angew Chem Int Ed 52:3858–3862. https://doi.org/10.1002/anie.201300256
90. Nagai A, Guo Z, Feng X et al (2011) Pore surface engineering in covalent organic frameworks. Nat Commun 2:1–8. https://doi.org/10.1038/ncomms1542
91. Nandi S, Singh SK, Mullangi D et al (2016) Low band gap benzimidazole COF supported Ni_3N as highly active OER catalyst. Adv Energy Mater 6:1601189. https://doi.org/10.1002/aenm.201601189
92. Narayanam JMR, Stephenson CRJ (2011) Visible light photoredox catalysis: applications in organic synthesis. Chem Soc Rev 40:102–113. https://doi.org/10.1039/b913880n
93. Pachfule P, Acharjya A, Roeser J et al (2019) Donor-acceptor covalent organic frameworks for visible light induced free radical polymerization. Chem Sci 10:8316–8322. https://doi.org/10.1039/c9sc02601k
94. Pachfule P, Kandambeth S, Díaz Díaz D, Banerjee R (2014) Highly stable covalent organic framework-Au nanoparticles hybrids for enhanced activity for nitrophenol reduction. Chem Commun 50:3169–3172. https://doi.org/10.1039/c3cc49176e
95. Philp D, Fraser Stoddart J (1996) Self-assembly in natural and unnatural systems. Angew Chem (Int Ed Engl) 35:1154–1196. https://doi.org/10.1002/anie.199611541
96. Preet K, Gupta G, Kotal M et al (2019) Mechanochemical synthesis of a new triptycene-based imine-linked covalent organic polymer for degradation of organic dye. Cryst Growth Des 19:2525–2530. https://doi.org/10.1021/acs.cgd.9b00166
97. Pyles DA, Crowe JW, Baldwin LA, McGrier PL (2016) Synthesis of benzobisoxazole-linked two-dimensional covalent organic frameworks and their carbon dioxide capture properties. ACS Macro Lett 5:1055–1058. https://doi.org/10.1021/acsmacrolett.6b00486
98. Rao MR, Fang Y, de Feyter S, Perepichka DF (2017) Conjugated covalent organic frameworks via michael addition-elimination. J Am Chem Soc 139:2421–2427. https://doi.org/10.1021/jacs.6b12005
99. Ren S, Bojdys MJ, Dawson R et al (2012) Porous, fluorescent, covalent triazine-based frameworks via room-temperature and microwave-assisted synthesis. Adv Mater 24:2357–2361. https://doi.org/10.1002/adma.201200751
100. Ritchie LK, Trewin A, Reguera-galan A et al (2010) microporous and mesoporous materials synthesis of COF-5 using microwave irradiation and conventional solvothermal routes. Microporous Mesoporous Mater 132:132–136. https://doi.org/10.1016/j.micromeso.2010.02.010
101. Rogge SMJ, Bavykina A, Hajek J et al (2017) Metal-organic and covalent organic frameworks as single-site catalysts. Chem Soc Rev 46:3134–3184. https://doi.org/10.1039/c7cs00033b
102. Ross GJ, Watts JF, Hill MP, Morrissey P (2000) Surface modification of poly(vinylidene fluoride) by alkaline treatment: 1. The degradation mechanism. Polymer 41:1685–1696. https://doi.org/10.1016/S0032-3861(99)00343-2
103. Ross GJ, Watts JF, Hill MP, Morrissey P (2001) Surface modification of poly(vinylidene fluoride) by alkaline treatment Part 2. Process modification by the use of phase transfer catalysts. Polymer 42:403–413. https://doi.org/10.1016/S0032-3861(00)00328-1
104. Segura JL, Mancheño MJ, Zamora F (2016) Covalent organic frameworks based on Schiff-base chemistry: Synthesis, properties and potential applications. Chem Soc Rev 45:5635–5671. https://doi.org/10.1039/C5CS00878F
105. Segura JL, Royuela S, Mar Ramos M (2019) Post-synthetic modification of covalent organic frameworks. Chem Soc Rev 48:3903–3945. https://doi.org/10.1039/c8cs00978c
106. Seo W, White DL, Star A (2017) Fabrication of holey graphene: catalytic oxidation by metalloporphyrin-based covalent organic framework immobilized on highly ordered pyrolytic graphite. Chem Eur J 23:5652–5657. https://doi.org/10.1002/chem.201605488
107. Sharma RK, Yadav P, Yadav M et al (2020) Recent development of covalent organic frameworks (COFs): synthesis and catalytic (organic-electro-photo) applications. Mater Horiz 7:411–454. https://doi.org/10.1039/c9mh00856j

108. Shunmughanathan M, Puthiaraj P, Pitchumani K (2015) Melamine-based microporous network polymer supported palladium nanoparticles: a stable and efficient catalyst for the Sonogashira coupling reaction in water. ChemCatChem 7:666–673. https://doi.org/10.1002/cctc.201402844
109. Stegbauer L, Schwinghammer K, Lotsch B, v. (2014) A hydrazone-based covalent organic framework for photocatalytic hydrogen production. Chem Sci 5:2789–2793. https://doi.org/10.1039/c4sc00016a
110. Stegbauer L, Zech S, Savasci G et al (2018) Tailor-made photoconductive pyrene-based covalent organic frameworks for visible-light driven hydrogen generation. Adv Energy Mater 8:1703278. https://doi.org/10.1002/aenm.201703278
111. Stoddart JF (2009) Thither supramolecular chemistry? Nat Chem 1:14–15. https://doi.org/10.1038/nchem.142
112. Sun Q, Aguila B, Perman J et al (2017) postsynthetically modified covalent organic frameworks for efficient and effective mercury removal. J Am Chem Soc 139:2786–2793. https://doi.org/10.1021/jacs.6b12885
113. Szczesniak B, Borysiuk S, Choma J, Jaroniec M (2020) Mechanochemical synthesis of highly porous materials. Mater Horiz 7:1457–1473. https://doi.org/10.1039/D0MH00081G
114. Tilford RW, Gemmil RW et al (2006) Facile synthesis of a highly crystalline, covalently linked porous boronate network. Chem Mater 18:5296–5301. https://doi.org/10.1021/cm061177g
115. Uribe-Romo FJ, Hunt JR, Furukawa H et al (2009) A crystalline imine-linked 3-D porous covalent organic framework. J Am Chem Soc 131:4570–4571. https://doi.org/10.1021/ja8096256
116. Uribe-Romo FJ, Doonan CJ, Furukawa H, et al (2011) Crystalline covalent organic frameworks with hydrazone linkages. J Am Chem Soc 133:11478–611481. https://doi.org/10.1021/ja204728y
117. Vardhan H, Hou L, Yee E et al (2019) Vanadium docked covalent-organic frameworks: an effective heterogeneous catalyst for modified Mannich-type reaction. ACS Sustainable Chemistry and Engineering 7:4878–4888. https://doi.org/10.1021/acssuschemeng.8b05373
118. Wang X, Ma R, Hao L et al (2018) Mechanochemical synthesis of covalent organic framework for the efficient extraction of benzoylurea insecticides. J Chromatogr A 1551:1–9. https://doi.org/10.1016/j.chroma.2018.03.053
119. Wang H, Zeng Z, Xu P et al (2019) Recent progress in covalent organic framework thin films: fabrications, applications and perspectives. Chem Soc Rev 48:488–516. https://doi.org/10.1039/c8cs00376a
120. Wei H, Chai S, Hu N et al (2015) The microwave-assisted solvothermal synthesis of a crystalline two-dimensional covalent organic framework with high CO_2 capacity. Chem Commun 51:12178–12181. https://doi.org/10.1039/c5cc04680g
121. Wei PF, Qi MZ, Wang ZP et al (2018) Benzoxazole-linked ultrastable covalent organic frameworks for photocatalysis. J Am Chem Soc 140:4623–4631. https://doi.org/10.1021/jacs.8b00571
122. Xiong C, Ning H, Jia G et al (2014) Towards covalent organic frameworks with predesignable and aligned open docking sites. Chem Commun 50:6161–6163. https://doi.org/10.1039/c4cc01825g
123. Xu H, Chen X, Gao J et al (2014) Catalytic covalent organic frameworks via pore surface engineering. Chem Commun 50:1292–1294. https://doi.org/10.1039/c3cc48813f
124. Xu H, Gao J, Jiang D (2015) Stable, crystalline, porous, covalent organic frameworks as a platform for chiral organocatalysts. Nat Chem 7:905–912. https://doi.org/10.1038/nchem.2352
125. Xue R, Gou H, Zheng Y et al (2020) A new squaraine-linked triazinyl-based covalent organic frameworks: preparation, characterization and application for sensitive and selective determination of Fe^{3+} cations. ChemistrySelect 5:10632–10636. https://doi.org/10.1002/slct.202002232
126. Yaghi OM (2019) Reticular chemistry in all dimensions. ACS Cent Sci 5:1295–1300. https://doi.org/10.1021/acscentsci.9b00750

127. Yaghi OM, O'Keeffe M, Ockwig NW et al (2003) Reticular synthesis and the design of new materials. Nature 423:705–714. https://doi.org/10.1038/nature01650
128. Yu W, Gu S, Fu Y et al (2018) Carbazole-decorated covalent triazine frameworks: novel nonmetal catalysts for carbon dioxide fixation and oxygen reduction reaction. J Catal 362:1–9. https://doi.org/10.1016/j.jcat.2018.03.021
129. Yusran Y, Fang Q, Valtchev V (2020) Electroactive covalent organic frameworks: design, synthesis, and applications. Adv Mater 32:2002038. https://doi.org/10.1002/adma.202002038
130. Yusran Y, Li H, Guan X et al (2020) Covalent organic frameworks for catalysis. EnergyChem 2. https://doi.org/10.1016/j.enchem.2020.100035
131. Zhang P, Dai S (2017) Mechanochemical synthesis of porous organic materials. Journal of Materials Chemistry A 5:16118–16127. https://doi.org/10.1039/C7TA04829G
132. Zhang W, Jiang P, Wang Y et al (2015) Bottom-up approach to engineer two covalent porphyrinic frameworks as effective catalysts for selective oxidation. Catal Sci Technol 5:101–104. https://doi.org/10.1039/c4cy00969j
133. Zhang FM, Sheng JL, di Yang Z et al (2018) Rational design of MOF/COF hybrid materials for photocatalytic H2 evolution in the presence of sacrificial electron donors. Angew Chem Int Ed 57:12106–12110. https://doi.org/10.1002/anie.201806862
134. Zhang M, Zhang M, Zhang M et al (2020) electron beam irradiation as a general approach for the rapid synthesis of covalent organic frameworks under ambient conditions. J Am Chem Soc 142:9169–9174. https://doi.org/10.1021/jacs.0c03941
135. Zhang M, Lu M, Lang ZL, et al (2020a) Semiconductor/covalent-organic-framework Z-scheme heterojunctions for artificial photosynthesis. Angew Chemie Int Ed 59:12106–12110. https://doi.org/10.1002/anie.202000929
136. Zhao S, Dong B, Ge R et al (2016) Channel-wall functionalization in covalent organic frameworks for the enhancement of CO_2 uptake and CO_2/N_2 selectivity. RSC Adv 6:38774–38781. https://doi.org/10.1039/c6ra04859e
137. Zhao F, Liu H, Mathe SDR et al (2018) Covalent organic frameworks: from materials design to biomedical application. Nanomaterials 8:15. https://doi.org/10.3390/nano8010015
138. Zhi Y, Li Z, Feng X et al (2017) Covalent organic frameworks as metal-free heterogeneous photocatalysts for organic transformations. J Mater Chem A 5:22933–22938. https://doi.org/10.1039/c7ta07691f
139. Zhu J, Yang C, Lu C et al (2018) Two-dimensional porous polymers: from sandwich-like structure to layered skeleton. Acc Chem Res 51:3191–3202. https://doi.org/10.1021/acs.accounts.8b00444
140. Zhu D, Zhang Z, Alemany LB et al (2021) Rapid, ambient temperature synthesis of imine covalent organic frameworks catalyzed by transition-metal nitrates. Chem Mater 33:3394–3400. https://doi.org/10.1021/acs.chemmater.1c00737

MOFs as Heterogeneous Catalysts

Metal–Organic Frameworks (MOFs) as Heterogeneous Catalysts: An Overview

Sushma Yadav and Priti Malhotra

Contents

1 Introduction 356
2 Synthesis of Metal–Organic Frameworks 357
3 Conventional Solvothermal Synthesis of MOF 357
4 Catalysis by MOF 358
5 Photocatalytic Role of MOF-Based Heterogeneous Catalyst 360
5.1 Electrocatalysis by MOF-Based Molecules 362
6 Conclusion and Perspective 363
Abbreviations 364
References 364

Abstract Researchers experience great thrills when they can design their catalysts to synthesize their target materials and such opportunities are possible by metal–organic frameworks (MOFs). These are hybrid materials demolishing the demarcation between organic and inorganic molecules. The inherent problems encountered with homogeneous catalysts have been effortlessly overcome using heterogeneous catalysts in the form of MOFs. Their intrinsic thermocatalytic features and ability to act as perfect hosts for metal nanoparticles (NPs) and as templates for designing new nanocatalysts are their assets that can be explored for various heterogeneous catalytic processes. The engineered MOFs possess programmable catalytic profiles which can be exploited for a wide range of reactions. The underlying chapter critically reviews the heterogeneous catalytic performances of MOFs in view of their dynamic bonds between the metal centers and organic molecules which offer them as highly localized and effective sites for suitable catalysis.

P. Malhotra (✉)
Department of Chemistry, Daulat Ram College, University of Delhi, Delhi 110007, India

S. Yadav
Department of Chemistry, University of Delhi, Delhi 110007, India

S. Gulati (ed.), *Metal-Organic Frameworks (MOFs) as Catalysts*,
https://doi.org/10.1007/978-981-16-7959-9_13

Graphical Abstract

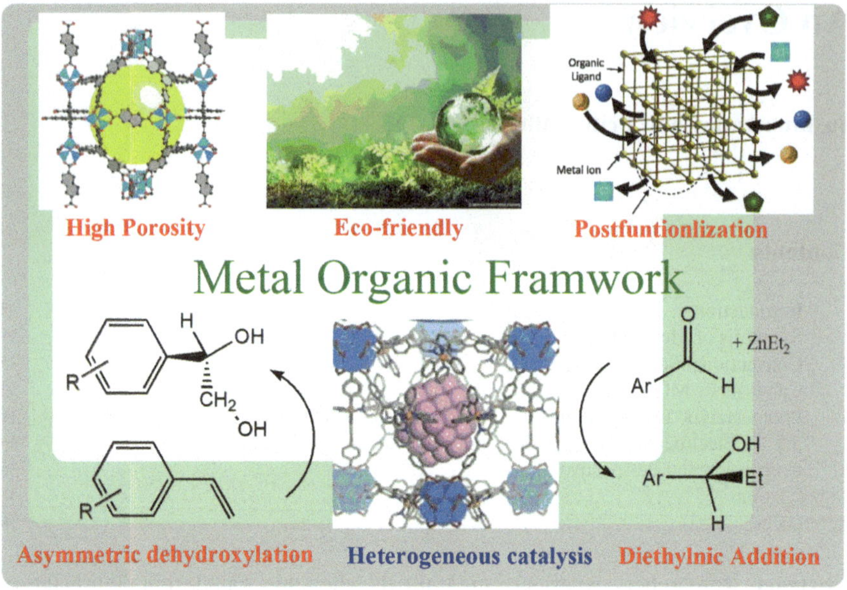

Keywords Heterogeneous · Metal–organic frameworks (MOFs) · Catalyst

1 Introduction

Metal–organic frameworks (MOFs) have gained enormous significance in material sciences because of their extraordinary properties such as large surface area, high porosity, and modifiable pore volumes [1]. All of which enable the MOFs to possess enhanced absorption capability of the matrices in the UV–Visible range thereby meeting the requirements of a broad range of applications. Research areas in the past decade have largely been built around MOFs and scientists have been able to trap its ubiquitous advantages for their fruitful utilization into several fields. The most premier research in the area was done by Robson and co-workers [1–4] which was later continued by Kitagawa et al. [5–7] Yaghi and co-workers [8, 9], and Ferry et al. [10]. Since the crystalline MOFs are derived from both organic and inorganic building units forming clusters of metal ions coordinated to rigid and stable organic molecules leading to two- or three-dimensional pores in the resulting structures. They possess a higher grade of porosity as compared to inorganic zeolites and exhibit exceptional flexibility. Both their thermal stability and chemical stabilities surpass the individual stability of the host or the guest molecules.

From the past 15 years, MOFs have emerged to be recognized as a rising generation of catalysts especially significant as heterogeneous catalysts. They are acknowledged as superior catalysts as compared to zeolites because of their extraordinarily fascinating properties namely abundance of catalytic surface and a greater number of catalytic sites, tuneable assembly, and adjustable pore volume. Scientists have contributed their great efforts and have succeeded in converting organic compounds and various gas molecules such as CO_2 into high value-added chemicals linking them with MOFs and also nanoscaled MOFs.

The explosion in the MOF centered research is due to their extensive catalytic efficiency mainly heterogeneous catalysis. Their versatility is accountable by not only their hybrid nature utilizing the synergism of the characteristics of the mixed components but also the stability and rigidity imparted by their interactions. A manifold of interest is aroused among engineers and scientists of material sciences pertinent to the MOF systems and as the diversity in their combination of constituents is growing their utilization advantages are opening further novel research gates. For example, as a useful consequence of their permanent porosity and functionality, MOFs based on carboxylates and imidazolates were exploited for not only catalytic applications [11, 12] but also for drug delivery [2, 3], carbon-sequestering [11, 13] hydrogen gas storage [14] and separation of impurities [11, 15, 16].

2 Synthesis of Metal–Organic Frameworks

Synthesis of metal–organic frameworks follows varied methods ranging from conventional solvothermal synthesis, dry-gel conversion, microwave-assisted synthesis, electrochemical, sonochemical, mechanochemical, microfluidic synthesis, ionothermal conversion methods. [17]. These methods are prevalent for commonly used metals such as Cu^{2+}, Zn^{2+}, Fe^{3+}, Al^{3+}, Cr^{3+}, and Zr^{4+} especially with commonly employed organic ligand namely 1,3,5-benzenetricarboxylic acid and functionalized 1,4-benzenedicarboxylic acid [18].

3 Conventional Solvothermal Synthesis of MOF

Conventional electrical heating is commonly employed for MOFs synthesis at small-scale solvothermally (Fig. 1). One of the widely employed techniques in microwave synthesis using the hydrothermal Fig. 1 [19]. It bears the advantage of phase selectivity and tuneable size and morphology control [20–22]. The microfluidic method of synthesizing MOF is commercially viable as it is a rapid and continuous process and suitable for industrial needs [23].

The electrochemical synthesis methodology of MOFs involves the anodic discharge of metal ions into the solution continuously and its interaction with the organic molecule in the solution leading to the formation of electrochemically driven

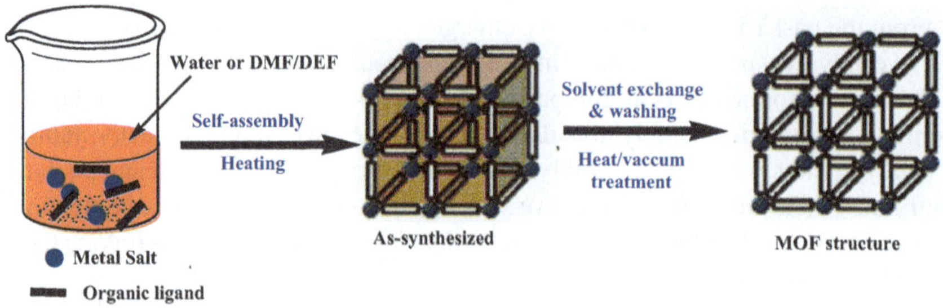

Fig. 1 Conventional solvothermal synthesis of MOF

MOF which is also a continuous process [24]. Another method of synthesis of MOF is the mechanochemical synthesis scheme where bonds are broken mechanically, followed by chemical transformations [24, 25].

The distinguishing characteristics of MOFs which are currently being explored by scientists for various applications include:

1. High-adsorption capacity
2. High-grade porosity
3. High crystallinity
4. High-thermal stability and (>400 °C) and chemical stability
5. Magnetic properties induced by metal and metal oxide components
6. Semiconducting nature
7. Tuneable compatibility with both organic and inorganic materials
8. Possibility of synthesizing biocompatible frameworks
9. Exceptional response to external stimuli including UV–Visible light.

4 Catalysis by MOF

MOFs and their derivatives from the past two decades have been attracting loads of attention toward their catalytic behaviour as their role in several organic reactions is well recognized. Attributed to their active functionality linked to both metallic and organic centers, their versatility as heterogeneous catalysts is quite fascinating [26–28]. However, as compared to conventional pristine MOFs and their derived molecules, their nanoscaled counterparts or their nanocomposites seem to be superior and show greater benefits in catalytic performances. The reason being that conventional MOFs have a longer path for diffusion and so the mass transfer and exposure to active sites get limited. The nano-supported MOFs reap the benefits of nanoscaled materials and the stability of organic framework both and lead to greater efficiency of these nanocatalysts showing integrated advantages. Their facile workability, enhanced stability, and reusability endow them with exceptional qualities as heterogeneous catalysts, a few of which are cited as follows:

1. Zeolitic imidazolate frameworks (ZIFs) are commonly used and consist of imidazolate linked to metal ions in a bivalent state and its tetrahedral structure mimics the aluminosilicate zeolite and so they demonstrate the structural and chemical advantages of both MOFs and zeolites leading to enhanced surface properties, modifiable porosity, variable functionality, and high-grade stability both thermal and chemical [29]. Nanoscaled MOFs such as ZIF and bimetallic CoZn-ZIF supported on MgAl–LDH (layered double hydroxide) were employed as a versatile heterogeneous catalyst for the Knoevenagel condensation reaction and reduction of 4-nitrophenol to 4-aminophenol. These MOF nanocomposites can be used both as catalysts and as catalyst precursors because of the dispersing and directing potential of LDH.
2. Synergistic photocatalytic application of MOF and its derivatives: Photocatalysis is a significant category of catalyst among heterogeneous catalysts and is recognized as the most important methodology for sustainable activity. Utilizing solar energy for cleaning the environment has multifold advantages that can fruitfully be harvested for sustainability. For example, photocatalytic reduction of organic contaminants of wastewater which includes all organic dyes and pharmaceutical wastes, CO_2 reduction for the production of fuels, fixation of CO_2 for energy storage as in lithium batteries and H_2 production, etc. The environmentally benign nature of such activities, cost-effectiveness, long-term stability, and high efficiency are their major highlights and puts them into the brightest spotlight for environmentalists and material scientists. In recent research, photocatalytic efficacy of MOFs has been widely explored [30, 31]. After irradiation of light, MOFs undergo recombination of electron holes at their active sites whose efficacy gets augmented at the heterojunction of the nanocomposites [32]. Several studies on the heterogeneous catalytic applications have been conducted so far such as numerous MOFs which include MOF-5 [33] and MOF-253-Pt [34]. Other similar molecules are UiO-66, NTU-9, UiO-67, MIL-53. Similar to many other photocatalysts, the great advantage of MOF lies in its tuneable nature for light adsorption apart from tuneable pore size and escalated surface area.
3. Catalytic performance of ultra-fine CrPd NPs anchored on a MIL-101-NH_2 system: Gao et al. [35] designed through a chemically wet approach and created CrPd NPs@ MIL-101-NH_2 which showed excellent catalytic behaviour for generating hydrogen from formic acid in an aqueous medium. The CrPd NPs were effectively dispersed on MOFs and their catalytic efficiency paved way for innovations on many metallic combinations which could be functionalized and supported MOFs for various other catalytic events in an aqueous medium [35].
4. Azolium based MOFs with metal complexes as catalysts: Wu et al. proposed that few of the ruthenium complexes such as $RuCl_3$, $[RuCp^*Cl_2]_2$ (Cp^* 1/4 pentamethylcyclopentadienyl) and $[Ru(C_6Me_6)Cl_2]_2$ (C_6Me_6 1/4 hexamethylbenzene) could be effectively immobilized getting supported on an azolium based MOF and the resulting heterogeneous ruthenium-based catalyst was designated as Ru_x-NHC-MOF (x1/4 1, 2, 3; NHC 1/4N-heterocyclic carbine).

This heterogeneous catalyst after getting retrieved had enhanced the electron-donating capacity of the ligand C_6Me_6 thereby showing high catalytic ability toward hydrogenation of CO_2 to give rise to formic acid [36]. The catalyst recovery was also high for subsequent usage.

5. Iron carbonyl-based MOF as a recyclable heterogeneous catalyst VNU-21 ($Fe_3(BTC)(EDB)_2$ 12.27 H_2O) were engaged for catalyzing the preparation of quinazolinones in a dual step method in the presence of oxygen. Employing VNU-21 the decarboxylation process of phenylacetic acid was carried out in the initial step which was later followed by the oxidative type of cyclization scheme using 2- amino benzamides leading to the formation of quinazolinones by the metal-free initiation process in the final step. VNU-21 showed excellent heterogeneous catalytic efficiency for quinazolinones synthesis and have proved to be beneficially for industries and material scientist [37].
6. MOF supported catalyst for Frischer Tropsch reaction: Sun et al. observed that their novel MOF-supported MIL-53(Al) (Al (OH)- $[O_2C\text{-}C_6H_4\text{-}CO_2]$) were highly stable thermally and provided extraordinary porosity in its cobalt-based matrix for catalyzing Fischer Tropsch reaction which was enabled by the MOF-based immobilization of Co NPs. The Co/MIL-53(Al) hybrids were able to catalyze the reaction such that the maximum yield of diesel and gasoline is obtained [38].
7. The catalytic activity of the Bronsted-Lewis acidic $[(CH_2COOH)_2IM]HSO_4$@H-UiO-66: Ye et al. attempted the combination of $[(CH_2COOH)_2IM]HSO_4$, Zr ions, and H-UiO-66. To give rise to the catalyst $[(CH_2COOH)_2IM]HSO_4$@H-UiO-66 which was used for the esterification reaction of oleic acid and methanol with more than 90% yield of biodiesel and high-grade reusability [39].
8. Shape selective catalyst MIL-101-NH_2 for effective catalysis. Novel shape-selective catalysts were developed by Chong et al. anchoring on MIL-101-NH_2 nanocavities bridging through acylamino groups to create IL (OAc-)-MIL-101-NH_2 which solved the purpose of immobilizing the ionic liquid into the MOF. The catalyst gave outstanding outcomes for the formation of 3-aryl-2-oxazolidinones as their shape-selective nature is enabled by MOF confinement along with the mixing diverse characteristics of its components. Thus, an efficient heterogeneous catalyst evolved for facile synthesis of 3-aryl-2-oxazolidinones by encompassing ILs as the organic catalytic molecule into MOF [40].

5 Photocatalytic Role of MOF-Based Heterogeneous Catalyst

MOFs are the recently recognized photocatalysts in the promising field of photocatalysis in order to harness solar energy and convert it to chemical energy. Photocatalysis by MOFs is attributed to their inherent structural features providing large surface area, organized porosity, and majorly their via modulation of metal clusters interwoven with organic molecules which facilitates multiple activities such as charge separation, activation of reactants through adsorption, and highly significant light adsorption. Collectively, it leads to the emergence of high-grade photocatalysts addressing many solutions and applications of the current era giving rise to a promising new technology [41].

To highlight the photocatalytic versatility of MOFs (Fig. 2), few examples are cited in the following content:

1. Photocatalytic production of a hydrogen-A new class of 2D- 2D hybrid interface was structured by Cao et al. for the quantum production of hydrogen photocatalytically. UNiMOF/g-C_3N_4 (UNG) was designed which was found to be suitable for the production of hydrogen following the self-assembly pathway electrostatically [42]. In another scientific report, Bibi et al. have proposed the design of NH_2-MIL-125/TiO_2/CdS which bore MOF in H-TiO_2/CdS cavities and the resulting heterojunctions were capable of generating the highest amount of hydrogen in just one hour exposure to visible light [43].

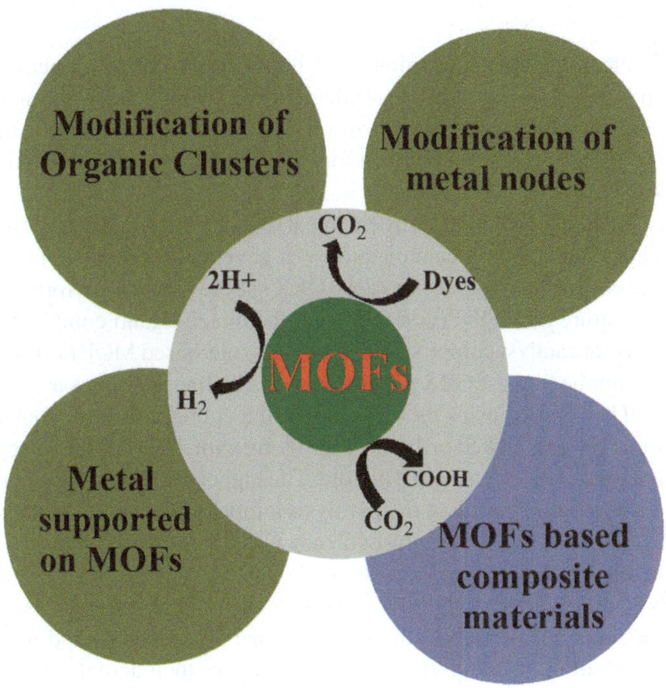

Fig. 2 Photocatalytic versatility of MOFs

2. Zn, Au, Pd and Pt-based MOF nano shuttles for photocatalytic conversion of CO_2: Han et al. have structured metal NPs and mixed metal NPs in the MOF-74 matrix such as Pt/MOF-74, Au@Pd@MOF-74, and Pt/ Au@Pd@MOF-74 for catalytic conversion of CO_2 in reverse water-gas shift reaction among which Au@Pd@MOF-74 show 100% selectivity for CO. Another catalyst Pt/Au@Pd@MOF-74 initiates the conversion of CO_2 to CH_4 at a bare minimum concentration of catalyst in addition to the 99.6% selectivity of CO during the change of CO_2 to CO [44].
3. Photo-induced oxidation of water employing Ti-MOF @ Fe_2O_3 nanorods: Li et al. managed to enhance the photochemical performance of Fe_2O_3 nanorods multifold by linking it with Ti-MOF as an even and stable $Ti_xFe_1{-}xOy$ shell were formed on Fe_2O_3 nanorods which enabled higher photon density, wider charge separation between the electrodes when used photo electrochemically and also facilitating the transfer of charges between the porous holes and the electrolytes thereby promoting the oxidation of water molecule [45].

5.1 *Electrocatalysis by MOF-Based Molecules*

Research in electrocatalysts has pinned all hopes on creating inexpensive, facile, effective, and low-emission catalytic products. Their overall goal is to innovate such effective catalysts which are stable, cost-effective, and are able to release oxygen through oxygen release reactions meant for metal-air batteries and catalytic splitting of water [46, 47].

In fuel cells, the reaction for electrochemical reduction of oxygen takes place at the cathode and is a highly significant reaction as it governs the efficacy of the fuel cell. The electrochemical reduction of oxygen (ORR) at the cathode performs according to the hybrid nature of MOF depending upon the metal–ligand combination. It has been reported that catalyst comprising of Co-imidazole-based MOF (ZIF-67) (Fig. 3) is able to exhibit the highest ORR performance in alkaline as well as acidic nature of electrolytes. The catalyst owes its versatility to the coordination of aromatic ligand which contains N and is linked to the Co^{3+} ion along with ZIF-67 thereby generating a highly active site for its electrocatalytic role. The highest performance of the catalyst is achieved when optimization of the pyrolysis temperature and the pathway of acid leaching is done such that the catalyst efficacy turns out to be comparable to that of Pt/C in acid and alkaline environment both [48].

Metal NPs and mixed metal NPs encapsulated in MOF nanosheets are highly effective electrochemical catalysts that demonstrate bifunctionality and are employed for energy generation. It is important to further enhance their activity, durability, and stability so that oxygen evolution reaction and oxygen reduction reaction can both be easily carried out electrochemically. For example, Pt NPs @MOFs demonstrating low over potential gaps (665 mV E_{gap}) were employed in Zn-air batteries for potential results in the evolution of oxygen [49]. It was also observed that apart from low over potential gaps, high faradaic efficiency up to 99.5% for electrostatic oxygen

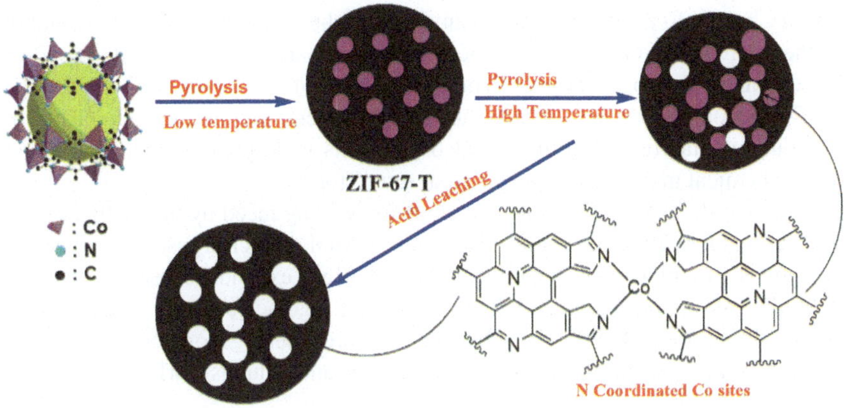

Fig. 3 Scheme for preparing and optimizing Co @ZIF-67

generation was attained in which there was also a substantial contribution of the locally centered crystallinity and disorder of long-range in the various bimetallic MOFs as mentioned by Li et al. [50].

Another electrochemical catalyst MoS_2/Co-MOF nanocomposite has been reported by Zhu et al. which provides high efficiency in catalyzing hydrogen evolution reaction (HER). Its major advantage is that it is environmentally friendly and highlights clean energy production. The nano hetero junction endorses synergism of Co-MOF along with MoS_2 not only enhancing the surface area but also the redox potential of the electrochemical catalytic activity [51].

6 Conclusion and Perspective

Out of the various strategies for nanoscaled catalysis, the MOF-based nanocomposites were one of the most successful strategies to enhance their catalytic performance as they could be fabricated according to the desired behaviour. The improvements can be manipulated by modifying the electronic state through the metals incorporated, the pore volume and its environment, the organic linkage, and post-fabrication modifications on the active sites of the skeleton. The overall benefits of MOF-based structures arising from the inherently outstanding organic units and diverse topological features have paved the way for many newly fabricated MOFs and their successful applications in numerous catalytic performances. MOFs are a fast-emerging class of compounds to govern the world of heterogeneous catalysis and truly the next generation of heterogeneous catalysis enabling large-scale synthetic processes for the production of value-added compounds giving enormous boost to both the scientific world and industrial growth. They are surely the future substitute for homogeneous catalysis as their heterogeneous catalytic events are slated for major advantages and benefits creating a world of merger organic and inorganic systems. The industry banking on fine chemicals can see a bright future of MOF-based catalytic synthesis of various chemicals to generate scaled-up and escalated yield of products thriving on the heterogeneous catalytic performance of MOFs, MOF-based NPs, and MOF nanocomposites. MOFs facilitate the implementation of green chemistry not only due to catalytic versatility but also due to its widespread photochemical and electrochemical marvels stamping a great future for them.

However, there are a few practical challenges being faced by the methodologies used during fabrication and application events of MOFs which arouse more interest of the scientist to explore novel techniques in this field and expand the research prospects of MOFs. As opportunities and challenges exist together even for heterogeneous catalysts by MOFs and MOF-based hybrids, they are arousing more focus of scientific research as their capabilities entice many interests and arduous work are expected in near future in this area.

Additionally, nanoscale MOFs-based heterogeneous catalysts have generated special enthusiasm of researchers and a keen interest to explore novel methodologies, discoveries, and phenomena.

Abbreviations

MOF	Metal organic framework
ZIFs	Zeolitic imidazolate frameworks
LDH	Layered double hydroxides
ORR	Reduction of oxygen
NPs	Nanoparticles

References

1. Lin Z, Bian W, Zheng J, Cai Z (2015) Magnetic metal-organic framework nanocomposites for enrichment and direct detection of small molecules by negative-ion matrix-assisted laser desorption/ionization time-of-flight mass spectrometry. Chem Commun 51:8785–8788. https://doi.org/10.1039/c5cc02495a
2. Fu YY, Yang CX, Yan XP (2013) Incorporation of metal-organic framework UiO-66 into porous polymer monoliths to enhance the liquid chromatographic separation of small molecules. Chem Commun 49:7162–7164. https://doi.org/10.1039/c3cc43017k
3. Chang CL, Qi XY, Zhang JW, Qiu YM, Li XJ, Wang X, Bai Y, Sun JL, Liu HW (2015) Facile synthesis of magnetic homochiral metal-organic frameworks for "enantioselective fishing." Chem Commun 51:3566–3569. https://doi.org/10.1039/c4cc09988e
4. Shih YH, Chien CH, Singco B, Hsu CL, Lin CH, Huang HY (2013) Metal-organic frameworks: new matrices for surface-assisted laser desorption-ionization mass spectrometry. Chem Commun 49:4929–4931. https://doi.org/10.1039/c3cc40934a
5. Lu G, Hupp JT (2010) Metal-organic frameworks as sensors: a ZIF-8 based fabry-pérot device as a selective sensor for chemical vapors and gases. J Am Chem Soc 132:7832–7833. https://doi.org/10.1021/ja101415b

6. Banerjee R, Phan A, Wang B, Knobler C, Furukawa H, O'Keeffe M, Yaghi OM (2008) High-throughput synthesis of zeolitic imidazolate frameworks and application to CO2 capture. Science (80-) 319:939–943. https://doi.org/10.1126/science.1152516
7. Zheng J, Cheng C, Fang WJ, Chen C, Yan RW, Huai HX, Wang CC (2014) Surfactant-free synthesis of a Fe_3O_4@ZIF-8 core-shell heterostructure for adsorption of methylene blue. CrystEngComm 16:3960–3964. https://doi.org/10.1039/c3ce42648c
8. Sonderegger H, Rameshan C, Lorenz H, Klauser F, Klerks M, Rainer M, Bakry R, Huck CW, Bonn GK (2011) Surface-assisted laser desorption/ionization-mass spectrometry using TiO_2-coated steel targets for the analysis of small molecules. Anal Bioanal Chem 401:1963–1974. https://doi.org/10.1007/s00216-011-5255-1
9. Nitta S, Kawasaki H, Suganuma T, Shigeri Y, Arakawa R (2013) Desorption/ionization efficiency of common amino acids in surface-assisted laser desorption/ionization mass spectrometry (SALDI-MS) with nanostructured platinum. J Phys Chem C 117:238–245. https://doi.org/10.1021/jp308380z
10. Chen G, Ding X, Cao Z, Ye J (2000) Determination of melatonin and pyridoxine in pharmaceutical preparations for health-caring purposes by capillary electrophoresis with electrochemical detection. Anal Chim Acta 408:249–256. https://doi.org/10.1016/S0003-2670(99)00809-0
11. Ma R, Lu M, Ding L, Ju H, Cai Z (2013) Surface-assisted laser desorption/ionization mass spectrometric detection of biomolecules by using functional single-walled carbon nanohorns as the matrix. Chem A Eur J 19:102–108. https://doi.org/10.1002/chem.201202838
12. Pan C, Xu S, Hu L, Su X, Ou J, Zou H, Guo Z, Zhang Y, Guo B (2005) Using oxidized carbon nanotubes as matrix for analysis of small molecules by MALDI-TOF MS. J Am Soc Mass Spectrom 16:883–892. https://doi.org/10.1016/j.jasms.2005.03.009
13. Walton BL, Verbeck GF (2014) Soft-landing ion mobility of silver clusters for small-molecule matrix-assisted laser desorption ionization mass spectrometry and imaging of latent fingerprints. Anal Chem 86:8114–8120. https://doi.org/10.1021/ac5010822
14. Nordström A, Apon JV, Uritboonthai W, Go EP, Siuzdak G (2006) Surfactant-enhanced desorption/ionization on silicon mass spectrometry. Anal Chem 78:272–278. https://doi.org/10.1021/ac051398q
15. Min Q, Zhang X, Chen X, Li S, Zhu JJ (2014) N-Doped graphene: an alternative carbon-based matrix for highly efficient detection of small molecules by negative ion MALDI-TOF MS. Anal Chem 86:9122–9130. https://doi.org/10.1021/ac501943n
16. Zhang H, Cha S, Yeung ES (2007) Colloidal graphite-assisted laser desorption/ionization MS and MS n of small molecules. 2. Direct profiling and MS imaging of small metabolites from fruits. Anal Chem 79:6575–6584. https://doi.org/10.1021/ac0706170
17. Safaei M, Foroughi MM, Ebrahimpoor N, Jahani S, Omidi A, Khatami M (2019) A review on metal-organic frameworks: synthesis and applications. TrAC—Trends Anal Chem 118:401–425. https://doi.org/10.1016/j.trac.2019.06.007
18. Lee YR, Kim J, Ahn WS (2013) Synthesis of metal-organic frameworks: a mini review. Korean J Chem Eng 30:1667–1680. https://doi.org/10.1007/s11814-013-0140-6
19. Park SE, Chang JS, Young KH, Dae SK, Sung HJ, Jin SH (2004) Supramolecular interactions and morphology control in microwave synthesis of nanoporous materials. Catal Surv from Asia 8:91–110. https://doi.org/10.1023/B:CATS.0000026990.25778.a8
20. Jhung SH, Chang JS, Hwang JS, Park SE (2003) Selective formation of SAPO-5 and SAPO-34 molecular sieves with microwave irradiation and hydrothermal heating. Microporous Mesoporous Mater 64:33–39. https://doi.org/10.1016/S1387-1811(03)00501-8
21. Jhung SH, Chang JS, Hwang YK, Grenèche JM, Férey G, Cheetham AK (2005) Isomorphous substitution of transition-metal ions in the nanoporous nickel phosphate VSB-5. J Phys Chem B 109:845–850. https://doi.org/10.1021/jp046188g
22. Kang KK, Park CH, Ahn WS (1999) Microwave preparation of a titanium-substituted mesoporous molecular sieve. Catal Lett 59:45–49. https://doi.org/10.1023/A:1019004101326
23. Schoenecker PM, Belancik GA, Grabicka BE, KSW, (2012) Kinetics Study and crystallization process design for scale-up of UiO-66-NH2 synthesis. AIChE J 59:215–228. https://doi.org/10.1002/aic

24. Stock N, Biswas S (2012) Synthesis of metal-organic frameworks (MOFs): routes to various MOF topologies, morphologies, and composites. Chem Rev 112:933–969. https://doi.org/10.1021/cr200304e
25. Pichon A, Lazuen-Garay A, James SL (2006) Solvent-free synthesis of a microporous metal-organic framework. CrystEngComm 8:211–214. https://doi.org/10.1039/b513750k
26. Gascon J, Aktay U, Hernandez-Alonso MD, van Klink GPM, Kapteijn F (2009) Amino-based metal-organic frameworks as stable, highly active basic catalysts. J Catal 261:75–87. https://doi.org/10.1016/j.jcat.2008.11.010
27. Opanasenko M, Dhakshinamoorthy A, Hwang YK, Chang JS, Garcia H, Čejka J (2013) Superior performance of metal-organic frameworks over zeolites as solid acid catalysts in the prins reaction: green synthesis of nopol. Chemsuschem 6:865–871. https://doi.org/10.1002/cssc.201300032
28. Chughtai AH, Ahmad N, Younus HA, Laypkov A, Verpoort F (2015) Metal-organic frameworks: versatile heterogeneous catalysts for efficient catalytic organic transformations. Chem Soc Rev 44:6804–6849. https://doi.org/10.1039/c4cs00395k
29. Li P, Zeng HC (2016) Immobilization of metal-organic framework nanocrystals for advanced design of supported nanocatalysts. ACS Appl Mater Interfaces 8:29551–29564. https://doi.org/10.1021/acsami.6b11775
30. Zhang T, Lin W (2014) Metal-organic frameworks for artificial photosynthesis and photocatalysis. Chem Soc Rev 43:5982–5993. https://doi.org/10.1039/c4cs00103f
31. Wang S, Wang X (2015) Multifunctional metal-organic frameworks for photocatalysis. Small 11:3097–3112. https://doi.org/10.1002/smll.201500084
32. Hong J, Chen C, Bedoya FE, Kelsall GH, O'Hare D, Petit C (2016) Carbon nitride nanosheet/metal-organic framework nanocomposites with synergistic photocatalytic activities. Catal Sci Technol 6:5042–5051. https://doi.org/10.1039/c5cy01857a
33. Llabre FX, Corma A, Garcia H, Valencia D, Vera C De (2007) <2007_JPC_applications for metal organic frameworks as quantum dot Semiconductors.pdf>. 80–85
34. Zhou T, Du Y, Borgna A, Hong J, Wang Y, Han J, Zhang W, Xu R (2013) Post-synthesis modification of a metal-organic framework to construct a bifunctional photocatalyst for hydrogen production. Energy Environ Sci 6:3229–3234. https://doi.org/10.1039/c3ee41548a
35. Gao D, Wang Z, Wang C, Wang L, Chi Y, Wang M, Zhang J, Wu C, Gu Y, Wang H, Zhao Z (2019) CrPd nanoparticles on NH2-functionalized metal-organic framework as a synergistic catalyst for efficient hydrogen evolution from formic acid. Chem Eng J 361:953–959. https://doi.org/10.1016/j.cej.2018.12.158
36. Wu C, Irshad F, Luo M, Zhao Y, Ma X, Wang S (2019) Ruthenium complexes immobilized on an Azolium based metal organic framework for highly efficient conversion of CO_2 into formic acid. ChemCatChem 11:1256–1263. https://doi.org/10.1002/cctc.201801701
37. To TA, Vo YH, Nguyen HTT, Ha PTM, Doan SH, Doan TLH, Li S, Le HV, Tu TN, Phan NTS (2019) Iron-catalyzed one-pot sequential transformations: synthesis of quinazolinones via oxidative Csp3–H bond activation using a new metal-organic framework as catalyst. J Catal 370:11–20. https://doi.org/10.1016/j.jcat.2018.11.031
38. Sun B, Tan H, Liu S, Lyu S, Zhang X, Zhang Y, Li J, Wang L (2019) Novel Cobalt catalysts supported on metal-organic frameworks MIL-53(Al) for the Fischer-Tropsch synthesis. Energy Technol 7. https://doi.org/10.1002/ente.201800802
39. Ye C, Qi Z, Cai D, Qiu T (2019) Design and synthesis of ionic liquid supported hierarchically porous Zr metal-organic framework as a novel Brønsted-Lewis acidic catalyst in biodiesel synthesis. Ind Eng Chem Res 58:1123–1132. https://doi.org/10.1021/acs.iecr.8b04107
40. Chong SY, Wang TT, Cheng LC, Lv HY, Ji M (2019) Metal-organic framework MIL-101-NH 2 -supported acetate-based butylimidazolium ionic liquid as a highly efficient heterogeneous catalyst for the synthesis of 3-Aryl-2-oxazolidinones. Langmuir 35:495–503. https://doi.org/10.1021/acs.langmuir.8b03153
41. Li Y, Xu H, Ouyang S, Ye J (2016) Metal-organic frameworks for photocatalysis. Phys Chem Chem Phys 18:7563–7572. https://doi.org/10.1039/c5cp05885f

42. Cao A, Zhang L, Wang Y, Zhao H, Deng H, Liu X, Lin Z, Su X, Yue F (2019) 2D–2D Heterostructured UNiMOF/g-C 3 N 4 for enhanced photocatalytic H2 production under visible-light irradiation. ACS Sustain Chem Eng 7:2492–2499. https://doi.org/10.1021/acssuschemeng.8b05396
43. Bibi R, Huang H, Kalulu M, Shen Q, Wei L, Oderinde O, Li N, Zhou J (2019) Synthesis of amino-functionalized Ti-MOF derived yolk-shell and hollow heterostructures for enhanced photocatalytic hydrogen production under visible light. ACS Sustain Chem Eng 7:4868–4877. https://doi.org/10.1021/acssuschemeng.8b05352
44. Han Y, Xu H, Su Y, Xu Z, Liang, Wang K, Wang W (2019) Noble metal (Pt, Au@Pd) nanoparticles supported on metal organic framework (MOF-74) nanoshuttles as high-selectivity CO_2 conversion catalysts. J Catal 370:70–78. https://doi.org/10.1016/j.jcat.2018.12.005
45. Li CH, Huang CL, Chuah XF, Senthil Raja D, Hsieh CT, Lu SY (2019) Ti-MOF derived TixFe1−xOy shells boost Fe_2O_3 nanorod cores for enhanced photoelectrochemical water oxidation. Chem Eng J 361:660–670. https://doi.org/10.1016/j.cej.2018.12.097
46. Lajevardi A, Tavakkoli Yaraki M, Masjedi A, Nouri A, Hossaini Sadr M (2019) Green synthesis of MOF@Ag nanocomposites for catalytic reduction of methylene blue. J Mol Liq 276:371–378. https://doi.org/10.1016/j.molliq.2018.12.002
47. Yi X, He X, Yin F, Chen B, Li G, Yin H (2019) Co-CoO-Co3O4/N-doped carbon derived from metal-organic framework: the addition of carbon black for boosting oxygen electrocatalysis and Zn-Air battery. Electrochim Acta 295:966–977. https://doi.org/10.1016/j.electacta.2018.11.142
48. Wang X, Zhou J, Fu H, Li W, Fan X, Xin G, Zheng J, Li X (2014) MOF derived catalysts for electrochemical oxygen reduction. J Mater Chem A 2:14064–14070. https://doi.org/10.1039/c4ta01506a
49. Xia Z, Fang J, Zhang X, Fan L, Barlow AJ, Lin T, Wang S, Wallace GG, Sun G, Wang X (2019) Pt nanoparticles embedded metal-organic framework nanosheets: a synergistic strategy towards bifunctional oxygen electrocatalysis. Appl Catal B Environ 245:389–398. https://doi.org/10.1016/j.apcatb.2018.12.073
50. Li J, Huang W, Wang M, Xi S, Meng J, Zhao K, Jin J, Xu W, Wang Z, Liu X, Chen Q, Xu L, Liao X, Jiang Y, Owusu KA, Jiang B, Chen C, Fan D, Zhou L, Mai L (2019) Low-crystalline bimetallic metal-organic framework electrocatalysts with rich active sites for oxygen evolution. ACS Energy Lett 4:285–292. https://doi.org/10.1021/acsenergylett.8b02345
51. Zhu M, Ma Q, Ding SY, Zhao YZ, Song WQ, Ren HP, Song XZ, Miao ZC (2019) A molybdenum disulfide and 2D metal-organic framework nanocomposite for improved electrocatalytic hydrogen evolution reaction. Mater Lett 239:155–158. https://doi.org/10.1016/j.matlet.2018.12.108

Recent Trends of Metal–Organic Frameworks in Heterogeneous Catalysis

Manoj Trivedi, Sanjay Kumar, Aryan Arora, and Kartika Goyal

Contents

1 Introduction ... 369
2 Designing MOFs with Intrinsic Catalytic Activity ... 370
2.1 MOFs Containing Open Metal Sites and Non-Decorated Structures ... 372
2.2 Catalysis Based on MOF Defects ... 373
2.3 Catalysis Based on Decorated MOFs ... 375
2.4 CO_2 Activation ... 376
2.5 Carbon-Hydrogen Bond Functionalization ... 377
2.6 Other Reactions Catalyzed by MOFs ... 378
3 MOFs Containing Metal Nanoparticles ... 378
3.1 MOFs Synthesis ... 379
3.2 Catalysis ... 383
4 Outlook, Challenges, and Future Perspectives ... 385
References ... 386

Abstract In the current scenario, various catalytic processes are used for manufacturing majority of chemical products. An efficient catalytic material is a very dynamic research field and is in great demand for the development of society. In the past decade, huge number of publications appeared on the designs of new catalytic solids such as MOFs have significantly increased. These materials have potential application in multidisciplinary fields. The advantage of heterogeneous catalysis based on MOFs includes easy product separation, recyclability, and stability. In this chapter, we have summarized the current heterogeneous aspects of MOFs in catalysis having intrinsic catalytic activity, metal nanoparticles, and precursors.

1 Introduction

Today, metal–organic frameworks (MOFs) represent a novel class of polymeric materials in material chemistry [1–3]. All over the world about 10,000 products is synthesized by the use of catalysts globally which accounts for about 15% of total world

M. Trivedi (✉) · S. Kumar · A. Arora · K. Goyal
Department of Chemistry, Sri Venkateswara College, University of Delhi, New Delhi 110021, India

S. Gulati (ed.), *Metal-Organic Frameworks (MOFs) as Catalysts*,
https://doi.org/10.1007/978-981-16-7959-9_14

GDP. The market value for industrial catalyst used in various catalytic processes is found to be around $16 billion [4]. MOF materials are promising heterogeneous catalytic systems due to their unique features: (i) diverse structures, (ii) hybrid chemical nature, (iii) presence of easily accessible vacant sites, (iv) tailorability, and lastly (v) an organized porosity. In fact, being these characteristics attracted lots of world researchers to work on catalysis engineering and their application in catalysis. The important structural characteristics of MOFs from catalytic aspects are shown in Fig. 1: (1) by the use of metal nodes with empty coordination sites; (2)using linker containing organocatalytic center; (3) photocatalytic activation; (4) as a catalytic template and their ability to encapsulate additional catalytic sites; (5) modifications of the MOF materials at the nanolevel; (6) as precursors that can be decomposed in a controlled way to yield nanoparticles; and (7) by the combination of the method mentioned above. During the past decade, we have witnessed that a huge amount of literatures on MOF catalysis have been published [5–12]. Therefore, we focused on the current trends of MOF catalysis with highlighting the energy and environmental applications.

The chapter consists of three sections that can be discussed along (i) Designing various MOF having intrinsic catalytic activity, (ii) MOFs containing metal nanoparticles, (iii) heterogeneous catalysis based on MOF materials. Overall, we believe that this chapter would be very helpful for beginner as well as experienced researchers in thrilling MOF catalysis research area.

2 Designing MOFs with Intrinsic Catalytic Activity

Since their revelation, MOFs have been considered as the best alternative of homogeneous catalysts for the catalysis community. Incontrovertibly, MOFs are represented as molecules arranged in a crystalline lattice that leads to solid material having intrinsic catalytic activity. In this regard, different MOFs have been synthesized which acted as single-site catalysts [13–16]. Currently, three methods are used to get materials having intrinsic catalytic activity as shown in Fig. 2: (i) by generating open metal sites, (ii) by defects in structure, and (iii) using different linkers. Following the same order, we have discussed the recent reports on this field. Earlier, the MOF industry has vigorously fixated on the utilization of MOFs robust predicated on metals with valency of 3 or 4, along with elongating the boundaries MOFs catalyzed reactions.

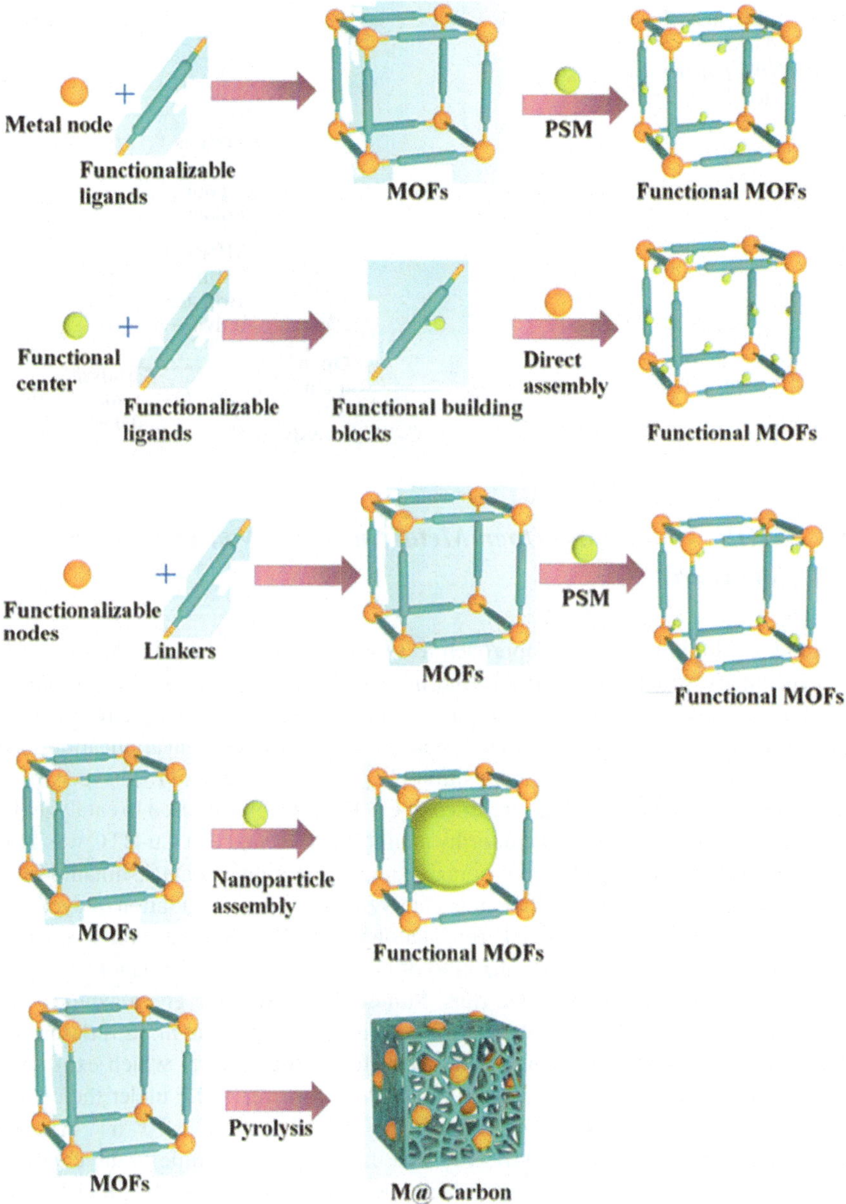

Fig. 1 Schematic representation for functionalization of various MOF materials

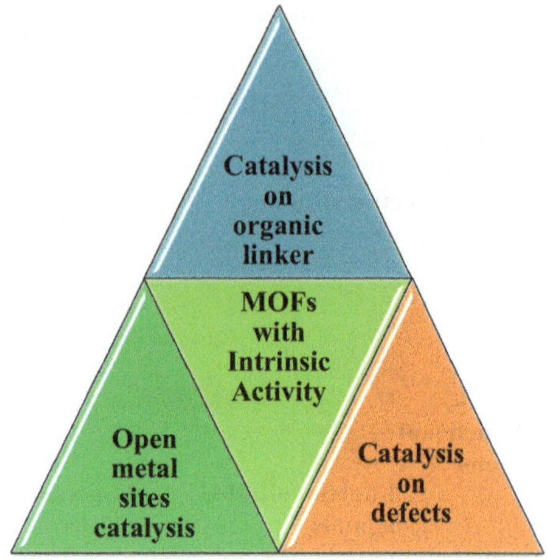

Fig. 2 Different methods were used to create catalytic active MOFs with intrinsic catalytic activity

2.1 MOFs Containing Open Metal Sites and Non-Decorated Structures

Tu et al. explored the one-pot approach to synthesize copper-based metalloligands and secondary structures based on Zinc in 2017 [17]. Properties of MOFs can be tuned by incorporating different inorganic building blocks into a single framework. The resulting MOFs (FDM-3-7) were tested as catalysts for different application such as CO oxidation and H_2O_2 decomposition [17]. Nikseresht et al. [18] have synthesized analogues of tacrine in the presence of Cu-BTC which was used to catalyze the Friedländer reaction. Dhakshinamoorthy et al. [19, 20] found that Cu-BTC was efficient in catalyzing the synthesis of borasiloxanes. Nagarjun and Dhakshinamoorthy [21] found that the FeBTC is efficient in catalyzing the aerobic oxidation of cyclooctane. Han et al. [22] synthesized iron analogue of MIL-100 that exhibited good catalytic performance in the acetalization of benzaldehyde with methanol and recycled its catalytic activity upto five runs. Sun et al. reported the green synthesis of MIL-100(Cr) under HF- and solvent-free conditions [9] Wang et al. [23] described the application of iron analogue of MIL-100 for ozone removal which exhibits a long-lasting ozone conversion efficiency of 100% for over 100 h under the conditions of a relative humidity of 45% and space velocity of 1.9×10^5 h^{-1} at room temperature. Iron analogue of MIL-100 was used for the first time in the catalytic transformation of hexose sugars into lactic acid (LA) as reported by Huang et al. [24] with 32% yield as compared to other catalysts like Cu-BTC and MIL-100(Cr). Rostamnia and Alamgholiloo [25] synthesized Iron MOF and used for Michael reaction. Han et al. successfully evacuated iron analogue of MIL-100 under various temperatures which utilized as catalysts for selective ethylene tetramerization in the presence of co-initiators [26]. MIL-100(Fe) has been subjected to evacuate at 250 °C. The resulting MOF contained high concentration of Fe^{2+} sites as well as porosity with the maximum catalytic activity. Oligomerization performance was affected by

Scheme 1 Schematic representation of Knoevenagel condensation reaction

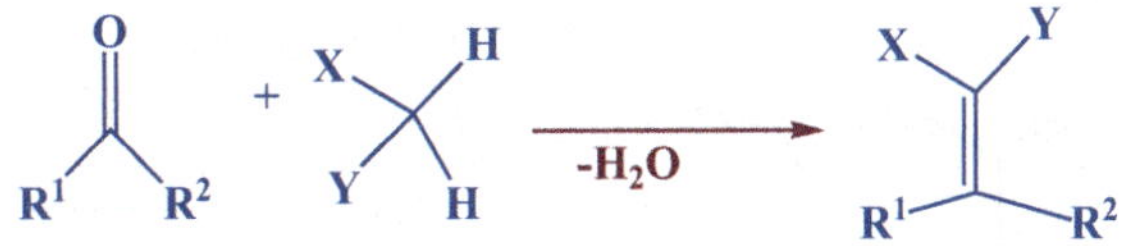

the cocatalysts via scavenging and the presence of additional open sites. Liu et al. utilized the same concept to synthesize the chromium analogue of MIL-100 [27] with moderate catalytic activities and high selectivities. Rivera-Torrente et al. [28] used a similar catalytic system in the presence of different cocatalyst. Comparative catalytic activities of MIL-100 and MIL-101 were examined. It was found that MIL-101 showed better catalytic activity compared to MIL-100. Santiago-Portillo et al. [29] synthesized 6 isostructural MIL-101-X materials by the postsynthetic treatment and used for the epoxide ring-opening, acetalization, and Prinscoupling. Yu et al. [30] synthesized some Fe-MOFs and used them in ozonation reaction. Among all Fe-MOFs, MIL-53 showed the catalytic performances owing to the presence of large Lewis acid sites and porosity [30]. Islamoglu et al. [31] synthesized a framework of Ce-BDC which proved to be effective in nerve agent simulant and soman hydrolysis. This showed faster hydrolysis as compared to UiO-66 [32]. The same group was extending their work by the use NU-901 as a catalyst and found that it also showed faster DMNP hydrolysis [33]. Shaabani et al. [34] synthesized heterocycle scaffolds by utilizing UiO-66 as a Lewis acid solid. As compared to other catalysts, this catalyst showed higher activity because of large number of vacant sites. UiO-66 was also used as catalyst in oxidative desulfurization by Zhang et al. [35] and it showed 100% conversion of ODS into dibenzothiophene. Dalapti et al. [36] synthesized a Ce-mixed valence-based MOF which helped in oxidizing the chromogenic peroxidase. Rojasec-Buzo et al. reported that Hf-based MOFs that exhibit high catalytic activity [37]. It was found that Hf-based MOFs showed a higher efficiency than Zr-based MOFs. For example, the Knoevenagel condensation reaction (Scheme 1).

Mistry et al. [38] synthesized two MOFs consisting of Cd-sulfoisophthalate layers. These layers had bipyridyl linkers forming three-dimensional structures. It also showed good catalytic activities for Knoevenagel condensation. Zhang et al. [39] used nanosheets of very thin Ni-MOF as a heterogeneous catalysis in a condensation reaction. Abdollahi and Morsali [40] synthesized a Zn MOF (TMU-41) which utilized in condensation reaction.

2.2 *Catalysis Based on MOF Defects*

Multiple deformation in MOFs has been produced by the displacement reaction by a number of different groups [41]. Molecular simulations provide a valuable information in understanding the structure and catalytic property of the MOF [14, 42]. Elemental and thermogravimetric analyses were used to quantified the defects [43].

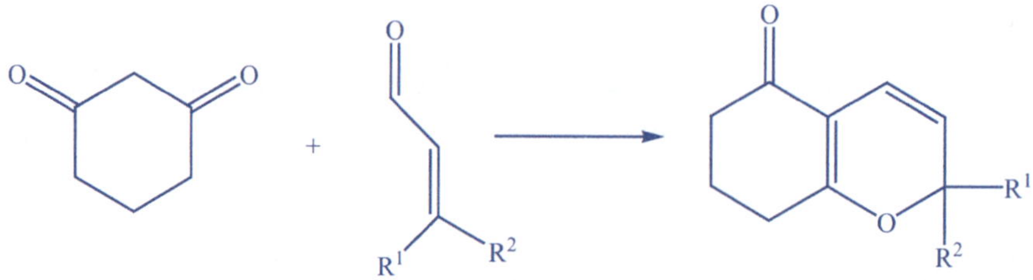

Scheme 2 [3 + 3] cycloaddition reaction

The demolition of the UiO-66 occurred in two steps, dihydroxylation of the framework and weight loss at around 400 °C that results in the formation of UiO. The techniques such as TGA, PXRD, UV-vis, NMR, and high-resolution TEM are used to get further information regarding defects [44, 45]. These defects significantly improve the activity of UiO-66 [46]. The properties of UiO-66 can be easily tuned by applying postsynthetic treatments [47, 48]. This concept has largely been utilized and reported [49–51]. Comparative study of UiO-66 and amino-functionalized-UiO-66 toward the Fischer esterification have been examined by Caratelli et al. [52] Hajek et al. [53] used defective UiO-66 to study the Oppenauer oxidation of primary alcohols. The hydrated material was found to be more active compared to dehydrated material for the conversion of prenol with furfural. Cai and Jiang synthesized hierarchical pore (HP) MOFs [54]. These MOFs have been used in cycloaddition reactions (Scheme 2). HP-UiO-66 showed better activity than the conventional UiO-66. Dissegna et al. [55] used the water adsorption measurements as a key parameter to determine the defects in acetic acid and trifluoroacetic acid-modulated UiO-66 MOFs.

Ye et al. [56] reported the mechano-synthesis of UiO-66 with larger defects and found that the number of missing linkers per $Zr_6O_4(OH)_4(BDC)_6$ reached around 2.16. As a result, increased defects in UiO-66 exhibited higher catalytic activities in various oxidation reactions. Kuwahara et al. [57] explored the use of sulfonic-acid-functionalized UiO-66 for the hydrogenation between levulinic acid and its esters to produce γ-valerolactone. The high yield of γ-valerolactone was obtained. It did not show any loss on prolonged use and was reusable for about four cycles. Yang and co-workers [58] investigated the chemical behavior of Zr-nodes in UiO-66 and UiO-67 MOF through IR and NMR spectroscopy. They demonstrated the role of nodes in catalysis for ethanol dehydration leading to diethyl ether selectively at 473–523 K. Zwoliński and Chmielewski [59] isolated a series of TEMPO-appended UiO-66 and UiO-67 using a method developed by Katz et al. [60] and used for aerobic oxidation of alcohols. Zhuang et al. [61] fabricated UiO-68-TEMPO with the use of H_2tpdc-TEMPO ligand with varied missing linker defects.

The researchers concluded that the synergistic effect between them was responsible for high catalytic activity of UiO-68-TEMPO [61]. Benzoic acid was used to tune the vacant linker defects in UiO-68-TEMPO. The defective microcrystal showed high catalytic activity as compared to the single-crystalline UiO-68-TEMPO. Zhao et al. [62] isolated MIL-101(Cr) hydrothermally with the use of phenylphosphonic acid. The role of phenylphosphonic acid was produced defects in structure and larger porosity in MIL-101(Cr). The three Zr-based MOFs, namely MOF-525, PCN-222, and PCN-224, have been synthesized by Epp et al. [63]. They were used in the synthesis of propylene carbonate. Mousavi et al. [64] studied MOF catalysts for CO_2 cycloaddition reaction. Four different MOFs based on DABCO (M=Zn, Co, Ni, Cu) were synthesized solvothermally. Zn-DABCO showed excellent activity and 100% selectivity. The synthesis of Ce-MOF (BIT-58) was reported by Chen et al. [65] which has suitable size for the catalytic conversion of benzaldehyde to 2-benzylidenemalononitrile [65]. The catalytic application of ZIF-8 MOF toward ring-opening polymerization was explored by Luo et al. in the absence of solvents or cocatalysts [66]. Two different synthetic strategies such as the solvothermal approach and the spray drying process were used for synthesizing ZIF-8. Chaemchuen et al. [67] explored the quantitative determination of defects in three MOFs. The structural defects increased the activity of MOFs in various reactions.

2.3 Catalysis Based on Decorated MOFs

Lots of significant attention have been paid toward tuning of MOFs at the postsynthetic level for various catalytic applications [68, 69]. Cao et al. [70] have explored the multivariate functionalities of a MOF as shown in Fig. 3. Normally, a multivariate MOF has been prepared by mixed linker single-step synthesis or by modification at postsynthetic level. Latter one is usually used for MOFs containing two or more

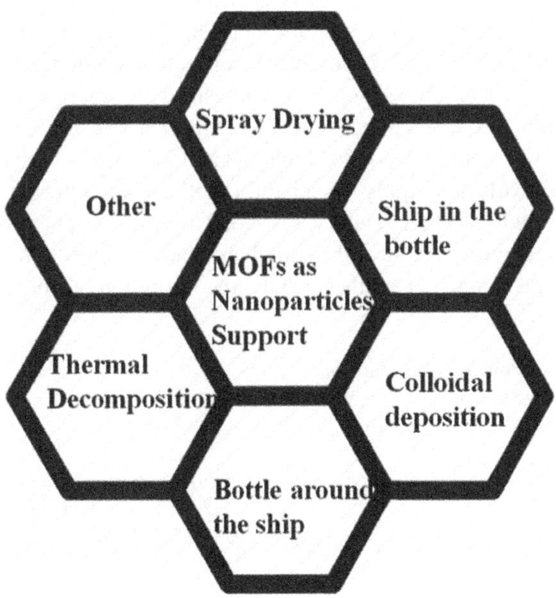

Fig. 3 Schematic representation of different methods used for the formation of nanoparticle-MOF composites

different functionalities. Zhou et al. [71] described an isoreticular family of catalysts based on the multicomponent metal-organic framework MUF-77 that exhibited several unique features such as simultaneous enhancements in reactivity and stereochemical selectivity for aldol reactions, the ability to catalyze Henry reactions that cannot be accomplished by homogeneous analogues, and discrimination between different reaction pathways (Henry vs aldol) that compete for a common substrate.

2.4 CO_2 Activation

The synthesis of cyclic carbonates is widely studied in MOF catalysis (Scheme 3).

Cyclic carbonate is the most stable thermal product of this reaction along with a by-product like polycarbonates. The reaction is facilitated in the presence of a cocatalyst such as tetraalkylammonium halide. The vital role of cocatalyst is believed to generate an alkoxide intermediate through nucleophilic attack which further undergoes addition with carbon dioxide and forming the cyclic carbonate and regenerates the cocatalyst. Various groups have worked on improving the CO_2 sorption capacity of MOF to achieve the high activity. Imidazoles and/or ionic liquids have proved to be providing alternate solutions in keeping the view in mind for their high carbon dioxide capture and solubility [72]. Bhin et al. [73] synthesized the chloro-functionalized ZIF-95 MOF which showed efficient storage and coupling CO_2. Although this reaction occurs at elevated temperatures, cocatalyst allows the reaction to occur at lower temperatures for higher conversions. Demir et al. [74] have been successfully synthesized Zr-MOFs and applied on the same reaction. Authors have synthesized MOF-53(VCl_3) and MOF-53(VCl_4) from Zr-based MOF. The yield of cyclic carbonates was 79.6% in case of MOF-53. Liang et al. reported a bifunctional imidazolium functionalized MOF. He used the isoreticular synthesis and postsynthetic modifications [75]. The same group also synthesized chromium-based MOF (FJI-C10) containing Cr^{3+} and free halogen ions [76]. Ding et al. [77] tuned the UiO-67 solvothermally with imidazolium decorated biphenyl and Zr^{IV} ions which exhibited adsorption of carbon dioxide and highly selective for methane and nitrogen. Kurisingal et al. [78] synthesized U6N@ILA and U6N@ILB by the modification of the UiO-66-amino with methylimidazole and methylbenzimidazole through a condensation reaction. Ding and Jiang [79] synthesized the incorporation of imidazole-based poly(ionic liquid)s

Scheme 3 Cyclic carbonate from coupling of CO_2 with epoxide

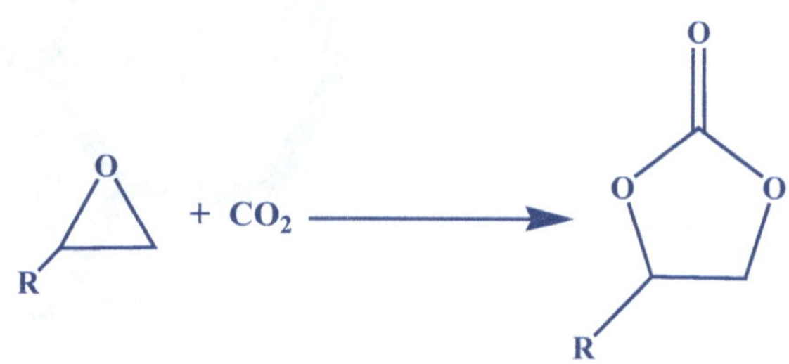

into a MIL-101. Aguila et al. [80] synthesized MIL-101(Cr)-IP, which was used in the fixation of carbon dioxide into epoxides at 50 °C. The increased catalytic activity was due to the interaction between halide ions and the Cr^{III} Lewis acid sites. Gao et al. [81] developed Zr-phosphonate framework that shows dual catalytic behavior. Tang et al. [82] synthesized Au/Zn MOF. Synergic effect of the two metallic components and hollow structures led to high catalytic activity [82]. Lately, numerous MOF-based systems have been devised focusing on CO_2 cycloaddition such as zinc-based ZIF-23 [83], NH_2-MIL-101(Al) [84], Zr-based Cu-VPI-100 and Ni-VPI-100 [85], and Zn MOFs [86], as well as Cu^{II} MOF [87] and Cd^{II} MOF [88]. Xiong et al. [89] reported two Gd and Cu cluster-based MOFs. The resulting MOFs were used to catalyze CO_2 carboxylation at room pressure and 80–100 °C. A porous silver polymer used for a similar reaction was investigated by Zhou et al. [90]. It gave 92% yields. Ji et al. [91] encapsulated single and triatomic Ru atoms clusters on ZIF-8 and found to be very effective catalyst for the semihydrogenation of alkynes. Absolute regioselectivity of terminal alkynes was obtained using the ZIF-8 shell. Carboxylation of propargylic alcohol was investigated by Hou et al. [92] The authors synthesized cluster-based heterometallic MOFs $[(NH_2C_2H_6)_{0.75}[Cu_4I_4{\cdot}(L)_3{\cdot}(In)_{0.75}]{\cdot}DMF{\cdot}H_2O]$ and isolate more than 90% yields of product. The other method for utilization of CO_2 focused on its fixation as reported by Padmanaban et al. [93] Kang et al. [94] synthesized three 3-D MOFs based on Co, Mn, Ni. All compounds showed efficient catalysis for the cycloaddition reaction with CO_2 under mild conditions. Zhang et al. [95] explored the successful use of UiO-68 for CO_2 hydrosilylation reaction. Very few works till date have been reported on single-site-MOF for carbon dioxide conversion. An et al. [96] successfully isolated the immobilization of iridium complexes into a UiO-type MOF. Li et al. [97] successfully performed the encapsulation of the ruthenium complex in UiO-66. The resulting material [Ru]@UiO-66 was used for the hydrogenation of CO_2 to formate in DMF/DBU mixtures.

2.5 *Carbon-Hydrogen Bond Functionalization*

Carbon-hydrogen bond activation is an important popular reaction between MOF investigators. [98] Wang et al. [99] synthesized a Mn-based MOF which was used as a catalyst for the direct amination of C–H bonds. Hoang et al. [100] reported the heterogeneous catalyst Cu-CPO-27 for C–H amination in 89% yields. Copper-based MOF (VNU-18) was used as an efficient catalyst for the direct coupling of carbonyls and N–H amines as explored by Tran et al. [101]. A iron-based MOF (VNU-20) was synthesized solvothermally by Pham et al. [102] which showed an efficient catalyst for the functionalization of coumarins with N,N-dimethylanilines. An iron MOF (VNU-21) was reported by To et al. [103] which was used for single-step synthesis of quinazolinones. Xu et al. [104] reported the use of Cu-TPPB MOF in the C–H activation of amidines. The advantage of this system was that it occurred in air atmosphere and utilized over a broad substrate range. Amidines without ortho substituents which were derived from aryl nitriles formed products in greater yields

[104]. Ikuno et al. [105] were the first to generate Cu-MOF (Cu-NU-1000) which was used for the formation of CH_3OH selectively. Incoporation of Cu sites on MOF use for the conversion of CH_4 to CH_3OH was also explored by Baek et al. [106] Osadchii et al. [107] synthesized a high-spin Fe MOF by electrocatalytic or hydrothermal method which was used for C–H bond functionalization of CH_4.

2.6 Other Reactions Catalyzed by MOFs

Two postsynthetic strategies were utilized for attachment of oxovanadium salt to UiO-66-NH_2 by Pourkhosravani et al. [108] Firstly the free NH_2 groups of UiO-66-NH_2 were directly reacted with salicylaldehyde, followed by reaction with oxovanadium salt to form UiO-66-SI/VO(acac). In other method, UiO-66-NH_2 was reacted with oxovanadium salt to form UiO-66-NH_2/VO$(acac)_2$. Both catalysts explored the epoxidation of geraniol with complete conversion with complete selectivity. Zhang et al. [109] described a method for the metalation of a UiO-67 with an iridium(III) compound which was used as a methane borylation catalyst. The UiO-66-supported nickel catalysts were synthesized by Li et al. [110] by atomic layer deposition (ALD) which was used in ethylene hydrogenation. Zhang et al. [111] have synthesized zirconium-based MOFs that showed oxidation of cyclohexene. Zhang et al. [112] synthesized a NH_2CONH_2-UiO-68 which was very effective for C–C bond formation. This MOF exhibited excellent catalytic activity compared to others analogue urea derivatives. An immobilized azide-functionalized UiO-66 MOF was synthesized by Elumalai et al. [113] which was used for C–C coupling reaction under mild conditions. It was used over a wide range and gave yields up to 93%. Oozeerally et al. [114] synthesized UiO-66-MSBDC that can be used as a heterogeneous catalyst for fructose and HMF from glucose. A phosphate-modified NU-1000 was also used as a catalyst for the conversion of glucose into fructose and 5-hydroxymethylfurfural (HMF) as discussed by Yabushita et al. [115]. The same group utilized NU-1000 as a promising material for further catalyst development [116]. NU-1000 consists a trihexagonal tiling using eight connected $Zr_6(\mu_3\text{-}O)_4(\mu_3\text{-}OH)_4(H_2O)_4(OH)_4$ nodes at each vertex and 1,3,6,8-(*p*-benzoate)pyrene linkers positioned at each edge [117]. This contained Zr_6 nodes as Lewis site and channels that can facilitate activity which was the attractive for the design of catalyst [118]. Kim and coworkers [119] reported the incorporation of single atoms and clusters of platinum on the NU-1000 which exhibited greater activity for hydrogenation of ethylene.

3 MOFs Containing Metal Nanoparticles

Currently, much attention has been paid toward metal nanoparticles (MNPs) as supporters that act as important catalytic sites. Metal-supported nanoparticles have been widely employed in various industrial processes. However, with supported catalysts, structure sensitivity has most often related to the performance of catalysis. The overall goal of the catalyst engineer is to design a catalyst with maximize productivity and selectivity. MOFs have been widely used as support for metal nanoparticles since the early days of MOF catalysis [120–123] but currently, various MOF-based materials have been developed rely on catalysis. All centers can be refined to improve metal-support interactions [124–126] and their use as supports for nanoparticles. Numerous excellent reviews related to synergistic effect between MOFs and MNPs exist but this phenomenon is still unexplored. However, without doubt, this phenomenon has played an important role in catalytic processes. These resulting materials show a better performance in catalysis than do MNPs immobilized on conventional materials. Typically, such porous and crystalline composites showed high dispersion of nanoparticles within the pores and easy way of active site for substrates made them efficient materials in catalyzing various reactions. The different roles of MOFs in these composites are: they (i) To provide uniform distribution and stability of MNPs within pores and (ii) selectivity of the reaction (iii) the synergistic interactions between MOF and an MNP to modify the electronic properties of MNPs (iv) catalyze one-pot tandem reactions. ICP, IR, and UV-vis spectroscopy and NMR are common techniques to characterize the MNPs within MOF structures. Powder XRD technique is used to determine the particle's size. XPS technique is used to determine the oxidation state of the metal. TEM and SEM are powerful techniques to determine the morphology of NPs' and their dispersion in lattice. Finally, nitrogen physisorption gave information about porosity prior to and after nanoparticle deposition. Herein, we discussed the different methods used for the synthesis of MNP@MOF and MOF@MNP composite materials as shown in Fig. 3 and their recent development in heterogeneous catalysis.

3.1 MOFs Synthesis

The composite material plays a pivotal role in catalytic performance as most of the metal nanoparticles catalyzed reactions are structure sensitive. Ning et al. [127] used different methods such as impregnation-reduction-H_2, impregnation-reduction-$NaBH_4$, deposition–precipitation-carbonization, deposition–precipitation-H_2, and colloid-immobilization to synthesize a series of MOF supported Au nanoparticles(Au@UiO-66-X) (Au@UiO-66, Au@UiO-66-NH_2, Au@UiO-66-NO_2, Au@UiO-66-COOH, and Au@UiO-66-NH_3Cl). Out of these, it is found that the impregnation-reduction-H_2 method is the best where a 100% selectivity of methyl-2-furoate with a complete conversion was obtained using Au@UiO-66 as catalyst. TEM images of the Au@UiO- 66 displayed a large size (10–25 nm) and irregular dispersion as synthesize by impregnation-reduction-$NaBH_4$, deposition–precipitation-carbonization, and deposition–precipitation-H_2 methods. However, methods such as colloid-immobilization and impregnation-reduction-H_2 yielded Au nanoparticles having a size of 2–3 nm and uniform distribution. Authors have found that the crystal structure of UiO-66 remained stable during synthesis while using the colloid-immobilization and impregnation-reduction-H_2 methods, whereas the other methods led to disintegrate of the framework.

Catalytic activities of Au@UiO-66 synthesized by colloid-immobilization and impregnation-reduction-H_2 methods were confirmed by small size of the Au nanoparticles and the well-preserved pore structure of UiO-66. TEM and XRD study of Au@UiO-66 catalyst revealed that no agglomeration of Au particle could be observed and UiO-66 retained its crystalline structure even after five catalytic cycles prepared by impregnation-reduction-H_2 method. Butson et al. discussed the importance of the various synthetic methods [128] and inferred that MOFs having the pore breathing effect could responsible for a preferential phase behavior for nanoparticles isolated by different reduction methods. MIL-53 has known for its breathing phenomenon.

Synthetic Strategy for MNP@MOF Composite

It was also known as "ship in bottle." It was prepared by the assembly of the active species within the pores of a MOF support. The advantage of this method is to prevent MNPs from aggregation. However, this method is more challenging because of various factors such as the wettability of the MOF's surface, interactions and pore environment, thermal, chemical, and mechanical stabilities of the framework affected their synthesis. The common techniques used for the deposition of metal precursors into MOFs are chemical vapor deposition, solution impregnation, double-solvent impregnation, and one-pot synthesis.

CVD

In this method, activated MOF is reacted to a volatile metal precursor in vacuum. During the evacuation, the metal precursor diffused into MOF channels. The precursor turned into nanoparticles after the reduction or thermal treatment. There is large fluctuation of particle size distribution while using this method. The volatile complexes $[(\eta^5\text{-}C_5H_5)Pd(\eta^3\text{-}C_3H_5)]$, $[(\eta^5\text{-}C_5H_5)Cu(PMe_3)]$, and $[(CH_3)Au\text{-}(PMe_3)]$ compounds were introduced on MOF-5 as reported by Hermes et al. [129] The Pd/Au/Cu@MOF was isolated by the reduction. The size of Au particles was in a range from 5 to 20 nm, whereas the average size range was 1 to 2 nm in the case of palladium and copper. This method has been hardly ever using for the preparation of MNP@MOF composite materials for catalysis because of very expensive method [130]. The advantage of this method is to allow the synthesis of very small particles within MOFs.

Solid Grinding

It involves the mixing of MOF powder with a volatile metal precursor. The volatile metal precursor entered into MOF channels and subsequently undergoes the reduction which gave the composite material. Ishida et al. [131] introduced the first deposition of a metal precursor on an MOF by this method.

Solution Impregnation

This method involved mixing a presynthesized MOF powder with dissolved metal precursors. The metal precursor solution entered in MOF pores because of the capillary effect. Here the precursor's undergoes reduction to form metal nanoparticles. The disadvantage of this method is the poor particle distribution on both the interior and exterior surfaces of MOFs. Sabo et al. [132] employed firstly this method to form metal nanoparticles within MOFs. MOF-5 was stirred with a $Pd(acac)_2$ solution followed by the slow evaporation in inert atmosphere. The resulting product was reduced under H_2 atmosphere and heated to 200 °C for 1 h. Finally, the Pd^{2+} was reduced into Pd. The isolated catalyst was not stable in air at room temperature because of lower hydrothermal stability of MOF-5 support. This problem can be overcome by the use of $-NH_2$ or-bpy moieties which facilitate the metal precursor tend to prefer coordination to these groups, thus allowing for greater location control. Hwang et al. [133] used this method first time for the synthesis of MIL-101 which is known for easy generation of coordinatively unsaturated metal sites by the replacement of terminal water molecules. Goswami et al. [134] reported the channel-templated growth of AuNPs within a stable zirconium-based metal–organic framework, NU-1000, via channel-anchored monometallic precursors which were used as efficient catalyst for condensed-phase hydrogenation of 4-nitrophenol to 4-aminophenol. Rivera-Torrente et al. [135] compared MIL-100(Fe) and disordered Basolite® F300 with identical iron 1,3,5-benzenetricarboxylate composition exhibit very divergent properties when used as a support for Pd nanoparticle deposition. MIL-100(Fe) shows a regular MTN-zeotype crystal structure with two types of cages, whereas Basolite® F300 lacks long-range order beyond 8 Å and has a single-pore system. The medium-range configurational linker-node disorder in Basolite® F300 results in a reduced number of Lewis acid sites, yielding more hydrophobic surface properties compared to hydrophilic MIL-100(Fe). The hydrophilic/hydrophobic nature of MIL-100(Fe) and Basolite® F300 impacts the amount of Pd and particle size distribution of Pd nanoparticles deposited during colloidal synthesis and dry impregnation methods, respectively. Authors suggested that polar (apolar) solvents/precursors attractively interact with hydrophilic (hydrophobic) MOF surfaces, allowing tools at hand to increase the level of control over, for example, the nanoparticle size distribution. Yang et al. [136] synthesized trimetallic NiFePd nanoparticles anchored on MIL-101(Cr) via a simple impregnation method which was highly active catalyst for the complete dehydrogenation of $N_2H_4BH_3$ and decomposition of $N_2H_4 \cdot H_2O$ with 100% hydrogen selectivity even after five recycles [136].

Double Solvent Approach

This is a common method for the synthesis of MNP@MOF composites which allows the uniform distribution of NPs on the internal surface of the pores. The capillary effect was the main driving force for this method. MOF is dispersed in a hydrophobic solvent such as hexane then added a small amount of the aqueous solution of the metal precursor and allow for stirred dispersion. Small metal clusters were encapsulated within the MOF matrix after the reduction. Aijaz et al. [137] used this method for the synthesis of well-dispersed and size-controlled metal nanoparticles inside the pores of a MOF. Sun et al. [138] used this method for the synthesis of a non-noble trimetallic system (Cu@Co@Ni). This system showed efficient activity for the tandem hydrogenation of nitroarenes by the hydrogen from ammonia borane dehydrogenation and CO oxidation. Recently, some literatures related to this method for the preparation of MNP@MOF composite materials were published [139–141].

Colloidal Deposition

This method is excellent technique for deposition of particles on the exterior surface of a MOF. MOF and metal nanoparticles were mixed and stirred in a solvent. If the size of NPs' allowed for its penetration into a MOF, the particles are distributed on both surfaces. If not then the NPs are attached on the external surface due to physical adsorption and/or electrostatic interactions.

Templated Synthesis

This method allowed the introduction of presynthesized MNPs into the MOF precursor solution. Surfactants such as PVP, cetyltrimethylammonium bromide (CTAB), and others are used to prevent MNPs from agglomeration. The disadvantage of this method is that binders are difficult to wash away from the framework completely, and these binders limit the access to MNPs so the catalytic performance decreases. However, this method allows for MNP@MOF growth with controlled NP size and shape [142]. A sacrificial template synthesis is used to avoid this problem.

Thermal Decomposition

This method involved the partial thermal decomposition of MOFs which allows for uniform distribution of metal oxide particles inside hierarchical pores.

Spray Drying

This method was also used for the modification of MOFs [143, 144]. This method involved the agglomeration of MOF nanocrystals to form spherical particles by fast evaporation of a solvent from the sprayed droplet. The MOF particles are converted into spherical beads (average size = 3.4 ± 1.8 μm) by maintaining the initial size of nanoparticles.

3.2 *Catalysis*

MNP@MOF composites have been known as excellent catalytic material over wide range of organic transformation reactions. MIL, UiO, and ZIF families are widely used as MOF supports. Here we described the application of MNP@MOF in heterogeneous catalysis.

Co-related Chemistry

Catalytic CO oxidation to CO_2 has been chosen as a model reaction [145]. Tsumori et al. [145] synthesized a "Quasi-MOF" in which MIL-101 has an interaction between the node and the Au nanoparticles that result in boosting the catalytic activity. A solution impregnation technique was used for the inclusion of Au nanoparticles (3 nm) over MOF. The partial framework decomposition of Cr-MIL-101 occurs at 573 K that releases CO_2 and form Cr–O sites that are necessary condition for the adsorption and activation of O_2 species in the CO oxidation reaction. Au@Cr-MIL-101 was used in CO oxidation in the range of temperature from 193 to 52 K. The catalyst showed 100%CO conversion at temperatures from 193 to 298 K but showed several lifetimes up to 6500 min.

CO_2 Utilization

This is a powerful method for the valorization of carbon dioxide because the main product CO can be further converted into valuable chemicals and fuels. Han et al. [146] reported various composite materials based on MOF-74(Zn) such as Au@Pd@MOF-74, Pt@MOF-74, and Pt/Au@Pd@MOF-74. The best catalytic performance was shown by Pt@MOF-74 in 1:1 ratio of H_2/CO_2 at 400 °C and 2 MPa of CO_2. Encapsulation of Pt/Au nanoparticles in UiO-66 [147] and $[Co_2(oba)_4(3\text{-}BPDH)_2]\cdot 4H_2O_2$ [148] were explored under the same reaction conditions. Both the MOF showed decomposition of the framework at such a temperature which was the main issue for their application as supports in high-temperature catalysis. The methanation of CO_2 occurred at lower temperature by using Ni catalysts. Zhao et al. [149] synthesized a Ni@UiO-66 catalyst which was used for the reduction of CO_2 to CH_4

under 1 MPa of H_2/CO_2 at a temperature from 200 to 340 °C. This catalyst showed a superior performance as compared to conventional systems such as Ni/ZrO_2 and Ni/SiO_2. Another potential application of CO_2 is the carboxylation reaction. Dutta et al. [150] synthesized a new MOF which was used for the carboxylation of terminal alkynes under mild conditions of 1 atm and 60 °C, and the reaction catalyst could be recovered and used again with an average 10% loss in yield for every further catalytic run.

Dehydrogenation

Hydrogen-rich compounds are the good source of hydrogen production and hydrogen storage. This approach has been utilized by many research groups because of its solid nature, stability, safer use at room temperature and high hydrogen content. MOF-supported nanoparticles were proved as perspective catalysts for on-demand hydrogen generation from these compounds [151, 152]. Li et al. [153] synthesized the immobilization of $Ni_{0.5}Fe_{0.5}$–CeO_x particles on MIL-101 which exhibited 100% conversion and 100% H_2 with a TOF of 351.3 h^{-1} and better catalyst as compared to $Ni_{0.5}Fe_{0.5}$MIL-101. The catalyst was stable throughout five cycles. Furthermore, $Ni_{0.5}Fe_{0.5}$–CeO_x/ZIF-67, ZIF-8, and UiO-66 were also used as a catalyst for the same reaction which showed complete decomposition with TOF values of 361.5, 192.3, and 300 h^{-1}, respectively.

Hydrogenation Reactions

Redfern et al. [154] reported the selective acetylene semihydrogenation catalyzed by Cu nanoparticles supported inside a Zr-based NU-1000 MOF. The size of copper NPs was of <1. The Cu@NU-1000 showed a TOF of 100 ± 20 h^{-1} as compared to Cu/ZrO_2 with a 1:1 H_2/C_2H_2 ratio with ethane, 1-butene, and 1,3-butadiene as by-products. The high acetylene conversion was achieved leading to the increase in the amount of C4 up to 15 mol % with <2% of ethane present in the mixture. Whereas, through Cu@NU-1000 catalyst, the full utilization of acetylene was achieved with 99.5% selectivity toward ethylene and less than 0.5 mol % of ethane, and C4 products with the industrially used mixture of 97% ethylene and 3% acetylene. The reduction of olefins has been studied over different MNP/MOF composite materials [155]. Meng et al. [156] used impregnation method to synthesize a Pt@UIO-66-NH_2 catalyst followed by annealing to fabricate a mesoporous MOF (Pt@UiO-66-NH_2-2 h). The TOF was found to 30 times more for the mesoporous solid which is due to the increased diffusion rate of the reagents. Chen et al. [157] used the solution impregnation method to synthesize MOF140-AA by the use of Ni^{2+} as a metal source and squaric acid which catalyzed the phenol hydrogenation. The catalytic performances were compared to Pd/SBA-15, Pd/ZrO_2, Pd/Al_2O_3, and Pd/SiO_2 and found that MOF140-AA was showing the best performance. The reduction of nitroarenes was chosen as a model hydrogenation reaction along with others industrially important reaction [158–163]. Yuan et al. [164] prepared MIL-101@Pt@FePCMP which showed a high catalytic activity for the selective reduction of the C=O bond as compared to MIL-101, and FeP-CMP. Lin et al. [165] introduced the deposition of Ru NPs on the surface of MIL-101-SO_3H which was used as an efficient catalyst for the conversion of methyl levulinate (ML) to γ-valerolactone (GVL).

Carbon–Carbon Coupling Reactions

This is an important reaction in synthetic organic chemistry. The NP@MOF catalysts were utilized as efficient catalyst for Suzuki–Miyaura coupling [166]. Xiong et al. [167] synthesized new La-MOF containing Fe_3O_4 nanoparticles (NPs), proved to be an advantageous strategy to produce a superior heterogeneous catalyst for Suzuki–Miyaura C–C bond formation. This showed very active, quite stable, magnetically recoverable, and reusable in coupling of bromo- or iodoaryls with arylboronic acids affording practically quantitative yields of biaryls (>99%) and high TONs and TOFs. The catalyst has been recycled up to 12 times without significant loss of the catalytic activity. Tan and Zeng [168] prepared a series of composites such as Pd/M-HKUST-1-R which were used as efficient catalyst for the oxidation of benzyl alcohol to benzyl aldehyde and the Knoevenagel condensation.

4 Outlook, Challenges, and Future Perspectives

The catalytic applications based on metal–organic framework are growing at are markable rate. During the past decade, we have witnessed that MOFs find their place in catalysis based on various methods used for bulk scale synthesis of MOFs or MOF-derived materials. Although a lot of MOF materials have been synthesized, these systems are still further explored with respect to it utilization and chemical and thermal properties. Many applications-based MOFs have been developed for utilizing in different organic transformations. Though a lot of effort has been devoted to this field, still comparison based on stability is still unexplored. The use of MOFs precursors is very successful for the synthesis of advanced catalytic materials. Thermal decomposition of MOFs has opened a new era for the synthesis of highly advanced heterogeneous catalysts. The interesting properties that possess by MOFs have inspired the researchers to apply these materials in a variety of catalytic reactions. We believe that MOFs field will grow continuously to receive extensive attention in the near future and help the catalysis community to solve all these issues arouse in this field.

References

1. Batten SR, Hoskins BF, Robson R (1995) J Am Chem Soc 117:5385–5386
2. Kitagawa S, Kawata S, Nozaka Y, Munakata M (1993) J Chem Soc Dalton Trans 1399–1404
3. Ahmad N, Chughtai AH, Younus HA, Verpoort F (2014) Coord Chem Rev 280:1–27
4. Hagen J (2015) Economic importance of catalysts. In: Hagen J (ed) Industrial catalysis, a practical approach. Wiley: 2015; Chapter 17, pp 459–462
5. Yang D, Gates BC (2019) ACS Catal 9:1779–1798
6. Hall JN, Bollini P (2019) React. Chem Eng 4:207–222
7. Dhakshinamoorthy A, Li Z, Garcia H (2018) Chem Soc Rev 47:8134–8172
8. Zhu L, Liu X-Q, Jiang H-L, Sun L-B (2017) Chem Rev 117:8129–8176
9. Mao Y, Qi H, Ye G, Han L, Zhou W, Xu W, Sun Y (2019) Microporous Mesoporous Mater 274:70–75
10. Wang R, Kapteijn F, Gascon J (2019) Chem-Asian J 14:3452–3461
11. Oar-Arteta L, Wezendonk T, Sun X, Kapteijn F, Gascon J (2017) Mater Chem Front 1:1709–1745
12. Rogge SMJ, Bavykina A, Hajek J, Garcia H, Olivos-Suarez AI, Sepúlveda-Escribano A, Vimont A, Clet G, Bazin P, Kapteijn F, Daturi M, Ramos-Fernandez EV, Llabres i Xamena FX, VanSpeybroeck V, Gascon J (2017) Chem Soc Rev 46:3134–3184
13. Sun Y, Du Q, Wang F, Dramou P, He H (2021) New J Chem 45:1137–1162
14. Rojas-Buzo S, Bohigues B, Lopes CW, Meira DM, Boronat M, Moliner M, Corma A (2021) Chem Sci 12:10106–10115
15. Drake T, Ji P, Lin W (2018) Acc Chem Res 51:2129–2138
16. Kang Y-S, Lu Y, Chen K, Zhao Y, Wang P, Sun W-Y (2019) Coord Chem Rev 378:262–280
17. Tu B, Pang Q, Xu H, Li X, Wang Y, Ma Z, Weng L, Li Q (2017) J Am Chem Soc 139:7998–8007
18. Nikseresht A, Ghasemi S, Parak S (2018) Polyhedron 151:112–117
19. Dhakshinamoorthy A, Asiri AM, Concepcion P, Garcia H (2017) Chem Commun 53:9998–10001
20. Nagaraj A, Amarajothi D (2017) J Colloid Interface Sci 494:282–289
21. Nagarjun N, Dhakshinamoorthy A (2019) Mol Catal 463:54–60
22. Han L, Qi H, Zhang D, Ye G, Zhou W, Hou C, Xu W, Sun Y (2017) New J Chem 41:13504–13509
23. Wang H, Rassu P, Wang X, Li H, Wang X, Feng X, Yin A, Li P, Jin X et al (2018) Angew Chem Int Ed 57:16416–16420
24. Huang S, Yang K-L, Liu X-F, Pan H, Zhang H, Yang S (2017) RSC Adv 7:5621–5627
25. Rostamnia S, Alamgholiloo H (2018) Catal Lett 148:2918–2928
26. Han Y, Zhang Y, Zhang Y, Cheng A, Hu A, Wang Z (2018) Appl Catal A 564:183–189
27. Liu S, Zhang Y, Han Y, Feng G, Gao F, Wang H, Qiu P (2017) Organometallics 36:632–638
28. Rivera-Torrente M, Pletcher PD, Jongkind MK, Nikolopoulos N, Weckhuysen BM (2019) ACS Catal 9:3059–3069
29. Santiago-Portillo A, Navalón S, Concepción P, Álvaro M, García H, ChemCatChem 9:2506–2511
30. Yu D, Wu M, Hu Q, Wang L, Lv C, Zhang L (2019) J Hazard Mater 367:456–464
31. Islamoglu T, Atilgan A, Moon S-Y, Peterson GW, DeCoste JB, Hall M, Hupp JT, Farha OK (2017) Chem Mater 29:2672–2675
32. Katz MJ, Mondloch JE, Totten RK, Park JK, Nguyen ST, Farha OK, Hupp JT (2014) Angew Chem Int Ed 53:497–501
33. Chen Z, Islamoglu T, Farha OK, Appl ACS (2019) Nano Mater 2:1005–1008
34. Shaabani A, Mohammadian R, Hooshmand SE, Hashemzadeh A, Amini MM (2017) ChemistrySelect 2:11906–11911
35. Zhang X, Huang P, Liu A, Zhu MA (2017) Fuel 209:417–423
36. Dalapati R, Sakthivel B, Ghosalya MK, Dhakshinamoorthy A, Biswas SA (2017) CrystEng-Comm 19:5915–5925

37. Rojas-Buzo S, García-García P, Corma A (2018) Chemsuschem 11:432–438
38. Mistry S, Sarkar A, Natarajan S (2019) Cryst Growth Des 19:747–755
39. Zhang X, Chang L, Yang Z, Shi Y, Long C, Han J, Zhang B, Qiu X, Li G, Tang Z (2019) Nano Res 12:437–440
40. Abdollahi N, Morsali A (2019) Polyhedron 159:72–77
41. Dissegna S, Epp K, Heinz WR, Kieslich G, Fischer RA (2018) Adv Mater 30:1704501
42. Vandichel M, Hajek J, Ghysels A, De Vos A, Waroquier M, Van Speybroeck V (2016) CrystEngComm 18:7056–7069
43. Nasalevich MA, Hendon CH, Santaclara JG, Svane K, Van der Linden B, Veber SL, Fedin MV, Houtepen AJ, Van derVeen MA, Kapteijn F (2016) Sci Rep 6:23676
44. Ren J, Ledwaba M, Musyoka NM, Langmi HW, Mathe M, Liao S, Pang W (2017) Coord Chem Rev 349:169–197
45. Liu L, Chen Z, Wang J, Zhang D, Zhu Y, Ling S, Huang K-W, Belmabkhout Y, Adil K, Zhang Y (2019) Nat Chem 11:622–628
46. Chakarova K, Strauss I, Mihaylov M, Drenchev N, Hadjiivanov K (2019) Microporous Mesoporous Mater 281:110–122
47. Dhakshinamoorthy A, Santiago-Portillo A, Asiri AM, Garcia H (2019) ChemCatChem 11:899–923
48. Taddei M (2017) Coord Chem Rev 343:1–24
49. Fang Z, Bueken B, De Vos DE, Fischer RA (2015) Angew Chem Int Ed 54:7234–7254
50. Ling S, Slater B (2016) Chem Sci 7:4706–4712
51. Wu H, Chua YS, Krungleviciute V, Tyagi M, Chen P, Yildirim T, Zhou W (2013) J Am Chem Soc 135:10525–10532
52. Caratelli C, Hajek J, Cirujano FG, Waroquier M, Llabrés FX, Xamena I, Van Speybroeck V (2017) J Catal 352:401–414
53. Hajek J, Bueken B, Waroquier M, De Vos D, VanSpeybroeck V (2017) ChemCatChem 9:2203–2210
54. Cai G, Jiang H-L (2017) Angew Chem 129:578–582
55. Dissegna S, Hardian R, Epp K, Kieslich G, Coulet M-V, Llewellyn P, Fischer RA (2017) CrystEngComm 19:4137–4141
56. Ye G, Zhang D, Li X, Leng K, Zhang W, Ma J, Sun Y, Xu W, Ma S, Appl ACS (2017) Mater Interfaces 9:34937–34943
57. Kuwahara Y, Kango H, Yamashita H (2017) ACSSustainable. Chem Eng 5:1141–1152
58. Yang D, Ortuño MA, Bernales V, Cramer CJ, Gagliardi L, Gates BC (2018) J Am Chem Soc 140:3751–3759
59. Zwoliński KM, Chmielewski MJ (2017) ACS Appl Mater.Interfaces 9:33956–33967
60. Katz MJ, Brown ZJ, Colón YJ, Siu PW, Scheidt KA, Snurr RQ, Hupp JT, Farha OK (2013) Chem Commun 49:9449–9451
61. Zhuang J-L, Liu X-Y, Zhang Y, Wang C, Mao H-L, Guo J, Du X, Zhu S-B, Ren B, Terfort A, Appl ACS (2019) Mater Interfaces 11:3034–3043
62. Zhao T, Dong M, Yang L, Liu Y (2018) Catalysts 8:394
63. Epp K, Semrau AL, Cokoja M, Fischer RA (2018) ChemCatChem 10:3506–3512
64. Mousavi B, Chaemchuen S, Moosavi B, Zhou K, Yusubov M, Verpoort F (2017) ChemistryOpen 6:674–680
65. Chen Y, Zhang S, Chen F, Cao S, Cai Y, Li S, Ma H, Ma X, Li P, Huang X (2018) J. Mater. Chem. A 6:342–348
66. Luo Z, Chaemchuen S, Zhou K, Verpoort F (2017) Chemsuschem 10:4135–4139
67. Chaemchuen S, Luo Z, Zhou K, Mousavi B, Phatanasri S, Jaroniec M, Verpoort F (2017) J Catal 354:84–91
68. Cohen SM (2017) J Am Chem Soc 139:2855–2863
69. Islamoglu T, Goswami S, Li Z, Howarth AJ, Farha OK, Hupp JT (2017) Acc Chem Res 50:805–813
70. Cao C-C, Chen C-X, Wei Z-W, Qiu Q-F, Zhu N-X, Xiong Y-Y, Jiang J-J, Wang D, Su C-Y (2019) J Am Chem Soc 141:2589–2593

71. Zhou T-Y, Auer B, Lee SJ, Telfer SG (2019) J Am Chem Soc 141:1577–1582
72. Wang Y, Guo L, Yin L (2019) Catal Lett 149:985–997
73. Bhin KM, Tharun J, Roshan KR, Kim DW, Chung Y, Park D-W (2017) J CO_2 Util 17:112–118
74. Demir S, Usta S, Tamar H, Ulusoy M (2017) MicroporousMesoporous Mater 244:251–257
75. Liang J, Chen R-P, Wang X-Y, Liu T-T, Wang X-S, Huang Y-B, Cao R (2017) Chem Sci 8:1570–1575
76. Liang J, Xie Y-Q, Wang X-S, Wang Q, Liu T-T, Huang Y-B, Cao R (2018) Chem Commun 54:342–345
77. Ding L-G, Yao B-J, Jiang W-L, Li J-T, Fu Q-J, Li Y-A, Liu Z-H, Ma J-P, Dong Y-B (2017) Inorg Chem 56:2337–2344
78. Kurisingal JF, Rachuri Y, Pillai RS, Gu Y, Choe Y, Park D-W (2019) Chemsuschem 12:1033–1042
79. Ding M, Jiang H-L (2018) ACS Catal 8:3194–3201
80. Aguila B, Sun Q, Wang X, O'Rourke E, Al-Enizi AM, Nafady A, Ma S (2018) Angew Chem Int Ed 57:10107–10111
81. Gao C-Y, Ai J, Tian H-R, Wu D, Sun Z-M (2017) Chem Commun 53:1293–1296
82. Tang L, Zhang S, Wu Q, Wang X, Wu H, Jiang Z (2018) J Mater Chem A 6:2964–2973
83. Ryu H, Roshan R, Kim M-I, Kim D-W, Selvaraj M, Park D-W (2017) Korean J Chem Eng 34:928–934
84. Senthilkumar S, Maru MS, Somani RS, Bajaj HC, Neogi S (2018) Dalton Trans 47:418–428
85. Zhu J, Usov PM, Xu W, Celis-Salazar PJ, Lin S, Kessinger MC, Landaverde-Alvarado C, Cai M, May AM, Slebodnick C (2018) J Am Chem Soc 140:993–1003
86. Agarwal RA, Gupta AK, De D (2019) Cryst Growth Des 19:2010–2018
87. Guo F (2019) Inorg Chem Commun 101:87–92
88. Zhang X, Liu H-L, Zhang D-S, Geng L (2019) Inorg Chem Commun 101:184–187
89. Xiong G, Yu B, Dong J, Shi Y, Zhao B, He L-N (2017) Chem Commun 53:6013–6016
90. Zhou Z, He C, Yang L, Wang Y, Liu T, Duan C (2017) ACS Catal 7:2248–2256
91. Ji S, Chen Y, Zhao S, Chen W, Shi L, Wang Y, Dong J, Li Z, Li F, Chen C (2019) Angew Chem Int Ed 58:4271–4275
92. Hou S-L, Dong J, Jiang X-L, Jiao Z-H, Zhao B (2019) Angew Chem Int Ed 58:577–581
93. Padmanaban S, Kim M, Yoon S (2019) J Ind Eng Chem 71:336–344
94. Kang X-M, Shi Y, Cao C-S, Zhao B (2019) Sci China: Chem 62:622–628
95. Zhang X, Sun J, Wei G, Liu Z, Yang H, Wang K, Fei H (2019) Angew Chem 131:2870–2875
96. An B, Zeng L, Jia M, Li Z, Lin Z, Song Y, Zhou Y, Cheng J, Wang C, Lin W (2017) J Am Chem Soc 139:17747–17750
97. Li Z, Rayder TM, Luo L, Byers JA, Tsung C-K (2018) J Am Chem Soc 140:8082–8085
98. Liu M, Wu J, Hou H (2018) Chem- Eur J 25:2935–2948
99. Wang L, Agnew DW, Yu X, Figueroa JS, Cohen SM (2018) Angew Chem Int Ed 57:511–515
100. Hoang TT, To TA, Cao VTT, Nguyen AT, Nguyen TT, Phan NTS (2017) Catal. Commun 101:20–25
101. Tran TV, Le HTN, Ha HQ, Duong XNT, Nguyen LHT, Doan TLH, Nguyen HL, Truong T (2017) Catal. Sci Technol 7:3453–3458
102. Pham PH, Doan SH, Tran HTT, Nguyen NN, Phan ANQ, Le HV, Tu TN, Phan NTS (2018) Catal. Sci Technol 8:1267–1271
103. To TA, Vo YH, Nguyen HTT, Ha PTM, Doan SH, Doan TLH, Li S, Le HV, Tu TN, Phan NTS (2019) J Catal 370:11–20
104. Xu F, Kang W-F, Wang XN, Kou H-D, Jin Z, Liu C-S (2017) RSC Adv 7:51658–51662
105. Ikuno T, Zheng J, Vjunov A, Sanchez-Sanchez M, Ortuño MA, Pahls DR, Fulton JL, Camaioni DM, Li Z, Ray D (2017) J Am Chem Soc 139:10294–10301
106. Baek J, Rungtaweevoranit B, Pei X, Park M, Fakra SC, Liu Y-S, Matheu R, Alshmimri SA, Alshehri S, Trickett CA (2018) J Am Chem Soc 140:18208–18216
107. Osadchii DY, Olivos-Suarez AI, Szécsényi Á, Li G, Nasalevich MA, Dugulan IA, Crespo PS, Hensen EJM, Veber SL, Fedin MV (2018) ACS Catal 8:5542–5548

108. Pourkhosravani M, Dehghanpour S, Farzaneh F, Sohrabi S (2017) React Kinet Mech Catal 122:961–981
109. Zhang X, Huang Z, Ferrandon M, Yang D, Robison L, Li P, Wang TC, Delferro M, Farha OK (2018) Nat Catal 1:356–362
110. Li Z, Peters AW, Liu J, Zhang X, Schweitzer NM, Hupp JT, Farha OK (2017) Inorg Chem Front 4:820–824
111. Zhang X, Vermeulen NA, Huang Z, Cui Y, Liu J, Krzyaniak MD, Li Z, Noh H, Wasielewski MR, Delferro M (2018) ACSAppl Mater. Interfaces 10:635–641
112. Zhang H, Gao X-W, Wang L, Zhao X, Li Q-Y, Wang X-J (2019) CrystEngComm 21:1358–1362
113. Elumalai P, Mamlouk H, Yiming W, Feng L, Yuan S, Zhou H-C, Madrahimov ST, Appl ACS (2018) Mater Interfaces 10:41431–41438
114. Oozeerally R, Burnett DL, Chamberlain TW, Walton RI, Degirmenci V (2018) ChemCatChem 10:706–709
115. Yabushita M, Li P, Islamoglu T, Kobayashi H, Fukuoka A, Farha OK, Katz A (2017) Ind Eng Chem Res 56:7141–7148
116. Rimoldi M, Howarth AJ, DeStefano MR, Lin L, Goswami S, Li P, Hupp JT, Farha OK (2017) ACS Catal 7:997–1014
117. Planas N, Mondloch JE, Tussupbayev S, Borycz J, Gagliardi L, Hupp JT, Farha OK, Cramer CJ (2014) J Phys Chem Lett 5:3716–3723
118. Wang TC, Vermeulen NA, Kim IS, Martinson ABF, Stoddart JF, Hupp JT, Farha OK (2016) Nat Protoc 11:149–162
119. Kim IS, Li Z, Zheng J, Platero-Prats AE, Mavrandonakis A, Pellizzeri S, Ferrandon M, Vjunov A, Gallington LC, Webber TE (2018) Angew Chem Int Ed 57:909–913
120. Yang Q, Xu Q, Jiang H-L (2017) Chem Soc Rev 46:4774–4808
121. Xiang W, Zhang Y, Lin H, Liu C-J (2017) Molecules 22:2103
122. Yu J, Mu C, Yan B, Qin X, Shen C, Xue H, Pang H (2017) Mater Horiz 4:557–569
123. Li B, Ma J-G, Cheng P (2019) Small 15:1804849
124. Garibay SJ, Cohen SM (2010) Chem Commun 46:7700–7702
125. Deria P, Mondloch JE, Karagiaridi O, Bury W, Hupp JT, Farha OK (2014) Chem Soc Rev 43:5896–5912
126. Yin Z, Wan S, Yang J, Kurmoo M, Zeng M-H (2019) Coord Chem Rev 378:500–512
127. Ning L, Liao S, Liu X, Guo P, Zhang Z, Zhang H, Tong X (2018) J Catal 364:1–13
128. Butson JD, Kim H, Murugappan K, Saunders M, Buckley CE, Silvester DS, Szilágyi PA (2019) ChemPhysChem 20:745–751
129. Hermes S, Schröter M-K, Schmid R, Khodeir L, Muhler M, Tissler A, Fischer RW, Fischer RA (2005) Angew Chem Int Ed 44:6237–6241
130. Friedrich M, Klarner M, Hermannsdörfer J, Kempe R (2018) Polyhedron 155:441–446
131. Ishida T, Nagaoka M, Akita T, Haruta M (2008) Chem- Eur J 14:8456–8460
132. Sabo M, Henschel A, Fröde H, Klemm E, Kaskel S (2007) J Mater Chem 17:3827–3832
133. Hwang YK, Hong D-Y, Chang J-S, Jhung SH, Seo Y-K, Kim J, Vimont A, Daturi M, Serre C, Férey G (2008) Angew Chem Int Ed 47:4144–4148
134. Goswami S, Noh H, Redfern LR, Otake K-i, Kung C-W, Cui Y, Chapman KW, Farha OK, Hupp JT (2019) Chem Mater 31:1485
135. Rivera-Torrente M, Filez M, Hardian R, Reynolds E, Seoane B, Coulet M-V, Oropeza-Palacio FE, Hofmann JP, Fischer RA, Goodwin AL (2018) Chem.-Eur. J. 24:7498–7506
136. Yang K, Yang K, Zhang S, Luo Y, Yao Q, Lu Z-H (2018) J. AlloysCompd. 732:363–371
137. Aijaz A, Karkamkar A, Choi YJ, Tsumori N, Rönnebro E, Autrey T, Shioyama H, Xu Q (2012) J Am Chem Soc 134:13926–13929
138. Sun J-L, Chen Y-Z, Ge B-D, Li J-H, Wang G-M, Appl ACS (2019) Mater Interfaces 11:940–947
139. Chen W-J, Cheng B-H, Sun Q-T, Jiang H (2018) ChemCatChem 10:3659–3665
140. Zheng X, Song X, Wang X, Zhang Z, Sun Z, Guo Y (2018) New J Chem 42:8346–8350
141. Zhao Y, Li Y, Pang H, Yang C, Ngai T (2019) J Colloid Interface Sci 537:262–268

142. Sugikawa K, Furukawa Y, Sada K (2011) Chem Mater 23:3132–3134
143. Carné-Sánchez A, Imaz I, Cano-Sarabia M, Maspoch D (2013) Nat Chem 5:203–211
144. Garzón-Tovar L, Rodríguez-Hermida S, Imaz I, Maspoch D (2017) J Am Chem Soc 139:897–903
145. Tsumori N, Chen L, Wang Q, Zhu Q-L, Kitta M, Xu Q (2018) Chem 4:845–856
146. Han Y, Xu H, Su Y, Xu Z-L, Wang K, Wang W (2019) J Catal 370:70–78
147. Zheng D-Y, Zhou X-M, Mutyala S, Huang X-C (2018) Chem- Eur J 24:19141–19145
148. Zhao X, Xu H, Wang X, Zheng Z, Xu Z, Ge J, Appl ACS (2018) Mater Interfaces 10:15096–15103
149. Zhao Z-W, Zhou X, Liu Y-N, Shen C-C, Yuan C-Z, Jiang Y-F, Zhao S-J, Ma L-B, Cheang T-Y, Xu A-W (2018) Catal. Sci Technol 8:3160–3165
150. Dutta G, Jana AK, Singh DK, Eswaramoorthy M, Natarajan S (2018) Chem-Asian J 13:2677–2684
151. Bakuru VR, Velaga B, Peela NR, Kalidindi SB (2018) Chem.-Eur. J. 24:15978–15982
152. Baguc IB, Ertas IE, Yurderi M, Bulut A, Zahmakiran M, Kaya M (2018) Inorg Chim Acta 483:431–439
153. Li S-J, Wang H-L, Wulan B-R, Zhang X-B, Yan J-M, Jiang Q (2018) Adv Energy Mater 8:1800625
154. Redfern LR, Li Z, Zhang X, Farha OK, Appl ACS (2018) Nano Mater 1:4413–4417
155. Kar AK, Srivastava R (2018) New J Chem 42:9557–9567
156. Meng F, Zhang S, Ma L, Zhang W, Li M, Wu T, Li H, Zhang T, Lu X, Huo F (2018) Adv Mater 30:1803263
157. Chen H, He Y, Pfefferle LD, Pu W, Wu Y, Qi S (2018) ChemCatChem 10:2558–2570
158. Augustyniak AW, Sadakiyo M, Navarro JAR, Trzeciak AM (2018) ChemistrySelect 3:7934–7939
159. Sadeghzadeh SM, Zhiani R, Emrani S (2018) New J Chem 42:988–994
160. Kaur R, Chhibber M, Mahata P, Mittal SK (2018) ChemistrySelect 3:3417–3425
161. Yin D, Li C, Ren H, Liu J, Liang C (2018) ChemistrySelect 3:5092–5097
162. Chen S, Lucier BEG, Luo W, Xie X, Feng K, Chan H, Terskikh VV, Sun X, Sham T-K, Workentin MS, Appl ACS (2018) Mater Interfaces 10:30296–30305
163. Yan R, Zhao Y, Yang H, Kang X-J, Wang C, Wen L-L, Lu Z-D (2018) Adv Funct Mater 28:1802021
164. Yuan K, Song T, Wang D, Zhang X, Gao X, Zou Y, Dong H, Tang Z, Hu W (2018) Angew Chem Int Ed 57:5708–5713
165. Lin Z, Luo M, Zhang Y, Wu X, Fu Y, Zhang F, Zhu W (2018) Appl Catal A 563:54–63
166. Azad M, Rostamizadeh S, Nouri F, Estiri H, Fadakar Y (2019) Mater Lett 236:757–760
167. Xiong G, Chen X-L, You L-X, Ren B-Y, Ding F, Dragutan I, Dragutan V, Sun Y-G (2018) J Catal 361:116–125
168. Tan YC, Zeng HC (2018) Nat Commun 9:4326

MOFs as Sensors

Metal–Organic Frameworks (MOFs) as Sensors for Environmental Monitoring

Sushma Yadav and Priti Malhotra

Contents

1 Introduction 394
2 Water Sensors Based on MOFs 395
  2.1 Luminescent Technology 396
  2.2 Sensing of Heavy Metal Pollutants 397
  2.3 Sensing of Organic Molecules Present as Contaminants 398
3 Electrochemical Sensors 399
4 Different Other Sensors in Aqueous Medium 400
5 Environmental Gas Sensors Based on MOF 401
6 Conclusion 404
Abbreviations 405
References 405

Abstract The uniqueness of metal–organic frameworks is demonstrated in their high surface area, highly ordered self-assembled structure, tailorable pore size, high density of active sites, and high catalytic activity which enable them to act as efficient sensing materials. MOFs provide magnificent platforms for detecting environmental contaminants which include detection of heavy metals, anions, organic pollutants, and gases both in water and air. In recent researches, huge progress has been achieved in utilizing special optical, electrochemical properties and field-effect transistor sensing abilities of MOFs. Their sensing capabilities could be further improved through MOF-based composites obtained by incorporating varied functional materials. The nanoscaled pore size of MOFs permits the diffusion of small molecules into the pores or through the MOF layer while excluding the larger molecules subjective of the size, shape, and conformation of MOFs. The MOFs also exhibit a dynamic response to external stimuli which include pressure changes, light, pH, temperature, and various other guest molecules showing the high grade of selectivity toward different analyze molecules. The ongoing chapter reviews and summarizes the outcomes of current findings and challenges of MOFs while

P. Malhotra (✉)
Department of Chemistry, Daulat Ram College, University of Delhi, Delhi 110007, India

S. Yadav
Department of Chemistry, University of Delhi, Delhi 110007, India

S. Gulati (ed.), *Metal-Organic Frameworks (MOFs) as Catalysts*,
https://doi.org/10.1007/978-981-16-7959-9_15

being used as highly sensitive and efficient electronic sensors for environmental contaminants.

Graphical Abstract

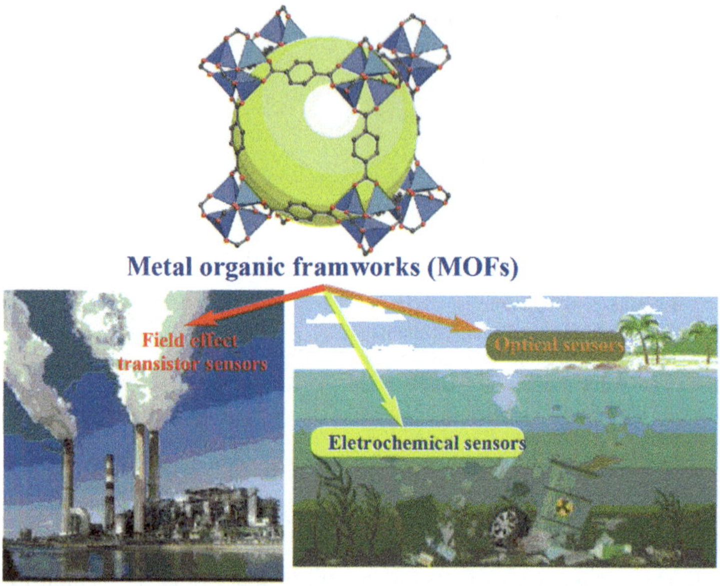

Keywords MOFs · Sensors · Environment · Monitoring · Contaminants

1 Introduction

There is constantly increasing demand for controlling environmental pollution and finding solutions for global warming, climate change, and other damaging causes in today's world. The compelling urge for designing methods to reduce the impact of pollution urgently requires an assessment of the elements of pollution and destruction. For effective monitoring, a comprehensive system is required for analyzing and evaluating the contaminants and pollutants precisely and accurately. Various types of sensors have been proposed in the past by numerous scientists among whom MOF sensors and MOF-based hybrid sensors have currently emerged as popular tools which are capable of assisting the scientists in first evaluating the pollutants and then carrying out the remediation exercise. Even the minutest number of pollutants need to be determined precisely for their complete removal and also to work out measures for curtailing and controlling them. The accuracy in monitoring the pollutants can be possible only if highly sensitive, reliable, and selective devices are available, and many novel techniques are employed for both analyzing and implementing damage control pathways for the environment. In the recent past, MOFs are growing popular as such coordination polymers which are endowed with high order tunable porosity and crystallinity. The presence of organic molecules linked to metal ions clusters contribute heavily to their extraordinary characteristics leading to the enhanced surface area, a large volume of modifiable pores, and fascinating catalytic efficacy attributed to dense active sites. A battery of MOFs and MOF-based NPs has been fabricated in the past which are employed as excellent sensors for detecting the contaminants present in the surrounding air and water bodies. Detection of pollutant heavy metal ions, hazardous organic molecules, and gases are performed by MOF-based environment sensors owing to their unique optical and electrochemical versatility and sensitivity to field-effect which enables them to heave as accurate sensors. MOF composites show improved properties as environment sensors and have initiated a novel class of sensor molecules providing ubiquitous platforms for measuring and signaling both water and gas contaminants. A large number of MOF-based sensors are reported, and a new world entity has motivated a promising area of research and given a new direction to the scientific community and custodians of environmental remediation and sustainability [1].

The extraordinary ability of MOF for reversible adsorption, high catalytic performance, the diverse functionality, and tunable structural features contributes largely to make them suitable chemical sensors [2]. The whole event of sensing while absorption of the target guest molecules takes place engages various structural, physical, and chemical changes on the surface of MOFs. In the current past, MOFs have exhibited distinguished characteristics especially for detecting and monitoring environmental contaminants such as poisonous heavy metals, hazardous chemicals, and gases of toxic nature [3, 4]. The general principles on which MOFs and MOF-based materials work are an optical phenomenon, electrochemical propositions, and FET theory of sensors.

2 Water Sensors Based on MOFs

The most probable substances which pollute water include industrial wastes containing heavy metals, organic effluents, pharmaceutical wastes such as antibiotics and bacteria's which are harmful to human and animal health and the ecosystem in the environment. Thus, high resolution, reliable, and sensitive sensors are in urgent need for determining the type and extent of contamination in water. MOFs have played a significant and promising role in sensing and detecting the chemicals in aqueous solutions from various sources and have exhibited their ability in their reversible capture and adsorption. Sensors based on MOFs function based on signals obtained through luminescent technology, electrochemical phenomenon, and colorimetric mechanism thereby emerging as the most versatile candidates for detecting and monitoring the target pollutant molecules present in water.

2.1 Luminescent Technology

Substantial advances have taken place in the past related to chemical sensing performances of MOFs which are based on their luminescence property [4, 5]. The process of luminescence involves the returning of electrons from the excited singlet state to the ground state through the emission of photons in the whole process [6]. The complete mechanism is known as "turn- off" mechanism [7] which originates when absorption of molecules of analytes occurs leading to the quenching and attenuation of the phenomenon of luminescence on the MOF surfaces thereby causing changes in the emissions which can be easily detected. Another, "turn-on" mechanism has also been reported in the past literature [7] suggesting that luminescent progression plays a vital role in the ability of MOF in detecting both inorganic and organic components and thus providing a multifunctional rostrum capable of emitting luminescent signals. The contribution of both the organic skeleton and inorganic matrix of MOF composites is equally responsible for their luminescent marvel. It has been observed that charge transfer between metal and ligand can also cause luminescence within MOFs through diverse functionalities. Either varied interaction with the guest molecules or the inherent changes in the MOF functionalities, both can equally influence luminescence and induce signals for sensing [6, 8–12].

Sensing of Inorganic anions in water by diverse MOF sensors using luminescent technology has gained importance as it not only resulted in evaluating the common nutrients present in groundwater such as phosphates, but it could also manage selective sensing of the degraded products of the nutrients as reported by Qian et al. [13]. Based on the selective nature of photoluminescence, MOF based on Tb (III) ions TbNTA1 (NTA = nitrilotriacetate) were studied, and it was observed that their highly specific and selective response toward PO_4 ions was not impacted by other anions such as NO^{3-}, $CO_3{}^{2-}$, $SO_4{}^{2-}$, $NO_2{}^{-}$, and halide ions. The signals of quenching effect of changes in bond strength of Tb-O generate luminescence which is transformed to signals as sensors [13]. Another group reported the detection of phosphate ions where MOF-5 was used in combination with ZnO quantum dots [14], and its interaction with phosphate groups involved electron transfer reaction giving rise to the ZnO QD quenching leading to its fluorescence recovery whose sensitivity had a fairly good limit of detection (~53 nM), however, the luminescence-based sensors cannot be reused. Recently another MOF was reported which provides reusable sensing for phosphate ion which was 3D-MOF Eu-BTB (based on H_3BTB = 1,3,5-benzenetribenzoate and could be reused for five subsequent runs [15].

Drinking water often consists of ClO^- ions, and there should be an optimum level of ClO^- ion present in it as its very low level cannot destroy pathogens, whereas very high levels can form degraded products which cause health damage. Lu and co-workers have shown in a recent study that a novel MOF-based material can detect free chlorine using NH_2-MIL-53(Al) which effectively provided the surface for fluorescent sensing. It provided fluorescent Al nanoplates as a sensing platform in conjugation with the hybrid MOF [16] and detected free chlorine. The NH_2-MIL-53(Al) was highly soluble in water, showed exceptional stability, and its wider benefits were seen in evaluating the difference between tap water and swimming pool water. In another experiment CN^- ion which is also a water contaminant ion and was sensed by a MOF ZIF-90 based hybrid selectively at two-micrometer levels [17] and gave an important method to monitor the CN^- ion levels within permissible limits of WHO.

2.2 Sensing of Heavy Metal Pollutants

Heavy metal ions are currently most worrisome pollutants of water and also the environment as a whole because it is one of the non-biodegradable substances and causes serious damages. Some heavy metals ions which include lead (Pb), copper (Cu), mercury (Hg), Cadmium (Cd), chromium (Cr), and arsenic (As) are a few of the high-level toxic elements and cause hazards to human health even if they are present in a minute or trace amount. Thus, it is highly pertinent and significant to employ MOF-linked sensors for the determination of heavy metal ions in water. Lin et al. [18] reportedly used poly (ethylenimine)-capped CQDs (BPEI-CQDs) for sensing Cu^{2+} based on MOF-linked fluorescent technology. The branched poly (ethylenimine)-capped CQDs (BPEI-CQDs) embedded inside ZIF-8 possess powerful fluorescent (40% quantum yield) and high-grade selectivity during the detection of Cu^{2+} ions. The BPEI-CQDs/ZIF-8 composites can both detect and adsorb Cu^{2+} ions as they command both fluorescent activity, and adsorption properties have been used for ultrasensitive and highly selective copper ion sensing. ZIF-8 not only exhibits excellent fluorescent activity and selectivity derived from CQDs but can also accumulate Cu^{2+} owing to its high adsorption property. The accumulation effect of MOFs can amplify the sensing signal. The fluorescent intensity of BPEI-CQDs/ZIF-8 was quenched with the presence of Cu^{2+}. This sensing platform can detect Cu^{2+} in a wide concentration range of 2–1000 nM and a lower LOD of 80 pM. Compared with other fluorescent sensors without the amplifying function of MOF or the introduction of a guest luminophore, this sensing platform has a much lower detection limit of approximately two orders of magnitude. This sensor was also applied in real water sample tests and showed good performance. This study indicated that novel sensing platforms can be designed and applied in heavy metal ion detection by incorporating MOFs with fluorescent nanostructures.

Detection of mercury ion (Hg^{2+}) is of prime importance because it is the most toxic substance and has the highest degree of health damages to both animals and humans. Hg^{2+} based substances pose highly poisonous and greater risks to human health. An expeditious and effective strategy for sensing the presence of mercury ions in a highly selective manner is needed to overcome this challenge to prevent damage to mankind. Chi group have reported that Ru-MOFs [19] are employed with luminescent $Ru(bpy)_2{}^{3+}$ and in conjugation provide an excellent sensor and adsorbent for mercury ions. Normally, Ru-MOFs get precipitated as a yellow powder in an aqueous solution which emits a red color under UV light, but in the presence of Hg^{2+}, the Ru-MOFs get decomposed fast by Hg^{2+} ions and were able to release a substantial amount of luminescent guest molecules into the water, *i.e.*, $Ru(bpy)_2{}^{3+}$

generating powerful signals due to fluorescence or electro-chemiluminescence. The efficient working of the sensor is subject to the concentration of Hg^{2+} and showed excellent results in the range of 25 pM–50 nM.

Chen et al. [20] were able to design lanthanide-based structure of MOF nanocomposites with a turn-on mechanism of fluorescent sensor which could manage the detection of Hg^{2+} ions *via* internal filter effect. Eu-isophthalate MOF NPs were fluorescent and were synthesized such that their diameter was in the range of 400 nm and was made to link to IDA. The coordination of IDA with the MOF particle was able to quench the MOF-based emission because the imidazole showed a strong absorbance in the region of MOF excitation, and upon coordination of Hg^{2+} with IDA, the latter gets released from the surface of MOF thereby causing a change in the emission. The interaction which is sensitive to light was also selective having LOD at 2 nm and either did not respond to the presence of other ions such as Pb^{2+}, Mg^{2+}, Ag^{+}, K^{+}, and Na^{+} or showed negligible sensitivity.

2.3 Sensing of Organic Molecules Present as Contaminants

VOCs are another category of water pollutants that are a threat to the environment as their toxicity is also a menace and needs to be addressed. Such VOCs which are poisonous are mainly derivatives of benzene and other aromatic compounds and are generally present in the environment in industrial effluents and are substantively toxic resulting in long-term destruction of the ecosystem [21]. MOF seems to be an obvious choice because their organic component can easily recognize the organic molecules through feasible reactions and bond formation thereby converting the changes to identifiable and readable signals. In recent times, a new MOF with Eu^{3+} within the matrix of MOF $Zr_6(u_3–O)4(OH)_4(bpy)_{12}$ was proposed whose fluorescent characteristics was able to analyze VOC giving exceptional results [22]. In this process, the MOF and VOCs combination can be easily recognized as a signal as the fluorescence properties of MOF nanocomposite are majorly dependent upon the VOCs binding. Nitroaromatic molecules were probed by Morsali et al. [23] after exploring MOFs TMU-31 and TMU-32 when they were mixed with aromatic compounds containing nitro groups facilitating the interaction of urea entity inside the MOF pores with the nitro group of the analyte. Various nitroaromatics such as nitrobenzene, 1,3-dinitrobenzene, 2,4,6-trinitrotoluene, and 2, 4-dinitrotoluene were probed using fluorescence emission spectra and quenching potential of TMU-31 and TMU-32 combination MOFs.

Pharmaceutical wastes are also one of the commonly found organic contaminants and cause biological damages to the ecosystem. Antibiotics such as TC is among such organic adulterants of water, and their degradation is a difficult process. MOF based on zirconium (PCN-128Y) was tested by the Cuan group, and they succeeded in the removal and detection of TC with luminescent MOF (PCN-128Y) [24] as the accumulated TC within the pores of MOF further enhanced the detection response.

The main center for sensing undergoes the rise in luminescence quenching at PCN-128Y owing to the joint effect of electron movement from PCN-128Y to TC which is induced by photons. The enhanced absorption ability of the MOF PCN-128Y further attenuates the detection abilities for which strong bonding between Zr and TC is also responsible.

3 Electrochemical Sensors

Quantification of contaminants using recently proposed advanced methods is focused on electrochemical sensor systems which involve amperometric, voltammetric, and impedimetric techniques. The electrochemical methods based on nanostructures and MOF nanocomposites are on the rise currently and have shown great promise as environmental sensors. The working of electrochemical sensors is based on the principle of redox reactions and is generally conducted through a three-electrode system comprising of a reference electrode, working electrode, and a counter electrode. Measuring the various changes in electrical signals, electrical potential, and the current for various analytes can detect and analyze them precisely [25, 26]. MOFs show potential as electrochemical sensing surface modifiers because of their high surface area and pore volume, good absorbability, and high catalytic activity [27, 28]. Attributed to their high surface area, high catalytic performance [29, 30], large pore volume, and efficient adsorption ability, MOFs show extraordinary potential for being effective electrochemical sensors. Many stable MOFs and MOF nanocomposites have been fabricated to date which can be employed as electrochemical sensors ubiquitously. Although MOFs have lower conductivity and because of their reversible nature in an aqueous medium, their stability also decreases. Yet it is possible to develop MOF as efficient electrode material by conjugation with diverse functionality and with nanomaterials having high redox potential. Their association with materials of high electrical conductivity can also help to design MOFs which can be effectively employed as electrochemical sensors [31, 32]

Pb^{2+} causes maximum damage to the aquatic life and has been successfully probed by electrochemical method with high sensitivity as even trace amounts are harmful and require precise monitoring and detection. Researchers [33] designed streptavidin-modified reduced graphene oxide-tetraethylene pentamine-gold nanoparticle (rGO-TEPA-Au) for sensing Pb^{2+}, and it was observed that this MOF-NPs- RGO composite was an excellent sensor for DNAzyme immobilization. During the process, the DNAzyme breaks and gets free from the catalyst forming a single strand of DNA linked to the sensor which later gets hybridized with Fe-MOFs/PdPt NPs attaching to the surface of the electrode. Both Fe-MOFs and PbPt NPs perform differently, and they show the activity of peroxidase and catalytic reduction of hydrogen peroxide, respectively. The hydrogen peroxide reduction is the main activity which is measured through its reduction current to assess Pb^{2+} ion within a 2 pM limit of detection the range of sensing being 0.005–1000 nM.

Pond water generally has enhanced nitrite NO^{2-} ion which is another contaminant of water that causes health damages. Using a MOF/rGO composite for making electrodes and then determining the nitrite levels electrochemically was important research conducted by Mobin et al. [34]. The combination sensor Cu-MOF/rGO proved to be active in sensing and determining NO^{2-} ion concentration through the electrochemical oxidation process. Sensing of organic compounds by MOFs *via* electrochemical process is also commonly encountered. Various derivatives of phenol such as resorcinol, catechol, and hydroquinone commonly exist in the environment as harmful pollutants [35]. Employing a MOF-based electrode, these dihydroxybenzene molecules could be detected and analyzed. The process involves the coating of chitosan on the electrode in the initial step which catalyzes the reduction of GO (graphene oxide) to rGO, and the resulting ERGO/CS upon getting linked to the MOF $Cu_3(BTC)_2$ forms an efficient electroactive platform that acts as a marker and analyzer for the pollutant organic molecules [35]. With enhanced conductivity provided by ERGO/CS, there was a subsequent rise in the response to the current flowing thereby showing raised sensing abilities with a limit of detection ranging from 0.44 to 0.33 M for various dihydroxybenzene molecules. This way many such novel MOFs came to be recognized as effective electrochemical sensors for evaluating harmful phenol derivatives which are present as contaminants in real-time water samples. Another class of organic molecules that cause water pollution is chlorinated phenols which pose serious harms to human health such as 2,4 dichlorophenol whose minute concentrations are also quite damaging [36]. Dong group [37] designed 1,3,5-benzenetricarboxylic acid copper ($Cu_3(BTC)_2$) based MOFs which exhibited extraordinary efficacy as sensors in the effective range of 0.04 to 1.0 l M and a good limit of detection, *i.e.*, ~9 nM.

4 Different Other Sensors in Aqueous Medium

The versatility of MOF-based electrochemical sensors built on $Cu_3(BTC)_2$ was attributed to high adsorption ability due to enhanced surface area and effective charge transfer. Sensors based on colorimetric measurements are also prevalent and often used for monitoring and detecting the pollutants of water by tracking chromogenic changes, *i.e.*, color changes [38, 39] in solutions. Gu group of researchers observed the changes in Ru complex-mixed MOFs (RuUiO-67) and tracked their response as probing molecules for the detection of Hg^{2+}through colorimetric signals and responses [40]. It was observed that dyes consisting of thiocyanate in $Ru(H_2bpydc)(bpy)(NCS)_2$ (H_2L) complex exhibited strong interaction with Hg^{2+} because of the complex formation between thiocyanate and Hg^{2+} ions.. Presence of Hg^{2+} ions produced signals with RuUiO-67 which offered a platform for recognizing and producing signal indications. Hg^{2+} produced a red to yellow change in color in the solution meant for detection, and the sensitivity of the complex was high. The limit of detection of the probe is 0.51 M for Hg^{2+} proving the good capability of the sensor in colorimetric suspension.

SERS has also proven to be a promising spectroscopic tool that demonstrates high potential as it is capable of responding to even a single molecule providing high-resolution spectroscopic results. Both biological probes and chemical sensing are performed by this technique due to its dependence on the distance between the target analyte and the nanocomposites getting highly reduced, its sensitivity gets enhanced, and it yields unique benefits [41]. The evaluation of organic molecule p-phenylenediamine which is a water contaminants was detected by Li et al. [42] employing Au NPs linked MOFs Au NPs/MIL-101 through SERS analysis. Their combined effect of Au NPs and the MOF matrices not only raised the adsorption of the analyte molecules but also enhanced the metal magnetic fields response giving precise results by SERS. This technique promoted a novel method for evaluating p-phenylenediamine as a pollutant and a biomarker for tumor cells in human blood serum namely alpha-fetoprotein.

5 Environmental Gas Sensors Based on MOF

MOFs have been endowed with the marvel of possessing high-level sustainable porosity, having the free and flexible volume in size along with tunable and workable inherent surface area [43–45]. The super level crystallinity augments its strength of interaction between the host molecule and the guest entity [46]. The MOFs contribute to the suitability of the environment of the matrix in such a way, so that it is conducive for the dwelling of the guest molecule thereby raising the sensitivity of the signals produced by strong interactions between the guest (analyte) and the host molecules [46]. Therefore, apart from acting as an important sensor for water pollutants, MOFs and their nanocomposites also play a vital role as gas sensors and are equally significant in detecting gas pollutants in the environment. Both the physical and chemical properties of materials based on MOFs are complimentary to their efficacy as gas sensors.

The FET-based analysis and detection of various gases generally utilize two electrodes, designed as the source electrode and drain electrode, coupled with a gate electrode, channel material, and gate oxide as shown in Fig. 1 [47]. The main active component in the system is the channel material which assists in monitoring the fluctuation in the conductance of the sensor due to the adsorption of analyte gases which get adsorbed physically on the host. The sensing capabilities are subject to the bandgap, threshold energy, and charge mobility of the channel material, and all these properties are those inherent in the material itself. They decide the sensitivity of the sensor and its overall precision in detecting and monitoring. Depending on the nature of gases whether they are oxidizing or reducing, and the semiconducting nature of the material, the interaction between the sensor and the gases are reflected in the conductance observed in the channel material. Thus, the overall change in the conductance produces readable signals and enables the detection of gases [48]. The use of FET sensing is highly probable during the applications of MOF-based gas sensors because it is chemical-free and can have instant responses even to the minute

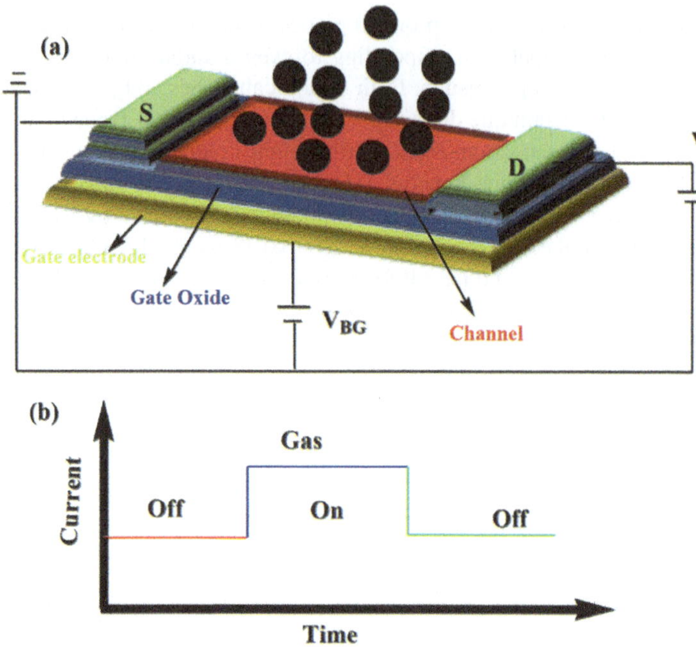

Fig. 1 **a** Schematic illustration of FET sensor with source and drain electrodes, channel material, gate oxide, and the gate electrode. **b** Sensor current change after gas adsorption and desorption

concentration of pollutant gases. For further enhancement in the sensing exercise for gases other gas sensing materials can also be coordinated with the MOF materials [49–51], and novel composite materials of MOF can be used for further upgradation of the sensing activity [30, 52]

Dinca's group has worked on the sensing of ammonia gas using MOF-linked FET sensors and observed excellent sensing behavior with high precision [53]. He and his co-workers explored noble metal@MOF hybrids which gave a good performance in sensing VOCs also gave high-level efficiency for the detection and monitoring of formaldehyde gas at room temperature and low concentrations which were as low as 0.25 ppm [54]. Employing transition metal oxides and noble metals nanostructures along with MOF led to many FET-based gas sensors nanocomposites whose performances were high grade and they showed good overall sensitivity and selectivity. Examples are Pd–ZnO/$ZnCo_2O_4$ [55] meant to be acetone sensor, Au@ZnO@ZIF-8 [56], removes VOCs along with gas sensing, and Au@MOF-5 ($Zn_4O(BDC)_3$) [57] for monitoring various other gases (Fig. 2).

Recently, gas sensing MOF-based films have also contributed to the large focus of the scientific community owing to their vast surface in contact with the gas molecules, escalated stability, and rapid flow of gas to the exposed area [53]. A study has revealed that a ZnO@ZIF-8 matrix nanorod film (thickness of 100 nm) was fabricated for analyzing and detecting H_2 over CO_2 using a solution deposition method [58].

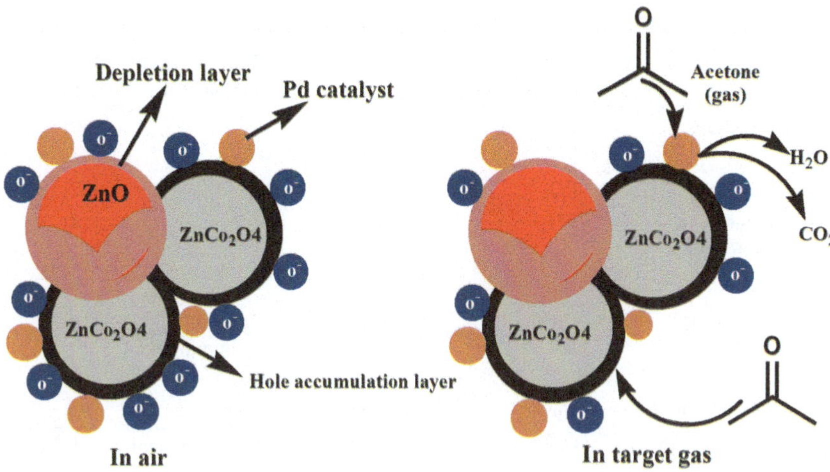

Fig. 2 Schematic of acetone sensing mechanism with Pd–ZnO/$ZnCo_2O_4$ hollow spheres

MOF sensors coupled with other sensing devices have also been explored for gas detection (Table 1). Salama et al. proposed a unique sensor for the detecting and monitoring of sulfur dioxide (SO_2) at normal temperature [66]. The sensor was based on a layer of Indium on MOF (MFM-300), which was further plated on the electrode and was able to exhibit extraordinary sensitivity toward SO_2 gas even at its very low concentration, *i.e.*, 75 ppm. The MFM-300 sensor was also able to perform exceptionally for detecting SO_2 in the presence of CH_4, NO_2, CO_2, and H_2 and showed high selectivity as well.

Dou stak worked on a luminescent MOF film sensor [MIL-100(In).Tb^{3+}) and found them to show rapid response to O_2 levels [67]. It was revealed in a study by Zhang group incorporated Eu^{3+} ions in MIL-124 which was then able to detect very low concentrations of NH_3 present in ambient air, and it exhibited a high level of sensing abilities [68]. Eddaoudi et al. observed that thin film of Re, the rare-earth metal-based MOFs developed into an interesting platform for detecting hydrogen sulfide (H_2S) gas as a pollutant at room temperature [69].

The RE-MOF had good performance for H_2S detection for very minimal concentrations as low as 100 ppb with a detection limit of 5.4 pp. Therefore, MOF-based materials and MOF nanocomposites have been recognized as versatile gas sensors which follow a facile approach for detection of various gases and command extraordinary performance. Their excellence lies in their rapid detection response, easy recovery high-order sensitivity and selectivity, and simple protocol of working.

Table 1 Gas and water contamination detection sensors based on MOFs

Sensing material	Target contaminant	LOD	Environmental sample test	References
Eu(III)@UMOFs	Hg^{2+}, Ag^{+}, and S^{2-}	–	–	[59]
UiO-66-NH_2	$PO_4{}^{3-}$	1.25 μM	–	[32]
APTMS-ZnO QDs@MOF-5	$PO_4{}^{3-}$	53 nM	–	[14]
UiO-66-NH_2	Hg^{2+}	17.6 nM	–	
NH_2-MIL-53(AI)	ClO^{-}	0.04 μM	Tap and swimming pool water	[16]
CDs@Eu-DPA MOFs	Cu^{2+}	26.3 nM	Real water	[60]
Eu-UiO-66(Zr)-$(COOH)_2$	Cd^{2+}	0.06 μM	Environmental water	[61, 62]
Zn_3(TDPAT)-$(H_2O)_3$	Nitrobenzene	50 ppM		[61]
$Zn_4O(BDC)_3$	Volatile organic compounds	4.9 nM	Real water	[63]
Cu-MOF/rGO	Pb^{2+}	33 nM	Pond water	[34]
UiO-66-NH_2	NO^{2-}	0.01 μM		[64]
$Cu_3(BTC)_2$	NO^{2-}	9 nM	Reservoir raw water	[65]
Cu-MOF-199/SWCTs	Cu^{2+}	0.08 and 1 μM	River water	[37]
RuUiO_2-67	Hg^{2+}	0.5 μM		[40]
$Tb_{1.7}Eu0_{.3}(BDC)_3{\cdot}(H_2O)_4$	Cd^{2+}	0.25 mM	Lead-polluted water samples	
Pd-ZnO/$ZnCo_2O_4$	Acetone	0.4–5 ppm		[55]
$Cu_3(HITP)_2$	Ammonia	0.5–10 ppm		[30]
ZnO@ZIF-8	Formaldehyde	10–200 ppm		[58]
ZIF-67	Formaldehyde	5–500 ppm		[54]

6 Conclusion

MOF-based sensors have been developed on various physical phenomena such as electrochemical, optical, FET signals, and SERS signals for sensing environmental pollutants. MOFs have been widely utilized as sensing and evaluating materials for heavy metal, organic compounds, anions, and gas detection as they demonstrate the extraordinary structure and unique physical and chemical properties. Their vast surface area and manipulative porosity, special catalytic performance, reversible behavior, and desirable adaptation of chemical functionalization make them special candidates for sensing environment contaminants. As cited above, MOF-based hybrids have exhibited exceptional performance as sensors, and there is enough scope for improvement in the future for overcoming the challenges faced in this field. One of the future exercises would be to attempt at various combinations of functionality with MOFs and MOF-based hybrid materials to attain further competence in this area. Attempting at different combinations with materials of high conductivity such as carbon-based nanomaterials, GO, and rGO with MOFs is to aim at higher stability, sustainability, and electroconductivity. It is also significant that to enable them to interact and recognize various types of target analytes rapidly and to generate an appropriate response fast and precisely, novel functional groups, metal NPs, and composites may be incorporated, so that new heights are attained. Though a bright and promising future holds for MOF and the platform provided by its hybrids, optimization and fruitful utilization still need to be addressed with more linkages and amalgamation with new materials.

Abbreviations

MOF	Metal-organic framework
NPs	Nanoparticles
FET	Field-effect transistor
NTA	Nitrilotriacetate
CQDs	Carbon quantum dots
BPEI	Branched poly (ethylenimine)
LOD	Limit of detection
IDA	Imidazole-4,5- dicarboxylic acid
VOCs	Volatile organic compounds
TC	Tetracycline
rGO	Reduced graphene oxide
TEPA	Tetraethylene pentaamine
GO	Graphene oxide
SERS	Surface-enhanced Raman scattering

References

1. Fang X, Zong B, Mao S (2018) Metal-organic framework-based sensors for environmental contaminant sensing. Nano-Micro Lett 10:1–19
2. Chidambaram A, Stylianou KC (2018) Electronic metal-organic framework sensors. Inorg Chem Front 5:979–998
3. Stavila V, Talin AA, Allendorf MD (2014) MOF-based electronic and opto-electronic devices. Chem Soc Rev 43:5994–6010
4. Lustig WP, Mukherjee S, Rudd ND, Desai AV, Li J, Ghosh SK (2017) Metal-organic frameworks: functional luminescent and photonic materials for sensing applications. Chem Soc Rev 46:3242–3285

5. Kumar V, Kim KH, Kumar P, Jeon BH, Kim JC (2017) Functional hybrid nanostructure materials: advanced strategies for sensing applications toward volatile organic compounds. Coord Chem Rev 342:80–105
6. Cui Y, Yue Y, Qian G, Chen B (2012) Luminescent functional metal-organic frameworks. Chem Rev 112:1126–1162
7. Zhang M, Feng G, Song Z et al (2014) Two-dimensional metal-organic framework with wide channels and responsive turn-on fluorescence for the chemical sensing of volatile organic compounds. J Am Chem Soc 136:7241–7244
8. Zhang SY, Shi W, Cheng P, Zaworotko MJ (2015) A mixed-crystal lanthanide zeolite-like metal-organic framework as a fluorescent indicator for lysophosphatidic acid, a cancer biomarker. J Am Chem Soc 137:12203–12206
9. Zhou J, Li H, Zhang H, Li H, Shi W, Cheng P (2015) A bimetallic lanthanide metal-organic material as a self-calibrating color-gradient luminescent sensor. Adv Mater 27:7072–7077
10. Li H, Shi W, Zhao K, Niu Z, Li H, Cheng P (2013) Highly selective sorption and luminescent sensing of small molecules demonstrated in a multifunctional lanthanide microporous metal-organic framework containing 1D honeycomb-type channels. Chem A Eur J 19:3358–3365
11. Zhao B, Cheng P, Chen X, Cheng C, Shi W, Liao D, Yan S, Jiang Z (2004) Design and synthesis of 3d–4f metal-based zeolite-type materials with a 3D nanotubular structure encapsulated "Water" Pipe. J Am Chem Soc 126:3012–3013
12. Wu S, Lin Y, Liu J, Shi W, Yang G, Cheng P (2018) Rapid detection of the biomarkers for carcinoid tumors by a water stable luminescent lanthanide metal-organic framework sensor. Adv Funct Mater 28:1–10
13. Xu H, Xiao Y, Rao X, Dou Z, Li W, Cui Y, Wang Z, Qian G (2011) A metal-organic framework for selectively sensing of PO43- anion in aqueous solution. J Alloys Compd 509:2552–2554
14. Zhao D, Wan X, Song H, Hao L, Su Y, Lv Y (2014) Metal-organic frameworks (MOFs) combined with ZnO quantum dots as a fluorescent sensing platform for phosphate. Sens Actuators B Chem 197:50–57
15. Xu H, Cao C-S, Zhao B (2015) A water-stable lanthanide-organic framework as recyclable luminescent probe for detecting pollutant phosphorus anions. Hang J Mater Chem C3:10715–10722
16. Lu T, Zhang L, Sun M, Deng D, Su Y, Lv Y (2016) Amino-functionalized metal-organic frameworks nanoplates-based energy transfer probe for highly selective fluorescence detection of free chlorine. Anal Chem 88:3413–3420
17. Karmakar A, Kumar N, Samanta P, Desai AV, Ghosh SK (2016) A post-synthetically modified MOF for selective and sensitive aqueous-phase detection of highly toxic cyanide ions. Chem—A Eur J 22:864–868
18. Lin X, Gao G, Zheng L, Chi Y, Chen G (2014) Encapsulation of strongly fluorescent carbon quantum dots in metal-organic frameworks for enhancing chemical sensing. Anal Chem 86:1223–1228
19. Lin X, Luo F, Zheng L, Gao G, Chi Y (2015) Fast, sensitive, and selective ion-triggered disassembly and release based on tris(bipyridine)ruthenium(II)-functionalized metal-organic frameworks. Anal Chem 87:4864–4870
20. Li Q, Wang C, Tan H, Tang G, Gao J, Chen CH (2016) A turn on fluorescent sensor based on lanthanide coordination polymer nanoparticles for the detection of mercury(II) in biological fluids. RSC Adv 6:17811–17817
21. Kesselmeier J, Staudt M (1999) An overview on emission, physiology and ecology. J Atmos Chem 33:23–88
22. Zhou Y, Yan B (2016) A responsive MOF nanocomposite for decoding volatile organic compounds. Chem Commun 52:2265–2268
23. Azhdari Tehrani A, Esrafili L, Abedi S, Morsali A, Carlucci L, Proserpio DM, Wang J, Junk PC, Liu T (2017) Urea metal-organic frameworks for nitro-substituted compounds sensing. Inorg Chem 56:1446–1454
24. Zhou Y, Yang Q, Zhang D, Gan N, Li Q, Cuan J (2018) Detection and removal of antibiotic tetracycline in water with a highly stable luminescent MOF. Sens Actuators B Chem 262:137–143

25. Chen X, Wang Y, Zhang Y, Chen Z, Liu Y, Li Z, Li J (2014) Sensitive electrochemical aptamer biosensor for dynamic cell surface N -glycan evaluation featuring multivalent recognition and signal amplification on a dendrimer-graphene electrode interface. Anal Chem 86:4278–4286
26. Fang X, Liu J, Wang J, Zhao H, Ren H, Li Z (2017) Dual signal amplification strategy of Au nanopaticles/ZnO nanorods hybridized reduced graphene nanosheet and multienzyme functionalized Au@ZnO composites for ultrasensitive electrochemical detection of tumor biomarker. Biosens Bioelectron 97:218–225
27. Wang Y, Hou C, Zhang Y, He F, Liu M, Li X (2016) Preparation of graphene nano-sheet bonded PDA/MOF microcapsules with immobilized glucose oxidase as a mimetic multi-enzyme system for electrochemical sensing of glucose. J Mater Chem B 4:3695–3702
28. Zhang C, Wang X, Hou M, Li X, Wu X, Ge J (2017) Immobilization on metal-organic framework engenders high sensitivity for enzymatic electrochemical detection. ACS Appl Mater Interfaces 9:13831–13836
29. Wu XQ, Ma JG, Li H, Chen DM, Gu W, Yang GM, Cheng P (2015) Metal-organic framework biosensor with high stability and selectivity in a bio-mimic environment. Chem Commun 51:9161–9164
30. Campbell MG, Sheberla D, Liu SF, Swager TM, Dincə M (2015) Cu3(hexaiminotriphenylene)2: an electrically conductive 2D metal-organic framework for chemiresistive sensing. Angew Chemie Int Ed 54:4349–4352
31. Wang X, Wang Q, Wang Q, Gao F, Gao F, Yang Y, Guo H (2014) Highly dispersible and stable copper terephthalate metal-organic framework-graphene oxide nanocomposite for an electrochemical sensing application. ACS Appl Mater Interfaces 6:11573–11580
32. Xu Z, Yang L, Xu C (2015) Pt@UiO-66 heterostructures for highly selective detection of hydrogen peroxide with an extended linear range. Anal Chem 87:3438–3444
33. Yu Y, Yu C, Niu Y, Chen J, Zhao Y, Zhang Y, Gao R, He J (2018) Target triggered cleavage effect of DNAzyme: Relying on Pd-Pt alloys functionalized Fe-MOFs for amplified detection of Pb2+. Biosens Bioelectron 101:297–303
34. Saraf M, Rajak R, Mobin SM (2016) A fascinating multitasking Cu-MOF/rGO hybrid for high performance supercapacitors and highly sensitive and selective electrochemical nitrite sensors. J Mater Chem A 4:16432–16445
35. Perry DA, Razer TM, Primm KM, Chen TY, Shamburger JB, Golden JW, Owen AR, Price AS, Borchers RL, Parker WR (2013) Surface-enhanced infrared absorption and density functional theory study of dihydroxybenzene isomer adsorption on silver nanostructures. J Phys Chem C 117:8170–8179
36. Huang S, Qu Y, Li R, Shen J, Zhu L (2008) Biosensor based on horseradish peroxidase modified carbon nanotubes for determination of 2,4-dichlorophenol. Microchim Acta 162:261–268
37. Li J, Xia J, Zhang F, Wang Z, Liu Q (2018) An electrochemical sensor based on copper-based metal-organic frameworks-graphene composites for determination of dihydroxybenzene isomers in water. Talanta 181:80–86
38. Kong B, Zhu A, Luo Y, Tian Y, Yu Y, Shi G (2011) Sensitive and selective colorimetric visualization of cerebral dopamine based on double molecular recognition. Angew Chemie Int Ed 50:1837–1840
39. Chen L, Chan L, Fu X, Lu W (2013) Highly sensitive and selective colorimetric sensing of Hg2+ based on the morphology transition of silver nanoprisms. ACS Appl Mater Interfaces 5:284–290
40. Wang Z, Yang J, Li Y, Zhuang Q, Gu J (2018) Zr-Based MOFs integrated with a chromophoric ruthenium complex for specific and reversible Hg2+ sensing. Dalt Trans 47:5570–5574
41. Yang S, Slotcavage D, Mai JD, Guo F, Li S, Zhao Y, Lei Y, Cameron CE, Huang TJ (2014) Electrochemically created highly surface roughened Ag nanoplate arrays for SERS biosensing applications. J Mater Chem C 2:8350–8356
42. Hu Y, Liao J, Wang D, Li G (2014) (hu 2014) Fabrication of gold nanoparticle-embedded metal−organic raman scattering detection.pdf
43. Xia H, Zhang J, Yang Z, Guo S, Guo S, Xu Q (2017) 2D MOF nanoflake-assembled spherical microstructures for enhanced supercapacitor and electrocatalysis performances. Nano-Micro Lett. https://doi.org/10.1007/s40820-017-0144-6

44. Li B, Wen HM, Zhou W, Chen B (2014) Porous metal-organic frameworks for gas storage and separation: what, how, and why? J Phys Chem Lett 5:3468–3479
45. Sun CY, Wang XL, Zhang X et al (2013) Efficient and tunable white-light emission of metal-organic frameworks by iridium-complex encapsulation. Nat Commun 4:1–8
46. Zhang Y, Bai X, Wang X, Shiu KK, Zhu Y, Jiang H (2014) Highly sensitive graphene-Pt nanocomposites amperometric biosensor and its application in living cell H_2O_2 detection. Anal Chem 86:9459–9465
47. Mao S, Chang J, Pu H, Lu G, He Q, Zhang H, Chen J (2017) Two-dimensional nanomaterial-based field-effect transistors for chemical and biological sensing. Chem Soc Rev 46:6872–6904
48. Mao S, Lu G, Chen J (2014) Nanocarbon-based gas sensors: progress and challenges. J Mater Chem A 2:5573–5579
49. Lee J-H, Houk RTJ, Robinson A, Greathouse JA, Thornberg SM, Allendorf MD, Hesketh PJ (2010) Investigation of microcantilever array with ordered nanoporous coatings for selective chemical detection. Micro- Nanotechnol Sens Syst Appl II 7679:767927
50. Klein N, Herzog C, Sabo M, Senkovska I, Getzschmann J, Paasch S, Lohe MR, Brunner E, Kaskel S (2010) Monitoring adsorption-induced switching by 129Xe NMR spectroscopy in a new metal-organic framework Ni2(2,6-ndc)2(dabco). Phys Chem Chem Phys 12:11778–11784
51. Venkatasubramanian A, Lee JH, Stavila V, Robinson A, Allendorf MD, Hesketh PJ (2012) MOF @ MEMS: design optimization for high sensitivity chemical detection. Sens Actuators B Chem 168:256–262
52. Campbell MG, Liu SF, Swager TM, Dincə M (2015) Chemiresistive sensor arrays from conductive 2D metal-organic frameworks. J Am Chem Soc 137:13780–13783
53. Liu B (2012) Metal-organic framework-based devices: separation and sensors. J Mater Chem 22:10094–10101
54. Chen EX, Yang H, Zhang J (2014) Zeolitic imidazolate framework as formaldehyde gas sensor. Inorg Chem 53:5411–5413
55. Koo WT, Choi SJ, Jang JS, Kim ID (2017) Metal-organic framework templated synthesis of ultrasmall catalyst loaded $ZnO/ZnCo_2$ O_4 hollow spheres for enhanced gas sensing properties. Sci Rep 7:1–10
56. Wang D, Li Z, Zhou J, Fang H, He X, Jena P, Bin ZJ, Wang WN (2018) Simultaneous detection and removal of formaldehyde at room temperature: Janus au@zno@zif-8 nanoparticles. Nano-Micro Lett 10:1–11
57. Liu Y, He L, Liu Y, Liu J, Xiong Y, Zheng J, Tang Z (2013) Core-shell noble-metal@metal-organic-framework nanoparticles with highly selective sensing property. Angew Chemie Int Ed 52:3741–3745
58. Wu X, Xiong S, Mao Z, Hu S, Long X (2017) A designed ZnO@ZIF-8 core-shell nanorod film as a gas sensor with excellent selectivity for H_2 over CO. Chem A Eur J 23:7969–7975
59. Xu XY, Yan B (2017) Intelligent molecular searcher from logic computing network based on Eu(III) functionalized UMOFs for environmental monitoring. Adv Funct Mater 27:1–11
60. Hao J, Liu F, Liu N, Zeng M, Song Y, Wang L (2017) Ratiometric fluorescent detection of Cu2+ with carbon dots chelated Eu-based metal-organic frameworks. Sens Actuators B Chem 245:641–647
61. Ma D, Li B, Zhou X, Zhou Q, Liu K, Zeng G, Li G, Shi Z, Feng S (2013) A dual functional MOF as a luminescent sensor for quantitatively detecting the concentration of nitrobenzene and temperature. Chem Commun 49:8964–8966
62. Xu XY, Yan B (2016) Eu(III) functionalized Zr-based metal-organic framework as excellent fluorescent probe for Cd2+ detection in aqueous environment. Sens Actuators B Chem 222:347–353
63. Lajevardi A, Tavakkoli M, Masjedi YA, Nouri A, Hossaini Sadr M (2019) Green synthesis of MOF@Ag nanocomposites for catalytic reduction of methylene blue. J Mol Liq 276:371–378
64. Peterson GW, Mahle JJ, Decoste JB, Gordon WO, Rossin JA (2016) Extraordinary NO_2 removal by the metal-organic framework UiO-66-NH2. Angew Chemie Int Ed 55:6235–6238
65. Yang Y, Wang Q, Qiu W, Guo H, Gao F (2016) Covalent immobilization of cu3(btc)2 at chitosan-electroreduced graphene oxide hybrid film and its application for simultaneous detection of dihydroxybenzene isomers. J Phys Chem C 120:9794–9803

66. Chernikova V, Yassine O, Shekhah O, Eddaoudi M, Salama KN (2018) Highly sensitive and selective SO_2 MOF sensor: The integration of MFM-300 MOF as a sensitive layer on a capacitive interdigitated electrode. J Mater Chem A 6:5550–5554
67. Dou Z, Yu J, Cui Y, Yang Y, Wang Z, Yang D, Qian G (2014) Luminescent metal-organic framework films as highly sensitive and fast-response oxygen sensors. J Am Chem Soc 136:5527–5530
68. Zhang J, Yue D, Xia T, Cui Y, Yang Y, Qian G (2017) A luminescent metal-organic framework film fabricated on porous Al2O3 substrate for sensitive detecting ammonia. Microporous Mesoporous Mater 253:146–150
69. Yassine O, Shekhah O, Assen AH, Belmabkhout Y, Salama KN, Eddaoudi M (2016) H_2S sensors: fumarate-based fcu-MOF thin film grown on a capacitive interdigitated electrode. Angew Chemie Int Ed 55:15879–15883

Metal–Organic Frameworks for Pesticide Sensing: Trend in the Recent Years

Navin Kumar Mogha and Dhanraj T. Masram

Contents

1 Introduction 412
2 Synthetic Methodologies of MOFs 414
2.1 Diffusion-Based Synthesis 414
2.2 Conventional Solution-Based Method 415
2.3 Solvo/hydrothermal Method 415
2.4 Microwave-Assisted Synthesis 416
2.5 Sonochemical Method 416
2.6 Electrochemical Synthesis 417
2.7 Mechanochemical Method 417
3 Applications in Organophosphorus Pesticides Sensing 418
3.1 Luminescent Sensing Mechanism 418
3.2 Electrochemical Sensing System 419
4 Conclusion and Future Aspects 421
Abbreviations 422
References 422

Abstract Pesticides play a significant part in ensuring steady and sustained growth with increased production in agriculture throughout the world. It is very well-documented by the Food and Agriculture Organization of the United Nations (FAO) that more than 30% of agricultural output has been restored by the use of pesticides. Though the non-judicious use of pesticides is a cause of concern due to its association with the environment and human health hazards, it is resulted due to more than 90% of the total pesticide input (~4.6 million tons) getting redistributed in the ecological cycle via runoff into the environment and the presence of residual pesticides in agricultural products. Inept pesticides usage can be accounted for different types of environmental problems, such as nonpoint pollution, degradation of soil, eutrophication, development of resistance in pathogen and pest, pesticide bioaccumulation leading to biodiversity damage. One of the larger problems associated with the inefficient use of conventional pesticides formulation is off-target loss, as only 0.1% is

N. K. Mogha
Pesticide Laboratory, Analytical Science Division, Shriram Institute for Industrial Research, Delhi, India

N. K. Mogha · D. T. Masram (✉)
Department of Chemistry, University of Delhi, Delhi, India

S. Gulati (ed.), *Metal-Organic Frameworks (MOFs) as Catalysts*,
https://doi.org/10.1007/978-981-16-7959-9_16

available for the biological uptake after leaching and dust drift. Additionally, the toxic solvents and ingredients used for conventional formulations of pesticides are bound to leach in water or soil causing serious health and environmental concerns. Hence, it is imperative to develop sensitive and easy-to-use sensors for the on-the-spot detection of pesticides to minimize the limitations associated with the use of conventional methodologies, such as the requirement of skilled manpower, high cost, and inefficient logistics of instruments. Incidentally, research on metal–organic frameworks (MOFs) is drawing a lot of interest as an alternative functional material to conventional nanomaterials owing to their plentiful advantages over them. This chapter brings a review of different modes of synthesis of MOFs and recent advances in their applications for the detection of different pesticides because of their tailorable physicochemical properties. Furthermore, present challenges in developing MOFs-based detection systems and their future perspectives are also well-discussed.

Keywords MOFs · Pesticides · Nanomaterials · Environment pollution · Sensors

1 Introduction

Specific sophisticated materials are required to perform any particular applications, and the invention of these types of materials is being led by advancements in nanotechnology. Metal–organic frameworks (MOFs) are among these materials having applications in nearly every field of science and technology. MOFs are a group of innovative porous functional material, made up of bridge organic linkers together with different metal ions called units, forming a crystalline framework with stable porosity [1]. In contrast to other porous materials, for instance, porous carbon nanomaterials, metal complexes, and their hybrids, MOFs offer novel topological architecture, plentiful functional groups having specific catalytical properties along with tunable porosity. In the last decade, based upon their size, functional flexibility, and geometric shape, more than 20,000 different types of MOFs are reported. MOFs have a surface area falling in the range of 1000 to 10,000 m^2/g which is way higher than the conventional porous materials like porous carbon and zeolites [2]. A wide range of porosity and array of different types of MOFs have made them a model material for their applications in catalysis and sensing [3].

Primarily, MOFs are synthesized by generating certain inorganic building blocks without decomposing the organic linkers used. Moreover, the nature and the properties of the MOFs formed greatly depend upon the method of synthesis. Furthermore, applications of the synthesized MOFs are also reliant on their major properties, for instance, morphological characters, porosity, size of the particles, and distribution. Henceforth, a detailed knowledge of the MOFs synthetic methodology plays a significant role in choosing diverse structures of MOFs having preferred physical and chemical properties necessary for sensing applications. Though MOFs synthesis using different organic ligands and metal ions appears easy, however, attaining the desired morphological structure with specific properties is relatively challenging.

Therefore, parameters such as time, temperature, pressure, choice of solvent, pH, organic linker, the concentration of metal ion, and linkers also play a very important role in MOFs synthesis. In the literature, various methods are reported to carry out the synthesis of MOFs; however, only a few of them are mostly used due to their importance for achieving desired MOFs. These methods include conventional solution-based synthesis, solvothermal synthesis, sonochemical method, microwave method, electrochemical method, mechanochemical method, etc.

Approximately 30% of the worldwide agricultural output is dependent upon the usage of pesticides; consequently, they play a very important role in modern-day agriculture. However, the use of the disproportionate and non-judicial pesticide is the cause of concern due to the associated health issues with them for instance dizziness, nausea, irritation of the eye and skin. Moreover, chronic diseases such as neurological disorders, asthma, diabetes, and even cancer are also associated with pesticide use [4–6]. Organophosphorus pesticides (OPs) are phosphorus-containing organic compounds, accounting for more than one-third of the total usage of pesticides across the globe [2–6]. As a result of their extremely toxic nature, they not only eradicate the agricultural pests and pathogens but due to the presence of their residues in water, soil, and agricultural products are an environmental and human health safety concern [4, 7–10]. The primary mode of action of OPs is the irreversible inhibition of the acetylcholinesterase (AChE) enzyme, which could lead to serious damages to our respiratory, reproductive as well as nervous systems. Moreover, the immunotoxic nature of the OPs can also increase the risk of cancers in human beings [11–15]. Nevertheless, pesticides including OPs cannot be barred due to the associated social and economic developments of the countries involved in agriculture. Henceforth, a quick, dependable, and sustainable process for the quantification of pesticides in different matrices is urgently required for the sake of public health safety and security. Traditionally, pesticides are detected with the use of the conventional methods and instruments such as high-performance liquid chromatography (HPLC) [16], gas chromatography (GC) [16], potentiometry [17], capillary electrophoresis [16], and flow injection spectrophotometry [18]. However, due to reoccurring disadvantages associated with these methods for instance time-consuming sample preparation and sample pretreatment, high cost of operation, maintenance of sophisticated instruments [18]. Consequently, demand for the use of quick, reliable, and efficient detection methods for pesticides is constantly growing for the past many years. Development of the chromogenic, luminescent chemosensors, immune, and enzymatic biosensors are trying to meet this demand through partially [7, 19–21]due to a lot of research and development is still required for the development of a fully functional sensor for the recognition of various OPs. The recent development in the nanotechnology and use of functional materials such as graphene, graphene oxide, carbon nanotubes, nanoparticles, and quantum dots has provided many required wings to these sensing techniques. This current progress has paved the way for the growth of more effective pesticide sensors with much lesser response time with increased accuracy and precision [22, 23], which can be used in both solid and liquid matrices [18]. On the other hand, with new opportunities comes to the drawbacks also, for example, such innovative pesticide sensors can also have some shortcomings, like complicated synthetic

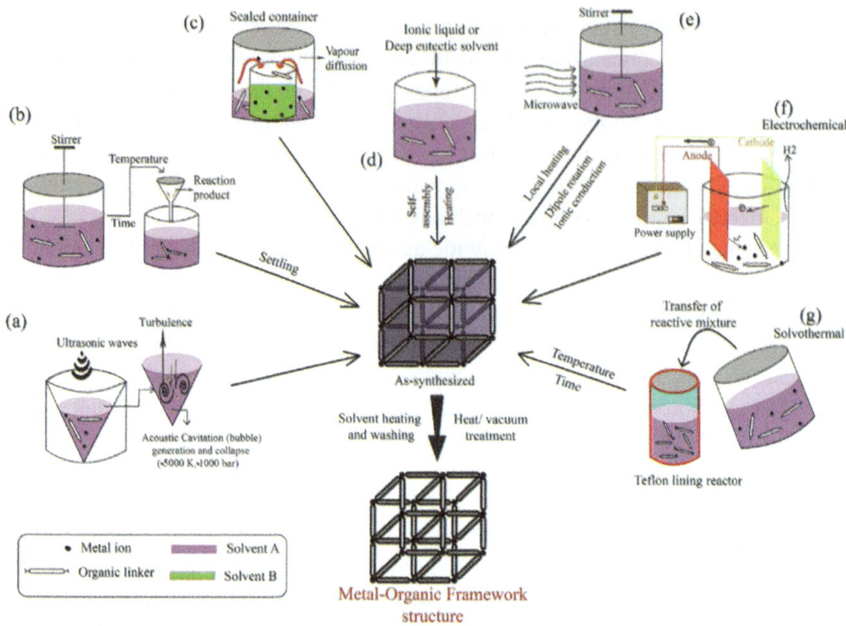

Fig. 1 Some basic methods adopted for the synthesis of MOFs. **a** Sonochemical method, **b** conventional solution-based method, **c** diffusion-based synthesis, **d** ionothermal method, **e** Microwave-assisted synthesis, **f** electrochemical synthesis, and **g** solvo/hydrothermal method. Adopted with reprinting permission from Ref. [24]

procedures, insufficient stability, molecular organization paucities, and interference by other analytes.

Herein this current chapter, we describe the diverse methodologies (Fig. 1) for the synthesis of MOFs, followed by, different sensing applications of MOFs for the detection of OPs in recent years. This chapter will offer novel understandings into the development as well as applications of MOFs in the field of a sensing system for the efficient detection of pesticides for their better environmental monitoring and human health safety.

2 Synthetic Methodologies of MOFs

2.1 *Diffusion-Based Synthesis*

The basic principle of diffusion synthesis is gradual mixing of the various species for their interaction. In one method, termed solvent liquid diffusion, where two layers of different density solvents are made out of which one is precipitating solvent while the other containing the product dissolved in the solvent; however, it is necessary that both the layers must be separated by solvent, and at that interface crystal of MOFs grows. In another method, physical barriers such as two different size containers are used and reactants are gradually diffused across this barrier. Furthermore, in this method, a gel can also be used as a medium for diffusion and crystallization, so as to prevent fast diffusion rates and precipitation of bulk material. Diffusion techniques are best for achieving the best quality crystals which can respond to the X-ray diffractions in particular if the product so formed is insoluble in most solvents [25–31]. The major advantage of this method is mild synthesizing conditions but there are some disadvantages of this method as well such as prolonged synthesis time [32]. One of the research groups [33] employed the diffusion method for the synthesis of UiO-661 by mixing reactants and solvent till a transparent solution is obtained and crystallizing the product using a precipitant solvent.

2.2 Conventional Solution-Based Method

Here in this method, organic linkers, concerned metal salts, and remaining reactants are dissolved in the concerned solvent, and reaction is performed at a fixed temperature for a specified time period and followed by filtration to isolate the reaction products, and MOFs are allowed to crystal out from the solvent gradually due to solvent evaporation [34, 35]. However, the reactions conditions must be strictly followed in this type of synthesis. A research group has synthesized ceramic-based MOFs using this method [36].

2.3 Solvo/hydrothermal Method

The solvothermal synthetic method has been in use for many years in material chemistry, where reactants are dissolved in the solvent and sealed in a Teflon-lined container, and heated at a certain temperature for a fixed time period. However, when water is used as the solvent instead of organic solvent then this type of synthesis is termed hydrothermal synthesis [37]. It was originally used for the synthesis of zeolites; however, recently it is applicable in the MOFs' synthesis also. MOFs synthesis using this method includes mixing all reactants in a solvent in certain proportions in a Teflon-lined container, followed by heating at a fixed temperature for a definite time. Afterward, the completion of the reaction, it is allowed to cool till room temperature to obtain pure MOFs crystals, which are washed and dried before further use [38, 39]. The primary advantage of this type of synthetic methodology is that by modifying the reaction condition crystal growth of MOF with the rate of nucleation can be controlled easily [40]. Moreover, due to recent advancements in the synthesis and applications, ionic liquids are also being used as solvents for the synthesis of MOFs; for example, choline chloride was used as the co-solvent for the synthesis of HUSK-1 [41].

2.4 Microwave-Assisted Synthesis

MOFs synthesis using microwave-assisted synthesis method is based upon the chief principle that the interaction between electromagnetic radiation and the mobile electric charge produces oscillations in the molecules with different polarities, increasing the temperature of the system to many folds as the end product. This method is termed as one of the best methods due to its high energy, shorter time period, cost-effectiveness, and ease of regulation [42]. Cr_MIL-100 was synthesized using this method in 4 h with 44% yield which was equivalent to the chemically synthesized MOFs. A similar MOF based on the iron was also synthesized using microwave-assisted synthesis only in 30 min [43]. Similarly, different groups have reported the nanoscale synthesis of Cr_MIL-101 by employing this method [44, 45]. Furthermore, IRMOF-1 when synthesized by microwave-assisted method yields higher CO_2 adsorption properties with better quality crystals as compared to conventional chemical synthesis [46]. Similarly, the microwave-assisted synthesis of IRMOF-16 $(OH)_2$ has helped in improving the crystal quality as well as the porosity of the material with respect to an increment in the irradiation time [47]. Some other examples of microwave-assisted synthesis include synthesis of imidazolate-based ZIF-8 having crystal shape of rhombic dodecahedral the with larger specific surface area in 4 h [43], triazolate-based MOFs, namely $[Zn_5Cl_4(BBTA)_3]$ and $Zn_5Cl_4(BTDD)_3$ [48], in 10 and 30 min, respectively.

2.5 Sonochemical Method

Sonochemistry is the employing high-energy ultrasound waves to carry out an organic transformation. In this method, sonication causes the formation and collapse of the bubble called acoustic cavitation due to the generation of high pressure and temperature, which results in faster heating and cooling rates of more than 1010 K/s, producing fine crystallites [24, 49, 50]. Furthermore, homogeneity and enhanced nucleation reduce the crystallization time and also particle size for MOFs, making this method more sustainable to employ [49]. The primary objective for using sonochemical methods in MOF science is because this method is quick, energy-efficient, environmentally benign, and easy to operate, and ambient working temperature. Fast reactions by the sonochemical method are paving way for its use in scaling up reactions for the synthesis of different MOFs [51]. The first MOF synthesized using the sonochemical method was Zn-carboxylates [52]while some other examples include HUSK-1, Mg-MOF-74 [53], IRMOF-9 [54], and MOF-5 [55].

2.6 Electrochemical Synthesis

Electrochemical synthesis of MOFs is mostly used in industries to produce MOFs powder. Its important advantages over other methods include ambient temperature, exclusion of anions like nitrates from the metal salts, and very fast reaction times. However, it has a limitation of producing bulk materials rather than crystals in scale-up reactions. However, in this method two different approaches are described in the literature, one is anodic deposition while the other is cathodic deposition. Anodic deposition includes the formation of MOF film on the respective metal anode which is in contact with the solution of ligand, resulting in the anodic dissolution of metal. On the other hand, in cathodic deposition, the cathodic surface is in contact with a solution comprising probase, ligand/organic linkers, and metal ions. As the pH increases near the cathodic surface, film formation takes place over the cathode where a local base is generated as a result of electrochemical reduction the which leads to subsequent ligand deprotonation, thus inducing formation of MOF [56]. The important parameter which needed to be taken care of during electrochemical synthesis of MOFs is choice of electrolyte, range of temperature, voltage–current density and structural/textural properties of the electrode, etc. BASF researcher in 2005 first reported the electrochemical synthesis of MOF HKUST-1 [57] producing both powder and films. Similarly, another group deposited a thin layer of HKUST-1 over copper mesh as anode using the electrochemical method of MOFs synthesis [58]. Some examples of MOFs synthesized by the electrochemical method include ZIF-8, HKUST-13, Al-MIL-53-NH2, and Al-MIL-100 [59].

2.7 Mechanochemical Method

The mechanochemical synthesis method includes the chemical transformation after breaking the intramolecular bonds employing mechanical forces. Mechanochemical transformations are being used for a very long time in synthetic chemistry, and its application includes the co-crystal formation of pharmaceutical actives, multicomponent synthesis, polymer sciences, solid-state chemistry, etc. [25]. In the field of MOFs synthesis, the first use of the mechanochemical method first came into light in 2006 [60]. The use of the mechanochemical method for synthesis of MOFs is strongly advocated by many research groups due to its being environmentally benign, use of no or scarce use of organic solvents, ambient temperature range of synthesis [61]; moreover, products can be obtained in good quantitative yields within a short span of time [62, 63]. The synthesis of HKUST-1 utilizing mechanochemical method has been comprehensively explored where it was synthesized by the process of ball milling with the very high specific surface area of 1713 m^2/g and porosity due to the presence of copper dimeric paddlewheel unit [64]. Furthermore, a modified technique known as liquid-assisted grinding (LAG) has demonstrated that the addition of a small quantity of solvent can accelerate the reaction by enhancing the molecular mobility of the reactants. Moreover, solvents can display a structure-directing influence in the synthesis of MOFs. For instance, MOFs based upon the imidazolate were synthesized using mechanochemical method however resulted only in non-porous MOFs with ZnO or Zinc chloride [65, 66]. However, a significant increase in the yields was obtained when used in combination with a small amount of DMF with hydoxy-imidazole and hydroxymethanol imidazole to obtain ZIF-4 and ZIF-8 respectively. Recently, this technique has been further modified to ion- and liquid-assisted grinding for the specific synthesis of pillared-layered MOFs in high efficiency, validating the structure-directing effect of ions and solvents [65, 67].

3 Applications in Organophosphorus Pesticides Sensing

3.1 Luminescent Sensing Mechanism

Recently, MOFs have arose as one of the most encouraging advance functional materials with application in the detection of gases [68, 69], solvents [70, 71], small molecules [72, 73], and explosives [74, 75], etc. Pesticide sensing strategies involving MOFs are primarily classified into two categories luminescent and electrochemical. In the case of luminescent MOFs (LMOFs), the detection capabilities depend upon the analyte–MOF interaction as a result of which LMOFs are termed as the best candidate for application in pesticide sensing [76, 77]. LMOF-based pesticide detection is primarily dependent upon three important points (1) intermolecular distances deviation between organic linkers and the metal ion, (2) chemical interaction between metal and the analyte molecule, and (3) interaction between the organic linker and analyte molecule based on the host–guest relationship [76]. Quenching in the optical intensity is used as the furthermost preferred method for signal transduction in LMOFs [78]. This quenching behavior is resulted due from an overlay between the electron donor and acceptor [79]. However, in some case it is the deviations in the oxidoreduction potential of the moieties present in MOFs that bring the quenching response [23]. Remarkably, an uncommon phenomenon where due to the interaction between LMOFs and analyte can increase the luminescence intensity, using which analyte can be quantified [23, 78, 79]. Interaction between the analyte and the MOFs can bring changes in the physicochemical properties of the system, and sometimes, these changes can be perceived directly by naked eyes also. Consequently, an extremely high specificity and adsorption capacity of the MOFs makes them the ideal candidate for the development of MOFs-based pesticide sensors. However, this field requires some more research inputs so as to understand the different variations in quenching rate, quenching pathway, and effect of porosity. A group reported the synthesis of LMOF, ZnPO-MOF with a high surface area of 4073.9 $m^2\ g^{-1}$ having outstanding adsorption capacity and sensing capabilities for the quick and selective detection of methyl parathion, where LOD was found to be 0.12 ppb by a luminescence quenching pathway [80]. Similarly, in 2019, another research group is shown

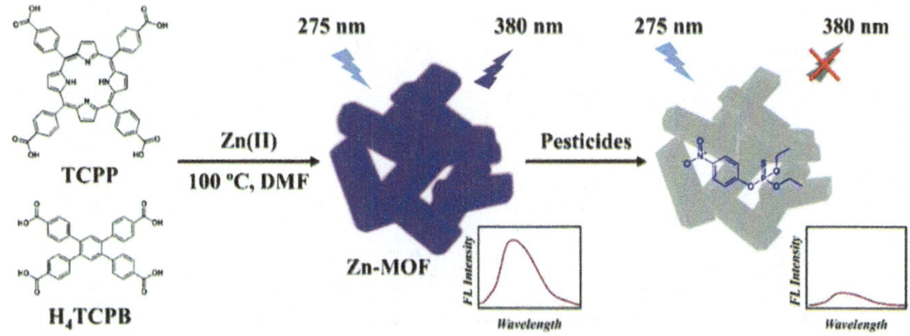

Fig. 2 An illustration depicting the preparation of Zn-MOF and its application in parathion sensing. Adopted with reprinting permission from Ref. [82]

that a water-stable 3D Zr-MOF is appropriate for rapid and sensitive detection of OPs in particular, methyl parathion, using luminescence quenching pathway with a quenching efficiency of as high as 66.4% and very low reported LOD of 0.115 ppb [81]. Moreover, another research group [82] reported a highly luminescent zinc metal–organic framework (Zn-MOF) which emitted violet fluorescence under ultraviolet irradiation. Furthermore, it showed selectivity toward parathion as compared to other OPs. Quenching efficiency was found to be stable for 15 days with LOD of 1.950 ppb (Fig. 2).

Additionally, another group [83, 84] reported two LMOFs, Cd-MOF, and Ln-MOF, where both the MOFs have shown quenching of fluorescence azinphos-methyl other OPs. A series of isostructural Ln-MOFs was also reported for the detection of OPs [85] and was shown excellent sensing behavior for the detection of chlorpyrifos, another OPs, having a very LOD of 0.14 ppb, with excellent recyclability, where even after the five cycles the quenching efficiency does not change significantly.

In a very interesting work [85], some researchers identified several pesticides rather than one using LMOFs with different quenching efficiencies. In this work (Fig. 3), a 3D cadmium-based MOF was synthesized and the quenching efficiency for chlorpyrifos was 52%, while for azinphos-methyl it was found to be 90%.

3.2 *Electrochemical Sensing System*

In recent years, the scientific community has witnessed a boom in the advance of electrochemical sensors for the sensing of various analytes including different pesticides. Owing to their high specificity and sensitivity makes electrochemical sensors a prime candidate for the efficient and sustainable detection of various pesticides. Though the research and development on the MOFs-based electrochemical sensors are in the very nascent stage, the primary reason behind this is the poor electrical conductivity ($<10 - 10$ S cm^{-1}) of pristine MOFs [86]. Consequently, this limitation can be overcome by modifying MOFs with different functional materials

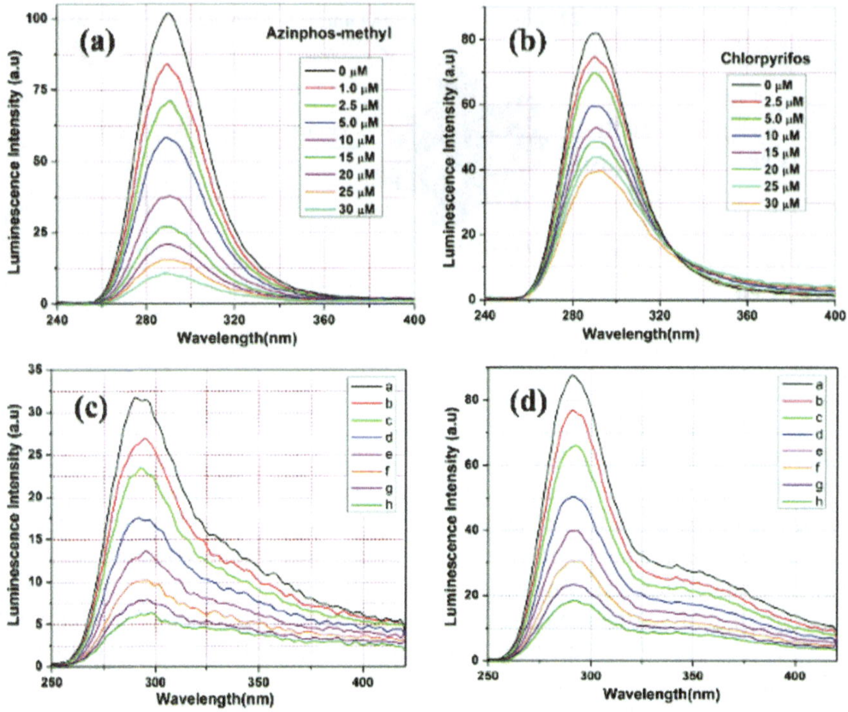

Fig. 3 Luminescence intensity of 3D cadmium-based MOF dispersed in water with increasing concentrations of (**a**) azinphos-methyl and (**b**) chlorpyrifos, the luminescence intensity of 3D cadmium-based MOF dispersed in apple (**c**) and tomato (**d**) extracts in the presence of different concentrations of azinphos-methyl. Adapted with reprinting permission from Ref. [85]

such as nanoparticles, nanotubes, and graphene. Henceforth, such modifications are suggested to improve the conductivity of MOFs so as to use them in electrochemical sensors more efficiently. In one of the modifications, the role of charge carriers can be played by doped guest molecules inside the porous structure of the MOF and can easily induce required conductivity [86]. Other modification techniques to induce conductivity to MOFs are by introducing heterobimetallic structures proton doping, redox-active linkers, and metallic clusters specific transition metals series. Moreover, different methodologies to alter the electrode surface and use of MOFs with heightened conductivity have generated a greater push in the field of MOFs-based electrochemical sensors for monitoring, detection, and identification of different pesticide molecules such method was explored for the detection of glyphosate [86], using MIP-MOF thin films functionalized by gold nanoparticles. The as-prepared MIP-MOF electrochemical sensor showed excellent sensing of glyphosate having a very low limit of quantification (0.8 pg/L), with the recovery of real samples ranging from 102.6 to 98.7%.

In a similar work, trace-level detection of the Methyl Parathion was performed, which was based on an absorptive reaction using Cd-based MOFs synthesized from the precursors [87]. The MOF/GCE electrochemical sensor exhibited a high-quality response (LOD-0.006 μg/mL) for methyl parathion, which was very efficient, sensitive, selective, and reproducible [88].

Another class of electrochemical sensors is electrochemical biosensors, and these detections of pesticides by biosensors take place utilizing specific enzymes producing electroactive substances [88]. These types of biosensors usually employ enzymes which include butyrylcholinesterase, acetylcholinesterase, tyrosine, cholesterol oxidase, organophosphorus hydrolase (OPH), and organophosphorus acid anhydrolase (OPAA), different types of antibodies, and nucleic acids [6]. In electrochemical biosensors, the chemical reaction taking place over the electrode is converted proportionally to electrochemical signals and their efficiency is controlled by the loading amount of the bioactive species, and hence, here MOFs can play an important role due to their enormous porosity. Biosensors based upon the OPH hydrolysis have been reported for the detection of OPs [16]. Similarly, OPAA can detect fluorine-based OPs, because its primary site of action is P–F bonds, for example, the detection of nerve gases sarin and soman [89].

4 Conclusion and Future Aspects

This chapter highlights MOF as an advanced functional material for applications in the detection of organophosphorus pesticides. Conventionally numerous different approaches were used for the efficient detection of pesticides and their residues; however, their practical limitations were the main drawbacks, inhibiting their day-to-day uses. Consequently, due to the high surface area, controllable porosity, modifiable luminescent qualities, and interaction with the guest species, MOF can be a perfect platform for the detection of pesticides. MOFs have exceptional chromic, luminescent, and optical properties which can be fine-tuned using various synthetic methodologies, as per the requirement for sensing applications. Choice of metal source, type of organic linker used, and the other reactants also play a very important part in determining the application of MOFs. Electrochemical sensing of pesticides employing MOFs is a promising field but still underdeveloped due to the non-conductive behavior of pristine MOFs; however, when in composite with different functionalities, and metallic particles synergistic behavior helps in the development of desired sensors.

MOF-based sensors are still very new in particular electrochemical, so a more balanced approach in the future could be useful for the development of MOFs-based sensors to make them easy to deploy in the field, financially cost-effective, adaptable, subtle, and quick in nature.

Abbreviations

MOFs	Metal−organic frameworks
FAO	Food and Agriculture Organization of the United Nations
Ops	Organophosphorus pesticides
AChE	Acetylcholinesterase
HPLC	High-performance liquid chromatography
GC	Gas chromatography
BBTA	1*H*,5*H*-Benzo(1,2-d:4,5-d′)bistriazole
BTDD	Bis-(1*H*-1,2,3-triazolo-[4,5-b],[4c′,5′-i])dibenzo-[1,4]-dioxin
LAG	Liquid-assisted grinding
LMOFs	Luminescent MOFs
LOD	Limit of detection
OPH	Organophosphorus hydrolase
OPAA	Organophosphorus acid anhydrolase

References

1. Zhou HC, Long JR, Yaghi OM (2012) Introduction to metal-organic frameworks. Chem Rev 112(2):673–674. https://doi.org/10.1021/cr300014x
2. Farha OK et al (2012) Designing higher surface area metal-organic frameworks: are triple bonds better than phenyls? J Am Chem Soc 134(24):9860–9863. https://doi.org/10.1021/ja302623w
3. Furukawa H, Cordova KE, O'Keeffe M, Yaghi OM (2013) The chemistry and applications of metal-organic frameworks. Science (80-) 341(6149). https://doi.org/10.1126/science.1230444
4. Kim KH, Kabir E, Jahan SA (2017) Exposure to pesticides and the associated human health effects. Sci Total Environ 575:525–535. https://doi.org/10.1016/j.scitotenv.2016.09.009
5. Evenset A et al (2016) Seasonal variation in accumulation of persistent organic pollutants in an arctic marine benthic food web. Sci Total Environ 542(Pt A):108–120. https://doi.org/10.1016/j.scitotenv.2015.10.092
6. Vikrant K, Tsang DCW, Raza N, Giri BS, Kukkar D, Kim KH (2018) Potential utility of metal-organic framework-based platform for sensing pesticides. ACS Appl Mater Interfaces 10(10):8797–8817. https://doi.org/10.1021/acsami.8b00664
7. Zhu X et al (2013) Metal-organic framework (MOF): a novel sensing platform for biomolecules. Chem Commun 49(13):1276–1278. https://doi.org/10.1039/c2cc36661d
8. Wang A, Costello S, Cockburn M, Zhang X, Bronstein J, Ritz B (2011) Parkinson's disease risk from ambient exposure to pesticides. Eur J Epidemiol 26(7):547–555. https://doi.org/10.1007/s10654-011-9574-5
9. Cui HF, Wu WW, Li MM, Song X, Lv Y, Zhang TT (2018) A highly stable acetylcholinesterase biosensor based on chitosan-TiO_2-graphene nanocomposites for detection of organophosphate pesticides. Biosens Bioelectron 99:223–229. https://doi.org/10.1016/j.bios.2017.07.068
10. Obare SO et al (2010) Fluorescent chemosensors for toxic organophosphorus pesticides: a review. Sensors 10(7):7018–7043. https://doi.org/10.3390/s100707018
11. Thakur S, Kumar P, Reddy MV, Siddavattam D, Paul AK (2013) Enhancement in sensitivity of fluorescence based assay for organophosphates detection by silica coated silver nanoparticles using organophosphate hydrolase. Sens Actuat B Chem 178:458–464. https://doi.org/10.1016/J.SNB.2013.01.010

12. Hulse EJ, Davies JOJ, Simpson AJ, Sciuto AM, Eddleston M (2014) Respiratory complications of organophosphorus nerve agent and insecticide poisoning: implications for respiratory and critical care. Am J Respir Crit Care Med 190(12):1342–1354. https://doi.org/10.1164/rccm.201406-1150CI
13. Carullo P, Cetrangolo GP, Mandrich L, Manco G, Febbraio F (2015) Fluorescence spectroscopy approaches for the development of a real-time organophosphate detection system using an enzymatic sensor. Sensors (Switzerland) 15(2):3932–3951. https://doi.org/10.3390/s150203932
14. Peter JV, Sudarsan TI, Moran JL (2014) Clinical features of organophosphate poisoning: a review of different classification systems and approaches. Ind J Crit Care Med 18(11):735–745. https://doi.org/10.4103/0972-5229.144017
15. Mogha NK, Sahu V, Sharma M, Sharma RK, Masram DT (2016) Biocompatible ZrO_2-reduced graphene oxide immobilized AChE biosensor for chlorpyrifos detection. Mater Des 111:312–320. https://doi.org/10.1016/j.matdes.2016.09.019
16. Kumar P, Kim KH, Deep A (2015) Recent advancements in sensing techniques based on functional materials for organophosphate pesticides. Biosens Bioelectron 70:469–481. https://doi.org/10.1016/j.bios.2015.03.066
17. Fernando JC et al (1993) Rapid detection of anticholinesterase insecticides by a reusable light addressable potentiometric biosensor. J Agric Food Chem 41(3):511–516. https://doi.org/10.1021/jf00027a031
18. Zheng YH, Hua TC, Sun DW, Xiao JJ, Xu F, Wang FF (2006) Detection of dichlorvos residue by flow injection calorimetric biosensor based on immobilized chicken liver esterase. J Food Eng 74(1):24–29. https://doi.org/10.1016/j.jfoodeng.2005.02.009
19. Hu S et al (2017) Facile synthesis of Fe_3O_4/g-C_3N_4/HKUST-1 composites as a novel biosensor platform for ochratoxin A. Biosens Bioelectron 92:718–723. https://doi.org/10.1016/j.bios.2016.10.006
20. Li Q et al (2014) Graphene/graphene-tube nanocomposites templated from cage-containing metal-organic frameworks for oxygen reduction in Li-O_2 batteries. Adv Mater 26(9):1378–1386. https://doi.org/10.1002/adma.201304218
21. Wei X et al (2012) Sensitive fluorescence biosensor for folate receptor based on terminal protection of small-molecule-linked DNA. Chem Commun 48(49):6184–6186. https://doi.org/10.1039/c2cc31979a
22. Jung S, Kim Y, Kim SJ, Kwon TH, Huh S, Park S (2011) Bio-functionalization of metal-organic frameworks by covalent protein conjugation. Chem Commun 47(10):2904–2906. https://doi.org/10.1039/c0cc03288c
23. Kumar P, Deep A, Kim KH (2015) Metal organic frameworks for sensing applications. TrAC Trends Anal Chem 73:39–53. https://doi.org/10.1016/j.trac.2015.04.009
24. Sharanyakanth PS, Radhakrishnan M (2020) Synthesis of metal-organic frameworks (MOFs) and its application in food packaging: a critical review. Trends Food Sci Technol 104:102–116. https://doi.org/10.1016/j.tifs.2020.08.004
25. Safaei M, Foroughi MM, Ebrahimpoor N, Jahani S, Omidi A, Khatami M (2019) A review on metal-organic frameworks: Synthesis and applications. TrAC Trends Anal Chem 118:401–425. https://doi.org/10.1016/j.trac.2019.06.007
26. Gao T, Dong BX, Pan YM, Liu WL, Teng YL (2019) Highly sensitive and recyclable sensing of Fe^{3+} ions based on a luminescent anionic $[Cd(DMIPA)]^{2-}$ framework with exposed thioether group in the snowflake-like channels. J Solid State Chem 270:493–499. https://doi.org/10.1016/j.jssc.2018.12.008
27. Moradi E, Rahimi R, Safarifard V (2019) Sonochemically synthesized microporous metal-organic framework representing unique selectivity for detection of Fe^{3+} ions. Polyhedron 159:251–258. https://doi.org/10.1016/j.poly.2018.11.062
28. Yoo J, Ryu UJ, Kwon W, Choi KM (2019) A multi-dye containing MOF for the ratiometric detection and simultaneous removal of Cr_2O_{72}—In the presence of interfering ions. Sensors Actuators, B Chem. 283:426–433. https://doi.org/10.1016/j.snb.2018.12.031

29. Oveisi M, Alinia Asli M, Mahmoodi NM (2019) Carbon nanotube based metal-organic framework nanocomposites: Synthesis and their photocatalytic activity for decolorization of colored wastewater. Inorganica Chim Acta 487:169–176. https://doi.org/10.1016/j.ica.2018.12.021
30. Bakhtiari N, Azizian S (2018) Nanoporous carbon derived from MOF-5: a superadsorbent for copper Ions. ACS Omega 3(12):16954–16959. https://doi.org/10.1021/acsomega.8b02278
31. Abazari R, Mahjoub AR, Shariati J (2019) Synthesis of a nanostructured pillar MOF with high adsorption capacity towards antibiotics pollutants from aqueous solution. J Hazard Mater 366:439–451. https://doi.org/10.1016/j.jhazmat.2018.12.030
32. Wang FX, Wang CC, Wang P, Xing BC (2017) Syntheses and applications of UiO series of MOFs. Chin J Inorg Chem 33(5):713–737. https://doi.org/10.11862/CJIC.2017.105
33. Shearer GC et al (2013) In situ infrared spectroscopic and gravimetric characterisation of the solvent removal and dehydroxylation of the metal organic frameworks UiO-66 and UiO-67. Top Catal 56(9–10):770–782. https://doi.org/10.1007/s11244-013-0027-0
34. Liu W, Chen H (2018) 1D energetic metal-organic frameworks: synthesis and properties. Cailiao Daobao/Materials Rev 32(1):223–227. https://doi.org/10.11896/j.issn.1005-023X.2018.02.013
35. Dai J et al (2018) Synthesis of novel microporous nanocomposites of ZIF-8 on multiwalled carbon nanotubes for adsorptive removing benzoic acid from water. Chem Eng J 331:64–74. https://doi.org/10.1016/j.cej.2017.08.090
36. Ye JW et al (2018) Room-temperature sintered metal-organic framework nanocrystals: a new type of optical ceramics. Sci China Mater 61(3):424–428. https://doi.org/10.1007/s40843-017-9184-1
37. Yang F et al (2015) Proton conductivities in functionalized UiO-66: tuned properties, thermogravimetry mass, and molecular simulation analyses. Cryst Growth Des 15(12):5827–5833. https://doi.org/10.1021/acs.cgd.5b01190
38. Ming Liu J, Liu T, Chen Wang C, Hui Yin X, Hu Xiong Z (2017) Introduction of amidoxime groups into metal-organic frameworks to synthesize MIL-53(Al)-AO for enhanced U(VI) sorption. J Mol Liq 242:531–536. https://doi.org/10.1016/j.molliq.2017.07.024
39. Guo H, Zheng Z, Zhang Y, Lin H, Xu Q (2017) Highly selective detection of Pb^{2+} by a nanoscale Ni-based metal-organic framework fabricated through one-pot hydrothermal reaction. Sens Actuat B Chem 248:430–436. https://doi.org/10.1016/j.snb.2017.03.147
40. Li B et al (2012) A strategy toward constructing a bifunctionalized MOF catalyst: post-synthetic modification of MOFs on organic ligands and coordinatively unsaturated metal sites. Chem Commun 48(49):6151–6153. https://doi.org/10.1039/c2cc32384b
41. Anumah A et al (2019) Metal-organic frameworks (MOFs): recent advances in synthetic methodologies and some applications. Chem Methodol 3(3):283–305. https://doi.org/10.22034/CHEMM.2018.139807.1067
42. Klinowski J, Almeida Paz FA, Silva P, Rocha J (2011) Microwave-assisted synthesis of metal-organic frameworks. Dalt Trans 40(2):321–330. https://doi.org/10.1039/c0dt00708k
43. Chansi, Bhardwaj R, Hadwani K, Basu T (2019) Role of metal-organic framework (MOF) for pesticide sensing. Nanosci Sustain Agric 75–99. https://doi.org/10.1007/978-3-319-97852-9_4
44. Khan NA, Kang IJ, Seok HY, Jhung SH (2011) Facile synthesis of nano-sized metal-organic frameworks, chromium-benzenedicarboxylate, MIL-101. Chem Eng J 166(3):1152–1157. https://doi.org/10.1016/j.cej.2010.11.098
45. Jhung SH, Lee JH, Yoon JW, Serre C, Férey G, Chang JS (2007) Microwave synthesis of chromium terephthalate MIL-101 and its benzene sorption ability. Adv Mater 19(1):121–124. https://doi.org/10.1002/adma.200601604
46. Lu CM, Liu J, Xiao K, Harris AT (2010) Microwave enhanced synthesis of MOF-5 and its CO_2 capture ability at moderate temperatures across multiple capture and release cycles. Chem Eng J 156(2):465–470. https://doi.org/10.1016/j.cej.2009.10.067
47. Kim DO et al (2011) Synthesis of MOF having functional side group. Inorganica Chim Acta 370(1):76–81. https://doi.org/10.1016/j.ica.2011.01.030
48. Denysenko D et al (2011) Elucidating gating effects for hydrogen sorption in MFU-4-type triazolate-based metal-organic frameworks featuring different pore sizes. Chem Eur J 17(6):1837–1848. https://doi.org/10.1002/chem.201001872

49. Gedanken A (2004) Using sonochemistry for the fabrication of nanomaterials. Ultrason Sonochem 11(2):47–55. https://doi.org/10.1016/j.ultsonch.2004.01.037
50. Stock N, Biswas S (2012) Synthesis of metal-organic frameworks (MOFs): routes to various MOF topologies, morphologies, and composites. Chem Rev 112(2):933–969. https://doi.org/10.1021/cr200304e
51. Bang JH, Suslick KS (2010) Applications of ultrasound to the synthesis of nanostructured materials. Adv Mater 22(10):1039–1059. https://doi.org/10.1002/adma.200904093
52. Stock N (2010) High-throughput investigations employing solvothermal syntheses. Microporous Mesoporous Mater 129(3):287–295. https://doi.org/10.1016/j.micromeso.2009.06.007
53. Kim J, Yang ST, Choi SB, Sim J, Kim J, Ahn WS (2011) Control of catenation in CuTATB-n metal-organic frameworks by sonochemical synthesis and its effect on CO_2 adsorption. J Mater Chem 21(9):3070–3076. https://doi.org/10.1039/c0jm03318a
54. Li ZQ et al (2009) Ultrasonic synthesis of the microporous metal-organic framework $Cu_3(BTC)_2$ at ambient temperature and pressure: an efficient and environmentally friendly method. Mater Lett 63(1):78–80. https://doi.org/10.1016/j.matlet.2008.09.010
55. Son WJ, Kim J, Kim J, Ahn WS (2008) Sonochemical synthesis of MOF-5. Chem Commun 47:6336–6338. https://doi.org/10.1039/b814740j
56. Mueller U et al (2005) Method for electrochemical production of a crystalline porous metal organic skeleton material, p. WO 2005/049892
57. Mueller U, Schubert M, Teich F, Puetter H, Schierle-Arndt K, Pastré J (2006) Metal-organic frameworks—Prospective industrial applications. J Mater Chem 16(7):626–636. https://doi.org/10.1039/b511962f
58. Van Assche TRC, Desmet G, Ameloot R, De Vos DE, Terryn H, Denayer JFM (2012) Electrochemical synthesis of thin HKUST-1 layers on copper mesh. Microporous Mesoporous Mater 158:209–213. https://doi.org/10.1016/j.micromeso.2012.03.029
59. Martinez Joaristi A, Juan-Alcañiz J, Serra-Crespo P, Kapteijn F, Gascon J (2012) Electrochemical synthesis of some archetypical Zn^{2+}, Cu^{2+}, and Al^{3+} metal organic frameworks. Cryst Growth Des 12(7):3489–3498. https://doi.org/10.1021/cg300552w
60. Pichon A, Lazuen-Garay A, James SL (2006) Solvent-free synthesis of a microporous metal-organic framework. CrystEngComm 8(3):211–214. https://doi.org/10.1039/b513750k
61. Kumar G, Mogha NK, Masram DT (2021) Zr-based metal-organic framework/reduced graphene oxide composites for catalytic synthesis of 2,3-Dihydroquinazolin-4(1H)-one derivatives. ACS Appl Nano Mater 4(3):2682–2693. https://doi.org/10.1021/acsanm.0c03322
62. Kaupp G (2009) Mechanochemistry: the varied applications of mechanical bond-breaking. CrystEngComm 11(3):388–403. https://doi.org/10.1039/b810822f
63. Beldon PJ, Fábián L, Stein RS, Thirumurugan A, Cheetham AK, Friščić T (2010) Rapid room-temperature synthesis of zeolitic imidazolate frameworks by using mechanochemistry. Angew. Chemie 122(50):9834–9837. https://doi.org/10.1002/ange.201005547
64. Klimakow M, Klobes P, Thünemann AF, Rademann K, Emmerling F (2010) Mechanochemical synthesis of metal-organic frameworks: a fast and facile approach toward quantitative yields and high specific surface areas. Chem Mater 22(18):5216–5221. https://doi.org/10.1021/cm1012119
65. Fujii K et al (2010) Direct structure elucidation by powder X-ray diffraction of a metal-organic framework material prepared by solvent-free grinding. Chem Commun 46(40):7572–7574. https://doi.org/10.1039/c0cc02635b
66. Willans CE et al (2011) Tripodal imidazole frameworks: Reversible vapour sorption both with and without significant structural changes. Dalt Trans 40(3):573–582. https://doi.org/10.1039/c0dt01011a
67. Yuan W, Friščič T, Apperley D, James SL (2010) High reactivity of metal-organic frameworks under grinding conditions: parallels with organic molecular materials. Angew Chemie Int Ed 49(23):3916–3919. https://doi.org/10.1002/anie.200906965
68. Li W et al (2016) MOF-derived hierarchical hollow ZnO nanocages with enhanced low-concentration VOCs gas-sensing performance. Sens Actuat B Chem 225:158–166. https://doi.org/10.1016/j.snb.2015.11.034

69. Assen AH, Yassine O, Shekhah O, Eddaoudi M, Salama KN (2017) MOFs for the sensitive detection of ammonia: deployment of fcu-MOF thin films as effective chemical capacitive sensors. ACS Sens 2(9):1294–1301. https://doi.org/10.1021/acssensors.7b00304
70. Liu XJ et al (2016) A water-stable metal-organic framework with a double-helical structure for fluorescent sensing. Inorg Chem 55(15):7326–7328. https://doi.org/10.1021/acs.inorgchem.6b00935
71. Zhao SS, Yang J, Liu YY, Ma JF (2016) Fluorescent aromatic tag-functionalized MOFs for highly selective sensing of metal ions and small organic molecules. Inorg Chem 55(5):2261–2273. https://doi.org/10.1021/acs.inorgchem.5b02666
72. Tomar K, Gupta M, Gupta AK (2016) A novel 2D porous Cd(II) MOF with a (4,4) connected binodal network: synthesis and photoluminescence sensing of small molecules. Inorg Chem Commun 64:16–18. https://doi.org/10.1016/j.inoche.2015.12.009
73. Wang S, Shan L, Fan Y, Jia J, Xu J, Wang L (2017) Fabrication of Ln-MOFs with color-tunable photoluminescence and sensing for small molecules. J Solid State Chem 245:132–137. https://doi.org/10.1016/j.jssc.2016.10.006
74. Wang K, Liu T, Liu Y, Tian X, Sun J, Zhang Q (2016) Fluorescent heterometallic MOFs: tunable framework charges and application for explosives detection. CrystEngComm 18(42):8301–8308. https://doi.org/10.1039/c6ce01818a
75. Xu H, Liu F, Cui Y, Chen B, Qian G (2011) A luminescent nanoscale metal-organic framework for sensing of nitroaromatic explosives. Chem Commun 47(11):3153–3155. https://doi.org/10.1039/c0cc05166g
76. Liu B (2012) Metal-organic framework-based devices: separation and sensors. J Mater Chem 22(20):10094–10101. https://doi.org/10.1039/c2jm15827b
77. Kuppler RJ et al (2009) Potential applications of metal-organic frameworks. Coord Chem Rev 253(23–24):3042–3066. https://doi.org/10.1016/j.ccr.2009.05.019
78. Cui Y, Yue Y, Qian G, Chen B (2012) Luminescent functional metal-organic frameworks. Chem Rev 112(2):1126–1162. https://doi.org/10.1021/cr200101d
79. Allendorf MD, Bauer CA, Bhakta RK, Houk RJT (2009) Luminescent metal-organic frameworks. Chem Soc Rev 38(5):1330–1352. https://doi.org/10.1039/b802352m
80. Xu X et al (2018) Sensitive detection of pesticides by a highly luminescent metal-organic framework. Sens Actuat B Chem 260:339–345. https://doi.org/10.1016/j.snb.2018.01.075
81. He K et al (2019) A water-stable luminescent metal-organic framework for rapid and visible sensing of organophosphorus pesticides. ACS Appl Mater Interfaces 11(29):26250–26260. https://doi.org/10.1021/acsami.9b06151
82. Tang J, Ma X, Yang J, Feng DD, Wang XQ (2020) Recent advances in metal-organic frameworks for pesticide detection and adsorption. Dalt Trans 49(41):14361–14372. https://doi.org/10.1039/d0dt02623a
83. Singha DK, Majee P, Mondal SK, Mahata P (2019) Detection of pesticide using the large stokes shift of luminescence of a mixed lanthanide co-doped metal–organic framework. Polyhedron 158:277–282. https://doi.org/10.1016/j.poly.2018.10.066
84. Singha DK, Majee P, Mondal SK, Mahata P (2017) Highly selective aqueous phase detection of azinphos-methyl pesticide in ppb level using a cage-connected 3D MOF. ChemistrySelect 2(20):5760–5768. https://doi.org/10.1002/slct.201700963
85. Wei W, Wang J, Bin Tian C, Du SW, Wu KC (2018) A highly hydrolytically stable lanthanide organic framework as a sensitive luminescent probe for DBP and chlorpyrifos detection. Analyst 143(22):5481–5486. https://doi.org/10.1039/c8an01606b
86. Sun L, Campbell MG, Dincə M (2016) Electrically conductive porous metal-organic frameworks. Angew Chemie Int Ed 55(11):3566–3579. https://doi.org/10.1002/anie.201506219
87. Kempahanumakkagari S, Kumar V, Samaddar P, Kumar P, Ramakrishnappa T, Kim KH (2018) Biomolecule-embedded metal-organic frameworks as an innovative sensing platform. Biotechnol Adv 36(2):467–481. https://doi.org/10.1016/j.biotechadv.2018.01.014

88. Wen LL et al (2010) Efficient detection of organophosphate pesticide based on a metal-organic framework derived from biphenyltetracarboxylic acid. Cryst Growth Des 10(7):2835–2838. https://doi.org/10.1021/cg1004352
89. Simonian AL et al (2001) Enzyme-based biosensor for the direct detection of fluorine-containing organophosphates. Anal Chim Acta 442(1):15–23. https://doi.org/10.1016/S0003-2670(01)01131-X

MOFs as Catalysts for the Capture and Degradation of Chemical Warfare Agents

Computational Approach Toward Identification and Catalytic Degradation of Chemical Warfare Agents Using MOFs

Richa Arora and Upasana Issar

Contents

1 Introduction 432
2 Why Use MOFs for Adsorption and Detoxification of CWAs? 435
3 Detoxification of Nerve Agents and Their Simulants by Zirconium-Based MOFs 436
4 Other MOFs for Capturing and Catalytic Degradation of CWAs 442
5 How Good Are the Surrogate Molecules in Mimicking the Adsorption Behavior of Real CWAs? 443
6 Conclusions 445
Abbreviations 446
References 447

Abstract Chemical warfare agents (CWAs) are toxic synthetic chemical weapons used for mass destruction in wars. Despite the prohibition on their use by several countries, some unscrupulous groups still use them. The most toxic and commonly used CWAs are nerve agents which are organophosphate-derived compounds. These are categorized as tabun (GA), sarin (GB), soman (GD), cyclosarin (GF), V-series nerve agents such as VR, VX, and VM. These agents attack the central nervous system of humans depending on the duration of time of exposure. To minimize the threat caused by CWAs, it is imperative to identify and destroy them. Several metal oxides nanoparticles, mixed-metal oxides, nanocomposites, etc. have been explored over the years as catalysts for the decomposition of CWAs. One such family is metal–organic frameworks (MOFs). MOFs are a highly stable and porous hybrid of inorganic metal and organic framework that have many applications in the field of catalysis and even as sensors. Numerous experimental and theoretical studies have shown the identification, adsorption, and detoxification of nerve agents on the surface of MOFs. Computational and theoretical studies not only help in understanding the geometrical and electronic properties of the structures/complexes involved but also prove to be beneficial in elucidating the mechanism of action. Therefore, the chapter aims to highlight the recent work on the catalytic degradation and adsorption of CWAs using MOFs from a computational modeling standpoint.

R. Arora
Department of Chemistry, Shivaji College, University of Delhi, Delhi, India

U. Issar (✉)
Department of Chemistry, Kalindi College, University of Delhi, Delhi, India

S. Gulati (ed.), *Metal-Organic Frameworks (MOFs) as Catalysts*,
https://doi.org/10.1007/978-981-16-7959-9_17

Keywords Nerve agents · Chemical warfare agents · Metal–organic frameworks · Adsorption · Detoxifications · Molecular simulations · DFT

1 Introduction

Chemical warfare agents (CWAs) are toxic chemical substances that cause an extremely hazardous effect on the human body, even death when deployed in the system. Historically, the first use of the chemical substance in any battlefield occurred in 1915 during World War-I (WW-I), when Germany released chlorine gas as a chemical weapon upon two French colonial divisions at Ypres, Belgium [72]. Later, with the discovery and usage of mustard gases, millions of soldiers were killed during WW-I [72]. Since then, these weapons were invariably being used for mass destruction by warring parties. CWAs can be classified based on the nature of chemical species used, their mechanism of action on the human body (target organ or tissue), and on the extent of destruction caused (degree of lethality) [68]. Figure 1 displays major classes of CWAs along with common examples.

Blister agents cause burning and blistering of skin; choking agents block respiratory tract and causes respiratory failure; blood agents inhibit the absorption of oxygen in the bloodstream and causes death due to lack of oxygen, and riot-control agents being the least harmful of all, causes temporary irritation in eyes, mouth, throat, lungs, and skin [68]. Out of all, nerve agents are considered the deadliest form of CWAs. When exposed, these hazardous agents being in the liquid/gaseous phase can be readily inhaled or absorbed through the skin. Once inside the body, these agents disrupt the functioning of nerves by hindering the mechanism by which these nerves transfer information signals to various organs. This eventually causes the breakdown of the central nervous system. Nerve agents are broadly categorized as G-series and V-series nerve agents, based on their place of origination. G-series nerve agents came into existence during World War II (1936–1949) in Germany [4]. These comprise tabun (GA), sarin (GB), soman (GD), and cyclosarin (GF). While, V (venomous)-series agents were developed around the 1950s in the United Kingdom. These consist of VE, VG, VM, VR, and VX [26]. Structurally, these nerve agents

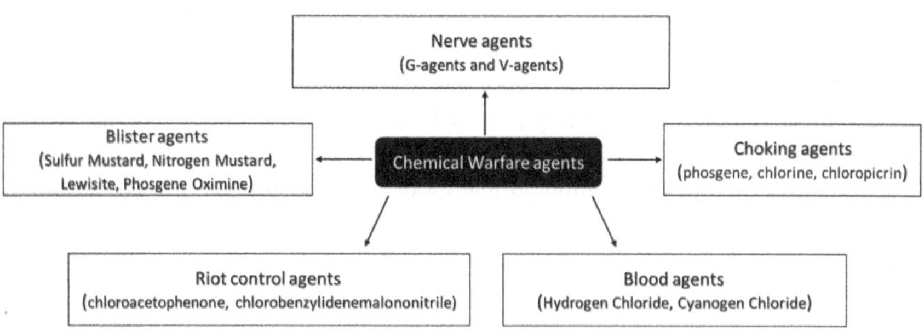

Fig. 1 Classification of chemical warfare agents

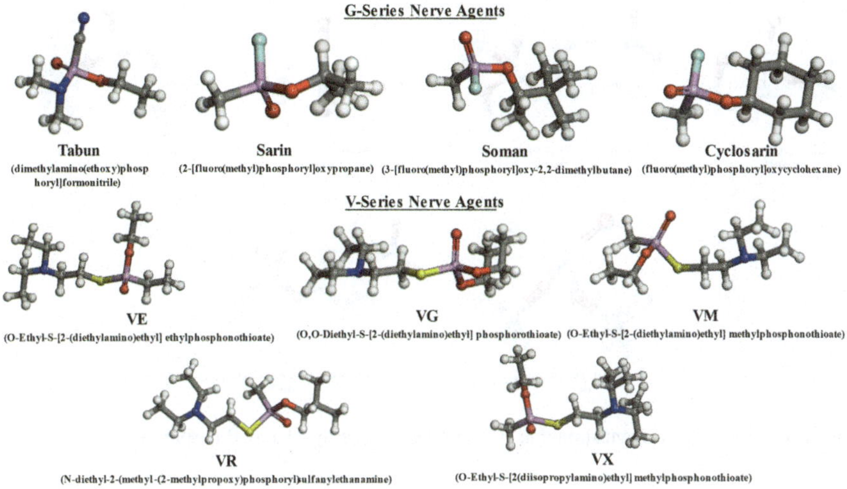

Fig. 2 G-series and V-series nerve agents along with IUPAC names (Color scheme: C, gray; O, red; P, purple; N, blue; S, yellow; F, cyan, and H, white)

contain organophosphorus group with variations in the phosphoric acid ester unit. Figure 2 shows structure of G-series and V-series warfare agents along with their IUPAC names.

Nerve agents are proven to affect the functioning of the enzyme called acetylcholinesterase (AChE) present in postsynaptic neuromuscular junctions in nerves. The role of this enzyme is to hydrolyze a naturally occurring neurotransmitter, acetylcholine, into choline and acetic acid [42]. Organophosphorus compounds irreversibly bind with this enzyme by phosphorylating its serine 203-OH residue, resulting in the accumulation of acetylcholine in receptors in large quantities [3]. This accumulation leads to the high concentration of acetylcholine in the synaptic cleft, resulting in the nonstop simulations of glands and muscles, ultimately causing asphyxiation [45]. Also, prolonged exposure to these nerve agents has shown long-term mental and behavioral defects [14].

Considering all the threat and lethal damages that these nerve agents cause on humanity, their development, production, acquisition, stockpiling, transfer, and usage have been banned under Chemical Weapons Convention, 1997 (United Nations Office for Disarmament Affairs. https://www.un.org/disarmament/wmd/chemical/). Sadly, despite all these efforts, there have been recent evidence of the use of CWAs in wars [74]. It is then imperative for a scientific community to come up with a plausible and effective solution to the destruction of CWAs. Due to the extreme lethality of nerve agents, their direct use for scientific research is avoided, rather surrogate molecules that share similar characteristics with less toxicity are employed. These agents are called nerve agent simulants (Fig. 3).

There are two ways to address the release of CWAs. One is the irreversible capturing of nerve agents onto the surface of any material. The second and the

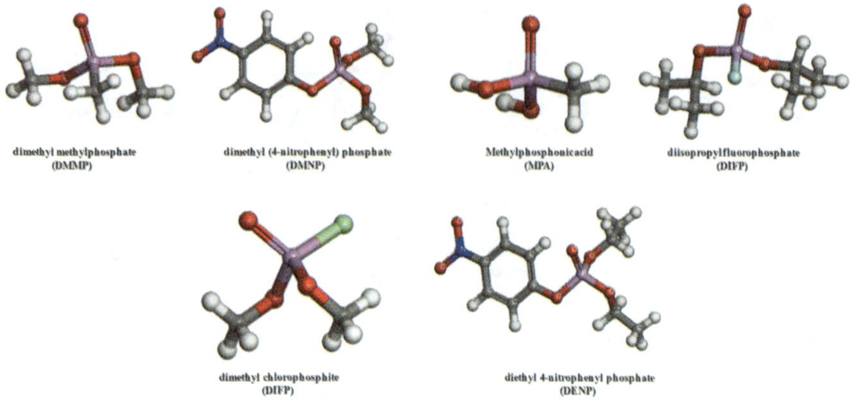

Fig. 3 Common nerve agent simulants (Color scheme: C, gray; O, red; P, purple; N, blue; Cl, green; F, cyan, and H, white)

most effective way is the degradation of CWAs by any catalyst to get non-toxic products. Organophosphorus compounds undergo hydrolysis via nucleophilic attack by water or hydroxide ion on its phosphate triester unit converting it to its corresponding diester derivative [15, 85, 86]. Scheme 1 displays the general hydrolytic pathway of organophosphorus ester in presence of water and hydroxide ions.

Routes adopted in both ways are different (Scheme 1). In base catalyzed degradation, P=O bond is cleaved by OH^- group while in a neutral or acidic medium C-O bond breaking occurs. There are many enzymes available in nature that help in destroying the activity of nerve agents by hydrolyzing their phosphate ester bonds [36, 87],however, it is not practical to utilize them at large scale or in real-life situations.

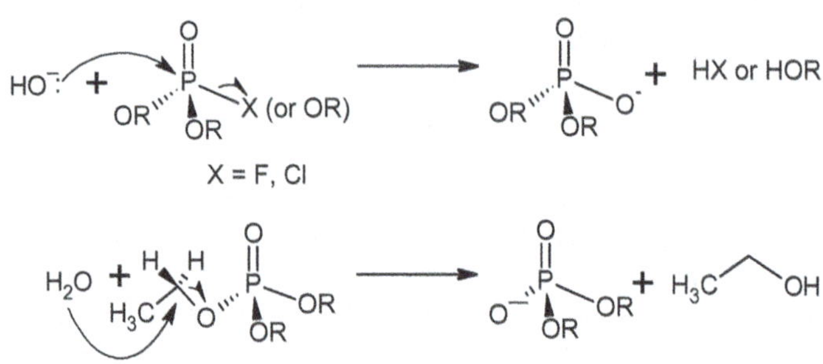

Scheme 1 Hydrolysis routes of organophosphorus esters in presence of OH^- and water

2 Why Use MOFs for Adsorption and Detoxification of CWAs?

Metal–organic frameworks (MOFs), a term coined almost two decades ago [84], have gained immense popularity in the last decade [17, 21] as potential heterogeneous catalysts.

MOFs comprise a uniform array of metal-containing secondary building units (SBU) separated by organic linkers. This creates a permanent porous geometric structure which enhances their potential use in storing of gas, separation of gas, and catalysis [17]. These are the crystalline-ordered molecules containing metallic inorganic nodes and organic linkers as connectors. Given the endless possibilities of combinations of nodes and linkers, thousands of stable MOFs exist. These molecules have a large surface area [40] and high permanent porosity [59], making an ideal surface for reaction or for adsorption of molecules. Sturluson et al. have reviewed the importance of molecular modeling and simulations in the discovery as well as deployment of MOFs for various applications. The motto is to highlight reliable computational techniques that can be used to get the most robust MOFs material for adsorption-based engineering applications [71].

MOF $Zn_2Ca(BTC)_2(H_2O)_2](DMF)_2$, made from zinc and 1,3,5-benzenetricarboxylic acid (BTC), was the first to show capturing of nerve agent simulant, MPA [88]. After that, a lot of studies have been reported that involve variations in metallic center, organic linker, or SBU of MOFs to capture CWAs, and their simulants. Apart from capturing of CWAs, some MOFs having Lewis-acidic metals are also able to display hydrolysis of nerve agents [29–31, 65, 66, 78]. The mechanism of hydrolysis of the nerve agent on the surface of MOFs can be thought of as being similar to the mechanism involved in its enzymatic hydrolysis by enzymes such as phosphotriesterase (PTE) [82], human serum paraoxonase 1 [13], and diisopropylfluorophosphatase (DFPase) [83]. The molecular simulations followed by density functional theory (DFT) calculations on the catalytic hydrolysis of simulant DMNP (also called methyl paraoxan) by PTE showed that the first step is the binding of reactant molecule with one of the Zn atoms in binuclear zinc enzyme, replacing the associated water. The next step is the S_N2 type nucleophilic attack of –OH present in the active site of PTE on DMNP, making a five-coordinate intermediate. Since the intermediate is short lived, the phosphodiester bond in DMNP breaks to yield non-toxic products [7]. On similar lines, MOFs having metallic nodes easily mimic the enzymatic system and hence can be an ideal choice for the detoxification of CWAs. The MOFs family attracts special importance for detoxification due to the presence of a highly periodic structure (active site) that can be tuned in many ways to adsorb CWAs.

The adsorptive capture of sarin on selective hydrophobic MOFs was examined under bulk mixture solvent conditions using molecular modeling techniques. The detoxification of sarin occurs only after its selective adsorption at MOFs. Quantitatively, the adsorption value can get affected if one changes the intrinsic flexibility of MOFs [58].

Although a lot of work has happened in the past decade on the catalytic destruction of CWAs into non-toxic products by MOFs, most of the available literature is exhaustively from the experimental point of view. However, there is an upper edge of using computational studies in the present case, as it avoids all the experimental hassles associated with dealing with toxic nerve agents. Therefore, this chapter aims to draw light on the advancement of molecular modeling and simulations on the capturing and destruction of CWAs and their simulants by MOFs as catalysts.

3 Detoxification of Nerve Agents and Their Simulants by Zirconium-Based MOFs

Among the most promising and extensively investigated candidates as catalysts for the detoxification of CWAs are zirconium-based MOFs. The research group led by Hupp and Farha deserves a special mention as they have done exhaustive research on Zr-based MOFs and have contributed a lot in understanding the mechanistic aspect of interactions between Zr-based MOFs and CWAs over the past decade [33].

These MOFs contain a strong Zr^{IV} Lewis-acidic metal center periodically arranged in a crystalline geometry, making Zr_6 nodes. Zr-based MOFs display strong thermal and chemical stability owing to the presence of very strong Zr-oxo node-linker bonds [38]. Most of the studies involving MOFs are based on the variations in the organic connector in MOFs and how these linkers bind to SBU. The SBU however remains identical in these MOFs. Three Zr-based MOFs, namely UiO-66 [6], NU-1000 [48], and MOF-808 [16] have been popularly adopted in a majority of the studies. These MOFs contain identical zirconium oxide/hydroxide core having $Zr_6(\mu_3\text{-}O)_4(\mu_3\text{-}OH)_4$ SBU, but differ in organic linkers and connectivity around SBU (Fig. 4. Six zirconium atoms are placed at the vertices of the octahedron with four hydroxylated faces and four oxide group faces. In UiO-66, SBUs are linked by twelve organic linkers; in NU-1000, it is eight organic linkers, while in MOF-808 six organic linkers are connected by SBUs [51, 52].

This variation creates a substantial difference in the pore size of these MOFs and thereby access to the catalytic center Zr_6-SBUs [77]. Early experimental studies displayed fast degradation of nerve agent simulant on UiO-66 [30], and later, a substantial improvement in the rate was observed on simply functionalizing the linker group. On introducing amine group on terephthalate in UiO-66, the overall efficiency of proton transfer increases. Another MOF, UiO-67 which has the same SBU as UiO-66 with only a difference in the size of the linker also displays a faster reaction rate due to more accessibility to the catalytic site [29, 31].

Researchers continue to work on optimizing the structure of MOFs by modulating the functional groups on linkers and pore size in an attempt to get the best moiety as a catalyst to degrade nerve agents [61]. The catalytic efficiency of MOFs for the hydrolytic reaction of nerve agents can be improved in different ways. One can introduce the defects sites or change the metal at nodes or by functionalizing the

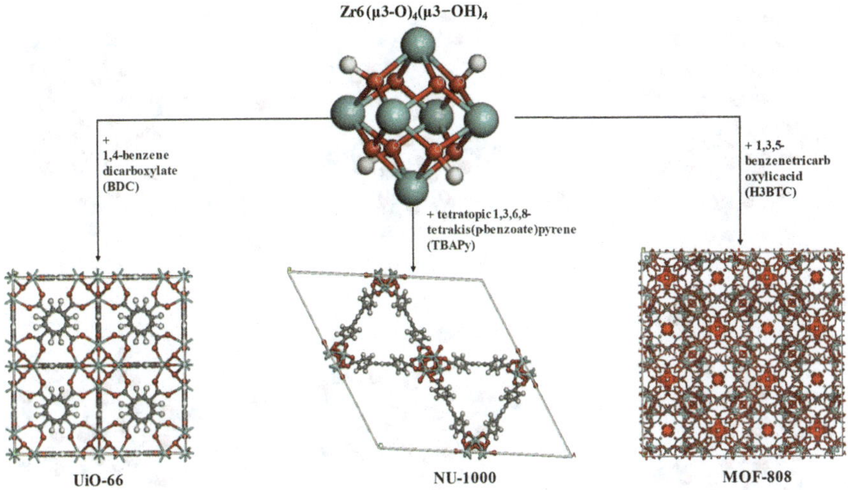

Fig. 4 Representation of core $Zr_6(\mu_3\text{-}O)_4(\mu_3\text{-}OH)_4$ along with organic connectors BDC, TBAPy, and H3BTC forming UiO-66, NU-1000, and MOF-808, respectively (Color scheme: C, gray; O, red; Zr, cyan; and H, white)

organic linker present in MOFs [27]. Meanwhile, it is also important to understand the mechanism behind this degradation process. Computational studies have shed some light in this respect and have been able to propose the degradation reaction pathway.

The nodes of MOF catalytically hydrolyze the phosphoester bond in nerve agents. This is because the Zr–OH–Zr moieties existing in Zr_6 nodes along with linker in UiO-66 actually resembles Zn–OH–Zn present in the active site of phosphotriesterase enzyme responsible for degrading G-type nerve agent. It has been reported that the rate of hydrolysis reaction is dependent on the pore size of MOFs. If the size of the pore is small, then the nerve agent sits only on the surface of MOF, and the hydrolytic reaction becomes slow [12].

To explore this further, Mondloch and co-workers performed experimental as well as computational studies on the decomposition of DMNP and GD by NU-1000 [49]. NU-1000 has fewer organic linkers compared to UiO-66 which makes its pore diameter bigger. This allows the nerve agent molecule to permeate better inside the MOF, proof of which can be seen from the fact that NU-1000 hydrolyses CWA approximately 30 times faster as compared to UiO-66. DFT calculations at M06-L level suggested two modes of binding for DMNP at node $Zr_6(\mu_3\text{-}O)_4(\mu_3\text{-}OH)_4(H_2O)_4(OH)_4$ of NU-1000. One, in which DMNP interacts with water ligated node, and the other, in which terminal water was removed from the node, and therefore, DMNP interacts directly with Lewis-acidic Zr^{IV} sites. Figure 5 displays the free energies of binding predicted for the interaction between DMNP and GD with MOF. The first step is the formation of a hydrogen bond between simulant and node-ligated water/hydroxide group (Fig. 5a). The interaction is a favorable process

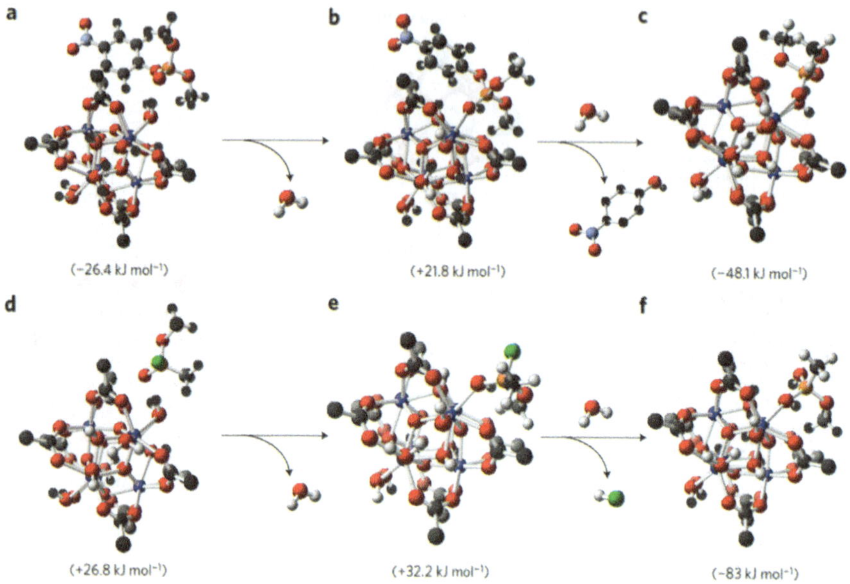

Fig. 5 Binding free energies of the hydrolysis of simulant DMNP (**a–c**) and nerve agent GD (**d–f**) on node of NU-1000. Reprinted with permission from Mondloch et al. [49]. Copyright 2015 Springer Nature Limited

(negative free energy) not only because of the stabilization due to hydrogen bonding but also because of the favorable π-π stacking interactions among aromatic rings of DMNP and organic linker. However, the same step becomes unfavorable for GD (Fig. 5d) owing to the absence of these stabilizing forces. The next step is the removal of the terminal water molecule of NU-1000 by nerve agent or simulant (Fig. 5b, e). This is unfavorable for both DMNP and GD, but in the last step where hydrolysis takes place, the reaction becomes favorable for both (Fig. 5c, f) forming stable products. In the second type of binding mode, electrostatic interactions play a pivotal role. Stronger electrostatic interactions exist as P=O unit of DMNP interacts directly with Zr in the node of MOFs [49]. The results of this computation and simulations are in accordance with the enzymatic hydrolysis of DMNP [7].

In another study, Chen et al. [9] utilize DFT calculations to understand the role of NU-1000 in the hydrolysis reaction of DMNP. They performed the hydrolysis of DMNP by OH^- nucleophile in the absence of any catalyst and studied the reaction profile. The same reaction is then carried out in presence of NU-1000 as a catalyst. The study inferred that the dehydration from the node must occur first in order to have effective DMNP binding on Zr atom. To gain more clarity on the role of NU-1000, all the transition state structures were analyzed. The presence of hydroxyl (–OH) group on the node of MOF facilitates the dehydration step and thus reduces the barrier height for the reaction [9]. It has also been established experimentally that the MOFs undergo structural distortions at high temperatures. In the distorted structure, two sets of distances are observed between two zirconium atoms, making the node structure asymmetric [63]. Even though these distortions do not change the long-range order crystalline structure of MOFs, but still these unconventional defects are shown to affect the properties of MOFs. Utilizing this, Chen and co-workers also studied the interaction between distorted NU-1000 and DMNP through DFT calculations. They reported that in the distorted MOFs, DMNP directly attacks the Zr atom at the node, and there is no essential requirement of dehydration step to occur first [9].

Momeni and the group utilized both periodic and cluster modeling to study the catalytic hydrolysis of sarin on Zr-based MOFs. The reaction profile suggested that the steps involving displacement of water and attack of nucleophile are dependent on the number of organic linkers in MOFs as well as on the functional group substituents on these organic linkers in the SBUs. A decrease in the activation energies for these steps is observed when the substitution is done by the amino group at the *ortho* position of the organic linker. This means that amino-substituted UiO-66 and MOF-808 display better hydrolytic activity [47]. The role of amino functionalities on MOFs for the hydrolysis of DMNP is further investigated using the DFT study. The authors proposed that the presence, as well as proximity of amino group to the nerve agent, play a significant role in its detoxification [23, 24]. In MOFs, organic linker connectivity and metal identity at hexanuclear M^{IV} cluster note can be modulated in many ways so as to come up with new promising materials.

Recently, Mendonca et al. utilized DFT studies to examine these factors associated with MOFs in detail and their role in the degradation of CWAs and their simulants. A thorough examination revealed that no single metal or node can be considered as the optimal choice for the entire range of organophosphates and hence, should be selected as per the agent/simulant chosen. Based on the large amount of data obtained from the DFT calculations, the authors also generated a quantitative structure-activity relationship (QSAR) model to explain the variations observed in binding parameters for molecules (nerve agents, simulants, and hydrolytic by-products) across several MOFs node topologies, connecting species and metals. The results showed the cumulative role of molecular and node properties for effective binding. In conclusion, the developed QSAR model could pave way for predicting the binding energetics of any nerve agent on the hexacoordinated metal node [43].

Although the hydrolysis of the nerve agents by Zr-MOFs is a highly favorable process in basic media with very small half-lives for the hydrolysis of CWAs, yet there are few problems that need to be addressed. The catalytic efficiency of Zr-based MOFs is dependent on buffers for pH moderation and regeneration of its active site [23, 24, 34, 39, 49, 57]. The methodology can be successfully utilized to destroy stockpiles of CWAs but show little inefficiency in instant applications such as in gas masks or in protective fabrics [8]. Therefore, researchers are focusing more on the gas-phase hydrolysis of CWAs on MOFs so as to come up with a robust material that can be used under any condition [80].

To address the issue, one can take the help of computational studies which can effectively propose the possible mechanism under any desired conditions. Once all the species involved in the pathway are known along with transition energies, adequate steps can be made to modulate the MOFs so as to increase the reaction rate and thereby improving the overall kinetics of the adsorption process. One such study suggested a complete reaction mechanism for the decomposition of GB on Zr-based MOF (UiO-66 and MOF-808) via molecular modeling. The reaction with UiO-66 requires a linker defect in which two Zr atoms are low coordinated. Under the dry condition, the first step in the decomposition requires the attachment of GB on Zr atom in SBU. This attachment takes place via sp^2 hybridized oxygen of GB [29, 31]. After this, nucleophilic addition of the OH group of zirconium occurs perpendicularly with GB forming a penta-coordinated phosphorus intermediate. The last step is the elimination of hydrogen fluoride or isopropanol forming products. Now, the formation of products depends on the way nucleophilic addition of hydroxide moiety takes place on GB. GB being a chiral molecule can attack the MOF via different orientations (facing different tetrahedron faces). The final products could be either HF and isopropyl methyl phosphonic acid (IMPA)-MOF or isopropyl alcohol and methyl phosphonofluoridic acid (MPFA)-MOF. The same reaction steps can be obtained with MOF-808 [76].

The above study suggests a decent pathway for the detoxification of nerve agents, but if one is interested in finding the catalytic turnover, then it is non-favorable. This is because the hydrolytic product IMPA binds very strongly with the zirconium nodes of MOFs making it very difficult to recover the catalyst for the next cycle. This is called product inhibition on the catalyst. It is more pronounced if the hydrolytic product binds in a bidentate manner onto the node (eg., using two adjacent Zr-oxo bonds for binding on MOFs) instead of monodentate binding (with one Zr atom). Subsequent experimental studies also support the theoretical findings that stronger the binding forces between product and node, the greater is the product inhibition, and the lesser efficient will be the destruction path for the nerve agent [64, 79].

One solution to this problem is to block some sites on the nodes of MOFs such that only a few sites are available for the nerve agent, and hence, monodentate binding could take place more favorably. This task can be achieved using atomic layer deposition or solvothermal deposition techniques where metal atoms are deposited as single-atom catalysts on nodes of desired MOFs [2, 22, 25, 32, 37, 48, 54, 56, 80].

NU-1000 (Fig. 4) can be considered as a perfect moiety for metalation as it contains a large pore size, enabling proper diffusion of metal ions. Zr_6 nodes have reactive terminal groups that further facilitate the easy insertion of metal atoms. Experimental and theoretical studies have confirmed successfully single-atom deposition for metals like vanadium, molybdenum, chromium, and copper on NU-1000 nodes [19, 55]. In this regard, Mendonca and Snurr utilized DFT calculations to thoroughly investigate various single-atom transition metals deposited on NU-100. The performance of these metal-loaded MOFs was then tested for the hydrolysis of sarin in gas-phase.

A total of 36 metal-loaded NU-1000 systems were examined. The oxidation state of metal deposited varied from +2 to +4. Out of these, M^{IV}NU-1000 systems showed a strong affinity of sarin, while M^{II}NU-1000 systems displayed strong binding with IMPA. To shed more light on catalytic hydrolysis of sarin, Ti^{IV}-NU-1000 system was chosen (Scheme 2). Titanium being in the same oxidation state as that of zirconium, Ti^{IV}-loaded NU-1000 becomes an ideal choice. The authors report an interesting catalytic cycle for detoxification. In the first step, O (from P=O) of sarin attacks

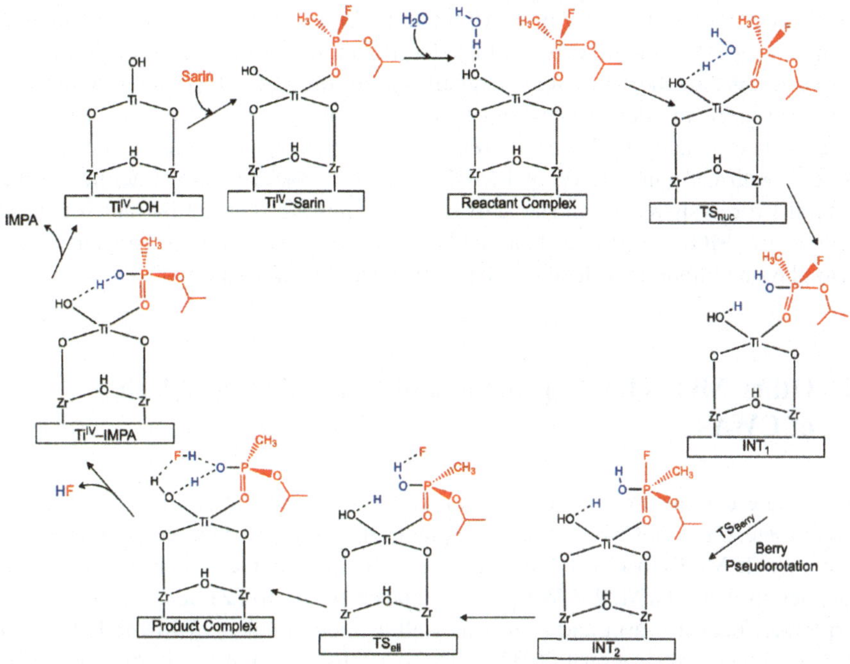

Scheme 2 Proposed mechanism of catalytic hydrolysis of sarin on Ti^{IV}-NU-1000 cluster. Reprinted (adapted) with permission from Mendonca and Snurr [44]. Copyright {2020} American Chemical Society

Ti^{IV}. Then, water from a humid environment forms a hydrogen bond with OH-Ti making a reactant complex. The nucleophilic addition of OH from water takes place at sarin forming a transition state (TS_{nuc}) which then leads to the formation of penta-coordinated intermediate (INT_1). In trigonal bipyramidal geometry, the most electronegative atom favors axial position, and elimination is also favored at that position. The intermediate INT_1, thus undergoes Berry pseudorotation such that -F atom comes at an axial position and forms INT_2 [11, 69]. In the elimination step, proton transfer takes place from –OH (attached to P) forming a new transition state TS_{eli}. The next step is the formation of a product complex stabilized by various hydrogen bonds. The desorption of HF then takes place from this product complex forming the IMPA-Ti^{IV}-NU-1000 species where the IMPA is linked in the monodentate mode with MOF. Elimination of IMPA takes place in the last step, thereby, regenerating the catalyst. In all, the authors suggested that single-atom metal deposited systems, force the products to bind in the monodentate manner showing better turnover frequencies of catalyst [44].

With the aim to get the most robust zirconium-based MOFs for the degradation of sarin, Zr_6, Zr_{12} and Ce^{IV} substituted MOFs were investigated through periodic boundary conditions at PBE level [60] calculations with D3 dispersion correction [20]. It is not mandatory that the defects in MOFs always lead to enhanced activity of catalyst. Rather, the activity of MOFs to hydrolyze nerve agent is the function of how and where the defects are introduced and also their relative position with respect to one another. It is in fact a challenge for the research community to carefully introduce the defect sites and hence make highly engineered materials for heterogeneous catalysis. The study revealed that defects introduced by substitution of Zr^{IV} with Ce^{IV} in UiO-66 and MOF-808 reduce the activation energies in the overall hydrolysis reaction. The activation energies are reduced by half for these bi-defective MOFs. Hydrogen bonding between metal oxide node and organic linker also plays an important role during catalytic degradation reaction [47].

4 Other MOFs for Capturing and Catalytic Degradation of CWAs

Systematic computational studies comprising of a combination of Grand Canonical Monte Carlo (GCMC) simulations and DFT were carried out on a series of water-stable MOFs to adsorb CWAs and their simulants to predict the energetics and interactions between MOF/CWA pairs. A total of twenty-five Zr- and Ti-MOFs were explored to adsorb sarin and soman as well as their simulants DMMP, DIFP, and pinacolyl methylphosphonate (PMP). These MOFs vary in pore size/topology and display diverse adsorption sites. The study reports excellent adsorption enthalpies (as high as -100 kJ mol^{-1}) for some MOF/CWA pairs. The Monte Carlo simulations revealed that MOFs are best in class porous materials displaying a high affinity for nerve agents even at low loading. DFT analysis suggests mostly van der Waals types of interactions and rules out the possibility of any charge transfer between MOF and CWA. The study also suggests that though DMMP and DIFP can be used as good surrogates of sarin; extra attention is required for choosing a simulant of soman. PMP is reported to be a better alternative simulant for soman compared to DMMP and DIFP [70].

The hydrolysis of CWAs can also be carried out using Ti-MOFs. Soares and co-workers performed the first-ever DFT calculations at the periodic level to study the degradation of nerve agent GB, and its simulants DMMP and DIFP on newly discovered and highly stable Ti-MOF and MIP-177(Ti) which is carboxylate Ti-MOF [70]. The reaction pathway and transition energies involved between species during the detoxification process obtained from DFT studies were analyzed carefully and the results were also compared with Zr-based MOFs. The results showed that MIP-177 (Ti) display adsorption capacity equivalent to well-established Zr-based MOFs. The mechanism of the degradation of sarin on MIP-177 is similar to its adsorption on Zr-based MOFs. O (of P=O bond) of sarin attacks the titanium node, and then, elimination of HF occurs resulting in the bidentate binding of IMPA onto the node. Lastly, IMPA-MIP-177 (Ti) complex is stable enough that a high amount of energy is required to desorb IMPA [70].

5 How Good Are the Surrogate Molecules in Mimicking the Adsorption Behavior of Real CWAs?

Since most of the experimental research focuses on the adsorption of simulant molecules rather than real nerve agents on MOFs, one cannot be 100% sure of the effectiveness of the methodology. As there are thousands of MOFs available, a high-throughput computational screening has been established to be the most beneficial tool for molecular simulations [10]. GCMC simulations can be another effective way of predicting the hydrophobicity and adsorption capacity of MOFs [18], but one cannot screen a large number of materials using this technique. This is because GCMC simulations usually have long equilibration times and hence are computationally too expensive [46].

To make things clear, Agrawal and co-workers made use of molecular simulations to systematically examine whether these simulants mimic the real agents effectively or not. Four simulants DMMP, DCP, DIFP, DMNP, and two real nerve agents, sarin, and soman are tested using molecular simulations on a set of 2969 experimentally MOFs crystal structures (obtained from CoRE Computation-Ready, Experimental MOF Database) in the limit of dilute adsorption. Each system was tested with both a generic force field and a DFT-derived force field. The enthalpy of adsorption and Henry's constant is the key parameters in predicting the best available simulant. The heat of adsorption data of each molecule within MOF showed that DCP and DMMP correlate well with sarin, while DMNP can be said to be the only feasible option for soman out of all other simulants. DMNP does not display an as strong correlation with soman as DCP or DMMP display with sarin. These results are also in coherence with the obtained Henry's constants and therefore can be well utilized in the low adsorbent concentrations. The study reported no substantial change in the adsorptive capacity of nerve agents and their simulants on MOFs on changing the force fields. Hence, the adsorbate-adsorbent interactions can be said to be independent of specific details of the force field [1].

One more important factor in governing the efficiency of the adsorption process of CWA on MOFs is the presence of atmospheric moisture or in the general hydrophilic environment during adsorption. It is the adsorption of water from the atmosphere that hinders the capturing performance of CWA on MOFs. A hydrophobic [$Zn_4(\mu_4$-O)-(μ_4-4-carboxy-3,5-dimethyl-4-carboxy-pyrazolato$)_3$] MOF resembling the same crystal structure as that in MOF-5, selectively captures sarin and mustard gas even under ambient moisture conditions. A comparison with hydrophilic HKUST-1 ([$Cu_3(BTC)_2$]) also affirms that the efficiency of CWA enhances in dry conditions. This happens because the coordinatively unsaturated metal sites (preferred binding sites) on MOFs become ineffective in presence of water. Therefore, pore size and surface hydrophobicity are key factors in the CWA-MOF recognition process [50]. Keeping this in view, the identification of the best porous material for the adsorption of CWA was computationally done using a systematic screening strategy. High-throughput screening was performed on 2969 MOFs available in the database to select 1647 MOFs that had appropriate pore sizes easily accessible by CWA. Monte Carlo simulations in the canonical ensemble (CMC) using the Widom test particle method [81] was then employed on these 1647 MOFs to finally shortlist 156 hydrophobic MOFs. Since the Widom method significantly reduces the simulation time for water adsorption, water is no more a competing species with CWAs during the adsorption process. GCMC simulations at various pressures were running on each of these 156 selected MOFs and CWAs pair. Sarin, soman, mustard gas, and water were chosen as adsorbates. Out of 156 MOFs, three hydrophobic MOFs were identified as the best adsorbents having good adsorbing/loading capacity >4 mol kg^{-1} for capturing sarin, soman, mustard gas. To verify these computational results, experimental adsorption studies of diethyl sulfide (DES), a simulant of mustard gas was carried out on one of the three best MOFs, viz. $[Ni_3(BTP)_2]$ under humid conditions. In this, MOF metallic node is joined with 1,3,5-tris (1*H*-pyrazol-4-yl) benzene (H_3BTP). Exceptional adsorption selectivity and stability of DES on $[Ni_3(BTP)_2]$ assures that the chosen computational strategy can be utilized in making suitable materials for effectively capturing CWAs [53]

UiO can be considered to be the most effective class of MOFs. The properties of UiO can be modulated by introducing various functional groups keeping their inherent stability intact [5, 28, 35]. There are comparatively fewer studies of the adsorption of CWAs or CWAs simulants on functionalized MOFs [62]. This could be because of the fact CWAs interact effectively with SBU rather than the attached linker [51, 52, 79]. So, the relationship between changing the functional group on organic linkers in MOFs, and their absorptive capacity is quite an unexplored field. However, it has been established that the interaction between guests and MOFs increases in presence of defects in MOFs such as missing linkers [51, 52, 73, 75, 79]. This suggests that the properties of defect-free MOFs can certainly be tuned by changing the functional groups on linkers. Ruffley et al. made use of these results and performed DFT studies on the adsorption of DMMP in functionalized UiO-67. Different substituents such as $-H$, $-CH_3$, $-SH$, $-NH_2$, $-N_3$, $-NO_2$, $-Br$, and $-Cl$ are introduced at the 2-position of linker 1,1′-biphenyl-4,4′-dicarboxylate (BPDC). The interaction between each functionalized MOFs and DMMP was carried out using DFT. Based on the DFT calculations and GCMC simulations, it is suggested that amino- and methyl-functionalized linkers in UiO-67 display strong binding affinities with DMMP as compared to unfunctionalized biphenyl linkers. These MOFs were then synthesized and experimentally tested for DMMP. The results found from the temperature-programmed desorption and IR studies also confirm the obtained trend of adsorption affinities for DMMP as predicted in theoretical simulations. Hence, such ab initio calculations can give good guidance in designing better MOFs to be used in experiments [67].

In another study, the detoxification of sarin and its surrogates, DIFP, and DMMP were tested onto two newly discovered titanium-based MOFs, viz. Ti-MOF and MIP-177(Ti) via computational techniques. The study revealed that DIFP mimics sarin better than DMMP onto these MOFs. The degradation mechanism as well as the reaction energies for DIFP-MOF interactions came out to be very close to that of the sarin-MOF pair. Hence, DIFP can be treated as a reliable surrogate of sarin [70].

Matito-Martos and co-workers performed high-throughput computational screening on a set of 2932 MOFs taken from the CoRE database to capture sarin, soman, and mustard gas (also known as sulfur mustard) and their simulants. Point charges were assigned for understanding the electrostatic interactions between CWA-MOF pairs. The first filtration criterion is the pore size of MOFs. It is required for the proper permeation of CWAs. On this basis 1275 MOFs were rejected as their pore limiting diameter was less than 3.72 Å. After that, Widom insertions were used to screen the remaining 1647 MOFs to calculate Henry's coefficient and enthalpy of adsorption. The heat of adsorption data showed that DES is a good surrogate for mustard gas. Moving to the next step, the authors found Henry's constant of water in 1647 MOFs and selected 156 hydrophobic MOFs. To further investigate the storage capacity of these 156 MOFs, GCMC simulations were performed. $[Ni_3(BTP)_2]$ MOF (CSD code: UTEWOG) was found to be the best for adsorbing DES selectively under humid conditions [41].

6 Conclusions

Over the last decade, MOFs have arisen as revolutionary heterogeneous catalysts for the degradation of extremely toxic CWAs. Zirconium-based MOFs have topped in the category and have been exploited in much detail for this purpose. This is because of the presence of Lewis-acidic Zr^{IV} active sites that ensures rapid hydrolysis of phosphodiester bond of nerve agents. The catalytic performance of these materials can be tuned by changing the node topology, metal, or organic linkers. Functional group substitution and defects also alter the nature and reactivity of these MOFs. With particular reference to nerve agents, it has been shown experimentally and validated computationally by a number of researchers that catalytic performance of MOFs for hydrolysis of CWAs is largely a function of three parameters. First and foremost is the ease with which CWAs or their simulants access the MOF's active site. This is probably the main factor in determining the kinetics of the hydrolytic reaction. The rate of reaction also increases by decreasing the connectivity in MOFs (reducing the organic linkers) thereby increasing the pore size diameter. Moreover, the catalytic performance of MOFs has been shown to increase by increasing the defect density in these materials. In addition to these modifications, the activity of MOFs can also be altered by altering the coordination site around the M_6 node and by introducing functionalization in the organic linkers. Amine functionalization on MOFs improves the rate of degradation of CWAs. Computational and experimental results coherently established the mechanistic pathways of detoxification and suggested all the probable reactive and non-reactive species formed during the hydrolysis reaction. There are enough studies that validate the use of surrogates such as DMNP for real nerve agents as these simulants demonstrate the exact same steps of hydrolysis as CWAs.

The high-throughput screening of the MOFs available in the database can be tested for the adsorption of any nerve agent. This can be followed by molecular simulations and modeling on selected CWA-MOF pairs. Based on the energetics of the reaction, one can predict the exact mechanism of the adsorption process and various species involved. Moreover, alterations in the metal/node/organic linker can also be done depending on the adsorbate or adsorbent. Therefore, by utilizing the combination of numerous computational tools, the best MOFs for the degradation of CWAs can be found which can be suggested to chemists for the synthesis and experimental studies.

Abbreviations

AChE	Acetylcholinesterase
BTC	1,3,5-Benzenetricarboxylic acid
CoRE	Computation-Ready, Experimental MOF Database
CWAs	Chemical warfare agents
DENP	Diethyl 4-nitrophenyl phosphate
DES	Diethyl sulfide
DFPase	Diisopropylfluorophosphatase
DFT	Density Functional Theory
DIFP	Diisopropyl fluorophosphate
DIFP	Dimethyl chlorophosphite
DMMP	Dimethyl methylphosphate
DMNP	Dimethyl (4-nitrophenyl) phosphate
GA	Tabun
GB	Sarin
GCMC	Grand Canonical Monte Carlo
GD	Soman
GF	Cyclosarin
H3BTP	1,3,5-Tris(1H-pyrazol-4-yl)benzene
IMPA	Isopropyl methyl phosphonic acid
INT	Intermediate
MOFs	Metal–organic frameworks
MPA	Methylphosphonic acid
MPFA	Methyl phosphonofluoridic acid
PMP	Pinacolyl methylphosphonate
PTE	Phosphotriesterase
QSAR	Quantitative structure-activity relationship
SBU	Secondary building units
TS	Transition state
WW-I	World War-I

References

1. Agrawal M, Sava Gallis DF, Greathouse JA, Sholl DS (2018) How useful are common simulants of chemical warfare agents at predicting adsorption behavior? J Phys Chem C 122:26061–26069
2. Ahn S, Thornburg NE, Li Z, Wang TC, Gallington LC, Chapman KW, Notestein JM, Hupp JT, Farha OK (2016) Stable metal-organic framework-supported niobium catalysts. Inorg Chem 55:11954–11961
3. Allgardsson A, Berg L, Akfur C, Hörnberg A, Worek F, Linusson A, Ekström FJ (2016) Structure of a prereaction complex between the nerve agent sarin, its biological target acetylcholinesterase, and the antidote HI-6. Proc Natl Acad Sci U S A 113:5514–5519
4. Arduini F, Palleschi G (2012) Disposable electrochemical biosensor based on cholinesterase inhibition with improved shelf-life and working stability for nerve agent detection. In: Nikolelis D (ed) Portable chemical sensors: weapons against bioterrorism. Springer, Dordrecht, pp 261–278
5. Biswas S, Van Der Voort P (2013) A general strategy for the synthesis of functionalised UiO-66 frameworks: characterisation, stability and CO_2 adsorption properties. Eur J Inorg Chem 2013:2154–2160
6. Cavka JH, Jakobsen S, Olsbye U, Guillou N, Lamberti C, Bordiga S, Lillerud KP (2008) A new zirconium inorganic building brick forming metal organic frameworks with exceptional stability. J Am Chem Soc 130:13850–13851
7. Chen S-L, Fang W-H, Himo F (2007) Theoretical study of the phosphotriesterase reaction mechanism. J Phys Chem B 111:1253–1255
8. Chen Z, Islamoglu T, Farha OK (2019) Toward base heterogenization: a zirconium metal-organic framework/dendrimer or polymer mixture for rapid hydrolysis of a nerve-agent simulant. ACS Appl Nano Mater 2:1005–1008
9. Chen H, Liao P, Mendonca ML, Snurr RQ (2018) Insights into catalytic hydrolysis of organophosphate warfare agents by metal-organic framework NU-1000. J Phys Chem C 122:12362–12368
10. Colon YJ, Snurr RQ (2014) High-throughput computational screening of metal-organic frameworks. Chem Soc Rev 43:5735–5749
11. Daniel KA, Kopff LA, Patterson EV (2008) Computational studies on the solvolysis of the chemical warfare agent VX. J Phys Org Chem 21:321–328
12. DeCoste JB, Peterson GW (2014) Metal-organic frameworks for air purification of toxic chemicals. Chem Rev 114:5695–5727
13. Fairchild SZ, Peterson MW, Hamza A, Zhan CG, Cerasoli DM, Chang WE (2011) Computational characterization of how the VX nerve agent binds human serum paraoxonase 1. J mol model 17:97–109
14. Figueiredo TH, Apland JP, Braga M, Marini AM (2018) Acute and long-term consequences of exposure to organophosphate nerve agents in humans. Epilepsia, 59 Suppl 2(Suppl 2):92–99
15. Florián J, Warshel A (1998) Phosphate ester hydrolysis in aqueous solution: associative versus dissociative mechanisms. J Phys Chem B 102:719–734
16. Furukawa H, Gándara F, Zhang YB, Jiang J, Queen WL, Hudson MR, Yaghi OM (2014) Water adsorption in porous metal-organic frameworks and related materials. J Am Chem Soc 136:4369–4381
17. Furukawa H, Cordova KE, O'Keeffe M, Yaghi OM (2013) The chemistry and applications of metal-organic frameworks. Science (New York, NY) 341(6149):1230444
18. Ghosh P, Kim KC, Snurr RQ (2014) Modeling water and ammonia adsorption in hydrophobic metal-organic frameworks: single components and mixtures. J Phys Chem C 118:1102–1110
19. Goetjen TA, Zhang X, Liu J, Hupp JT, Farha OK (2019) Metal-organic framework supported single site chromium(III) catalyst for ethylene oligomerization at low pressure and temperature. ACS Sustain Chem Eng 7:2553–2557
20. Grimme S, Antony J, Ehrlich S, Krieg H (2010) A consistent and accurate ab initio parametrization of density functional dispersion correction (DFT-D) for the 94 elements H-Pu. J Chem Phys 132:154104

21. Howarth AJ, Majewski MB, Farha OK (2019) 6—Metal-organic frameworks for capture and detoxification of nerve agents. Iin: Ghosh SKBT (ed) Elsevier, pp 179–202
22. Ikuno T, Zheng J, Vjunov A, Sanchez-Sanchez M, Ortuño MA, Pahls DR, Fulton JL, Camaioni DM, Li Z, Ray D, Mehdi BL, Browning ND, Farha OK, Hupp JT, Cramer CJ, Gagliardi L, Lercher JA (2017) Methane oxidation to methanol catalyzed by Cu-Oxo clusters stabilized in NU-1000 metal-organic framework. J Am Chem Soc 139(30):10294–10301
23. Islamoglu T, Ortuño MA, Proussaloglou E, Howarth AJ, Vermeulen NA, Atilgan A, Asiri AM, Cramer CJ, Farha OK (2018a) Presence versus proximity: the role of pendant amines in the catalytic hydrolysis of a nerve agent simulant. Angew Chemie 130:1967–1971
24. Islamoglu T, Ortuño MA, Proussaloglou E, Howarth AJ, Vermeulen NA, Atilgan A, Asiri AM, Cramer CJ, Farha OK (2018b) Presence versus proximity: the role of pendant amines in the catalytic hydrolysis of a nerve agent simulant. Angew Chem Int Ed Engl 57:1949–1953
25. Islamoglu T, Goswami S, Li Z, Howarth AJ, Farha OK, Hupp JT (2017) Post synthetic tuning of metal−organic frameworks for targeted applications. Acc Chem Res 50:805–813
26. Jang YJ, Kim K, Tsay OG, Atwood DA, Churchill DG (2015) Update 1 of: destruction and detection of chemical warfare agents. Chem Rev 115:PR1–PR76
27. Kalaj M, Momeni MR, Bentz KC, Barcus KS, Palomba JM, Paesani F, Cohen SM (2019) Halogen bonding in UiO-66 frameworks promotes superior chemical warfare agent simulant degradation. Chem Commun 55:3481–3484
28. Kandiah M, Nilsen MH, Usseglio S, Jakobsen S, Olsbye U, Tilset M, Larabi C, Quadrelli EA, Bonino F, Lillerud KP (2010) Synthesis and stability of tagged UiO-66 Zr-MOFs. Chem Mater 22:6632–6640
29. Katz MJ, Klet RC, Moon S-Y, Mondloch JE, Hupp JT, Farha OK (2015a) One step backward is two steps forward: enhancing the hydrolysis rate of UiO-66 by decreasing [OH^-]. ACS Catal 5:4637–4642
30. Katz MJ, Mondloch JE, Totten RK, Park JK, Nguyen ST, Farha OK, Hupp JT (2014) Simple and compelling biomimetic metal-organic framework catalyst for the degradation of nerve agent simulants. Angew Chemie Int Ed 53:497–501
31. Katz MJ, Moon S-Y, Mondloch JE, Beyzavi MH, Stephenson CJ, Hupp JT, Farha OK (2015b) Exploiting parameter space in MOFs: a 20-fold enhancement of phosphate-ester hydrolysis with UiO-66-NH_2. Chem Sci 6:2286–2291
32. Kim IS, Li Z, Zheng J, Platero-Prats AE, Mavrandonakis A, Pellizzeri S, Ferrandon M, Vjunov A, Gallington LC, Webber TE, Vermeulen NA, Penn RL, Getman RB, Cramer CJ, Chapman KW, Camaioni DM, Fulton JL, Lercher JA, Farha OK, Hupp JT, Martinson ABF (2018) Sinter-resistant platinum catalyst supported by metal-organic framework. Ange Chem Int Ed 57:909–913
33. Kirlikovali KO, Chen Z, Islamoglu T, Hupp JT, Farha OK (2020) Zirconium-based metal-organic frameworks for the catalytic hydrolysis of organophosphorus nerve agents. ACS Appl Mater Interfaces 12:14702–14720
34. de Koning MC, van Grol M, Breijaert T (2017) Degradation of paraoxon and the chemical warfare agents VX, Tabun, and Soman by the metal-organic frameworks UiO-66-NH_2, MOF-808, NU-1000, and PCN-777. Inorg Chem 56:11804–11809
35. Kutzscher C, Nickerl G, Senkovska I, Bon V, Kaskel S (2016) Proline functionalized UiO-67 and UiO-68 type metal−organic frameworks showing reversed diastereoselectivity in aldol addition reactions. Chem Mater 28:2573−2580
36. Li W-S, Lum KT, Chen-Goodspeed M, Sogorb MA, Raushel FM (2001) Stereoselective detoxification of chiral sarin and soman analogues by phosphotriesterase. Bioorg Med Chem 9:2083–2091
37. Li Z, Peters AW, Bernales V, Ortuño MA, Schweitzer NM, DeStefano MR, Gallington LC, Platero-Prats AE, Chapman KW, Cramer CJ, Gagliardi L, Hupp JT, Farha OK (2017) Metal-organic framework supported cobalt catalysts for the oxidative dehydrogenation of propane at low temperature. ACS Cent Sci 3:31–38
38. Liu Y, Howarth AJ, Vermeulen NA, Moon SY, Hupp JT, Farha OK (2017) Catalytic degradation of chemical warfare agents and their simulants by metal-organic frameworks. Coord Chem Rev 346:101–111

39. Liu Y, Moon S-Y, Hupp JT, Farha OK (2015) Dual-function metal-organic framework as a versatile catalyst for detoxifying chemical warfare agent simulants. ACS Nano 9:12358–12364
40. Mason JA, Veenstra M, Long JR (2014) Evaluating metal-organic frameworks for natural gas storage. Chem Sci 5:32–51
41. Matito-Martos I, Moghadam PZ, Li A, Colombo V, Navarro JAR, Calero S, Fairen-Jimenez D (2018) Discovery of an optimal porous crystalline material for the capture of chemical warfare agents. Chem Mater 30:4571–4579
42. McHardy SF, Wang HYL, McCowen SV, Valdez MC (2017) Recent advances in acetylcholinesterase Inhibitors and reactivators: an update on the patent literature (2012–2015). Expert Opin Ther Pat 27:455–476
43. Mendonca ML, Ray D, Cramer CJ, Snurr RQ (2020) Exploring the effects of node topology, connectivity, and metal identity on the binding of nerve agents and their hydrolysis products in metal-organic frameworks. ACS Appl Mater Interfaces 12:35657–35675
44. Mendonca ML, Snurr RQ (2020) Computational screening of metal-organic framework-supported single-atom transition-metal catalysts for the gas-phase hydrolysis of nerve agents. ACS Catal 10:1310–1323
45. Mercey G, Verdelet T, Renou J, Kliachyna M, Baati R, Nachon F, Jean L, Renard PY (2012) Reactivators of acetylcholinesterase inhibited by organophosphorus nerve agents. Acc Chem Res 45:756–766
46. Moghadam PZ, Fairen-Jimenez D, Snurr RQ (2016) Efficient identification of hydrophobic MOFs: application in the capture of toxic industrial chemicals. J Mater Chem A 4:529–536
47. Momeni MR, Cramer CJ (2018) Dual role of water in heterogeneous catalytic hydrolysis of sarin by zirconium-based metal-organic frameworks. ACS Appl Mater Interfaces 10:18435–18439
48. Mondloch JE, Bury W, Fairen-Jimenez D, Kwon S, DeMarco EJ, Weston MH, Sarjeant AA, Nguyen ST, Stair PC, Snurr RQ, Farha OK, Hupp JT (2013) Vapor-phase metalation by atomic layer deposition in a metal-organic framework. J Am Chem Soc 135:10294–10297
49. Mondloch JE, Katz MJ, Ghosh WC, Liao P, Bury W, Wagner G, Hall MG, DeCoste JB, Peterson GW, Snurr RQ, Cramer CJ, Hupp JT, Farha OK (2015) Destruction of chemical warfare agents using metal-organic frameworks. Nat Mater 14:512−516
50. Montoro C, Linares F, Quartapelle Procopio E, Senkovska I, Kaskel S, Galli S, Masciocchi N, Barea E, Navarro JAR (2011) Capture of nerve agents and mustard gas analogues by hydrophobic robust MOF-5 type metal-organic frameworks. J Am Chem Soc 133:11888–11891
51. Moon S-Y, Liu Y, Hupp JT, Farha OK (2015) Instantaneous hydrolysis of nerve-agent simulants with a six-connected zirconium-based metal-organic framework. Angew Chemie Int Ed 54:6795–6799
52. Moon S-Y, Wagner GW, Mondloch JE, Peterson GW, DeCoste JB, Hupp JT, Farha OK (2015) Effective, facile, and selective hydrolysis of the chemical warfare agent VX using Zr6-based metal-organic frameworks. Inorg Chem 54:10829–10833
53. Navarro JAR (2018) Impact of defects on pyrazolate based metal organic frameworks. Isr J Chem 58:112–1118
54. Noh H, Cui Y, Peters AW, Pahls DR, Ortuño MA, Vermeulen NA, Cramer CJ, Gagliardi L, Hupp JT, Farha OK (2016) An exceptionally stable metal-organic framework supported molybdenum(VI) oxide catalyst for cyclohexene epoxidation. J Am Chem Soc 138:14720–14726
55. Noh H, Kung C-W, Otake K, Peters AW, Li Z, Liao Y, Gong X, Farha OK, Hupp JT (2018) Redox-mediator-assisted electrocatalytic hydrogen evolution from water by a molybdenum sulfide-functionalized metal-organic framework. ACS Catal 8:9848–9858
56. Otake K, Cui Y, Buru CT, Li Z, Hupp JT, Farha OK (2018) Single-atom-based vanadium oxide catalysts supported on metal-organic frameworks: selective alcohol oxidation and structure-activity relationship. J Am Chem Soc 140:8652–8656
57. Palomba JM, Credille CV, Kalaj M, DeCoste JB, Peterson GW, Tovar TM, Cohen SM (2018) High-throughput screening of solid-state catalysts for nerve agent degradation. Chem Commun 54:5768–5771

58. Park J, Agrawal M, Sava Gallis DF, Harvey JA, Greathouse JA, Sholl DS (2020) Impact of intrinsic framework flexibility for selective adsorption of sarin in non-aqueous solvents using metal-organic frameworks. Phys Chem Chem Phys 22:6441–6448
59. Peng Y, Krungleviciute V, Eryazici I, Hupp JT, Farha OK, Yildirim T (2013) Methane storage in metal-organic frameworks: current records, surprise findings, and challenges. J Am Chem Soc 135:11887–11894
60. Perdew JP, Burke K, Ernzerhof M (1996) Generalized gradient approximation made simple. Phys Rev Lett 77:3865–3868
61. Peterson GW, Moon S-Y, Wagner GW, Hall MG, DeCoste JB, Hupp JT, Farha OK (2015) Tailoring the pore size and functionality of UiO-type metal-organic frameworks for optimal nerve agent destruction. Inorg Chem 54:9684–9686
62. Peterson GW, Destefano MR, Garibay SJ, Ploskonka A, McEntee M, Hall M, Karwacki CJ, Hupp JT, Farha OK (2017) Optimizing toxic chemical removal through defect-induced UiO-66-NH_2 metal–organic framework. Chem Eur J 23:15913–15916
63. Platero-Prats AE, Mavrandonakis A, Gallington LC, Liu Y, Hupp JT, Farha OK, Cramer CJ, Chapman KW (2016) Structural transitions of the metal-oxide nodes within metal-organic frameworks: on the local structures of NU-1000 and UiO-66. J Am Chem Soc 138:4178–4185
64. Plonka AM, Wang Q, Gordon WO, Balboa A, Troya D, Guo W, Sharp CH, Senanayake SD, Morris JR, Hill CL, Frenkel AI (2017) In situ probes of capture and decomposition of chemical warfare agent simulants by Zr-based metal organic frameworks. J Am Chem Soc 139:599–602
65. Roy A, Srivastava AK, Singh B, Mahato TH, Shah D, Halve AK (2012) Degradation of sulfur mustard and 2-chloroethyl ethyl sulfide on Cu–BTC metal organic framework. Microporous Mesoporous Mater 162:207–212
66. Roy A, Srivastava AK, Singh B, Shah D, Mahato TH, Srivastava A (2012) Kinetics of degradation of sulfur mustard and sarin simulants on HKUST-1 metal organic framework. Dalt Trans 41:12346–12348
67. Ruffley JP, Goodenough I, Luo TY, Richard M, Borguet E, Rosi NL, Johnson JK (2019) Design, synthesis, and characterization of metal-organic frameworks for enhanced sorption of chemical warfare agent simulants. J Phys Chem C 123:19748–19758
68. Schwenk M (2018) Chemical warfare agents. Classes and targets. Toxicol Lett 293:253–263
69. Šečkute J, Menke JL, Emnett RJ, Patterson EV, Cramer CJ (2005) Ab initio molecular orbital and density functional studies on the solvolysis of sarin and O, S-dimethyl methylphosphonothiolate, a VX-like compound. J Org Chem 70:8649–8660
70. Soares CV, Leitão AA, Maurin G (2019) Computational evaluation of the chemical warfare agents capture performances of robust MOFs. Microporous Mesoporous Mater 280:97–104
71. Sturluson A, Huynh MT, Kaija AR, Laird C, Yoon S, Hou F, Feng Z, Wilmer CE, Colón YJ, Chung YG, Siderius DW, Simon CM (2019) The role of molecular modelling and simulation in the discovery and deployment of metal-organic frameworks for gas storage and separation. Mol Simul 45:1082–1121
72. Szinicz L (2005) History of chemical and biological warfare agents. Toxicology 214:167–181
73. Taddei M (2017) When defects turn into virtues: the curious case of zirconium-based metal-organic frameworks. Coord Chem Rev 343:1–24
74. Trapp R (2017) The use of chemical weapons in Syria: Implications and consequences. In: Friedrich B, Hoffmann D, Renn J, Schmaltz F, Wolf M (eds) One hundred years of chemical warfare: research, deployment, consequences. Springer, Cham, pp 363–375
75. Trickett CA, Gagnon KJ, Lee S, Gándara F, Bürgi H-B, Yaghi OM (2015) Definitive molecular level characterization of defects in UiO-66 crystals. Angew Chemie Int Ed 54:11162–11167
76. Troya D (2016) Reaction mechanism of nerve-agent decomposition with Zr-based metal organic frameworks. J Phys Chem C 120:29312–29323
77. Vermoortele F, Vandichel M, Van de Voorde B, Ameloot R, Waroquier M, Van Speybroeck V, De Vos DE (2012) Electronic effects of linker substitution on lewis acid catalysis with metal-organic frameworks. Angew Chemie Int Ed 51:4887–4890
78. Wang S, Bromberg L, Schreuder-Gibson H, Hatton TA (2013) Organophophorous ester degradation by chromium(III) terephthalate metal-organic framework (MIL-101) chelated

to *N*,*N*-dimethylaminopyridine and related aminopyridines. ACS Appl Mater Interfaces 5:1269–1278
79. Wang G, Sharp C, Plonka AM, Wang Q, Frenkel AI, Guo W, Hill C, Smith C, Kollar J, Troya D, Morris JR (2017) Mechanism and kinetics for reaction of the chemical warfare agent simulant, DMMP(g), with zirconium(IV) MOFs: an ultrahigh-vacuum and DFT study. J Phys Chem C 121:11261–11272
80. Wang X, Zhang X, Li P, Otake KI, Cui Y, Lyu J, Krzyaniak MD, Zhang Y, Li Z, Liu J, Buru CT, Islamoglu T, Wasielewski MR, Li Z, Farha OK (2020) Vanadium catalyst on isostructural transition metal, lanthanide, and actinide based metal-organic frameworks for alcohol oxidation. J Am Chem Soc 141:8306–8314
81. Widom B (1963) Some topics in the theory of fluids. J Chem Phys 39:2808
82. Wong K-Y, Gao J (2007) The reaction mechanism of paraoxon hydrolysis by phosphotriesterase from combined QM/MM simulations. Biochemistry 46:13352–13369
83. Wymore T, Field MJ, Langan P, Smith JC, Parks JM (2014) Hydrolysis of DFP and the nerve agent (S)-sarin by DFPase proceeds along two different reaction pathways: implications for engineering bioscavengers. J Phys Chem B 118:4479–4489
84. Yaghi OM, Li H (1995) Hydrothermal synthesis of a metal-organic framework containing large rectangular channels. J Am Chem Soc 117:10401–10402
85. Yang Y-C (1999) Chemical detoxification of nerve agent VX. Acc Chem Res 32:109–115
86. Yang YC, Baker JA, Ward JR (1992) Decontamination of chemical warfare agents. Chem Rev 92:1729–1743
87. Yang S, Doetschman D, Schulte J, Sambur J, Kanyi CW, Fox JD, Kowenje C, Jones BR, Sherma ND (2006) Sodium X-type faujasite zeolite decomposition of dimethyl methylphosphonate (DMMP) to methylphosphonate: nucleophilic zeolite reactions I. Microporous Mesoporous Mater 92:56–60
88. Zou R, Zhong R, Han S, Xu H, Burrell AK, Henson N, Cape JL, Hickmott DD, Timofeeva TV, Larson TE, Zhao Y (2010) A porous metal-organic replica of α-PbO_2 for capture of nerve agent surrogate. J Am Chem Soc 132:17996–17999

Metal–Organic Frameworks (MOFs) as Versatile Detoxifiers for Chemical Warfare Agents (CWAs)

Laishram Saya and Sunita Hooda

Contents

1 Introduction 454
2 Classification of CWAs and Their Characteristic Properties 456
 2.1 Nerve Agents 457
 2.2 Vesicants 458
 2.3 Blood Agents 459
 2.4 Choking Agents 460
 2.5 Riot-Control Agents 460
 2.6 Psychomimetic Agents 461
3 Toxic Effects of CWAs 463
 3.1 Nerve Agents 463
 3.2 Vesicants 463
 3.3 Blood Agents 464
 3.4 Riot-Control Agents 464
 3.5 Psychomimetic Agents 464
4 Model CWA Simulants and Their Significance 465
5 Strategic Routes for Fabrication of MOFs for Effective Sequestration of CWAs 466
 5.1 Tuning the Pore Properties and Surface Hydrophobicity 467
 5.2 Partial Oxidation Approach 468
 5.3 Post-Synthetic Functionalization with Amine Groups 468
 5.4 Inducing Lewis Acidity Through Missing-Linker Defect 469
6 MOFs as Catalysts for Detoxification of CWAs 469
 6.1 Degradation of Simulants of Real CWAs 469
 6.2 Degradation of Real Nerve Agents 476
7 MOFs for Adsorption of CWAs 479
8 MOF-Modified Fabrics for CWA Adsorption 481
9 Conclusions and Future Aspects 482
Abbreviations 483
References 484

L. Saya
Department of Chemistry, Sri Venkateswara College (University of Delhi), Dhaula Kuan, New Delhi 110021, India

S. Hooda (✉)
Department of Chemistry, Acharya Narendra Dev College (University of Delhi), Govindpuri, Kalkaji, New Delhi 110019, India
e-mail: sunitahooda@andc.du.ac.in

S. Gulati (ed.), *Metal-Organic Frameworks (MOFs) as Catalysts*,
https://doi.org/10.1007/978-981-16-7959-9_18

Abstract Chemical warfare agents (CWAs) are considered as one of the most fatal weapons potentially strong to cause extreme toxicity and disastrous effects to a large population. They were used as weapons for the first time in 1915 during World War I (WWI) when Ypres, a Belgian city, was attacked by the German military. Sulfur mustard, a dreadful chemical warfare agent, which was used in the subsequent battles became the major cause of chemical casualties in WWI. These chemicals imposed harsh after-effects even years after they were deployed. Nerve agents and vesicants are particularly known to be extremely harmful, among the various classes of CWAs; even short-term exposure to these chemicals can lead to severe after-effects. Above all, CWAs also release various volatile organic compounds (VOCs), which comprise an important group of air pollutants, which can potentially cause serious health effects to mankind including mutagenesis and carcinogenesis. In view of these consequences, capture and subsequent degradation of these agents to less or completely non-toxic by-products are of paramount importance. Being highly toxic, degradation of hazardous CWAs through catalytic reactions such as hydrolysis, methanolysis, and oxidation has been proved to be one of the best methods that can eventually transform them into less-toxic products. Research communities throughout the globe have been making relentless attempts on developing novel catalytic materials in this field. Metal-organic frameworks (MOFs), being specifically designed making use of organic linkers and inorganic nodes, offer scope for fabrication of a versatile range of materials with great diversity in structural and chemical properties, characterized by their high stability, crystalline, and ordered nature with significantly large surface areas, high porosity, and free volume. The presence of freely available metal sites and/or numerous functional moieties on the surface of the MOFs allows adsorption or capture of certain toxic CWAs with high selectivity and efficiency via various interactions which may be either H-bonds, ionic or Coulombic interactions, coordination bonds, Π-Π* interactions, etc. or a combination of these. Moreover, further functionalization with coordinating or conjugating agents also imparts them good catalytic properties. The pore properties along with the specificity of the functional groups in the MOFs together ascribe to the subsequent catalytic degradation of highly toxic CWAs and their simulants.

Keywords Nerve agents · Vesicants · Simulants · Degradation · Half-life · Adsorption

1 Introduction

Since prehistoric times, the use of dangerous materials from chemical and biological sources has been an indispensable part of evolution, and their utility as weapons for homicide was quite common [1]. Even though the use of these hazardous substances in warfare dates back to ancient Greek and Roman times, their after-effects had been relatively less known due to the lack of adequate knowledge. Rapid advances in various fields of chemistry and the establishment of chemical industries, during the nineteenth century, were often accompanied by several accidents with toxic and hazardous chemicals [2]. The severity of these accidents gave insight into the harsh toxicological effects and the possible manufacture of these chemicals on a large scale for utilization as weapons. These chemicals, which became popularly known as the chemical warfare agents (CWAs), were weaponized for the first time in 1915 during World War I (WWI) when Ypres, a Belgian city, was attacked by the German military releasing 168 metric tons of chlorine gas which killed an estimated 5000 troops [3]. The same battlefields witnessed the development of sulfur mustard, a dreadful chemical warfare agent, two years later, that became the major cause of chemical casualties in WWI. Since then, the use of CWAs in warfare has become a trend with at least 12 conflicts including both the first and second Persian Gulf Wars witnessing the use of CWAs on a large scale. The most dangerous part is, even years after these chemical weapons were deployed, civilians remain inadvertently exposed to these chemicals. For instance, a great number of fishermen have burned accidentally during WWI when more than 50,000 metric tons of mustard shells were released.

CWAs are categorized into various types based on volatility, chemical structure, or most commonly on the nature of toxic effects produced. The mode of classification, properties, and toxic effects produced by each of the CWAs is described in the following sections. Of all the various types, nerve agents and vesicants are the most common. Soman and Sarin are some examples of nerve agents, while vesicants are best represented by sulfur mustard groups. Even minimal exposure to these chemicals can lead to severe effects including blisters on the skin, suffocation or choking, and irritation in the eyes and respiratory tract [4]. Above all, the use of CWAs also releases another major group of air pollutants commonly referred to as the volatile organic compounds (VOCs), which comprise chemicals with a vapor pressure above 10 Pa at 293 K, and can be the potential cause of photochemical smog and result in carcinogenesis, teratogenesis, and mutagenesis [5]. In view of these toxic effects, it is imperative to realize that detection, capture, and subsequent degradation of these agents to less or completely non-toxic by-products is of paramount importance which has indeed attracted a great deal of research attention. However, treatment of CWAs involves taking care of major safety issues and their extreme toxicity which makes it important to preferably devise techniques that are associated with toxicity-free products. Degradation through catalytic reactions involving hydrolysis, methanolysis, and oxidation can eventually transform CWAs into simpler less-toxic products; hence, they are considered as one of the best methods.

Research communities throughout the globe have been making continuous attempts on developing novel catalytic materials in this field [6, 7]. A wide range of materials has been developed for the catalytic detoxification or degradation of CWAs. Nevertheless, each of them is associated with one or the other restrictions like poor storage stability, corrosive properties, slow degradation rates, loss of catalytic activity due to poisoning, and low surface area. [8–10]. In this regard, metal–organic frameworks (MOFs) have been proved to be potential candidates for the annihilation of a wide range of CWAs through adsorption or catalytic degradation. MOFs are highly porous and crystalline two- or three-dimensional complex structures with plenty of metal centers, and that are catalytically active [11, 12]. These frameworks are normally fabricated using organic ligands, with desired functional groups, amalgamated with secondary building units (SBUs), preferably inorganic metal nodes [13, 14]. Among various MOFs, Zr-based MOFs have been most extensively investigated. One of the main reasons is their extreme structural stability toward a versatile range of guest molecules, which makes it a preferred candidate as a catalyst for the degradation of CWAs as well as for capturing highly persistent and volatile CWAs. Another important structural aspect is the close similarity that exists between the Zr–OH–Zr bonds of Zr_6 clusters and the Zn–OH–Zn active sites in the phosphotriesterase enzyme. These act as efficient catalytic sites capable of destructing organophosphates (OP), which are the major functional components of all CWAs as well as model simulants [15–17]. Zr-based MOFs are particularly known for their high efficacy toward catalytic degradation of CWAs being characterized by their large surface area and outstanding physicochemical properties such as mechanical, thermal, and hydrolytic stabilities. UiO-66-NH_2, an amino-functionalized form of UiO-66, one of the most common MOFs, behaves as a biomimetic catalyst with excellent catalytic activity toward numerous CWAs [18]. The proximal amino groups play a key role in modulating the catalytic environment around the active sites of the Zr_6 node [19]. In recent times, smart textiles, with good nerve agent detoxifying capacity, have also been developed, wherein MOFs are decorated or incorporated onto fabrics or textiles via electrospinning technique [20, 21]. This is one of the simple practical applications of powdery MOFs for detoxifying CWAs through degradation. In addition, physical spraying [22], self-assembly [23], and in situ growths [24, 25] are some other applications of these MOFs for the same purpose. In this line, a new class of materials based on carbon networks have emerged which are referred to as porous coordination networks (PCN) possessing characteristic sheet-like exploited structure that imparts a large surface area favorable for catalytic application. Some of the MOFs which are commonly investigated for the degradation of real and simulant CWAs are depicted in Fig. 1.

2 Classification of CWAs and Their Characteristic Properties

All CWAs are characterized by specific chemical, physicochemical, and physiological properties [26]. They can be classified into various categories on the basis of many parameters. Based on volatility, CWAs can be classified as persistent (less volatile) agents like sulfur mustard and VX and non-persistent (more volatile) agents such as hydrogen cyanide and chlorine, phosgene. CWAs can also be either organophosphorus (OP), organofluorine, organosulfur compounds, or arsenicals, depending on their chemical structure. However, classifying them on the basis of the type of harmful effects caused to humans is the most preferable. The various types of CWAs along with their characteristic properties on this basis are described below:

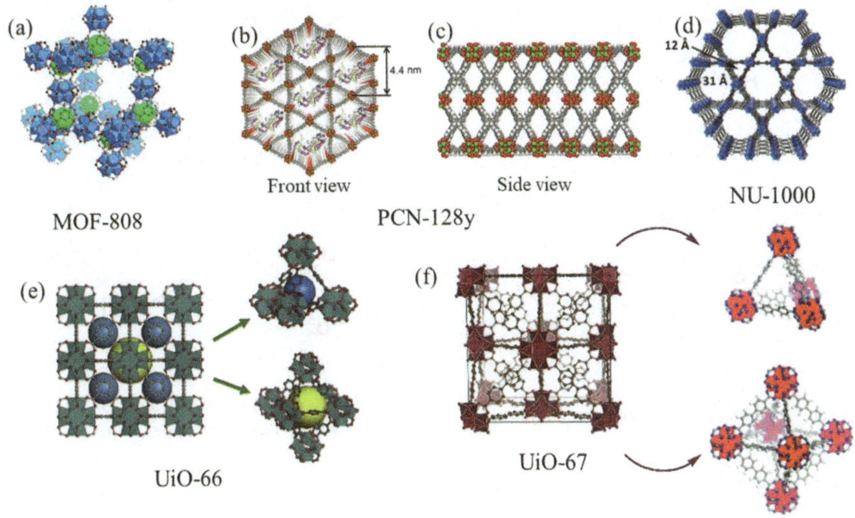

Fig. 1 Structures of common MOFs: **a** MOF-808, **b** front view and **c** side view of PCN-128y, **d** NU-1000, **e** UiO-66, **f** UiO-67. Reprinted with permission from Refs. [70, 86, 102–104]

2.1 Nerve Agents

Nerve agents are purely synthetic chemicals belonging to the class of OP compounds. They are so-called due to their effect on the nervous system. Tabun (GA) is the first-ever nerve agent developed, in the 1930s, by Gerhard Schrader, a German chemist after which a series of nerve agents, popularly called the G-agents were developed [27]. Sarin (GB) and Soman (GD) are some of the common G-agents. In an attempt to increase their potential and environmental persistence, more stable versions of the G-agents, called the V-agents, were developed in the 1960s. The best example is VX, a sulfur-containing OP, that is known for higher potency and structural stability with less volatility, and less solubility in water than Sarin. VX bears high environmental persistence even up to several weeks after release and acts directly through skin contact.

All the nerve agents appear as colorless liquids in the pure state. The G-agents have a fruity smell, while the V-agents have a characteristic amine-like smell. Sarin has an extremely high solubility in water; Soman is sparingly soluble, while tabun and VX show intermediate behavior in this respect. The reactions of all G-agents with alkali end up with the generation of non-toxic phosphonic acid as these nerve agents undergo reactions through the breakage of the P-X bond. Chemical structures of common nerve agents are shown in Fig. 2.

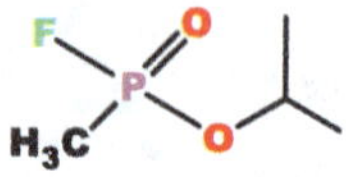

Isopropyl methylphosphonofluoridate
Sarin (GB)

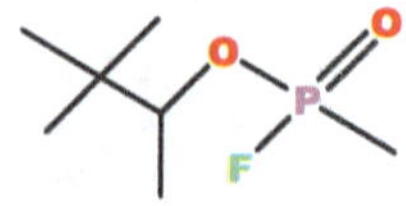

O-1,1,2-trimethylpropylmethylphosphonofluoridate
Soman (GD)

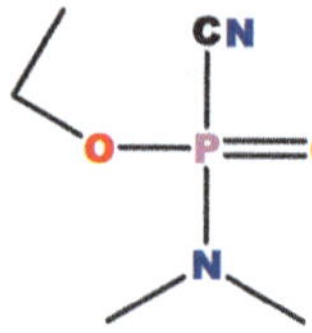

O-Ethyl N,N dimethyl-phosphoramidocyanate
Tabun (GA)

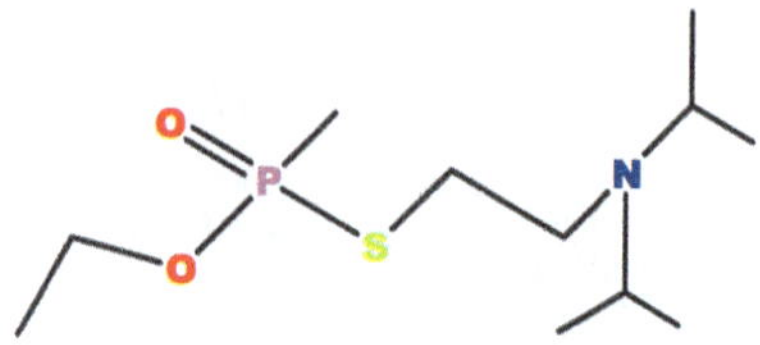

S-{2-[Di(propan-2-yl)amino]ethyl} O-ethyl methylphosphonothioate (VX)

Fig. 2 Chemical structures of common nerve agents

2.2 *Vesicants*

Vesicants, commonly known as blistering agents, comprise a class of extremely dangerous compounds that can cause burn-like skin injuries [28–31]. On inhalation, vesicants initially attack the upper respiratory tract, gradually travel to the lungs, and cause pulmonary inflammation. They can also cause severe eye injuries. Mustards and arsenicals are the two main classes of vesicants. Sulfur mustard, nitrogen mustards, and lewisites are some important members of vesicants, among which sulfur mustard is the most common and is, therefore, sometimes referred to as the king of all CWAs.

Sulfur mustard and nitrogen mustards are both colorless and odorless liquids in their pure forms. However, characteristic mustard or garlic-like smell develops when sulfur mustard becomes impure. It is less volatile, highly soluble in organic solvents but sparingly soluble in water, with high storage stability. Nitrogen mustards are even less volatile, less soluble, and show better resistance to oxidizing agents compared to sulfur mustard. On the other hand, pure lewisites bear a metallic odor with solubility almost similar to that of sulfur mustard in water and relatively less stable. Storage stability is one of the most basic requirements for a CWA which is lacking in nitrogen mustards and lewisites which make them less popular as CWAs compared to sulfur mustard. Chemical structures of the common vesicants are shown in Fig. 3.

Sulfur mustard got its prominence from its large-scale use in WWI. However, nitrogen mustards, which were first developed in the 1930s, were not synthesized further in large amounts for warfare purposes due to the reason mentioned above. Lewisites are non-flammable, and their toxicity is almost similar to that of sulfur mustard due to which they were synthesized in 1918 for warfare purposes. Mechlorethamine (HN2, Mustargen) has found more peaceful applications as a cancer chemotherapeutic agent than as warfare agents; therefore, it has been used as the standard compound for this purpose for several years.

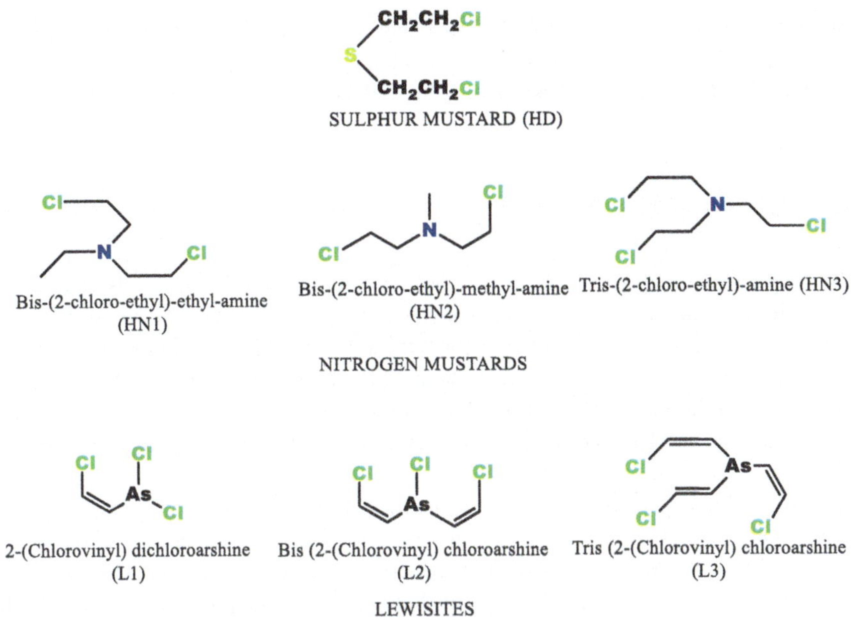

Fig. 3 Chemical structures of common vesicants

2.3 *Blood Agents*

Cyanogenic agents, popularly known as the blood agents [32–35] contain a cyanide group of chemicals as the main component that can cause an immediate effect on the normal metabolism by stopping the body tissues to utilize oxygen. Hydrogen cyanide (HCN) and cyanogen chloride (CNCl) are the most popular members that belong to this class. These agents attack the inner walls of the mitochondria by interrupting the electron transport chain; thus, they interfere with the production of blood components. Hence, the name "blood agent" is given. However, cyanogenic agents are not known to cause direct effects on the blood. Since these agents interrupt the functioning of certain specific enzymes, they are sometimes referred to as systemic agents also. Due to their extremely high volatility, blood agents are less useful as CWAs. However, depending on their efficacy, they can be used as warfare weapons, if used in enclosed space.

Even much before the potential of hydrogen cyanide (HCN), as a warfare agent, was realized during WWI; it was discovered long back in 1872, by a Swedish chemist and was used on a large scale for commercial purposes. It was first utilized in the form of a warfare weapon in the battle of Somme in 1916, by the French. The second most common blood agent, cyanogen chloride was used commercially in many industrial applications during WWI, rather than as a warfare agent. One of the advantages of cyanogen chloride over hydrogen cyanide is that it is a heavier gas and can potentially cause severe effects even with low doses of exposure. It has the greatest penetrating capacity of all other agents and can even penetrate readily into the filter materials of a gas mask. Thus, the introduction of cyanogen chloride was helpful in overcoming many of the disadvantages faced in handling HCN.

2.4 Choking Agents

Choking agents, also known as the pulmonary agents, as the name itself, indicates, affect the respiratory system starting from the nose, going down the throat, and finally attacking the lungs. In critical conditions, these agents cause death due to choking that results from swelling of the respiratory tract membranes followed by filling of the lungs with liquid thereby depriving the lungs of oxygen. These types of fatalities are also known as "dry-land drownings"[36]. Chlorine, perfluoro isobutylene (PFIB), phosgene, diphosgene, nitric oxide, etc. belong to this class of CWAs, out of which, chlorine and phosgene are the most common.

Choking agents were synthesized and used on a large scale during WWI as an effective class of CWA. Apart from being used as a warfare agent, chlorine as well as phosgene finds enormous applications in chemical industrial processes. Even though these compounds were used on a large-scale in the design of anti-human weapons by terrorists [37], control on such an illegal use was difficult.

Chlorine and phosgene are generally heavier than air. Chlorine appears as a yellowish-green gas with a characteristic pungent smell, at room temperature and is liquefiable under moderate pressure. Phosgene was used in such large quantities in WWI that almost 80% of the deaths due to chemicals was considered to be caused by phosgene alone [37]. It can be hardly distinguished by its odor even at toxic levels, and death is caused after a significant delay of up to 24 h following exposure. Diphosgene, which is basically a combination of phosgene and chloroform, is another CWA that is relatively easier to handle than phosgene as it is a liquid at room temperature. Diphosgene has the ability to penetrate through canisters and is more persistent than both chlorine and phosgene. Perfluoro isobutylene (PFIB) is another CWA that is about ten times as toxic as phosgene. It is generated during the synthesis of polytetrafluoroethylene (Teflon) as a by-product on overheating. Similar to phosgene, PFIB has a considerable latent period of symptom development after exposure.

2.5 Riot-Control Agents

Riot-control agents are also referred to as lachrymators, eye irritants, or harassing agents. They are more popularly known as the so-called "tear gases" among the general public. They are used as CWAs due to their tear stimulating capabilities. These agents can cause tearing and blepharospasm or immediate temporary irritation in the eyes along with irritation in the upper respiratory tract [38]. On exposure, these

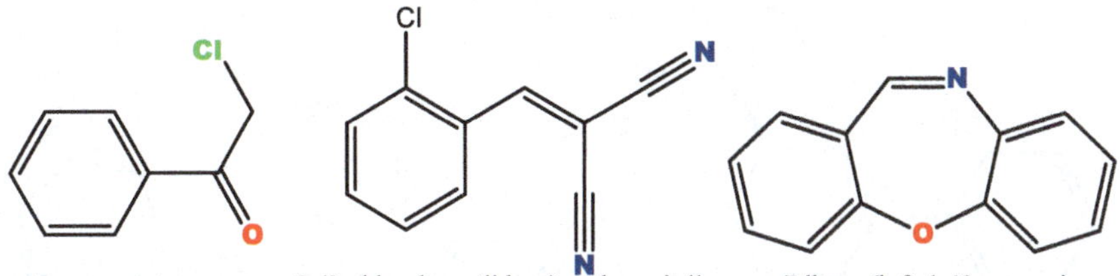

Fig. 4 Chemical structures of common riot-control agents

chemicals can cause immediate secretion of tears, pain in the eyes, and even irritation on the skin to such a degree that causes victims to behave irrationally leading to uncoordinated activities. Based on the nature of toxic effects produced, riot-control agents can be classified into the following types:

1. Lachrymators: These cause lachrymation and irritation primarily in the eyes.
2. Vomiting agents: These induce vomiting.
3. Sternutators: These basically induce sneezing and irritation in the upper respiratory tract.

CN, CS, and CR are the three most significant chemicals among the list of riot-control agents; chemical structures along with their IUPAC names are given in Fig. 4. They exist in the form of either solid or get dispersed as aerosols at room temperature. These chemicals show relatively low solubility in water but significantly high solubility in most of the organic solvents.

2.6 *Psychomimetic Agents*

These chemical agents are a little different from the above-mentioned classes as they are not known to cause any severe effect or damage to the autonomic nervous system or any serious physical disabilities. Rather, they affect the psychological behavior of the victims persistently causing alterations in their thoughts, mood, and perception leading to abnormalities in their behavior [39, 40]. When administered in mild doses (<10 mg), these chemicals can induce symptoms related to abnormal functioning of the central nervous system that includes numbness leading to partial paralysis, hallucinations, and other psychotic disorders.

Psychomimetic agents, extracted from plants, have been used as incapacitating agents since ancient times. They were used for the first time by Solon's soldiers, as a warfare agent, during a war in 600 BC. They threw hellebore roots, which have good incapacitating capability, into the streams supplying water to the enemy troops. The years that followed also witnessed the use of such psychomimetic agents, extracted from a variety of plants, as warfare agents. For instance, belladonna plants were used for creating disorientation in the enemy troops by Hannibal's army in 184 BC. Various

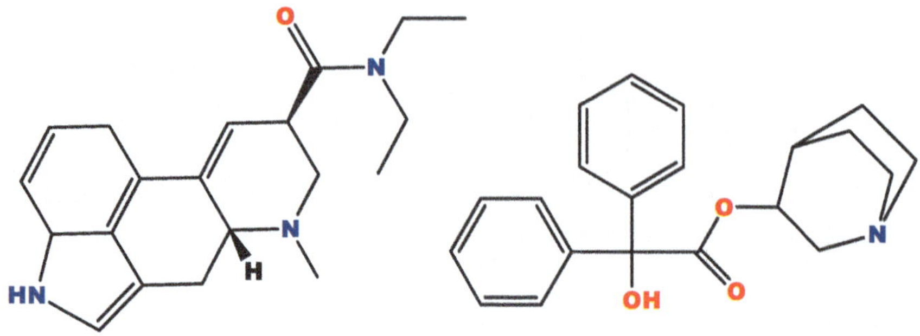

lysergic acid diethylamide
(LSD)

3-quinuclidinyl benzilate
(BZ)

Fig. 5 Chemical structures of LSD and BZ

psychomimetic agents containing the indole group were discovered during World War II (WWII). Lysergic acid diethylamide (LSD) and marijuana derivatives are some of them which are relatively non-lethal incapacitating agents but can induce psychotic effects in victims (Fig. 5). Apart from this, several other drugs of similar type were investigated during the time by the United States Army for military purposes. These include ketamine and phencyclidine which are known as dissociative drugs, other psychedelic drugs like tetrahydrocannabinol (THC), potent opioids like fentanyl, and many glycolate anticholinergics [41]. 3-quinuclidinyl benzilate is one of the most important anticholinergic agents which were synthesized especially to be used in warfare weapons on a large scale in the 1960s during which it even received the NATO code name "BZ" [42]. BZ shows such a high potency which is almost 25-times that of atropine. Serious effects such as acute brain-related syndrome, associated with delirium that can last for 2–3 days, may be caused due to the intake of even less than 1 mg of BZ.

Under the category of psychedelics, LSD has been the most popular, which has attracted much attention and interest since the 1950s. LSD has been proved to induce complete incapacitation with the intake of an oral dose as low as 2.5 μg/kg, causing unpredictable behavior which is mainly due to emotions, intense thoughts, and abnormal perceptions [43]. High dosages of LSD can even result in critical conditions like visual and auditory hallucinations [44]. LSD toxicity is associated with typical symptoms like dilation of the pupils, increase in blood pressure as well as body temperature. Affected individuals are not even able to follow a simple sequence of instructions given to them or pay concentration to a simple task. It was observed from studies that administration of a minimal dose of <200 μg can even lead to complete disorganization of well-trained simulated military units which makes it efficient as warfare agents. LSD can also be used in aerosol form. Aerosol LSD has an estimated ID_{50e} of 6 μg/kg and has 50% incapacitation potential among exposed individuals [45].

3 Toxic Effects of CWAs

3.1 Nerve Agents

These agents directly attack the muscarinic and nicotinic receptors within the central nervous system. They irreversibly inhibit the action of acetylcholinesterase (AChE), an important enzyme responsible for the hydrolysis of acetylcholine (ACh), a neurotransmitter released at the nerve synapses, nerve-muscle, and nerve-gland junctions [46, 47]. This results in the accumulation of ACh eventually causing overexcitation or paralysis. Typical symptoms associated with poisoning like constriction of the pupil, increased secretion of saliva, perspiration, running nose, urination, muscle cramps, cardiac arrhythmias, bronchoconstriction, decrease in heart rate and blood pressure start appearing immediately as the nerve agents enter the system. In most critical conditions, respiratory muscles get paralyzed, inhibiting the basic function of the respiratory center, finally resulting in death. An immediate death most likely occurs in the case of high concentrations of the nerve agent.

3.2 Vesicants

Being lipophilic, mustards can easily penetrate the cellular membranes of skin, most textiles, and rubber, where it gets converted to sulphonium ions, which are highly reactive. It damages the DNA, RNA as well as proteins through irreversible alkylation, and ultimately the cell succumbs to death [48]. Lewisites show effects relatively much faster; they get absorbed onto the skin almost ten times faster where they directly bind to the sulfhydryl groups thereby inactivating them. As a result, immediate pain along with irritation is felt in the affected organ subsequently followed by more severe systematic symptoms.

On the other hand, mustards are normally characterized by a specific latent period in the development of symptoms post-exposure [49]. The mode of exposure, environmental conditions, and amount of exposure decides the severity of the symptoms as well as the length of the latent period. Mustard, be in the gaseous or aerosol form, directly attacks the skin, eyes, and the respiratory tract, followed by chemical damage within 1–2 min of exposure. However, pain is felt only after 4–6 h. Mild exposure causes erythema and swelling in the nose and larynx, even in the trachea which may later lead to respiratory obstruction. In critical cases, even a mere bacterial infection in the lungs may get developed into bronchopneumonia which is the real reason behind death following mustard exposure. In severe cases, it may even lead to leucopenia owing to bone marrow damage. Nausea, vomiting, diarrhea, and pain are common symptoms of the consumption of foods or water containing liquid mustard. There is even danger of vomiting and cardiac disorders only due to exposure to the skin. Coagulation necrosis and sun burn-like erythema to vesicles are the major signs of skin injuries that later merge into blisters. The toxic effect on the eyes includes conjunctivitis and may be as extreme as corneal opacification, ulceration, and rupture.

3.3 Blood Agents

Cyanide, due to its high affinity for metal ions, readily binds with Fe^{3+} ions of the cytochrome oxidase enzyme, on entering the biological system, depriving the body tissues of oxygen, the most vital requirement. This eventually leads to death due to respiratory failure. Mild dose exposure to HCN causes symptoms like headache, giddiness, weakness, confusion, accompanied by nausea and vomiting sometimes. Whereas, high levels of exposure cause serious symptoms such as difficulty and pain in breathing, uncoordinated body movement, respiratory failure, cardiac irregularities, and even coma leading to death [34, 35].

Phosgene is again a highly reactive chemical. On entering the body system, it can combine with numerous functional groups such as –SH, $-NH_2$, and –OH groups present in many of the biological macromolecules, including enzymes, which are vital for the normal functioning of the body. It causes acylation of various tissue elements of the lungs that increases the permeability of the alveolar membrane ultimately resulting in poisoning. This leads to pulmonary edema eventually causing death. Low-dose inhalation causes breathing difficulty, bradycardia, and hypotension, increased salivation, nausea, micturition, etc. Exposure to higher concentrations affects the vital lung functions that cause pulmonary edema, eventually followed by death [51–53].

3.4 Riot-Control Agents

These agents first attack the eyes causing pain along with conjunctivitis, lachrymation, and blepharospasm, with symptoms being felt within 10–30 s [38]. A direct chemical action on sensory receptors in skin and mucosa is considered as the possible mechanism for the quick action of these agents. It probably involves an enzymatic process dependent on nicotinamide adenine dinucleotide hydrogenase (NADH).

3.5 Psychomimetic Agents

The common symptoms that develop after ingestion of psychomimetic agents include restlessness, dizziness, failure to obey orders, lack of body coordination, vomiting, mouth dryness, increased body temperature, dilated pupils causing blurred vision, stammering, hallucinatory, and psychotic behavior like inappropriate reactions, irrational fear, perceptual distortions, and phobias [43–45]. The sympathetic nervous system, a part of the autonomic nervous system, is particularly stimulated by LSD which is responsible for symptoms like dilated pupils, increased body temperature, and a rise in the blood sugar level. LSD can also block serotonin, a hormone-like substance that plays a key role in the propagation of impulses in various nerves. Such a blockage disturbs the normal biochemistry of psychic functions. LSD is also known to affect neurophysiologic functions related to dopamine activity. Following ingestion or inhalation of psychedelics, clinical effects can start showing up as fast as within 30 min or as long as 20 h.

4 Model CWA Simulants and Their Significance

As discussed in previous sections, real CWAs such as Sarin, Soman, and VX are known for their extreme toxic and hazardous effects even at minimal doses. Hence, handling these chemicals for investigation purposes is associated with major risks and requires special care and permission to carry out in specific laboratories. In order to minimize this risk, most of the research on CWAs is eventually carried out with simulant molecules rather than the real CWAs. Model simulant CWAs are basically functional analog molecules that can effectively mimic the real CWA molecules in their chemical reactions, as a result of similarity in their chemical and physical properties. This makes it possible to carry out adsorption or degradation studies of real CWAs with the help of these simulants with relatively lesser risk or danger of toxicity. Even though a variety of simulants are developed and reported in the literature, identification of the most accurate simulant which can sufficiently mimic a particular CWA is a challenging task. A simulant should desirably show similarity in various physicochemical properties, their behavior in the environment as well as physiological effects caused by the real CWAs [50]. As far as the fate of the real CWAs in the environment is concerned, several other properties such as bioavailability, volatility and ease of hydrolysis also need to be analyzed properly [51]. Some of the common simulants that can mimic reals CWAs are dimethyl 4-nitrophenyl phosphate (DMNP), 2-chloroethyl ethyl sulfide (CEES), *p*-nitrophenyl diphenyl phosphate (PNPDPP), diisopropylfluorophosphate (DIFP), and diethyl chloro phosphonate (DCP), chemical structures of which are shown in Fig. 6 The physiochemical properties of nerve agents such as Soman, sulfur mustard, and VX are similar to those of OP pesticides [52, 53]. Hence, DMNP, a popular pesticide consisting of a phosphate-ester bond, is a suitable simulant that can adequately represent nerve agents particularly, the G-series agents. On the other hand, DIFP and DCP can be suitable representatives for the cholinesterase family that actually plays the role of inhibitors in organophosphate pesticides [54]. Due to major similarities in the above-mentioned aspects with the real CWAs, these model simulants are mainly employed for monitoring the catalytic activity of a material in the hydrolysis, oxidation, or methanolysis of real nerve agents.

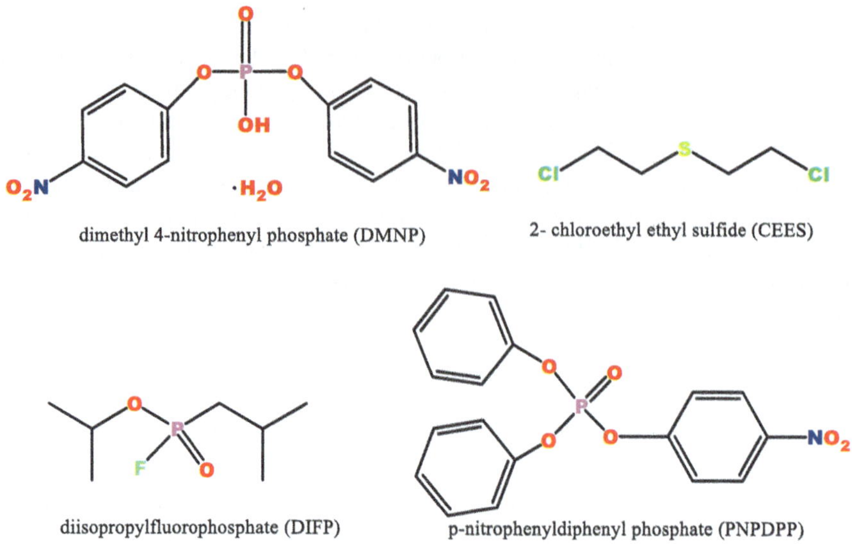

Fig. 6 Chemical structures of common model CWA simulants

5 Strategic Routes for Fabrication of MOFs for Effective Sequestration of CWAs

Elucidating the mechanism of catalytic degradation of CWAs by MOFs requires a clear knowledge of the structural properties of the MOFs as well as the target material, i.e., CWAs. The morphology of the MOF surface and its surface area, pore properties such as pore size, total pore volume, and most importantly the choice of the metal, ligands and the organic linkers are some important structural aspects that are required to be taken care of during the fabrication of an MOF for targeted catalytic degradation of CWAs. The specific requirements related to the structure of the MOF can be achieved through two possible approaches. The first is called the pre-synthesis process; wherein, the MOF has to be fabricated keeping in mind the degradation potential of the specific CWA. The second one comprises various post-synthetic approaches through which the already existing MOF framework is modified with the introduction of specific functional groups that may desirably alter the properties of the MOF. It can be achieved either by the insertion of Bronsted groups such as metal alkoxides into the existing frameworks, utilization of the coordinated or free or defect-free moieties of the parent MOF, or ion exchange of the original ligands with multiple ligands, increasing the acidic character of the original materials [55]. The important strategic routes adopted in the structural modification MOFs for the annihilation of CWAs through various reactions are discussed in the following sections.

5.1 Tuning the Pore Properties and Surface Hydrophobicity

In order to capture the CWAs efficiently and for their subsequent degradation, adequate knowledge is required on the various parameters which control their interactions with the MOF. The recognition, as well as the interactions between the MOF, and the CWA molecules are mainly driven and monitored by the surface morphology parameters that include the pore size, total pore volume, and surface hydrophobicity. Hence, tuning these pore properties of the MOF during the synthesis is critical for achieving the effective degradation of a specific CWA. For instance, morphology analysis of [Zn_4O(3,5-dimethyl-4-carboxy-pyrazolato)$_3$], which is a Zn-MOF, shows the presence of narrow window-like porous structures having diameters of 6 Å and a pore volume of 47% [56]. Crystal structure of the MOF possesses cubic Fm-3 m space group comprising of Zn_4O^{6+} groups linked to the *N*-site of the ligand (*N*,*N′*,O,O′-exo-tetradentate 3,5-dimethyl-4-carboxy-pyrazolate) [56] (Fig. 7a). Such a structure imparts remarkable mechanical, thermal, and chemical stability to the MOF as required for practical applications of selectively capturing harmful CWAs. Moreover, kinetic diameters of the selected CWA simulants, i.e., DES and DIFP are quite close to the pore size of this specific MOF. This makes it distinguishable and more efficient as a catalyst from other compounds with larger molecular diameters.

Another important property that is affected by the pore size is the gas–liquid partitioning properties of different compounds. This can be explained on the basis of Henry's law constants, and the heat of adsorption values associated with different CWAs (adsorbates). Another important parameter is the partition coefficients of H_2O-CWA system which can be obtained through reverse gas chromatography experiments [56]. The observed Henry's law constant values for DES, DIFP, benzene, cyclohexane, and water were 2850, 71.2, 1090, 0.07, and 0.26, respectively. Further comparison of these results with other known MOFs such as molecular sieve active carbon carboxen [$Cu_3(btc)_2$] led to the conclusion that the coordinatively unsaturated metal sites are ineffective in capturing the MOFs. It gives insight into the fact

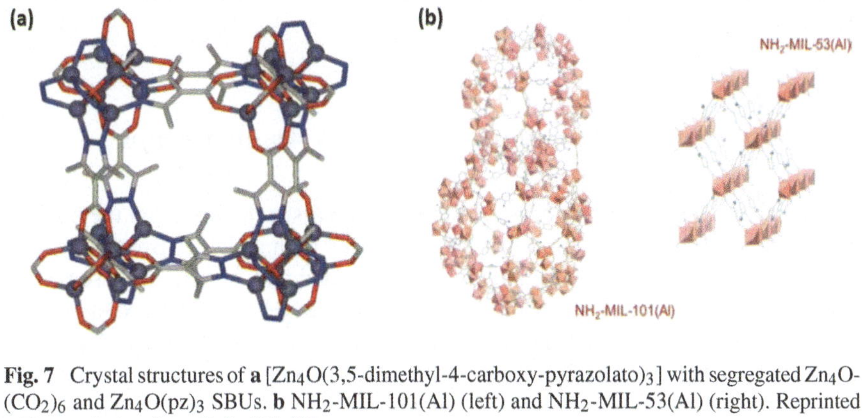

Fig. 7 Crystal structures of **a** [Zn_4O(3,5-dimethyl-4-carboxy-pyrazolato)$_3$] with segregated Zn_4O-$(CO_2)_6$ and $Zn_4O(pz)_3$ SBUs. **b** NH_2-MIL-101(Al) (left) and NH_2-MIL-53(Al) (right). Reprinted with permission from Refs. [56, 59]

that a wise and proper selection of metal and ligand pairs which will give a close to accurate pore size, and preferably, a desired morphology in the resulting MOF is paramount for the effective capturing and degradation of selective CWAs as well as their simulants.

5.2 *Partial Oxidation Approach*

Conventional degradation techniques like methanolysis and hydrolysis are associated with certain technical drawbacks such as complicated experimental procedures, extended reaction times, and above all, causes complete oxidation thereby generating extremely toxic by-products. Some of them are even more toxic than the original CWA. Hence, there have been continuous attempts to explore alternative approaches to avoid or in the least, minimize the release of hazardous secondary products during the degradation process especially in the complete oxidation of CWAs. One of the effective ways is to restrict the degradation up to only partial oxidation. Toxic nerve agents, like HD, can be subjected to partial oxidation via a photosensitization method using singlet oxygen ($^1O_2$), which can be generated, for instance, by using an organic chromophore like pyrene. Pyrene as well as its analogs has a low energy triplet state along with the possibility of an easy and efficient intersystem crossing between the singlet and the triplet state making them good sources of $^1O_2$. The best example is MOF NU-1000 which is usually fabricated using Zr ions and 4,4′,4″,4‴-(pyrene-1,3,6,8-tetrayl) tetrabenzoic acid (H_4TBAPy) [57]. NU-1000 exists in the form of 3D channel-like structures, which are self-assembled, with diameters of 31 and 12 Å. Such a typical morphology favors the diffusion of the CWA molecules as well as the degradation products throughout the MOF [58]. Furthermore, the accumulation of the pyrene molecules on the MOF surface is minimized by the wider channel diameter. This factor again enhances the generation of $^1O_2$ [70, 71]. Unlike complete oxidation, partial oxidation of NU-1000 MOF through photosensitization with UV light-emitting diode (LED) forms bis(2-chloroethyl) sulfoxide (CEESO) which is non-toxic.

5.3 *Post-Synthetic Functionalization with Amine Groups*

The amine groups present in the MOF structure can assist in improving the efficacy of the MOF toward degradation of CWA simulants such as DIFP. Bromberg et al. demonstrated the covalent attachment of amino-functional, aluminum-based MOF particles to reactive adhesives, resulting in super-stable NH_2-MIL-101(Al) and NH_2-MIL-53-Al MOFs, whose crystal structures are shown in Fig. 7b [59]. The deposition of these adhesives can be performed on a variety of surfaces. It is followed by a further covalent modification of the attached MOFs making use of a highly nucleophilic alkylamino pyridine moiety like 4-methylaminopyridine (4-MAP). Attachment of such a nucleophile can facilitate the hydrolytic degradation of the CWA [64]. Moreover, 4-MAP has properties similar to common catalytic nucleophiles like dimethyl aminopyridine (DMAP) [60, 61]. It is interesting to note that in the later step of conjugation with the additional -NH_2 group, 4-MAP selectively modifies only the surface -NH_2 groups and those within the porous structure. However, the rest of the -NH_2 groups are not affected and can interact freely with the isocyanate layers present on the adhesives [59]. These amine-functionalized MOFs apparently cleave the P-F bond through a nucleophilic attack on the phosphorous ions causing efficient degradation of DIFP molecules. It is also suggested that the primary amino groups present in the MOF can further be incorporated with other functionally active moieties, including OP degrading enzymes, reactive polymers, polyoxometalates, or organometallic catalysts which can give rise to efficient detoxifying materials for various types of CWAs.

5.4 Inducing Lewis Acidity Through Missing-Linker Defect

The flexibility of tuning the Lewis acidity of MOFs makes these materials one of the most advantageous for diverse applications. Zr-based MOFs such as UiO-66 possess Lewis acidic nature at the Zr^{4+} centers which act as favorable sites for the introduction of basic hydroxide residues. A typical post-synthetic approach adopted for the effective capture, and subsequent degradation of CWA simulants is the insertion of acidic and/or basic sites (e.g., metal alkoxide groups) and generation of missing-linker defects which can further enhance the phosphotriesterase activity of UiO-66. For example, the catalytic activity of defect-free UiO-66 as well as missing-linker defect [$Zr_6O_4(AcO)_6$(bdc)] [UiO66@AcO] was compared through the incorporation of Bronsted acidic sites in [$Zr_6O_4(AcO)_4(bdc)_5(SO_4H)_2$] ([UiO-66@$SO_4$H]) [55] and lithium alkoxides in [$Zr_6O_6(bdc)_6(LiOtBu)_{0.3}$] ([UiO-66@LiOtBu]), [$Zr_6O_8(bdc)_4(LiOtBu)_{0.3}$] ([UiO66@AcOLiOtBu]) and [$Zr_6O_6(bdc)_6(LiOEt)_{0.25}$] ([UiO66@LiOEt]) [62, 63]. The missing organic linker gives rise to unsaturated coordination around the Zr_6 nodes and is found to show an induced synergistic effect due to the alkoxide basicity combined with the Lewis acidity of the metal centers that eventually imparts an improved phosphotriesterase activity to UiO-66. This overall enhances the degradation performance of UiO-66 toward diverse nerve agents.

6 MOFs as Catalysts for Detoxification of CWAs

6.1 Degradation of Simulants of Real CWAs

Treatment of CWAs involves taking care of major safety issues and their extreme toxicity. Most of the time, direct investigation of the degradation profiles of toxic

CWAs is extremely dangerous. Therefore, model CWA simulants are generally used as clones for real CWAs for the analysis of the degradation performance of various catalysts such as MOFs. Since these simulants have chemical structures very similar to those of the real CWAs, they can significantly cut down the quantum of efforts in carrying out a number of experiments with real dangerous CWAs, provided the models are sufficiently accurate.

Degradation of DMNP

Among all MOFs, zirconium-based ones (Zr-MOFs), being known for their unique properties such as extended structural stability, large surface area, tunable pore sizes are most extensively used for the degradation of DMNP. The degradation efficiency of various Zr-MOFs certainly depends on the functional groups present in the MOF framework. For instance, the extent of basicity of the MOF influences the overall catalytic performance. Peterson et al. found that among the isoreticular series of Zr-MOFs each consisting of Zr (IV)-carboxylate cluster, e.g., UiO-66, UiO-67, and UiO-66-NH_2; the amine-containing framework, UiO-66-NH_2, can exhibit higher degradation capacity as compared to -OH functional group-containing MOFs (e.g., UiO-66 and UiO-67) [66]. This was evident from the degradation efficiency values of 95, 96, and 98% obtained for UiO-66, UiO-67, and UiO-66-NH_2, respectively, within a duration of 1 h. The amine groups (-NH_2) present in the chemical structure of UiO-66-NH_2 enhance the basicity of the MOF as a catalyst, which in turn increases the catalytic efficacy of the MOF toward DMNP. The pore aperture size of the MOF relative to the kinetic diameter of DMNP also apparently plays an important role in the degradation behavior of the MOF. UiO-67 has a pore dimension of 8–11.5 Å, which is very close to the kinetic diameter (11 4.5 Å) of DMNP. As a result, UiO-67 was found to be more efficient as a catalyst compared to UiO-66 with a pore dimension of 6 Å [64, 65]. The nature of the solvent can also influence the catalytic behavior as it affects the chemical stability of the MOFs [66]. The amino group in the UiO-66-NH_2 sometimes acts as an active electron-rich site instead of taking part in hydrogen bonding with the guest molecules. In some specific solvents, similar electronic transitions probably occur between the amino group of UiO-66-NH_2 and the solvent as was in the case with guest molecules. Such a phenomenon is favorable for inducing stability in the MOF in presence of such solvents particularly. In order to avoid the formation of aggregates of MOF particles in the reaction medium [67], which happens commonly, there has been a recent trend of using these materials in the form of a thin film immobilized or fixed on a functional surface. This can minimize the matrix effect and proves convenient with more satisfactory results in large-scale applications.

One of the recent studies by Xiong et al. demonstrated the synthesis of defective UiO-66-NH_2, possessing significantly large pore volume as well as surface area by modulating the amounts of acetic acid added [68]. The acid modulation strategy proves beneficial in a way that it results in increased pore space along with defects in the resulting MOF. Consequently, a greatly enhanced performance is observed in the catalytic properties of such acid-modulated defective UiO-66-NH_2 MOFs toward the detoxification of CWA simulants. This is evident from the considerably shortened half-life (0.76 min) of DMNP degradation. Figure 8 gives a description of the synthesis route, their SEM images at varying magnifications, and characterization parameters of defective UiO-66-NH_2.

NU-1000 is another common Zr-based MOF used for the degradation of DMNP that shows a much faster degradation attaining $t_{1/2}$ in 15 min as compared to 25 min for UiO-66 [69]. Pristine and dehydrated NU-1000 show degradation efficiencies of 80 and 100%, respectively, within 60 min of reaction time. Moreover, dehydrated NU-1000 also exhibited much faster degradation relative to the pristine form. The

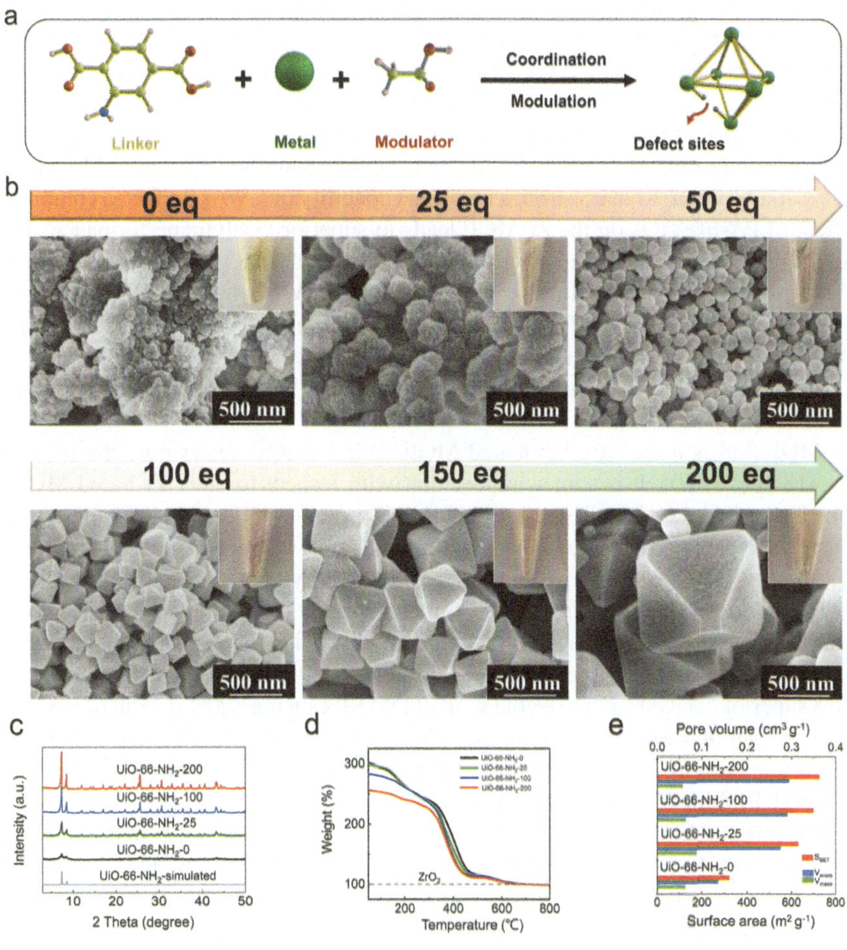

Fig. 8 **a** Pictorial representation for the synthesis of defective UiO-66-NH_2. **b** SEM images, **c** XRD plots, **d** thermogravimetric data and **e** pore properties of acetic acid-modulated UiO-66-NH_2-x (x = 0, 25, 50, 100, 150, 200). Reprinted with permission from Ref. [68]

catalytic process is enhanced by hydrogen bond formation between DMNP molecule and the hydroxyl groups in the ligand terminals of NU-1000. Apart from hydrogen bonding, electrostatic interactions between the P=O bond of DMNP and Zr of the dehydrated NU-1000 were also observed [69]. Hence, it can favorably interact with the DMNP molecules through direct coordination thereby resulting in enhanced catalytic performance with a much shorter $t_{1/2}$ value.

A smart composite material, (PA-6@TiO_2@UiO-66/UiO-66-NH_2/UiO-67), is reported which is capable of 90% degradation of DMNP to p-nitro phenoxide. It was obtained by coating polyamide-6 nanofiber (PA-6) with titanium dioxide (TiO_2) via atomic layer deposition (ALD) method which is followed by incorporation with MOF materials. The later step was achieved using the solvothermal growth method [67]. A number of similar materials using different starting materials were synthesized such as PA-6@TiO_2@UiO-66, PA-6@TiO_2@UiO-66-NH_2, PA-6@TiO_2@UiO-67, PA-6@TiO_2, and PA-6. Analysis of their degradation behavior showed $t_{1/2}$ values of 135, 7.3, 7.4, 1170, and 3950 min, respectively, for the synthesized materials. These findings revealed that the efficacy for these materials is enhanced as compared to their pristine forms, which may be due to the amalgamation of PA with TiO_2, whose sheet-like structures have a strong capturing capability of CWAs. In fact, a combined effect of PA-6@TiO_2 on the Zr-MOF leads to superior catalytic performance. The addition of oxygen atoms enhances the gap between the TiO_2 sheets from 5.5 Å to 7.5 Å, which again favors the interactions with the CWAs whose kinetic diameter is approximately 4.5 Å [64]. This leads to a better packing of host and guest molecules thereby enhancing the degradation efficiency. Moreover, the matrix structure of PA-6 acts as an added factor in supporting the diffusion of CWAs into the porous structure of MOF and TiO_2 [67].

MOF-808 is a recently fabricated MOF with a much greater pore diameter of 4.8–18 Å [52, 70]. It is somewhat peculiar with respect to other Zr-based MOFs, like UiO-66 and NU-1000 because MOF-808 has only six coordinated sites being occupied out of the available 12 or 8 coordinated sites [71]. Solvent molecules initially fill these vacant sites, which are easily replaceable by guest molecules like CWAs [87]. Moreover, due to the presence of uncoordinated ligand molecules, the CWAs can move freely and undergo interactions readily with the freely available metal centers through crystal-to-crystal transformation [88]. Hence, MOF-808 proves to be a superior catalyst for the degradation of DMNP causing 100% destruction within a short span of 4 min of reaction time. The calculated turnover frequency (TOF) of MOF-808 is above1.4, a value which is 10 times and 350 times those of UiO-66-NH_2 and NU-1000-dehydrated, respectively [70].

The pH conditions of the solvent medium also play a notable role in the catalytic activity of MOFs toward DMNP. A buffering reagent, N-ethyl morpholine is normally used during the hydrolysis of these CWAs that is assumed to cause deprotonation of water molecules and form acidic by-products such as HF [72]. In order to achieve similar buffering action in solid form, Moon et al., via a post-synthetic pathway, designed a solid catalyst by incorporating NU-1000 with polyethyleneimine (PEI) [73]. This was observed to have degradation efficiency similar to that of *N*-ethyl morpholine buffered NU-1000 material. The fabricated NU-1000/PEI (1.5 mmol of

NU-1000 per 0.003 mmol of PEI) showed a $t_{1/2}$ value of 8 min in converting DMNP into p-nitro phenoxide, a result which is similar to that of NU-1000 buffered with *N*-ethyl morpholine (1.5 mmol of NU-1000 in 0.39 mmol of N-ethyl morpholine solution).

Degradation of CEES

Among other methods, partial oxidation is the most preferably adopted as one for the effective degradation of CEES. It can be carried out in the presence of specific solvents or suitable sources of light for sensitization like UV light, which may be in the form of light-emitting diodes (LEDs) using specially designed catalysts [74–77]. Zr-based MOFs again prove to be good catalysts for the purpose. However, certain factors play a significant role in determining the rate of the degradation process such as the presence of Lewis acid sites, basic alkoxide units, and the efficiency of the ligands present in the MOF catalyst as well as the oxidizing agent and type of light being used for sensitization. For example, UiO-66@LiOtBu, being incorporated with metal alkoxide units, is found to show a much faster oxidative degradation of CEES to 2-chloroethyl ethyl sulfoxide (CEESO) and is associated with a rate constant value of 0.17 min^{-1} and $t_{1/2}$ of 3 min as compared to UiO-66 and UiO-66@SO_4H [22]. The presence of tert-Bu ion gives rise to an alkoxide basicity which works together with the Lewis acidity due to the metal nodes which amounts to an induced synergistic effect in UiO-66-LiOtBu. This is responsible for the faster oxidation rate in the case of UiO-66@LiOtBu [22, 78]. On the other hand, the absence of such Lewis acid sites in UiO-66 can cause accumulation of the by-products which prevents the interaction between the MOF and CEES thereby resulting in a shielding effect [22]. This ultimately hinders the degradation process. In the case of UiO-66@SO_4H, the presence of only the Lewis acid decreases the reaction pH, inducing a poisoning effect that again retards the oxidation phenomenon. In an experimental attempt to justify the poisoning effects by adding HCl to the MOFs, it was shown that UiO-66@LiOtBu exhibited the least poisoning effects because of the basic alkoxide (tert-Bu) units which take part in maintaining the pH of the reaction medium by neutralizing the extra acidity produced by the added HCl. On the other hand, UiO-66 completely loses its catalytic property with the addition of HCl [22]. This might be most possibly due to the structural collapse that results from the minimization of pore properties of UiO-66 under highly acidic conditions.

Catalytic poisoning can be normally avoided using strong oxidizing agents like hydrogen peroxide and tert-butyl hydro peroxide. However, the use of such strong oxidants may lead to the generation of toxic sulfone by-products like 2-chloroethyl ethyl sulfone ($CEESO_2$) which are even more toxic than the parent material [79]. Photocatalytic oxidation that takes place in the presence of a suitable light source is a good alternative that generates no harmful secondary by-product. The catalytically active moiety in the MOF gets photosensitized by the light which generates $^1O_2$

species that acts as the catalyzing agent to convert CEES to CEESO [80]. Porphyrin-based MOF, PCN-222/MOF-545, is a good photocatalyst for the oxidative degradation of CEES. $^1O_2$ can be generated under white, blue, and red LED light. The type of light source used apparently plays a prominent role in deciding the degradation kinetics. For instance, degradation of CEES using PCN-222/MOF-545 as a photocatalyst, was observed to be the most efficient using blue light, achieving 100% degradation within 25 min with a short $t_{1/2}$ value of 13 min. In the presence of blue light, Π-Π* photosensitized transitions take place in the porphyrin molecules of the MOF, which can effectively energize the triplet oxygen to get promoted to $^1O_2$; this attributes to the enhanced performance of the MOF in blue light.

The ligand molecules in the MOF also play a significant role as photosensitizers in the overall photocatalytic degradation process. Liu et al., in their study, made a comparative analysis of the degradation behavior of NU-1000 MOF and separate pyrene-based linker molecules toward the oxidation of CEES resulting in interesting outcomes. The $t_{1/2}$ values observed were 6.2 and 10.5 min, respectively, for the MOF and the linker species [58]. The presence of the six coordinated Zr ions as well the incorporation of the photosensitizer groups into the MOF structure probably enhances the oxidation reaction. The acceleration in the degradation process can be attributed to the heterogeneity of the MOF surface which keeps the pyrenes separated from the rigid MOF framework. This ultimately reduces the chances of deactivation through aggregation or accumulation of the by-products. Hence, it is worthy to conclude that the synergistic effect of both the ligand and metal nodes ultimately causes an enhanced degradation of CEES, with a careful selection of suitable ligand so as to achieve maximum $^1O_2$ generation [58].

Degradation of DIFP

All the [UiO-66@X] materials, where X is a metal alkoxide, are found to be active catalysts for the degradation of DIFP through hydrolysis, unlike the hydrophobic MOF counterparts [81]. These materials have been proved to be even more effective than the "state-of-the-art" activated carbon Blcher-101408 [22]. Incorporation of LiOtBu to UiO-66 results in a highly active, reusable heterogeneous catalyst, [UiO-66@LiOtBu] that exhibits such a short half-life of 5 min with a TOF of 0.13 min^{-1}, and complete hydrolysis of the P-F bond is achieved in 30 min. Almost similar results are obtained for [UiO-66@LiOEt] also but with slower kinetics, i.e., a half-life of 30 min and TOF of 0.017 min^{-1} and complete degradation in 250 min. The difference in the hydrolysis rates may be due to the higher basicity of t-BuO compared to that of EtO. These findings once again justify the strong synergistic effect resulting from the alkoxide basicity (t-BuO, EtO, etc.) and Lewis's acidity due to the presence of metal centers in the MOFs. Such an induced effect improves the phosphotriesterase activity of UiO-66 to a great extent. Another intriguing observation is that UiO-66 loses its catalytic activity after an approximate 75% degradation of the DIFP, which is indicative of catalytic poisoning due to the accumulation of the products. The catalytic poisoning was confirmed by adding equimolar quantities of various acids like methyl phosphonic acid or phosphoric acid which showed considerable retardation in the catalytic performance of UiO-66. Probably, poisoning of the active catalytic sites present on the surface of UiO-66 further followed by incorporation of -HSO_4 active acidic sites on the surface of UiO-66 leads to a significant loss of catalytic activity in UiO-66. In contrast, the catalytic performance of [UiO66@LiOtBu] remains unaffected.

Some specific enzymes like organophosphorus acid anhydrolase (OPPA) are known to be useful in the catalytic hydrolysis of bonds like P–F, P–O, PCN, and P-S bonds [82] which are the common bonds targeted to be cleaved in many of the CWAs and their simulants. Hence, immobilization of these enzymes by encapsulation on MOFs is also found to decrease the catalytic efficiency [83–85]. For instance, immobilization of MOF PCN-128y with (OPPA) causes a slow interparticle mass transfer diffusion of the DIFP molecules into the MOF surface resulting in much lower catalytic efficiency for OPPA@PCN-128y compared to free OPPA [86]. Moreover, the enzyme encapsulation causes a significant shrinking of the pore size resulting in a significant decrease in pore volumes from 0.19 to 0.14 $cm^3\ g^{-1}$ (for the triangular channels) and 0.52–0.13 $cm^3\ g^{-1}$ (for the hexagonal channels) in PCN-128y. This restricts the intra-particle diffusion of DIFP in the MOF pores. All these cumulatively bring about retardation in the overall catalytic performance of the MOF. In fact, only micropores are available on the surface of the MOFs for interaction with DIFP molecules after being immobilized by the enzyme [87]. Moreover, enzyme activity is usually affected by temperature; an increase in the activation temperature is found to be accompanied by an enhanced catalytic efficiency of OPPA@PCN-128y.

Degradation of PNPDPP

PNPDPP can be degraded through methanolysis using MOFs such as UiO-66 [64] and Al (III)-porphyrin containing MOFs [88, 89] as catalysts. A series of modularly prepared metalloporphyrin dimers is shown to catalyze the methanolysis of PNPDPP with must higher rates as compared to the uncatalyzed reactions. The methanolysis step induced by methoxide groups combined with activation by the Lewis acid localized at MOF cavities can accelerate the degradation phenomenon of PNPDPP up to 1300 times. According to a study by Tottel et al. wherein a porous organic polymer (POP) was incorporated with an aluminum (III)-porphyrin associated with axial ligands, it was observed that the removal of these ligands renders a great extent of microporosity to the POP. These aluminum-porphyrin-blended organic polymers are highly active and reusable catalysts that prove to be efficient in the methanolysis of PNPDPP. Interestingly, the best catalytic performance could be achieved through supercritical CO_2 processing of the POP [88]. This technique, in contrary to the conventional way of activation through heating, can potentially impart greater pore size and pore volume to the POPs, which can significantly assist in enhancing the substrate accessibility ultimately accelerating the degradation process. However,

UiO-66 is comparatively found to show superior performance than these Al (III)-porphyrin containing MOFs. UiO-66 shows a $t_{1/2}$ value of 45–50 min with a much faster initial rate of 5106 s^{-1} [64] compared to Al (III)-porphyrin containing MOFs that shows 2–10 times slower rate kinetics. Again, the reason for this enhanced degradation activity is probably the suitable pore diameter of UiO-66 (approx. 6 Å) as compared to that of PNPDPP (approx. 4.5 Å), along with a relatively greater surface area of 1450 m^2 g^{-1}. All these factors prove favorable for the degradation of PNPDPP.

6.2 *Degradation of Real Nerve Agents*

CWAs are highly toxic to be handled in laboratories and investigation of the degradation and toxicity profiles of such real CWAs actually require specific permissions and special care and attention to carry out. This has put restrictions on the scope of research in this field, in spite of its extreme importance. Due to this, only a limited number of research has been performed so far on the removal of real CWAs either through degradation or adsorption using MOFs or any other materials. However, researchers have managed to carry out research and experiments making use of specific MOFs in the degradation of some selected real nerve agents. Functionalization of MOFs is usually followed as a general trend that is expected to bring about effective degradation of most of the nerve agents; at the same time, it serves the purpose of reducing the generation of toxic by-products [66, 67]. Certain parameters such as the nature of the functional groups both in the MOF as well as the target CWAs, pore size, and volume, and the morphology especially the mesoporosity of the MOFs have been proved to play a key role in the degradation behavior. The following sections give a description of the use of MOFs in the detoxification of some common real CWAs.

Degradation of Sulfur Mustard

Partial oxidation is more commonly adopted for the degradation of sulfur mustard because of the highly toxic end product sulfone being formed after complete oxidation. The majority of the degradation mechanisms involve the generation of $^1O_2$ which has the potential for selective and partial oxidation of HD. Liu et al. experimented on NU-1000, a pyrene-based MOF which showed excellent photocatalytic degradation efficiency toward real sulfur mustard as well as its simulant CEES [58]. Pyrene is one of the most commonly used organic chromophores which has attracted much research attention and is extensively investigated owing to its alluring photophysical properties. As mentioned in Sect. 5.2, the presence of a triplet state with a relatively low energy combined with the possibility of the occurrence of an efficient and easy intersystem crossing between the two spin states in pyrene, and its derivatives makes them best sources for the generation of singlet oxygen, $^1O_2$ [90–92].

NU-1000 MOF was irradiated with a light-emitting diode from a UV source that successfully generated $^1O_2$ which in turn attacks the CEES molecules leading to its partial oxidation to non-toxic CEESO. This undergoes further elimination yielding 2-chloroethyl vinyl sulfoxide [58]. This method was found to be equally effective when examined on real HD wherein, it undergoes successful oxidation to non-toxic products. The process is not associated with the generation of any toxic by-product like sulfones. Comparison of the degradation profiles of CEES and HD reveals that the degradation of HD involves a much longer $t_{1/2}$ value (33 min) than the partial oxidation of CEES (6.2 min) [58]. A possible reason for such a huge difference may be the higher solubility of CEES in methanol that accelerates the degradation process, compared to the real CWA. CEES has a solubility of 1392 mg L^{-1} in water that is much higher compared to HD, which is only 684 mg L^{-1} in water. This makes the degradation mechanism through $^1O_2$ highly effective.

Degradation of Soman

UiO-based MOF materials have been claimed to be extraordinarily effective catalysts for the degradation of Soman nerve agents. UiO-66 and UiO-67 along with their functionalized analogs have been examined for the degradation of Soman to pinacolyl methyl phosphate [66]. The trends in the degradation profiles for Soman are interesting when compared to those of its simulant, DMNP. Pristine UiO-66 exhibited much better performance than its functionalized analogs toward degradation of the real nerve agent. It is important to note that amino-functionalized analogs of both MOFs such as UiO-66-NH_2 and UiO-67-NH_2 exhibit almost similar trends of percentage conversion for degradation of Soman. A complete conversion of Soman after 8 min was achieved for both the MOFs. On the contrary, in the case of degradation of DMNP, the amino-functionalized analog, UiO-66/67-NH_2, showed a superior degradation performance compared to the original MOF materials, i.e., UiO-66 and UiO-67 [66]. A complete degradation of DMNP was achieved at 60, 4, 20, 12, and 9.5 min, respectively, with UiO-66, UiO-66-NH_2, UiO-67, UiO-67-NH_2, and UiO-67-$N(Me)_2$, respectively. Such a notable variation arises from the difference in the specific bond through which hydrolysis takes place in Soman as well as DMNP. Hydrolysis of Soman first initiates with the cleavage of the P–F bond while DMNP with the P–O bond. The P-F bonds are relatively easier to cleave compared to P–O bonds. The chemical structure of UiO-66 favors the cleavage of the P–F bond in Soman; hence, it is a more efficient catalyst among the other selected MOFs. Indeed, only a slight improvement is observed even with amine functionalization in UiO-66-NH_2. The increased pore size also acts as an added advantage in accelerating Soman degradation. It is noteworthy that, unlike with the simulant DMNP, amine functionalization apparently does not have much effect on the speed of degradation of Soman as is evident from the half-life of UiO-67-NH_2 which is only slightly longer than that of UiO-67. The effect is much more pronounced in the case of its simulant, DMNP. Results show 52 and 97% conversion of Soman to PMPA by UiO-66 and UiO-67, respectively, which were achieved within a reaction time of 3.5 min. On the contrary, percentage conversions of 92 and 95%, respectively, within 3.5 min were observed using UiO-67-NH_2 and UiO-67-$N(Me)_2$ [66]. Such a prominent difference in the degradation performance toward degradation of Soman arises from the difference in the pore parameters between UiO-66 and UiO-67.

As described in Sect. 6.1.1, MOF-functionalized nanofibers, PA-6@ TiO_2@UiO-67, synthesized by Zhao et al. [67] show ultra-fast degradation and excellent detoxifying capacity toward real nerve agent Soman (GD). The half-life of Soman degradation by these nanofibers is as short as 2.3 min compared to 3 min and 3.7 min for other UiO-66 analogs, namely PA-6@TiO_2@UiO-66 and PA-6@ TiO_2@UiO-66-NH_2, respectively [67]. A greater degradation efficiency was observed as compared to the pristine UiO-66 as well as methyl incorporated analogs, for which larger pore size is the major contributing factor that allows a faster penetration of guest molecules into the catalyst [66, 67]. The catalytic efficiency of the MOFs can also be enhanced through the detachment of the coordinated water molecules from the MOF framework which in turn strengthens the interactions between the active metal nodes of the MOF and the organophosphate sites of the nerve agent. For instance, pristine and dehydrated NU-1000 exhibited such a prominent difference in their $t_{1/2}$ values: 3 min and 36 min, respectively, toward the catalytic conversion of Soman [69]. Mesoporous channels of the MOF framework, significant Vander Waals interactions between the linkers in the MOF framework and alkyl groups in Soman, along with the availability of eight coordinated nodes together attribute to the excellent hydrolysis of the dehydrated forms.

NU-1000 with PEI (linear-polyethyleneimine) buffer composite is also an efficient catalyst for hydrolytic degradation of GD through the cleavage of the P-O bond similarly in the case of DMNP. $t_{1/2}$ value of 1.8 min is obtained which is much faster as compared to 36 min for pristine NU-1000 and 3 min for dehydrated NU-1000 [69]. OPPA@PCN-128y is another MOF modified via a post-synthetic route which was experimented for degradation of Soman [86]. A minimal amount of 3.75 mg of OPPA@PCN-128y could effectively hydrolyze 3 mM of Soman, which implicates almost 90% conversion in 60 min at 25 °C. However, several milligrams of traditional catalyzing materials are required to defluorinate a small quantity of GD.

Degradation of VX

Unlike HD and GD, the chemical structure of VX contains both P–O and P–S bonds where hydrolysis can take place. This makes hydrolytic degradation of VX relatively difficult. The hydrolysis is usually associated with two by-products, namely ethyl methyl phosphonic acid (EMPA) and diisopropylamino ethyl mercaptan (DESH), which is greatly influenced by the pH of the medium. As discussed in the above sections, post-synthetic modification through selective functionalization of the MOFs proves to be an efficient method devised for enhancing the degradation efficiency of the catalyst, without causing a significant disturbance in the existing parent framework. For instance, De Coste et al. reported the treatment of Zr-based MOFs, UiO-66, and UiO-66-NH_2 with tetrafluoromethane (CF_4) and hexafluoroethane (C_2F_6)

plasmas, as a post-synthetic treatment [93]. This results in strong interactions between the Zr–O bonds of the MOF and the fluoride radicals generated by the perfluoro alkane plasma which in turn accounts for the development of mesoporosity. A hierarchical pore structure is finally obtained which permits the binding and subsequent diffusion of the by-products, as well, onto the MOF surface away from the reaction medium. This further expedites the degradation process as a whole. Thus, the post-synthetic modulation of the pore parameters favorably leads to a reduction in the half-life from 653 min for UiO-66-NH_2 to 257 min for P-UiO-66-NH_2-CF_4-60, which amounts to almost 60% reduction. This causes considerable variation in the degradation performance of VX. Moon et al. demonstrated the effect of amine functionalization of UiO-67 through the synthesis of UiO-67-NH_2 and UiO-67-N$(Me)_2$ which again proved to be significantly selective and effective catalysts for the hydrolysis of VX [22]. In a medium of pH 10 buffered solution of *N*-ethyl morpholine, MOFs like UiO-67, UiO-67-NH_2, and UiO-67-N$(Me)_2$ are observed to potentially hydrolyze the P–S bond in VX with the generation of relatively less-toxic EMPA and DESH. The hydrolysis efficacy of UiO-67-N$(Me)_2$ for VX in the above condition is quite high which is evident from such a small $t_{1/2}$ of 1.8 min compared to other well-known materials in the field. Most interestingly, the catalytic hydrolysis was also observed to work well in the absence of buffer. NU-1000-dehydrated/PEI composite has also been examined for the hydrolysis of VX [73] and was found to be able to break the P–S bond within a $t_{1/2}$ of 12.7 min reaching 80% conversion within only 60 min. Apparently, a large number of amine groups is required for the hydrolysis of VX that plays a key role in maintaining the appropriate pH required for the protonation of the tertiary amine group that has a pKa value of 8.6 [94, 95]. The desired number of amine groups can be manipulated again through a post-synthetic modification strategy.

7 MOFs for Adsorption of CWAs

Besides degradation, adsorption can also prove to be one of the efficient techniques for the removal and detoxification of CWAs from the environment. MOFs, owing to their high porosity, large surface area, and tunable reactivity with plenty of functional groups, are excellent adsorbents for such chemicals. Moreover, MOFs can be specifically designed or tailored to impart a high adsorption capability of CWAs on their surface permitting the degradation of these toxic chemicals into less or completely non-toxic substances. Son et al. fabricated a series of hexanuclear Zr-based MOFs namely MOF-808, NU-1008, MOF-525, NU-1200, UiO-66, and its modified forms such as defective UiO-66 (UiO-66-defective), NU-1000 and its analogs like dehydrated NU-1000 (NU-1000-Dehyd), benzoic acid-capped NU-1000 (NU-1000-BA), NU-901, which were analyzed for the adsorption of GB, and CEES [96]. These MOFs show considerable variations in parameters such as surface area, pore functionalization, pore volume, connectivity of the SBU units, and nature of the open metal sites. Careful observation of the CWA adsorbed on the surface of the MOFs series proved helpful in elucidating the influence of each parameter in trapping the CWAs. Large surface areas and pore volumes along with extensive hydrogen-bonding interactions result in a more facile diffusion of CWA molecules into the MOF and hence greater access to the binding sites. This ultimately results in superior adsorption of both GB and CEES by these MOFs. Compared to the rest of the above-mentioned MOFs, UiO-66, defective UiO-66, and MOF-808 showed the highest reactivities toward GB compared. This is attributed to the presence of a higher number of active sites on the MOF surface per unit volume. Moreover, the SBU identity seems to predominantly govern the behavior of CWA uptake. MOFs with aqua and/or hydroxyl functional groups on the node exhibited a greater uptake capacity of GB and CEES, most likely due to extensive hydrogen-bonding interactions. It was also evident that MOFs with open metal sites (OMS) showed better uptake performance under dry conditions while in moisture, the OMS interact more strongly with water, hence decreased CWA loading. UiO-66, in spite of having relatively lower uptake capacity, exhibited superior retention or lesser desorption of loaded GB, most probably due to its microporosity. In the case of CEES, the highest retention capacity could be observed for UiO-66 and NU-1000.

Asha et al., also investigated the adsorption behavior of NU-1000 and UiO-67 toward CWA simulants, CEES and DMNP, respectively, from an aqueous medium [97]. The adsorption phenomenon is proved to follow pseudo-second-order kinetics model with chemical interactions between the MOFs and the CWAs. 2-CEES contains thioether and chloro groups which get involved in hydrogen-bonding interactions with the Zr–OH groups present at the nodes of these MOFs, which is apparently responsible for the chemical interactions. Additionally, the strong binding of DMNP to the MOF is due to the strong interaction of the Zr nodes with the phosphate groups in the DMNP. NU-1000 exhibited excellent adsorptive performance toward 2-CEES and DMNP with high uptake capacities of 4.197 and 1.70 mmol g^{-1}, respectively, and recyclability up to 3 cycles. On the other hand, UiO-67 showed comparatively poor results with an uptake of 4.0 and 0.90 mmol g^{-1} of 2-CEES and DMNP, respectively.

The capture of harmful VOCs as well as model simulants of Sarin and mustard gas was also examined by Montoro et al. using [Zn_4(μ4-O)-(μ4-4-carboxy-3,5-dimethyl-4-carboxy-pyrazolato)$_3$], a Zr-based MOF, whose crystal structure resembles that of a well-known MOF, [Zn_4O(1,4-benzenedicarboxylato)$_3$] system (MOF-5) [56, 98]. In addition to its remarkable thermal, mechanical, and chemical integrity, comparison of the capture performance of this MOF with well-known adsorbent materials such as $Cu_3(btc)_2$] and carboxen (also referred to as molecular sieve active carbon), leads to interesting conclusions. The unsaturated metal sites which are the primary guest-binding sites are found to be relatively ineffective in the capture of VOCs in ambient moisture conditions. Rather, the pore size and surface hydrophobicity are considered to play a pivotal role in the VOC-MOF recognition process.

8 MOF-Modified Fabrics for CWA Adsorption

For easy applications of MOFs in the real world for degradation of CWAs, various techniques have been adopted to incorporate various MOFs, like Zr-based and Cu-based MOFs, into fabric form or textiles, which is made feasible at the minute level through electrospinning [99]. Giannakoudakis et al. synthesized smart textiles comprising of cotton, Cu-BTC MOF, and oxidized graphitic carbon nitride (g-C_3N_4-ox) nanospheres which proved to be good nerve agent detoxifiers as well as colorimetric detectors [100]. The deposition of MOFgCNox, a highly porous nanocomposite of Cu-BTC and g-C_3N_4-ox, onto cotton textiles results in a stable advanced hybrid textile with an excellent photocatalytic degradation ability for dimethyl chlorophosphite, a nerve gas simulant. The whole detoxification process is accompanied by a distinct visible color change which makes possible the selective detection of CWAs and subsequent monitoring of their penetration into the protective layer. The introduction of the visible light-sensitive g-C_3N_4 nanospheres into the MOF imparts significant mesoporosity to the surface of the nanocomposite that results in a tremendous increase in the surface reactivity. Moreover, the active phase is so well dispersed on the matrix of the cotton textile which altogether leads to excellent adsorption of not only the CWA simulant molecule but also its degradation products. The model simulant of nerve agents can be oxidized to non-toxic compounds. Maya et al. developed advanced self-detoxifying adsorbents of CWAs taking advantage of improved phosphotriesterase catalytic activity of lithium alkoxide doped Zr-MOFs which were integrated into the surface of silk fibroin fibers [63]. The favorable air-permeation properties of the textiles work together with the self-detoxifying properties of the MOF material thereby resulting in superior smart fabrics. The performance of these UiO-66 modified materials toward the degradation of common CWA simulants like DIFP, DMMP, CEES, and diethyl sulfide (DES) was quite promising as compared to other conventionally known materials including metal oxides, activated carbons, and their composites and other known hydrophobic MOFs.

A mechanically strong, filtration efficient polyacrylonitrile (PAN) nanofibers incorporated with PIM-1 fibers have also been fabricated. Subsequent amalgamation with UiO-66-NH_2 particles leads to further enhancement of the adsorption capacity at the same time maintaining good mechanical strength, and breathability [101]. The resulting PIM/PAN/MOF composite fiber mat is found to exhibit remarkable integrated properties, which prove beneficial toward sequestration of CWA simulants. A comparative analysis with a neat PIM-1 fiber mat reveals that PIM/PAN/MOF composite shows more than 99% filtration efficiency, almost 70 times increased tensile strength, a promisingly high water–vapor transmissibility rate of 1013 g/m^2 with a large surface area of 574 m^2/g. A schematic demonstration showing blocking penetration of toxic vapor, aerosol, and liquid by this composite fiber mat is shown in Fig. 9.

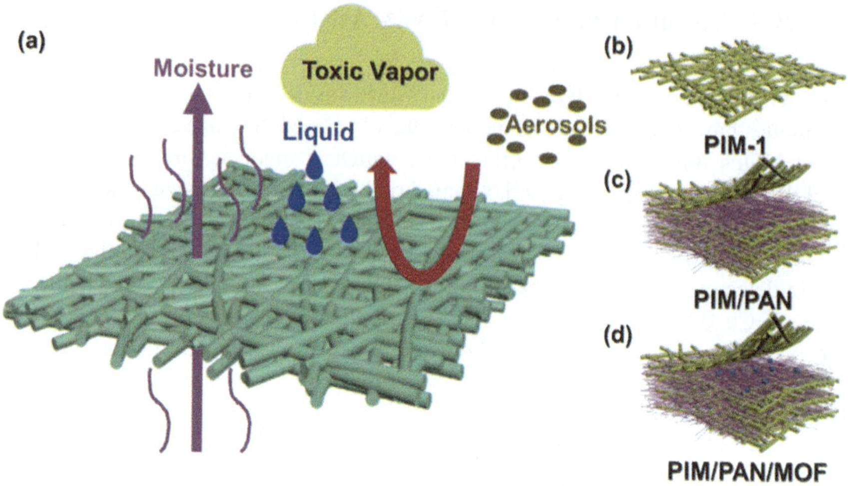

Fig. 9 **a** Highly permeable and selective fiber mat preventing the penetration of toxic materials, **b** PIM-1 fiber web, **c** PIM/PAN fiber web with integrated PAN layers in PIM-1 matrix. **d** PIM/PAN/MOF composite fiber web. Reprinted with permission from Ref. [101]

9 Conclusions and Future Aspects

In view of the prolonged hazardous effects of CWAs, devising innovative techniques simultaneously with the exploration and design of sustainable materials for the elimination of these toxic chemicals from the environment is of paramount importance. However, it is a matter of grave concern that only a few institutions are able to carry out research in this field, most likely due to the extreme toxicity associated with it as well as lack of adequate safety measures. The limited number of literature in the field is conclusive that MOFs, especially Zr-based MOFs, are potential candidates for the capture and degradation of a wide range of CWAs including real CWAs such as GD, HD, and VX as well as their model simulants. Specific fabrication strategies based on pore dimensions and surface morphology, post-synthetic modifications, missing-linker defects, etc. have been adopted for enhancing the degradation efficiency of the MOFs. Hydrolysis, methanolysis, and partial oxidation being the common reactions involved in the degradation process; it is generally observed that various parameters like pore dimensions of the MOF relative to the kinetic diameter of the target CWA, nature of the solvent medium and its pH, nature of the ligand units as well the functional moieties present in the MOF play significant roles in the catalytic efficiency. Apart from degradation, adsorption on MOF surfaces is also one of the simplest techniques adopted for the removal of CWAs, which is favored by the highly porous and large surface area of MOFs. In recent times, various smart textiles comprising of natural/synthetic fabrics incorporated with MOFs have also been designed which also prove to be potential detoxifiers for various CWAs. In spite of their superior removal performance, real-life applications of MOFs, on a large scale is still anonymous probably due to difficulties in achieving similar structural properties because of the rigorous specific conditions being required. Hence, there is still scope for further research and investigations in the field in terms of certain aspects such as simplifying the fabrication routes, exploring ways for carrying out research on real CWAs with minimum toxicity, and most importantly discovering MOFs which are applicable for multiple types of CWAs.

Acknowledgements The authors are thankful to Principal, Sir Venkateshwara College, University of Delhi and Principal, Acharya Narendra Dev College, University of Delhi for their valuable cooperation and guidance.

Abbreviations

4-MAP	Methylaminopyridine
ACh	Acetyl choline
AchE	Acetyl cholinesterase
ALD	Atomic layer deposition
CEES	2-Chloroethyl ethyl sulfide
CEESO	2-Chloroethyl ethyl sulfoxide
$CEESO_2$	2-Chloroethyl ethyl sulfone
CN	2-Chloroacetophenone
CR	Dibenz (b,f)-1,43-oxazepine
CS	2-(2-Chlorobenzylidene) malononitrile
CWAs	Chemical Warfare Agents
DCP	Diethyl chloro phosphonate
DESH	Diisopropylamino ethyl mercaptan
DIFP	Diisopropylfluorophosphate
DMNP	Dimethyl 4-nitrophenyl phosphate
EA-2192	S-[2(diisopropyla mino) ethyl]methyl phosphonic acid
EMPA	Ethyl methyl phosphonic acid
GB	Sarin
GD	Soman
H_4TBAPy	4,4′,4″,4‴-(Pyrene-1,3,6,8-tetrayl) tetrabenzoic acid
HD	Sulfur mustard
HN1	Bis-(2-chloroethyl)-ethyl-amine
HN2	2-Chloroethyl)-methyl-amine
HN3	Tris-(2-chloroethyl)-amine
ID_{50}	Lethal Dose 50
LED	Light-emitting diode
LSD	Lysergic acid diethylamide
MOFs	Metal-organic Frameworks
NADH	Nicotinamide adenine dinucleotide hydrogenase

OMS	Open metal sites
OP	Organophosphorus
PA-6	Polyamide-6 nanofiber
PAN	Polyacrylonitrile
PCN	Porous coordination networks
PEI	Polyethyleneimine
PFIB	Perfluoro isobutylene
PIM	Polymers of intrinsic microporosity
PMPA	Pinacolyl methyl phosphonic acid
PNPDPP	P-nitrophenyl diphenyl phosphate
POPs	Porous organic polymers
SBUs	Secondary building units
TA	Tabun
THC	Tetrahydrocannabinol
TOF	Turn Over Frequency
VOCs	Volatile Organic Compounds
VX	S-{2-[Di(propan-2-yl) amino] ethyl} O-ethyl methylphosphonothioate
WW I	World War I
WW II	World War II
Zr-MOFs	Zirconium-based Metal–Organic Frameworks

References

1. Szinicz L (2005) History of chemical and biological warfare agents. Toxicology 214:167–181
2. Coleman K (2005) A history of chemical warfare. Palgrave Macmillan, New York
3. Bajgar J, Fusek J, Kassa J, Kuca K, Jun D (2020) Chapter 3—Global impact of chemical warfare agents used before and after 1945. In: Gupta RC (ed) Handbook of toxicology of chemical warfare agents. Academic Press, San Diego, pp 17–24
4. Kuca K, Pohanka M (2010) Chemical warfare agents. EXS 100:543–558
5. Barea E, Montoro C, Navarro JA (2014) Toxic gas removal—metal-organic frameworks for the capture and degradation of toxic gases and vapours. Chem Soc Rev 43:5419–5430
6. Huang NY, Gu J, Chen D, Xu Q (2021) MOF/hydrogel catalysts for efficient nerve-agent degradation. Chem Catalysis 1(3):502–504
7. Wagner GW, Peterson GW, Mahle JJ (2012) Effect of adsorbed water and surface hydroxyls on the hydrolysis of VX, GD, and HD on titania materials: the development.of self-decontaminating paints. Ind Eng Chem Res 51:3598–3603
8. Bandosz TJ, Laskoski M, Mahle J et al (2012) Reactions of VX, GD, and HD with $Zr(OH)_4$: near instantaneous decontamination of VX. J Phys Chem C 116:11606–11614
9. Dong J, Hu J, Chi Y, et al (2017) A polyoxoniobate-polyoxovanadate double-anion catalyst for simultaneous oxidative and hydrolytic decontamination of chemical warfare agent simulants. Angew Chem Int Ed 56:4473–4477
10. Hou Y, An H, Zhang Y et al (2018) Rapid destruction of two types of chemical warfare agent simulants by hybrid polyoxomolybdates modified by carboxylic acid ligands. ACS Catal 8:6062–6069
11. Bobbitt NS, Mendonca ML, Howarth AJ et al (2017) Metal-organic frameworks for the removal of toxic industrial chemicals and chemical warfare agents. Chem Soc Rev 46:3357–3385

12. Liu Y, Howarth AJ, Vermeulen NA et al (2017) Catalytic degradation of chemical warfare agents and their simulants by metal-organic frameworks. Coord Chem Rev 346:101–111
13. Yaghi OM, Keeffe MO, Ockwig NW et al (2003) Reticular synthesis and the design of new materials. Nature 423:705–714
14. Bai Y, Dou Y, Xie LH, Rutledge W et al (2016) Zr-based metal-organic frameworks: design, synthesis, structure, and applications. Chem Soc Rev 45:2327–2367
15. Larsson A, Qvarnström J, Lindberg S (2021) In vitro human skin decontamination efficacy of MOF-808 in decontamination lotion following exposure to the nerve agent VX. Toxicol Lett 339:32–38
16. Wu H, Yildirim T, Zhou W (2013) Exceptional mechanical stability of highly porous zirconium metal-organic framework UiO-66 and its important implications. J Phys Chem Lett 4:925–930
17. Monsdloch JE, Katz MJ, Planas N et al (2014) Are Zr6-based MOFs water stable? Linker hydrolysis versus capillary-force driven channel collapse. Chem Commun 50:8944–8946
18. Katz MJ, Moon SY, Mondloch JE et al (2015) Exploiting parameter space in MOFs: a 20-fold enhancement of phosphate-ester hydrolysis with UiO-66-NH_2. Chem Sci 6:2286–2291
19. Isl1amoglu T, Ortuno MA, Proussaloglou E, et al (2018) Presence versus proximity: the role of pendant amines in the catalytic hydrolysis of a nerve agent simulant. Angew Chem Int Ed 57:1949–1953
20. Dwyer DB, Dugan N, Hoffman N et al (2018) Chemical protective textiles of UiO-66-integrated PVDF composite fibers with rapid heterogeneous decontamination of toxic organophosphates. ACS Appl Mater Interfaces 10:34585
21. Yao A, Jiao X, Chen D et al (2019) Photothermally enhanced detoxification of chemical warfare agent simulants using bioinspired core-shell dopamine-melanin@ metal-organic frameworks and their fabrics. ACS Appl Mater Interfaces 11:7927–7935
22. Kalaj M et al (2019) Spray-coating of catalytically active MOF–polythiourea through postsynthetic polymerization. Angew Chem Int Ed 58(8):2336
23. Lee DT, Zhao J, Peterson GW et al (2017) Catalytic "MOF-cloth" formed via directed supramolecular assembly of UiO-66-v crystals on atomic layer deposition-coated textiles for rapid degradation of chemical warfare agent simulants. Chem Mater 29:4894–4903
24. Zhao J, Lee DT, Yaga RW et al (2016) Ultra-fast degradation of chemical warfare agents using MOF-nanofiber kebabs. Angew Chem Int Ed 55:13224–13228
25. Lee DT, Zhao J, Oldham CJ et al (2017) UiO-66-NH_2 metal-organic framework (MOF) nucleation on TiO_2, ZnO, and Al_2O_3 atomic layer deposition-treated polymer fibers: role of metal oxide on MOF growth and catalytic hydrolysis of chemical warfare agent simulants. ACS Appl Mater Interfaces 9:44847–44855
26. Ganesan K, Raza SK, Vijayaraghavan R (2010) Chemical warfare agents. J Pharm Bio Allied Sci 2:166–178
27. López-Muñoz F, Alamo C, Guerra JA et al (2008) The development of neurotoxic agents as chemical weapons during the National Socialist period in Germany. Rev Neurol 47:99–106
28. Rowell M, Kehe K, Balszuweit F, Thiermann H (2009) The chronic effects of sulfur mustard exposure. Toxicology 263:9–11
29. Young RA, Bast CB (2009) Chapter 8: mustards and vesicants. Gupta RC (ed) Handbook of toxicology of chemical warfare agents. Academic Press,San Diego, p 69
30. Smith WJ (2009) Therapeutic options to treat sulfur mustard poisoning—the road ahead. Toxicology 263:70–73
31. Kehe K, Thiermann H, Balszuweit F et al (2009) Acute effects of sulfur mustard injury—Munich experiences. Toxicology 263:3–8
32. Jyothi MS, Nagarajan V, Chandiramouli R (2020) Benzyl alcohol and 2-methyldecalin vapor adsorption studies on ß-bismuthene sheets—A DFT outlook. Chem Phys Lett 755
33. Bartlett JG, Sifton DW, Kelly GL (eds) (2002) PDR guide to biological and chemical warfare response, 1st edn. Thompson Healthcare Publications, Montvale, NJ, pp 1–404
34. Borowitz JL, Isom GE, Baskin SI (2001) Chemical warfare agents: toxicity at low levels. CRC Press, Boca Raton, p 305
35. Cummings TF (2004) The treatment of cyanide poisoning. Occup Med 54:82–85

36. Zellner T, Eyer F (2020). Choking agents and chlorine gas—History, pathophysiology, clinical effects and treatment. Toxicol Lett 320:73–79
37. Hoenig SL (2007) Choking agents. Compendium of chemical warfare agents. Springer
38. Beswick FW (1983) Chemical agents used in riot control and warfare. Hum Toxicol 2:247
39. Fusek J, Bajgar J, Kassa J et al (2009) Psychotomimetic agent BZ(3-quinuclidinyl benzilate). In: Gupta RC (ed) Handbook of toxicology of chemical warfare agents. Elsevier, London, pp 135–142
40. Costa LG, Furlong, C, Gupta RC (eds) (2009) Handbook of toxicology of chemical warfare agents. Elsevier-Academic Press, Amsterdam, pp 1023–1031
41. Ketchum JS (2006) Chemical warfare: secrets almost forgotten. ChemBooks, Santa Rosa, CA
42. Goodman E (2010). Historical contributions to the human toxicology of atropine: behavioural effects of high doses of atropine and military uses of atropine to produce intoxication. Eximdyne, Wentzville, Missouri, p 62
43. Nichols DE (2004) Hallucinogens. Pharmacol Ther 101:131–181
44. Holze F, Vizeli P, Ley L, Müller F, Dolder P, Stocker M et al (2021) Acute dose-dependent effects of lysergic acid diethylamide in a double-blind placebo-controlled study in healthy subjects. Neuropsychopharmacology 46(3):537–544
45. Kinston W, Rosser R (1974) Disaster effects on mental and physical state. J Psychosom Res 18:437–456
46. Bajgar J, Fusek J, Kassa J et al (2009) Chemical aspects of pharmacological prophylaxis against nerve agent poisoning. Curr Med Chem 16:2977–2986
47. Bajgar J (2005) Complex view on poisoning with nerve agents and organophosphates. Acta Medica (Hradec Kralove) 48:3–21
48. Shakarjian MP, Heck DE, Gray JP et al (2010) Mechanisms mediating the vesicant actions of sulfur mustard after cutaneous exposure. Toxicol Sci 114:5–19
49. Kehe K, Szinicz L (2005) Medical aspects of sulphur mustard poisoning. Toxicology 214:198–209
50. Lavoie J, Srinivasan S, Nagarajan R (2011) Using cheminformatics to find simulants for chemical warfare agents. J Hazard Mater 194:85–91
51. Bartelt-Hunt SL, Knappe DRU, Barlaz MA (2008) A review of chemical warfare agent simulants for the study of environmental behaviour. Crit Rev Environ Sci Technol 38:112–136
52. Moon SY, Liu Y, Hupp JT et al (2015) Instantaneous hydrolysis of nerve-agent simulants with a six-connected zirconium-based metal-organic framework. Angew Chem Int Ed Engl 54:6795–6799
53. Climent E, Biyikal M, Gawlitza K, Dropa T, Urban M, Costero AM, Martínez-Máñez R, Rurack K, et al (2016) A rapid and sensitive strip-based quick test for nerve agents Tabun, Sarin, and Soman Using BODIPY-modified silica materials. Chem Eur J 22:11138–11142
54. Gupta RC (2006) In: Gupta RC (ed) Toxicology of organophosphate and carbamate compounds. Elsevier, Burlington, pp 103–160
55. Wu H, Chua YS, Krungleviciute V et al (2013) Unusual and highly tunable missing-linker defects in zirconium metal-organic framework UiO-66 and their important effects on gas adsorption. J Am Chem Soc 135:10525–10532
56. Montoro C, Linares F, Procopio EQ et al (2011) Capture of nerve agents and mustard gas analogues by hydrophobic robust MOF-5 type metalorganic frameworks. Am Chem Soc 133:11888
57. Stylianou KC, Heck R, Chong SY et al (2010) A guest-responsive fluorescent 3D microporous metal-organic framework derived from a long-lifetime pyrene core. J Am Chem Soc 132:4119–4130
58. Liu Y, Buru CT, Howarth AJ et al (2016) Efficient and selective oxidation of sulfur mustard using singlet oxygen generated by a pyrene-based metal-organic framework. J Mater Chem A 4:13809–13813
59. Bromberg L, Klichko Y, Chang EP et al (2012) Alkylaminopyridine-modified aluminum aminoterephthalate metal-organic frameworks as components of reactive self-detoxifying materials. ACS Appl Mater Interfaces 4:4595–4602

60. Price KE, Mason BP, Bogdan AR et al (2006) Microencapsulated linear polymers: "Soluble" heterogeneous catalysts. J Am Chem Soc 128:10376–10377
61. Gil-San-Millan R, López-Maya E, Hall M et al (2017) Chemical warfare agents detoxification properties of zirconium metal-organic frameworks by synergistic incorporation of nucleophilic and basic sites. ACS Appl Mater Interfaces 9:23967–23973
62. Meng Q, Doetschman DC, Rizos AK et al (2011) Adsorption of organophosphates into microporous and mesoporous NaX zeolites and subsequent chemistry. Environ Sci Technol 45:3000–3005
63. López-Maya E, Montoro C, Rodríguez-Albelo LM et al (2015) Textile/metal-organic-framework composites as self-detoxifying filters for chemical-warfare agents. Angew Chem Int Ed Engl 54:6790–6794
64. Katz MJ, Mondloch JE, Totten RK et al (2014) Simple and Compelling biomimetic metal-organic framework catalyst for the degradation of nerve agent simulants. Angew Chem Int Ed Engl 53:497–501
65. Cavka JH, Jakobsen S, Olsbye U et al (2008) A new zirconium inorganic building brick forming metal organic frameworks with exceptional stability. J Am Chem Soc 130:13850–13851
66. Peterson GW, Moon SY, Wagner GW et al (2015) Tailoring the pore size and functionality of UiO-type metal-organic frameworks for optimal nerve agent destruction. Inorg Chem 54:9684–9686
67. Zhao J, Lee DT, Yaga RW et al (2016) Ultra-fast degradation of chemical warfare agents using MOF–nanofiber kebabs. Angew Chem 128:13418–13422
68. Xiong J, Wang L, Qin X et al (2021) Acid-promoted synthesis of defected UiO-66-NH_2 for rapid detoxification of chemical warfare agent simulant. Mater Lett 302:130427.
69. Mondloch JE, Katz MJ, Isley WC et al (2015) Destruction of chemical warfare agents using metal-organic frameworks. Nat Mater 14:512–516
70. Jiang J, Gándara F, Zhang YB et al (2014) Superacidity in sulfated metal-organic framework-808. J Am Chem Soc 136:12844–12847
71. Liang W, Chevreau H, Ragon F et al (2014) Metal-organic frameworks as media for the catalytic degradation of chemical warfare agents. Cryst Eng Comm 16:6530–6533
72. DeCoste JB, Peterson GW (2014) Metal-organic frameworks for air purification of toxic chemicals. Chem Rev 114:5695–5727
73. Moon SY, Proussaloglou E, Peterson GW et al (2016) Detoxification of chemical warfare agents using a Zr_6-based metal-organic framework/polymer mixture. Chem Eur J 22:14864–14868
74. Liu Y, Moon SY, Joseph T et al (2015) Dual-function metal-organic framework as a versatile catalyst for detoxifying chemical warfare agent simulants. ACS Nano 9:12358–12364
75. Ringenbach CR, Livingston SR, Kumar D et al (2005) Vanadium-doped acid-prepared mesoporous silica: synthesis, characterization, and catalytic studies on the oxidation of a mustard gas analogue. Chem Mater 17:5580–5586
76. Wagne GW, Yang YC (2002) Rapid nucleophilic/oxidative decontamination of chemical warfare agents. Ind Eng Chem Res 41:1925–1928
77. Boring E, Geletii YV, Hill CL (2001) A homogeneous catalyst for selective O_2 oxidation at ambient temperature. diversity-based discovery and mechanistic investigation of thioether oxidation by the Au (III)Cl_2NO_3(thioether)/O_2 system. J Am Chem Soc 123:1625–1635
78. Ameloot R, Aubrey M, Wiers BM et al (2013) Ionic conductivity in the metal-organic framework UiO-66 by dehydration and insertion of lithium tert-butoxide. Chem Eur J 19:5533–5536
79. Carniato F, Bisio F, Psaro R et al (2014) Niobium(V) saponite clay for the catalytic oxidative abatement of chemical warfare agents. Angew Chem Int Ed Engl 53:10095–10098
80. Liu Y, Howarth AJ, Hupp JT et al (2015) Catalytic degradation of chemical warfare agents and their simulants by metal-organic frameworks. Angew Chem 127:9129–9133
81. Padial NM, Procopio EQ, Montoro C et al (2013) Highly Hydrophobic isoreticular porous metal-organic frameworks for the capture of harmful volatile organic compounds. Angew Chem Int Ed 52:8290–8294

82. Daczkowski CM, Pegan SD, Harvey SP (2015) Engineering the organophosphorus acid anhydrolase enzyme for increased catalytic efficiency and broadened stereospecificity on Russian VX. Biochemistry 54:6423–6433
83. Lykourinou V, Chen Y, Wang XS et al (2011) Immobilization of MP-11 into a mesoporous metal-organic framework, MP-11@mesoMOF: a new platform for enzymatic catalysis. J Am Chem Soc 133:10382–10385
84. Feng D, Liu TF, Su J et al (2015) Wang. Stable metal-organic frameworks containing single-molecule traps for enzyme encapsulation. Nat Commun 6:5979
85. Chen Y, Lykourinou V, Hoang T et al (2012) Size-selective biocatalysis of myoglobin immobilized into a mesoporous metal-organic framework with hierarchical pore sizes. Inorg Chem 51:9156–9158
86. Li P, Moon SY, Guelta MA et al (2016) Encapsulation of a nerve agent detoxifying enzyme by a mesoporous zirconium metal-organic framework engenders thermal and long-term stability. J Am Chem Soc 138:8052–8055
87. Cheng T, Harvey SP, Chen GL (1996) Cloning and expression of a gene encoding a bacterial enzyme for decontamination of organophosphorus nerve agents and nucleotide sequence of the enzyme. Appl Environ Microbiol 62:1636–1641
88. Totten RK, Kim YS, Weston MH et al (2013) Enhanced catalytic activity through the tuning of micropore environment and supercritical CO_2 processing: Al(porphyrin)-based porous organic polymers for the degradation of a nerve agent simulant. J Am Chem Soc 135:11720–11723
89. Totten RK, Ryan P, Kang B et al (2012) Enhanced catalytic decomposition of a phosphate triester by modularly accessible bimetallic porphyrin dyads and dimers. Chem Commun 48:4178–4180
90. Dinçalp H, Kızılok S, Içli S et al (2014) Targeted singlet oxygen generation using different DNA-interacting perylene diimide type photosensitizers. J Fluorescene 24:917–924
91. Okamoto M, Tanaka F (2002) Quenching by oxygen of the lowest singlet and triplet states of pyrene and the efficiency of the formation of singlet oxygen in liquid solution under high pressure. J Phys Chem A 10:3982
92. Grossman JN, Stern AP, Kirich ML et al (2016) Anthracene and pyrene photolysis kinetics in aqueous, organic, and mixed aqueous-organic phases. Atmos Environ 128:158
93. DeCoste JB, Rossin JA, Peterson GW (2015) Hierarchical pore development by plasma etching of Zr-based metal-organic frameworks. Chem Eur J 21:18029–18032
94. Yang YC (1999) Chemical detoxification of nerve agent VX. Acc Chem Res 32:109–115
95. Yang YC, Szafraniec LL, Beaudry WT et al (1996) Autocatalytic hydrolysis of V-type nerve agents. Org Chem 61:8407–8413
96. Son FA, Wasson MC, Islamoglu T et al (2020) Uncovering the role of metal-organic framework topology on the capture and reactivity of chemical warfare agents. Chem Mater 32:4609–4617
97. Asha P, Sinha M, Mandal S (2017) Effective removal of chemical warfare agent simulants using water stable metal-organic frameworks: mechanistic study and structure—Property correlation. RSC Adv 7:6691–6696
98. Deng H, Doonan CJ, Furukawa H et al (2010) Multiple functional groups of varying ratios in metal-organic frameworks. Science 327:846
99. Phadatare A, Kandasubramanian B (2020) Metal organic framework functionalized fabrics for detoxification of chemical warfare agents. Ind Eng Chem Res 59:569–586
100. Giannakoudakis DA, Hu Y, Florent M et al (2017) Smart textiles of MOF/g-C_3N_4 nanospheres for the rapid detection/detoxification of chemical warfare agents. Nanoscale Horiz 2:356–364
101. Wang S, Pomerantz N, Dai Z et al (2020) Polymer of intrinsic microporosity (PIM) based fibrous mat: combining particle filtration and rapid catalytic hydrolysis of chemical warfare agent simulants into a highly sorptive, breathable, and mechanically robust fiber matrix. Mater Today Adv 8:100085
102. Hao L, Hurlock MJ, Li X, Ding G, Kriegsman KW, Guo X, Zhang Q (2019) Efficient oxidative desulfurization using a mesoporous Zr-based MOF. Catal Today 350:64–70

103. Zou, D, Liu D (2019) Understanding the modifications and applications of highly stable porous frameworks via UiO-66. Mater Today Chem 12:139–165
104. Vahabi AH, Norouzi F, Sheibani E et al (2021) Functionalized Zr-UiO-67 metal-organic frameworks: Structural landscape and application. Coord Chem Rev 445:214050

Chiral MOFs for Asymmetric Catalysis

Chiral Metal-Organic Frameworks for Asymmetrical Catalysis

Bhavna Dwivedi, Pawan K. Mishra, and Vivek Mishra

Contents

1 Introduction 494
2 Application of MOFs as Catalyst in Asymmetric Reactions 495
 2.1 Chiral MOF Catalyzed Asymmetric Michael Addition 495
 2.2 Enantioselective Ring-Opening of Epoxides/aziridines Using MOF Catalysts 497
 2.3 Chiral MOFs -Catalyzed Alkene Functionalization 497
 2.4 Asymmetric Dihydroxylation 500
 2.5 Chiral MOF Catalyzed Asymmetric Carbonyl-ene Reactions 501
 2.6 Asymmetric Aldol Reaction Using MOF Catalyst 502
 2.7 Asymmetric Hydrosilation and Hydroboration of Carbonyl Using Chiral MOFs 503
 2.8 Asymmetric Domino Reaction 503
 2.9 Asymmetric Diels–Alder Reaction 504
 2.10 Homochiral MOF-Mediated Asymmetric Cyanosilylation 505
 2.11 Asymmetric Henry Reaction 508
 2.12 Asymmetric Friedel–Crafts Reactions 509
 2.13 Asymmetric Tandem Reactions 510
3 Conclusion 510
List of Abbreviations 511
References 511

Abstract Due to the vital importance of optically pure organic entities, asymmetric catalysis always plays an extensive role in this asymmetric organic transformation. In asymmetric synthesis, metal-organic frameworks (MOFs) have emerged as a viable heterogeneous catalytic systems for asymmetric catalytic organic transformations due to their exclusive properties by way of crystalline highly porous structure, structural tunability, and specific catalytic sites. They also serve as a valuable

---

B. Dwivedi
School of Physical Sciences, Jawaharlal Nehru University, New Delhi, India

P. K. Mishra
School of Chemistry, Cardiff University, Cardiff, UK

V. Mishra (✉)
Amity Institute of Click Chemistry Research and Studies, Amity University, Noida, Uttar Pradesh, India
e-mail: vmishra@amity.edu

S. Gulati (ed.), *Metal-Organic Frameworks (MOFs) as Catalysts*,
https://doi.org/10.1007/978-981-16-7959-9_19

platform for studying superficial mechanistic knowledge and designing catalysts. This chapter provides an instant of CMOF advances for asymmetric catalysis. The designs, compositions, and topologies of these porous coordination materials, as well as their catalytic performance, are addressed. This chapter is expected to shed light on heterogeneous asymmetric catalysis using CMOFs and inspire future studies in this interesting subject.

Keywords Asymmetric catalysis · Chiral MOFs · Organic transformation · Porous materials

1 Introduction

Metal-organic frameworks have used one of the most important crystalline porous materials in organic synthesis, material science, and industries because of their remarkable applications in various areas such as separation of isomers, gas adsorption, intelligent sensors, catalysis, and many more [1–5]. In particular, MOFs have been widely engaged as homogenous and heterogeneous catalysts for various asymmetric reactions due to their local chiral environments, which may be produced by exploiting optically active organic linkers or assembling helical shapes [6–8].

Chiral MOFs provide a possible opportunity, presumably utilizing heterogeneous and homogeneous catalysts, such as those that are easily separated and handled [9–11]. To design chiral MOFs, depending on the nature of coordinating ligands, metals, synthesizing coordinating ligands always receive significant attention to determine the properties of MOFs in terms of better stereocontrol, hydrolytic stability tolerance in presence of acid and base, and utility as a catalyst toward asymmetric reaction [12]. Every metal-organic framework must have at least these three essential components for successful application in symmetric reaction: (1) stable porous structure, (2) the existence of active catalytic centers in the inner region, and (3) the chiral surrounding of the porous structure. MOFs provide more surface area and more prosperous opportunities to modify their structure for catalytic design centers and chiral restraints compared to different porous materials, e.g., zeolites. However, by default, organic ligands are used as a chiral auxiliary that was synthesized using developed synthetic protocols in organic synthesis. Metal cations or secondary building units (SBUs) are worked as active catalytic sites in their respective MOFs.

Most of the enantiopure MOFs or complex metal-organic framework-based catalysts have been synthesized via one of the following general methodologies that are shown in Fig. 1.

In the chapter, the authors provide a concise overview of the catalytic application of chiral metal-organic frameworks (CMOFs) in various types of asymmetric reactions all over the previous decade.

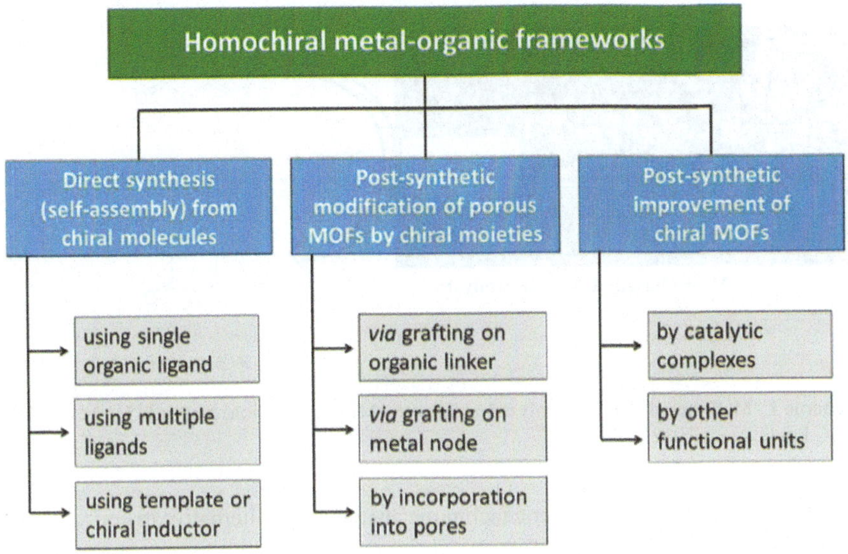

Fig. 1 Representation of various synthetic routes for access of homochiral MOFs (picture taken from [9]. © Copyright Elsevier 2021

2 Application of MOFs as Catalyst in Asymmetric Reactions

Metal-organic frameworks, MOFs have emerged as a potential catalytic system for several asymmetric organic reactions including asymmetric Aldol reaction, Michael addition, Henry reaction, Cyanosilylation, asymmetric epoxide ring-opening, cycloaddition, etc.

2.1 Chiral MOF Catalyzed Asymmetric Michael Addition

The Michael reaction, which is frequently employed in pharmaceutical and natural product synthesis, is one of the most effective techniques for creating C–C bonds [13, 14]. In recent years, significant progress has been made in the development of homochiral MOF catalyst and their application in asymmetric Michael addition reactions. Zhu and co-workers developed a range of new MOF-based organocatalyst CDMILs by utilizing commonly available starting materials MIL-10 and a chiral amine, (1R, 2R)-1,2-diphenylethylenediamine (Scheme 1). These catalysts showed excellent catalytic reactivity in the asymmetric Michael addition [15]. Thus, using this chiral heterogeneous catalyst, the asymmetric synthesis of (S)-warfarin with significant enantioselectivity may be performed on a gram-scale (2.8 g) with high

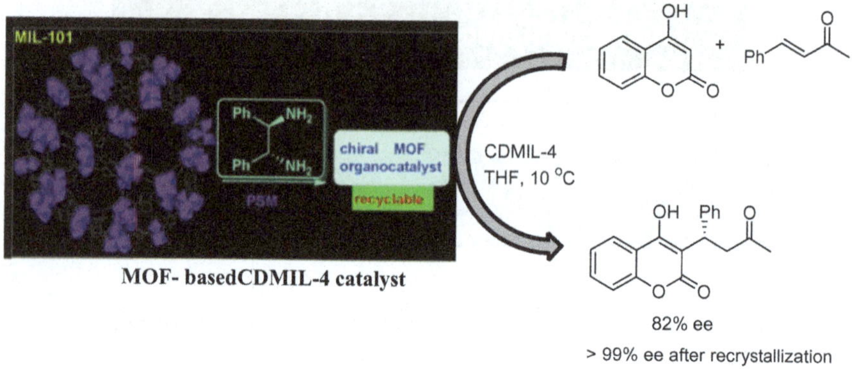

Scheme 1 MOF-based CDMIL catalyzed asymmetric Michael addition reaction [15]. © Copyright RSC Publications 2012

yield (92%), making the synthetic technique an appealing alternative to conventional methods.

Another, 8-connected zirconium and 2,5-thiophenedicarboxylate-based CMOFs, DUT-67 via post-modification through solvent-assisted linker incorporation using L-proline have demonstrated by Nguyen et. al. in Scheme 2 [16]. The resultant CMOF, DUT-67, could serve as an effective heterogeneous catalyst for the asymmetric Michael addition of trans-nitrostyrene to cyclohexanone with good to outstanding yields (~96%) and enantioselectivity when compared to L-proline in homogenous catalysis with lower ee, ~38%. Additionally, Zr-MOF has been recycled and reused up to five steps without losing their catalytic activity and crystallinity and could be easily separated from the products. As a result, chiral DUT-67-pro looks to be a suitable catalyst in asymmetric Michael addition reaction of unmodified ketones.

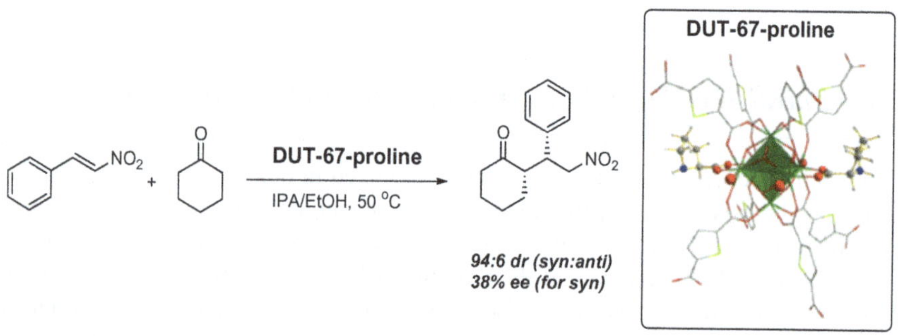

Scheme 2 DUT-67-pro-catalyzed asymmetric Michael addition of cyclohexanone to trans-nitrostyrene. © Copyright ACS Publications 2018

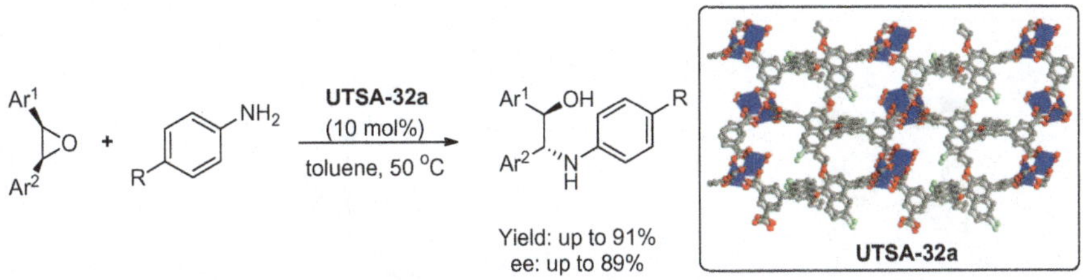

Scheme 3 UTSA-32a catalyzed asymmetric ring-opening of *meso*-epoxide with aromatic amines [20]. © Copyright RSC Publications 2013

2.2 *Enantioselective Ring-Opening of Epoxides/aziridines Using MOF Catalysts*

Chiral epoxides and aziridines are prevalent structural motif that underwent an asymmetric synthetic transformation to access other optically active functionalities [17]. Chiral MOFs have been used as heterogeneous catalysts for these reactions due to their excellent catalytic activity, site-selectivity, and high enantiomeric excess. Several asymmetric ring-openings (ARO) of epoxide with various nucleophiles have been reported [18, 19]. Zhao and co-workers have reported a binaphthyl framework based on a novel chiral catalyst viz., UTSA-32a. This homochiral catalyst was produced via the solvothermal treatment of $Zn(NO_3)_2.6H_2O$ and chiral ligand H4L at 110 °C in the co-solvent of DMA and EtOH for two days. Further, its catalytic activity was investigated for the enantioselective ring-opening reaction [20]. This catalyst is good and efficient for the enantioselective ring-opening reaction of meso-epoxide with amines to deliver corresponding β-amino alcohol as a product in good to excellent yields and up to 89% ee value (Scheme 3).

By employing (R)-2,2′dihydraxyl-1,10′-binaphthalene-5,5′-dicarboxylic acid as chiral ligand, two homochiral metal-organic frameworks (MOFs) for examples, (R)-CuMOF-1 and (R)-ZnMOF-1 have been reported by Tanaka and fellow-workers [21]. This catalyst is efficiently employed in ARO reaction between *meso*-epoxide rings with arylamines. The reaction provides better selectivity, good to excellent yields, and high enantioselectivity of related products. Moreover, retention of catalytic activity of these catalysts after use makes more efficient for this reaction (Scheme 4).

2.3 *Chiral MOFs -Catalyzed Alkene Functionalization*

Alkene functionalization catalyzed by chiral MOFs has emerged as a promising platform for accessing a wide range of organic compounds with complex structural designs [22]. Several reactions have been well investigated using chiral metal-organic framework catalysts, including asymmetric dihydroxylation, aziridination, amination, hydroboration, and epoxidation. Lin and colleagues [23] employed chiral

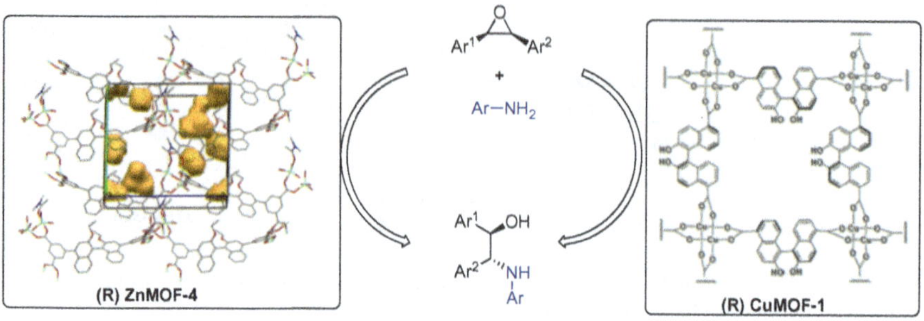

Scheme 4 (R)-CuMOF-1 or (R)-ZnMOF-4 catalyzed ARO of meso-epoxide with aromatic amines [21]. © Copyright RSC Publications 2018

MOFs catalyst generated from $[Zn_4(4\text{-}O)(O_2CR)_6]$, secondary building unit, and Mn-(salen) derived dicarboxylic acid to demonstrate a sequential asymmetric epoxidation of alkene followed via ring-opening to deliver the corresponding product in good yield and excellent enantioselectivity (Scheme 5). This catalyst shows high efficiency in synthesizing complex compounds with excellent regio- or stereo control mode.

In 2017, Liu et al. [24] have demonstrated a series of Zn MOFs having chiral linkers of metal-salen complexes (MLX) of metals including Cu, V Cr, Mn, Fe, and Co

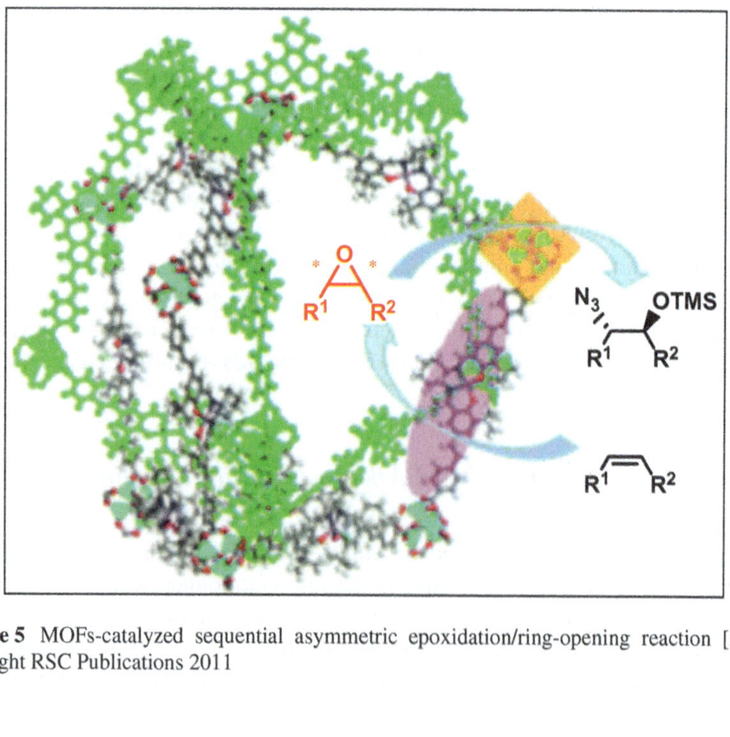

Scheme 5 MOFs-catalyzed sequential asymmetric epoxidation/ring-opening reaction [23]. © Copyright RSC Publications 2011

(Scheme 6a). This synthetic methodology can produce binary and ternary combinations of MOFs by varying ratios of different metals in salen complexes, for example, MOFs-1^{Cu}, MOFs-1^{CuV}, MOFs-1^{CuCr}, MOFs-1^{CuFe}, MOFs-1^{CuCo}, MOFs-1^{CuMnCr}, MoFs-1^{CuMnCo} revealing its catalytic activity (Scheme 6b). This binary MOFs-CuMn was an efficient heterogeneous catalyst for asymmetric epoxidation of 2,2-dimethyl-2H-chromene using 2-^{t}BuSO$_2$PhIO and iodosyl mesitylene, respectively. Additionally, these catalysts are successfully employed for sequential epoxidation/ring-opening reaction in the presence of various nucleophiles such as arylamines, thiols, trimethylsilylazide, ether, and water to produce desired products in good yields with excellent enantioselectivity (up to 99%) (Scheme 6c). These catalysts were reused for at least 3–4 sequential catalytic reactions with a decrease of 1–2% conversion.

The Cui group developed a three-dimensional robust homochiral porous metal – organic framework using chiral Cu(salen) ligand linked with chiral dicarboxylic acid functionality (Scheme 7). This catalyst is an effective heterogeneous catalyst for

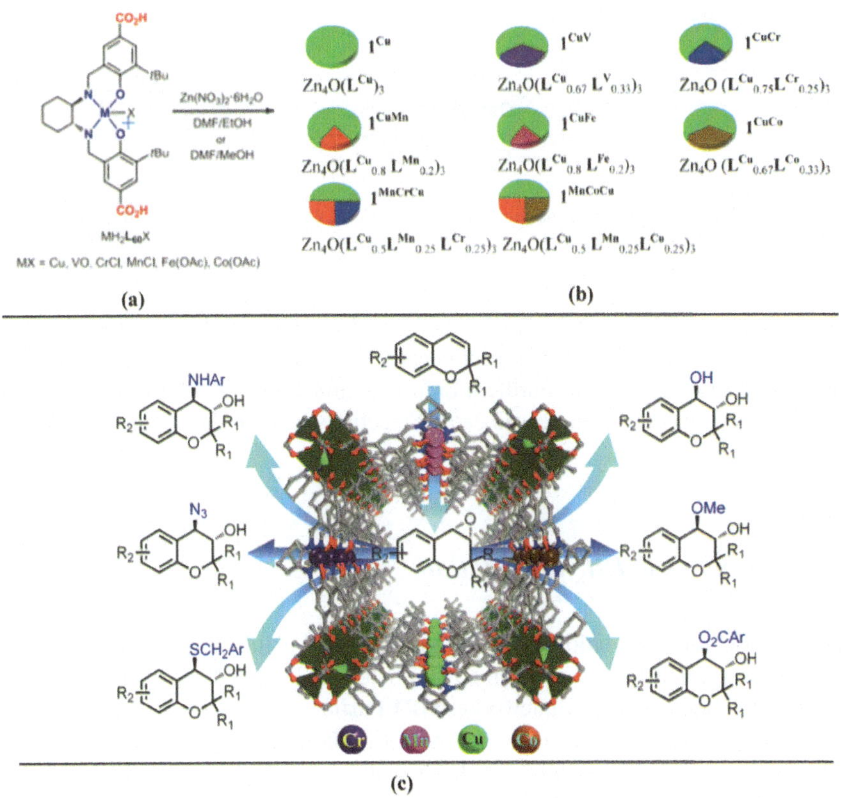

Scheme 6 **a** various M(salen) complexes; **b** MOFs catalysts with binary and trinary combination of metal-salen complexes; **c** examples of MOFs-catalyzed asymmetric epoxidation/ring-opening reactions [24]. © Copyright ACS Publications 2017

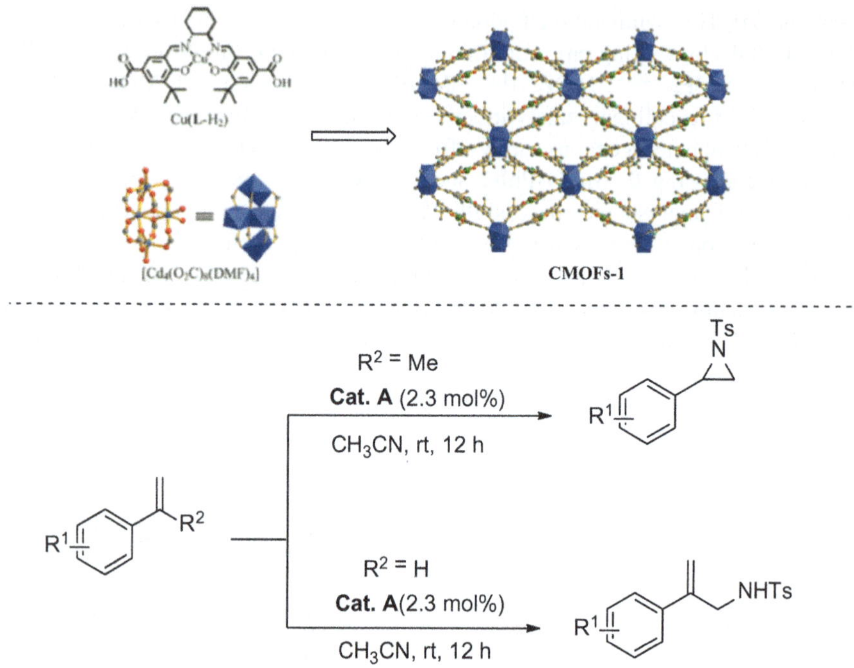

Scheme 7 Aziridination and amination of alkene using chiral Cu(salen)-based MOFs catalyst [25].

olefins functionalization such as aziridination and allylic amination reactions. In addition, the catalyst remains more useful because of easy to separate from the product and reusability for reaction. Furthermore, the chiral framework limit may result in substrate size selectivity, increased catalytic activity, and product enantioselectivity [25].

2.4 *Asymmetric Dihydroxylation*

Asymmetric dihydroxylation of alkene using a chiral metal-organic framework has been demonstrated by Duan and his teammates [26]. They have synthesized two asymmetric metal-organic frameworks, **NiPYI** using a material $Ni_2H[BW_{12}O_{40}]$ and 4,4-bipyridine with L-tert-butoxycarbonyl-2-(imidazole)-1-pyrrolidine (LBCIP) via solvothermal reaction with 60% yield. The subsequent MOFs, (S)-NiPYI and (R)-NiPYI, were employed as a catalyst to access asymmetric 1,2-diols with opposite absolute configuration via asymmetric dihydroxylation of styrene derivatives along with 15% H_2O_2 in good yield with excellent ee (up to 95%) (Scheme 8).

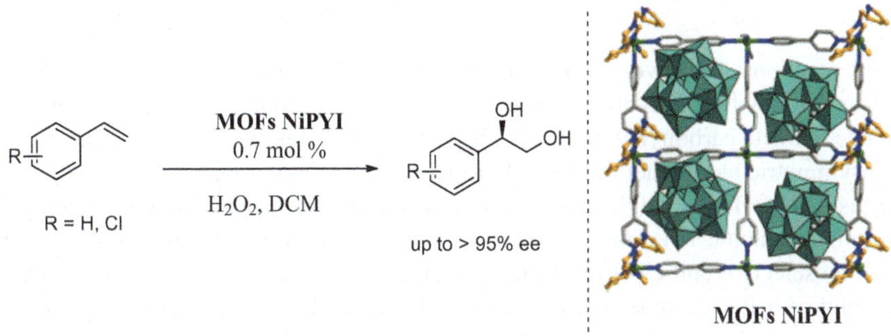

Scheme 8 NiPYI-catalyzed asymmetric 1,2-Dihydroxylation [26]. © Copyright ACS Publications 2013

2.5 *Chiral MOF Catalyzed Asymmetric Carbonyl-ene Reactions*

This is a chemical reaction between an alkene and allylic hydrogen in which compound (enophiles) has multiple bonds, i.e., formation of new C–C σ-bond due to 1,5-hydrogen shift and ene double bond migration. The reaction is also termed as Alder-Ene reaction [27].

Kim, Jeong, and co-workers reported four types of Zn/Ti-based MOFs catalysts that were employed as enantioselective catalysts in the asymmetric carbonyl ene-reaction. During the reaction, the more stoichiometric quantity of Zn/(S)-KUMOF-1 catalyst is needed that confirms substrate size plays an important role for reaction deficiency, on the other hand, Ti/(S)-KUMOF-1 catalyst performed a catalytic reaction on the crystal surface (Scheme 9) [28].

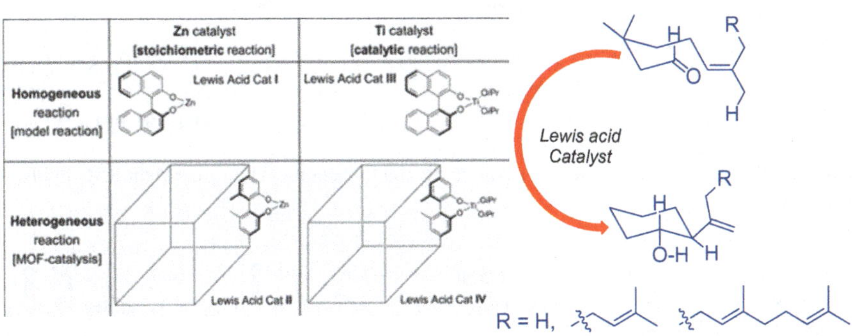

Scheme 9 KUMOF catalyzed asymmetric carbonyl-ene reactions [28]. © Copyright ACS Publications 2019

2.6 Asymmetric Aldol Reaction Using MOF Catalyst

Aldol reactions are a well-known reaction for C–C bond formation in which β-hydroxy aldehyde or β-hydroxy ketone is obtained as a product via reaction of enol or enolate ion with carbonyl compounds [29]. Chiral MOF has been used as the catalyst for asymmetric aldol reactions [30]. For example, Liu and groups reported a highly porous MIL-101-PP1 MOF catalyst developed utilizing Lewis acid and co-catalyst L-proline (Scheme 10). The aldol product was formed in 91% yield with excellent diastereoselective and enantioselective (ee approximately 92%) when this catalyst was applied in the asymmetric aldol reaction. Compared to heterogeneous and homogeneous catalysis, the asymmetric direct aldol synthesis showed high diastereo- and enantioselectivities in this reaction [31].

The diastereoselective aldol reaction was performed between 4-nitrobenzaldehyde and acyclic/cyclic ketones using a heterogenous Zr-based metal-organic frameworks catalyst for examples, UiO-66, Zr-NDC, and UiO-67 that have been reported by Voort and co-workers (Scheme 11) [32]. After screening various solvents, the catalyst UiO-66 revealed better reactivity in DCM solvent with 100% conversion, while the highest diastereoselectivity was in methanol solvent. Furthermore, these modified Zr-MOFs displayed excellent catalytic activity with diastereoselectivity in a lesser reaction time.

Using the chiral (S)-5-[1-(tert-butoxycarbonyl)pyrrolidine-2-carboxamido]isophthalic acid (IPA-L-Pro-BOC) and BTC ligands, Chen and

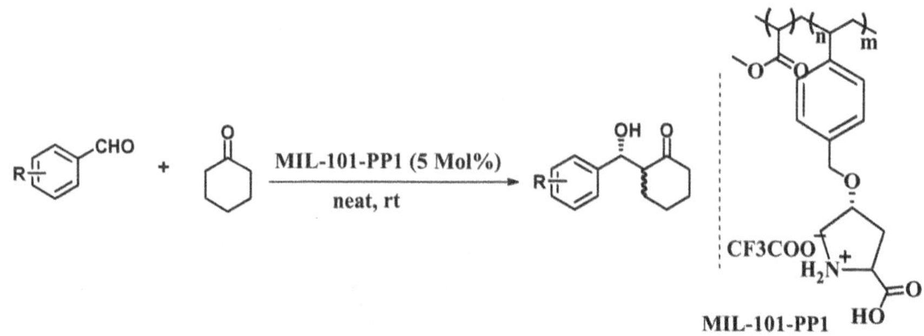

Scheme 10 MIL-101-PP1-catalyzed asymmetric aldol reaction [31]. © Copyright RSC Publications 2018

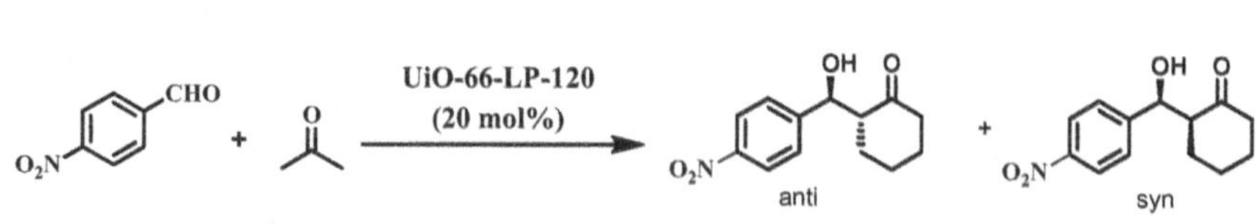

Scheme 11 Asymmetric aldol reactions using zirconium-based MOFs, UiO-66-LP-120 catalyst [32]. © Copyright RSC Publications 2018

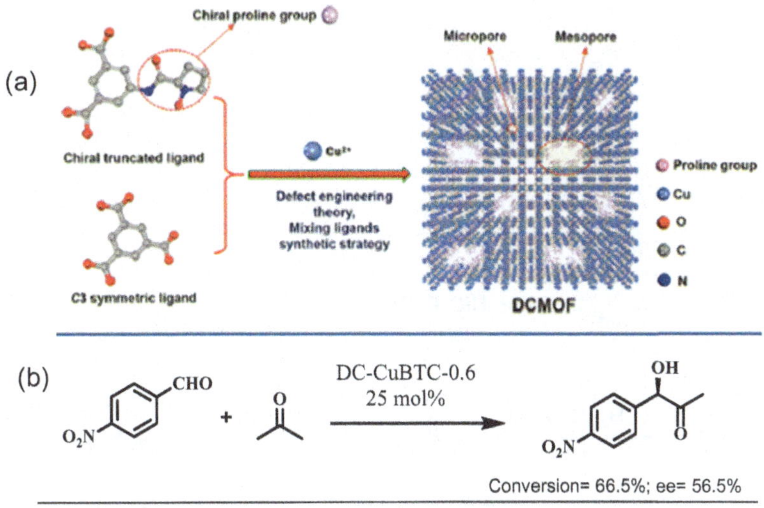

Scheme 12 **a** Synthesis of DCMOFs; **b** Asymmetric aldol reaction using DCMOFs as the heterogeneous catalyst [33]. © Copyright ACS Publications 2021

groups developed a defective DCMOFs catalyst that showed good reactivity and reusability for the aldol reaction at 25 °C (Scheme 12) [33].

2.7 Asymmetric Hydrosilation and Hydroboration of Carbonyl Using Chiral MOFs

A series of chiral MOFs, L-val-UiO-FeCl, L-ala-UiO-FeCl, L-leu-UiO-FeCl, vol-UiO-68-FeCl with naturally occurring asymmetric amino acids have been developed by Manna and co-workers (Scheme 13) [34, 35]. These catalysts provided high reactivity and enantioselectivity (>99%) as compared to their homogenous control catalyst when employed as a catalyst for the hydrosilylation and hydroboration reaction. The developed Fe-based MOF catalysts offered high turnover numbers (up to 10,000) and better recyclability and reusability (>15 times) without loss of enantioselectivity.

2.8 Asymmetric Domino Reaction

A similar and highly stable porous chiral heterogenous MOFs, UiO-67 was reported by Li and co-workers that modified by α,α-L-diaryl prolinol through immobilization of a chiral prolinol side chain into 4,4′-biphenyldicarboxylic acid (Scheme 14). [36]

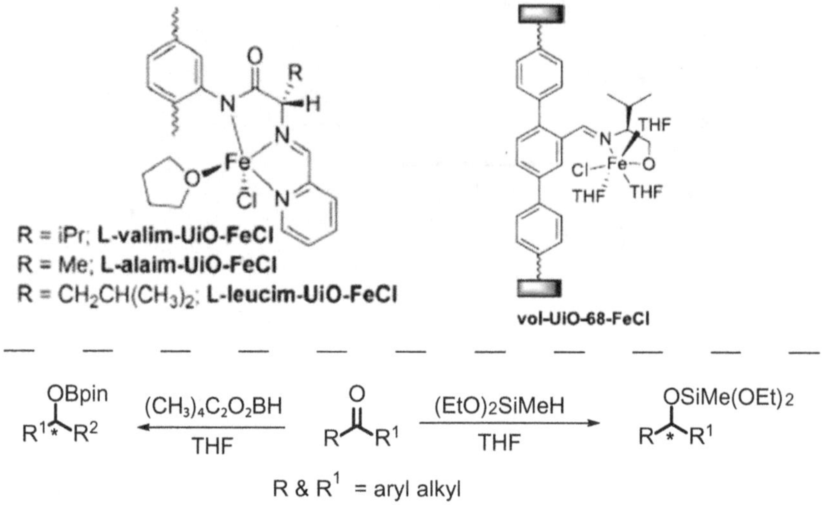

Scheme 13 UiO-Fe-catalyzed asymmetric hydrosilylation and hydroboration of ketones [34, 35]. © Copyright Wiley and ACS Publications 2021

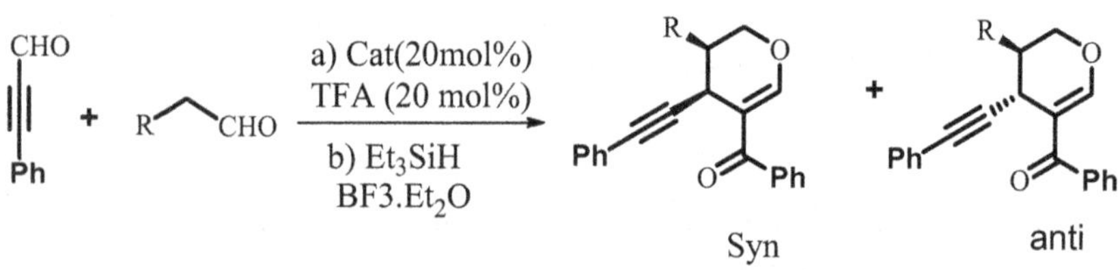

Scheme 14 MOF catalyzed asymmetric domino reaction [36]. © Copyright Elsevier 2021

The UiO-67, used as a catalyst first time for asymmetric domino reaction, i.e., asymmetric hydration/aldol/oxa-Diels–Alder cascade reaction with high yields (88%) and exceptional enantioselectivity up to 80%.

2.9 Asymmetric Diels–Alder Reaction

Diels–Alder reactions are the most versatile tools in organic synthesis for carbon–carbon bond formation [37]. In addition, the asymmetric variant of this reaction has much-gained attention by using chiral MOF as a catalyst in recent years because of capturing the reactants and assemble them in a controlled chiral region with Lewis acidic metal sites. In this regard, Tanaka and co-workers [38] have reported a chiral MOF synthesizing by solvothermal reaction between $Cu(NO_3)_3.3H_2O$ and chiral ligand, (R)-2,2′-dihydroxy-1,1′-binaphthyl-4,4′-dibenzoic acid in the mixture

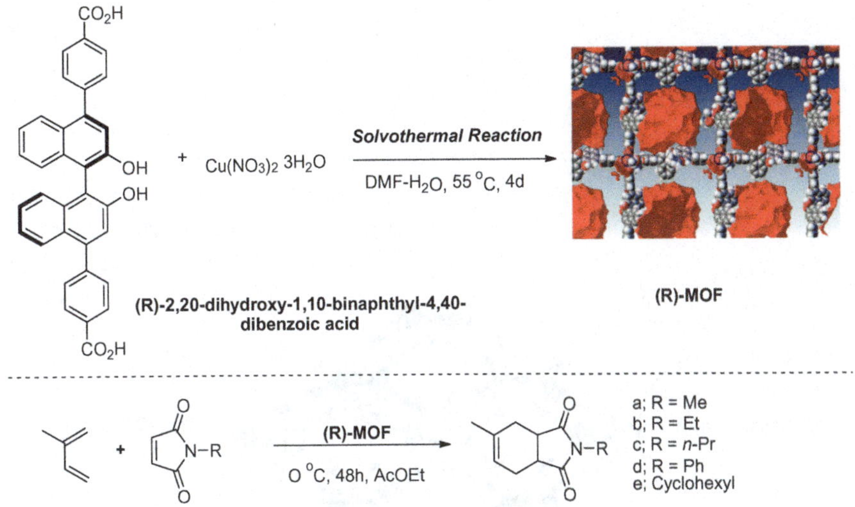

Scheme 15 (R)-MOF catalyzed asymmetric domino reaction [38]. © Copyright RSC Publications 2016

of dimethylformamide and water (Scheme 15). This catalyst employed asymmetric D.A. reaction to produce the [4+2] cycloadduct product in excellent yield and up to 75% ee, employing isoprene and *N*-alkyl substituent of maleimide as substrates. In addition, the enantioselectivity of the product depends on the variation of length of N-alkyl substituent of maleimide.

Liu, Cui and co-workers reported a more stereoselective chiral Cr-Salen-functionalized MOF by heating of chiral Cr-Salen complex, cadmium iodide (CdI_2) and sodium acetate in the mixture of DMF and MeOH [39]. Screening of this catalyst for various asymmetric [4+2] cycloaddition reactions showed high reactivity and enantioselectivity (>99%) (Scheme 16). An advantageous feature of catalyst is several times recyclability and reusability for the reaction without loss of catalytic efficacy and enantioselectivity.

2.10 Homochiral MOF-Mediated Asymmetric Cyanosilylation

Cyanosilylation reaction is one most studied nucleophilic addition reactions in carbonyl compounds. Due to the remarkable properties of the metal-organic framework, MOFs have been extensively used as a highly efficient heterogeneous catalyst for cyanosilylation of various aldehyde or ketone with TMSCN. The silylated cyanohydrins smoothly converted to various valuable organic compounds including

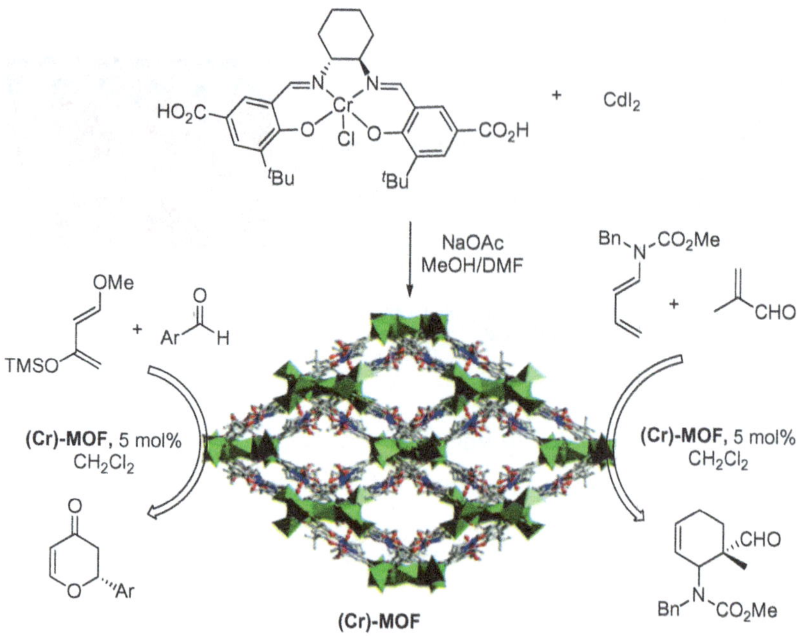

Scheme 16 (Cr)-MOF catalyzed [4+2] cycloaddition reaction [39]. © Copyright RSC Publications 2016

β-amino alcohol, cyanohydrins, etc. Cui and co-workers have reported an enantiopure chiral MOF of BINOL ligand [40]. This MOF becomes the more efficient and recyclable catalyst for asymmetric cyanation of aldehyde after exchanging one proton of the dihydroxyl group for Li(I) ions. The synthetic usefulness of cyanation was proven with 98% ee in the access of a nonselective adrenoceptor blocking drug, (S)-bufuralol (Scheme 17a). Voort groups synthesized a vanadium–salen based chiral Cd-bpdc MOF using (R,R)-1,2-cyclohexanediamino-N,N'-bis(3-tert-butyl-5-(4-pyridyl)salicylidene) under solvothermal conditions (Scheme 17b). This MOF has a high BET surface area of 574 m^2g^1 and is inherently microporous. Compared to previous metallosalen-based MOFs, this framework has a greater CO_2 uptake capacity at 273 K and 1 bar. This chiral material was used for asymmetric cyanosilylation of carbonyl compounds in the absence of solvent [41]. The catalyst is recyclable and reusable, and it performed well in terms of conversion and ee. Further, Jiang and co-workers reported a porphyrin–salen-based chiral metal-organic framework (ps-CMOF) that could serve as the heterogeneous catalyst for enantioselective cyanosilylation of aldehydes and provide the silylated cyanohydrin product in good yield with promising enantioselectivity. However, ketones showed poor reactivity toward this catalytic system (Scheme 17c) [42].

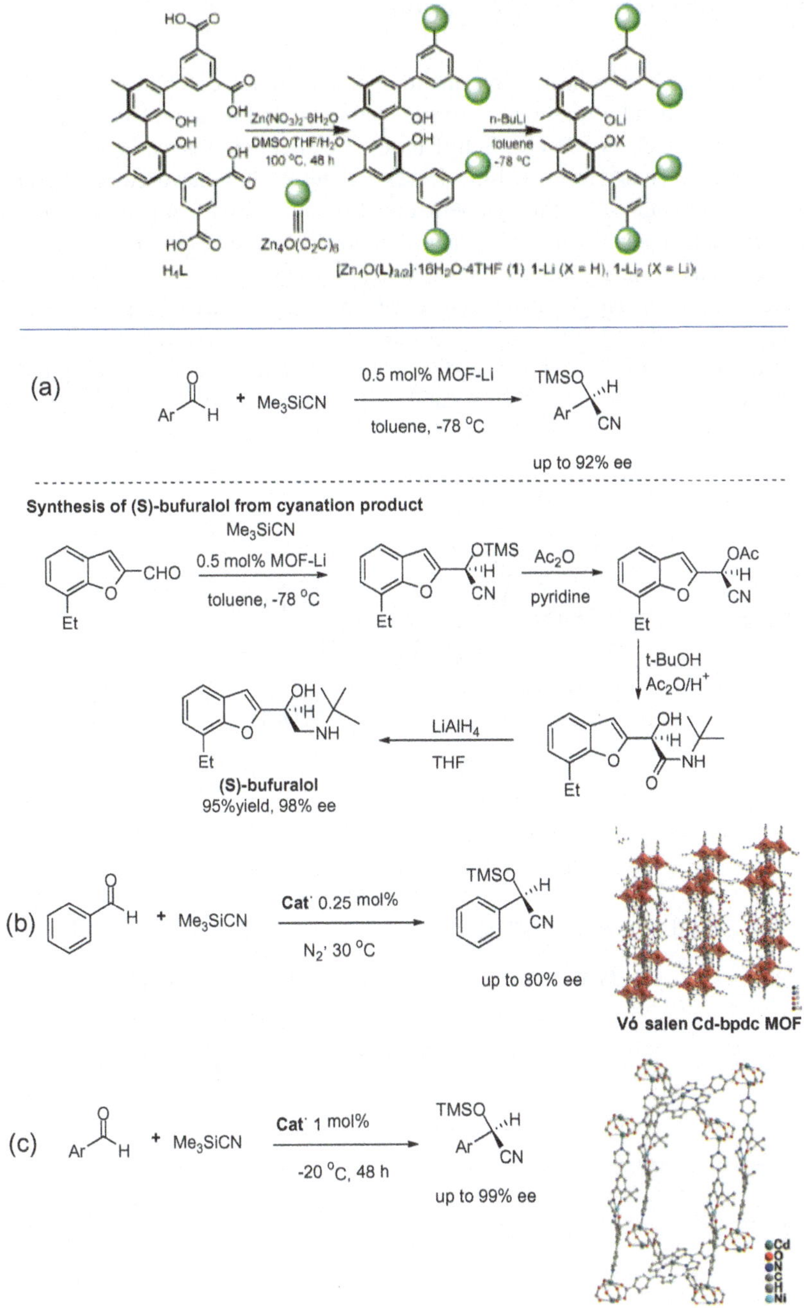

Scheme 17 MOF catalyzed asymmetric cyansilylation of aldehydes [41, 42]. © Copyright RSC Publications 2016 and ACS Publications 2014

2.11 *Asymmetric Henry Reaction*

Henry reaction is one of the most promising synthetic strategies for the construction of carbon–carbon bond formation in organic chemistry. In literature, various homogenous catalysts such as BINOL, Schiff bases, salen complex have been developed for asymmetric Henry reaction [43]. Recently, chiral MOF has used an alternative heterogeneous catalyst for this synthetic transformation. Ren, Jiang, and co-workers have reported an enantiopure (homochiral) Cu(salen)-based MOF used as a heterogeneous catalyst for asymmetric Henry reaction (Scheme 18) [44]. The nitroaldol product was obtained in good yield with excellent enantioselectivity as compared to a homogenous catalyst. In the case of bulky aldehyde substrate, activity was decreased with similar enantioselectivity. In addition, easy separation, and reusability for five (05) following runs without loss of catalytic activity proved more efficient of this developed MOF.

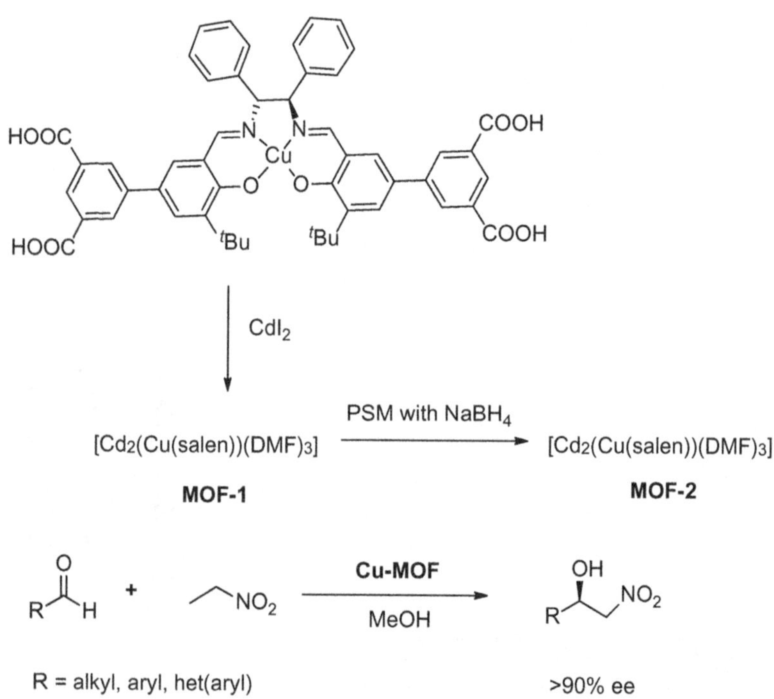

Scheme 18 CMOF-mediated asymmetric Henry reaction [44]. © Copyright ACS Publications 2018

2.12 Asymmetric Friedel–Crafts Reactions

In recent years, homochiral MOFs emerged as an efficient catalyst for asymmetric Friedel–Crafts reaction due to the presence of potential Lewis acid metal sites with the chiral environment of ligands [45]. Liu et al. prepared Zr-MOF that efficiently catalyzed enantioselective Friedel–Crafts (FC) reaction employing indole and N-sulfonyl aldimines in asymmetric FC reaction in reasonable yields with good to excellent enantioselectivity [46]. Later, three Mn-CMOFs that was synthesized from 1,1-bihenol-phosphonocarboxylate ligands having 3,5- bis(trifluoromethyl)-, bismethyl-, and bisfluoro-phenyl substituents at the 3,3-position [47]. Further, to evaluate the catalytic efficiency of these MOFs, asymmetric Friedel–Crafts alkylation was performed between various derivatives of indole and β, γ-unsaturated α-keto esters. The catalyst was also efficient in the case when asymmetric F. C. alkylation carried pyrrole with nitroalkene. The yield of corresponding products was good to excellent with excellent enantioselectivity up to 97%. Moreover, these catalysts showed outstanding recyclability and reusability of more than ten successive runs without losing their catalytic efficiency (Scheme 19).

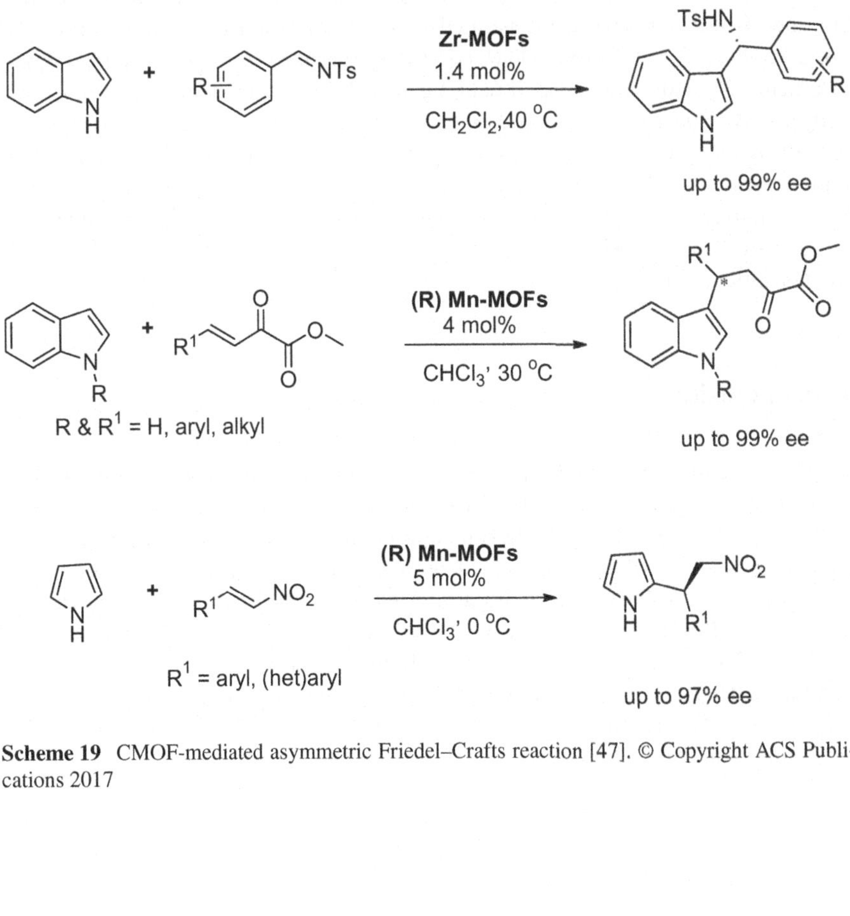

Scheme 19 CMOF-mediated asymmetric Friedel–Crafts reaction [47]. © Copyright ACS Publications 2017

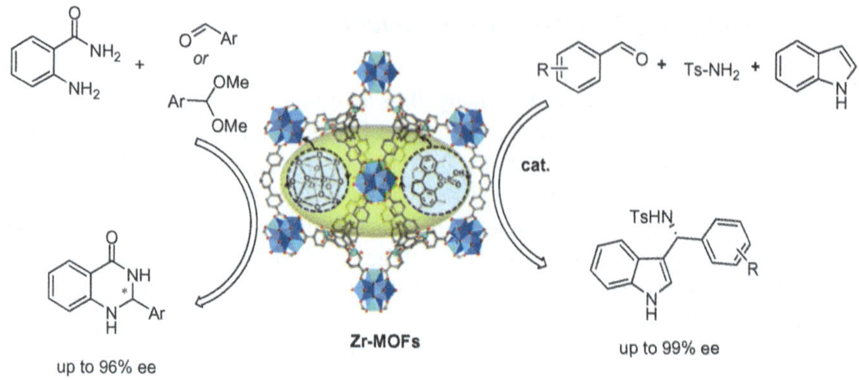

Scheme 20 (Zr)-MOF catalyzed asymmetric tandem reaction [46]. © Copyright ACS Publications 2019

2.13 *Asymmetric Tandem Reactions*

MOFs become a promising catalyst for asymmetric tandem reactions. Liu and Cui and co-workers have demonstrated a couple of chiral crystalline porous Zr-based MOFs, $[Zr_6O_4(OH)_8(H_2O)_4(L)_2]$, was synthesized from enantiopure 4,4,6,6-tetra(benzoate) and -tetra(2-naphthoate) ligands of 1,1-spirobiindane-7,7-phosphoric acid in DMF and formic acid (scheme 20) [46]. These materials showed an efficient catalytic system for condensation/cyclization reaction using aldehydes with 2-aminobenzamide. These CMOF could also serve as a heterogeneous catalyst toward three-component reactions. Three-component reactions between indole, aldehyde, and *para*-toluene sulfonamide have furnished corresponding in good yields and in good enantioselectivity.

3 Conclusion

Because optically pure organic entities are so essential, asymmetric catalysis, such as asymmetric Michael addition, aldol reaction, Henry reaction, cyanosilation, asymmetric epoxide ring-opening, and cycloaddition, plays a key role in this asymmetric organic transition. MOFs have arisen as a potential catalytic system for asymmetric catalytic organic transformations due to their exclusive properties, e.g., highly regular crystalline porous structure, catalytic sites, specific properties, and structural tunability. They also provide an excellent foundation for basic mechanistic understanding and catalyst design. This chapter is anticipated to emphasize heterogeneous asymmetric catalysis utilizing chiral MOFs as well as stimulate more research in this promising field.

List of Abbreviations

CMOF	Chiral metal-organic framework
F.C. alkylation	Friedel Crafts alkylation
ARO	Asymmetric Ring Opening
D.A. reactions	Diels Alder reactions
TMSCN	Trimethylsilyl cyanide

References

1. Furukawa H, Cordova KE, O'Keeffe M, Yaghi OM (2013) The chemistry and applications of metal-organic frameworks. Science 341:1230444
2. Bétard A, Fischer RA (2012) Metal-organic framework thin films: from fundamentals to applications. Chem Rev 112:1055–1083
3. Xuan W, Zhu C, Liu Y, Cui Y (2012) Mesoporous metal-organic framework materials. Chem Soc Rev 41:1677–1695
4. Dandan Li, Hai-Qun Xu, Long J, Hai-Long J (2019) Metal-organic frameworks for catalysis: state-of-the-art, challenges, and opportunities. Energy Chem 1:100005
5. Jiang HL, Xu Q (2011) Porous metal-organic frameworks as platforms for functional applications. Chem Commun 47:3351–3370
6. Corma A, García H, Llabrés i Xamena FX (2010) Engineering metal organic frameworks for heterogeneous catalysis. Chem Rev 110:4606–4655
7. Farrusseng D, Aguado S, Pinel C (2009) Metal-organic frameworks: opportunities for catalysis. Angew Chem Int Ed 48:7502–7513
8. Huang YB, Liang J, Wang XS, Cao R (2017) Multifunctional metal-organic framework catalysts: synergistic catalysis and tandem reactions. Chem Soc Rev 46:126–157
9. (a) Dybtsev DN, Bryliakov KP (2021) Asymmetric catalysis using metal-organic frameworks. Coord Chem Rev 437:213852 (b) Liu Y, Xuan W, Cui Y (2010) Engineering homochiral metal-organic frameworks for heterogeneous asymmetric catalysis and enantioselective separation. Adv Mater 22:4112−4135
10. Yoon M, Srirambalaji R, Kim K (2012) Homochiral meta-organic frameworks for asymmetric heterogeneous catalysis. Chem Rev 112:1196–1231
11. Ma L, Falkowski JM, Abney C, Lin W (2010) A series of isoreticular chiral metal-organic frameworks as a tunable platform for asymmetric catalysis. Nat Chem 2:838–846
12. Cohen SM (2012) Postsynthetic methods for the functionalization of metal-organic frameworks. Chem Rev 112:970–1000
13. Zhang Y, Wang W (2013) C−C bond formation by michael reaction. In: Rios Torres R (ed) Stereoselective organocatalysis. Wiley, Hoboken, NJ, pp 147−203
14. Takemoto Y, Stadler M (2012) C−C bond formation: michael reaction. In: Carreira EM, Yamamoto H (eds) Comprehensive chirality, vol 6. Elsevier Science, Oxford, pp 37−68
15. Shi T, Guo Z, Yu H, Xie J, Zhong Y, Zhu W (2013) Atom-economic synthesis of optically active warfarin anticoagulant over a chiral MOF organocatalyst. Adv Synth Catal 355:2538–2543
16. Nguyen KD, Kutzscher C, Drache F, Senkovska I, Kaskel S (2018) Chiral functionalization of a zirconium metal-organic framework (dut-67) as a heterogeneous catalyst in asymmetric Michael addition reaction. Inorg Chem 57:1483–1489
17. Sweeney JB (2002) Aziridines: epoxides' ugly cousins? Chem Soc Rev 31:247–258
18. Nielsen LPC, Jacobsen E N (2006) Catalytic asymmetric epoxide ring-opening chemistry. In: Yudin AK (ed) Aziridines and epoxides in organic synthesis. Wiley-VCH, Weinheim, Chapter 7, p 229

19. Jacobsen EN (2000) Asymmetric catalysis of epoxide ring-opening reactions. Acc Chem Res 33:421–431
20. Regati S, He Y, Thimmaiah M, Li P, Xiang S, Chen JB, Zhao CG (2013) Enantioselective ring-opening of meso-epoxides by aromatic amines catalyzed by a homochiral metal-organic framework. Chem Commun 49:9836–9838
21. Tanaka K, Kinoshita M, Kayahara J, Uebayashi Y, Nakaji K, Morawiak M, Urbanczyk-Lipkowska Z (2018) Asymmetric ring-opening reaction of meso -epoxides with aromatic amines using homochiral metal-organic frameworks as recyclable heterogeneous catalysts. RSC Adv 8:28139–28146
22. Xia Q, Yuan C, Li, Y, Cui Y (2019) Design and assembly of a chiral composite metal-organic framework for efficient asymmertric sequential transformation of alkenes to amino alcohols Chem Commun 55(2019):9136–9139
23. Song F, Wang C, Lin W (2011) A chiral metal-organic framework for sequential asymmetric catalysis. Chem Commun 47:8256–8258
24. Xia Q, Li Z, Tan C, Liu Y, Gong W, Cui Y (2017) Multivariate metal-organic frameworks as multifunctional heterogeneous asymmetric catalysts for sequential reactions. J Am Chem Soc 139:8259–8266
25. Liu Y, Li Z, Yuan G, Xia Q, Yuan C, Cui Y (2016) Chiral Cu(salen)-based metal-organic framework for heterogeneously catalyzed aziridination and amination of olefins. Inorg Chem 55:12500–12503
26. Han Q, He C, Zhao M, Qi B, Niu J, Duan C (2013) Chiral Cu(salen)-based metal-organic framework for heterogeneously catalyzed aziridination and amination of olefins. J Am Chem Soc 135:10186–10189
27. Hilt G, Treutwein J (2007) Cobalt-catalyzed Alder-ene reaction. Angew Chem Int Ed 46:8500–8502
28. Han J, Lee MS, Thallapally PK, Kim M, Jeong N (2019) Identification of reaction sites on metal-organic framework-based asymmetric catalysts for carbonyl-ene reactions. ACS Catal 9:3969–3977
29. Mahrwald R (2004). Modern Aldol reactions, vols 2. Wiley, KGaA, Weinheim, Germany, pp 1218–23. ISBN 978-3-527-30714-2.
30. Demuynck ALW, Goesten MG, Ramos-Fernandez EV, Dusselier M, Vanderleyden J, Kapteijn F, Gascon J, Sels BF (2014) Induced chirality in a metal-organic framework by postsynthetic modification for highly selective asymmetric Aldol reactions. ChemCatChem 6:2211–2214
31. Dong XW, Yang Y, Che JX, Zuo J, Li XH, Gao L, Hua YZ, Liu XY (2018) Heterogenization of homogeneous chiral polymers in metal-organic frameworks with enhanced catalytic performance for asymmetric catalysis. Green Chem 20:4085–4093
32. Feng X, Sekhar JH, Leus K, Wang G, Ouwehand J, Voort VDP (2018) L-proline modulated zirconium metal organic frameworks: simple chiral catalysts for the aldol addition reaction. J Catal 365:36–42
33. Chen Z, Yan X, Li M, Wang S, Chen C (2021) Defect-engineered chiral metal-organic frameworks for efficient asymmetric Aldol reaction. Inorg Chem 60:4362–4365
34. Newar R, Akhtar N, Antil N, Kumar A, Shukla S, Begum W, Manna K (2021) Amino acid-functionalized metal-organic frameworks for asymmetric base-met catalysis. Angew Chem Int Ed 60:10964–10970
35. Antil N, Akhtar N, Newar R, Begum W, Kumar A, Chauhan M, Manna K (2021) Chiral iron(ii)-catalysts within valinol-grafted metal-organic frameworks for enantioselective reduction of ketones. ACS Catal 11:10450–10459
36. Ren H, Cheng L, Yang J, Zhao K, Zhai Q, Li Y (2021) Recyclable and reusable chiral α, α-L-diaryl prolinol heterogeneous catalyst grafting to UiO-67 for enantioselective hydration/aldol/oxa-Diels Alder domino reaction. Catal Commun 149:106–249
37. Kagan' HB, Riant O (1992) Catalytic asymmetric Diels-Alder reactions. Chem Rev 92:1007–1019
38. Tanaka K, Nagase S, Anami T, Wierzbicki M, Urbanczyk-Lipkowska Z (2016) Enantioselective Diels–Alder reaction in the confined space of homochiral metal-organic framework RSC Adv 6:111436–111439

39. Xia Q, Liu Y, Li Z, Gong W, Cui YA (2016) Cr(salen)-based metal-organic framework as a versatile catalyst for efficient asymmetric transformations Chem Commun 52:13167–13170.
40. Mo K, Yang Y, Cui Y (2014) A Homochiral metal-organic framework as an effective asymmetric catalyst for cyanohydrin synthesis. J Am Chem Soc 136:1746–1749
41. Bhunia A, Dey S, Moreno JM, Diaz U, Concepcion P, Hecke KV, Janiak C, Voort PVD (2016) A homochiral vanadium–salen based cadmium bpdc MOF with permanent porosity as an asymmetric catalyst in solvent-free cyanosilylation. Chem Commun 52:1401–1404
42. Li J, Ren Y, Qi C, Jiang H (2017) The first porphyrin–salen based chiral metal-organic framework for asymmetric cyanosilylation of aldehydes. Chem Commun 53:8223–8226
43. Boruwa J, Gogoi N, Saikia PP, Barua NC (2006) Catalytic asymmetric Henry reaction Tetrahedron. Asymmetry 17:3315–3326
44. Fan Y, Ren Y, Li J, Yue C, Jiang H (2018) Enhanced activity and enantioselectivity of henry reaction by the postsynthetic reduction modification for a chiral Cu(salen)-based metal-organic framework. Inorg Chem 57:11986–11994
45. Zhang Y, Liu X, Lin L, Feng X (2018) Recent advance in catalytic asymmetric friedel-crafts reactions. Progr Chemistry 30:491–504
46. Gong W, Chen X, Jiang H, Chu D, Cui Y, Liu Y (2019) Highly stable Zr(IV)-based metal-organic frameworks with chiral phosphoric acids for catalytic asymmetric Tandem reactions. J Am Chem Soc 141:7498–7508
47. Chen X, Jiang H, Hou B, Gong W, Liu Y, Cui Y (2017) Boosting chemical stability, catalytic activity, and enantioselectivity of metal-organic frameworks for batch and flow reactions. J Am Chem Soc 136:1746–1749

MOFs as Catalysts for the Storage of Methane

Metal–Organic Frameworks: Promising Materials for Methane Storage

Pooja Rani, Ahmad Husain, and Girijesh Kumar

Contents

1 Introduction 518
2 Basic Terminology Used in CH_4 Storage 519
2.1 Excess, Absolute and Total Adsorption 519
2.2 Gravimetric and Volumetric Capacity 520
2.3 Working Capacity or Deliverable Capacity 520
3 CH_4 Storage in MOFs 521
3.1 Requirements for MOFs as ANG Adsorbent 521
3.2 Characterizations Required for MOFs to Investigate Their CH_4 Adsorption Capacity 522
4 Design and Synthesis of MOFs for CH_4 Storage 523
4.1 Rigid MOFs for CH_4 Storage 524
4.2 Flexible MOFs for CH_4 Storage 540
5 Conclusion and Outlook 544
Abbreviations 544
References 545

Abstract Porous crystalline materials are also known as metal–organic frameworks (MOFs) represents the ultimate chemical platform to the scientific community not only because of their captivating topological features but also due to their remarkable and wide range of beneficial applications such as magnetism, molecular recognition, biomedicine, fluorescence, heterogeneous catalysis and particularly gas storage and separation (Rani et al. in Cryst Growth Des 20:7141–7151, 2020 [1]; Kumar and Das in Inorg Chem Front 4:202–233, 2017 [2]; Kumar and Gupta in Chem Soc Rev 42:9403–9453, 2013 [3]; Husain et al. in CrystEngComm 22:5980–5986, 2020 [4]; Kumar et al. in Inorg Chem Front 8:1334–1373, 2021 [5]; Fang et al. in Chem Soc Rev 49:3638–3687, 2020 [6]; Ghasemzadeh et al. in Green Chem 22:7265–7300, 2020 [7]; Goetjen et al. in Chem. Commun 56:10,409–10,418, 2020 [8]; He et al.

P. Rani · G. Kumar (✉)
Department of Chemistry and Centre for Advanced Studies in Chemistry, Panjab University Chandigarh, Chandigarh, India
e-mail: gkumar@pu.ac.in

A. Husain (✉)
Department of Chemistry, DAV University, Jalandhar, Punjab 144012, India
e-mail: ahmad10620@davuniversity.org

S. Gulati (ed.), *Metal-Organic Frameworks (MOFs) as Catalysts*,
https://doi.org/10.1007/978-981-16-7959-9_20

in Chem. Soc. Rev 43:5657–5678, 2014 [9]). The storage and separation of natural gases particularly CH_4 are some of the most demanding and challenging tasks for the petrochemical industry. Natural gas (NG), whose main component is methane (CH_4) has been considered as an alternative energy resource for transportation fuel because of its high abundance and relatively low CO_2 emission (Makal et al. in Chem Soc Rev 41:7761–7779, 2012 [10]). However, the current challenge with the widespread use of CH_4 is its low volumetric energy storage density. The storage and transportation of CH_4 require high-pressure compression (200–300 bar) which is potentially risky and costly (Wen et al. in Chem Commun 56:13,117–13,120, 2020 [11]; Li et al. in Energy Environ Sci 8:2504–2511, 2015 [12]). Nevertheless, this problem could be resolved using the adsorbent, which is employed to increase the storage capacity as compared to a compressed gas tank (Makal et al. in Chem Soc Rev 41:7761–7779, 2012 [10]; Kundu et al. in Chem Mater 31:2842–2847, 2019 [13]; Wu et al. in J Phys Chem C 123:8550–8559, 2019 [14]). Therefore, adsorbed natural gas systems will be better suited in terms of both cost and safety concerns. From the past few decades, great efforts had been devoted to developing highly efficient, cost-effective, stable, reusable as well as environment-friendly porous adsorbent materials. Among all the porous adsorbent materials, MOFs are considered the most promising storage adsorbents due to their tunable structure and high porosity (Li et al. in Energy Environ Sci 8:2504–2511, 2015 [12]). MOFs have various advantages over conventional adsorbent materials, such as high surface area, easily tunable pore size, shape, and desired functionalities that suit CH_4 storage (Zhang et al. in Chem Eur J 24:7866–7881, 2018 [15]).

Keywords Metal–organic frameworks · Methane storage · ANG adsorbent · Working and deliver capacity · Natural gas (NG)

1 Introduction

Owing to the rapid development of industrialization, consumption of fossil fuels such as coal, diesel, petrol, and natural gases are rising enormously, therefore causing not only challenges to the sustainability of oil reserves but also increasing serious environmental and economic issues [16, 17]. Therefore, the development of alternate resources for conventional petroleum-based transportation fuels is strongly encouraged [18]. Among the various clean energy fuels, hydrogen (H_2) is considered the most ideal cleaner energy source because of its pollution-free burning, high chemical energy production, and water being the only by-product [19]. In addition to hydrogen, natural gas (comprising mainly 95% methane, CH_4), is also considered an important fuel because of its high research octane number (RON), relative abundance, high thermal efficiency, and reasonably low CO_2 emissions [20, 21].

However, the comparatively low volumetric energy storage density of natural gas challenges its excessive use for the transportation of vehicles. To overcome this issue, three main strategies have been established. In the first approach, liquefied natural gas (LNG) was used wherein CH_4 is stored under cryogenic conditions, which compresses CH_4 to 0.16 vol % with an energy density of approximately 25 MJ L^{-1} compared to gaseous CH_4. But the main disadvantage of this approach is to cool down the whole storage tank to −162 °C, which raises the operational cost [22]. In the second method compressed natural gas (CNG) is used, wherein CH_4 is stored under high-pressure of 200–250 bar in fuel tanks. Wherein, CH_4 is compressed to 1 vol % and therefore, increasing its energy density to 10 MJ L^{-1}. However, this protocol has a lot of limitations in respect of its design and development and is limited to few countries like Europe, Asia, and South America due to its serious safety concerns in transport and flammable nature. The third protocol also well-known as adsorbed natural gas (ANG) comprises the utilization of porous adsorbent materials to store the CH_4 under much lower pressure compared to CNG [23].

Therefore, adsorption technology using porous materials offers a safer, simpler, and cost-effective method for the storage of NG under ambient temperature and pressure. In the past few decades, a wide range of porous materials such as zeolites, activated carbon, porous silica, porous polymers, carbon nanotubes, covalent organic frameworks (COFs), porous organic polymers (POPs), and MOFs have been explored [10, 24]. Among these, MOFs are considered as the potential porous materials having all features required to become a perfect candidate for ANG technology. MOFs are an important class of crystalline nanoporous materials that are constructed from metal ions/clusters connected by multitopic organic ligands. Furthermore, MOFs offer high surface area and porosity, tunable pore volume, open metal sites, and functionalized as well as well-ordered pore surfaces [25]. These topological and tailored features facilitate the host–guest interactions between MOFs and CH_4 and make it a perfect adsorbent material for the safe storage of CH_4 [10, 15]

This chapter covers a comprehensive list of remarkable examples of MOFs that have been used as porous adsorbent materials for the storage of CH_4 and thus, providing a perspective to the contemporary scientists to develop an alternative as a replacement for conventional fuels and cope with the problem of sustainability of oil reserves, economic and environmental issues.

2 Basic Terminology Used in CH_4 Storage

2.1 Excess, Absolute and Total Adsorption

The basic term such as excess, absolute and total uptake has been widely used in the literature during the determination of CH_4 storage capacity from the adsorption isotherms. Excess adsorption (n_{ex}) is defined as the amount of adsorbed gas molecules having interactions with the pore surface of the adsorbent [18]. Similarly, absolute adsorption is defined as the sum of gas molecules exhibit interactions with the pore surface of the adsorbent as well as those that do not have gas–solid interactions however staying in the adsorbed region [26]. But, it is impossible to determine the adsorbed region experimentally, therefore, absolute adsorption cannot be provided directly and thus, calculated using high-pressure CH_4 adsorption measurements [27]. Moreover, the term total uptake (n_{tot}) is defined as the total amount of gas molecules present inside a material's pores and can be calculated using the following equation [28].

$$n_{tot} = n_{ex} + \rho_{bulk}(P, T) \times V_p,$$

where ρ_{bulk} is obtained from the NIST Refprop [29] and V_p is generally determined from a N_2 adsorption isotherm at 77 K.

2.2 Gravimetric and Volumetric Capacity

The CH_4 adsorption capacity is usually measured with the help of either gravimetric or volumetric protocol [30]. In the gravimetric method, CH_4 uptake is defined as the mass of CH_4 adsorbed per unit mass of adsorbents. On the other hand, in the volumetric methodology uptake is defined as the volume of CH_4 adsorbed under standard pressure and temperature and divided by the volume of adsorbents. It is important to mention that the volumetric capacity plays an important role compared to the gravimetric capacity in the case of vehicular applications because passenger vehicles have limited space for gas tanks. Therefore, in order to calculate volumetric capacity, the density of adsorbent is required. Notably, for several tested MOFs the ideal crystallographic density has been extensively used to calculate the maximum possible volumetric uptake of CH_4 [30]. Theoretically, the real packing density of a particular MOF would be much lower than its crystallographic density [31, 32]. Thus, it is very important and essential to specify the types of density used to calculate the volumetric capacity. As there are different types of density such as crystallographic, bulk, wafer, and pellet are used to report the volumetric storage capacity [18].

2.3 Working Capacity or Deliverable Capacity

The working capacity which is also known as deliverable capacity is defined as the amount of CH_4 released from the porous adsorbent when the pressure is released from high-pressure (usually 35 bar or 65 bar) to the lower service pressure which is nearly 5.8 bar [9, 24, 33]. Notably, the working capacity of a particular MOF is more important than their CH_4 storage capacity in terms of their practical applicability, as it governs the driving range of natural gas vehicles (NGVs) [18]. Moreover, the working capacity of a specific MOF depends upon the choice of the upper adsorption pressure (normally 35 and 65 bar) and lower working pressure which is generally in between 5 and 10 bar. The working capacity is always less than the total CH_4 storage capacity. In order to improve the working capacity, it is necessary to increase the storage capacity at 35/65 bar and simultaneously decrease the amount of adsorbed CH_4 at 5/5.8 bar [9, 24]. Furthermore, the working capacity of particular adsorbents toward the vehicular applications is directly related to the internal thermal effects of the ANG storage system as well as adsorbents themselves. As it is well-known that adsorption is exothermic, whereas desorption is an endothermic process [30], a slight charge or discharge of CH_4 on the adsorbent can cause a sudden rise or drop in the temperature over the entire adsorbent bed, and thus, one can conclude that all the aforementioned factors significantly decrease the working capacity of adsorbents. Hence, to avoid this problem, one can choose the adsorbent having low heat adsorption and high heat capacity.

3 CH_4 Storage in MOFs

3.1 Requirements for MOFs as ANG Adsorbent

For the CH_4 storage in ANG application, the employed MOF should not only possess a high uptake capacity but more importantly, it must have a high delivery capacity (working capacity). As it is the volumetric CH_4 delivery capacity that determines the driving range of an ANG vehicle, not the volumetric uptake capacity. In the utilization of porous adsorbent material for the storage as well as the release of CH_4, the guest adsorption ability of the concerned adsorbent is mainly considered. Which is defined as the difference between the storage capacity and amount of left CH_4 in adsorbent during desorption process at 35/65 and 5.8 bar pressure, respectively [15]. In this regard, the US department of energy (DOE) has set up some targets for the onboard storage of H_2 and CH_4, and therefore, the MOFs must meet these targets to become the best porous adsorbent materials. These target values for the H_2 storage are found to be, 5.5 wt % on a gravimetric basis and 40 g L^{-1} on the volumetric basis at an operational temperature of −40 to 60 °C and a pressure of less than 100 bar. On the other hand, the target values for the onboard CH_4 storage are 0.5 g g^{-1} of total gravimetric uptake capacity at 65 bar and 350 cm^3 (STP) volumetric deliverable capacity in between 5.8 and 65 bar at 298 K [25].

Based on the aforementioned target values set by DOE, USA, it is important to exceed these parameters so that the low packing density of the MOFs can be compensated as it directly affects the volumetric storage capacity [24]. Apart from the high volumetric CH_4 and working capacity, there are some other important parameters that should be kept in mind while synthesizing a porous MOF material for ANG applications:

1. The material must have high thermal, chemical as well as mechanical stability because other than CH_4, NG also contains some non-CH_4 components like CO_2, H_2S, H_2O, and so on as an impurity. Sometimes, these impurities significantly affect the metal–ligand coordination bonds and disrupt the structural integrity

of the MOF material. However, to avoid this problem, the guard bed methodology can be used so that these non-CH_4 impurities can be eliminated before filling the storage tank while using the MOF material in the ANG application. Despite its remarkable utility, the guard bed protocol not only makes the fuelling system expensive but also complicates the fuelling system. Therefore, the development of a chemically and thermally stable porous MOF material is necessary to ensure its recyclability and reusability while using it in practical ANG applications. Furthermore, as mentioned in para 2.3 of this chapter that improvement of the volumetric uptake capacity of MOFs is necessary, which could be accomplished by compressing the framework structure in order to reduce the number of voids and improve the bulk density. However, the compression of MOFs can seriously damage the structure of MOF and thus, its storage capacity. Hence, high mechanical stability is also necessary for the concerned MOFs along with their thermal stability to help them to be compressed at high-pressure without destructing the structural integrity and gas adsorption ability.

2. MOF adsorbent material should have a low affinity toward the strongly adsorbing species such as ethane, propane, and butane other components of NG fuel. In addition, due to their larger size and high polarizability compare to CH_4, these components exhibit stronger binding affinity toward the adsorbent resulted in a poor CH_4 storage capacity.
3. The adsorbent should have low adsorption heat and high heat capacity.
4. The MOF adsorbent material should be easy to synthesize at an industrial scale and must be cost-effective, wherein the design of the organic linkers plays a crucial role. One of the main reasons for the high-cost synthesis of MOFs are the multistep synthesis of the organic linkers [33]. Therefore, the synthesis of organic linkers from easily available cheap substrates which are based on renewable sources is highly recommended. In addition, while designing a particular MOF for gas sorption properties, the topology, pore surface area, and pore volume should be considered as they are the main factors that directly relate to the CH_4 uptake capacity of a particular MOF [24].

3.2 Characterizations Required for MOFs to Investigate Their CH_4 Adsorption Capacity

The adsorption isotherm can give detailed information regarding adsorption capacity and isosteric heat of adsorption, however, it does not trace the adsorption sites [15]. Therefore, to get in-depth knowledge about the host–guest interactions and CH_4 storage capability of MOFs, several techniques such as single-crystal X-ray diffraction (SCXRD), neutron diffraction, powder X-ray diffraction (PXRD), solid-state NMR, DFT calculations, and so on can be employed. Crystallographic and diffraction methods demonstrated a detailed analysis of the CH_4 adsorption in MOFs [34]. Among all the investigated techniques neutron diffraction is found to be more useful to identify the MOF adsorption sites. SCXRD is also an important characterization technique used to get detailed structural information of MOF material. Whereas, PXRD technique is helpful to define the structural integrity of MOF material before and after the adsorption of CH_4 [35]. Besides, PXRD combined with Rietveld refinement can also provide basic structural information of MOF. Solid-state NMR is a sensitive technique that can help in the study of adsorbed guest molecules in the pore along with their interactions with the host MOFs [36].

Calculation of isosteric heat of adsorption (Q_{st}). To know the efficiency and affinity of a particular MOF toward a specific gas, it is very important to calculate the Q_{st} value as it is an important indicator to designate the strength of interactions between host (adsorbent) and guests (adsorbate) [36, 37]. In literature, the following form of virial type expression is widely used to fit and calculate the total CH_4 adsorption from the concern isotherm data at 273 and 298 K.

$$\ln P = \ln N + 1/T \sum_{i=0}^{m} aiNi + \sum_{i=0}^{n} biNi$$

where P = pressure in bar, N = amount of adsorbed CH_4 in cm^3 (STP) cm^{-3}, T(K) = temperature, a_i and b_i represent the virial coefficients. Whereas, m, n signifies = number of coefficients required to effectively describe the isotherms. Finally, virial coefficients (a_0 through a_m) values have been used in the following expression to estimate the Q_{st}.

$$Q\text{st} = -R \sum_{i=0}^{m} aiNi$$

Here, Q_{st} is the coverage-dependent isosteric heat of adsorption and R is the universal gas constant of 8.3147 J K^{-1} $mol^{-1.}$

Notes Reports are also available in literature wherein, coverage-dependent Q_{st} is also calculated using the Clausius-Clapeyron equation [38].

$$Q_{st} = \frac{RT1T2}{T2 - T1} \ln \frac{P1}{P2}$$

where, R = molar gas constant (8.314 J/K/mol) and P and T stands for pressure and temperature, respectively.

4 Design and Synthesis of MOFs for CH_4 Storage

The design and construction of porous coordination polymers (PCPs) also known as MOFs attracted considerable attention over the past few decades not only because of their intriguing structural features but also due to their wide range of application in molecular recognition, drug delivery, shape- and size-selective heterogeneous catalysis, particularly the sorption of greenhouse gases. In this regards, Kitagawa and co-workers have published the very first report in 1997, wherein they have prepared three porous framework containing materials with the common formula $\{[M_2(4,4'\text{-bpy})_3(NO_3)_4].(H_2O)x\}_n$ (where M = Co, $x = 2$ (**1a**); Ni, $x = 4$ (**1b**), Zn, $x = 4$ (**1c**) [39]. The single-crystal X-ray analysis revealed that **1a** exhibits a unique three-dimensional (3D) network architecture having the channelling cavities with dimensions of about 3 × 6 and 3 × 3 Å, respectively along with the *a* and *b* axes. The structure of **1b** and **1c** are determined with the help of PXRD analysis. These cavities are filled with the lattice occluded water molecules having no binding interactions with the crystal framework. Finally, the authors have employed **1a** as CH_4 sorption material and noticed a low uptake with a value of 2.3 mmol per 1.0 g of anhydrous material. Nevertheless, these findings opening the door to future researchers working in the domain to develop new porous materials for sorption applications. Since then, a lot of research has been conducted in this area and some of the porous MOFs have shown remarkable CH_4 sorption properties and conveyed a message that the selection of the organic linker played a crucial role in the sorption study. This chapter compiles some of the recent examples of MOFs which are investigated for CH_4 sorption properties. These MOFs are categorically divided into two categories depending upon the nature of the organic linker used in the synthesis of MOF architecture and their structural diversity and adjustability; (i) rigid MOF, (ii) flexible MOF.

4.1 Rigid MOFs for CH_4 Storage

The CH_4 storage capacities of a particular MOF is affected by several factors such as pore size and their distribution, selection of organic linker, open metal sites, framework densities, and so on [40–44]. In 1999, Williams and co-workers [45] prepared a chemically functionalized nanoporous framework material with formula $[Cu_3(TMA)_2(H_2O)_3]_n$ (**HKUST-1; 2a**) (where TMA = benzene-1,3,5-tricarboxylate) having well-defined 1 nm-size channels with fourfold symmetry. Further, these nanochannels interconnect to provide a 3D-connected honeycomb network having pores with large hexagonal-shaped windows. Their porosity was measured based on N_2 adsorption/desorption isotherm and found Brunauer–Emmett–Teller (BET) surface area of 692.2 $m^2\ g^{-1}$ along with the Langmuir surface area of 917.6 $m^2\ g^{-1}$, and total pore volume 0.333 $cm^3\ g^{-1}$. These findings not only provide hope to future researchers but also laid down the first stone at the beginning of the gas sorption study in the porous framework containing materials. Later on, in 2008 Kaskel and co-workers [31] examined (**2a**) along with $Zn_2(bdc)_2dabco$ (**2b**), and $Cr_3F(H_2O)_2O(bdc)_3$ (**2c**) MOFs toward the high-pressure CH_4 adsorption and observed the highest excess adsorption of 15.7 wt% at 303 K, whereas the effective volumetric storage capacity was found to be 28 $m^3\ m^{-3}$ at 150 bar.

In thrust of development of porous functional framework materials, in the year 2015, Chen and co-workers [12] have reported three isoreticular

NOTT-101 MOFs having the formulas; $Cu_2(\mathbf{L^2})$(ZJU-5; **3a**), $[Cu_2\mathbf{L^3}(H_2O)_2]\cdot 3DMF\cdot 5H_2O$ (UTSA-75; **3b**) and $[Cu_2\mathbf{L^4}(H_2O)_2]\cdot 5DMF\cdot 3H_2O$ (UTSA-76; **3c**) where the phenyl ring of the linker $\mathbf{L^1}$ (see Chart 1 for details) is replaced with pyridine, pyridazine, and pyrimidine groups. Furthermore, the authors have also prepared three multivariate MOFs using a mixed ligand approach and with the formulas; $[Cu_2(\mathbf{L^1})_{0.8}(\mathbf{L^4})_{0.2}(H_2O)_2]\cdot 4DMF\cdot 4H_2O$ (UTSA-77; **3d**), $[Cu_2(\mathbf{L^1})_{0.33}(\mathbf{L^4})_{0.67}(H_2O)_2]\cdot 4DMF\cdot 5H_2O$ (UTSA-78; **3e**) and $[Cu_2(\mathbf{L^1})_{0.2}(\mathbf{L^4})_{0.8}(H_2O)_2]\cdot 4DMF\cdot 4H_2O$ (UTSA-79; **3f**). These multivariate MOFs contain different ratios of linkers $\mathbf{H_4L^1}$–$\mathbf{H_4L^4}$ [46–48]. Single-crystal X-ray analysis revealed that these MOFs comprised of dinuclear paddlewheel $Cu_2(COO)_4$ secondary building units (SBUs) and concerned organic linkers, which are linked together to construct 3D NbO-type structures. This framework structure contains two types of cages; the first one is a cuboctahedral cage having a diameter of 10.2 Å, whereas, second is a larger irregular cage of about 9.6 × 22.3 Å^2. After the incorporation of functional organic groups, the pore surfaces within these isoreticular MOFs become slightly different. The BET surface area of all the MOFs was found to be in the range of 2800–2880 m^2g^{-1} (see Table 1 for details).

Finally, the authors have employed these MOFs toward the CH_4 storage capacities and the results were compared with NOTT-101 and observed the lowest total volumetric CH_4 capacity of 237 cm^3 (STP) at 65 bar and RT in comparison to other studied MOFs. Also, the incorporation of the pyridine ring into the framework enhances the CH_4 storage capacity of **3a** to 249 cm^3 (STP) cm^{-3} (Fig. 1). Interestingly, the incorporation of pyridazine and pyrimidine functionalities into the NOTT-101 enhances the CH_4 storage capacity of **3b** to 251 cm^3 (STP) and 257 cm^3 (STP) for **3c**. Similarly, they have also calculated the CH_4 storage working capacity for **3a–3c** and found 188 cm^3 (STP) working capacity for **3a,** which is higher than NOTT-101 (181 cm^3, STP). On the other hand, the functionalized MOFs **3b** and **3c** exhibit 192 and 197 cm^3 (STP) working capacity, respectively.

The authors sincerely investigated the reason for this enhanced uptake of CH_4 and argued that the enhanced CH_4 storage mainly depends on the secondary adsorption sites and general pore surface. It is observed that for functionalized MOF having better secondary adsorption surfaces, the CH_4 uptake at higher pressure would be higher due to the enhanced Vander Walls interactions between CH_4 molecules and Lewis basic N sites of the organic linker. These outcomes were supported by computational analyses which indicated that the Lewis basic N sites and the freedom of the functionalized organic linkers were the two main reasons for this remarkably improved volumetric CH_4 storage and working capacities. Thus, these findings open a new door for contemporary research scientists to design new porous MOFs with more improved CH_4 storage as well as working capacity.

Later on, to improve the CH_4 storage and working capacities, He and co-workers [49] constructed a new NbO-type MOF with formula; $[Cu_2\mathbf{L^5}(H_2O)_2]\cdot 6DMF\cdot 2H_2O]$ (ZJNU-53; **4a**) utilizing a new organic linker $\mathbf{H_4L^5}$ obtained from the slight modification of linker $\mathbf{H_4L^6}$ (see Chart 1 for details) employed to synthesize ZJNU-50 [50]. The aim of the employment of naphthalene ring in place of the benzene ring in the spacer is to increase the CH_4 uptake at high-pressure of 65 bar than that

Ligands:

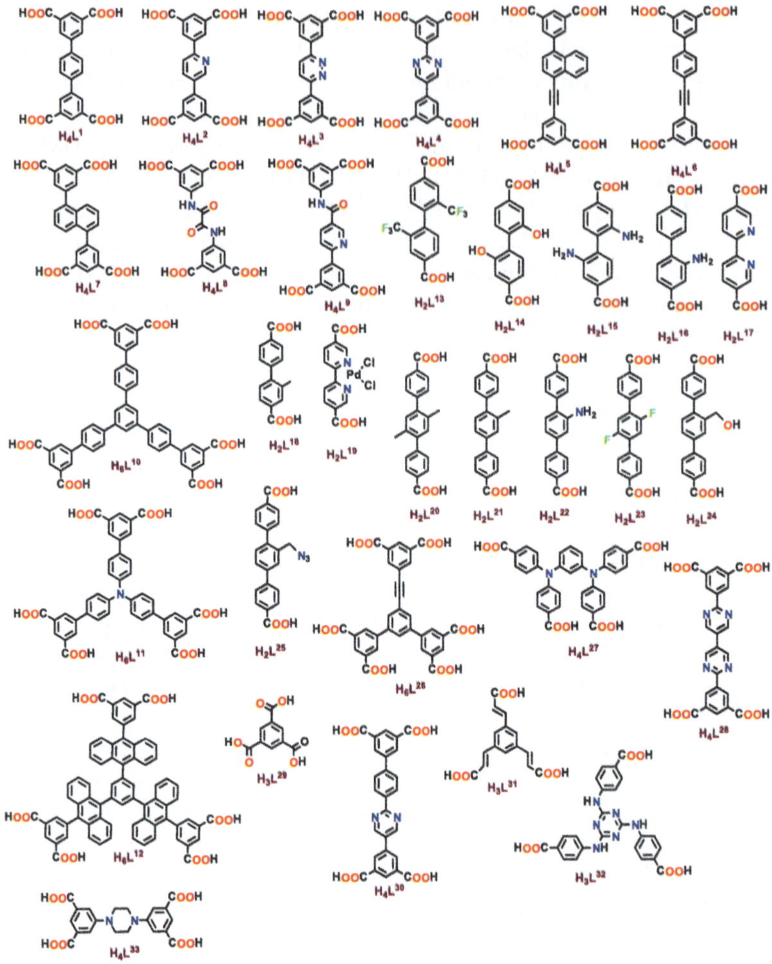

Co-ligands:

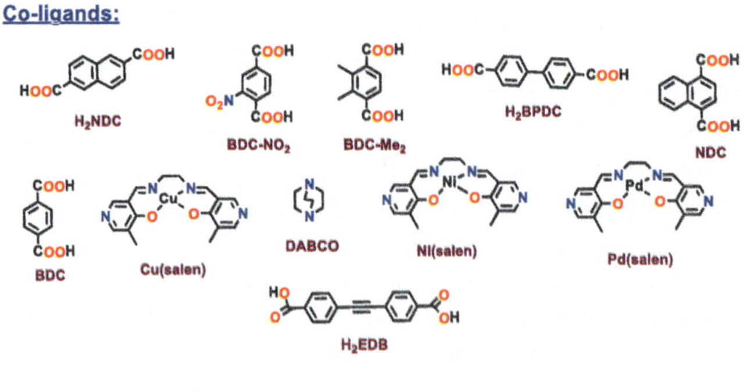

◂**Chart 1** Ligands and co-ligands used in the preparation of diverse range of MOFs discussed in this chapter. The abbreviation used to denote these organic linkers are provided herewith, with their details, Ligands: $\mathbf{H_4L^1}$ = [1,1′:4′,1″-terphenyl]-3,3″,5,5″-tetracarboxylic acid; $\mathbf{H_4L^2}$ = 5,5′-(pyridine-2,5-diyl)diisophthalic acid; $\mathbf{H_4L^3}$ = 5,5′-(pyridazine-2,5-diyl)diisophthalic acid; $\mathbf{H_4L^4}$ = 5,5′-(pyrimidine-2,5-diyl)diisophthalic acid; $\mathbf{L^5}$ = 5,5′-(naphthalene-1,4-diyl-ethyne-1,2-diyl)diisophthalic acid; $\mathbf{L^6}$ = 5,5′-(benzene-1,4-diyl-ethyne-1,2-diyl) diisophthalic acid; $\mathbf{L^7}$ = 5,5′-(naphthalene-1,5-diyl)diisophthalic acid; $\mathbf{L^8}$ = 5,5′-(oxalylbis(azanediyl))diisophthalic acid; $\mathbf{L^9}$ = 5-(5-((3,5-dicarboxyphenyl)carbamoyl)pyridin-2-yl)isophthalic acid; $\mathbf{L^{10}}$ = 1,3,5-tris(3′,5′-dicarboxy[1,1′-biphenyl]-4-yl)benzene; $\mathbf{H_4L^{11}}$ = 4′,4‴,4⁗-nitrilotris(([1,1′-biphenyl]-3,5-dicarboxylic acid; $\mathbf{H_4L^{12}}$ = 5,5′,5″-(benzene-1,3,5-triyltris(anthracene-10,9-diyl))triisophthalic acid; $\mathbf{H_4L^{13}}$ = 2,2′-Bis(trifluoromethyl)-4,4’-biphenyldicarboxylic acid; $\mathbf{L^{14}}$ = 2′-Dihydroxy-1,1′-biphenyl-4,4′-dicarboxylic acid; $\mathbf{L^{15}}$ = 2,2′-Diamino-1,1′-biphenyl-4,4′-dicarboxylic acid; $\mathbf{L^{16}}$ = 2-amino-[1,1′-biphenyl]-4,4′-dicarboxylic acid; $\mathbf{L^{17}}$ = [2,2′-bipyridine]-5,5′-dicarboxylic acid; $\mathbf{L^{18}}$ = 2-methyl-[1,1′-biphenyl]-4,4′-dicarboxylic acid; $\mathbf{L^{19}}$ = Pd(H_2bpydc)Cl_2; $\mathbf{L^{20}}$ = 2, 5-dimethyl-[1,1′: 4′,1′ terphenyl]-4,4″-dicarboxylic acid; $\mathbf{L^{21}}$ = 2′-methyl-[1,1′: 4′,1′ terphenyl]-4,4″-dicarboxylic acid; $\mathbf{L^{22}}$ = 2’-amino-[1,1′: 4′,1′ terphenyl]-4,4″-dicarboxylic acid; $\mathbf{L^{23}}$ = 2, 5-difluoro-[1,1′: 4′,1′ terphenyl]-4,4″-dicarboxylic acid; $\mathbf{L^{24}}$ = 2′-hydroxymethyl-[1,1′: 4′,1′ terphenyl]-4,4″-dicarboxylic acid; $\mathbf{L^{25}}$ = 2′-azidomethyl-[1,1′: 4′,1′ terphenyl]-4,4″-dicarboxylic acid; $\mathbf{L^{26}}$ = 5,5′,5″-Benzene-1,3 diylbis(2-isophthalate)-5-yl(1-ethynyl-2-isophthalate; $\mathbf{L^{27}}$ = 4,4′,4″,4‴-(1,3 phenylenebis(azanetriyl)) tetrabenzoate; $\mathbf{H_6L^{28}}$ = 5,5′-([5,5′-bipyrimidine]-2,2′-diyl)diisophthalic acid; $\mathbf{L^{29}}$ = 1,3,5-Benzenetricarboxylic acid; $\mathbf{L^{30}}$ = 4′-(5-(3,5-dicarboxyphenyl)pyrimidin-2-yl)-[1,1′-biphenyl]-3,5-dicarboxylic acid; $\mathbf{L^{31}}$ = benzene-1,3,5-tri-β-acrylic acid; $\mathbf{L^{32}}$ = 4,4′,4″-s-triazine-1,3,5-triyltri-p-aminobenzoic acid; $\mathbf{L^{33}}$ = 5,5′-(piperazine-1,4-diyl)diisophthalic acid. Coligands: **DABCO** = 1,4-diazabicyclo[2.2.2]octane; **salen** = 5,5′-((1E,1′E)-(ethan-1,2-diylbis(azanylylidene))bis(methanylylidene))bis(3-methylpyridine-4-ol); $\mathbf{H_2BDC}$ = 1,4-Benzene dicarboxylic acid; $\mathbf{H_2NDC}$ = 2,4-Napthalene dicarboxylic acid; $\mathbf{BDC\text{-}Me_2}$ = 2,5-dimethylterephthalic acid; $\mathbf{BDC\text{-}NO_2}$ = 2-nitroterephthalic acid; $\mathbf{H_2EDB}$ = 4,4′-ethynylenedibenzoate; $\mathbf{H_2BPDC}$ = 4,4′-biphenyldicarboxylate;

of low-pressure of 5 bar by reducing the pore size by allocating the large cavities into smaller ones and augmenting the interactions between the methane molecules and framework. Further, it is well-known fact that to attain the high CH_4 delivering capacity for a particular MOF, the CH_4 uptake capacity should be simultaneously enhanced and diminished at 65 and 5 bar, respectively. Single-crystal X-ray analysis reveals that **4a** exhibits an overall 3D non-interpenetrated 4,4-connected NbO-type network with a Schläfli topological symbol of $\{6^4{\cdot}8^2\}$ [51, 52]. The 3D framework architecture of **4a**, contains two different types of nano-cages, associated by sharing dinuclear paddlewheel SBUs, and ordered in an alternating manner running parallel to crystallographic *c*-axis (Fig. 2a). The spherical cage has a diameter of 14 Å and is comprised of 12 ligands and 6 SBUs, whereas the shuttle-shaped is composed of six ligands and 12 SBUs. The shuttle-shaped cage in **4a** was further categorized into three cages owing to the naphthalene moiety directing inward as compared to MOF ZJNU-50, thereby, enhancing the gas adsorption at higher pressure.

Finally, the authors sincerely utilized **4a** for the CH_4 sorption studies and noticed a gravimetric CH_4 uptake of 0.206 g g^{-1}, equivalent to a volumetric uptake of 188 cm^3 (STP) cm^{-3} at 35 bar and 298 K, which is quite better than DOE's old target of 180 cm^3 (STP) cm^{-3} for CH_4 storage under identical conditions. Notably, at 65 bar,

Table 1 A comprehensive list of MOFs showing the CH_4 uptake at different working capacities such as 35, 65 and 80 bar and 25 °C

MOFs	BET surface area (m^2 g^{-1})	V_p (g cm^{-3})	CH_4 adsorption at room temperature and 35/65/80 bar (cm^3 cm^{-3})		Q_{sv}(KJ mol^{-1})	References
			Total uptake capacity [cm^3 (STP) cm^{-3}	CH_4 working capacity [cm^3 (STP) cm^{-3}		
VNU-22	1030	0.41	132/155/NA	77/101/NA	16.5	[72]
Fe-NDC	1240	0.54	134/160/NA	83/108/NA	16.6	[72]
DUT-4	1308	0.68	122/167/NA	82/124/NA	NA	[72]
Ni-MOF-74	1350	0.51	228/251/267	106/129/152	NA	[48]
VNU-21	1440	0.58	142/182/NA	102/140/153	15.7	[72]
DUT-5	1613	0.81	114/134/NA	94/114/NA	NA	[72]
LIFM-82	1624	0.71	196/246/271	143/193/218	17.5	[62]
LIFM-83	1715	0.72	192/241/265	140/189/213	16.59	[62]
HKUST-1	1850	0.78	227/267/277	150/190/200	17	[60]
PCN-14	2000	0.85	195/230/250	128/157/178	NA	[55]
MAF-38	2022	0.808	226/263/273	150/187/197	21.6 ± 1.8	[92]
NJU-Bai 41	2370	0.92	204/245/NA	NA/172/NA	17.77	[54]
MOF-519	2400	0.938	200/259/279	151/210/230	14.6	[93]
pbz-MOF-1	2415	0.99	140/192/210	110/162/180	NA	[92]
MFM-132a	2466	1.06	162/201/213	109/150/162	NA	[57]
NOTT-101a	2805	1.08	194/237/NA	138/181/NA	NA	[11]
UTSA-77	2807	1.08	202/249/NA	141/188/NA	15.04	[12]
UTSA-76	2820	1.092	211/257/NA	151/197/NA	15.44	[34]
ZJU-5	2829	1.08	201/249/NA	140/188/NA	14.15	[47]
NJU-Bai 42	2830	1.07	203/247/NA	141/193/NA	14.49	[54]
UTSA-75	2838	1.06	205/251/NA	146/192/NA	14.93	[12]
UTSA-78	2840	1.09	207/252/NA	146/191/NA	15.09	[12]
UTSA-79	2877	1.08	207/255/NA	145/193/NA	14.65	[12]
Co(bdp)	2911	1.02	161/203/NA	155/197/NA	NA	[94]
ZJNU-53	3034	1.084	188/241/NA	135/190/NA	15.02	[49]
NJU-Bai 43	3090	1.22	202/254/NA	146/198/NA	14.45	[54]
NU-125	3120	1.29	182/232/NA	133/183/NA	15.1	[41]
NU-800	3150	1.34	140/197/NA	110/167/NA	12.5	[95]
UTSA-110	3241	1.263	187/241/NA	136/190/NA	NA	[72]
UTSA-111	3252	1.229	NA/234/NA	NA/183/NA	14.5	[11]
MOF-520	3290	1.277	162/215/231	125/178/194	13.6	[94]

(continued)

Table 1 (continued)

MOFs	BET surface area (m^2 g^{-1})	V_p (g cm^{-3})	CH_4 adsorption at room temperature and 35/65/80 bar (cm^3 cm^{-3})		Q_{sv}(KJ mol^{-1})	References
			Total uptake capacity [cm^3 (STP) cm^{-3}	CH_4 working capacity [cm^3 (STP) cm^{-3}		
MOF-5	3320	1.38	126/188/NA	104/166/NA	NA	[72]
MOF-905-NO_2	3380	1.29	132/185/NA	107/160/NA	10.7	[72]
MFM-115a	3394	1.38	186/238/256	138/191/208	NA	[57]
MOF-950	3440	1.30	145/195/209	109/159/174	11.9	[23]
MOF-905	3490	1.34	145/206/228	120/181/203	11.7	[23]
MOF-905-Me_2	3640	1.39	138/192/NA	111/165/NA	10.3	[72]
FDM-8	3643	1.54	NA/193/NA	NA/167/NA	10.4	[11]
MFM-112a	3800	1.62	162/218/236	125/181/200	NA	[57]
Cu-tbo-MOF-5	3971	1.12	151/199/216	110/158/175	20.4	[96]
PCN-66	4000	1.63	136/187/NA	101/152/NA	17.73	[65]
NU-140	4300	1.97	138/200/NA	108/170/NA	14	[97]
BUT-22	4380	2.01	NA/182/NA	NA/152/NA	12.0	[11]
MOF-205	4460	2.16	120/183/NA	101/164/NA	NA	[72]
MOF-177	4500	1.89	122/NA/185	102/NA/185	NA	[98]
MOF-210	6240	3.6	82/141/166	69/128/154	NA	[98]
NU-111	4930	2.09	138/206/NA	111/179/NA	14.2	[60]
DUT-49	5476	2.91	112/176/NA	92/156/NA	NA	[99]
Al-soc-MOF-1	5585	2.3	122/197/222	101/176/201	11	[100]
NU-1501-Fe	7140	2.90	NA/168/NA	NA/147/NA	NA	[62]
NU-1501-Al	7310	2.91	NA/163/NA	NA/143/NA	9.7	[11]

the volumetric CH_4 uptake stretched 241 cm^3 (STP) cm^{-3}, which is (Fig. 2b) almost up to the DOE's CH_4 storage target of 263 cm^3 (STP) cm^{-3} [53]. Under these circumstances, the density of CH_4 stored in the micropore was 0.2432 g cm^{-3}, corresponding to the density of compressed CH_4 at 383 bar and 298 K. Apart from the high CH_4 storage capacities, **4a** also displayed high CH_4 deliverable capacities (135 cm^3 (STP) cm^{-3}), from 35 to 5 bar at 298 K, which further increased to 190 cm^3 (STP) cm^{-3} if 5 and 65 bar lower and upper-pressure limits, respectively. The authors argue that this improved CH_4 storage and working capacities of **4a** might be due to the presence of additional adsorption surface in the MOF **4a**, by virtue of ligand modification to generate the smaller pores and facilitates the interactions between the framework and CH_4 molecules.

In the year 2017, Bai and co-workers [54] have prepared a family of isomorphic MOFs based upon PCN-14 MOF with formulas; $[Cu_2(\mathbf{L^7})(H_2O)_2]_n$ (NJU-Bai 41;

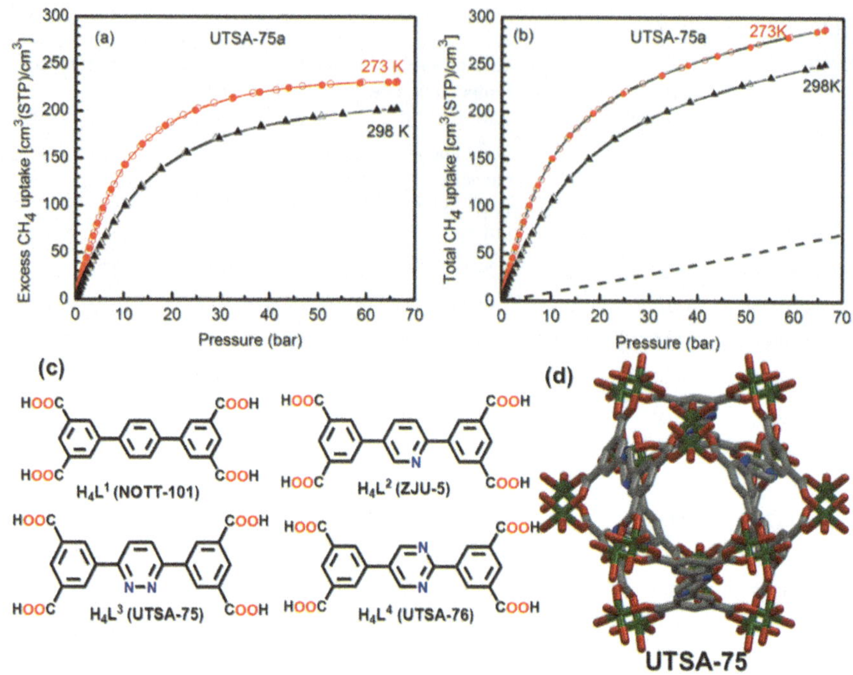

Fig. 1 Excess (**a**) and total (**b**) high-pressure CH_4 sorption isotherms of UTSA-75a at 273 K (red) and 298 K (black). The filled and open symbols represent adsorption and desorption, respectively. For comparison, data of pure CH_4 gas stored in a high-pressure gas tank is represented as a black line in (**b**). **c** Schematic structure of the organic ligands $H_4\mathbf{L}^1$–$H_4\mathbf{L}^4$ that serve as linkers in NOTT-101; ZJU-5 (**3a**), UTSA-75, and UTSA-76, respectively. **d** Crystal structures packing of UTSA-75 displaying pores and cavity. Figure 1a, b are reproduced with the permission of Ref. [12]. Copyright © 2015, the Royal Society of Chemistry. *Note* Only selected images for the CH_4 sorption are shown throughout the entire chapter for clarity and better understanding

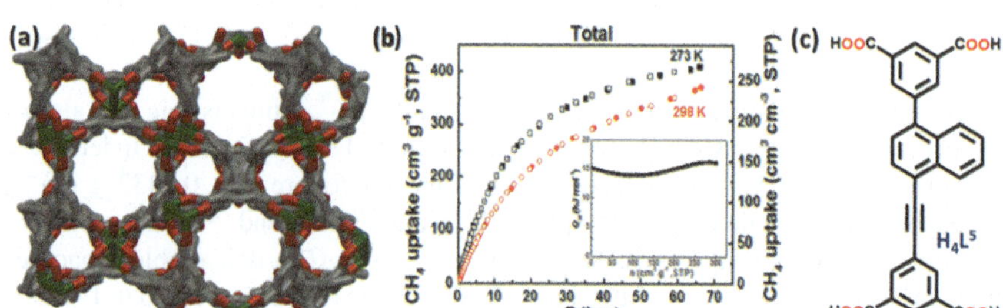

Fig. 2 **a** Crystal packing displaying pores and cavity of ZJNU-53. **b** Total volumetric CH_4 adsorption isotherms of ZJNU-53. **c** Schematic structure of the organic linker for ZJNU-53. Figure 2b is reproduced with the permission of Ref. [49]. Copyright @ 2016 the Royal Society of Chemistry

5a), $[Cu_2(\mathbf{L^8})(H_2O)_2]_n$ (NJU-Bai 42; **5b**) and $[Cu_2(\mathbf{L^9})(DMF)_2]_n$ (NJU-Bai 43; **5c**) (see Chart 1 for $\mathbf{L^7}$–$\mathbf{L^9}$). As it is well-known that PCN-14 is a member of MOF-505 family comprises of a shuttle-shaped cage (6, 10 and 6 Å) and a noble candidate for the fine-tuning of pore size and exhibits BET surface area of 1753 m^2/g, pore volume of 0.87 cm^3/g and Q_{st} of 30 kJ/mol. It also shows an absolute CH_4 adsorption capacity of 230 v/v, which is 28% higher than the DOE target (180 v/v) for CH_4 storage [55, 56]. Keeping these fascinating and remarkable results of MOF, PCN-14, and in an effort to increase the diameter of shuttle-shaped cages, Bai groups replaced the anthracene ring of PCN-14 with naphthalene ring and prepared **5a**. Notably, **5a**, exhibits an improved diameter of 8, 12, and 8 Å. Further, to decrease the interactions between the MOF frameworks with CH_4 molecules, **5b** was prepared by replacing the naphthalene ring with a cylindrical oxamide group. The framework architecture of **5b** displays a partition-less shuttle-shaped cage (13 × 27 Å). Further, authors have also replaced a half of the oxamide moiety with pyridyl moiety to get the structure of **5c** having a larger shuttle-shaped cage of 13 × 30 Å (Fig. 3a–c). Finally, the authors have exploited all these MOFs toward CH_4 sorption analysis along with MOF-505 and PCN-14. It was observed that the volumetric CH_4 uptake capacity of **5a** enhanced to 245 cm^3 (STP) compare to 230 cm^3 (STP) of PCN-14, whereas working capacity improved to 172 cm^3 (STP) in comparison to PCN-14, 157 cm^3 (STP).

The enhanced volumetric and working capacity of **5a** can be accredited to the larger pore size which resulted in the efficient utilization of space. Moreover, installed slender oxamide moieties in **5b** created a hydrophilic cage with a large pore size and

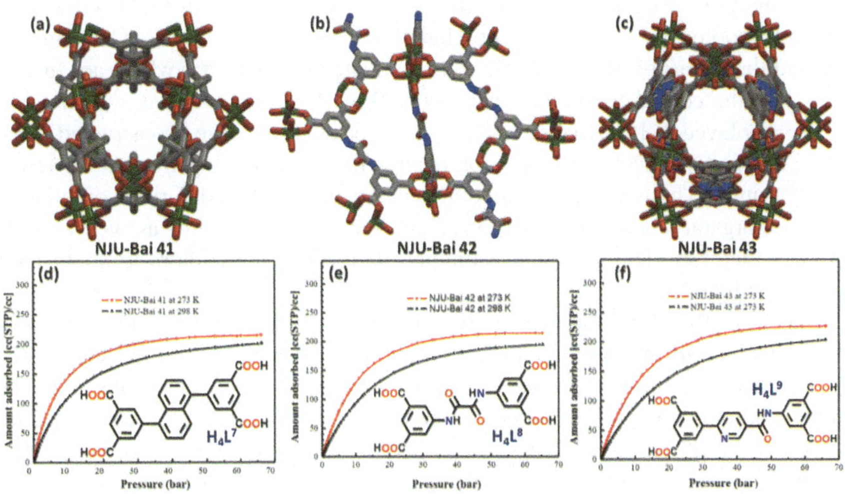

Fig. 3 Crystal structures packing of NJU-Bai 41 **(a)**, NJU-Bai 42 **(b)**, and NJU-Bai 43 **(c)** displaying pores and cavity. Excess volumetric CH_4 adsorption isotherms of NJU-Bai 41 **(d)**, NJU-Bai 42 **(e)**, and NJU-Bai 43 **(f)** at 273 K and 298 K. The inset of Fig. 2d–f shows the respective ligand chemical structure from which the MOF has been constructed. Figure 2d–f is reproduced with the permission of Ref. [54]. Copyright @ 2017 Wiley–VCH

therefore decreased the CH_4-framework interactions, resulted in high volumetric uptake (247 cm^3, STP) and working (193 cm^3, STP) capacity. Similarly, **5c** exhibited the enhanced volumetric uptake capacity of 254 cm^3 (STP) along with a volumetric working capacity of 198 cm^3 (STP). These results proved that the fine-tuning of pore chemistry and structure can be an effective approach for boosting the volumetric and gravimetric CH_4 uptakes as well as working capacity. In addition, the effect of temperature on the CH_4 sorption was also studied and observed that the storage working capacity of **5a** decreased from 172 to 161 cm^3 (STP), when the temperature is decreased from 298 to 273 K at a pressure between 65 and 5 bar. In contrast to **5a**, the volumetric uptake and working capacity of **5b** do not alter. In the case of **5c**, the working capacity increased from 198 cm^3 (STP) to 205 cm^3 (STP), which is ascribed to the higher pore volume and internal surface area. From the aforementioned findings, it is almost clear that the fine-tuning of MOF-505 analogues decreased the low-pressure CH_4 uptake and increased the high-pressure CH_4 uptake.

In the same year (2017), Schröder and co-workers [57] have reported a set of three (3,24)-connected MOFs having the formulas; $[Cu_3(\mathbf{L^{10}})(H_2O)_3)]$·8DMSO·15DMF·3$H_2O$ (MFM-112; **6a**), $[Cu_3(\mathbf{L^{11}})(H_2O)_3]$·5 DMF ·10 H_2O (MFM-115, **6b**) and $[Cu_3(\mathbf{L^{12}})(H_2O)_3]$·9DMF (MFM-132; **6c**) (see Chart 1 for L^{10}–L^{12}). The authors functionalized the framework architecture of newly synthesized MOFs **6a–6c** by incorporating a central phenyl ring, a nitrogen center at the core of the hexacarboxylate, and anthracene moiety, respectively. The functionalization of the central part of the hexacarboxylate ligands provided the different functionalities and pore geometries to the concerned MOF. Single-crystal X-ray analysis revealed that these MOFs are isostructural with (3,24)-connected network topology, wherein the isophthalate units of the hexacarboxylate ligands bridged the $[Cu_2(O_2CR)_4]$ paddlewheels to create cuboctahedra, which are further linked by the central triangular ligand core (Fig. 4a–c). In addition, the structure of **6c** displayed a $[Cu_{24}(\text{isophthalate})_{24}]$ cuboctahedron cage, constructed from 12 $\{Cu_2\}$ paddlewheels and 24 isophthalates linkers stemming from 24 different $(\mathbf{L^{10}})^{6-}$ units. The anthracene functionalization in **6c**, offered a rich compilation of metal-organic coordination cages of different sizes, geometry as well as pore surface, therefore demonstrating a unique platform to the authors to study their host-guest interactions.

Subsequently, MOFs **6a–6c** were investigated for CH_4 sorption and found that at 35 bar, **6a** and **6b** showed high total gravimetric uptakes of 322 and 304 cm^3 (STP) g^{-1}, respectively, compared to those measured for the best behaving MOFs under the same conditions [58–61]. In addition, **6b** showed a high volumetric CH_4 capacity of 186 v/v at 35 bar and 298 K. In contrast, **6a** exhibits a lower total volumetric uptake of 162 v/v at 35 bar. Though, **6c** shows lower total gravimetric CH_4 uptake of 249 cm^3 (STP) g^{-1} compare to **6a**, however, exhibits the same volumetric uptake of 162 v/v at 35 bar and 298 K owing to its higher crystal density. At 80 bar, pressure, **6a** and **6b** exhibit a remarkable increase in their total CH_4 adsorption capacities of 469 and 419 cm^3 g^{-1}, and volumetric uptakes of 256 and 236 v/v, respectively.

Under identical condition **6c** displays a lower uptake of 213 v/v in comparison to **6a** and **6b**, might be because of its lower surface area and pore volume. Further,

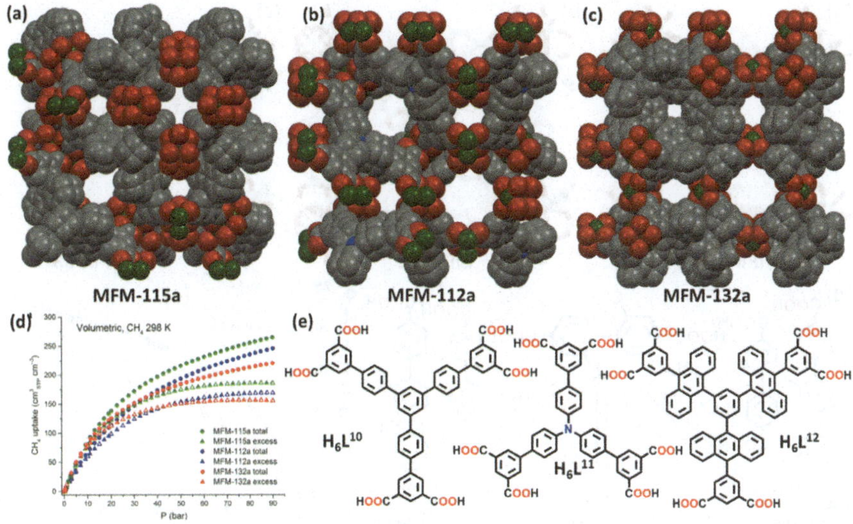

Fig. 4 **a–c** Crystal packing displaying pores and cavity of MFM-115a, MFM-112a, and MFM-132a, respectively. **d** High-pressure volumetric CH_4 adsorption isotherms for MFM-115a, MFM-112a, and MFM-132a in the pressure range 0–90 bar at 298 K. **e** Schematic structure of the organic ligands H_6**L**^{8}–H_6**L**^{10} that serve as linkers for MFM-115a, MFM-112a, and MFM-132a, respectively. Figure 4d is reproduced with the permission of Ref. [57], © 2017 American Chemical Society

authors have also investigated the deliverable CH_4 capacities of **6a–6c** and noticed high deliverable capacities of 181 v/v and 191 v/v between 65 and 5 bar at 298 K for that **6a** and **6b**, respectively. In addition, **6a** and **6b** also displayed a remarkably high deliverable CH_4 capacity of 200 and 208 v/v between 80 and 5 bar and at room temperature. Compared to **6a** and **6b**, **6c** exhibited lower deliverable CH_4 capacities of 150 and 162 v/v at 65 and 80 bar (to 5 bar), respectively. Finally, the Schröder groups established an argument that the pore environment created by the rotation of phenylene rings in organic linker affected significantly the CH_4 sorption and might be a controlling factor of the gas adsorption in host materials.

Su and co-workers [62] have adopted a new strategy called swing or multirole strategy and prepared a multifunctional Zr-MOF (LIFM-28; **7a**). The basic requirements for multirole MOF synthesis are (i) flexible functionalization with different modules and (ii) easy alteration between different functional varieties. In this regard, Su groups designated **7a** as the parent framework and prepared a series of fully functionalized MOFs LIFM-70 – 86 (**7a–7q**) by using linkers **L**$^{13-25}$ (see Chart 1 for details) and which are constructed by replacing the binding sites of the parent linker **L**^{13}. Single-crystal X-ray analysis of all the functionalized MOFs, **7a–7q** reveals their isomorphous structures, however having different functional units. Further, the parent MOF **7a** exhibits an 8-connected Zr_6 cluster having four pairs of replaceable H_2O terminals along the crystallographic *c*-axis (A site) and *a/b* axis (B site) and generated two types of pockets namely A and B (Fig. 5a-c) [62]. Moreover, the authors have employed all these MOFs for a diverse range of applications. For

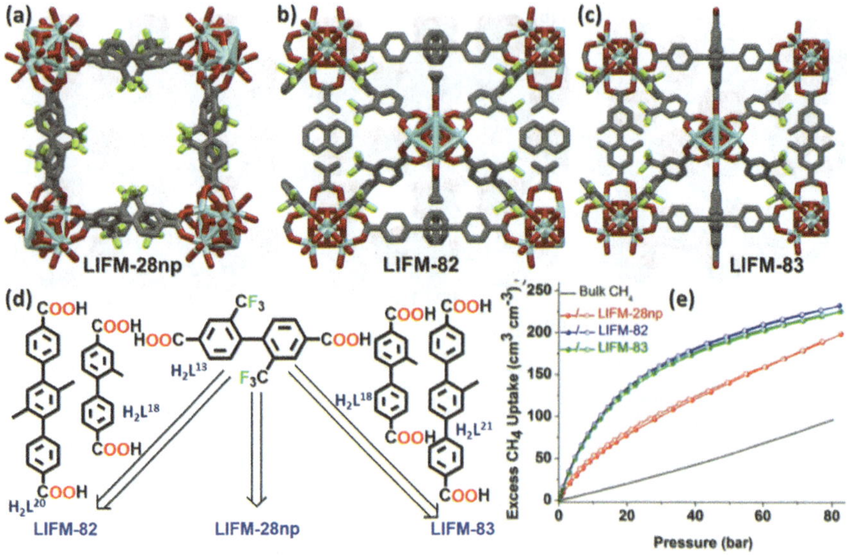

Fig. 5 **a–c** Crystal packing displaying pores and cavity of LIFM-28np, LIFM-82, and LIFM-83, respectively. **d** Schematic structure of the organic ligands that serve as linkers in respective MOFs. **e** The total methane volumetric uptake of LIFM-28np, LIFM-82, and LIFM-83. The bulk density of CH_4 is represented as a black curve. Figure 5e is reproduced with the permission of Ref. [62], © 2017 American Chemical Society

instance, MOFs **7 h** and **7j** were employed for CO_2 sorption properties because of the presence of amine groups in their framework architecture. Similarly, MOFs **7 m** and **7n** were employed for CH_4 sorption properties due to the presence of methyl groups, which are supposed to favor the CH_4 adsorption. MOFs **7f**, **7 h,** and **7j** were examined for fluorescence behavior and observed an enhancement in the fluorescent compare to it was seen **7a**. Whereas, MOF **7p** was employed for click reaction by the author.

It is important to mention that other aspects of applications of all these investigated MOFs are beyond the scope of this chapter so let us briefly discuss the CH_4 adsorption properties of MOFs **7 m** and **7n.** The CH_4 adsorption data implied that the total volumetric uptake capacity for **7 m** and **7n** was found to be 271 and 265 cm^3 (STP) cm^{-3} at 80 bar, respectively. Whereas, CH_4 working capacity for **7 m** and **7n** was found to be 218 and 213 cm^3 (STP) cm^{-3} at 5–80 bar and 298 K (Fig. 5e). These sorption values appreciably surpassed the values noticed for the standard MOF like HKUST-1 under similar conditions [23]. Finally, authors augured that these types of reversible and dynamic processes of spacer installation and removal guarantee easy multifunctional switching of **7a** and provides the framework with different functionalities for targeted applications, such as gas sorption and separation, catalysis, click reaction, luminescence sensing, and extraordinary CH_4 storage via the introduction of different functional units.

In the same year, Chen and co-workers [63] have synthesized a new **ntt**-type MOF UTSA-61 with the formula $[Cu_3\mathbf{L^{26}}(H_2O)_{2.5}]\cdot xG$ (**8a**) (where G = non-coordinated solvent molecules; see Chart 1 for $\mathbf{L^{26}}$). The MOF **8a** was synthesized using a solvothermal protocol and characterized by PXRD and Rietveld refinement using the GSAS package [64]. The PXRD analysis revealed that the $(\mathbf{L^{28}})^{6-}$ is coordinated to six Cu_2 paddlewheels wherein each Cu_2 paddlewheel is further connected to four linkers (Fig. 6a). Each Cu center displays a square-pyramidal geometry where the axial positions are occupied by solvent molecules. Out of the four metal sites, two are present on special positions which are involved in a -$[Cu_2(O_2CR)_4H_2OCu_2(O_2CR)_4]$-connection with their symmetry-related counterpart. The dendritic hexacarboxylate linker $\mathbf{L^{26}}$ might support three cuboctahedral cages where two cuboctahedral cages are placed next to each other. The 3D framework architecture of **8a** consists of four Cu_2 paddlewheels and one large-$[Cu_2(O_2CR)_4H_2O\text{-}Cu_2(O_2CR)_4]$- cluster, which are connected with the help of $(\mathbf{L^{26}})^{6-}$ linker. The overall framework of **8a** is parallel to other **ntt**-type frameworks and features three types of cages, (i) cuboctahedral cage (cubOh), (ii) truncated tetrahedral cage (T-Td), and truncated octahedral cage (TOh). The cubOh cage is constructed from 24 isophthalate units. The T-Td cage has four $(\mathbf{L^{26}})^{6-}$ linkers on the four hexagonal faces, whereas eight $(\mathbf{L^{26}})^{6-}$ linkers lie on the hexagonal surfaces of TOh cage.

Further, the author examined BET surface area and pore volume of **8a** and observed 2171 m^2g^{-1} and 0.968 cm^3g^{-1}, respectively. The high surface area, as well as the porosity, motivated them to exploit **8a** toward the CH_4 sorption studies and obtained storage capacity of 194 cm^3 (STP) cm^{-3} at 35 bar and 298 K, which reached 224 cm^3 (STP) cm^{-3} when pressure is increased to 65 bar (Fig. 6b). These values are the highest among all the reported **ntt**-type MOFs [65–67]. Furthermore, **8a** exhibits the gravimetric uptake values of 0.241 g g^{-1}, which is larger than HKUST-1 (0.216 g g^{-1}), Ni-MOF-74 (0.148 g g^{-1}), PCN-14 (0.204 g g^{-1}) and so on [18]. MOF **8a** also displayed the working capacity of 176 cm^3 (STP) cm^{-3} at 298 K, which is comparable to the best **ntt**-type MOFs, such as NU-111 (179 cm^3 (STP)

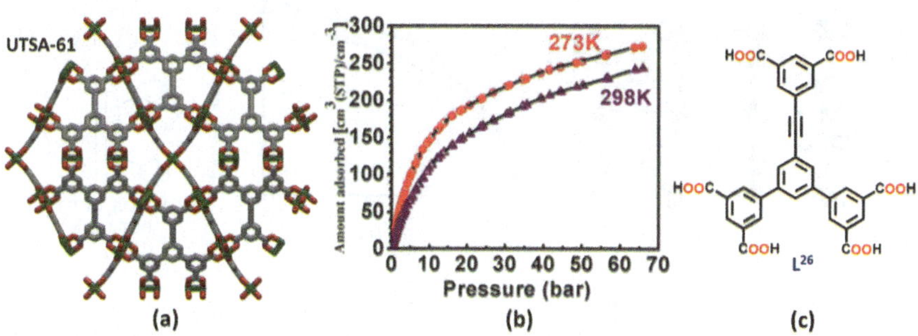

Fig. 6 **a** Crystal packing displaying pores and cavity of UTSA-61. **b** Total volumetric CH_4 adsorption isotherms of UTSA-61 at 273 and 298 K. **c** Schematic structure of the organic linker for UTSA-61. Figure 6b is reproduced with the permission of Ref. [63], © 2017 American Chemical Society

cm^{-3}) [68] and NU-125 (183 cm^3 (STP) cm^{-3}) [41] and better than other reported **ntt**-type MOFs (refer to reference 63 for details). The Chen groups further evaluated the Q_{st} value for **8a** and found to be 17.5 kJ mol^{-1} which is a little higher than that of NU-111 (14.2 kJ mol^{-1}) and NU-125 (15.1 kJ mol^{-1}). This better Q_{st} value for **8a** is primarily ascribed to the comparatively high concentration of open metal sites. In addition, at 5 bar of pressure, **8a** displayed volumetric CH_4 uptake of 68 cm^3 (STP) cm^{-3} in comparison to NU-111 [27 cm^3 (STP) cm^{-3}] and NU-125 [50 cm^3 (STP) cm^{-3}]. Finally, the authors have argued that when CH_4 loading was increased, the open metal sites as well as the cages are occupied by the CH_4 molecules, resulted in an increment of $CH_4 \cdots CH_4$ interactions, which leads to increases in the Q_{st} value at higher concentration. Therefore, **8a** can be considered as one of the best performing **ntt**-type MOFs for CH_4 storage.

Kaskel and co-workers [69] have also built a MOF with formula, $\{Cu_4(\mathbf{L^{27}})_2\}$ (DUT-71; **9a**) (see Chart 1 for $\mathbf{L^{27}}$), and later on functionalized it via post-synthetic modification using M(salen) [M=Cu, Ni, Pd] and DABCO [DABCO = 1,4-diazabicyclo[2.2.2]octane]. The functionalization afforded a series of porous materials having a common formula, $[M_4(mpbatb)_2(Cu(salen))_{0.5}(DABCO)_x]$ (where, M=Cu **{9b}**, Ni **{9c}**, Pd **{9d}**) and collectively named as DUT-117(M) (Fig. 7). The structure of parent MOF **9a,** contains two types of cages having the diameter of 21.4 × 11.0 Å (larger cage) and 22.7 × 5.2 Å (smaller cage). Notably, the structure of **9a** exhibits 76.2% solvent-accessible void per unit cell volume, however, the framework broke down during the desolvation process. Therefore, Kaskel groups used linear salen derivatives and DABCO as a cross-linker to stabilize the Cu-paddlewheel structure of **9a.** Besides, some other reasons also exist for the introduction of the cross-linker in the framework of **9a**; (i) to decrease the pore size, (ii) to generate more surface area, (iii) installation of more open metal sites in structure and expedite the metalation. Generally, two types of cross-linkers have been used by the author to attain the maximum degree of cross-linking (i) salen derived metalloligand is used

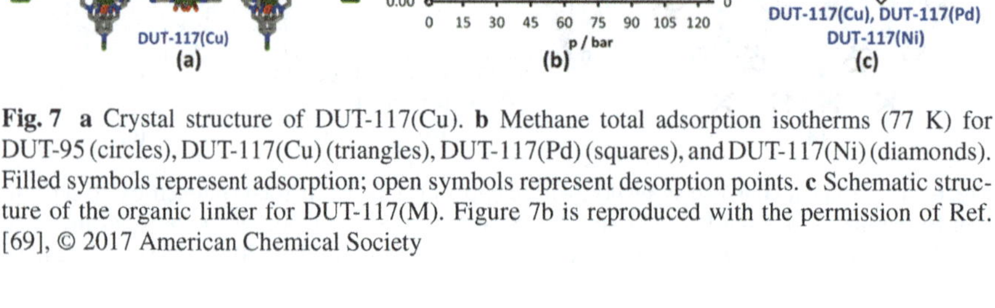

Fig. 7 **a** Crystal structure of DUT-117(Cu). **b** Methane total adsorption isotherms (77 K) for DUT-95 (circles), DUT-117(Cu) (triangles), DUT-117(Pd) (squares), and DUT-117(Ni) (diamonds). Filled symbols represent adsorption; open symbols represent desorption points. **c** Schematic structure of the organic linker for DUT-117(M). Figure 7b is reproduced with the permission of Ref. [69], © 2017 American Chemical Society

to cross-link the larger pores, whereas (ii) DABCO is used to cross-link the smaller pore.

The PXRD analysis revealed that **9b–d** could be desolvated without losing its structural integrity and porosity. PXRD studies were performed using Le Bail analysis and observed that unit cell volume decreased up to 5% upon activation. The author claimed that if the distance between the paddle wheels is less than 11.2 Å, the material is quite rigid against solvent removal. Whereas, longer Cu-Cu spacings induce some flexibility in the framework architecture. Finally, Kaskel groups employed these functionalized MOFs **9b–d,** toward the various gas adsorption such as H_2 (at -196 K), CH_4 (298 K), and CO_2 (between −78 to 25 °C). Moreover, MOFs **9b–d** exhibits BET surface area of 2637, 2939, and 2737 m^2g^{-1}, whereas, pore volume was found to be 1.04, 1.15, and 1.07 cm^3 g^{-1}, respectively. MOF **9a** also contains pores, however, their size is too large for the ideal storage of gases at moderate pressures. The author successfully reduced the pore size with the help of the installation of pillaring ligands, which resulted in an optimal value (Fig. 7b). Further, the incorporation of salen based linker creates additional adsorption sites and metal centers in MOFs **9b–d**. The CH_4 adsorption isotherms of **9c** exhibit the highest capacity up to 240 mg g^{-1} at 65 bar, whereas **9b** and **9d** exhibited gravimetric uptake of 201 mg g^{-1} and 229 mg g^{-1}, respectively. Under identical conditions, **9b** displays the volumetric uptake capacity of 214 cm^3 cm^{-3} at 65 bar with the working capacity of 171 cm^3 $cm^{-3.}$ Though, **9b** and **9d** offering additional metal sites, their working capacities are found to be lower than that of DUT-95 (164 cm^3 cm^{-3}) [70].

As discussed above as well in literature several MOFs are known for their individually high volumetric or high gravimetric CH_4 uptake but only a few examples of MOFs are available which simultaneously exhibit high gravimetric as well as volumetric CH_4 uptake. In this regard, in 2018 Chen and co-workers [71] synthesized a new MOF with the formula, $[Cu_2(\mathbf{L^{28}})(H_2O)_2]_n$, (UTSA-110; **10**) (see Chart 1 for $\mathbf{L^{28}}$) using the solvothermal protocol. The ligand $\mathbf{L^{28}}$ utilized for the preparation of **10** consists of N-functional sites and therefore, it is expected that the resulted **10** would have high gravimetric and volumetric uptake. Single-crystal X-ray analysis revealed that **10** is isoreticular to MOF, NOTT-102 and comprises of a dinuclear paddlewheel $Cu_2(COO)_4$ SBUs connected by the $O_{carboxylate}$of linker $(\mathbf{L^{27}})^{4-}$ to build 3D NbO-type structures (Fig. 8a). The framework architecture of **10** exhibits two types of cages; the first one is a cuboctahedral cage having a dimension of about 10.5 × 14.7 Å^2, whereas, the second one is a large irregular elongated cage with the dimension of ≈ 9.6 × 28.6 Å^2. Notably, due to the larger size of the linker the size of the cages in **10** are much larger than the NOTT-101 (11.2 and 20.6 Å) along a particular crystallographic axis. Finally, the Chen groups exploited the activated sample of **10** (noted as UTSA-110a; **10a,** hereafter) toward the sorption studies and observed 3241 m^2/g BET surface area and 1.263 cm^3/g pore volume. These significant results provide the advantage to the author to record the CH_4 adsorption isotherms of **10a.** The activated sample of MOF **10a** shows one of the highest gravimetric uptakes of about 402 cm^3 (STP) g^{-1} at 65 bar and 298 K (Fig. 8b). Whereas, exhibits total volumetric uptake of 241 cm^3 (STP) cm^{-3} at RT and 65 bar with a working capacity of 190 cm^3 (STP) cm^{-3}. The authors argued that these results are quite better than that

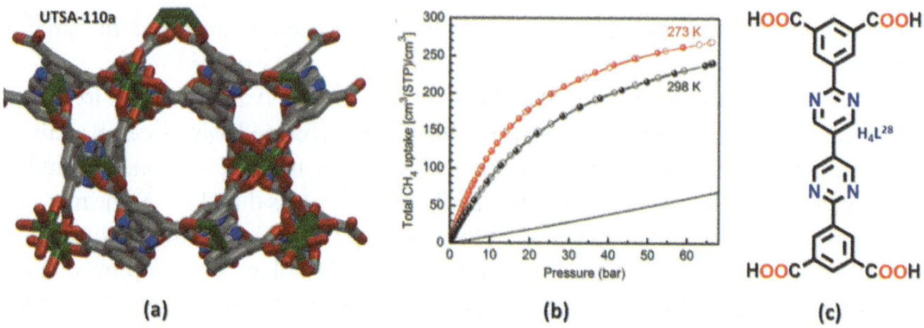

Fig. 8 **a** Crystal packing displaying pores and cavity of UTSA-112a. **b** Total volumetric CH_4 adsorption isotherms of UTSA-112a. **c** Schematic structure of the organic linker for UTSA-112a. Figure 8b is reproduced with the permission of Ref. [71], © 2018 WILEY–VCH Verlag GmbH & Co. KGaA, Weinheim

of NOTT-102, therefore, indicating that newly installed functional N sites might have a positive response toward CH_4 storage capacities. Further, the volumetric working capacity of 190 cm^3 (STP) cm^{-3} shown by the **10a** is the highest so far described for robust MOFs and thus, opens the door for contemporary researchers to develop new MOFs having higher gravimetric/volumetric working capacities simultaneously.

Later on, Tu and co-workers [72] have reported three iron-based MOFs with formulas; $[Fe_3(\mathbf{L^{29}})(EDB)_2{\cdot}12.27H_2O]$ (VNU-21; **11a**) [73], and $\{[Fe_3(\mathbf{L^{29}})(BPDC)_2]{\cdot}11.97H_2O\}$ (VNU-22; **11b**) [74] and $\{[Fe_3O(NDC)_2SO_4(HCO_2)(H_2O)_2]{\cdot}0.6H_2O\}$ (Fe-NDC; **11c**) (see Chart 1 for linkers $\mathbf{L^{29}}$, EDB^{2-}, $BPDC^{2-}$ and NDC^{2-} details). Both **11a** and **11b** are constructed by adapting the mixed linker strategy, in which the sinusoidal $[Fe_3(CO_2)_7]_\infty$ iron-rod SBUs were linked through the tritopic $(\mathbf{L^{29}})^{3-}$ and ditopic EDB^{2-}linkers and exhibits rectangular channels of 7 × 12.4 and 6.8 × 8.7 Å^2, respectively. Whereas, the framework of **11c** consists of new $[Fe_3O(CO_2)_5(SO_4)(H_2O)_2]_\infty$ rod SBUs sewed by the ditopic NDC^{2-} linkers to build the 3D architecture decorated by channels of dimension 9 × 9 Å^2 along O_z axis (Fig. 9a–c). Further, these MOFs were investigated for CH_4 sorption and observed that **11b** and **11c** exhibits volumetric uptakes of 155/160 cm^3 (STP) cm^{-3} at 65 bar and 164/167 cm^3 (STP) cm^{-3} at 80 bar, respectively. Whereas, **11a** shows poor CH_4 uptake compare to **11c** and **11b** at pressures below 20 bar (Fig. 9e). Notably, the authors have observed a significant improvement after increasing the pressure and obtained a value of 142, 182, and 194 cm^3 (STP) cm^{-3} at 35, 65, and 80 bar. The values at 65 and 80 bar are not only better than those of **11b** and **11c**, but also higher than that of the traditional MOFs with large 1D channels, for instance, DUT-4 (164 cm^3 (STP) cm^{-3}) and DUT-5 (134 cm^3 (STP) cm^{-3}) [75]. The quite better performance of **11a** over **11b**, **11c,** and standard MOFs can be attributed to its narrow channels (7.0 × 12.4 Å^2), which provide a perfect fitting space to pack the CH_4 molecules with strong binding energy.

In continuation of the development of MOFs having both high volumetric and gravimetric uptake capacities, very recently, Chen and co-workers [11] synthesized

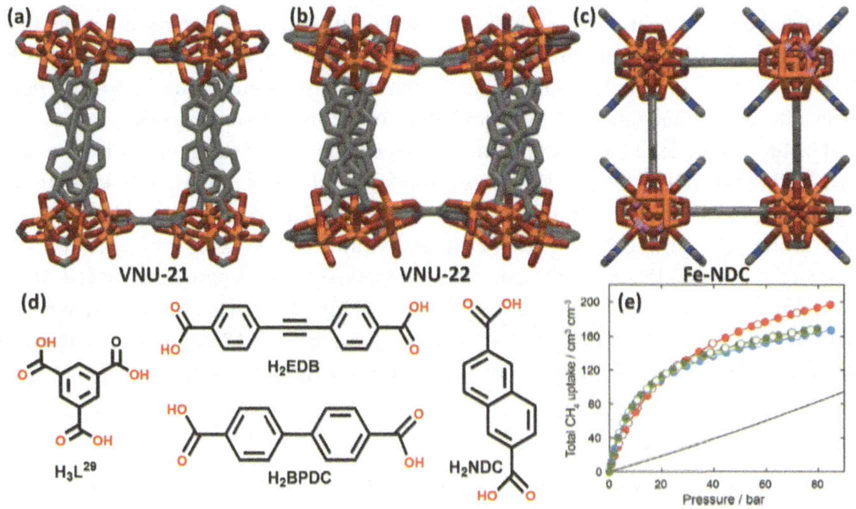

Fig. 9 **a–c** Crystal packing displaying pores and cavity of VNU-1, VNU-2, and Fe-NDC, respectively. **d** Schematic structure of the organic ligands that serve as linkers. **e** The total methane volumetric uptake of VNU-21 (red), VNU-22 (blue), and Fe-NDC (green). The bulk density of methane is represented as a black curve. Figure 9e is reproduced with the permission of Ref. [72], © the Partner Organisations 2019

a new MOF with formula, $[Cu_2(\mathbf{L^{30}})(H_2O)_2]_n$ (UTSA-111; **12**) (see Chart 1 for $\mathbf{L^{30}}$) by incorporating a pyrimidine ring into the organic linker of MOF NOTT-101 and extended the linker with functional N sites [12]. A Single-crystal X-ray diffraction analysis revealed that the framework of **12** comprised of dinuclear paddlewheel $Cu_2(COO)_4$ SBUs linked by the $O_{carboxylates}$ of the linker $(\mathbf{L^{30}})^{4-}$ to form 3D NbO-type structure (Fig. 10a). The framework architecture of **12** features one cuboctahedral cage having a dimension of about 10.5 × 14.7 Å^2 and another large irregular

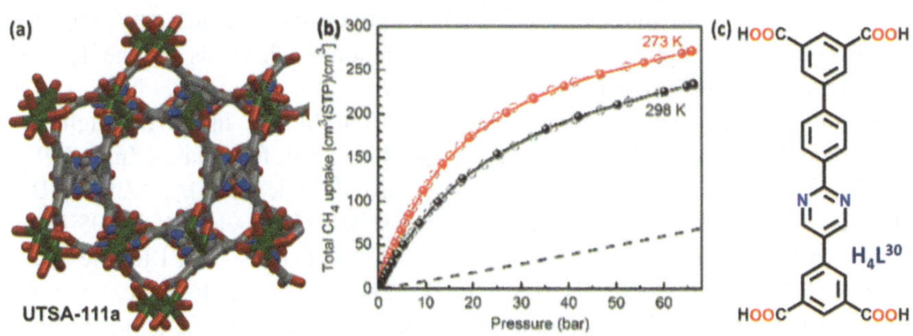

Fig. 10 **a** Crystal packing displaying pores and cavity of UTSA-111a. **b** Total volumetric CH_4 adsorption isotherms of UTSA-111a. **c** Schematic structure of the organic linker for UTSA-111a. Figure 10b is reproduced with the permission of Ref. [11], © The Royal Society of Chemistry 2020

elongated cage of nearly 9.6 × 28.6 Å^2. The incorporation of pyrimidine rings into the organic linkers causes the elongation of the cages in **12**, compared to NOTT-101 (10.2 and 20.6 Å). The authors argue that the instantaneous tuning of the pore sizes and functionalization of the framework of **12** might facilitate the gravimetric and volumetric CH_4 storage capacities. Further, the activated sample of **12** noted as **12a** (UTSA-101a), hereafter, shows remarkably high gravimetric uptake without sacrificing its volumetric performance (at 65 bar and RT) compared to NOTT-101a. Therefore, **12a** showed high gravimetric and volumetric working capacities of 309 cm^3 (STP) g^{-1} and 18 cm^3 (STP) cm^{-3} simultaneously, in comparison to NOTT-101a (250 cm^3 (STP) g^{-1} and 172 cm^3 (STP) cm^{-3}) (Fig. 10b). This unique balance of both high gravimetric and volumetric working capacities in **12a** outperformed some of the standard MOFs available in the literature (see Table 1 for details).

4.2 Flexible MOFs for CH_4 Storage

As discussed above, that the thermal and chemical robustness of MOFs, comprising their structural rigidity during reversible adsorption/desorption phenomenon, are the prime features in terms of their probable industrial applications in the domain of heterogeneous catalysis and gas storage/separation and thus, MOFs becomes the interesting porous materials, which are explored well in last decades [35, 76, 77]. On the other hand, the advantages and possible application of flexible materials were known very early. The flexible MOFs also known as Soft Porous Crystals (SPC) exhibits reversible stretching of the structure under the influence of external stimuli such as host–guest interactions, photochemical, thermal, mechanical, adsorption–desorption, and so on [78]. Further, the flexibility in a MOF can appear in various ways such as rearrangement of the metal-carboxylate cluster constituting the SBU, the rotation of the benzene ring of the organic ligand, and the expansion/contraction of the cell volume [79]. The different flexibility behaviors shown by a MOF includes breathing [80–82], expansion [83], linker rotation [84] or subnetwork displacement [85, 86]. The distinguishing flexible behavior of MOFs is not only scientifically important but also practically applicable for various applications such as gas storage/delivery, separation, catalysis, and sensing.

Keeping the fascinating importance of flexible MOFs, in 2016, Yaghi and co-workers [23] have prepared a series of Zn-MOFs with formulas; $\{Zn_4O(\mathbf{L^{31}})_2\}$ (MOF-950; **11a**), $\{Zn_4O(\mathbf{BDC})(\mathbf{L^{31}})_{4/3}\}$ (MOF-905; **11b**), $\{Zn_4O(\mathbf{BDC\text{-}Me_2})(\mathbf{L^{31}})_{4/3}\}$ (MOF-905-Me$_2$; **11c**), $\{Zn_4O(\mathbf{NDC})(\mathbf{L^{31}})_{4/3}\}$ (MOF-905-Naph; **11d**), $\{Zn_4O(\mathbf{BDC\text{-}NO_2})(\mathbf{L^{31}})_{4/3}\}$ (MOF-905-NO$_2$; **11e**) (see Chart 1 for the details of linker $\mathbf{L^{31}}$ and co-ligands $\mathbf{H_2BDC}$ and $\mathbf{H_2NDC}$). These MOFs were well characterized by different analytical and spectroscopic techniques including SCXRD and PXRD. The structural analysis reveals, that similar to other **ith-d** net containing Zn-MOFs [87, 88] in this structure also, each $Zn_4O(-CO_2)_6$ unit is connected to six carboxylates, four equatorial stemming from $\mathbf{L^{31}}$ and two axial coming from BDC, resulted in the generation of two types of micropores in the framework; (i) a dodecahedral pore having the diameter of 18 Å, constructed from four BDC and eight **L**[31] linkers intersecting 12 zinc-based SBUs, and (ii) a tetrahedral pore with the diameter of 6 Å, consists of four zinc-based SBUs linked by two BDC and four **L**[31] linkers (Fig. 11a). The functionalized **11b** MOFs exhibits larger cages ranges from 15.3 to 17.6 Å (for MOF-905-Naph and -Me_2, respectively), whereas, the diameter of smaller cages is still greater than that of the expanded version, MOF-205. Finally, the authors have examined all these Zn-MOFs, **11a–e**, toward the CH_4 adsorption properties and observed the high CH_4 volumetric work capacity of 203 cm^3 (STP) cm^{-3} at 80 bar and 298 K for **11b** (Fig. 11b). Functionalized MOF **11b** also exhibits the highest CH_4 uptake of 310 cm^3 g^{-1} followed by **11a** (297 cm^3 g^{-1}), **11c** (294 cm^3 g^{-1}), **11d** (289 cm_3 g^{-1}), and **11e** (261 cm^3 g^{-1}). The maximum excess CH_4 uptake for **11b** in volumetric units is 167 cm^3 cm^{-3} at 80 bar and 298 K, which exceeds **11a** (153 cm^3 cm^{-3}) and **11d**, **11c** and **11e** (160, 151, and 144 cm^3 cm^{-3}), respectively.

Later on, Zhang and co-workers [89] have prepared four mesoporous MOFs with formulas; $\{(Zn_4O)_3(\mathbf{L^{32}})_4(\mathbf{BDC})_3\}$ (ST-1; **12a**), $\{(Zn_4O)_3(\mathbf{L^{32}})_4(\mathbf{NDC})_3\}$ (ST-2; **12b**), $\{(Zn_4O)_3(\mathbf{L^{32}})_4(BDC)(BPDC)_2\}$ (ST-3; **12c**) and $\{(Zn_4O)_5(\mathbf{L^{32}})_6(\mathbf{BPDC})_6\}$ (ST-4; **12d**) with the help of $Zn_4O(\text{-}CO_2)_6$ cluster and suitable organic linkers (see Chart 1 for details). The basic strategy was to mix tritopic linker **L**[32] with ditopic linkers such as **H_2BDC**, **H_2NDC**, **H_2BPDC**, and further combination of **H_2BDC** and **H_2BPDC** to achieve ultrahigh porosity without lengthening the organic struts (Fig. 12). The authors accomplished the pore geometries in these ST MOFs, **12a–d** and found similar compositions, however, have different topologies. For instance, **12a** exhibits **muo** net and composed of both 0-D cages [$\{4^3.5^6\}$; inner diameter: 2.2 × 2.2 × 1.5 nm^3] and 1D channels (inner diameter: 3.2 nm). Whereas, **12b** shows umt net and contains several types of cavities such as (i) elongated polyhedral cages [$\{4^3.5^{18}\}$; inner diameter: 3.3 × 3.2 × 2.3 nm^3]; (ii) spherical icosahedral cages [$\{5^{12}\}$; inner diameter: 2.4 × 2.4 × 2.3 nm^3]; (iii) prolate polyhedral cages [$\{4^3.5^6\}$;

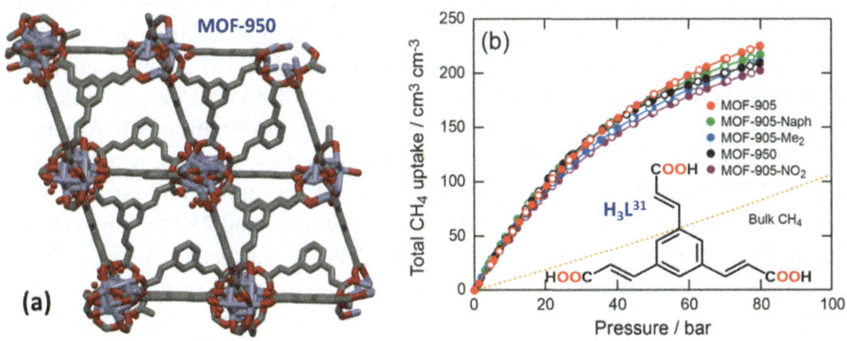

Fig. 11 **a** Crystal structure of **MOF-950**. **b** Methane total adsorption isotherms (298 K). Filled symbols represent adsorption; open symbols represent desorption points. The schematic structure of the organic linker (H_3**L**[31]) for MOF-950 is provided in the inset. Figure 11b is reproduced with the permission of Ref. [23], © 2016 American Chemical Society

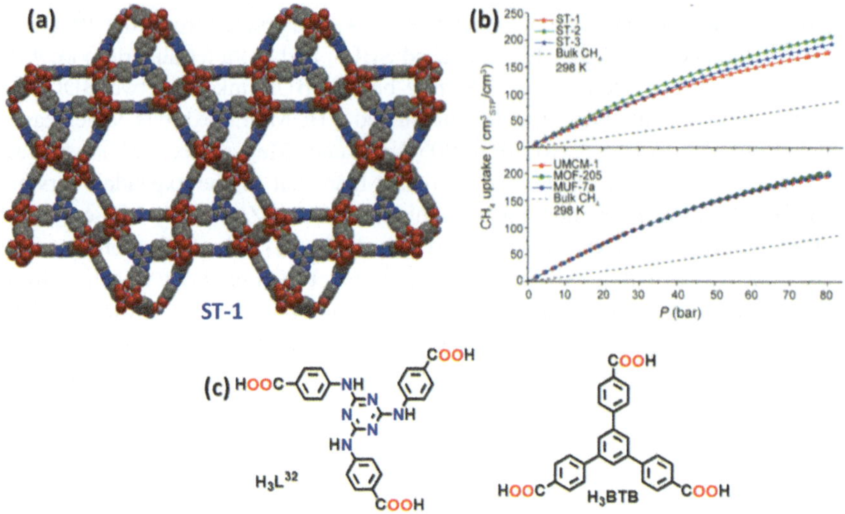

Fig. 12 **a** Crystal structure of **ST-1**. **b** High-pressure (0–80 bar) methane uptake isotherms for the ST MOFs and their analogues at 298 K. **c** Schematic structure of the organic linker. Figure 12b is reproduced with the permission of Ref. [89], © 2017 American Chemical Society

inner diameter: 2.3 × 2.3 × 1.5 nm^3), and (iv) adamantane-like cages [$\{5^4\}$; inner diameter: 0.8 nm). On the other hand **12c** offer a unique **ith-d** net with an elongated pore [$\{5^{12}\}$; inner diameter: 3.2 × 3.0 × 2.7 nm^3] having a BPDC vs BDC linker ratio of 2:1. Furthermore, as an exceptional case, **12d** features **ott** net with colossal cages [$\{4^6.5^{24}.6^8\}$, 3.0 nm + 6 × 2.0 nm) terminated with dangling ligands, spherical mesoporous cages ([$5^{12}.6^4$]; inner diameter: 3.0 nm), cubic cages [$\{4^6\}$; inner diameter: 1.8 nm], and microporous adamantane-like cages [$\{5^4\}$; inner diameter: 1.0 nm].

Finally, Zhang's groups exploited all these ST MOFs, **12a–d** toward the ultrahigh CH_4 sorption via Grand canonical Monte Carlo (GCMC) simulation, and observed higher deliverable CH_4 uptakes at 298 K and 5–250 bar (Fig. 12). Significantly, these ST MOFs show enhanced CH_4 uptakes in the range of 80–250 bar, which can be accredited to the engineered pore geometries and energy distribution [90]. Among all these investigated ST MOFs, **12b** is recognized as the best material, which exhibits 324 cm^3 STP/cm^3 CH_4 uptake at 298 K and 250 bar, along with the highest deliverable capacity of 305 cm^3 STP/cm^3 at 298 K and in between 5 and 250 bar. Importantly, the ultrahigh deliverable capacity of 289 cm^3 STP/cm^3 has been realized at 298 K and 5−200 bar, which can deliver 31 and 12% more than the CNG process at 200 bar and 250 bar, respectively. The author compare their results with other reported CH_4 uptake data and claimed that this performance is a new record in MOFs and surpasses the previous record holder of activated carbons.

In the same year (2017), Zhang and co-workers [91] have prepared a first cycloaliphatic ring functionalized MOF with formula, $[Cu_2(\mathbf{L^{33}})(DMF)_2]\cdot 3DMF\cdot 7H_2O$ (NJU-Bai 19; **13**) using a flexible organic linker H_4L^{33} (see Chart 1) and copper salt under solvothermal conditions. The author incorporated the aliphatic fragments and N sites in the form of piperazine to replace the central benzene ring of the linker used in the preparation of NOTT-101. This replacement aims to just increase the uptake and working capacity of the MOF toward CH_4 gas. Single-crystal X-ray analysis reveals that the structure of **13** is composed of binuclear Cu(II) paddlewheel nodes, which are bridged by the tetratopic linker $(L^{33})^{4-}$ to build a 3D (4,4)-connected NbO-type net (Fig. 13a). Further, the framework structure of **13** features two types of cages. The first one is a smaller spherical cage having a diameter of 9.7 Å and the second one is a larger shuttle-shaped cage with a dimension of about 9.3 × 22.2 Å. Besides, the structure of **13** also consists of triangular channels having a diameter of around 5 Å viewing along the c-axes, whereas, triangular windows can be observed along the a/b axis, wherein the size is controlled by the length of the spaces. Notably, MOF **13** exhibits an accessible pore volume of 71.5% per unit cell volume.

Finally, Zhang groups exploited functionalized MOF **13** for the permanent porosity and found 2803 and 3007 m^2g^{-1} BET surface area with a pore volume of 1.063 cm^3 g^{-1}, which are comparable to that of NOTT-101 (2805 and 3045 m^2g^{-1}, respectively). This comparable porosity of **13** with that of NOTT-101 encouraged the author to study its methane storage capacities. The results of this study are shown in Fig. 13b of this chapter, which shows a total volumetric CH_4 storage capacity of 200.4 cm^3 (STP) cm^{-3} at 298 K and 35 bar, thus surpassing the DOE's earlier target of 180 cm^3 (STP) cm^{-3}. Further, at 65 bar, the volumetric CH_4 uptake of **13** enhanced to 246.4 cm^3 (STP) cm^{-3}, which is consistent with the amount of CH_4 stored in a CNG tank at 249 bar. Importantly, the volumetric CH_4 working capacity of **13** remained unaffected (from 185 to 189 cm^3 (STP) cm^{-3}) at 273 K temperature and in between 65 and 5 bar, working pressures, as some of the recognized MOFs having high Q_{st} exhibited a drop in their volumetric CH_4 working capacity with the decrease in temperature. This might be ascribed to the relatively low Q_{st} value

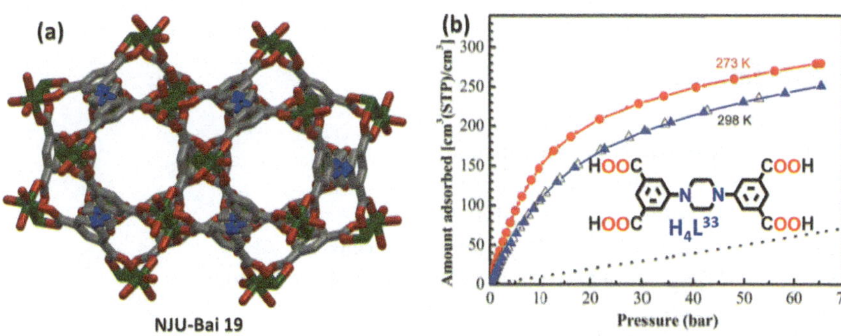

Fig. 13 **a** Crystal structure of **NJU-Bai 19.** **b** The total high-pressure CH_4 adsorption isotherms for NJU-Bai 19 at 273 K and 298 K (filled and open symbols represent adsorption and desorption, respectively). Data of pure CH_4 stored in the high-pressure gas tank are presented as a black line. The schematic structure of the organic linker (H_4L^{33}) for **NJU-Bai 19** is provided in the inset. Figure 13b is reproduced with the permission of Ref. [91], © The Royal Society of Chemistry 2017

for CH_4 adsorption. Moreover, at 65 bar pressure and RT, the full tank CH_4 deliver capacity of **13** is approximately four times greater than the unfilled tank. Therefore, the author argued that the introduction of the aliphatic piperazine ring to the linker opened a new door for contemporary scientists to construct the MOFs with enhancing the volumetric and working capacity of CH_4 using a cycloaliphatic backbone.

5 Conclusion and Outlook

The potential application of metal-organic frameworks (MOFs) in natural gas-(NG) based applications are still in their infancy stages and have essentially been restricted to adsorptive storage. These initial outcomes can be further expanded potentially, provided the various applications of NG by using MOFs with advanced adsorbents. Several MOFs have already been in use for selective gas adsorption in the presence of various gaseous impurities. The most important challenge is the development of the synthetic route for the large-scale synthesis of high purity MOFs at a relatively lower cost. The next major challenge is the stability of the concerned MOF in harsh environments for various applications. The recyclability of MOFs to be periodically regenerated is also a key area of consideration for future developments of MOFs. The MOFs are incredibly sought potential materials for the high storage capacity of NG owing to astonishing tunability of the surface and high porosity. We strongly believe that this book chapter would provide a detailed insight into the challenging area of methane gas storage and its uses in ambient environments. We hope this chapter will act as a stepping stone for this vision.

Acknowledgements G Kumar thanks SERB, Government of India for the financial assistance in the form of SERB Research Scientist Support Grant with Ref. No. SB/SRS/2020-21/42/CS. PR thanks UGC for the award of a Senior Research Fellowship (Ref. No. 19/06/2016(i)EU-V).

Abbreviations

VNU	Vietnam National University
NDC	2,6-Naphthalene dicarboxylate
DUT	Dresden University of Technology
LIFM	Lehn Institute of Functional Materials
HHKUST	Hong Kong University of Science and Technology
PCN	Porous Coordination Network
NJU-Bai	Nanjing University Bai's group
MFM	Manchester Framework Material
NOTT	University of Nottingham
UTSA	University of Texas at San Antonio
ZJU	Zhejiang University

bdp	1,4-benzenedipyrazolate
ZJNU	Zhejiang Normal University
NU	Northwestern University
FDM	Fudan materials
BUT	Beijing University of Technology
NA	Stands for not available

References

1. Rani P, Gauri HA, Bhasin KK, Kumar G (2020) A doubly interpenetrated Cu^{II} metal-organic framework for selective molecular recognition of nitroaromatics. Cryst Growth Des 20:7141–7151
2. Kumar G, Das SK (2017) Coordination frameworks containing compounds as catalysts. Inorg Chem Front 4:202–233
3. Kumar G, Gupta R (2013) Molecularly designed architectures–the metalloligand way. Chem Soc Rev 42:9403–9453
4. Husain A, Rani P, Alisha SA, Mondal T, Saha S, Bhasin KK, Trivedi M, Kumar G (2020) Luminescent Cd II metal-organic frameworks based on isoniazid using a mixed ligand approach. CrystEngComm 22:5980–5986
5. Kumar G, Kumar G, Gupta R (2021) Effect of pyridyl donors from organic ligands *versus* metalloligands on material design. Inorg Chem Front 8:1334–1373
6. Fang R, Dhakshinamoorthy A, Li Y, Garcia H (2020) Metal-organic frameworks for biomass conversion. Chem Soc Rev 49:3638–3687
7. Ghasemzadeh MA, Mirhosseini-Eshkevari B, Tavakoli M, Zamani F (2020) Metal-organic frameworks: advanced tools for multicomponent reactions. Green Chem 22:7265–7300
8. Goetjen TA, Liu J, Wu Y, Sui J, Zhang X, Hupp JT, Farha OK (2020) Metal-organic framework (MOF) materials as polymerization catalysts: a review and recent advances. Chem Commun 56:10409–10418
9. He Y, Zhou W, Qian G, Chen B (2014) Methane storage in metal-organic frameworks. Chem Soc Rev 43:5657–5678
10. Makal TA, Li J-R, Lu W, Zhou HC (2012) Methane storage in advanced porous materials. Chem Soc Rev 41:7761–7779
11. Wen H-M, Shao K, Zhou W, Li B, Chen B (2020) A novel expanded metal-organic framework for balancing volumetric and gravimetric methane storage working capacities. Chem Commun 56:13117–13120
12. Li B, Wen H-M, Wang H, Wu H, Yildirim T, Zhou W, Chen B (2015) Porous metal-organic frameworks with Lewis basic nitrogen sites for high-capacity methane storage. Energy Environ Sci 8:2504–2511
13. Kundu T, Shah BB, Bolinois L, Zhao D (2019) Functionalization-induced breathing control in metal-organic frameworks for methane storage with high deliverable capacity. Chem Mater 31:2842–2847
14. Wu X, Xiang S, Su J, Cai W (2019) Understanding quantitative relationship between methane storage capacities and characteristic properties of metal-organic frameworks based on machine learning. J Phys Chem C 123:8550–8559
15. Zhang Y, Lucier BEG, Fischer M, Gan Z, Boyle PD, Desveaux B, Huang Y (2018) A multifaceted study of methane adsorption in metal-organic frameworks by using three complementary techniques. Chem Eur J 24:7866–7881
16. Férey G (2008) Hybrid porous solids: past, present, future. Chem Soc Rev 37:191–214
17. Horike S, Shimomura S, Kitagawa S (2009) Soft porous crystals. Nat Chem 1:695–704

18. Li B, Wen H-M, Zhou W, Xu JQ, Chen B (2016) Porous metal-organic frameworks: promising materials for methane storage. Chem 1:557–580
19. Ma S, Zhou HC (2010) Gas storage in porous metal-organic frameworks for clean energy applications. Chem Commun 46:44–53
20. Service RF (20140 Stepping on the gas. Science 346:538–541
21. Yeh S (2007) An empirical analysis on the adoption of alternative fuel vehicles: the case of natural gas vehicles. Energy Policy 35:5865–5875
22. Santos JC, Lima JA, Gurgel JM, Marcondes F (2019) Improvement of methane storage capacity in activated carbon bed with bidisperse packing. Braz J Chem Eng 36:831–843
23. Jiang J, Furukawa H, Zhang YB, Yaghi OM (2016) High methane storage working capacity in metal-organic frameworks with acrylate links. J Am Chem Soc 138:10244–10251
24. Konstas K, Osl T, Yang Y, Batten M, Burke N, Hill AJ, Hill MR (2012) Methane storage in metal-organic framework. J Mater Chem 22:16698–16708
25. Hu Z, Kundu T, Wang Y, Sun Y, Zeng K, Zhao D (2020) Modulated hydrothermal synthesis of highly stable MOF-808(Hf) for methane storage. ACS Sustain Chem Eng 8:17042–17053
26. Zhou W, Wu H, Hartman MR, Yildirim T (2007) Hydrogen and methane adsorption in metal−organic frameworks: a high-pressure volumetric study. J Phys Chem C 111:16131–16137
27. Dincă M, Dailly A, Liu Y, Brown CM, Neumann DA, Long JR (2006) Hydrogen storage in a microporous metal−organic framework with exposed Mn^{2+} coordination sites. J Am Chem Soc 128:16876–16883
28. Poirier E, Dailly A (2008) Investigation of the hydrogen state in IRMOF-1 from measurements and modeling of adsorption isotherms at high gas densities. J Phys Chem C 112:13047–13052
29. See the NIST Refprop database at http://webbook.nist.gov/chemistry/fluid/
30. Mason JA, Veenstra M, Long JR (2014) Evaluating metal-organic frameworks for natural gas storage. Chem Sci 5:32–51
31. Senkovska I, Kaskel S (2008) High-pressure methane adsorption in the metal-organic frameworks $Cu_3(btc)_2$, $Zn_2(bdc)_2$dabco, and $Cr_3F(H_2O)_2O(bdc)_3$. Microporous Mesoporous Mater 112:108–115
32. Seki K (2001) Design of an adsorbent with an ideal pore structure for methane adsorption using metal complexes. Chem Commun 1496−1497
33. He Y, Chen F, Li B, Qian G, Zhou W, Chen B (2018) Porous metal-organic frameworks for fuel storage. Coord Chem Rev 373:167–198
34. Li W, Liu J, Zhao D (2016) Mesoporous materials for energy conversion and storage devices. Nat Rev Mater 1:16023
35. Furukawa H, Cordova KE, O'Keeffe M, Yaghi OM (2013) The chemistry and applications of metal-organic frameworks. Science 341:1230444
36. Kim D, Ahn Y-H, Lee H (2015) Phase equilibria of CO_2 and CH_4 hydrates in intergranular meso/macro pores of MIL-53 metal-organic framework. J Chem Eng Data 60:2178–2185
37. Czepirski L, JagieŁŁo J (1989) Virial-type thermal equation of gas-solid adsorption. J Chem Eng Sci 44:797–801
38. Wang K, Huang H, Liu D, Wang C, Li J, Zhong C (2016) Covalent triazine-based frameworks with ultramicropores and high nitrogen contents for highly selective CO_2 capture. Environ Sci Technol 50:4869–4876
39. Kondo M, Yoshitomi T, Matsuzaka H, Kitagawa S, Seki K (1997) Three-dimensional framework with channeling cavities for small molecules: $\{[M_2(4,4'\text{-bpy})_3(NO_3)_4]\cdot xH_2O\}_n$ (M · Co, Ni, Zn). Angew Chem Int Ed 36:1725–1727
40. Guo Z, Wu H, Srinivas G, Zhou Y, Xiang S, Chen Z, Yang Y, Zhou W, O'Keeffe M, Chen B (2011) Angew Chem Int Ed 50:3178–3181
41. Wilmer CE, Farha OK, Yildirim T, Eryazici I, Krungleviciute V, Sarjeant AA, Snurr RQ, Hupp JT (2013) Gram-scale, high-yield synthesis of a robust metal-organic framework for storing methane and other gases. Energy Environ Sci 6:1158–1163
42. Wilmer CE, Leaf M, Lee CY, Farha OK, Hauser BG, Hupp JT, Snurr RQ (2012) Large-scale screening of hypothetical metal-organic frameworks. Nat Chem 4:83–89

43. Kong G-Q, Han Z-D, He Y, Ou S, Zhou W, Yildirim T, Krishna R, Zou C, Chen B, Wu C-D (2013) Expanded organic building units for the construction of highly porous metal-organic frameworks. Chem Eur J 19:14886–14894
44. Pang J, Jiang F, Wu M, Yuan D, Zhou K, Qian J, Su K, Hong M (2014) Coexistence of cages and one-dimensional channels in a porous MOF with high H_2 and CH_4 uptakes. Chem Commun 50:2834–2836
45. Chui SS-Y, Lo SM-F, Charmant JPH, Orpen AG, Williams ID (1999) A chemically functionalizable nanoporous material $[Cu_3(TMA)_2(H_2O)_3]_n$. Science 283:1148–1150
46. Li B, Wen H-M, Wang H, Wu H, Tyagi M, Yildirim T, Zhou W, Chen B (2014) A porous metal-organic framework with dynamic pyrimidine groups exhibiting record high methane storage working capacity. J Am Chem Soc 136:6207–6210
47. Rao X, Cai J, Yu J, He Y, Wu C, Zhou W, Yildirim T, Chen B, Qian G (2013) A microporous metal-organic framework with both open metal and Lewis basic pyridyl sites for high C_2H_2 and CH_4 storage at room temperature. Chem Commun 49:6719–6721
48. Wu H, Zhou W, Yildirim T (2009) High-capacity methane storage in metal–organic frameworks M2(dhtp): the important role of open metal sites. J Am Chem Soc 131:4995–5000
49. Song C, Liu H, Jiao J, Bai D, Zhou W, Yildirim T, He Y (2016) High methane storage and working capacities in a NbO-type metal-organic framework. Dalton Trans 45:7559–7562
50. Song C, Ling Y, Feng Y, Zhou W, Yildirim T, He Y (2015) A NbO-type metal-organic framework exhibiting high deliverable capacity for methane storage. Chem Commun 51:8508–8511
51. Li M, Li D, O'Keeffe M, Yaghi OM (2014) Topological analysis of metal-organic frameworks with polytopic linkers and/or multiple building units and the minimal transitivity principle. Chem Rev 114:1343–1370
52. He Y, Li B, O'Keeffe M, Chen B (2014) Multifunctional metal-organic frameworks constructed from meta-benzenedicarboxylate units. Chem Soc Rev 43:5618–5656
53. See the DOE MOVE program at https://arpa-e-foa.energy.gov/
54. Zhang M, Zhou W, Pham T, Forrest KA, Liu W, He Y, Wu H, Yildirim T, Chen B, Space B, Pan Y, Zaworotko MJ, Bai J (2017) Fine-tuning of MOF-505 analogues to reduce low-pressure methane uptake and enhance methane working capacity. Angew Chem Int Ed 129:11584–11588
55. Ma S, Sun D, Simmons JM, Collier CD, Yuan D, Zhou HC (2008) Metal-organic framework from an anthracene derivative containing nanoscopic cages exhibiting high methane uptake. J Am Chem Soc 130:1012–1016
56. Chen B, Ockwig NW, Millward AR, Contreras DS (2005) High H_2 adsorption in a microporous metal-organic framework with open metal sites. Angew Chem Int Ed 44:4745–4749
57. Yan Y, Kolokolov DI, Silva ID, Stepanov AG, Blake AJ, Dailly A, Manuel P, Tang CC, Yang S, Schröder M (2017) Porous metal-organic polyhedral frameworks with optimal molecular dynamics and pore geometry for methane storage. J Am Chem Soc 139:13349–13360
58. Yan Y, Lin X, Yang S, Blake AJ, Dailly A, Champness NR, Hubberstey P, Schröder M (2009) Exceptionally high H_2storage by a metal-organic polyhedral framework. Chem Commun 1025–1027
59. Yan Y, Blake AJ, Lewis W, Barnett SA, Dailly A, Champness NR, Schröder M (2011) Modifying cage structures in metal-organic polyhedral frameworks for H_2 storage. Chem Eur J 17:11162–11170
60. Peng Y, Krungleviciute V, Eryazici I, Hupp JT, Farha OK, Yildirim T (2013) Methane storage in metal-organic frameworks: current records, surprise findings, and challenges. J Am Chem Soc 135:11887–11894
61. He Y, Zhou W, Yildirim T, Chen B (2013) A series of metal-organic frameworks with high methane uptake and an empirical equation for predicting methane storage capacity. Energy Environ Sci 6:2735–2744
62. Chen C-X, Wei Z-W, Jiang J-J, Zheng S-P, Wang H-P, Qiu Q-F, Cao C-C, Fenske D, Su C-Y (2017) Dynamic spacer installation for multirole metal-organic frameworks: a new direction toward multifunctional MOFs achieving ultrahigh methane storage working capacity. J Am Chem Soc 139:6034–6037

63. Xu G, Li B, Wu H, Zhou W, Chen B (2017) Construction of ntt-type metal-organic framework from C2-symmetry hexacarboxylate linker for enhanced methane storage. Cryst Growth Des 17:4795–4800
64. Larson AC, Dreele VRB (1994) General structure analysis system, Report LAUR 86–748; Los Alamos National Laboratory: New Mexico USA
65. Yuan D, Zhao D, Sun D, Zhou HC (2010) An isoreticular series of metal-organic frameworks with dendritic hexacarboxylate ligands and exceptionally high gas-uptake capacity. Angew Chem Int Ed 49:5357–5361
66. Guo Z, Wu H, Srinivas G, Zhou Y, Xiang S, Chen Z, Yang Y, Zhou W, O'Keeffe M, Chen B (2011) A metal-organic framework with optimized open metal sites and pore spaces for high methane storage at room temperature. Angew Chem Int Ed 50:3178–3181
67. Li B, Zhang Z, Li Y, Yao K, Zhu Y, Deng Z, Yang F, Zhou X, Li G, Wu H, Nijem N, Chabal YJ, Lai Z, Han Y, Shi Z, Feng S, Li J (2012) Enhanced binding affinity, remarkable selectivity, and high capacity of CO_2 by dual functionalization of a rht-type metal-organic framework. Angew Chem Int Ed 51:1412–1415
68. Peng Y, Srinivas G, Wilmer CE, Eryazici I, Snurr RQ, Hupp JT, Yildirim T, Farha OK (2013) Simultaneously high gravimetric and volumetric methane uptake characteristics of the metal-organic framework NU-111. Chem Commun 49:2992–2994
69. Müller P, Bon V, Senkovska I, Getzschmann J, Weiss MS, Kaskel S (2017) Crystal engineering of Phenylenebis(azanetriyl))tetrabenzoate based metal-organic frameworks for gas storage applications. Cryst Growth Des 17:3221–3228
70. Müller P, Wisser FM, Bon V, Grünker R, Senkovska I, Kaskel S (2015) Postsynthetic paddle-wheel cross-linking and functionalization of 1,3-Phenylenebis(azanetriyl)tetrabenzoate-based MOFs. Chem Mater 27:2460–2467
71. Wen H-M, Li B, Li L, Lin R-B, Zhou W, Qian G, Chen B (2018) A metal-organic framework with optimized porosity and functional sites for high gravimetric and volumetric methane storage working capacities. Adv Mater 30:1704792
72. Tu TN, Nguyen HTD, Tran NT (2019) Tailoring the pore size and shape of the one-dimensional channels in iron-based MOFs for enhancing the methane storage capacity. Inorg Chem Front 6:2441–2447
73. To TA, Vo YH, Nguyen HTT, Ha PTM, Doan SH, Doan TLH, Li S, Le HV, Tu TN, Phan NTS (2019) Iron-catalyzed one-pot sequential transformations: synthesis of quinazolinones via oxidative Csp3 AH bond activation using a new metal-organic framework as catalyst. J Catal 370:11–20
74. Doan SH, Tran NKQ, Pham PH, Nguyen VHH, Nguyen NN, Ha PTM, Li S, Le HV, Le NTH, Tu TN, Phan NTS (2019) A new synthetic pathway to Triphenylpyridines via cascade reactions utilizing a new iron-organic framework as a recyclable heterogeneous catalyst. Eur J Org Chem 2382–2389
75. Senkovska I, Hoffmann F, Fröba M, Getzschmann J, Böhlmann W, Kaskel S (2009) New highly porous aluminium based metal-organic frameworks: Al(OH)(ndc) (ndc = 2,6-naphthalene dicarboxylate) and Al(OH)(bpdc) (bpdc = 4,4′-biphenyl dicarboxylate). Microporous Mesoporous Mater 122:93–98
76. Li J-R, Sculley J, Zhou H-C (2012) Metal-organic frameworks for separations. Chem Rev 112:869–932
77. Schneemann A, Henke S, Schwedler I, Fischer RA (2014) Targeted manipulation of metal-organic frameworks to direct sorption properties. ChemPhysChem 15:823–839
78. Schneemann A, Bon V, Schwedler I, Senkovska I, Kaskel S, Fischer RA (2014) Flexible metal–organic frameworks. Chem Soc Rev 43:6062–6096
79. Lee JH, Jeoung S, Chung YG, Moon HR (2019) Elucidation of flexible metal-organic frameworks: research progresses and recent developments. Coord Chem Rev 389:161–188
80. Férey G (2004) A rationale for the large breathing of the porous aluminum terephthalate (MIL-53) upon hydration. Chem Eur J 10:1373–1382
81. Sanselme M, Grenèche JM, Riou-Cavellec M, Férey G (2002) $[Fe_2(C_{10}O_8H_2)]$: An antiferromagnetic 3D iron(II) carboxylate built from ferromagnetic edge-sharing octahedral chains (MIL-62). Chem Commun 2172–2173

82. Volkringer C, Loiseau T, Guillou N, Férey G, Elkaïm E, Vimont A (2009) XRD and IR structural investigations of a particular breathing effect in the MOF-type gallium terephthalate MIL-53(Ga). Dalton Trans 2241–2249
83. Mellot-Draznieks C, Serre C, Surblé S, Audebrand N, Férey G (2005) Very large swelling in hybrid frameworks: a combined computational and powder diffraction study. J Am Chem Soc 127:16273–16278
84. Fairen-Jimenez D, Moggach SA, Wharmby MT, Wright PA, Parsons S, Düren T (2011) Opening the gate: framework flexibility in ZIF-8 explored by experiments and simulations. J Am Chem Soc 133:8900–8902
85. Bureekaew S, Sato H, Matsuda R, Kubota Y, Hirose R, Kim J, Kato K, Takata M, Kitagawa S (2010) Control of interpenetration for tuning structural flexibility influences sorption properties. Angew Chem Int Ed 49:7660–7664
86. Maji TK, Matsuda R, Kitagawa S (2007) A flexible interpenetrating coordination framework with a bimodal porous functionality. Nat Mater 6:142–148
87. Klein N, Senkovska I, Gedrich K, Stoeck U, Henschel A, Mueller U, Kaskel S (2009) A mesoporous metal-organic framework. Angew Chem Int Ed 48:9954–9957
88. Liu L, Telfer SG (2015) Systematic ligand modulation enhances the moisture stability and gas sorption characteristics of quaternary metal-organic frameworks. J Am Chem Soc 137:3901–3909
89. Liang C-C, Shi Z-L, He C-T, Tan J, Zhou H-D, Zhou H-L, Lee Y, Zhang Y-B (2017) Engineering of pore geometry for ultrahigh capacity methane storage in mesoporous metal-organic frameworks. J Am Chem Soc 139:13300–13303
90. Lee Y, Barthel SD, Dłotko P, Moosavi SM, Hess K, Smit B (2017) Quantifying similarity of pore-geometry in nanoporous materials. Nat Commun 8:15396
91. Zhang M, Chen C, Wang Q, Fu W, Huang K, Zhou W (2017) A metal-organic framework functionalized with piperazine exhibiting enhanced CH_4 storage. J Mater Chem A 5:349–354
92. Alezi D, Spanopoulos I, Tsangarakis C, Shkurenko A, Adil K, Belmabkhout Y, O'Keeffe M, Eddaoudi M, Trikalitis PN (2016) Reticular chemistry at its best: directed assembly of hexagonal building units into the awaited metal-organic framework with the intricate polybenzene topology, pbz-MOF. J Am Chem Soc 138:12767–12770
93. Gándara F, Furukawa H, Lee S, Yaghi OM (2014) High methane storage capacity in aluminum metal-organic frameworks. J Am Chem Soc 136:5271–5274
94. Mason JA, Oktawiec J, Taylor MK, Hudson MR, Rodriguez J, Bachman JE, Gonzalez MI, Cervellino A, Guagliardi A, Brown CM, Llewellyn PL, Masciocchi N, Long JR (2015) Methane storage in flexible metal-organic frameworks with intrinsic thermal management. Nature 527:357–361
95. Gomez-Gualdron DA, Gutov OV, Krungleviciute V, Borah B, Mondloch JE, Hupp JT, Yildirim T, Farha OK, Snurr RQ, Computational design of metal-organic frameworks based on stable zirconium building units for storage and delivery of methane. Chem Mater 26:5632–5639
96. Spanopoulos I, Tsangarakis C, Klontzas E, Tylianakis E, Froudakis G, Adil K, Belmabkhout Y, Eddaoudi M, Trikalitis PN (2016) Reticular synthesis of HKUST-like tbo-MOFs with enhanced CH_4 storage. J Am Chem Soc 138:1568–1574
97. Barin G, Krungleviciute V, Gomez-Gualdron DA, Sarjeant AA, Snurr RQ, Hupp JT, Yildirim T, Farha OM (2014) Isoreticular series of (3,24)-connected metal-organic frameworks: facile synthesis and high methane uptake properties. Chem Mater 26:1912–1917
98. Furukawa H, Ko N, Go YB, Aratani N, Choi SB, Choi E, Yazaydin AO, Snurr RQ, O'keeffe M, Kim J, Yaghi OM (2010) Ultrahigh Porosity in Metal-Organic Frameworks. Science 329:424-428
99. Stoeck U, Krause S, Bon V, Senkovska I, Kaskel S (2012) A highly porous metal-organic framework, constructed from a cuboctahedral super-molecular building block, with exceptionally high methane uptake. Chem Commun 48:10841–10843
100. Alezi D, Belmabkhout Y, Suyetin M, Bhatt PM, Weseliński ŁJ, Solovyeva V, Adil K, Spanopoulos I, Trikalitis PN, Emwas A-H, Eddaoudi M (2015) MOF crystal chemistry paving the way to gas storage needs: aluminum-based soc-MOF for CH_4, O_2, and CO_2 storage. J Am Chem Soc 137:13308–13318

Photocatalysis by MOFs

Photocatalysis by Metal–Organic Frameworks (MOFs): An Overview

Ankur Srivastava, Manju Yadav, Bani Mahanti, Arun Kumar, and Mrituanjay D. Pandey

Contents

1 Introduction 554
2 MOF-Based Photocatalysis 554
2.1 Enhancing Light-Harvesting 555
2.2 Enhancing Electron–Hole Separation 556
3 MOFs as Photothermal Catalysis 557
3.1 Photothermal Effect of Plasmonic Metals 558
3.2 Photothermal Effect of Both MOFs and Plasmonic Metals 559
4 Conclusion 559
References 560

Abstract Metal–organic frameworks (MOFs) are porous crystalline materials with sequential unique characteristics, such as wide surface vicinity, high content of transition metals, porosity, and tunable physical properties after synthesis, and this makes itself in the group of heterogeneous catalysts. The MOFs composites tagged with featured metals and nanoparticles consist of eminent prospects in many applications such as photocatalytic reduction of harmful oxides of nitrogen and carbon, H_2 generation, and environmental debris treatment. Furthermore, various strategies have been designed to amend the MOFs for photocatalytic performance enhancements, such as the mixed-metal/linker approach, metal ion/ligand immobilization approach, loading metal nanoparticles, and magnetic recycling. In addition, light-mediated catalysis, viz photothermal catalysis, and photocatalysis, show vital efforts in the conversion of solar energy to chemical/thermal energy through the interaction of light with matter. This catalytic effort and the interrelated recent researches against the challenge of insufficient solar light utilization have been presented here briefly in a few headings.

A. Srivastava · M. Yadav · M. D. Pandey (✉)
Department of Chemistry, Institute of Science, Banaras Hindu University, Varanasi, Uttar Pradesh 221005, India
e-mail: mdpandey.chem@bhu.ac.in

B. Mahanti
Chemistry Section, MMV, Banaras Hindu University, Varanasi, Uttar Pradesh 221005, India

A. Kumar
Department of Chemistry School of Chemical Sciences, National Institute of Science Education and Research Bhubaneswar, P.O. Jatni, Khurda, Odisha 752050, India

S. Gulati (ed.), *Metal-Organic Frameworks (MOFs) as Catalysts*,
https://doi.org/10.1007/978-981-16-7959-9_21

The work toward obtaining a stable technology, low-cost applications, and future development on account are also presented.

Keywords Metal–organic framework · Photocatalysis · Photothermal catalysis

1 Introduction

Today, the world is facing two major challenges: the scarcity of fossils energy and environmental toxic wastages. Eliminating poisonous organic–inorganic pollutants with strong degradation properties from the atmosphere and wastewater has become a burning research topic. For example, the atmospheric CO_2 loadings have increased from pre-industrial levels of about 280 ppm to about 390 ppm now. Researchers suggest that at the current top-up rate within the next couple of decades, the average global temperatures will rise about 2 °C much before the end of the century. This degree of global warming is sufficient to cause massive worldwide economic and political upheaval. Conventional methods and industrial technologies employing liquid or solid sorbents for coagulation or adsorption currently exist, generally experience worse and high operating expenses and still releasing other derived irreversible pollutant products, and otherwise require a high cost to regenerate the trapping agents for reuse. To avoid enormous global disruptions in human society in the future, a substantial cut in anthropogenic pollutants replacing by alternate sustainable energy and a collaborated effort to decrease the existing pollutant levels are a must to do.

Photocatalytic degradation of pollutants paves the way for high-order performance green application. In a variety of renewable energy resources, solar energy is a rich and safe option which leads to touching the scale of ever-increasing global energy demand. Solar light-based technologies have attracted great attention for environmental remediation [1]. Light-mediated catalysis, i.e., photothermal catalysis, and photocatalysis serve a prominent role in obtaining fuels, chemicals, and heat. In the context of the researches in semiconducting materials, the photocatalysis-enabled [2] MOFs are being considered as the most prominent semiconductors, capable of replacing the traditional inorganic semiconductors. The design and synthesis of efficient MOFs for the purpose has extremely enlarged applications in a wide variety of research fields.

2 MOF-Based Photocatalysis

MOFs are a special class of organic compounds possessing a series of distinctive properties owing to their porous and crystalline structures. The rich topology, large surface area (over 6000 m^2g^{-1}), and tunable porous structure of MOFs provide exceptional opportunities for designing numerous application-based compounds. For instance, in an abundance of metals (particularly transitions metals), the MOFs can easily be manipulated for intrinsic catalytic activity (heterogeneous catalysts) in the photo- and electro-catalytic processes. The rapid development of the MOF materials for photocatalysis is one of the great works in the field of environmental remediation. The photocatalysts are semiconductors, and photocatalysis is a phenomenon where the photocatalysts materials affect the rate of a chemical reaction on exposure to light. The MOF constituents, metal nodes (ions/oxide/cluster), and organic linkers (photoactive) provide a large number of selection choices making it feasible to obtain exceptionally tuned and relevant photocatalysts capable of removing pollutants, bacterial disinfection, CO_2 fixation, etc. Unlike traditional inorganic semiconductors, which are recognized as delocalized conduction bands or valance bands, these MOF photocatalyst are recognized as organized molecules in a crystalline lattice with superior properties of both inorganic and organic constituents which can serve efficiently in microelectronic devices [3]. Ti- and Zr-based MOFs gained significant interest in photocatalysis; some examples of the types are UiO-66 ($Zr_6O_4(OH)_4(BDC)_6$ and MIL-125 ($Ti_8O_8(OH)_4$ $(BDC)_6$, [UiO: University of Oslo, MIL: Materials Institute Lavoisier BDC: 1,4-benzene dicarboxylate)]. With the development of understanding between demand and applicability, a variety of MOFs composites are prepared recently. The photocatalytic elimination of unwanted aqueous organic pollutants such as dyes, phenols, insecticides, pharmaceuticals, and personal care products provides an interesting hope. Similarly, the removal of Cr(VI) and radiative U(VI) can be achieved by its photo-reduction into trivalent states suitable for easy precipitation and separation. Inactivation of harmful bacteria, photo-oxidation of some gaseous pollutants (NO, toluene) into harmless products by MOF-based composites shows enormously tunable photocatalytic properties using solar light. Many reviews appeared in the literature describing different features of MOFs, involving environmental applications [1]. Usually, the photocatalytic mechanism includes three successive reaction steps: (1) The absorption of light in photocatalysts gives rise to exciting excitons; (2) transfer of excitons to catalytic centers and breaking to release of electron and protons photoactive sites; then lastly (3) the reaction of the electron and proton active sites with reactants previously adsorbed on the catalytic surface to perform the chemical reactions and provides electrical neutrality [4].

2.1 Enhancing Light-Harvesting

The solar energy spectrum is comprised of ~5% UV light, ~50% near-infrared (NIR) light, and 42–45% visible light. It is most important to put out the photocatalysts absorption of the NIR region to UV and visible region for getting an upgrade in the solar energy application. In photoactive MOFs, a vast majority of organic linkers function as light-harvesting antennas. Many methods are employed to upgrade the MOF's light absorption by modulating the linkers, such as by trapping $-NH_2$ or other groups with aromatic carboxylate ligands [2]. The strategy is much helpful and reveals to expand the effective light response in the visible range up to ~600 nm. Such MOFs self-assemblies with distinguished structures and physiochemical characteristics [5, 6] can cleverly be used as essential photocatalytic moieties.

The bandgap value plays a key role in semiconducting materials for their particular applications (e.g., photocatalysts), which in MOFs, can be tuned by varying metal cluster size (or type) and functionalizing the organic linker. The MOFs architectural networks consist of manageable unfasten channels which help in tuning the bandgap and its effectiveness [7]. After all the photocatalytic reactions of semiconductor, photocatalysts begin by absorption of incident photons of equal energy or greater than their energy associated with bandgap [8]. Heterogeneous photocatalysis reaction of inorganic semiconductors and MOFs involves four steps: (1) dispersion of reactant from major phases (gas or liquid) to the photocatalysts surface, (2) produces electron/hole pairs through absorption of light by photocatalysts, (3) on the surface of photocatalysts, redox reactions take place between reactive species and adsorbed reactants, and (4) from the catalyst surface adsorption/desorption of intermediate products occurs. It is relevant to mention here that the optoelectronic feature of MOFs promotes the photo-redox reactions through the excitation of metal nodes and organic ligands [4]. Further, MOFs bi-functionality, by increasing the number of reactive sites as well as life values of photogenerated electron pairs, is the way to overcome major drawbacks of conventional semiconductor-type photocatalysts [9, 10].

The hydrogen generation from the catalytic splitting of water molecules by harvesting solar energy would be a sustainable energy source with high energy-yielding hydrogen for upcoming days. Although inorganic semiconductors are being in large use now, organic semiconductors are being established as high efficient photovoltaic molecular semiconductors. The suitable reason behind this is the formation of the tunable bandgap, the lightweight of the molecules, and their flexibility. Moreover, doing fabrication in the vigorous sites appeared the realistic factors for the development of high efficient photovoltaic materials [11].

2.2 Enhancing Electron–Hole Separation

Comparatively weak linker bonds between the metal nodes and organic linker, inadequate bandgaps, and the poor semiconducting properties of MOFs were the reason for not being directly employed in many photocatalytic applications. However, several tunable structural properties qualify them as unique platform materials for semiconductors. In photocatalysis, photogenerated electron–hole pairs are formed, whereas their separation is a vital key for performance. Simultaneously basic knowledge of charge separation as well as migration of electron–hole pairs in MOFs is important to describe the different types of photocatalysts. MOFs photocatalysts, like others, after being exposed to light, undergo an electron–hole reassimilation process, which results in a remarkable decrease in catalytic efficiency. To overcome, other metal species and semiconductors are generally added into MOFs to assimilate as co-catalysts as the best strategy is to create MOF-based combinations. However, it remains under the limit to encourage electron–hole separation for many [2]. In traditional nano-doped semiconductor photocatalysts, the metal nanoparticles generally play an important role in creating electron–hole separation. Likewise, MOF-metal nanoparticles combinations, in which metal nanoparticles behave as an electron acceptor or as co-catalysts, become an essential part for increasing electron–hole separation. Additionally inherited advantage giving by the porosity of MOFs, the size and location of the co-catalyst play important roles because these components influence directly the relocation path of charge carriers and application ability of active centers [2].

3 MOFs as Photothermal Catalysis

A significant amount of research has been done to harvest solar energy into several formats and is being in the application too, still, its storage and usage offer a big challenge because of the limited conversion efficiency. More research work is required to empower solar energy harvesting.

In addition to MOF-photocatalysis, photothermal catalysis can also play a promising role in solar to chemical/thermal energy conversion. The photothermal catalysis over MOFs-based materials is in a growing phase and offers a different proposal for well-organized solar energy transformations. Upon light irradiation, the photoexcitation process occurs on semiconducting materials by photothermal effect. In this process, thermal energy (heat) can be produced [2]. The photoexcitation of semiconducting materials (e.g., carbon-based materials) is brought by local heating of the lattice which promotes endothermic reactions [12] (Fig. 1).

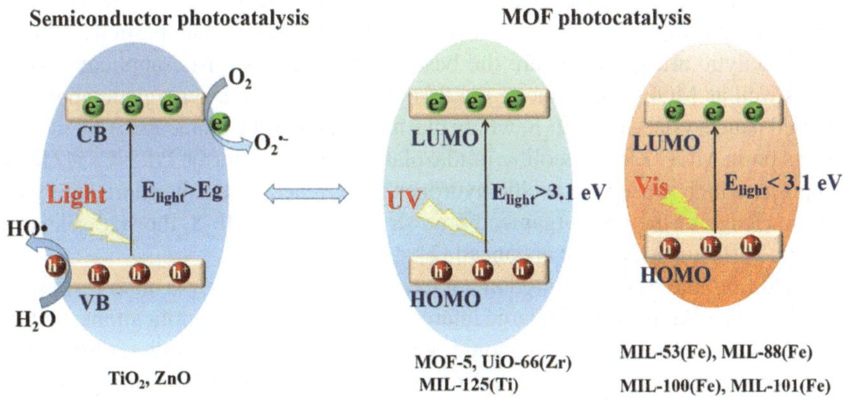

Fig. 1 Mechanism for semiconductor photocatalysis (left); comparison of band gaps and light source (UV or visible light) between representative MOFs (right) [12]

The earlier studies for the development of oxide derivatives using MOFs found the advanced characteristics in the above materials which are observed by enhanced porosity and high surface area in the variety of catalytic activity. It is worth mentioning here that a specific light-promoted effect has been demonstrated to promote catalytic action of UV–visible light, inducing the oxygen vacancies, which is found little different from photocatalytic theory [13].

3.1 Photothermal Effect of Plasmonic Metals

The plasmons are best described as an oscillation of electron density concerning the fixed positive ions in a metal. Plasmonics, in a broad sense, deals with surface plasmon resonance (SPR)-related science and technology and is capable of facilitating interesting applications in the wavelength ranging from ultraviolet up to infrared. Recently, surface plasmons (SPs) have attracted growing interest due to their applications in plasmon catalysis, surface-plasmon-driven photochemistry, and many other specific areas [14]. Generally, the plasmonic materials use the surface plasmon resonance effects to get special optical properties not seen in nature.

To understand further, the SPR arises from the interaction of light with a free-electron pool at a metal–dielectric interface, where the energy of the photon is transferred to collective excitations of the electrons, which occurs at a specific wavelength of light; exactly at the matching momentum of the photon with the plasmon. The recent researches involving plasmonic nanomaterials to harvest the light at the nanoscale has now become a hot topic of the field. As a result, many application-based plasmonic nanomaterials are being researched and developed swiftly.

Furthermore, the plasmonic nanomaterials (e.g., Ag, Au), upon visible-light irradiation, can be excited enough to generate hot carriers which can be exploited easily in combination with the MOF-semiconductor boost efficient electron–hole separation. Here, the highly tunable structure of MOFs qualifies them for efficient carriers for plasmonic catalysis by incorporating plasmonic nanoparticles (NPS). These catalytic arrangements are the best example of synergistic applications for NPs as well as MOFs. One among the first attempts of such a synergistic approach is the Pd nanocrystals and ZIF-8, where plasmonic palladium nanocubes (NCs) after dispersion in MOF, ZIF-8 (zeolitic imidazolate framework) forms Pd, NCs@ZIF-8, composite which is used in specific hydrogenations reaction. Furthermore, applying the knowledge of the well-organized pore structure of the ZIF-8, the size-selective catalytic performance can be enhanced [2].

In addition, the nanostructured metal having discrete surface plasmon resonance (SPR) property has been drawn large interest due to its extensive applications. Unlike most dyes or pigments, the plasmonic nanoparticles carry a specific color that is highly connected with their size and morphology. Interestingly by tuning the size and shape, the performance and optical properties of plasmonic nanoparticles can be tuned without altering the chemical composition. These nanoparticles SPR are very strong absorbers and scatters of light and used recently in lateral flow diagnostics and surface-enhanced spectroscopy, etc. The Pd and Ag hybrid nanocomposites are reported to be strong SPR absorbers studied for cancer remedy, whereas a layer of silica-coated Pd–Ag nanocomposites is found to be fairly biocompatible [15].

3.2 Photothermal Effect of Both MOFs and Plasmonic Metals

In NPS-MOF composite, the SPR of NPS transforms light energy into local heat, which thermally accelerates the catalytic reactions. This high solar-to-heat conversion capability of the NPS opens a green route for temperature-mediated photocatalytic reactions. The plasmon NPs along with photoactive MOFs with certain catalytic functions, can boost significantly photothermal transformations. There are many polymeric agents which exhibit photothermal effects, such as carbon-based polymeric agents and porphysomes (porphysomes are self-assembly motifs made from porphyrin-lipid conjugates (photonic nanoparticles), carry both the multifunctional properties of porphyrin building-blocks and the properties of nanostructure itself). Combination of noble metals and MOFs; such as Pt/PCN-224(Zr) composites (Zr-carboxylate MOFs with metalloporphyrins, PCN stands for porous coordination network) show synergistic photothermal effect capable of activating O_2 to $^1O_2$ under light irradiation. Here, light exposure not only converts light energy into heat but also induces oxygen activation. In the Zn analog, the oxygen activation was controlled by competitive electron transfer between Pt surface and PCN-224(Zn), which was brought in by adjusting the light intensity [16]. The study is quite helpful and provides the idea of tuning the light intensity for photothermal excitement for the required amount of electrons in metal-MOF composites and metal surface to enhanced the catalysis process [2].

4 Conclusion

Photocatalysts' nature of MOFs has signified their excellent capability in a different area. The possible application of the metal–organic framework in environmental remedies and their various problems have been observed, which can be overcome by reducing costs by improving high yield and systematically working on structural controls along with the synthesis process. Despite MOFs photocatalysis and especially MOFs photothermal catalysis are in the young stage, the researchers are leading swiftly to a certain level of advancement. Hence, more attempts have to be done in the path of growth continued, and some different challenges also have to be addressed.

Acknowledgements AS and MY are acknowledge Banaras Hindu University, Varanasi, for providing the facility for doing the research work. BM and MDP also acknowledge BHU for providing a seed grant and incentive grant under MoE Govt. India, Institute of Eminence (IoE) scheme.

References

1. Wang Q, Gao Q, Al-enizi AM (2019) Recent advances in MOF-based photocatalysis : environmental remediation under visible light. Inorganic Chem Front 22–26. https://doi.org/10.1039/c9qi01120j
2. Xiao JD, Jiang HL (2018) Metal-organic frameworks for photocatalysis and photothermal catalysis. Acc Chem Res 4. https://doi.org/10.1021/acs.accounts.8b00521
3. Kolobov N, Goesten MG, Gascon J (2021) Metal–organic frameworks: molecules or semiconductors in photocatalysis. Angew Chem Int Edn 14. https://doi.org/10.1002/anie.202106342
4. Younis SA, Kwon EE, Qasim M, Kim KH, Kim T, Kukkar D, Dou X, Ali I (2020) Metal-organic framework as a photocatalyst: progress in modulation strategies and environmental/energy applications. Prog Energy Combust Sci 81:100870 https://doi.org/10.1016/j.pecs.2020.100870
5. Rossin A, Tuci G, Luconi L, Giambastiani G (2017) Metal-organic frameworks as heterogeneous catalysts in hydrogen production from lightweight inorganic hydrides. ACS Catal 7(8):5035–5045. https://doi.org/10.1021/acscatal.7b01495
6. Zhang Y, Ying JY (2015) Main-chain organic frameworks with advanced catalytic functionalities. ACS Catal 5(4):2681–2691. https://doi.org/10.1021/acscatal.5b00069
7. Nasalevich MA, Van Der Veen M, Kapteijn F, Gascon J (2014) Metal-organic frameworks as heterogeneous photocatalysts: advantages and challenges. CrystEngComm 16(23):4919–4926. https://doi.org/10.1039/c4ce00032c
8. Meng X, Yun N, Zhang Z (2019) Recent advances in computational photocatalysis: a review. Can J Chem Eng 97(7):1982–1998. https://doi.org/10.1002/cjce.23477
9. Luo J, Zhang S, Sun M, Yang L, Luo S, Crittenden JC (2019) A critical review on energy conversion and environmental remediation of photocatalysts with remodeling crystal lattice, surface, and interface. ACS Nano 13(9):9811–9840. https://doi.org/10.1021/acsnano.9b03649
10. Khan MM, Adil SF, Al-Mayouf A (2015) Metal oxides as photocatalysts. J Saudi Chem Soc 19(5):462–464. https://doi.org/10.1016/j.jscs.2015.04.003
11. Nangia AK, Desiraju GR (2019) Crystal engineering: an outlook for the future. Angew Chemie Int Edn 58(13):4100–4107. https://doi.org/10.1002/anie.201811313
12. Qi Wang Q, Gao Q, Al-Enizi AM, Nafady A, Ma S (2020) Recent advances in MOF-based photocatalysis: environmental remediation under visible light Inorg. Chem Front 7:300–339. https://doi.org/10.1039/C9QI01120J
13. Chen X, Li Q, Zhang M, Li J, Cai S, Chen J, Jia H (2020) MOF-templated preparation of highly dispersed Co/Al_2O_3 composite as the photothermal catalyst with high solar-to-fuel efficiency for CO_2 methanation. ACS Appl Mater Interfaces 12(35):39304–39317. https://doi.org/10.1021/acsami.0c11576
14. Gutiérrez Y, Brown AS, Moreno F, Losurdo M (2020) Plasmonics beyond noble metals: exploiting phase and compositional changes for manipulating plasmonic performance. 128, 080901. https://doi.org/10.1063/5.0020752
15. Huang X, Tang S, Liu B, Ren B, Zheng N (2011) Enhancing the photothermal stability of plasmonic metal nanoplates by a core-shell architecture. Adv Mater 23(30):3420–3425. https://doi.org/10.1002/adma.201100905
16. Chen Y-Z, Wang ZU, Wang H, Lu J, Yu S-H, Jiang H-L (2017) Singlet oxygen-engaged selective photo-oxidation over Pt nanocrystals/porphyrinic MOF: the roles of photothermal effect and Pt electronic state. J Am Chem Soc 139:2035–2044

Metal Organic Frameworks as Photocatalyst for Water Purification

Naveen Goyal and Pragya Arora

Contents

1 Introduction 562
2 Water Purification 563
  2.1 Methods for Water Purification 564
3 Photocatalysis and Its Requirement for Water Purification 565
  3.1 Mechanism of Photocatalysis 566
  3.2 Factors Affecting Photocatalysis 568
  3.3 Kinetics of Photocatalysis 569
4 Photocatalytic Reactors for Treatment of Water 571
5 General Photocatalytic Materials 572
6 Metal–Organic Frameworks (MOFs) 572
  6.1 Synthesis of MOFs 575
  6.2 Characterization of MOFs 576
7 MOFs as Photocatalyst 578
  7.1 Metal–Organic Frameworks as Photocatalysts for Dye Degradation 579
  7.2 Metal–Organic Frameworks as Photocatalysts for Primary Pollutants 584
  7.3 Metal–Organic Frameworks as Photocatalysts for Photoreduction of Heavy Metals 586
8 Summary and Future perspectives 587
Abbreviations 588
References 589

Abstract Earth's surface is covered by 71% of water, out of which 97% is present in oceans and 3% in rivers as a freshwater source. Water is essential for life, and one cannot underestimate its role and necessity in the biosphere. On the other hand, emerging problems like rising population and pollution due to fast industrialization have led to its scarcity and disturbed the natural water cycle. Thus, it is the need of the hour to address the issue with measures like appropriate decomposition of the industrial and other biological waste. Several researchers have been attempted to design

N. Goyal (✉)
Materials Research Centre, Indian Institute of Science, Bangalore, India
e-mail: naveengoyal@iisc.ac.in

P. Arora
Department of Chemistry, Indian Institute of Technology, Kanpur, India
e-mail: pragya20@iitk.ac.in

S. Gulati (ed.), *Metal-Organic Frameworks (MOFs) as Catalysts*,
https://doi.org/10.1007/978-981-16-7959-9_22

various techniques to clean the polluted water by degrading these hazardous pollutants using various methods, including physical, biological, and chemical processes. Among these processes, heterogeneous photocatalysis has been considered as the greener, ecological, and sustainable method as it holds the potential of utilizing solar energy as its energy source. Fast depleting energy resources and rising pollution, both major problems could be managed using photocatalysis for the purification of polluted water. This chapter describes the role of metal–organic frameworks (MOFs) as efficient photocatalyst for the photodegradation of contaminants present in the water. MOFs are a new class of hybrid porous materials with inorganic and organic components. The uniformly distributed pores in MOFs provide an opportunity to separate harmful pollutants being dumped in the water. This chapter discussed the photocatalytic process in detail and the use of MOFs as photocatalysts for the photodegradation of various pollutants. Although having good adsorption tendency and photocatalytic activity, the stability of MOFs in water is a significant concern. Various strategies such as functionalization, doping of metal atoms, and decoration of metal nanoparticles to further improve MOFs' stability and photocatalytic activity have been presented.

Keywords Water purification · Photocatalysis · Metal–organic frameworks · Dye degradation · Photoreduction of heavy metals

1 Introduction

The increasing population has led to a rise in water demands. With 70% of the earth's area occupied with water resources, still the access to freshwater is limited. Fast industrialization with huge waste dumping into the rivers has led to a further scarcity of freshwater. Thus, it is the need of the hour to take certain measures to control the emerging water pollution. One method to get pure water from polluted water is its purification. Purification of polluted water is a challenging and essential task to meet the rising demands [1].

The literature contains a huge number of methods involving various materials that have been used for water purification [2]. The materials, like zeolite's, activated carbon, and silica gel, have been opted for the purpose that works on the adsorption principle and does separate the pollutants from water [3]. But all these materials show certain disadvantages like the saturation of adsorbent pores and limited selectivity to adsorb all the pollutants present in the water.

From the last decade, an alternative approach has been used for water purification which is photocatalysis. It includes photodegradation of contaminants present in the water [4]. Semiconductor materials like titanium oxide and zinc oxide are being extensively used as a photocatalyst to remove organic and inorganic pollutants from water. These materials work efficiently in ultraviolet light but had a limited tendency to work efficiently in the visible range of the solar spectrum. Thus, there is a necessity to develop materials that could work efficiently and effectively in solar light. MOFs are one such sort of material that have been the center of study in recent times. These materials because of their large surface area and high cavity volume offer a plethora of opportunities for both adsorption and photocatalysis of pollutants present in water [5, 6]. These materials contain metal nodes and organic linkers to construct 1D, 2D, and 3D frameworks. Functionalization of either metal nodes or organic linkers has led to an increase in their catalytic activity for photo-degradation of dyes and also for the photoreduction of heavy metals present in the aqueous medium.

This chapter describes the need for the purification of water and the processes required to purify it. The photocatalysis process including its mechanism and kinetics is discussed in detail. Some insight is given in the synthesis methodologies and characterization of MOFs. The last part of the chapter reviewed the application of MOFs as a catalyst for photo-degradation of various dyes present in water and photoreduction of heavy metals. More emphasis has been given to improve the photocatalytic activity of MOFs by functionalization in terms of metal atom doping and heterostructure formation along with its stability in polar solvents.

2 Water Purification

Water purification is a process that involves a chain of processes where various chemical and biological contaminants, solid suspended particles, and harmful gases are separated from polluted water [1]. The water that we receive at home does undergo different purification processes as shown in Fig. 1.

Pre-treatment—In this process, the polluted water is made to pass through different beds of sands, sieves, and micro-filters. This step provides raw quality water and includes different mediums for filtration.

Purification—In this process, the raw quality water received from pre-treatment is purified with a combination of nano-filtration, UV light, and reverse osmosis.

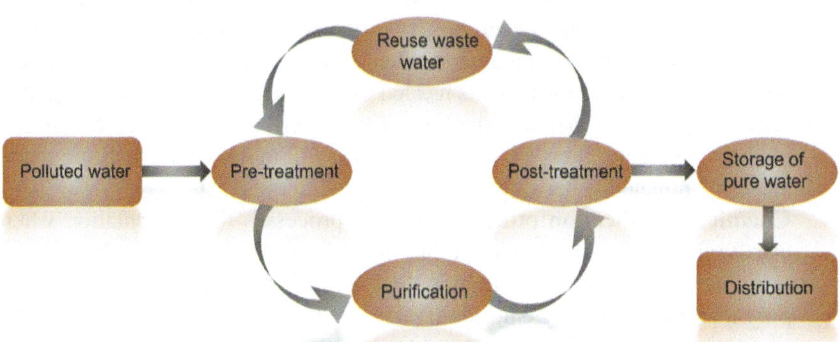

Fig. 1 Water purification cycle

After-treatment—This process is based on the requirement of taste and level of purity demanded by the receiver. It includes further purification with UV light, chlorination, activated carbon, remineralization, and additives.
Reuse waste water—The wastewater discarded from the first three steps is either made to run through the initial steps again or is used for non-drinkable water consumption purposes.
Storage—The purified water is stored for further consumption.
Distribution—It is necessary to distribute the purified water through clean suitable piping to keep it free from contaminants.

The former four steps are essential to remove the pollutants and make the water pure enough for consumption.

2.1 Methods for Water Purification

Various methods have been adopted to clean the polluted water in order to balance rising demands that include different physical, biological, and chemical processes [7] as shown in Fig. 2:

i. **Physical processes** like sedimentation are the primary step where the contaminated water is made to stand still, so that pollutants could settle at the bottom via gravitational forces. While other processes like filtration involve the passing of the polluted water through membranes containing zeolites, which further separate out the pollutants depending on the size of pores present in the membrane. Distillation, on the other hand, provides pure water in the form of vapors where the contaminants are separated from the residue left in the other compartment [8]. Apart from these, materials like active charcoal with high surface area being used to adsorb the pollutants from water or impurities present in water.
ii. **Biological purification** processes utilize biological activated sludge treatment where small microorganisms like bacteria degrade the organic waste present in water in aerobic and anaerobic conditions. It is considered as one of the economic methods as compared to other chemical processes. But continuous dumping of antimicrobial agents has led to a decrease in the efficiency of such bacteria and makes them resistant over longer periods. Other than this, simple methods like sand filters are also used where 1–2 m deep sand beds are made, and contaminated water is made to seep through them [2].
iii. **Chemical purification** processes include processes like flocculation where clarifying agents like aluminum chloride, aluminum sulfate, bentonite clay, ferric chloride, ferrous sulfate monohydrate, poly-aluminum-chloride (PAC), poly-aluminum hydroxy chloride (PAHC), etc., are added that cause colloidal particles to flocculate and thus are separated [2]. While processes like chlorination involve the addition of chlorine or hypochlorite that kill microbes present in the water and purify it. However, these processes are not environment friendly as they release harmful by-products which are sometimes even carcinogenic.

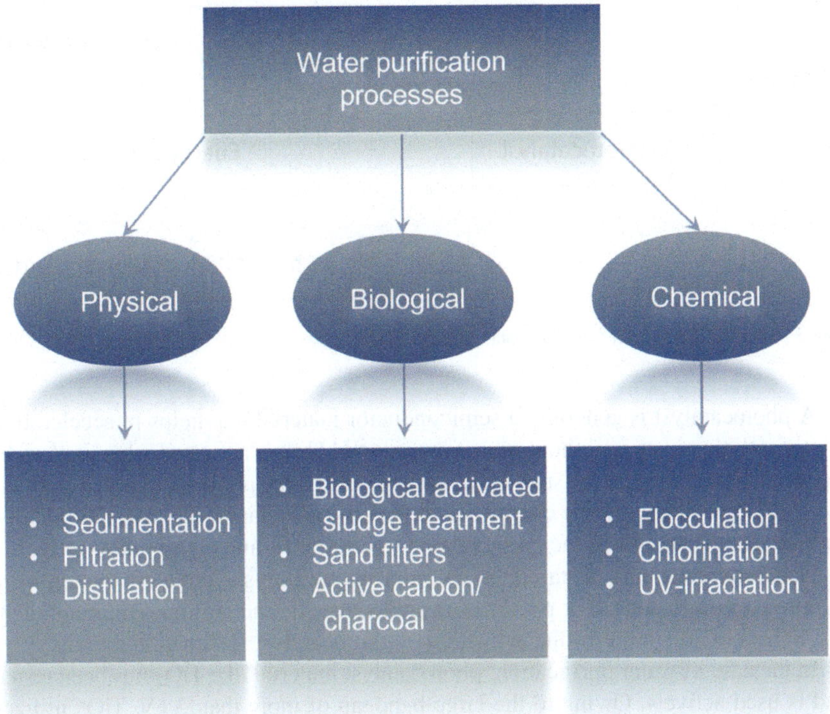

Fig. 2 Schematic showing water purification processes

Another chemical process that is extensively used for water purification is photocatalysis by virtue of UV irradiation [9]. This process includes chemicals that generate (in-situ) powerful oxidants which leads to the formation of reactive oxygen species (ROSs) like OH° radicals, which oxidize the pollutants through the chain reactions. Details of photocatalysis, its mechanism, and kinetics along with its application in water purification are discussed in the subsequent section.

3 Photocatalysis and Its Requirement for Water Purification

Photocatalysis is the process where a photon source (electromagnetic radiation) and a photocatalyst are utilized to create ROSs [10]. ROSs are powerful oxidants and are generated because of the charge separation, i.e., formation of a hole in the valence band and transfer of an excited electron to the conduction band of a photocatalyst material.

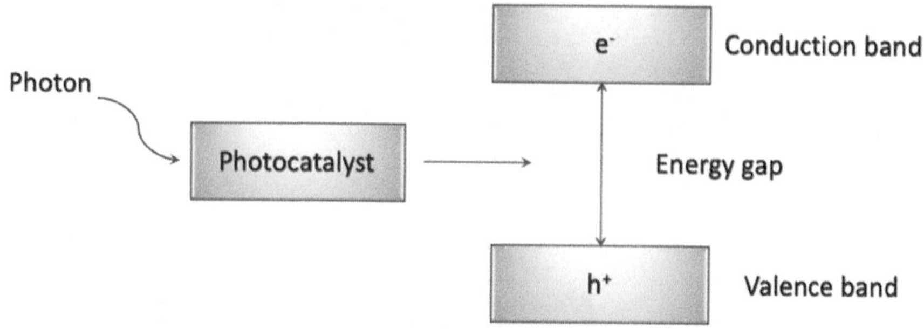

Fig. 3 Schematic diagram for photocatalyst activation

A photocatalyst is generally a semiconductor material that helps in accelerating photocatalytic reaction without its consumption [11]. It is activated when it absorbs a photon. The condition for such materials to undergo the photoexcitation process is the consistency between the energy source and the bandgap of the material. Ideally, a photocatalyst should be inexpensive, non-toxic, stable, and highly photoactive.

Generally, for a photocatalytic reaction or photocatalysis, three components are required (i) Light source or photon, (ii) photocatalyst, and (iii) an oxidizing agent [5]. Schematic diagram for the activation of photocatalyst is shown in the Fig. 3.

In the case of water purification, photocatalyst material like TiO_2 has been tested and is used actively. Owing to the large bandgap of more than 3 eV, TiO_2 utilizes only UV light for its activation. Its activity in visible light and sunlight is very less, and therefore, it cannot be graded as a greener catalyst for water purification.

Recently, there has been a huge development in the field of photocatalytic purification of water, where the goal is to develop such photocatalysts like metal–organic frameworks (MOFs) that could utilize solar energy as their energy source.

Photocatalysis offers certain advantages over other conventional methods: It is a green method as it decomposes the organic pollutants into non-harmful simple molecules like water, simple mineral acids, and carbon dioxide. While on the other hand, chemical methods like flocculation and chlorination degrade the organic pollutants into secondary contaminants. Photocatalysis is relatively simple and affordable, as it offers the advantage of performing the process at room temperature and with the ability to choose solar energy as its energy source [12].

3.1 Mechanism of Photocatalysis

Mechanism of photocatalysis schematically shown in Fig. 4. List of processes involved in photocatalytic degradation of pollutants [13] to purify water are given below.

i. Transfer of pollutants on photocatalysts' surface from the water
ii. Adsorption of pollutants on the photocatalyst surface

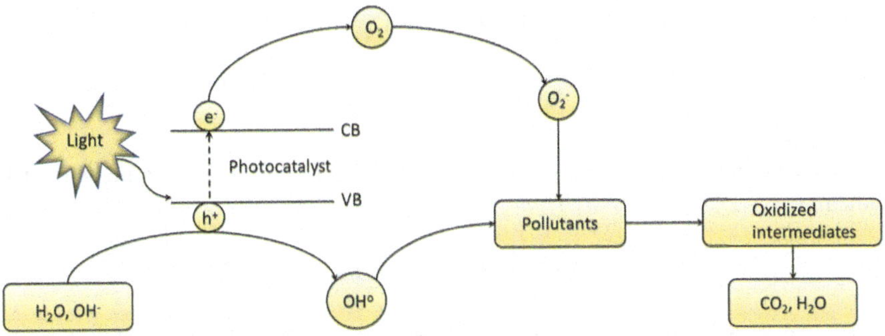

Fig. 4 Schematic showing mechanism of photocatalysis

iii. Photocatalytic degradation of pollutants on the surface of photocatalyst by means of photochemical reactions
iv. Desorption process
v. Removal of products from the liquid phase.

When a photon of light, having energy greater than or equal to the bandgap, irradiates on the photocatalyst, molecular excitation takes place. This excitation causes the formation of electrons and holes in the conduction and valence band of the material, respectively [5, 13, 14].

$$\text{Photocatalyst} \xrightarrow{\text{hv}} e^- + h^+$$

Then, the hole (h^+) generated in the photon irradiation step gets utilized to oxidize different species adsorbed on the catalyst surface like adsorbed water (H_2O_{ad}), adsorbed organic substrate (RX_{ad}), or the adsorbed $OH^-{}_{ad}$ ion.

$$h^+ + RX_{ad} \rightarrow RX^+_{ad}$$

$$h^+ + H_2O_{ad} \rightarrow OH^\circ_{ad} + H^+$$

$$h^+ + OH^-_{ad} \rightarrow OH^\circ_{ad}$$

The OH°_{ad} radicals formed are very reactive and are used to oxidize the adsorbed hydrocarbon pollutants (RX_{ad}), which is given by the reaction-

$$OH^\circ_{ad} + RX_{ad} \rightarrow \text{Intermediates}$$

Further, the molecular oxygen dissolved in water acts as an oxidizing agent by scavenging electrons from the conduction band of photocatalyst and forms reactive species like superoxide which can also degrade the pollutants [15, 16].

$$e^- + O_2 \rightarrow O_2^-$$

$$O_2^- + RX_{ad} \rightarrow \text{Intermediates}$$

To accelerate the photocatalytic reaction, H_2O_2 is also used to generate extra OH° radicals.

$$H_2O_2 + hv \rightarrow 2\ OH^\circ$$

However, there are chances of the recombination of the hole and the electron, which tends to limit the efficiency of the photocatalysis process [5].

Currently, research is more focused on developing methods to increase the efficiency of the photocatalytic process. Thus, one needs to understand the factors affecting this process.

3.2 *Factors Affecting Photocatalysis*

The photocatalytic reaction rate is affected by various factors as shown in Fig. 5 and explained below.

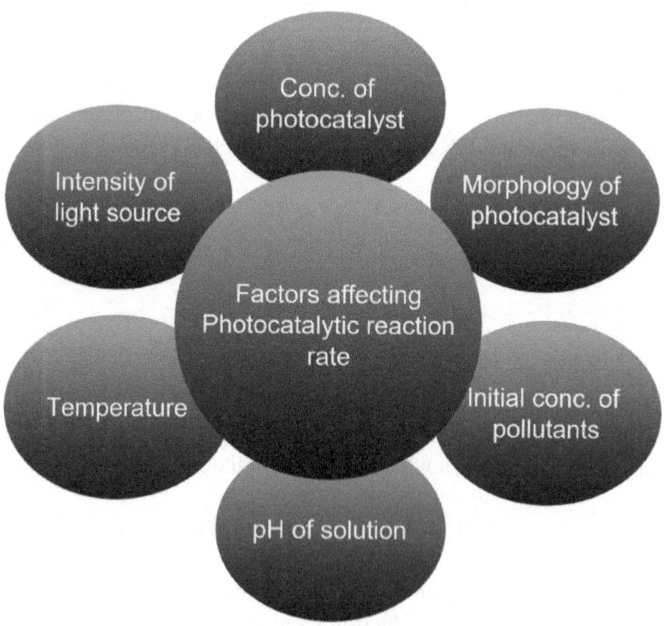

Fig. 5 Factors affecting photocatalysis reaction rate

i. **Concentration and morphology of photocatalyst**: It has been observed that with an increase in the concentration of photocatalyst, photocatalytic reaction rate increases. This is because of the increase in the amount of light absorbed by photocatalyst, which leads to the formation of a greater number of reactive radicals. However, after a certain optimal concentration of photocatalyst, the reaction rate starts decreasing because of light scattering or light blocking by suspended photocatalyst in water. The morphology of photocatalysts also affects the reaction rate. It has been observed that small particle-sized photocatalysts have a higher photocatalytic reaction rate in the decomposition of organic pollutants as compared to large particle-sized photocatalysts [13, 17].
ii. **The initial concentration of pollutants**: Initial concentration of pollutants has significant effects on the photocatalytic reaction rate. Increasing the concentration of pollutants leads to a reduction in reaction kinetics. This is due to a decrease in photonic efficiency as the surface of a photocatalyst becomes saturated with the increased concentration of adsorbed pollutants [18].
iii. **The intensity of light source**: On increasing the light intensity, the photocatalytic reaction rate increases. This is due to the increase in the number of photons absorbed by photocatalysts [17]. However, as the photocatalyst surface becomes saturated, an increase in light intensity does not have much effect on the reaction rate. Also, photocatalytic reaction rate depends on the type of illumination source used whether it is ultraviolet (UV), visible, or solar lamp.
iv. **pH of solution**: The pH of a solution also affects the photocatalytic reaction kinetics. pH affects the photocatalyst surface charge which leads to self-aggregation of the photocatalyst. Charging of the photocatalyst surface affects the process of adsorption of pollutants on the photocatalyst surface which eventually changes the reaction rate [19].
v. **Temperature**: Varying temperature has a significant effect on photocatalytic reaction rates. It has also been shown that in some cases, with increase in temperature of reaction temperature above 80 °C, the reaction rates get decreased. This is because of the increase in recombination of charges generated and also due to the desorption of pollutants from the surface of the photocatalyst [20]. Increase in temperature also influences the dissolved oxygen amount. Dissolved oxygen concentration decreases with increasing temperature. Since dissolved oxygen is required to generate ROSs which is used to degrade pollutant therefore with increase in temperature, the overall reaction rate decreases [21].

3.3 Kinetics of Photocatalysis

Photocatalytic degradation of pollutant present in water involve species like pollutants and O_2 molecules. Since O_2 molecules are present in excess, so the whole reaction can be described using pseudo-first-order reaction with respect to the pollutant concentration [5, 22] and follows the equation:

$$r = -\frac{d[C_{ad}(t)]}{dt}$$

$$r = k \cdot [C_{ad}(t)]$$

In the equation, r is the rate of reaction, $[C_{ad}(t)]$ is the concentration of pollutant adsorbed on the photocatalyst surface at a time 't,' and k is rate constant.

The total concentration of pollutant can be written as the sum of concentration of pollutant in the fluid phase and concentration of pollutant adsorbed on the photocatalyst [15].

$$[C(T)] = [C(t)] + [C_{ad}(t)]$$

where $[C(T)]$ is the total concentration of pollutant, $[C(t)]$ is the concentration of pollutant in the fluid phase, and $[C_{ad}(t)]$ is the concentration of pollutant adsorbed on photocatalyst surface.

Considering that adsorption equilibrium is achieved at every time, and following the Langmuir adsorption–desorption isotherm, and the equation for the rate of reaction can be described as:

$$r = \frac{k \cdot k_L \cdot [C(t)]}{1 + k_L \cdot [C(t)]}$$

where k_L is Langmuir constant and $[C(t)]$ is the concentration of pollutant in fluid at a time 't.'

In general, the concentration of pollutants is very less in water, and therefore, the denominator of the rate equation can be modified as

$$\{1 + k_L[C(t)]\} \sim 1.$$

Hence, the rate equation becomes,

$$r = k \cdot k_L \cdot [C(t)]$$

$$r = k_{app} \cdot [C(t)]$$

where $k_{app} = k \cdot k_L$ and known as apparent rate constant.

Value of apparent constant can be determined by using integration method. Integration of above equation in the limits 0 to t will give

$$\ln \frac{[C_0]}{[C_t]} = k_{app} \cdot t$$

where $[C_0]$ is the initial concentration of pollutants and $[C_t]$ is the concentration of a pollutant at time t. The slope of the graph between $\ln([C_0]/[C_t])$ versus time will give the value k_{app}.

4 Photocatalytic Reactors for Treatment of Water

Photocatalytic reactors are the devices used for the photocatalytic treatment of water. These can be classified in the following manner depending on various factors:

i. **State of the photocatalyst**: Photocatalyst in a reactor is mainly used in two ways, either in the suspended form or by attaching it to support. Based on this, reactors are of two types—(a) Slurry reactors, (b) Immobilized reactors [23, 24]. Photocatalysts are freely suspended in water in the slurry reactors, while photocatalysts are attached to fixed support in immobilized reactors. The difference between the slurry reactor and the immobilized reactor is given in Table 1.
ii. **Type of illumination source**: The irradiation source used in the design of photocatalytic reactors could be (a) solar light or (b) UV polychromatic lamps. Under the solar illumination source, photocatalytic reactors are categorized under two categories, namely concentrating reactors and non-concentrating reactors, where concentrating reactors require small reactor volume and uses direct light, while a non-concentrating reactor requires a large reactor volume and uses both direct and diffused light, which leads to less or negligible optical loss [15].
iii. **Irradiation source position**: The position of the irradiation source is an important feature of a photocatalytic reactor. Based on the position of lamp/illuminating source, reactors are distinguished as:

Table 1 Difference between the slurry reactor and immobilized reactor

Slurry reactors	Immobilized reactors
Photocatalytic surface area-to-reactor volume ratio is high	Photocatalytic surface area-to-reactor volume ratio is low
After the photocatalytic process, filtration is required to remove the photocatalyst	No need for any additional filtration step to remove photocatalyst
Photocatalyst is uniformly suspended in the reactor	Photocatalyst is continuously operated
Pressure drop through the reactor is low	Pressure drop through the reactor is significantly high

(a). reactor having immersed light source in which lamp is placed within the unit.
(b). reactor using an external light source. The light source is placed outside the unit.
(c). reactors having distributed light sources in which reflectors are being used to transport light from source to reactor [15].

5 General Photocatalytic Materials

Materials like metal oxides, metal sulfides, and metal phosphides of Ti, Zn, W, Ce, Fe, In, Bi, etc., have been studied deeply as photocatalysts [25]. The morphological modifications and electronic modifications to these materials show enhanced catalytic results. Morphologically modified materials involve the formation of heterojunctions with other materials, while electronic modifications include doping of other metal atoms. But there are certain problems that have been observed with these photoactive materials like metal oxides which offer low electronic conductivity and poor apparent quantum efficiency, metal phosphides have a fast recombination rate of charge carriers, and metal sulfides possess poor stability. Hence, it is required to look for alternate materials as photocatalysts.

One such material is metal–organic frameworks, MOFs. The material is a combination of organic and inorganic constituents, already known for its excellent adsorption and separation applications owing to its robust size, ultra-high surface area, large pore volume, and size [26]. The material particularly is favorable to degrade pollutants in water first because of its adsorption capacity and second because of its photocatalytic activity. The combination of these properties makes MOFs a superior candidate for the degradation of pollutants present in the water. The purification of water using MOFs is started long back, but, in this chapter, some recent progress in the synthesis of MOFs and their photocatalytic activity has been discussed.

6 Metal–Organic Frameworks (MOFs)

MOFs are a new class of hybrid crystalline 3D porous materials having cage-like structures. These materials are a combination of two components—(i) metal clusters or metal nodes also termed as secondary building units (SBUs), and (ii) organic linkers. Both metal nodes and organic linkers combine to construct 1D, 2D, and 3D networks [6]. Metals ions commonly used for the synthesis of MOFs are Zn^{2+}, Cu^{2+}, Al^{3+}, Zr^{4+}, Cr^{3+}, and Fe^{3+}, while organic linkers used for binding metal nodes are anions of di or tricarboxylic acid, sulfates, and phosphates [27] as shown in Fig. 6. The arrangement of these constituents defines the physicochemical properties of MOFs and plays a pivotal role in tuning their properties [28]. General reaction scheme for the synthesis of MOFs is shown in Fig. 7.

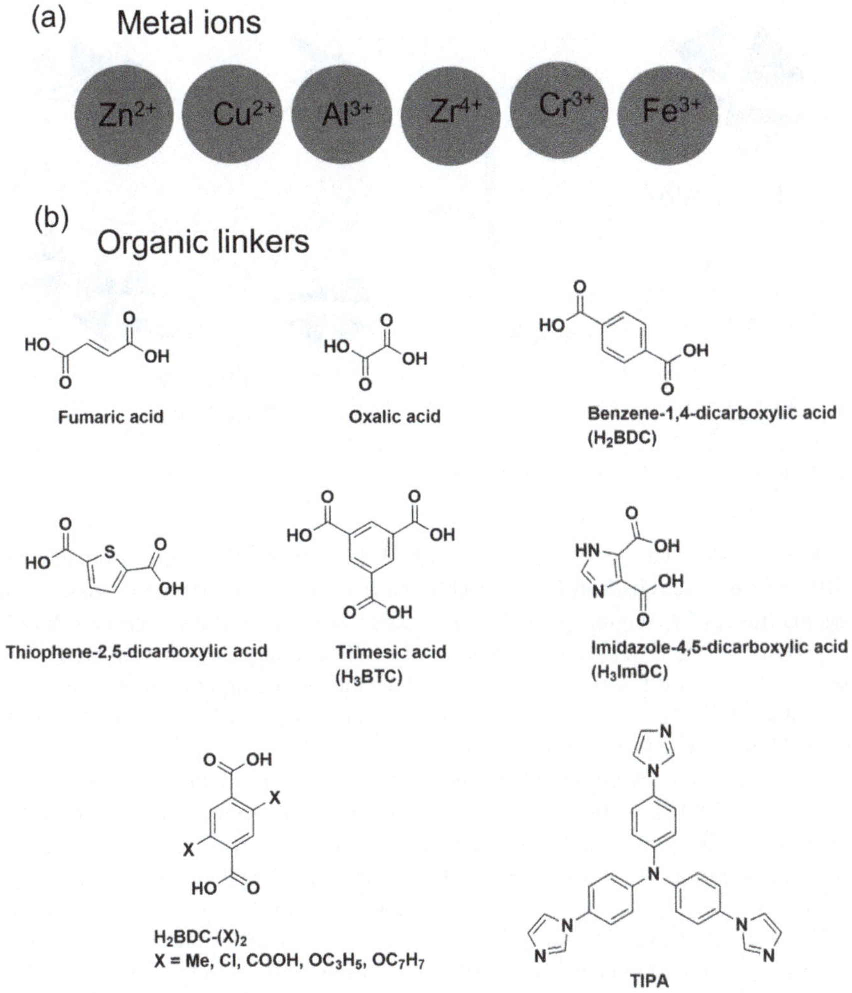

Fig. 6 Commonly used metal ions (**a**), and organic linkers (**b**) in the synthesis of MOFs

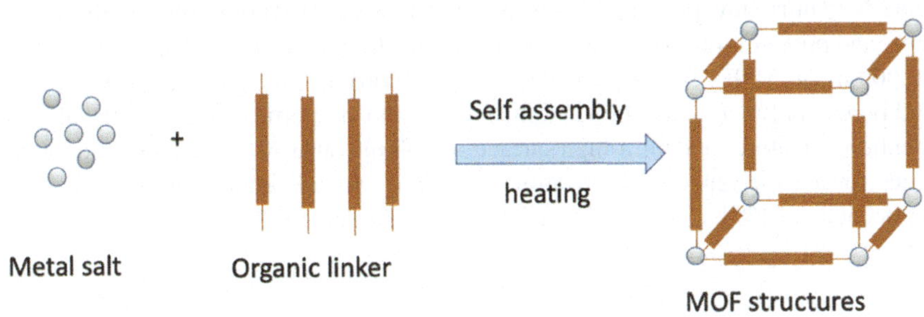

Fig. 7 General reaction scheme of MOFs

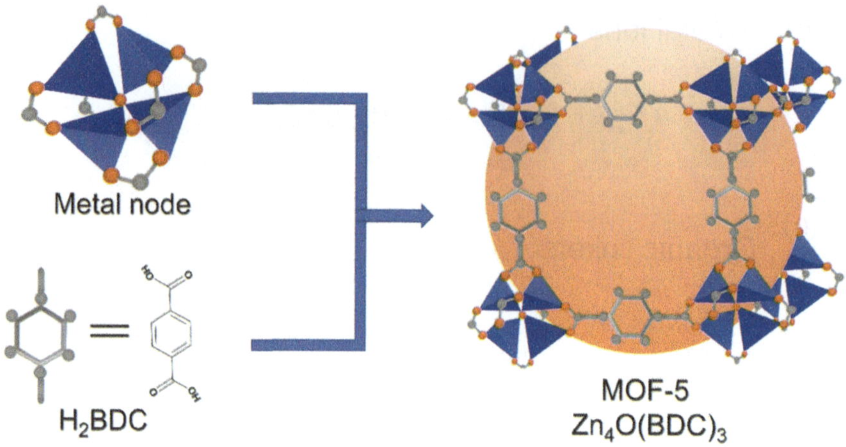

Fig. 8 Reaction scheme for synthesis of MOF-5

The first MOF has been synthesized in 1999 [29], by a combination of octahedra SBUs of Zn_4O tetrahedron (metal node) and benzene-1,4-dicarboxylic acid as an organic linker [28]. named as MOF-5. A reaction scheme for the synthesis of MOF-5 is shown in Fig. 8. After the synthesis of MOF-5, a large number of MOFs has been synthesized till now by modifying either metal ion, organic ligand, or metal node and can be classified under different categories such as UiO based MOFs, MIL based MOFs, and ZIF based MOFs [30].

SBUs in MOF is important as their coordination geometries determine the topology of MOFs. Various approaches have been attempted to tune the SBUs including exchanging metal ions, generating defects, transforming the oxidation states. Furthermore, by adjusting the geometry, ratio, and length of the organic linkers and by introducing versatile functional groups, the morphology and properties of MOFs could be tuned. Overall, it is the combination of SBUs and organic ligands (as linkers) that determine the final framework. SBUs restricts linkers movement and secures the position of metal centers [5, 31].

The unique arrangement of SBUs and organic linkers give rise to the stable framework with intriguing properties such as uniformly separated pores, ultra-high surface area, and pore volume. MOFs have been reported to have a surface area of more than 10,000 m^2/g. MOFs further provides an opportunity to tune the pore size, surface, and pores via functionalization and thus tailor their properties. Owing to the abovementioned features and ultra-high surface-to-volume ratio, MOFs are suitable candidates for a wide range of applications including gas adsorption and storage [32], photocatalysis [33], electrocatalysis [34], and drug delivery [35].

6.1 Synthesis of MOFs

Various approaches that have been used to synthesize metal–organic frameworks (MOFs) are summarized in Table 2. MOFs have been conventionally prepared using solvothermal (Hydrothermal: when the solvent is water) synthesis technique. This technique involves the use of conventional electric heating with control over the temperature [36, 37]. Generally, in solvothermal reactions high boiling point solvents are used. However, solvothermal or hydrothermal reactions require quite a large reaction time [38]. Therefore, to accelerate the reaction time for synthesizing MOFs, and to obtain high-quality crystals, alternate synthesis techniques have been employed including microwave-assisted method, sonochemical method, electrochemical synthesis, and mechanochemical methods as shown in Fig. 9.

Microwave-assisted synthesis methods have been extensively used for rapid crystallization of MOFs along with the control over morphology [39]. This synthesis technique involves the interaction of polar solvent molecules or ions present in the solution

Table 2 Summary of various synthesis modes of MOFs

Synthesis method	Energy	Time (h)	Temperature (K)
Solvothermal	Thermal energy	50–90	353–453
Microwave	Microwave radiation	0.5–4	303–373
Sono-chemical	Ultrasonic radiation	0.5–2	273–313
Electrochemical	Electric energy	< 1	273–303
Mechanochemical	Mechanic energy	0.5–2	298
Slow evaporation	No external energy	> 168	298

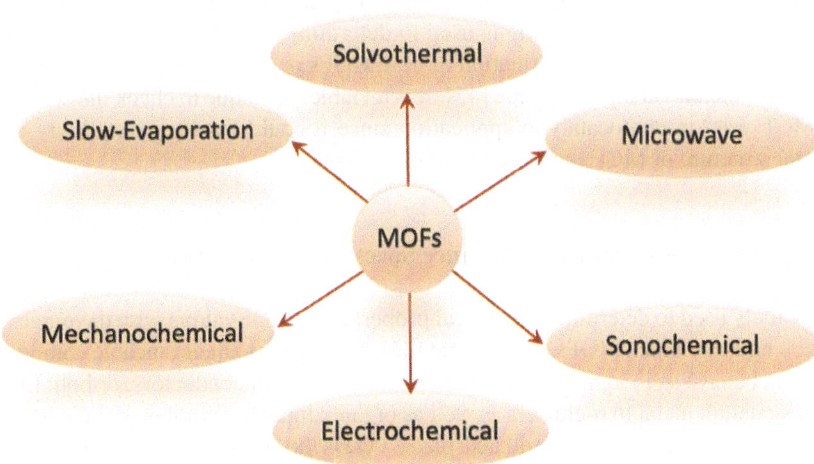

Fig. 9 Synthesis method used for MOFs

mixture with the incident electromagnetic wave. It requires a lower synthesis time than the conventional solvothermal technique. On the other hand, the sonochemical synthesis technique requires the use of ultrasonic radiation (20,000 Hz–10 MHz) [40]. Ultrasonic radiations produce acoustic cavitation, which leads to physical or chemical transformations. The process provides high temperature (~5000 K) and pressure (~1000 bar), causing rapid heating and cooling which further leads to the production of very fine crystals. Apart from these, the synthesis of MOFs via electrochemical methods uses anode as a metal ion source. The dissolution of metal ions from the anode during electrolysis reacts with organic molecules dispersed in conducting electrolytes present in the reaction mixture to form MOFs [41]. Mechanochemical technique involves breaking the intramolecular bonds and subsequent chemical transformation. Mechanochemical synthesis does not require any solvent and can occur at room temperature [42]. Slow evaporation method does not require any external energy source, but it requires a longer synthesis time than any other technique. Apart from these methods, ionothermal method [43], spray-drying methods [44], and diffusion methods are also popular for synthesizing MOFs.

6.2 *Characterization of MOFs*

MOFs synthesized from above-mentioned synthesis methods are characterized by various techniques described below.

Powder X-Ray Diffraction (PXRD)

It is used to determine crystal structure, polymorphic forms, and crystallinity of as-synthesized MOFs. Also, PXRD is used to determine crystallite size with the help of Scherrer's equation. The formation of new MOFs or composite or heterojunction can be predicted via this technique [45]. It is a viable technique to check the stability of MOF used in photocatalytic application since it could test the crystallinity and crystal structure of MOF recovered after usage [31].

Ultraviolet–Visible Diffuse Reflectance Spectroscopy (UV–Vis DRS)

It is widely used to determine the optical properties of MOFs along with the range of wavelengths in which MOFs absorb light [46]. MOFs due to their tuneable composition can behave as inorganic semiconductors, organic semiconductors, or both [33]. MOFs contain metal oxo-clusters as well as organic ligands. So, when light is irradiated, mainly two transitions could take place. One is the electronic transition between the valance band to the conduction band of the metal oxo-cluster, and the second is the electronic transition from HOMO to LUMO in organic ligand as shown in Fig. 10.

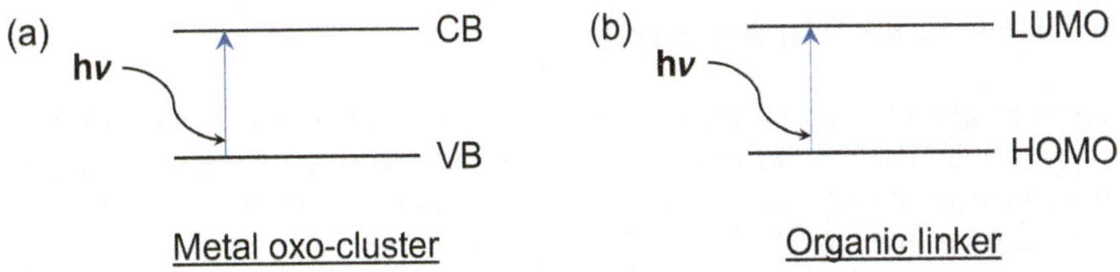

Fig. 10 **a** Electronic transition in metal oxo-cluster and **b** electronic transition occur in organic linkers

Bandgap energy of these semiconductors can be determined by UV–Vis DRS technique by employing Tauc plot approximation [47].

Tauc Equation

$$(\alpha \text{hv})^{1/n} = A(\text{hv} - E_g)$$

where h is Planck's constant,

ν refers to the frequency of electromagnetic radiation,

E_g is the optical band gap,

A is a constant,

n is also constant having a value of 2 or 1/2 depending on the nature of band gaps. For direct bandgap, $n = 1/2$, and for indirect bandgap materials $n = 2$.

The bandgap value is determined from the graph between $(\alpha h\nu)^{1/n}$ versus hν and extrapolating the curve to the x-axis.

Scanning Electron Microscopy (SEM) and Transmission Electron Microscopy (TEM)

SEM is employed to determine the topology of MOFs. TEM is used to determine the particle size and microstructures. TEM is also used to acquire crystallographic information including crystal structure, plane indices, and dislocations [48]. Owing to the porous structure, MOFs could have varied morphology such as cubes, rhombohedral, and octahedral. Homogeneity of elements in as-prepared MOFs can be determined by energy dispersive spectroscopy (EDS) coupled with these microscopes. An elemental map obtained from these techniques helps in determining the chemical composition of MOFs.

Brunauer–Emmett–Teller (BET)

BET is employed to measure the surface area of the MOFs. Adsorption isotherms acquired in this method can also give information regarding the pore size and volume along with homogeneity of MOFs [49].

Photoluminescence (PL) Spectroscopy

This technique is used to determine the electronic properties of as-prepared MOFs. PL spectra are used to obtain the bandgap of the materials. PL spectra can give information about quantum efficiency and the recombination rate of photogenerated charge carriers present in MOFs [5].

Thermogravimetry Analysis (TGA)

TGA is most commonly employed to measure the thermal stability of as-prepared MOFs. In this method, the weight loss of the sample is plotted against temperature. The TGA curve is also helpful in estimating the solvent-accessible pore volume [5].

Another important parameter for characterization of MOFs is aqueous stability test. Aqueous stability tests are performed to check the chemical stability of MOFs in water for their use in water purification using photocatalytic reactions.

To check the repeatability and stability of MOFs, post-characterization is also performed to see if the morphology of MOFs is maintained [50].

7 MOFs as Photocatalyst

MOFs are basically a hybrid of organic and inorganic materials. Intriguing features of semiconducting MOFs not only make them an efficient adsorbent material but also make them work as a photocatalyst [51]. MOFs having semiconductor property show optical transition and lead to the generation of electron and hole pairs upon irradiation of light which is the required property of a material to be photocatalyst. As compared to conventional semiconductor photocatalyst (such as TiO_2), MOFs offer many advantages because of their porous and robust structure [52]. The porous robust structure further allows encapsulating different chromophores into its voids, which helps in absorption of more amount of light and generation of a large number of electron and hole pair, making it even more active as a photocatalyst [53, 54]. In addition to that doping of heteroatom or metal nanoparticles, decoration over the surface of MOFs promotes facile charge transfer and helps in reducing the recombination rate of electron–hole pair. Another added advantage of MOFs as photocatalysts is the functionalization of organic linkers with groups like NH_2 which increases the delocalization in the aromatic ring and shows absorption band in the visible region and make MOFs a suitable catalyst in the visible region as well [55].

Tunable optical properties of MOFs make them an excellent candidate for their usages as photocatalysts materials. These materials can work efficiently in degrading various organic and inorganic pollutants available in the water. The aqueous stability of MOF is an important parameter for photocatalytic water purification. The metal-linker bond in MOF is a weak coordination bond and can be easily cleaved by water; thus, it must be modified to resist its breaking. Various modifications strategies have been attempted for the design and synthesis of stable MOFs. One such example is forming a stronger bond considering the HSAB principle. The bond between the metal ions and linkers is stronger if the linkers are strong Lewis base and the metal ions are in their high valent oxidation state, so that they act as strong Lewis acid. Thus, by HSAB theory, the interaction will eventually be stronger, and water might not cause the bond cleavage [56]. Another factor like the length of the bond between metal ions and organic linkers is also needed to be considered for synthesizing water-resistant MOFs. The longer the bond, the more will be the possibility of attack by a water molecule. Increasing the steric hindrance and thus, appropriate modifications to the organic groups can further prevent the water attack to metal-linker bond [57].

Following are some case studies where these materials have been used as a photocatalyst to decompose organic contaminants like organic dyes and inorganic contaminants like toxic heavy metal ions.

7.1 *Metal–Organic Frameworks as Photocatalysts for Dye Degradation*

Dyes such as rhodamine 6 G (R6G), methylene blue (MB), rhodamine B, and methyl orange are considered to be pollutants because the presence of these dyes in water harms aquatic species. Structure of these dyes are shown in Fig. 11. These dyes are mainly discharged from the textile and paper industries. Due to the presence of aromatic rings, these dyes are stable at ambient conditions and non-degradable in water. Hence, removal of these dyes is of utmost importance to purify water. Till now, various research groups investigated their photodegradation performance using

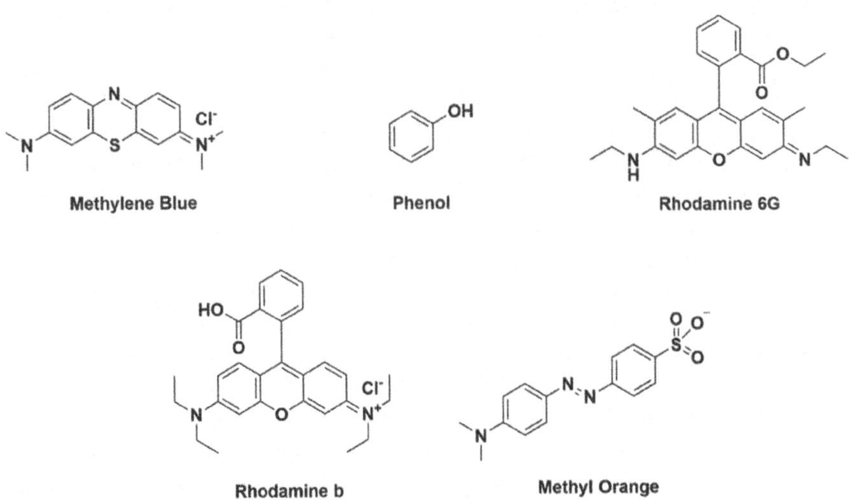

Fig. 11 Representative structure of organic pollutants studied for photodegradation

MOFs. Researchers also compared the photodegradation rate of these dyes using different illumination sources such as UV lamps, visible light, and solar light.

Table 3 summarizes the photodegradation of various pollutants using MOFs. Jing Du et al. in 2011, carried out a study of photodegradation of methylene blue dye under both UV and visible light illumination sources [58]. They tested the photodegradation activity of MIL-53(Fe) as a photocatalyst. Their result predicted that MIL-53(Fe) exhibits lower photocatalytic activity toward methylene blue dye under both light sources. However, they found a huge increment in photocatalytic activity on the addition of H_2O_2, since H_2O_2 (as shown in Sect. 3.1) leads to the generation of a greater number of OH^o radicals which is required for degrading the dyes and also H_2O_2 has been shown to reduce the recombination rate of generated electron–hole pairs. They have also investigated the photodegradation rate of methylene blue dye. In presence of H_2O_2, MIL-53(Fe) under UV–Vis light reduces 99% of dye in 20 min; however, in absence of H_2O_2, only 11% reduction in dye concentration was observed even after 40 min [58].

It has been seen that complete degradation of dye is a problem in visible light. Several Fe based MOFs have been tested as photocatalyst [59], but none of them could achieve 100% degradation rates. MIL-88, MIL-100, and MIL-101 MOFs were used for the photodegradation of rhodamine 6G dye [60]. MIL-88 shows better activity compared to others but could achieve a 100% degradation rate only in presence of H_2O_2.

Most of the studies of dye degradation using MOFs have been carried out by taking UV or visible light source. From an environmental point of view, more focus should be on the usage of a greener approach by taking sunlight as an irradiation source. Xu et al. [61] in 2016 studied the photodegradation of methylene blue and rhodamine B dyes using MOFs containing metal nodes of Cu and organic linker of imidazole in sunlight. Their results show that the methylene blue degraded approximately 92% in 32 h and rhodamine B degraded approximately 91.4% in 24 h. However, none of the dye managed to degrade completely in solar light.

A higher degradation rate of dyes could be achieved using such organic linker which has more number of electrons in delocalization. More delocalized electrons favor the charge transfer mechanism and show a higher degradation rate. A notable example is shown by Zhang et al. in 2014, they used MOFs containing tetrazole as an organic linker and metal node containing Ag for the photodegradation of rhodamine 6G and methylene blue and achieved a higher degradation rate.

The choice of metal ion also affects the photodegradation rate of dyes. In 2013, Wang et al. compared the degradation rate of methylene blue dye using two MOF-[M(4-bpah) (1,3-bdc) (H_2O)], (M = Cd, Co), one containing Co metal node and other containing Cd metal node. Photodegradation rate of methylene blue using [Co(4-bpah) (1,3-bdc) (H_2O)] MOF was found to be 25% in 4 h, while photodegradation rate of methylene blue using [Cd(4-bpah) (1,3-bdc) (H_2O)] MOF was found to be 40% in 4 h [62]. However, a slight increment in photodegradation rate was observed by Wen et al. [63] by changing metal ion to Ni. Photodegradation rate of methylene blue using [Ni(4-bpah) (1,3-bdc) (H_2O)] was found to be 55% in 4 h.

Table 3 Application of MOFs for photodegradation of targeted dyes

MOFs	Pollutant	Irradiation source used	Reference
MIL-53(Fe)	Methylene blue	UV–visible	Du et al. [58]
MIL-53(Fe)	Methylene blue	UV–visible + H_2O_2	Du et al. [58]
MIL-53(Fe)	Methylene blue	Visible	Du et al. [58]
MIL-53(Fe)	Methylene blue	Visible + H_2O_2	Du et al. [58]
Cu (II)-imidazole	Methylene blue	Visible	Xu et al. [61]
Zn (II)-imidazole	Methylene blue	Visible	Xu et al. [61]
Cu (II)-imidazole	Methylene blue, rhodamine B	Solar light	Xiao et al. [66]
Cu (II)-pyrazine	Methylene blue	Solar light	Wang et al. [67]
Cu (II)-pyrazole	Methylene blue, methyl orange	UV + H_2O_2	Bala et al. [68]
Cd (II)-imidazole	Methylene blue, methyl orange	UV	Liu et al. [69]
Ag (I)-MOF	Rhodamine 6G, methyl orange	UV	Zhang et al. [5]
Cu(dm-bim)	Rhodamine B, methyl orange	Visible	Wen et al. [70]
MIL-88B, MIL-100 (Fe)	Rhodamine 6G	Visible	Laurier et al. [60]
Cd(4-bpah)(1,3-bdc)(H_2O)	Methylene blue	UV	Wang et al. [62]
Co(4-bpah)(1,3-bdc)(H_2O)	Methylene blue	UV	Wang et al. [62]
Ni(4-bpah)(1,3-bdc)(H_2O)	Methylene blue	UV	Wang et al. [62]
NTU-9 (Ti)	Rhodamine B, methylene blue	Visible	Gao et al. [64]

(continued)

Table 3 (continued)

MOFs	Pollutant	Irradiation source used	Reference
UiO-66 (Zr-Ti)	Methylene blue	Solar light	Wang et al. [65]

where, dm-bim = 5,6-dimethyl-benzimidazolate, 4-bpah = N, N'-bis(4-pyridinecarboxamide)-1,2-cyclohexane, 1,3-bdc = 1,3-benzenedicarboxylate

Gao et al. [64] in 2014 successfully synthesized MOFs based on titanium (Ti) named NTU-9, similar to TiO_2, NTU-9 acts as an efficient photocatalyst and degraded the rhodamine B and methylene blue dyes in a very short time span (20–80 min) using visible light as the irradiation source. However, recent research also focussed on the synthesis of more Ti-based MOFs which can perform as efficient photocatalysts in sunlight. Wang et al. [65] in 2016 successfully synthesized UiO-66(Zr-Ti) from UiO-66(Zr) by exchanging some of the Zr^{4+} by Ti^{4+}. They performed a photocatalytic reaction in sunlight and found that UiO-66(Zr-Ti) was able to degrade 80% of methylene blue under sunlight.

Till now, it is very clear that the activity of MOFs in photocatalysis is highly dependent on the metal ion, organic linker, and type of irradiation source used. It is worthwhile to mention here that despite huge research on MOFs as a photocatalyst, the activity of MOFs as photocatalysts is limited mainly in sunlight. Another drawback of MOFs is their stability in water. Therefore, to achieve better photocatalytic activity, certain structural modifications have been performed by various research groups to functionalize MOFs.

Type of structural modifications has been performed to increase the photocatalytic activity of MOFs.

i. Amine linkage
ii. Doping of metal
iii. Decoration of metal nanoparticles
iv. Heterostructures formation

Amine linkage: Introduction of the amino group (-NH_2) in the organic linker tends to increase the delocalization in the aromatic ring which shifts the absorption band to a higher wavelength, toward the visible region. The amino group also takes part in lone pair transport from organic linker to metal clusters [14]. It has been observed in certain cases that organic linker in MOFs containing NH_2 group shows better photocatalytic activity. Jianchuan He et al. [71] in 2018 synthesized NH_2-MIL-88B (Fe) MOFs by microwave method and found that 98% of methylene blue dye photo-degraded by synthesized photocatalyst in 45 min under visible light.

Doping of metal: Doping of metal in the MOFs leads to an increase in the photodegradation rate of dyes, and this is because of the generation of a new energy level that favors metal to metal charge transfer. Li et al. [72] in 2014 synthesized Ni-doped ZIF-8 (Zn) named BIT-11 and observed the complete photodegradation of methylene blue dye in 30 min using a visible light source. One another interesting example of complete photodegradation of methylene blue dye was observed by Abdelhameed et al. in sunlight by using Cr^{3+} and Ag^{+} doped NH_2-MIL-125 MOFs. They also observed that the introduction of silver atoms suppresses the recombination rate of electron and hole pairs and also increases its stability in water [73].

Decoration of metal nanoparticles: Another approach to increase the photocatalytic activity of MOFs is to decorate metal nanoparticles on the surface of MOFs. Liang et al. [74] synthesized the metal nanoparticle decorated MOFs. They loaded Pt, Pd, and Ag metal nanoparticles over MIL-100 MOFs and compared the activity of each one for photodegradation of methyl orange in visible light. They predicted that Pt-loaded MIL-100 showed the best catalytical activity among all and degraded methyl orange almost 100% in 40 min.

Heterostructures formation: Heterostructures consist of two or more materials and generally lead to improvement in an already existing application. A similar approach has been done by various researchers to improve the photocatalytic activity of MOFs. Since TiO_2, ZnO asis a well-known photocatalyst, a combination of oxides with MOFs could generate novel photocatalyst for dye degradation. Keeping that in mind, Liu et al. [75] synthesized heterostructure of TiO_2 nanosheet with MIL-100 (Fe) and found that synthesized heterostructure is efficient in degrading methylene blue under visible light, but it also requires the use of H_2O_2. Similarly, the approach has been done by Feng et al. [76], and they synthesized heterostructure of Fe_2O_3-MIL-53 (Fe). Although as-synthesized heterostructure is not able to degrade 100% methylene blue because of its paramagnetic character (due to the presence of Fe_2O_3), it was able to recover from the aqueous solution and found very stable in water.

Another interesting example of a heterostructure for photocatalysis is given by Sha et al. [77], who combined the oxide of bismuth and tungsten with MOFs and synthesized Bi_2WO_6/UiO-66 (Zr). Their synthesized heterostructure was able to degrade 100% of rhodamine B in 180 min using a visible light source. However, later, they modified the heterostructure to BiOBr/UiO-66 (Zr) and found increment in photocatalytic activity. BiOBr/UiO-66 (Zr) heterostructure cap completely degrade rhodamine B dye in just 15 min using a visible light source [78]. They predicted that increment inactivity could be due to the generation of more reactive radicals or due to reducing the recombination rate of electron and hole pairs. Summary of photocatalytic activity of functionalized MOFs is given in Table 4.

Many research groups focussed on developing heterostructure because of the increase in photocatalytic activity. Recently, Huang et al. [79] synthesized heterostructure based on reduced graphene oxide named rGO/NH_2-MIL-125 which they found efficient for complete photodegradation of methylene blue dye in 30 min. Wu et al. [80] also synthesized reduced graphene-based heterostructure which shows excellent photocatalytic activity for complete photodegradation of rhodamine B and methylene blue dyes in sunlight within 20 min.

Table 4 Application of functionalized MOFs for photodegradation of targeted dyes

MOFs		Pollutant	Irradiation source used	Reference
Metal-doped MOFs	UiO-66 (Ti-doped)	Methylene blue	Solar light	Wang et al. [65]
	ZIF-8 (Ni-doped)	Methylene blue	Visible	Li et al. [72]
	NH_2-MIL-125 (Cr and Ag doped)	Methylene blue	Solar light	Abdelhameed et al. [73]
Metal nanoparticle decorated MOFs	Pt-MIL-100	Methyl orange	Visible	Liang et al. [74]
	Pd-MIL-100	Methyl orange	Visible	Liang et al. [74]
	Ag-MIL-100	Methyl orange	Visible	Liang et al. [74]
Heterostructures based on MOFs	TiO_2-MIL-100 (Fe)	Methylene blue	Visible + H_2O_2	Liu et al. [75]
	Fe_2O_3-MIL-53 (Fe)	Methylene blue	Visible	Feng et al. [76]
	Bi_2WO_6-UiO-66	Rhodamine B	Visible	Sha et al. [77]
	BiOBr-UiO-66	Rhodamine B	Visible	Sha et al. [78]
	rGO-NH_2-MIL-125	Methylene blue	Visible	Huang et al. [79]
	rGO/MIL-88 (Fe)	Rhodamine B, methylene blue	Solar light	Wu et al. [80]

7.2 Metal–Organic Frameworks as Photocatalysts for Primary Pollutants

Primary contaminants are polycyclic aromatic hydrocarbons (PAH), volatile organic compounds, pesticides, herbicides, phenol, alkylphenol, and many more as predicted by the Environmental Protection Agency, USA [81]. Presence of these substances in water makes the life of aquatic organisms unbearable. So, removal of these substances from water is need of the hour [82]. Among them, phenol-related compounds are studied the most as target pollutants for water purification. Alvaro et al. [83] in 2007 studied the photodegradation of phenol using MOF-5 for phenol degradation under UV light as an illumination source. They found that the activity of MOF-5 is comparable to the well-known photocatalyst TiO_2 suggest the high efficiency of MOFs for phenol degradation. Phenol degrades in series of reactions as shown in Fig. 12 [84]. Mechanism of photodegradation of phenol is similar to as described in Sect. 3.1 above.

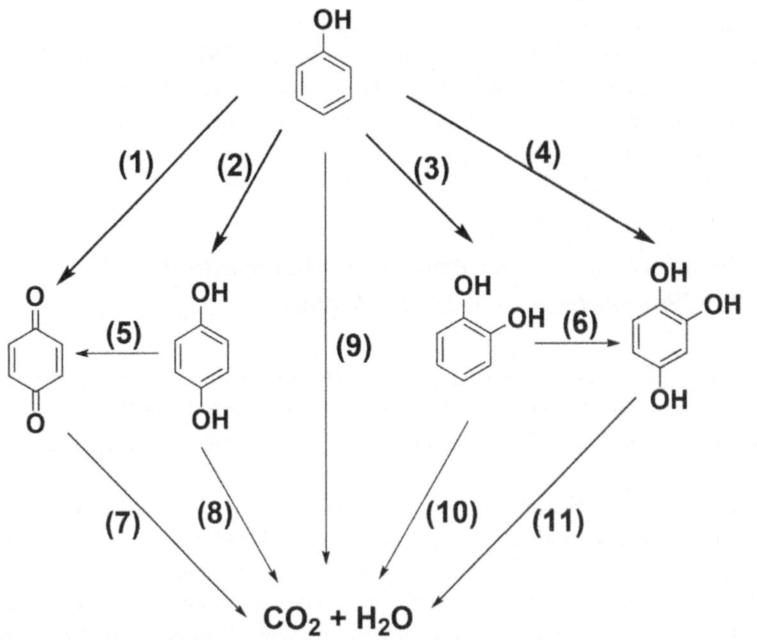

Fig. 12 Reaction map for photodegradation of phenol

Reactions shown in Fig. 12 are as follows:

Partial Oxidation Reactions

1. Phenol to para-benzoquinone
2. Phenol to para-hydroquinone
3. Phenol to catechol
4. Phenol to hydroxyquinol
5. Quinol to para-benzoquinone
6. Catechol to hydroxyquinol.

Total Oxidation Reactions

1. Hydroxyquinol to carbon dioxide and water
2. Quinol to carbon dioxide and water
3. Phenol to carbon dioxide and water
4. Catechol to carbon dioxide and water
5. Hydroxyquinol to carbon dioxide and water.

Certain advancement has been done to improve the photocatalytic activity of MOFs for photodegradation of phenol and phenol like compounds. In 2017, Surib et al. [85] demonstrated photodegradation of chlorophenol using metal ions intercalated Cd-MOF in sunlight as an illumination source. They found that photocatalytic activity of intercalated MOFs is far better than the general photocatalyst TiO_2. However, the use of MOFs for photodegradation of primary pollutants has not been explored much in literature.

7.3 Metal–Organic Frameworks as Photocatalysts for Photoreduction of Heavy Metals

Heavy metals like lead, mercury, arsenic, chromium, cadmium, etc., are toxic, bioaccumulative, and persist in the environment for longer periods [86]. All these factors make them hazardous and an urgent matter of concern. Naturally, these metals are released into the environment from weathering of metal-containing rocks and volcanic eruptions, while anthropogenically they are released from industrial waste, mining, tanneries, etc. Intake of these heavy metals by plants and animals could be poisonous as they will disrupt the whole food chain. Poisoning from these metals could occur through both direct and indirect exposure, improperly packaged food, medicines, etc. Upon prolonged exposure to lower levels of heavy metals, one may suffer from symptoms like headache, weakness, muscle pain, constipation, while exposure to higher levels of heavy metals may cause serious effects on the nervous and immune systems [87].

There are many ways to treat heavy metal pollutants in wastewater resources, like adsorption, electrolysis, membrane filtration, biological treatment, photocatalysis, etc. Adsorption technology has many advantages as it is an easy, efficient, conventional technique and can be effectively used to adsorb various pollutants from wastewater [88]. Earlier, materials like metal oxides, zeolites, carbon nanotubes, activated carbon, organic resins, etc., were used, but these materials offer certain disadvantages in adsorbing heavy metals. The smaller pores of activated carbon to trap heavy metals, the slow adsorption capacity of metal oxides, and the reusability of organic resins are few drawbacks of conventional materials used for capturing heavy metals.

Moreover, it is difficult for a particular material with defined properties to adsorb heavy metals; however, heavy metals can be converted from a high valence state (toxic form) to a lower valence state (less-toxic form). Materials like MOFs are an appropriate choice for these applications since, these are not only efficient adsorbent due to its highly porous and robust structure, but they also act as a photocatalyst [87]. MOFs contain organic likers which upon adequate functionalization could work as antennas in order to trap energy from the source. The energy can be easily transferred to the active site via the organic linker channels where the degradation of the heavy metal would occur. The photocatalysis of heavy metals in a high oxidation state could convert them to a low-oxidation state which is a less-toxic form. Hence, the combined adsorption and photocatalyst activity make MOFs the potential, efficient, and effective candidate to trap heavy metal impurities [89].

Photocatalysis of chromium metal has been investigated by various research groups with various modified and unmodified MOFs. UiO-66 with amine functionalities does the photocatalytic reduction of Cr (VI) to Cr (III) in presence of water [90]. This amine functionalization on UiO-66(NH_2) extends the range of this MOF to absorb visible light photons, while unfunctionalized UiO-66 is inactive for this process in visible light. Further, in the same line when this UiO-66(NH_2) was loaded with Ag nanoparticles, literature shows that silver facilitates the Cr (VI) reduction with stable rates. Not only is the zirconium oxide cluster present in UiO-66, but iron-based MOFs are also efficient for the purpose. Rather Fe based MOFs like MIL-100(Fe) show better efficiency for photocatalysis of Cr (VI) [52, 91].

Liang et al. [74] synthesized metal atom decorated MOFs and achieved 100% photoreduction of chromium metal ions from higher valance state to lower valance state in just 8 min. They also compared the activity of different metal atoms decorated on MOFs. For this purpose, they synthesized Au, Pd, and Pt decorated MIL-100 (Fe) and found that the best results were given by Pt decorated MIL-100 (Fe).

On a similar line, as discussed in dye degradation, the heterostructure of MOFs with other semiconductors tends to improve the photoreduction efficiency of chromium metal. One notable example is given by Huang et al. [92], who synthesized the heterostructure of g-C_3N_4 with MIL-53 (Fe) and found the complete photoreduction of Cr (VI) to Cr (III) in 3 h. Shen et al. [93] synthesized heterostructure of reduced graphene oxide with UiO-66 (NH_2) and found that 100% photoreduction of chromium metal ion takes place in 100 min. Further improvement in photoreduction efficiency is achieved by Liang et al. [94], who observed 100% photoreduction of Cr (VI) to Cr (III) in just 80 min by using rGO/MIL-53 (Fe).

From the above examples, it is noteworthy that huge development of photocatalyst has been done for the photocatalysis of chromium metal. But a lot of work is still needed for the photoreduction of other heavy metals.

8 Summary and Future perspectives

In this chapter, considering the importance of water, its rising demand, and emerging scarcity issue, we have discussed various methods for the treatment of wastewater. A brief introduction of the methods like physical processes involving sedimentation, distillation, filtration, etc., biological processes including biological sludge treatment, etc., followed by chemical processes like flocculation and chlorination has been discussed. Their disadvantages such as releasing secondary pollutants into the environment are also presented. As an alternative to the above-mentioned methods, photocatalysis has been described as a greener, ecological, and sustainable technique to purify water. The photocatalytic mechanism, factors affecting the efficiency of photocatalysis, and the kinetics of photocatalysis have been discussed in detail. This is followed by few conventional photocatalysts including metal oxides, metal sulfides, and metal phosphides which are effective for wastewater treatment are listed. Owing to the disadvantages like poor conductivity, and fast recombination of charge carriers, the application of conventional materials for photocatalysis is limited. In this regard, MOFs being porous, high surface area material, and hybrid of organic and inorganic constituents that offer properties of a semiconductor have been explored as photocatalysts in addition to their adsorption and separation applications. Thus, for being a photocatalyst in addition to the high adsorption tendency, porosity, and robust design, their potential to degrade toxic organic–inorganic pollutants from wastewater resources has been discussed in this chapter. Since the working environment is aqueous, their aqueous medium stability has been presented along with the methods to improve it. Several strategies to improve the aqueous medium stability of MOFs including modifications in the organic linker design, the choice of metal ions used, and creating steric hindrance to the water molecules from attacking the weaker bonds have been presented.

At last, we have described several case studies to reveal the application of MOFs as photocatalysts for the degradation of primary pollutants, organic dyes, and heavy metal ions from wastewater. In the case studies, several methods like an encapsulation of chromophores into the porous framework and incorporation of light-absorbing organic linkers to modify the photocatalytic activity of MOFs have been discussed.

There is still a long way to go for understanding and implying those modifications to enhance the efficiency and capacity of MOFs as photocatalysts. Also, there is a lot of lagging with the stability and recyclability of MOFs. In the case of degradation of dyes, it has been observed that a lot of heteroatoms doping and heterostructure incorporation into the MOFs is required to make it more photocatalytic active. But the disadvantage is lying in terms of their cost and lengthy synthesizing process. Therefore, a large room for development is required in terms of cost-effective processes to synthesize MOFs with high yield and enhanced photocatalytic activity. Also, considering a greener approach, more focus should be on synthesizing those MOFs that could work as photocatalysts in sunlight. Similarly, very few MOFs are known that can degrade the heavy metal ions from high valent toxic form to low valent less toxic form. Mostly, work has been done to degrade high valent ions of chromium metal. Thus, more advancements are required to photo-catalyze other metal ions like nickel, cadmium, lead, mercury, and arsenic using MOFs. Having proper research on the above-mentioned challenges would make MOFs the most efficient photocatalyst for future generations.

Abbreviations

MOFs	Metal-organic frameworks
PAC	Poly-aluminum chloride
PAHC	Poly-aluminum hydroxy chloride
ROSs	Reactive oxygen species
PXRD	Powder X-ray diffraction

UV	Ultraviolet
DRS	Diffuse reflectance spectroscopy
HOMO	Highest occupied molecular orbital
LUMO	Lowest unoccupied molecular orbital
SEM	Scanning electron microscope
TEM	Transmission electron microscope
EDS	Energy dispersive spectroscopy
BET	Brunauer–Emmett–Teller
PL	Photoluminescence
TGA	Thermogravimetry analysis
R6G	Rhodamine 6G
MB	Methylene blue
PAH	Polycyclic aromatic hydrocarbons

References

1. Shannon MA et al (2008) Science and technology for water purification in the coming decades. Nature 452:301–310
2. Samer MSE-M (2015) Biological and chemical wastewater treatment processes. IntechOpen. https://doi.org/10.5772/61250
3. Yousef R, Qiblawey H, El-Naas MH (2020) Adsorption as a process for produced water treatment: a review. Processes 8
4. Loeb SK et al (2019) The technology horizon for photocatalytic water treatment: sunrise or sunset? Environ Sci Technol 53:2937–2947
5. Bedia J et al (2019) A review on the synthesis and characterization of metal organic frameworks for photocatalytic water purification. Catalysts 9
6. Furukawa H, Cordova KE, O'Keeffe M, Yaghi OM (2013) The chemistry and applications of metal-organic frameworks. Science 341
7. Kunduru KR et al (2017) Nanotechnology for water purification: applications of nanotechnology methods in wastewater treatment. In: Grumezescu AMBT-WP (ed) Water purification. Academic Press, pp 33–74. https://doi.org/10.1016/B978-0-12-804300-4.00002-2
8. Sharma S, Bhattacharya A (2017) Drinking water contamination and treatment techniques. Appl Water Sci 7:1043–1067
9. Mohamed A, Salama A, Nasser WS, Uheida A (2018) Photodegradation of Ibuprofen, Cetirizine, and Naproxen by PAN-MWCNT/TiO_2–NH_2 nanofiber membrane under UV light irradiation. Environ Sci Eur 30:47
10. Zhang J, Chen F, He B (2018) Photocatalysis
11. Lee S-Y, Park S-J (2013) TiO_2 photocatalyst for water treatment applications. J Ind Eng Chem 19:1761–1769
12. Kumar S et al (2014) Nanotechnology-based water treatment strategies. J Nanosci Nanotechnol 14:1838–1858
13. Russo V et al (2020) Applications of metal organic frameworks in wastewater treatment: a review on adsorption and photodegradation. Front Chem Eng 2:15
14. Zango ZU et al (2020) A critical review on metal-organic frameworks and their composites as advanced materials for adsorption and photocatalytic degradation of emerging organic pollutants from wastewater. Polymers 12
15. De Lasa HI, Serrano B, Salaices M (2005) Photocatalytic reaction engineering. Springer

16. Regmi C, Joshi B, Ray SK, Gyawali G, Pandey RP (2018) Understanding mechanism of photocatalytic microbial decontamination of environmental wastewater. Front Chem 6:33
17. Curco D, Gimenez J, Addardak A, Cervera-March S, Esplugas S (2002) Effects of radiation absorption and catalyst concentration on the photocatalytic degradation of pollutants. Catal Today 76:177–188
18. Araña J et al (2004) Photocatalytic degradation of formaldehyde containing wastewater from veterinarian laboratories. Chemosphere 55:893–904
19. Malato S, Fernández-Ibáñez P, Maldonado MI, Blanco J, Gernjak W (2009) Decontamination and disinfection of water by solar photocatalysis: recent overview and trends. Catal Today 147:1–59
20. Gaya UI, Abdullah AH (2008) Heterogeneous photocatalytic degradation of organic contaminants over titanium dioxide: a review of fundamentals, progress and problems. J Photochem Photobiol C Photochem Rev 9:1–12
21. Chong MN, Lei S, Jin B, Saint C, Chow CWK (2009) Optimisation of an annular photoreactor process for degradation of Congo Red using a newly synthesized titania impregnated kaolinite nano-photocatalyst. Sep Purif Technol 67:355–363
22. Hu A et al (2012) Adsorption and photocatalytic degradation kinetics of pharmaceuticals by TiO_2 nanowires during water treatment. Waste Biomass Valorization 3
23. Janssens R, Mandal MK, Dubey KK, Luis P (2017) Slurry photocatalytic membrane reactor technology for removal of pharmaceutical compounds from wastewater: towards cytostatic drug elimination. Sci Total Environ 599–600:612–626
24. Zheng X, Shen Z-P, Shi L, Cheng R, Yuan D-H (2017) Photocatalytic membrane reactors (PMRs) in water treatment: configurations and influencing factors. Catalysts 7
25. Chong MN, Jin B, Chow CWK, Saint C (2010) Recent developments in photocatalytic water treatment technology: a review. Water Res 44:2997–3027
26. Wen M et al (2019) Metal–organic framework-based nanomaterials for adsorption and photocatalytic degradation of gaseous pollutants: recent progress and challenges. Environ Sci Nano 6:1006–1025
27. Sharmin E, Zafar F (2016) Introductory chapter: metal organic frameworks (MOFs). In: Metal-organic frameworks. IntechOpen
28. Yaghi OM, Li G, Li H (1995) Selective binding and removal of guests in a microporous metal–organic framework. Nature 378:703–706
29. Li H, Eddaoudi M, O'Keeffe M, Yaghi OM (1999) Design and synthesis of an exceptionally stable and highly porous metal-organic framework. Nature 402:276–279
30. Chui SS-Y, Lo SM-F, Charmant JPH, Orpen AG, Williams ID (1999) A chemically functionalizable nanoporous material $[Cu_3(TMA)_2\ (H2O)_3]_n$. Science 283:1148–1150
31. Howarth AJ et al (2017) Best practices for the synthesis, activation, and characterization of metal-organic frameworks. Chem Mater 29:26–39
32. Eddaoudi M et al (2002) Systematic design of pore size and functionality in isoreticular MOFs and their application in methane storage. Science 295:469–472
33. Li Y, Xu H, Ouyang S, Ye J (2016) Metal–organic frameworks for photocatalysis. Phys Chem Chem Phys 18:7563–7572
34. Mukhopadhyay S, Basu O, Nasani R, Das SK (2020) Evolution of metal organic frameworks as electrocatalysts for water oxidation. Chem Commun 56:11735–11748
35. Lawson HD, Walton SP, Chan C (2021) Metal-organic frameworks for drug delivery: a design perspective. ACS Appl Mater Interfaces 13:7004–7020
36. Stock N, Biswas S (2012) Synthesis of metal-organic frameworks (MOFs): routes to various MOF topologies, morphologies, and composites. Chem Rev 112:933–969
37. Dey C, Kundu T, Biswal BP, Mallick A, Banerjee R (2014) Crystalline metal-organic frameworks (MOFs): synthesis, structure and function. Acta Crystallogr Sect B, Struct Sci Cryst Eng Mater 70:3–10
38. Lee Y-R, Kim J, Ahn W-S (2013) Synthesis of metal-organic frameworks: a mini review. Korean J Chem Eng 30:1667–1680

39. Ni Z, Masel RI (2006) Rapid production of metal–organic frameworks via microwave-assisted solvothermal synthesis. J Am Chem Soc 128:12394–12395
40. Son W-J, Kim J, Kim J, Ahn W-S (2008) Sonochemical synthesis of MOF-5. Chem Commun 6336–6338. https://doi.org/10.1039/B814740J
41. Martinez Joaristi A, Juan-Alcañiz J, Serra-Crespo P, Kapteijn F, Gascon J (2012) Electrochemical synthesis of some archetypical Zn^{2+}, Cu^{2+}, and Al^{3+} metal organic frameworks. Cryst Growth Des 12:3489–3498
42. Klimakow M, Klobes P, Thünemann AF, Rademann K, Emmerling F (2010) Mechanochemical synthesis of metal–organic frameworks: a fast and facile approach toward quantitative yields and high specific surface areas. Chem Mater 22:5216–5221
43. Parnham ER, Morris RE (2007) Ionothermal synthesis of zeolites, metal-organic frameworks, and inorganic-organic hybrids. Acc Chem Res 40:1005–1013
44. Carné-Sánchez A, Imaz I, Cano-Sarabia M, Maspoch D (2013) A spray-drying strategy for synthesis of nanoscale metal-organic frameworks and their assembly into hollow superstructures. Nat Chem 5:203–211
45. Gándara F, Bennett TD (2014) Crystallography of metal-organic frameworks. IUCrJ 1:563–570
46. Capano G et al (2020) On the electronic and optical properties of metal-organic frameworks: case study of MIL-125 and MIL-125-NH_2. J Phys Chem C 124:4065–4072
47. Tauc J (1970) Absorption edge and internal electric fields in amorphous semiconductors. Mater Res Bull 5:721–729
48. Tsuruoka T et al (2016) Morphology control of metal-organic frameworks based on paddle-wheel units on ion-doped polymer substrate using an interfacial growth approach. Langmuir 32:6068–6073
49. Thommes M (2010) Physical adsorption characterization of nanoporous materials. Chem Ing Tech 82:1059–1073
50. Farha OK, Hupp JT (2010) Rational design, synthesis, purification, and activation of metal–organic framework materials. Acc Chem Res 43:1166–1175
51. Zeng L, Guo X, He C, Duan C (2016) Metal-organic frameworks: versatile materials for heterogeneous photocatalysis. ACS Catal 6:7935–7947
52. Shi L et al (2015) An amine-functionalized iron(III) metal-organic framework as efficient visible-light photocatalyst for Cr(VI) reduction. Adv Sci 2:1500006
53. Wang Z, Cohen SM (2009) Postsynthetic modification of metal–organic frameworks. Chem Soc Rev 38:1315–1329
54. Pidko EA, Hensen EJM (2013) Computational approach to chemical reactivity of MOFs. Met Org Fram Heterog Catal 209–234
55. Lerma-Berlanga B et al (2021) Effect of linker distribution in the photocatalytic activity of multivariate mesoporous crystals. J Am Chem Soc 143:1798–1806
56. Aliyu Mohammed H, Ariffin A, Wibowo A (2020) Cation exchange in metal-organic frameworks (MOFs): the hard-soft acid-base (HSAB) principle appraisal. Inorg Chim Acta 511:119801
57. Ding M, Jiang H-L (2021) Improving water stability of metal–organic frameworks by a general surface hydrophobic polymerization. CCS Chem 3:2740–2748
58. Du J-J et al (2011) New photocatalysts based on MIL-53 metal–organic frameworks for the decolorization of methylene blue dye. J Hazard Mater 190:945–951
59. Ran J-W, Liu S-W, Wu P, Pei J (2013) Efficient photocatalytic properties of a dinuclear iron complex with bis[2-hydroxybenzaldehyde]hydrazonate ligand. Chin Chem Lett 24:373–375
60. Laurier KGM et al (2013) Iron(III)-based metal-organic frameworks as visible light photocatalysts. J Am Chem Soc 135:14488–14491
61. Xu X-Y, Chen Q-C, Yu Y-D, Huang X-C (2016) Ligand induced anionic cuprous cyanide framework for cupric ion turn on luminescence sensing and photocatalytic degradation of organic dyes. Inorg Chem 55:75–82
62. Wang X-L et al (2013) Structural diversities and fluorescent and photocatalytic properties of a series of cuii coordination polymers constructed from flexible bis-pyridyl-bis-amide ligands with different spacer lengths and different aromatic carboxylates. Cryst Growth Des 13:3561–3576

63. Wen L et al (2012) Visible-light-driven photocatalysts of metal-organic frameworks derived from multi-carboxylic acid and imidazole-based spacer. Cryst Growth Des 12:1603–1612
64. Gao J et al (2014) A p-type Ti(iv)-based metal–organic framework with visible-light photo-response. Chem Commun 50:3786–3788
65. Wang A et al (2016) Titanium incorporated with UiO-66(Zr)-type Metal-Organic Framework (MOF) for photocatalytic application. RSC Adv 6:3671–3679
66. Ren S et al (2019) Synthesis, structures and properties of six lanthanide complexes based on a 2-(2-carboxyphenyl)imidazo(4,5-: F)-(1,10)phenanthroline ligand. RSC Adv 9:3102–3112
67. Zhang J-Y, Chang Z-H, Wang X-L, Wang X, Lin H-Y (2021) Different Anderson-type polyoxometalate-based metal–organic complexes exhibiting –OH group-directed structures and electrochemical sensing performance. New J Chem 45:3328–3334
68. Bala S et al (2014) Designing functional metal-organic frameworks by imparting a hexanuclear copper-based secondary building unit specific properties: structural correlation with magnetic and photocatalytic activity. Cryst Growth Des 14:6391–6398
69. Liu L et al (2014) Polynuclear CdII polymers: crystal structures, topologies, and the photodegradation for organic dye contaminants. Cryst Growth Des 14:3035–3043
70. Wen T, Zhang D-X, Liu J, Lin R, Zhang J (2013) A multifunctional helical Cu(i) coordination polymer with mechanochromic, sensing and photocatalytic properties. Chem Commun 49:5660–5662
71. He J, Zhang Y, Zhang X, Huang Y (2018) Highly efficient Fenton and enzyme-mimetic activities of NH_2-MIL-88B(Fe) metal organic framework for methylene blue degradation. Sci Rep 8:5159
72. Li R et al (2014) Nickel-substituted zeolitic imidazolate frameworks for time-resolved alcohol sensing and photocatalysis under visible light. J. Mater. Chem. A 2:5724–5729
73. Abdelhameed RM, Simões MMQ, Silva AMS, Rocha J (2015) Enhanced photocatalytic activity of MIL-125 by post-synthetic modification with CrIII and Ag nanoparticles. Chem Eur J 21:11072–11081
74. Liang R, Jing F, Shen L, Qin N, Wu L (2015) M@MIL-100(Fe) (M = Au, Pd, Pt) nanocomposites fabricated by a facile photodeposition process: Efficient visible-light photocatalysts for redox reactions in water. Nano Res 8:3237–3249
75. Liu X et al (2017) A sandwich-like heterostructure of TiO_2 nanosheets with MIL-100(Fe): a platform for efficient visible-light-driven photocatalysis. Appl Catal B Environ 209
76. Feng X, Chen H, Jiang F (2017) In-situ ethylenediamine-assisted synthesis of a magnetic iron-based metal-organic framework MIL-53(Fe) for visible light photocatalysis. J Colloid Interface Sci 494:32–37
77. Sha Z, Sun J, On Chan HS, Jaenicke S, Wu J (2014) Bismuth tungstate incorporated zirconium metal–organic framework composite with enhanced visible-light photocatalytic performance. RSC Adv 4:64977–64984
78. Sha Z, Wu J (2015) Enhanced visible-light photocatalytic performance of BiOBr/UiO-66(Zr) composite for dye degradation with the assistance of UiO-66. RSC Adv 5:39592–39600
79. Huang L, Liu B (2016) Synthesis of a novel and stable reduced graphene oxide/MOF hybrid nanocomposite and photocatalytic performance for the degradation of dyes. RSC Adv 6:17873–17879
80. Wu Y, Luo H, Wang H (2014) Synthesis of iron(iii)-based metal–organic framework/graphene oxide composites with increased photocatalytic performance for dye degradation. RSC Adv 4:40435–40438
81. Gasperi J, Garnaud S, Rocher V, Moilleron R (2008) Priority pollutants in wastewater and combined sewer overflow. Sci Total Environ 407:263–272
82. Hossain MF (2019) Chapter 10: Water pollution. In Hossain MF (Ed) Sustainable development for mass urbanization. Elsevier, pp 151–267. https://doi.org/10.1016/B978-0-12-817690-0.00010-5
83. Alvaro M, Carbonell E, Ferrer B, Llabrés i Xamena FX, Garcia H (2007) Semiconductor behavior of a metal-organic framework (MOF). Chem Eur J 13:5106–5112
84. Salaices M, Serrano B, de Lasa HI (2004) Photocatalytic conversion of phenolic compounds in slurry reactors. Chem Eng Sci 59:3–15

85. Surib NA et al (2017) Ag^+, Fe^{3+} and Zn^{2+}-intercalated cadmium(ii)-metal–organic frameworks for enhanced daylight photocatalysis. RSC Adv 7:51272–51280
86. Tchounwou PB, Yedjou CG, Patlolla AK, Sutton DJ (2012) Heavy metal toxicity and the environment. Exp. Suppl. 101:133–164
87. Kobielska PA, Howarth AJ, Farha OK, Nayak S (2018) Metal–organic frameworks for heavy metal removal from water. Coord Chem Rev 358:92–107
88. Sarkar S, Adhikari S (2018) Adsorption technique for removal of heavy metals from water and possible application in wastewater-fed aquaculture. In: Wastewater management through aquaculture, pp 235–251. https://doi.org/10.1007/978-981-10-7248-2_12.
89. Feng M, Zhang P, Zhou H-C, Sharma VK (2018) Water-stable metal-organic frameworks for aqueous removal of heavy metals and radionuclides: a review. Chemosphere 209:783–800
90. Shen L, Wu W, Liang R, Lin R, Wu L (2013) Highly dispersed palladium nanoparticles anchored on UiO-66(NH_2) metal-organic framework as a reusable and dual functional visible-light-driven photocatalyst. Nanoscale 5:9374–9382
91. Wang C-C et al (2016) Photocatalytic Cr(VI) reduction in metal-organic frameworks: a mini-review. Appl Catal B Environ 193:198–216
92. Huang W, Liu N, Zhang X, Wu M, Tang L (2017) Metal organic framework g-C_3N_4/MIL-53(Fe) heterojunctions with enhanced photocatalytic activity for Cr(VI) reduction under visible light. Appl Surf Sci 425
93. Shen L, Liang S, Wu W, Liang R, Wu L (2013) Multifunctional NH_2-mediated zirconium metal–organic framework as an efficient visible-light-driven photocatalyst for selective oxidation of alcohols and reduction of aqueous Cr(vi). Dalton Trans 42:13649–13657
94. Liang R, Shen L, Jing F, Qin N, Wu L (2015) Preparation of MIL-53(Fe)-reduced graphene oxide nanocomposites by a simple self-assembly strategy for increasing interfacial contact: efficient visible-light photocatalysts. ACS Appl Mater Interfaces 7:9507–9515

Metal-Organic Framework as a Photocatalyst: Recent Growth in Environmental Applications

Anita Yadav and Rakesh Kumar Sharma

Contents

1 Introduction 596
2 Possible Photocatalytic Applications of MOFs 600
2.1 Reduction of Carbon Dioxide (CO_2) on MOFs 600
2.2 Hydrogen Production via Water Splitting on MOFs 614
2.3 Photoelectric Conversion/ Photoelectrochemical Applications 615
2.4 Reduction of N_2 (Nitrogen Fixation) with MOFs 616
2.5 Photocatalytic Degradation of Organic Contaminants 618
2.6 Photocatalytic Application in Organic Synthesis and Transformation Reactions 623
2.7 Photocatalytic Reduction of Heavy Metal Ions for Detoxification 624
2.8 Photocatalytic Antibacterial Activity 625
2.9 Photocatalytic Removal of Toxic Gases 626
3 The Common Mechanism Involved in Photocatalysis by MOFs 626
4 Conclusion and Future Perspectives 630
Abbreviations 631
References 634

Abstract Photocatalysts are considered as an encouraging and environment-friendly technology for solar energy conversion to thermal/chemical energy via interactions between light matters. Metal–organic frameworks (MOFs) are considered recently as innovative photocatalysts because of their immense coordination between active metal centers and organic linkers. This property offers unique structural characteristics that contribute to the large surface area, ordered structure, highly porous nature, and ultrahigh structural tunability. The highly porous nature is responsible for concentrating the organic pollutants in the cavity of MOFs to gain enhanced activity, tunable characteristic allows the productive charge separation, and generation of reactive oxygen species capacity is responsible for the speedy decomposition. Such advancement extends enhanced chances for improving their efficiency in several photocatalytic applications such as CO_2/H_2 generation and environmental remediation. MOFs furnish well-defined advantages in comparison with any other conventional photocatalysts under visible, UV, and NIR light irradiation. Some of the challenges conventional photocatalysts face include inadequate usage of sunlight

A. Yadav · R. K. Sharma (✉)
Nanotechnology Lab, Department of Chemistry, University of Delhi, Delhi, India

S. Gulati (ed.), *Metal-Organic Frameworks (MOFs) as Catalysts*,
https://doi.org/10.1007/978-981-16-7959-9_23

or restricted active sites. MOFs-based photocatalysts tend to solve several practical problems at an industrial scale. The research on MOF photocatalyzed processes is well in demand, and various mechanistic judgments have been calculated so that their potential can be utilized. Hence, it is substantial, to sum up, this field of research supplies thorough awareness regarding its development and future aspects. This chapter proposes a comprehensive and critical analysis regarding photocatalysis and historical development, synthesis, and modifications in MOF networks. Further, we will also represent the present status of MOFs involved in photocatalytic processes like electron–hole pair, formation of reactive species, and influence on water chemistry in detail. Later on, we will also emphasize the advancements which have been accomplished to date on the photocatalytic performances of MOFs in energy and environmental remediations via experimental investigations.

Keywords Metal–organic frameworks (MOFs) · Functionalization · Photocatalysis · Light · Degradation · Contaminants · Advanced oxidation process

1 Introduction

Photocatalysis is a catalytic reaction that is persuaded by light energy [1]. It is the potential of harvesting solar energy which is easily available to human beings [2]. It is an important chemical method by which energy and environmental crises can be handled by converting solar energy to chemical potential [3] and requires the irradiation of light of a particular wavelength onto the semiconductor material [4]. This is an advanced oxidation process (AOP) [5] for proficient degradation of harmful organic contaminants present in the water by photocatalytic materials [6]. AOP is the conversion of light energy to chemical energy together with the formation of free radicals, like hydroxyl radicals, which possess the tendency to break down the harmful organic pollutants into non-toxic by-products [7]. Around a billion people in developing countries do not get drinking water, and nearly 3 billion people are deficient in proper sanitation. The contamination of water resources with chemicals, pesticides, industrial wastes, and pharmaceuticals is a rising concern in both developing and the developed countries [2, 8]. The noteworthy research initiated after 1970 when the splitting of water was established by utilizing near UV light with the help of a platinum cathode and a titania photoanode [9]. The advantages of photocatalysis include their potential to degrade pollutants to non-toxic by-products within a short period under normal conditions. A perfect photocatalyst is stable in aqueous as well as in organic solvents under acidic or alkaline pH, should stand strong light irradiation, easily available, low cost, and reusable, and also it must possess high porosity to enhance its catalytic activity [10]. Therefore, scientists came across various mesoporous and microporous materials due to their size, morphology, pore volume, etc. Photocatalysis involves the formation of electron–hole pairs, e^- recombination, photocatalytic electron transport, oxygen reduction, and oxidation by holes present in the valence band [4]. Out of several materials available like silicates, graphene oxide, metal-oxide nanoparticles, and other carbon materials, metal–organic frameworks (MOFs) show lots of potentials [10].

MOFs are theoretically assembled three-dimensional crystalline porous materials and assist as an exciting stage for the study and designing of simulated systems [11–13]. These form an innovative class of inorganic–organic hybrid materials comprised of metal ions/clusters (e.g., secondary building units (SBU)) and organic ligands (e.g., unidentate, bidentate, multidentate bridging ligands) [14] bonded via coordinate bond [15, 16]. The name metal–organic framework was invented first time in 1992 [17]. In 1990, Hoskins and Robson reported scaffold-like material having Cu(I) as a center and tetracyano-tetra phenylmethane as a ligand [17, 18]. Later on, in 1999 Yaghi and his co-workers extended the metal framework and formed a porous crystalline structure called MOF-5 [17]. Over 70,000 MOFs have been discovered till now [19, 20]. MOFs have become an active field in coordination chemistry and came out to be an emerging class of research due to their ultimate diversity in structure, super high surface area (over 6000 m^2g^{-1}) [21], well-defined porous structure, manageable pore size, and adjustable functional groups [22]. Its easily tailorable characteristics, i.e., unsaturated metal sites, provide distinctive features for several applications which include catalysis, gas storage, magnetism, sensing, carbon dioxide capture, electronics, electrochemistry, and biomedical applications [23]. Furthermore, MOFs are widely utilized as heterogeneous catalysts under irradiation of UV/visible light for H_2 generation, CO_2 reduction, and also for the breakdown of harmful organic contaminants utilizing photogenerated electrons and holes [24–26]. MOFs were used as a catalyst in 1994 for the first time for cyanosilylation of aldehydes [27]. MOFs have the potential to tackle the limitations of conventional photocatalysts such as their low stability, surface area and reusability [19]. These are arranged in the crystalline lattice which is distinct from inorganic semiconductors having delocalized valence and conduction band [21]. The activity of photocatalysis by MOFs can be magnified by the incorporation of photoactive ligands into MOFs.

The design of charge transfer pathways in MOFs offers improvement in semiconductor properties and encourages their use in photocatalysis under sunlight, UV, and visible light irradiation. The charge transfer mechanisms can be modified using chromophoric ligands, metal centers, and also the encapsulated molecule present in the MOF composites [28]. Several synthetic procedures to modify the crystalline properties of MOFs include solvothermal approaches (e.g., microwave-assisted, sonochemical, conventional heating) and non-solvothermal methods (e.g., mechanochemical, electrochemical, spray drying, and flow chemistry) [14, 29]. The most commonly used solvothermal method takes place in a closed reaction vessel under pressure (autogenous) at a temperature greater than the boiling point of the solvent used and allows the self-assembly of MOF crystals. The most commonly used solvents in this process are *N*, *N*-dimethylformamide (DMF), methanol, and ethanol, where the reaction is performed at a temperature lower than 220 °C and left for crystallization for several hours or days. These different methods help in controlling and modifying the size, morphology, and functionalization of crystals [17, 30]. In comparison, non-solvothermal methods form at a lower temperature than the boiling point of the solvent [31]. Using these synthetic methods, the rate of growth of MOFs can be managed to alter the coordination geometry, surface area, and porosity of formed MOF. Hence, having significant knowledge of the coordination environment between metal centers and organic linkers it is important to utilize the synthetic method best suited for the fabrication of the appropriate MOFs photocatalysts. Various techniques and instrumental methods are used for the characterizations of the MOFs (like porosity, crystallinity, optical, electronic properties, and morphology). The commonly used are High-Resolution Transmission Electron Microscopy (HR-TEM), N_2 adsorption/desorption isotherms, Single-Crystal or Fourier Transform Infrared Spectroscopy (FTIR), Field Emission Scanning Electron Microscopy (FESEM), Powder X-ray Diffraction (XRD), Raman Spectroscopy, X-ray Photoelectron Spectroscopy (XPS), Nuclear Magnetic Resonance (NMR), UV–visible Diffuse Reflectance Spectroscopy (UV–Vis DRS), Surface Plasmon Resonance (SPR), Thermogravimetric Analysis (TGA), Photocurrent Responses, Inductively Coupled Plasma Optical Emission Spectroscopy (ICP-OES), Electrochemical Impedance Spectroscopy (EIS), and Photoluminescence (PL) Spectroscopy. Out of these FTIR, Raman, single-crystal XRD, NMR, and neutron diffraction are utilized to elucidate the formation and structural defects in MOFs [32]. The analytical characterizations when combined with the quantum chemical approach (e.g., molecular dynamics, Monte Carlo simulations, and quantum mechanics) provide data related to electronic, reactivity, and mechanical features of MOFs [33]. The studies provide an understanding of the synthesis and functionalization of MOFs via ex situ/in situ non-solvothermal/solvothermal methods [14, 34].

Alvaro et al. (2007) were the ones who created the concept of using MOFs as photocatalysts while exploring the semiconducting properties of MOF-5 ($Zn_4O(BDC)_3$, BDC^{2-}=1,4-benzodicarboxylate). In their study, they found that the ligand used in the solution form in fabricating MOF undergoes some changes which lead to the concept of electron ejection from the organic ligand molecule [35]. Upon light irradiation, MOFs show the properties of semiconductors. The organic ligand harvests light from various sources and leads to the activation of the metal center from ligand to metal charge transition (LMCT). It is an important criterion for photocatalytic activities. The porous nature of MOFs contributes broadly to the photocatalytic processes. Several MOFs having metal centers Fe, Cr, Zr, and Ti display stability in water and tend to harvest solar energy [36]. Such MOFs have a small bandgap; hence, excitation is eased in the presence of visible light [10]. These new MOFs-based heterogeneous photocatalysts have several restrictions which limit their applications. Generally, due to the large bandgap of MOFs, they collect only about 5% of total sunlight energy imparted. Also, quick recombination of electron–hole pairs occurs which leads to unsatisfactory results. Several MOFs have poor stability, low reproducibility, high cost, and kinetic instabilities [37], and hence they agglomerate easily [14, 38] and are unable to harness sunlight energy. To overcome these limitations, MOFs have been functionalized with metal, metal-oxide nanoparticles, graphene, graphene oxide, zeolites, carbon nanotubes (CNT), combination with conductive materials, etc. Various techniques used for the synthesis of functionalized MOFs are surface functionalization, deposition, impregnation, and fabrication. The functionalization leads to reduced bandgap, successful separation of electron–hole pairs (photogenerated), and low photocurrent and photoluminescence response to harness solar light energy effectively [39]. Such composites have a higher absorption of visible light, greater stability in the environment, and more reusability than normal unfunctionalized MOFs [10]. The functionalized MOFs broaden the area of utilization of MOFs in photocatalytic reactions [40]. The activities of MOF-based composites have been classified based on three types of active sites such as type I, where the SBU behaves as semiconductor dot photocatalysts in MOF crystal, type II where organic ligands have chromophores in their structure, and type III is when a photocatalytic species is encapsulated within the pores of MOF [14]. However, the functionalization and encapsulation within MOF frameworks sometimes reduce the tendency of metal sites to absorb photons due to the blockage of the pores available in MOF. The blocking of pores also encourages steric hindrance to the reactant molecules, which obstruct the transfer of mass from reaction media to catalyst sites [14, 41]. Hence, designing molecules without altering the properties is desirable for an effectual gathering of sunlight energy. The classification of synthesis, modification strategies, properties, and applications of MOFs are shown in Fig. 1.

Photostability of MOFs—MOFs possess an enhanced photocatalytic rate compared with other photocatalysts due to their distinctive characteristics of high surface area, tailorable physical–chemical properties, and desired topology which facilitate speedy transport and accommodation of the desired molecule. The theoretical bandgap of MOFs for the photocatalytic application depends upon the metal ion and organic ligands and varies between 1.0 and 5.5 eV [42]. The photocatalytic phenomenon of MOFs can be suggestively amended via altering modulation of the composition and structure of MOFs [14]. The photocatalytic rate also depends upon the richness in the number of metal nodes acting as active catalyst sites, easy encapsulation of photosensitizers, and functional organic linkers along adjustable bandgap

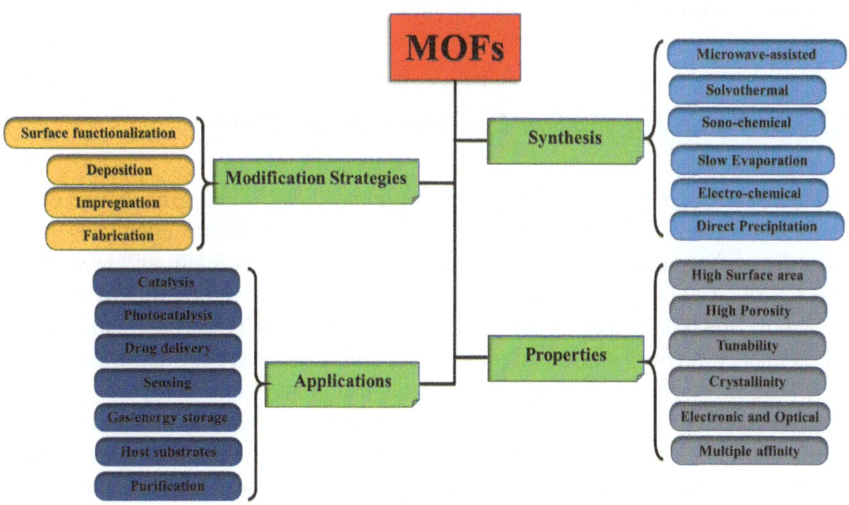

Fig. 1 Schematic categorization of synthesis, modifications, properties, and applications of MOFs

[13, 43]. MOFs are photoresponsive and tend to absorb light via a metal ion center and an organic ligand. First of all, the structure of MOFs should be evaluated and characterized to gain knowledge of the mechanism involved. Furthermore, the porous nature and tailorable characteristics of MOFs lead to the adjustment of the desired molecule through open framework structures and, hence, provide an advantage in the removal of toxic pollutants from water [15]. Moreover, MOFs possess high stability and, hence, do not agglomerate easily. Last but not the least, response toward visible light is highly achieved by the ligands which act as organic chromophores [23, 44].

So far, abundant techniques have been utilized to treat contaminated water resources and to generate energy sources. This chapter summarizes the photocatalytic application of MOFs in the direction of removal of harmful organic pollutants, i.e., dyes, pesticides, antibiotics, volatile organic compounds, heavy metal ions, etc., from the water. Photocatalytic CO_2 reduction, H_2 generation, and some photoelectrochemical applications have also been included. The objective is to provide the up-to-date knowledge and references for photocatalytic activities of MOFs. In this respect, all the MOFs and functionalized MOFs which have been reported are summarized along with the mechanisms involved. Lastly, the limitations and future aspects have also been discussed.

2 Possible Photocatalytic Applications of MOFs

The utilization of MOFs for photocatalytic applications has gained massive attention in the past few years, owing to the rise in awareness and realization of the unique physical and chemical properties of MOFs. Some of the photocatalytic applications which we have discussed in the chapter have been illustrated in Fig. 2, and MOFs reported for various photocatalytic activities are summarized in Table 1.

2.1 Reduction of Carbon Dioxide (CO_2) on MOFs

The emission of CO_2 results in global warming causing the greenhouse effect. The CO_2 conversion method using solar energy into valuable chemicals is the most demanding nowadays to lessen the energy shortage problem and greenhouse effect [128]. The photocatalytic reduction of CO_2 produces carbon monoxide (CO), methanol (MeOH), formic acid (HCOOH), and methane (CH_4). These end products act as energy carriers and are significant for industrial practical applications [95, 128]. The shortage of energy and the greenhouse effect have become the topic of concern because of the excess depletion of fossil fuels. Because of the thermodynamic stability of carbon dioxide (CO_2), its reduction effectively via a photocatalytic system has become a major task [129]. The various photocatalytic systems have been studied till now including inorganic semiconductors and metal complexes which are active only in the UV region and show limited application due to low adsorption

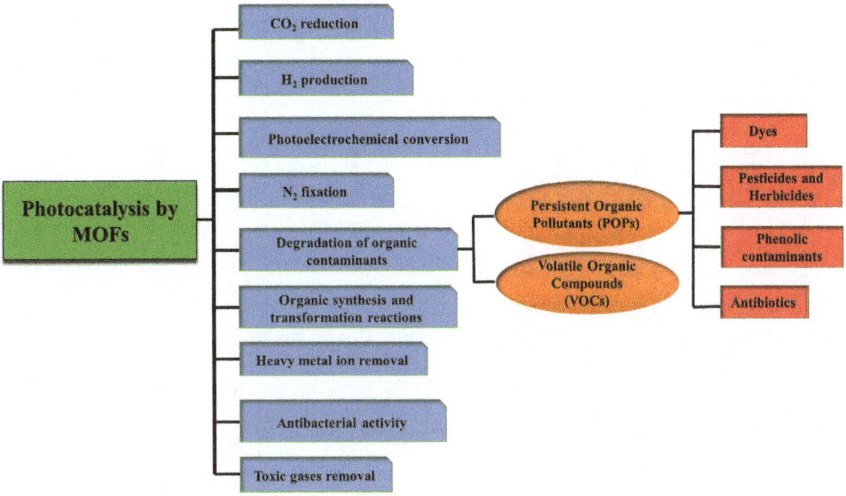

Fig. 2 Schematic representation of photocatalytic applications of MOFs

capacity, inadequate light collection, and requirement of photosensitizer and a sacrificial agent [24, 46]. There is demand for the development of photocatalyst which shows convincing adsorption capacity, the high separation efficiency of electron–hole pairs, recyclability, and outstanding photosensitivity [46, 48]. In the past few years, scientists have developed a strong interest in utilizing the MOFs as photocatalysts (adsorbent) mimicking natural photosynthesis [37] for the conversion of CO_2 (adsorbate) and is a favorable solution for the environmental problem and for producing renewable energy in high density [130, 131]. Researchers are using solar and electrical energies to activate CO_2 and its conversion into fuels with the help of MOFs as catalysts [43].

Han et al. [46] fabricated a stable gas–solid system using TPVT-MOFs crystals via a solvothermal approach and then combined these crystals with g-C_3N_4. The TPVT-MOFs@g-C_3N_4 composite thus obtained was used directly, without sacrificial agent and photosensitizer. The production rate of CO (56.4 $\mu molg^{-1}h^{-1}$) by TPVT-MOFs@g-C_3N_4-10 was found to be 3.2 times greater than pure g-C_3N_4 upon irradiation of visible light. The interaction provides the favorable condition of the valence band (VB) and conduction band (CB) for the reduction of CO_2 and oxidation of H_2O photocatalytically. Wang et al. [48] synthesized a series of ZIF-67_X ($X = 1, 2, 3$) co-catalysts nanocrystals possessing distinct morphologies via a solvent-induced approach. ZIF-67_3 having morphology leaf-like displays greater photocatalytic reduction of CO_2 into CO. Their work involves the use of MOFs by regulating their morphologies for CO_2 capture and, hence, further enhancing their photocatalytic performances. A study of CO_2 conversion photocatalytically to CO by covalently attached photocatalyst $Re^I(CO)_3$(BPYDC)Cl to UiO-67 (MOF) forming (Re_n-MOF) was performed by Choi and co-authors [47]. By controlling the density in the framework, they developed a series of compounds ($n = 0$ to 3, 5, 11, 16, and 24

Table 1 MOFs reported for various photocatalytic applications

Photocatalytic application	MOF used	Synthetic method	Shape and size	Surface area	Light irradiated	Ref.
CO_2 reduction	MIL-X(Fe) ($X = 101, 53, 88B$)	Hydrothermal	–	–	Visible	[24]
	$\{Cd_3[RuL_1]_2{\cdot}2(Me_2NH_2){\cdot}solvent\}_n$ $\{Cd[Ru\text{-}L_2]{\cdot}3(H_2O)\}_n$	Solvothermal	–	–	–	[45]
	TPVT-MOFs	Solvothermal	TPVT-MOFs deposited on the surface of g-C_3N_4 (thin nanosheet)	37.2 m^2/g	Visible	[46]
	$ReI(CO)_3(BPYDC)Cl$	–	–	–	Visible	[47]
	ZIF-67_X ($X = 1, 2, 3$)	Solvent induced	ZIF-67_1-rhombic dodecahedral (edge length-250 nm) ZIF-67_2-pitaya-like (particle size-2.5 mm) ZIF-67_3-2D leaf-like (height-180 nm)	(1698.8–16.245) m^2/g	Visible	[48]
	MOF-X (X = Ni, Co. Cu)	–	–	–	visible	[49]
	Bis(4′-(4-carboxyphenyl)-terpyridine)ruthenium(II)	–	–	50 m^2/g	Visible	[50]
H_2 production	CuNWs/ ZIF-8 CZX ($X = 1$–4 depending upon the wt% ratio of Zn and ZIF-8)	Microwave irradiation	ZIF-8 (uniform decahedral nanocrystals) attached to the CuNWs	(1044–89) m^2/g	–	[51]
	TPVT-MOFs	Solvothermal	TPVT-MOFs deposited on the surface of g-C_3N_4 (thin nanosheet)	37.2 m^2/g	Visible	[46]

(continued)

Table 1 (continued)

Photocatalytic application	MOF used	Synthetic method	Shape and size	Surface area	Light irradiated	Ref.
	Pd/UiO-66 X ($X = 1, 3, 5\%$ i.e., amount of H_2PdCl_4 added)	Solvothermal followed by impregnation	Pd/MOF-smooth UiO-66 surface with a lot of particulate matter	(843–838.9) m^2/g	Visible	[52]
	$Mo_3S_{13}2-$ /MIL-125-NH_2 1 T-MoS_2/MIL-125-NH_2	–	–	–	Visible	[25]
	Nickel phosphonate MOF	Hydrothermal	–	–	Sunlight	[53]
	MIL-167/MIL-125-NH_2 heterojunctions	Nucleation growth	MIL-167 (20–60 μm) large crystals are covered by smaller crystals of MIL-125-NH_2 (300–400 nm)	–	Visible	[54]
	1D titanium phosphonate MOF	Hydrothermal	Nanowires with an average diameter ≈80 nm	160 m^2/g	Visible	[55]
	Ni-BDC Ni/Cu-BTC	Solvothermal	Agglomerates of macro-sized loosely packed sheets Double-sided pyramidal micro-crystals	15 m^2/g 1450 m^2/g	Visible	[56]
N_2 fixation	XAu@UiO-66 ($X = 0.9, 1.9, 3.0, 4.1, 5.2$ wt%)	Impregnation and reduction	Octahedral shape with a size of 146 ± 14 nm	(1220–893) m^2/g	Visible	[57]
	UiO-66-fresh UiO-66-UV–vis	UiO-66-UV–vis- by ligand exchange process	UiO-66-fresh and UiO-66-UV–vis- octahedron shape (diameters-200 nm)	995 m^2/g 819 m^2/g	Visible	[58]
	MIL-101(Fe) MIL-101(Cr)	Solvent thermal	Octahedral morphology (particle size ~200–600 nm)	1413 m^2/g 2941 m^2/g	Xenon lamp (300 W)	[59]

(continued)

Table 1 (continued)

Photocatalytic application	MOF used	Synthetic method	Shape and size	Surface area	Light irradiated	Ref.
	NH_2-MIL-125 (Ti)	Solvothermal	Regular moon cake-shaped crystallites (diameter ~800 nm and thickness ~400 nm)	1146 m^2/g	visible	[60]
	Nano-MOF-74(Zn)@DF-C_3N_4	Solvothermal followed by ultrasonic suspension	Nano-MOF-74 is well dispersed on DF-C_3N_4 with a nano-size of <20 nm	–	Xe lamp (300 W)	[61]
Degradation of persistent organic pollutants (POPs)	MOF-5 derived ZnO and ZnO@C composites (for RhB degradation)	Heating under atmospheric conditions	Cubic structure	>500 m^2/g	UV	[62]
	UiO-66(nTi) ($n = 0.25, 0.5, 0.75, 1, 1.25, 1.5$) (for MB degradation)	Facial modified post grafting method	Spherical agglomerations	(710.4–590.4) m^2/g	Sunlight	[63]
	MIL-68-In MIL-68-In derived In_2S_3 nanorods (for MO degradation)	Sulfide under solvothermal conditions	Homogeneous hexagonal prisms hexagonal nano rod-like	321.4 m^2/g	Visible	[64]
	NH_2-MIL-125(Ti) rGO–NMTi (for MB degradation)	Solvothermal	Snowflake Sandwich-like structure	1233 m^2/g 750.3 m^2/g	Visible	[65]
	A@FeBTC (for RhB degradation)	Hydrothermal	Compact spheres dispersed over FeBTC (cubic—0.1–0.3 μm)	–	Visible	[66]
	NH_2-MIL-125(Ti) (A) CTAB/NH_2-BDC (X)(B) ($X = 0.5$–2.0) (for RhB degradation)	Solvothermal	A smooth circular plate having rough defects and pores	1298m^2/g (1258–1133) m^2/g	Visible	[67]

(continued)

Table 1 (continued)

Photocatalytic application	MOF used	Synthetic method	Shape and size	Surface area	Light irradiated	Ref.
	$\{[Co(L)]0.0.5(CH_3CN)0.5(H_2O)\}_n$ (HPU-5) $\{[Mn_2(L)_2(H_2O)_2]0.2(CH_3OH)0.3(H_2O)0.2(DMA)\}$ (HPU-6) (for MO, RhB degradation)	Solvothermal	–	–	UV	[68]
	MOF-X (*X*-1, 2) (for MO, CR, Orange II sodium salt degradation)	Solvothermal	Homogeneous surface (MOF-1 = 2.6 mm and MOF-2 = 3.4 mm)	(101.66–110.80) m^2/g	Normal	[69]
	MIL-100-X (*X* = 1, 2, 3) (for BB41 degradation)	Hydrothermal	Nanoporous	–	–	[70]
	[Mn(DMTDC)(DMF)] (for MB degradation)	Solvothermal	–	320 m^2/g	UV	[71]
	UiO-66-X (*X* = NH_2, CNT, RGO) (for AY, AO degradation)	Solvothermal	UiO-66-NH_2-similar to UiO-66 (average size ~50 nm) UiO-66-CNT- layer-like	(814.2–166.8) m^2/g	UV, Visible	[72]
	Cu-H_3-btc–Ag_2O-NPs Cu-H_2-bdc–Ag_2O-NPs (for OG degradation)	Sonochemical-assisted hydrothermally	Irregular plate-like (30–80 nm) A hierarchical spherical particle with rough surfaces (2–10 nm)	–	Blue LED	[73]
	Copper(I) 3,5-diphenyltriazolate MOF (CuTz-1) (for MO, methyl blue, MB, RhB degradation)	Hydrothermal	–	215.5 m^2/g	Xe lamp, natural sunlight	[74]
	Pillared layer NNU-36 (for RhB, MB, R6G degradation)	Solvothermal	Plate-shaped	–	Visible (H_2O_2)	[75]

(continued)

Table 1 (continued)

Photocatalytic application	MOF used	Synthetic method	Shape and size	Surface area	Light irradiated	Ref.
	MIL-53(Fe)·X Ni/Fe-MOF-Y ($X = H_2O$, DMF and $Y = 0.3.H_2O$, 0.3.DMF) (for RhB degradation)	Solvothermal	Ni^{2+}/Fe^{3+}(0.1)-a mixture of octahedral and hexagonal bipyramids Ni^{2+}/Fe^{3+}(0.3–0.7)-octahedral and hexagonal bipyramids with cracks on a crystal surface	(158–480) m^2/g	Visible	[76]
	ZIF-8 nanocrystals (for RhB degradation)	Simple mixing	Uniform nanocrystals (average diameter—88.8 ± 12.8 nm)	1698 m^2/g	Visible or UV	[77]
	$MgFe_2O_4$@MOF (for RhB, R6G degradation)	Solvothermal	Mono-dispersed (diameter—20–50 nm)	519.86 m^2/g		[78]
	MIL/CNT(X) ($X = 0.01, 0.03$) (for RB5 degradation)	Hydrothermal	MIL/CNT(X)-MOF crystal is grown on CNT	499 m^2/g		[79]
	CuO NPs deposited over ZIF-8 (for R6G degradation)	Nucleation and growth process	Ultra-small colloidal CuO NPs (~5 nm) on the surface of cubic ZIF-8 (~80 nm)	65.40 m^2/g	Sunlight	[80]
	Ag_2CrO_4-AgCl/Ag-HKUST-1 (for AB, OG degradation)	Hydrothermal	Small spherical nanoparticles (100–200 nm)	–	UV–vis	[81]
	Ag_2WO_4@MIL-125-NH_2 Ag_3VO_4@MIL-125-NH_2 (for RhB, MB degradation)	Doping	Spherical nanoparticles	185.2 m^2/g 129.5 m^2/g	UV, visible	[82]
	Ni-MOF (for CV degradation)	Microwave irradiation	1D-wire like structure	–	Simulated solar light	[83]
	$[Zn(L)(H_2O)]\cdot H_2O$ (for RhB, MO degradation)	Solvothermal	–	–	300 W Xe lamp	[84]

(continued)

Table 1 (continued)

Photocatalytic application	MOF used	Synthetic method	Shape and size	Surface area	Light irradiated	Ref.
	MIL-53-Fe MOF/magnetic magnetite/biochar composites (for RhB degradation)	Solvothermal	Octahedral rod-like (diameter ~0.5–1.1 μm and length ~07–1.5 μm)	687 m^2/g	UV	[85]
	$ZnO/Ni_{0.9}Zn_{0.1}O$ (for MB degradation)	Calcination	Nano rod-like		UV	[86]
	M-ZnO-X ($X = 450, 500, 550$) (for RhB degradation)	Calcination	Cubic structure (side length of ~25 μm)	(760–9.28) m^2/g	Solar simulated	[87]
	$\{[Cd(PA)(4,4'bpy)_2](H_2O)\}_n$ (for RhB, MO degradation)	Solvothermal	Fibrous morphology with rod-shaped crystals	342.14 m^2/g	Visible	[88]
	La-BDC MOF La-NDC MOFs (for RhB degradation)	Solvothermal	La-MOFs- rods like (diameters-1–20 μm)	442.19m^2/g 29.774m^2/g	Visible	[89]
	MOF/$CuWO_4$ heterostructure (for MB degradation)	Facile synthesis	MOF/$CuWO_4$-well-dispersed $CuWO_4$ particles on rod-like MOF	801 m^2/g	LED	[90]
	2D/2D MOF-5/LTH hybrid (for MB degradation)	Hydrothermal	2D/2D plate like with smooth surfaces	–	UV–visible	[91]
	Fe_3O_4@Sn-MOF (for AR3R degradation)	Hydrothermal	Uniform	59.949 m^2/g	Simulated light source	[26]
	$[(Cu(H_2L)(4,4'\text{-bipy})0.5(H_2O)]$ $[Co(C_{14}H_{14}O_6.5P_2)(4,4'bipy)0.5(H_2O)_2]\cdot H_2O$ (for MB degradation)	Hydrothermal	–	–	UV	[92]

(continued)

Table 1 (continued)

Photocatalytic application	MOF used	Synthetic method	Shape and size	Surface area	Light irradiated	Ref.
	La-PTC MOF (for MB, RhB, MO degradation)	Solvothermal	Rod shape	22.23 m^2/g	Visible	[93]
	Zr-porphyrin MOFs (3D) Zr-porphyrin MOFs (2D) (for CO, DR23, RB5, MB, RhB degradation)	Solvothermal	Uniform hexagonal prism (length ~300 – 500 nm and diameter ~500 – 800 nm)	1485.5 m^2/g 431.6 m^2/g	Visible	[94]
	Phosphonate-based MOFs; STA-12(M) (M = Mn, Fe, Co, Ni) (for RhB, MO degradation)	–	–	–	Visible	[95]
	[Zn(bpe)(fdc)]0.2DMF (BUT-206) (for CV, RhB degradation)	Solvothermal	–	–	UV-Hg source (80 W)	[96]
	MOFx/P–TiO_2 (for RhB degradation)	Self-assembly	MOF1.0/P–TiO_2-small cuboids of TiO_2 scattered on the surface	50.60m^2/g	Visible	[97]
	$Cu_3(BTC)_2$ (for RhB degradation)	Hydrothermal	Pyramid-shaped	32.16 m^2/g	visible light	[98]
	[Cd(bpy)(H_2O)L]n(BUC-66) $\{Co(bpy)(H_2O)L]\_H_2O\}_{2n}$ (for MO degradation)	Hydrothermal	–	–	UV	[99]
	UiO-67 (for GP, GF degradation)	Solvothermal	Irregular shape (in µm)	2172 m^2/g	Normal	[100]
	$Ag_3PO_4/BiPO_4$–MOF–graphene (for atrazine degradation)	Solvothermal	Irregular nano-cocoons (size-50 to 500 nm)	–	Visible, UV	[101]
	3-D CaFu MOF (for imidacloprid degradation	solvothermal	Uniform, Regular truncated octahedral (average particle size ~30 nm)	–	–	[102]

(continued)

Table 1 (continued)

Photocatalytic application	MOF used	Synthetic method	Shape and size	Surface area	Light irradiated	Ref.
	FeMOF-X ($X = 71, 31, 11, 13, 17$) Fe(BDC)(DMF,F) (for phenol degradation)	Solvothermal	FeMOF-71-bulky irregular The crystal size increases from FeMOF-31 to FeMOF-11 FeMOF-13, FeMOF-17 and Fe(BDC)(DMF,F)-big triangular prisms	–	–	[103]
	g-C_3N_4/PDI@MOF (for BPA, PNP degradation)	In situ growth	spindle-shaped NH_2-MIL-53(Fe) are in close contact with the surface of the g-C_3N_4/PDI	–	Visible	[104]
	MOF/$CuWO_4$ heterostructure (for 4-nitrophenol degradation)	Facile synthesis	MOF/$CuWO_4$-well-dispersed $CuWO_4$ particles on rod-like MOF	801 m^2/g	LED	[90]
	MOF [Zn(BDC)(DMF)] (for 4-nitrophenol degradation)	Solvothermal	Regular brick shaped	575.6 m^2/g	–	[105]
	UMOFNs/Ag_3PO_4 (for 2-chlorophenol degradation)	Sonication	2D UMOFNs wrap the surface of tetrahedral Ag_3PO_4	1.60 m^2/g	Visible	[106]
	Fe-MIL-101 ($X = 101, 100, 53$) (for tetracycline degradation)	Hydrothermal	Fe-MIL-101-regular octahedron (500 nm) Fe-MIL-100-granules (500 nm–1 μm) Fe-MIL-53-bundle (diameter-80 μm)	(252.5–21.42) m^2/g	Visible	[107]

(continued)

Table 1 (continued)

Photocatalytic application	MOF used	Synthetic method	Shape and size	Surface area	Light irradiated	Ref.
	MIL-100(Fe)/PANI composite (for tetracycline degradation)	Ball milling	PANI (fibrous structure) was uniformly coated on MIL-100(Fe) (regular polyhedrons or octahedrons)	–	White	[108]
	BiOBr/UiO-66 (for tetracycline degradation)	Hydrothermal	Octahedral (average grain size ~100 nm)	–	Visible	[109]
	MIL-68-In MIL-68-In derived In_2S_3 nanorods (for tetracycline degradation)	Sulfide under solvothermal conditions	Homogeneous hexagonal prisms hexagonal nano rod-like	321.4 m^2/g	Visible	[64]
	MOF-derived core/shell C-TiO_2/$CoTiO_3$ type II heterojunction (for CIP degradation)	Direct calcination	C-TiO_2/$CoTiO_3$-nanocake shape	–	Visible	[110]
	AgI/$Bi_{12}O_{17}Cl_2$ heterojunction (for sulfamethazine degradation)	hydrothermal precipitation	AgI (diameters—50 to 200 nm) is deposited on the surface of the $Bi_{12}O_{17}Cl_2$nanosheets (width and length ranges of 50–100 nm and 0.2–1.0 μm)	–	Visible	[111]
	g-C_3N_4/PDI@MOF (for tetracycline, carbamazepine degradation)	In situ growth	Spindle-shaped NH_2-MIL-53(Fe) in close contact with g-C_3N_4/PDI	–	Visible	[104]
Degradation of volatile organic compounds (VOCs)	Ni-MOF/NF (for toluene removal)	Solvothermal	Ni-MOF spherical (size in nm to μm) composed of agglomerated nanosheets and uniformly growth on Ni foam	22.07 m^2/g	Visible	[112]

(continued)

Table 1 (continued)

Photocatalytic application	MOF used	Synthetic method	Shape and size	Surface area	Light irradiated	Ref.
	FeMOF-0.5 ($X = 0.5, 0.25, 0$) (for toluene, benzene, p-xylene removal)	Solvothermal	FeMOF-0.5-spindle (7.5 μm long and 3 μm wide) FeMOF-0.5-spindle rods (14 μm long and 2 μm wide) FeMOF-0.5-short spindle rods (13 μm long and 3 μm wide)	(1800–1660) m^2/g	Simulated solar light	[113]
Photocatalytic organic synthesis and transformations	X-MIL-125(Ti) ($X = NH_2$, MR) (for oxidation of benzyl alcohol to benzaldehyde)	Solvothermal followed by functionalization	–	(1500–1250) m^2/g	Visible	[114]
	Ti-MOF@Pt@DM-LZU1 (for hydrogenation of olefins)	Hydrothermal	Octahedron	144 m^2/g	Visible	[115]
	MIL-100(Fe)/α-Fe_2O_3-X ($X = 15, 30$) (for oxidation of O-xylene)	Hydrothermal	Small (lateral dimension ~5 nm)	763 m^2/g 1138 m^2/g	Visible	[116]
	MIL-125-NH_2 (for oxidation of benzyl alcohol to benzaldehyde)	Hydrothermal	–	–	Visible	[117]
	Bi-TATB Bi-BTC (for oxidation of benzyl alcohol to benzaldehyde)	Solvothermal	Thick sheet morphology irregular bulk morphology	151.73 m^2/g 7.27 m^2/g	UV–vis	[118]
	Ce-UiO66-X (X = H, Br, NO_2, 4F, PDC) Ce-MIL140A-4F (for oxidation of substituted benzylic alcohols to benzaldehydes)	Solvent free	Irregular octahedral shape (size ~30–150 nm)	(827–547 m^2/g) 279 m^2/g	Near-ultraviolet	[119]

(continued)

Table 1 (continued)

Photocatalytic application	MOF used	Synthetic method	Shape and size	Surface area	Light irradiated	Ref.
Removal of toxic metal ions (Cr(IV) and Cr(VI)	MIL-125(Ti) NH_2-MIL-125(Ti)	Solvothermal	Well-crystallized analogous microcosmic (400–600 nm)	1343.9 m^2/g	Visible	[120]
	g-C_3N_4/MIL-53(Fe) CMFe-X ($X = 1, 3, 5, 7$)	Solvothermal	g-C_3N_4 coated on MIL-53(Fe) (polyhedron microrods ~7–20 μm)	(19.4–13.8) m^2/g	Visible	[121]
	BUC-21/titanate nanotube (BT-X)	Ball milling of 2D MOF, BUC-21, and titanate nanotubes (TNTs)	BT-1-nanotubular structure of TNTs (outer diameter—9 nm and tube length—50 to 300 nm) distributed on the surface of BUC-21 (irregular block)	7.85 m^2/g	UV	[122]
	UiO-66-NH_2(Zr/Hf) metal–organic framework	Seeding method on the α-Al_2O_3 substrate	Octahedrons (300–500 nm)	–	Simulated sunlight	[123]
	Cd_TIPA	Solvothermal	1D hexagonal nanotube-like channels	8.158 m^2/g	Visible	[124]
	$\{[Zn(PA_2-)(4,4'\text{-bpy})]\text{-}(H_2O)\}_n$	Solvothermal	Interconnected flakes	–	Sunlight	[125]
	Zr-based MOFs	hydrothermal	Uio-66-$(OH)_2$ uniform particles (size of 70–120 nm)	561.51 m^2/g	Visible	[126]
	Bimetallic Mn–Fe MOFs	–	–	–	Visible	[127]
	Pillared layer NNU-36	Solvothermal	Plate shaped	–	Visible (H_2O_2)	[75]
	$[Cd(bpy)(H_2O)L]n$(BUC-66) $\{Co(bpy)(H_2O)L]\_H_2O\}_{2n}$	Hydrothermal	–	–	UV	[99]
	$[(Cu(H_2L)(4,4'\text{-bipy})0.5(H_2O)]$ $[Co(C1_4H_{14}O_6.5P_2)(4,4'\text{bipy})0.5(H_2O)_2]\cdot H_2O$	Hydrothermal	–	–	UV	[92]
	$\{[Cd(PA)(4,4'\text{bpy})_2](H_2O)\}_n$	Solvothermal	Fibrous with rod-shaped crystals	342.14 m^2/g	Visible	[88]

(continued)

Table 1 (continued)

Photocatalytic application	MOF used	Synthetic method	Shape and size	Surface area	Light irradiated	Ref.
	MOFx/P–TiO_2	Self-assembly	MOF1.0/P–TiO_2 small cuboids of TiO_2 scattered on the surface	$50.60 m^2/g$	Visible	[97]
	MIL-100(Fe)/PANI composite	ball-milling	PANI (fibrous structure) was uniformly coated on MIL-100(Fe) (regular polyhedrons or octahedrons)	–	White	[108]
	Phosphonate-based MOFs; STA-12(M) (M = Mn, Fe, Co, Ni)	–	–	–	Visible	[95]

complexes present per unit cell), and Re_3-MOF shows the highest activity. Further, Ag $\subset$ Re_3-MOF formed by coating of Re_3-MOF over Ag nanocubes enhances the photocatalytic conversion upon visible light irradiation by seven folds having the stability of 48 h.

2.2 *Hydrogen Production via Water Splitting on MOFs*

Sunlight-driven hydrogen production from water splitting acts as a favorable solution for environmental pollution and energy crises [51] and has gained a lot of attention in recent years. But it is limited as it requires highly stable and active photocatalysts at a low cost. Hydrogen (H_2) is an ideal future fuel and a potential source for the generation of power due to an energy content of 140 $MJkg^{-1}$. It is recognized as an attractive source of energy from both economic and environmental standpoints. Hydrogen is a valuable fuel for purposes like chemical reactions, fuel cell vehicles, and petroleum refining [132, 133]. The products of water splitting (H_2 and O_2) are pollution-free combustion products, and the hydrogen acts as a fuel leaving behind water vapors, a by-product, which can be recycled naturally or industrially. Moreover, it also acts as an oxygen scavenger and can remove oxygen traces and hence reduce oxidative corrosion. Several countries are shifting to hydrogen-based economies from hydrocarbons in the past few decades [133]. The catalytic efficacy of the evolution of hydrogen is dependent upon the electron generation and transfer, stability of the catalyst, reactant adsorption, etc. [37]. Several photocatalysts like TiO_2, NiO, CdS, NiS, CuS, and C_3N_4 were advanced for H_2 production. Research work done using metal nanoparticles till now shows poor catalytic activity in H_2 production despite having a large surface area because of the lacking of H^+ and electron transfer channels between metal and support [51]. Other challenges in H_2 production include insufficient separation efficiency of charge carriers and susceptibility to oxidation, many photocatalysts are made from rare and expensive components, and also doping is required for the activity enhancement of some of the catalysts [55]. Metal–organic frameworks have exceptional features which attracted remarkable attention as catalysts for photocatalytic H_2 production, by water splitting, as an energy carrier for future sustainable energy. The MOFs as heterogeneous catalysts are very exciting and encouraging to provide better performance [37, 134]. Due to large surface area, more active sites are available for better contact with the incoming molecule, and its porous structure provides a path for migration of electrons and also facilitates separation of the charge carrier under normal conditions [52, 135]. For the efficient production of hydrogen, coupling of MOFs with co-catalyst (<10 wt%) like platinum NPS is done which attracts the photogenerated electrons and acts as an electron donor to the holes to protect the degradation of MOF [76].

Zhang et al. [51] synthesized CuNW/ZIF-8 using the microwave antenna method. The coordination of Zn^{2+} with 2-methylimidazole is induced by high temperature into ZIF-8, followed by the formation of assembly with copper nanowires (CuNWs). It reveals high activity and stability in hydrogen production out of hydrolysis of

NH_3BH_3 due to the cooperative effect between CuNWs and ZIF-8 for the adsorption of reactant in large capacity, dissociation and activation, transfer of an electron and also provides the way for the rapid transfer of H^-/H^+ ions to copper active sites. Some other 1D threaded MOFs were also prepared to show the uniqueness of the mentioned strategy. Kampouri et al. [54] constructed the MIL-167/MIL-125-NH_2 MOF/MOF heterojunction which shows improved optoelectronic properties, photocatalytic activity (H_2 production), and effective charge separation. The heterojunction formed showed better performance (455 μmol $h^{-1}g^{-1}$) as compared to its individual components MIL-167 (0.8 μmol $h^{-1}g^{-1}$) and MIL-125-NH_2 (51.2 μmol $h^{-1}g^{-1}$), upon irradiation with visible light, without any co-catalyst. Morshedy et al. [56] fabricated nickel–benzene dicarboxylic acid (Ni-BDC) and nickel/copper–benzene tri-carboxylic acid (Ni/Cu-BTC) via solvothermal method having mono and binary central metal ion, respectively. Ni-BDC shows sheet-like nonporous and possesses 15 m^2/g surface area while Ni-Cu-BTC is porous with bipyramidal morphology having a surface area of 1450 m^2/g. The Ni-MOF shows hydrogen purity of 50% and productivity of 200 mmol/h while Ni/Cu-MOF showed 100% hydrogen storage because of two metal cations.

2.3 Photoelectric Conversion/ Photoelectrochemical Applications

The need for sustainable energy sources, mainly solar energy, is constantly rising for the past few decades. Even though energy provided by sun's radiation is roughly 10,000 times more than the total consumption of energy in a year, still the efficiency of the photovoltaic devices is not adequate [17]. The most efficient technology to date for solar energy conversion is silicon-based devices [17]. Ahead of host–guest chemistry, recently MOFs have been explored for the application of energy harvesting and energy-storing devices [136] and are gaining tremendous attention due to their porous nature and huge surface area [137]. The great diversity and choice in the selectivity of organic linkers can combine to form the required MOF with suitable properties. However, a solar cell entirely builds on MOFs for collecting solar energy and is rarely inspected. Due to boundless flexibility in the metal ions and ligands coordination, an endless amount of final products is possible [138]. This enhances the capturing of solar irradiation by organic ligand (also called an antenna) which in turn affects photocatalytic activity and efficiency of power conversion of MOF-dependent devices. Many MOFs show insulator properties, and only a few of them have some special ligands in their structure, which are responsible for the conducting properties [139]. Therefore, MOFs with better electronic properties for photovoltaic applications are still very challenging. The conductivity of MOFs is significantly enhanced via molecular doping of iodine which overcomes resistance in charge transfer in the MOF-dependent photovoltaic devices [137]. The thin films of MOFs have also been prepared to collect radiations of solar energy in TiO_2-based solar cells. This solar cell has I_{sc} of 1.25 mA cm^2, V_{oc} of 0.49 V, FF of 0.43, and power conversion efficiency of 0.26%. When supported with carbon nanotubes, the power conversion efficiency was increased to 0.46% [140]. Although the figure is less than the already available devices, there is always a scope for further improvement in their performances [140].

Lee et al. [140] had constructed thin films of ruthenium-based MOF via the layer-by-layer method (LbL) and explored it as a sensitizer in the solar cell. The iodine doping is necessary for the transference of electrons to TiO_2 from the MOF framework. The HOMO–LUMO energy gap of the frameworks was suitable for its use as a sensitizer for TiO_2. A type of solar cell having a liquid junction is constructed when the LBL thin film of the framework is combined with TiO_2 mesoporous film. This solar cell has I_{sc} of 2.56 mA cm^{-2}, V_{oc} of 0.63 V, FF of 0.63, and E_{ff} of 1.22%. Li et al. [137] synthesized a thin layer of copper(II) benzene-1,3,5-tricarboxylate MOF by LbL technique and explored it as a layer that absorbs light in TiO_2-dependent solar cells. The HOMO and LUMO levels of MOF films were found to have an energy of −5.37 and −3.82 eV, respectively, with respect to vacuum. When this solar cell is doped with iodine, it has I_{sc} of 1.25 mAcm^{-2} and E_{ff} of 0.26% under sunlight irradiation of 1 sun (100 mW cm^{-2}). While non-iodine-doped solar cell shows conversion efficiency of 0.008%, the iodine-doped MOFs act as an active layer and iodine doping enhances the conductivity and considerably reduces the charge transfer resistance across TiO_2/MOF/electrolyte interface. Zhan et al. [141] suggested a self-template method to synthesize ZnO@ZIF-8 heterostructures. Here, ZnO is not only acting as a template but helps in the formation of ZIF-8 by providing Zn^{2+} ions too. The reaction temperature and solvent composition are important for the fabrication of ZnO@ZIF-8 heterostructures. These nanorods arrays present selective photoelectrochemical responses toward hole scavengers like H_2O_2 and ascorbic acid due to the presence of ZIF-8. These nanorods were applied for the detection of H_2O_2 and hence act as a promising material in electronic devices like sensors having molecule size selectivity.

2.4 *Reduction of N_2 (Nitrogen Fixation) with MOFs*

As an essential source of energy, ammonia (NH_3) is essential for life and to industrial processes as well. NH_3 not only acts as a precursor for the synthesis of various chemicals but also acts as an important medium for energy storage [57]. It is also known as the other hydrogen in the fuel world. Although nitrogen consists of 78% of the earth's atmosphere, its fixation is kinetically complex and requires high activation barrier. Due to negative electron affinity, it cannot be used directly for the synthesis of other nitrogenous compounds and hence contributes a lot to environmental pollution [142]. N_2 molecules are highly inert due to the presence of triple bond in N_2, which shows poor proton affinity and hence hampering the transport of electrons [143]. The goal of fixing N_2 artificially is to advance sustainable catalytic processes to meet the increasing demands for the protection of the environment and energy efficiency.

In the Haber–Bosch process, high pressure and temperature are required to reduce N_2 and also lead to the emission of 2.3 tons of greenhouse gases. To deal with such problems, nitrogen fixation photocatalytically has gained a lot of attention recently. When nitrogen fixation was achieved in stark under nitrogen atmosphere at ambient pressure in contradiction to the conventional Haber–Bosch process, this diminishes the need for high energy and hence is a process that is worth studying [59]. The quest for NH_3 production leads to the progress of Photocatalytic Nitrogen Reduction Reaction (PNRR) at atmospheric pressure and room temperature using solar energy in an eco-friendly manner [143]. The efficiency of PNRR depends upon the well-organized coupling between electrons, photons, and dinitrogen which is affected due to the requirement of high kinetic energy barriers for the activation of N_2 molecules [143]. To date, abundant photocatalysts are reported for PNRR such as BiOBr, graphite carbon, TiO_2, metal oxides, and metal sulfides[58]. MOFs are studied widely in recent years, and they upon the combination with catalytic materials form new materials with enhanced activity. MOFs adsorb water-soluble nitrogen and lessen the distance between the catalytic site and nitrogen which weakens the triple bond in N_2 [142].

Chen et al. [143] fabricated gas-permeable Au@MOF membranes, consisting of interconnected nanoreactors, the MOF acts as an ideal assembly for the dispersion of gold nanoparticles (AuNPs). Upon irradiation of the sample with visible light, the electrons were generated on the gold nanoparticles which were then directly combined with the N_2 molecules already adsorbed over gold surfaces. The Au@MOF membrane eases the transfer of mass of N_2 molecules and protons, hence raise the reaction at the gas/membrane/solution interface. The N_2 produced was allowed to react with H_2 to produce ammonia with an evolution rate of 18.9 mmol $g_{Au}{}^{-1}h^{-1}$ having a quantum efficiency of 1.54%. Gao and co-authors [58] utilized UiO-66 with linker and cluster defects (photo-excited) for the reduction of nitrogen. The production rate of NH_4^+ was 196 $\mu molg^{-1}h^{-1}$ upon UV irradiation and 68 $\mu molg^{-1}h^{-1}$ upon irradiation with visible light in the air without any sacrificial agent. The linker defect is more responsible for the enhancement of the performance than cluster defects. Shang et al. [57] synthesized a porphyrin-based MOF (PMOF) having Fe as an active metal center for N_2 photocatalytic reduction reaction (NPRR) under normal conditions. Porphyrin acts as a photosensitizer, and the Fe atom provides a favorable binding site through π-back bonding for N_2. The PMOF incorporated with aluminum (Al) to provide high stability and already present Fe (and thus named as Al-PMOF(Fe)) is dispersed and resides at porphyrin ring promotes adsorption as well as activation of N_2. Al-PMOF(Fe) shows an 82% enhancement in the yield of NH_3 (635 μgg^{-1}) and about a 52% increase in the rate of NH_3 production (127 $\mu gh^{-1}g^{-1}$) as compared to PMOF only. The catalyst was reusable for up to three cycles. This porphyrin-based MOF study for the PNRR will create a way for the designing of catalysts for artificial photosynthesis.

2.5 *Photocatalytic Degradation of Organic Contaminants*

Water contamination is a major hazard in the whole world with increasing industrialization. Statics calculations predicted that over 1000 million people will suffer from water scarcity in 2021 [19, 144]. The rising contamination in the environment has put the environmental moieties in danger such as human and animals health, freshwater bodies, and the atmosphere to a much greater extent worldwide. Out of several types of pollutants in water, POPs are raising concern worldwide as these are highly toxic, semi-volatile, and bio-accumulative and exhibit long-range transport via environmental sources [145, 146]. POPs are carcinogenic, damage the central nervous system, and also affect the ecological balance [19]. These are further categorized mainly as (a) dyes, (b) pesticides and herbicides, (c) phenols, and (d) antibiotics. Even at a very low concentration (ng/L-μg/L), these contaminants are proven to be fatal to human health as well as aquatic life. Due to the similarities in the structure of POPs, their removal or degradation by adsorption or photocatalysis is an effective fashion [19]. POPs commonly have aromatic rings or azo fragments and hence may be adsorbed via π-complexation, electron donator–acceptor interactions, π-π interaction, electrostatic interaction, polar–polar interaction, hydrogen bonding, etc. [19]. Several conventional methods were used to treat wastewater, but these were not effective and hence lead to persistent scum in water [147]. Scientists are continuously using several methods to eliminate these harmful contaminants. Metal–organic frameworks (MOFs) due to tailorable features and various active sites can photocatalytically remove these harmful agents from water [146]. MOFs act as an adsorbent owing to their porosity, high surface area, and tunability which gives access via adsorption sites to entrap the harmful contaminants.

Persistent Organic Pollutants (POPs)

Dyes

Synthetic dyes are widely used in areas like the paper industry, textile industry, agriculture industry, and food industry. Eleven million tons of dyes containing wastewater have been discharged every year [94]. The removal of dyes from water is important as the quality of water is intensely affected by the color, and even minute concentration is visible and causes high toxicity and danger to water organisms [19]. In the phenomenon of photocatalysis, highly reactive radical species (e.g., ·OH and $\cdot O_2$) are generated which helps in the conversion of organic dyes into harmless products, i.e., CO_2 and H_2O. Hence, it acts as a promising method, highly efficient and economical as well [96]. Zhao et al. [94] constructed a self-cleaning membrane of hydrophilic Zr-porphyrin MOFs of two and three dimensions embedded in the polyamide active layer having better separation performance. Out of the two, 3D Zr-porphyrin MOF shows outstanding activity upon visible light irradiation. The uniform 1D acts as a membrane for dye degradation on the membrane surface and also acts as an additional transport channel. The optimum membrane displays enhanced water permeate flux (110.4 $Lm^{-2}h^{-1}$) which is three times better than a pristine TFC membrane. It shows the dye removal efficiency close to 100% for direct red 23, Congo red, reactive black 5, rhodamine B, and methylene blue. Zeng et al. developed a phosphate-based MOF/P–TiO_2 composite (MOF_x/P–TiO_2) via step-by-step self-assembly. The loading of Cd-MOF in the MOF_x/P–TiO_2 influences the degradation (97.6% in 25 min) of pollutants, rhodamine B (RhB). The activity of MOF_x/P–TiO_2 is due to the enhanced light absorption capacity and effectual charge separation. The band structure of the catalyst is responsible for inducing a superfast interfacial charge transfer path. The catalysts formed possess good stability and are reusable even after five cycles [97]. Emam et al. [82] doped silver vanadate (Ag_3VO_4) and silver tungstate (Ag_2WO_4) nanoparticles in Ti-containing MOF (MIL-125-NH_2) to enhance the photocatalytic degradation of methylene blue (MB) and rhodamine B (RhB) dyes under UV and visible light. The dye degradation rate increased by a factor of 1.6–2.1 and 2.1–4.9 when doped with Ag_2WO_4 and Ag_3VO_4, respectively. Due to the low bandgap, Ag_3VO_4@MIL-125-NH_2 exhibits high activity and was reusable for up to five regeneration cycles. A new MOF [Zn(bpe)(fdc)]·2DMF (BUT-206) is reported by Alamgir et al. [96] and utilized for the degradation of crystal violet (CV) and rhodamine B photocatalytically. BUT-206 shows a degradation efficiency of 92.5% under UV light irradiation within 120 min without photosensitizer or co-catalyst. The dye degradation follows pseudo-first-order kinetics. BUT-206 shows good stability and was regenerated up to five cycles.

Pesticides and Herbicides

Pesticides and herbicides are toxic, persistent in water, cause extreme damage to the ecosystem, and threaten the health of humans and aquatic organisms. These are used to control pests and to promote plant growth [19]. These cannot be eliminated effectively from an aqueous system. Imidacloprid is commonly used for bugs, pests, and termite control. Upon exposure, this affects thyroid, liver, and body weight in addition to hematological and cardiovascular effects [102]. Organophosphorus pesticides (OPs) are highly toxic and are widely used for agriculture production. It binds with acetylcholinesterase (AChE), resulting in paralysis and organ failure [100]. Herbicides such as S-triazines are also widely used in agriculture areas. Atrazine is used for controlling grass and annual broad weeds [101]. Hence, the removal of toxic materials is one of arduous challenges [101]. Singh et al. [102] tested the adsorption of imidacloprid from the aqueous ecosystem by developing a single crystal of calcium fumarate. The sorption was affected by solution pH, and imidacloprid showed the greatest adsorption at pH 6.5. The removal efficiency of imidacloprid was 98.3% in 70 min, and the adsorption capacity was 467.23 mgg^{-1} by CaFu MOFs. The equilibrium adsorption data fit very well in Langmuir model. Zhu et al. [100] showed the removal of glyphosate (GP) and glufosinate (GF) using Zr-based MOFs of UiO-67. Due to adequate pore size and strong affinity, the capacity of adsorption approaches 3.18 $mmolg^{-1}$ and 1.98 $mmolg^{-1}$ for GP and GF, respectively. The reaction kinetics is described by pseudo-second order, and adsorption data fit well in the Langmuir model than in the Freundlich model. For the very first time, Mohaghegh et al. [101] reported a novel $Ag_3PO_4/BiPO_4$ (AB)–graphene-based nanocomposite and evaluated its photodegradation ability for atrazine under visible and UV light irradiation. The degradation efficiency is better for graphene (GR)-based nanocomposites than for pure AB. GR nanocomposite displays effective separation of electron–hole pair's accounts for the improved efficiency.

Phenolic Contaminants

Phenols are hazardous and carcinogenic and can cause severe impact even at a very low concentration. Phenolic contaminants are mainly obtained from petrochemicals, plastics, synthetic fibers, resins, oil refineries, and coking industries [19]. Kusutaki et al. fabricated highly efficient tetrahedral core–shell ultrathin MOF nanosheet (UMOFNs)/Ag_3PO_4 composite. The photocatalytic activity for degradation of 2-chlorophenol (2-CP) was found to be higher for UMOFNs/Ag_3PO_4 composite than Ag_3PO_4 upon irradiation with visible light. The photocatalyst shows complete degradation within 7 min, and the rate of reaction was higher than pure Ag_3PO_4. The Ag nanoparticles formed lead to the efficient separation of electron–hole pairs [106]. Li et al. [104] developed g-C_3N_4/PDI@MOF by in situ growth of NH_2-MIL-53(Fe) onto g-C_3N_4/PDI layer. It was utilized for the removal of phenolic pollutants with H_2O_2 by irradiating visible LED light with an efficiency of almost 100% within 10 min and 30 min for bisphenol A (BPA) and p-nitrophenol (PNP), respectively. The photocatalytic activity is accredited to the interface contact and electronic structure match between NH_2-MIL-53(Fe) and g-C_3N_4/PDI. It is favorable for charge separation and expedites the photodegradation process. The catalysts possess good stability and are reusable. Sun et al. [103] synthesized a series of materials, i.e., MIL-53(Fe)-X (X = ratios of n($FeCl_3$)/n($FeCl_2$)) to Fe(BDC)(DMF,F). The crystal morphology changes to a big triangular prism from a small irregular shape. These catalysts were utilized for phenol degradation with H_2O_2 at neutral pH under mild conditions where Fe(BDC)(DMF, F) shows the best photocatalytic efficiency.

Antibiotics

Antibiotics are extensively used to support the health of the human and endorse the development of fauna and flora [148]. The antibiotic drugs mainly come from animal feed, pharmaceutical plants and hospitals, human excretion, and aquaculture. Approximately, 200,000 tons of antibiotics are consumed every year [149]. Antibiotics even at trace concentration lead to antibiotic resistance and harm the ecosystem, human health, and aquatic wildlife [19]. Tetracycline is the extensively (second largest) used antibiotic in agriculture, bacterial infection, and livestock products as well. It accounts for almost one-third of the antibiotics consumption and production [150] and is discovered in river water, surface water, and underground water [109]. Due to the widespread use of tetracyclines, it is continuously released in the environment at a considerable level. This accumulation of antibiotics has resulted in the growth of antibiotic-resistant pathogens, hence causing danger to the health of humans, harm the ecological balance, and poor metabolic transformation in vivo [107, 109, 151]. The antibiotics upon accumulation in the body induce mutations, affect kidneys, and have teratogenic and mutagenic effects [152]. The tetracycline existence results in affecting the activity and composition of microorganisms, and hence, thereby affecting the ecological balance of the microbial flora. It causes toxicity to the algae species and crustaceans as well [107]. Due to their stable structures and recalcitrance to degradation, antibiotics cannot be removed effectively. Therefore, it is necessary to find a method for the effective removal of antibiotics to treat wastewater. Wang and co-authors [107] facilitated Fe-based MOFs, i.e., Fe-MIL-X ($X = 101$, 100, and 53) for the removal of harmful antibiotics like tetracycline. Fe-MIL-101 showed a removal efficiency of 96.6% when the initial concentration of tetracycline taken was 50 mg/L. Fe-MIL-100 and Fe-MIL-53 showed removal efficiency of 57.4% and 40.6% under the same circumstances. The effect of concentration, adsorption time, and tetracycline concentration upon degradation was inspected. Fe-MIL-101 showed great stability and reusability even after three photodegradation cycles. The active species involved in photocatalytic degradation as proved by trapping experiments and ESR studies were ·O, ·OH, and H^+. The invaginated hexagonal In_2S_3 nanorods structure was synthesized by sulfidation of In-MOF (MIL-68-In) by Fang et al. in 2018 [64]. The In_2S_3–8 h displays extraordinary photocatalytic performance for tetracycline degradation upon visible light irradiation. The crystallinity and optical properties promote an effectual separation of electron–hole pairs. A study by Lin et al. [110] showed the development of a C-TiO_2/$CoTiO_3$ type II core–shell heterojunction synthesized by direct calcination using MIL-125/Co as a precursor and sacrificial template. The formed composite can gather visible light and conquer the recombination of electron–hole pairs, resulting in noteworthy photocatalytic degradation activity for ciprofloxacin with an efficiency of 99.6% upon visible light irradiation for 2 h.

Volatile Organic Compounds (VOCs)

The volatile organic compounds (VOCs) like aldehydes, ketones, chlorinated hydrocarbons, and aromatics (ethylbenzene, benzene, toluene, and xylenes) [113] because of their toxicity and their tendency to persist in the environment have created a bad effect on human health as well as the ecology of the environment. The most commonly used VOC group, i.e., aromatics, which accounts for ~20–50% [116] is produced by industries, solvent use, petrochemical processing, paints, and vehicle exhaust, and so on and has created concern worldwide [153]. Even in small concentrations, they can cause cancer, teratogenesis, headaches, etc. They speed up the photochemical smog and formation of secondary organic aerosols [112]. For the purification of the environment, removal of these aromatics and other VOCs is significant, and therefore, several methods have been explored for the same. Two of the commonly used mechanism to eliminate VOCs based on their functional groups include (a) absorptive removal of VOCs by materials and (b) decomposition of VOCs into non-harmful products like CO_2 and H_2O using catalytic materials [154]. The efficiency of adsorption is dependent on the surface area, tunability, and porosity of the material. Several adsorbents which are reported so far include zeolites, activated carbon, silica, etc. The problem with adsorption is that when the adsorbent becomes saturated, the VOCs molecules are released into the external environment with the changing temperature and pressure conditions [103]. Also, adsorption generates secondary products which are highly hazardous. Hence, the decomposition of VOCs into harmless products using catalytic material is more promising [113]. Photocatalytic oxidation (PCO) has the potential to convert hazardous VOCs into harmless end products under mild conditions upon solar or visible light illumination [155]. TiO_2-based photocatalysts for the photocatalytic oxidation of VOCs came into notice due to their low cost, easy accessibility, and high stability [156]. Regardless of several advantages, conventional semiconductors catalysts have major drawbacks including insufficient PCO activity, poor affinity, and selectivity, less stability because of structural characteristics, fast recombination of electron–hole pairs, low surface area [157], etc., which significantly hinder their application under solar or visible light irradiation [112]. Hence, for effectual elimination of VOCs, substitutes or auxiliary materials having good affinity and extraordinary PCO activity are desirable [112, 116].

MOFs offer opportunities for efficient photocatalytic degradation of VOCs. Owing to the remarkable high surface area, ultrahigh porosity-ordered pore structure, MOFs are a better candidate for the trapping of gaseous molecules such as O_3, CO_2, and VOCs over traditional catalysts. MOFs possess marvellous LMCT property and hence act as an auspicious favorable candidate for the PCO of organic contaminants [116]. Li et al. [113] had reported some isostructural iron-based MOF (Fe-MOF) for the efficient adsorption and degradation of VOCs under normal conditions. The Fe-MOF shows a removal efficiency of almost 100% for toluene at a starting concentration of 460 ppm. The Fe-MOF is reusable for up to 100 cycles, and toluene gets converted to harmless products like CO_2 under 100 mW/cm^2 solar irradiation. This group of Fe-MOFs acts as an adsorbent as well as a photocatalyst for toluene trusting the high porosity and catalytic activity. Ding et al. [112] prepared Ni-MOF nanosheets (2D) vertically grown on foam (nickel) to form Ni-MOF/NF via the solvothermal method. Ni foam is acting as a skeleton to provide vertical support to Ni-MOF and as a sacrificial template to offer Ni ions for the proper growth of MOF. The Ni-MOF/NF offers an electron transport channel and hence benefitting the e^- transport channels. The mineralization efficiency of 86.6% could be achieved at 98.1% of ethyl acetate removal which cannot be achieved by 3D bulk Ni-MOF counterparts. Chen et al. [116] fabricated MIL-100(Fe)/α-Fe_2O_3 catalysts via a hydrothermal method which proves to be an effective composite for o-xylene adsorption. The MIL-100(Fe)/α-Fe_2O_3 shows 100% and 90% removal efficiency of o-xylene upon irradiation with 250 W Xe lamp and visible light, respectively.

2.6 Photocatalytic Application in Organic Synthesis and Transformation Reactions

The organic synthetic and transformation reaction by photocatalysis is an efficient technique for the synthesis of various molecules like aldehydes, ketones, alcohols, etc. The traditionally used methods include the use of hazardous oxidizing reagents like potassium dichromate and potassium permanganate, also producing harmful side products. Therefore, the utilization of photocatalytic systems acts as an alternative and is proven to be a more sustainable and greener approach [158]. Recently, photocatalytic MOFs have emerged as a powerful tool for the synthesis of complex molecules. Visible light photocatalyst shows fascinating applications due to their easy synthesis and environmental friendliness [159]. Zhu et al. [159] reported the synthesis of a composite of MOF and CdS nanoparticle (CdS/MOF) by a solvothermal method which shows the selectivity in oxidation of benzyl alcohol to benzaldehyde. The MOF acts as support while cadmium acetate acts as a precursor for the CdS nanoparticles. The CdS/MOF shows appreciable enhancement in the photocatalytic efficiency in comparison with pure CdS at room temperature. This may be ascribed to the increased light absorption and enhanced surface area. A study reported by Long and co-authors [160] utilized amine-functionalized zirconium MOF (UiO-66-NH_2) for the oxygenation of several organic compounds like alkanes, alcohols, and olefins and showed high selectivity and efficiency of the photocatalyst. The cavities of UiO-66-NH_2 (in nm) act as a magnificent microphotoreactor. It is believed to follow the mechanism commenced by the photogenerated electron transfer during the organic transformations. Sun and Kim [115] fabricated Ti-MOFs@Pt@DM-LZU1 having manageable surface wettability. The Pt nanoparticles were encapsulated in the interfacial pores constructed between Ti-MOF@Pt and DM-LZU1. The Pt eases the separation of charge over photoactive Ti-MOF and DM-LZU1 shell which is hydrophobic and encourages reactant enrichment. The styrene was fully converted into ethylbenzene (selectivity > 99%) within 40 min under visible light irradiation with time of flight (TOF) value of 577 h^{-1} using Ti-MOF@Pt@DM-LZU1 while no activity was observed without catalyst or under nitrogen atmosphere revealed the catalyzation of reaction by H_2 as a proton source. In the dark, with catalyst, only 41% conversion was observed. Interfacial pores facilitate mass and electron transport between reactant and Pt NPs. The catalyst was regenerated up to three cycles without any loss of activity. The Ti-MOF upon excitation with visible light irradiation can transfer electrons to Pt NPs, resulted in an enhanced density of electrons and hence increased hydrogenation performance. Wei et al. [161] developed 3D In-MOF with absorption in the visible light region and further employed this for the oxidation of benzylamines. The prepared MOF was proved to be highly selective for the transformation of a substrate to imines upon irradiation with visible light. Hydrogen peroxide enhances the rate of the reaction more effectively than those under heating. The study exhibits the high potential of photoactive MOFs for organic transformations.

2.7 *Photocatalytic Reduction of Heavy Metal Ions for Detoxification*

With the growth of industrialization, heavy metal contamination results in a severe effect on health and has become a major concern [122]. Chromium is a toxic contaminant that arises from industries such as steel fabrication, leather tanning, printing, electroplating, photography, textile manufacturing, cement, pigments, smelting, and so on. In contrast to cadmium or lead, chromium metal exists in several oxidation states (Cr(IV), Cr(VI), and Cr(III)) [162]. Cr(VI) and Cr(IV) are highly toxic and mutagenic and are carcinogenic for the environment and human beings causing liver damage, bronchitis, ulcers, and dermatitis [122, 163] while Cr(III) is non-toxic, environment friendly, and also an essential metal in human nutrition [121]. Hence, the reduction of Cr(VI) to Cr(III) before discharging water into the environment is an effective remedy for wastewater treatment. Out of the several methods like chemical reduction, precipitation, ion exchange, microbial reduction, coagulation, membrane separation, and electro-reduction, photocatalytic reduction has been considered to have high performance and is cost effective as well [120]. Photocatalysis presents a sustainable technology with some benefits such as complete mineralization, low cost, mild conditions, and free of any hazardous chemical substances [122]. The Cr(III) formed during the procedure is precipitated by the addition of lime or NaOH to form $Cr(OH)_3$, which requires further costing and complex types of equipment. Therefore, the development of methods for Cr(VI) reduction is highly desired [122]. Several nano-sized photocatalysts, like, TiO_2, CdS, ZnO, CdS, and g-C_3N_4, were utilized for the reduction of Cr(VI) into Cr(III) upon irradiation under UV light or visible light. But unfortunately, such photocatalysts suffer from poor reduction capacity, fast electron–hole recombination, etc.

Metal–organic frameworks (MOFs) act as emerging photocatalyst and are effectively utilized to reduce Cr(VI) into Cr(III). The most important factor responsible for the removal of harmful toxic metal ions in water is the water stability of MOFs. If the MOF is water sensitive, it will lead to structural degradation of the material [164]. MOFs having water stability can preserve the structure as well as functional stability during the removal of harmful contaminants from water [165]. Zhao et al. [75] constructed a novel MOF photocatalyst, i.e., NNU-36 via controllable pillared layer method using visible light. The prepared complex has good stability and exhibits an extended range of visible light absorption. NNU-36 has been proven to be highly efficient to reduce Cr(VI) by photocatalytic experiments upon visible light irradiation. pH plays a vital role to reduce Cr(VI). The hole scavengers (i.e., methanol) enhance the catalytic efficiency to a greater extent. Hence, efficient photocatalysts can be synthesized for environmental remediation by utilizing solar energy. Zeng et al. [92] fabricated two MOF photocatalysts (i.e., $[(Cu(H_2L)(4,4'\text{-bipy})0.5(H_2O)]$ and $\{[Co(C_{14}H_{14}O_{6.5}P_2)(4,4'\text{-bipy})0.5(H_2O)_2]\cdot H_2O\}$ via hydrothermal method. Both the complexes exhibit active performance for the photoreduction of Cr(VI) to Cr(III) under UV light irradiation. As demonstrated by the control experiments, pH value plays an important role in Cr(VI) reduction. Both the photocatalysts were reusable and stable and hence a potential candidate for environment remediation. Kaur et al. [125] utilized 2D zinc-based MOF $\{[Zn(PA^{2-})(4,4'\text{-bpy})]\text{-}(H_2O)\}_n$ via the solvothermal method and radiate green fluorescence. The bright green fluorescence of Zn-MOF is quenched when it interacts with $Cr_2O_7^{2-}$ ion, and hence it acts as a Cr(VI) detector via fluorescence signaling with the limit of the detection value of 4.12 μM. This MOF not only works as a sensor but also as a photocatalyst to reduce Cr(VI) to Cr(III) under natural sunlight. The Zn-MOF has good stability and is reusable and shows almost complete Cr(VI) reduction. Garazhian et al. [127] synthesized bimetallic MOFs (i.e., STA-12-Mn–Fe), and almost complete removal of Cr(VI) was accomplished under natural sunlight irradiation and fluorescent lamp (40 W) irradiation at pH 2. The reaction was initiated with an initial 20 mgL^{-1} Cr(VI) concentration and photocatalyst (10 mg) within 30 min. The rate constant of 0.132 min^{-1} was observed for pseudo-first-order kinetics. The bimetallic MOFs show improved photocatalytic performance due to the cooperative effect between the two metal ions in comparison with the single metal MOFs upon sunlight irradiation.

2.8 Photocatalytic Antibacterial Activity

The contamination of food, medical equipment, and the environment by bacteria is threatening lives of the people all over the world. There is an emergent need for the development of efficient antibacterial agents. Nanoparticles, metal oxides, and metallic substances constitute first-generation, second-generation, and third-generation antibacterial agents, respectively. Metal–organic frameworks (MOFs) fall under the category of third-generation antibacterial agents. MOFs possess photocatalytic bactericidal properties which help in controlling air and water pollution owing to their high specific surface area, tunable properties, and controlled release [166]. They also act as a reservoir for antibacterial agents [167]. Hence, they are a good candidate for application in medical diagnosis, wastewater treatment, antibacterial coating, and food preservation [166]. Most of the air purifiers available have good removal capacity but poor biocidal effect. A study was reported by Li and co-authors [168] for the filtration of air for passive pollution control using a series of MOFs and test their potential. ZIF-8 (Zinc-imidazole metal–organic framework) shows >99.99% inactivation of *Escherichia coli* upon solar irradiation for 2 h. The mechanistic approach indicates that photoelectrons trapped at charge trapping metal (Zn^+) center via LMCT cause ROS production. The ZIF-8 incorporated air filters possess astonishing performance for pollution control, having >99.99% photocatalytic bacterial killing efficiency within 30 min. Mohaghegh et al. [169] developed a multifunctional MIL-88B(Fe)–Ag/TiO_2 nanotubes/Ti plates by dip coating and further utilized it as an antibacterial agent for the degradation of *Escherichia coli* (*E. coli*). The rate of degradation by the synthesized catalyst was found to be four times than that of TiO_2 nanotubes/Ti upon UV irradiation. The composite demonstrated good antibacterial properties against E. coli by killing all the bacterial colonies upon photocatalytic treatment for 60 min on MIL-88B(Fe)–Ag/TiO_2 nanotubes/Ti plates. The premise shows the cooperative effect between the MIL-88B(Fe) and Ag/TiO_2 nanotubes/Ti plates. The removal of such pollutants with high efficiency is an effortless approach for the treatment of polluted water using MIL-88B(Fe)–Ag/TiO_2 nanotubes/Ti plate.

2.9 *Photocatalytic Removal of Toxic Gases*

The release of anthropogenic pollutants in the environmental matrixes includes harmful industrial gases, vapors, and products of chemical reactions, and also the intended release of chemical warfare agents is causing risk at a higher level and is a growing concern. The chief origin of these toxic gases NO_x, NH_3, SO_x, H_2S, CO, and mustard gas is anthropogenic, is a significant concern for air pollution, and has become a hot topic of research. Fossil fuel burning causes the emission of SO_2, CO, and NO_2, etc. NO_x and SO_x involve in the formation of photochemical smog, H_2S is a corrosive, odorous gas, and NH_3 acts as a contaminant in several chemical and pharmaceutical industries such as fermentation agents, fertilizer, refrigerant, and antimicrobial agents which cause a great threat to the health and environment [170, 171]. Keeping in view the harmful effects of these gases, their capture and degradation are of utmost priority for the safety of the environment and the health of living beings. Several already available technologies like incineration, oxidation, and membrane separation are there for environmental remediation, but they suffer from their low efficiency, high cost, and formation of secondary pollutants. In this aspect, MOFs appear to be an alternative and have shown appealing applications in adsorption and degradation due to cost-effective technology and environmentally benign nature [170–172]. Liu et al. [173] fabricated heterogeneous catalyst porphyrin-based MOF (PCN-222/MOF-545) and used it as a photosensitizer for the generation of singlet oxygen by the use of already available LEDs. The singlet oxygen generated under ambient conditions can oxidize mustard gas simulant (2-chloroethyl ethyl sulfide) selectively to non-toxic product (2-chloroethyl ethyl sulfoxide) without the formation of sulfone product which is highly toxic. It is a more convenient and realistic approach for the detoxification of mustard gas than the already available conventional methods.

3 The Common Mechanism Involved in Photocatalysis by MOFs

There are mainly four fundamental processes [174] shown by MOF-based photocatalysts as follows:

Photoexcitation: The absorption of light might happen via metal ions or clusters or by organic ligands (e.g., unidentate, bidentate, multidentate ligands). MOFs may possess few broad absorption bands often seen in the UV–Vis region based on the type of chromophore present in MOF. These absorption bands are attributed to metal-to-ligand (MLCT) or ligand-to-metal (LMCT) charge transfer transitions or π-π^* transitions of aromatic rings present in the organic ligand. Hence, band theory has certain limitations in explaining the absorption of light and successive transitions in MOF materials. Therefore, it is prominent to use molecular orbital theory (MOT), where the highest occupied molecular orbitals (HOMO) and lowest unoccupied molecular orbital (LUMO) peaks decide the 'band gaps' of MOF. The inorganic secondary building units and organic ligands owing discrete molecular orbitals and MOFs have numerous distinct absorbance bands, and the band having the longest wavelength (lowest energy) defines the band gaps. LMCT effects are necessary in those cases where MOFs have chromophore units in the organic ligands. Hence, the knowledge of the position of the orbitals is a necessary condition to understand the mechanism. The photogenerated electrons and holes have the same energy levels as the HOMO and LUMO orbitals of MOFs.

Charge separation, transfer, and transport processes: The process of charge separation, transfer, and transport is instigated and continued via the excessive absorption of light energy, which results in the separation of charged excitons which are then transferred to the photocatalyst surface. The type of material, particle size, and material crystallinity are responsible for the efficiency of these processes. Structural deficiency present in chromophore centers may serve as the recombination centers where charge carriers can be trapped and leads to a considerable loss of energy through the release of heat. The nanocrystals having small size particles commonly hold the reasonable transfer of charge performance. If the size of particles is very small, then improved surface defects are the reason for recombination.

Charge carrier recombination: In this process, the energy which was absorbed during excitation gets released in the form of heat upon recombination of electrons and holes. The energy emitted is captured and translated via photoluminescence (PL) spectroscopy. The charge carrier's recombination is the largest loss of energy in such systems. The recombination occurs at the surface of the photocatalyst as well as in the bulk; hence, both recombinations are taken into account. The recombination can be quenched by decreasing the particle size or by applying an external source of energy.

Charge utilization (Surface electrocatalysis): The charge carriers, i.e., electrons and holes, migrate to the surface of the photocatalyst. These are persuaded to engage in chemical reactions. The rate of the chemical reaction impacts the charge transfer and transition [174]. All the four processes involved in the mechanism of photochemical reaction are illustrated schematically in Fig. 3.

Therefore, the efficiency of photocatalysis is based on the cumulative effects [175] of the following phenomena represented in Eq. (A):

$$\eta_c = \eta_{abs} \times \eta_{cs} \times \eta_{cr} \times \eta_{cu} \tag{A}$$

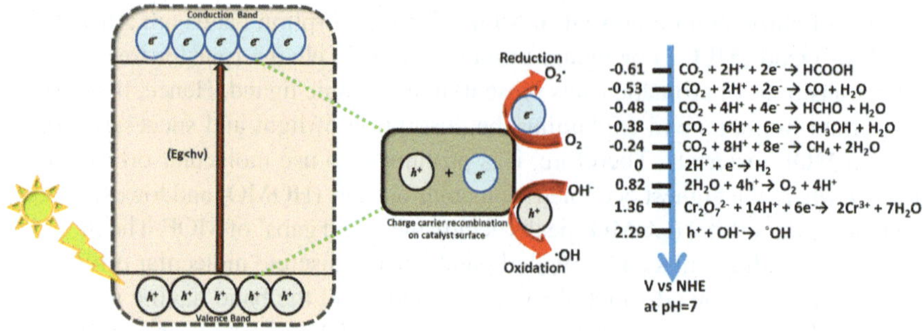

Fig. 3 Fundamental processes involved in the photocatalysis by MOFs and redox potential of different species formed during photocatalysis

where η_c is the solar energy conversion efficiency, η_{abs} is light absorption efficiency, η_{cs} is charge excitation/separation, transfer, transport efficiency, η_{cr} is charge recombination efficiency, and η_{cu} is charge utilization efficiency for the photocatalytic reactions [176].

Taking into account all four processes, Liu et al. have explained that photocatalysis is a complex catalytic process and is controlled by various factors. A wide range of studies has been carried out to analyze the aspects and mechanisms responsible for the photocatalytic processes. The reactive radical species plays a prominent role in such processes. In a particular system of MOFs, the electrons present in the valence band (VB) are excited when a photocatalyst absorbs UV or visible light with energy beyond its corresponding bandgap energy (Eg) and instinctively jump into the conduction band (CB) leaving a hole (h^+), resulting in the generation of photo-excited electron–hole pairs (Eq. 1). The photogenerated electron (e^-) acts as a strong reductant, and the hole (h^+) acts as a strong oxidant. The photogenerated electron and hole then transfer to the photocatalyst surface and participate in redox reactions with the species which are adsorbed and lead to the formation of hydroxyl radical ($^{\cdot}OH$) (Eqs. 2, 3) and superoxide radical ($\cdot O_2{}^-$) (Eq. 4). Hydroxyl radical is a powerful and non-selective oxidizing radical under neutral and acidic conditions having redox potential of 1.83 and 2.65 V, respectively, versus NHE [177–180].

$$\text{Catalyst} + \text{hv} \rightarrow \text{Catalyst}\left(e^-_{CB} + h^+_{VB}\right) \tag{1}$$

$$h^+ + OH^- \rightarrow \cdot OH \tag{2}$$

$$h^+ + H_2O \rightarrow \cdot OH + H^+ \tag{3}$$

$$e^- + O_2 \rightarrow \cdot O_2^- \tag{4}$$

Three cases can be considered to explain the mechanistic approach further as follows:

(a) When the photocatalyst is utilized for the photocatalytic degradation of organic contaminants (like dyes, antibiotics, pesticides, etc.), these are degraded by ·OH generated on the valence band (Eq. 5), or it can be directly oxidized in the conduction band by photo-excited h^+ (Eqs. 6, 7) represent that the electrons can be utilized for the reduction of highly toxic metal ions, e.g., Cr(VI) into Cr(III) (low toxic form).

$$\text{Dye} + \cdot\text{OH} \rightarrow \text{Degradation products} \tag{5}$$

$$\text{Dye} + h^+ \rightarrow \text{Degradation products} \tag{6}$$

$$Cr_2O_7^{2-} + 14H^+ + 6e^- \rightarrow 2Cr^{3+} + 7H_2O,\ E_0 = 1.36\ V \tag{7}$$

(b) When the photocatalyst is utilized for water splitting (for production of H_2), the value of the valence band of the photocatalyst must lie below the energy of the redox couple, i.e., O_2/H_2O, and the value of conduction band must lie above the energy of the H^+/H_2 redox couple. The production of O_2 and H_2 may follow these Eqs. (8) and (9), respectively. The holes in the valence band are involved in the water oxidation (Eq. 8)

$$2H_2O + 4h^+ \rightarrow O_2 + 4H^+ \tag{8}$$

$$2H^+ + e^- \rightarrow H_2,\ E_0 = 0 \tag{9}$$

(c) When the photocatalyst is utilized for CO_2 reduction, the only electrons having sufficient reduction potential versus NHE are consumed at pH = 7 (Eqs. 10–14) [179].

$$CO_2 + 2H^+ + 2e^- \rightarrow HCOOH,\ E_0 = -0.61\ V \tag{10}$$

$$CO_2 + 2H^+ + 2e^- \rightarrow CO + H_2O,\ E_0 = -0.53\ V \tag{11}$$

$$CO_2 + 4H^+ + 4e^- \rightarrow HCHO + H_2O,\ E_0 = -0.48\ V \tag{12}$$

$$CO_2 + 6H^+ + 6e^- \rightarrow CH_3OH + H_2O,\ E_0 = -0.38\ V \tag{13}$$

$$CO_2 + 8H^+ + 8e^- \rightarrow CH_4 + 2H_2O,\ E_0 = -0.24\ V \quad (14)$$

It is found that H_2O_2 is responsible for increasing the rate of formation of hydroxyl radical (·OH) [179] in various ways:

(a) It acts as an electron acceptor through oxygen, which inhibits the recombination of the photogenerated electrons and holes; the reduction of H_2O_2 at the CB produces hydroxyl radical (Eq. 15),

$$e^-_{CB} + H_2O_2 \rightarrow OH^- + \cdot OH \quad (15)$$

(b) it accepts an electron from superoxide to give rise to the hydroxyl radical (Eq. 16); and

$$\cdot O_2^- + H_2O_2 \rightarrow OH^- + \cdot OH + O_2 \quad (16)$$

(c) self-decomposition upon illumination (Eq. 17).

$$H_2O_2 + h\nu \rightarrow 2\cdot OH \quad (17)$$

4 Conclusion and Future Perspectives

The MOFs appear to be an emergent photocatalytic material, have attracted ample research interest, and demonstrate a promising future. This chapter is the outcome of the well-ordered and evaluative review, which covers the several studies reported on the existing situation of MOFs and modified MOFs nanocomposites in the photocatalytic performance of CO_2 reduction, H_2 production, adsorption, and degradation of numerous organic contaminants (i.e., pesticides, dyes, and antibiotics) during the past few years. The sharp increasing trend shows the enthusiasm and interest of the research society in this field. Some perceptions of the mechanism of action have also been gained. The admirable photocatalytic actions of certain MOFs are ascribed due to (a) their extremely porous structure which proficiently fixates the harmful pollutants inside the cavity and is advantageous for the study and characterization of photocatalysts, (b) enormous structural tunability leads to intensified solar harnessing and worthwhile charge separation, (c) the unceasing production of reactive oxygen species (ROS) to attack the harmful organic pollutants and beneficial for expeditious decomposition, and (d) combination of photocatalytic and adsorbing properties of MOFs can create material having high catalytic efficiency and improved visible light response. MOFs act as an exceptional contender to fabricate capable heterojunction photocatalysts, empowering MOFs recognizable advantages over other conventional semiconductor photocatalysts. These are also considered semiconductors established on their optical transition properties, electrochemical, and photochemical activities.

MOFs themselves suffer from poor stability in an aqueous medium, and also their reusability is a crucial challenge. Except for few types of MOFs, the rest are unstable in water results in low recovery and even cause secondary pollution due to metal leaching. The stability can be enhanced via the introduction of metal of high valence state and multidentate ligands in the MOFs. The modifications with functionalities like metal-oxo clusters, hybridized metal ions, and Π-conjugated ligands result in enhanced photoactive nature and efficient degradation of organic pollutants than the conventional photocatalysts. The adsorption and photocatalysis should be combined simultaneously for the effectual treatment with minimal consumption of energy using MOFs. In all, we have confidence that in the future, MOFs may serve as the most powerful and perfect choice for light harvesting to accomplish the photocatalytic activities as it tends to surpass the current materials for novel applications.

Acknowledgements The authors acknowledge the University of Delhi for providing financial assistance in terms of the Research and Development (R&D) Grant. We are also thankful to the University Grant Commission for providing research fellowships.

Abbreviations

AB	Amido black
AChE	Acetylcholinesterase
Ag_3VO_4	Silver vanadate
Ag_2WO_4	Silver tungstate
AO	Acid orange
AOP	Advanced oxidation process
AR3R	Acid red 3R
AuNPs	Gold nanoparticles
AY	Acid yellow
BB41	Basic blue 41
BiOBr	Bismuth oxybromide
BPA	Bisphenol A
BPYDC	2,2′-Bipyridine-5,5′-dicarboxylate
CB	Conduction band
CdS	Cadmium sulfide
CH_4	Methane
CIP	Ciprofloxacin
CNT	Carbon Nanotubes
2-CP	2-Chlorophenol
CO	Carbon monoxide

CO_2	Carbon dioxide
Cr	Chromium
CR	Congo red
CuNWs	Copper nanowires
CuS	Copper sulfide
CV	Crystal violet
1D	One dimensional
2D	Two dimensional
3D	Three dimensional
DMF	N, N-dimethyl formamide
DR23	Direct red 23
E. coli	*Escherichia coli*
E_{ff}	Power conversion efficiency
EIS	Electrochemical impedance spectroscopy
FESEM	Field emission scanning electron microscopy
FF	Fill factor
FTIR	Fourier transform infrared spectroscopy
$g\text{-}C_3N_4$	Graphitic carbon nitride
GF	Glufosinate
GP	Glyphosate
GR	Graphene
HCOOH	Formic acid
H_2	Hydrogen
H_2O	Water
H_2O_2	Hydrogen peroxide
HOMO	Highest occupied molecular orbital
HR-TEM	High-resolution transmission electron microscopy
ICP-OES	Inductively coupled plasma optical emission spectroscopy
I_{sc}	Short-circuit current
LbL	Layer-by-layer
LED	Light-emitting diode
LMCT	Ligand to metal charge transition
LUMO	Lowest unoccupied molecular orbital
MB	Methylene blue
MeOH	Methanol
MIL	Matériaux de l'Institut Lavoisier
MLCT	Metal to ligand charge transfer
MO	Methyl orange
MOFs	Metal–organic frameworks
MOF-5	$Zn_4O(BDC)_3$
MOT	Molecular orbital theory
$NaBH_4$	Sodium borohydride
NaOH	Sodium hydroxide
NH_3	Ammonia
NH_3BH_3	Ammonia borane

Ni-BDC	Nickel–benzene dicarboxylic acid
Ni/Cu-BTC	Nickel/copper–benzene tri-carboxylic acid
NiO	Nickel oxide
NIR	Near infrared
NiS	Nickel sulfide
NMR	Nuclear magnetic resonance
N_2	Nitrogen
4-NP	4-Nitrophenol
NPs	Nanoparticles
OG	Orange dye
O_2	Oxygen
O_3	Ozone
PCO	Photocatalytic oxidation
PL	Photoluminescence spectroscopy
PMOF	Porphyrin-based metal−organic framework
PNP	P-nitrophenol
PNRR	Photocatalytic nitrogen reduction reaction
POP	Persistent organic pollutants
R6G	Rhodamine 6G
RB5	Reactive black 5
RhB	RhodamineB
ROS	Reactive oxygen species
SBU	Secondary building units
SPR	Surface plasmon resonance
TC	Tetracycline hydrochloride
TIPA	Tris-(4-imidazolylphenyl)amine
TiO_2	Titanium dioxide
TPVT	2,4,6-Tris(2-(pyridine-4-yl)vinyl)-1,3,5-triazine
UiO-67	Zr 6 (m 3-O) 4 (m 3-OH) 4 (BPDC) 12
UV	Ultraviolet–visible
UV−Vis DRS	UV−visible diffuse reflectance spectroscopy
VB	Valence band
V_{oc}	Open-circuit voltage
VOCs	Volatile organic compounds
Xe	Xenon
XPS	X-ray photoelectron spectroscopy
XRD	X-ray diffraction
ZIF	Zeolitic imidazolate framework (ZIF)
ZnO	Zinc oxide

References

1. Ravelli D, Dondi D, Fagnoni M, Albini A (2009) Photocatalysis. A multi-faceted concept for green chemistry. Chem Soc Rev 38:1999–2011. https://doi.org/10.1039/B714786B
2. Kalkan AK, Chopra N (2015) Harnessing sunlight by photocatalysis: a sustainable pathway for renewable fuels and clean water. Nanomaterials Energy 2:114–116. https://doi.org/10.1680/NME.13.00012
3. Zeng L, Guo X, He C, Duan C (2016) Metal-organic frameworks: versatile materials for heterogeneous photocatalysis. ACS Catal 6:7935–7947. https://doi.org/10.1021/ACSCATAL.6B02228
4. Ying C (2021) Metal-organic framework as a new photocatalyst for environmental pollutant treatment. IOP Conf Ser Earth Environ Sci 631:012021. https://doi.org/10.1088/1755-1315/631/1/012021
5. Sharma A, Kumar A, Li C et al (2020) Microencapsulated UV filter@ZIF-8 based sunscreens for broad spectrum UV protection. RSC Adv 10:34254–34260. https://doi.org/10.1039/D0RA05828A
6. Yadav A, Sharma A, Sharma RK (2019) Mesoporous iron gallate nanocomplex for adsorption and degradation of organic dyes. Colloids Surf A Physicochem Eng Asp 579. https://doi.org/10.1016/J.COLSURFA.2019.123694
7. He J, Zhang Y, Zhang X, Huang Y (2018) Highly efficient Fenton and enzyme-mimetic activities of NH2-MIL-88B(Fe) metal organic framework for methylene blue degradation. Sci Rep 81(8):1–8. https://doi.org/10.1038/s41598-018-23557-2
8. Scerpella DL, Bouranis NG, Webster MJ et al (2021) Using geographic information systems (GIS) for targeted national recruitment of community-dwelling caregivers managing dementia-related behavioral and psychological symptoms: a recruitment approach for a randomized clinical trial. J Geogr Inf Syst 13:302–317. https://doi.org/10.4236/JGIS.2021.133017
9. Fujishima A, Honda KI, Kikuchi SI (1969) Photosensitized electrolytic oxidation on semiconducting n-type TiO_2 electrode. J Soc Chem Ind Japan 72:108–113. https://doi.org/10.1246/NIKKASHI1898.72.108
10. Zango ZU, Jumbri K, Sambudi NS et al (2020) A critical review on metal-organic frameworks and their composites as advanced materials for adsorption and photocatalytic degradation of emerging organic pollutants from wastewater. Polymers 12:2648. https://doi.org/10.3390/POLYM12112648
11. Sharma A, Kumar A, Li C et al (2021) A cannabidiol-loaded Mg-gallate metal–organic framework-based potential therapeutic for glioblastomas. J Mater Chem B 9:2505–2514. https://doi.org/10.1039/D0TB02780D
12. Kumar A, Sharma A, Chen Y et al (2021) Copper@ZIF-8 core-shell nanowires for reusable antimicrobial face masks. Adv Funct Mater 31:2008054. https://doi.org/10.1002/ADFM.202008054
13. Wang CC, Du XD, Li J et al (2016) Photocatalytic Cr(VI) reduction in metal-organic frameworks: a mini-review. Appl Catal B Environ 193:198–216. https://doi.org/10.1016/J.APCATB.2016.04.030
14. Younis SA, Kwon EE, Qasim M et al (2020) Metal-organic framework as a photocatalyst: progress in modulation strategies and environmental/energy applications. Prog Energy Combust Sci 81:100870. https://doi.org/10.1016/J.PECS.2020.100870
15. Li Z, Wang L, Qin L et al (2021) Chemosphere recent advances in the application of water-stable metal-organic frameworks : adsorption and photocatalytic reduction of heavy metal in water. Chemosphere 285:131432. https://doi.org/10.1016/j.chemosphere.2021.131432
16. Li J-R, Kuppler RJ, Zhou H-C (2009) Selective gas adsorption and separation in metal–organic frameworks. Chem Soc Rev 38:1477–1504. https://doi.org/10.1039/B802426J
17. Yildirim O, Bonomo M, Barbero N et al (2020) Application of metal-organic frameworks and covalent organic frameworks as (photo)active material in hybrid photovoltaic technologies. Energies 13:1–48

18. Hoskins BF, Robson R (1990) Design and construction of a new class of scaffolding-like materials comprising infinite polymeric frameworks of 3D-linked molecular rods. A reappraisal of the Zn(CN)2 and Cd(CN)2 structures and the synthesis and structure of the diamond-related frameworks [N(CH3)4][CuIZnII(CN)4] and CuI[4,4',4″,4‴-tetracyanotetraphenylmethane]BF4•xC6H5NO2. J Am Chem Soc 112:1546–1554. https://doi.org/10.1021/JA00160A038
19. Pi Y, Li X, Xia Q et al (2018) Adsorptive and photocatalytic removal of persistent organic pollutants (POPs) in water by metal-organic frameworks (MOFs). Chem Eng J 337:351–371. https://doi.org/10.1016/J.CEJ.2017.12.092
20. Moghadam PZ, Li A, Wiggin SB et al (2017) Development of a Cambridge structural database subset: a collection of metal-organic frameworks for past, present, and future. Chem Mater 29:2618–2625. https://doi.org/10.1021/ACS.CHEMMATER.7B00441
21. Wang Q, Gao Q, Al-Enizi AM et al (2020) Recent advances in MOF-based photocatalysis: environmental remediation under visible light. Inorg Chem Front 7:300–339. https://doi.org/10.1039/C9QI01120J
22. Yuan N, Gong X, Sun W, Yu C (2021) Advanced applications of Zr-based MOFs in the removal of water pollutants. Chemosphere 267. https://doi.org/10.1016/j.chemosphere.2020.128863
23. Zhang X, Wang J, Dong XX, Lv YK (2020) Functionalized metal-organic frameworks for photocatalytic degradation of organic pollutants in environment. Chemosphere 242:125144. https://doi.org/10.1016/J.CHEMOSPHERE.2019.125144
24. Wang D, Huang R, Liu W et al (2014) Fe-based MOFs for photocatalytic CO_2 reduction: role of coordination unsaturated sites and dual excitation pathways. ACS Catal 4:4254–4260. https://doi.org/10.1021/CS501169T
25. Nguyen TN, Kampouri S, Valizadeh B et al (2018) Photocatalytic hydrogen generation from a visible-light-responsive metal-organic framework system: stability versus activity of molybdenum sulfide cocatalysts. ACS Appl Mater Interfaces 10:30035–30039. https://doi.org/10.1021/ACSAMI.8B10010
26. Yue L, Cao Y, Han Y et al (2021) Preparation of core-shell structured Fe3O4@Sn-MOF composite and photocatalytic performance. J Alloys Compd 870:159339. https://doi.org/10.1016/j.jallcom.2021.159339
27. Fujita M, Kwon YJ, Washizu S, Ogura K (2002) Preparation, clathration ability, and catalysis of a two-dimensional square network material composed of cadmium(II) and 4,4′-Bipyridine. J Am Chem Soc 116:1151–1152. https://doi.org/10.1021/JA00082A055
28. Zhang T, Lin W (2014) Metal–organic frameworks for artificial photosynthesis and photocatalysis. Chem Soc Rev 43:5982–5993. https://doi.org/10.1039/C4CS00103F
29. Tao AR, Habas S, Yang P (2008) Shape control of colloidal metal nanocrystals. Small 4:310–325. https://doi.org/10.1002/SMLL.200701295
30. Rubio-Martinez M, Avci-Camur C, Thornton AW et al (2017) New synthetic routes towards MOF production at scale. Chem Soc Rev 46:3453–3480. https://doi.org/10.1039/C7CS00109F
31. Rabenau A (1985) The role of hydrothermal synthesis in preparative chemistry. Angew Chemie Int Ed English 24:1026–1040. https://doi.org/10.1002/ANIE.198510261
32. Feng Y, Chen Q, Jiang M, Yao J (2019) Tailoring the properties of UiO-66 through defect engineering: a review. Ind Eng Chem Res 58:17646–17659. https://doi.org/10.1021/ACS.IECR.9B03188
33. Odoh SO, Cramer CJ, Truhlar DG, Gagliardi L (2015) Quantum-chemical characterization of the properties and reactivities of metal-organic frameworks. Chem Rev 115:6051–6111. https://doi.org/10.1021/CR500551H
34. Stock N, Biswas S (2011) Synthesis of metal-organic frameworks (MOFs): routes to various MOF topologies, morphologies, and composites. Chem Rev 112:933–969. https://doi.org/10.1021/CR200304E
35. Alvaro M, Carbonell E, Ferrer B et al (2007) Semiconductor behavior of a metal-organic framework (MOF). Chem A Eur J 13:5106–5112. https://doi.org/10.1002/CHEM.200601003

36. Liang R, Jing F, Shen L et al (2015) MIL-53(Fe) as a highly efficient bifunctional photocatalyst for the simultaneous reduction of Cr(VI) and oxidation of dyes. J Hazard Mater 287:364–372. https://doi.org/10.1016/J.JHAZMAT.2015.01.048
37. Zhang Y, Mao F, Wang L et al (2020) Recent advances in photocatalysis over metal-organic frameworks-based materials. Sol RRL 4:1900438. https://doi.org/10.1002/SOLR.201900438
38. Kumar P, Kim KH, Kim YH et al (2016) A review of metal organic resins for environmental applications. J Hazard Mater 320:234–240. https://doi.org/10.1016/J.JHAZMAT.2016.08.037
39. Yang Z, Xu X, Liang X et al (2017) Fabrication of Ce doped UiO-66/graphene nanocomposites with enhanced visible light driven photoactivity for reduction of nitroaromatic compounds. Appl Surf Sci 420:276–285. https://doi.org/10.1016/J.APSUSC.2017.05.158
40. Wang Z, Huang J, Mao J et al (2020) Metal–organic frameworks and their derivatives with graphene composites: preparation and applications in electrocatalysis and photocatalysis. J Mater Chem A 8:2934–2961. https://doi.org/10.1039/C9TA12776C
41. Zhu QL, Li J, Xu Q (2013) Immobilizing metal nanoparticles to metal-organic frameworks with size and location control for optimizing catalytic performance. J Am Chem Soc 135:10210–10213. https://doi.org/10.1021/JA403330M
42. Shen L, Liang R, Wu L (2015) Strategies for engineering metal-organic frameworks as efficient photocatalysts. Cuihua Xuebao/Chin J Catal 36:2071–2088. https://doi.org/10.1016/S1872-2067(15)60984-6
43. Wang S, Wang X (2015) Multifunctional metal-organic frameworks for photocatalysis. Small 11:3097–3112. https://doi.org/10.1002/SMLL.201500084
44. Zhou T, Du Y, Borgna A et al (2013) Post-synthesis modification of a metal–organic framework to construct a bifunctional photocatalyst for hydrogen production. Energy Environ Sci 6:3229–3234. https://doi.org/10.1039/C3EE41548A
45. Zhang S, Li L, Zhao S et al (2015) Construction of interpenetrated ruthenium metal-organic frameworks as stable photocatalysts for CO_2 reduction. Inorg Chem 54:8375–8379. https://doi.org/10.1021/ACS.INORGCHEM.5B01045
46. Han Z, Fu Y, Zhang Y et al (2021) Metal-organic framework (MOF) composite materials for photocatalytic CO_2 reduction under visible light. Dalt Trans 50:3186–3192. https://doi.org/10.1039/d1dt00128k
47. Choi KM, Kim D, Rungtaweevoranit B et al (2016) Plasmon-enhanced photocatalytic CO_2 conversion within metal-organic frameworks under visible light. J Am Chem Soc 139:356–362. https://doi.org/10.1021/JACS.6B11027
48. Wang M, Liu J, Guo C et al (2018) Metal–organic frameworks (ZIF-67) as efficient cocatalysts for photocatalytic reduction of CO_2: the role of the morphology effect. J Mater Chem A 6:4768–4775. https://doi.org/10.1039/C8TA00154E
49. Wang XK, Liu J, Zhang L et al (2019) Monometallic catalytic models hosted in stable metal-organic frameworks for tunable CO_2 photoreduction. ACS Catal 9:1726–1732. https://doi.org/10.1021/ACSCATAL.8B04887
50. Mahmoud ME, Audi H, Assoud A et al (2019) Metal-organic framework photocatalyst incorporating Bis(4′-(4-carboxyphenyl)-terpyridine)ruthenium(II) for visible-light-driven carbon dioxide reduction. J Am Chem Soc 141:7115–7121. https://doi.org/10.1021/JACS.9B01920
51. Zhang D, Liu P, Xiao S et al (2016) Microwave-antenna induced in situ synthesis of Cu nanowire threaded ZIF-8 with enhanced catalytic activity in H_2 production. Nanoscale 8:7749–7754. https://doi.org/10.1039/C5NR07505J
52. Jin Z (2017) Yang H (2017) Exploration of Zr–metal–organic framework as efficient photocatalyst for hydrogen production. Nanoscale Res Lett 121(12):1–10. https://doi.org/10.1186/S11671-017-2311-6
53. Salcedo-Abraira P, Vilela SMF, Babaryk AA et al (2020) Nickel phosphonate MOF as efficient water splitting photocatalyst. Nano Res 142(14):450–457. https://doi.org/10.1007/S12274-020-3056-6
54. Kampouri S, Ebrahim FM, Fumanal M et al (2021) Enhanced visible-light-driven hydrogen production through MOF/MOF heterojunctions. ACS Appl Mater Interfaces 13:14239–14247. https://doi.org/10.1021/acsami.0c23163

55. Zhu Y-P, Yin J, Abou-Hamad E et al (2020) highly stable phosphonate-based MOFs with engineered bandgaps for efficient photocatalytic hydrogen production. Adv Mater 32:1906368. https://doi.org/10.1002/ADMA.201906368
56. Morshedy AS, El SHMA, El NAMA, Zaki T (2020) Hydrogen production and in situ storage through process of water splitting using mono/binary metal-organic framework (MOF) structures as new chief photocatalysts. Energy Fuels 34:11660–11669. https://doi.org/10.1021/ACS.ENERGYFUELS.0C01559
57. Shang S, Xiong W, Yang C et al (2021) Atomically dispersed iron metal site in a porphyrin-based metal-organic framework for photocatalytic nitrogen fixation. ACS Nano 15:9670–9678. https://doi.org/10.1021/ACSNANO.0C10947
58. Gao W, Li X, Zhang X et al (2021) Photocatalytic nitrogen fixation of metal–organic frameworks (MOFs) excited by ultraviolet light: insights into the nitrogen fixation mechanism of missing metal cluster or linker defects. Nanoscale 13:7801–7809. https://doi.org/10.1039/D1NR00697E
59. Li G, Li F, Liu J, Fan C (2020) Fe-based MOFs for photocatalytic N2 reduction: key role of transition metal iron in nitrogen activation. J Solid State Chem 285. https://doi.org/10.1016/J.JSSC.2020.121245
60. Huang H, Wang XS, Philo D et al (2020) Toward visible-light-assisted photocatalytic nitrogen fixation: a titanium metal organic framework with functionalized ligands. Appl Catal B Environ 267:118686. https://doi.org/10.1016/j.apcatb.2020.118686
61. Ding Z, Wang S, Chang X et al (2020) Nano-MOF@defected film C3N4Z-scheme composite for visible-light photocatalytic nitrogen fixation. RSC Adv 10:26246–26255. https://doi.org/10.1039/d0ra03562a
62. Yang SJ, Im JH, Kim T et al (2011) MOF-derived ZnO and ZnO@C composites with high photocatalytic activity and adsorption capacity. J Hazard Mater 186:376–382. https://doi.org/10.1016/J.JHAZMAT.2010.11.019
63. Wang A, Zhou Y, Wang Z et al (2016) Titanium incorporated with UiO-66(Zr)-type metal-organic framework (MOF) for photocatalytic application. RSC Adv 6:3671–3679. https://doi.org/10.1039/C5RA24135A
64. Fang Y, Zhu SR, Wu MK et al (2018) MOF-derived In2S3 nanorods for photocatalytic removal of dye and antibiotics. J Solid State Chem 266:205–209. https://doi.org/10.1016/J.JSSC.2018.07.026
65. Huang L, Liu B (2016) Synthesis of a novel and stable reduced graphene oxide/MOF hybrid nanocomposite and photocatalytic performance for the degradation of dyes. RSC Adv 6:17873–17879. https://doi.org/10.1039/C5RA25689E
66. Araya T, Chen CC, Jia MK et al (2017) Selective degradation of organic dyes by a resin modified Fe-based metal-organic framework under visible light irradiation. Opt Mater (Amst) 64:512–523. https://doi.org/10.1016/J.OPTMAT.2016.11.047
67. Hu S, Liu M, Li K et al (2016) Surfactant-assisted synthesis of hierarchical NH2-MIL-125 for the removal of organic dyes. RSC Adv 7:581–587. https://doi.org/10.1039/C6RA25745C
68. Li H, Li Q, He Y et al (2017) Two novel porous MOFs with square-shaped cavities for the removal of toxic dyes: adsorption or degradation? New J Chem 41:15204–15209. https://doi.org/10.1039/C7NJ02904G
69. Ezugwu CI, Asraf MA, Li X et al (2018) Selective and adsorptive removal of anionic dyes and CO_2 with azolium-based metal-organic frameworks. J Colloid Interface Sci 519:214–223. https://doi.org/10.1016/J.JCIS.2018.02.003
70. Mahmoodi NM, Abdi J, Oveisi M et al (2018) Metal-organic framework (MIL-100 (Fe)): synthesis, detailed photocatalytic dye degradation ability in colored textile wastewater and recycling. Mater Res Bull 100:357–366. https://doi.org/10.1016/J.MATERRESBULL.2017.12.033
71. Fang XD, Yang LB, Dou AN et al (2018) Synthesis, crystal structure and photocatalytic properties of a Mn (II) metal-organic framework based on a thiophene-functionalized dicarboxylate ligand. Inorg Chem Commun 96:124–127. https://doi.org/10.1016/J.INOCHE.2018.08.017

72. Lin KY, Yang H, Hsu FK (2018) Zr-Metal organic framework and derivatives for adsorptive and photocatalytic removal of acid dyes. Water Environ Res 90:144–154. https://doi.org/10.2175/106143017X15054988926604
73. Tabatabaei N, Dashtian K, Ghaedi M et al (2018) Novel visible light-driven Cu-based MOFs/Ag_2O composite photocatalysts with enhanced photocatalytic activity toward the degradation of orange G: their photocatalytic mechanism and optimization study. New J Chem 42:9720–9734. https://doi.org/10.1039/C7NJ03245E
74. Liu C-X, Zhang W-H, Wang N et al (2018) Highly efficient photocatalytic degradation of dyes by a copper–triazolate metal–organic framework. Chem A Eur J 24:16804–16813. https://doi.org/10.1002/CHEM.201803306
75. Zhao H, Xia Q, Xing H et al (2017) Construction of pillared-layer MOF as efficient visible-light photocatalysts for aqueous Cr(VI) reduction and dye degradation. ACS Sustain Chem Eng 5:4449–4456. https://doi.org/10.1021/ACSSUSCHEMENG.7B00641
76. Nguyen VH, Nguyen TD, Bach LG et al (2018) Effective photocatalytic activity of mixed Ni/Fe-base metal-organic framework under a compact fluorescent daylight lamp. Catalysts 8:487. https://doi.org/10.3390/CATAL8110487
77. Chin M, Cisneros C, Araiza SM et al (2018) Rhodamine B degradation by nanosized zeolitic imidazolate framework-8 (ZIF-8). RSC Adv 8:26987–26997. https://doi.org/10.1039/C8RA03459A
78. Tian H, Peng J, Lv T et al (2018) Preparation and performance study of $MgFe_2O_4$/metal–organic framework composite for rapid removal of organic dyes from water. J Solid State Chem 257:40–48. https://doi.org/10.1016/J.JSSC.2017.09.017
79. Oveisi M, Alinia Asli M, Mahmoodi NM (2019) Carbon nanotube based metal-organic framework nanocomposites: synthesis and their photocatalytic activity for decolorization of colored wastewater. Inorganica Chim Acta 487:169–176. https://doi.org/10.1016/J.ICA.2018.12.021
80. Chakraborty A, Islam DA, Acharya H (2019) Facile synthesis of CuO nanoparticles deposited zeolitic imidazolate frameworks (ZIF-8) for efficient photocatalytic dye degradation. J Solid State Chem 269:566–574. https://doi.org/10.1016/J.JSSC.2018.10.036
81. Jalali S, Rahimi MR, Dashtian K et al (2019) One step integration of plasmonic Ag2CrO4/Ag/AgCl into HKUST-1-MOF as novel visible-light driven photocatalyst for highly efficient degradation of mixture dyes pollutants: Its photocatalytic mechanism and modeling. Polyhedron 166:217–225. https://doi.org/10.1016/J.POLY.2019.03.045
82. Emam HE, Ahmed HB, Gomaa E et al (2019) Doping of silver vanadate and silver tungstate nanoparticles for enhancement the photocatalytic activity of MIL-125-NH2 in dye degradation. J Photochem Photobiol A Chem 383:111986. https://doi.org/10.1016/J.JPHOTOCHEM.2019.111986
83. Devarayapalli KC, Vattikuti SVP, TVM S et al (2019) Facile synthesis of Ni-MOF using microwave irradiation method and application in the photocatalytic degradation. Mater Res Express 6:1150h3. https://doi.org/10.1088/2053-1591/AB5261
84. Dong J-P, Shi Z-Z, Li B, Wang L-Y (2019) Synthesis of a novel 2D zinc(II) metal–organic framework for photocatalytic degradation of organic dyes in water. Dalt Trans 48:17626–17632. https://doi.org/10.1039/C9DT03727F
85. Navarathna CM, Dewage NB, Karunanayake AG et al (2020) Rhodamine B adsorptive removal and photocatalytic degradation on MIL-53-Fe MOF/magnetic magnetite/biochar composites. J Inorg Organomet Polym Mater 30:214–229. https://doi.org/10.1007/S10904-019-01322-W
86. Zhong M, Qu S-Y, Zhao K et al (2020) Bimetallic metal-organic framework derived ZnO/Ni0.9Zn0.1O nanocomposites for improved photocatalytic degradation of organic dyes. ChemistrySelect 5:1858–1864. https://doi.org/10.1002/SLCT.201904629
87. Wang Y, Ge S, Cheng W et al (2020) Microwave hydrothermally synthesized metal-organic framework-5 derived C-doped ZnO with enhanced photocatalytic degradation of rhodamine B. Langmuir 36:9658–9667. https://doi.org/10.1021/ACS.LANGMUIR.0C00395

88. Kaur H, Kumar R, Kumar A et al (2019) Trifunctional metal–organic platform for environmental remediation: structural features with peripheral hydroxyl groups facilitate adsorption, degradation and reduction processes. Dalt Trans 48:915–927. https://doi.org/10.1039/C8DT04180F
89. Buhori A, Zulys A, Gunlazuardi J (2020) Synthesis of Lanthanum metal-organic frameworks (La-MOFs) as degradation photocatalyst of Rhodamine-B. AIP Conf Proc 2242:040033. https://doi.org/10.1063/5.0013010
90. Ramezanalizadeh H, Manteghi F (2018) Synthesis of a novel MOF/CuWO4 heterostructure for efficient photocatalytic degradation and removal of water pollutants. J Clean Prod 172:2655–2666. https://doi.org/10.1016/J.JCLEPRO.2017.11.145
91. Bhuvaneswari K, Palanisamy G, Pazhanivel T et al (2021) In-situ development of metal organic frameworks assisted ZnMgAl layered triple hydroxide 2D/2D hybrid as an efficient photocatalyst for organic dye degradation. Chemosphere 270:128616. https://doi.org/10.1016/j.chemosphere.2020.128616
92. Zeng T, Wang L, Feng L et al (2019) Two novel organic phosphorous-based MOFs: synthesis, characterization and photocatalytic properties. Dalt Trans 48:523–534. https://doi.org/10.1039/C8DT04106G
93. Zulys A, Adawiah A, Gunlazuardi J, Yudhi MDL (2021) Light-harvesting metal-organic frameworks (MOFs) La-PTC for photocatalytic dyes degradation. Bull Chem React Eng Catal 16:170–178. https://doi.org/10.9767/bcrec.16.1.10309.170-178
94. Zhao P, Wang J, Han X et al (2021) Zr-porphyrin metal-organic framework-based photocatalytic self-cleaning membranes for efficient dye removal. Ind Eng Chem Res 60:1850–1858. https://doi.org/10.1021/acs.iecr.0c05583
95. Farrokhi A, Bivareh F, Dejbakhshpour S, Moghaddam AZ (2020) Insight into the photocatalytic properties of phosphonate-based metal–organic frameworks for reduction of Cr (VI) and synergistic elimination of organic dyes under natural sunlight. Appl Organomet Chem 34. https://doi.org/10.1002/AOC.5938
96. Alamgir TK, Wang YJ et al (2021) Construction of a mixed ligand MOF as “green catalyst” for the photocatalytic degradation of organic dye in aqueous media. RSC Adv 11:23838–23845. https://doi.org/10.1039/d1ra02994k
97. Zeng T, Shi D, Cheng Q et al (2020) Construction of novel phosphonate-based MOF/P–TiO_2 heterojunction photocatalysts: enhanced photocatalytic performance and mechanistic insight. Environ Sci Nano 7:861–879. https://doi.org/10.1039/C9EN01180C
98. Samuel MS, Savunthari KV, Ethiraj S (2021) Synthesis of a copper (II) metal–organic framework for photocatalytic degradation of rhodamine B dye in water. Environ Sci Pollut Res
99. Yi XH, Wang FX, Du XD et al (2018) Highly efficient photocatalytic Cr(VI) reduction and organic pollutants degradation of two new bifunctional 2D Cd/Co-based MOFs. Polyhedron 152:216–224. https://doi.org/10.1016/J.POLY.2018.06.041
100. Zhu X, Li B, Yang J et al (2014) Effective adsorption and enhanced removal of organophosphorus pesticides from aqueous solution by Zr-based MOFs of UiO-67. ACS Appl Mater Interfaces 7:223–231. https://doi.org/10.1021/AM5059074
101. Mohaghegh N, Tasviri M, Rahimi E, Gholami MR (2015) Comparative studies on Ag3PO4/BiPO4–metal-organic framework–graphene-based nanocomposites for photocatalysis application. Appl Surf Sci Complete 216–224. https://doi.org/10.1016/J.APSUSC.2015.05.135
102. Singh S, Kaushal S, Kaur J et al (2021) CaFu MOF as an efficient adsorbent for simultaneous removal of imidacloprid pesticide and cadmium ions from wastewater. Chemosphere 272
103. Sun Q, Liu M, Li K et al (2016) Controlled synthesis of mixed-valent Fe-containing metal organic frameworks for the degradation of phenol under mild conditions. Dalt Trans 45:7952–7959. https://doi.org/10.1039/C5DT05002B
104. Li Y, Fang Y, Cao Z et al (2019) Construction of g-C3N4/PDI@MOF heterojunctions for the highly efficient visible light-driven degradation of pharmaceutical and phenolic micropollutants. Appl Catal B Environ 250:150–162. https://doi.org/10.1016/J.APCATB.2019.03.024

105. Samuel MS, Bhattacharya J, Parthiban C et al (2018) Ultrasound-assisted synthesis of metal organic framework for the photocatalytic reduction of 4-nitrophenol under direct sunlight. Ultrason Sonochem 49:215–221. https://doi.org/10.1016/J.ULTSONCH.2018.08.004
106. Kusutaki T, Katsumata H, Tateishi I et al (2019) Tetrahedral UMOFNs/Ag3PO4 core-shell photocatalysts for enhanced photocatalytic activity under visible light. ACS Omega 4:15975–15984. https://doi.org/10.1021/ACSOMEGA.9B02042
107. Wang D, Jia F, Wang H et al (2018) Simultaneously efficient adsorption and photocatalytic degradation of tetracycline by Fe-based MOFs. J Colloid Interface Sci 519:273–284. https://doi.org/10.1016/J.JCIS.2018.02.067
108. Chen DD, Yi XH, Zhao C, et al (2020) Polyaniline modified MIL-100(Fe) for enhanced photocatalytic Cr(VI) reduction and tetracycline degradation under white light. Chemosphere 245:125659. https://doi.org/10.1016/J.CHEMOSPHERE.2019.125659
109. Li X, Zhang D, Bai R et al (2020) Zr-MOFs based BiOBr/UiO-66 nanoplates with enhanced photocatalytic activity for tetracycline degradation under visible light irradiation. AIP Adv 10:125228. https://doi.org/10.1063/5.0030228
110. Lin B, Li S, Peng Y et al (2021) MOF-derived core/shell C-TiO_2/$CoTiO_3$ type II heterojunction for efficient photocatalytic removal of antibiotics. J Hazard Mater 406:124675. https://doi.org/10.1016/j.jhazmat.2020.124675
111. Zhou C, Lai C, Xu P et al (2018) In situ grown AgI/Bi12O17Cl2 heterojunction photocatalysts for visible light degradation of sulfamethazine: efficiency, pathway, and mechanism. ACS Sustain Chem Eng 6:4174–4184. https://doi.org/10.1021/ACSSUSCHEMENG.7B04584
112. Ding X, Liu H, Chen J et al (2020) In situ growth of well-aligned Ni-MOF nanosheets on nickel foam for enhanced photocatalytic degradation of typical volatile organic compounds. Nanoscale 12:9462–9470. https://doi.org/10.1039/D0NR01027H
113. Li P, Kim S, Jin J et al (2020) Efficient photodegradation of volatile organic compounds by iron-based metal-organic frameworks with high adsorption capacity. Appl Catal B Environ 263:118284. https://doi.org/10.1016/J.APCATB.2019.118284
114. Nasalevich MA, Goesten MG, Savenije TJ et al (2013) Enhancing optical absorption of metal–organic frameworks for improved visible light photocatalysis. Chem Commun 49:10575–10577. https://doi.org/10.1039/C3CC46398B
115. Sun D, Kim D-P (2020) Hydrophobic MOFs@metal nanoparticles@COFs for interfacially confined photocatalysis with high efficiency. ACS Appl Mater Interfaces 12:20589–20595. https://doi.org/10.1021/ACSAMI.0C04537
116. Chen L, Wang X, Rao Z et al (2021) In-situ synthesis of Z-scheme MIL-100(Fe)/α-Fe_2O_3 heterojunction for enhanced adsorption and Visible-light photocatalytic oxidation of O-xylene. Chem Eng J 416:129112. https://doi.org/10.1016/j.cej.2021.129112
117. Chambers MB, Wang X, Ellezam L et al (2017) Maximizing the photocatalytic activity of metal-organic frameworks with aminated-functionalized linkers: sub-stoichiometric effects in MIL-125-NH2 maximizing the photocatalytic activity of metal-organic frame-works with aminated-functionalized linkers: sub-stoichiometric effects in MIL-125-NH2. J Am Chem Soc 139. https://doi.org/10.1021/jacs.7b02186
118. Zhang R, Liu Y, Wang Z et al (2019) Selective photocatalytic conversion of alcohol to aldehydes by singlet oxygen over Bi-based metal-organic frameworks under UV–vis light irradiation. Appl Catal B Environ 254:463–470. https://doi.org/10.1016/J.APCATB.2019.05.024
119. Campanelli M, Del GT, De AF et al (2019) Solvent-free synthetic route for Cerium(IV) metal-organic frameworks with UiO-66 architecture and their photocatalytic applications. ACS Appl Mater Interfaces 11:45031–45037. https://doi.org/10.1021/ACSAMI.9B13730
120. Wang H, Yuan X, Wu Y et al (2015) Facile synthesis of amino-functionalized titanium metal-organic frameworks and their superior visible-light photocatalytic activity for Cr(VI) reduction. J Hazard Mater 286:187–194. https://doi.org/10.1016/J.JHAZMAT.2014.11.039
121. Huang W, Liu N, Zhang X et al (2017) Metal organic framework g-C3N4/MIL-53(Fe) heterojunctions with enhanced photocatalytic activity for Cr(VI) reduction under visible light. Appl Surf Sci 425:107–116. https://doi.org/10.1016/J.APSUSC.2017.07.050

122. Wang X, Liu W, Fu H et al (2019) Simultaneous Cr(VI) reduction and Cr(III) removal of bifunctional MOF/Titanate nanotube composites. Environ Pollut 249:502–511. https://doi.org/10.1016/J.ENVPOL.2019.03.096
123. Du XD, Yi XH, Wang P et al (2019) Robust photocatalytic reduction of Cr(VI) on UiO-66-NH2(Zr/Hf) metal-organic framework membrane under sunlight irradiation. Chem Eng J 356:393–399. https://doi.org/10.1016/J.CEJ.2018.09.084
124. Niu G, Si C, Jiao J et al (2020) An electron-rich metal-organic framework for highly efficient photocatalytic reduction of Cr(VI). J Alloys Compd 830. https://doi.org/10.1016/J.JALLCOM.2020.154696
125. Kaur H, Sinha S, Krishnan V, Koner RR (2020) Photocatalytic reduction and recognition of Cr(VI): new Zn(II)-based metal-organic framework as catalytic surface. Ind Eng Chem Res 59:8538–8550. https://doi.org/10.1021/ACS.IECR.9B06417
126. Xie H, Ma D, Liu W et al (2020) Zr-Based MOFs as new photocatalysts for the rapid reduction of Cr(VI) in water. New J Chem 44:7218–7225. https://doi.org/10.1039/D0NJ00457J
127. Garazhian Z, Farrokhi A, Rezaeifard A et al (2021) The enhanced visible-light-induced photocatalytic activities of bimetallic Mn-Fe MOFs for the highly efficient reductive removal of Cr(vi). RSC Adv 11:21127–21136. https://doi.org/10.1039/d1ra01986d
128. Wang H, Zhu QL, Zou R, Xu Q (2017) Metal-organic frameworks for energy applications. Chem 2:52–80. https://doi.org/10.1016/J.CHEMPR.2016.12.002
129. Liu X, Inagaki S, Gong J (2016) Heterogeneous molecular systems for photocatalytic CO_2 reduction with water oxidation. Angew Chem Int Ed Engl 55:14924–14950. https://doi.org/10.1002/ANIE.201600395
130. Zhao S-N, Wang G, Poelman D, Voort P Van Der (2018) Metal organic frameworks based materials for heterogeneous photocatalysis. Molecules 23:2947. https://doi.org/10.3390/MOLECULES23112947
131. Burtch NC, Jasuja H, Walton KS (2014) Water stability and adsorption in metal-organic frameworks. Chem Rev 114:10575–10612. https://doi.org/10.1021/CR5002589
132. Pullen S, Fei H, Orthaber A et al (2013) Enhanced photochemical hydrogen production by a molecular diiron catalyst incorporated into a metal-organic framework. J Am Chem Soc 135:16997–17003. https://doi.org/10.1021/JA407176P
133. El Naggar AMA, Kazak C (2016) Preparation and characterization of novel nano-structured porous nickel alloy composite induced by electroless deposition and its performance in the hydrogen separation. Sep Purif Technol 160:73–80. https://doi.org/10.1016/J.SEPPUR.2016.01.015
134. Wang C, Liu D, Lin W (2013) Metal-organic frameworks as a tunable platform for designing functional molecular materials. J Am Chem Soc 135:13222–13234. https://doi.org/10.1021/JA308229P
135. Du P, Schneider J, Jarosz P, Eisenberg* R (2006) Photocatalytic generation of hydrogen from water using a platinum(II) terpyridyl acetylide chromophore. J Am Chem Soc 128:7726–7727. https://doi.org/10.1021/JA0610683
136. Silva CG, Corma A, García H (2010) Metal–organic frameworks as semiconductors. J Mater Chem 20:3141–3156. https://doi.org/10.1039/B924937K
137. Lee DY, Shinde DV, Yoon SJ et al (2013) Cu-based metal-organic frameworks for photovoltaic application. J Phys Chem C 118:16328–16334. https://doi.org/10.1021/JP4079663
138. Gascon J, Hernández-Alonso MD, Almeida AR et al (2008) Isoreticular MOFs as efficient photocatalysts with tunable band gap: an operando FTIR study of the photoinduced oxidation of propylene. ChemSusChem 1:981–983. https://doi.org/10.1002/CSSC.200800203
139. Zeng MH, Wang QX, Tan YX et al (2010) Rigid pillars and double walls in a porous metal-organic framework: single-crystal to single-crystal, controlled uptake and release of iodine and electrical conductivity. J Am Chem Soc 132:2561–2563. https://doi.org/10.1021/JA908293N
140. Lee DY, Kim E-K, Shin CY et al (2014) Layer-by-layer deposition and photovoltaic property of Ru-based metal–organic frameworks. RSC Adv 4:12037–12042. https://doi.org/10.1039/C4RA00397G

141. Zhan WW, Kuang Q, Zhou et al (2013) Semiconductor@metal-organic framework core-shell heterostructures: a case of ZnO@ZIF-8 nanorods with selective photoelectrochemical response. J Am Chem Soc 135:1926–1933. https://doi.org/10.1021/JA311085E
142. Ding Z, Wang S, Chang X et al (2020) Nano-MOF@defected film C3N4 Z-scheme composite for visible-light photocatalytic nitrogen fixation. RSC Adv 10:26246–26255. https://doi.org/10.1039/D0RA03562A
143. Chen L-W, Hao Y-C, Guo Y et al (2021) Metal-organic framework membranes encapsulating gold nanoparticles for direct plasmonic photocatalytic nitrogen fixation. J Am Chem Soc 143:5727–5736. https://doi.org/10.1021/JACS.0C13342
144. Gleick PH (2009) A look at twenty-first century water resources development. Water Int 25:127–138. https://doi.org/10.1080/02508060008686804
145. Wang C-C, Li J-R, Lv X-L et al (2014) Photocatalytic organic pollutants degradation in metal–organic frameworks. Energy Environ Sci 7:2831–2867. https://doi.org/10.1039/C4EE01299B
146. Rasheed T, Ahmad N, Nawaz S, Sher F (2020) Photocatalytic and adsorptive remediation of hazardous environmental pollutants by hybrid nanocomposites. Case Stud Chem Environ Eng 2:100037. https://doi.org/10.1016/J.CSCEE.2020.100037
147. Rasheed T, Hassan AA, Bilal M et al (2020) Metal-organic frameworks based adsorbents: a review from removal perspective of various environmental contaminants from wastewater. Chemosphere 259:127369. https://doi.org/10.1016/J.CHEMOSPHERE.2020.127369
148. Marson JW, Baldwin HE (2019) An overview of acne therapy, part 1: topical therapy, oral antibiotics, laser and light therapy, and dietary interventions. Dermatol Clin 37:183–193. https://doi.org/10.1016/J.DET.2018.12.001
149. Baran W, Adamek E, Ziemiańska J, Sobczak A (2011) Effects of the presence of sulfonamides in the environment and their influence on human health. J Hazard Mater 196:1–15. https://doi.org/10.1016/J.JHAZMAT.2011.08.082
150. Wammer KH, Slattery MT, Stemig AM, Ditty JL (2011) Tetracycline photolysis in natural waters: loss of antibacterial activity. Chemosphere 85:1505–1510. https://doi.org/10.1016/J.CHEMOSPHERE.2011.08.051
151. Zhou Y, Liu X, Xiang Y et al (2017) Modification of biochar derived from sawdust and its application in removal of tetracycline and copper from aqueous solution: adsorption mechanism and modelling. Bioresour Technol 245:266–273. https://doi.org/10.1016/J.BIORTECH.2017.08.178
152. Qi N, Wang P, Wang C, Ao Y (2018) Effect of a typical antibiotic (tetracycline) on the aggregation of TiO_2 nanoparticles in an aquatic environment. J Hazard Mater 341:187–197. https://doi.org/10.1016/J.JHAZMAT.2017.07.046
153. Tan Z, Lu K, Jiang M et al (2019) Daytime atmospheric oxidation capacity in four Chinese megacities during the photochemically polluted season: a case study based on box model simulation. Atmos Chem Phys 19:3493–3513. https://doi.org/10.5194/ACP-19-3493-2019
154. Meng J, Wang X, Yang X et al (2019) Enhanced gas-phase photocatalytic removal of aromatics over direct Z-scheme-dictated H3PW12O40/g-C3N4 film-coated optical fibers. Appl Catal B Environ 251:168–180. https://doi.org/10.1016/J.APCATB.2019.03.063
155. Boyjoo Y, Sun H, Liu J et al (2017) A review on photocatalysis for air treatment: from catalyst development to reactor design. Chem Eng J 310:537–559. https://doi.org/10.1016/J.CEJ.2016.06.090
156. Fujishima A, Honda K (1972) Electrochemical photolysis of water at a semiconductor electrode. Nature 238(5358):37–38. https://doi.org/10.1038/238037a0
157. Mahmood A, Wang X, Shi G et al (2020) Revealing adsorption and the photodegradation mechanism of gas phase o-xylene on carbon quantum dots modified TiO_2 nanoparticles. J Hazard Mater 386. https://doi.org/10.1016/J.JHAZMAT.2019.121962
158. López-Magano A, Jiménez-Almarza A, Alemán J, Mas-Ballesté R (2020) Metal–organic frameworks (MOFs) and covalent organic frameworks (COFs) applied to photocatalytic organic transformations. Catalysts 10:720. https://doi.org/10.3390/CATAL10070720

159. Ke F, Wang L, Zhu J (2015) Facile fabrication of CdS-metal-organic framework nanocomposites with enhanced visible-light photocatalytic activity for organic transformation. Nano Res 86(8):1834–1846. https://doi.org/10.1007/S12274-014-0690-X
160. Long J, Wang S, Ding Z et al (2012) Amine-functionalized zirconium metal–organic framework as efficient visible-light photocatalyst for aerobic organic transformations. Chem Commun 48:11656–11658. https://doi.org/10.1039/C2CC34620F
161. Wei H, Guo Z, Liang X et al (2019) Selective photooxidation of amines and sulfides triggered by a superoxide radical using a novel visible-light-responsive metal-organic framework. ACS Appl Mater Interfaces 11:3016–3023. https://doi.org/10.1021/ACSAMI.8B18206
162. Hsu C-L, Wang S-L, Tzou* Y-M, (2007) Photocatalytic reduction of Cr(VI) in the presence of NO3- and CI- electrolytes as influenced by Fe(III). Environ Sci Technol 41:7907–7914. https://doi.org/10.1021/ES0718164
163. Xia Q, Huang B, Yuan X et al (2018) Modified stannous sulfide nanoparticles with metal-organic framework: toward efficient and enhanced photocatalytic reduction of chromium (VI) under visible light. J Colloid Interface Sci 530:481–492. https://doi.org/10.1016/J.JCIS.2018.05.015
164. Wang C, Liu X, Demir NK et al (2016) Applications of water stable metal–organic frameworks. Chem Soc Rev 45:5107–5134. https://doi.org/10.1039/C6CS00362A
165. Li Z, Wang L, Qin L et al (2021) Recent advances in the application of water-stable metal-organic frameworks: adsorption and photocatalytic reduction of heavy metal in water. Chemosphere 285:131432. https://doi.org/10.1016/J.CHEMOSPHERE.2021.131432
166. Shen M, Forghani F, Kong X et al (2020) Antibacterial applications of metal–organic frameworks and their composites. Compr Rev Food Sci Food Saf 19:1397–1419. https://doi.org/10.1111/1541-4337.12515
167. Wyszogrodzka G, Marszałek B, Gil B, Dorożyński P (2016) Metal-organic frameworks: mechanisms of antibacterial action and potential applications. Drug Discov Today 21:1009–1018. https://doi.org/10.1016/J.DRUDIS.2016.04.009
168. Li P, Li J, Feng X et al (2019) Metal-organic frameworks with photocatalytic bactericidal activity for integrated air cleaning. Nat Commun 101(10):1–10. https://doi.org/10.1038/s41467-019-10218-9
169. Mohaghegh N, Faraji M, Abedini A (2018) Highly efficient multifunctional Ag/TiO_2 nanotubes/Ti plate coated with MIL-88B(Fe) as a photocatalyst, adsorbent, and disinfectant in water treatment. Appl Phys A 1251(125):1–10. https://doi.org/10.1007/S00339-018-2324-8
170. Barea E, Carmen Montoro R, Navarro JA (2014) Toxic gas removal—metal–organic frameworks for the capture and degradation of toxic gases and vapours. Chem Soc Rev 43:5419–5430. https://doi.org/10.1039/C3CS60475F
171. Wen M, Li G, Liu H et al (2019) Metal–organic framework-based nanomaterials for adsorption and photocatalytic degradation of gaseous pollutants: recent progress and challenges. Environ Sci Nano 6:1006–1025. https://doi.org/10.1039/C8EN01167B
172. Furukawa H, Cordova KE, O'Keeffe M, Yaghi OM (2013) The chemistry and applications of metal-organic frameworks. Science 341. https://doi.org/10.1126/SCIENCE.1230444
173. Liu Y, Howarth AJ, Hupp JT, Farha OK (2015) Selective photooxidation of a mustard-gas simulant catalyzed by a porphyrinic metal-organic framework. Angew Chemie Int Ed 54:9001–9005. https://doi.org/10.1002/ANIE.201503741
174. Alshammari A, Jiang Z, Cordova KE (2016) Metal organic frameworks as emerging photocatalysts. Semicond Photocatal Mater Mech Appl. https://doi.org/10.5772/63489
175. Li X, Yu J, Jaroniec M (2016) Hierarchical photocatalysts. Chem Soc Rev 45:2603–2636. https://doi.org/10.1039/C5CS00838G
176. Li X, Yu J, Low J et al (2015) Engineering heterogeneous semiconductors for solar water splitting. J Mater Chem A 3:2485–2534. https://doi.org/10.1039/C4TA04461D
177. Gligorovski S, Strekowski R, Barbati S, Vione D (2015) Environmental implications of hydroxyl radicals (·OH). Chem Rev 115:13051–13092. https://doi.org/10.1021/CR500310B

178. Wen Y, Feng M, Zhang P et al (2021) Metal organic frameworks (MOFs) as photocatalysts for the degradation of agricultural pollutants in water. ACS ES&T Eng 1:804–826. https://doi.org/10.1021/ACSESTENGG.1C00051
179. Liu Y, Liu Z, Huang D et al (2019) Metal or metal-containing nanoparticle@MOF nanocomposites as a promising type of photocatalyst. Coord Chem Rev 388:63–78. https://doi.org/10.1016/J.CCR.2019.02.031
180. Low J, Cheng B, Yu J et al (2017) Surface modification and enhanced photocatalytic CO_2 reduction performance of TiO_2: a review. ApSS 392:658–686. https://doi.org/10.1016/J.APSUSC.2016.09.093

Metal–Organic Frameworks (MOFs) Catalyzed Photodegradation of Organic Dyes: Syntheses, Designing Strategies, and Mechanism

Ayushi Singh and Abhinav Kumar

Contents

1 Introduction 645
2 Synthetic Approaches, Designing, and Structures of Photosensitive MOFs 648
2.1 Synthetic Approaches 648
2.2 Designing 650
2.3 Structures and Photocatalyses of MOFs 652
3 Photocatalytic Mechanism 666
3.1 Direct Photodegradation Process 667
3.2 Sensitization-Mediated Degradation Process 669
4 Conclusions 669
Abbreviations 670
References 671

Abstract Metal–organic frameworks (MOFs) or coordination polymers (CPs) are an assorted class of materials finding applications in areas, viz. sensing and absorption/separation of gas/small molecules, storage of gas, etc. In addition, MOF integrating nanoparticles exhibit display varied photocatalytic behavior. Nevertheless, the application of MOFs as photocatalysts had recently been investigated and explored. The presented chapter lays attention on synthetic protocols, designing stratagems, and architectures of photosensitive MOFs. Also, it will be providing details about the mechanistic pathway employing which MOFs execute photocatalytic degradation of organic dyes mostly existing in industrial wastewater.

Keywords Metal–organic frameworks · Photocatalysis · Organic dyes · Mechanism

1 Introduction

During the past twenty years, increasing interest had been witnessed to investigate a new hybrid variety of nanoporous, crystalline materials "metal–organic frameworks"

A. Singh · A. Kumar (✉)
Department of Chemistry, Faculty of Science, University of Lucknow, Lucknow 226007, India
e-mail: kumar_abhinav@lkouniv.ac.in

S. Gulati (ed.), *Metal-Organic Frameworks (MOFs) as Catalysts*,
https://doi.org/10.1007/978-981-16-7959-9_24

(MOFs). These are known as hybrid materials because they comprise metallic nodes connected through organic linkers [1]. These emerging materials find application in a plethora of application areas [2–12]. Additionally, these materials offer a tuneable framework, large surface area per unit volume as well as porosity. This allows swift mass transport as well as interactions with substrate molecules. Apart from that, they have coordinatively unsaturated metal centers and exposed sites which induce catalytic properties in them [13, 14]. Also, a plethora of MOFs on irradiation has exhibited semiconducting behavior which in turn induces photocatalytic properties in them. Because of these peculiar features associated with these materials, a range of MOFs has been deployed as photocatalysts for safe and sustainable degradation of aromatic pollutants, as the catalyst for the evolution of hydrogen gas, for CO_2 reduction, water cleavage, etc. [15–17]. These materials lay an apt platform to develop heterogeneous catalysts for photocatalyzed degradation reactions. This is because these materials are (1) highly efficient, (2) can operate under ambient temperature, (3) offer good reproducibility and recyclability, (4) easy to handle, and (5) mineralize toxic contaminants to simple non-toxic compounds [18].

The aromatic entity of ligands in MOFs on irradiation generates charge-separated states, which induces light-harvesting properties in them and makes them a photocatalyst. MOF-based photocatalysts offer several advantages over conventional photocatalysts: (a) Their suitable pore dimensions enable the movement of organic pollutant and concomitant movement of the mineralized products through their open channels; (b) the molecular framework as well as the electron communication features of MOFs can be easily regulated; (c) this class of materials could be synthesized by many synthetic protocols which lead to higher crystallinity and diverse morphology in them [19]. Additionally, the stability of MOF is decided by several factors such as nature of metal ions, coordination numbers of metal cations and connectivity of metal clusters, its basic character, and hydrophobicity of ligands. However, the existence of stable framework with porosity with open secondary building unit (SBU) facilitates adsorption of organic dyes in/on the MOF which then could be decomposed [20–22]. Apart from this, using post-synthetic modification strategy, additional functionalization can be introduced in MOFs which uplifts their efficiency to photochemically decompose the organic dyes, [23, 24] a prominent organic contaminants existing in water bodies [25]. Presently, over 1000,000 commercial dyes are produced from industries [26–28] which are chemically stable and remain unaffected by light, heat, and oxidizing agents. Thus, the removal of these carcinogenic organic dyes is highly required. Common dye substrates for photodegradation by MOFs are presented in Scheme 1.

Hence, for a clean environment as well as for the well-being of humans and animals, the removal of these organic dyes is of utmost importance where MOFs can play a vital and promising role as photocatalysts. The chapter presented herein provides a comprehension of the recent development in the use of MOFs as photocatalysts for the decomposition of toxic aromatic dyes. Apart from that, the synthetic protocols deployed for MOFs suitable for photocatalysis are presented. Also, the MOFs comprising main group metals, transition metals, and *f*-block metals with

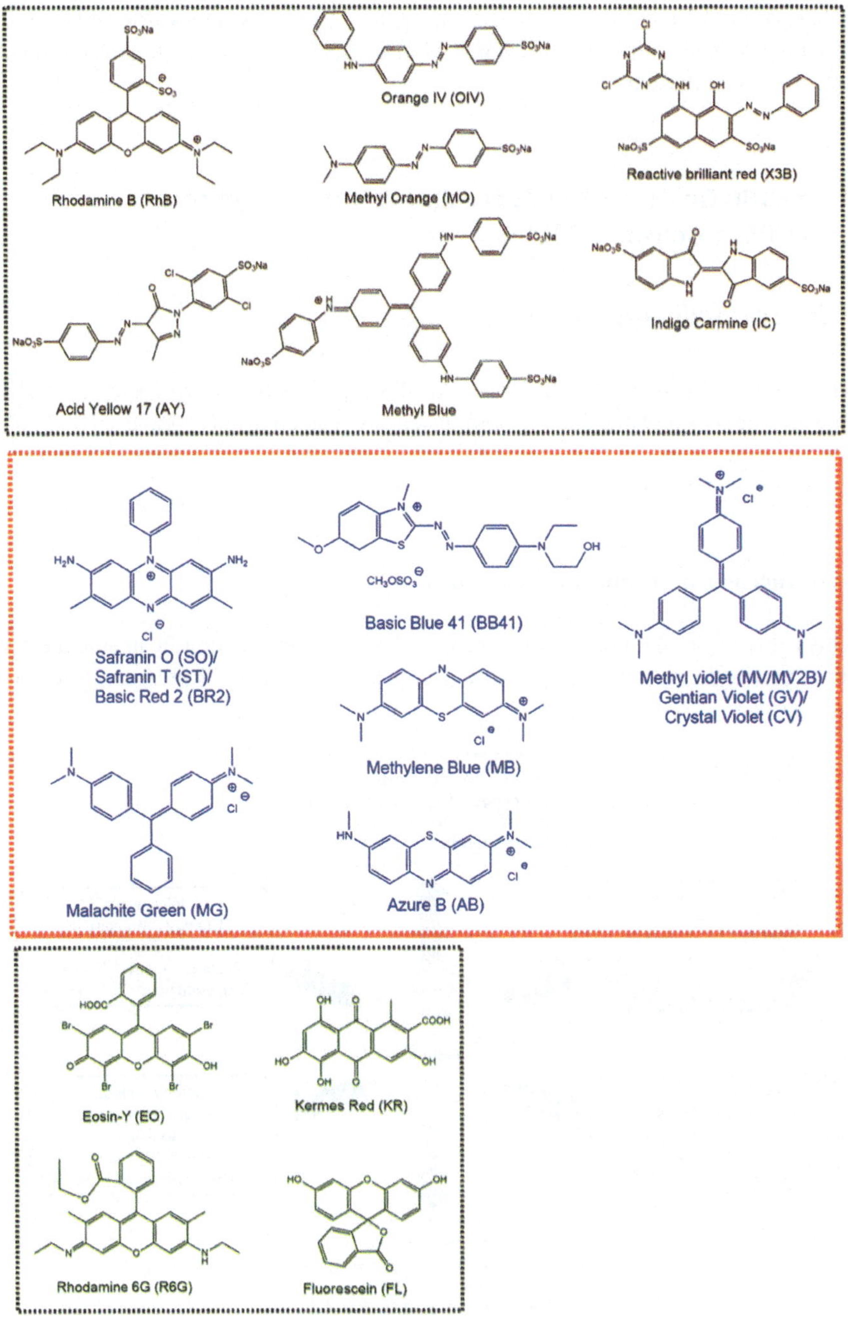

Scheme 1 Chemical structures of some widely used dyes

enhanced performances as photocatalysts against dye's decomposition and eventually mechanistic pathway using which MOFs usually execute photodegradation of dyes had been discussed.

2 Synthetic Approaches, Designing, and Structures of Photosensitive MOFs

2.1 *Synthetic Approaches*

The liquid-phase syntheses are conventionally deployed for the syntheses of MOFs, in which ligands and metal ion nodes/clusters are reacted in appropriate solvent for a specific time period. Figure 1 recapitulates various approaches as well as conditions for MOF syntheses [29, 30].

Solvothermal/Hydrothermal Approach

This is the most commonly used approach that executes reaction in an autoclave at the prescribed controlled temperature and/or under autogenous pressure. Solvents,

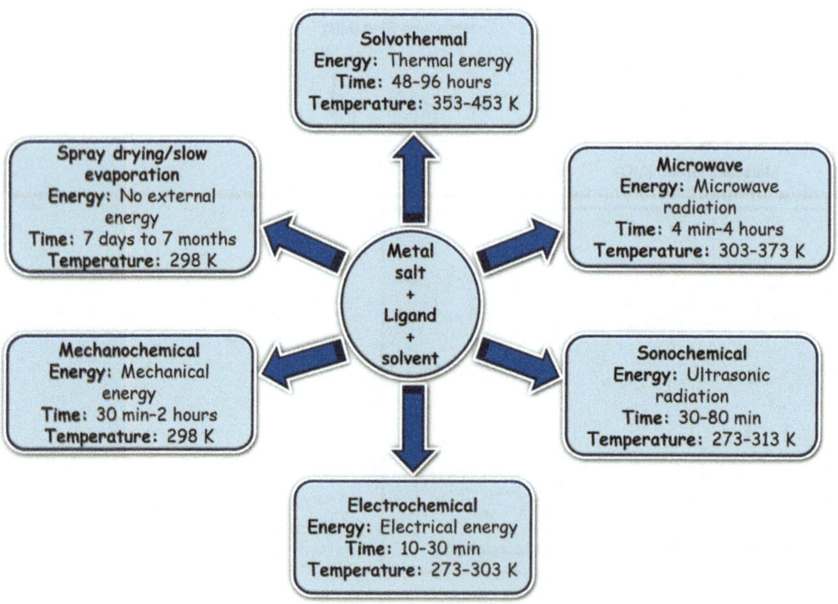

Fig. 1 Various protocols and synthetic conditions used for the preparation of MOF

such as dimethylformamide, dimethylformamide, and acetonitrile, which have high boiling points are used in this approach [31, 32]. Apart from this, solvent mixtures are employed to sort out the problem of solubility. But, this protocol suffers from some limitations like extensive reaction period, elevated temperature, and expensive solvents. Hence, hydrothermal synthetic protocol utilizing water renders a more eco-compatible and economic method.

Microwave Approach

This is a fast synthetic protocol in which microwave (MW) irradiation/heating results in better crystallinity, development of perforated architecture, reduced particle size with better control over the morphology of MOFs [31–33]. Also, for this synthetic protocol effect of variation in power, temperature, solvent concentration, and substrate, irradiation time had been explained by Klinowski et al. [32]. Also, well-known MOFs were synthesized deploying solvothermal method and have also been synthesized using this method [34–38].

Sonochemical Approach

This protocol using sound waves assists in the growth of consistent nucleation centers and speeds up the crystallization than the solvothermal protocol. Using this approach, many MOFs had been synthesized, e.g., MOF-5 having 5–25 mm crystals were isolated sonochemically using 1-methyl-2-pyrrolidinone in half an hour [39, 40]. Here also, how variation in preparation conditions/power, etc., affect structure and porosity of MOFs was explained [40].

Electrochemical Approach

This process is used for the fabrication of MOF-based thin films over the substrate's surface in mild conditions that decreases the possibility of film cracking normally encountered in other protocols. Direct electro-syntheses can be performed through anodic solubilization of metal cations and/or deprotonation of organic ligand [41, 42]. In contrast to other approaches, this method does not necessitate the use of metal salts as precursors as the metal ions generated by oxidizing electrodes. Also, this procedure could be used for industrial-scale production of MOFs [41–43].

Mechanochemical Approach

In principle, three mechanochemical methods such as (1) grinding under neat conditions, (2) grinding with the assistance of liquid, and (3) ion/liquid-assisted grinding are used [44–47]. Neat grinding is a solvent-free method but liquid-assisted grinding needs a small amount of liquid while the second methodology needs a small quantity of liquid and salt in traces. There is an upsurge in the interest in mechanochemical syntheses as it requires a very less quantity of solvent and eco-friendly nature. Also, less energy is required, and thus it can be regarded as a highly efficient and simple protocol [44–48].

Slow Evaporation Approach

This method operates at room temperature and does not need additional energy. But the significant crystallization period is the chief disadvantage associated with this method.

Apart from the abovementioned methods, a few less-explored methods used for the syntheses of MOFs are (1) diffusion, [49] (2) ionothermal, [50] and (3) spray-drying approaches [51]. The second approach utilizes ionic liquids as a solvent and also engenders template and electro-neutralizing groups. Its utilization offers several advantages, like low melting, good heat stability, and a broad liquifiable range. Additionally, by changing IL composition, the MOF's architecture could be altered which results in alteration in physicochemical properties. The range of MOFs prepared using various methods, exhibited important variation in yield, size, surface area, morphology, and porosity using the IL method [52–67].

The final stage of MOF synthesis is their purification and/or activation, which removes co-crystallized linkers, metal oxides, which are untrapped in apertures of the main material. This step can be performed using "sieving" operating based on the differences in the densities of the main material, and impurities or purification can be accomplished by the elevated temperature solvent treatment. The MOF activation is performed by removing guest molecules from the pores by low pressure (vacuum) drying, supercritical drying, and/or solvent exchange [31, 32].

2.2 Designing

To develop an appropriate MOF that could be utilized as a photocatalyst, the essential requirements are its (1) high stability in different types of solvents, (2) stability toward variation in pH, and (3) stability under strong radiation. Additionally, MOF-based photocatalysts should have low cost and could be synthesized and utilized easily. As mentioned earlier, the rational assembly of SBU and organic linkers engenders MOFs. Hence, the alteration like organic linkers as well as SBUs results in a varied class of frameworks. To control over porosity and required properties, appropriate MOF designing is needed [67]. To design a targeted MOF that could be utilized as a photocatalyst, the presence of a light-harvesting center, reaction sites, and rational charge/electron transport are important requirements. To introduce a light-absorbing center in the targeted MOF, the strong absorption of visible/IR radiation is needed for enhanced capturing of sunlight [68]. Hence, ligands with extended π-conjugation improve electron delocalization and lead to a required bathochromic shift in the UV–Vis spectrum. Further, to introduce catalytic centers, the presence of sole metal or metal oxide centers where catalytic reaction generates is required. Thus, introducing extensive conjugation in ligand and adding electron-rich metal cluster decrease bandgap and enhance the photochemical reaction. Apart from this, in situ functionalization is another stratagem that further improves photochemical properties. Also, post-synthetic modification can be employed which alters MOF's structure [23, 24]. Such modification reduces the electron/hole recombination and uplifts charge mobility. Additionally, the photocatalysts having small particle size also decline the electron/hole recombination and uplift catalytic performance [69]. Apart from this, to design the desired photocatalysts, the energy parameter of the light-capturing center as well as reaction sites must be analogous for easy electron transfer.

Hence, overall it could be concluded that to construct MOF-based photocatalysts, rational designing of the single organic linker is pivotal as it executes varied functions such as assembling a framework by coordinating at least two-reaction centers, light-harvesting, electrons, and holes transfer. The nature of antenna, anchors, and extent of conjugation in aromatic ligand considerably affect the kinetics of charge transfer [70].

Antennae

In a specific organic entity behaving as a ligand, the antenna acts as a light/photon harvester. In such a ligand system, varied functionalities act as electron donors to π-bonded electron cloud of the organic linker which in turn lowers band-gap energy of MOF [15, 71–80].

Anchoring Groups

The anchoring tendency of the ligand plays a pivotal role in the transfer of electrons, and it also stabilizes the MOF's architecture. Apart from that, charge transfer rates at the time of photocatalysis are also controlled by these anchors [81]. Hence, the nature of anchor plays a key role in uplifting the performance of any MOF employed as a photocatalyst. For example, the –COOH group exhibits relatively faster electron transfer than the $-PO_3H_2$ group. This could be attributed to the smaller size and weight of carbon as compared to phosphorus and strong double bonding over a single bond. This observation was corroborated by using non-adiabatic MD simulation techniques. Also, quantum chemical calculations suggested that the –COOH group possesses relatively more electron density than the $-PO_3H_2$ group. Hence, as compared to the $-PO_3H_2$ group, –COOH groups are a good option as an anchor in MOF being employed as photocatalysts.

Charge Transportation Phenomenon

The charge transportation phenomenon takes place through ligand or via space. This phenomenon relies on supramolecular architecture and the extent of the conjugation ligand. In long-range transport, charge recombination and relaxation phenomena enhance the utilization of charge carriers. This recombination process could be decreased by curtailing the chain/bridge length of the ligand. But, this leads to concomitant suppression in the rate of charge transfer because of the shift in the electronic absorption spectrum toward the UV region [82]. However, incorporation of an antenna/electron–donor extends organic linker's conjugation which enhances light-capturing capacity with subsequent acceleration in charge transfer kinetics, and reverse phenomenon operates with the incorporation of the electron-withdrawing group [83]. Hence, linkers play a vital role in the electron transport pathway as electrons might be restricted at the organic connector for a small time before arriving in catalytic sites and thereby facilitating the catalytic reaction [84].

2.3 Structures and Photocatalyses of MOFs

Metal centers from all four blocks of the periodic table have been employed for the design and fabrication of MOF-based photocatalysts. The most commonly used ligands employed for the syntheses of these MOFs include poly-/multi-topic carboxylate and nitrogen-containing heterocyclic ligands. These ligands are extensively conjugated and had been used as connectors/linkers for MOFs [85]. On the basis of the nature of metal ions, in this section MOFs had been classified into the main group, transition, lanthanides, and actinide metals, and their utility as photocatalysts have been summarized in the presented section.

Main Group Metal-Based MOFs

The use of main group metal center-based MOFs as photocatalysts for the degradation of dyes had not been investigated comprehensively. However, Li et al. reported an indium-based MOF $\{[In_3OCl(H_2O)_2(\mathbf{L^1})_2]\}_n$ (where $\mathbf{L^1}$ = 4,4′,4″-s-triazine-1,3,5-triyl-tri-*p*-aminobenzoate) which degraded 94.6% rhodamine B (RhB) in 70 min and ~93.1% methylene blue (MB) in 190 min under 300 W xenon lamp irradiation [86].

Another anionic indium-based MOF, $[(CH_3)_2(NH_2)^+][In(H_4\mathbf{L^2})]\cdot(DMF)_5$ ($\mathbf{L^2}$ = 1,3,6,8-tetrakis(3,5-isophthalate)pyrene) abbreviated as JUC-138 was reported by Zhao et al. having E_g at ~3.34 eV (Fig. 2) which photodegraded ~90% AB in 240 min in UV light (Fig. 2) [87]. The MOF **In-I**, $\{[(CH_3)_2NH_2]_3(In_3(\mathbf{L^3})_4)\}\cdot(solvent)_x$) 5′-(5-carboxy-1H-benzo[d]imidazol-2-yl)[1,1′:3′,1″-terphenyl]-4,4″-dicarboxylate ($\mathbf{L^3}$) with doubly interpenetrated framework degraded MB and MO separately and mixture of both MB and MO [88]. The MOF under UV irradiation decomposed 95.89% of MB in 1 h and 92.80% MO 2 h.

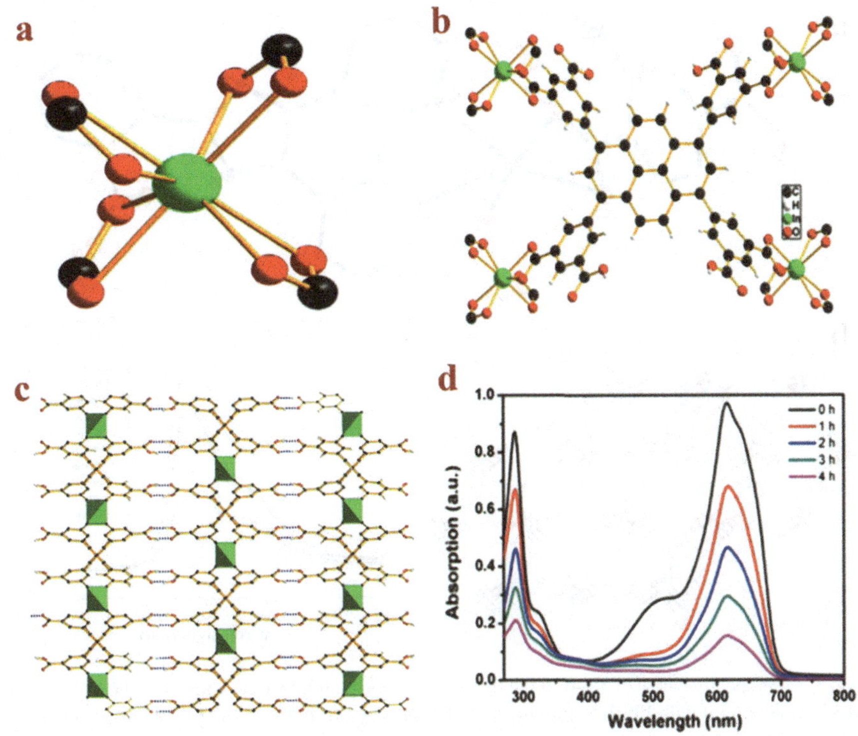

Fig. 2 **a** Coordination style of In^{3+} in **JUC-138**. **b** Coordination style of H_8**L**^{2} in **JUC-138**. **c** Hydrogen bonds constructed 3D supramolecular structure. Photodegradation of AB **d** dye with **JUC-138** as the catalyst. Reproduced with permission from Ref. [87]. Copyright 2017 American Chemical Society

Among lead(II), the **Pb-1,** [Pb(**L**^{4})(H_2O)]$_n$) (**L**^{4} = 4,4′-(1H-1,2,4-triazol-1-yl)methylenebis(benzoate)) decomposed 74.2% of MB in 125 min [89]. Another series of four MOFs [[Pb(**L**^{5})(**L**^{6})]$_n$ (**Pb-2**), [Pb_2(**L**^{5})$_2$(**L**^{7})]$_n$·2nH_2O (**Pb-3**), [Pb_2(**L**^{5})$_2$(**L**^{7})]$_n$ (**Pb-4**) [Pb(**L**^{5})(**L**^{8})$_{0.5}$]$_n$ (**Pb-5**) and [Pb_2(**L**^{5})$_2$(**L**^{9})]$_n$ (**Pb-6**) having 2-(2-carboxylicphenyl)imidazo(4,5-f)-(1,10)phenanthroline (**L**^{5}) and carboxylate ligands pyridine-4-carboxylate (**L**^{6}), 2-amino-1,4-benzenedicarboxylate (**L**^{7}), 1,4-benzenedicarboxylate (**L**^{8}), and 2-hydroxy-1,4-benzenedicarboxylate (**L**^{9}) [90] decomposed MB up to 53.1% with **Pb-2**; 99.3% with **Pb-4**, 84.6% with **Pb-5** and 72.1% with **Pb-6** in 4.6 h. This indicated that **Pb-4** was the best photocatalyst among all and this was ascribed to the presence of –NH_2 in ligand (Fig. 3c). A 3D MOF **Pb-7**, {[Pb_{11}(**L**^{10})$_7$(μ_4-O)$_3$(μ_3-OH)$_2$]·(H_2O)}$_n$ was fabricated employing 2-methyl-4,4′-biphenyldicarboxylate (**L**^{10}) [91] which along with sodium persulfate under visible light photodegraded 63.5% of MO in time span of 150 min. A holodirected Pb(II) 3D-MOF, viz. SCP 2, [Pb(**L**^{11})$_2$]$_n$ (where **L**^{11} = Nicotinate) comprising hexagonal bi-pyramid geometry around Pb under UV and sunlight irradiation degraded 94.27% of MB in 1 h in UV- and 91.11% after 2 h in sunlight in concomitant presence of

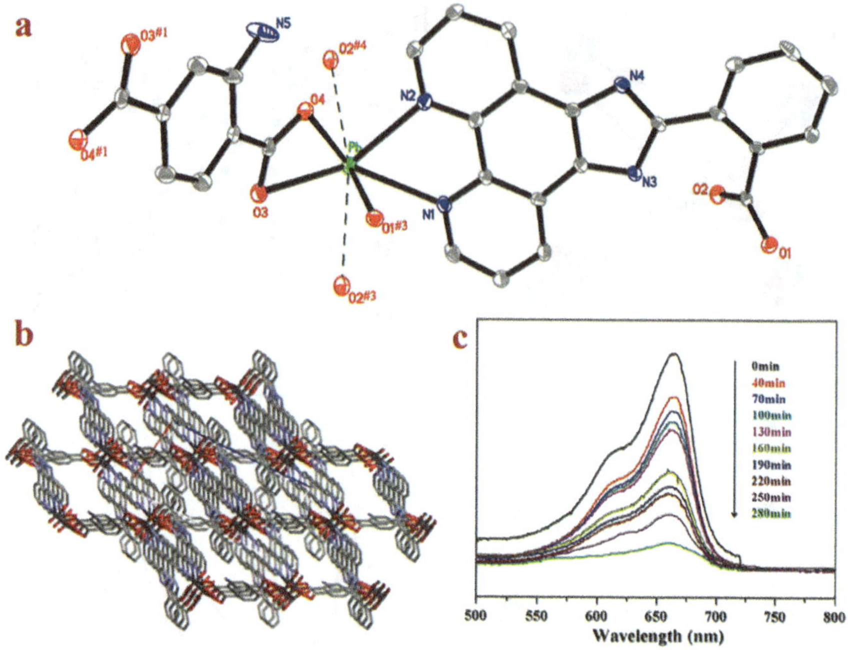

Fig. 3 **a** Perspective view of **Pb-4,** **b** the three-dimensional architecture, **c** the periodic UV–Vis of MB in visible light with **Pb-4**. Reproduced with permission from Ref. [90]. Copyright 2018 Royal Society of Chemistry

H_2O_2 [92]. Another Pb(II) MOF, **Pb-8**, $[Pb_2\mathbf{L^{12}}(H_2O)]_n{\cdot}3nH_2O{\cdot}nDMF$ ($\mathbf{L^{12}}$ = 5-(bis(4-carboxybenzyl)amino)isophthalate) degraded 64.2% of MB in 250 min [93]. The observed differences in photocatlytic properties of these Pb(II) MOFs could be ascribed to the alteration in networks from 1D chain → 2D sheet → 3D framework due to the variation in coordination modes of linkers.

Also, organometallic complex of bismuth(V) **Bi-I** (where **Bi-I** = $Ph_3Bi(\mathbf{L^{13}})_2$ and $H\mathbf{L^{13}}$ = 4-fluorobenzoic carboxyl) [94] under UV irradiation for 1 h, degraded 91.8% RhB, 79.8% MB, and 76.3% MV and also in visible light degraded 90.7% RhB in 8 h.

Transition Metals Based MOFs

This class of MOFs is most widely studied for their application as photocatalysts for dye photodegradation.

Among manganese-based MOFs, an important photocatalyst was SC1 with composition $[Mn(\mathbf{L^{14}})_4(NCS)_2]$ and obtained using ethyl isonicotinate ligand ($\mathbf{L^{14}}$), which degraded 72.62% eosin-Y (EO) in sunlight in 3 h, 93.91% under UV irradiation in 130 min [95]. Recently, a highly stable Fe-MOF, **JOU-22** with formula

$[Fe_2(OH)_2(\mathbf{L^{15}})]0.2DMF.2H_2O$ (JOU-22, $\mathbf{L^{15}}$ = biphenyl-3,3′,4,4′-tetracarboxylate, DMF = *N,N*-dimethylformamide) had been reported, which displayed efficient visible light-induced photodegradation of dyes such as MB, MO, BG, and Rh B [96]. Around 63.1% MB was degraded in 80 min under visible light irradiation. While with hydrogen peroxide, the 95.5% MB underwent photodegradation in the same time period (Fig. 4). This photocatalyst also degraded dyes, viz. MO, RhB, and BG dyes, in the presence of hydrogen peroxide (Fig. 4).

Unlike manganese and iron, cobalt is the most commonly employed for fabrication of MOF-based photocatalysts. A dinuclear pillar-layered-like MOF, $[Co(\mathbf{L^{16}})(\mathbf{L^{17}})_2]_n$ (**Co-1**) (where $\mathbf{L^{16}}$ = tris(4-pyridin-4 ylphenyl)amine and $\mathbf{L^{17}}$ = 4,4′-sulfonyldibenzoate) had been reported which degraded ~94% of RhB under UV light in 1 h using hydrogen peroxide as initiator [56]. An anionic MOF, SCNU-Z2, obtained using heterotopic tripodal ligand $\mathbf{L^{18}}$ = 5,5′-(3-1H-imidazol-1yl-1,5 phenylene)bis-2H-tetrazolate was the first example among MOFs which degraded 97% MB in dark in time of 15 min [97].

Another cobalt MOF **Co-2**, $\{[Co_3(\mathbf{L^{19}})_2(\mathbf{L^{20}})_2(H_2O)_2]\cdot 14H_2O\}_n$ ($\mathbf{L^{19}}$ = 1-aminobenzene-3,4,5-tricarboxylate; 4,4′-bipyridyl ($\mathbf{L^{20}}$) degraded 96.28% MB,

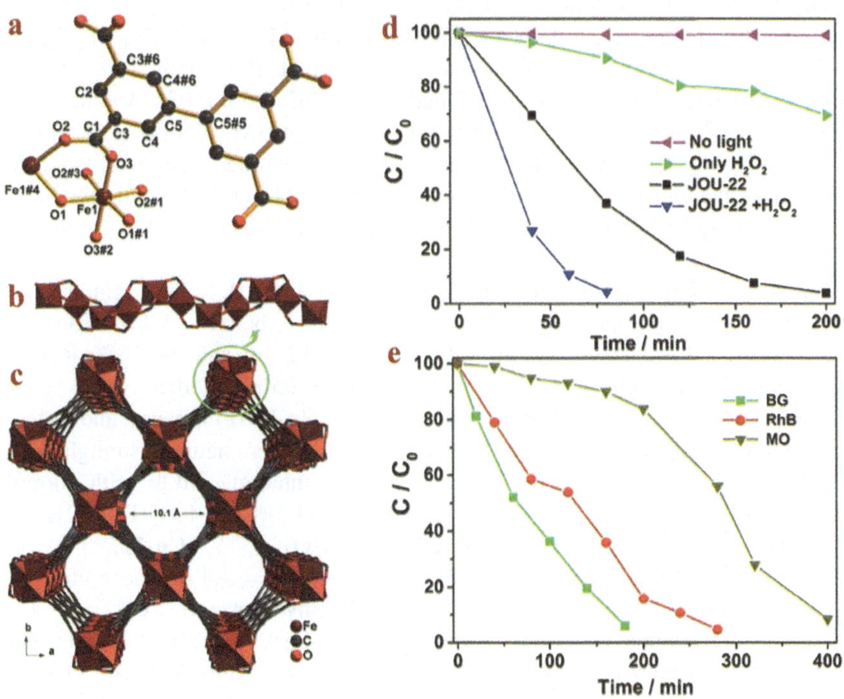

Fig. 4 **a** The geometry around Fe in **JOU-22**. **b** view of Fe–O chain. **c** 3D framework with 1D channels. **d** reaction kinetics of degradation of MB in varied conditions. **e** photodegradation kinetics of BG, RhB, and MO in visible light with **JOU-22** and H_2O_2. Reproduced with permission from Ref. [96]. Copyright 2021 Royal Society of Chemistry

95.79% RhB, 89.63% GV, 81.13% MO, 78.26% fluorescein, and 74.15% R6G, in 60 min in visible light and H_2O_2 [98]. Also, MOF **Co-3**, $\{[Co(\mathbf{L^{21}})_2]SO_4\}_n$ 1,3,5-tris(imidazol-1-ylmethyl)benzene (**L**[21]) had been reported, and its photocatalytic performance was assessed with the variation in pH [99]. The pH variation revealed that the efficiency of this MOF improved at alkaline pH and ~97.77% MB degraded in 25 min at pH 12.2. This is because, in alkaline solution, the **Co-3** acquired a negative charge due to adsorption of OH^- ions and attracted positively charged MB dye.

Another MOF, **Co-4**, $[Co_3(\mathbf{L^{22}})_2(DMF)(\mathbf{L^{23}})]\cdot DMF$ ($H_3\mathbf{L^{22}}$ = biphenyl-3,4′,5-tricarboxylic acid and $\mathbf{L^{23}}$ = 1,3-bis(4-pyridyl)propane) degraded nearly 90% of RhB in 2 h in with H_2O_2 [100]. While, **Co-5**, $[Co(\mathbf{L^{24}})(\mathbf{L^{8}})_{1/2}]_n$ ($H_2\mathbf{L^8}$ = 1,4-benzenedicarboxylic acid, $\mathbf{L^{24}}$ = 3,5 bis(pyridin-4-ylmethoxy)benzoate) having Co(II) paddle wheel SBU on UV irradiation with H_2O_2, photodegraded 95% for methyl blue and 84% for MB, and ST degraded completely degraded [101]. A green synthetic protocol had been adopted for the preparation of MOF **Co-6**, $[\{Co_{0.5}(\mathbf{L^{25}})_{0.5}(\mathbf{L^{26}})_{1.25}\}]_\infty$ ($\mathbf{L^{25}}$ = 5-azidoisophthalate; $\mathbf{L^{26}}$ = 1,2-bis(4-pyridyl)ethylene) [102]. This MOF was converted to nanoscale MOF (NMOF) by grinding which photodegraded 92% MB and 90% RhB in visible light in time span of 2 h in the presence of H_2O_2. A porous MOF comprising square cavities $\{[Co(\mathbf{L^{27}})]\cdot 0.5(CH_3CN)\cdot 5(H_2O)\}_n$ (HPU-5) (where $\mathbf{L^{27}}$ = (5-(3,5-Di-pyridin-4-yl-[1,2,4]triazol-1-ylmethyl)-isophthalate)) [103] which along with H_2O_2 photodegraded 83% degradation of MO in 2 h and 99.3% RhB 2.5 h [103]. Another MOF, **Co-7** $\{[Co(\mathbf{L^{28}})(\mathbf{L^{29}})]\}_n$ (where $\mathbf{L^{28}}$ = 6-(4-carboxyphenyl) nicotinate and $\mathbf{L^{29}}$ = 1,4-bis(4-pyridyl)benzene) with 3D, threefold interpenetrated framework degraded 100% MO dye under UV irradiation in the presence of H_2O_2 [104]. Additionally, it too decomposed ST, neutral red, RhB, methyl blue, and MB up to 78%, 74%, 71%, 67%, and 53%, respectively.

Another MOF **Co-8**, $\{[Co(\mathbf{L^{30}})(H_2O)_2](\mathbf{L^8})\cdot(H_2\mathbf{L^8})\}_n$, (1,1,2,2-tetrakis[4-(1H-1,2,4-triazol-1-yl)phenyl]ethylene (**L**[30]) and benzene-1,4-dicarboxylate (**L**[8])) degraded 92.48% MB in 135 min irradiation and 90.56% RhB in 3 h in visible light irradiation and H_2O_2 [105]. Another MOF **Co-9**, $[Co(\mathbf{L^{32}})(\mathbf{L^{31}})_{1.5}]_n$ (where $\mathbf{L^{31}}$ = 4,4′-bis(imidazole-1-yl)-biphenyl and $\mathbf{L^{32}}$ = tetrabromoterephthalate) degraded 85.83% MB under natural sunlight irradiation in 8 h [106]. A series of four cobalt containing MOF with formula $\{[Co_4(\mathbf{L^4})_4(H_2O)]\cdot 2CH_3CN\cdot 4H_2O\}_n$ (**Co-10**), $\{[Co(\mathbf{L^4})(\mathbf{L^{33}})]\cdot 1.5CH_3OH\cdot 2H_2O\}_n$ (**Co-11**), $\{[Co_2(\mathbf{L^4})_2(\mathbf{L^{34}})]\cdot 6CH_3CH_2OH\cdot 4H_2O\}_n$ (**Co-12**), and $\{[Co(\mathbf{L^4})(\mathbf{L^{35}})(H_2O)]\cdot DMA\}_n$ (**Co-13**) (where $\mathbf{L^{33}}$ = 4,4′-bis(benzoimidazo-1-yl)biphenyl, $\mathbf{L^{34}}$ = 1,4-bis(imidazol-1-ylmethyl)benzene, $\mathbf{L^{35}}$ = 4,4′-bis(imidazolyl)biphenyl) were synthesized using 4,4′-(1H-1,2,4-triazol-1-yl)methylenebis(benzoate) ($H_2\mathbf{L^4}$) [107] used as catalysts against MB, MO, OIV, and MV, and these exhibited decent photocatalytic performance. Another two MOFs, viz. $\{[Co(\mathbf{L^{36}})_{0.5}(\mathbf{L^{37}})]\cdot 4H_2O\}_n$ (**Co-14**) and $\{[Co_5(\mathbf{L^{36}})_4(\mathbf{L^{38}})_4]\cdot 2NO_3\cdot 2DMA\cdot 2H_2O\}_n$ (**Co-15**), were designed using tripodal linker 1,3,5-tris(2-methylimidazol-1-yl)benzene (**L**[36]) and isophthalate linkers ($\mathbf{L^{37}}$ = 5-aminoisophthalate and $\mathbf{L^{38}}$ = 5-hydroxyisophthalate) and in 70 min, degraded 94.36% MB (**Co-14**) and 89.57%

(**Co-15**) in visible light with H_2O_2 [108]. Another MOF **Co-16**, [Co(H**L**^{39})(**L**^{40})]$_n$ (**L**^{39} = 4-(4-carboxylphenylmethylthio)benzoate and **L**^{40} = 1,3-bis(1-imidazo-1-ly)benzoate) exhibited good photocatalytic activity with 95.0%, 94.7%, and 94.1% photodegradation of MB, MV, and RhB, respectively, in 2 h in UV light [109].

In addition to cobalt-based MOFs, nickel(II)-based MOFs had also been utilized as photocatalysts for dye degradation. Two Ni(II)-based MOFs with formula [Ni(**L**^{41})$_{1.5}$(**L**^{42})(H_2O)]$_n$ (**Ni-1**) and [Ni(**L**^{34})(**L**^{43})]$_n$ (**Ni-2**) (5-substituted isophthalate (5-nitroisophthalic acid (**L**^{42}), 5-methoxyisophthalate (**L**^{43}), 1,4-bis(1-imidazolyl)benzene (**L**^{41}), and 1,4-bis(imidazol-1-ylmethyl)benzene (**L**^{34})) were used as photocatalysts against MO solution and degraded 94% (**Ni-1**) and 95% (**Ni-2**) in visible light and H_2O_2 in time span of 70 min [110]. Three MOFs, viz. {[Ni(**L**^{44})(**L**^{38})]·H_2O}$_n$ (**Ni-3**), [Ni(**L**^{44})(**L**^{45})]$_n$ (**Ni-4**) and [Ni(**L**^{44})(**L**^{46})]$_n$ (**Ni-5**) (1,4-bis(benzimidazol-1-yl)-2-butylene (**L**^{44}), 5-hydroxyisophthalate (**L**^{38}) and 2,6-naphthalenedicarboxylate (**L**^{45}), and 4,4'-oxybis(benzoate) (**L**^{46})) were reported [111]. **Ni-3–Ni-5** exhibited good photocatalysis against degradation of MB 91.5% (**Ni-3**), 93.4% (**Ni-4**), and 87.0% (**Ni-5**) in 2 h in UV light. A series of nickel MOFs comprising three dimensional architectures, viz. [Ni(**L**^{47})(**L**^{20})(H_2O)$_2$] (**Ni-6**), [Ni$_4$(**L**^{47})$_4$(H_2O)$_8$] (**Ni-7**), [Ni$_2$(**L**^{47})$_2$(**L**^{48})(H_2O)] (**Ni-8**), and [Ni(**L**^{47})(**L**^{41})$_{0.5}$(H_2O)]·H_2O (**Ni-9**) (**L**^{47} = 5-(4′-carboxylphenyl)nicotinate; **L**^{20} = 4,4'-cipyridyl; **L**^{48} = 4,4′-bis(imidazol-1-ylmethyl)biphenyl; **L**^{41} = 1,4-bis(1-imidazoly)-benzene) were used photocatalysts [112] which degraded MO up to 84.3%, 80.6%, 76.6%, and 68.5% **Ni-6**, **Ni-7**, **Ni-8**, and **Ni-9**, respectively, after 75 min in visible light.

Another MOF, **Ni-10** with formula [Ni(**L**^{49})(**L**^{50})]$_n$ (where **L**^{49} = 1,6-bis(benzimidazol-1-yl)hexane, H_2**L**^{50} = isophthalate) [113]. The photocatalytic properties of both bulk- and nano-structured MOFs were estimated for the photodegradation of MB in UV light. For bulk MOF 85.6% and for nano-MOF 90.6% degradation of MB was observed in 2.5 h. A new ternary MOF **Ni-11**, {[Ni$_2$(**L**^{51})$_2$(**L**^{52})$_2$(H_2O)]·8.5H_2O}$_n$ (where **L**^{51} = 1,4-bis(5,6-dimethylbenzimidazol-1-yl)-2-butylene and **L**^{52} = 1,4-naphthalenedicarboxylate) having three-dimensional 3,6T13 topology was used for the degradation of MB in UV irradiation, and 92.7% photodegradation of MB was observed in 2 h [114]. A new MOF **Ni-12**, {[Ni(**L**^{53})(**L**^{20})(H_2O)$_2$]·(H_2O)$_5$}$_n$ 4,5-bis(pyrazol-1-yl) phthalate (**L**^{53}) was also used as photocatalyst for degradation of MB which 91.0% after 120 min and for dye MO, the degradation ratio after 180 min as 82.1% under UV light irradiation (Fig. 5c,d) [115].

Copper-based MOFs are also studied as photocatalysts for dye degradation. A copper-based MOF **Cu-1**, {[Cu(**L**^{54})(**L**^{55})]·H_2O}$_n$ (**L**^{54} = tris(4-(1,2,4-triazol-1-yl)phenyl)amine and **L**^{55} = pimelate) [116] exhibited 90.3% of MB in 75 min and 90.1% RhB in 2.5 h in visible light and H_2O_2, and the MOF was reusable for minimum five cycles. Another copper paddlewheel MOF **Cu-2**, [Cu$_4$(**L**^{56})$_2$(H_2O)$_4$·5DMF]$_n$ (H_4**L**^{56} = 3,5-di(3,5-dicarboxylatephenyl)nitrobenzene) degraded 81.3% of MV in 45 min under UV irradiation [117]. While, **Cu-3**, [Cu$_2$(**L**^{56})·5DMF]$_n$ exhibiting 3D framework comprising 1D hexangular channels degraded 70% MV in 45 min irradiation [118]. Another MOF, **Cu-4**,

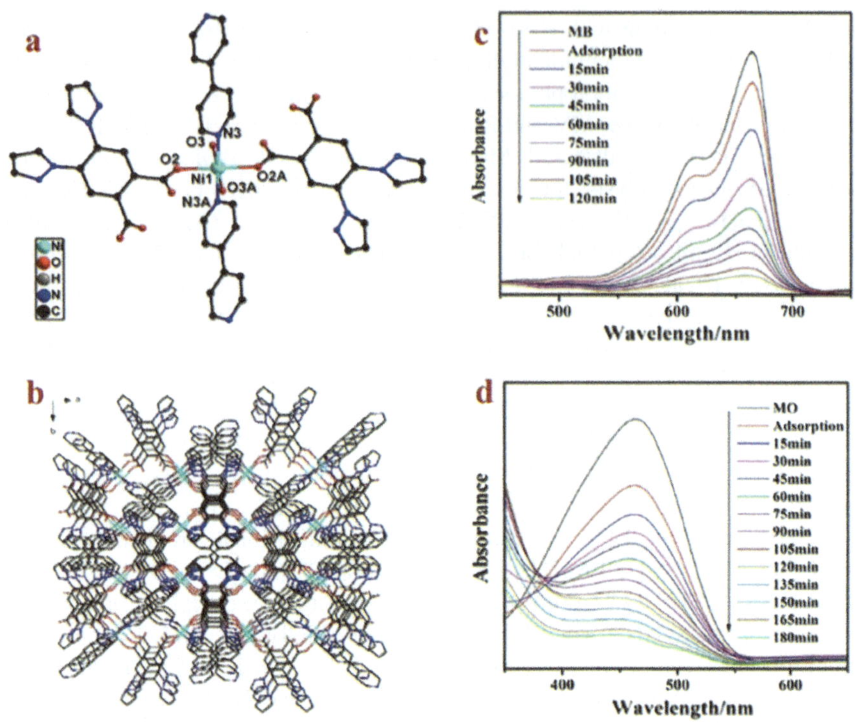

Fig. 5 **a** The geometry around Ni(II) in **Ni-12**, **b** the view of the framework along the *c* axis. Variation in UV–Vis of MB (**c**) and MO (**d**) in the presence of **Ni-12** and H_2O_2 in UV light. Reproduced with permission from Ref. [115]. Copyright 2021 Wiley Online Library

$[Cu(H_2\mathbf{L}^{57})(\mathbf{L}^{58})]_n$ (where $\mathbf{L}^{57}$ = terphenyl-3,3′,5,5′-tetracarboxylate and $\mathbf{L}^{58}$ = 1,4-bis(1-imidazol-yl)-2,5-dimethyl benzene) degraded 97.3% MB in 120 min with H_2O_2 in UV light [119]. A peculiar semiconducting 3D framework (**Cu-5**), $\{[Cu^ICu^{II}_2(\mathbf{L}^{59})_2]NO_3\cdot 1.5DMF\}_n$ (4′-(3,5-dicarboxylatephenyl)-4,2′:6′,4″-terpyridine ($\mathbf{L}^{59}$)) degraded 80% MB in very small time span of 30 min [120]. The MOF-199 synthesized hydrothermally degraded 99% of BB41 in 180 min in UV light [121]. The MOF **Cu-6**, $[Cu(\mathbf{L}^{20})Cl]_n$ [122] enhancement in photocatalytic properties in presence of H_2O_2 photodegraded 93.93% MB in 2.5 h. An efficient panchromatic active **Cu-7**, $[Cu_{1.5}(\mu_2\text{-}O)(H_2\mathbf{L}^{60})]$ ($\mathbf{L}^{60}$ = 4,4′-[6-(dimethylamino)-1,3,5-triazine-2,4-diyl]bis(azanediyl)dibenzoate) photodegraded ~99% of MB in 70 min in UV and 86% MB in 90 min in visible light [123]. The MOF **Cu-8**, $\{[Cu(\mathbf{L}^{61})(\mathbf{L}^{54})]\cdot 2H_2O\}_n$ ($\mathbf{L}^{54}$ = tris[4-(1,2,4-triazol-1-yl)phenyl]amine; $\mathbf{L}^{61}$ = succinate) [124] degraded 92.6% MB in 75 min and 90.3% RhB in 2 h with the assistance of H_2O_2. The MOF **Cu-9**, $\{[Cu(\mathbf{L}^{62})(H_2O)]\cdot 0.25(CH_3CN)\}_n$ ($\mathbf{L}^{62}$ = 4,4′-dicarboxylate-4″-nitrotriphenylamine) photodegraded 97.2% of CV, 92.1% of R6G, 87.8% of CR, and 70.2% of BR2 dyes in 40 min with H_2O_2 [125].

Another self-catenating MOF **Cu-10** $[Cu(\mathbf{L^{63}})_{0.5}(\mathbf{L^{64}})]_n$ ($\mathbf{L^{63}}$ = 1,4-bis(2-methyl-imidazol-1-yl-methyl)benzene and $\mathbf{L^{64}}$ = glutarate) photodegraded 90.3% Rh B in 165 min with H_2O_2 in UV light [126]. Another series of Cu-based MOFs, viz. $\{[Cu_4(OH)_2(\mathbf{L^{63}})_2(\mathbf{L^{65}})_2(H_2O)_2]\cdot 3H_2O\}_n$ (**Cu-11**), $[Cu_2(\mathbf{L^{63}})(\mathbf{L^{50}})_2]_n$ (**Cu-12**), $[Cu_2(\mathbf{L^{63}})(\mathbf{L^{66}})_2]_n$ (**Cu-13**), $\{[Cu_2(\mathbf{L^{63}})(\mathbf{L^{67}})_2]\cdot H_2O\}_n$ (**Cu-14**) (where $\mathbf{L^{65}}$ = 1,2,4-benzenetricarboxylate, $\mathbf{L^{50}}$ = isophthalate, $\mathbf{L^{66}}$ = 5-methyl-isophthalate, $\mathbf{L^{67}}$ = 1,4-benzenediacetate) degraded MB up to 94.0% (**Cu-11**), 90.7% (**Cu-12**), 89.1% (**Cu-13**), and 91.6% (**Cu-14**) in 1.5 h [127]. These MOFs also degraded RhB up to 93.0% (**Cu-11**), 84.7% (**Cu-12**), 81.3% (**Cu-13**), and 88.6% (**Cu-14**) in 3 h irradiation. Apart from these two dyes, they had been used as photocatalysts against MO and photodegraded this dye up to 92.7% (**Cu-11**), 85.1% (**Cu-12**), 77.2% (**Cu-13**), and 87.3% (**Cu-14**) in long time span of 5 h. The MOF **Cu-15** $\{[Cu(\mathbf{L^{68}})(\mathbf{L^{50}})][Cu_4(OH)_2(\mathbf{L^{68}})(\mathbf{L^{50}})_3]\cdot 4H_2O\}_n$ (where $\mathbf{L^{68}}$ = 1-imidazol-1-yl-3-(1,2,4-triazol-4-yl)propane,) degraded MB, RhB, and MO upto 90.5%, 89.2%, and 87.8% after 1.5 h, 3 h, and 4 h, respectively, with H_2O_2 [128]. Another MOF **Cu-16**, $[Cu_6(CN)_7(\mathbf{L^{69}})_2(OH_3)]_3{}^{\infty}$, ($\mathbf{L^{69}}$ = Hexamethylenetetramine) degraded CR, MV, and MB dyes up to 96.86, 98.92, and 99.31% in time span of 8, 25, and 20 min, respectively, in UV light [129]. In ultrasonic conditions, the same MOF in presence of hydrogen peroxide decomposed 97.09, 98.96, and 99.69% CR, MV, and MBin 6, 18, and 15 min, respectively.

Zinc(II)-based MOFs have also been synthesized and utilized as photocatalysts for dye degradation. Among varied Zn(II) MOFs, **Zn-1**, $[Zn(\mathbf{L^{70}})(H_2O)]\cdot H_2O$ comprising 4 (pyridine-4-yl) phthalate) ($H_2\mathbf{L^{70}}$) as photocatalyst degraded 98.5% RhB as well as 83.8% MO in 120 min in visible light for five consecutive cycles [130]. Another MOF, NNU-36 $[Zn_2(\mathbf{L^{71}})(\mathbf{L^{72}})_2]\cdot 2DMF$ ($\mathbf{L^{71}}$ = 9,10-bis(4′-pyridylethynyl)-anthracene and $\mathbf{L^{72}}$ = 4,4′-biphenyldicarboxylate) exhibited photodegradation of RhB [131]. Its photocatalytic properties were uplifted from 46.6 to 96% on adding H_2O_2. Another MOF, **Zn-2**, $\{[Zn(\mathbf{L^{73}})]\cdot 2.7\ DMF\}$ ($\mathbf{L^{73}}$ = 1,4 bis(triazol-1-yl)terephthalate) three-dimensional uninodal tetra-connected $CdSO_4$ topology decomposed 88.8% RhB in UV light [132]. Another, Zn-based MOF **Zn-3**, $\{[Zn_5(\mathbf{L^{74}})_2(DMF)_2(\mu_3\text{-}H_2O)]\cdot 2DMF\}$ synthesized using symmetrical pentacarboxylic acid ligand, 5-di(3′,5′ dicarboxylphenyl)benzoate ($\mathbf{L^{74}}$), degraded 73.1% MV under UV irradiation [133]. A three-dimensional MOF **Zn-4,** $\{[Zn_4(\mathbf{L^{52}})_{3.5}(\mu^4\text{-}OH)(DMF)]\cdot 1.7DMF\}_n$ ($\mathbf{L^{52}}$ = 1,4-naphthalenedicarboxylate) degraded 78.3% of MV in 40 min [134]. A biocompatible MOF **Zn-5**, $\{[Zn_2(\mathbf{L^{75}})_2]\cdot 0.5H_2O\}_n$, synthesized using phytochemical ferulate ($\mathbf{L^{75}}$), photodegraded 88% RhB in 100 min in UV light [135]. Another zinc MOF **Zn-6**, $\{[Zn_2(\mathbf{L^{57}})(DMF)_3]\cdot 2DMF\cdot 2H_2O\}$ having terphenyl-3,3″,5,5″-tetracarboxylate ($\mathbf{L^{57}}$) ligand photo-decomposed 72.5% and 92.8% against MV and RhB in UV light, respectively [136]. Another MOF, **Zn-7** $[Zn(\mathbf{L^{75}})(\mathbf{L^{20}})_4\cdot 2H_2O]_n$ synthesized using 1,4-bis(3-carboxylatebenzyl)piperazine acid ($\mathbf{L^{75}}$) ligand and ancillary ligand 4,4′-bipyridyl ($\mathbf{L^{20}}$) having threefold interpenetrating 3D net with *pcu* topology degraded 77.36% MV in 40 min [137].

Another MOF **Zn-8**, $[Zn(\mathbf{L^{76}})(\mathbf{L^{77}})_2]_n$ (where $\mathbf{L^{76}}$ = 1,5-naphthalenedisulfonate, and $\mathbf{L^{77}}$ = 2,4,5-tri(4-pyridyl)-imidazolate) exhibited 98.1% degradation of RhB in 2 h [138]. Two Zn-MOFs, viz. $\{[Zn(\mathbf{L^{78}})(\mathbf{L^{79}})]_2\}_n$ (**Zn-9**) and $\{[Zn(\mathbf{L^{78}})(\mathbf{L^{61}})]_2\}_n$

(**Zn-10**) (where HL^{78} = 1,3-bis(5,6-dimethylbenzimidazol-1-yl)-2-propanol, $\mathbf{L^{79}}$ = *trans,trans*-muconate and $\mathbf{L^{61}}$ = succinate) in 2 h degraded 97.5% and 97.0% of MB, respectively, in UV light [139]. MOF **Zn-11**, $[Zn_2\mathbf{L^{12}}]_n$ ($\mathbf{L^{12}}$ = 5-(bis(4-carboxybenzyl)amino)isophthalate) in 250 min, photodegraded 91.4% MV [93].

Very recently, a Zn(II)-MOF, **Zn-12** $[Zn_4(\mathbf{L^{80}})_2(\mathbf{L^{81}})_4{\cdot}H_2\mathbf{L^{81}}]{\cdot}4H_2O$ ($\mathbf{L^{80}}$ = 3,5-bis(3,4- dicarboxylphenoxy)benzoate, $\mathbf{L^{81}}$ = 3,3′,5,5′ -tetramethyl-4,4′-bipyrazole) had been synthesized and used as photocatalyst against MB and photodegraded this dye up to 91.33% (Fig. 6c) [140]. Additionally, this MOF was used as photocatalyst against other dyes, viz. MO, CR, AO, and ACB, and exhibited percentage photodegradation up to 74.86%, 26.10%, 61.29%, and 17.25%, respectively, in 100 min under UV light. The catalytic efficiency of **Zn-12** followed the order MB > MO > AO > CR > ACB (Fig. 6d).

Some silver MOFs find applications as photocatalysts for dye degradation. Three sandwich-like MOFs, $Ag_4(\mathbf{L^{20}})_4(\mathbf{L^{37}})_2{\cdot}11H_2O$ (**Ag-1**), $Ag_2(\mathbf{L^{26}})_2(\mathbf{L^{45}}){\cdot}2H_2O$ (**Ag-2**), and $[Ag_2(\mathbf{L^{26}})_{1.5}(\mathbf{L^{82}})_{0.5}](\mathbf{L^{82}})_{0.5}{\cdot}7H_2O$ (**Ag-3**) (where $\mathbf{L^{20}}$ = 4,4′-bipyridine, $\mathbf{L^{37}}$ = 5-aminoisophthalate, $\mathbf{L^{26}}$ = 1,2-di(4-pyridyl)ethylene, $\mathbf{L^{45}}$ = 2,6-naphthalenedicarboxylate, and $H_2\mathbf{L^{82}}$ = stilbenedicarboxylate) [141] exhibited degradation of MB up to 98.2% (**Ag-1**), 99.8% (**Ag-2**), and 99.9% (**Ag-3**) in 3 h under

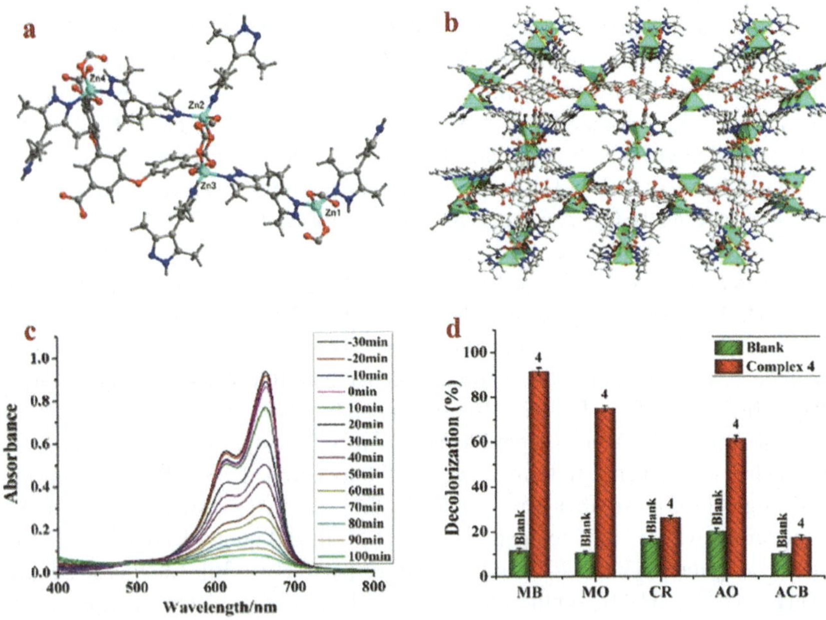

Fig. 6 **a** Coordination environment around Zn1 and Zn2 in **Zn-12**, **b** the 3D framework, **c** periodic UV–Vis spectra of MB in presence of **Zn-12**, and **d** comparative *k* values for different dyes. Reproduced with permission from Ref. [140]. Copyright 2021 Elsevier

UV light. Another MOF, **Ag-4**, $[Ag_2(en)_2(\mathbf{L^{83}})]_n$ (en = ethylenediamine, $\mathbf{L^{83}}$ = 5,5′-azotetrazolate) displayed 90.8% degradation of R6G in UV light [142]. The MOF **Ag-5**, $[Ag(H\mathbf{L^{50}})(\mathbf{L^{84}})]_n$ ($\mathbf{L^{50}}$ = isophthalate, $\mathbf{L^{84}}$ = 1,6-bis(5,6-dimethylbenzimidazol-1-yl)hexane)) degraded ~ 95.9% MB in 150 min [143]. A three-dimensional MOF **Ag-6**, $Ag_2(\mathbf{L^{26}})_2(\mathbf{L^{85}})](H_2O)_{5.4}$ ($\mathbf{L^{26}}$ = 1,2-di(4-pyridyl)ethylene, and $\mathbf{L^{85}}$ = 1,4-phenylenediacetate) [144] degraded 96% MB in 75 min in UV light. A highly efficient Ag-based CP **Ag-7**, $[Ag_2(H_2\mathbf{L^{86}})(\mathbf{L^{20}})_2{\cdot}2H_2O]$ ($H_4\mathbf{L^{86}}$ = 1,3-bis(3,5-dicarboxyphenoxy)benzene, $\mathbf{L^{20}}$ = 4,4′-bipyridyl which photodegraded MB and Rh B up to 90.24% and 93.56%, respectively, in 100 min under UV light (Fig. 7f,g) [145].

Among Cd-based MOF, **Cd-1**, $[Cd(\mathbf{L^{7}})(\mathbf{L^{20}})]$ (where $H_2\mathbf{L^{7}}$ = 2-aminoterephthalate and $\mathbf{L^{20}}$ = 4,4′-bipyridyl) exhibited good photocatalytic activity and degraded 82% MB in 90 min in visible and 72% MB in UV irradiation [146]. Another two cadmium MOFs synthesized employing 1,3-di((2′,4′-dicarboxylatephenyl)benzene ($\mathbf{L^{87}}$) ligand imidazole linkers

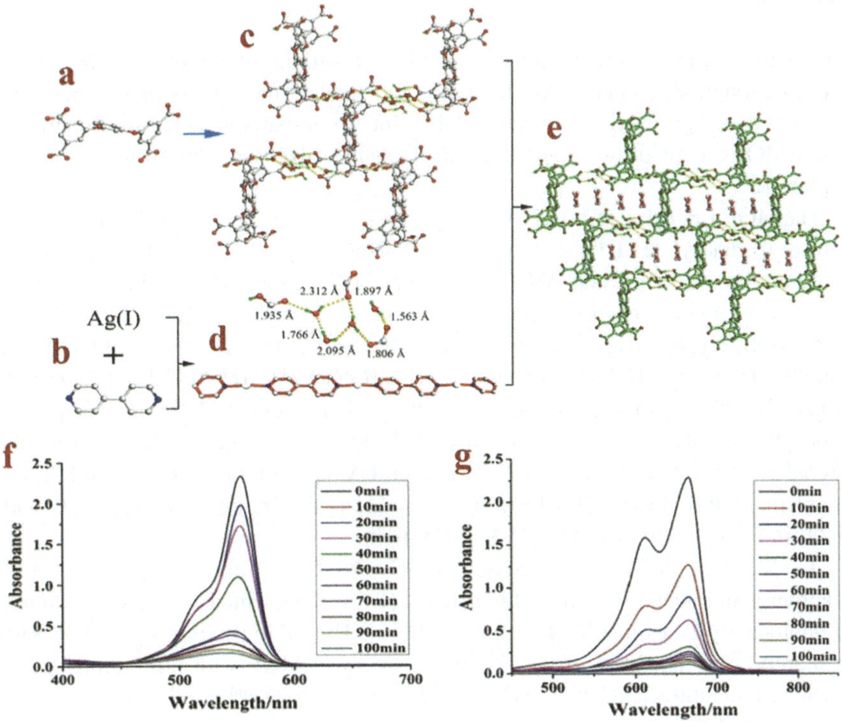

Fig. 7 **a** The view of $\mathbf{L}^{86\,4-}$, **b, d** self-assembly of $\mathbf{L}^{86\,4-}$ and its 1D chain, **c** 2D supramolecular sheet by the H-bonding interactions between $\mathbf{L}^{86\,4-}$ and water molecules, and **e** the 2D sheet incorporating 1D chain. Photodegradation of **f** Rh B and **g** MB. Reproduced with permission from Ref. [145]. Copyright 2021 Elsevier

1,3-bis(2-methylimidazol-1-ylmethyl)benzene (**L**[88]) and 1,3-bis(imidazol-1-ylmethyl)benzene (**L**[89]) with compositions $[(Cd_2(\mathbf{L^{87}})(\mathbf{L^{88}})(H_2O)_{0.5})\cdot H_2O]_n$ (**Cd-2**) and $[(Cd_2(\mathbf{L^{87}})(\mathbf{L^{89}}))\cdot H_2O]_n$ (**Cd-3**) degraded 99.5% (**Cd-4**) and 99.8% (**Cd-5**) of dye MO in 1 h with H_2O_2 in UV light [147]. Two three-dimensional multinodal MOFs $[Cd_3(\mathbf{L^{90}})(\mathbf{L^{20}})_{0.5}(H_2O)_4]\cdot 5H_2O$ (**Cd-6**) and $[(CH_3)_2NH_2]_2[Cd_{11}(\mathbf{L^{90}})_4(DMF)_4(H_2O)_8]\cdot 4H_2O$ (**Cd-7**) (**L**[90] = 3,4-di(3,5-dicarboxylphenyl)phthalate) photodecomposed 88.6% RhB using **Cd-6**, 91.5% in presence of **Cd-7** [148]. Two MOFs, viz. $\{[Cd(\mathbf{L^{91}})(\mathbf{L^{66}})]\cdot 0.33H_2O\}_n$ (**Cd-8**) and $\{[Cd(\mathbf{L^{92}})(\mathbf{L^{93}})]\cdot 1.35H_2O\}_n$ (**Cd-9**) (where **L**[66] = 5-methylisophthalate; **L**[91] = 1,5-bis(benzimidazol-1-yl)pentane; **L**[93] = tetrachloroterephthalate and **L**[92] = 1,2-bis(2-methylbenzimidazol-1-yl)benzene), [149] decomposed 89% MB using **Cd-8** as photocatalyst and 91.2% MB using **Cd-9** in 2 h. The MOF **Cd-10**, $[Cd(\mathbf{L^{95}})_2(NO_3)_2]$ ((**L**[95] = 1,10-phenanthroline) degraded IC dye 77.83% and 98.06% in sunlight and UV light irradiation, respectively, in less than 1 h [150].

Lanthanide-Based MOFs

The 4*f* inner transition metal ions display high coordination numbers with variable coordination modes and due to which they had created interest among the inorganic chemists to design and develop their MOFs. But, in comparison with transition metal-based MOFs, their utility as photocatalysts against dye degradation had been less explored.

The MOF **La-1** $[La_2(\mathbf{L^{96}})_3(H_2O)_2]_n$ synthesized using (2E,2′E)-3,3′-(anthracene-9,10-diyl) diacrylate (**L**[96]) degraded 91.2% RhB within 2 h in visible light [151]. Another three-dimensional MOF **La-2** $[La(H\mathbf{L^{97}})_3]_n$ ($H_2\mathbf{L^{97}}$ = chalcone dicarboxylic acid ligand) comprising honeycomb-like cavities photodegraded 79 and 74%, MB and RhB, respectively, in UV light for 3 h (Fig. 8e,f) [152]. Another two MOFs, namely $[La(\mathbf{L^{37}})_{1.5}(H_2O)]_n$ (**La-3**) and $\{[La(\mathbf{L^{37}})(\mathbf{L^{98}})_{0.5}(H_2O)_2](H_2O)\}_n$ (**La-4**) (where $H_2\mathbf{L^{98}}$ = oxalic acid and 5-aminoisophthalic acid ($H_2\mathbf{L^{37}}$), also displayed good photocatalytic properties against RhB [153]. These newly designed MOFs degraded 81.6% (**La-3**) and 69.7% (**La-4**) in UV light in a long period of 8 h. An interesting MOF **La-5**, $\{[La(\mathbf{L^{6}})_3(H_2O)_2]\}_n$ [154] decomposed 80.55% MB with H_2O_2 in 2 h in sunlight and 95.33% in UV light.

A visible light responsive MOF NNU-15(Ce) comprising 4,4′-(diethynylanthracene-9,10-diyl)dibenzoate (**L**[99]) represented the first example of photocatalysts against RhB with H_2O_2 [155]. The MOF decomposed ~99% RhB in merely 12 min with *k* of 0.2397 min^{-1}. A novel MOF with three-dimensional porous architecture, $[Pr(\mathbf{L^{100}})_{1.5}(H_2O)]\cdot 8H_2O\cdot 3.5DMF]$ (**L**[100] = 2,3,5,6 tetramethyl-1,4-benzenedicarboxylate) under irradiation of xenon lamp, degraded Congo red up to ~100% in 24 h [156].

Using molecular building blocks (MBBs) strategy europium-based MOF, **Eu-1** $[(CH_3)_2NH_2]_2[Eu_6(OH)_8(\mathbf{L^{101}})_6(H_2O)_6].(DMF)_{15}$ (**L**[101] = 4,4′-(9,10-anthracenediyl)dibenzoate) had been designed which possessed *fcu* topology [157]. In presence of H_2O_2, the MOF photodegraded 99% MB in 30 min and 98%

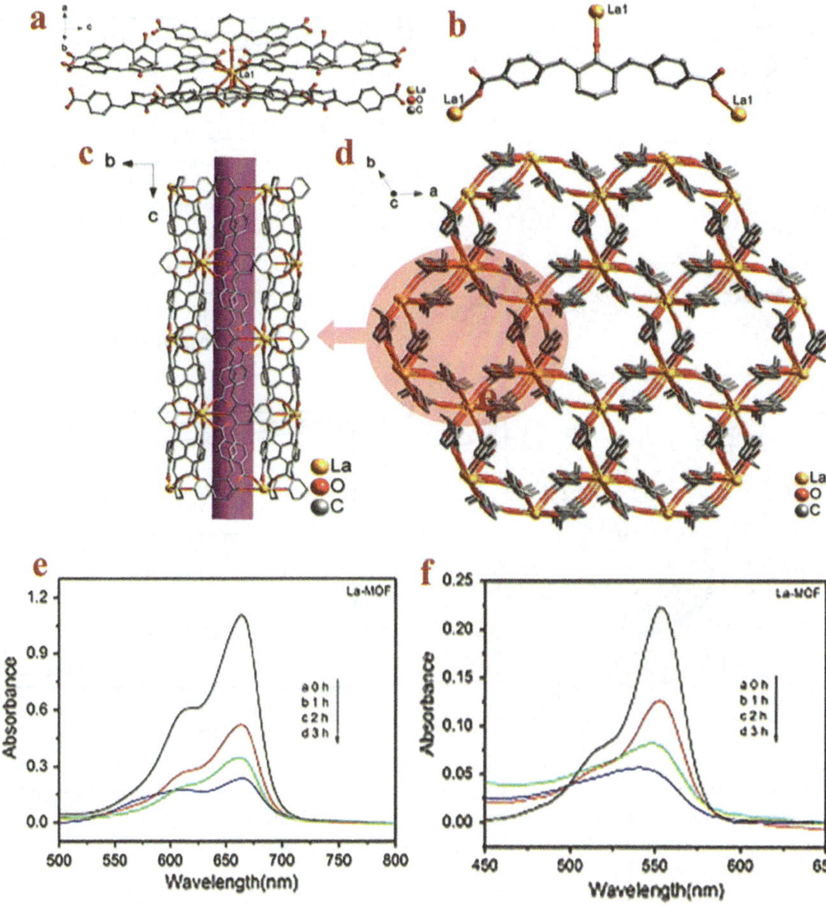

Fig. 8 **a** Geometry around La^{3+} in **La-2**, **b** the coordination mode of $\mathbf{L}^{124\,2-}$, **c** one-dimensional tube like structure, and **d** porous framework along *c* axis. UV–Vis spectra of **e** MB and **f** RhB dyes with **La-2**. Reproduced with permission from Ref. [152]. Copyright 2019 Elsevier

RhB in 36 min (Fig. 9d, e). However, the chalcone dicarboxylic acid analog, **Eu-2,** [Eu(H$\mathbf{L}^{97}$)$_3$] (chalcone dicarboxylic acid ligand (H$_2$$\mathbf{L}^{97}$)) exhibited merely 72% MB degradation in time span of 180 min [152]. The observed differences in the photodegradation ability in both were due to the presence of anthracene ligand in **Eu-1** which was absent in **Eu-2**.

An efficient gadolinium-MOF **Gd-1**, {[Gd($\mathbf{L}^{102}$)$_{1.5}$(H_2O)$_2$]·2H_2O}$_n$ having *N,N′*-dioxide-3,3′-benzocinnoline dicarboxylate ($\mathbf{L}^{102}$) decomposed ~95% RhB in 45 min under UV light, [158] while **Gd-2**, Gd(H($\mathbf{L}^{97}$)$_3$) decomposed only 72% RhB in 180 min under similar reaction conditions. The observed differences maybe because of the presence of different ligand systems which had altered the architecture and in turn bandgaps of both the photocatalyst.

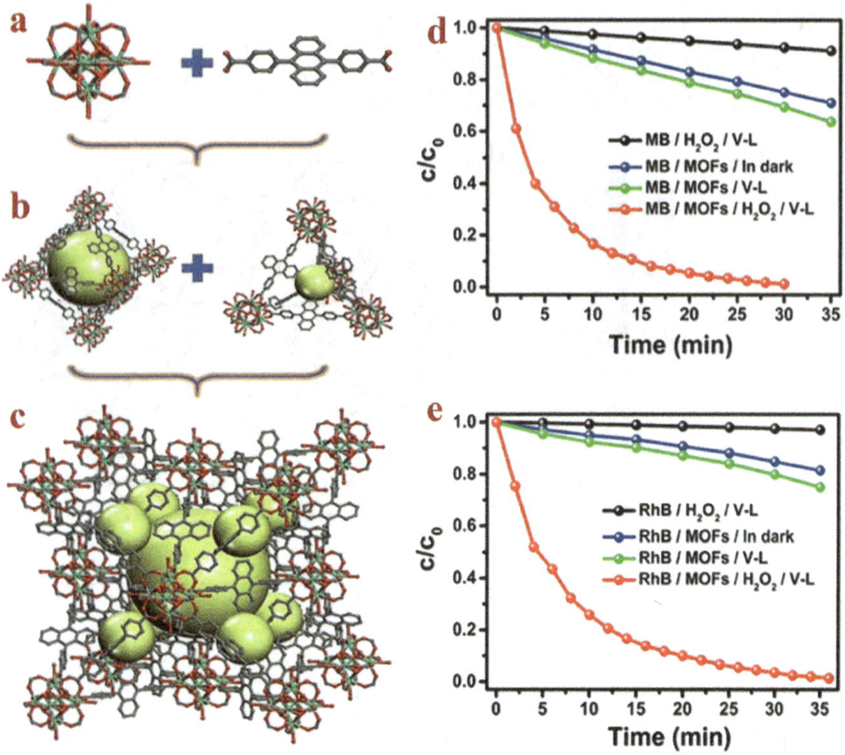

Fig. 9 **a** Molecular building blocks comprising Eu_6-cluster as well as the linear aromatic ligand, **b** octahedral and tetrahedral cages, **c** the fcu net of **Eu-1**. Color scheme: Eu (cyan), C (gray), and O (red). Photodegradation of **d** MB, **e** RhB. Reproduced with permission from Ref. [157]. Copyright 2018 Royal Society of Chemistry

The MOF NNU-15(Tb) = {[Tb(H_2**L**^{99})$_{1.5}$(H_2O)$_2$(DMF)]·(DMF)$_2$} (**L**^{99} = 4,4′-(diethynylanthracene-9,10-diyl)dibenzoate) photodegraded ~90% RhB in 1 h in visible light and H_2O_2 [155]. Another MOF **Tb-1** [Tb(**L**^{5})(**L**^{103})]$_n$ (**L**^{5} = 2-(2-carboxylatephenyl)imidazo(4,5-f)-(1,10)phenanthroline and **L**^{103} = adipate) photodegraded 81.2% MB degradation in 370 min irradiation period [159]. Two Tb(III)-based MOFs, viz. [Tb_2(**L**^{52})$_{0.5}$(**L**^{52})$_{0.5}$(**L**^{52})$_{0.5}$(**L**^{52})$_{0.5}$(**L**^{52})$_{0.5}$(**L**^{52})$_{0.5}$(DMA)]$_n$ (**Tb-2**) and [Tb(**L**^{104})$_{0.5}$(**L**^{104})$_{0.5}$(**L**^{104})$_{0.5}$(DMF)]$_n$ (**Tb-3**) having **L**^{52} = naphthalene-1,4-dicarboxylate, **L**^{104} = anthracene-9,10-dicarboxylate [160] photodegraded RhB up to 92.8% (**Tb-2**) and 89.2% (**Tb-3**).

The dysprosium-based MOF, NNU-15(Dy) = [Dy(H_2**L**^{99})$_{1.5}$(H_2O)$_2$(DMF)]·(DMF)$_2$ also had been used photocatalyst against RhB degradation with the assistance of H_2O_2 and visible light and degraded ~95% RhB in 25 min [155].

Three Yb(III)-based MOFs, viz. [Yb(**L**^{8})$_{1.5}$(DMF)]$_n$ (**Yb-1**), [Yb(**L**^{105})$_{0.5}$(NO_3)(DMF)$_2$]$_n$ (**Yb-2**), and [Yb(**L**^{104})$_{0.5}$(**L**^{104})$_{0.5}$(NO_3)(DMF)(H_2O)]$_n$

(**Yb-3**), with ligands of different bulkiness terephthalate ($\mathbf{L^{8}}$); 5,5′-(diazene-1,2-diyl)diisophthalate ($\mathbf{L^{105}}$) and anthracene-9,10-dicarboxylate ($\mathbf{L^{104}}$) are also reported [161]. These MOFs decomposed RhB ~85.8% (**Yb-1**), 96.5% (**Yb-2**), and 96.9% (**Yb-3**) in UV light and in time period of 6 h. Using **Yb-2**, the 93.9% RhB decomposed in 180 min. The observed differences in their photocatalytic performances resulted due to the difference in their topologies [162].

Actinide-Based MOFs

Although many reports dealing with the photocatalytic properties of actinide MOFs for dye degradation are not investigated comprehensively, significant attention had been devoted to the uranyl-based MOFs comprising unique U=O double bonds to display photocatalytic activity. In this regard, a MOF **U-1**, $[(UO_2)_5(H_2O)_3(H_3O)(\mathbf{L^{106}})(\ H\mathbf{L^{106}})]\cdot 8H_2O$ comprising flexible hexapodal ligand 1,3,5-triazine-2,4,6-triaminohexaacetate ($\mathbf{L^{106}}$) synthesized and used as a photocatalyst to degrade RhB up to 92.9% dye in 2 h (Fig. 10c) [163].

Two UO_2 complexes, $(UO_2)_2(\mathbf{L^{107}})(DMA)_2$ (**U-2**) and $[(UO_2)_2(\mathbf{L^{67}})(\mu_3OH_2)]0.2[HN(CH_3)_2].H_2O$ (**U-3**) (where $H_4\mathbf{L^{107}}$ = 1,2,4,5-benzenetetracarboxylic acid) comprising 3D and 2D architectures, respectively, were also used as photocatalysts against RhB [164]. In presence of **U-2**, the aqueous solution of RhB became almost colorless in 130 min under UV irradiation. However, **U-3** decomposed merely half of the RhB solution till 160 min. The observed differences in photocatalytic properties of the two compounds were ascribed to the structural differences between the two MOFs. **U-2** was a mononuclear compound while **U-3** possess a multi-nuclear structure. Another uranyl MOF, **U-4**, $[(UO_2(\mathbf{L^{108}})(H_2O)]\cdot H_2O$ (where $H_2\mathbf{L^{108}}$ = 2,5-pyridinedicarboxylic acid) [165] exhibited degradation of MB up to 84.4, 92.4, and 100% on the irradiation with different light sources viz. 165 W, 330 W, and 500 W Hg-lamp, respectively, within 90 min. The experiments suggested that the photocatalytic efficiency of MOFs was uplifted with the rise in the power of Hg lamp.

Overall it could be concluded that the *d*-block especially the 3D transition metal-based MOFs finds significant application as photocatalysts for dye degradation. This could be ascribed to their low-cost production, high thermal, aqueous, and chemical stability. Apart from these features, their topology can be controlled and tuned during synthesis which results in appropriate and desired architectures with suitable bandgaps to induce semiconducting and in turn photocatalytic properties in them. Their reasonable architectural stability and integrity had further complemented their properties and because of which they could be recycled and reused for several cycles without appreciable loss in the catalytic properties.

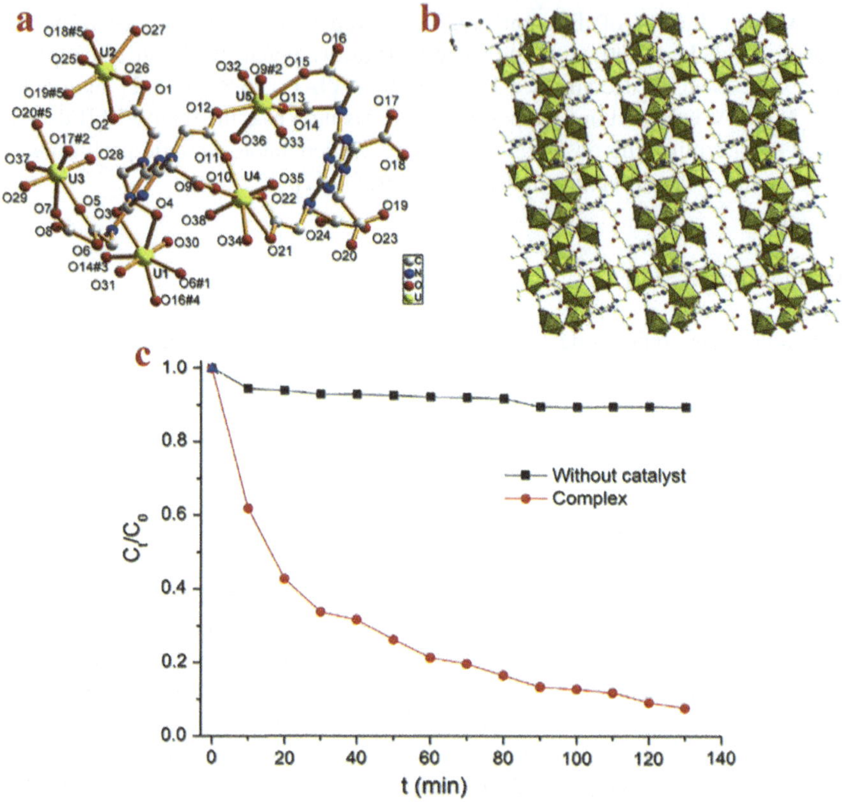

Fig. 10 **a** Structure of **U-1**, hydrogens was omitted for clarity, **b** extensive 3D network and the view of uncoordinated carboxyl group, **c** concentration change in RhB without **U-1** and with **U-1**. C_0 and C_t stand for the concentrations of RhB before and after irradiation. Reproduced with permission from Ref. [163]. Copyright 2019 Elsevier

3 Photocatalytic Mechanism

The efficiency of MOF-assisted photocatalysis depends on three important factors (1) separation of charge, (2) recombination of charge, and (3) interfacial of electrons. Also, small bandgap parameters of the MOFs facilitate electron transition and are beneficial for photocatalysis. The two basic mechanistic pathways using which MOFs execute the photodegradation of dyes are (a) direct photodegradation sometimes referred to as advanced oxidation process and (b) sensitization-mediated degradation process.

3.1 Direct Photodegradation Process

The MOFs utilize the direct photodegradation process for the photodecomposition of dyes through different pathways (Fig. 11). In such processes, MOFs display LMCT by transferring the electron from the photoexcited organic ligand to metal using the antenna effect. On irradiation, the photoexcited MOF generates electron (e^-)–hole (h^+) pairs, which in all cases is the first step in the dye photodegradation. The electron–hole pairs are separated as photoinduced electrons excited to the conduction band, leaving an equal quantity of holes (h^+) in the valence band (Eq. 1) [90, 107]. In order to revert back to the ground state, the excited electrons and holes react with OH^- to yield ·OH radical to assist oxidation (Eq. 2) and also oxidation of water molecules takes place to yield hydroxyl radicals (·OH) (Eq. 3), which thereafter executes direct oxidation/mineralization of organic dyes to CO_2, H_2O (Eq. 9).

$$\mathrm{MOF} + h\nu \rightarrow h^+_{VB} + e^-_{CB} \quad (1)$$

$$h^+_{VB} + OH^- \rightarrow \cdot OH \quad (2)$$

$$h^+_{VB} + H_2O \rightarrow \cdot OH + H^+ \quad (3)$$

In another pathway, photoelectrons residing in conduction band are trapped by O_2 which gets reduced to $\cdot O_2^-$ radical (Eq. 4), and then it reacts with H^+ to engender ·OH radicals (Eqs. 5–8). Both ·OH and $\cdot O_2^-$ oxidize dyes to CO_2 and H_2O and few smaller compounds (Eq. 9).

$$e^-_{CB} + O_2 \rightarrow \cdot O_2^- \quad (4)$$

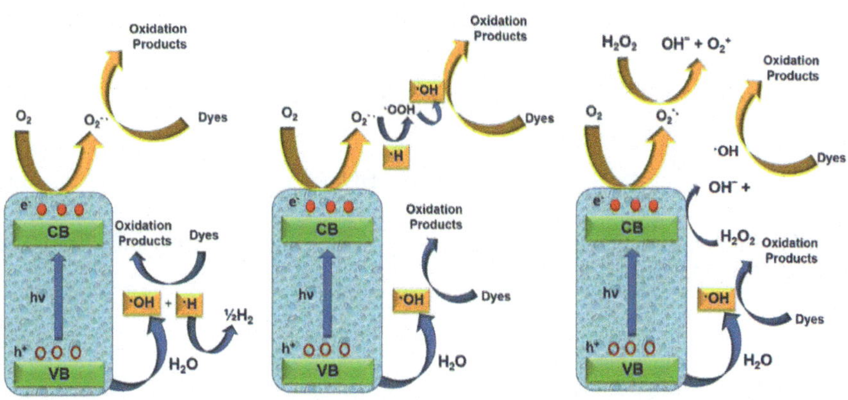

Fig. 11 Diagram showing the generation of ROS and degradation of dye through different AOPs using MOF-based photocatalysts

$$\cdot O_2^- + 2H^+ \rightarrow \cdot OH + OH^- \quad (5)$$

$$\cdot O_2^- + H^+ \rightarrow \cdot OOH \quad (6)$$

$$2\cdot OOH \rightarrow O_2 + H_2O_2 \quad (7)$$

$$H_2O_2 + h\nu \rightarrow 2\cdot OH \quad (8)$$

$$h_{VB}^+/^\cdot OH/^\cdot O_2^- + \text{Dye} \rightarrow \text{oxidized products } (CO_2 + H_2O + \text{small compounds}) \quad (9)$$

Further, to uplift the photocatalytic ability of MOFs, H_2O_2 is added which delays electron–hole pair recombination and increases the photodegradation rate (Eqs. 10–12). Also, the holes undergo a reaction with H_2O_2 molecules to yield ·OH radicals.

$$H_2O_2 + h\nu \rightarrow 2\cdot OH \quad (10)$$

$$e_{CB}^- + H_2O_2 \rightarrow \cdot OH + OH^- \quad (11)$$

$$\cdot O_2^- + H_2O_2 \rightarrow \cdot OH + OH^- + O_2 \quad (12)$$

But, several MOFs are unstable in H_2O_2; therefore, ultrasonic irradiation [)))] can be used in which ·OH radicals are generated through water splitting (Eqs. 13–15) [95, 129, 166].

$$H_2O +))) \rightarrow \cdot H + \cdot OH \quad (13)$$

$$\cdot OH + \cdot OH \rightarrow H_2 + O_2 \quad (14)$$

$$\cdot H + O_2 \rightarrow \cdot HO_2 \quad (15)$$

Additionally, some MOFs degrade dyes by employing both H_2O_2 and ultrasound (Eq. 16, 17) [95, 129, 166].

$$H_2O_2 \rightarrow 2 \cdot OH \quad (16)$$

$$H_2O_2 + \cdot OH \rightarrow \cdot HO_2 + H_2O \quad (17)$$

Also, H_2O_2 and $Na_2S_2O_8$ combinations had been used which served as electron acceptors to increase degradation efficiency [91]. The MOFs convert persulfate ($S_2O_8{}^{2-}$) to $SO_4^{\cdot}$ radicals using photo-Fenton-like process (Eqs. 18–20):

$$M^{2+} + S_2O_8^{2-} \rightarrow M^{3+} + SO_4^{2-} + SO_4^{\cdot-} \quad (18)$$

$$M^{3+} + S_2O_8^{2-} \rightarrow M^{2+} + 2SO_4^{\cdot-} \quad (19)$$

$$SO_4^{\cdot-} + \text{Dye} \rightarrow \text{degradation products} \quad (20)$$

3.2 Sensitization-Mediated Degradation Process

Some MOFs do not absorb visible light due to their large bandgap. Such MOFs photocatalyze the degradation of dyes through the photosensitization process [167–169]. In this process, dye absorbs visible light which excites its electron from HOMO → LUMO. If LUMO of dye lies at a higher energy scale than CB of MOFs, the photoexcited electrons of dyes are transferred to CB of MOFs. Thereafter, the MOF executes the photodegradation of dyes like the direct degradation process (vide supra) (Fig. 12).

4 Conclusions

From the presented discussion, it could be concluded that stability, cost economy, porosity, and photoactivity are some important prerequisites that must be associated with MOFs for their real-world application as photocatalysts for dye degradation. Hence, when appropriate rigid, semi-rigid, and flexible organic ligands with appropriate antennae and suitable co-ligand to complement photoactivity when coordinated to the main group, transition and inner transition metal centers can engender targeted MOFs that can display superior photocatalytic properties. Also, to enhance photocatalytic activity, the designing and syntheses of nanosized MOFs are gaining attention that can exhibit photocatalysis in the absence of promoters such as hydrogen peroxide. Among the plethora of MOF-based photocatalysts, the transition metal-based MOFs supersede well beyond the main group and inner transition metal-based MOFs. Hence, the utility of the main group and inner transition metal-based MOFs requires further investigation.

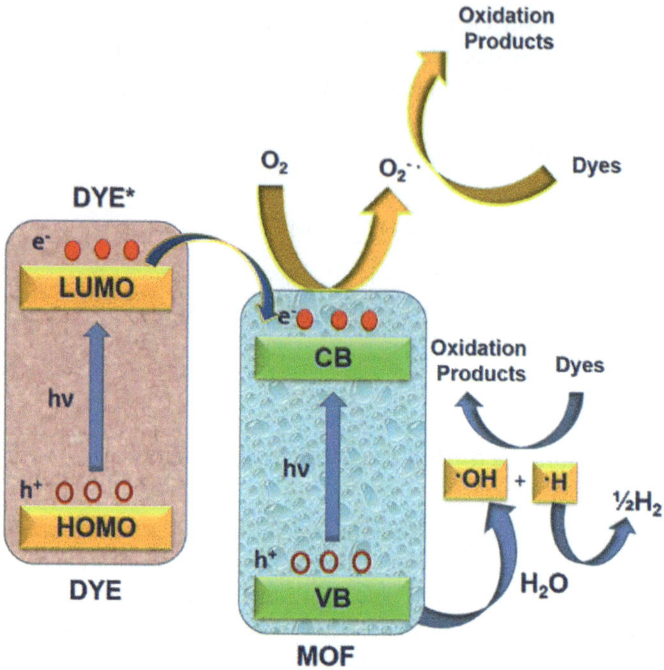

Fig. 12 Sensitization-mediated superoxide radicals ($\cdot O_2^-$) generation and dye degradation

Acknowledgements AS and AK are grateful to CSIR, New Delhi, for fellowship (file no. 09/107(0412)/2020-EMR-I) and project (01(2899)/17/EMR-II), respectively.

Abbreviations

AO	Ammonium oxalate
AOPs	Advanced oxidation processes
BQ	Benzoquinone
DMA	*N,N*-Dimethylacetamide
DMF	*N,N*-Dimethylformamide
DMPO	5,5-Dimethyl-1-pyrroline N-oxide
EDTA-2Na	Ethylenediaminetetraacetic acid
Eg	Energy bandgap
en	Ethylenediamine
ESR	Electron spin resonance
HOMO	Highest occupied molecular orbital
IR	Infrared
LMCT	Ligand to metal charge transfer

LUMO	Lowest unoccupied molecular orbital
MOCs	Metal–organic complexes
MOF	Metal–organic framework
MW	Microwave
NMOFs	Nanosized MOFs
SBU	Secondary building unit
SC	Supramolecular complex
TEOA	Triethanolamine
UV	Ultraviolet
VB	Valence band
CB	Conduction band

References

1. Zhao D, Timmons DJ, Yuan D, Zhou H-C (2011) Tuning the topology and functionality of metal−organic frameworks by ligand design. Acc Chem Res 44:123–133
2. Getman RB, Bae Y-S, Wilmer CE, Snurr RQ (2012) Review and analysis of molecular simulations of methane, hydrogen, and acetylene storage in metal–organic frameworks. Chem Rev 112:703–723
3. Wu H, Gong Q, Olson DH, Li J (2012) Commensurate adsorption of hydrocarbons and alcohols in microporous metal organic frameworks. Chem Rev 112:836–868
4. Kreno LE, Leong K, Farha OK, Allendorf M, Van Duyne RP, Hupp JT (2012) Metal–organic framework materials as chemical sensors. Chem Rev 112:1105–1125
5. Tan LL, Li H, Zhou Y, Zhang Y, Feng X, Wang B, Yang YW (2015) Zn^{2+}-triggered drug release from biocompatible zirconium MOFs equipped with supramolecular gates. Small 11:3806
6. Wen HM, Li B, Wang H, Wu C, Alfooty K, Krishna R, Chen B (2015) A microporous metal–organic framework with rare lvt topology for highly selective C_2H_2/C_2H_4 separation at room temperature. Chem Commun 51:5610–5613
7. Kim SN, Kim J, Kim HY, Cho HY, Ahn WS (2013) Adsorption/catalytic properties of MIL-125 and NH_2-MIL-125. Catal Today 204:85–93
8. Kim H, Yang S, Rao SR, Narayanan S, Kapustin EA, Furukawa H, Umans AS, Yaghi OM, Wang EN (2017) Water harvesting from air with metal-organic frameworks powered by natural sunlight. Science 356:430–434
9. Asha KS, Bhattacharjee R, Mandal S (2016) Complete transmetalation in a metal–organic framework by metal ion metathesis in a single crystal for selective sensing of phosphate ions in aqueous media. Angew Chem Int Ed 55:11528–11532
10. Lu XF, Liao PQ, Wang JW, Wu JX, Chen XW, He CT, Zhang JP, Li GR, Chen XM (2016) An alkaline-stable, metal hydroxide mimicking metal–organic framework for efficient electrocatalytic oxygen evolution. J Am Chem Soc 138:8336–8339
11. Comito RJ, Fritzsching KJ, Sundell BJ, Schmidt-Rohr K, Dinca M (2016) Single-site heterogeneous catalysts for olefin polymerization enabled by cation exchange in a metal-organic framework. J Am Chem Soc 138:10232−10237
12. Ramaswamy P, Wong NE, Shimizu GKH (2014) MOFs as proton conductors – challenges and opportunities. Chem Soc Rev 43:5913–5932
13. Loera-Serna S, Ortiz E (2016) Catalytic applications of metal-organic frameworks. In: Norena KE, Wang J-A (eds) Advanced catalytic materials-photocatalysis and other current trends. InTech

14. Yang QH, Xu Q, Jiang HL (2017) Metal–organic frameworks meet metal nanoparticles: synergistic effect for enhanced catalysis. Chem Soc Rev 46:4774–4808
15. Xiao J-D, Jiang H-L (2019) Metal–organic frameworks for photocatalysis and photothermal catalysis. Acc Chem Res 52:356–366
16. Dhakshinamoorthy, Li Z, García H (2018) Catalysis and photocatalysis by metal organic frameworks. Chem Soc Rev 47:8134–8172
17. Partha M, Giridhar M, Srinivasan N (2006) Novel photocatalysts for the decomposition of organic dyes based on metal-organic framework compounds. J Phys Chem B 110:13759–13768
18. Wang C-C, Li J-R, Lv X-L, Zhang Y-Q, Guo G (2014) Photocatalytic organic pollutants degradation in metal–organic frameworks. Energy Environ Sci 7:2831–2867
19. (a) Wang JL, Wang C, Lin WB (2012) Metal–organic frameworks for light harvesting and photocatalysis. ACS Catal 2:2630–2640; (b) Ding M, Cai X, Jiang H-L (2019) Improving MOF stability: approaches and applications. Chem Sci 10:10209–10230; (c) Subudhi S, Mansingh S, Swain G, Behera A, Rath D, Parida K (2019) HPW-anchored UiO-66 metal–organic framework: a promising photocatalyst effective toward tetracycline hydrochloride degradation and H_2 evolution via Z-scheme charge dynamics. Inorg Chem 58:4921–4934; (d) Swain G, Sultana S, Parida K (2019) One-pot-architectured Au-nanodot-promoted $MoS_2/Znln_2S_4$: a novel p–n heterojunction photocatalyst for enhanced hydrogen production and phenol degradation. Inorg Chem 58:9941–9955; (e) Subudhi S, Mansingh S, Tripathy SP, Mohanty A, Mohapatra P, Rath D, Parida K (2019) The fabrication of Au/Pd plasmonic alloys on UiO-66-NH_2: an efficient visible light-induced photocatalyst towards the Suzuki Miyaura coupling reaction under ambient conditions. Catal Sci Technol 9:6585–6597; (f) Subudhi S, Paramanik L, Sultana S, Mansingh S, Mohapatra P, Parida K (2020) A type-ii interband alignment heterojunction architecture of cobalt titanate integrated UiO-66-NH_2: a visible light mediated photocatalytic approach directed towards Norfloxacin degradation and green energy (Hydrogen) evolution. J Colloid Interface Sci 568:89–105
20. Mon M, Bruno R, Soria JF, Armentano D, Pardo E (2018) Metal–organic framework technologies for water remediation: towards a sustainable ecosystem. J Mater Chem A 6:4912–4947
21. Zango ZU, Jumbri K, Sambudi NS, Ramli A, Bakar NHHA, Saad B, Rozaini MNH, Isiyaka HA, Jagaba AH, Aldaghri O, Sulieman A (2020) A critical review on metal-organic frameworks and their composites as advanced materials for adsorption and photocatalytic degradation of emerging organic pollutants from wastewater. Polymers (Basel) 12:2648
22. Khan MS, Khalid Md, Shahid M (2020) What triggers dye adsorption by metal organic frameworks? The current perspectives. Mater Adv 1:1575–1601
23. Subudhi S, Tripathy SP, Parida K (2021) Highlights of the characterization techniques on inorganic, organic (COF) and hybrid (MOF) photocatalytic semiconductors. Catal Sci Technol 11:392–415
24. Subudhi S, Rath D, Parida KM (2018) A mechanistic approach towards the photocatalytic organic transformations over functionalised metal organic frameworks: a review. Catal Sci Technol 8:679–696
25. Suffet H, Malaiyandi M (eds) (1986) Organic pollutants in water, vol 214. American Chemical Society, Washington, DC
26. Shaheen SM, Kwon EE, Biswas JK, Tack FMG, Ok YS, Rinklebe J (2017) Arsenic, chromium, molybdenum, and selenium: geochemical fractions and potential mobilization in riverine soil profiles originating from Germany and Egypt. Chemosphere 180:553–563
27. Garvasis J, Prasad AR, Shamsheera KO, Jaseela PK, Joseph A (2020) Efficient removal of Congo red from aqueous solutions using phytogenic aluminum sulfate nano coagulant. Mater Chem Phys 251:123040
28. Mohan SV, Babu VL, Sarma PN (2007) Anaerobic biohydrogen production from dairy wastewater treatment in sequencing batch reactor (AnSBR): Effect of organic loading rate. Enzyme Microb Technol 41:506–515

29. Dey C, Kundu T, Biswal BP, Mallick A, Banerjee R (2014) Crystalline metal-organic frameworks (MOFs): synthesis, structure and function. Acta Crystallogr B 70:3–10
30. Sharanyakanth PS, Radhakrishnan M (2020) Synthesis of metal-organic frameworks (MOFs) and its application in food packaging: a critical review. Trends Food Sci Technol 104:102–116
31. Bedia J, Muelas-Ramos V, Peñas-Garzón M, Gómez-Avilés A, Rodríguez JJ, Belver C (2019) A review on the synthesis and characterization of metal organic frameworks for photocatalytic water purification. Catalysts 9:52
32. Klinowski J, Paz FAA, Silva P, Rocha J (2011) Microwave-assisted synthesis of metal–organic frameworks. Dalton Trans 40:321–330
33. Jhung SH, Lee JH, Chang JS (2005) Microwave synthesis of a nanoporous hybrid material, chromium trimesate. Bull Korean Chem Soc 26:880–881
34. Ni Z, Masel RI (2006) Rapid production of metal−organic frameworks via microwave-assisted solvothermal synthesis. J Am Chem Soc 128:12394–12395
35. Jhung SH, Lee JH, Yoon JW, Serre C, Ferey G, Chang JS (2007) Microwave synthesis of chromium terephthalate MIL-101 and its benzene sorption ability. Adv Mater 19:121–124
36. Choi JS, Son WJ, Kim J, Ahn WS (2008) Metal–organic framework MOF-5 prepared by microwave heating: factors to be considered. Microporous Mesoporous Mater 116:727–731
37. Seo YK, Hundal G, Jang IT, Hwang YK, Jun CH, Chang JS (2009) Microwave synthesis of hybrid inorganic–organic materials including porous $Cu_3(BTC)_2$ from Cu(II)-trimesate mixture. Microporous Mesoporous Mater 119:331–337
38. Jhung SH, Lee JH, Forster PM, Ferey G, Cheetham AK, Chang JS (2006) Microwave synthesis of hybrid inorganic–organic porous materials: phase-selective and rapid crystallization. Chem Eur J 12:7899–7905
39. Son W-J, Kim J, Kim J, Ahn W-S (2008) Sonochemical synthesis of MOF-5. Chem Commun 6336–6338
40. Vaitsis C, Sourkouni G, Argirusis C (2019) Metal Organic Frameworks (MOFs) and ultrasound: a review. Ultrason Sonochem 52:106–119
41. Joaristi M, Juan-Alcañiz J, Serra-Crespo P, Kapteijn F, Gascon J (2012) Electrochemical synthesis of some archetypical Zn^{2+}, Cu^{2+}, and Al^{3+} metal organic frameworks. Cryst Growth Des 12:3489–3498
42. Al-Kutubi H, Gascon J, Sudhölter EJR, Rassaei L (2015) Electrosynthesis of metal–organic frameworks: challenges and opportunities. Chem Electro Chem 2:462–474
43. Schlesinger M, Schulze S, Hietschold M, Mehring M (2010) Evaluation of synthetic methods for microporous metal–organic frameworks exemplified by the competitive formation of $[Cu_2(btc)_3(H_2O)_3]$ and $[Cu_2(btc)(OH)(H_2O)]$. Microporous Mesoporous Mater 132:121–127
44. Klimakow M, Klobes P, Thünemann AF, Rademann K, Emmerling F (2010) Mechanochemical synthesis of metal−organic frameworks: a fast and facile approach toward quantitative yields and high specific surface areas. Chem Mater 22:5216–5221
45. Stolar T, Užarević K (2020) Mechanochemistry: an efficient and versatile toolbox for synthesis, transformation, and functionalization of porous metal–organic frameworks. Cryst Eng Comm 22:4511–4525
46. Szczęśniak B, Borysiuk S, Choma J, Jaroniec M (2020) Mechanochemical synthesis of highly porous materials. Mater Horiz 7:1457–1473
47. Chen D, Zhao J, Zhang P, Dai S (2019) Mechanochemical synthesis of metal–organic frameworks. Polyhedron 162:59–64
48. Pichon, Lazuen-Garay A, James SL (2006) Solvent-free synthesis of a microporous metal–organic framework. CrystEngComm 8:211–214
49. Bian Y, Xiong N, Zhu G (2018) Technology for the remediation of water pollution: a review on the fabrication of metal organic frameworks. Processes 6:122
50. Parnham ER, Morris RE (2007) Ionothermal synthesis of zeolites, metal–organic frameworks, and inorganic–organic hybrids. Acc Chem Res 40:1005–1013
51. Rubio-Martinez M, Avci-Camur C, Thornton AW, Imaz I, Maspoch D, Hill MR (2017) New synthetic routes towards MOF production at scale. Chem Soc Rev 46:3453–3480

52. Díaz-García M, Mayoral Á, Díaz I, Sánchez-Sánchez M (2014) Nanoscaled M-MOF-74 materials prepared at room temperature. Cryst Growth Des 14:2479–2487
53. Yang D-A, Cho H-Y, Kim J, Yang S-T, Ahn W-S (2012) CO_2 capture and conversion using Mg-MOF-74 prepared by a sonochemical method. Energy Environ Sci 5:6465–6473
54. Wu X, Bao Z, Yuan B, Wang J, Sun Y, Luo H, Deng S (2013) Microwave synthesis and characterization of MOF-74 (M = Ni, Mg) for gas separation. Micropor Mesopor Mater 180:114–122
55. Haque E, Jhung SH (2011) Synthesis of isostructural metal–organic frameworks, CPO-27s, with ultrasound, microwave, and conventional heating: Effect of synthesis methods and metal ions. Chem Eng J 173:866–872
56. Jung DW, Yang DA, Kim J, Kim J, Ahn WS (2010) Facile synthesis of MOF-177 by a sonochemical method using 1-methyl-2-pyrrolidinone as a solvent. Dalton Trans 39:2883–2887
57. Furukawa H, Miller MA, Yaghi OM (2007) Independent verification of the saturation hydrogen uptake in MOF-177 and establishment of a benchmark for hydrogen adsorption in metal–organic frameworks. J Mater Chem 17:3197–3204
58. Tranchemontagne DJ, Hunt JR, Yaghi OM (2008) Room temperature synthesis of metal-organic frameworks: MOF-5, MOF-74, MOF-177, MOF-199, and IRMOF-0. Tetrahedron 64:8553–8557
59. Masoomi MY, Bagheri M, Morsali A, Junk PC (2016) High photodegradation efficiency of phenol by mixed-metal–organic frameworks. Inorg Chem Front 3:944–951
60. Masoomi MY, Bagheri M, Morsali A (2017) Porosity and dye adsorption enhancement by ultrasonic synthesized Cd(II) based metal-organic framework. Ultrason Sonochem 37:244–250
61. Razavi SAA, Masoomi MY, Morsali A (2018) Morphology-dependent sensing performance of dihydro-tetrazine functionalized MOF toward Al(III). Ultrason Sonochem 41:17–26
62. Gordon J, Kazemian H, Rohani S (2012) Rapid and efficient crystallization of MIL-53(Fe) by ultrasound and microwave irradiation. Micropor Mesopor Mater 162:36–43
63. Chalati T, Horcajada P, Gref R, Couvreur P, Serre C (2011) Optimisation of the synthesis of MOF nanoparticles made of flexible porous iron fumarate MIL-88A. J Mater Chem 21:2220–2227
64. Krawiec P, Kramer M, Sabo M, Kunschke R, Fröde H, Kaskel S (2006) Improved hydrogen storage in the metal-organic framework $Cu_3(BTC)_2$. Adv Eng Mater 8:293–296
65. Schlesinger M, Schulze S, Hietschold M, Mehring M (2010) Evaluation of synthetic methods for microporous metal–organic frameworks exemplified by the competitive formation of $[Cu_2(btc)_3(H_2O)_3]$ and $[Cu_2(btc)(OH)(H_2O)]$. Micropor Mesopor Mater 132:121–127
66. Seo Y-K, Hundal G, Jang IT, Hwang YK, Jun C-H, Chang J-S (2009) Microwave synthesis of hybrid inorganic–organic materials including porous $Cu_3(BTC)_2$ from Cu(II)-trimesate mixture. Micropor Mesopor Mater 119:331–337
67. Li Z-Q, Qiu L-G, Xu T, Wu Y, Wang W, Wu Z-Y, Jiang X (2009) Ultrasonic synthesis of the microporous metal–organic framework $Cu_3(BTC)_2$ at ambient temperature and pressure: An efficient and environmentally friendly method. Mater Lett 63:78–80
68. Chen X, Burda C (2008) The electronic origin of the visible-light absorption properties of C-, N- and S-doped TiO_2 nanomaterials. J Am Chem Soc 130:5018–5019
69. Takanabe K (2017) Photocatalytic water splitting: quantitative approaches toward photocatalyst by design. ACS Catal 7:8006–8022
70. Bag PP, Sahoo P (2020) Designing metal-organic frameworks based photocatalyst for specific photocatalytic reactions: a crystal engineering approach. In: Rajendran S, Naushad M, Ponce LC, Lichtfouse E (eds) Green photocatalysts for energy and environmental process, environmental chemistry for a sustainable world, vol 36. Springer Nature, Switzerland AG, pp 141–186
71. Hendon CH, Tiana D, Fontecave M, Sanchez C, D'arras L, Sassoye C, Rozes L, Mellot-Draznieks C, Walsh A (2013) Engineering the optical response of the titanium-MIL-125 metal–organic framework through ligand functionalization. J Am Chem Soc 135:10942–10945

72. Wang J, Rao CY, Lu L, Zhang SL, Muddassir M, Liu JQ (2021) Efficient photocatalytic degradation of methyl violet using two new 3D MOFs directed by different carboxylate spacers. Cryst Eng Comm 23:741–747
73. Hu ML, Safarifard V, Doustkhah E, Rostamnia S, Morsali A, Nouruzi N, Beheshti S, Akhbari K (2018) Taking organic reactions over metal-organic frameworks as heterogeneous catalysis. Micropor Mespor Mat 256:111–127
74. Liu KG, Rouhani F, Gao XM, Abbasi-Azad M, Li JZ, Hu XD, Wang W, Hu ML, Morsali A (2020) Bilateral photocatalytic mechanism of dye degradation by a designed ferrocene-functionalized cluster under natural sunlight. Catal Sci Technol 10:757–767
75. Liu J, Liu G, Gu C, Liu W, Xu J, Li B, Wang W (2016) Rational synthesis of a novel 3,3,5-c polyhedral metal–organic framework with high thermal stability and hydrogen storage capability. J Mater Chem A 4:11630–11634
76. Zhong YY, Li XS, Chen JH, Wang XX, Wei LT, Fang LQ, Kumar A, Zhuang SZ, Liu JQ (2020) Recent advances in MOF-based nanoplatforms generating reactive species for chemodynamic therapy. Dalton Trans 49:11045–11058
77. Kinik FP, Ortega-Guerrero A, Ongari D, Ireland CP, Smit B (2021) Pyrene-based metal organic frameworks: from synthesis to applications. Chem Soc Rev 50:3143–3177
78. Deria P, Mondloch JE, Karagiaridi O, Bury W, Hupp JT, Farha OK (2014) Beyond post-synthesis modification: evolution of metal–organic frameworks *via* building block replacement. Chem Soc Rev 43:5896–5912
79. Yin H-Q, Wang X-Y, Yin X-B (2019) Rotation restricted emission and antenna effect in single metal–organic frameworks. J Am Chem Soc 141:15166–15173
80. Zhao Y, Li D (2020) Lanthanide-functionalized metal–organic frameworks as ratiometric luminescent sensors. J Mater Chem C 8:12739–12754
81. Akimov V, Asahi R, Jinnouchi R, Prezhdo OV (2015) What makes the photocatalytic CO_2 reduction on N-doped Ta_2O_5 efficient: insights from nonadiabatic molecular dynamics. J Am Chem Soc 137:11517–11525
82. Abrahamsson M, Johansson PG, Ardo S, Kopecky A, Galoppini E, Meyer GJ (2010) Decreased interfacial charge recombination rate constants with N3-type sensitizers. J Phys Chem Lett 1:1725–1728
83. Karnahl M, Kuhnt C, Ma F, Yartsev A, Schmitt M, Dietzek B, Rau S, Popp J (2011) Tuning of photocatalytic hydrogen production and photoinduced intramolecular electron transfer rates by regioselective bridging ligand substitution. Chem Phys Chem 12:2101–2109
84. Gao C, Wang J, Xu H, Xiong Y (2017) Coordination chemistry in the design of heterogeneous photocatalysts. Chem Soc Rev 46:2799–2823
85. Lu W, Wei Z, Gu Z-Y, Liu T-F, Park J, Park J, Tian J, Zhang M, Zhang Q, Gentle T III, Bosch M, Zhou H-C (2014) Tuning the structure and function of metal–organic frameworks *via* linker design. Chem Soc Rev 43:5561–5593
86. Li H-P, Dou Z, Chen S-Q, Hu M, Li S, Sun H-M, Jiang Y, Zhai Q-G (2019) Design of a multifunctional indium–organic framework: fluorescent sensing of nitro compounds, physical adsorption, and photocatalytic degradation of organic dyes. Inorg Chem 58:11220–11230
87. Zhao N, Sun F, Zhang N, Zhu G (2017) Novel pyrene-based anionic metal–organic framework for efficient organic dye elimination. Cryst Growth Des 17:2453–2457
88. Li Q, Xue D-X, Zhang Y-F, Zhang Z-H, Gao Z, Bai J (2017) A dual-functional indium–organic framework towards organic pollutant decontamination *via* physically selective adsorption and chemical photodegradation. J Mater Che A 5:14182–14189
89. Zong Z, Fan C-B, Zhang X, Meng X-M, Jin F, Fan Y-H (2019) Synthesis, crystal structures and dye removal properties of a series of metal-organic frameworks based on *N*-heterocyclic carboxylic acid ligands. Microporous Mesoporous Mater 282:82–90
90. Li Z-M, Qiao Y, Liu C-B, Zhou Y-F, Wang X-Y, Charpentier PA, Che G-B, Xu WZ, Liu L-H, Zhu E-W (2018) Syntheses, crystal structures, adsorption properties and visible photocatalytic activities of highly stable Pb-based coordination polymers constructed by 2-(2-carboxyphenyl)imidazo(4,5-*f*)-(1,10)phenanthroline and bridging linkers. Dalton Trans 47:7761–7775

91. Xiao J-X, Ma D-Y (2018) Syntheses, structures, luminescent and catalytic properties of two 3D metal-organic frameworks. Inorg Chim Acta 483:6–11
92. Etaiw SE-DH, Abd El-Aziz DM, Marie H, Ali E (2018) Cd (II) and holodirected lead (II) 3D-supramolecular coordination polymers based on nicotinic acid: Structure, fluorescence property and photocatalytic activity. Solid State Sci 79:15–22
93. Qiao Y, Ma Y, Jiang W, Wang X, Guan W, Che G, Li W, Qin F (2018) A series of metal–organic frameworks constructed by a rigid-flexible 5-(bis(4-carboxybenzyl)amino)isophthalic acid: syntheses, crystal structures and physical properties. Cryst Eng Comm 20:7782–7794
94. Zhang X-Y, Wu R-X, Bi C-F, Zhang X, Fan Y-H (2018) A new organobismuth (V) complex with fluorobenzoic ligands: Synthesis, crystal structure, photodegradation properties. Inorg Chim Acta 483:129–135
95. Etaiw SE-DH, Marie H (2019) Sonochemical nanostructure of Mn(II) supramolecular complex: X-ray structure, sensing and photocatalytic properties. Sens Actuat B 290:631–639
96. Xie A-D, Hu M-G, Luo Y-H, Zhu X-G, Wang Z-H, Geng W-Y, Zhang H, Zhang D-E, Zhang H (2021) Synthesis of a stable iron(III)–organic framework for a visible light induced simultaneous photocatalytic reduction of Cr(VI) and the degradation of organic dyes in water. New J Chem 45:13406–13414
97. Deng S-Q, Miao Y-L, Tan Y-L, Fang H-N, Li Y-T, Mo X-J, Cai S-L, Fan J, Zhang W-G, Zheng S-R (2019) An anionic nanotubular metal–organic framework for high-capacity dye adsorption and dye degradation in darkness. Inorg Chem 58:13979–13987
98. Hou H, Shao CZ, Han X, Liu Y, Xu W, Wu Q, Xie Q, Zhao Y (2019) Metal-dependent photocatalytic activity and magnetic behaviour of a series of 3D Co–Ni metal organic frameworks. Dalton Trans 48:6191–6197
99. Shang Q, Zeng T, Gao K, Liu N, Cheng Q, Liao G, Pan Z, Zhou H (2019) A novel nitrogen heterocyclic ligand-based MOF: synthesis, characterization and photocatalytic properties. New J Chem 43:16595–16603
100. Shao Z, Huang C, Han X, Wang H, Li A, Han Y, Li K, Hou H, Fan Y (2015) Two (3,6)-connected porous metal–organic frameworks based on linear trinuclear $[Co_3(COO)_6]$ and paddlewheel dinuclear $[Cu_2(COO)_4]$ SBUs: gas adsorption, photocatalytic behaviour, and magnetic properties. Dalton Trans 44:12832–12838
101. Das D, Mandal A, Ganguly S, Mukherjee S (2019) Photodegradation of some organic dyes over two metal–organic frameworks with especially high efficiency for safranine T. Dalton Trans 48:13869–13879
102. Qin L, Hu Q, Zheng Q, Dou Y, Yang H, Zheng H (2020) Green synthesis of nanoscale cobalt(II)-based MOFs: highly efficient photo-induced green catalysts for the degradation of industrially used dyes. Cryst Eng Comm 22:2327–2335
103. Xu Z, Li H, Li Q, He Y, Tang Q (2017) Two novel porous MOFs with square-shaped cavities for the removal of toxic dyes: adsorption or degradation? New J Chem 41:15204–15209
104. Wang Z-X, Tian H-X, Ding J-G, Li B-L, Wu B (2020) Syntheses, crystal structures, dye degradation and luminescence sensing properties of four coordination polymers. Acta Cryst C C76:23–29
105. Xiao Q-Q, Song Z-W, Li Y-H, Cui G-H (2019) A Co-MOF with a (4,4)-connected binodal two-dimensional topology: synthesis, structure and photocatalytic properties. J Solid State Chem 276:331–338
106. Zong Z-A, Fan C-B, Zhang X, Meng X-M, Jin F, Fan Y-H (2019) Two difunctional Co(II) coordination polymers for natural sunlight photocatalysis of methylene blue and selective fluorescence sensing of Cr(VI) ion in water media. Cryst Eng Comm 21:673–686
107. Wang J, Chen N-N, Zhang C, Jia L-Y, Fan L (2020) Four Co(II) coordination polymers based on 4,4′-(1*H*-1,2,4-triazol-1-yl)methylenebis(benzoic acid): syntheses, structural diversity, magnetic properties, dye adsorption and photocatalytic properties. Cryst Eng Comm 22:811–820
108. Fan C, Zong Z, Zhang X, Su B, Zhang X, Zhu Z, Bi C, Fan Y (2019) Functional group induced structural diversities and photocatalytic, magnetic and luminescence sensing properties of four cobalt(II) coordination polymers based on 1,3,5-tris(2-methylimidazol-1-yl)benzene. Polyhedron 170:515–522

109. Liang F, Ma D (2020) Syntheses, structural diversity and photo-degradation and dye adsorption properties of novel Ni(II)/Co(II) coordination polymers modulated by 4-(4-carboxylphenylmethylthio)benzoic acid ligand. J Mol Struct 1208:127814
110. Qu Y-H, Yang Y-J, Dong G-Y (2020) Luminescence and catalytic properties of two nickel(II)-organic frameworks constructed by 5-substituted isophthalate and N-donor mixed ligands. Polyhedron 180:114431
111. Wang Z, Ren Y, Cao J, Tang L, Zhang M, Zhou S (2018) Synthesis, structures, and photocatalytic properties of three new nickel(II) coordination polymers containing bis(benzimidazole) ligands with different coordination architectures. New J Chem 42:17991–18000
112. Li J-X, Liu D, Qin Z-B, Dong G-Y (2019) Structural assembly from 1D to 3D motivated by the linear co-ligands, and the magnetic and photocatalytic properties of five Ni^{II} coordination polymers with 5-(4′-carboxylphenyl)nicotinic acid. Polyhedron 160:92–100
113. Hao Z-C, Wang S-C, Yang Y-J, Cui G-H (2020) Sonochemical synthesis of two nano-sized nickel(II) coordination polymers derived from flexible bis(benzimidazole) and isophthalic acid ligands. Polyhedron 181:114466
114. Wen L, Zhao J, Lv K, Wu Y, Deng K, Leng X (2012) Syntheses, structural diversities and photocatalytic properties of three nickel(II) coordination polymers based semi-bis(benzimidazole) and aromatic dicarboxylic acid ligands. Cryst Growth Des 12:1603–1612
115. Yang Y, Zhu C, Li R, Gu X, She S, Anorg Z (2021) Two new metal-organic frameworks constructed by 4,5-bis(pyrazol-1-yl) phthalic acid for photocatalytic properties. Allg Chem 647:1560–1567
116. Wang F, Liu W, Li S, Gable RW, Zhong H, Xu Q, Huang G, Singh A, Liu J, Kumar A (2018) Synthesis, structure, spectral characteristic and photocatalytic degradation of organic dyes of a copper metal-organic framework based on tri(triazole) and pimelate. Inorg Chem Commun 97:109–112
117. Pan Y, Liu W, Liu D, Ding Q, Liu J, Xu H, Trivedi M, Kumar A (2019) A 3D metal-organic framework with isophthalic acid linker for photocatalytic properties. Inorg Chem Commun 100:92–96
118. He X, Fang K, Guo XH, Han J, Lu XP, Li MX (2015) A homochiral Cu(I) coordination polymer based on achiral precursors and its photocatalytic properties. Dalton Trans 44:13545–13549
119. Wu Z-L, Wang C-H, Zhao B, Dong J, Lu F, Wang W-H, Wang W-C, Wu G-J, Cui J-Z, Cheng P (2016) Coligand syntheses, crystal structures, luminescence and photocatalytic properties of five coordination polymers based on rigid tetracarboxylic acids and imidazole linkers. Angew Chem Int Ed 55:4938–4942
120. Mahmoodi NM, Abdi J (2019) A semi-conductive copper–organic framework with two types of photocatalytic activity. Microchem J 144:436–442
121. Zhang M, Wang L, Zeng T, Shang Q, Zhou H, Pan Z, Cheng Q (2018) Nanoporous metal-organic framework (MOF-199): Synthesis, characterization and photocatalytic degradation of Basic Blue 41. Dalton Trans 47:4251–4258
122. Wei SY, Song G, Bai F, Shi Z, Xing Y-H (2015) Two pure MOF-photocatalysts readily prepared for the degradation of methylene blue dye under visible light. Chem Plus Chem 80:1007–1013
123. Wang X, Luan J, Lin H, Xu C, Liu G, Zhang J (2013) Large tripodal spacer ligands for the construction of microporous metal–organic frameworks with diverse structures and photocatalytic activities. Cryst Eng Comm 15:9995–10006
124. Yang X-F, Chen X-L, Zhu H-B, Shen Y (2018) Synthesis, structure and photocatalytic degradation of organic dyes of a copper(II) metal–organic framework (Cu–MOF) with a 4-coordinated three-dimensional $CdSO_4$ topology. Polyhedron 157:367–373
125. Qian L-L, Han S-S, Zheng L-Y, Yang Z, Li K, Li B-L, Wu B (2019) An acid–base resistant paddle-wheel Cu(II) coordination polymer for visible-light-driven photodegradation of organic dyes. Polyhedron 162:303–310
126. Shi L-L, Zheng T-R, Li M, Qian L-L, Li B-L, Li H-Y (2017) Syntheses, structures and properties of structural diversity of 3D coordination polymers based on bis(imidazole) and dicarboxylate. RSC Adv 7:23432–23443

127. Zheng T-R, Blatov VA, Qian L-L, Tang D-Y, Zhang Y-Q, Wang Z-X, Li B-L, Wu B (2018) A series of five-coordinated copper coordination polymers for efficient degradation of organic dyes under visible light irradiation. Polyhedron 148:81–87
128. Etaiw SEH, Fayed TA, Aziz DMAE, Khatab HM (2019) An unusual (4,6)-coordinated copper(II) coordination polymer: High efficient degradation of organic dyes under visible light irradiation and electrochemical properties. Appl Organometal Chem 34:e5301
129. Dong J-P, Shi Z-Z, Li B, Wang L-Y (2019) Ultrasound-assisted nanoscaled supramolecular coordination polymer as an efficient recyclable catalyst for photocatalytic degradation of dye pollutants. Dalton Trans 48:17626–17632
130. Jiang J, Furukawa H, Zhang Y-B, Yaghi OM (2016) Synthesis of a novel 2D zinc(II) metal–organic framework for photocatalytic degradation of organic dyes in water. J Am Chem Soc 138:10244–10251
131. Jin J, Wu J, Liu W, Ma A, Liu J, Singh A, Kumar A (2018) Construction of pillared-layer MOF as efficient visible-light photocatalysts for aqueous Cr(VI) reduction and dye degradation. New J Chem 42:2767–2775
132. Cheng J, Hu T, Li W, Chang Z, Sun C (2020) A new Zn(II) metal–organic framework having 3D $CdSO_4$ topology as luminescent sensor and photocatalyst for degradation of organic dyes. J Solid State Chem 282:121125
133. Pan Y, Ding Q, Xu H, Shi C, Singh A, Kumar A, Liu J (2019) Luminescent sensing and photocatalytic degradation properties of an uncommon (4,5,5)-connected 3D MOF based on 3,5-di(3′,5′-dicarboxylphenyl)benzoic acid. Cryst Eng Comm 21:4578–4585
134. Zhou E-H, Li B-H, Chen W-X, Luo Z, Liu J, Singh A, Kumar A, Jin J-C (2017) A new Zn(II)-based 3D metal–organic framework with uncommon *sev* topology and its photocatalytic properties for the degradation of organic dyes. J Mol Struct 1149:352–356
135. Liu J, Kumar A, Li B, Prasad R, Batten SR, Wu J, Jin J, He Y (2017) Photocatalytic degradation of organic dyes by a stable and biocompatible Zn(II) MOF having ferulic acid: Experimental findings and theoretical correlation. Cryst Eng Comm 19:6464–6472
136. Yuan F, Lia Y, Ling X-Y, Yuan C-M, Zhou C-S, Wang J, Singh A, Kumar A, Chen F-Y (2019) A 3D luminescent Zn(II) MOF for the detection of high explosives and the degradation of organic dyes: an experimental and computational study. Inorg Chem Commun 109:107576
137. Wang J-J, Zhang D-J, Cao Y-L, Li X-R, Wang Y-R, Li Y-A, Tian L-L, Hu T-L (2018) A new 3D three-interpenetration metal–organic framework and its photocatalytic property: A combined experimental and theoretical investigation. Inorg Chim Acta 482:447–453
138. Song W-C, Liang L, Cui X-Z, Wang X-G, Yang E-C, Zhao X-J (2018) Selective fluorescent sensing and photocatalytic properties of Zinc(II) and Cadmium(II) coordination architectures with naphthalene-1,5-disulfonate and 2,4,5-tri(4-pyridyl)-imidazole. Cryst Eng Comm 20:668–678
139. Etaiw SE-DH, Marie H (2018) Syntheses, crystal structures and photocatalytic properties of three zinc (II) coordination polymers constructed by mixed ligands. J Photochem Photobiol A 364:478–491
140. Wang G-L, Wang J, Zhou L, Zhou M, Wang X, Zhou S-H, Lu L, Trivedi M, Kumar A (2021) Structural diversity in four Zn(II)/Cd(II) coordination polymers tuned by flexible pentacarboxylate and N-donor coligands: Photocatalysts for enhanced degradation of dyes. Dyes Pigm 195:109695
141. Chorab MW, Patroniak V, Kubicki M, Kądziołka G, Przepiórski J, Michalkiewicz B (2012) Three silver-based complexes constructed from organic carboxylic acid and 4,4′-bipyridine-like ligands: Syntheses, structures and photocatalytic properties. J Catal 291:1–8
142. Li A-L, Hao Z-C, Han C, Cui G-H (2020) Structures, photoluminescence and photocatalytic properties of two novel metal–organic frameworks based on tetrazole derivatives. Appl Organometal Chem 34:e5313
143. Kulovi S, Dalbera S, Dey SK, Maiti (Choudhury) S, Puschmann H, Zangrando E, Dalai S (2018) Cobalt(II) and silver(I) coordination polymers containing flexible bis (benzimidazol-1-yl)hexane ligands: synthesis, crystal structures, sensing and photocatalytic properties. Chemistry Select 3:5233–5242

144. Chen J-M, Zhou Q-K, Hou Y-X, Liu D (2017) Hemocompatible 3D silver(I) coordination polymers: synthesis, X-ray structure, photo-catalytic and antibacterial activity. Acta Cryst C C73:1017–1023
145. Shi C, Nie Z-H, Zhao L, Lu L, Cheng F, Chen X, Tan G, Liu Q-Q, Wang J, Chauhan R, Kumar A (2021) Efficient degradation of dyes in water by two Ag-based coordination polymers containing 1,3-bis(3,5-dicarboxylphenoxy)benzene and N-donor linkers. Polyhedron 207:115362
146. Gu J-Z, Cai Y, Wen M, Shi Z-F, Kirillov AM (2018) A metal–organic framework based multifunctional catalytic platform for organic transformation and environmental remediation. Dalton Trans 47:14327–14339
147. Wu Y-P, Wu X-Q, Wang J-F, Zhao J, Dong W-W, Li D-S, Zhang Q (2016) Assembly of a series of d^{10} coordination polymers based on W-shaped 1,3-di(2′,4′-dicarboxyphenyl)benzene: from syntheses, structural diversity, luminescence, to photocatalytic properties. Cryst Growth Des 16:2309–2316
148. Sun YQ, Zhong JC, Ding L, Chen YP (2015) Assembly of two novel $Cd_3/(Cd_3 + Cd_5)$-cluster-based metal–organic frameworks: structures, luminescence, and photocatalytic degradation of organic dyes. Dalton Trans 44:11852–11859
149. El-bendary MM, Rüffer T, Arshad MN, Asiri AM (2019) Multicomponent self-assembly of two Cd(II)-based coordination polymers: synthesis, structures and photocatalytic properties. J Mol Struct 1192:230–240
150. Ren W, Han F, Lu J-K, Hu Y-C (2020) Synthesis and structure characterization of Pt(IV) and Cd(II) 1,10-phenanthroline complexes; fluorescence, antitumor and photocatalytic property. J Inorg Organomet Polym Mater 30:1782–1789
151. Pan W, Gong C, Zeng X, Hub C, Zhang Y, Zhu D-R, Xu H, Guo H, Zhang J, Xie J (2019) A new La(iii)-MOF for efficient dye photodegradation and protective effect on exercise pain after total knee arthroplasty by reducing nicotinic acetylcholine receptors expression. Polyhedron 169:24–31
152. Wang J-J, Si P-P, Yang J, Zhao S-S, Li P-P, Li B, Wang S-Y, Lu M, Yu S-X (2019) Assembly of porous lanthanide metal–organic frameworks constructed by chalcone dicarboxylic acid and exploration of their properties. Polyhedron 162:255–262
153. Xia Q, Yu X, Zhao H, Wang S, Wang H, Guo Z, Xing H (2017) La(III)-based MOFs with 5-aminoisophthalic acid for optical detection and degradation of organic molecules in water. Cryst Growth Des 17:4189–4195
154. Das MC, Xu H, Wang Z, Srinivas G, Zhou W, Yue Y-F (2011) Ultrasonic synthesis of 1D-Zn(II) and La(III) supramolecular coordination polymers nanoparticles, fluorescence, sensing and photocatalytic property. Chem Commun 47:11715–11717
155. Du X, He H, Du L, Li W, Wang Y, Jiang Q, Yang L, Zhang J, Guo S (2019) Syntheses of novel lanthanide metal–organic frameworks for highly efficient visible-light-driven dye degradation. Polyhedron 171:221–227
156. Liu X, Liu B, Li G, Liu Y (2018) Porous Pr(III)-based organic framework for dye-adsorption and photo degradation with (4,5)-c net. J Mater Chem A 6:17177–17185
157. Wu W, Liu W, Shen X, Zhong H, Xu J, Li B, Liu J-Q, Singh AK, Singh A, Kumar A (2018) Two anthracene-based metal–organic frameworks for highly effective photodegradation and luminescent detection in water. Inorg Chem Commun 95:104–106
158. Ren S, Jiang W, Wang Q, Li Z, Qiao Y, Che G (2019) A new 3D Gd-based metal-organic framework with paddle-wheel unit: structure and photocatalytic property. RSC Adv 9:3102–3112
159. Wang J-J, Si P-P, Liu M-J, Chen Y, Yu S-X, Lu M, Wang S-Y, Li B, Li P-P, Zhang R-C (2018) Synthesis, structures and properties of six lanthanide complexes based on a 2-(2-carboxyphenyl)imidazo(4,5-*f*)-(1,10)phenanthroline ligand. Polyhedron 157:63–70
160. Wang J-J, Chen Y, Liu M-J, Fan R-Y, Si P-P, Yang J, Pan Y-Y, Chen Y, Zhao S-S, Xu J (2018) Selective fluorescent sensing and photodegradation properties of Tb(III)-based MOFs with different bulky backbone ligands. Polyhedron 154:411–419

161. Liang L-L, Hu Y-Q, Zhao J-S (2020) Yb(III)-based MOFs with different bulky backbone ligands for optical detection and degradation of organic molecules in wastewater. J Solid State Chem 282:121085
162. Nguyen HL, Gandara F, Furukawa H, Doan TLH, Cordova KE, Yaghi OM (2016) Two uranium coordination polymers constructed by a polycarboxylic acid: Structural variation, photoluminescent and photocatalysis properties. J Am Chem Soc 138:4330–4333
163. Hou Y-N, Xing Y-H, Bai F-Y, Guan Q-L, Wang X, Zhang R, Shi Z (2014) Synthesis, crystal structure, photoluminescence property and photoelectronic behavior of two uranyl-organic frameworks constructed from 1, 2, 4, 5-benzenetetracarboxylic acid as ligand. Spectrochim Acta A 123:267–272
164. Si Z-X, Xu W, Zheng Y-Q (2016) Synthesis, structure, luminescence and photocatalytic properties of an uranyl-2,5-pyridinedicarboxylate coordination polymer. J Solid State Chem 239:139–144
165. Daiguebonne C, Deluzet A, Camara M, Boubekeur K, Audebrand N, Gerault Y, Baux C, Guillou O (2003) Ultrasonic degradation of aqueous phenolsulfonphthalein (PSP) in the presence of nano-Fe/H_2O_2. Cryst Growth Des 3:1015–1020
166. Wang DK, Huang RK, Liu WJ, Sun DR, Li ZH (2014) Mechanistic insights into photodegradation of organic dyes using heterostructure photocatalysts. ACS Catal 4:4254–4260
167. Bibi R, Shen Q, Wei L, Hao D, Li N, Zhou J (2018) Hybrid BiOBr/UiO-66-NH_2 composite with enhanced visible-light driven photocatalytic activity toward RhB dye degradation. RSC Adv 8:2048–2058
168. Shinde DR, Tambade PS, Chaskar MG, Gadave KM (2017) Photocatalytic degradation of dyes in water by analytical reagent grades ZnO, TiO_2 and SnO_2: a comparative study. Drink Water Eng Sci 10:109–117
169. Molla MAI, Tateishi I, Furukawa M, Katsumata H, Suzuki T, Kaneco S (2017) Evaluation of reaction mechanism for photocatalytic degradation of dye with self-sensitized TiO_2 under visible light irradiation. Open J Inorg Non-Met Mater 7:1–7

Role of MOFs in Bio-catalysis

Multifaceted Metal–Organic Frameworks: An Emerging Platform for Biocatalytic Reactions

Shikha Gulati, Kartika Goyal, Nandini Sharma, Sanjay Kumar, and Kanchan Batra

Contents

1 Introduction ... 684
2 Structural Morphology and Potential of Mofs Over Conventional Bio Catalysts ... 685
3 Fundamental Properties Contributing Towards Biocatalysis ... 689
4 Strategies for Enzyme Immobilization ... 691
4.1 Surface Immobilization ... 692
4.2 Covalent Binding ... 693
4.3 Cage Inclusion ... 694
4.4 In Situ MOF Formation and Enzyme Immobilization ... 695
5 Outlook to Applications of MOFs as Immobilized Biocatalysts ... 696
5.1 Biosensing and Detection ... 696
5.2 MOFs as Host for Biomimetic Catalysis ... 697
5.3 Enzyme-MOF for Digestion of Proteins ... 698
6 Conclusion and Future Outlook ... 698
Abbreviations ... 699
Important Websites ... 699
References ... 700

Abstract Since the nineteenth century, the scientific community has experienced huge advancements and as a result, they are now delving into newer and innovative technological approaches to tackle present-day problems. One such emergent technology is metal–organic frameworks (MOFs), which belong to a category of coordination polymers, synthesized from ion or metal aggregate nodes and functional organic ligands that are united by coordinate bonds. They have garnered a lot of scientific curiosity in the previous decade due to their multimodal structures and wide-ranging functionalities exemplified by their surface area, easily adjustable porosity, electronic and optical properties, crystallinity, etc. These special abilities render them

S. Gulati · K. Goyal · S. Kumar
Department of Chemistry, Sri Venkateswara College, University of Delhi, Delhi 110021, India

N. Sharma
Department of Biological Sciences, Sri Venkateswara College, University of Delhi, Delhi 110021, India

K. Batra (✉)
Department of Zoology, Kalindi College, University of Delhi, East Patel Nagar, New Delhi 110008, India

S. Gulati (ed.), *Metal-Organic Frameworks (MOFs) as Catalysts*,
https://doi.org/10.1007/978-981-16-7959-9_25

superior in many fields of application like biosensing, bioimaging, cancer therapy, drug delivery and most importantly, in biocatalysis. Biological macromolecules and enzymes are produced by living cells and exhibit a wide array of important functions like biochemical synthesis by increasing the efficacy and rate of such living reactions. But these enzymes also possess some inherent limitations like low tolerance to temperature and pH variations, easy deactivation and when they are taken out of living systems, their handling proves to be quite costly, making them unstable and futile in industrial catalysis. Because of these drawbacks and the requirement of immobilization of natural enzymes during catalytic reactions, the use of MOFs as biocatalytic platforms is being explored thoroughly and is discussed in this chapter. This chapter aims to highlight the multifunctional aspects of MOFs which make them efficient bioimmobilization candidates in assisting enzymatic catalysis reactions on an industrial scale. Along with this, the synthesis and new surface functionalization strategies for making MOFs suitable for enzyme immobilization in biocatalysis are explained and future prospects of this technology are explored in the current chapter.

Keywords Biocatalysis · Metal–organic frameworks (MOFs) · Enzymes · Enzyme immobilization · Catalysis · Porous host matrix

1 Introduction

Nanotechnology has been opening a variety of doors in numerous fields of importance. One such highly utilized and revolutionary prospect, which will be discussed in this chapter, is metal–organic frameworks or MOFs. Metal–organic frameworks are essentially inorganic/organic materials (which are crystalline in nature) that display a high aptitude for porosity, which is also one of their characteristic properties. The structure of MOFs consists of a systematic array of positively charged metal ions composed in the form of a three-dimensional (3D) network, surrounded by bridging ligands that are organic and function as 'linkers' to the metal ion network. Such a network of metal ions is also referred to as secondary building units [1]. These metal ions function as nodes that bind the arms of the organic linkers together to result in the formation of a methodical, repeating, cage-like structure. This attribute provides MOFs, the ability to possess a large inner surface area, along with many other unique properties. These diverse properties include high surface area, structural diversity (making them particularly tunable materials), crystallinity, high porosity, ordered structure, adjustable and easily tweaked topography, both hydrophobicity as well as hydrophilicity, electronic transduction, optical affinity and many more [2]. MOFs have a large scale of structural and functional diversity, making them efficient and highly applicable in a myriad of fields, namely biomedical imaging, gas separation, drug delivery, carbon dioxide or CO_2 capture and most importantly biocatalysis, which will be highlighted in this chapter [3].

It is widely known that enzymes are macromolecules that are undoubtedly the foremost biological catalysts in the biosphere. These enzymes subsequently have far-reaching applications in pharmaceuticals, sensing, chemical and drug synthesis, food as well as energy processing, etc. The efficiency of enzymes over other types of traditional catalysts is because of their high activity including turnover numbers, selectivity (region-, chemo-, stereo-) and even specificity, which give them an edge in pertinent and sophisticated reactions required in industries [4]. Although enzymes prove to be better than artificial catalysts, their industrial applications are detrimentally affected because of their fragile and easily tampered biological nature including poor chemical, thermal, operational and even storage stability [5]. To combat this disability, a new low-cost alternative of trapping the required enzyme in a distinct special structure with intact enzymatic activity was devised which was named 'enzyme immobilization'. This technique is said to improve and enhance the stability, recovery and recyclability of important biocatalysts and to achieve this, scientists have found that materials having high porous strength emerge as viable biocatalytic platforms. During the exercise of enzyme immobilization, it is vital to comprehend the interactions of matrix and enzymes so as to maximize the activity and stability, without compromising the chemical and structural dexterity of the biocatalyst [6].

Thus, the above-mentioned and discussed structure, topology and properties of MOFs have a high potential of being compatible and efficient matrices for enzyme immobilization, essential in biocatalysis. The importance of MOFs in biocatalysis, their advantages and functionalization strategies to become methodical biocatalytic platforms for numerous applications is the points of discussion in this chapter.

2 Structural Morphology and Potential of Mofs Over Conventional Bio Catalysts

As discussed in the above introduction, metal–organic frameworks or MOFs are unique coordination compounds that have both inorganic and organic components linked together by strong bonds. Their distinguished properties because of flexibility and variations in the constituents' size, geometry and functionality have led the scientists to develop, study and report beyond 20,000 different types of MOFs in the past decade. These porous structures owe the uniqueness of their properties to their mode of synthesis as it involves certain procedures which end up providing the MOFs their indispensable attributes and are important factors in affecting their structural and functional aspects. In overview, the process of MOF synthesis can be achieved in quite a number of ways, namely direct precipitation, evaporation, sonochemical, electrochemical, solvothermal and microwave-assisted, a few of which are succinctly summarized in Table 1 [7].

During the synthesis of MOFs, many properties can be realized by minor changes in processing conditions, which can be summarized as follows.

1. In the course of forming the microemulsion, which is a spontaneous mixture formed by water, surfactant, low molecular weight alcohol (or co-surfactant)

Table 1 Some conventional synthetic methods of MOFs with their advantages and disadvantages are listed below

S.no	Method of synthesis	Advantages	Disadvantages	References
1	Slow diffusion	– Possible in ambient conditions (temperature and pressure, etc.) – Useful for X-ray diffraction studies, as single large crystals of MOFs, are normally formed	– Highly time-consuming – The resultant material formed is usually in very low quantities	[8]
2	Electrochemical	– It is a quick and clean approach – Applied to industrially produce MOF named HKUST-1 or Hong Kong University of Science and Technology-1	– Apart from HKUST-1, no other MOF has been produced using this synthetic method	[9]
3	Solvothermal	– Possesses a large functional temperature range (80–250 °C) – Can perform cooling and heating temperature ramps for increasing crystal growth – Has easy industrial transposition	– High energy consumption – Reaction time in days – High-cost requirements in purchase	[10]
4	Heating assisted by microwave	– Simplistic and fuel-efficient method – Time required in crystallization is reduced with yield improvement – Morphology, particle distribution and phase selectivity can be controlled during synthesis – Easy variation in MOFs possible with proper control of parameters	– Isolation of single large crystals is tedious – Its industrial implementation is not easy and quick but quite difficult	[11]

(continued)

Table 1 (continued)

S.no	Method of synthesis	Advantages	Disadvantages	References
5	Sonochemical	– Fundamental for isolating phase-pure substances – Particle size and structure can be homogenized easily and quickly – Is highly suitable for nano-MOFs synthesis	– Ultrasonic waves used can break the crystalline integrity and hinder the formation of large single crystals	[12]

and the hydrophobic element, if the organic component acts like the continuous phase whilst the water counterpart behaves similar to the dispersed phase, then an automatic reverse microemulsion can be formed. This in turn affects the extent of properties of MOFs.

2. Initially, the crystallization is achieved by collisions between droplets of the aqueous medium, and the complexity of such droplets can be regulated by modifying the arrangement of either/both continuous phase as well as a surfactant, to give MOFs with adjustable structure and size, as result.
3. Another processing condition that plays an important role is the temperature. It has been observed that in MOFs, temperature affects the rate of crystal growth and relative nucleation, which is a key factor in helping scientists to fabricate a variety of such frameworks with different crystal structures and sizes by controlling the temperature.
4. Along with this, it has been proved that the introduction of certain compatible additives with required qualities can generate a specific crystal size with every synthesis. It can be exemplified by capping agents and blocking agents, which are known to stop and slow down the crystal growth and formation, respectively.
5. Similarly, the overall morphology of MOFs, including their size and structure, can also be tweaked and monitored by certain modulators. Altogether, certain conditions of processing play a categorical role in forming distinguished crystal structures of MOFs which are largely temperature, pressure, time, external additives, temperature source, etc. [13].

After understanding the nuances of synthesis of MOFs, it is important to relate it with their existing properties which give them a better vantage point in the application of biocatalysis as compared to other conventional biocatalysts like zeolites, carbonaceous materials, hydrogels, graphene oxides, mesoporous silica, carbon nanotubes, other polymers, etc. [14]. As discussed in the introduction, such conventional platforms are known to have disabling limitations in the form of low enzymatic loading, restricted mass transfer, enzyme denaturation, etc. To overcome this, MOFs have been introduced in the field of biocatalysis because of their multimodal practicalities and profitable properties.

The biggest advantage of MOFs stems from their inherent structural morphology where they have one organic branch consisting of ditopic or polytopic carboxylates (or some other negatively charged organic group) that acts as 'linkers' to metal-containing inorganic units, which result in the formation of architecturally sturdy crystalline MOFs. These have a prototypical porosity of almost 50% greater than their own actual crystal volume. This is achieved mainly because of the presence of tunable functional groups that are organic in nature, along with the use of transition metals with varying geometry like tetrahedral, linear, square planar, etc., which confer wide-ranging functionalities to the MOFs. The biocatalytic nature of MOFs aids in a multitude of reactions like oxidation, ring-opening reactions, aldol condensation, epoxidation reactions, etc. Another major benefit of MOFs is their surface area value which ranges typically from 1000 to 10,000 m^2/g, which is much more than the traditional porous frameworks used in biocatalysis like zeolites and carbonaceous materials. Presently, these MOFs are the only class of porous materials with permanent porosity having the highest degree of extensive variability and multiplicity, thus making them the most sought-after models for biocatalysis. Their multifaceted functionalities stem from the following uniquely advantageous attributes-

- Ordered and methodical structural integrity present in these biocatalytic platforms
- Tunable porosity due to option of adjusting pore size of crystals
- The diameter of frameworks can be fine-tuned according to the requirement of enzymes
- The high degree of stability as they show resistance towards external forces causing the change in their intrinsic morphology
- Also, portray resistance towards metal aggregation in crystalline structures
- The processing conditions of MOFs are economical as well as convenient and can be easily controlled in order to get the required results
- Their sample collection is more or less effortless and very simplistic
- MOFs also show opposition towards architectural disintegration caused by any external agents so as to keep the enzyme immobilized and viable for a long time
- Ability to accept attachment of other foreign substances on or around the pores and their surface
- MOFs also portray tremendous thermal and chemical stability making them susceptible and highly receptive towards post-synthetic functionalization
- These biocatalytic platforms also possess the capability to vary their properties like size and nature without compromising the principal topology that has in turn given rise to the isoreticular principle which has been applied to create, MOFs containing the features of largest aperture and lowest density of 98 Å and 0.13 g/cm^3, respectively [15].

The multiple advantages and properties of MOFs as better biocatalytic platforms over conventional biocatalysts are summarized in Fig. 1.

Apart from these wide-ranging advantages of MOFs, there are certainly distinctive and fundamental particularities that play an important role in exhibiting biocatalytic function of such MOFs, which are, hybrid nature, enantioselectivity and linker functionality which are discussed in detail in the following sections.

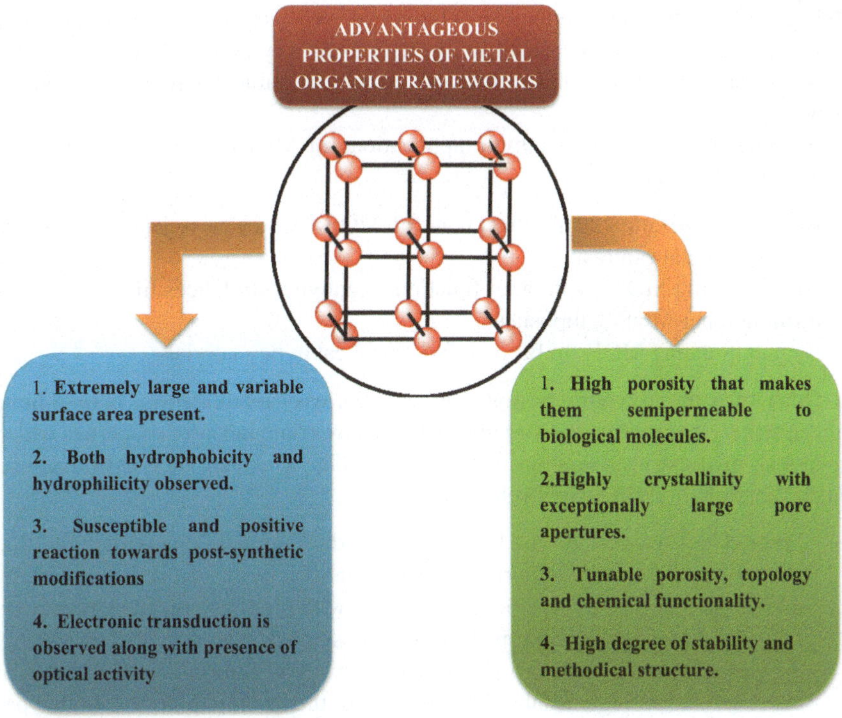

Fig. 1 Advantageous properties of MOFs that enhance their biocatalytic functionality

3 Fundamental Properties Contributing Towards Biocatalysis

To recapitulate, MOFs or porous coordination polymers (PLCs) are crystalline, highly porous, tunable platforms comprising of clusters of metal ions in a network joined by organic linkers, possessing a high degree of flexibility, making them much better and efficient than classic porous substances. Their unique and presently unparalleled characteristics increase their applicability in a diversified area of employment, fluctuating from gas storage, CO_2 capture, photoluminescence to drug delivery, electronic supercapacitors, biocatalysis, etc. In this chapter, so far, we have covered the elements of MOFs, their synthesis, properties and inherent structural and morphological advantages, which make them a superior choice as catalysts, especially in biological reactions. In biocatalytic reactions, the major disability which is faced during the industrial application of this process is the sensitivity and susceptibility of enzymes to become denatured and get destroyed by even slight ambient changes. Thus, scientists came up with the approach of enzyme immobilization, to maximize the stability as well as the functionality of such enzymes. This is where MOFs play an important role as their irreplaceable topology and properties, tunable metal nodes and ligands, ability to show interactions like van der Waals forces, hydrogen bonding, coordinate bonding and covalence make them extremely potential and compatible matrices for enzyme immobilization and creating a stable microenvironment for enzymes [16].

In order to achieve the catalytic applications of MOFs, three basic ways are employed, which are-

- By functionalization of the backbone of MOFs, which is organic linkers and inorganic metal centres,
- By integrating MOFs with other functional catalytic substances like enzymes or forming other MOF composites,
- By manipulating MOFs to become functional catalytic materials.

Other than the exceedingly high surface area, easy modification, pore-volume and extremely ordered crystal structure, MOFs have some inherent properties which govern all their extrinsic and intrinsic functionalities, which are their hybrid nature, linker functionality and enantioselectivity-

1. **Hybrid Nature**—MOFs are known to show dual nature in their structures, that is, they have the ability to portray both hydrophobicity as well as hydrophilicity due to the presence of organic linkers along with inorganic metal centres, with only a few minor tweaks during the process of synthesis. This ability is extremely useful as it opens a lot of doors with respect to interactions with various kinds of chemical and biological substances and makes them compatible with both types of affinities, water-loving as well as water-repelling. Apart from this, their affinities are not only changeable but also tunable in their respective wavelengths, allowing them to portray a variable range in hydrophobic and hydrophilic nature. Such a property allows MOFs to form bonds and interact with a myriad of substances which is especially useful during biocatalytic reactions as they involve wide-ranging nature of materials [17].
2. **Enantioselectivity**—This unique property allows MOFs to act as extremely potential catalysts in both heterogeneous and homogenous catalysis as well as matrices for other biocatalysts like enzymes. Enantioselectivity can be described as the propensity of catalysts to show preference in enantiomers by chiral selector carbon and when it is employed in catalytic reactions, it is called enantioselective catalysis. Because of the mild changes in synthetic conditions, even very delicate functional variations can be incorporated in MOFs and because of this ability, they can show the property of enantioselectivity. For instance, either metal complexes or enantiopure chiral ligands can be assimilated into MOF structures allowing them to become efficient and proper asymmetric catalysts. However, this cannot be achieved through traditional catalysts like microporous crystalline materials or even zeolites, thus making MOFs a better platform for catalysis [18].
3. **Linker Functionality**—The organic linkers present the structure of MOFs can undergo a multitude of variations with respect to their origin, type of molecule used, compatibility with others, etc. As a result, these linker groups become a

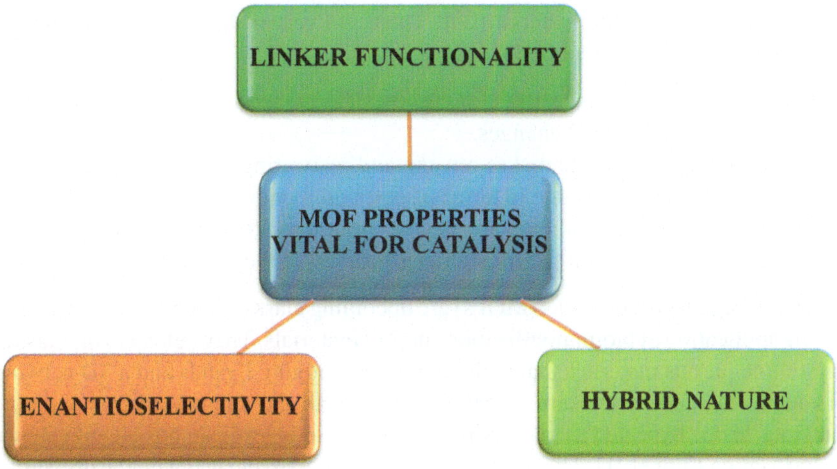

Fig. 2 A schematic representation of fundamental catalytic properties of MOFs

means to provide distinctive functionalities to the MOFs and allow the formation of a variety of structures by simple changes in these linkers. In the ideal conditions, judicious selection of organic linkers and metal ion aggregates along with precisely tailored conditions can help scientists achieve predetermined MOFs, fulfilling exact requirements. However, this is not possible as realizing such a well-educated design is extremely difficult and requires total control on all parameters. If the organic linkers are declared rigid, then assembly of various metal ions around them to form MOFs can be predicted to a sharper degree and this approach is referred to as reticular synthesis. Here, variable orientations, bonding to metal ions and numerous geometries can be achieved with the help of organic linker functionality, making them better potential candidates for catalytic activity [19].

The three fundamental properties of MOFs helping in catalytic activity are visualized in the following schematic (Fig. 2).

4 Strategies for Enzyme Immobilization

Enzymes are sophisticated biomacromolecules that are broadly used in plethora of domains, for example, biosensors, pharmaceuticals and biofuel cells. The high activity and specificity of enzymes make them the useful catalyst in basic fundamental reactions over conventional catalysts. For the purpose of commercialization and industrialization of enzymes, the different practices of enzyme immobilization have been thoroughly studied in recent years to enhance the practical performance under different reaction conditions. An immobilized enzyme is one that is attached to an inert and insoluble material [20]. There are certain benefits for the enzyme immobilization that are reported below-

- Immobilizing an enzyme allows for improved resistance against reaction variables like temperature and pH changes.
- Immobilizing an enzyme makes reusing and recovering easier that further saves money and leads to an edge in terms of cost and money.
- In certain cases, immobilizing an enzyme minimizes the reaction time.
- It leads to an increase in the functional efficiency of the enzyme.

Metal–organic frameworks (MOFs) are upcoming class of future novel candidates for the application of bioimmobilization support materials. They belong to the class of crystalline porous materials that include organic bridging ligands and a 3D network of metal ions. The disclosure of MOFs holds forth the strategy of immobilizing enzymes and is expected as an important platform to investigate hosted enzymes in the upcoming future [21]. The following key attributes associated with the utilization of MOFs make them a perfect choice over conventional host materials-

- Their exceptional strengths of high surface area
- Large yet tunable pore sizes
- Pore walls susceptible to functionalization
- Diversified architectural structure

Additionally, MOFs possess the ability to provide a stable and compatible surroundings for immobilized enzymes, which effectively lessen the loss of enzyme conformation under denaturation conditions. Enzyme-MOF composites are known for their reusability and recyclability, which makes them a greener alternative against traditional supporting materials like silica, collagen and cellulose [22].

To date, various strategies have been proposed by scientists for immobilizing enzymes of great interest using pristine or functionalized MOFs. The idea is typically derived from similar immobilization techniques reported in the literature for other available supporting materials like biopolymers, mesoporous oxides, nanotubes and layered doubled hydroxides. Figure 3 visualizes a schematic demonstrating the various immobilization techniques to encapsulate enzymes using MOFs.

4.1 Surface Immobilization

Surface immobilization stands for fitting of biocatalysts on the surface of material making use of electrostatic, hydrophobic or van der Waals interaction. MOFs are known to exhibit comparatively larger surface area that offers an increased loading capacity of enzyme molecules through surface immobilization. The functionalization of MOFs makes the stable linkages between MOF and the enzyme.

Recently, Zhong et al. an enzyme named glucose oxidase (GOx) was combined with a metal–organic framework MOF-545(Fe), which owned peroxidase-like activity. The as-constructed multi-enzyme system showed higher efficiency of 92%

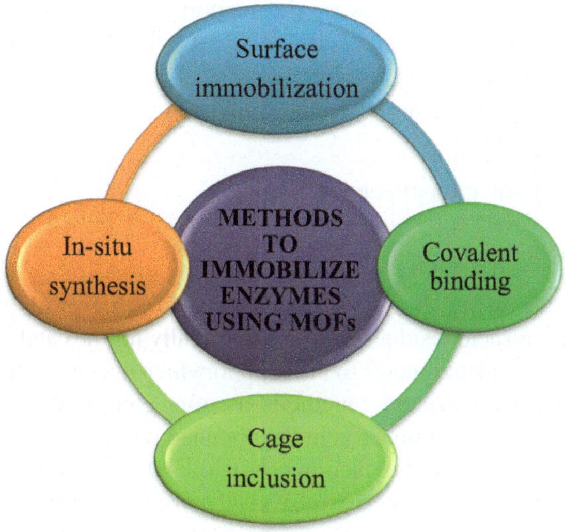

Fig. 3 A visual representation of different immobilization strategies to encapsulate enzyme using MOFs

in contrast to only 40% activity of the free enzyme for tandem catalysis. Besides, MOF-545(Fe) not only served as a support material for the enzyme immobilization but also supported in the task of cooperating with natural enzymes for the cascade reactions [23].

In another such study conducted by scientists, Cu-BTC (BTC: 1,3,5-benzene tricarboxylate)-based hierarchically porous MOF matrix was administered successfully to immobilize the *Bacillus subtilis* lipase (BSL2). As result, immobilized BSL2 demonstrated improved enzymatic activity and ideal reusability during the esterification reaction. Even after the passage of 10 cycles, the immobilized BSL2 appeared to exhibit 90.7% of its earlier enzymatic activity and 99.6% of its earlier conversion [24].

Besides, other matrixes such as Zr/Al/Fe/Cr-based MOFs are also widely utilized to immobilize various enzymes through the surface immobilization technique, which is perfectly documented by Liang and coworkers in their recent study [25]. These results gave the assurance that the porous MOF materials have enormous potential for utilization as effective biological hybrid materials for catalysis.

4.2 Covalent Binding

Although various enzymes have effectually connect to the surface of MOFs by surface immobilization, still, adsorbed enzymes held by weak van der Waal forces that revealed poor stability under certain conditions. In recent studies, the multipoint covalent coupling has been postulated as one of the robust chemical bonds utilized for the conjugation of biomolecules or enzymes. As a result of multiple covalent bond formation between the support carrier and enzyme conformational flexibility was diminished and additionally, it fixes the enzyme such that it prevents protein denaturation or unfolding [26, 27].

There are various methods under covalent binding-

- Diazoation
- Formation of peptide bond
- Group activation
- Polyfunctional reagents

However, it is worth noting that to achieve a high level of bound activity, the amino acid residues required essentially for the catalytic activity must not indulge in the covalent linkage to the support which may pose difficulties under certain circumstances. Thus, these methods are often employed when there is a strict requirement of enzyme elimination in the obtained products [28].

In a study, the nano/microscale UiO-66-NH_2 metal–organic framework (MOF) materials were successfully synthesized and soybean epoxide hydrolase (SEH), an important hydrolase for the preparation of vicinal diols, was systematically immobilized for the first time onto the prepared UiO-66-NH_2 MOF. Results predicted that SEH@UiO-66-NH_2 possessed more than 17.6 U activity after two hours of incubation at 45 °C, whereas for free, SEH activity of 10.1 U was obtained under the same condition [29].

4.3 *Cage Inclusion*

Cage inclusion often referred as encapsulation is the method that offers the inclusion of enzyme molecules within cages of MOFs through the process of diffusion. The encapsulated enzymes are known to exhibit enhanced stability even under typical environments [30]. The caging of enzyme can be achieved by any of the following methods:

- By inclusion of enzyme within a highly cross-linked polymer matrix.
- By dissolution of enzyme in a non-aqueous phase.
- By separating the enzyme from a bulk solution utilizing a semipermeable microcapsule.

In a study, it was reported that encapsulation of enzymes significantly decreased the leaching of enzymes from carrier support. Due to a diverse array of properties, MOFs can be employed as potential host matrices. Feng et al. reported the development of ultra-large mesoporous stable MOFs and used it for encapsulation of three enzymes, namely horseradish peroxidase (HRP), cytochrome complex (Cyt c) and microperoxidase-11 (MP-11). The carrier-encapsulated HRP was portrayed to exhibit higher catalytic potential in volatile organic solvents as compared to the free derivative [31].

Microperoxidase-11 was immobilized into a mesoporous MOF consisting of nanoscopic cages that showed superior catalytic performance for the oxidation of small substrates combined with considerable reusability and stability. Besides, MOF-encapsulated immobilized enzymes were reusable for multiple rounds with no significant leaching from the carrier [32].

4.4 In Situ MOF Formation and Enzyme Immobilization

The above-discussed methods make the potential possibility of MOFs with large pores to host enzymes. However, this entails two major problems-

- Only a few MOFs possess sufficient large mesopores.
- Secondly, these are restricted for smaller dimension enzymes.

However, these shortcomings are eliminated with the in situ synthesis method. In this method, MOF formation and enzyme immobilization are carried out simultaneously. The MOF building blocks and enzymes are mixed in the solution that results in the crystallization of MOF particles with embedded enzymes at the particle surface site. Like others, this technique also offers a 3D microenvironment for the biomolecules without causing any leaching. This method is generally carried out at lower temperatures to prevent the enzyme from denaturation [33, 34].

This approach is said to bring in a set of noteworthy advantages of encapsulation of a larger number of enzymes owing to the highly uniform and methodical nanosizes, greater specific areas and tunable pore sizes of MOFs [15].

In a recent study by a group of researchers, an 'indirect' in situ MOF fabrication technique was reported for the immobilization of functional biomolecules. The received enzyme–MOF composites were found to be extremely microporous, strongly-built and provided a straightforward way to encapsulate enzymes for size-selective and recyclable biocatalytic systems [35]. Table 2 lists down some of the recently reported MOF–enzyme biocatalytic platforms.

Table 2 List of some MOF-enzyme composite and corresponding strategies used for immobilization

S.no	MOF	Enzyme	Type of immobilization	Reference
1	Fe-BTC MOF	Laccase and lipase	In situ or post-synthesis methodology	[36]
2	La-MOF-NH_2	Acetylcholinesterase	Encapsulation	[37]
3	Fe-MOF-NH_2,	Acetylcholinesterase	Encapsulation	[37]
4	Fe-MOF	Alcohol dehydrogenase, lipase and glucose oxidase	Co-precipitation	[38]
5	NH_2 MOF	Glucose oxidase	Covalent binding	[39]

5 Outlook to Applications of MOFs as Immobilized Biocatalysts

Immobilized enzymes are actively utilized in economic reuse and for the processing of different products. Immobilized enzymes have potential applications in biosensing, catalysis and other industrial applications. The recent progress in this domain lies in the fact that they are economical, environmentally friendly and easy to use in comparison with other parallel technologies. Figure 4 illustrates the useful enzyme properties that were obtained after immobilization in MOFs. In the further section, the recent growth and applications of enzyme immobilized biocatalysts are discussed.

5.1 Biosensing and Detection

Biosensors have gained higher interest in the past decade in the arena of quality control, food safety and the diagnosis of cancer. These are basically enzyme-immobilized sensing and detection devices where MOFs are known to create a crucial impact by hosting enzymes and protecting them against perturbations. This activity is owing to the possibility of tuning the pore size and easy functionalization of pore walls, which as a result favours the specific interactions between enzymes and porous walls of MOFs.

Recently, smart porous Cu-hemin MOFs were prepared by combing Cu^{2+} with hemin for electrochemical detection of glucose. The resulting nanocomposites with

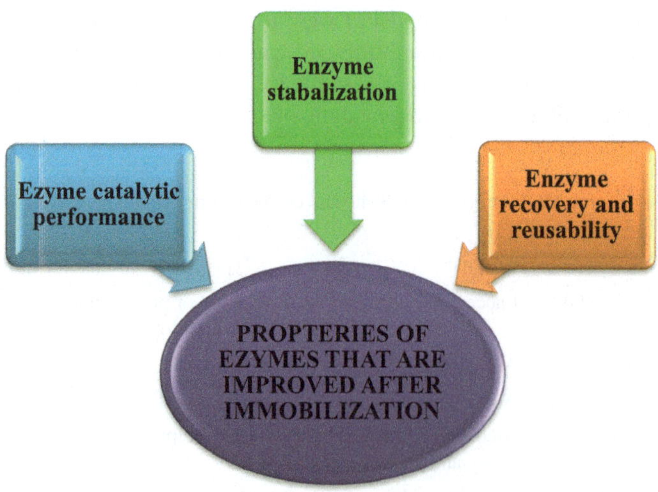

Fig. 4 Improved properties obtained after immobilization of enzymes

a larger surface area facilitated the easy loading of GOD/glucose oxidase molecules in the pores to retain their bioactivity [21].

In another study, two lanthanide zeolite-like metal–organic frameworks (Ln-ZMOFs) were synthesized. Further with varying the stoichiometric ratio of Tb^{3+} and Eu^{3+} during synthesis, three mixed crystal MZMOFs with varying Eu: Tb stoichiometry was obtained. Fluorescent studies showed that a methanol suspension of $Eu_{0.6059}Tb_{0.3941}$-ZMOF-3 (MZMOF-3) exhibited selective detection of lysophosphatidic acid (LPA), a biomarker for ovarian cancer and other gynaecologic cancers [40].

Such proposed approaches were proven to be very effective after the inclusion of MOFs which portrays that these porous structures can be the future promising candidates for application in fields like biosensing and detection.

5.2 MOFs as Host for Biomimetic Catalysis

Enzymes are potential biocatalysts owing to their large availability, non-toxicity, high catalytic efficiency and environmental friendliness. Nevertheless, these advantages are enough to convince their practical implications yet their high cost of formation and complex purification of heterogeneous catalysts from these enzymes is a main limitation for their practical use. To look after these shortcomings, a significant attention has been given to exploration of biomimetic catalysts that possess the potential to sustain under typical reaction conditions like high temperatures, strong acid or base and in different organic solvents including methanol, ethanol, DMF or dimethylformamide and DCM or dimethyl carbonate with improved stability than that of their enzyme equivalents. Biomimetic catalysis is a domain of biomimetic chemistry that includes chemical catalysis mimicking the important characteristics of enzymes.

Porous materials that comprise either organic or inorganic compounds are widely attempted for biomimetic catalysis, but they exhibit inherent limitations. In addition, organic compounds are usually amorphous wherein they lack crystalline structures, on the contrary, inorganic compounds are lacking structural flexibility. Explicitly, well-designed MOFs can largely expand the repertoire of porous materials by combining the useful properties of both organic and inorganic materials into one system. Moreover, with numerous extraordinary properties like large specific surface area, exceptional porosity and structural versatility, MOFs hold a distinctive position for the development of biomimetic catalysts in comparison with other porous materials [41].

Qin et al. put on a hemin molecule into a newer type of host material, i.e. amino-containing MOF (MIL-101(Al)-NH_2) for the very first time. Using H_2O_2, the as-prepared Hemin@MIL-101 displayed peroxidase-like activity via catalytic oxidation of the substrate 3,3,5,5-tetramethylbenzidine (TMB) [42].

5.3 Enzyme-MOF for Digestion of Proteins

Trypsin is an often-used proteolytic enzyme that catalyzes protein digestion and transforms into peptides for proteomics research and industrial production. However, its practical implications are often hampered due to the long-running time (18–24 h) and self-digestion in the reaction mixture. To this context, surface immobilization of trypsin on MOFs prevails over the self-digestion and improves their recycling capacity [43].

The efficiency of digestion of the trypsin-FITC@MOF bioreactor was discovered to enhance to a higher extent even in repeated utilization, over the use of free enzyme-assisted digestion or nanoparticle immobilized enzyme reactors reported so far.

A group of researchers has been fortunate in immobilizing the proteolytic enzyme upon MOFs [MIL-101(Cr), MIL-88B (Cr) and MIL-88B-NH_2 (Cr)] activated by the presence of dicyclohexylcarbodiimide (DCC) via the formation of peptide linkage. The resultant trypsin–MOFs composite was utilized for the digestion of BSA with the help of ultra-sonication that was carried out for two minutes. The developed trypsin–MOF conjugate was reusable and recoverable [44].

6 Conclusion and Future Outlook

Metal–organic frameworks or MOFs have prudentially been declared as the emerging and highly applicable branch of material science which has wide-ranging significance in numerous fields, be it their relevance in supercapacitors, gas storage and separation, CO_2 capture or upcoming use in drug delivery, biosensing, bioimaging and biocatalysis. Their exceptional porosity and unprecedented surface area combined with other significant properties have provided them with well-deserved attention in many scientific areas, especially in the domain of biocatalysis. The emergence of MOFs as viable host matrices for enzyme immobilization and their individual catalytic potential has opened a variety of opportunities in the industrial development of this aspect. Currently, this territory of MOFs application is in its preliminary phase, where many functionalization strategies have been designed and are being tested to become novel support materials in biocatalysis which are stable, tough, recoverable and cost-effective biocatalytic platforms. There is a fairly prominent way to go so as to achieve the full potential of MOFs, for instance, precision in the control of overall morphology and assembly of these frameworks is yet to be achieved which can open newer areas in this field. Along with this, the formation of yields like multivariate frameworks can allow synergistic and other types of functional results with enzymes and similar materials. There is still plenty of research and innovation required in this field to overcome the shortcomings of MOFs which disable them to achieve their maximum potential, and we hope that active scientific research and aptitude be applied to realize the prospective applications of MOFs in biocatalysis.

Abbreviations

MOFs	Metal–organic frameworks
HKUST-1	Hong Kong University of Science and Technology-1
PLCs	Porous coordination polymers
GOx	Glucose oxidase
BTC	1,3,5-benzene tricarboxylate
BSL2	*Bacillus subtilis* lipase
SHE	Soybean epoxide hydrolase
HRP	Horseradish peroxidase
Cyt c	Cytochrome complex
MP-11	Microperoxidase-11
Ln-ZMOFs	Lanthanide zeolite-like metal–organic frameworks
GOD	Glucose oxidase
LPA	Lysophosphatidic acid
DMF	Dimethylformamide
DCM	Dichloromethane
TMB	Tetramethylbenzidine

Important Websites

- https://www.nanowerk.com/mof-metal-organic-framework.php
- https://agscientific.com/blog/2021/01/importance-of-enzyme-immobilization/
- https://biocyclopedia.com/index/biotechnology/microbial_biotechnology/enzyme_technology/biotech_enzyme_immobilization.php
- https://www.eurekalert.org/news-releases/740946
- https://www.bnl.gov/nsls2/news/highlights/news.php?a=217175
- https://blog.novomof.com/metal-organic-frameworks-as-nanocarriers-for-therapeutic-agents-delivery
- https://www.chemistryworld.com/news/mof-squeezes-catalyst-into-shape-for-switching-reactivity/3008458.article
- https://www.materialstoday.com/biomaterials/news/novel-biological-mof-can-lock-dna/
- https://cordis.europa.eu/project/id/209241
- https://www.sciencedirect.com/topics/chemistry/metal-organic-framework

References

1. Gaigneaux E (2002) Scientific bases for the preparation of heterogeneous catalysts. In: Proceedings of the 8th international symposium, Louvain-la-Neuve, Belgium, 9–12 Sept 2002, p 1129
2. Furukawa H, Cordova KE, O'Keeffe M, Yaghi M (2013) The chemistry and applications of metal-organic frameworks. Science 341(6149). https://doi.org/10.1126/science.1230444
3. Mueller U, Schubert M, Teich F, Puetter H, Schierle-Arndt K, Pastré J (2006) Metal–organic frameworks—prospective industrial applications. J Mater Chem 16(7):626–636. https://doi.org/10.1039/B511962F
4. Sun H, Zhang H, Ang EL, Zhao H (2018) Biocatalysis for the synthesis of pharmaceuticals and pharmaceutical intermediates. Bioorg Med Chem 26(7):1275–1284. https://doi.org/10.1016/J.BMC.2017.06.043
5. Zhou Z, Hartmann M (2013) Progress in enzyme immobilization in ordered mesoporous materials and related applications. Chem Soc Rev 42(9):3894–3912. https://doi.org/10.1039/C3CS60059A
6. Mohamad NR, Marzuki NH, Buang NA, Huyop F, Wahab RA (2015) An overview of technologies for immobilization of enzymes and surface analysis techniques for immobilized enzymes. Biotechnol Biotechnol Equip 29(2):205–220. https://doi.org/10.1080/13102818.2015.1008192
7. Meek ST, Greathouse JA, Allendorf MD (2011) Metal-organic frameworks: a rapidly growing class of versatile nanoporous materials. Adv Mater 23(2):249–267. https://doi.org/10.1002/ADMA.201002854
8. Springuel G, Robeyns K, Norberg B, Wouters J, Leyssens T (2014) Cocrystal formation between chiral compounds: how cocrystals differ from salts. Cryst Growth Des 14(8):3996–4004. https://doi.org/10.1021/CG500588T
9. Campagnol N, Souza ER, De Vos DE, Binnemans K, Fransaer J (2014) Luminescent terbium-containing metal–organic framework films: new approaches for the electrochemical synthesis and application as detectors for explosives. Chem Commun 50(83):12545–12547. https://doi.org/10.1039/C4CC05742B
10. Zhang Y, Bo X, Nsabimana A, Han C, Li M, Guo L (2014) Electrocatalytically active cobalt-based metal–organic framework with incorporated macroporous carbon composite for electrochemical applications. J. Mater. Chem. A 3(2):732–738. https://doi.org/10.1039/C4TA04411H
11. Phang WJ, Lee WR, Yoo K, Ryu DW, Kim B, Hong CS (2014) pH-Dependent proton conducting behavior in a metal-organic framework material. Angew Chemie Int Ed 53(32):8383–8387. https://doi.org/10.1002/ANIE.201404164
12. Jin L-N, Liu Q, Sun W-Y (2014) An introduction to synthesis and application of nanoscale metal–carboxylate coordination polymers. CrystEngComm 16(19):3816–3828. https://doi.org/10.1039/C3CE41962B
13. Zou K-Y, Li Z-X (2018) Controllable syntheses of MOF-derived materials. Chem A Eur J 24(25):6506–6518. https://doi.org/10.1002/CHEM.201705415
14. Li M et al (2020) Fabricating covalent organic framework capsules with commodious microenvironment for enzymes. J Am Chem Soc 142(14):6675–6681. https://doi.org/10.1021/JACS.0C00285
15. Salunkhe RR, Kaneti YV, Yamauchi Y (2017) Metal-organic framework-derived nanoporous metal oxides toward supercapacitor applications: progress and prospects. ACS Nano 11(6):5293–5308. https://doi.org/10.1021/acsnano.7b02796
16. Majewski MB, Howarth AJ, Li P, Wasielewski MR, Hupp JT, Farha OK (2017) Enzyme encapsulation in metal–organic frameworks for applications in catalysis. CrystEngComm 19(29):4082–4091. https://doi.org/10.1039/C7CE00022G
17. Silva P, Vilela SMF, Tomé JPC, Paz FAA (2015) Multifunctional metal–organic frameworks: from academia to industrial applications. Chem Soc Rev 44(19):6774–6803. https://doi.org/10.1039/C5CS00307E

18. Long J, Yaghi O, Ma L, Abney C, Lin W (2009) Metal-organic frameworks issue Reviewing the latest developments across the interdisciplinary area of metal-organic frameworks from an academic and industrial perspective Enantioselective catalysis with homochiral metal-organic frameworks. https://doi.org/10.1039/b807083k
19. Lu W et al (2014) Tuning the structure and function of metal-organic frameworks via linker design. https://doi.org/10.1039/c4cs00003j
20. Wang X, Lan PC, Ma S (2020) Metal-organic frameworks for enzyme immobilization: beyond host matrix materials. ACS Cent Sci 6(9):1497–1506. https://doi.org/10.1021/acscentsci.0c00687
21. Bilal M, Adeel M, Rasheed T, Iqbal HMN (2019) Multifunctional metal–organic frameworks-based biocatalytic platforms: recent developments and future prospects. J Mater Res Technol 8(2):2359–2371. https://doi.org/10.1016/J.JMRT.2018.12.001
22. Sirisha VL, Jain A, Jain A (2016) Enzyme immobilization: an overview on methods, support material, and applications of immobilized enzymes, 1st edn, vol 79. Elsevier Inc.
23. Zhong X, Xia H, Huang W, Li Z, Jiang Y (2020) Biomimetic metal-organic frameworks mediated hybrid multi-enzyme mimic for tandem catalysis. Chem Eng J 381(June 2019):122758. https://doi.org/10.1016/j.cej.2019.122758
24. Cao Y, Wu Z, Wang T, Xiao Y, Huo Q, Liu Y (2016) Immobilization of: Bacillus subtilis lipase on a Cu-BTC based hierarchically porous metal-organic framework material: a biocatalyst for esterification. Dalt. Trans. 45(16):6998–7003. https://doi.org/10.1039/c6dt00677a
25. Liang S, Wu XL, Xiong J, Zong MH, Lou WY (2020) Metal-organic frameworks as novel matrices for efficient enzyme immobilization: an update review. Coord Chem Rev 406:213149. https://doi.org/10.1016/j.ccr.2019.213149
26. Cui J, Ren S, Sun B, Jia S (2018) Optimization protocols and improved strategies for metal-organic frameworks for immobilizing enzymes: current development and future challenges. Coord Chem Rev 370:22–41. https://doi.org/10.1016/j.ccr.2018.05.004
27. Mateo C, Palomo JM, Fernandez-Lorente G, Guisan JM, Fernandez-Lafuente R (2007) Improvement of enzyme activity, stability and selectivity via immobilization techniques. Enzyme Microb Technol 40(6):1451–1463. https://doi.org/10.1016/j.enzmictec.2007.01.018
28. Guisan JM (2013) Immobilization of enzymes and cells in series editor. Immobil Enzyme Cells 1051:1–375. https://doi.org/10.1007/978-1-62703-550-7
29. Cao SL et al (2016) Novel nano-/micro-biocatalyst: soybean epoxide hydrolase immobilized on UiO-66-NH2 MOF for efficient biosynthesis of enantiopure (R)-1, 2-octanediol in deep eutectic solvents. ACS Sustain Chem Eng 4(6):3586–3595. https://doi.org/10.1021/acssuschemeng.6b00777
30. Mehta J, Bhardwaj N, Bhardwaj SK, Kim KH, Deep A (2016) Recent advances in enzyme immobilization techniques: metal-organic frameworks as novel substrates. Coord Chem Rev 322:30–40. https://doi.org/10.1016/j.ccr.2016.05.007
31. Feng D et al (2015) Stable metal-organic frameworks containing single-molecule traps for enzyme encapsulation. Nat Commun 6:1–8. https://doi.org/10.1038/ncomms6979
32. Lykourinou V et al (2011) Immobilization of MP-11 into a mesoporous metal-organic framework, MP-11@mesoMOF: a new platform for enzymatic catalysis. J Am Chem Soc 133(27):10382–10385. https://doi.org/10.1021/ja2038003
33. Gkaniatsou E, Sicard C, Ricoux R, Mahy JP, Steunou N, Serre C (2017) Metal-organic frameworks: a novel host platform for enzymatic catalysis and detection. Mater Horizons 4(1):55–63. https://doi.org/10.1039/c6mh00312e
34. Gascón V, Castro-Miguel E, Díaz-García M, Blanco RM, Sanchez-Sanchez M (2017) In situ and post-synthesis immobilization of enzymes on nanocrystalline MOF platforms to yield active biocatalysts. J Chem Technol Biotechnol 92(10):2583–2593. https://doi.org/10.1002/jctb.5274
35. Hou C et al (2015) Facile synthesis of enzyme-embedded magnetic metal-organic frameworks as a reusable mimic multi-enzyme system: Mimetic peroxidase properties and colorimetric sensor. Nanoscale 7(44):18770–18779. https://doi.org/10.1039/c5nr04994f

36. Nobakht N, Faramarzi MA, Shafiee A, Khoobi M, Rafiee E (2018) Polyoxometalate-metal organic framework-lipase: an efficient green catalyst for synthesis of benzyl cinnamate by enzymatic esterification of cinnamic acid. Int J Biol Macromol 113(2017):8–19. https://doi.org/10.1016/j.ijbiomac.2018.02.023
37. Dong S, Peng L, Wei W, Huang T (2018) Three MOF-templated carbon nanocomposites for potential platforms of enzyme immobilization with improved electrochemical performance. ACS Appl Mater Interfaces 10(17):14665–14672. https://doi.org/10.1021/acsami.8b00702
38. Gascón V, Carucci C, Jiménez MB, Blanco RM, Sánchez-Sánchez M, Magner E (2017) Rapid in situ immobilization of enzymes in metal-organic framework supports under mild conditions. ChemCatChem 9(7):1182–1186. https://doi.org/10.1002/cctc.201601342
39. Tudisco C et al (2016) Covalent immobilization of glucose oxidase on amino MOFs via post-synthetic modification. RSC Adv 6(109):108051–108055. https://doi.org/10.1039/c6ra19976c
40. Zhang X et al (2015) Preparation of M@$BiFeO_3$ nanocomposites (M = Ag, Au) bowl arrays with enhanced visible light photocatalytic activity. J Am Ceram Soc 98(7):2255–2263. https://doi.org/10.1111/jace.13543
41. Serhan M et al (2019) Total iron measurement in human serum with a smartphone. In: Proceedings of annual meeting conference on AIChE, vol 2019, Nov 2019. https://doi.org/10.1039/x0xx00000x
42. Qin FX, Jia SY, Wang FF, Wu SH, Song J, Liu Y (2013) Hemin@metal-organic framework with peroxidase-like activity and its application to glucose detection. Catal Sci Technol 3(10):2761–2768. https://doi.org/10.1039/c3cy00268c
43. Wang Q, Liu S, Tai W, Li J, Ye M, Yang L (2018) Purification, chemical structure and antioxidant activity of active ingredient (LPT-3d) separated from Lachnum sp. Process Biochem 71:166–174. https://doi.org/10.1016/j.procbio.2018.04.012
44. Shih YH et al (2012) Trypsin-immobilized metal-organic framework as a biocatalyst in proteomics analysis. ChemPlusChem 77(11):982–986. https://doi.org/10.1002/cplu.201200186

MOFs Encapsulated Metal Nanoparticles as Catalysts

Metal–Organic Frameworks (MOFs) Encapsulated Nanoparticles: Potential Catalysts for Diverse Organic Reactions

Gunjan Purohit and Diwan S. Rawat

Contents

1 Introduction 706
2 Encapsulation and Its Importance 706
3 Metal NPs Encapsulated in MOFs for Hydrogenation Reactions 708
4 MNPs@MOFs in C–C Cross-Coupling Reactions 718
Abbreviations 721
References 722

Abstract This chapter summarizes the developing interdisciplinary area of metal–organic frameworks (MOFs) encapsulated metal nanoparticles and their application in various organic transformations viz., hydrogenation-dehydrogenations, CO oxidation (liquid/gas phased), multicomponent tandem, and C–C cross-coupling reactions, etc. The encapsulated/anchored metal nanoparticles to MOFs i.e., MNPs@MOFs are one of the promising preeminent catalytic systems that have shown selectivity, reactivity, and recyclability that may open up a new dimension in industrial processes. The enhanced reactivity of MNPs@MOFs in heterogeneous catalysis is attributed to the synergistic effect from both components i.e., metal nanoparticles and MOFs. A variety of examples shown in the present chapter concludes the superiority of MOFs as supporting material to host metal NPs in their confined space leading to a tailorable and attractive system for catalysis. Efforts will be made to critically summarize the literature in the area of MOFs encapsulated metal nanoparticles as catalysts during the last 10 years.

Keywords Encapsulated/anchored metal nanoparticles to MOFs · Hydrogenation-dehydrogenations · Multicomponent/tandem reactions · C–C cross-coupling reactions

G. Purohit · D. S. Rawat (✉)
Department of Chemistry, University of Delhi, Delhi 110007, India

S. Gulati (ed.), *Metal-Organic Frameworks (MOFs) as Catalysts*,
https://doi.org/10.1007/978-981-16-7959-9_26

1 Introduction

The unique and astonishing properties of metal–organic frameworks (MOFs) are well known in the literature which in turn makes it an excellent candidate for metal nanoparticles encapsulation [1–6]. The nanoporosity of MOFs which is mainly due to the structured network of multi-functional organic moieties linked together with inorganic units possessing a predictable/regular geometry makes it an attractive contender for catalysis [7, 8]. The MOFs are also known as porous coordination polymers (PCPs) [6, 9]. In this chapter we will be comprehensively discussing the metal NPs encapsulated on MOFs with a special focus placed on designing, developing a nanomaterial for its potential application in the area of catalysis. Metals, in general, are chemically inert whereas, the nano-dimensioned metal nanoparticles become highly active including an array of noble metal NPs viz. palladium (Pd), platinum (Pt), gold (Au), silver (Ag), etc. Generally, the naked metal nanoparticles show a peculiar tendency to agglomerate or aggregation leading to decreased reactivity. Such nanoparticles when anchored on supporting systems namely zeolites, silicates, carbons, alumina, etc., or using surfactants can be handled easily followed by retaining their chemical reactivity and activity [10]. The highly exposed volume to surface area ratio as well as its inherent surface energies of MOFs makes it an excellent as well as exciting supporting material for metal nanoparticles encapsulation [11]. The porous nature of MOFs having higher pore volumes provides greater accessibility for metal nanoparticles to be encapsulated within during catalysis. Moreover, the fine-tuning of shapes, sizes/size-distribution of MOFs by adding chelating groups in the process of functionalization makes it easier to bound metal nanoparticles to it thereby making the system more stabilized. So, in either way by selecting the suitable MOF substrate, numerous metal/metal oxide NPs viz., Cu, Pt, Pd, Au, Ag, Ni, Ru and titania-oxide can be embedded and immobilized successfully into MOFs by following suitable chemical/synthetic techniques. [12, 13].

In heterogeneous catalysis, the cage-like structure of MOFs enables the incorporation of metal nanoparticles in a confined space thereby maintaining the dimensions of few nanometers to be best suited for the organic transformations viz., hydrogenation-dehydrogenations, CO oxidation (liquid/gas phased), multicomponent tandem, and C–C cross-coupling reactions, etc.

2 Encapsulation and Its Importance

The intrinsic crystallinity of MOFs allows one to design the uniform and single-site active catalytic species. The tailormade MOFs make an attractive catalytic system possessing a higher surface area in comparison to counterparts such as metal oxides, zeolites, and other porous materials. The porosity in MOFs can be a source of a special cavity to provide an environment wherein the reactants or substrates can interact efficiently to enhance the chemical activity and selectivity. Such sites are the

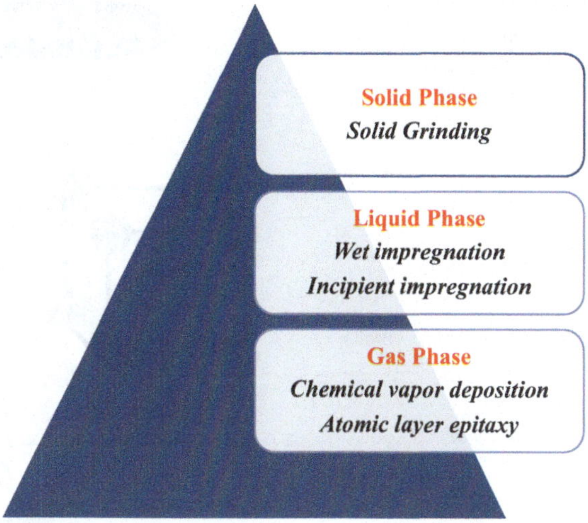

Fig. 1 Graphical representation of synthetic routes to prepare metal-encapsulated MOFs

perfect destination for metal nanoparticles to encapsulate within the system of MOFs. There are numerous reports available that depict the chemical synthetic routes namely liquid, gas, and solid impregnation methods for the preparation of metal-encapsulated MOFs as can be seen from Fig. 1.

The most widely accepted liquid impregnation method is quite simple as it involves the immersion of porous support in some solvent-containing metal precursor, then removal of solvent molecules followed by reacting it with reducing agents in the presence of molecular hydrogen or hydrogen source such as sodium borohydride, hydrazine hydrate, etc. This liquid impregnation method is further classified into wet impregnation and dry impregnation method (as discussed thoroughly in earlier chapters). The simple grinding of metal salts to MOFs of interest with any solvent can result in metal-encapsulated MOFs which is commonly known as the solid impregnation method. In 2008, Haruta and coworkers employed this simple strategy of solid impregnation method to host gold nanoparticles to MOFs precursors such as MIL-53, MOF-5, ZIF-8, HKUST-1, etc. [14].

In the gas impregnation method, the volatile organometallic precursors which are in the gaseous phase can be encapsulated in MOFs by following the photochemical/thermal-decomposition that resulted in metal-encapsulated MOFs NPs. This method was again devoid using a solvent system and high loading of metal NPs can be entrapped within the MOFs cavity. In the year 2005, Fischer et al., encapsulated various iron, nickel-based organometallic complexes within the cavities of MOF-5 by making use of the gas impregnation method [15].

Conclusively, by designing and making a fusion of the aforementioned methods one can derive a synthetic methodology for making metal-encapsulated MOFs which is consecutively known as the "Assembly method". In this type of strategy, one can either entrap the metal NPs into the pre-designed MOFs or can make use of the

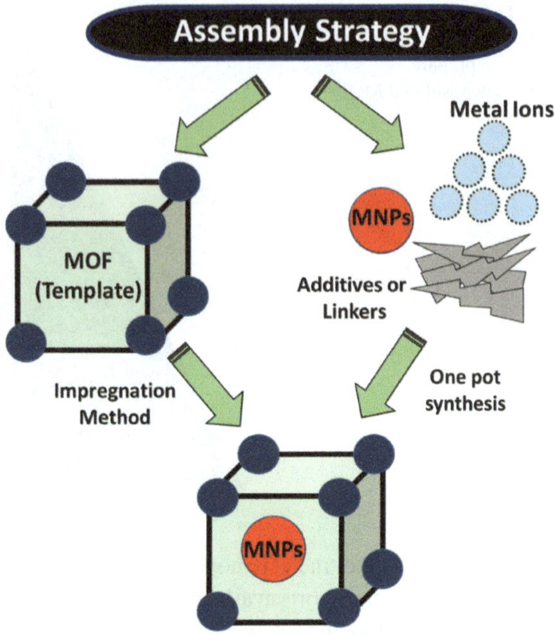

Fig. 2 Pictorial representation of assembly method to synthesize MNPs@MOFs

one-pot synthetic methodology. Figure 2 illustrates the pictorial representation of the assembly method for making MNPs@MOFs.

3 Metal NPs Encapsulated in MOFs for Hydrogenation Reactions

The infinite network structures in metal–organic frameworks (MOFs) make it a new generation of hybrid materials wherein the bridging organic ligands provide an excellent structural site to imbibe metal nanoparticles to it. Such characteristics of MOFs have allowed their usage in an efficient heterogeneous catalysis application [16–21]. Catalytic hydrogenations employing heterogeneous catalysis comprising of reduction of nitro compounds, carbonyls, unsaturated hydrocarbons, CO_2, etc., is one of the prominent research areas which finds its wide pertinency in the arena of industrial applications namely fine chemicals, agrochemicals, pharmaceuticals, food products, coal processing, fragrances/perfumes, etc. [22–26]. Generally, the typical hydrogenation reaction process involves the presence of molecular hydrogen which is accompanied by a catalyst that may be or may not be recovered from the reaction mixture as shown in Fig. 3.

The reaction process depending on the optimized reaction conditions may possess high TON or TOF values which in turn justifies the efficiency of the catalytic system. The noble, as well as non-noble metal nanoparticles such as palladium (Pd), platinum

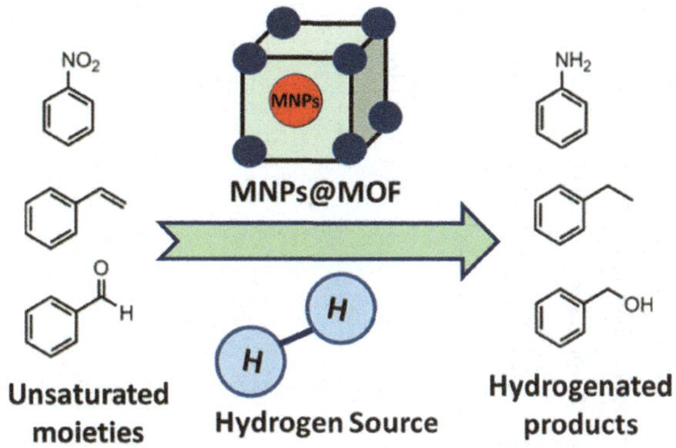

Fig. 3 Schematic representation of hydrogenation of unsaturated carbonyl, arenes, olefins

(Pt), rhodium (Rh), ruthenium (Ru), gold (Au), iron (Fe), cobalt (Co), nickel (Ni) etc., based catalytic systems which can be either supported on solid supports or unsupported ones, are well known in literature which shows typical selective hydrogenation reactivity [27–30]. It is interesting to note that depending on the supporting material, the chemical reactivity of embedded metal nanoparticles can be fine-tuned leading to enhanced reactivity and stability. MOF, being an interesting material possessing high porosity and tuneable structural networks finds its way to be an excellent host to imbibe noble metal nanoparticles [31]. In recent times, the metal nanoparticles encapsulated on MOFs show interesting and excellent applicability for the reduction/hydrogenations of nitro compounds, CO_2, carbonyl, unsaturated hydrocarbons, chemo-selective hydrogenations, etc., which we will be discussing elaborately by highlighting such examples.

Back in 2005, Li et al. demonstrated catalytic direct hydrogenation of nitrobenzene in ethanol using Pt encapsulated on MOFs, i.e., Pt@MIL-101 NPs in water at room temperature using 4.0 MPa H_2 (entry 2, Table 1) [32]. Within a time frame of 30 min, a complete hydrogenated product could be obtained showing a higher TOF value of 25,438 h^{-1}. All the substituted nitroarenes, be it *-ortho, -meta,* or *-para* could afford good selectivity as well as reactivity. The higher values of TOFs may be accounted for the hydrophobic nature of the support MIL-101. It should be noteworthy to mention that Pt@MIL-101 could also be hydrogenated the substituted benzaldehydes to their respective alcohols under mild conditions (RT, 40 atm H_2, 1.0 h) resulting in values of TOF around 2500–5147 h^{-1}. When the substituted benzaldehydes with strong electronegative groups such as –F were used, 100% selectivity was observed. Although, the lower reactivity of Pt@MIL-101 catalyst can be seen if we compare it with Pt supported on carbon nanotubes [33]. In comparison with the industrial catalyst such as Raney nickel (Ni) or Pt/Pd@Al_2O_3 which are widely used for the reduction of substituted nitroarenes requires higher temperature reaction conditions, the Pt@MIL-101 catalyst shows higher reactivity as well as selectivity. Table 1 summarizes the different metal-encapsulated MOFs catalytic systems which show promising results for reduction reactions of substituted nitroarenes [29, 32, 34–57].

Table 1 MNPs@MOF derived catalytic systems for the hydrogenation of substituted nitroarenes

Entry	Catalytic system	Reaction conditions	MOF synthesis method	Product yield/conversion	Year	Ref.
1	Au/MPC or Pd/MPC	$NaBH_4$, H_2O, RT	Calcination of Fe-MIL-88A	1×10^{-2} s^{-1} or 1.2×10^{-2} s^{-1}	2014	[34]
2	Pt/MIL-101	EtOH 4.0 MPa H_2 298 K, 30 min	Hydrothermal followed by impregnation	TOF: 25,438 h^{-1}	2015	[32]
3	Co@C-N-900-15 h	^{i}PrOH, 150 °C, 80 h	Pyrolysis of Co-MOF	80%	2015	[35]
4	Pd@ Zn-BDC	$NaBH_4$, RT, 3.0 min	Hydrothermal followed by impregnation	>99%	2015	[36]
5	Co@Pd/NC (MOF Templated)	0.1 MPa H_2, EtOAc, RT, 45 min	Thermolysis followed by impregnation	>98%	2015	[37]
6	Pd@MOF or Pd@ZIF-8	$NaBH_4$, RT, 25.0 min	Pre-synthesized Pd NPs and encapsulation on ZIF	>99%	2015	[38]
7	γ-Fe_2O_3 NPs@C (from Fe-MIL-88A)	N_2H_4.H_2O, EtOH, 85 °C, 1 h	Solvothermal followed by pyrolysis	100%	2016	[39]
8	Pd@MIL-101	$NaBH_4$, RT, 25.0 min	double solvents	>99%	2012	[40]
9	Co–CoO@N (From ZIF-67)	NH_3BH_3, CH_3OH, H_2O RT, 1.5 h	Pyrolysis	100%	2016	[29]
10	Cu-MIL-101(Cr)	Microwave, $NaBH_4$	Impregnation	>99%	2013	[41]
11	Ni@C650 (Ni-MOF)	EtOH, 0.5 MPa H_2,160 °C	direct thermolysis	>99%	2017	[42]
12	Au–Ag@ZIF-8	$NaBH_4$, RT, H_2O	Sequential deposition reduction	>99%	2011	[43]
13	Cu/Co@NPCC-600	$NaBH_4$, RT, H_2O	Carbonization	>99%	2017	[44]
14	Au@MIL-100(Fe)	$NaBH_4$, RT, H_2O, 15 min	Step-by-step assembly strategy	>99%	2013	[45]

(continued)

Table 1 (continued)

Entry	Catalytic system	Reaction conditions	MOF synthesis method	Product yield/conversion	Year	Ref.
15	$Zn_{0.3}Co_{2.7}$@NC	$NaBH_4$, RT, H_2O, 5 min	Self-template strategy	100%	2017	[46]
16	Cobalt-terephthalic acid MOF@C-800-	THF—H_2O, 20 bar H_2, 120 °C, 20 h	MOF–carbon template followed by pyrolysis	98%	2018	[47]
17	Co_2P/CNx	THF—H_2O, 5.0 MPa H_2, 60 °C, 6 h	Thermal treatment	>99%	2018	[30]
18	Ru/UiO-66	HCOOH: $^iPrOH/H_2O$, 150 °C, 3.0 h	Calcination followed by encapsulation	100	2018	[48]
19	Co@CN-800	H_2O, 2.0 MPa H_2, 70 °C, 2 h	Pyrolysis of ZIF-9	100	2019	[49]
20	Au-Fe_3O_4@MIL-100(Fe)	$NaBH_4$, RT,	Layer-by-layer assembly	>99%	2015	[50]
21	Pt/NPC-900 from Cu-MOF	KBH_4, H2O, RT	Annealing followed by chemical reduction	0.200 s^{-1} g^{-1} L or >98%	2018	[51]
22	Co_3S_4 from ZIF-67	Na_2S/Na_2SO_3 H_2O, RT, 6 h	Solvothermal followed by the thermal annealing process	99.8%	2017	[52]
23	NTR@ZIF-8 (Biocatalyst)	PBS buffer (pH 7.4), NADH, 37 °C, 2 h	Enzyme (NTR) immobilization on ZIF-8	>90%	2021	[53]
24	$Cu_{1.96}S$/NSC	$NaBH_4$, RT, H_2O, 3 min	Solvothermal followed by vulcanization	100%	2021	[54]
25	Fe_3O_4@SiO_2-NH_2-CoCu@UiO-66	^iPrOH, 110 °C	Hydrothermal synthesis (Half way injection)	>99%	2021	[55]

More recently, in 2019 Hua et al., reported an excellent methodology wherein, the functionalized Co-MOF prepared by single-step solvothermal strategy followed by high-temperature pyrolysis under nitrogen atmosphere of zeolitic ZIF-9 MOF moiety which leads to the formation of Co@CN NPs [49]. The Co@CNs exhibit good reactivity toward the hydrogenation of nitro-aromatics (entry 19, Table 1). The milder reaction conditions of 70 °C, 2.0 MPa H_2 pressure make the procedure more viable. The catalytic system could tolerate a variety of substituents, and also the cobalt core makes this system magnetically active; thereby the catalyst can be retrieved efficiently for further cycles [49]. Such approaches are of great importance in developing an economical and practical synthetic process. The transfer hydrogenation synthetic approach also works in a similar fashion wherein instead of using high/1.0 atm pressure molecular hydrogen, a hydrogen source such as isopropanol, ethanol, borohydride, etc., are used. A complete overview of different synthetic approaches for hydrogenating nitroarenes are illustrated in Table 1 [29, 32, 34–55].

The reduction reactions of nitrophenol to corresponding aminophenols are of industrial importance as it is a key intermediate to synthesize paracetamol which is an antipyretic/analgesic drug [57]. Moreover, 4-nitrophenol is an industrial toxic which has to be removed from the environment so that it doesn't harm the wastewater streams/atmosphere (Ref: entry 1, Table 1) [34]. Moreover, reduced amines such as aniline are prominently used in the polymer industry i.e., to fabricate polyurethane, colors-pigments, or dyes [57]. The UV–*vis* spectroscopy is a reliable characterizing tool that depicts the UV absorbance peaks of both the compounds i.e., the 4-nitrophenol and 4-aminophenol in the UV spectra [33]. In the past few years much, many reports have been furnished which states the incorporation of metal nanoparticles (noble or non-noble metals) to MOFs cavities for the reduction of nitro-functionalities to their corresponding amines (entry 19–25, Table 1) [49–55]. It is interesting to know that not only metal NPs can be immobilized within the cage-like structures of MOFs, enzymes can also be entrapped. Recently 2021, B. Shen et al., reported first time the incorporation of nitroreductase (NTR) into the zeolitic structure of ZIF-8 [53]. It was the first of its own type "a biocatalyst" which not only exhibits excellent chemical reactivity but also tolerates the harsh reaction conditions of high temperature, an organic solvent system and strong alkali/acid treatment. This biocatalyst "NTR@ZIF-8" proves to be recyclable for five successive cycles for the bio-catalyzed reduction of 4-nitrophenyl to 4-aminophenol. (Entry 23, Table 1) [53].

MNPs@MOFs exhibit a significant reactivity toward the reduction of unsaturated moieties such as alkene/alkynes etc., in more or less a similar fashion. The concept remains the same that the MNPs@MOFs will activate the olefins wherein hydrogen will react to form a corresponding saturated product. In 2018, Liang et al. reported the use of non-noble metal "Ni/NiO@C" a core–shell catalytic system that selectively could hydrogenate the phenylacetylene to the corresponding ethylbenzene in the presence of 0.1 MPa H_2 at 50 °C in ethanol as a solvent (entry 1, Table 2) [58]. When Ni-MOF-74 calcined at 500 °C under argon atmosphere it forms a mesoporous and uniform sized Ni/NiO@C NPs [58]. Kaskel et al., reported the size-selective hydrogenation of olefins using Pd@MIL-101 (Cr) as catalyst (entry 4, Table 2) [62]. A variety of Pd-based catalytic systems were compared such as palladium on charcoal

Table 2 MNPs@MOF derived catalytic systems for the hydrogenation of Olefins

Entry	Catalytic system	Reaction conditions	MOF synthesis method	Product yield/conversion	Year	Ref.
1	Ni/NiO@C (Ni-MOF-74)	EtOH, 50 °C, 1.0 MPa H_2, 1.0 h	Hydrothermal followed by Pyrolysis	97%	2018	[58]
2	Pt@ZIF-8	EtOH, RT, 24 h, 1 atm H_2	Assembly on unprotected Pt NPs	95% (TOF = 2403 h^{-1})	2013	[59]
3	Ni/SiO_2 and Co/SiO_2	8 MPa H_2, 100 °C, 1.8 h	Liquid impregnation	>99%	2016	[60]
4	Pt@MIL-101 (Cr)	35 °C, 1.5 bar H_2	Wet Impregnation	>99%	2013	[62]
5	Ni@C/Ni-MOF-74	EtOH, 30 °C, 0.5 h	Pyrolysis	>97%	2018	[63]
6	Pd@MDPC	EtOH, 25 °C, 1.0 atm H_2	carbonization of ZIF-67	>98%	2016	[61]
7	Ru-MOF-Mt	6.0 MPa H_2, 160 °C, 2.5 h	Solvothermal	>99.9%	2016	[64]
8	Pt@SALEM-2	EtOAc, RT, 1 bar H_2	Solvent-Assisted Linker Exchange	30 (7)	2016	[65]
9	Cu@ZIF-8	Toluene, 1.5 MPa H_2, 130 °C, and 10 h	Controlled pyrolysis	>99%	2021	[66]
10	(Ru SAs/N–C)	Toluene, 35.0 bar H_2, 100 °C	coordination-assisted strategy	98.5–99.4%	2017	[67]
11	Pd@ZIF-8	EtOAc, 35 °C, 24 h, 1.0 bar H_2	Solid grinding method	84.1	2016	[68]
12	Pd@MOF-5	35 °C, 1.0 atm H_2	Wet impregnation	>99%	2017	[69]
13	Pd@MSS@ZIF-8	EtOAc, 35 °C, 1.0 atm H_2	supramolecular templating (Presynthesis)	>99%	2014	[70]
14	C@Pd@ZIF-8	80 °C, 20.0 atm H_2	Wet impregnation followed by carbonization	95.1%	2015	[71]
15	Pd/SiO_2@ZIF-8	EtOAc, 35 °C, 1.0 bar H_2	Growing of ZIF-8 on pre-synthesized Pd@SiO_2	>99%	2014	[72]

(continued)

Table 2 (continued)

Entry	Catalytic system	Reaction conditions	MOF synthesis method	Product yield/conversion	Year	Ref.
16	Pd@UiO-67	THF, 25 °C, 1.0 atm H_2	Co-precipitation	>99%	2007	[73]
17	Ni@MesMOF-1	1.0 atm H_2, RT, 4 h	Gas phased impregnation	>99%	2010	[74]
18	Pd nanocubes@ZIF-8	1.0 atm H_2, RT, 90 min, 100 mW cm^{-2} visible hν	Plasmon-Driven Photothermal Conversion	>99%	2016	[75]
19	Pd@ZIF-8	1-Hexene, 25 °C, 0.1–1.0 MPa H_2, 4–24 h	Dispersion polymerization	>99%	2020	[76]
20	Pd@ZIF-L	1-Hexene, 1 atm H_2	Pre-synthesized Pd NPs to ZIF	100%	2016	[77]

or Pd-entrapped-MOF-5 NPs which concludes the superiority of Pd@MIL-101 as it shows 100% conversion of styrene to ethylbenzene under viable reaction conditions (entry 4, Table 2) [62]. The pore size of MIL-101 makes an approachable room to react and provides stability in an aqueous medium. Olefin hydrogenation is well known in literature but mostly with the complete reduction parameters. From the organic chemistry perspective, a well-designed catalytic system should not only hydrogenate the substrate but also should retain its chemical reactivity and selectivity.

The size-selective olefin hydrogenation is attributed to the fact accounted for the dependence of the nano pore size of MOFs. For larger substrates, no such reactivity or activity can be seen. In principle, the generalized trend of reactivity follows the order of cyclooctene < cyclohexene < 1-hexene. Evidently, when larger substrates possessing expanded pore sizes such as Ir@ZIF-8 or Pd@ZIF-L, an appreciable reactivity can be observed for olefins like styrene, phenylacetylene, cyclohexene, etc. as depicted in Table 2 [58–77]. In 2016, Xue et al., used Pd@ZIF-L catalytic system which successfully hydrogenate the 1-hexene with 100% conversion attributing the fact that the layered zeolitic 2-D framework of ZIF-L could incorporate the Pd NPs in it (entry 20, Table 2) [77]. Although, with heavier sized tetraphenylacetylene, no such conversion was observed under 1 atmospheric hydrogen pressure [77].

Leus et al., obtained better reactivity with Pt@MIL-101 (Cr) wherein linear/cyclic olefins hydrogenated to respective products with full conversion at room temperature. Substrates like cyclooctene or cyclohexene were smoothly hydrogenated at room temperature under a hydrogen pressure of 6 atmospheres with a conversion of 98% in 2–6-h time [78]. Moreover, with Pd@MDPC substrates like 1-octyne/4-pentyn-1-ol were successfully hydrogenated to corresponding alkenes (entry 6, Table 2) [61]. B. Han et al., in 2016 successfully synthesized Ni@SiO_2 NPs using liquid impregnation method followed by thermolysis for the selective hydrogenation of benzene

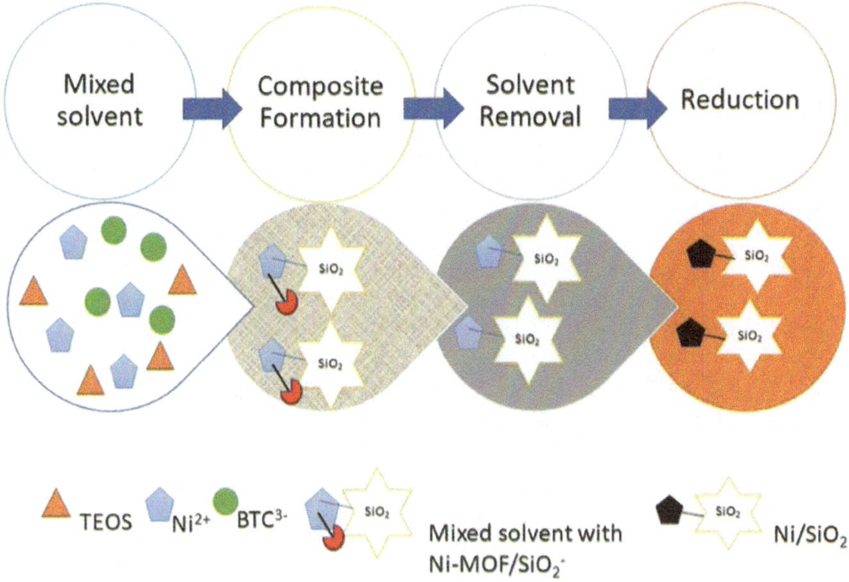

Fig. 4 Schematic representation of hydrogenation of benzene to cyclohexane using Ni/SiO_2 NPs derived from Ni-MOF/SiO_2

to cyclohexene in just 1.8 h time (entry 3, Table 2) [60]. Fig. 4 shows a pictorial representation for the synthesis of Ni@SiO_2 NPs from its MOF template (entry 3, Table 2) [60]. Table 2, summarizes the effective and popular catalytic systems which could hydrogenate the olefins with much higher success rates along with maintaining the reactivity and selectivity of MNPs@MOFs. Much research is still going on as with the higher/larger olefins complete conversion with shorter reaction time is still not reported with other nanoparticles.

Apart from nitro-aromatics or olefin reduction, the other popular MOF derived catalytic systems includes a variety of organic substrate such as carbonyls (be it aldehyde or ketone), diols, phenols, etc. All these substrates are an important part of natural products or fine chemicals. It is an important aspect of an MNPs@MOFs catalyst that it not only selectively reduces the carbonyl/diols/phenols but also retains the other structural part unharmed.

The use of microwave-assisted chemical-vapor deposition "MWCVD" is a powerful synthetic methodology which was used by Zhang et al., in 2018 to prepare M_2Si@C nano-catalyst from M-MOF-74 (M = Co, Fe, Ni) where silicates are embedded in the porous carbon support (entry 4, Table 3) [82]. This nanomaterial i.e., M_2Si@C of nano dimension (8–12 nm) were then used as a catalyst for the hydrogenation of α,β-unsaturated aldehyde (cinnamaldehyde). Both Nickel and cobalt-based catalytic systems showed very good selectivity and reactivity depending on the C=C or C=O bonds. With a Co-based system, 60% selectivity was observed leading to the formation of cinnamyl alcohol exclusively while the nickel-based

Table 3 MNPs@MOF derived catalytic systems for the hydrogenation of aldehydes, ketones diols, phenol, etc.

Entry	Catalytic system	Reaction conditions	MOF synthesis method	Product yield/conversion	Year	Ref.
1	Pt@UiO-66-NH_2	Cinnamaldehyde, 25 °C, 44 h, MeOH, 40 bar H_2	Wetness/capillary impregnation	98.7%	2014	[79]
2	Co@CN-900	Cinnamaldehyde, *n*-hexanol, 80 °C, 48 h	Thermolysis	> 98%	2017	[80]
3	Pd@MIL-101(Cr)-NH_2	Cinnamaldehyde, EtOAc/Toluene/Acetone, 20 °C, 1.0 atm H_2	Wetness/capillary impregnation	> 99%	2016	[81]
4	M_2Si@C (M = Fe, Co, Ni)	Cinnamaldehyde, 1.0 MPa H2, EtOH, 60 °C, 1.5 h	Microwave-assisted chemical vapor deposition (MWCVD)	90%	2018	[82]
5	Pt@MIL-100(Fe)@MIL-100(Fe)	Cinnamaldehyde, ^{i}PrOH, RT, 1.0 atm H_2	Homoepitaxial growth on pre-synthesized Pt-MOF Core	95%	2016	[83]
6	Pt_3Co/PCT	Cinnamaldehyde, 0.2 MPa H_2, EtOH, 80 °C, 3.0 h	Amino acid-assisted solvothermal strategy	96%	2019	[84]
7	Pd@UiO-66-NH_2	Crotanaldehyde, ^{i}PrOH, 30 bar H_2, 70 °C, 18 h	Impregnation	74.8%	2021	[85]
8	Co_3O_4/N-Gr/Fe_3O_4-800–1	Cinnamaldehyde, Toluene, 30.0 bar H_2, 110 °C	Wet impregnation	> 99%	2021	[86]
9	Rh@MIL-101 (Cr)	Cinnamaldehyde EtOH, 30 °C, 1 MPa H_2, 3.0–12 h	Double solvent method	> 98%	2020	[87]
10	Pt/$Ni_xFe_{1-x}Al_2O_{4+\delta}$	Cinnamaldehyde 2 MPa H_2 ^{i}PrOH + H_2O, 90 °C, 0.5 h	Sol gel followed by impregnation	92.2%	2021	[88]
11	Pd-PVP@ZIF-8	1,4-butynediol, H_2O, 50 °C, 2 MPa H_2	PVP-assisted synthetic method	> 99%	2014	[89]

(continued)

Table 3 (continued)

Entry	Catalytic system	Reaction conditions	MOF synthesis method	Product yield/conversion	Year	Ref.
12	Pd@MIL-101(Cr)	1,4-butynediol, H_2O, 0.5 MPa H_2, 50 °C	Metal–organic chemical vapor deposition (MOCVD)	> 94%	2017	[90]
13	Pt@ZIF-8	1,4-butynediol, H_2O, 92 20 atm H_2, 120 °C, 1.5 h.03	Chemical Reduction	100%	2016	[91]
14	Pd@MOF-5/Pd@IRMOF-3	1,4-butynediol, MeOH, 1.0 atm H_2, 20 °C	Direct Mixing	> 95%	2013	[92]
15	Co@C-N-900–15 h	Acetophenone, ^{i}PrOH, 80 °C. 17 h	Co-containing MOF as sacrificial template	94%	2015	[93]
16	Pd@UiO-66 or Pd@UiO-67	Aceto or Benzo-phenone, ^{i}PrOH, 5 bar H_2, 60 °C	CVD “Chemical vapour Deposition”	99%	2015	[94]
17	Ru@ZIF-8	Acetophenone, EtOH, 40 atm H_2, 40 °C, 1.0 h	Wetness/capillary impregnation	98.8%	2013	[95]
18	NiPd@MIL-101	3-heptanone, 20 bar H_2, 60 °C, 20 h	CVD “Chemical vapour Deposition”	75	2012	[96]
19	Ni@MOF-5	Acetophenone, ^{i}PrOH, autoclave, 83 °C, 1.5 h	Impregnation methods	>99%	2020	[97]
20	Rh@S-MIL-101	Phenol, H_2O, 50 °C, 5.0 bar H_2	Wetness/capillary impregnation	>99%	2015	[98]
21	Co–Ni@CN-600	Phenol, 0.8 MPa H_2, ^{i}PrOH, 100 °C, 12 h	Doped MOF-templated strategy	>99%	2017	[99]
22	Pd@MIL-53(Al)	Phenol, H_2O, 5 h, 19–50 °C	Wetness/capillary impregnation	>99%	2014	[100]
23	Ru@MIL-101	Phenol, H_2O, 50 °C, 5 bar H_2	Gas phased impregnation	>90%	2016	[101]
24	Pd/MOF140-AA	Phenol, H_2O, 260 °C, 2 MPa H_2 2.0 h	Incipient wetness impregnation followed by Thermolysis	45.1%	2018	[102]

system was chemo-selective for C=C bonds with 90% selectivity (entry 4, Table 3) [82]. As shown in Table 3 entries 1–10 [79–88], a variety of Pd, Pt, Co, Rh, cobalt oxide-based catalytic system shows exclusive chemo-selectivity for reducing the cinnamaldehyde to corresponding alcohol or reduced C–C products. The selective hydrogenation of unsaturated diols (1,4-butynediol to 1,4-butenediol) is of industrial importance aspect as it finds its applications in the diverse areas of production of vitamins, endusulfan, etc. [104]. Conclusively, Table 3, summarizes the effective and efficient catalytic system for chemoselective and active hydrogenations of carbonyls, phenols, diols, etc. [79–102].

One of the industrially important intermediates is "cyclohexanone" as it is used in the synthesis of nylon-6 via caprolactam or nylon-66 [105]. So if one can do selective hydrogenation of phenol it will lead to the formation of cyclohexanone. Rather than using a two-step synthetic strategy, it is always preferred to have a one-pot synthesis using metals such as Ni, Pd, Pt, etc., which is cost-effective, time-saving, and simple [103]. In recent times and as shown in Table 3, entries 20–24 [98–102], the MNPs@MOF shows excellent selectivity for the hydrogenation of phenol leading to cyclohexanone as a product exclusively. Chen et al., in 2018 demonstrated the synthesis of Pd/MOF140-AA using incipient wetness impregnation followed by thermolysis [102]. The catalytic system Pd/MOF140-AA was found to be an efficient catalyst that could successfully hydrogenate the phenol to cyclohexanol completely with an overall product yield of 45.1% (entry 24, Table 3) [102]. In 2016, Ertas et al. reported the synthesis of a recyclable Ru/MIL-101 catalyst for the aqueous-phase mediated catalytic transformation phenol to cyclohexanone at a 5.0 bar initial H_2 pressure with a selectivity of >90% (entry 23, Table 3) [101].

Conclusively, MNPs@MOF prove to be a facile and good catalytic system for the hydrogenations of various substrates using either molecular hydrogen or transfer hydrogenation at ambient reaction conditions. If we compare the industrially used catalyst, MNPs@MOFs are promising candidates in terms of selectivity, reactivity, and activity. However, one should devoid the use of multistep synthetic protocols for making MNPs@MOFs. The catalytic mechanism involved in the chemistry of MOFs-derived catalysts is still unexplored which limits its applications to some extent. But with the advancement of theoretical modeling studies and techniques, the fate of MNPs@MOF can be understood fairly.

4 MNPs@MOFs in C–C Cross-Coupling Reactions

To construct biaryl units, the palladium-catalyzed C–C coupling reactions such as Sonogashira, Suzuki–Miyaura, Stille, Negishi, Heck reactions are of great use as it possesses diverse importance in the industrial as well as an academic arena [106, 116, 118]. Numerous pharmaceutical products such as anticancer, antibiotics or fertilizers, pesticides are synthesized using the concept of C–C coupling reactions [106, 107]. In principle, the phosphines ligands and their analogs or *N*-heterocycle carbene containing palladium-salts or palladium-based organometallics have been thoroughly explored for C–C coupling reactions. But all these ligands or salts are air/moisture sensitive and hence maintaining an inert atmosphere or use of dry solvents makes the process troublesome as the removal of catalyst from the reaction mixture is tedious work. So, to overcome all these impediments Pd nanoparticles are extensively used. As the metal when goes into nanoforms becomes highly energized or active which usually leads to agglomeration which in turn is responsible for the decrement in its chemical reactivity. Later, the problem of agglomeration took care of by using appropriate supporting materials such as zeolites, polymers, graphene, carbon spheres, silica, alumina, metal oxides, or stabilizers [108, 111–113]. Even with such an advancement in the area of palladium-based nanomaterials, when it comes to industrial purposes it is palladium on charcoal (Pd/C) which is widely accepted and used thoroughly. However, if we look into the MOF-based Pd nanomaterials it can be generalized that much less attention has been paid to this as can be clearly seen in Table 4.

Recently, in 2021 Han et al., reported the synthesis and catalytic role of Pd@HKUST-1 in Suzuki coupling (entry 6, Table 4) [116]. Double solvent approach was used to make the catalytic system which gave good yields of coupled products. Earlier in 2014, Zadehahmadi et al., demonstrated a double ion exchange followed by a reduction approach to synthesize Pd@UiO-66-NH_2 system for Suzuki-Miyamura coupling reactions (entry 1, Table 4) [114]. A reaction between aryl halide and phenylboronic acid gave the desired C–C coupled (biaryl) product with as high a yield of 85%. The optimized reaction scheme did well when various substituents of phenylboronic acids and aryl halides were employed. Shang et al., in 2014 reported the successful entrapment of ultrafine Pd nanoparticles in the cage-like cavities of MIL-101 by employing the double solvent approach [120]. The synthesized Pd@MIL-101 showed minimum aggregation thus leading to a stable and well dispersed catalytic system. The Suzuki and Heck coupling reaction was performed using Pd@MIL-101 NPs showing good recyclability and 90–99% yield of desired C–C coupled products under ambient reaction conditions (entry 5, Table 4) [120].

In the C–C coupling reaction role of MOF is restricted to act as a supporting material; as such it does not have any catalytic role. The overall reactivity of catalysts is only attributed to the embedded Pd(0) NPs. Functionalized MOF provides more stability and durability by fine-tuning encapsulated naked Pd NPs. Overall, there is much more to explore the usage of MOF specifically for C–C coupling reactions.

Recently Jeong et al., reported the facile and enantioselective synthesis of embedment of Zn or Ti on MOF support i.e., Zn/(*S*)-KUMOF-1 or Ti/(*S*)-KUMOF-1 [122]. This was the first of its own type wherein Zn/(*S*)-KUMOF-1 catalyst was used for the enantioselective carbonyl-ene reaction. It is interesting to know that the active reactive sites of MOF support i.e., KUMOF-1 can either be at outer or inner side depending on the interaction of reactant with it which is kind of contradictory to the previous reports in literature according to which the chiral environment of reaction site depends on its location [122].

Table 4 MNPs@MOF derived catalytic systems for C–C coupling reactions

Entry	Catalytic system	Reaction conditions	MOF synthesis method	Product yield %	Year	Ref.
1	Pd@UiO-66-NH_2	Suzuki–Miyaura coupling between aryl halides and phenylboronic acid, K_2CO_3, DMF/H_2O, 60 °C, air	direct ion-exchange method followed by reduction	85–95	2014	[114]
2	Cu/Cu_2ONPs@rGO	Sonogashira coupling between phenylacetylene and iodobenzene, Cs_2CO_3, DMF, 80 °C, 8 h	Solvothermal reduction	91	2018	[115]
3	Pd@MOF-5	Suzuki–Miyaura coupling between bromobenzene & phenylboronic acid,	Chemical vapour deposition	81	2012	[117]
4	a-Cu@C from BDC-MOF	C–N coupling reaction, KOH, DMSO, 110 °C, 24 h	deflagrated pyrolysis	74	2019	[119]
5	Pd@MIL-101	Suzuki–Miyaura coupling between aryl halides and phenylboronic acid, K_2CO_3, EtOH-H_2O, RT, 1–12 h	Double solvent approach	90–99	2014	[120]
6	Pd@HKUST-1	Suzuki homocoupling between phenylboronic acid & elemental I_2, base, EtOH, N_2 (1 bar), 70 °C, 7 h	Capillary impregnation reduction	28–91	2021	[121]

Conclusions and Outlook

In this chapter, the use of metal-encapsulated MOFs i.e., MNPs@MOFs were highlighted as an efficient heterogeneous catalytic system in hydrogenations of nitroaromatics, olefins, arenes, aldehydes, ketones, diols, phenols, and C–C coupling reactions. Though its a relatively new area, MOFs have gained much attention in past few years. The metal–organic frameworks i.e., MOFs appeared to be an excellent host to immobilize metal NPs within its highly porous and active cavities. The three important aspects of MOFs are (i) higher surface areas, porosity, fine-tuning of pore sizes, (ii) possibility of extended functionalization of MOFs, (iii) control on nucleation growth of metal NPs because of size-dependent cavities. These remarkable intrinsic properties of MOFs, cease the possibility of NPs aggregation. The stable cavities or pores of MOFs allow minimum leaching effect thereby leading to better reactivity and activity. Not only metal NPs but also their corresponding metal oxides or *bi/tri* metallic systems, enzymes can be easily incorporated with the cages of MOFs showing better synergism for the enhanced catalytic activity. Although, there is much more theoretical as well as practical evidences, are required with which one can understand the reason behind its anomalous activity of MNPs@MOFs are good when the chemo-selectivity is limited but when in the case of enantioselectivity, this system somewhat lacks behind its homogeneous counterparts.

Henceforth we believe, if one can modify the nodes, linkers, dependency of reactants to the pores, the enhancement in enatioselectivity may be obtained. A successful catalyst that is industrially viable has the characteristics of showing higher TON/TOF values. Therefore, achieving higher TOF/TON values are still one the main future goal for MNPs@MOFs. For example, a catalyst can show consecutive recyclable cycles but the area of concern is how fast and how smooth reaction can go? Is it stable or doesn't show the effect of leaching? If, such problems are addressed in near future the fate of MNPs@MOFs is definitely welcoming for industrial as well as academic purposes.

Abbreviations

NPs	Nanoparticles
MNPs	Metal nanoparticles
MOFs	Metal–organic frameworks
MNPs@MOFs	Encapsulated/anchored metal nanoparticles to MOFs
PCPs	Porous coordination polymers
TON	Turn over numbers
TOF	Turn over frequency
NTR	Nitroreductase
^{i}PrOH	Isopropanol
EtOH	Ethanol
H_2O	Water
UV–*vis*	Ultra violet visible
MPa	Mega pascal
atm	Atmospheric pressurem
EtOAc	Ethyl acetate
H_2	Molecular hydrogen
THF	Tetrahydrofuran
$NaBH_4$	Sodium borohydride
HCOOH	Acetic acid

BTC	Bis(trichloromethyl)carbonate
RT	Room temperature
K	Kelvin
$N_2H_4.H_2O$	Hydrazine hydrate
NH_3BH_3	Borane-ammonia complex
CH_3OH or MeOH	Methanol
KBH_4	Potassium tetrahydroborate
Na_2S	Sodium sulfide
Na_2SO_3	Sodium sulfite
PBS buffer	Phosphate-buffered saline buffer
NADH	Nicotinamide adenine dinucleotide
°C	Degree Celsius
TEOS	Tetraethyl orthosilicate
Ni	Nickel
Pd	Palldium
Pt	Platinum
Fe	Iron
Cu	Copper
Au	Gold
Ag	Silver
Zn	Zinc
Ti	Titanium
Ru	Ruthenium
Rh	Rhodium
Co	Cobalt
CO	Carbon monoxide
CO_2	Carbon dioxide
K_2CO_3	Potassium carbonate
DMF	Dimethylformamide
Cs_2CO_3	Caesium carbonate
DMSO	Dimethyl sulfoxide
I_2	Elemental Iodine
N_2	Molecular nitrogen
MWCVD	Microwave-assisted chemical-vapour deposition
Pd/C	Palladium on charcoal

References

1. Natarajan S, Mandal S (2008) Open-framework structures of transition-metal compounds. Angew Chem Int Ed 120:4876–4907
2. Morris RE, Wheatley PS (2008) Gas storage in nanoporous materials. Angew Chem Int Ed 120:5044–5059

3. Maspoch D, Ruiz-Molina D, Veciana J (2007) Old materials with new tricks: multifunctional open-framework materials. Chem Soc Rev 36:770–818
4. Janiak C (2003) Engineering coordination polymers towards applications. Dalton Trans 2781–2804
5. Mueller U, Schubert M, Teich F, Puetter H, Schierle-Arndt K, Pastre J (2006) Metal–organic frameworks—prospective industrial applications. J Mater Chem 16:626–636
6. Ferey G (2008) Hybrid porous solids: past, present, future. Chem Soc Rev 37:191–214
7. Kitagawa S, Kitaura R, Noro SI (2004) Functional porous coordination polymers. Angew Chem Int Ed 43:2334–2375
8. Uemura T, Kitaura R, Ohta Y, Nagaoka M, Kitagawa S (2006) Nanochannel-promoted polymerization of substituted acetylenes in porous coordination polymers. Angew Chem Int Ed 45:4112–4116
9. Rowsell JLC, Yaghi OM (2004) Metal–organic frameworks: a new class of porous materials. Microporous Mesoporous Mater 73:3–14
10. Corma A, Garcia H (2008) Supported gold nanoparticles as catalysts for organic reactions Chem. Soc Rev 37:2096–2126
11. Cushing BL, Kolesnichenko VL, O'Connor CJ (2004) Recent advances in the liquid-phase syntheses of inorganic nanoparticles. Chem Rev 104:3893–3946
12. Moon HR, Limb DW, Suh MP (2013) Fabrication of metal nanoparticles in metal–organic frameworks. Chem Soc Rev 42:1807–1824
13. Dhakshinamoorthya A, Garcia H (2012) Catalysis by metal nanoparticles embedded on metal–organic frameworks. Chem Soc Rev 41:5262–5284
14. Ishida T, Nagaoka M, Akita T, Haruta M (2008) Deposition of gold clusters on porous coordination polymers by solid grinding and their catalytic activity in aerobic oxidation of alcohols. Chem Eur J 14:8456–8460
15. Hermes S, Schroter MK, Schmid R, Khodeir L, Muhler M, Tissler A, Fischer RW, Fischer RA (2005) Angew Chem Int Ed 44:6237–6241
16. Moulton B, Zaworotko MJ (2001) From molecules to crystal engineering: supramolecular isomerism and polymorphism in network solids. Chem Rev 101:1629–1658
17. Evans OR, Lin W (2002) Crystal engineering of NLO materials based on metal−organic coordination networks. Acc Chem Res 35:511–522
18. Ferey G, Mellot-Draznieks C, Serre C, Millange F (2005) Crystallized frameworks with giant pores: are there limits to the possible? Acc Chem Res 38:217–225
19. Yaghi OM, O'Keeffe M, Ockwig NW, Chae HK, Eddaoud M, Kim J (2003) Reticular synthesis and the design of new materials. Nature 423:705–714
20. Hill RJ, Long DL, Champness NR, Hubberstey P, Schroder M (2005) New approaches to the analysis of high connectivity materials: design frameworks based upon 4^4- and 6^3-subnet tectons. Acc Chem Res 38:335–348
21. Bradshaw D, Warren JE, Rosseinsky MJ (2007) Reversible concerted ligand substitution at alternating metal sites in an extended solid. Science 315:977–988
22. Seh ZW, Kibsgaard J, Dickens CF, Chorkendorff I, Nørskov JK, Jaramillo TF (2017) Combining theory and experiment in electrocatalysis: insights into materials design. Science 355:4998
23. Kozachuk O, Luz I, Xamena F, Noei H, Kauer M, Albada HB, Bloch ED, Marler B, Wang YM, Muhler M (2014) Multifunctional, defect-engineered metal-organic frameworks with ruthenium centers: sorption and catalytic properties. Angew Chem Int Ed 53:7058–7062
24. Zhong J, Yang X, Wu Z, Liang B, Huang Y, Zhang T (2020) State of the art and perspectives in heterogeneous catalysis of CO_2 hydrogenation to methanol. Chem Soc Rev 49:1385–1413
25. Ding K, Uozumi Y (2008) Handbook of asymmetric heterogeneous catalysis. Wiley-VCH Verlag GmbH & Co. KGaA. ISBN: 9783527623013. https://doi.org/10.1002/9783527623013
26. Nerozzi F (2012) Heterogeneous catalytic hydrogenation. Platin Met Rev 56:236
27. Sun X, Olivos-Suarez AI, Oar-Arteta L, Rozhko E, Osadchii D, Bavykina A, Kapteijn F, Gascon J (2017) Metal-organic framework mediated Cobalt/Nitrogen-doped carbon hybrids as efficient and chemoselective catalysts for the hydrogenation of nitroarenes. ChemCatChem 9:1854–1862

28. Wang X, Zhong W, Li Y (2015) Nanoscale co-based catalysts for low-temperature CO oxidation, Catal. Sci Technol 5:1014–1020
29. Ma X, Zhou YX, Liu H, Li Y, Jiang HL (2016) A MOF-derived Co–CoO@N-doped porous carbon for efficient tandem catalysis: dehydrogenation of ammonia borane and hydrogenation of nitro compounds. Chem Commun 52:7719–7722
30. Yang SL, Peng L, Oveisi E, Bulut S, Sun DT, Asgari M, Trukhina O, Queen WL (2018) MOF-derived cobalt phosphide/carbon nanocubes for selective hydrogenation of nitroarenes to anilines. Chem Eur J 24:4234–4238
31. Rosler C, Fischer RA (2015) Metal–organic frameworks as hosts for nanoparticles. CrystEngComm 17:199–217
32. Pan H, Li X, Yu Y, Li J, Hu J, Guan Y, Wu P (2015) Pt nanoparticles entrapped in mesoporous metal–organic frameworks MIL-101 as an efficient catalyst for liquid-phase hydrogenation of benzaldehydes and nitrobenzenes. J Mol Catal A: Chem 399:1–9
33. Liao YT, Chen JE, Isida Y, Yonezawa T, Chang WC, Alshehri SM, Yamauchi Y, Wu KCW (2016) ChemCatChem 8:502–509
34. Dong Z, Le X, Liu Y, Dong C, Ma J (2014) Metal organic framework derived magnetic porous carbon composite supported gold and palladium nanoparticles as highly efficient and recyclable catalysts for reduction of 4-nitrophenol and hydrodechlorination of 4-chlorophenol. J Mater Chem A 2:18775–18785
35. Long J, Zhou Y, Yingwei L (2015) Transfer hydrogenation of unsaturated bonds in the absence of base additives catalyzed by a cobalt based heterogeneous catalyst. Chem Commun 51:2331–2334
36. Zhu Y, Wang YM, Liu P, Wu YL, Wei W, Xia CK, Xie JM (2015) Cage-like pores of a metal-organic framework for separations and encapsulation of Pd nanoparticles for efficient catalysis. New J Chem 39:2669–2674
37. Shen K, Chen L, Long JL, Zhong W, Li YW (2015) MOFs-templated Co@Pd core-shell NPs embedded in N-doped carbon matrix with superior hydrogenation activities. ACS Catal 5:5264–5271
38. Wang C, Zhang HY, Feng C, Gao ST, Shang NZ, Wang Z (2015) Multifunctional Pd@MOF core–shell nanocomposite as highly active catalyst for p-nitrophenol reduction. Catal Commun 72:29–32
39. Li Y, Zhou YX, Ma X, Jiang HL (2016) Metal-organic framework-templated synthesis of γ-Fe_2O_3 nanoparticles encapsulated in porous carbon for efficient and chemoselective hydrogenation of nitro compounds. Chem Commun 52:4199–4202
40. Yadav M, Aijaz A, Xu Q (2012) Highly catalytically active palladium nanoparticles incorporated inside metal-organic framework pores by double solvents method. Funct Mater Lett 5:1250039
41. Wu F, Qiu LG, Ke F, Jiang X (2013) Copper nanoparticles embedded in metal–organic framework MIL-101(Cr) as a high performance catalyst for reduction of aromatic nitro compounds. Inorg Chem Commun 32:5–8
42. Tang B, Song WC, Yang EC, Zhao XJ (2017) MOF-derived Ni-based nanocomposites as robust catalysts for chemoselective hydrogenation of functionalized nitro compounds. RSC Adv 7:1531–1539
43. Jiang HL, Akita T, Ishida T, Haruta M, Xu Q (2011) Synergistic catalysis of Au@Ag core-shell nanoparticles stabilized on metal-organic framework. J Am Chem Soc 133:1304–1306
44. Li H, Yue F, Yang C, Xue P, Li NN, Zhang Y, Wang JD (2017) Structural evolution of metal−organic framework and derived hybrids composed of metallic cobalt and copper encapsulated in nitrogen-doped porous carbon cubes with high catalytic performance. CrystEngComm 19:64–71
45. Ke F, Zhu JF, Qiu LG, Jiang X (2013) Controlled synthesis of novel Au@MIL-100(Fe) core–shell nanoparticles with enhanced catalytic performance. Chem Commun 49:1267–1269
46. Xu XJ, Li H, Xie HT, Ma YH, Chen TX, Wang JD (2017) Zinc cobalt bimetallic nanoparticles embedded in porous nitrogen-doped carbon frameworks for the reduction of nitro compounds. J Mater Res 32:1777–1786

47. Murugesan K, Senthamarai T, Sohail M, Alshammari AS, Pohl MM, Beller M, Jagadeesh RV (2018) Cobalt-based nanoparticles prepared from MOF–carbon templates as efficient hydrogenation catalysts. Chem Sci 9:8553–8560
48. Yang Q, Zhang HY, Wang L, Zhang Y, Zhao J (2018) Ru/UiO-66 catalyst for the reduction of nitroarenes and tandem reaction of alcohol oxidation/knoevenagel condensation. ACS Omega 3:4199–4212
49. Hua A, Lua X, Caia D, Pana H, Jinga R, Xiaa Q, Zhoua D, Xia Y (2019) Selective hydrogenation of nitroarenes over MOF-derived Co@CN catalysts at mild conditions. Mol Catal 472:27–36
50. Ke F, Wang LH, Zhu JF (2015) Multifunctional Au-Fe_3O_4@MOF core–shell nanocomposite catalysts with controllable reactivity and magnetic recyclability. Nanoscale 7:1201–1208
51. Wu XQ, Zhao J, Wu YP, Dong WW, Li DS, Li JR, Zhang Q (2018) Ultrafine Pt nanoparticles and amorphous nickel supported on 3D mesoporous carbon derived from Cu-metal–organic framework for efficient methanol oxidation and nitrophenol reduction. ACS Appl Mater Interfaces 10:12740–12749
52. Xu Y, Lv XJ, Chen Y, Fufu W (2017) Highly selective reduction of nitroarenes to anilines catalyzed using MOF-derived hollow Co3S4 in water under ambient conditions. Catal Commun 101:31–35
53. Shen B, Ding R, Dai J, Ji Y, Wang Q, Wang Y, Huang H, Zhang X (2021) Encapsulating nitroreductase into metal-organic framework: Boosting industrial performance for the reduction of nitro-aromatics. Green Synth Catal. https://doi.org/10.1016/j.gresc.2021.09.001
54. Wang D, Zhang C, Zeng F, Hu X, Li C, Su Z (2021) Efficient catalytic reduction of 4-nitrophenol by Cu1.96S/NSC hybrid material derived from metal-organic framework. In: E3S web conf 252, 02065
55. Li Y, Li YN, Zheng JW, Dong XY, Guo RX, Wang YM, Hu ZN, Ai Y, Liang Q, Sun HB (2021) Metal-organic framework-encapsulated CoCu nanoparticles for the selective transfer hydrogenation of nitrobenzaldehydes: engineering active armor by the half-way injection method. Chem Eur J 27:1080–1087
56. Zhou YG (2007) Asymmetric hydrogenation of heteroaromatic compounds. Acc Chem Res 40:1357–1366
57. Pollak P (2011) Fine chemicals, 2nd edn. Wiley, Hoboken, New Jersey
58. Guo XL, Chen X, Su DS, Liang CH (2018) Preparation of Ni/C core-shell nanoparticles through MOF pyrolysis for phenylacetylene hydrogenation reaction. Acta Chim Sinica 76:22–29
59. Wang P, Zhao J, Li XB, Yang Y, Yang QH, Li C (2013) Assembly of ZIF nanostructures around free Pt nanoparticles: efficient size-selective catalysts for hydrogenation of alkenes under mild conditions. Chem Commun 49:3330–3332
60. Kang X, Liu H, Hou M, Sun X, Han H, Jiang T, Zhang Z, Han B (2016) Synthesis of supported ultrafine non-noble subnanometer-scale metal particles derived from metal-organic frameworks as highly efficient heterogeneous catalysts. Angew Chem Int Ed 55:1080–1084
61. Li X, Zhang W, Liu Y, Li R (2016) Palladium nanoparticles immobilized on magnetic porous carbon derived from ZIF-67 as efficient catalysts for the semi hydrogenation of phenylacetylene under extremely mild conditions. ChemCatChem 8:1111–1118
62. Khajavi H, Stil HA, Kuipers HPCE, Gascon J, Kapteijn F (2013) Shape and transition state selective hydrogenations using egg-shell Pt-MIL-101(Cr) catalyst. ACS Catal 3:2617–2626
63. Nakatsuka K, Yoshii T, Kuwahara Y, Mori K, Yamashita H (2018) Controlled pyrolysis of Ni-MOF-74 as a promising precursor for the creation of highly active Ni nanocatalysts in size-selective hydrogenation. Chem Eur J 24:898–905
64. Chen DC, Huang M, He S, He SL, Ding LP, Wang Q, Yu SM, Miao SD (2016) Ru-MOF enwrapped by montmorillonite for catalyzing benzene hydrogenation. Appl Clay Sci 119:109–115
65. Stephenson CJ, Hupp JT, Farha OK (2016) Post assembly transformation of a catalytically active composite material, Pt@ZIF-8, via solvent-assisted linker exchange. Inorg Chem 55:1361–1363

66. Ling JL, Chen K, Wu CD (2021) Interwrapping distinct metal-organic frameworks in dual-MOFs for the creation of unique composite catalysts. Research 2021:9835935
67. Wang X, Chen W, Zhang L, Yao T, Liu W, Lin Y, Ju H, Dong J, Zheng L, Yan W, Zheng X, Li Z, Wang X, Yang J, He D, Wang Y, Deng Z, Wu Y, Li Y (2017) Uncoordinated amine groups of metal-organic frameworks to anchor single Ru sites as chemoselective catalysts toward the hydrogenation of quinoline. J Am Chem Soc 139:9419–9422
68. Ding SS, Yan Q, Jiang H, Zhong ZX, Chen RZ, Xing WH (2016) Fabrication of Pd@ZIF-8 catalysts with different Pd spatial distributions and their catalytic properties. Chem Eng J 296:146–153
69. Xu D, Pan Y, Zhu LK, Yusran Y, Zhang DL, Fang QR, Xue M, Qiu SL (2017) Simple coordination complex-derived Ni NP anchored N-doped porous carbons with high performance for reduction of nitroarenes. CrystEngComm 19:6612–6619
70. Zhang T, Li B, Zhang XF, Qiu JS, Han W, Yeung KL (2014) Pd nanoparticles immobilized in a microporous/mesoporous composite ZIF-8/MSS: a multifunctional catalyst for the hydrogenation of alkenes. Microporous Mesoporous Mater 197:324–330
71. Zhou WQ, Zou BH, Zhang WN, Tian DB, Huang W, Huo FW (2015) Synthesis of stable heterogeneous catalysts by supporting carbon-stabilized palladium nanoparticles on MOFs. Nanoscale 7:8720–8724
72. Lin L, Zhang T, Zhang XF, Liu HO, Yeung KL, Qiu JS (2014) New Pd/SiO_2@ZIF-8 core–shell catalyst with selective, antipoisoning, and antileaching properties for the hydrogenation of alkenes. Ind Eng Chem Res 53:10906–10913
73. Sabo M, Henschel A, Froede H, Klemm E, Kaskel S (2007) Solution infiltration of palladium into MOF-5: synthesis, physisorption and catalytic properties. J Mater Chem 17:3827–3832
74. Park YK, Choi SB, Nam HJ, Jung DY, Ahn HC, Choi K, Furukawa H, Kim J (2010) Catalytic nickel nanoparticles embedded in a mesoporous metal–organic framework. Chem Commun 46:3086–3088
75. Yang QH, Xu Q, Yu SH, Jiang HL (2016) Pd nanocubes@ZIF-8: integration of plasmon-driven photothermal conversion with a metal–organic framework for efficient and selective catalysis. Angew Chem Int Ed 55:3685–3689
76. Zhao Y, Ni X, Ye S, Gu ZG, Li Y, Ngai T (2020) A smart route for encapsulating Pd nanoparticles into a ZIF-8 hollow microsphere and their superior catalytic properties. Langmuir 26:2037–2043
77. Xue SL, Jiang H, Zhong ZX, Low ZX, Chen RZ, Xing WH (2016) Palladium nanoparticles supported on a two-dimensional layered zeolitic imidazolate framework-L as an efficient size-selective catalyst. Microporous Mesoporous Mater 221:220–227
78. Leus K, Dendooven J, Tahir N, Ramachandran RK, Meledina M, Turner S, Tendeloo GV, Goeman JL, Eycken VD, Detavernier C, Voort PVD (2016) Atomic layer deposition of pt nanoparticles within the cages of MIL-101: a mild and recyclable hydrogenation catalyst. Nanomaterials (Basel) 6
79. Guo ZY, Xiao CX, Maligal-Ganesh RV, Zhou L, Goh TW, Li XL, Tesfagaber D, Thiel A, Huang WY (2014) Pt Nanoclusters confined within metal–organic framework cavities for chemoselective cinnamaldehyde hydrogenation. ACS Catal 4:1340–1348
80. Liu XM, Cheng SJ, Long JL, Zhang W, Liu XH, Wei DP (2017) MOFs-Derived Co@CN bi-functional catalysts for selective transfer hydrogenation of α, β-unsaturated aldehydes without use of base additives. Mat Chem Front 1:2005–2012
81. Nagendiran A, Pascanu V, Gomez AB, Miera GG, Tai CW, Verho O, Martín-Matute B, Backvall JE (2016) Mild and selective catalytic hydrogenation of the C=C bond in α, β-unsaturated carbonyl compounds using supported palladium nanoparticles. Chem Eur J 22:7184–7189
82. Zhang LL, Chen X, Peng ZJ, Liang CH (2018) Chemoselective hydrogenation of cinnamaldehyde over MOFs-derived M2Si@C (M = Fe Co, Ni) silicides catalysts. Mol Catal 449:14–24
83. Liu HL, Chang LN, Chen LY, Li YW (2016) Nanocomposites of platinum/metal–organic frameworks coated with metal–organic frameworks with remarkably enhanced chemoselectivity for cinnamaldehyde hydrogenation. ChemCatChem 8:946–951

84. Gu Z, Chen L, Li X, Chen L, Zhang Y, Duan C (2019) NH_2-MIL-125(Ti)-derived porous cages of titanium oxides to support Pt–Co alloys for chemoselective hydrogenation reactions. Chem Sci 10:2111–2117
85. Lo WS, Chou LY, Young AP, Ren C, Goh TW, Williams BP, Chen L, Ismail SYMN, Huang W, Sung CK (2021) Probing the interface between encapsulated nanoparticles and metal−organic frameworks for catalytic selectivity control. Chem Mater 33:1946−1953
86. Shaikh MN, Abdelnaby MM, Hakeem AS, Nasser GA, Yamani ZH (2021) Co_3O_4/Nitrogen-doped graphitic carbon/Fe_3O_4 nanocomposites as reusable catalysts for hydrogenation of quinoline, cinnamaldehyde, and nitroarenes. ACS Appl Nano Mater 4:3508–3518
87. Liu Q, Li Y, Fan Y, Su CY, Li G (2020) Chemoselective hydrogenation of α, β-unsaturated aldehydes over Rh nanoclusters confined in a metal–organic framework. J Mater Chem A 8:11442–11447
88. Xin H, Zhang W, Xiao X, Chen L, Wu P, Li X (2021) Selective hydrogenation of cinnamaldehyde with $Ni_xFe_{1-x}Al_2O_{4+\delta}$ composite oxides supported Pt catalysts: C=O versus C=C selectivity switch by varying the Ni/Fe molar ratios. J Catal 393:126–139
89. Zhang MM, Yang YB, Li C, Liu Q, Williams CT, Liang CH (2014) PVP–Pd@ZIF-8 as highly efficient and stable catalysts for selective hydrogenation of 1,4-butynediol, Catal. Sci Technol 4:329–332
90. Yina D, Lia C, Rena H, Shekhahb O, Liu J, Liang C (2017) Efficient Pd@MIL-101(Cr) hetero-catalysts for 2-butyne-1,4-diol hydrogenation exhibiting high selectivity. RSC Adv 7:1626–1633
91. Li C, Zhang MM, Di X, Yin DD, Li WZ, Liang CH, Chin (2016) One-step synthesis of Pt@ZIF-8 catalyst for the selective hydrogenation of 1,4-butynediol to 1,4-butenediol. J Catal 37:1555–1561
92. Isaeva VI, Tkachenko OP, Afonina EV, Kozlova LM, Kapustin GI, Grunert W, Solov'eva SE, Antipin IS, Kustov LM (2013) 2-Butyne-1,4-diol hydrogenation over palladium supported on Zn^{2+}-based–MOF and host–guest MOF/calix[4]arene materials. Microporous Mesoporous Mater 166:167–175
93. Long J, Zhou Y, Li Y (2015) Transfer hydrogenation of unsaturated bonds in the absence of base additives catalyzed by a cobalt-based heterogeneous catalyst. Chem Commun 51:2331–2334
94. Luz I, Rosler C, Epp K, Xamena FXLI, Fischer RA (2015) Pd@UiO-66-type MOFs prepared by chemical vapor infiltration as shape-selective hydrogenation catalysts. Eur J Inorg Chem 2015:3904–3912
95. Liu MM, Fan BB, Shi XF, Li RF (2013) Ru/ZIF-8 with a chiral modifier for asymmetric hydrogenation of acetophenone. Catal Commun 42:20–24
96. Hermannsdorfer J, Friedrich M, Miyajima N, Albuquerque RQ, Kummel S, Kempe R (2012) Ni/Pd@MIL-101: synergistic catalysis with cavity-conform Ni/Pd nanoparticles. Angew Chem Int Ed 51:11473–11477
97. Chelavi AN, Shahabadi VZ, Sayyahi S, Ardakani HA (2020) Optimization of the transfer hydrogenation reaction of acetophenone on Ni@MOF-5 nanoparticles using response surface methodology. Res Chem Intermed 46:445–458
98. Ertas IE, Gulcan M, Bulut A, Yurderi M, Zahmakiran MJ (2015) Rhodium nanoparticles stabilized by sulfonic acid functionalized metal-organic framework for the selective hydrogenation of phenol to cyclohexanone. Mol Catal A: Chem 410:209–220
99. Li AQ, Shen K, Chen JY, Li Z, Li YW (2017) Highly selective hydrogenation of phenol to cyclohexanol over MOF-derived non-noble Co-Ni@NC catalysts. Chem Eng Sci 166:66–76
100. Zhang DM, Guan YJ, Hensen EJM, Xue T, Wang YM (2014) Tuning the hydrogenation activity of Pd NPs on Al–MIL-53 by linker modification, Catal. Sci Technol 4:795–802
101. Ertas IE, Gulcan M, Bulut A, Yurderi M, Zahmakiran M (2016) Metal-organic framework (MIL-101) stabilized ruthenium nanoparticles: highly efficient catalytic material in the phenol hydrogenation. Microporous Mesoporous Mater 226:94–103
102. Chen H, He Y, Pfefferle LD, Pu W, Wu Y, Qi S (2018) Phenol catalytic hydrogenation over palladium nanoparticles supported on metal-organic frameworks in aqueous phase. ChemCatChem 10:2558–2570

103. Fridman VZ, Davydov AA (2000) Dehydrogenation of cyclohexanol on copper-containing catalysts: I. the influence of the oxidation state of copper on the activity of copper sites. J Catal 195:20–30
104. Chaudhari RV, Jaganathan R, Kolhe DS, Emig G, Hofmann H (1987) Kinetic modelling of hydrogenation of butynediol using 0.2% Pd/C catalyst in a slurry reactor. Appl Catal 29:141–159
105. Musser MT (2011) Cyclohexanol and cyclohexanone. Ullmann's Encyclopedia of Industrial Chemistry
106. Miyaura N, Suzuki A (1995) Palladium-catalyzed cross-coupling reactions of organoboron compounds. Chem Rev 95:2457–2483
107. Wolfe JP, Singer RA, Yang BH, Buchwald SL (1999) Highly active palladium catalysts for suzuki coupling reactions. J Am Chem Soc 121:9550–9561
108. Evangelisti C, Panziera N, D'Alessio A, Bertinetti L, Botavina M, Vitulli G (2010) New monodispersed palladium nanoparticles stabilized by poly-(N-vinyl-2-pyrrolidone): preparation, structural study and catalytic properties. J Catal 272:246–252
109. Han J, Kim S, Lee MS, Kim M, Jeong N (2020) Development of heterogeneous enantioselective catalysts using chiral metal-organic frameworks (MOFs). J Vis Exp 155
110. Han J, Lee MS, Thallapally PK, Kim M, Jeong N (2019) Identification of reaction sites on metal-organic framework-based asymmetric catalysts for carbonyl-ene reaction. ACS Catal 9:3969–3977
111. Mondal J, Modak A, Dutta A, Basu S, Jha SN, Bhattacharyya D, Bhaumik A (2012) One-pot thioetherification of aryl halides with thiourea and benzyl bromide in water catalyzed by Cu-grafted furfural imine-functionalized mesoporous SBA-15. Chem Commun 48:8000–8002
112. Islam SM, Mondal P, Roy AS, Mondal S, Hossain D (2010) Heterogeneous Suzuki and copper-free Sonogashira cross-coupling reactions catalyzed by a reusable palladium(II) complex in water medium. Tetrahedron Lett 51:2067–2070
113. Biffis A, Zecca M, Basato M, Eur J (2001) Metallic palladium in the Heck reaction: active catalyst or convenient precursor? Inorg Chem 2001:1131–1133
114. Kardanpour R, Tangestaninejad S, Mirkhani V, Moghadam M, Mohammadpoor-Baltork I, Khosropour AR, Zadehahmadi FJ (2014) Highly dispersed palladium nanoparticles supported on amino functionalized metal-organic frameworks as an efficient and reusable catalyst for Suzuki cross-coupling reaction. Organomet Chem 761:127–133
115. Sun W, Gao L, Sun X, Zheng G (2018) A novel route with a Cu(ii)-MOF-derived structure to synthesize Cu/Cu_2O NPs@graphene: the electron transfer leads to the synergistic effect of the Cu(0)–Cu(I) phase for an effective catalysis of the Sonogashira cross-coupling reactions. Dalton Trans 47:5538–5541
116. Hillier AC, Grasa GA, Viciu MS, Lee HM, Yang C, Nolan SP (2002) Catalytic cross-coupling reactions mediated by palladium/nucleophilic carbene systems. J Organomet Chem 653:69–82
117. Zhang MM, Guan JC, Zhang BS, Su DS, Williams CT, Liang CH (2012) Chemical vapor deposition of Pd(C3H5)(C5H5) to synthesize Pd@MOF-5 catalysts for suzuki coupling reaction. Catal Lett 142:313–318
118. Yousaf M, Zahoor AF, Akhtar R, Ahmad M, Naheed S (2020) Development of green methodologies for Heck, Chan-Lam, Stille and Suzuki cross-coupling reactions. Mol Divers 24:821–839
119. Nguyen-Sorenson AHT, Anderson CM, Balijepalli SK, McDonald KA, Matzger AJ, Stowers KJ (2019) Highly active copper catalyst obtained through rapid MOF decomposition. Inorg Chem Front 6:521–526

120. Shang NZ, Gao ST, Zhou X, Feng C, Wang Z, Wang C (2014) Palladium nanoparticles encapsulated inside the pores of a metal–organic framework as a highly active catalyst for carbon–carbon cross-coupling. RSC Adv 4:54487–54493
121. Tang H, Yang M, Lia X, Zhoua ML, Bao YS, Cui XY, Zhao K, Zhang YY, Han ZB (2021) Synthesis of biaryl compounds via Suzuki homocoupling reactions catalyzed by metal organic frameworks encapsulated with palladium nanoparticles. Inorg Chem Commun 123:108368
122. Han J, Kim S, Lee MS, Kim M, Jeong Na (2020) Development of heterogeneous enantioselective catalysts using chiral Metal-Organic Frameworks (MOFs). J Vis Exp 155

Recent Progress in the Synthesis and Electrocatalytic Application of Metal–Organic Frameworks Encapsulated Nanoparticle Composites

Naveen Goyal and Rajeev Kumar Rai

Contents

1 Introduction 732
2 Challenges Associated with MOFs for Catalysis 735
2.1 Synthetic Challenge 735
2.2 Chemical and Thermal Stability 736
2.3 Electrical Conductivity 737
3 Synthesis of NP Encapsulated MOFs 737
3.1 Impregnation Approach 738
3.2 Encapsulation Approach 739
3.3 One-Pot Synthetic Approach 741
4 Reaction Mechanism of ORR, OER, and HER 741
4.1 Mechanism of Oxygen Reduction Reaction (ORR) 742
4.2 Mechanism of Oxygen Evolution Reaction (OER) 743
4.3 Mechanism of Hydrogen Evolution Reaction (HER) 744
5 Parameters to Evaluate the Electrochemical Performances 744
5.1 Parameters to Evaluate HER/OER Performance 744
5.2 Parameters to Evaluate ORR Performance 746
6 MOF and MOF-Based Electrocatalysts 747
6.1 MOF as Electrocatalyst 747
6.2 MOF Encapsulated Nanostructures for Electrocatalysis 749
6.3 MOF-Derived Electrocatalysts for ORR, OER, and HER 751
7 Summary and Future Perspectives 758
Abbreviations 759
References 760

Abstract Owing to uniformly distributed pores with tunable sizes and ultrahigh surface area, metal–organic frameworks (MOFs) have gained enormous interest in the rational design of electrocatalysts. However, the semiconducting or insulating nature of MOFs limits their electrocatalytic properties. Considerable efforts have been attempted to improve the catalytic properties of MOFs such as increasing the conductivity of MOFs, formation of composites, or using MOFs as the template

N. Goyal · R. K. Rai (✉)
Materials Research Centre, Indian Institute of Science, Bangalore 560012, India
e-mail: rajeevrai@iisc.ac.in

N. Goyal
e-mail: naveengoyal@iisc.ac.in

S. Gulati (ed.), *Metal-Organic Frameworks (MOFs) as Catalysts*,
https://doi.org/10.1007/978-981-16-7959-9_27

for MOF-derived nanostructures. MOFs also provide an opportunity to encapsulate metal and semiconductors nanoparticles (NPs), which show enhanced electrocatalytic activity compared to individual components because of the synergistic effect. Uniform and large pores in MOFs promote the facile mass transfer, diffusion of the redox-active species, and further reduce the aggregation of metal nanoparticles and hence, enhance their structural and catalytic stability.

This chapter presents a review of the recent progress in MOFs, MOF encapsulated catalytically active NPs and MOF-derived nanostructures for electrochemical applications. To begin with, some key challenges related to MOFs for electrocatalysis are presented, followed by the state-of-the-art advances in the synthesis of MOF encapsulated NPs. Electrochemical performances of MOFs are discussed. Further, owing to the synergistic effect and utilizing accessible metal sites of MOFs along with encapsulated NPs for enhancement in electrocatalytic activity is discussed. MOFs acting as sacrificial templates to synthesize various carbon-based electrocatalysts are also discussed. In last, a summary and future perspective of MOFs and MOF-based electrocatalysts are proposed.

Keywords Metal–organic frameworks · Encapsulated nanoparticle · Hydrogen evolution reaction (HER) · Oxygen reduction reaction (ORR) · Oxygen evolution reaction (OER)

1 Introduction

Considering the growing population and technological development, Prof. Smalley has pointed out the ongoing energy demand as one of the major challenges for the upcoming years and named it as "terawatt challenge" [1, 2]. He estimated that by 2050, the world would require 60 terawatts of energy to comfortably support the world's population, which is equivalent to burning 10 billion barrels of oil per day. To supply so much energy, we need a "new oil" in the form of a renewable and clean energy source. In this regard, electrochemical energy devices, including water-splitting [3, 4], fuel cells [5, 6], and rechargeable metal-air batteries [7], have been recognized as potential future energy sources due to their ultrahigh energy density and environment-friendly character. These electrochemical devices require coupled electrochemical half-reactions, namely hydrogen evolution reaction (HER), oxygen reduction reaction (ORR), oxygen evolution reaction (OER), and hydrogen oxidation reaction (HOR). The slow kinetics of these reactions have limited the development of electrochemical energy devices and demands electrocatalysts to lower the energy barrier and catalyze these reactions. To date, various precious metals such as Pt, Ru, and Ir have remained the "holy grail" for these electrochemical reactions [8, 9]. The usage of scarce and costly noble metals as catalysts in electrochemical energy devices has led to the design of advanced, low-cost, and earth-abundant materials. In this regard, metal–organic frameworks (MOFs) have emerged as promising electrochemical catalysts for HER, OER, and ORR. [10].

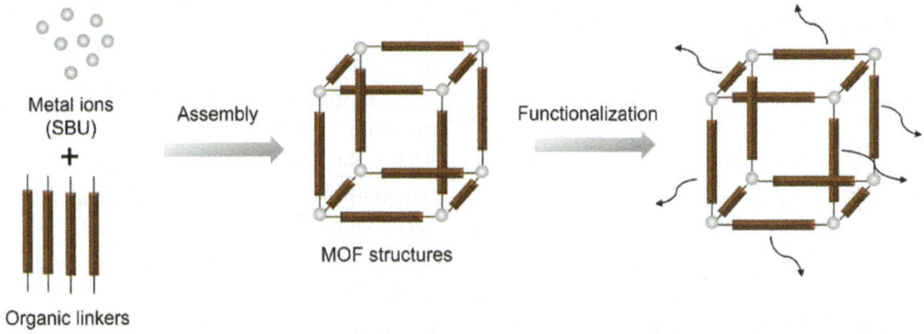

Fig. 1 Representative scheme showing the assembly of metal ions (SBU) and organic linkers to form MOF and further functionalization to modify the properties

Metal–organic frameworks (MOFs) with the modular structure are hybrid organic–inorganic crystalline compounds and consist of periodically arranged inorganic nodes linked by organic linkers (Fig. 1) [11]. The two components of MOFs, namely inorganic nodes, also known as secondary building units (SBU), are metal ions or clusters of transition metals [12], p block metals or lanthanides, and organic linkers such as ditopic or polytopic linkers including carboxylate [12], sulfates [13], nitrogen-donor groups: Azoles or heterocyclic compounds are coordinative bonded and give rise to crystalline structure with porous texture [14]. Coordination number of SBU and nature of organic ligands determine the MOFs geometry. Metals in MOFs can have an octahedral, trigonal prism, square paddlewheel, and trigonal geometries. Since the first discovery of MOF-5 by Yaghi et al. in 1999 [11], various combinations of SBUs and organic linkers provide an opportunity to synthesize MOFs with diverse functionalities. So far, more than 20,000 MOFs have been reported in the literature [15] and classified under different categories such as UIO-based MOFs, MIL, and ZIF-based MOFs.

A combination of inorganic and organic subunits with synergistic effects holds a promising potential for endless possibilities. The coordination of long-chain organic linkers and SBUs results in an open framework with ample storage space, permanent and uniform pores, stable frameworks, and high surface area in contrast to other conventional porous materials [16]. MOFs provide added advantage to the atomically uniform structure of pores. Owing to the ultrahigh surface area (up to 10,000 m^2/g), linker dependent adjustable pore size and shape, the various possible combinations of organic linkers and surface functionalization, MOFs have emerged as interesting materials for heterogeneous catalysis. SBUs introduce the Lewis acidity and undergo the redox reaction during the catalytic processes. Since SBUs are highly ordered and therefore shows similar catalytic activity which is difficult to achieve in conventional catalysts. Tuning the chemical environment around the pores in MOFs to exhibit hydrophobic or hydrophilic properties can further tailor their catalytic properties. The unique, diverse, and tuneable properties of MOFs have been exploited for various applications including gas sensing and storage [17], light-harvesting [18], and drug delivery [19, 20]. Few applications of MOFs are listed in Table 1.

Table 1 Few examples of MOFs with their applications

MOFs	Metal used	Organic linker	Application	Reference
MOF-5	Zn	H_2BDC	Methane storage and separation, electrocatalysis	Eddaoudi et al. [16]
MIL-101	Cr	H_3BTC	Drug delivery, electrocatalysis	Horcajada et al. [20]
IRMOF-3	Zn	H_2BDC	Gas storage and separation	Rowsell et al. [29]
IRMOF-9	Zn	Bpdc	Adsorption and storage	Rowsell et al. [29]
HKUST-1	Cu	H_3BTC	Gas storage and adsorption	Rowsell et al. [29]
MOF-199	Cu	H_3BTC	Antibacterial	Rodrı́guez et al. [19]
MIL-53	Fe	H_2BDC	Photocatalysis, adsorption	Liang et al. [18]
MIL-100	Fe, Cr	Tri/tetra-carboxylate	Photocatalysis, delivery of nitric oxide	Laurier et al. [30]
Co-TDM	Co	H_8TDM	High potent bactericidal activity	Zhuang et al. [31]
MIL-88	Fe	H_2BDC	Photocatalysis	Laurier et al. [30]
MIL-125	Ti	H_2BDC	Photocatalysis, adsorption	George et al. [32]
UiO-66	Zr	H_2BDC	Dye degradation	Pu et al. [33]
ZIF-8	Zn	HMeIm	Photocatalysis, adsorption, gas storage	Jing et al. [34]
UU-100	Co	H_2dcpg	Electrocatalysis	Roy et al. [35]
CTGU-5	Co	Bib	Electrocatalysis	Wu et al. [36]

Where H_2BDC = 1,4-benzenedicarboxylic acid,
H_3BTC = 1,3,5-benzenetricarboxylic acid
Bpdc = 4,4′-biphenyldicarboxylate
H_8TDM = tetrakis [(3,5-dicarboxyphenyl)-oxamethyl] methane
HIm = imidazole
HMeIm = 2-methyl imidazole
H_3BPDC = 4,4′-biphenyldicarboxylic acid
H_2dcpg = diphenyglyoxime-4,4′-dicarboxylic acid
Bib = 1,4-bis(imidazole) butane

Intriguing features of MOFs make them emerge as new generation heterogeneous catalysts for electrochemical reactions [21, 22]. However, the low catalytic activity due to limited active metal sites, low electrical conductivity, low thermal, and chemical stability limits the MOFs to attain their true potential. To further, enhance the catalytic activity and achieve the realistic catalytic property, MOFs can be integrated with different functional materials. MOFs can be integrated with metal nanoparticles (NP), metal oxides, quantum dots of polyoxometalates (POMs) to form a composite nanostructure [23–27]. Combined properties of MOFs and functional materials have been exploited for catalysis with enhanced activity, selectivity, and stability. Fischer et al. have synthesized the first MOF-metal nanoparticle composite in 2005 [28]. They have combined the properties of Pd nanoparticles and MOF-5 to significantly enhance the catalytic properties. The synergistic effect originating from the interaction of metal and organic linkers with the active components enhances its catalytic properties. Owing to the small size, high surface energy NPs tend to agglomerate and thus loses their intrinsic catalytic property. Encapsulation of NP by MOFs further prevent the agglomeration and retain the catalytic activity [23, 24]. Porous structure and possibility of pore's surface functionalization in MOF further facilitate the diffusion of the reactant and product from the electrochemical active sites (NP). MOFs can also act as a sacrificial template to form carbon-based MOF-derived catalysts.

Considering the recent growing interest in the MOFs and MOF-NP composites for catalytic applications, a summary of current state-of-the-art catalysts is presented in this chapter. To begin with, some key challenges associated with MOF are discussed followed by various synthesis methods for the preparation of MOF-NP. Next, we have discussed the electrochemical mechanism of ORR, OER, and HER followed by parameters to evaluate the performance of electrocatalysts. Furthermore, we have summarized the catalytic performances of various MOFs, MOF-NP composite, and MOFs-derived nanostructures in the field of electrocatalysis. At last, we present the summary and future perspective of MOF-based electrocatalysts.

2 Challenges Associated with MOFs for Catalysis

2.1 Synthetic Challenge

Understanding the structural and compositional complexity of MOFs to design novel frameworks with the distinct and desired property is one of the compelling areas of research in MOFs. Controlling the MOFs' pore size, shape, and volume is the critical parameters for highly shape-selective catalysis. Such shape-controlled catalysis is well explored for zeolites; however, such reports are not available for MOFs. Furthermore, the ability to encapsulate the condensed phase nanostructures in the MOFs provide an opportunity for novel physical property investigation. Owing to the various possibility of organic linkers, the distance, and angle between metal nodes and linkers, and thus, subsequent control over the MOF topology is unparallel and should be investigated for energy-related applications. An added advantage of MOFs to control the internal surface chemistry of pores through the incorporation of organic and inorganic functional groups along with tuneable topology and composition should be synthesized and exploited for various applications. Figure 2 summarizes the possible modification that can be done during the synthesis of MOFs.

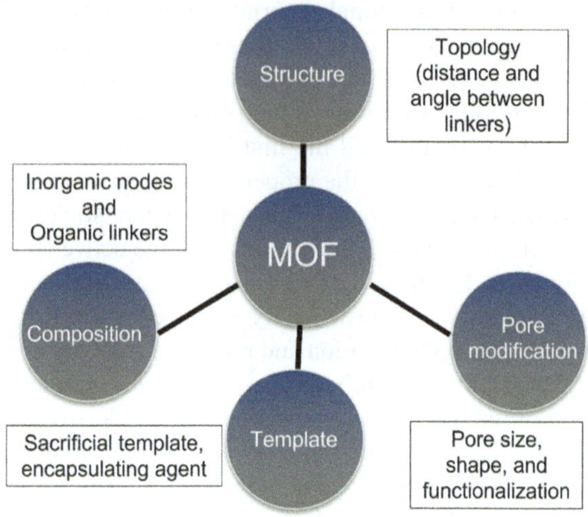

Fig. 2 Summary of various tuneable parameters of MOFs

2.2 *Chemical and Thermal Stability*

MOFs' thermal and chemical stability possess a major roadblock for their application in catalysis. MOFs tend to degrade under harsh reaction conditions such as polar solvents or high temperatures. In MOFs, metal ions are weakly coordinatively bonded with organic linkers, which get disrupted in polar solvents, especially in water [11]. The stronger polar-polar interaction of solvents with SBUs tends to destroy the coordination bond between the metal node and organic linkers. Another issue related to the structural stability of MOFs is the leaching of metal ions (active sites) under polarization in an acidic medium leading to the loss the catalytic activity. Similarly, MOFs are stable up to 150–350 °C and can withstand their frameworks with any structural changes. However, the prolonged exposure to the high temperature can decompose the MOFs into metal oxide and carbon. Xia et al. have illustrated cobalt imidazolate framework CIF as ORR and reported that prolong exposure of CIF to the high temperature leads to the formation of amorphous carbon [37].

Various approaches have been explored to increase the chemical and thermal stability of MOFs. One such approach is using the hard-soft acid–base (HSAB) principle to strengthen the bonding between metal and organic linkers [38]. Highly acidic metals can readily react with highly basic organic linkers to form a robust metal–organic bond [39, 40]. Another approach is to shield the metal and organic linkers bond by functionalization. Functionalization of the bond can decrease the interaction of the exogenous molecules with the metal–organic linker bonds.

2.3 Electrical Conductivity

The uniform and permanent porous structure of MOFs are ideal candidates for electrochemical redox reactions with fast diffusion kinetics of electron and redox species. However, their low electrical conductivity due to poor overlap between orbitals of metal ions and linkers possess a major challenge for their application. The charge transfer and subsequent electrical conductivity in MOFs are best explained in terms of the redox hoping mechanism which depend on the separation of active sites and density of states between them.

To date, various approaches have been used to increase electrical conductivity. The creation of defects by using mixed oxidation state metal ions has been shown to increase the electrical conductivity of MOFs. Hole formation by incorporating Fe^{3+} in place of Fe^{2+} in Fe-based MOFs has shown to increase the electrical conductivity by five times and decreases the charge activation energy [41]. The increase in electrical conductivity can be attributed to the hole delocalization due to the $Fe^{2+/3+}$ mixed-valence state. Similarly, the formation of organic holes has also been shown to increase the electrical conductivity in tetrathiafulvalene-tetrabenzoate-based MOFs [42]. Second approach to increase the electrical conductivity is to increase the interaction between the linkers and metals ions. For example, Sulfur-based bridging ligands have been shown to increase the conductivity by strongly interacting with metal ions. Owing to effective overlap between the metal ions and ligands, Cu-HITP-based MOF has shown high electrical conductivity up to 20 S m^{-1} [43].

Another strategy used to increase the functionality of MOFs is the functionalization of pores by incorporating redox-active species. Incorporation of conjugated and redox-active molecule 7,7,8,8-tetracyanoquinodimethane in HKUST-1 has been shown to increase the electrical conductivity by 6 times [44]. Along with the electrical conductivity, another factor that needs to be considered is ionic conductivity through the pores. A less packed MOFs provide better conduction of ions during the electrochemical processes which can be further increased by introducing a network of functional groups. Umeyama et al. have shown an intrinsic proton conductor using a network like structure formation in Zn-based MOF due to hydrogen bonding between the layers. [45].

Unique, diverse, and tuneable property of MOFs provide an opportunity to design MOFs with different functionalities. The properties can be further tailored by incorporating nanoparticles in MOFs due to synergistic interaction between the metals and organic linkers have been widely explored for electrocatalysis.

3 Synthesis of NP Encapsulated MOFs

Synthesis methods of MOF encapsulated nanoparticles can be broadly classified into three categories as shown in Fig. 3:

1. Impregnation approach

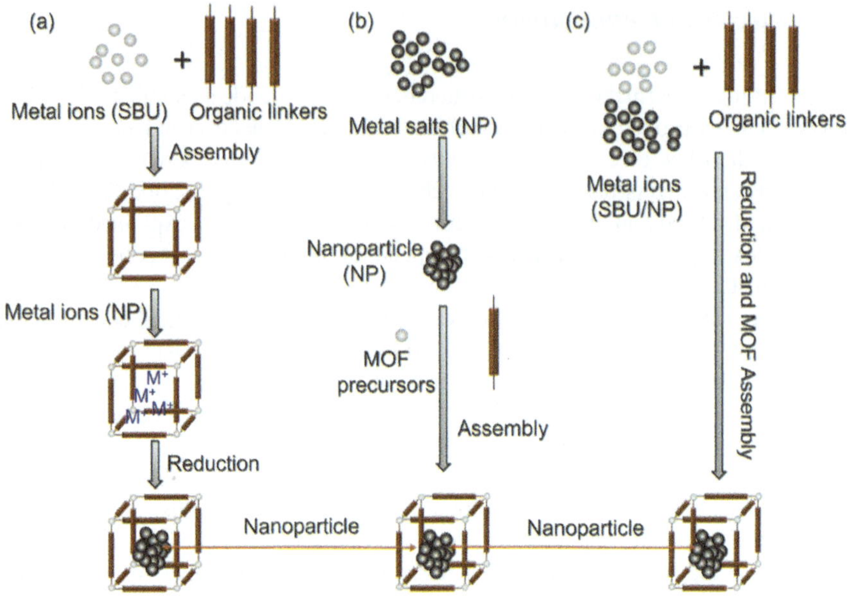

Fig. 3 Reaction schemes for the synthesis of MOF-NP composites **a** impregnation approach, **b** encapsulation approach, and **c** one-pot synthetic approach

2. Encapsulation approach
3. One-pot synthetic approach

3.1 Impregnation Approach

The impregnation approach also refers to the "ship in a bottle" approach, where ship refers to metal or metal oxide nanoparticles while bottle refers to MOFs [46]. This approach includes three steps as shown in Fig. 3a. The first step is a synthesis of MOFs which includes heating of metal salt and organic linkers in suited solvent via either solvothermal or microwave techniques. The second step includes the incorporation of metal precursors into the synthesized MOFs. For incorporation of the metal precursor into the pores/cavity of MOFs, the double solvent approach is mostly used. The double solvent approach prevents the agglomeration of NPs on the external surface of MOFs. The third step is the reduction of metal precursors into the channels of MOFs to produce metal nanoparticles inside the pores of MOFs. $NaBH_4$ is mostly used as a reducing agent.

Xu et al. synthesized the first MOF encapsulated nanoparticle composite by this approach. They formed Pt@MIL-101 (Cr) by using H_2PtCl_6 solution as a metal precursor and a double solvent approach to infilter Pt precursor into the cavity of

MIL-101 (Cr) [47]. Similar method has been used to synthesize Pd@NH_2-UiO-66 (Zr) by Sun et al. by using mild reduction condition [48].

This method can be employed to synthesize alloy and core–shell nanoparticles inside MOFs. Xu et al. synthesized alloy nanoparticles of AuNi inside the pores of MIL-101 (Cr) [49]. $HAuCl_4$ was used as Au precursor, and $NiCl_2$ was taken as Ni precursor and co-reduced using $NaBH_4$ as a reducing agent. They observed that some of the alloy nanoparticles agglomerated on the external surface of MOFs. To overcome this issue, they increased the concentration of reducing agent $NaBH_4$ which led to the generation of small-sized nanoparticles and found that the external surface of MOFs become completely free of nanoparticles. Jiang et al. also observed the effect of reducing agent and synthesized core–shell nanoparticles of PdCo inside the voids of MIL-101 (Cr) by using mild reducing agent NH_3BH_3 while alloy nanoparticles of PdCo have been formed inside the voids of MIL-101 (Cr) by using strong reducing agent $NaBH_4$ [50].

A similar approach has been used to encapsulate metal oxide nanoparticles and a mixture of metal nanoparticles and metal oxide nanoparticles into the cavity of MOFs. Han et al. synthesized CoOx/MIL-101 (Cr) composite using a double solvent approach [51]. They used $Co(NO_3)_2$ solution as Co precursor and added to the synthesized MOFs MIL-101 (Cr) and dispersed in hexane using ultrasonication. After that, they calcined the whole mixture to form CoOx/MIL-101 (Cr). An et al. synthesized Cu/ZnO nanoparticles inside the MOFs. Cu and ZnO nanoparticles were produced inside the pores of MOFs by step wise reduction method using H_2 as reducing agent [52].

The impregnation approach was used to synthesize a huge number of metal nanoparticle-MOF composites. However, control of the size and morphology of NPs in the pore of MOFs is still a challenge. Also, this method has one more disadvantage in terms of aggregation of NPs on the external surface of MOFs.

3.2 *Encapsulation Approach*

The encapsulation approach is also referred to as "bottle around the ship." This approach consists of two steps as shown in Fig. 3b. The first step is the formation of metal nanostructures with controlled morphology by reducing suitable metal salt using a reducing agent. The second step involves the growth of MOFs around the synthesized nanostructures. This method has an advantage over the impregnation method since morphology-controlled nanostructures could be synthesized by selecting proper capping agents and reaction conditions. For example, shape-controlled gold nanomaterials could be synthesized as nanoparticles, nanowires, nanorods, and nanocubes. Once morphology-controlled nanostructure has been synthesized, the only step that remains an encapsulation of these nanomaterials into the pores of MOFs.

Kitagawa et al. synthesized Pd nanocubes surrounded by HKUST-1 (Cu) [53]. Pd nanocubes were synthesized using Na_2PdCl_4 as Pd precursor, L-ascorbic acid as a mild reducing agent, bromide ions as capping agent, and water as solvent. Synthesized Pd nanocubes were added into the solution containing $Cu(NO_3)_2$ and trimesic acid and stirred for 48 h at room temperature to form Pd@HKUST-1 (Cu). However, they do not found uniformity of Pd nanocubes in the product.

An alternate approach was developed by Tang et al. by introducing a layer of surfactant in between the MOFs and nanostructures to form uniform core–shell nanocomposites. They successfully synthesized Pd@IRMOF-3 (Zn) core–shell nanocomposite [54]. PVP was used as a stabilizing agent to stabilize Pd nanoparticles also was used as a bridging layer in between Pd nanoparticles and IRMOF-3 (Zn) to form uniform nanocomposites. A similar research group also synthesized nanowires surrounded by MOFs. First, they synthesized PVP-capped silver nanowires and then added these synthesized Ag nanowires to the reaction mixture containing $Zn(NO_3)_2{\cdot}6H_2O$ and 2-methylimidazolate to form Ag@ZIF-8 (Zn) core–shell nanocomposites. In this case also, PVP act as a bridging layer between nanowires and MOFs [55].

Not only organic molecules like PVP but ionic surfactants can also be used as a bridging ligand. Tsung et al. synthesized Pd@ZIF-8 (Zn) core–shell nanocomposites using cetyltrimethylammonium bromide (CTAB) as a bridging ligand [56]. First, they synthesized Pd nanocubes capped by CTAB, because of the presence of a hydrophobic chain in CTAB, it interacted with a non-polar group of organic ligands of ZIF-8 and facilitated control alignment of MOF around Pd nanocubes. The various research group also tried to incorporate more than one type of nanoparticle into the pores of MOFs. One notable example is shown by Zhang et al. by synthesizing Au/Pt@ZIF-8 (Zn) core–shell nanocomposites.

Another interesting approach to synthesize MOF encapsulated nanoparticles is to sandwich the nanoparticle layer in between the layers of MOFs. Tang et al. synthesized Pt nanoparticles sandwiched inside the MIL-101 (Fe) [57]. Synthesis of such nanocomposite includes the synthesis of MIL-101 (Fe) core first, followed by adsorption of already synthesized Pt nanoparticles and then the growth of MIL-101(Fe) shell. In this case also, PVP act as a bridging ligand to sandwich Pt nanoparticles inside the MOFs.

The introduction of a bridging ligand sometimes diminishes the catalytic activity of core–shell nanocomposite. Therefore, various research groups tried to focus on an alternative approach to synthesize core–shell nanocomposite by using a hard template that could be removed after the whole synthesis procedure. Su et al. synthesized yolk-shell nanocomposite of Au@ZIF-8 (Zn) by hard template method. In this method, SiO_2 was used as a hard template [58]. The whole synthesis procedure includes four steps: The first step is a synthesis of morphology control nanoparticles. The second step is a coating of as-synthesized nanostructure with a hard template which is SiO_2 to form Au@SiO_2 core–shell nanostructure. The third step is the introduction of the core–shell of Au@SiO_2 to the reaction mixture of ZIF-8 (Zn) to synthesize Au@SiO_2@ZIF-8 (Zn). Forth step is the removal of the hard template. NaOH solution is used to etch out layers of SiO_2 to form a yolk-shell nanocomposite of Au@ZIF-8 (Zn). However, due to the presence of a limited number of hard templates, more synthesis strategies are required to synthesize uniform core–shell nanocomposite containing MOF encapsulated nanoparticles.

3.3 One-Pot Synthetic Approach

In the impregnation and encapsulation approach described above, two or more two steps are required to synthesize MOF-nanoparticle composites. It is always desirable to reduce the timing and number of steps of the reaction for the facile synthesis. Recently, research groups are more focused on synthesizing MOF-nanoparticle composites in a single step as shown in Fig. 3c, but the challenge is to balance the agglomeration of nanoparticles with nucleation and growth timing of core of nanoparticle and shell of MOFs.

Tang et al. synthesized the first core–shell nanoparticle-MOFs composite via one-pot reaction scheme. They synthesized Au core and MOF-5 (Zn) shell composite by mixing all the required precursors such as $HAuCl_4$, zinc nitrate, and benzene-1,2-dicarboxylic acid organic ligand in the reaction mixture along with solvent ethanol and capping agent PVP [59]. Capping agent plays a crucial role in synthesizing core–shell composites. In this case, the growth of MOF-5 takes place on the PVP-capped Au nanoparticles. A similar method has been used to synthesize Pd-MOF core–shell nanocomposites by Li et al. They used $Pd(NO_3)_2$ as Pd precursor, $ZrCl_4$, organic linker as 2,2′-bipyridine-5,5′-dicarboxylic acid, reducing agent as NH_3BH_3 in DMF solvent. They obtained uniform 3.2 nm Pd nanoparticles surrounded by UiO-67 (Zr) [60].

Although a one-pot reaction scheme is a less time consuming and easy way to synthesize core–shell nanocomposites but not many reports are there in the literature by this method. This is because of difficultly in tuning the morphology and size of nanoparticles along with the growth of MOFs over the nanoparticle. Therefore, it is need of the hour to develop simple, facile one-pot chemical methods to synthesize these composites.

4 Reaction Mechanism of ORR, OER, and HER

HER/OER/ORR are the major electrochemical processes being used in water-splitting, fuel cells, and metal-air batteries as shown in Fig. 4. These electrochemical processes involve multiple steps with intermediates being adsorbed on the active sites of the electrocatalysts. Thus, the electrocatalytic activity of catalysts greatly depends on the interaction of the active sites with the intermediates, which is measured as the adsorption energy of various intermediates on the redox-active sites of the electrocatalysts. The interaction depends on the electronic structure of the active sites, which in turn can be manipulated through the structural and compositional variation in the electrocatalyst. These electrochemical processes can have a different mechanism

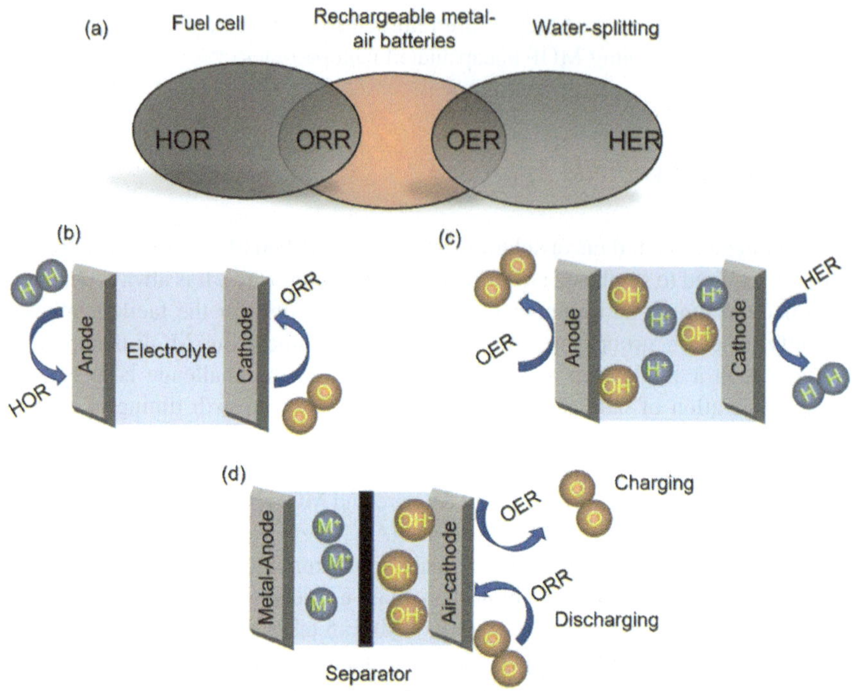

Fig. 4 **a** Representative schematic showing the various electrochemical processes involved in fuel cells, rechargeable batteries, and water-splitting; **b** electrochemical process in fuel cells; **c** electrochemical process in water-splitting; and **d** electrochemical process in rechargeable metal-air batteries

and their accepted reaction mechanism in acidic and alkaline mediums is explained below [61].

4.1 Mechanism of Oxygen Reduction Reaction (ORR)

ORR is one of the fundamental reactions in fuel cells and metal-air batteries where O_2 is being reduced to form H_2O or OH^- and require an equilibrium potential of 1.23 V with respect to RHE. However, the other kinetic barriers increase the required equilibrium potential. The most acceptable ORR mechanism with high efficiency in acid and alkaline medium is shown below:

Acidic medium:

$$\text{Overall reaction}: O_2 + 4H^+ + 4e^- \rightarrow H_2O$$

$$\text{Steps}: O_2 + 2H^+ + 2e^- \rightarrow H_2O_2$$

$$H_2O_2 + 2H^+ + 2e^- \rightarrow 2H_2O$$

Alkaline medium:

$$\text{Overall reaction}: O_2 + 2H_2O + 4e^- \rightarrow 4OH^-$$

$$\text{Steps}: O_2 + 2H_2O + 2e^- \rightarrow OOH^- + OH^-$$

$$OOH^- + H_2O + 2e^- \rightarrow 3OH^-$$

4.2 *Mechanism of Oxygen Evolution Reaction (OER)*

OER is the reverse of ORR where O_2 gas is evolved from oxidation of OH^- or H_2O. It is another crucial fundamental reaction in fuel cells and metal-air batteries. ORR and OER both are generally $4e^-$ processes and have various intermediate steps where oxygenated species undergo various transitions. The most acceptable OER mechanism in acid and alkaline medium is shown below.

Acidic medium:

Overall reaction: $2H_2O \rightarrow O_2 + 4H^+ + 4e^-$
Steps: $\theta + H_2O \rightarrow OH^\theta + H^+ + e^-$
$OH^\theta \rightarrow O^\theta + H^+ + e^-$
$O^\theta + H_2O \rightarrow OOH^\theta + H^+ + e^-$
$OOH^\theta \rightarrow O^\theta{}_2 + H^+ + e^-$
$O^\theta{}_2 \rightarrow \theta + O_2$

Alkaline medium:

Overall reaction: $4OH^- \rightarrow O_2 + 2H_2O + 4e^-$
Steps: $\theta + OH^- \rightarrow OH^\theta + e^-$
$OH^- + OH^\theta \rightarrow H_2O + O^\theta + e^-$
$O^\theta + OH^- \rightarrow OOH^\theta + e^-$
$OOH + OH^- \rightarrow O^\theta{}_2 + e^-$
$O^\theta{}_2 \rightarrow \theta + O_2$

θ is used to show the active site of the catalyst.

4.3 Mechanism of Hydrogen Evolution Reaction (HER)

Hydrogen is a promising environment-friendly energy source and possesses high energy density. Hydrogen gas is one of the feeds in fuel cells and can be produced by the electrolysis of water. The reaction mechanism of HER in acidic and alkaline mediums is presented below.

Acidic medium:

$$\text{Overall equation}: 2H^+ + 2e^- \rightarrow H_2$$

- Volmer step: $\theta + H^+ + e^- \rightarrow H^{\theta}$
- Tafel step: $H^{\theta} + H^{\theta} \rightarrow H_2$
- Heyrovsky step: $H^+ + e^- + H^{\theta} \rightarrow H_2$

Alkaline medium:

$$\text{Overall equation}: 2H_2O + 2e^- \rightarrow H_2 + 2OH^-$$

- Volmer step: $H_2O + e^- \rightarrow H^{\theta}+OH^-$
- Tafel step: $H^{\theta} + H^{\theta} \rightarrow H_2$
- Heyrovsky step: $H_2O+ e^- + H^{\theta} \rightarrow H_2 + OH^-$

θ is used to show the active site of the catalyst.

Unlike OER and ORR, hydrogen is the adsorption species in the intermediate steps of HER and the equilibrium potential for HER is 0 V. For a good electrocatalyst, the adsorption energy of H (ΔG_{H*}) should be 0 eV, i.e., hydrogen should not be adsorbing too strongly or too weakly. Any steps in the HER mechanism can be the rate-determining steps, and Tafel slope values can predict the rate-determining step.

5 Parameters to Evaluate the Electrochemical Performances

5.1 Parameters to Evaluate HER/OER Performance

The performance of any catalyst for HER/OER is evaluated by various parameters as shown in Fig. 5.

Parameters are briefly described as follows

1. **Overpotential**: Overpotential is the foremost important parameter for evaluating the performance of an electrocatalyst. It is defined as the difference between the values of applied potential (E) and the equilibrium potential (E_{eq}). The equilibrium potential of any electrochemical reaction is thermodynamically

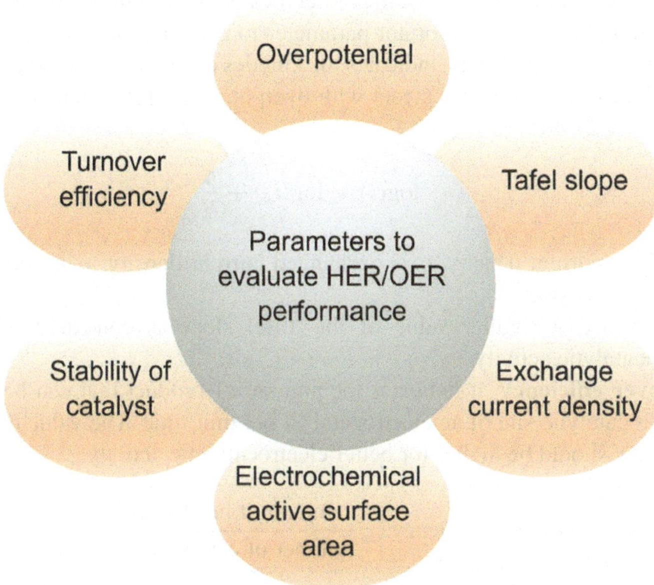

Fig. 5 Different parameters to evaluate the HER and OER performance of electrocatalysts

determined potential. Overpotential is denoted by "η." In general, the lower the value of overpotential, the better will be the performance of that material as electrocatalysts for HER/OER.

$$\eta = E - E_{\text{eq}}$$

2. **Electrochemical active surface area (ECSA)**: It is defined as the electrochemically active area available on the surface of any catalyst, which eventually indicates the availability of active sites on the catalyst surface where adsorption of species could take place. The higher value of ECSA is suggested for better electrocatalyst.
3. **Exchange current density**: Exchange current density (i_o) is defined as the exchange current (j_o) per unit area of the electrode (A). Its magnitude is used to describe charge transfer interactions between reactant and the used electrocatalyst.

$$i_o = \frac{j_o}{A}$$

Generally, the high value of i_o is desirable for better electro catalytical activity. This is evident from the fact that Pt has a higher value of i_o and is a superior HER electrocatalyst.

4. **Tafel slope**: Tafel slope is used to predict the mechanism of the electrochemical reaction and hence an important parameter to evaluate the performance of an electrocatalyst. Tafel slope indicates the kinetics of the reaction and predicts the increment in the value of current with overpotential. Tafel equation is written as

$$\log(i) = \log(i_o) + \frac{\eta}{b}$$

where i is current density, i_o is exchanged current density, η is overpotential, and b is Tafel slope.

   Generally, a smaller value of the Tafel slope is considered for better electrocatalytic activity.

5. **Turnover efficiency**: It indicates the amount of product that can be obtained per unit catalytic site of an electrocatalyst per unit time. The value of turnover efficiency should be higher for better electrocatalytic activity.

$$\text{Turnover efficiency} = \frac{\text{amount of product molecules}}{\text{number of catalytic site} \times \text{time}}$$

6. **Stability of catalyst**: Stability of catalyst is another foremost important parameter since it determines the usability of that catalyst on large scale. The catalyst must be stable and should not change its morphology and property after test cycles.

5.2 *Parameters to Evaluate ORR Performance*

The performance of an electrocatalyst used for ORR is evaluated by various parameters as described below

1. **Mass activity**: For any material to act as a catalyst for ORR should have a high value of the mass activity. Mass activity is described as the product of kinetic current density (i_k) to the geometric area of the electrode (A) being used divide by the mass of catalyst (m) loaded on the electrode.

$$\text{Mass activity} = i_k \times \frac{A}{m}$$

2. **Specific activity**: Specific activity is another important parameter to evaluate the performance of ORR. The specific activity represents how many active sites available on the catalyst surface. More will be the number of active sites, the better will be the performance of an electrocatalyst because of the increase in electrochemical active surface area.

3. **Onset and half-wave potential**: Both onset potentials and half-wave potential are used to determine the properties of catalyst used for ORR. Both parameters were determined by performing linear sweep voltammetry (LSV).

6 MOF and MOF-Based Electrocatalysts

The advantage of MOF having well-separated metal nodes, large pore channels, and ultrahigh surface area has been exploited for electrocatalysis. However, as previously mentioned MOFs have low electronic conductivity, hence various strategies have been explored to increase the conductivity, stability and subsequently enhance the electrochemical activity. The following sections are going to discuss the advancement in MOFs, MOF-NP composite, and MOF-derived electrocatalyst.

6.1 MOF as Electrocatalyst

Pristine MOFs with engineered microstructures including choice of metal centers, ligand modification have been designed for the catalytic application. Examples of MOFs as electrocatalysts for OER, ORR, and HER are presented in Table 2. Tripathy et al. synthesized a Co-MOF ([$Co_4(BTC)_3$ $(BIM)_6$] [solvent]) through a solvothermal method and investigated its electrocatalytic properties for ORR and OER [62]. Similarly following hydrothermal method, a Fe-MOF was prepared on nickel foam was reported by Zhang et al. The MOF was investigated for OER in 1 M KOH solution. The catalyst requires an overpotential of 240 mV and a Tafel slope of 72 mV/dec [63].

Several strategies have been explored to improve the catalytic property of MOFs. Lin et al. performed the theoretical studies on the TM-ZIF-based catalysts for OER and ORR and have revealed that several ligands and coordination numbers affect their catalytic properties, and unsaturated metal ions have better catalytic activity [64]. Further, the emergence of conductive MOFs has boosted electrochemical performances. Huang et al. have used hexaiminohexaazatrinaphthalene (HAHATN) as a conjugated ligand to synthesize bimetallic sited conductive MOF [65]. The synthesized MOF has an extra $M\text{-}N_2$ moiety, and DFT studies have shown that the $M\text{-}N_2$ moiety has a higher degree of unsaturation and thus better catalytic property for HER. It has been also shown that bimetal and multimetal nodes in the MOF can enhance the electrochemical properties. Tuning the electronic states of the reactive species Co by introducing Fe in Co-MOF has been shown to enhance the OER activity. CoFe-MOF required an overpotential of 265 mV to attain $j = 10$ mA/cm^2 and has a Tafel slope of 44 mv/dec [66]. Other multimetallic nodes (Co, Fe, Ni)-MOF based on MIL and MOF-74 have been prepared which show the better electrochemical activity as compared to monometallic-MOF. [67, 68].

Table 2 Summary of various MOFs for electrocatalytic OER, ORR, and HER applications

Catalyst	Application	Condition	Performance		Reference
			Overpotential (mV)	Tafel slope (mV/dec)	
$Ni_3(Ni_3{\bullet}HAHATN)_2$	HER	0.1 M KOH	115	45.6	Huang et al. [65]
CoFe-MOF	OER	1 M KOH	265	44	Zou et al. [66]
2D $Co_6O(dhbdc)_2$	Total water-splitting	0.1 M KOH	211	-	Huang et al. [71]
TMOF-4 nanosheets	OER	1 M KOH	318	54	Song et al. [72]
Co_xFe_{1-x}-MOF-74	OER	1 M KOH	280	56	Zhao et al. [67]
Co-MOF	OER ORR	1 M KOH	280 E_{onset} = 0.85 V	51	Tripathy et al. [62]
Fe-MOF-NF	OER	1 M KOH	240	72	Zhang et al. [63]
Ni-PC-MOF	OER	1 M KOH	<250	-	Jia et al. [73]
$Co(Hpcyz)_4.H_2O$	HER	0.5 M H_2SO_4	E_{onset} = 101	121	Zhao et al. [70]

Redox reaction at the metal center changes the coordination state and thus destruct the MOF framework. Various approaches have been employed to stabilize the MOF in a redox reaction environment. One such approach was shown by Khrizanforova et al., a redox stable bimetallic Co/Ni-MOF was prepared by using ferrocenyl diphosphinate and 4,4′-bipyridine [69]. The introduction of bipyridine in MOF increases the structural stability in organic or in aqueous medium and better activity in comparison with the ferrocenyl diphosphinate-based polymer. Structures of MOFs also play a crucial role in electrocatalytic activity. Various efforts have been done to study the structure–property correlation of MOFs. One such example was reported by Zhao et al. who synthesized a 3D Co-MOF showing eightfold 6^6 interpenetrating networked frameworks. The Co-MOF when tested for HER in 0.5 M H_2SO_4 shows low onset potential and low Tafel slope of 101 mV and 121 mV/dec, respectively [70]. Replacing the metal centers with redox-active polyoxometalate ions have also shown promising electrocatalytic activity.

The dimensionality of MOFs plays a crucial role in regulating the catalytic activity of MOFs. In bulk MOFs, most the electrochemical active sites (SBUs) are well within the structure and thus, not accessible to redox species. Thus, 2D MOFs have been explored with the advantage of rapid mass and electron transfer, higher surface area, making the active sites to be exposed completely, and also acts as a model for structure–property relation. An in situ electrochemical exfoliation of redox-active catechol functionalised ligand-based pillared-layer MOF to 2D MOF was designed for total water-splitting reaction and requires 211 mV overpotential to achieve 10 mA/cm^2 current [71]. Song et al. reported a 2D MOF based on the 1D chain of $[Co_4(OH)_2]^{6+}$ from exfoliation of 3D MOF [72]. The exfoliated 2D MOF shows a robust OER catalyst with an overpotential of 318 mV and a Tafel slope of 54 mV/dec. 2D conductive MOFs with high surface area MOFs can be directly synthesized. Jia and co-workers synthesized a large π-conjugated conductive linker phthalocyanine (Pc)-based Ni-Pc MOF via a solution method and explored for OER [73].

6.2 *MOF Encapsulated Nanostructures for Electrocatalysis*

As mentioned above, the catalytic activity of the MOFs can be improved by incorporating electrochemically active nanoparticles. Small nanoparticles have higher surface energy and coarsen during the catalytic processes. Encapsulation of inorganic nanoparticles in MOF nanocages with strong interaction between them prevent agglomeration and improve catalytic stability. Few examples of MOF encapsulated nanostructures for electrochemical applications are presented in Table 3.

Encapsulated redox-active Keggin type polyoxometalate (POM) ion $[CoW_{12}O_{40}]^{6+}$ in the voids of ZIF-8 was reported by Mukhopadhya and co-worker for water oxidation reaction (OER).[23] The POM@ZIF-8 was in situ synthesized. The choice of Keggin is crucial for the encapsulation of POM. $[CoW_{12}O_{40}]^{6+}$ is small enough to be accommodated in the nanocages of ZIF-8 but large enough to prevent its leaching out from the void. The synthesized POM@ZIF-8 shows excellent stability, even after 1000 catalytic cycles, current values were stable, and exhibit high Faradaic efficiency of 96% in the neutral medium. Controlled electrochemical experiments depict that the formation of CoO_x on the surface of POM-ZIF-8 is the actual active site for the reactions. Similarly, Abdelkader-Fernandez et al. synthesized a Co-based POM $(SiW_9CO_3(H_2O)_3O_{37}]^{10-}$ in ZIF-8 and ZIF-67 nanocages [24]. The synergistic effect originating from the electron transfer between SiW_9CO_3/ZIF-67 results in the activation of POM active sites as well as the creation of the unsaturated metal center in ZIF-67, which was not observed in SiW_9CO_3/ZIF-8 nanostructure.

Along with the incorporation of POMs, metal nanoparticles can also be integrated within the pores of MOF and facilitate electrochemical activity. Zheng et al. reported tiny Pd nanoparticles encapsulated in the pores of electrically conducting MOF-74 [25]. The catalyst was synthesized using the impregnation method, where Pd salt was added to MOF-74-Co followed by reduction. The Pd-MOF-74-Co shows better electrochemical performance for HER and ORR in both acidic and alkaline mediums. Pd-MOF-Co-74-3 (highest loading) shows $E_{onset} = 0.938$ V and half-wave potential of 0.798 V in 0.1 M KOH which was equal to commercial Pt/C. Decreasing the coarsening tendency of Pt quantum dots by enclosing them into the MOF is one of the promising tendencies to exploit the full catalytic efficiency of Pt and reduce the loading of noble metals. Ye et al. designed a one-step synthesis of PtQDs@Fe-MOF

Table 3 MOF encapsulated nanostructures for electrochemical applications

Catalysts	Application	Condition	Performance		Reference
			Potential (V)	Tafel slope (mV/dec)	
POM@ZIF-8	OER	0.1 M Na_2SO_4	–	–	Mukhopadhyay et al. [23]
SiW_9CO_3/ZIF-67	OER	0.1 M KOH	η_{10} = 0.47	113.6	Abdelkader-Fernandez et al. [24]
Pd-MOF-Co-74–3	ORR HER	01 M KOH 0.1 M H_2SO_4	E_{onset} = 0.938 E_{onset} = -0.4	57	Zheng et al. [25]
$Ru_{50}Rh_{50}$-UiO-66-NH_2	HER	0.5 M H_2SO_4 1 M PBS 1 KOH	η_{10} = 0.077 η_{10} = 0.114 η_{10} = 0.177	79 93.4 11.8	Ding et al. [27]
CoO_x-ZIF-67	OER	1 M KOH	E_{10} = 1.548	–	Dou et al. [76]
PtQDs@Fe-MOF	HER OER Total water-splitting	1 M KOH	η_{10} = 0.033 η_{10} = 0.144 E_{10} = 1.47	28.6	Ye et al. [26]
PtCNW/PCN-222	HER	1 M PBS	η_{10} = 0.114	78.6	Xiao et al. [75]
(Ni-Cu@Ni-Cu-MOF-74)	HER	1 M KOH	–	98	Ma et al. [74]
$Pt_{0.5\%}$/Cu-MOF-74	ORR	0.1 M KOH	$E_{1/2}$ = 0.8	–	Anand Prakash [77]
NiSO-BDC	OER	1 M KOH	η_{10} = 0.298	58.6	Wang et al. [78]

on nickel foam [26]. This core@shell structure has been studied for HER, OER, and total water-splitting reactions.

Bimetallic RuRh alloys with various atomic percentages were encapsulated using impregnation method in UiO-66-NH_2 [27]. Equiatomic RuRh alloy ($Ru_{50}Rh_{50}$-UiO-66-NH_2) shows overpotential of 77, 114, and 177 mV to reach a current density of 10 mA/cm^2 in acidic, neutral, and alkaline medium, respectively, which are comparable to Pt/C and Ru/C. In terms of stability, the encapsulated catalyst has outperformed the commercial catalysts. In situ Cu-Ni alloy nanoparticles were encapsulated in Cu-Ni-MOF-74 [74]. Encapsulated nanoparticle (Ni-Cu@Ni-Cu-MOF-74) shows better electrical conductivity and thus better OER performance in 1 M KOH. The catalysts also show better stability in comparison with individual components.

Host and guest nanostructures affect the property of hybrid materials. Xiao et al. investigated the host–guest effect by growing Pt/C nanowires (PtCNW) and Pt/C nanodots (PtCND) into the channel and cages of PCN-22 and PCN-221, respectively [75]. Time-resolve photoluminescence study reveals the faster electron transfer kinetics in the case of nanowires and when tested for HER under light illumination, PtCNW/PCN-222 shows better performance with an overpotential of 115 mV to reach the current density of 10 mA/cm^2.

Dou et al. designed another synthetic strategy to synthesize atomically dispersed catalytic active sites in the MOF and reported a dispersed CoO_x-ZIF-67 by oxygen plasma treatment of ZIF-67 [76]. Oxygen plasma treatment on the MOF breaks the Co–N bond and subsequently react with plasma to form CoO_x. During OER studies, CoO_x-ZIF-67 required a potential of 1.548 V to reach a current density of 10 mA/cm^2 which is comparable to commercial RuO_2. Not only encapsulating the nanomaterials but integration on the surface of MOFs can also enhance the properties. Anand Prakash reported a Pt nanoparticle anchored on Cu-MOF for ORR with different loading of the metal nanoparticle. $Pt_{0.5\%}$/Cu-MOF-74 shows the onset potential of 0.97 V and $E_{1/2} = 0.84$ versus RHE with better long-term stability [77].

Not only metal nanoparticles but even encapsulation of transition metal chalcogenides in MOF have also been shown as cost-effective catalysts with improving catalytic activity. For example, Nickel sulfides show poor catalytic activity toward OER. However, heterogeneous nanostructures show improved activity. Similarly, Wang et al. [78] synthesized Ni-MOF modified Ni_3S_2/NiS for OER application. A two-step method was employed to synthesize the catalyst. First, a hollow Ni_3S_2/NiS nanoparticle was synthesized by sulfurization of Ni nanoparticle and then a thin layer of Ni-MOF was coated on the surface of Ni_3S_2/NiS. The Ni metal was provided from the nanostructure, and terephthalic acid was added as a linker to form the MOF coating. The OER activity was investigated in a 1 M KOH solution and requires an overpotential of 248 mV to reach the current density of 10 mA/cm^2.

The above discussion with examples demonstrates the incorporation of nanostructures into the voids/ pores of MOFs is an effective way to improve their catalytic property. The synergistic effect originating with either charge transfer or conductive support facilitate the catalysis.

6.3 MOF-Derived Electrocatalysts for ORR, OER, and HER

Over the past two decades, pristine MOFs have been extensively explored for heterogeneous catalysis, but the intrinsic challenges of low electrical conductivity and stability under electrocatalytic conditions possess a challenge for their applications. However, MOFs can also act as a sacrificial element to overcome the above-mentioned challenges. They can be converted to graphitic carbon support or N-doped carbon films, or heteroatom doped carbon materials with excellent electrical and ionic conductivity, porous materials having high surface area, exhibit better electrochemical activity for various reactions as shown in Fig. 6. The well-separated

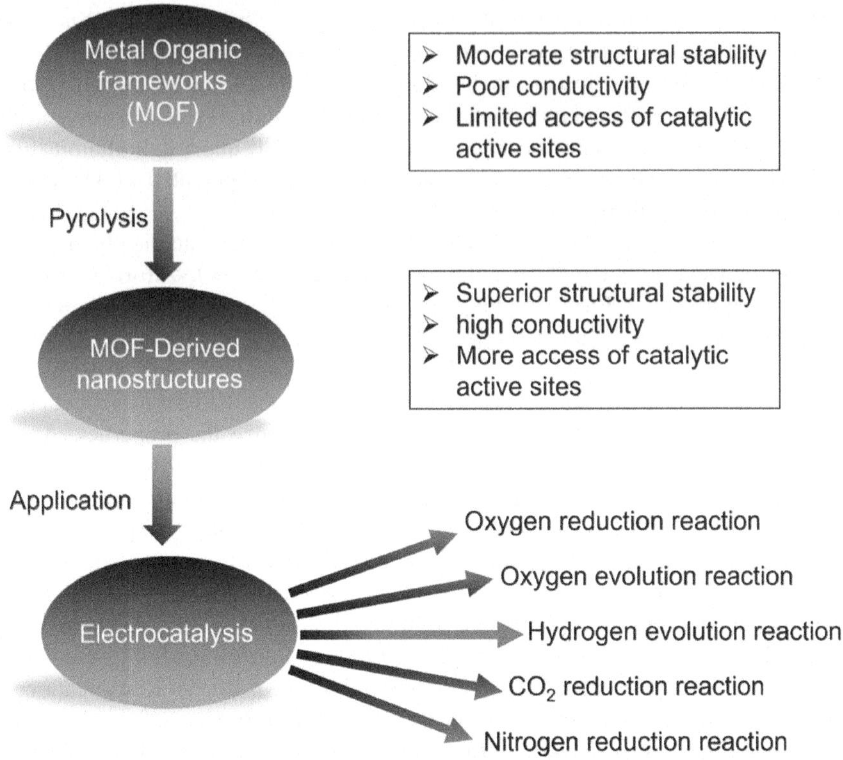

Fig. 6 Representative scheme of MOF to MOF-derived nanostructure along with property comparison and possible electrochemical applications

metal nodes in MOFs can be easily converted to highly dispersed metal clusters over carbon support with enhanced electrocatalytic activity due to uniform distribution of electrochemical active metal centers and better electrical conductivity.

Thermal treatment of MOFs in the different atmospheres is the most common strategy to synthesize metal incorporated MOF-derived catalysts. Calcination of MOFs in the air will convert organic ligands to carbonaceous gases leaving behind the metal nodes as metal oxides. However, pyrolysis in an inert atmosphere such as Ar and N_2 can lead to the formation of metal decorated/incorporated carbon support (graphitic or amorphous carbon or carbon nanotubes). In the following subsections, we will briefly discuss the MOF-derived electrocatalysts for ORR, OER, and HER.

MOF-Derived Electrocatalysts for ORR

Metal-free heteroatom doped MOF-derived carbon matrices are being developed as high-performance and low-cost electrocatalysts for ORR/OER/HER. Carbon matrices with single element doping or multi-element doping have been investigated as a catalyst. Carbon matrices provide facile support with high electrical conductivity to boost the catalytic function of MOF-derived catalysts for HER, OER, and ORR. Owing to the electronegativity differences between N and C which results in surface polarity and enhancement of the conductivity, N-doped carbon matrix synthesized after pyrolysis and removing the metal traces from the MOF has shown to enhance the ORR electrocatalytic activity. N can be doped with different coordination and can activate different electrochemical reactions. For example, pyridinic N acts as active sites for ORR in acidic medium [79] and OER in alkaline medium [80]. However, N in graphitic coordination is active sites for ORR in alkaline medium [80] and acidic condition acts as HER catalyst [81]. Zhao et al. reported nitrogen-doped MOF-derived carbon by pyrolysis of amine-functionalized MOF (amino-MIL-53(Al)) [82]. Wu et al. further showed the activation of MOF-derived N-doped nanocarbon matrix by treating with NH_3 to regulate the N coordination [83]. The catalyst shows the half-wave potential of 28 mV higher than the Pt/C in an alkaline medium.

Other heteroatoms can be further doped along with N such as P, B, and S to enhance the electrocatalytic properties. Synergistic effects originating from the electronegativity difference, and charge redistribution has been shown to tune the adsorption of intermediates and electron transfer process to improve the ORR. N, S co-doping in ZIF-8-derived nanocarbon-based electrocatalyst has shown to enhance the ORR activity [22]. Using MOF-5 as a template, Li et al. have shown co-doping of P, N, and S in the carbon matrix [84]. The doped nanostructure has shown better cathodic current density as compared to commercial Pt/C.

Metal precursors can also be easily incorporated in the pores of MOFs before pyrolysis to construct metal-MOF-derived composite. Owing to the stronger interaction between metal and carbon film, metal particles can be uniformly dispersed over carbon support. Single-atom transition metals incorporation in N-doped MOF-derived nanostructures has also been investigated for ORR. In such catalysts, TM-N_x sites are considered to be electrochemically active sites. TM can be introduced in the nanostructure intrinsically or by providing the external source of metals. Proietti et al. have designed Fe-based ORR catalyst by pyrolysis of ZIF-8, 1, 10-phenanthroline as N-source and Iron(II) acetate as Fe source [85]. ZIF-8 acts as a host for phenanthroline and Fe salt which was subsequently heated to form the catalyst. The catalyst shows comparable efficiency as commercial Pt/C. Zhang et al. investigated the size effect of Fe-based catalyst derived from Fe doped ZIF, and the best catalytical activity was found to be of 50 nm-sized catalyst [86]. Enhanced catalytic property of this catalyst was associated with Fe-N_4 -C coordination. Owing to the absence of aggregation of Fe nanoparticles during the catalysis, the catalyst shows improved stability. Pyrolysis of Co-MOF leads to the formation of cobalt nanoparticles and a compound-based electrocatalyst on the carbon support [87, 88]. Along with the doping, MOF-derived framework can also act as a substrate to incorporate ORR active metal nanoparticles. Wang et al. have shown a highly ordered Pt_3Co intermetallic nanoparticle derived from MOF as a stable ORR catalyst [89]. Co and N co-doped support were first synthesized followed by deposition of Pt nanocrystals. On thermal treatment, the Co atom diffuses into the Pt crystal to form Pt_3Co. The intermetallic nanocrystal shows

Table 4 Summary of MOF-derived electrocatalysts for ORR

Catalyst (MOF-derived nanocrystals)	Application	Performances		References
		E_{onset} (V)	$E_{1/2}$ (V)	
N-doped NH_2-MIL-53(Al) MOF	ORR	–0.13 (vs. Ag/AgCl)	–	Zhao et al. [82]
NH_3 functionalized N-doped NH_2-MIL-53(Al) MOF	ORR	0.96	0.84	Wu et al. [83]
N, S co-doped ZIF-8 carbon matrix	ORR	−0.13 (vs Ag/AgCl)	–	Song et al. [22]
P, N, S co-doped MOF-5 carbon matrix	ORR	–0.006 (vs, Ag/AgCl)	–	Li et al. [84]
Fe and N incorporated ZIF-8 matrix	ORR	–	–	Proietti et al. [85]
Fe doped ZIF-8-based nanocrystal	ORR	-	–	Zhang et al. [86]
MOF-derived Pt_3Co intermetallic nanoparticle	ORR	–	0.92	Wang et al. [89]
Zn-MOF-derived B, N-doped porous carbon matrix	ORR	–	0.894	Qian et al. [21]

half-wave potential of 0.92 V (vs. RHE) with excellent stability. A summary of above discussed MOF-derived electrocatalysts for ORR is presented in Table 4.

MOF-Derived Electrocatalysts for OER

The challenges associated with OER process, such as oxidation of carbon materials and stability issue, limit the application of metal-free MOF-derived catalysts. However, doping of heteroatoms in the carbon matrix has been investigated for OER. N-doping in carbon matrix in electron-withdrawing pyridinic coordination has been shown to improve the OER process [80]. Lei et al. have reported an N-doped carbon material as bifunctional HER and OER electrocatalyst which was synthesized by pyrolyzing ZIF-8 followed by cathodic polarization treatment [90]. The improved catalytic property was attributed to various functional group formations during cathodic polarization treatment. Multiatom doping in carbon matrices is another method to improve the OER activity of MOF-derived metal-free electrocatalyst for OER. Qian et al. have shown co-doping of B and N in Zn-MOF-derived metal-free porous carbon as a bifunctional electrocatalyst for OER/ORR [21]. The nanostructure shows high ORR activity having the onset potential, $E_{onset} = 0.894$ V versus RHE, with better long-term durability than Pt/C. For the OER process, the catalyst requires a 1.55 V (vs RHE) potential to reach a current density of 10 mA/cm^2

in a 6 M KOH solution. N-doping with the presence of positively charge C atoms along with B doping facilitates the adsorption of OH^- and H_2O to enhance the OER activity.

Precious (Ru, Ir, Pt) and non-precious TM (Fe, Ni, Co) can also be incorporated in the carbon matrix which shows improved stability and activity as compared to metal-free MOF-derived electrocatalysts for OER. During the OER process, several metal-based materials can undergo partial oxidation which can further improve their catalytic property. However, in metal-MOF-derived OER catalyst, the major part of the catalyst is still carbon-based which still possess the challenge of stability. Li et al. have reported a Co-MOF-derived bifunctional HER/OER catalyst which was prepared by pyrolysis of ZIF-67 in an inert atmosphere [91]. The catalyst has been investigated for water-splitting reactions and requires a 1.64 V potential to achieve 10 mA/cm^2. Multi metallic doping in MOF-derived nanocarbon has also been explored for OER. Yang et al. have shown that multimetallic doping in carbon framework can tune the electronic structure of the material and subsequently alter the catalytic activity [92]. They synthesize FeCoNi doped MOF-derived catalysts have reported remarkable HER and OER catalytic performance for FeCo doped and FeCoNi doped catalysts.

Incorporation of metals such as Ru and Ir into the MOF has also been shown to enhance the catalytic property. Chen et al. have synthesized Ru/Cu-doped RuO_2 complex integrated in carbon matrix as catalyst. The catalyst was synthesized through thermolysis of Ru-modified Cu-1,3,5-benzenetricarboxylic acid (BTC). The catalyst acts as a highly efficient HER and OER catalyst in alkaline conditions [93]. Through a galvanic replacement between $IrCl_3$ and Co-NC, which was synthesized by thermolysis of ZIF-67, Ir-based electrocatalyst Co@Ir-NC was reported [94]. This catalyst has shown superior long-term stability as compared to commercial Pt/C and RuO_2 for HER and OER, respectively.

MOF-Derived Electrocatalysts for HER

Although various carbon-based nanostructures have been investigated for HER, metal-free MOF-derived electrocatalysts have not been very well explored and future exploration is desired. Lei et al. have shown the N-doped MOF-derived carbon nanostructure as HER and OER catalyst [90]. The catalyst requires 155 mV versus RHE to attain 10 mA/cm^2 of current and having a Tafel slope of 54.7 mV/dec. Another example of metal-free MOF-derived catalysts has been reported by Lin et al. [95] where a novel and unique dual graphitic N-doped in carbon hex-ring was synthesized by pyrolysis of Cu-BTC precursor followed by NH_4OH treatment to remove Cu and facilitate the N-doping. The catalyst requires only 57 mV of potential to reach a current density of 10 mA/cm^2 exhibit Tafel slope of 44.6 mV/dec. Further, DFT reveals that carbon bonded to the dual graphitic N is responsible for the improved catalysis and has the least ΔG_{H^*} value of 0.01 eV.

Pt group metal (PGM)-based electrocatalysts have been extensively studied for the HER. However, the agglomeration of PGM during the electrochemical process reduces the catalytic activity. However, incorporation of PGM into MOF-derived carbon nanostructures with better conductivity and strong interaction between them can prevent agglomeration and enhance stability and activity. Ying et al. synthesized a highly dispersed Pt nanoparticle encapsulated in an N-doped hollow carbon framework by pyrolysis of tannic acid-coated Pt-encapsulated ZIF-8 [96]. The porous nature of the carbon framework facilitates the mass transport and provides strong anchoring support to the Pt nanoparticle, thus displays better HER catalysis as a comparison with commercial Pt/C with enhanced stability. Metal alloys such RuCo and IrCo nanoparticle decorated N-carbon have been shown to tune the catalytic activity of C atoms for HER [97, 98]. Another precious metal (Ru)-MOF-derived catalyst with better activity and stability was synthesized by selective etching of Cu nanoparticle form Cu/Ru-MOF-derived carbon nanostructure [99].

Non-precious metals such as Ni, W have also been incorporated into the MOF-derived carbon framework [100]. The atomically dispersed metals help in utilizing the complete efficiency of metals electrochemical activity. Chen et al. reported atomically dispersed W atoms on MOF-derived carbon skeleton and tested for HER. The catalyst was synthesized by pyrolysis of WCl_5/UIO-66-NH_2. [100]. The catalyst requires overpotential close to Pt/C to achieve 10 mA/cm^2 current and exhibit excellent stability. Metal nanoparticle decorated MOF-derived carbon support has also been synthesized. Pyrolysis of Ni-MOF under different conditions leads to the formation of Ni nanoparticles with surface nitridation on the carbon support [101]. The surface-modified Ni-catalyst attains a current of 20 mA/cm^2 at a low overpotential of 88 mV. The metal nanoparticle of Co is also encapsulated in MOF-derived carbon nanostructures such as graphene oxide and carbon nanotubes for HER [102]. Examples of MOF-derived nanostructures for HER and OER are presented in Table 5.

MOF-Derived Electrocatalysts for Electrochemical Reduction Reaction of CO_2 (CORR) and N_2 (NRR)

MOFs act as a sacrificial template for various carbon-based nanostructures having highly dispersed metals, metal oxides, and heteroatom doped. MOF-derived nanostructures are also explored as electrochemical catalysts for nitrogen reduction to NH_3 and CO_2 reduction to various fuels. The growing pollution and global warming demand the conversion of gases such as CO_2 to value-added chemicals. Another energy demanding industrial process is a synthesis of NH_3 from N_2. The porous structure of MOF facilitates the mass transport of CO_2 or N_2 to the catalytically active sites. Examples of MOF-derived electrocatalysts for electrochemical conversion of CO_2 and N_2 are given in Table 6.

Depending upon the reaction pathways, CO_2 can be either converted to CO or other hydrocarbons. ZIF-derived catalysts generally tend to catalyze the CO_2 to CO. However, the presence of Cu as an active site promotes multi-carbon products. Ion-exchange of Zn^{2+} in ZIF-8 by the adsorbed Ni nanoparticles leads to the formation of a single-atom Ni-based MOF catalyst for CO_2 reduction to CO with

Table 5 Summary of MOF-derived nanostructures for HER and OER electrocatalytic applications

Catalyst (MOF-derived nanocrystals)	Application	Performances		References
		Potential at 10 mV/cm^2	Tafel slope (mV/dec)	
ZIF-8 pyrolyzed N-doped carbon matrix	OER HER	476 mV 155 mV	78.5 54.7	Lei et al. [90]
Zn-MOF-derived B, N co-doped porous carbon matrix	OER	1.38 V	117	Qian et al. [21]
Co-MOF-derived matrix	OER HER	370 mV 298 mV	76,131	Li et al. [91]
Fe, Co, Ni-doped MOF-derived matrix	OER HER	288 mV 149 mV	60 77	Yang et al. [92]
Cu-BTC-derived Ru/Co doped complex	OER HER	204 mV 28 mV	56 35	Chen et al. [93]
ZIF-67-derived Co@Ir nanocrystals	OER HER	280 mV 121 mV	– 41.9	Li et al. [94]
Pt encapsulated ZIF-8	HER	57 mV	27	Ying et al. [96]
RuCo decorated N-doped carbon matrix	HER	28 mV	31	Su et al. [98]
IrCo decorated N-doped carbon matrix	HER	24 mV	23	Jiang et al. [97]
W decorated MOF-derived carbon matrix	HER	85 mV	53	Chen et al. [100]
Ni decorated MOF-derived carbon matrix	HER	88 mV (at 20 mV cm^{-2})	71–83	Wang et al. [101]

ultrahigh turnover frequency (5273 h^{-1}) and 71% faradic frequency [103]. Nam et Al. designed Cu-HKUST-based catalysts to promote the formation of multi-carbon products during electroreduction of CO_2 [104]. The symmetric structure of Cu-dimer (SBU of HKUST) was distorted to the undercoordinated asymmetric structure by thermal annealing. The catalyst shows faradic efficiency of 45% for the formation of C_2H_4 as the product. Another Co-based MOF-derived catalyst has been reported by Wang et al. [105] Nitrogen coordination around Co was found to regulate the catalytic activity and Co coordinated with two N atoms was found to be the best catalyst for CO_2 reduction to CO. Recently, Yao et al. have synthesized a Cu@Cu_xO nanoparticle from calcination of Cu-based MOF (HKUST-1) [106]. The presence of Cu^+/Cu

Table 6 Summary of catalyst used for electrocatalytic conversion of CO_2 and N_2 to other carbon fuels and NH_3, respectively

Catalyst	Product	Performance (Faradic efficiency % or yield)	Reference
ZIF-8-derived Ni SAs/N–C	CO	71.9	Zhao et al. [103]
Calcined HKUST-1 (HKUST-CuAs)	CO C_2H_4	45 45	Nam et al. [104]
ZIF-67-derived Co/N–C	CO	94	Wang et al. [105]
Cu@Cu_xO	Ethylene C_{2+} product	51 70	Yao et al. [106]
N-doped porous carbon	NH_3	–	Liu et al. [107]
C@NiO@Ni microtubes	NH_3	10.9	Luo et al. [110]
N, P co-doped carbon matrix	NH_3 NH_4OH	1.08 μgh^{-1} mg^{-1} 5.77×10^{-4} μgh^{-1} mg^{-1}	Song et al. [109]
Ru SAs/N–C	NH_3	29.6 120.9 μgh^{-1} mg^{-1}	Geng et al. [111]

interface in the catalyst promotes the *CO–CO adsorption and thus favors the formation of C_{2+} product and suppresses the mono carbon products. The Faradic efficiency of ethylene and C_{2+} products was reported to be 51% and 70%, respectively.

NH_3 is another industrial important compound that is generally synthesized by the Haber–Bosch process. The electrochemical synthesis of NH_3 from N_2 is one of the promising alternatives to the Born-Haber process. However, the formation of NH_3 from N_2 is energy demanding as the inert N_2 must be adsorbed, activated, and N–N triple bond must be broken, thus requires electrocatalyst. The competitive HER as a side reaction presents another challenge in the selective electrochemical synthesis of NH_3. Co-doping of heterogeneous atoms in the carbon skeleton has been exploited for NRR. N-doped porous carbon matrix derived from ZIF-8 has been shown as NRR catalyst, exhibiting the yield of 1.44 $mmolg^{-1}$ h^{-1} [107]. The pyridinic and pyrrolic N-doping create the actives site in the catalyst [107, 108]. Other heteroatoms co-doping in the carbon matrix has also been reported as a catalyst for NRR [109]. Recently, Luo et al. designed a MOF-derived C@NiO@Ni microtubes for NRR with a yield of 43.15 $\mu g\, h^{-1}$ μg^{-1} catalyst and faradic efficiency of 10.9% [110]. Ru single-atom decorated on N-doped carbon was synthesized for NRR from pyrolysis of Ru-MOF precursor [111].

7 Summary and Future Perspectives

In this chapter, we have presented the latest progress in exploiting MOFs and MOF-based nanostructures for electrochemical processes such as HER, OER, and ORR which are the key process in water-splitting, fuel cells, and metal-air batteries. MOFs have a highly porous structure with a large surface area, but at the same time, they have certain drawbacks which have been discussed in this chapter including poor conductivity and their chemical and thermal stability. Followed by a brief mechanism of ORR, OER and HER are presented with parameters used to evaluate the catalysts performances. A summary of MOFs, MOFs encapsulated nanoparticles, and MOF-derived nanostructures for electrocatalytic application is also discussed.

Although, the potentials of MOFs are being explored but still there exists large room for development. From a synthetic point of view, there are only a few MOFs that are stable under harsh conditions. Thus there is a need to explore the synthesis methods for designing a robust and stable MOF for electrocatalysis. Also, only limited metal nodes such as Ni, Co, Cr, and Fe are explored for electrocatalysis, which results in a limited number of electrocatalysts. The intrinsic poor conductivity of MOFs, and thus poor charge transfer, also possess a challenge for electrocatalytic applications. Hence, there is still a need to synthesize MOFs with large electrical conductivity and stable frameworks. A better electrocatalyst should have low overpotential and high current, thus engineering the MOF structure to expose a greater number of redox-active sites, with easy accessibility, and better mass transport is required. The synergistic effect originating from the interaction of MOFs and nanoparticles can enhance the catalytic performances. The advantage of interface engineering between nanoparticles and MOFs should be exploited in future. By considering the above-mentioned properties and challenges, efficient MOFs with better electrochemical performances should be synthesized.

Abbreviations

MOFs	Metal-organic frameworks
HER	Hydrogen evolution reaction
OER	Oxygen evolution reaction
ORR	Oxygen reduction reaction
CORR	Carbon dioxide reduction reaction
NRR	Nitrogen reduction reaction
SBU	Secondary building unit
POMs	Poly oxometalates
NP	Nanoparticle
HSAB	Hard-soft acid–base
CIF	Cobalt imidazolate framework
CTAB	Cetyl trimethyl ammonium bromide
ECSA	Electrochemical active surface area
LSV	Linear sweep voltammetry
PGM	Pt group metal

References

1. Smalley RE (2005) Future global energy prosperity: the terawatt challenge. Mrs Bull 30:412–417
2. Kurtz SR et al (2020) Revisiting the Terawatt challenge. MRS Bull 45:159–164
3. Vojvodic A, Nørskov JK (2011) Optimizing perovskites for the water-splitting reaction. Science 334:1355–1356
4. Yan Y, Xia BY, Zhao B, Wang X (2016) A review on noble-metal-free bifunctional heterogeneous catalysts for overall electrochemical water splitting. J Mater Chem A 4:17587–17603
5. Zhu C, Li H, Fu S, Du D, Lin Y (2016) Highly efficient nonprecious metal catalysts towards oxygen reduction reaction based on three-dimensional porous carbon nanostructures. Chem Soc Rev 45:517–531
6. Debe MK (2012) Electrocatalyst approaches and challenges for automotive fuel cells. Nature 486:43–51
7. Cook TR et al (2010) Solar energy supply and storage for the legacy and nonlegacy worlds. Chem Rev 110:6474–6502
8. Rao RR et al (2017) Towards identifying the active sites on $RuO_2(110)$ in catalyzing oxygen evolution. Energy Environ Sci 10:2626–2637
9. Tian X et al (2019) Engineering bunched Pt-Ni alloy nanocages for efficient oxygen reduction in practical fuel cells. Science (80-.) 366:850–856
10. Yang L, Zeng X, Wang W, Cao D (2018) Recent progress in MOF-derived, heteroatom-doped porous carbons as highly efficient electrocatalysts for oxygen reduction reaction in fuel cells. Adv Funct Mater 28:1704537
11. Li H, Eddaoudi M, O'Keeffe M, Yaghi OM (1999) Design and synthesis of an exceptionally stable and highly porous metal-organic framework. Nature 402:276–279
12. Serre C, Millange F, Surblé S, Férey G (2004) A route to the synthesis of trivalent transition-metal porous carboxylates with trimeric secondary building units. Angew Chemie Int Ed 43:6285–6289
13. Shimizu GKH, Vaidhyanathan R, Taylor JM (2009) Phosphonate and sulfonate metal organic frameworks. Chem Soc Rev 38:1430–1449
14. Yaghi OM et al (2003) Reticular synthesis and the design of new materials. Nature 423:705–714
15. Furukawa H, Cordova KE, O'Keeffe M, Yaghi OM (2013) The chemistry and applications of metal-organic frameworks. Science (80-.) 341
16. Eddaoudi M et al (2002) Systematic design of pore size and functionality in isoreticular MOFs and their application in methane storage. Science (80-.) 295:469–472
17. Mueller U et al (2006) Metal–organic frameworks—prospective industrial applications. J Mater Chem 16:626–636
18. Liang R, Jing F, Shen L, Qin N, Wu L (2015) MIL-53 (Fe) as a highly efficient bifunctional photocatalyst for the simultaneous reduction of Cr (VI) and oxidation of dyes. J Hazard Mater 287:364–372
19. Rodríguez HS, Hinestroza JP, Ochoa-Puentes C, Sierra CA, Soto CY (2014) Antibacterial activity against Escherichia coli of Cu-BTC (MOF-199) metal-organic framework immobilized onto cellulosic fibers. J Appl Polym Sci 131
20. Horcajada P et al (2006) Metal-organic frameworks as efficient materials for drug delivery. Angew Chem Int Ed Engl 45:5974–5978
21. Qian Y et al (2017) A metal-free ORR/OER bifunctional electrocatalyst derived from metal-organic frameworks for rechargeable Zn-Air batteries. Carbon N Y 111:641–650
22. Song Z et al (2017) Origin of the high oxygen reduction reaction of nitrogen and sulfur co-doped MOF-derived nanocarbon electrocatalysts. Mater Horizons 4:900–907
23. Mukhopadhyay S, Debgupta J, Singh C, Kar A, Das SK (2018) A Keggin polyoxometalate shows water oxidation activity at neutral pH: POM@ZIF-8, an efficient and robust electrocatalyst. Angew Chemie Int Ed 57:1918–1923

24. Abdelkader-Fernández VK, Fernandes DM, Balula SS, Cunha-Silva L, Freire C (2020) Oxygen evolution reaction electrocatalytic improvement in POM@ZIF nanocomposites: a bidirectional synergistic effect. ACS Appl Energy Mater 3:2925–2934
25. Zheng F et al (2019) Immobilizing Pd nanoclusters into electronically conductive metal-organic frameworks as bi-functional electrocatalysts for hydrogen evolution and oxygen reduction reactions. Electrochim Acta 306:627–634
26. Ye B et al (2019) Pt (1 1 1) quantum dot engineered Fe-MOF nanosheet arrays with porous core-shell as an electrocatalyst for efficient overall water splitting. J Catal 380:307–317
27. Ding Z et al (2019) RhRu alloyed nanoparticles confined within metal organic frameworks for electrochemical hydrogen evolution at all pH values. Int J Hydrogen Energy 44:24680–24689
28. Hermes S et al (2005) Metal@ MOF: loading of highly porous coordination polymers host lattices by metal organic chemical vapor deposition. Angew Chemie Int Ed 44:6237–6241
29. Rowsell JLC, Yaghi OM (2006) Effects of functionalization, catenation, and variation of the metal oxide and organic linking units on the low-pressure hydrogen adsorption properties of metal-organic frameworks. J Am Chem Soc 128:1304–1315
30. Laurier KGM et al (2013) Iron (III)-based metal–organic frameworks as visible light photocatalysts. J Am Chem Soc 135:14488–14491
31. Zhuang W et al (2012) Highly potent bactericidal activity of porous metal-organic frameworks. Adv Healthc Mater 1:225–238
32. George P, Dhabarde NR, Chowdhury P (2017) Rapid synthesis of titanium based metal organic framework (MIL-125) via microwave route and its performance evaluation in photocatalysis. Mater Lett 186:151–154
33. Pu S, Xu L, Sun L, Du H (2015) Tuning the optical properties of the zirconium–UiO-66 metal–organic framework for photocatalytic degradation of methyl orange. Inorg Chem Commun 52:50–52
34. Jing H-P, Wang C-C, Zhang Y-W, Wang P, Li R (2014) Photocatalytic degradation of methylene blue in ZIF-8. Rsc Adv 4:54454–54462
35. Roy S et al (2019) Electrocatalytic hydrogen evolution from a cobaloxime-based metal-organic framework thin film. J Am Chem Soc 141:15942–15950
36. Wu Y-P et al (2017) Surfactant-assisted phase-selective synthesis of new cobalt MOFs and their efficient electrocatalytic hydrogen evolution reaction. Angew Chemie Int Ed 56:13001–13005
37. Xia BY et al (2016) A metal–organic framework-derived bifunctional oxygen electrocatalyst. Nat Energy 1:15006
38. Wells AF (2012) Structural inorganic chemistry. Oxford university press
39. Cavka JH et al (2008) A new zirconium inorganic building brick forming metal organic frameworks with exceptional stability. J Am Chem Soc 130:13850–13851
40. Zhang M et al (2014) Symmetry-guided synthesis of highly porous metal-organic frameworks with fluorite topology. Angew Chemie Int Ed 53:815–818
41. Sun L et al (2017) Is iron unique in promoting electrical conductivity in MOFs? Chem Sci 8:4450–4457
42. Park SS et al (2015) Cation-dependent intrinsic electrical conductivity in isostructural tetrathiafulvalene-based microporous metal-organic frameworks. J Am Chem Soc 137:1774–1777
43. Sun L, Campbell MG, Dincă M (2016) Electrically conductive porous metal-organic frameworks. Angew Chemie Int Ed 55:3566–3579
44. Talin AA et al (2014) Tunable electrical conductivity in metal-organic framework thin-film devices. Science (80-.) 343:66–69
45. Umeyama D, Horike S, Inukai M, Itakura T, Kitagawa S (2012) Inherent proton conduction in a 2D coordination framework. J Am Chem Soc 134:12780–12785
46. Luan Y, Qi Y, Gao H, Zheng N, Wang G (2014) Synthesis of an amino-functionalized metal–organic framework at a nanoscale level for gold nanoparticle deposition and catalysis. J Mater Chem A 2:20588–20596

47. Aijaz A et al (2012) Immobilizing highly catalytically active Pt nanoparticles inside the pores of metal-organic framework: a double solvents approach. J Am Chem Soc 134:13926–13929
48. Sun D, Li Z (2016) Double-solvent method to Pd nanoclusters encapsulated inside the cavity of NH_2–UiO-66 (Zr) for efficient visible-light-promoted suzuki coupling reaction. J Phys Chem C 120:19744–19750
49. Zhu Q-L, Li J, Xu Q (2013) Immobilizing metal nanoparticles to metal–organic frameworks with size and location control for optimizing catalytic performance. J Am Chem Soc 135:10210–10213
50. Chen Y-Z, Xu Q, Yu S-H, Jiang H-L (2015) Tiny Pd@Co core-shell nanoparticles confined inside a metal-organic framework for highly efficient catalysis. Small 11:71–76
51. Han J et al (2015) Metal–organic framework immobilized cobalt oxide nanoparticles for efficient photocatalytic water oxidation. J Mater Chem A 3:20607–20613
52. An B et al (2017) Confinement of ultrasmall Cu/ZnO_x nanoparticles in metal-organic frameworks for selective methanol synthesis from catalytic hydrogenation of CO_2. J Am Chem Soc 139:3834–3840
53. Li G et al (2014) Hydrogen storage in Pd nanocrystals covered with a metal-organic framework. Nat Mater 13:802–806
54. Zhao M et al (2014) Core-shell palladium nanoparticle@metal-organic frameworks as multifunctional catalysts for cascade reactions. J Am Chem Soc 136:1738–1741
55. Liu X et al (2015) Solar-light-driven renewable butanol separation by core-shell Ag@ZIF-8 nanowires. Adv Mater 27:3273–3277
56. Hu P et al (2014) Surfactant-directed atomic to mesoscale alignment: metal nanocrystals encased individually in single-crystalline porous nanostructures. J Am Chem Soc 136:10561–10564
57. Zhao M et al (2016) Metal-organic frameworks as selectivity regulators for hydrogenation reactions. Nature 539:76–80
58. Wang S et al (2016) Nanoreactor based on macroporous single crystals of metal-organic framework. Small 12:5702–5709
59. He L et al (2013) Core–shell noble-metal@ metal-organic-framework nanoparticles with highly selective sensing property. Angew. Chemie 125:3829–3833
60. Chen L, Chen H, Li Y (2014) One-pot synthesis of Pd@MOF composites without the addition of stabilizing agents. Chem Commun 50:14752–14755
61. Radwan A, Jin H, He D, Mu S (2021) Design engineering, synthesis protocols, and energy applications of MOF-derived electrocatalysts. Nano-Micro Lett 13:1–32
62. Tripathy RK, Samantara AK, Behera JN (2019) A cobalt metal–organic framework (Co-MOF): a bi-functional electro active material for the oxygen evolution and reduction reaction. Dalt Trans 48:10557–10564
63. Zhang X, Liu Q, Shi X, Asiri AM, Sun X (2018) An Fe-MOF nanosheet array with superior activity towards the alkaline oxygen evolution reaction. Inorg Chem Front 5:1405–1408
64. Lin C-Y, Zhang J, Xia Z (2019) Coordination-dependent catalytic activity and design principles of metal-organic frameworks as efficient electrocatalysts for clean energy conversion. J Phys Chem C 123:214–221
65. Huang H et al (2020) Conductive metal-organic frameworks with extra metallic sites as an efficient electrocatalyst for the hydrogen evolution reaction. Adv Sci 7:2000012
66. Zou Z et al (2019) Expediting in-situ electrochemical activation of two-dimensional metal-organic frameworks for enhanced OER intrinsic activity by iron incorporation. ACS Catal 9:7356–7364
67. Zhao X et al (2018) Mixed-node metal-organic frameworks as efficient electrocatalysts for oxygen evolution reaction. ACS Energy Lett 3:2520–2526
68. Li F-L, Shao Q, Huang X, Lang J-P (2018) Nanoscale trimetallic metal-organic frameworks enable efficient oxygen evolution electrocatalysis. Angew Chemie Int Ed 57:1888–1892
69. Khrizanforova V et al (2020) 3D Ni and Co redox-active metal–organic frameworks based on ferrocenyl diphosphinate and 4,4′-bipyridine ligands as efficient electrocatalysts for the hydrogen evolution reaction. Dalt Trans 49:2794–2802

70. Zhou Y-C et al (2019) A new 3D 8-fold interpenetrating 6^6-dia topological Co-MOF: Syntheses, crystal structure, magnetic properties and electrocatalytic hydrogen evolution reaction. J Solid State Chem 279:120929
71. Huang J et al (2018) Electrochemical exfoliation of pillared-layer metal-organic framework to boost the oxygen evolution reaction. Angew Chem Int Ed Engl 57:4632–4636
72. Song X, Peng C, Fei H (2018) Enhanced electrocatalytic oxygen evolution by exfoliation of a metal-organic framework containing cationic one-dimensional $[Co_4(OH)_2]^{6+}$ chains. ACS Appl Energy Mater 1:2446–2451
73. Jia H et al (2018) A novel two-dimensional nickel phthalocyanine-based metal–organic framework for highly efficient water oxidation catalysis. J Mater Chem A 6:1188–1195
74. Ma X, Qi K, Wei S, Zhang L, Cui X (2019) In situ encapsulated nickel-copper nanoparticles in metal-organic frameworks for oxygen evolution reaction. J Alloys Compd 770:236–242
75. Xiao Y-H, Tian W, Jin S, Gu Z-G, Zhang J (2020) Host-guest thin films by confining ultrafine Pt/C QDs into metal-organic frameworks for highly efficient hydrogen evolution. Small 16:2005111
76. Dou S et al (2017) Atomic-scale CoOx species in metal-organic frameworks for oxygen evolution reaction. Adv Funct Mater 27:1702546
77. Parkash A (2020) Pt nanoparticles anchored on Cu-MOF-74: an efficient and durable ultra-low Pt electrocatalyst toward oxygen reduction reaction. ECS J Solid State Sci Technol 9:65021
78. Wang J, Zeng HC (2019) A hybrid electrocatalyst with a coordinatively unsaturated metal-organic framework shell and hollow Ni_3S_2/NiS core for oxygen evolution reaction applications. ACS Appl Mater Interfaces 11:23180–23191
79. Guo D et al (2016) Active sites of nitrogen-doped carbon materials for oxygen reduction reaction clarified using model catalysts. Science (80-.) 351:361 LP–365
80. Yang HB et al (2016) Identification of catalytic sites for oxygen reduction and oxygen evolution in N-doped graphene materials: development of highly efficient metal-free bifunctional electrocatalyst. Sci Adv 2:e1501122
81. Jiao Y, Zheng Y, Davey K, Qiao S-Z (2016) Activity origin and catalyst design principles for electrocatalytic hydrogen evolution on heteroatom-doped graphene. Nat Energy 1:16130
82. Zhao X et al (2014) One-step synthesis of nitrogen-doped microporous carbon materials as metal-free electrocatalysts for oxygen reduction reaction. J Mater Chem A 2:11666–11671
83. Wu M et al (2017) A facile activation strategy for an MOF-derived metal-free oxygen reduction reaction catalyst: direct access to optimized pore structure and nitrogen species. ACS Catal 7:6082–6088
84. Li J-S et al (2014) Heteroatoms ternary-doped porous carbons derived from MOFs as metal-free electrocatalysts for oxygen reduction reaction. Sci Rep 4:5130
85. Proietti E et al (2011) Iron-based cathode catalyst with enhanced power density in polymer electrolyte membrane fuel cells. Nat Commun 2:416
86. Zhang H et al (2017) Single atomic iron catalysts for oxygen reduction in acidic media: particle size control and thermal activation. J Am Chem Soc 139:14143–14149
87. Li Z et al (2016) Directed growth of metal-organic frameworks and their derived carbon-based network for efficient electrocatalytic oxygen reduction. Adv Mater 28:2337–2344
88. Zhou T et al (2016) Nitrogen-doped cobalt phosphate@nanocarbon hybrids for efficient electrocatalytic oxygen reduction. Energy Environ Sci 9:2563–2570
89. Wang XX et al (2018) Ordered Pt_3Co intermetallic nanoparticles derived from metal-organic frameworks for oxygen reduction. Nano Lett 18:4163–4171
90. Lei Y et al (2018) Metal-free bifunctional carbon electrocatalysts derived from zeolitic imidazolate frameworks for efficient water splitting. Mater Chem Front 2:102–111
91. Li X, Niu Z, Jiang J, Ai L (2016) Cobalt nanoparticles embedded in porous N-rich carbon as an efficient bifunctional electrocatalyst for water splitting. J Mater Chem A 4:3204–3209
92. Yang Y et al (2017) Tuning electronic structures of nonprecious ternary alloys encapsulated in graphene layers for optimizing overall water splitting activity. ACS Catal 7:469–479
93. Yang K et al (2018) Ultrasmall Ru/Cu-doped RuO_2 complex embedded in amorphous carbon skeleton as highly active bifunctional electrocatalysts for overall water splitting. Small 14:1803009

94. Li D et al (2018) Total water splitting catalyzed by Co@Ir core-shell nanoparticles encapsulated in nitrogen-doped porous carbon derived from metal-organic frameworks. ACS Sustain Chem Eng 6:5105–5114
95. Lin Z et al (2019) Dual graphitic-N doping in a six-membered C-ring of graphene-analogous particles enables an efficient electrocatalyst for the hydrogen evolution reaction. Angew Chem Int Ed Engl 58:16973–16980
96. Ying J et al (2017) Nitrogen-doped hollow porous carbon polyhedrons embedded with highly dispersed Pt nanoparticles as a highly efficient and stable hydrogen evolution electrocatalyst. Nano Energy 40:88–94
97. Jiang P et al (2018) Tuning the activity of carbon for electrocatalytic hydrogen evolution via an iridium-cobalt alloy core encapsulated in nitrogen-doped carbon cages. Adv Mater 30:1705324
98. Su J et al (2017) Ruthenium-cobalt nanoalloys encapsulated in nitrogen-doped graphene as active electrocatalysts for producing hydrogen in alkaline media. Nat Commun 8:14969
99. Qiu T et al (2019) Highly exposed ruthenium-based electrocatalysts from bimetallic metal-organic frameworks for overall water splitting. Nano Energy 58:1–10
100. Chen W et al (2018) Single tungsten atoms supported on MOF-derived N-doped carbon for robust electrochemical hydrogen evolution. Adv Mater 30:1800396
101. Wang T, Zhou Q, Wang X, Zheng J, Li X (2015) MOF-derived surface modified Ni nanoparticles as an efficient catalyst for the hydrogen evolution reaction. J Mater Chem A 3:16435–16439
102. Chen Z et al (2018) Ultrafine Co nanoparticles encapsulated in carbon-nanotubes-grafted graphene sheets as advanced electrocatalysts for the hydrogen evolution reaction. Adv Mater 30:1802011
103. Zhao C et al (2017) Ionic exchange of metal-organic frameworks to access single nickel sites for efficient electroreduction of CO_2. J Am Chem Soc 139:8078–8081
104. Nam D-H et al (2018) Metal-organic frameworks mediate Cu coordination for selective CO_2 electroreduction. J Am Chem Soc 140:11378–11386
105. Wang X et al (2018) Regulation of coordination number over single Co sites: triggering the efficient electroreduction of CO_2. Angew Chemie Int Ed 57:1944–1948
106. Yao K et al (2020) Metal–organic framework derived copper catalysts for CO_2 to ethylene conversion. J Mater Chem A 8:11117–11123
107. Liu Y et al (2018) Facile ammonia synthesis from electrocatalytic N_2 reduction under ambient conditions on N-doped porous carbon. ACS Catal 8:1186–1191
108. Mukherjee S et al (2018) Metal-organic framework-derived nitrogen-doped highly disordered carbon for electrochemical ammonia synthesis using N_2 and H_2O in alkaline electrolytes. Nano Energy 48:217–226
109. Song P, Kang L, Wang H, Guo R, Wang R (2019) Nitrogen (N), Phosphorus (P)-codoped porous carbon as a metal-free electrocatalyst for N_2 reduction under ambient conditions. ACS Appl Mater Interfaces 11:12408–12414
110. Luo S, Li X, Gao W, Zhang H, Luo M (2020) An MOF-derived C@NiO@Ni electrocatalyst for N_2 conversion to NH_3 in alkaline electrolytes. Sustain Energy Fuels 4:164–170
111. Geng Z et al (2018) Achieving a record-high yield rate of 120.9 μgNH_3 $mgcat.^{-1}h^{-1}$ for N_2 electrochemical reduction over Ru single-atom catalysts. Adv Mater 30

Concluding Remarks and Future Perspectives of MOFs as Catalysts

Concluding Remarks About Metal–Organic Frameworks (MOFs): From Properties to Potential Applications

Sanjay Kumar, Aryan Arora, Kartika Goyal, Shikha Gulati, and Manoj Trivedi

Contents

1 Introduction ... 768
2 Structural Properties of Metal–Organic Frameworks (MOFs) ... 769
  2.1 Stability of MOFs ... 769
  2.2 Pore Characteristics and Surface Area Properties of MOFs ... 770
3 Emphasis on the Functionalization of MOFs ... 771
  3.1 Synthesis Strategies for Functionalization ... 773
  3.2 Applications of Functionalized MOFs ... 774
4 Catalytic Role of MOFs in Chemical Reactions and Their Implications ... 775
  4.1 Metal or Metal Clusters as Lewis Acids or Bronsted Acids ... 777
  4.2 Metal Catalyzed CO_2 Conversions ... 780
  4.3 Metal Catalyzed Oxidation ... 780
  4.4 Metal Catalyzed Reduction ... 780
  4.5 Metal Catalyzed Cross-Coupling Reactions ... 781
5 Future Application Potential of MOFs ... 781
  5.1 Multivariate MOFs ... 781
  5.2 Defective MOFs ... 782
  5.3 2-dimensional MOFs ... 782
  5.4 Composites Based on MOFs ... 783
  5.5 Nanomaterials Based on MOFs ... 783
6 Challenges and Future Opportunities ... 784
7 Conclusion ... 784
Abbreviations ... 785
References ... 785

Abstract In previous few decades, the Metal Organic Frameworks (MOFs) field has undergone an instrumental change. They have grown into a material of central importance, particularly in inorganic and materials chemistry. Previously discussed chapters talk about the functionalization and main structural characters of MOFs. Unlike other conventional porous materials, MOFs have extremely ordered, rich and tunable structures. Researches around MOFs are primarily aimed at utilizing their structural information for different functionalization. The combination of MOFs with a variety of functional components, such as metal nanoparticles, oxides, graphene, etc. show superior properties to their individual counterparts. They find numerous

S. Kumar · A. Arora · K. Goyal · S. Gulati (✉) · M. Trivedi
Department of Chemistry, Sri Venkateswara College, University of Delhi, Delhi 110021, India

S. Gulati (ed.), *Metal-Organic Frameworks (MOFs) as Catalysts*,
https://doi.org/10.1007/978-981-16-7959-9_28

applications in different fields, because of their unique characteristics. Conventionally MOFs were used in gas storage and separation. Recent advancements include applications in catalysis, chemical sensors, proton conduction, and medicinal fields. MOFs possessing hierarchical structures are believed to have a high potential for applications including bulky molecules like biomass up-gradation. Even though MOFs are of great significance, there is still a lack of research in some specific areas. Researches in MOF faces numerous challenges like the obscuration of their precise growth mechanism, precise control of target structures, lack of new strategies for the construction of MOF material, and issues like cost, stability, etc. for industry use. This chapter aims to summarize and conclude all the structural properties, functionalization, and applications of MOFs discussed in the previous chapters. We will also discuss the future perspectives of MOFs in different fields, giving a fair idea of our present knowledge of MOFs and their true potential.

Keywords Metal organic frameworks · Functionalization · Porosity · Catalysis · Bio-catalyst

1 Introduction

Porous materials have drawn a great deal of attention in numerous scientific fields like physics, chemistry, and material science [1, 2]. They play a significant role in various applications like adsorption, catalysis, biomedicine, etc. Activated carbons and zeolites are some popularly used porous materials, but have certain shortcomings like irregular structure, pores, and absence of clear structure–property relationships. Hence, the discovery of a superior porous material happens to be a potent field of research.

Metal–organic frameworks (MOFs), or porous coordination polymers (PCPs), happen to be an interesting class of porous compounds. They are formed using organic linkers and metal ions or clusters and are a latest type of crystalline materials with quite huge surface areas (1000–10,000 m^2/g). In addition to this, they have high porosity, tunable structures, and are flexible unlike zeolites and activated carbon [3, 4]. They have high amount of metal nodes and an infinite number of organic linkers, making their compositions and structure easy to tune precisely. They are considered as a subclass of coordination polymers. MOFs have a variety of applications, like gas sorption and separation, chemical sensors, biomedicine, heterogeneous catalysis, etc. [4, 5]. Systematically distributed active sites increase their catalytic efficiency significantly. In 1994 Fujita et al. reported the first-ever MOF-based catalyst, a 3D Cd (II)-MOF with 4,4′-bipyridine [6, 7].

The strategies used to synthesize functional MOFs are broadly classified into three categories: (i) by creating open metal sites or unsaturated metal centers upon removing coordinated solvent molecules; (ii) introduction of guest-accessible functional groups into the pre-functionalized organic linkers; (iii) post-synthetic modification of MOF materials either by attaching new functional ligands to metal sites or introducing a functional material to MOF pores [8].

All the properties discussed above make MOFs a central topic of discussion. This chapter aims to summarize our existing knowledge of MOFs, which was discussed in the previous chapters. It will give an overview of the structure, characterization, and functionalization of MOFs. We will also discuss the role of MOFs as a catalyst in different reactions. Lastly, conclusions, perspectives for potential applications, and challenges posed to MOFs are indicated.

2 Structural Properties of Metal–Organic Frameworks (MOFs)

MOFs have diverse crystalline structures; this provides the basis for their multiple roles. They are prepared by combining metals or metal clusters with organic linkers, mostly carboxylic acid or nitrogen-containing ligands. The metals or metal clusters are known as secondary building units, SBUs. Structures of MOFs are decided by the geometry of SBUs and the structure of organic ligands, this helps in their tuning to a particular extent by deciding desired SBUs and linkers [2–17]. Pore surfaces of MOFs can be used to generate functional sites, for example, functional groups like pyridyl and amine can directly incorporate into the organic linkers. Post-synthetic modification can also introduce the functional groups incompatible with MOF synthesis. These can be tuned as per their specific application. Pore sizes of MOFs vary from tiny to large and can accumulate variety of species like single metal atoms, nanoparticles, metal complexes, organic dyes, polyoxometalates, polymers, and small enzymes [9, 10].

2.1 Stability of MOFs

Despite the large number of structures of MOFs reported in the past years, they generally have weak stability, like poor water, acid/base, thermal, and mechanical stability. Practical applications require the stability of MOFs; hence a number of methods are used to improve this [11, 12]. Direct methods to improve the stability involve strengthening of the coordination bonds between SBUs and organic linkers. Till date, few water-stable MOFs like chromium-based MIL-101 series, zeolitic imidazolate frameworks (ZIFs) [13], zirconium-based carboxylates [14], aluminum-based carboxylates, the pyrazole-based MOFs [15], have been extensively studied.

New strategies are being developed to increase the stability of MOFs, generating hydrophobic surfaces to enhance the water/moisture stability of MOFs is one such approach. A lot of researchers have reported different ways to achieve this stability. Nguyen and Cohen suggested a combination of medium to long alkyl groups within IRMOF-3, to guard the moisture-sensitive MOF by transforming it into a water repellant material [16]. Yang et al. discovered a series of fluorinated MOFs with excellent water stability [17]. Jiang et al. synthesized a novel and highly stable MOF (USTC-6), featuring high hydrophobicity [18]. Several similar approaches and materials were also reported by different researchers.

Furthermore, a lot of approaches are also targeted to improve thermodynamic stability. Zhu et al. reported that metal ions (Cu^{2+}, Cd^{2+}, or Fe^{2+}) doping into a gyroidal MOF, STU1A, resulted in high stability [19]. Unlike water/moisture stability, MOFs with good acid/base stability are quite difficult to construct. As per the HSAB principle, interactions between hard Lewis acids and bases, or vice-versa are much stronger than those between hard acids and soft bases, or vice-versa [20]. A common technique to form MOFs with acid/base stability is to integrate the high oxidation state metal ions (hard acids) with the carboxylate ligands (hard bases) to form strong bonds, unaffected by any chemical attack. Zr-based MOFs stand out among such MOFs [21].

With applications involving elevated temperatures, thermal stability becomes a key factor. Thermal stability is usually affected by the number of linkers connected to each metal node and the node linker bond strength. High valence metal ions often create thermally stable MOFs. Some MOFs like MIL-53 and UiO-66 [14] are exceptionally stable up to 773 K. Thermal stabilities has become necessary for emerging MOFs because it is a useful predictor of resistance to other stresses. From an engineering perspective, the mechanical stability of MOFs under vacuum or pressure is extremely important. Generally, high porosity and surface area make MOFs less mechanically stable, which can grow during mechanical loading [22]. Methods to improve this stability are yet to be explored and significant development is awaited [23].

2.2 *Pore Characteristics and Surface Area Properties of MOFs*

The permanent and highly ordered porosity of MOFs provide the grounds for their different functions. Most of the MOFs fabricated to date have been usually microporous with a pore diameter less than 2 nm. They have shown significantly high adsorption capacities for various gases like hydrogen and carbon dioxide. Narrow pores of MOFs usually don't allow the hosting of immense objects and anchoring molecular functionalities, limiting their usage in some applications. For some recent applications like catalysis and drug carrier, mesoporous MOFs (pore diameter 2–50 nm) and macroporous (pore diameter > 50 nm) MOFs are required. A number of strategies are used to construct and tune MOFs with enormous pore sizes, stretching of ligand's length happens to be one such method [24, 25]. The Brunauer–Emmett–Teller (BET) surface area and Langmuir surface area are also the key factors for the functional applications of MOFs [26]. Theoretically, maximum surface area of MOFs can reach around 14,600 m^2/g [27].

3 Emphasis on the Functionalization of MOFs

One of the prominent features of MOFs is their ability to be integrated with various kinds of functions, which can be incorporated into different parts of MOF structure. These parts can be metal ions/clusters, organic bridging ligands, and empty spaces inside the cavities, as shown in Fig. 1.

Organic functional groups are the most used groups for the construction of functional MOFs. This is because of their rich host–guest chemistry. The aim of this integration is to enrich the host–guest chemistry between MOFs as host and other small molecules as guests. This improves the efficiency of MOFs in different applications. Functionalization also affects structural properties such as crystallinity, porosity, flexibility, stability, and topology. Therefore, pre-designing of functional MOFs is used to get the desired functionality, stability, and porosity. Organic functional groups can

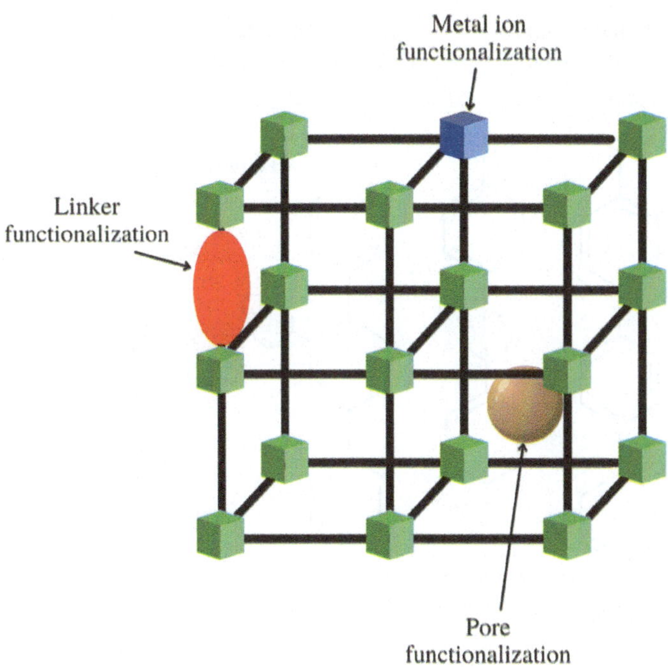

Fig. 1 Functionalizable parts of MOFs

be classified based on two approaches: on the basis of their chemical and structural properties and on the basis of their role in MOF structure.

Classification on the basis of chemical and structural properties can have four categories:

(i) Carbonyl-based functions (Imide, urea, amide, ketone, etc.)
(ii) Oxygen-based functions (Hydroxy, ether, N-oxide, Azoxy, Oxodiazole)
(iii) Sulfur-based functions (Thiol, sulfide, thiourea, sulfonate)
(iv) Nitrogen-based functions (Heterocyclic azines/azoles like pyridine, pyrazole)

Classification on the basis of role in MOF structure can be categorized as (Fig. 2):

(i) Functional groups as coordinating sites
(ii) Functional groups as guest-interactive sites

There are coordinating functional groups as well. They coordinate to metal ions during self-assembly construction MOFs and are intact. This group includes functions like carboxylate, sulfonate, phosphonate, and some of the heterocyclic groups like azoles, pyridine, and diazines. These functions are also used as guest-interactive sites. The functional groups discussed above can also be placed under two categories, for e.g., functions like carboxylate and azoles are applied to both coordinating and guest interactive sites.

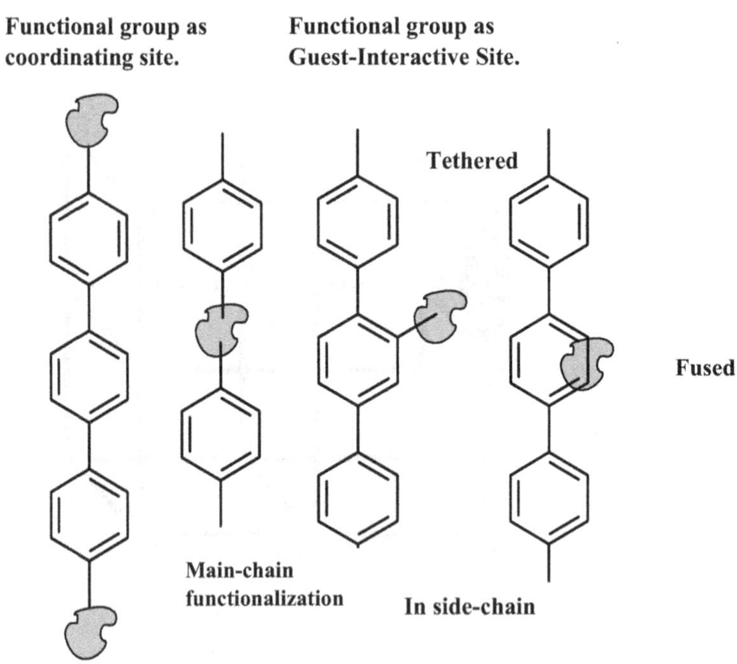

Fig. 2 Classification of organic functional groups based on their role and position in MOF framework

Functional groups also affect the structure of MOFs. These groups can be applied as coordinating sites or guest-interactive sites. Some parameters like the effect on crystallinity, porosity, topology, the orientation of coordination sites, etc., and synthesis conditions like a solvent, pH, etc. should be taken into consideration for the structure of FMOFs. For instance, if a function like a pyridine is applied to the structure of ligand with linear chains, the direction of carboxylate coordinating sites will not change but in the case of functions like urea, the structure is expected to change. In the case of functional ligands with three chains, urea does not change the orientation of carboxylate structures. As the secondary interactions between parts of frameworks are hard to predict, it is almost impossible to predict the structure-directing effects of the functions.

3.1 Synthesis Strategies for Functionalization

These strategies can be broadly categorized into three categories: (i) In-situ functionalization (ii) pre-synthetic modifications (iii) post-synthetic modifications [28].

In-Situ Functionalization

In this approach, the guests are introduced during the synthesis method either inside the pores or in the MOF matrix. This can be done using two ways either by trapping the molecules or embedding nanoparticles in MOF matrices. To trap the molecule the pore aperture of the MOF framework must be smaller than the kinetic diameter of the guest species. Secondly, the framework must possess binding sites. Under either of the two conditions, the molecule will be trapped [29]. The embedding of nanoparticles prevents the Ostwald ripening of the nanoparticles, resulting in MOFs with enhanced properties for different applications.

Pre-synthetic Modifications

Derivatization of organic linkers prior to synthesis can also help in introducing functionality. The substituents attached must not interfere in the formation of targeted MOF. Amino- or bipyridine-based linkers are generally used for this purpose. Post-synthetic deprotection is usually done to enhance the introduced functionalities [30].

Post-synthetic Modifications

These are the most effective modifications, the ones previously discussed undergo subtle changes in reaction parameters. They usually include ion-exchange reactions, solvent exchange reactions, and the incorporation of guest species in the pores [28]. They are classified as:

(i) Functionalization involving weak interactions
(ii) Functionalization involving string interactions
(iii) Functionalization involving covalent interactions

3.2 Applications of Functionalized MOFs

These functional MOFs find applications in numerous fields and they interact using different mechanisms. Some of them are listed below:

(i) Urea is used as a catalyst in Friedel-Craft Henry ring-opening and uses hydrogen bonding with electron-rich or anionic species. This is shown in Fig. 3 [5–36].

(ii) Oxalamide is used in gas adsorption of CO_2 and uses simultaneous oxalamide (O)…C(CO_2) and oxalamide (NH)…O(CO_2) interactions [31, 32].
(iii) Carbonyls are used in sensing organic solvents and use charge transfer between electroactive fluorenone and solvents [33].
(iv) Imide is used as a photocatalyst for C–H and C–C reduction. It uses light-induced electron transfer through π … π interactions [34].
(v) Diazines are used in the removal of Pb^{2+} and Hg^{2+} using pyrazine (N)…metal ion coordination interaction [35].
(vi) N-oxide is used in Li-storage using electrostatic interactions.
(vii) Amine is used in gas adsorption of H_2O using acid (gas molecule)…base (amine) interaction [36].

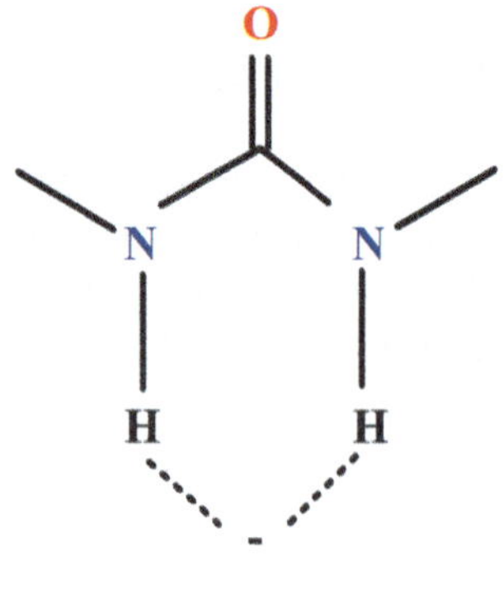

Fig. 3 H-bonding in urea with electron-rich species

Major groups used in gas adsorption include amine, heterocyclic azoles, amides, oxalamides, pyridine, and diazine. Heterocyclic azoles and amines show superior performance in CO_2 separation through their lewis basicity, polarity, and hydrogen bonding, whereas pyridines play an important role in methane storage. FMOFs as heterogeneous catalysts are specially used for CO_2 fixation and activation. The combination of catalytic activity of functional groups and the heterogeneous nature of MOFs provides a large number of unique properties for designing FMOF-based catalysts. Almost all functions show good results in sensing the analyte, since it requires different guest interactive sites. Different functions use different mechanisms, like chelation, hydrogen bonding, lewis acid/base interactions, etc. Some functions like amine, phosphonate, and sulfonate can adsorb photons, thus they can also act as antennas. They transfer these photons through energy or electron transfer to different parts of the structure especially metal clusters for photocatalysis applications.

4 Catalytic Role of MOFs in Chemical Reactions and Their Implications

Metal–organic frameworks have emerged as an interesting type of stable, tunable, and recyclable hybrid materials, consisting of organic bridging ligands and metal centers with properly confined geometries. MOFs have become one of the most promising candidates for heterogeneous catalysis applications [37]. To evaluate the behavior of such porous materials, two competing effects must be considered: activity is proportional to the number of sites active, but an increase in the number of active functions can also result in diffusional limitations. Stability is a vital parameter in heterogeneous catalysis, as it establishes the type of application the material is suited for [38]. MOFs have large surface areas available with apt microporous pores, and highly ordered crystalline structure, providing uniform and confined catalytic sites, size-, shape- and enantio-selectivities. MOFs can help in reducing the entropy loss and futher minimizes the transition state energy in reactions like enzymatic catalysis [39, 40]. They can also be utilized to eliminate the multimolecular catalyst deactivation path and extend the catalytic center's functional life owing to their periodic order and site isolation of catalytic struts.

MOFs are more suited for liquid-phase reactions. Such reactions are performed at 473 K where product diffusion plays an important role for the control of catalytic activity. Such conditions are typically used for the production of chemicals with low volatility. Therefore, MOFs are the optimal catalysts for such reactions [41, 42]. Unlike homogeneous catalysts, frameworks of MOFs can be controlled during the catalytic process. These can be easily recycled from the reaction systems, making them economically and environmentally feasible. Furthermore, the embedded, isolated, and proper active sites of MOFs are easily accessible via open channels. However, the diffusion of reactants and products is an impotant setback of MOFs catalysts compared to homogeneous catalysts.

Table 1 Comparison of Zeolites and MOFs

Properties	Zeolites	MOFs
Crystalline	Yes	Yes
Pore size (nm)	≥1	0.5–9.8
Surface area (m^2g^{-1})	<700	<10,400
Pore volume (cm^3)	0.1–0.5	>1
Thermal stability	High	Low to medium
Chemical versatility	Low	High
Chirality	Difficult to achieve	Easily obtained
Lewis acidity	High	Low to medium
Metal site density	Low	High

Comparison with other porous materials

(i) In comparison to heterogeneous catalysts like zeolites, MOFs are much unstable, imposing restricting to their practical application (Table 1).
(ii) Preparation of MOFs is relatively easy as reactions can be performed under mild conditions [43, 44].
(iii) MOFs show significant porosity with tunable cavities at the nanoscale, whereas it is relatively difficult to attain a size of 1.0 nm for cavities in zeolite structure.
(iv) The number of zeotype structures are definite because of the fixed tetrahedral SiO_4 and AlO_4 structural units, but MOFs can be prepared from infinite combinations of metals and functional ligands offering plethora of options [45, 46].

In MOFs structure metal centers as well as connecting ligands can act as catalytic centers. Here, three prominent methods are used to introduce such active sites into MOFs:

(i) Metal connecting nodes in unsaturated coordination surrounding can be used as active catalyst sites.
(ii) Bridging ligands offer active centers by introducing functional organic sites free from the construction of frameworks of MOFs, that can act as Lewis's acids or bases.
(iii) Catalytic sites can be immobilized by connecting nodes/bridging ligands within the MOFs by PSM, where free functional groups acts as "surface ligands" through coordination interactions. This approach is much more versatile.

Some of the catalytic reactions using MOF as a catalyst are discussed below.

4.1 Metal or Metal Clusters as Lewis Acids or Bronsted Acids

The catalytic metal centers are usually used as Lewis acids or bases, in order to retain the framework of MOFs in the process. Uniform channels promote a high dispersion of reactant molecules and allow tiny molecules to enter the interior surface and react with each metal site. MOFs also provide numerous active sites to absorb the reactant molecules and promote the reactions efficiently. Following are some examples of reactions successfully catalyzed by metal centers within the MOFs.

Cyanosilylation

A 2D square network catalyst $[Cd(bpy)_2](NO3)_2$ was first reported to successfully catalyze the cyanosilylation of aldehydes with the selectivity of shape [6]. The reaction occurs in pores and is dependent on the cavity size of the network. It was suggested that the lewis acidity of Cd(II) center catalyzes (Fig. 4) the imino nitrogen into an electron-donating substrate with selectivity because of the interaction between hydrophobic grid cavities of the catalyst and surface [47]. From here onwards a lot of similar MOF catalysts were reported by different researchers. Chiral Salen ligands for asymmetric catalysis have also been studied to improve the efficiency of a heterogeneous catalyst. In 2015, Cui et al. presented two chiral porous Ti(salen)-based MOFs for asymmetric cyanation of aldehydes.

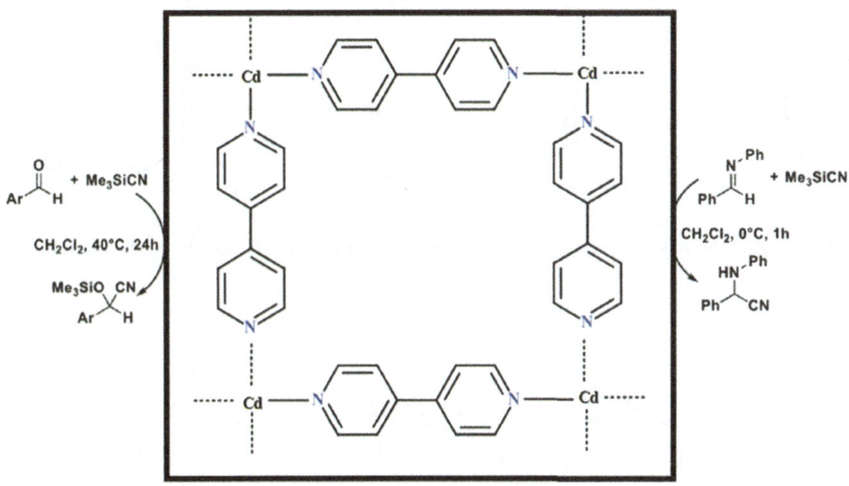

Fig. 4 The Cd(II) ion in the grid complex for cyanosilylation

Transesterification

Kim et al. reported the first transesterification utilizing a homochiral MOF $[Zn_3(\mu_3\text{-}O)\ (L\text{–}H)_6]0.2H_3O\cdot 12H_2O$ with enantiopure metal–organic clusters as SBUs, allowing the metal complexes in its pores enantioselectively [48]. Freshly prepared MOFs show ambient catalytic activity after being immersed in ethanol, to exchange water molecules and ethanol molecules inside the channel and dried up in the sun. The exposed pyridyl groups provide a catalytic opportunity and exposed Zn atoms also play a vital role in the catalytic procedure. Transesterifications catalyzed by Cd [49], Mn [50], and Cu [48] have not been reported in recent times.

Condensation Reactions

Gracia et al. synthesized an effective heterogeneous catalyst, Fe-MOF [Fe(BTC)], for the Claisen-Schmidt condensation (Fig. 5) reaction in toluene between acetophenone and benzaldehyde to selectively yield chalcone in the high amount [48]. Unlike other MOF analogs the activity of Fe(BTC) arises from acidic iron ions with coordinative vacancies and larger surface areas.

The catalyst is also able to synthesize various chalcone derivates selectively under mild conditions, which can be recycled easily. Koner's group also published a report on alkaline earth metal-based MOFs, Ba-MOF [51] and an Mg-MOF [52], which catalyzed aldol condensation reactions under environmentally friendly conditions (Fig. 6). The *s*-aldol products do not undergo further transformation to generate unsat-

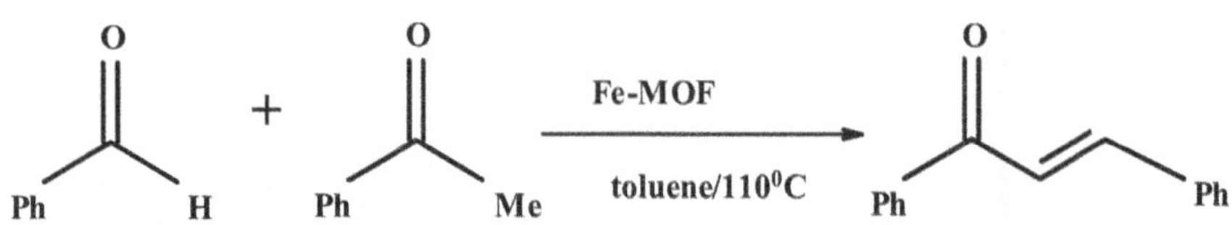

Fig. 5 Claisen-Schmidt condesnation of benzaldehyde with acetophenone. Reproduced from 3

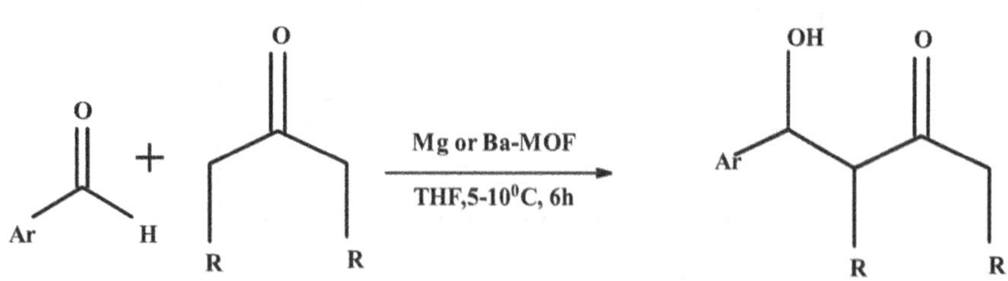

Fig. 6 Coordination environment of the catalytic centers for the Claisen-Schmidt condensation. Reproduced from 3

urated carbonyl compounds and the electron-donating groups afford lower conversions than those of electron-withdrawing groups because of negative inductive and mesomeric effects.

In 2016, Pomberio et al. reported three MOFs based on Cu, Zn, and Cd and an isophthalic acid, which catalyzed the Henry reaction of nitroethane with various aldehydes in an aqueous medium [53]. They were more environmental benignly than the ones used earlier.

Ring-Opening Reactions

Construction of homochiral MOFs through the introduction of chiral organic linkers to metal ions/clusters has attracted immense interest but the utilization of chiral MOFs in asymmetric catalytic reactions is rather limited [54, 55]. Tanka et al. reported a chiral heterogeneous catalyst, for the asymmetric ring-opening reaction of an epoxide with an amine under solvent-free conditions [56]. The MOF structure analysis reveals that each pair of Cu(II) ions in the catalyst serves as a Lewis acid with unsaturated open sites after desolvation. The structure of nucleophiles like amines and alcohols, also affect the stereoselectivity and the conversion of the products in the process of nucleophilic attack.

MOFs based on Lanthanides are also reported to work as catalysts for ring-opening reactions. In a Co-based Metallo ligand, without an open metal site, the Co ions, the lanthanide metal ion act as an active Lewis acid. In other similar MOFs, Farha et al. found that the number of defects is proportional to the catalytic activity like in styrene oxide ring-opening reaction [57].

Friedel–Crafts Reactions

Conventionally, acylation reactions between an aromatic substrate and an acyl component are catalyzed by stoichiometric amounts of strong Lewis acids and Bronsted acids. MOFs being recyclable present great potential in Friedel–Crafts reactions, given the economic and environmental reasons. Cu-MOF-74 has been reported lately with unsaturated metal sites as Lewis acid catalyst in the acylation of anisole [58]. Anisole conversion catalyzed by Cu-MOF-74 exceeds other MOF catalysts and conventional zeolites. Kim et al. reported sulfonated Zr-terephthalate MOF UiO-66-SO_3H, which showed excellent catalytic activity for acylation of p-xylene. Additional acidic sites enhanced the acidic character and activity of the catalyst.

4.2 Metal Catalyzed CO_2 Conversions

Carbon dioxide happens to be one of the most abundant, inexpensive, non-toxic, and bio-renewable carbon resources. The conversion of carbon dioxide into chemical feedstock is somewhat difficult owing to its thermodynamic stability and low reactivity [59]. Doped UiO-67 MOFs for the photocatalytic reduction of CO_2 to CO were first reported by Lin et al. [60]. In other research, Li et al. achieved other functionalized MOFs as photo-catalysts to reduce CO_2 to formate by introducing amino groups into MIL-125(Ti) or UiO-66(Zr) [61]. Amino functionality significantly improved the adsorption of light and CO_2. Such researches demonstrated the potential application of MOFs in artificial photosynthesis under mild conditions. Later in 2014, hydroboration of CO_2 was achieved by Sumby's group using N-heterocyclic carbine containing the fourfold interpenetrated 3D diamondoid MOF [62]. This was the first attempt to embed Cu(I) bis-NHC ligands as catalytic sites into MOF.

4.3 Metal Catalyzed Oxidation

Latest catalytic systems are inclined to use molecular oxygen as the cheapest oxidant for the oxidation of alkanes. A thermally stable 3D polymeric terephthalene lanthanide MOF, which catalyzed the oxidation of alkyl phenyl sulfides was reported by Ruiz-Valero et al. The catalyst was recovered by simple filtration without the obvious losses but reduced reactivity. The activity and selectivity were retained for around four cycles. Gracia et al. reported that MIL-100(Fe) selectively catalyzed the oxidation of thiophenol to diphenyl sulfide. Polyoxometallate (POM) MOF, like PMo-MOF and PW-MOF are efficient and reusable catalysts for the oxidation of numerous sulfides with high conversation and selectivity to sulfoxides [63].

4.4 Metal Catalyzed Reduction

Hatton et al. reported that aluminium (III) 2-aminoterephthalate MOFs were good supporters for aluminium isopropoxide (Al-*i*-Pro), dispersing the latter into pores via bond formation. The capacity of the composites to adsorb aldehydes from a saturated vapor atmosphere facilitated acetaldehyde dimerization [64]. Ethyl acetate is the major product of the composites, while the parent MIL-101 (Al)-NH_2 mainly affords crotonaldehyde via acetaldehyde self-condensation.

Lately, Stavila and the team made a detailed study on the IRMOF-74 series for the selective hydrogenolysis of the aryl-ether bond [65]. Mg-IRMOF-74 turned out to be an efficient catalyst, with outstanding stability for the reduction reaction. Additional activation sites through the insertion of Ti and Ni dopants into Mg-IRMOF-74 have led to further improvements.

4.5 *Metal Catalyzed Cross-Coupling Reactions*

These reactions include forming new C–C bonds and C-heteroatom bonds. Catalyzed by organometallic catalysts, these reactions are a vital part of the pharmaceutical industry and for the conjugated organic materials. Applications of homogeneous catalysts in this industry is fairly limited owing to their proneness to aggregation and the difficulty in recycling. Hence, MOFs are gradually applied in these organic transformations. Following are some examples of such reactions, which use MOFs as a catalyst.

(i) *Pd-catalysed cross coupling reactions*: Li et al. reported a mixed ligand strategy with H_2bpydc and bpdc to prepare Pd(II)@UiO-67 with the [$PdCl_2$(H_2-bpydc)] complex and bpdc ligand, which leads to high activity in the Heck and Suzuki–Miyaura coupling reactions of aryl chlorides.

(ii) *Cu-, Ag-, Au-catalysed cross-coupling reactions*: In 2009, Wu et al. discovered a reusable and efficient catalyst [CuL_2], for the cross-coupling reaction between imidazole and aryl boronic acid. The reaction featured the advantages of no additional additives, mild conditions, and high yields.

(iii) *Grignard Reactions*: Lin et al. reported CMOF with dative PSM by utilizing the BINOL-derived ligand (R)-6,6′-dichloro-2,2′-dihydroxy-1,1′-binaphthyl-4,4′-bypyridine. The BINOL ligand contains pyridyl donors designed for the assembly of the framework and an orthogonal, chiral 2,2′-binapthol group. The heterogeneous catalyst synthesized by this strategy for asymmetric reactions is ideal because the catalytic sites and their secondary environment are identical throughout the solid. CMOFs are used to catalyze the addition reaction of $ZnEt_2$ to aromatic aldehydes to produce aldehydes and chiral secondary alcohols.

5 Future Application Potential of MOFs

5.1 *Multivariate MOFs*

Multivariate MOFs are a totally different class of compounds. They contain multiple functionalities such as different metal ions and linkers. Introducing heterogeneity MOF structures leads to better functional performance than a simple combination of pure counterparts. MTV MOFs were first developed by Yaghi et al. He synthesized 18 MTV MOF-5 type structures consisting of up to eight distinct single-phased functionalities [66]. Intermingling functional groups also pose a difficulty to such MOFs, i.e., if they exist as random, alternating, or as clusters. Keeping this in mind, Kong et al. reported a methodology using a number of techniques with molecular simulations. Its effectiveness can be seen in the prediction and validation of the structures and functions of MTV MOFs. Spatial disorderness in ordered materials can be overcomed using this method. A one-pot method was reported by Sun et al. It aimed at including multi functionalities into stable Zr-MOFs through mixing ligands. Dong et al. reported a way to achieve programmable controlled release of ibuprofen, rhodamine B, and doxorubicin with MIL-101-based MTV MOFs. A physical mixture of single component MOF cannot be used for this purpose, making MTV MOFs [67].

5.2 *Defective MOFs*

Structural heterogeneity can also be achieved by introducing defects into the MOF structure. These defects provide new functional sites missing in pure MOFs, enhancing gas adsorption and separation, catalytic activity, and electrical and conductive properties [68, 69]. Missing linker defects were reported for the first time by Wu et al. They used acetic acid as a modulator in the synthesis of UiO-66. Variation of acetic acid concentration and reaction time was used to tune the vacancies. The defective UiO-66 formed showed a drastic increase in pore volume, surface area, and carbon dioxide adsorption. Non-defective "ideal" MOFs can also be prepared by carefully examining the reaction conditions. At a ratio of 2:1 for BDC:Zr, and 493 K temperaure, Shearer et al. fabricated an "ideal" UiO-66 which was thermally stable up to 723 K. A study has also confirmed that increased reaction temperatures lead to fewer defects.

A combination of mesoporous UiO-66 and phosphotungstic acid acted as a great catalyst for the methanolysis of styrene oxide. Modulator approach is generally used in defective MOFs. Jiang et al. reported the use of TFA to generate defective USTC-253-TFA. The formed compound had exposed metal centers, leading to an increase in CO_2 uptake, high selectivity, catalytic activity, and recyclability in the cycloaddition of CO_2 and epoxide at room temperature [70].

5.3 *2-dimensional MOFs*

MOFs can be fabricated into 2-dimensional nanosheets containing active sites. These sites are exposed on the surface. 2D MOFs often show a better performance in applications like surface-active separation, sensing, catalysis, and other applications [71]. Peng et al. fabricated a process to construct 1 nm thick molecular sieve nanosheets from the MOF $[Zn_2(bim)_4]_n$. The 2D MOF had an exceptionally high hydrogen selectivity [72]. The 2D nanosheets developed by Zhao et al. proved to be the best for sensing DNA. It had a threshold much lower than that for 3D MOF-based particles and was comparable to different 2D nanomaterials. Measurement of fluorescence quenching of the donors in 3D and 2D nanosheets, developed by Cao et al. revealed that 3D MOFs are more energy efficient than 2D MOFs. 2D MOFs can also be used in catalysis because of its exposed active sites. Shi et al. showed the modification of SBUs with monocarboxylic acids during synthesis of MOFs. It is also used in the tuning of hydrophobicity/hydrophilicity of the exposed active sites.

Low yield and lack of control of layer thickness are some hurdles posed in the formation of ultrathin 2D MOF nanosheets. Ding et al. proposed a technique to make 2D sheets using layered 3D MOF crystals. The 2D sheets made in this method were far superior heterogeneous photocatalysts than their 3D counterparts.

5.4 Composites Based on MOFs

Different sizes of pores and channels, combined with large surface areas of MOFs, provide ideal platforms for the construction of hybrid composite materials [73]. The pore spaces in MOFs are ideal to add functional groups for modified properties. Different functional sites have a synergic relationship among them. A locally confined environment provides stability for regular channels of MOFs which facilitate mass transport.

MOF-based composites can also be constructed using metal nanoparticles. This can be used for synergetic catalysis, like high activity for ammonia borane dehydrogenation and CO oxidation can be achieved using Pt nanoparticles which are confined into the pores of MIL-101 [74, 75]. Yang et al. also devised a Pd nanocubes@ZIF-8 composite material by encapsulating Pd nanocubes in ZIF 8 [75]. Pd@ZIF-8 can be used in efficient, selective, and recyclable catalytic hydrogenation of olefins. This was one of the early attempts to integrate plasmonic photothermal effects of metal nanocrystals with MOFs to enhance catalytic activity.

5.5 Nanomaterials Based on MOFs

MOF-based nanomaterials provide the ideal platform for the fabrication of numerous nanomaterials including carbons, metal oxides, metal chalcogenides, metal carbides, metal phosphides, and others. They are widely used in electrochemical energy storage and conversion and heterogeneous catalytic reactions. Thermolysis of the parent MOFs in the absence or presence of additional substances under an inert atmosphere can be used for the fabrication of such nanomaterials. The periodic structure ensures an even distribution of the active components in nanomaterials. Thermolysis improves the stability of parent MOF materials to give an ultra-stable nanostructure. Such structure can withstand harsh conditions in reactions like the Fischer–Tropsch process [76]. An improved electrical conductivity is also observed in MOF-derived carbon-based nanomaterials. This is beneficial for electrocatalysis.

Lately, MOF-derived nanomaterials with single atoms or small atom clusters have received a great deal of attention [77]. To date, reports on MOF-derived nanomaterials containing single atom are fairly limited.

6 Challenges and Future Opportunities

Recent interest and development in the domain of MOF studies are continued which is justified enough by vast applications of exceptional size dimensions, gas storage, and separation, owing to the inherent pores and their volume, to electronic applications, which demands precise control of the electronic structure. Despite the large potential applications, there are pertinent challenges that MOFs face in their conventional aspects and certain modern implications.

More detailed studies are required in various domains of MOFs to solve the current challenges. For instance, the porosity of most MOFs makes them potential agents for gas storage application but the size exclusion approach requires fine-tuning of pore size for separation of gases with similar kinetic diameters that is often a challenging task. Indeed, size exclusion is difficult for separating molecules of similar size and polarizability. Hence, this is one such area that needs future exploration by researchers across the globe.

Besides the edge of heterogeneous catalysts, various large-scale industrial processes still rely on homogeneous catalysis (e.g., Wacker oxidation, hydroformylation, ethylene oligomerization), owing to the lack of compositional and electronic control of heterogeneous catalysts. Heterogenizing molecular complexes through appendage to solid-state surfaces has been practiced successfully at molecular level control to solids, but more often this method leads to a significant reduction in activity or selectivity for the surface isolated complex. Thus, the question of finding heterogeneous catalysts for industrial processes is still open for further exploration. In this aspect, MOFs may possess exceptional opportunities due to their steric tunability.

These problems are not comprehensive but denote the subtleties that are significant to various other resulting applications of MOFs.

7 Conclusion

Here we summarized the key features, main structural characters, functionalization, and applications of MOFs in different fields. MOFs field has undergone explosive growth in the past few years. New methods are being devised for the functionalization of MOFs, which show superior properties to their parent counterparts. Coating MOF with functional materials is one such approach. Although some challenges still persists, for say like large-scale production methods to lower the cost of MOFs preparation. Industrial applications of MOFs are also being worked upon. There have been some significant contributions in the field of heterogeneous catalysis using MOFs, but it is still far behind as compared to available resources. A lot of new fabrications of MOFs are also being developed, for example, 2D MOFs, Quasi-MOFs, MOF-derived nanoparticles, etc. These discoveries do promise a bright future in the field of MOFs, given the diverse nature of the properties of MOFs. Collaborative efforts from industry experts and researchers from different fields can help in overcoming the obstacles and ensure advancement in the field of Metal Organic Frameworks.

Abbreviations

MOFs	Metal Organic Frameworks
SBUs	Secondary Building units
FMOFs	Functional metal organic frameworks
PSM	Post synthetic modifications
CMOF	Catalytic metal organic framework
MTV	Multi variate

References

1. Ruehle B, Clemens DL, Lee BY, Horwitz MA, Zink JI (2017) A pathogen-specific cargo delivery platform based on mesoporous silica nanoparticles. J Am Chem Soc 139:6663–6668
2. Hartmann M, Machoke AG, Schwieger W (2016) Catalytic test reactions for the evaluation of hierarchical zeolites. Chem Soc Rev 45:3313–3330
3. Allendorf MD, Stavila V (2015) Crystal engineering, structure-function relationships, and the future of metal-organic frameworks. CrystEngComm 17:229–246
4. Li B, Wen HM, Cui Y, Zhou W, Qian G, Chen B (2016) Emerging multifunctional metal-organic framework materials. Adv Mater 28:8819–8860
5. Doherty CM, Buso D, Hill AJ, Furukawa S, Kitagawa S, Falcaro P (2014) Using functional nano- and microparticles for the preparation of metal-organic framework composites with novel properties. Acc Chem Res 47:396–405
6. Fujita M, Yoon Jung Kwon st, Washizu S, Ogura K (1994) Preparation, clathration ability, and catalysis of a two-dimensional square network material composed of Cadmium(11) and 4,Y-Bipyridine
7. Kang YS, Lu Y, Chen K, Zhao Y, Wang P, Sun WY (2019) Metal–organic frameworks with catalytic centers: from synthesis to catalytic application. Coord Chem Rev 378:262–280
8. Yuan D, Zhao D, Sun D, Zhou HC (2010) An isoreticular series of metal-organic frameworks with dendritic hexacarboxylate ligands and exceptionally high gas-uptake capacity. Angew Chem Int Ed 49:5357–5361
9. Lu W, Wei Z, Gu ZY et al (2014) Tuning the structure and function of metal-organic frameworks via linker design. Chem Soc Rev 43:5561–5593
10. Tranchemontagne DJ, Tranchemontagne JL, O'keeffe M, Yaghi OM (2009) Secondary building units, nets and bonding in the chemistry of metal–organic frameworks. Chem Soc Rev 38:1257–1283
11. Bosch M, Zhang M, Zhou H-C (2014) Increasing the stability of metal-organic frameworks. Adv Chem 2014:1–8
12. Burtch NC, Jasuja H, Walton KS (2014) Water stability and adsorption in metal-organic frameworks. Chem Rev 114:10575–10612

13. Park KS, Ni Z, Cô AP, Choi JY, Huang R, Uribe-Romo FJ, Chae HK, O'keeffe M, Yaghi OM (2006) Exceptional chemical and thermal stability of zeolitic imidazolate frameworks
14. Cavka JH, Jakobsen S, Olsbye U, Guillou N, Lamberti C, Bordiga S, Lillerud KP (2008) A new zirconium inorganic building brick forming metal organic frameworks with exceptional stability. J Am Chem Soc 130:13850–13851
15. Colombo V, Galli S, Choi HJ, Han GD, Maspero A, Palmisano G, Masciocchi N, Long JR (2011) High thermal and chemical stability in pyrazolate-bridged metal-organic frameworks with exposed metal sites. Chem Sci 2:1311–1319
16. Nguyen JG, Cohen SM (2010) Moisture-resistant and superhydrophobic metal-organic frameworks obtained via postsynthetic modification. J Am Chem Soc 132:4560–4561
17. Yang C, Kaipa U, Mather QZ, Wang X, Nesterov V, Venero AF, Omary MA (2011) Fluorous metal-organic frameworks with superior adsorption and hydrophobic properties toward oil spill cleanup and hydrocarbon storage. J Am Chem Soc 133:18094–18097
18. Jiang ZR, Ge J, Zhou YX, Wang ZU, Chen D, Yu SH, Jiang HL (2016) Coating sponge with a hydrophobic porous coordination polymer containing a low-energy CF3-decorated surface for continuous pumping recovery of an oil spill from water. NPG Asia Mater. https://doi.org/10.1038/am.2016.22
19. Zhu X-W, Zhou X-P, Li D, Article type exceptionally water stable heterometallic gyroidal MOFs: tuning porosity and hydrophobicity by doping metal ions Received (in XXX, XXX) Xth XXXXXXXXX 20XX, Accepted Xth XXXXXXXXX 20XX. https://doi.org/10.1039/c0xx00000x
20. Pearson RG, Busch DH (1963) Hard and soft acids and bases
21. Bai Y, Dou Y, Xie LH, Rutledge W, Li JR, Zhou HC (2016) Zr-based metal-organic frameworks: design, synthesis, structure, and applications. Chem Soc Rev 45:2327–2367
22. Chapman KW, Halder GJ, Chupas PJ (2009) Pressure-induced amorphization and porosity modification in a metal-organic framework. J Am Chem Soc 131:17546–17547
23. Moosavi SM, Boyd PG, Sarkisov L, Smit B (2018) Improving the mechanical stability of metal-organic frameworks using chemical caryatids. ACS Cent Sci 4:832–839
24. Eddaoudi M, Kim J, Rosi N, Vodak D, Wachter J, O'Keeffe M, Yaghi OM (2002) Systematic design of pore size and functionality in isoreticular MOFs and their application in methane storage. Science 295:469–472
25. Deng H, Grunder S, Cordova KE et al (2012) Large-pore apertures in a series of metal-organic frameworks. Science 336:1018–1023
26. Senkovska I, Kaskel S (2014) Ultrahigh porosity in mesoporous MOFs: promises and limitations. Chem Commun 50:7089–7098
27. Farha OK, Eryazici I, Jeong NC, Hauser BG, Wilmer CE, Sarjeant AA, Snurr RQ, Nguyen ST, Yazaydin AÖ, Hupp JT (2012) Metal-organic framework materials with ultrahigh surface areas: is the sky the limit? J Am Chem Soc 134:15016–15021
28. Yaghi OM, Kalmutzki MJ, Diercks CS Introduction to reticular chemistry : metal-organic frameworks and covalent organic frameworks.
29. Yaghi OM, Kalmutzki MJ, Diercks CS (2019) 6 Functionalization of MOFs 6.1 introduction
30. Peng YL, Liu J, Zhang HF, Luo D, Li D (2018) A size-matched POM@MOF composite catalyst for highly efficient and recyclable ultra-deep oxidative fuel desulfurization. Inorg Chem Front 5:1563–1569
31. Alsmail NH, Suyetin M, Yan Y et al (2014) Analysis of high and selective uptake of CO_2 in an Oxamide-containing {Cu2(OOCR)4}-based metal-organic framework. Chem Eur J 20:7317–7324
32. Li XY, Li YZ, Yang Y, Hou L, Wang YY, Zhu Z (2017) Efficient light hydrocarbon separation and CO_2 capture and conversion in a stable MOF with oxalamide-decorated polar tubes. Chem Commun 53:12970–12973
33. Drache F, Bon V, Senkovska I, Adam M, Eychmüller A, Kaskel S (2016) Vapochromic luminescence of a zirconium-based metal-organic framework for sensing applications. Eur J Inorg Chem 2016:4483–4489

34. Zeng L, Liu T, He C, Shi D, Zhang F, Duan C (2016) Organized aggregation makes insoluble perylene diimide efficient for the reduction of aryl halides via consecutive visible light-induced electron-transfer processes. J Am Chem Soc 138:3958–3961
35. Rouhani F, Morsali A (2018) Fast and selective heavy metal removal by a novel metal-organic framework designed with in-situ ligand building block fabrication bearing free nitrogen. Chem Eur J 24:5529–5537
36. Ko N, Hong J, Sung S, Cordova KE, Park HJ, Yang JK, Kim J (2015) A significant enhancement of water vapour uptake at low pressure by amine-functionalization of UiO-67. Dalton Trans 44:2047–2051
37. Fadhel AZ, Pollet P, Liotta CL, Eckert CA (2010) Combining the benefits of homogeneous and heterogeneous catalysis with tunable solvents and nearcritical water. Molecules 15:8400–8424
38. Cruciani G (2006) Zeolites upon heating: factors governing their thermal stability and structural changes. J Phys Chem Solids 67:1973–1994
39. Murase T, Horiuchi S, Fujita M (2010) Naphtalene diels-alder in a self-assembled molecular flask. J Am Chem Soc 132:2866–2867
40. Yoshizawa M, Tamura M, Fujita M (2006) Diels-alder in aqueous molecular hosts: unusual regioselectivity and efficient catalysis. Science 312:251–254
41. Pal N, Bhaumik A (2015) Mesoporous materials: versatile supports in heterogeneous catalysis for liquid phase catalytic transformations. RSC Adv 5:24363–24391
42. Dhakshinamoorthy A, Opanasenko M, Čejka J, Garcia H (2013) Metal organic frameworks as heterogeneous catalysts for the production of fine chemicals. Catal Sci Technol 3:2509–2540
43. Anderson M, Terasaki O, Ohsuna T et al (1994) Structure of the microporous titanosilicate ETS-10. Nature 367:347–351. https://doi.org/10.1038/367347a0
44. Davis ME (1993) New vistas in zeolite and molecular sieve catalysis
45. Chen SS, Chen ZH, Fan J, Okamura TA, Bai ZS, Lv MF, Sun WY (2012) Synthesis and characterization of metal complexes with mixed 4-imidazole-containing tripodal ligand and varied dicarboxylic acid. Cryst Growth Des 12:2315–2326
46. Wilson ST, Lok BM, Messina CA, Cannan TR, Flanigen EM (1982) Aluminophosphate molecular sieves: a new class of microporous crystalline inorganic solids. J Am Chem Soc 104(3):1146–1147
47. Ohmori O, Fujita M (2004) Heterogeneous catalysis of a coordination network: cyanosilylation of imines catalyzed by a Cd(II)-(4,4'-bipyridine) square grid complex. Chem Commun 4:1586–1587
48. Hwang IH, Jo YD, Kim HY, Kang J, Noh JY, Hyun MY, Kim C, Kim Y, Kim SJ (2012) Novel Mn II coordination compounds constructed from benzoate and various bipyridyl ligands: magnetic property and catalytic activity. Polyhedron 42:282–290
49. 35010088
50. Eom GH, Kim JH, Jo YD, Kim EY, Bae JM, Kim C, Kim SJ, Kim Y (2012) Anion effects on construction of cadmium(II) compounds with a chelating ligand bis(2-pyridylmethyl)amine: Their photoluminescence and catalytic activities. Inorg Chim Acta 387:106–116
51. Dhakshinamoorthy A, Alvaro M, Garcia H (2010) Aerobic oxidation of benzyl amines to benzyl imines catalyzed by metal-organic framework solids. ChemCatChem 2:1438–1443
52. Maity T, Saha D, Das S, Koner S (2012) Barium carboxylate metal-organic framework—Synthesis, x-ray crystal structure, photoluminescence and catalytic study. Euro J Inorg Chem 4914–4920
53. Saha D, Maity T, Das S, Koner S (2013) A magnesium-based multifunctional metal-organic framework: synthesis, thermally induced structural variation, selective gas adsorption, photoluminescence and heterogeneous catalytic study. Dalton Trans 42:13912–13922
54. Karmakar A, Martins LMDRS, Hazra S, Guedes Da Silva MFC, Pombeiro AJL (2016) Metal-organic frameworks with pyridyl-based isophthalic acid and their catalytic applications in microwave assisted peroxidative oxidation of alcohols and Henry reaction. Cryst Growth Des 16:1837–1849
55. Tian LL, Wang C, Dawn S, Smith MD, Krause JA, Shimizu LS (2009) Macrocycles with switchable exo/endo metal binding sites. J Am Chem Soc 131:17620–17629

56. Ma L, Lin W (2008) Chirality-controlled and solvent-templated catenation isomerism in metal-organic frameworks. J Am Chem Soc 130:13834–13835
57. Tanaka K, Oda S, Shiro M (2008) A novel chiral porous metal-organic framework: Asymmetric ring opening reaction of epoxide with amine in the chiral open space. Chem Commun 820–822
58. Serhan M, Sprowls M, Jackemeyer D, Long M, Perez ID, Maret W, Tao N, Forzani E (2019) Total iron measurement in human serum with a smartphone. In: AIChE annual meeting, conference proceedings. https://doi.org/10.1039/x0xx00000x
59. Calleja G, Sanz R, Orcajo G, Briones D, Leo P, Martínez F (2014) Copper-based MOF-74 material as effective acid catalyst in Friedel-Crafts acylation of anisole. Catal Today 227:130–137
60. pnas.79.2.701
61. Wang C, Xie Z, Dekrafft KE, Lin W (2011) Doping metal-organic frameworks for water oxidation, carbon dioxide reduction, and organic photocatalysis. J Am Chem Soc 133:13445–13454
62. Fu Y, Sun D, Chen Y, Huang R, Ding Z, Fu X, Li Z (2012) An amine-functionalized titanium metal-organic framework photocatalyst with visible-light-induced activity for CO_2 reduction. Angew Chem Int Ed 51:3364–3367
63. Fei H, Sampson MD, Lee Y, Kubiak CP, Cohen SM (2015) Photocatalytic CO_2 reduction to formate using a Mn(I) molecular catalyst in a robust metal-organic framework. Inorg Chem 54:6821–6828
64. Haddadi H, Hafshejani SM, Farsani MR (2015) Selective and reusable oxidation of sulfides to sulfoxides with hydrogen peroxide catalyzed by organic-inorganic polyoxometalate-based frameworks. Catal Lett 145:1984–1990
65. Bromberg L, Su X, Hatton TA (2013) Aldehyde self-condensation catalysis by aluminum aminoterephthalate metal-organic frameworks modified with aluminum isopropoxide. Chem Mater 25:1636–1642
66. Stavila V, Parthasarathi R, Davis RW, el Gabaly F, Sale KL, Simmons BA, Singh S, Allendorf MD (2016) MOF-based catalysts for selective hydrogenolysis of carbon-oxygen ether bonds. ACS Catal 6:55–59
67. Deng H, Doonan CJ, Furukawa H, Ferreira RB, Towne J, Knobler CB, Wang B, Yaghi OM (2010) Multiple functional groups of varying ratios in metal-organic frameworks. Science 327:846–850
68. Wu H, Chua YS, Krungleviciute V, Tyagi M, Chen P, Yildirim T, Zhou W (2013) Unusual and highly tunable missing-linker defects in zirconium metal-organic framework UiO-66 and their important effects on gas adsorption. J Am Chem Soc 135:10525–10532
69. Ren J, Ledwaba M, Musyoka NM, Langmi HW, Mathe M, Liao S, Pang W (2017) Structural defects in metal–organic frameworks (MOFs): Formation, detection and control towards practices of interests. Coord Chem Rev 349:169–197
70. Fang Z, Bueken B, de Vos DE, Fischer RA (2015) Defektmanipulierte Metall-organische Gerüste. Angew Chem 127:7340–7362
71. Jiang ZR, Wang H, Hu Y, Lu J, Jiang HL (2015) Polar group and defect engineering in a metal-organic framework: synergistic promotion of carbon dioxide sorption and conversion. Chemsuschem 8:878–885
72. Zhao M, Lu Q, Ma Q, Zhang H (2017) Two-dimensional metal-organic framework nanosheets. Small Methods. https://doi.org/10.1002/smtd.201600030
73. Cao L, Lin Z, Shi W, Wang Z, Zhang C, Hu X, Wang C, Lin W (2017) Subscriber access provided by UB + Fachbibliothek Chemie | (FU-Bibliothekssystem) exciton migration and amplified quenching on two-dimensional metal-organic layers
74. Li S, Huo F (2015) Metal-organic framework composites: from fundamentals to applications. Nanoscale 7:7482–7501
75. Aijaz A, Karkamkar A, Choi YJ, Tsumori N, Rönnebro E, Autrey T, Shioyama H, Xu Q (2012) Immobilizing highly catalytically active Pt nanoparticles inside the pores of metal-organic framework: a double solvents approach. J Am Chem Soc 134:13926–13929

76. Yang Q, Xu Q, Yu S-H, Jiang H-L (2016) Pd Nanocubes@ZIF-8: integration of plasmon-driven photothermal conversion with a metal-organic framework for efficient and selective catalysis. Angew Chem 128:3749–3753
77. An B, Cheng K, Wang C, Wang Y, Lin W (2016) Pyrolysis of metal-organic frameworks to Fe_3O_4@Fe_5C_2 core-shell nanoparticles for fischer-tropsch synthesis. ACS Catal 6:3610–3618

The manufacturer's authorised representative in the EU is Springer Nature Customer Service Centre GmbH, Europaplatz 3, 69115 Heidelberg, Germany. If you have any concerns regarding our products, please contact ProductSafety@springernature.com

Printed and bound by CPI Group (UK) Ltd, Croydon, CR0 4YY
30/09/2026
02225333-0019